Energy Systems in Electrical Engineering

Series Editor

Muhammad H. Rashid, Florida Polytechnic University, Lakeland, USA

Energy Systems in Electrical Engineering is a unique series that aims to capture advances in electrical energy technology as well as advances electronic devices and systems used to control and capture other sources of energy. Electric power generated from alternate energy sources is getting increasing attention and supports for new initiatives and developments in order to meet the increased energy demands around the world. The availability of computer–based advanced control techniques along with the advancement in the high-power processing capabilities is opening new doors of opportunity for the development, applications and management of energy and electric power. This series aims to serve as a conduit for dissemination of knowledge based on advances in theory, techniques, and applications in electric energy systems. The Series accepts research monographs, introductory and advanced textbooks, professional books, reference works, and select conference proceedings. Areas of interest include, electrical and electronic aspects, applications, and needs of the following key areas:

- Biomass and Wastes Energy
- Carbon Management
- Costs and Marketing
- Diagnostics and Protections
- Distributed Energy Systems
- Distribution System Control and Communication
- Electric Vehicles and Tractions Applications
- Electromechanical Energy Conversion
- Energy Conversion Systems
- Energy Costs and Monitoring
- Energy Economics
- Energy Efficiency
- Energy and Environment
- Energy Management, and Monitoring
- Energy Policy
- Energy Security
- Energy Storage and Transportation
- Energy Sustainability
- Fuel Cells
- Geothermal Energy
- Hydrogen, Methanol and Ethanol Energy
- Hydropower and Technology
- Intelligent Control of Power and Energy Systems
- Nuclear Energy and Technology
- Ocean Energy
- Power and Energy Conversions and Processing
- Power Electronics and Power Systems
- Renewable Energy Technologies
- Simulation and Modeling for Energy Systems
- Superconducting for Energy Applications
- Tidal Energy
- Transport Energy

More information about this series at https://link.springer.com/bookseries/13509

Aashish Kumar Bohre · Pradyumn Chaturvedi ·
Mohan Lal Kolhe · Sri Niwas Singh
Editors

Planning of Hybrid Renewable Energy Systems, Electric Vehicles and Microgrid

Modeling, Control and Optimization

Springer

Editors
Aashish Kumar Bohre
Department of Electrical Engineering
National Institute of Technology
Durgapur, West Bengal, India

Pradyumn Chaturvedi
Department of Electrical Engineering
Visvesvaraya National Institute
of Technology
Nagpur, Maharashtra, India

Mohan Lal Kolhe
Faculty of Engineering and Science
University of Agder
Kristiansand, Norway

Sri Niwas Singh
Department of Electrical Engineering
Indian Institute of Technology
Kanpur, Uttar Pradesh, India

ISSN 2199-8582 ISSN 2199-8590 (electronic)
Energy Systems in Electrical Engineering
ISBN 978-981-19-0978-8 ISBN 978-981-19-0979-5 (eBook)
https://doi.org/10.1007/978-981-19-0979-5

This Springer imprint is published by the registered company Springer Nature Singapore Pte Ltd.
The registered company address is: 152 Beach Road, #21-01/04 Gateway East, Singapore 189721, Singapore

Preface

Nowadays, the electrical energy demand is growing worldwide very fast, and accordingly, the energy crises directly impact economics, society, and the development of any country. Modern energy systems are changing rapidly and reveal progressively many-sided features. Renewable energy sources such as solar, wind, and other renewable sources are capable enough to meet the growing load demand, which reduces the dependency on conventional fossil fuels. Alternatively, fossil fuel-based energy generation is the direct source of critical global warming issues. Therefore, promoting the use of renewable energy sources, for meeting electrical energy demand, is an important strategy to enhance the energy security of any country. In this connection, wind and solar-based electrical power generation has gained the attention of researchers all over the world. This enables high penetration of renewable sources with the main electrical grid. Power generation from these renewable energy resources is intermittent because of its dependence on environmental conditions. As a result, the power generation from solar and wind systems keeps on fluctuating and directly impacts the voltage magnitude, supply frequency, and waveform, and hence on the quality and quantity of supplying power to the interconnected grid system.

The electrical energy system is currently experiencing substantial transformations due to the increasing utilization of renewable energy systems and integrating electric vehicles in the current transportation system. Environmental compliance and energy saving are becoming increasingly crucial in the future. Today's need is to increase grid reliability and sustainability while considering customer services and grid operation. The changes are especially significant in the power distribution system, which must be changed into a smart grid or automated distribution system to effectively manage the system operation. The smart grid, also known as the next-generation/future grid, is a broadly spread automated energy delivery network that employs two-way electrical and information exchanges. The smart grid implementation will satisfy environmental standards to emphasize demand response events and efficient energy management. The smart grid supports plug-in electric vehicles, renewable distributed generation, and energy storage technologies, and these technologies have a significant impact on the system/grid. The integration of electric vehicles (EVs) provides opportunities and significantly impacts the microgrid/smart grid. EVs can be used as flexible loads in bi-directional mode either as grid-to-vehicle (G2V) for EV charging or vehicle-to-grid (V2G) to provide energy to the grid. The future plug-in-EVs infrastructure may be used to access the electricity from the smart grid to maximize the benefits of V2G mode.

The aim of this book is to provide an essential reference source, building on the available literature in the area of microgrid/smart-grid operation and control, providing further research opportunities in the field of renewable energy systems and electric vehicles. This specific text is expected to provide the primary and major resources necessary for researchers, academicians, students, faculties, and scientists across the globe to adopt and implement new inventions in the thrust area of renewable generation and their utilization. Therefore, the book provides knowledge on planning of hybrid renewable energy systems, electric vehicles and microgrids operation, control and optimization to share up-to-date scientific advancements in the core and related area. The main objectives of the book are as follows:

- Identify and explore the scope of different modeling, operation, and control methods for microgrids or smart grids with high share of renewable energy resources.
- Identify the opportunities of modern electric vehicle technology related to modeling, operation, and control and also explore its impacts on the grid.
- Identify and explore the possible configurations, operation, and control of hybrid renewable system for efficient energy management including a storage system.
- Identify the different modeling, operation, and control schemes/algorithms/approaches/techniques for implementation in future development of micro-grids/smart grids, electric vehicles technology, and hybrid renewable systems.

From recent trends, it can be observed that research is mainly focused on the planning of renewable energy systems, electric vehicles, and microgrids or smart grids based on the efficient modeling and control of microgrids in a smart grid environment to maximize technical, social, and economic benefits by applying the different advanced computational intelligence/optimization techniques. In the smart grid environment, applying different methods such as artificial intelligence has become a trend. This book will provide vast knowledge in the focused area to the researcher and it will also give common sources of information to the power engineers and academicians. Researchers are highly involved in this area because energy demand is exponentially growing, whereas energy resources are depleting day by day. This requires optimal allocation and utilization of different renewable sources in electrical power and energy system. This edition of the book will be a source of motivation to the researchers working in the area of renewable distributed generation, micro-grid, automation of power distribution, electric vehicle, and synchronized operation of the solar system, battery, and grid. This book will include chapters that focus on novel solution methodologies and current research in the focused area related to modeling, control, and optimization approaches. This book covers a wide range of diverse applications together with novel basics, modeling, control, and experimental results.

This book consists of 37 well-structured full chapters contributed by subject experts specialized in various topics addressed in this book. The included chapters in this book have been brought out after a rigorous review process in the broad areas of hybrid renewable systems, electric vehicles, and microgrids: modeling, control, and optimization. The main importance has been given to those chapters that offer novel approaches, experimental systems, and practical systems solutions in the recent era of renewable energy systems, EVs, and microgrids.

Durgapur, India — Aashish Kumar Bohre
Nagpur, India — Pradyumn Chaturvedi
Kristiansand, Norway — Mohan Lal Kolhe
Kanpur, India — Sri Niwas Singh

Acknowledgments

This book on "Planning of Hybrid Renewable Energy Systems, Electric Vehicles, and Microgrid: Modeling, Control, and Optimization" is an outcome of the inspiration and encouragement given by many individuals for whom these words of thanks are only a token of our gratitude and appreciation for them. Our sincere gratitude goes to the people who contributed their time and expertise to this book. We highly appreciate their efforts to achieve the goal of this book. The editors would like to acknowledge the help of all the people involved in this project and, more specifically, the editors would like to thank each one of the authors for their contributions and the editorial board/reviewers regarding the improvement of quality, coherence, and the content presentation of this book.

The editors would like to express their sincere thanks to Springer, Editorial team, for their continuous support and for giving us an opportunity to edit this book.

The editors are thankful to our family members for their prayers, encouragement, and care shown toward us during the completion of this book.

Thank you all.

We also express our gratitude to GOD for all the blessings.

About This Book

This book focuses on various challenges, solutions, and emerging technologies in the area of operation, control, design, optimization, and protection of microgrids in the presence of hybrid renewable energy sources and electric vehicles. The utilization of renewable energy sources (RES) offers a sustainable, economic, and eco-friendly solution for the modern power system network. This book provides an insight on the potential applications and recent development of different types of RESs including AC/DC microgrid, RES Integration issues with grid, electric vehicle technology, etc. Currently, the development of electric vehicles (EVs) and AC/DC smart microgrids are growing very fast worldwide due to numerous techno-socio-economic advantages within the system and meet the present global requirement. It is designed as an interdisciplinary platform for audiences working in the focused area to access information related to energy management, modeling, and control. It covers fundamental knowledge, design, mathematical modeling, application, and practical issues with sufficient design problems and case studies with detailed planning aspects. This book provides a guide for researchers, academicians, practicing engineers, professionals, and scientists, as well as graduate and postgraduate students working in the area of various applications of RES, electric vehicles, and AC/DC Microgrid.

Contents

Editors and Contributors

About the Editors

Aashish Kumar Bohre is an Assistant Professor in the Department of Electrical Engineering, National Institute of Technology (NIT), Durgapur, West Bengal, India. He has received his B.E. degree (2009) in Electrical and Electronics Engineering from UIT-RGPV Bhopal, M.P., India; M.Tech. degree (2011) in Power System Engineering and a Ph.D. degree (2016) in Electrical Engineering from Maulana Azad National Institute of Technology (MANIT), Bhopal, India. He has over six years of teaching experience and has supervised several Ph.D., M.Tech. and B.Tech. theses in different research areas. His research interest includes distribution system operation and planning, sustainable development of power system, distributed generation, power system optimization, operation and planning of transmission system, FACTS planning, smart grid, renewable energy, power system stability, transient analysis, power system reliability, electric vehicle and soft computing techniques, etc. Dr. Bohre has published several research papers in reputed journals and conferences and also authored four chapters and one research book in his research area.

Pradyumn Chaturvedi received Ph.D., M.E., and B.E. degrees in 2010, 2001, and 1996 from National Institute of Technology Bhopal (India), Rajiv Gandhi Proudyogiki Viswavidyalaya Bhopal (India), and Barkatullah University Bhopal (India) respectively. He is presently working as Assistant Professor (Grade-I) in the Department of Electrical Engineering, Visvesvaraya National Institute of Technology, Nagpur, India. He has more than 21 years of experience in teaching and research with more than 90 research papers published in International reputed journals and refereed International/National Conferences. He has also co-authored one book "Modeling and Control of Power Electronics Converter System for Power Quality Improvements", Academic Press, USA. Dr. Chaturvedi completed three sponsored research projects in the area of multilevel inverter application for renewable energy harvesting and currently working on DST-SERB Core Research Grant sponsored project on "Solid State Transformer for Off-Grid Rural Electrification". Two International Collaborative Projects are ongoing with him; One research project under India-Taiwan International Joint Research Program to develop Solid State Transformer for Low Voltage Distribution Network; and International Consortium Project under India-EU Horizon 2020 Program on Integrated Local Energy Systems is also ongoing. He is a Senior Member of IEEE. Dr. Chaturvedi is Associate Editor, IEEE IES Industrial Electronics Technology News, ITeN; and Chairperson, Resonant and Soft Switching Converter Subcommittee, IEEE IES Technical Committee of Power Electronics. Dr. Chaturvedi is the Founder & Chairperson, IEEE Bombay Section Joint Chapter of PELS-IES. He has successfully organized IEEE IAS/PELS/IES Sponsored International Conference on Smart Technology for Power, Energy and Control (STPEC) in 2020 and 2021 as Program Chair. His research interests include solid-state transformer, fault-tolerant multilevel converters, medium voltage DC grid, reduced device count multilevel inverters, renewable energy harvesting, and improved power quality converters.

Mohan Lal Kolhe is a full professor in smart grid and renewable energy at the Faculty of Engineering and Science of the University of Agder (Norway). He is a leading renewable energy technologist with three decades of academic experience at the international level and previously held academic positions at the world's prestigious universities, e.g., University College London (UK/Australia), University of Dundee (UK); University of Jyvaskyla (Finland); Hydrogen Research Institute, QC (Canada); etc. In addition, he was a member of the Government of South Australia's first Renewable Energy Board (2009–2011) and worked on developing renewable energy policies. Due to his enormous academic contributions to sustainable energy systems, he has been offered the posts of Vice-Chancellor of Homi Bhabha State University Mumbai (Cluster University of Maharashtra Government, India), full professorship(s)/chair(s) in 'sustainable engineering technologies/systems' and 'smart grid' from the Teesside University (UK) and Norwegian University of Science and Technology (NTNU) respectively.

Professor Kolhe is an expert evaluator of many prestigious international research councils (e.g., European Commission: Erasmus+ Higher Education—International Capacity Building, Royal Society London (UK), Engineering and Physical Sciences Research Council (EPSRC UK), Cyprus Research Foundation, etc.). Professor Kolhe has successfully won competitive research funding from the prestigious research councils (e.g., Norwegian Research Council, EU, EPSRC, BBSRC, NRP, etc.) for his work on sustainable energy systems. His research works in energy systems have been recognized within the top 2% of scientists globally by Stanford University's 2020, 2021 matrices. He is an internationally recognized pioneer in his field, whose top 10 published works have an average of over 175 citations each.

Sri Niwas Singh obtained his M.Tech. and Ph.D. in Electrical Engineering from Indian Institute of Technology Kanpur, in 1989 and 1995, respectively. He is presently Professor (HAG) in the Department of Electrical Engineering, Indian Institute of Technology Kanpur, India. Prof. Singh was the Vice-Chancellor of Madan Mohan

Malviya University of Technology, Gorakhpur during April 2017 to July 2020. Dr. Singh received several awards including the Young Engineer Award 2000 of Indian National Academy of Engineering, Khosla Research Award of IIT Roorkee, and Young Engineer Award of CBIP New Delhi (India), 1996. Prof Singh is a recipient of the Humboldt Fellowship of Germany (2005, 2007) and Otto-Monsted Fellowship of Denmark (2009–10). He received INAE Outstanding Teacher Award 2016, IEEE R10 Outstanding Volunteer Award 2016, NPSC 2020 Academic Excellence Award and 2021 IEEE Industry Application Society (IAS) Outstanding Educator/Mentor Award. His research interests include power system restructuring, FACTS, power system optimization & control, power quality, wind power, etc. Prof Singh has published over 225 journal publications, 2 textbooks, 8 edited books and 30 book chapters to his credit. Dr. Singh received 2013 IEEE Educational Activities Board Meritorious Award in Continuing Education, which is very prestigious award, first time by a person of R10 region (Asia-Pacific). Currently, He is IEEE R10 Vice-Chair, Technical Activities for 2019–2020 and India Council Chair 2019–2020.

Contributors

Parimal Acharjee Department of Electrical Engineering, National Institute of Technology Durgapur, Durgapur, India

Bimrew Tamrat Admasu Department of Mechanical Engineering, Bahir Dar Institute of Technology, Bahir Dar University, Bahir Dar, Ethiopia

Mohammad Rafiq B. Agrewale Automotive Engineering, ARAI Academy, Pune, India

Irfan Ahmed Department of Electrical Engineering, National Institute of Technology Durgapur (NITDgp), A-Zone, Durgapur, West Bengal, India

Getu Alemayehu Bahir Dar Energy Center, Bahir Dar Institute of Technology, Bahir Dar University, Bahir, Ethiopia

Syed Raahat Ara Department of Applied Sciences and Humanities, Faculty of Engineering, Jamia Millia Islamia, New Delhi, India

Arnab Ari National Institute of Technology Durgapur, Durgapur, West Bengal, India

Aadesh Kumar Arya Department of Electrical Engineering, College of Engineering Roorkee, Roorkee Haridwar, India

Anoop Arya Electrical Engineering Department, MANIT, Bhopal, India

Delele Worku Ayele Department of Chemistry, Bahir Dar University, Bahir Dar, Ethiopia

Faouzi Bacha Department of Electrical Engineering, University of Tunis, ENSIT, Tunis, Tunisia

El Manaa Barhoumi Department of Electrical and Computer Engineering, College of Engineering, Dhofar University, Salalah, Oman

Ikram Ben Belgacem Laboratoire de Génie Mécanique, Ecole Nationale d'Ingénieurs de Monastir, Université de Monastir, Monastir, Tunisie

Tushar Kanti Bera Department of Electrical Engineering, National Institute of Technology Durgapur (NITDgp), A-Zone, Durgapur, West Bengal, India

Jitendra Singh Bhadoriya Department of Electrical Engineering, NIT Kurukshetra, Kurukshetra, Haryana, India

Nandola Maitrey Bharatbhai Electrical Department, NIT Kurukshetra, Kurukshetra, India

Yogesh Krishan Bhateshvar ARAI Academy, Pune, India

Aniruddha Bhattacharya Department of Electrical Engineering, National Institute of Technology Durgapur (NITDgp), A-Zone, Durgapur, West Bengal, India

Sandeep Bhongade SGSITS, Indore, India

Partha Sarathee Bhowmik Department of Electrical Engineering, National Institute of Technology Durgapur (NITDgp), A-Zone, Durgapur, West Bengal, India

Aashish Kumar Bohre Department of Electrical Engineering, National Institute of Technology Durgapur (NITDgp), A-Zone, Durgapur, West Bengal, India

Ajoy Kumar Chakraborty Department of Electrical Engineering, National Institute of Technology Agartala, Agartala, India

Pradyumn Chaturvedi Visvesvaraya National Institute of Technology (VNIT) Nagpur, Nagpur, Maharashtra, India

Jagdish Chaudhari Nagpur Institute of Technology, Nagpur, India

Sourav Das Department of Electrical Engineering, National Institute of Technology Durgapur, Durgapur, India

Devendra Deshmukh Department of Mechanical Engineering, Indian Institute of Technology, Indore, India

Slah Farhani Laboratory of Computer for Industrial Systems, INSAT, University of Carthage, Carthage, Tunisia

S. P. Gawande Yeshwantrao Chavan College of Engineering, Nagpur, India

Suresh Kumar Gawre Department of Electrical Engineering, MANIT, Bhopal, India

Harsh Goud Indian Institute of Information Technology, Nagpur, Maharashtra, India

Govind Rai Goyal College of Engineering Roorkee, Roorkee, India

Atma Ram Gupta Department of Electrical Engineering, National Institute of Technology Kurukshetra, Kurukshetra, Haryana, India

Nigus Gabbiye Habtu Department of Chemical Engineering, Bahir Dar Institute of Technology, Bahir Dar University, Bahir Dar, Ethiopia

Sanjay Jain Electrical and Electronics Engineering Department, R.K.D.F. University, Bhopal, Madhya Pradesh, India

Siddharth Jain School of Electrical Engineering, Vellore Institute of Technology, Vellore, India

Prasad B. Joshi Yeshwantrao Chavan College of Engineering, Nagpur, India

Dong Wong Jung Department of Mechanical Engineering, Jeju National University, Jeju City, South Korea

Shubham V. Kadam Lokmanya Tilak College of Engineering, Navi Mumbai, India

Shilpa R. Kalambe Electrical Engineering Department, Dr. Babasaheb Ambedkar College of Engineering and Research, Nagpur, Maharashtra, India

Bhojraj N. Kale Mechanical Engineering Department, Dr. Babasaheb Ambedkar College of Engineering and Research, Nagpur, Maharashtra, India

C. Pearline Kamalini Saranathan College of Engineering, Trichy, Tamilnadu, India

Mohan Lal Kolhe University of Agder, Kristiansand, Norway

Mahesh T. Kolte Research Centre, Pimpri Chinchwad College of Engineering, Savitribai Phule Pune University, Pune, India

Aboli Kulkarni School of Electrical Engineering, Vellore Institute of Technology, Vellore, India

Prakash S. Kulkarni Visvesvaraya National Institute of Technology, Nagpur, India

Amit Kumar Department of Electrical Engineering, National Institute of Technology, Kurukshetra, Haryana, India

Nitin Kumar K. K. Wagh Institute of Engineering Education and Research, Nashik, Maharashtra, India

Rajeev Kumar School of Engineering, Indian Institute of Technology Mandi, Kamand, Himachal Pradesh, India

Shalendra Kumar Department of Mechanical Engineering, National Institute of Technology, Jamshedpur, Jharkhand, India

Satya Prasad Maddula GITAM (Deemed to be) University, Hyderabad, India

Saee P. Mane Lokmanya Tilak College of Engineering, Navi Mumbai, India

M. Marimuthu Saranathan College of Engineering, Trichy, Tamilnadu, India

Vijayakumar Mathaiyan Department of Mechanical Engineering, Jeju National University, Jeju City, South Korea

Sudhansu Mehera Department of Mechanical Engineering, Government College of Engineering, Keonjhar, Odisha, India

Shweta Mehroliya Electrical Engineering Department, MANIT, Bhopal, India

Arghya Mitra Visvesvaraya National Institute of Technology, Nagpur (M.S), India

Ravindra M. Moharil Yeshwantrao Chavan College of Engineering, Nagpur, India

Prateek Mundra Electrical Engineering Department, MANIT, Bhopal, India

Dipanshu Naware Visvesvaraya National Institute of Technology, Nagpur (M.S), India

Yogesh Kumar Nayak Electrical Engineering Department, Government College of Engineering, Keonjhar, Odisha, India

Akhil Nigam Chandigarh University, Mohali, Punjab, India

Karuna Nikum Department of Engineering Sciences and Humanities, TCET, Mumbai, India

Vasishta Bhargava Nukala Sreyas Institute of Engineering and Technology, Hyderabad, India

Paul C. Okonkwo Department of Mechanical and Mechatronics Engineering, College of Engineering, Dhofar University, Salalah, Oman

Chinmaya Prasad Padhy GITAM (Deemed to be) University, Hyderabad, India

Arnab Pal Department of Electrical Engineering, National Institute of Technology Agartala, Agartala, India

Rahul Pandey Shri Shankaracharya Group of Institution, Bhilai C.G., India

Santanu Paul Department of Electrical Engineering, Indian Institute of Science Bangalore, Bangalore, India

Aarti S. Pawar Research Centre, Pimpri Chinchwad College of Engineering, Savitribai Phule Pune University, Pune, India

Rajagopal Peesapati EEE Department, Raghu Engineering College, Vizag, India

Rajesh M. Pindoriya School of Computing and Electrical Engineering, Indian Institute of Technology Mandi, Kamand, Himachal Pradesh, India

Miska Prasad EEE Department, ACE Engineering College, Hyderabad, India

Ramendra Kumar Rai National Institute of Technology, Durgapur, India

Vijayanandh Raja Kumaraguru College of Technology, Coimbatore, India

Anup Kumar Rajak Department of Mechanical Engineering, Rewa Engineering College, Rewa, Madhya Pradesh, India

Bharat S. Rajpurohit School of Computing and Electrical Engineering, Indian Institute of Technology Mandi, Kamand, Himachal Pradesh, India

Md Tuhin Rana National Institute of Technology Durgapur, Durgapur, West Bengal, India

Zakir Hussain Rather Department of Energy Sc. and Engineering, Indian Institute of Technology Bombay, Mumbai, India

Kiran H. Raut G H Raisoni Institute of Engineering and Technology, Nagpur, India

Suchita Roy Shri Shankaracharya Group of Institution, Bhilai C.G., India

Gitimayee Sahu Lokmanya Tilak College of Engineering, Navi Mumbai, India

Rahul Samala Indian Institute of Technology Madras, Chennai, India

Yashwant Sawle Madhav Institute of Technology & Science, Gwalior, India

Nitin Kumar Saxena Department of Electrical and Electronics Engineering, KIET Group of Institutions, Ghaziabad, India

M. Senthilkumar Kumaraguru College of Technology, Coimbatore, India

Ashwani Kumar Sharma National Institute of Technology, Kurukshetra, India

Kamal Kant Sharma Chandigarh University, Mohali, Punjab, India

R. Shenbagalakshmi SKN Sinhgad Institute of Technology and Science, Lonavala, Maharashtra, India

Asha Shendge G H Raisoni Institute of Engineering and Technology, Pune, India

Pranay S. Shete Yeshwantrao Chavan College of Engineering, Nagpur, India

Rajesh Ranjan Shukla Electrical Engineering Department, Government College of Engineering, Keonjhar, Odisha, India

Rajkumar Singh Department of Mechanical Engineering, Rewa Engineering College, Rewa, Madhya Pradesh, India

Sri Niwas Singh Indian Institute of Technology (IIT) Kanpur, Kalyanpur, Kanpur, India

Sivaranjani Sivalingam Department of Mechanical Engineering, Jeju National University, Jeju City, South Korea

S. Srinivasamoorthy Department of Mechanical Engineering, Jeju National University, Jeju City, South Korea

Pankaj Swarnkar Maulana Azad National Institute of Technology, Bhopal, Madhya Pradesh, India

Biniyam Zemene Taye Bahir Dar Energy Center, Bahir Dar Institute of Technology, Bahir Dar University, Bahir, Ethiopia;
Faculty of Electrical and Computer Engineering, Bahir Dar Institute of Technology, Bahir Dar University, Bahir Dar, Ethiopia

Mohan P. Thakre K. K. Wagh Institute of Engineering Education and Research, Nashik, Maharashtra, India

Rishi K. Thakur School of Engineering, Indian Institute of Technology Mandi, Kamand, Himachal Pradesh, India

Ram Babu Thogaru Visvesvaraya National Institute of Technology, Nagpur (M.S), India

Shreastha Varun SGSITS, Indore, India

Swamy Naidu Neigapula Venkata National Institute of Technology, Raipur, India

R. Venugopal Saranathan College of Engineering, Trichy, Tamilnadu, India

S. Vignesh Automotive Engineering, ARAI Academy, Pune, India

S. Vijayalakshmi Saranathan College of Engineering, Trichy, Tamilnadu, India

K. C. Vora Automotive Engineering, ARAI Academy, Pune, India

Abhay Wagh Directorate of Technical Education, Mumbai, India

A. S. Werulkar Department of Electronics and Telecommunication Engineering, SVPCET, Nagpur, India

Ababay Ketema Worku Bahir Dar Energy Center, Bahir Dar Institute of Technology, Bahir Dar University, Bahir, Ethiopia

Temesgen Atnafu Yemata Department of Chemical Engineering, Bahir Dar Institute of Technology, Bahir Dar University, Bahir Dar, Ethiopia

Mohamed Zellagui Department of Electrical Engineering, University of Batna, Batna, Algeria

Chapter 1
Introduction

Aashish Kumar Bohre, Pradyumn Chaturvedi, Mohan Lal Kolhe, and Sri Niwas Singh

Abstract This chapter provides an overview of renewable energy, electric vehicles, microgrids or smart grids and their various applications. In the modern era, utilization of renewable energy sources is growing fast in different combinations of hybrid systems due to enormous availability and various technoeconomic advantages. A hybrid renewable energy system integrates different non-renewable and renewable sources along with storage systems to maintain system reliability and resiliency. Electric vehicles (EVs) are more efficient in energy saving, emission reduction and environmental protection than fuel-operated vehicles. As a result, EVs are becoming important with different applications in the transportation sector to reduce global warming. The adoption of EVs required sufficient charging infrastructure globally for the sustainable and regular operation of it's facilities. Therefore, the analysis of the impacts of EVs on the distribution grid/microgrid/smart grid is necessary for a reliable and economic operation of the system. The bidirectional electrical power flows with two-way digital control and communication capabilities have poised the energy producers and utilities to restructure the conventional power system into a robust smart distribution grid. These new functionalities and applications provide a pathway for clean energy technology. Further details in these areas are also presented here to get quick knowledge about advantages, utilization, and applications in these mentioned eras.

A. K. Bohre (✉)
National Institute of Technology Durgapur, Durgapur, India
e-mail: aashishkumar.bohre@ee.nitdgp.ac.in

P. Chaturvedi
Visvesvaraya National Institute of Technology (VNIT), Nagpur, India
e-mail: pradyumn.c@eee.vnit.ac.in

M. L. Kolhe
University of Agder, Kristiansand, Norway
e-mail: mohan.l.kolhe@uia.no

S. N. Singh
Indian Institute of Technology (IIT) Kanpur, Kanpur, India

A. K. Bohre et al. (eds.), *Planning of Hybrid Renewable Energy Systems, Electric Vehicles and Microgrid*, Energy Systems in Electrical Engineering,
https://doi.org/10.1007/978-981-19-0979-5_1

Keywords Renewable Energy · Hybrid Renewable Energy System (HRES) · Electric Vehicles (EVs) · Distribution grid/microgrid/smart grid

Nomenclature

RES	Renewable energy system
EVs	Electric vehicles
HRES	Hybrid renewable energy system
V2G	Vehicle to grid
V2H	Vehicle to home
V2B	Vehicle to building
ESS	Energy storage system
DSM	Demand-side management
SCADA	Supervisory control and data acquisition
AMI	Advanced metering infrastructure

1.1 Hybrid Renewable Energy System

The alternate renewable energy resource becomes a necessity due to depleting nature of conventional sources of energy and the continuous growth of energy demand. Also, the use of fossil fuels has contributed to climate change and global warming over the past decades and it has become essential to look towards environment-friendly energy sources. Renewable sources like solar energy, wind energy, biomass are seen as promising alternatives to conventional sources (Oladigbolu et al. 2020). However, they are not without drawbacks. In solar energy systems, generation drastically falls on cloudy days and at night; windmill output is low at low speeds and prone to breakdown at high speeds; biomass plant performance drops at low temperature, and so on. Compared to the conventional sources, their supply is irregular, uncertain and has low energy densities, making them less reliable. A hybrid renewable energy system (HRES) attempts to overcome these drawbacks by integrating more than one power generation system into a single unit. It usually combines renewable and conventional sources along with storage units to provide greener and more reliable energy (Sawle et al. 2021; Rezk et al. 2020). The general block diagram is shown in Fig. 1.1. Different storage systems can be used to ensure uninterrupted power supply when the power obtained from the sources is not enough to feed the load. Recent studies focus on the integration of multiple renewable energy systems such as solar, wind, biomass, micro-hydro and tidal and their various permutations and topologies, depending on the availability and feasibility (Khan et al. 2017; Sawle et al. 2018). Many research on hybrid wind and PV systems have been carried out in terms of HRES modelling, size optimization, reliability analysis, environmental and

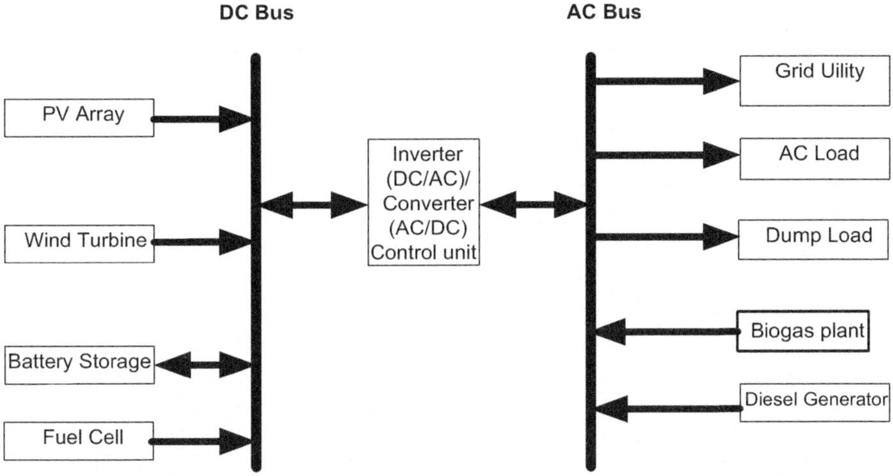

Fig. 1.1 Conceptual block diagram to develop HRES System

economic assessment. With the advent of new technologies, modern energy storage systems are cost-effective and more efficient. When integrated with renewable energy sources, an energy storage system plays a vital role as it absorbs energy during high generation and acts as a source during high demand. The storage device thus can be used to reduce the fluctuations in power generated by the renewable sources that are being exchanged with the grid. To improve the resiliency of the system/grid, it can also be used as an emergency backup to satisfy the critical loads in the presence of any disturbance. An array of energy storage options is available in the market, such as super-capacitors, superconducting magnetic energy storage, compressed air, pumped hydro storage, flywheels and rechargeable batteries.

The energy generation, control and storage elements used in HRES vary from application to application depending mainly upon availability, demand and affordability. After the selection of components, firstly, mathematical modelling for each component needs to be done, for further optimization of the study system. HRES are either grid-connected or off-grid. While grid-connected systems have the advantage of selling surplus power and drawing power when demand is high, off-grid systems are suitable for isolated remote locations (Khan et al. 2017; Sawle et al. 2018; Singh and Bansal 2018).

1.2 Electric Vehicles (EVs)

The electric vehicles are the future, not because they are better for the environment (which they are if you drive the car for any reasonable length of time) but because they are just better vehicles. They have better acceleration and torque, easy to handle, have more efficiency than the other vehicles powered by fuel, quieter, there is no smell, the investment is one time and it is very cheaper to run. As electric vehicle's usage is growing day by day the EVs in future becomes a reality (Silva et al. 2019). The time-taking method of charging an EV becomes a major problem to accept the electronic revolution of the automobile industry. Reducing the pollution in the environment is the most important topic of the present times. Pollution mainly arises from the burning of fossil fuel which emits CO_2 and other harmful gases in the environment. The transportation sector is introducing technologically advanced Electric Vehicles (EVs) which are ecofriendly. Developments in bidirectional capabilities of EVs batteries, charging technologies and communication infrastructure give rise to concepts such as Vehicle to Grid (V2G), Vehicle to Home (V2H), Vehicle to Building (V2B), etc. EVs charging will put significant load demand on the grid as well as uneven charging will create a disturbance in the grid (Rizvi et al. 2018). In order to reduce the burden on the grid, EVs can be charged by renewable energy sources. Since solar is the most widely available renewable energy source, it is considered for charging the EVs. Solar charging stations can be installed at homes, offices, parking lots, public areas and isolated areas. EV integration to the grid is the attraction of many researchers and engineers around the world due to the rapidly growing numbers of EVs in the global market. The use of EVs in the grid will play an important role in the future smart grid technology. The EVs will enable the participation of consumers in the energy sector (Moghaddam et al. 2018). Both utilities and the consumers will be benefited from the integration of vehicles and grids. The different types of charging methods are used for charging the EVs such as AC charging (slow), DC charging (fast), induction charging, battery swapping and smart charging.

1.3 Microgrids and Smart Grids

Microgrids and Smart grids are emerging as the latest trending aspect in power industries. The smart grid integrates the technology dealing with Information and Communication in almost all aspects of power systems starting from electricity generation till consumption in order to improve the reliability of energy consumption and service, minimize the environmental impact, enable active participation of the consumers, new products and markets, improves the efficiency leading to more safe and reliable energy production and distribution. The other benefits include reducing carbon emission, supporting the increased use of electric vehicles, and creating wider employment opportunities (Dileep 2020; Srinithi et al. 2021). Smart grids can be seen as a combination of microgrids and minigrids among which microgrid plays a major role in accomplishing authentic and more secure energy supply for the retail load as well as distributed generation. On the other hand, microgrid can be seen as a decentralized energy system comprising distributed energy sources with demand management, storage and generations with loads that are capable of operating either in parallel or independently.

The smart grid is an electrical network that integrates systems cost-efficiently. It connects all users, generators, consumers to ensure economically efficient, sustainable power systems with fewer losses, good quality and security of supply. The smart grid allows the integration of renewable technology, like solar and wind power production and also the adoption of plug-in electric vehicles in electrical grid. However, the implementation issues in the smart grid include interfacing renewable energy source (RES), implementing energy storage system (ESS), connect or disconnect the grid operations, integrating decentralized renewable power in distribution networks and developing demand-side management (DSM), when implementing smarter electricity networks. Regarding trends in development & implementation of smart grid, RE forecasting, RE scheduling, integrating supervisory control and data acquisition (SCADA) for RE generation and integrating RES through a power grid. Besides, installing advanced metering infrastructure (AMI), with price incentives for DSM, enlarging balancing areas, testing of different battery technologies, enhancing situational awareness, better visualization and stable operation of the grid are provided in this chapter (Madureira et al. 2011; Palizban et al. 2014; Ferraro et al. 2018). The advanced distribution management system (ADMS) approach has also addressed training, planning, optimizing, operating, analyzing and monitoring the distribution network. ADMS includes SCADA, DMS, OMS and DSM to analyze and manage DER. Besides, HEMT-based technology meets the needs for power distribution grid and end-user utilization. Therefore, new power electronic devices based on wide bandgap semiconductor devices such as SiC and GaN have described improving power density and power efficiency. Despite the benefits provided by smart grids, there are few technical limits in the renewable generation on the smart grid. Control, management and stability in power systems introduce several interesting issues for realizing the smart grid in the existing power network. Therefore, this chapter analyses the major issues associated with the integration of MW scale renewable energy into the weak distribution network and its solution methodologies. Moreover, high interfacing penetration of distributed energy resources (DER) with energy storage and microgrid control systems is an essential feature of future distribution grids for optimal utilization and management of DER. This feature helps distributed system operators follow the right path for transforming their classical grids into smart grids. High RES penetration to the existing power network can cause reliability and interoperability issues due to their intermittent nature resulting in uncertainty in the operation and control of the power system. In such an instance, ESS provides ancillary services in the power generation source to balance the mismatch between supply and demand. For this reason, ESS is a potential solution to support RES penetration for the smart grid. Concerning large-scale energy storage deployment in the power grid, a cost-effective storage system along with design aspects, operation and control of ESS in Distribution System Operator (DSO) for the smart grid are presented to smooth renewable generation output (Srinithi et al. 2021; Madureira et al. 2011; Palizban et al. 2014; Ferraro et al. 2018). Moreover, the impacts to large penetration of ESS include voltage regulation, active frequency damping control, wildfire impacts and transmission support are addressed to enhance the flexibility of distributed generation. Furthermore, frequency stability and control is a challenging problem in the design and operation of interconnected power networks. This chapter explores a conceptual block diagram of inverter interfaced control such as hierarchical control, virtual inertia control for maintaining power system frequency within the specified operating range in a power system. Synchronverter is another emerging technique discussed to overcome the deficit of inertia in the modern power system using proper control strategy. For enhancing the power system resilience, the end–end security life cycle and architecture of the resilient attack system for Wide-Area Monitoring, Protection, & Control (WAMPAC) are further presented. The potential solution is to form sustainable microgrids, which include RES and ESS. Sustainable microgrids with an advanced control strategy are also provided to enhance power system resiliency. With an increasing number of DG units integrated at both transmission and distribution systems, they have a crucial role in maintaining energy balance and power continuity as part of power quality (PQ) in future smart grids. Thus, the various power quality issues at the consumption as well as supply side, conventional and modern power quality control strategies are provided. Smart grid technologies such as volt/var management system (VVM), supervisory control and data acquisition (SCADA), power quality analyzer (PQA), geographic information system (GIS), distribution automation (DA), advanced metering infrastructure (AMI), and smart inverters are further presented to manage and improve power quality (Madureira et al. 2011; Palizban et al. 2014; Ferraro et al. 2018; John and Lam 2017). Simultaneously, solution methodologies with various control approaches are compared in terms of their respective merits and outcomes while providing optimal socio-economic benefits.

1.4 Conclusion

Renewable energy sources are the future of this world and converters are the heart of this system. We are entering a new era of renewable sources of energy. Our main supply will depend on a renewable resource; thus we have to prepare for this era and fix the quality issues. Wind energy, geothermal energy, hydro energy, tidal and biomass generate energy and battery used to store the power thus it needs to convert into AC/DC or vice-versa through a converter. In conversion and integration process of different devices, the system face some losses as well as the power qualities issues like voltage sag, voltage dip (because of faults). Nowadays, electric vehicles are growing very rapidly and it has become necessary to better the world today. The electric vehicle provides a pollution-free environment. The development of electric vehicles is not much satisfying with the construction of charging facilities. There are plenty of methods adopted in order to improvise the electric vehicle performance but there is only little concern about the charging infrastructure and charging methods. In the smart grid, the system deals with the recent developments, the advancements in technology, the threats to security and approaches for its protection. A major study to be made is about assessing the costs of protecting the grid against the threat of solar storms that send geomagnetic pulses slamming into vulnerable transformers and other critical grid equipment.

References

Dileep G (2020) A survey on smart grid technologies and applications. Renew Energy 1(146):2589–2625

Ferraro P, Crisostomi E, Shorten R, Milano F (2018) Stochastic frequency control of grid-connected microgrids. IEEE Trans Power Syst 33(5):5704–5713

John T, Lam SP (2017) Voltage and frequency control during microgrid islanding in a multi-area multi-microgrid system. IET Gener Transm Distrib 11(6):1502–1512

Khan MJ, Yadav AK, Mathew L (2017) Techno economic feasibility analysis of different combinations of PV-Wind-Diesel-Battery hybrid system for telecommunication applications in different cities of Punjab, India. Renew Sustain Energy Rev 76:577–607

Madureira AG, Pereira JC, Gil NJ, Lopes JP, Korres GN, Hatziargyriou ND (2011) Advanced control and management functionalities for multi-microgrids. Eur Trans Electr Power. 21(2):1159–1177

Moghaddam Z, Ahmad I, Habibi D, Phung QV (2018) Smart charging strategy for electric vehicle charging stations. IEEE Trans Transp Electrification 4(1):76–88. https://doi.org/10.1109/TTE.2017.2753403

Oladigbolu JO, Ramli MA, Al-Turki YA (2020) Feasibility study and comparative analysis of hybrid renewable power system for off-grid rural electrification in a typical remote village located in Nigeria. IEEE Access 8:171643–171663

Palizban O, Kauhaniemi K, Guerrero JM (2014) Microgrids in active network management—Part I: hierarchical control, energy storage, virtual power plants, and market participation. Renew Sustain Energy Rev 1(36):428–439

Rezk H, Al-Dhaifallah M, Hassan YB, Ziedan HA (2020) Optimization and energy management of hybrid photovoltaic-diesel-battery system to pump and desalinate water at isolated regions. IEEE Access 8:102512–102529

Rizvi SAA, Xin A, Masood A, Iqbal S, Ullah Jan M, Rehman H (2018) Electric vehicles and their impacts on integration into power grid: a review. In: Presented at the 2nd IEEE conference on energy internet and energy system integration (EI2), Beijing, China

Sawle Y, Gupta SC, Bohre AK (2018) Socio-techno-economic design of hybrid renewable energy system using optimization techniques. Renew Energy 119:459–472

Sawle Y, Jain S, Babu S, Nair AR, Khan B (2021) Prefeasibility economic and sensitivity assessment of hybrid renewable energy system. IEEE Access 9:28260–28271

Silva FC, Ahmed M, Martínez JM, Kim YC (2019) Design and implementation of a blockchain-based energy trading platform for electric vehicles in smart campus parking lots. Energies 12(24):4814

Singh R, Bansal RC (2018) Optimization of an autonomous hybrid renewable energy system using reformed electric system cascade analysis. IEEE Trans Industr Inf 15(1):399–409

Srinithi S, Vijayalakshmi S, Marimuthu M, Parathagan B (2021) Symmetric multilevel inverter using DC-DC zeta converter. In: IEEE 2021 2nd international conference for emerging technology (INCET) 2021 May 21, pp 1–5

Chapter 2
Planning Methodologies of Hybrid Energy System

Akhil Nigam and Kamal Kant Sharma

Abstract A hybrid system consists of conventional and nonconventional energy systems for the achievement of reliable operation to keep the balance between energy supply and load demand. Various methods have been employed for planning and sizing of the hybrid energy system to get optimal location. Due to weather conditions, some renewable energy sources such as solar and wind energy may be unable to provide continuous supply. In addition, stability is an important issue. This may be voltage stability, frequency stability, and rotor angle stability. Different optimization techniques have been developed for optimizing the parameters of the hybrid energy system. This manuscript deals with a review of different hybrid energy systems with optimization techniques to achieve their best optimal location and sizing. Some planning methods have been reviewed with in this manuscript and focused on the development of a new hybrid energy system with advanced techniques.

Keywords Renewable energy system · Solar system · Wind power · Particle swarm optimization · Genetic algorithm · Artificial bee colony algorithm

Abbreviations

PV	Photovoltaic
NPC	Net percent cost
RES	Renewable energy sources
MPPT	Maximum power point technique
PSO	Particle swarm optimization
HVAC	High voltage alternating current
ACO	Ant colony optimization
HVDC	High voltage direct current
GA	Genetic algorithm

A. Nigam · K. K. Sharma (✉)
Chandigarh University, Mohali, Punjab, India
e-mail: kamalkant.ee@cumail.in

A. K. Bohre et al. (eds.), *Planning of Hybrid Renewable Energy Systems, Electric Vehicles and Microgrid*, Energy Systems in Electrical Engineering,
https://doi.org/10.1007/978-981-19-0979-5_2

DCMG Direct current microgrid

2.1 Introduction

In the last decades, renewable energy source has increased unprecedented growth with globalized manner by using different energy sources such as solar PV cell, wind energy, hydropower, geothermal, biomass, etc. As per U.S. energy report, the energy consumption falls by 4% in the previous year of 2021 and will increase by 1.4% in year 2021. The report says that India will be the largest contributor in the field of renewable energy production by 2021 and overall total production throughout the world will be a double contributor as compared to 2020. These variable energy sources are different in various aspects than conventional energy sources. These renewable energy sources are required to produce electricity with increasing energy demand and with suitable efficieny. In comparison to conventional energy sources, various major aspects such as the small size of generators, location constrained, non-synchronous type generators, low-run cost fall under the category of renewable energy sources. Hence, these characteristics produce challenges in existing power systems. These challenges may include capacity of transmission grid and sufficiency of electricity generation. Renewable energy sources have taken considerable attraction throughout the world. All these things come due to insufficient amounts of fuels and discontinuity of supply. By using fossil fuels, energy consumption is low and overall efficiency is low due to large generation of gas emissions. In year 2019, overall consumption of energy by using fossil fuels was 82% by using 12 billion tons of fossil fuels.

Apart from this, numerous renewable energy sources have been implemented in different countries. Bioenergy production is carried out by using biomass to be used to generate renewable electricity and thermal electricity. All energy production goes under different processes like pyrolysis, hydrothermal liquefaction and gasification (Ali et al. 2020). In addition to energy production having intermittent availability of alternate energy sources such as wind and solar photovoltaic cells, sometimes we also need energy storage devices. The amalgamation of alternate energy sources depends upon weather conditions which impact on the reliability, stability and quality. After all few nonconventional energy generators have been coupled with transmission line systems with keeping superiority of similar generators interconnected with distributed generation systems. Through new regulation act, it requires renewable energy sources to perform similar to conventional energy sources and play a vital role in enhancing voltage and frequency stability. There are numerous renewable energy sources as classified below (Fig. 2.1).

From the diagram, renewable or nonconventional energy sources are such as wind power, biomass, solar energy, hydropower, and geothermal. Further, these sources have been classified into different types depending upon different modes of operation. For example, wind energy performs on ON-Shore and OFF-Shore mode and solar

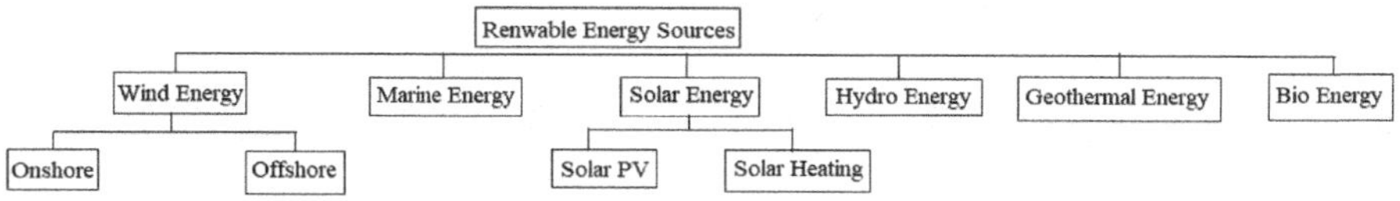

Fig. 2.1 Classification of renewable energy sources

energy can be performed as solar PV and solar thermal. Renewable energy sources have been considered as clean energy and critically important due to ecofriendly environmental parameters. These renewable energy sources cover domestic applications and provide energy service at the customer end with zero gas emission problems. Most of the researchers have deployed results in renewable energy sources and have focused on policy, financial challenges, and technical issues (Quazi et al. 2019).

Hence, the overall net capacity of renewable energy sources will be kept by 10% in years 2020 and 2021. The capacity of renewable energy can be increased by a new installation of energy sources with 85% net of all newly installed limits by the year 2017 from the energy report. Fossil fuel contributes around 74% for the production of electricity in the year 2017–18 and renewable energy sources have a contribution of 27% in electricity production (Report and IEA 2020).

This chapter includes the following sections:

- Section 2.1 describes introduction and classification of renewable energy sources.
- Section 2.2 describes about hybrid energy system.
- Section 2.3 describes the types of hybrid energy systems.
- Section 2.4 describes various technologies for hybrid energy system.
- Section 2.5 describes the planning methodologies of hybrid energy system.
- Section 2.6 describes different optimization techniques for obtaining an optimal location of energy system with the development of hybrid energy system using new techniques.

2.2 Hybrid Energy System

As per increasing energy demands, there is a need for environmental protection and secure electricity, then a new energy system has been introduced named as hybrid energy system. Hybrid energy system is an amalgamation of conventional and nonconventional energy sources through which electricity can be transferred to the customer side at an efficient cost. These sources perform well under all environmental conditions keeping all climate parameters as per requirement in establishing a hybrid energy system. Utilization of nonconventional energy sources reduce the cost of fossil fuels and draw more attraction. The energy report represents the use of RES and power production around 15% in the year 2020 (Guo et al. 2018). Sometimes, renewable energy sources may have a problem of low energy density and the energy production is low due to the use of only one source. To overcome all these problems, amalgamation of nonconventional energy sources has been introduced. Most of the research work has been done in the area of hybrid systems in distant areas applications. The installation of hybrid systems has been executed in different countries for the last few years. Overall hybrid systems have curtailed the total maintenance cost of standalone energy in different conditions.

Extensively, hybrid system consists of a standalone system which incorporates the features of traditional and alternate energy sources with having cells, controller, and power conditioning unit. The controller and power conditioning unit are used to maintain the power quality of the grid. Research work has been centralized on the performance of the system and the evolution of power converters (Nema et al. 2009). Hybridization of energy sources can increase reliability, however, there is an essential aspect of designing for such systems for achieving efficient performance. The hybrid energy system performs effectively by supplying electric load in case of OFF grid applications. These systems have an advantage in designing by taking one or more than one nonconventional energy sources with or without using traditional energy sources improve the system performance and provide reliability. However, there are many remote areas in which they are performing by using ordinary energy sources such as diesel, gas, etc., based generators to fulfill energy demand at the customer side.

For the implementation of alternate energy sources economically, it is required for designing an optimal model which is positioned on forecasting alternate energy sources by employing convenient methods. The modeling of hybrid energy system is a mosaic one which has the need of mathematical models of the alternate energy systems and then optimized. Many of the renewable energy sources have been characterized which involve a variation of output power. In addition to this, energy storage is also used to ensure the durability of power supply to load through amending power reliability (Anounea et al. 2018). As per government report, total renewable energy production is about to be 450 GW up to year 2030.

There are various advantages by using a hybrid energy system:

- Better utilization of renewable energy sources
- Low generation variability
- Better utilization of land for renewable energy sources
- Conclusive generation profile
- Enhancement of overall efficiency of system.

There are various schemes proposed by the Indian government such as installation of solar and other renewable energy sources with a capacity of 25,750 MW by the year 2022 and national solar mission up to 2022 with a production of 20GW. Many researchers have focused on developing hybrid energy systems by utilizing renewable energy sources into the grid. Hence through the contribution of energy storage systems, power-sharing regulation is required for the accomplishment of characteristics of energy sources (Babu et al. 2020).

2.3 Types of Hybrid System

There are discrete types of hybrid systems which consist numerous conventional and nonconventional energy sources. Depending upon environmental parameters, hybrid energy system performs under all considerations and provides reliable operation.

2.3.1 Solar Gas Turbine

A typical solar gas turbine structure consists of several components such as solar collector, solar receiver, combustion chamber, expander, and prime mover. The design of this hybrid system consisting of a solar tower associated with a gas turbine provides consistent energy with varying solar irradiance. The solar tower transforms solar energy into electricity. This system contains two types which focus sunlight into a single point having collectors such as heliostats and dish type. There is a system-integrated heat storage unit which performs in coordination with the steam line. Initially, storage chamber is to be considered empty when the size reaches up to 110 percent compressor outlet temperature (Kulor et al. 2021). This hybrid system has the advantage of solar panel fields and towers with combination of energy storage systems (Grangea et al. 2014).

2.3.2 Fuel Cell Gas Turbine

In this system, turbine avails fuel cells, for example, solid oxide fuel cell exhaust to generate compressed power. The gas turbine is used to extract thermal energy with high temperature for driving compressor in addition to provide pressurized air to the fuel cell. This hybrid system provides efficiency of fuel cell and power density for reliable operation. This system captures and oxidizes anode off-gas to excursion turbine machine and generate electricity. This method provides the requirement of airflow and heat to drive the turbine for the production of electricity (McLarty et al. 2014).

2.3.3 Solar Biomass Energy

In a solar biomass energy hybrid system, there are various components such as heliostats, parabolic troughs, solar PV panels, and hyperboloid reflectors. First, the condensed water is to be preheated by using bled steam from a turbine at a pressure of 0.5 MPa. After that, hot water flows into the heat exchanger and acquires heat energy from the solar PTC system when sunlight is available. Then runs a boiler of biomass where it is used to generate superheat at a pressure of 6 MPa and temperature of 772 K (Zhang et al. 2019). Then, superheated steam is used to build up electricity in turbine where sunlight is available. During the daytime, biomass boiler runs with low capacity and at nighttime, it runs with full load capacity. So due to this hybrid energy system, it is not only supplying but also generating clean energy and having the capacity to avoid unnecessary waste which can further be used for agricultural purposes. By using two-stage gasifier solar biomass, it will be better efficient utilization of energy produced (Bai et al. 2016).

2.3.4 Solar Geothermal Energy

This hybrid system is very clean and energy-saving to produce electricity by using solar or geothermal energy systems. In many countries, commercial solar and geothermal energy system has been established. The hybridization of solar and geothermal system consists of geothermal with organic rankine cycle, solar heating comprises of superheater, solar collector, and solar pump. The running fluid ORC is burnt by an evaporator, then superheated and moved to the turbine for electricity production. The exhausted steam from the turbine is condensed by using a condenser and run into a preheater. An evaporator completes the entire process for the generation of electricity. The overall cost of such type of hybrid system may be 1.5–3.5% than normal cost.

2.3.5 Solar Wind Energy

By using solar wind hybrid system in which wind energy is transformed into electricity by using wind turbine and for solar energy system, sunlight is transformed into electrical energy with the help of photovoltaic system. Hybrid solar-wind energy system combines solar PV array, wind turbine, controller, inverter, battery, generator, and another appliance. The charge controller and inverter are used to adjust output power as per load demand. This hybridization system performs for ON grid and OFF grid, both modes with their intermittent nature. Grid-connected systems are more economical rather than OFF grid mode connected system, since they do not require a battery bank. The suitable optimization method can be preferred for the measurement of net present value that helps to reduce future cost (González et al. 2015).

2.4 Technologies/Tools for Hybrid Energy System

There are various emerging technologies/tools for hybrid energy system operation which are highlighted on proper designing with the integration of multiple energy sources. It is also required for reducing the overall cost of system designing hence it can be met with using different software tools. Below, the description of different software applications for hybrid energy system have been described:

RET Screen: This simulation tool is excel-based clean management tool which incorporates in determining economical and technical viability. This is the most used simulation tool for reliability and feasibility studies in renewable energy sources and is free to download. It uses C language and was released in year 1988. It is attainable with higher number than 30 languages and have variants such as RET Screen 4 and RET Screen plus. RET screen plus includes monitoring the system performance with the integration of solar radiation data.

HOMER: HOMER tool is user-friendly and developed by National Renewable Energy Laboratory (NREL), USA. This application tool is used for designing and evaluating the performance for ON grid and OFF grid systems. In this simulation, virtual C++ language is required. It also represents the tables and charts for the comparison of different configuration and analyzes their performance by Net Percent Cost (NPC).

HYBRID2: It was developed by Renewable Energy Research Laboratory (RESL), USA in the year 1996. This tool can simulate both AC & DC distribution systems, renewable energy sources, energy storage, and converters. HYBRID 1 simulation tool was developed in the year 1994. It comprises of four major segments such as Simulation Module, Graphical User Interface, Graphical results Interface, and Economics Module which permit users to design projects and maintain all records.

HySim: It prefers to hydrological simulation tool and was developed by Sandia National Laboratory (SNL), 1987. It employs amalgamation system that combines solar photovoltaic, wind energy turbine, generator, and energy storages for a standalone system. This tool is having an economic analysis consisting of cost of fuel, energy, operational. and maintenance cost.

HybSim: This simulator tool was developed by Sandia National laboratories for the purpose of economic analysis of remote areas where energy demand is fulfilled by using renewable energy sources along with conventional energy sources. It requires a datasheet of load demand, solar irradiation, wind speed, etc.

TRNSYS: TRNSYS tool is developed by the University of Wisconsin and the University of Colorado, 1975. Initially, this tool was used only for thermal systems but after sometime, it was promoted for solar PV applications, diesel generator, and wind energy. It uses FORTRAN language which does not permit for optimization of energy sources. There are two versions released in years 2010 and 2012 as TRNSYS 17.0 and TRNSYS 17.1 respectively.

iHOGA: This tool is based on C++. It is used for solar PV, wind turbine, micro-hydro pumped, etc., designing hybrid system with single- or multi-objective function. It uses two versions such as PRO+ and EDU in which EDU is free.

INSEL: INSEL tool was designed by the University of Oldenburg, Germany. INSEL tool has more features for conventional simulation programs. It is utilized for the interpretation of operation and maintenance costs. It performs under the environment of MATLAB and simulink. This is not a free software tool.

SOMES: Simulation and Optimization model for Renewable Energy Systems tool has been developed at Utrecht University, Netherlands, 1987. This consists of a solar PV array, wind turbine, converters, and battery storage. This tool is used to evaluate performance on an hourly basis. It is coded in Turbo Pascal and runs on the windows platform.

SOLSTOR: It was developed by Sandia National Laboratory, 1970. This is used for simulation and optimization of integrated alternate energy systems including the peripherals such as solar photovoltaic, wind energy turbine, batteries, and converters. This minimizes total life cost by selecting the best size of equipments. Currently, this software tool is not under working.

2.5 Planning Methodologies of Hybrid Energy System

Traditionally remote areas are supplied energy by using diesel generators. Research has been carried out and shows the effects of diesel generators with negative results. Then, mostly used solution such as to deliver electricity by using renewable energy sources has been accepted. By keeping all environmental conditions such as solar irradiance and wind speed, there may be problems in the production of electricity continuously. Then, hybrid energy system has been preferred for continuity and reliable power supply. It is common to use, to implement conventional energy sources with alternate energy sources in addition with energy storage to design as microgrid (Emad et al. 2019).

There are different methodologies for the designing of hybrid energy systems with their parameters, power resources, and limitations. There are such as graphic construction method, analytical method, probabilistic method, and artificial intelligence method (Upadhyay and Sharma 2014) (Fig. 2.2).

2.5.1 Graphic Construction Method

This method is designed by Markvart (1996) for designing standalone solar photovoltaic and wind turbine configurations. This method is occupied on a condition in which the mean value of demand must be contented by taking the mean value of solar irradiance and speed of wind for fix capacity of solar photovoltaic generator

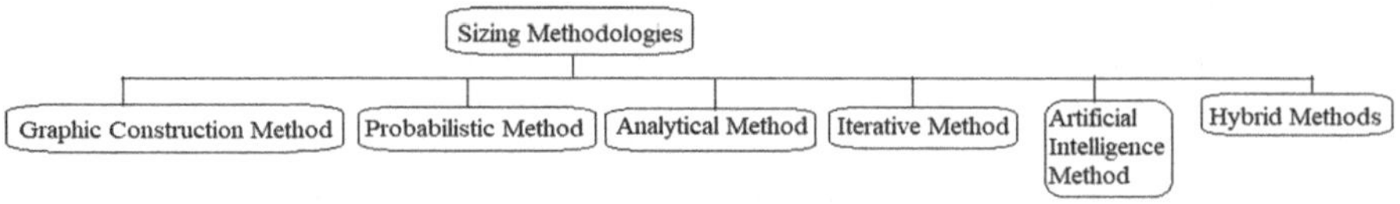

Fig. 2.2 Sizing methodologies of hybrid energy system

and wind energy turbine. On the basis of interpretation data, the trajectory is drawn between solar photovoltaic generator and wind energy turbine. The number of curves depends upon the number of times of data collected.

2.5.2 Analytical Method

In this method, hybrid energy models are shown by computational methods which outline the size of hybrid systems with its feasibility. The time taken in this method is less as compared with the Monte Carlo simulation method.

2.5.3 Probabilistic Method

It is one of the simplest methods for planning and sizing of hybrid energy system. The results obtained by this method may not be the best solution. In this method, few parameters are to be considered for optimization of hybrid energy system.

2.5.4 Artificial Intelligence Method

Many researchers have done work on artificial intelligence methods for finding out the best optimal solution for sizing of hybrid energy system. They have utilized genetic algorithm, particle swarm optimization on different platforms such as HOMER, ARENA 12.0 software, etc. These methods take less time as compared to conventional methods and provide the best solution.

2.5.5 Hybrid Method

These methods combine two or higher number of distinct methods which use positive consequences in order to obtain an optimal solution for the desired problem. The problems may be single objective or multi-objective depending upon the understanding of the importance of using different techniques. Sometimes, these methods may be complex but provide better results.

Table 2.1 represents different methods with their specifications for planning or sizing of hybrid energy systems (Bhandari et al. 2015):

Table 2.1 Sizing methods of hybrid energy systems

S. no	Methods	Parameters	Resources	Drawback
1	Graphic construction technique	Average solar radiation and wind speed data of each hour and month	PV/wind, PV/battery	Not comprising of slope PV module and height of wind tower
2	Logical technique	Average sun radiation and wind velocity data of each hour and month	Solar PV, wind, battery	Low flexibility
3	Probabilistic technique	Probabilistic access of calibrating of solar PV array and wind energy s	Solar PV, wind energy, battery	Not represent dynamic functioning of hybrid energy system
4	Frequent technique	Average sun radiation and wind velocity data of each hour and month	Solar PV, wind, battery	Not comprising of slope PV module and height of wind tower
5	Artificial intelligence technique	Average sun radiation & wind velocity data of each hour and month	Solar PV, wind, battery	Designing is very complex

2.6 Optimization Techniques

In this chapter section, optimization techniques have been discussed considering single objective and multi-objective functions. Many methods have been introduced for finding the best optimal solution of hybrid power system. For determining the reliability of hybrid power system, optimization is performed along with size and location of a component with achieving energy demand. For remote areas, artificial intelligence methods do not demand any requirement of environmental data for designing a unified energy system. There are many methods as described:

2.6.1 Particle Swarm Optimization

PSO method has several advantages over other methods as to reduce levelized cost of energy considering losses between demand and production sides. This method is very fast and reliable to obtain the optimal position. It is developed by Eberhart and Kennedy, 1995. It is based on the exchange of information between particles in the population preferred as a swarm. Each particle behaves as a point in hyperspace with essential properties as a memory of its own position and neighborhood's position. The performance of each particle is evaluated by the fitness function.

2.6.2 *Ant Colony Optimization*

ACO method is first published in the year 1996 which is influenced by the social act of insects in searching the precise direction for feed. Through this process, each information is shared by other ants by using the substance. This technique has been proposed for encountering the best optimal region and capacity of the system using renewable energy sources. Then it provides a better solution from genetic algorithm in less computational times (Abdmouleh et al. 2017).

2.6.3 *Artificial Bee Colony Optimization*

This method is inspired from a social act of honey bee swarm. Basically, there are three types of bees such as scout, worker, and onlooker. Worker bees are associated with food searchers and get back to their hive and dance. Onlooker bees watch worker bees' performance and select food sources from where worker bees carry food. At last, scout bees are independent on their own sources to search their food.

2.6.4 *Tabu Search Optimization*

It is meta-heuristic method discovered in 1996 to evaluate optimization problems. It is most identified for finding location and sizing of hybrid energy performance. It is having explicit memory and can be an application for continuous and discrete variables. The advantages of using this method are having many iterations and depending upon settings of parameters to obtain optimal location.

2.6.5 *Genetic Algorithm*

This method is inspired from natural selection and it is having different components such as chromosome encoding, selection, recombination, fitness function, and evolution. There are many choices for designing genetic algorithms for different applications. In the selection process, chromosomes are selected from the population and further used for reproduction. Members are selected based on performance and selected members to introduce in the next phase. In the crossover process, two chromosomes mix their genetic material and produce a new chromosome and the last step is mutation in which mutation of parents occur and develop a new chromosome.

Development of Hybrid Energy System Using New Techniques

For obtaining the performance of a hybrid energy system with its precise efficiency, few different combinations have been introduced. Padmanaban et al. (2019) have designed a photovoltaic–fuel cell hybrid grid with the integration of an ultracapacitor power system. To achieve the maximum power of this entire hybrid system, another maximum power tracking technique has been used such as Jaya Based MPPT (Padmanaban et al. 2019). This method is fast and effective with nil deviation near maximum power point. Hybrid optimization techniques such as combination of PSO and ABC for optimal location of energy system. Chakir et al. (2019) have proposed a hybrid PV battery energy system under TRNSYS environment (Chakir et al. 2019). For improvement of power quality of any hybrid power system, some AFACTS devices have been employed for reliable operation. Mostly used FACTS device such as DSTATCOM has been preferred for solar wind hybrid power system (Parija et al. 2019). The other concept has been taken of microgrid in order to decrease energy cost under different constraints. It can be configured under different converters between power sources and load. Hence, a methodology can be used for the purpose of energy-storing systems for charging and discharging (Kafazi et al. 2019). A hybrid system can be configured by using conventional and nonconventional energy sources such as solar and thermal hybrid system with PI controller and considering breaking point factors (Verma and Kori 2020). Similarly, for standalone systems, it is required to optimize the performance of energy system. Solar PV systems with different MPPT techniques can perform under sudden changes in weather conditions for the purpose of overcoming harmonic problems (Pradhan et al. 2020). There are different implementations of MPPT techniques for the operation of solar PV system such as Perturb and Observe and Incremental Conductance methods which perform by considering a few challenges such as steady-state oscillations, drift problem, designing problem, and lack of tracking steps. These all problems may be overcome by using such as drift-free method, variable step method, and prediction of parameters, etc. (Li et al. 2019).

For the optimization of hybrid energy system various techniques have been introduced in addition with multiple maximum power point techniques such as hill-climbing, fractional open-circuit voltage, fractional short-circuit current, etc. (Motahhir et al. 2019). Some of the heuristic techniques are preferred for optimization of hybrid renewable energy systems, for example, solar PV system in which partial shading is another issue which impacts the performance of energy system (Eltamaly and Farh 2019). There are more advantages of DC system over AC systems like DC is most compatible with renewable energy system hence, direct current microgrid has been introduced but having power management issue hence various control strategies have used such as centralized and decentralized. So, sizing of hybrid renewable energy system is an important factor from point of designing (Kumar et al. 2019). For the enhancement of power generation by using renewable energy system such as solar system requires the best MPPT technique, it can use hybridization of ANFIS and PSO methods (Priyadarshi et al. 2019).

There is a requirement for an improved power management control of a hybrid DCMG system which may be the combination of solar cell, fuel cell, and battery energy storage system. Based on the performance of BES, a better power management can be implemented by regulating modulation index and improving voltage stability (Senapati et al. 2019). Among renewable energy sources, wind is widely used energy source which may have integration with supercapacitor and lithium batteries for the establishment of hybrid energy storage wind power system (Yang et al. 2020). For maintaining the reliability of electricity under newly charging method combination with lithium-ion battery energy storage system has been introduced in which meta-heuristic algorithm has been used for the optimization of hybrid energy system (Trpovski et al. 2020). Energy storage system is used for maintaining continuous electricity which also faces the challenges such as best location and its capacity hence a lot of review work has been done by researchers (Ling Ai et al. 2019). There is a process such as distribution system asset which includes installation and optimal size of energy storage system containing other equipments such as transformer and feeders which operate on 33 bus systems (Masteri et al. 2018). For the next generation, flexible AC/DC hybrid energy system has been introduced because sometimes it is difficult for the certainty of generation and load hence this problem can be removed (Liu et al. 2019). Targeting as improving the efficiency of hybrid energy system, a new configuration has been introduced with wind and tidal energy sources. In this configuration, tidal power performs with the fluctuation of wind power under control strategy (Hu et al. 2017). Due to the special load such as EV charges point and new power generation system (solar and wind power generation system) can be taken for improvement of power quality of hybrid system (Luo and Huo 2020). Sometimes due to fluctuations in weather conditions, wind speed may be low or sunlight intensity may be low hence smoothening of such a hybrid system has been carried out with an energy storage system. But this may increase the maintenance cost of an entire hybrid system (Lamsal and Sreemal 2019).

From the point of stability, there are many issues which have to be resolved by configuring a hybrid system with advanced technologies in order to active power control, frequency control, and voltage control (Schinke and Elich 2018). Due to the use of power converters, few harmonics can be introduced specially in solar PV systems hence through the use of impedance transformation harmonics problem can be analyzed (Jingkai and Cao 2019). For hybridization system of HVAC and HVDC, probabilistic power flow can be used line current converter and voltage source converter (Shu et al. 2019). Hybrid energy system is an important element for finding the best optimal location with hybrid energy system planning moth-flame optimization method has been used (Ardabilli et al. 2020). For the minimization of operation cost of microgrid consisting of distributed energy resources combined heat and power generation unit can be installed to keep the balance between energy supply and load demand (Semero et al. 2020). To overcome the problem of switching state optimization hybrid method can be applied which uses DC programming and is compared with other heuristic methods (Schafer et al. 2020). The conventional designing of hybrid system takes a lot of time for simulation hence average models have been employed which may consist of bidirectional half-bridge converter and boost converter in addition to new power filter method for a smooth operation of hybrid energy system (Tang et al. 2020). Under islanding operations, some of the parameters of hybrid system may be in an unbalanced condition which causes instability of power system hence a time delay can be introduced in addition with some batteries (Hashimoto et al. 2019). Hybrid renewable energy source can be preferred as integrated renewable energy system so planning of these systems have been formulated by optimization, formulation, and configuration (Babatunde et al. 2020). Some of the hybrid energy system models have been configured by fuzzy logic method for proper selection of renewable energy sources and focuses on dynamic issues (Aikhuele et al. 2019). Due to intermittent changes in weather conditions, renewable energy sources may have a problem of instability such as small-signal stability hence proper location and sizing of hybrid energy system may overcome these instability problems (Ajeigbe et al. 2020). For the smooth operation of hybrid energy consisting of wind power may be configured with a pumped hydrosystem to improve the power quality of hybrid system by using a single value decomposition matrix (Hassan et al. 2019).

2.7 Conclusion

This chapter deals with a review of planning and sizing methodologies of hybrid system for the achievement of better operation. There are different hybrid energy systems that have been described with their parameters with their working. There are different optimization techniques that have been reviewed with their description and new techniques for the development of hybrid energy system. Hence, we can optimize and obtain the best optimal location and size of hybrid energy system using renewable energy sources.

2.8 Future Work

By hybridization, various energy sources can be integrated with electric vehicles. In this current era, electric vehicle is playing a major role as they are achieving momentum. There is advancement of battery system from lead acid to lithium ion. These batteries are having large potentials as compared with other batteries. Various types of electric vehicles are present which can be incorporated with hybrid energy system to keep a balance between energy supply and load demand.

2.9 Summary

In this manuscript, introduction of renewable energy system has been described with the description of hybrid energy system. As there is requirement of energy supply to fulfill energy demand. Hence through only one source, it is quiet difficult hence hybrid system has been introduced. There is complexity in hybrid system as it requires the best location and sizing for providing reliable energy. Hence different optimization techniques with operation tools for hybrid energy systems have been described. Further, hybrid energy systems can be utilized with electric vehicle technology interconnected with lithium-ion batteries and can be analyzed for their operation.

References

Abdmouleh Z, Gastli A, Ben-Brahim L, Haouari M, Al-Emadi NA (2017) Review of optimization techniques applied for the integration of distributed generation from renewable energy sources. Renew Energy 113:266–280. https://doi.org/10.1016/j.renene.2017.05.087

Aikhuele DO, Ighravwe DE, Akinyele D (2019) Evaluation of renewable energy technology based on reliability attributes using hybrid fuzzy dynamic decision-making model. Technol Econ Smart Grids Sustain Energy 4(16):1–7. https://doi.org/10.1007/s40866-019-0072-2

Ajeigbe OA, Munda JL, Hamam Y (2020) Towards maximizing the integration of renewable energy hybrid distributed generations for small signal stability enhancement: a review. Int J Energy Res 44(4):2379–2425. https://doi.org/10.1002/er.4864

Al-Shetwi AQ, Hannan MA, Jern KP, Mansur M, Mahila TMI (2020) Grid-connected renewable energy sources: Review of the recent integration requirements and control methods. J Cleaner Prod 253:1–17. https://doi.org/10.1016/j.jclepro.2019.119831

Anounea K, Bouyaa M, Astitob A, Abdellah AB (2018) Sizing methods and optimization techniques for PV-wind based hybrid renewable energy system: a review. Renew Sustain Energy Rev 93:652–673. https://doi.org/10.1016/j.rser.2018.05.032

Ardabilli NA, Jalal S, Shenvana S, Shayeghi H (2020) Optimization of a hybrid energy system architecture in island mode by a two-layer approach. In: 2020 10th smart grid conference, pp 1–6. https://doi.org/10.1109/SGC52076.2020.9335759

Babatunde OM, Munda JL, Hamam Y (2020) "Operations and planning of integrated renewable energy system: a survey. In: 2020 5th international conference on renewable energies for developing countries, pp 1–6. https://doi.org/10.1109/REDEC49234.2020.9163857

Babu TS, Vasudevan KR, Ramachandaramurthy VK, Sani SB, Chemud S, Lajim RM (2020) A comprehensive review of hybrid energy storage systems: converter topologies, control strategies and future prospects. IEEE Access 8:148702–148721. https://doi.org/10.1109/ACCESS.2020.3015919

Bai Z, Liu QB, Lei J, Hong H, Jin HG (2016) New solar-biomass power generation system integrated a two-stage gasifier. Appl Energy 194:310–319. https://doi.org/10.1016/j.apenergy.2016.06.081

Bhandari B, Lee K-T, Lee G-Y, Cho Y-M, Ahn S-H (2015) Optimization of hybrid renewable energy power systems: a review. Int J Precision Eng Manuf Green Technol 2(1):99–112. https://doi.org/10.1007/s40684-015-0013-z

Chakir S, Tabaa M, Moutaouakkil F, Medromi H, Alami K (2019) Architecture & methodology for a grid connected PV-battery hybrid system. In: 2019 7th international renewable & sustainable energy conference, pp 1–6. https://doi.org/10.1109/IRSEC48032.2019.9078174

Eltamaly AM, Farh HMH (2019) Dynamic global maximum power point tracking of the PV systems under variant partial shading using hybrid GWO-FLC. Sol Energy 177:306–316. https://doi.org/10.1016/j.solener.2018.11.028

Emad D, El-Hameed MA, Yousef MT, El-Fergany AA (2019) Computational methods for optimal planning of hybrid renewable microgrids: a comprehensive review and challenges. Arch Comput Methods Eng 27:1–23. https://doi.org/10.1007/s11831-019-09353-9

González A, Riba JR, Rius A, Puig R (2015) Optimal sizing of a hybrid grid-connected photovoltaic and wind power system. Appl Energy 154:752–762. https://doi.org/10.1016/j.apenergy.2015.04.105

Grangea BD, Falcoza C, Sirosb Q, Ferrierea F (2014) Simulation of hybrid solar gas-turbine cycle with storage integration. Energy Procedia 49:1144–1156. https://doi.org/10.1016/j.egypro.2014.03.124

Guo S, Liu Q, Sun J, Jin H (2018) A review on the utilization of hybrid renewable energy. Renew Sustain Energy Rev 91:1121–1147. https://doi.org/10.1016/j.rser.2018.04.105

Hashimoto S, Yamamoto T, Nara K, Tobaru N (2019) Capacity determination of the DC-side battery for hybrid batteries in PV generation system. In: 2019 IEEE innovative smart grid technologies—asia (ISGT Asia), pp 1745–1750. https://doi.org/10.1109/ISGT-Asia.2019.8881162

Hassan A, Hamam Y, Munda JL (2019) Minimizing the impact of intermittent wind power on multi-period power system operation with pumped hydro generation. Energies 12(18):1–22. https://doi.org/10.3390/en12183583

Hu E, Wang Z, Zhao H, Guo J, Yang H (2017) A novel control strategy to smooth power fluctuation of hybrid offshore wind and tidal power generation system. In: 2017 IEEE conference on energy internet and energy system integration, pp 1–4. https://doi.org/10.1109/EI2.2017.8245723

Jingkai W, Cao B (2019) Simulation analysis harmonic characteristics of photovoltaic power generation based on MATLAB. Energy Procedia 158:412–417. https://doi.org/10.1016/j.egypro.2019.01.125

El Kafazi I, Bannari R, Bouzi M (2019) Hybrid microgrid considering PV generator and energy storage systems. In: 2019 1st global power, energy and communication conference, pp 286–291. https://doi.org/10.1109/GPECOM.2019.8778615

Kulor F, Markus ED, Kanzumba K (2021) Design and control challenges of hybrid, dual nozzle gas turbine power generating plant: a critical review. Energy Rep 7:324–335. https://doi.org/10.1016/j.egyr.2020.12.042

Kumar J, Agarwal A, Agarwal V (2019) A review on overall control of dc microgrids. J Energy Storage 21:113–138. https://doi.org/10.1016/j.est.2018.11.013

Lamsal D, Sreemal V (2019) Output power smoothing control approaches for wind and photovoltaic generation system. Renew Sustain Energy Rev 113:1–22. https://doi.org/10.1016/j.rser.2019.109245

Li X, Wang Q, Wen H, Xiao W (2019) Comprehensive studies on operational principles for maximum power point tracking in photovoltaic systems. IEEE Access 7:121407–121420. https://doi.org/10.1109/ACCESS.2019.2937100

Ling Ai W, Ramachandaramurthy VK, Taylor P, Ekanayake J, Walker S, Sanjeevikumar P (2019) Review on the optimal placement, sizing and control of an energy storage system in the distribution network. J Energy Storage 21(02):489–504. https://doi.org/10.1016/j.est.2018.12.015

Liu X, Liu Y, Liu H, Yang Z (2019) Optimal planning of AC/DC hybrid system with renewable generations: an expansion planning modle. In: 2019 IEEE 3rd international conference on energy internet and energy system integration, pp 1236–1241. https://doi.org/10.1109/EI247390.2019.9061953

Luo W, Huo Q (2020) Research on modeling and planning method of distribution network with new and special energy load. In: 2020 15th international conference on industrial electronics and applications, pp 1696–1701. https://doi.org/10.1109/ICIEA48937.2020.9248264

Markvart T (1996) Sizing of hybrid photovoltaic-wind energy systems. Sol Energy 57(4):277–281. https://doi.org/10.1016/S0038-092X(96)00106-5

Masteri K, Venkatesh B, Freitas W (2018) A feeder investment model for distribution system planning including battery energy storage. Can J Electr Comput Eng 41(4):162–171. https://doi.org/10.1109/CJECE.2018.2865176

McLarty D, Brouwer J, Samuelsen S (2014) Hybrid fuel cell gas turbine system design and optimization. National Fuel Cell Research Center, Engineering Laboratory Facility, pp 1–12. https://doi.org/10.1115/1.4024569

Motahhir S, Hammoumi AE, Ghziza AE (2019) The most used MPPT algorithms: review and the suitable low-cost embedded board for each algorithm. J Clean Prod 246(1):1–38. https://doi.org/10.1016/j.jclepro.2019.118983

Nema P, Nema RK, Rangnekar S (2009) A current and future state of art development of hybrid energy system using wind and PV-solar: a review. Renew Sustain Energy Rev 13(8):2096–2113. https://doi.org/10.1016/j.rser.2008.10.006

Padmanaban SK, Priyadarshi N, Bhaskar MS, Holm-Nielsen JB, Hossain E, Azam F (2019) A hybrid photovoltaic-fuel cell for grid integration with Jaya based maximum power point tracking: experimental performance evaluation. IEEE Access 82978–82990. https://doi.org/10.1109/ACCESS.2019.2924264

Parija B, Behera S, Pattanayak R, Behera S (2019) Power quality improvement in hybrid power system using D-STATCOM. In: Proceedings of the third international conference on computing methodologies and communication, pp 564–567. https://doi.org/10.1109/ICCMC.2019.8819656

Pradhan C, Senapati MK, Malla SG, Nayak PK, Gjengedal T (2020) Coordinated power management and control of standalone PV-hybrid system with modified IWO based MPPT. IEEE Syst J 99:1–12. https://doi.org/10.1109/JSYST.2020.3020275

Priyadarshi N, Padmanaban S, Holm-Nielsen JB, Blaabjerg F, Bhaskar MS (2019) An experimental estimation of hybrid ANFIS–PSO-based MPPT for PV grid integration under fluctuating sun irradiance. IEEE Syst J 14(1):1218–1229. https://doi.org/10.1109/JSYST.2019.2949083

Quazi A, Hussain F, Rahim NA, Hardaker G, Alghazzawi D, Shaban K, Haruna K (2019) Towards sustainable energy: a systematic review of renewable energy sources, technologies and public opinions. IEEE Access 63837–63851. https://doi.org/10.1109/ACCESS.2019.2906402

Renewable Energy Market Report (2020) IEA, Outlook for year 2020 and 2021

Schafer F, Scheidier A, Braun M (2020) A hybrid optimization method combining network expansion planning and switching state optimization. IEEE Open Access J Power Energy 7:234–242. https://doi.org/10.1109/OAJPE.2020.3006344

Schinke A, Elich I (2018) Enhanced voltage and frequency stability for power system with high penetration of distribution photovoltaic generation. IFAC-Papers on Line 51(28):31–36. https://doi.org/10.1016/j.ifacol.2018.11.673

Semero YK, Zhang J, Zheng D (2020) Optimal energy management strategy in microgrids with mixed energy resources and energy storage system. IET Cyber-Phys Syst Theory Appl 5:80–84. https://doi.org/10.1049/iet-cps.2019.0035

Senapati MK, Pradhan C, Samantaray SR, Nayak PK (2019) Improved power management control strategy for renewable energy-based DC micro-grid with energy storage integration. IET Gener Transm Distrib 13(6):838–849. https://doi.org/10.1049/iet-gtd.2018.5019

Shu T, Lin X, Peng S, Du X, Chen H, Li F, Tang J, Li W (2019) Probabilistic power flow analysis for hybrid HVAC and LCC-VSC HVDC system. IEEE Access 7:142038–142052. https://doi.org/10.1109/ACCESS.2019.2942522

Tang X, Chen X, Zhang R, Li Q, Song Y, Zhang Y, Luo N (2020) 2020 10th international conference on power & energy systems, pp 189–194. https://doi.org/10.1109/ICPES51309.2020.9349726

Trpovski A, Banerjee P, Xu Y, Hamacher T (2020) A hybrid optimization method for distribution system expansion planning with lithium-ion battery storage systems. In: 2020 IEEE sustainable power and energy conference, pp 1–7. https://doi.org/10.1109/iSPEC50848.2020.9351208

Upadhyay S, Sharma MP (2014) A review on configurations, control and sizing methodologies of hybrid energy systems. Renew Sustain Energy Rev 38:47–63. https://doi.org/10.1016/j.rser.2014.05.057

Verma A, Kori AK (2020) Higher efficiency solar and thermal hybrid power plant based on PI controller. In: Proceedings of fourth international conference on computing methodologies and communication, pp 699–703. https://doi.org/10.1109/ICCMC48092.2020.ICCMC-000130

Yang M, Liu H, Zhao L (2020) optimal configuration method of hybrid storage system tracking the wind power plan. In: 2020 IEEE 3rd student conference on electrical machines and systems, pp 1–7. https://doi.org/10.1109/SCEMS48876.2020.9352355 .

Zhang C, Sun J, Ma J, Xu F, Qiu L (2019) Environmental assessment of a hybrid solar-biomass energy supplying system: a case study. Int J Environ Res Public Health 16(12):1–14. https://doi.org/10.3390/ijerph16122222

Chapter 3
Advanced Fault Diagnosis and Condition Monitoring Schemes for Solar PV Systems

Suresh Kumar Gawre

Abstract In the present era of smart technologies, the power sector has highly benefited as monitoring, supervision, and control have moved toward the intelligent power delivery. High-quality power estimation, self-healing, and machine-to-machine communication-based approaches have been appreciated to achieve more reliable and secured smart grids (SG). Renewable energy sources (RES), mainly solar photovoltaic (PV) systems are intermittent in nature and wisely utilized in power generation either as stand-alone or grid connected. Energy sector is focused on RES to reduce carbon footprint and affordable energy to all. Prediction, decision-making, and fast healing for recovery after faults in system, are prime objectives for fault diagnosis and condition monitoring of RES. Classical PV fault diagnosis schemes are available, which basically follow the general process of detection, feature extraction, and classification of fault data. Enormous data has to be handled by the processors either offline or online. In this chapter, fault detection schemes for handling preprocessing of raw data from various sensors through wire or wireless-based time domain or frequency domain methods like Fourier transform, Wavelet transform along with novel approaches based on the internet of things (IoT) have been rigorously reviewed. Traditional as well as advanced artificial intelligence (AI), machine learning (ML), and emerging approaches for PV fault classification and mitigation have been discussed thoroughly. Along with comprehensive and critical literature review, a smart PV fault classification scheme is proposed for the enhancement of the performance of solar PV systems.

Keywords Fault diagnosis · Renewable energy sources · Solar PV system (SPVS) · Fault detection · Fault classification · Internet of things (IoT) · Artificial intelligence (AI) · Machine learning (ML)

S. K. Gawre (✉)
Department of Electrical Engineering, MANIT, Bhopal, India
e-mail: sgawre28@gmail.com

A. K. Bohre et al. (eds.), *Planning of Hybrid Renewable Energy Systems, Electric Vehicles and Microgrid*, Energy Systems in Electrical Engineering,
https://doi.org/10.1007/978-981-19-0979-5_3

Nomenclature

I	PV Cell current, (A)
k	Boltzmann constant
R_{sh}	Shunt resistance of PV cell
R_{se}	Series resistance of PV cell
V	PV cell voltage
I_{ph}	Light generated current
I_o	Reverse saturation current for D diode
I_{o2}	Reverse saturation current for D_2 diode
V_T	Voltage due to temperature = $V_T(N_s{*}N{*}k{*}T)/q$
T_c	PV Cell temperature (operating)
N_1	Quality factor of D diode
N_2	Quality factor of the D_2 diode
V_{oc}	OC voltage of PV cell
V_m	Maximum voltage of PV cell
I_{sc}	SC current of PV cell
I_m	Max. current of PV cell
P_m	Max. power of PV cell
P_o	Ideal power of PV cell
P_{rad}	Solar radiant power
η	Power conversion Efficiency

3.1 Introduction

Non-conventional energy sources are widely accepted and adopted by most of the nations to reduce the carbon footprint and make affordable energy to all. The present green power generation movement has grabbed attention toward renewable energy sources (RES), especially solar photovoltaic systems (SPVS) (Tyagi et al. 2013). Global photovoltaic (PV) power addition is achieved for 2021 around 117 GW capacities installed, which is a 10% increase from 2020 and further expected 165 GW for 2023–25 (IEA 2020). Installed capacity of SPVS and the production always suffers by faults and mall operation. New norms and standards have been adopted, as classical monitoring parameters were not efficient, to provide relevant information for analysis and mitigation of faults (Yahyaoui and Segatto 2017). Technical progress in measurement and fault monitoring of devices, has drastically upgraded in the last few decades, which insures the accuracy and reliability of modern fault diagnosis and condition monitoring schemes. Practical approach of analysis along with smart fault monitoring techniques is the basic soft computing tools for intelligent power delivery in smart grid scenarios. SPVS are installed with the prediction of a well-operated system run up to 25 years. PV system component structure, mountings, and maintenance costs are included to estimate the payback period for the system

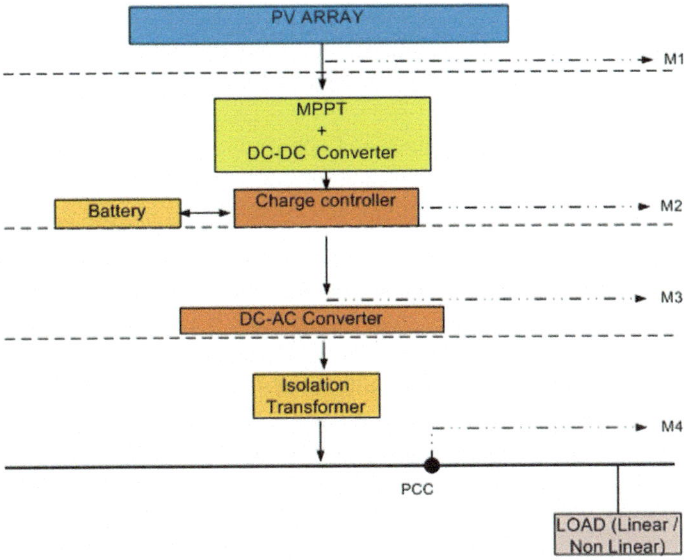

Fig. 3.1 Generalized location of fault monitoring sensors

installed. In this situation, a smart prediction of faults in essential parts of the PV system is desirable. Schematic diagram of the solar PV system with generalized fault monitoring sensors is shown in Fig. 3.1. It has different sections to be monitored at the solar PV module level, DC–DC converter level, DC to AC inverter level, charge controller level, and the point of common coupling (PCC) nearest to the load.

High-quality power standards are essential for accurate, errorless monitoring and real-time/online power estimation. Faults detection and classification in commercially available PV panels is discussed, where some focused on faults and partial shading analysis. Graph-based semi-supervised learning was proposed in research work (Alam et al. 2015; Hariharan et al. 2016; Hu et al. 2015; Kumar et al. 2018; Akram and Lotfifard 2015). Uncertainty and expandability of load always seek toward the smart decision-making for load shifting. Identification of specific faults in a particular area is always needed to estimate the severity (Pillai and Rajasekar 2018). Immediate data transfer to the data center for prompt action to ensure a more reliable and secure smart grid either offline or online, various solar PV fault detection and fault classification methods. Faults, defects, and shading conditions in PV array involve detection as a prime computational task. PV faults in solar PV array results significant power loss, lower reliability, very fast panel degradation, and further risk of fire (Gokmen et al. 2013). This chapter presents a comprehensive literature review along with a critical analysis of fault diagnosis and condition monitoring for solar PV systems. Major contributions are:

- Focus on novel signal processing detection methods.
- Provide research avenues in all detection and classification methods.

- Environment-based detection analysis.
- Internet of things (IoT) based approach for wireless fault detection.
- Artificial intelligence (AI), machine learning (ML), deep learning (DL) based review.
- For new researches, key challenges and opportunities have been discussed
- Mitigation and fault localization issues are covered for further course of action.
- Hybrid scheme is proposed using IoT and AI schemes.

In this chapter, fault analysis, classification, and mitigation techniques are thoroughly reviewed and organized in the following sections. Section 3.2 deals with solar PV cell modeling and fault analysis. In Sect. 3.3, basics of solar PV faults have been discussed along with the causes of faults and the location of faults. Section 3.4 describes various PV fault detection methods including online or offline detection using advanced signal processing tools, also focused on IoT-based and environmental-based detection methods as a modern approach. In Sect. 3.5, major advanced PV fault classification techniques have been included for review. Section 3.6 focused on fault mitigation techniques and solution-based approaches. Further in Sect. 3.7, key challenges, opportunities, and future research scope along with a proposed PV fault classification scheme with IoT application.

3.2 Solar PV Modeling and Fault Analysis

Solar PV cell model is required to analyze the behavior of the actual PV array. To build PV array voltage and current of desired value combination, series and parallel solar cells are used. Most of the researchers have proposed a PV cell model based on a single diode model whereas double diode model is more precise (Kumar et al. 2018; Akram and Lotfifard 2015). For representation of a practical PV cell, a light dependant current source is employed along with two parallel diodes as shown in Fig. 3.2.

Total output current of a solar PV cell is expressed in Eq. (3.1) as:

$$I = I_{ph} - I_0\left[exp\left(\frac{V - R_{se}I}{N_1 V_T}\right) - 1\right] - I_{02}\left[exp\left(\frac{V - R_{se}I}{N_2 V_T}\right)1\right] - \frac{V - R_{se}I}{R_{sh}} \quad (3.1)$$

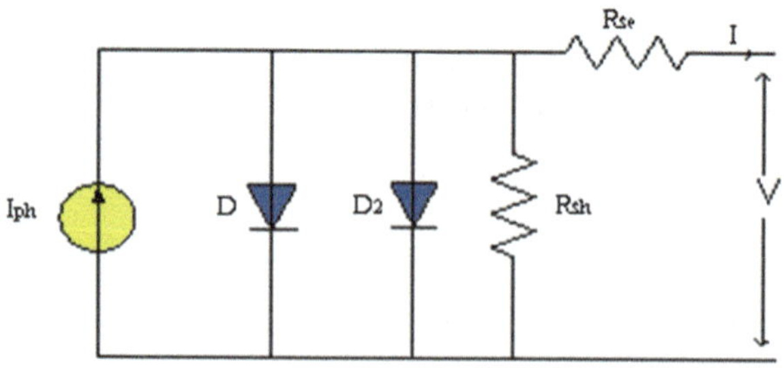

Fig. 3.2 Double diode model of PV cell

Output current of PV array depends on irradiance and temperature.

Basic electrical parameters generally used for fault analysis are Open Circuit Voltage (V_{oc}), Short Circuit Current (I_{sc}), Fill factor (FF), and Efficiency.

Open Circuit Voltage (V_{oc}) is the maximum output voltage that yields from a solar PV cell at its open terminals. Expression for V_{oc} is given in Eq. (3.2)

$$V_{oc} = \frac{kT_c}{q} ln\left(\frac{I_{ph}}{I_0} + 1\right) \tag{3.2}$$

Short Circuit Current (I_{sc}) is defined as the maximum current of solar PV cell when its output terminals are shorted i.e., (V = 0).

FF is the ratio of maximum PV power ($P_m = V_m * I_m$) of solar PV cell to the ideal PV power ($P_o = V_{oc} * I_{sc}$).

$$FF = \frac{V_m * I_m}{V_{oc} * I_{sc}} \tag{3.3}$$

Efficiency is the ratio of the PV power output (maximum power point P_m of a solar PV cell) to power input (solar radiation power). Expressed in Eq. (3.4) given below.

$$\eta = \frac{P_m}{P_{rad}} \tag{3.4}$$

A novel modeling PV systems method is proposed which uses information given from manufacturer's datasheet under standard-operating test conditions (STCs) and normal-operating cell temperature (NOCT) conditions (Akram and Lotfifard 2015). Intensive investigation of different fault causes, protection schemes, and issues of hidden faults in PV systems were discussed along with mitigation techniques (Pillai and Rajasekar 2018). Other papers exhibited the method which uses minimum sensors through operating voltage measurement of solar PV array and ambient temperature only (Gokmen et al. 2013; Jain et al. 2019; Khoshnami and Sadeghkhani 2018). In this paper, the author was proposed a technique for evaluation of PV panels by a computational strategy created for estimate and arrangement for PV fault analysis. This general strategy permits the stretching out of assessment in a multidimensional space, where all the present strategies have a slight precision, can effectively apply for an enormous zone of PV panels. Researchers described the most regularly utilized solar PV cell model and the generalized PV model utilizing MATLAB SIMULINK was created (Chen et al. 2016). The issue of model parameter assurance dependent on the four-parameter models is also addressed. At long last, improved logical formulae depicting the four-parameter model are introduced. This is trailed by the execution of a PV module MATLAB simulation comprising of arrangement of equal associated cells, appropriate for testing MPPT algorithm. Producing the results of solar light intensity and PV cell temperature, the qualities of the PV model are simulated using Wavelet-based detection, fast outlier detection with AI techniques

(Zhao et al. 2013, 2014; Yi and Etemadi 2016; Sreelakshmy et al. 2018; Roy et al. 2017). ANN technique with thermography-based MPPT methods were described in (Chine et al. 2016; Wang et al. 2015). Table 3.1 gives the summary of different modeling of PV system for MPPT, partial shading, fault diagnosis, and research avenues.

3.3 Faults in Solar PV Array

Faults can occur in the solar PV modules in several ways like extrinsic, i.e., defect created during manufacturing, failure occurring during the design, fabrication, or assembly process (Karmacharya and Gokaraju 2017; Harrou et al. 2019). Intrinsic that is a failure caused by a natural deterioration in the materials or by the manner in which the materials are combined during fabrication. External causes, i.e., transportation failure, connector failure, clamping, cable failure and lightning among which glass breakage is most often. Electrical faults due to insulation failure or equipment failure like bridging and short circuits (Kumar et al. 2018; Chine et al. 2016; Dobaria et al. 2018; Kajari-Schröder et al. 2012).

However, basic classifications of faults in solar panels are permanent, incipient, and intermittent as shown in Fig. 3.3. Intermittent faults include faults which are temporary in nature such as partial shading, leaf, bird drop, and adverse environmental reasons like dust, heavy snow accumulation, contamination, and humidity (Zaki et al. 2021; Pillai and Rajasekar 2018). Permanent PV faults include faults which can cause damage to PV modules like bridging diode faults, OC faults, SC faults, and interconnection damage. Incipient faults are slowly developing faults which may lead to permanent faults including faults such as cells degradation, partial damage, and corrosion in interconnections.

3.3.1 Permanent Faults

3.3.1.1 Line to Line Fault

At the time of sudden SC between two lines of different potential levels in array L-L fault occurs. Due to the sudden short circuit, a low impedance path is created through which high amount of current flows in PV array. The two lines involved can be either of different or same modules as shown in Fig. 3.4. Due to the high current flow, the voltage regulation becomes poor thus the output voltage is reduced. The sudden change in values of output voltage and current changes the I-V characteristic of solar PV array (Boggarapu et al. 2020; Wang et al. 2015). The immense high current can lead to the fire hazards in PV modules and may damage the complete system.

Table 3.1 Modeling fault and fault diagnosis

Modeling fault and fault diagnosis			
References	Techniques	Research avenues/Application remarks	Simulation Tool/Expemental setup
Kumar et al. (2018)	Tilt angle analysis	Partial shading	Experimental setup
Akram and Lotfifard (2015)	PV modeling, PNN	MPPT, AI application for detection and classification of OC, SC faults	Simulation
Gokmen et al. (2013)	Voltage and temperature measurement	Detection of OC and SC faults	Simulation and experimental setup
Chine et al. (2016)	Artificial neural networks	MPPT	Simulation
Hu et al. (2014)	Thermography-based virtual MPPT	MPPT, Shading fault	Software simulation and experimental setup
Wang et al. (2015)	Dynamic current–voltage characteristics	MPPT, Shading condition fault	Simulation and validation with experimental circuit
Maki and Valkealahti (2012), Jaraniya et al. (2020), Baccoli et al. (2021), Spataru et al. (2015), González-Longatt (2005), Chenni et al. (2007), Koutroulis and Blaabjerg (2012), Katoch, et al. (2018), Livera et al. (2020)	PV modeling	MPPT, fault due to partial shading	Simulation with experimental validation
Boutelhig et al. (2016), Planas et al. (2015)	Maximize discharge rate	Water pumping system	Experimental setup
Gomes et al. (2013), Fadil and Giri (2011)	Climatic sensor less technique	MPPT	MatLab/Simulink and dSPACE
Dierauf et al. (2013)	fault analysis	AD/DC microgrid	Simulation
Feaster and Wamsted (2020)	Diagnostic architecture	PV array fault	Simulation
Stember et al. (1982), Gong et al. (2020)	Fault analysis WT, PNN	Inverter	Simulation
Livera et al. (2019)	Failure, defect detection techniques	Data analysis of performance	Simulation

(continued)

Table 3.1 (continued)

Modeling fault and fault diagnosis			
References	Techniques	Research avenues/Application remarks	Simulation Tool/Expemental setup
Spataru et al. (2015)	Light I-V measurements photovoltaic systems based on	Sensors are required	Simulation and experimental setup

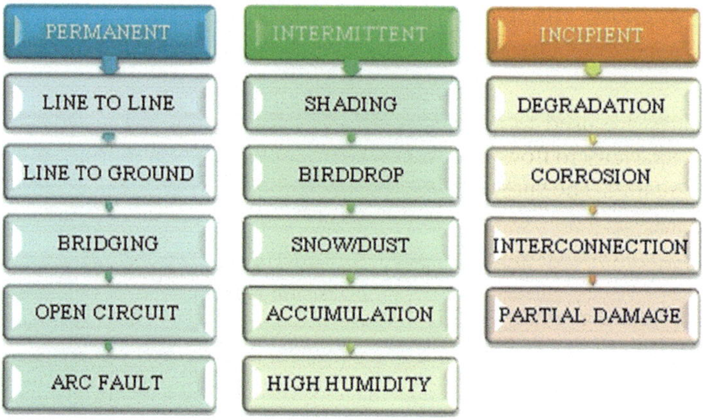

Fig. 3.3 Types of faults in PV Array

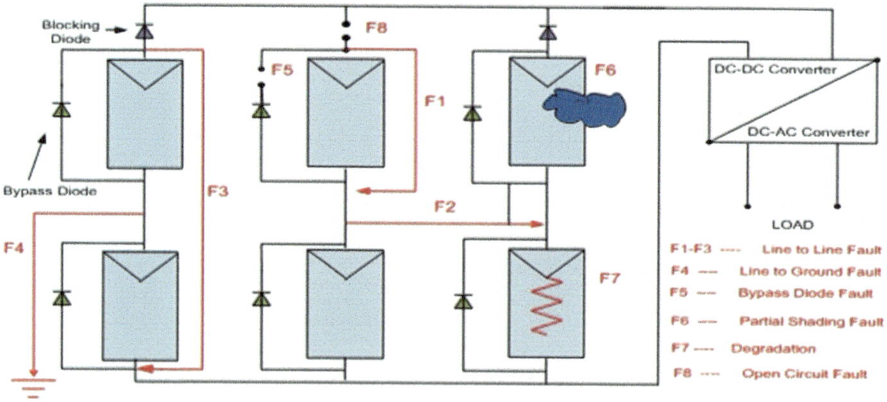

Fig. 3.4 Faults on PV array

3.3.1.2 Earth Fault

A ground fault occurs when the short circuit (as shown in Fig. 3.4) appears between any conductor and earth, causing a huge rise in the current through PV arrays and further potential damage to the PV modules. The output voltage is dropped which abruptly makes changes in the I-V characteristics of the solar PV array (Murtaza et al. 2019). Earth faults are very common in the PV arrays, it can occur due to any wire break or any insulation failure in the PV system (Saleh et al. 2017; Li et al. 2019).

3.3.1.3 Bypass Diode Fault

Bypass diodes are used to overcome the power loss caused by partial shading. It allows the electricity to flow around the cells when they are not providing any potential due to shading. The surge due to lightning can be one of the causes for this fault. If any fault occurs in this diode or it gets short-circuited, then it reduces the efficiency of the solar panels or in some cases leads to circulating current which can damage the solar array. Researches have successfully suggested methods to improve efficiency under partial shading condition and effective use of bypass diode (Gomes et al. 2013; Teo et al. 2020).

3.3.2 Intermittent Faults

3.3.2.1 Partial Shading

It is an abnormal condition in which some part of the PV array doesn't get enough solar irradiation to produce any potential across itself creating problem for solar cells in series in the array. It can virtually reduce the output of that particular string to zero as long as the shadow is present on the cell of that string. However, these days' arrays are provided with the bypass diode which minimizes the effects of partial shading. Mitigation techniques are provided by papers (Fadil and Giri 2011; Maki and Valkealahti 2012; Yahyaoui 2016).

3.3.2.2 Bubbles

Bubbles occur frequently in the PV modules due to difference in adhesion (generally, ethylene–vinyl acetate—EVA is used as adhesive). Due to the formation of bubbles, the heat dissipation rate of the solar cell reduces which eventually reduces the service life of the module. Due to the formation of bubbles, some parts of the solar panels did not get sufficient amount of irradiance due to which efficiency of the solar module decreases also the moisture can enter into the PV module, authors presented efficient analysis dust on solar PV array (Darwish et al. 2015).

3.3.3 Incipient Faults

3.3.3.1 Glass Breakage and Crack

This is very common due to hailstorms or due to the heavy burden on the glass of solar panels. Any breakage and crack in the glass of PV module can lead to contact with moisture and atmosphere in general thus leading to corrosion in the junction box and connections and may lead to internal arcing and other hazards in future. Glass breakage can be caused by snow or inappropriate mounting system glass chipping during installation and handling caused hot spots (Lu et al. 2016; Jaraniya et al. 2020)

3.3.3.2 Delamination

Occurs when the adhesion between the front panel and back fiber gets separated or the thermal properties of the plastics used are poor also. Occurs due to adhesion contamination (the adhesion used is ethylene–vinyl acetate or EVA) or due to environmental factors (Ramkiran et al. 2020). As a result, the moisture enters into the cell and leads to corrosion in the junction boxes and contacts which degrades the performance of the PV module (Fernández-Solas et al. 2021).

3.3.3.3 Burn Marks and Hotspots

Due to an uneven presence of irradiation or uneven heat exhaust or overloading, hotspots are created. These are places on the panels which become warmer than the other places. These can also occur due to manufacturing defects like badly soldered connections or a structural defect in solar cells. Due to bad soldering, the resistance to that particular part of the solar cell reduces in comparison to other parts. Hence, allows more current through it and leads to the formation of hotspots. If the hotspot persists for a longer time, it can cause short circuits and may lower the lifespan and performance of the PV module (Fernández-Solas et al. 2021). If hotspots persist for a longer time duration, they are converted to burn marks, burn marks depict physical damage through scrapes, tears, and scratches (Baccoli et al. 2021; Teo et al. 2020; Kajari-Schröder et al. 2012).

3.3.3.4 Front Panel Discoloration

Discoloration occurs in different parts of the solar panels due to mainly two reasons one is weathering in which the moisture ingresses in the module and due to corrosion, the outer panel of PV arrays deteriorates with time leads to discoloration and sometimes due to an additive reaction between oxygen and chemicals like adhesive present in solar panels leads to the formation of chromospheres which are not transparent and thus leads to discoloration (Fernández-Solas et al. 2021). Sometimes when the sensitivity of any cell is lower than the closest cells, it appears dull in comparison to the other cells. It does not lead to a failure but reduces the efficiency of the cell further improvement techniques are discussed by authors (Baccoli et al. 2021; Teo et al. 2020; Fathabadi 2015).

3.3.3.5 Snail Trails

The brownish or white lines on the solar panels or partial discoloration or of the front panel of the photovoltaic module called snail trails usually occur after a couple of years, have multiple causes like constant contact to moisture, poor level of fiber used in the front panels, and use of defective front metallization silver paste in the PV module manufacturing process (Pradeep Kumar et al. 2017; Boggarapu et al. 2020; Meyer et al. 2013). Snail trails were mostly located at the edge of the cell or near micro cracks.

3.3.3.6 Potential Induced Degradation (PID)

This fault occurs due to voltage difference developing between the earth and the panel. Usually, the frame of the solar panel is earthed which leads to the difference in voltage between PV modules and the earth produces a discharged voltage in the primary power circuit and leads to a leakage current to the earth (Fernández-Solas et al. 2021). The way to avoid this leakage due to PID is to choose a solar panel which has inbuilt PID resistance (Dierauf et al. 2013).

3.3.4 Arc Faults

Arc faults are the result of loose connections or any insulation damage. This type of fault can be minimal in the beginning but have the potential to create heavy fire hazards in the later stage. Arcs cannot be detected by the normal overcurrent relays because arcs flow through high impedance paths. It does not create a very heavy drop in the output potential or generate a heavy short circuit current, but can create heavy fire hazards which can cost millions (Alam et al. 2015; Zhao et al. 2013; Gokmen et al. 2012).

3.4 Detection of Faults in SPV System

During the whole life cycle of photovoltaic modules, it is very essential to regularly check the health of the PV devices and check whether they are satisfying certain parameters or not. During diagnosis of faults, it is not only to remove the faulty part from the healthy parts but also require the preventive diagnosis because in case of solar modules, it is very necessary to detect incipient faults or degradation at an early stage. Otherwise, these faults can slowly grow into large faults and can damage the solar modules. By detecting possible losses at an early stage, it is possible to reduce the economic losses, increase the reliability of the system and provide safe operation (Boggarapu et al. 2020) (Fig. 3.5).

3.4.1 Conventional Fault Detection Techniques

3.4.1.1 Visual Methods

This method can only be used when one is working on a small scale like for domestic purposes or mounted on a machine. On large scale, it is not possible to conduct visual inspections because it is very time taking to visually inspect each and every solar module and look for the faults. The person who is doing the inspection should be highly experienced otherwise, one might miss some faults. The National Renewable Energy Laboratory has made a checklist for visual inspection so that any fault may not be missed. Several defects may be detected like; cracked glass, delamination, corrosion, damaged wiring, bubbles, etc. Since it has high chances of human errors,

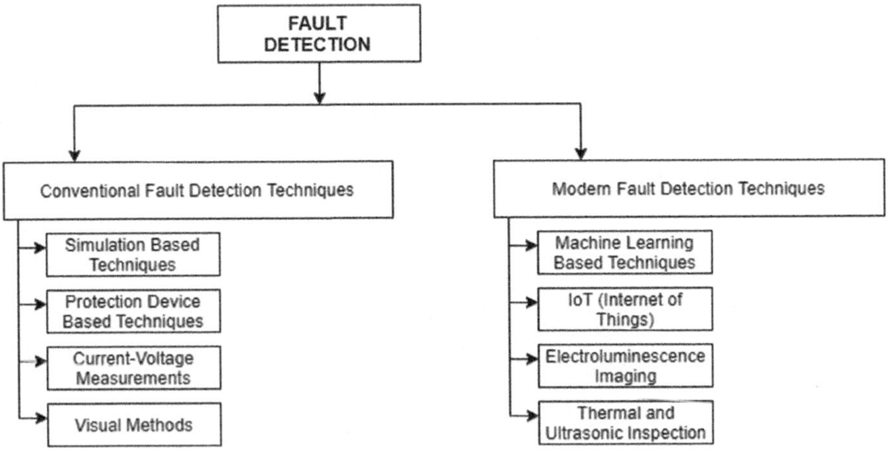

Fig. 3.5 Various fault detection techniques

this method is not reliable and it is very time-taking as well (Alam et al. 2015; Naveen Venkatesh and Sugumaran 2021).

3.4.1.2 Current–Voltage (I–V) Measurement Method

The principle of this technique is that whenever there is a fault in the PV string, majority fault always injects some noise or any abnormal voltage or current in the system which one can detect through measuring devices and by plotting I-V characteristic graph we can detect the faulty part or PV string of the system. By using this method, one can identify any kind of disconnection or partial shading because such defects reduce the output voltage of the string, which is easily detectable in this method (Boggarapu et al. 2020). However, it is not very accurate with advanced prediction of permanent fault by watching the degradation, also the exact location of the fault cannot be determined. Whole faulty string can be identified instead. Then, we can look for the exact location of fault by using some other techniques (Alam et al. 2015).

3.4.2 Modern Fault Detection Techniques

3.4.2.1 Thermal and Ultrasonic Inspection

The thermal image is captured through IR cameras; it captures the temperature differences in different PV cells and strings. This method can easily detect the faults like creation of burn marks and hotspots, snail trails, etc. This method is suitable for large-scale solar plants where inspection is not easy. In large-scale plants, the images are processed through the technology of image processing for fast and accurate results (Pillai and Rajasekar 2018). While the ultrasonic inspection is used to detect the glass breakage, void, or any other damage to the module, an ultrasonic transducer is used which transmits as well as receives the ultrasonic signals. In case of any fault, the signal received will have some disturbance which can be used to detect the fault (Zhao et al. 2014).

3.4.2.2 Electroluminescence Imaging Technique

In order to perform EL testing, it is necessary to provide PV strings with well-determined voltage so as to make a current flow and make cells emit electromagnetic emission invisible to naked eyes but visible to special cameras. The images are processed with the technology of image processing (if working on a large scale) or can be analyzed manually and faults can be diagnosed like PID, cracks, glass breakage, conversion efficiency error, etc. (Zhao et al. 2014; Yi and Etemadi 2016; Deitsch et al. 2012).

3.4.2.3 Detection with Wavelet Transform

Fault detection is a crucial process. Advanced signal processing using wavelet transform is more popular than Fourier transform for fault parameters analysis (Karmacharya and Gokaraju 2017, 2018). Continuous wavelet transform found very redundant, where the discrete wavelet transform (DWT) is represented by discrete scale and translated wavelet basis and is more suitable for real-time applications (Hu et al. 2015). DWT can be implemented by replacing a by ${a_0}^m$ and b by $nb_0{a_0}^m$ in Eq. (3.5).

$$DWT_{\Psi}x(m,n) = {a_0}^{-\frac{m}{2}}\int_{-\infty}^{\infty} x(t)\Psi^*\left(\frac{t - nb_0{a_0}^m}{{a_0}^m}\right)dt \tag{3.5}$$

where, m is scale and n is time-shift parameter.

Discrete wavelet transform is known as dyadic orthonormal wavelet transform. If $a_0 = 2$ and $b_0 = 1$ in Eq. (3.6), expressed as

$$DWT_{\Psi}x(m,n) = 2^{-\frac{m}{2}}\int_{-\infty}^{\infty} x(t)\Psi^*\left(\frac{t - n2^m}{2^m}\right)dt \tag{3.6}$$

It is very efficient algorithm to compute DWT with dyadic sampling of the time–frequency plane.

The WT can be implemented using multiresolution analysis (MRA) of a fault signal x(t), which is obtained by passing signal through a half-band digital low pass filter with impulse response. MRA technique is shown in Fig. 3.6.

A recorded time signal $c_O(n,)$ a sampled version of $x(t)$ of PV fault current, or voltage signal is decomposed into its detailed $d_1(n)$ and smoothed $c_1(n)$ signals with the application of filters $g(n)$ and $h(n)$. Here, $g(n)$ has a high-pass filter response which gives filtered signal $d_1(n)$, a detailed version of $c_O(n)$. Fault detection and classification have been successfully done using wavelet transform (Karmacharya and Gokaraju 2017). This research proposes a wavelet transform-based fault diagnosis technique for grid-connected solar systems. Defect detection for photovoltaic system is critical for PV power station management. The wavelet transformation with MRA is an effective method for detecting fault location and components fault analysis (Stember et al. 1982; Karmacharya and Gokaraju 2018; Gong et al. 2020). A thorough examination of the numerous flaws that cause PV modules failure has

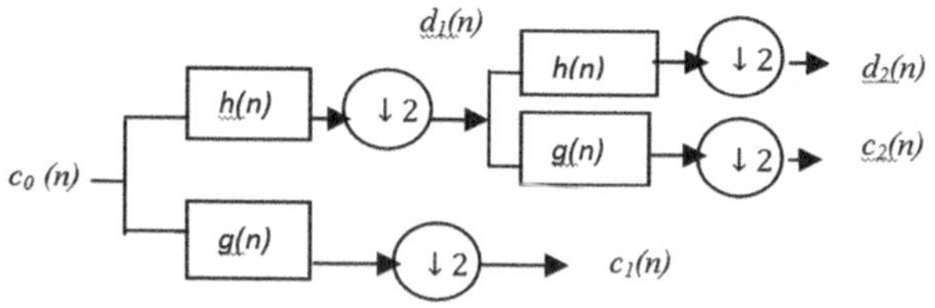

Fig. 3.6 Decomposition of c_O(n) (sample fault signal) into **2** scale

been explored. Following a review of relevant literature, a monitoring tool is developed combining thermography and artificial intelligence algorithms to detect various sorts of problems in PV modules while also filtering out non-significant anomalies (Fernández-Solas et al. 2021; Li et al. 2019).

Harrou et al. (2019) This paper employed k-nearest neighbor for fault detection. An ANN (artificial neural network) classifier is proposed to diagnose the defective PV module's transfer characteristics (I-V data), which uses multilayer perceptron (MLP) networks to determine the type and location of defects (Karmacharya and Gokaraju 2018). Further novel approaches are proposed for fault detection using deep convolution neural network (CNN) (Wu et al. 2019; Carletti et al. 2020; Pierdicca et al. 2018) us. Summary of advanced fault detection and localization techniques are given in Table 3.2

PV faults are subjected to various faults whether depending on climatic conditions, physically or any electrical faults. This is the root for the thesis as we are discussing about the classification of faults. Environment pollution and physical defect-based detection are equally important for the overall fault detection of PV array system. Further pollution detection techniques are suggested by authors for fault detection as given in Table 3.3.

3.4.2.4 IoT Based Detection Techniques

The values of current and voltages are gathered through sensors already mounted on PV modules. The data collected will be stored in the database using a Wi-Fi module or optical fiber than the values will be compared to current and voltages of healthy cells and previously recorded values at given illumination. If the value of voltage or current, is lower than the desired output at given illumination then the system will inform the control room about the fault as shown in the figure. The system is very useful for remote monitoring of fault detection and it constantly monitors the system for any kind of fault and keeps data for future use as well which increases its performance (Phoolwani et al. 2020). PV arrays are examined using IoT technology as shown in Fig. 3.7, which can considerably enhance PV array monitoring, performance, and maintenance (Hamied et al. 2018; SeyyedHosseini et al. 2020; Kekre and Gawre 2017).

An online fault supervision system of PV array using IoT is intended to implement effective monitoring system of PV operating conditions. Data is delivered to the data gateway over a dedicated IoT network (Hans and Tamhane 2020; Pereira et al. 2019; Kurukuru et al. 2019). Second, the data gateway takes data from the data collecting section, gathers fault signal data via the satellite, all before uploading them to the cloud-based website (Mellit et al. 2020). A variety of visualizations based unmanned aerial vehicle (UAV) applications for fault analysis discussed in various papers (Nie et al. 2020; Grimaccia et al. 2017; Akram et al. 2019) (Table 3.4).

Table 3.2 Fault detection and localization

Fault detection and localization		
References	Techniques	Research avenues/Application remarks
Yahyaoui and Segatto (2017), Alam et al. (2015), Hariharan et al. (2016), Kumar et al. (2018), Jain et al. (2019)	Digital twin approach	Fault modeling and analysis
Pillai and Rajasekar (2018), Jain et al. (2019)	Sensor optimization	Advanced fault detection
Karmacharya and Gokaraju (2017, 2018)	Wavelets and ANN	Advanced fault detection and classification
Hu et al. (2015)	MSD	Advanced fault detection
Khoshnami and Sadeghkhani (2018)	Sample entropy based	Fault detection
Chen et al. (2016), Zhao et al. (2013, 2014)	Fast outlier detection	Fast analysis
Yi and Etemadi (2016)	Multi-resolution signal decomposition and fuzzy	Advanced fault detection and classification
Sreelakshmy et al. (2018)	Maximal overlap discrete wavelet transform	Advanced fault detection
Roy et al. (2017)	Ground-fault detection	Fault detection
Harrou et al. (2019)	k-nearest neighbors	(EWMA) exponentially weighted moving average
Pradeep Kumar et al. (2017)	Wavelet packets	online
Zaki et al. (2021)	Fault classification	Deep learning
Pillai and Rajasekar (2018),	MPPT-based fault detection technique	Sensorless line–line and line–ground faults
Boggarapu et al. (2020), Murtaza et al. (2019)	Line to line fault	labVIEW
Seapan et al. (2020)	shading effect on I-V characteristics	MPPT
Platon et al. (2015)	Fault detection	Online
Silvestre et al. (2014), Livera et al. (2020)	Fault detection	PV systems for grid connected applications
Klise (2016)	Statistical detection methods	Fault detection
Mansouri et al. (2021)	Electroluminescence images analysis using Fourier image reconstruction	Defect detection of solar cells
Wu et al. (2019)	Improved deep learning network	Inverter
Talayero et al. (2020)	Performance monitoring	Solar plants

(continued)

Table 3.2 (continued)

Fault detection and localization		
References	Techniques	Research avenues/Application remarks
Chouder and Silvestre (2010)	Power losses analysis	Automatic supervision and fault detection
Zhang et al. 2013)	Application of independent component analysis for surface defect detection	Photovoltaic module
Carletti et al. (2020)	Deep CNN	UAV
Pierdicca et al. (2018)	Deep CNN for automatic detection	damaged photovoltaic cells
Li et al. (2019)	Hot spot detection	UAV
Nie et al. (2020)	Digital mapping of solar plant	UAV
Grimaccia et al. (2017)	Thermography, infrared images	Photovoltaic module

Table 3.3 Pollution detection techniques

References	Techniques	Research avenues/Application remarks
Darwish et al. (2015), Lu et al. (2016)	Photovoltaic performance analysis under dust pollution condition	Solar PV farms
Ramkiran et al. (2020)	PV modules with filters	Improve efficiency
Fernández-Solas et al. (2021)	Optical degradation impact on the spectral performance	Optical degradation
Dierauf et al. (2013)	Weather corrected performance ratio	Improve efficiency
Klein et al. (2016)	Electricity-based heating and cooling	Power balance
Dobaria et al. (2018)	Composite climate	PV plants
Meyer et al. (2013)	Root cause analysis of Snail trails	PV modules
Kajari-Schröder et al. (2012)	Criticality of cracks	PV modules
Bouraiou (2018)	Climatic conditions analysis under outdoor hot dry	Experimental setup to observe defects in PV modules
Ali et al. (2017)	Real-time detection of faults	light I-V measurement-based diagnosis
Li et al. (2019)	DL based defect analysis of PV module	PV farms at large scale

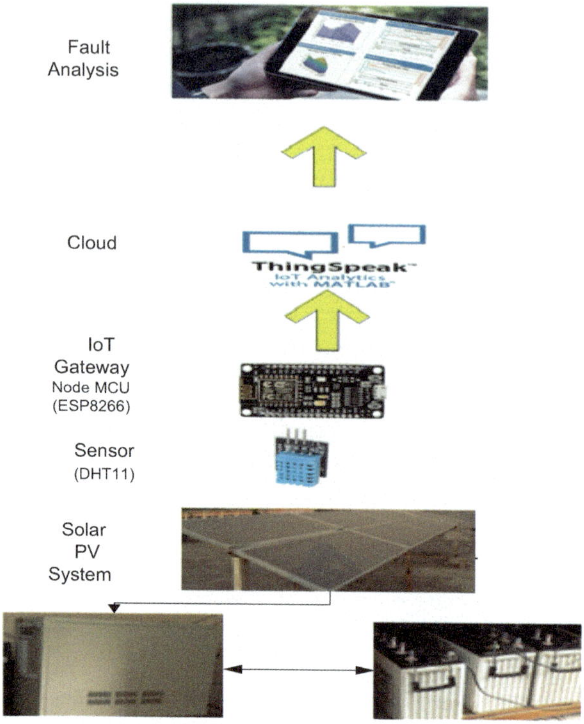

Fig. 3.7 IoT-based fault detection scheme

Table 3.4 IoT based fault detection techniques

IoT based detection techniques		
References	Advantages	Research avenues/Application remarks
Phoolwani et al. (2020)	Low cost	Thermal imaging
Hamied et al. (2018)	Experimental prototype	Controlling of PV systems with condition monitoring for micro inverter
SeyyedHosseini et al. (2020)	Multiagent-based monitoring for micro inverter	Fault monitoring and control
Kekre and Gawre (2017)	GPS based	Low cost system
Hans and Tamhane (2020)	Hybrid green energy driven	Street lighting system
Pereira et al. 2019)	Meteorological data	Photovoltaic module temperature monitoring
Nie et al. (2020), Grimaccia et al. (2017)	UAV for infrared images	Hotspots detection
Akram et al. (2019)	Processing of infrared images and thermograph	Detection of PV modules defect

3.5 Classification Based on AI Techniques

Step ahead after detection and feature extraction of fault signal is classification, further decision-making is a crucial part that will come into the picture when all the data sets are trained well for different output or fault conditions. Different classification techniques and the combination of different classifiers are discussed in this section for the proper categorization of faults in solar PV system for efficient mitigation of faults. Many classification approaches are available for fault classification under AI techniques like machine learning, k-nearest neighbor, random forest, ANN, CNN, etc.

3.5.1 Machine Learning-Based Technique

It is a technology which learns from the past experiences which is in the form of data and performs better day by day. The efficiency of a solar module depends upon many factors like irradiance, temperature, dust, shading, degradation, etc. So, it takes multiple inputs for a cell line reading of irradiance, current, voltage, temperature, etc., and then process it through an algorithm. There are many algorithms in the market like SVM (support vector machine), linear regression/classification, random forest and many more. This technology can also classify the fault on the basis of its past experience more accurately. Bigger the data set to train the algorithm or system more efficient will be the results (Boggarapu et al. 2020; Murtaza et al. 2019).

3.5.1.1 Support Vector Machines

The support vector machines (SVM) is popular and effective to the classification problems because it minimizes the generalization error using structural risk minimization. In most of the practical problem input data are non-separable, classification can as shown in Fig. 3.8

For given training set of SVM $\{(\mathbf{x}_i, y_i)\}_{i=1..n}$, $\mathbf{x}_i \in \mathbf{R}^d$, $y_i \in \{-1, 1\}$be clearly separated by a hyper plane with margin ρ. Then for each training sample $(\mathbf{x}_i, y_i)$:

$$\boldsymbol{w}^T\boldsymbol{x}_i + b \leq -\rho/2 \quad \text{if } y_i = -1 \tag{3.7}$$

$$\boldsymbol{w}^T\boldsymbol{x}_i + b \geq \rho/2 \quad \text{if } y_i = 1 \tag{3.8}$$

$$y_i\left(\boldsymbol{w}^T\boldsymbol{x}_i + b\right) \geq \rho/2 \tag{3.9}$$

here the minimization of cost function, given as

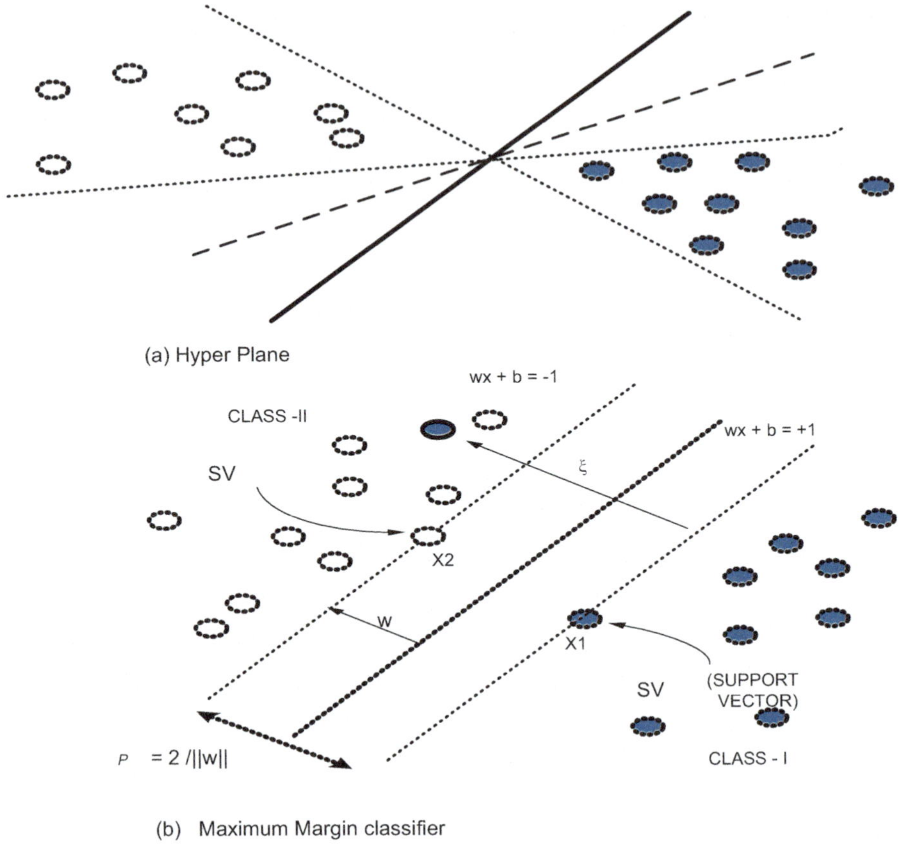

Fig. 3.8 SVM classifier

$$\min\phi(\mathrm{w}, \xi) = \frac{1}{2}w^T w + C\sum_{i=1}^{n}\xi_i \tag{3.10}$$

with constraints

$$yi(\mathbf{wTx}i + b) \geq 1 - \xi_{\mathrm{i}} \text{ and } \xi_{\mathrm{i}} \geq 0 \tag{3.11}$$

For $i = 1, 2, 3, \ldots, n$, where $C > 0$ is the regularization coefficient, and $\xi_i \geq 0$ is the stack variable (te distance between the margin and the samples xi lying on the wrong side of the margin).

Applying the Karush–Kuhn–Tucker conditions into Lagrange dual Problem

$$\mathrm{Q}(\alpha) = \sum_{i=1}^{n}\alpha_i - \frac{1}{2}\sum_{i,j=1}^{n}\alpha_i\alpha_j \mathrm{y}_i\mathrm{y}_j \mathrm{K}(\mathrm{x}_i\mathrm{x}_j) \tag{3.12}$$

With constraints

$$\sum_{i=1}^{n} \alpha_i y_i = 0 \text{ and } 0 \leq \alpha_i \leq C \quad (3.13)$$

The stack variable, $\boldsymbol{\xi}_\mathbf{i}$ and the weight vector, **w** do not exist in the dual form. Now, the decision function for soft margin support vector machines (SVM) classifier is

$$f(x) = sgn\left(\sum_{i=1}^{Ns} \alpha_i \mathrm{y}_i \mathrm{K}(\mathrm{x}_, \mathrm{x}_i) + b\right) \quad (3.14)$$

where, Ns is the no. of support vectors. $x_i, x_i \in \mathrm{SV}$,

$\mathrm{K}(\mathrm{x}_, \mathrm{x}_i) = \langle\phi(\mathrm{x}), \phi(\mathrm{x}_i)\rangle$ is the chosen kernel function.

Dealing with any number of feature, make SVM unique and simple to implement as a fault classifier (Zhao et al. 2013, 2014).

For feature extraction, a system identification approach based on wavelet and two-stage support vector machine (SVM) classifiers was used (Yi and Etemadi 2016). This detection approach simply takes data from the PV array's total voltage and current, as well as very less amount of labeled data for training of SVM classifier. The amount of energy generated from a PV array is governed by variables such as sunshine, array location, covered area, and solar panel quality. Each update increases the likelihood of an inaccuracy in the array. As a result, academics must emphasize rigorous study and the development of a procedure for quickly locating and fixing all types of errors. Machine learning (ML) is used along with voltage and current-based sensors to detect, identify, and classify common defects including open circuits, short circuits, and hot-spots (Kurukuru et al. 2019; Malhotra 2015).

3.5.2 Artificial Neural Network (ANN)

ANN is a computational model which resembles human brain behavior. Extremely complex problems can easily be dealt with ANN because of its ability of parallel computation and it is also adaptive to external disturbances which make it one of the promising computation techniques in power system (Boppana 2015; Gunda and Jones 2019). Authors have explained that ANN can adjust the data without any explicit specification of functional form for the underlying model. Further, the comparison of the performances of different classifiers on basis of features extracted by WT. In paper (Garoudja et al. 2017), authors provided a technique for fault detection and classification system which was based on WT and ANN with practical implementation (Hopwood et al. 2020).

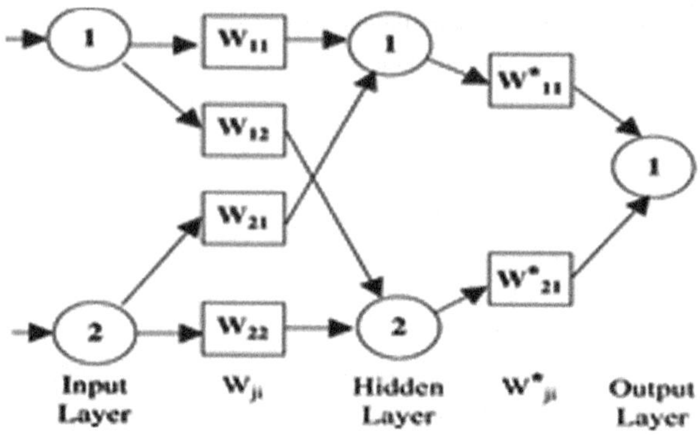

Fig. 3.9 Architecture of MFNN

3.5.2.1 Multilayer Feed Forward Network

In this technique, ANN with three layers was discussed where each layer has specific neurons with some weights where W_{ij} is weight of hidden layer neuron. After moving to the next iteration, weight of the neuron available in different layers gets updated final output as shown in Fig. 3.9. Different iterations result in the highest accuracy and better results will get from this method. To ensure accurate diagnosis and failure identification in PV systems, a fault classification technique is developed. This method takes into account the wavelet transform's feature extraction capabilities as well as the categorization attributes of radial basis function networks (RBFNs). The dynamic fusing of kernels is used to increase the effectiveness of the classification classifier (Lin et al. 2015; Garoudja et al. 2017; Hopwood et al. 2020). Further advanced deep CNN-based methods for fault classification have been proposed (Malhotra 2015; Aghaei et al. 2015; Tsanakas et al. 2015; Chouder and Silvestre 2010; Aziz et al. 2020; Rahman and Chen 2020; Bartler et al. 2018). AI technique-based classification is given in Table 3.5.

3.6 Fault Mitigation Techniques

3.6.1 Protection Device-Based Technique

Protection devices are basically used for the protection purposes like protection against over current, under voltage, etc. But nowadays, IED (intelligent electronic devices) are replacing the conventional electromechanical relays and IEDs are equipped with both alarm settings and relay settings (Boggarapu et al. 2020; Rahmann et al. 2016). This method is very useful in case of L-L and L-G fault as these faults are easily detectable. One can also install RCD (residual current device) which spots the difference in current passing through PV string terminals or PV array. In case of any fault definitely residual current will flow will lead to the actuation of RCD and thus

Table 3.5 PV Fault classification based on AI techniques

Classification based on AI techniques		
References	AI techniques	Research avenues/Application remarks
Hariharan et al. (2016)	Graph-based semi-supervised learning	Advanced AI technique suitable for
Hu et al.(2015)	SVM	Detection and classification
Lin et al. (2015)	ANN	fault diagnosis of PV array with online monitoring
Kurukuru et al. (2019), Aghaei et al. (2015), Tsanakas et al. (2015)	Machine learning techniques and thermography	Detection and classification
Gunda and Jones (2019)	WT and PNN	Fault detection and classification
Garoudja et al. (2017), Hopwood et al. (2020)	Neural network	PV panel fault analysis
Karmacharya and Gokaraju (2018)	Deep learning network	Inverter fault classification
Hooshyar et al. (2016)	Fault type classification	Microgrids including photovoltaic DGs
Deitsch et al. (2012)	Electroluminescence images	Photovoltaic module cells
Malhotra (2015)	Machine learning	fault prediction
Aziz et al. (2020), Rahman and Chen (2020), Bartler et al. (2018)	A novel CNN-based approach	Faults and defect detection

fault is detected (Yahyaoui et al. 2016), but it is a very expensive technique because it requires installation of a lot of devices and all the relays and protection devices should have optimum time setting and selectivity otherwise unwanted tripping can happen and which is not desirable (González-Longatt 2005).

3.6.2 Simulation-Based Technique

This method includes many subtechniques like power loss analysis method, earth capacitance measurement, and TDR (time domain reflector method) (González-Longatt 2005; Gunda and Jones 2019). In the power loss analysis method, the parameters of PV modules are fed to a software simulator to detect the power losses in the system and simulate the behavior of PV system then on the basis of comparison we can detect the fault in the system. Earth capacitance method is being used to detect the PID error in the module. In TDR (time domain reflector), a signal is injected into the system and the reverse signal is analyzed in the simulator to detect the level of detection (Mansouri et al. 2021).

3.6.3 Arc Fault Detection

Arc fault is responsible for fire in most of the cases. Whenever there is an arc, it injects some high-frequency noises in the system, which can be used to detect the arc fault in the system. So normal OC relays and IEDs cannot detect the arc fault because it is not a short circuit, arc is a flow of current through high impedance path. Thus, special type of devices like AFD (Arc fault detector) and AFCI (Arc fault circuit interrupter) are being used. These devices distinguish between arc noise and other noises (due to harmonics, etc.) and avoid false tripping (Zhao et al. 2013). Mitigation and solution are summarized in Table 3.6.

3.7 Key Challenges and Opportunities

Overall monitoring and mitigation is a challenging task as ground faults, L-L faults, and arc faults are prone to severe fire hazards and always suggested to have robust fault detection and mitigation techniques have been applied prevent fires (Pillai and Rajasekar 2018). Limitations persist in the conventional detection methods, due to undetected ground faults and cause severe damage to the PV array and equipments (Chen et al. 2016; Zhao et al. 2013, 2014). Some faults are instantaneous; some are repairable or needs to be replaced by the faulty portion/component. In a bigger canvas, every aspect of the PV fault must be taken into account while collecting data, so here opportunity is to deal with big data management, novel signal processing as WT is most suitable along with AI technique (Sreelakshmy et al. 2018; Pradeep Kumar et al. 2017; Lin et al. 2015). Moreover, the healing model or protection system has to be an effective and prompt response to faulty sections with ease of communication to the control center for instant action online or IoT-based novel approaches with suitable cyber security measures (Phoolwani et al. 2020; SeyyedHosseini et al. 2020; Grimaccia et al. 2017; Mellit et al. 2020). Power quality issues are directly related to faults and the frequency disturbance phenomenon (Hacke et al. 2018). Now the smart grids are more sensitive to nonlinear loads and equipped with modern monitoring and control devices, various IEEE standards, NEC (National Electrical Code), and IEC standards are available as a benchmark for the measurement of power quality issues related to faults penetration in SPVS (Pillai and Natarajan 2019). Major opportunities are-

- Highly automated SG will be required in future to have self-healing capability by prompt machine-to-machine communication-based approaches.

Table 3.6 Fault mitigation techniques

Fault mitigation techniques		
References	Techniques	Research avenues/Application remarks
Pillai and Rajasekar (2018)	PV system protection	Overall protection and PV fault analysis
Zhao et al. (2012)	L-L fault analysis	Fault protection challenges
Saleh et al. (2017)	Voltage-based protection scheme	Protection challenges for faults
Li et al. (2019)	Accurate detection of high-impedance L-L fault	Localization
Baccoli et al. (2021)	PV collector—reflector	Enhancing energy production
Teo et al. (2020)	Bypass diode	Partial shading loss
Fathabadi (2015)	PV array string and central based MPPT	Enhancement of performance
Rahmann et al. (2016)	Mitigation control against partial shading	PV system
Honrubia-Escribano et al. (2015)	Fault analysis	Power quality
Yahyaoui et al. (2016)	Control technique	Small scale single-phase grids
SETO in: 2020:a decade of progress, a promising future, DOE, New York, NY, USA (2010), Cristaldi et al. (2015), Hacke et al. (2018)	Improving photovoltaic inverter reliability	AC side
Jordan et al. (2020), Ristow et al. (2008), Pradeep Kumar and Fernandes (2017), Dumnic et al. (2018), Oprea et al. (2019), Spertino et al. (2019)	Equipment reliability, safety, and quality assurance protocols	Safety, maintenance
Mansouri et al. (2021)	Analysis and mitigation of faults	Adaptive neuro-fuzzy system
Golnas (2012), Sangwongwanich et al. (2018), Peters and Madlener (2017), Freeman et al. (2018)	System reliability and economic analysis	Inverter
Livera et al. (2020), Alsina et al. (2018)	Prevention reliability, machine learning	Industry

- Big data management with more accuracy, speed, and security are basic requirements while dealing with online—real-time communication between the service provider and customer, so data generation by utility related to faults and load parameters are equally important for the consumer end.
- Upgradation is crucial, while it matters to provide a secure and fast authentic system for close monitoring and further remedial actions.
- Before preprocessing data, monitoring points are required, at various levels and sections of RES-based systems. Optimal placement of monitoring points is a new area of research.
- Advanced AI to minimize distinguish procedure for fault analysis, which depends upon the most severely affected areas, whether it is a solar power-based system or wind generation-based system.
- Fast healing after faults in a smart grid should be objective for real-time diagnosis and condition monitoring.

Hybrid fault classification scheme is proposed where IoT-based detection and AI-based classifier is employed for SPVS as shown in Fig. 3.10. Here, IoT nodes fetch the fault signal information and send it to cloud (virtual data base), data can be retrieved remotely and further preprocessed using novel signal processing techniques (FT/WT/ST). Before classification, feature extraction is an important aspect to reduce the data and focus on specific features for the process of classification, here ANN or SVM-based methods are popular for the classification of faults. Further, a new deep learning CNN approach is suited for more efficient fault analysis.

3.8 Conclusion

Renewable energy sources are an essential part of the modern power grid and they always pose attention while considering replacing conventional energy sources in the future. For efficient maintenance and operation of SPVS robust and smart fault monitoring system is desirable. This chapter presents a comprehensive review on SPVS fault detection, classification, and mitigation techniques along with critical analysis. Signal processing-based approaches as well as experimental setup-based schemes along with online or offline fault diagnosis methods have been reviewed. It is observed that conventional/classical methods for SPVS fault diagnosis and condition monitoring are found durable and robust for classical power system model but when smart grid scenario is considered, wired, wireless communication devices with IoT and AI-based approaches are adopted heavily to make the system more flexible toward smart operation and control. A hybrid IoT and AI-based fault classification scheme for SPVS is proposed to enhance the performance of fault diagnosis. Future scope of the presented work is, to deal with cyber security issues, reconfigurable SPVS component and self-healing ability of renewable energy sources.

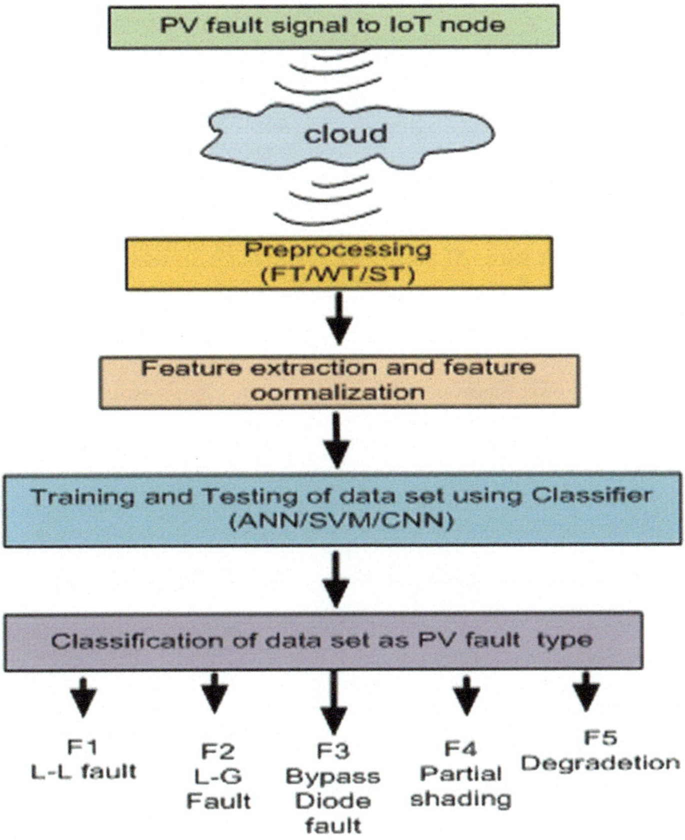

Fig. 3.10 Proposed PV fault classification scheme using IoT and advanced AI techniques

References

Aghaei M, Gandelli A, Grimaccia F, Leva S, Zich RE (2015) IR real time analyses for PV system monitoring by digital image processing techniques. In: Proceedings of international conference on event-based control, communication and signal processing (EBCCSP), pp 1–6

Akram MN, Lotfifard S (2015) Modeling and health monitoring of dc side of photovoltaic array. IEEE Trans Sustain Energy 6(4):1245–1253

Akram MW, Li G, Jin Y, Chen X, Zhu C, Zhao X, Aleem M, Ahmad A (2019) Improved outdoor thermography and processing of infrared images for defect detection in PV modules. Sol Energy 190:549–560

Alam MK, Khan F, Johnson J, Flicker J (2015) A comprehensive review of catastrophic faults in PV arrays: types, detection, and mitigation techniques. IEEE J Photovoltaics 5(3):982–997. https://doi.org/10.1109/JPHOTOV.2015.2397599

Ali MH, Rabhi A, El Hajjaji A, Tina GM (2017) Real time fault detection in photovoltaic systems. Energy Procedia 111(September):914–923

Alsina EF, Chica M, Trawiáski K, Regattieri A (2018) On the use of machine learning methods to predict component reliability from data driven industrial case studies. Int J Adv Manuf Technol 94(5–8):2419–2433
Aziz F, Ul Haq A, Ahmad S, Mahmoud Y, Jalal M, Ali U (2020) A novel convolutional neural network-based approach for fault classification in photovoltaic arrays. IEEE Access 8:41889–41904
Baccoli R, Kumar A, Frattolillo A, Mastino C, Ghiani E, Gatto G (2021) Enhancing energy production in a PV collector—reflector system supervised by an optimization model: experimental analysis and validation. Energy Convers Manag 229
Bartler A, Mauch L, Yang B, Reuter M, Stoicescu L (2018) Automated detection of solar cell defects with deep learning. In: Proceedings of 26th European signal processing conference (EUSIPCO), Sept 2018, pp 2035—2039
Boggarapu PK, Manickam C, Lehman B, Chilakapati N et al (2020) Identification of pre-existing/undetected line-to-line faults in PV array based on pre-turn on/off condition of the PV inverter. IEEE Trans Power Electron
Boppana S (2015) Outdoor soiling loss characterization and statistical risk analysis of photovoltaic power plants. Doctoral dissertation, Arizona State University
Bouraiou A et al (2018) Experimental investigation of observed defects in crystalline silicon PV modules under outdoor hot dry climatic conditions in Algeria. Sol Energy 159(December):475–487
Boutelhig A, Arab AH, Hanini S (2016) New approach to exploit optimally the PV array output energy by maximizing the discharge rate of a directly-coupled photovoltaic water pumping system (DC/PVPS). Energy Convers Manage 111:375–390
Carletti V, Greco A, Saggese A, Vento M (2020) An intelligent flying system for automatic detection of faults in photovoltaic plants. J Ambient Intell Hum Comput 11(5):2027–2040
Chen L, Li S, Wang X (2016) Quickest fault detection in photovoltaic systems. IEEE Trans Smart Grid 9(3):1835–1847
Chenni R, Makhlouf M, Kerbache T, Bouzid A (2007) A detailed modeling method for photovoltaic cells. Energy 32:1724–1730
Chine W, Mellit A, Lughi V, Malek A, Sulligoi G, Massi Pavan A (2016) A novel fault diagnosis technique for photovoltaic systems based on artificial neural networks. Renew Energy 90:501–512
Chouder A, Silvestre S (2010) Automatic supervision and fault detection of PV systems based on power losses analysis. Energy Convers Manage 51(10):1929–1937
Cristaldi L, Faifer M, Lazzaroni M, Khalil MMAF, Catelani M, Ciani L (2015) Diagnostic architecture: a procedure based on the analysis of the failure causes applied to photovoltaic plants. Measurement 67:99–107
Darwish ZA, Kazem HA, Sopian K, Al-Goul MA, Alawadhi H (2015) Effect of dust pollutant type on photovoltaic performance. Renew Sustain Energy Rev 41:735–744
Deitsch S, Christlein V, Berger S, Buerhop-Lutz C, Maier A, Gallwitz F, Riess C (2012) Automatic classification of defective photovoltaic module cells in electroluminescence images. Sol Energy 185:455–468
Dierauf T, Growitz A, Kurtz S, Cruz JLB, Riley E, Hansen C (2013) Weather corrected performance ratio. Contract 303:275–3000
Dobaria BV, Sharma V, Adeshara A (2018) Investigation of failure and degradation types of solar PV plants in a composite climate: abstract after 4–6, years of field operation. Lect Notes Electr Eng 435:227–235
Dumnic B, LiivikE, Milicevic D, Popadic B, Katic V, Blaabjerg F (2018) Fault analysis and field experiences of central inverter based 2 MW PV plant. In: Proceedings of 20th European conference on power electronics and applications, pp 1–5
El Fadil H, Giri F (2011) Climatic sensor less maximum power point tracking in PV generation systems. Contr Eng Pract 19:513–521
Fathabadi H (2015) Two novel techniques for increasing energy efficiency of photovoltaic-battery systems. Energy Convers Manage 105:149–166

Feaster S, Wamsted D (2020) IEEFA U.S.: utility-scale renewables top coal for the first quarter of 2020 [Online]. https://ieefa.org/ieefa-u-s-utility-scale-renewables-top-coal-for-thefirst-quarter-of-2020/
Fernández-Solas Á, Micheli L, Almonacid F, Fernández EF (2021) Optical degradation impact on the spectral performance of photovoltaic technology. Renew Sustain Energy Rev 141
Freeman JM, Klise GT, Walker A, Lavrova O (2018) Evaluating energy impacts and costs from PV component failures. In: Proceedings of IEEE 7th world conference on photovoltaic energy conversion, pp 1761–1765
Garoudja E, Harrou F, Sun Y, Kara K, Chouder A, Silvestre S (2017) Statistical fault detection in photovoltaic systems. Sol Energy 150:485–499
Gokmen N, Karatepe E, Celik B, Silvestre S (2012) Simple diagnostic approach for determining of faulted PV modules in string based PV arrays. Sol Energy 86:3364–3377
Gokmen N, Karatepe E, Silvestre S, Celik B, Ortega P (2013) An efficient fault diagnosis method for PV systems based on operating voltage-window. Energy Convers Manag 73:350–360
Golnas A (2012) PV system reliability: an operator's perspective. In: Proceedings of IEEE 38th photovoltaic specialists conference (PVSC), pp 1–6
Aureliano Gomes de Brito M, Galotto L, Poltronieri Sampaio L, Dazevedo Melo G, Canesin CA (2013) Evaluation of the main MPPT techniques for photovoltaic applications. IEEE Trans Industr Electron 60:1156–1167
Gong X, Wang N, Zhang Y, Yin S, Wang M, Wu G (2020) Fault diagnosis of micro grid inverter based on wavelet transform and probabilistic neural network. In: Proceedings of 39th Chinese control conference (CCC), pp 4078–4082
González-Longatt FM (2005) Model of photovoltaic module in Matlab. In: The proceedings of the second Ibero-American conference of electrical, electronics and computation students, pp 1–5
Grimaccia F, Leva S, Niccolai A (2017) PV plant digital mapping for modules' defects detection by unmanned aerial vehicles. IET Renew Power Gener 11(10):1221–1228
Gunda T, Jones CB (2019) Data-driven analysis of PV failures from O&M records. In: Proceedings of renewables O&M innovation workshop, Charlotte, NC, USA, Sandia National Lab. (SNL-NM)
Hacke P, Lokanath S, Williams P, Vasan A, Sochor P, TamizhMani G, Shinohara H, Kurtz S (2018) A status review of photovoltaic power conversion equipment reliability, safety, and quality assurance protocols. Renew Sustain Energy Rev 82:1097–1112
Hamied A, Mellit A, Zoulid M, Birouk R (2018) IoT-based experimental prototype for monitoring of photovoltaic arrays. In: 2018 international conference on applied smart systems (ICASS), Medea, Algeria, pp 1–5. https://doi.org/10.1109/ICASS.2018.8652014
Hans MR, Tamhane MA (2020) IoT based hybrid green energy driven street lighting system. In: 2020 fourth international conference on I-SMAC (IoT in Social, Mobile, Analytics and Cloud) (I-SMAC), Palladam, India, pp 35–41. https://doi.org/10.1109/I-SMAC49090.2020.9243365
Hariharan R, Chakkarapani M, Ilango GS, Nagamani C (2016) A method to detect photovoltaic array faults and partial shading in PV systems. IEEE J Photovoltaics 6(5):1278–1285. https://doi.org/10.1109/JPHOTOV.2016.2581478
Harrou F, Taghezouit B, Sun Y (2019) Improved k NN-based monitoring schemes for detecting faults in PV systems. IEEE J Photovoltaics 9(3):811–821
Honrubia-Escribano A, García-Sánchez T, Gómez-Lázaro E, Muljadi E, MolinaGarcía A (2015) Power quality surveys of photovoltaic power plants: characterisation and analysis of grid-code requirements. Renew Power Gener IET 9:466–473
Hooshyar A, El-Saadany EF, Sanaye-Pasand M (2016) Fault type classification in microgrids including photovoltaic DGs. IEEE Trans Smart Grid 7(5):2218–2229
Hopwood MW, Gunda T, Seigneur H, Walters J (2020) Neural network based classification of string-level IV curves from physically-induced failures of photovoltaic modules. IEEE Access 8:161480–161487
Hu Y, Cao W, Wu J, Ji B, Holliday D (2014) Thermography-based virtual MPPT scheme for improving PV energy efficiency under partial shading conditions. IEEE Trans Power Electron 29(11):5667–5672

Hu Y, Zhang J, Cao W, Wu J, Tian GY, Finney SJ, Kirtley JL (2015) Online two-section PV array fault diagnosis with optimized voltage sensor locations. IEEE Trans Ind Electron 62(11):7237–7246
IEA (2020) Renewables 2020. IEA, Paris. https://www.iea.org/reports/renewables-2020
Jain P, Poon J, Singh JP, Spanos C, Sanders SR, Panda SK (2019) A digital twin approach for fault diagnosis in distributed photovoltaic systems. IEEE Trans Power Electron 35(1):940–956
Jaraniya D, Nema RK, Gawre SK (2020) Design and simulation of power electronics interface for modified P & O maximum power point tracking under suddenly varying irradiance. In: 2020 IEEE international students' conference on electrical, electronics and computer science (SCEECS), pp 1–6
Jordan DC, Marion B, Deline C, Barnes T, Bolinger M (2020) PV field reliability status—analysis of 100 000 solar systems. Prog Photovoltaics Res Appl 28(8):739–754
Kajari-Schröder S, Kunze I, Köntges M (2012) Criticality of cracks in PV modules. Energy Procedia 27:658–663
Karmacharya IM, Gokaraju R (2017) Fault location in ungrounded photovoltaic system using wavelets and ann. IEEE Trans Power Del 33(2):549–559
Karmacharya IM, Gokaraju R (2018) Fault location in ungrounded photovoltaic system using wavelets and ANN. IEEE Trans Power Del 33(2):549–559
Katoch S et al (2018) Shading prediction, fault detection, and consensus estimation for solar array control. In: 2018 IEEE industrial cyber-physical systems (ICPS), St. Petersburg, Russia, 2018, pp 217–222. https://doi.org/10.1109/ICPHYS.2018.8387662
Kekre A, Gawre SK (2017) Solar photovoltaic remote monitoring system using IOT. In: 2017 international conference on recent innovations in signal processing and embedded systems (RISE), Bhopal, India, pp 619–623. https://doi.org/10.1109/RISE.2017.8378227
Khoshnami A, Sadeghkhani I (2018) Sample entropy-based fault detection for photovoltaic arrays. IET Renew Power Gener 12(16):1966–1976
Klein K, Langner R, Kalz D, Herkel S, Henning HM (2016) Grid support coefficients for electricity-based heating and cooling and field data analysis of present-day installations in Germany. Appl Energy 162:853–867
Klise KA (2016) Performance monitoring using Pecos. Sandia National Lab. (SNL-NM), Albuquerque, NM, USA, Tech. Rep. SAND2016–4303C
Koutroulis E, Blaabjerg F (2012) A new technique for tracking the global maximum power point of PV arrays operating under partial-shading conditions. IEEE J Photovolt 2:184–190
Kumar A, Gawre SK, Sarkar M, Gosula S (2018) A real-time comparative data analysis of different types of solar panels during partial shading with distinct tilt angles. In: 2018 15th IEEE India council international conference (INDICON), pp 1–6. https://doi.org/10.1109/INDICON45594.2018.8987115
Kurukuru VSB, Haque A, Khan MA, Tripathy AK (2019) Fault classification for photovoltaic modules using thermography and machine learning techniques. In: 2019 international conference on computer and information sciences (ICCIS), Sakaka, Saudi Arabia, pp 1–6. https://doi.org/10.1109/ICCISci.2019.8716442
Li Z, Yu Y, Wu C, Yang Z, Meng J (2019) Detection of high-impedance line-line fault in photovoltaic arrays based on voltage divider. In: 2019 IEEE sustainable power and energy conference (iSPEC), pp 786–791. IEEE
Li X, Yang Q, Lou Z, Yan W (2019) Deep learning based module defect analysis for large-scale photovoltaic farms. IEEE Trans Energy Convers 34(1):520–529
Lin H, Chen Z, Wu L, Lin P, Cheng S (2015) On-line monitoring and fault diagnosis of PV array based on BP neural network optimized by genetic algorithm. Multi-disciplinary trends in artificial intelligence. Springer International Publishing, pp 102–112
Livera A, Theristis M, Makrides G, Georghiou GE (2019) Recent advances in failure diagnosis techniques based on performance data analysis for grid-connected photovoltaic systems. Renew Energy 133:126–143

Livera A, Theristis M, Koumpli E, Theocharides S, Makrides G, Sutterlueti J, Stein JS, Georghiou GE (2020) Data processing and quality verification for improved photovoltaic performance and reliability analytics. Prog Photovoltaics Res Appl

Lu H, Lu L, Wang Y (2016) Numerical investigation of dust pollution on a solar photovoltaic (PV) system mounted on an isolated building. Appl Energy 180:27–36

Maki A, Valkealahti S (2012) Power losses in long string and parallel-connected short strings of series-connected silicon-based photovoltaic modules due to partial shading conditions. IEEE Trans Energy Convers 27:173–183

Malhotra R (2015) A systematic review of machine learning techniques for software fault prediction. Appl Soft Comput J 27:504–518

Mansouri MM, Hadjeri S, Brahami M (2021) New method of detection, identification, and elimination of photovoltaic system faults in real time based on the adaptive Neuro-fuzzy system. IEEE J Photovoltaics 11(3):797–805. https://doi.org/10.1109/JPHOTOV.2021.3051145

Mellit A, Hamied A, Lughi V, Pavan AM (2020) A low-cost monitoring and fault detection system for stand-alone photovoltaic systems using IoT technique. In: Zamboni W, Petrone G (eds) ELECTRIMACS 2019. Lecture notes in electrical engineering, vol 615. Springer, Cham. https://doi.org/10.1007/978-3-030-37161-6_26

Meyer S et al (2013) Snail trails: root cause analysis and test procedures. Energy Procedia 38:498–505

Murtaza AF, Bilal M, Ahmad R, Sher HA (2019) A circuit analysis based fault finding algorithm for photovoltaic array under LL/LG faults. IEEE J Emerg Sel Top Power Electron 1–1

Naveen Venkatesh S, Sugumaran V (2021) Fault diagnosis of visual faults in photovoltaic modules: a review. Int J Green Energy 18(1):37–50. https://doi.org/10.1080/15435075.2020.1825443

Nie J, Luo T, Li H (2020) Automatic hotspots detection based on UAV infrared images for large-scale PV plant. Electron Lett 56(19):993–995

Oprea S-V, Bâra A, Preoţescu D, Elefterescu L (2019) Photovoltaic power plants (PV-PP) reliability indicators for improving operation and maintenance activities. A case study of PV-PP Agigea located in Romania. IEEE Access 7:39142–39157

Pereira RIS, Jucá SCS, Carvalho PCM, Souza CP (2019) IoT network and sensor signal conditioning for meteorological data and photovoltaic module temperature monitoring. IEEE Lat Am Trans 17(06):937–944. https://doi.org/10.1109/TLA.2019.8896816

Peters L, Madlener R (2017) Economic evaluation of maintenance strategies for ground-mounted solar photovoltaic plants. Appl Energy 199:264–280

Phoolwani UK, Sharma T, Singh A, Gawre SK (2020) IoT based solar panel analysis using thermal imaging. In: 2020 IEEE international students' conference on electrical, electronics and computer science (SCEECS), Bhopal, India, pp 1–5. https://doi.org/10.1109/SCEECS48394.2020.114

Pierdicca R, Malinverni ES, Piccinini F, Paolanti M, Felicetti A, Zingaretti P (2018) Deep convolutional neural network for automatic detection of damaged photovoltaic cells. Int Arch Photogramm Remote Sens Spatial Inf Sci 42:893–900

Pillai DS, Rajasekar N (2018) An MPPT-based sensorless line–line and line–ground fault detection technique for PV systems. IEEE Trans Power Electron 34(9):8646–8659

Pillai DS, Rajasekar N (2018) A comprehensive review on protection challenges and fault diagnosis in PV systems. Renew Sustain Energy Rev 91. https://doi.org/10.1016/j.rser.2018.03.082

Pillai DS, Natarajan R (2019) A compatibility analysis on NEC, IEC, and UL standards for protection against line-line and line-ground faults in PV arrays. IEEE J Photovoltaics 9(3):864–871

Planas E, Andreu J, Gárate JI, de Alegría IM, Ibarra E (2015) AC and DC technology in microgrids: a review. Renew Sustain Energy Rev 43:726–749

Platon R, Martel J, Woodruff N, Chau TY (2015) Online fault detection in PV systems. IEEE Trans Sustain Energy 2015:1200–1207

Pradeep Kumar VVS, Fernandes BG (2017) A fault-tolerant single-phase grid-connected inverter topology with enhanced reliability for solar PV applications. IEEE J Emerg Sel Topics Power Electron 5(3):1254–1262

Pradeep Kumar B, Saravana Ilango G, Jaya Bharata Reddy M, Chilakapati N (2017) Online fault detection and diagnosis in photovoltaic systems using wavelet packets. IEEE J Photovoltaics 8(1):257–265
Rahman MRU, Chen H (2020) Defects inspection in polycrystalline solar cells electroluminescence images using deep learning. IEEE Access 8:40547–40558
Rahmann C, Vittal V, Ascui J, Haas J (2016) Mitigation control against partial shading effects in large-scale PV power plants. IEEE Trans Sustain Energy 7:173–180
Ramkiran B, Sundarabalan CK, Sudhakar K (2020) Performance evaluation of solar PV module with filters in an outdoor environment, Case Stud Therm Eng 21
Ristow A, Begovic M, Pregelj A, Rohatgi A (2008) Development of a methodology for improving photovoltaic inverter reliability. IEEE Trans Ind Electron 55(7):2581–2592
Roy S, Alam MK, Khan F, Johnson J, Flicker J (2017) An irradiance-independent, robust ground-fault detection scheme for PV arrays based on spread spectrum time-domain reflectometry (SSTDR). IEEE Trans Power Electron 33(8):7046–7057
Saleh KA, Hooshyar A, El-Saadany EF, Zeineldin HH (2017) Voltage-based protection scheme for faults within utility-scale photovoltaic arrays. IEEE Trans Smart Grid 9(5):4367–4382
Sangwongwanich A, Yang Y, Sera D, Blaabjerg F, Zhou D (2018) On the impacts of PV array sizing on the inverter reliability and lifetime. IEEE Trans Ind Appl 54(4):3656–3667
Seapan M, Hishikawa Y, Yoshita M, Okajima K (2020) Detection of shading effect by using the current and voltage at maximum power point of crystalline silicon PV modules. Sol Energy 211:1365–1372
SETO in 2020: a decade of progress, a promising future, DOE, New York, NY, USA (2010)
SeyyedHosseini M, Yazdian-Varjani A, Mohamadian M (2020) IOT based multi agent micro inverter for condition monitoring and controlling of PV systems. In: 2020 11th power electronics, drive systems, and technologies conference (PEDSTC), Tehran, Iran, pp 1–6. https://doi.org/10.1109/PEDSTC49159.2020.9088449
Silvestre S, Aires da Silva M, Chouder A, Guasch D, Karatepe E (2014) New procedure for fault detection in grid connected PV systems based on the evaluation of current and voltage indicators. Energy Convers Manage 86:241–249
Spataru S, Sera D, Kerekes T, Teodorescu R (2015) Diagnostic method for photovoltaic systems based on light I-V measurements. Sol Energy 119:29–44
Spertino F, Chiodo E, Ciocia A, Malgaroli G, Ratclif A (2019) Maintenance activity, reliability analysis and related energy losses in five operating photovoltaic plants. In: Proceedings of IEEE international conference on environment and electrical engineering and IEEE industrial and commercial power systems Europe, pp 1–6
Sreelakshmy J, Pradeep Kumar B, Saravana Ilango G, Nagamani C (2018) Identification of faults in PV array using maximal overlap discrete wavelet transform. In: 2018 20th national power systems conference (NPSC), pp 1–6. IEEE
Stember LH, Huss WR, Bridgman MS (1982) A methodology for photovoltaic system reliability & economic analysis. IEEE Trans Rel R–31(3):296–303
Talayero AP, Llombart A, Melero JJ (2020) Diagnosis of failures in solar plants based on performance monitoring. Renew Energy Power Qual J 18:33–128
Teo JC, Tan RHG, Mok VH, Ramachandaramurthy VK, Tan CK (2020) Impact of bypass diode forward voltage on maximum power of a photovoltaic system under partial shading conditions. Energy 191
Tsanakas JA, Chrysostomou D, Botsaris PN, Gasteratos A (2015) Fault diagnosis of photovoltaic modules through image processing and canny edge detection on field thermographic measurements. Int J Sustain Energy 34(6):351–372
Tyagi VV, Rahim NAA, Rahim NA, Selvaraj JAL (2013) Progress in solar PV technology: research and achievement. Renew Sustain Energy Rev 20:443–461. https://doi.org/10.1016/j.rser.2012.09.028

Wang W, Liu AC-F, Chung HS-H, Lau RW-H, Zhang J, Lo AW-L (2015) Fault diagnosis of photovoltaic panels using dynamic current–voltage characteristics. IEEE Trans Power Electron 31(2):1588–1599

Wu J, Yan Z, Sun Q (2019) Multiple faults detection of three-level NPC inverter based on improved deep learning network. In: Proceedings of international conference on applications and techniques in cyber intelligence. Springer, pp 1575–1583 [Online]. http://link-springer-com-443.webvpn.fjmu.edu.cn/chapter/10.1007%2F978-3-030-25128-4_195#citeas

Yahyaoui I (2016) Specifications of photovoltaic pumping systems in agriculture: sizing, fuzzy energy management and economic sensitivity analysis. Book. ISBN: 9780128120392. Elsevier

Yahyaoui I, Segatto MEV (2017) A practical technique for on-line monitoring of a photovoltaic plant connected to a single-phase grid. Energy Convers Manag 132. https://doi.org/10.1016/j.enconman.2016.11.031

Yahyaoui I, Tadeo F, Segatto MEV (2016) Control strategy for small-scale photovoltaic systems connected to single-phase grids. In: The proceeding of the international renewable energy congress. IREC, IEEE, pp 1–6

Yi Z, Etemadi AH (2016) Fault detection for photovoltaic systems based on multi-resolution signal decomposition and fuzzy inference systems. IEEE Trans Smart Grid 8(3):1274–1283

Zaki SA, Zhu H, Fakih MA, Sayed AR, Yao J (2021) Deep learning–based method for faults classification of PV system. IET Renew Power Gener 15:193–205

Zhang X, Sun H, Zhou Y, Xi J, Li M (2013) A novel method for surface defect detection of photovoltaic module based on independent component analysis. Math Problems Eng 2013

Zhao Y, De Palma J-F, Mosesian J, Lyons R, Lehman B (2012) Line–line fault analysis and protection challenges in solar photovoltaic arrays. IEEE Trans Ind Electron 60(9):3784–3795

Zhao Y, Lehman B, Ball R, Mosesian J, Palma J-F (2013) Outlier detection rules for fault detection in solar photovoltaic arrays. In: 2013 twenty-eighth annual IEEE applied power electronics conference and exposition (APEC), pp 2913–2920. IEEE

Zhao Y, Balboni F, Arnaud T, Mosesian J, Ball R, Lehman B (2014) Fault experiments in a commercial-scale PV laboratory and fault detection using local outlier factor. In: 2014 IEEE 40th photovoltaic specialist conference (PVSC), pp 3398–3403. IEEE

Chapter 4
Overview of Energy Management Systems for Microgrids and Smart Grid

Siddharth Jain, Aboli Kulkarni, and Yashwant Sawle

Abstract Global energy crisis and the ongoing transition toward green energy has necessitated the integration of renewable energy generation systems into the grid. This calls for an efficient and reliable energy management system that promotes a sense of balance between supply and demand of energy, reduces power loses and unanticipated peak loads, etc., for stable operation. Centralized optimization of energy resources is prevalent, but a demand response method is also proposed, wherein, the consumers adjust their power consumption to not exceed the generation. This book chapter gives a comprehensive idea about different energy management techniques and their requirement with an emphasis on the currently under process and more sophisticated energy-saving algorithms, that can benefit various stakeholders in the grid system.

Keywords Green energy · Centralized optimization · Demand response · Energy management techniques · Energy-saving algorithms

Nomenclature

EMS	Energy management system
AMI	Advanced metering infrastructure
CHP	Combined heat and power plants
BACS	Building automation and control system
PLC	Programmable Logic Controller
HMI	Human–Machine Interface
LP	Linear Programming
MILP	Mixed-integer Linear Programming

S. Jain · A. Kulkarni
School of Electrical Engineering, Vellore Institute of Technology, Vellore, India

Y. Sawle (✉)
Madhav Institute of Technology & Science, Gwalior, India
e-mail: yashsawle@gmail.com

A. K. Bohre et al. (eds.), *Planning of Hybrid Renewable Energy Systems, Electric Vehicles and Microgrid*, Energy Systems in Electrical Engineering,
https://doi.org/10.1007/978-981-19-0979-5_4

SCADA	Supervisory, data, and control acquisition systems
BMS	Building Management System

4.1 Introduction

Large-scale deployment of renewable energy resources is necessitated due to the global energy crisis, increasing carbon footprint, and ongoing green energy transition. The use of non-conventional sources of energy like solar, wind, tidal, geothermal, etc., help alleviate environmental pollutants and support sustainable development. Integrating these distributed renewable energy generation systems into the utility grid has brought into existence the microgrid concept. A microgrid acts as a self-sufficient system with two modes of operation: grid-connected mode and islanded mode of operation in case of grid failures. For the maximum utilization of the generated renewable energy, there has been considerable research in energy management systems for both the microgrid and smart grid. An ideal energy management system helps optimize the utilization of the generated renewable energy, insecure, efficient, reliable, well-coordinated, and intelligent ways. It also assists in monitoring and controlling the transmission, generation, and distribution system (Zia et al. 2018). This book chapter highlights:

- The necessity of an energy management system in a microgrid or smart grid
- The objectives of an energy management system and its constraints
- Various energy management techniques and energy-saving algorithms for microgrid
- Solutions to energy management problems in a microgrid (AI-based, metaheuristic approaches, etc.)
- Architecture of an energy management system in a smart grid
- Tools used in energy management systems in smart grid.

Therefore, a comprehensive review of the function, importance, constraints and barriers, etc., of an energy management system in a microgrid and a smart grid is given.

4.1.1 Microgrid

Several definitions of a microgrid are found in research works. A broad conclusion can be drawn that a microgrid is a group of distributed energy generation units and a cluster of connected electric loads with defined boundaries. They can be partially or completely controlled. A microgrid can also operate as an independent entity or in connection with the utility grid. These are known as islanded mode and grid-connected mode of operation, respectively. The concept of microgrids came into existence to ensure a reliable and economic energy supply. Any problem occurring in electrical energy collection can be solved more efficiently due to decentralized supervisory control, thus eliminating a more complex centralized system (Hirsch et al. 2018).

Several factors influence further research and development in microgrids. Primarily, the need for clean energy and the intermittencies that come with it is a major factor. Environmental emissions due to greenhouse gases associated with conventional energy sources have necessitated the integration of non-renewable energy sources in the grid. Further, microgrids are technically equipped to handle the variations in the generation of renewable energy with optimal energy scheduling and several energy storage systems. These systems include lithium-ion, nickel–cadmium, and lead-acid; regenerative fuel cells like vanadium redox and zinc-bromine; and highly efficient kinetic energy storage systems like flywheels.

Microgrids have also proven to be very economical in infrastructure, efficiency improvement methods, and ancillary services. Usage of absorption cooling technology in CHP (combined heat and power plants) has helped address increased load demand. Also, reduction in line losses and direct current distribution systems have improved grid efficiency greatly. A microgrid's services include control over voltage and reactive power generation, frequency regulation, compensation for harmonics in the generated voltage, power quality, consumer privacy, etc. Energy security is another important arena related to a microgrid that is greatly probed. Severe weather conditions, cyber-attacks, and cascading outages are the driving factors. In case of power outages in some areas due to adverse weather conditions, a microgrid can divert power to critical from their distributed generation units or energy storage systems to critical loads.

Further, microgrids enable segmented control over the grid. Therefore, a domino effect doesn't affect the entire primary grid in case of a power outage. Consumer information and communication of the central control with the localized control centers form microgrid operations. Physical and cyber-attacks on the grid compromise its reliability and affects consumer privacy. Therefore, significant research has been done in this area to mitigate the same.

Hence, microgrids' benefits include decarbonization, reduction in load congestion in the grid, improvement in electric utility grid operation, etc.

4.1.2 Smart Grid

The smart grid is an intelligent electric grid that allows the consumers to experience a sustainable, economical, efficient, and secure electrical energy supply. It integrates information technology for optimal energy supply and has a two-way communication between the grid operators and customers. Advanced metering infrastructure (AMI) allows real-time monitoring of energy consumption and ensures accurate energy pricing to its customers. Smart meters are an important aspect in this, enabling notifications in case of a power outage, power quality, and better energy management. Smart grids also allow dynamic pricing mechanisms subjected to load demand (Tuballa and Abundo 2016).

Smart grids having dispersed network infrastructure proves to be beneficial since it reduces maintenance costs and allows better control. They also have self-healing responses to any grid disturbances. It gives an active role to electricity consumers and provides demand flexibility. It also ensures a reduction of transmission and distribution power losses.

Therefore, the smart grid has revolutionized electrical energy generation and supply.

4.1.3 Literature Review

The electric current production and distribution by means of conventional grids that function for customers nowadays have been progressed over 100 years. Traditional grids have aided fine earlier; however, they cannot be suitable for the upcoming time to come (EUR22040 2006; Hamilton and Summy 2011) in requirements of their effectiveness and atmosphere friendliness. Moreover, fresh encounters are ascending from the growing user's requirements and the reduction of conventional sources of energy at a steady percentage (Birol 2006; Armaroli and Balzani 2007). A conventional power network comprises of central production of electric current, which is transferred and circulated to the users in one way (Yu et al. 2011). The key sources utilized to produce electricity globally are conventional sources: coal (43%) and natural gas (22.8%) (IEA 2015). However, predictions in (El-Hawary 2014) specify a fair of electric current and hybrid electric vehicles of 105.7 million until 2050, signifying 11% of the global electrical energy usage. They can be reviewed by a modern power system network, capable of increasing grid's productivity, dependability, and safety. It decreases peak load demand; suggests ecofriendly aids, tracing, and healing errors (Brown and Zhou 2013). To encounter non-conventional energy sources requirement to dodge forthcoming supply disasters, which offer an outstanding chance for the arrangement of smart grid tools (Heffner 2011). As peak demand of load is rising at a steady rate in all areas, smart grids placement may further help handle the growth in expected peak load (Gungor et al. 2011). Computerized controls, up-to-date communication structure, and sensing and energy management tools (Sawle et al. 2021) accomplish these characteristics. Smart grids have barricades that oppose advancing progress (Carvalho et al. 2012).

Moreover, smart grids are vital to incorporate numerous forms of non-conventional distributed production (Joskow 2012). The information gathered by smart electric meters delivers real-time info about electrical current usage of a utility service center (Shortfall, managing an electricity 2010). Administrations and utilities have met holes between supply and demand, guided to blackouts, load shedding, and production shortage (Sawle et al. 2018). Non-conventional sources of energy can lessen ecological and public influences (Urmee et al. 2009). Besides, it offers minimal charge options than non-renewable energy technologies offer for numerous applications (Pratt et al. 2010).

Moreover, it will enlarge the prevailing system's capacity and decrease greenhouse gas emissions, which are accountable for global warming and depletion of the ozone layer (Joskow and Wolfram 2012). However, the Energy Management System displays the real-time electricity tariffs, allowing consumers to schedule home smart appliances to off-peak periods when the electricity price is lower. It also allows the electric utility to send overload signals to smart homes to reduce electricity consumption to avoid interruptions and peak demand rates, assisting the grid balance (Meliopoulos 2002). The microgrid enriched with modern power electronic-based technology (Peng et al. 2009; Sawle et al. 2018) can offer higher dependability of service, better quality of power supply, and better efficiency of energy use by utilizing the available waste heat. The microgrid, an integrated form of DERs, is normally interfaced with load and utility grid by electronic power inverters (Olivares et al. 2014). In Fig. 4.1, we have proposed a framework for the chapter.

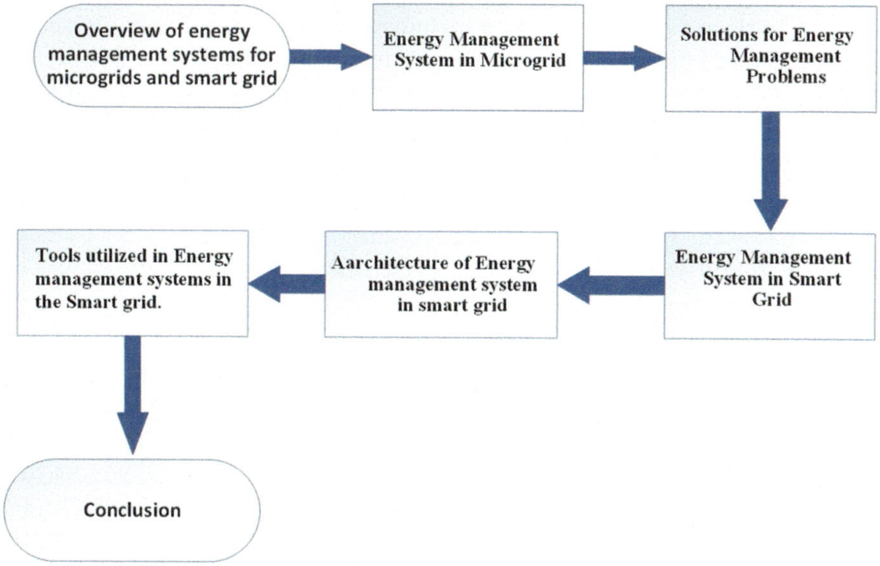

Fig. 4.1 Proposed framework

4.2 Energy Management Systems in Microgrid

4.2.1 Microgrid Energy Management

According to the International Organization for Standardization (ISO), the standard ISO 50001 emphasizes establishing, implementing, and maintaining an energy management system (EMS) that enables continual development in energy utilization, consumption, security, and efficiency to improve the overall energy performance. The incorporation assures an EMS's effective implementation with its strategies for decision-making of load forecasting modules, supervisory, data, and control acquisition systems (SCADA), and Human–Machine Interfaces (HMI). Further, the International Electrotechnical Commission standard IEC 61,970 pertaining to the application program interface of an energy management system to the power grid describes an EMS as a computer-aided system that consists of software which helps in information exchange to the primary control center, provides applications for the functionality required for the efficient operation of electrical generation, transmission and distribution facility, and promises minimal cost for energy supply. Functions of a microgrid EMS include analysis, monitoring, energy forecasting of distributed energy generation resources, reduction of operation costs, control over the market's energy prices, reduction of carbon dioxide emission, and a reliable energy supply and increase in the lifetime of the system components.

4.2.2 Objective Functions and Constraints of an Energy Management System

The objective functions are defined after considering various aspects such as the consumer preference, the equipment used in the grid, energy generation, storage, transmission and distribution, government rules and regulations, and tariff types. An EMS is primarily meant for reducing costs related to environmental factors, capital and operation, energy storage, etc. Due to greenhouse gas emissions and penalties for the same, costs are incurred under the environmental category. Allocation, production, maintenance, supply costs, purchase costs due to energy shortage, start-up and shutdown costs, battery degradation costs, fuel, and electrolyser costs form the capital and operational expenses. Expenditure due to charging/discharging, money spent on ultra-capacitors, hydrogen storage, batteries are the energy storage costs. Much work is done to propose several single and multiobjective optimization techniques to ensure the microgrid EMS's reliability. An EMS must function in the energy generated (minimum–maximum) limits of a distributed energy generation system for economic and safe operation. This is known as consumption constraint, and all kinds of loads like households, industries, etc., must operate within it. The physical capacity of the energy storage systems must also be considered since overcharging and discharging affect their health and efficiency. Reactive power support,

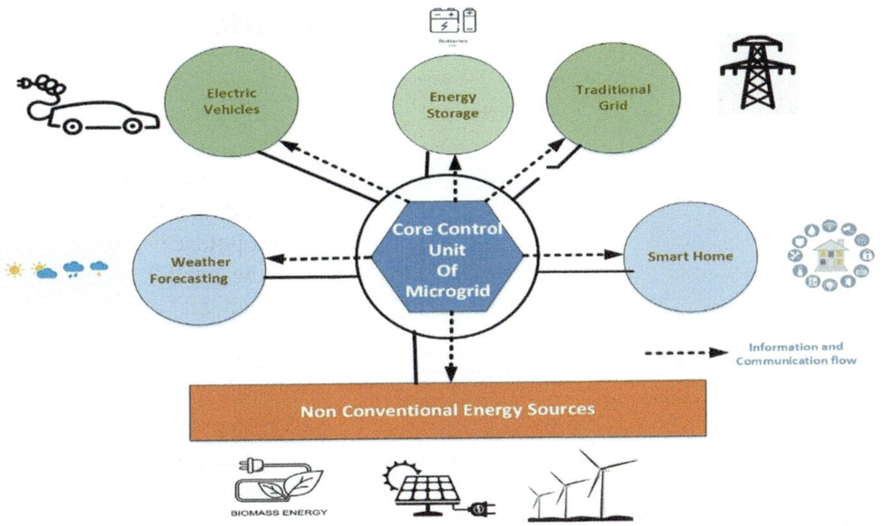

Fig. 4.2 The architecture of a microgrid

demand response, energy balancing, etc., are other constraints for the EMS to meet its objectives. Figure 4.2 demonstrates the architecture of the microgrid.

4.2.3 Optimization Techniques for Energy Management Systems

The supervisory, control, and data acquisition architecture for an EMS is either centralized or decentralized. In the centralized type of EMS SCADA, information such as the power generated by the distributed energy resources, the central controller of microgrid collects the consumers' power consumption, weather-related data, and cost-function. Further, the central control governs the optimal energy scheduling and passes on this information to the local control. However, in the decentralized type of EMS SCADA architecture, the central control of the microgrid sends and receives all the required information from the local control. The local control requests the central control for energy generation. With the integration of distributed renewable energy generation resources, electric vehicles, smart homes/buildings, and energy storage systems into the grid, the microgrid EMS strategies to achieve its objectives have diversified. They include the economical, reliable, and sustainable operation of the microgrid, economic dispatch, minimization of the system losses, controlling any intermittencies in the renewable energy generation resources, unit commitment, etc. Several optimization techniques have been proposed to deal with the problems faced by energy management systems and to ensure an efficient operation. They are discussed in the subsequent sections and shown in Fig. 4.3.

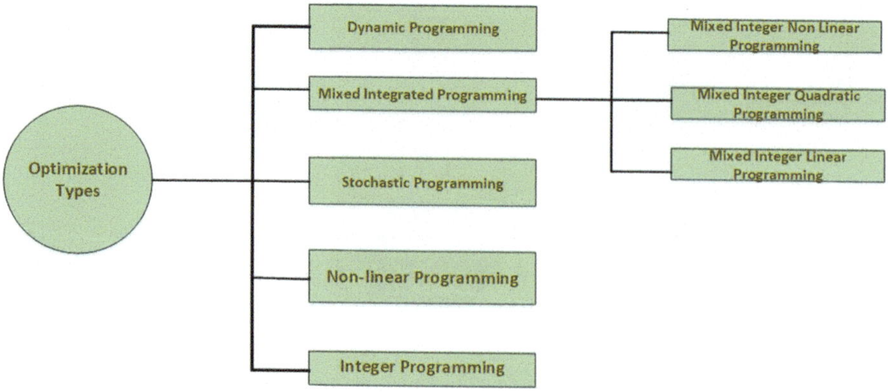

Fig. 4.3 Different optimization techniques

4.2.3.1 Linear Programming

One method proposed to minimize the objective functions is linear programming (L.P.) and mixed-integer linear programming (MILP). L.P. is used for the reduction of fluctuations in demand and also maintaining energy balance in microgrids with renewable energy generation systems (Davis and Thompson 2007). For minimal operating costs, certain energy management systems operate using mix-mode operation strategies. These include on and off mode, power-sharing mode, and continuous run mode. The expense functions of these strategies are minimized using L.P. and MILP (Sukumar 2017). Much research work is done wherein the MILP framework is utilized to solve several problems related to energy management. It proves to be a trade-off between low costs for operation and good energy services to the consumers. For example, in Vergara (2017), and EMS is proposed and developed as a nonlinear model. The costs for the operation of the microgrid are minimized by taking into consideration load-shedding and system outage. The nonlinear model is linearized into MILP using the Taylor series, estimated operation points, piecewise approximation, and auxiliary variable introduction. This model proves to be computationally efficient than the nonlinear one. The drawback is that the demand response and power losses are not taken into consideration. Further, the objective function in Tenfen and Finardi (2015) takes into consideration the curtailable and reschedulable loads in demand response and also costs pertaining to operation and management, start-up/shutdown, load shedding, etc., while proposing an optimal energy management strategy. In Tenfen and Finardi (2015), a load demand policy is determined to minimize operation costs due to technical and economic constraints. Microturbines with ambient temperature and ramps constraints and fuel cells with cycling and preheating costs constraints are modeled. The minimization of costs is done by using a piecewise linear function of diesel generators and battery sizing. Therefore, MILP is used to propose a cost minimization strategy.

These and several other research works use L.P. and MILP to carryout energy resource optimization in microgrids. However, significant work needs to be done to consider battery degradation, greenhouse gas emission, depth of discharge (DoD) on the battery's performance, system privacy, reliability, and efficiency issues.

4.2.3.2 Nonlinear Programming

The optimization approach used in several research works for a microgrid EMS is based on the nonlinear programming framework. In Sawle et al. (2017), a central controller is designed for a microgrid to operate optimization during an interconnected operation mode. The microgrid central controller aims to minimize the operational costs by considering the active power of the grid. The paper considers market policies, typical load profiles, current market prices, and renewable energy production during the design process. This approach is entirely based on nonlinear programming. The authors in Ornelas-Tellez and Jesus Rico-Melgoza (2013) propose an optimal control for minimizing a quadratic cost function and tracking the trajectory of a nonlinear system. The control law so developed comes from solving the Hamilton–Jacobi–Bellman (HJB) equation and is meant for ensuring efficiency in the power flow between sources in the grid. HJB equation is used primarily for state-dependent factorized non-linear systems. The intermittent nature of the renewable energy sources and the grid-connected and islanded modes of operation of the microgrid pose problems to optimization. For reliable and economic energy supply, the renewable energy sources' capacities have to be optimized, and their coordination with the grid should be managed. In Zhao et al. (2014), the authors proposed a probability of self-sufficiency (PSS) concept to depict that the microgrid can meet local demands quite self-sufficiently. Also, the problem of operation scheduling involving unit commitment (U.C.) and dispatch problems is solved. The U.C. problem is a mixed integer-based nonlinear programming optimization problem. Therefore, a nonlinear mixed integer-based optimization task solves various technical issues on both the supply and demand side like load management, electric vehicles, dynamic pricing, and on-site renewable energy generation.

4.2.3.3 Dynamic Programming

A dynamic programming approach is used to solve the problem related to energy consumption scheduling since it is based on a real-time pricing strategy. As seen from Ma et al. (2013), the energy consumption scheduling problem is cast off into a game wherein the consumers compete with each other to obtain energy at minimal costs. The non-cooperative energy consumption game has an optimal energy consumption solution, which is also known as the Nash equilibrium point. Also, in Sawle et al. (2016), for the integration of renewable energy generation-based microgrids in the main grid and low energy costs, the authors propose a dynamic model wherein the consumers are accountable for the complete schedule of energy demand also keep track of the energy prices. Furthermore, dynamic programming is also used to maximize the lifetime of the energy storage system used in the microgrid. The best way to achieve this is for the central controller of the microgrid to find out the most effective way of charging and discharging the energy storage system (e.g., battery) (Hooshmand et al. 2013). This can be determined by taking into consideration factors such as the levels of renewable energy generated, grid electricity rates at the time of use (TOU), and the load profile. The authors in Duan and Zhang (2013), for the regulation of the energy purchase of microgrids, have proposed a dynamic contract mechanism. Regulation of the energy is crucial since the inconsistencies in the renewable energy generation of the microgrid can, in turn, affect the utility grid. Therefore, the mechanism enables the microgrid to make time-specific commitments but also provides it with the flexibility of altering any future ones. These commitments are subjected to change in the case of current levels of energy in the energy storage system and load demands.

4.2.4 Stochastic Programming

The authors in Cau (2014) proposed a stochastic EMS for the control of a microgrid that has adopted battery and hydrogen energy storage systems. As opposed to the conventional EMS, which takes into consideration the state-of-charge of batteries, the consumers base this EMS on the uncertainty in the renewable energy produced and the demand. The authors believe in a stochastic approach to deal with these intermittencies. This EMS helps in the minimization of the utilization costs of the energy storage systems. Utilization costs include costs incurred due to maintenance, depreciation, operations, and replacement. The stochastic approach proves to be a much more effective option in cost reduction by 15% in the cases taken into consideration in Rezaei and Kalantar (2015). For an isolated droop-controlled microgrid, frequency is a key control variable. In Lasseter and Paigi (2004), an objective function dependent upon the same is formulated using mixed-integer linear programming, which aims at optimization of the frequency value that is subjected to deviation from its nominal value due to uncertainties in the generation and demand of energy. Scenario-based and reduction method is used to model these intermittencies, and a two-stage stochastic optimization approach is adopted to reduce these frequency deviations and the emission and operation costs.

Much research is done in this method of optimization, but the common drawbacks observed are no mechanisms for the mitigation of greenhouse gas emissions, choosing the higher depth of discharge values of the batteries, higher time of computation, and privacy issues among the consumers.

4.3 Solutions for Energy Management Problems

4.3.1 Introduction

Different research works have given several solution approaches to the problems encountered in the energy management systems of the microgrid. These problems are based on the optimization framework. The solutions make use of neural networks in the artificial intelligence-based approach, optimization algorithms like genetic, swarm in the meta-heuristic approach, etc. They are discussed thoroughly in subsequent subsections along with the relevant research work.

4.3.2 Artificial Intelligence-Based Approach

4.3.2.1 Neural Networks and Fuzzy Logic Based

Artificial neural networks are used for energy forecasting and prediction of load demand. The authors in Chaouachi et al. (2013) make use of artificial intelligence techniques for the intelligent energy management of a microgrid. This is done along with the usage of multiobjective-based linear programming optimization. The EMS takes into consideration the forecasted values of the renewable energy generated and the load demand for the minimization of operation and emission costs. The proposed machine learning model is characterized by improved learning and generalization capacities. Another important aspect of microgrid efficiency is the battery scheduling process, wherein the authors make use of fuzzy logic to deal with intermittencies.

Similarly, in Sawle and Gupta (2015), a two-step reinforcement learning algorithm-based neural network is used for the battery scheduling process. This also helps in increasing the rate of utilization of the battery during high load demand, minimizing the dependence on external sources for energy purchase, and maximizing the wind energy generation. The forecasted energy values feed the neural network to take optimum battery scheduling actions. The drawback of this process is that the demand response is not considered. A recurrent neural network is developed in Gamez Urias et al. (2015) to minimize energy purchase from the utility grid, maximize the utilization of the renewable energy generated, and for proposing an efficient energy management system for the microgrid. Kalman filter and hybrid wavelet functions help the neural network in achieving the same.

For maximization of the energy trading profits with the utility grid and managing the power flow in microgrid equipped with renewable energy generation systems and energy management systems, the authors in De Santis et al. (2017) have proposed a combination of fuzzy logic and genetic algorithm. A hierarchical genetic algorithm tuned rule base of a fuzzy inference system helps for this purpose. A poly-generation microgrid in the islanded mode of operation consisting of a fuzzy logic-based energy management system is designed in Kyriakarakos (2012). It helps in the minimization of the operation costs of the grid but does not consider the demand response and the computational time complexity. All of these neural networks and fuzzy logic approaches are meant for microgrid energy management systems with centralized supervisory control architecture. They take into consideration the intermittencies related to energy generation and load demand, with the major contribution being a reduction in the costs, pollutants, and ensuring proper energy sharing with the utility grid.

4.3.2.2 Game Theory and Markov Decision Process Based

The authors in Ma et al. (2016) have proposed a Stackelberg game approach in a microgrid consisting of combined heat and power (CHP) and photovoltaic (P.V.) prosumers to maximize profits and ensure fair profit distribution among them. This is ensured by maintaining Stackelberg equilibrium with the help of differential evolution-based heuristic algorithm and constrained nonlinear programming. In this game, the followers, i.e., the prosumers (CHP and P.V.) choose their own strategies in response to the strategies of the microgrid operators who is the leader. This approach shows a significant increase in the profit, along with an improvement in the load characteristics of the microgrid as a whole. Reference (Mohamed and Koivo 2011) gives an overview of a modified game theory based on nonlinear multiobjective optimization for deciding an optimum operation strategy for a microgrid. It also helps in minimization of the operational and maintenance costs and lower emission levels of greenhouse gases like CO_2, NO, NO_2, SO_2, etc. The objective function in this approach is the difference between the Pareto objectives and the normalized distance of the function from its worst value. A rolling horizon Markov decision process (MDP) is introduced in Sawle et al. (2018) for a grid-connected microgrid with multiple energy generation systems. To deal with the large states and decision space of MDP, a feasible base policy inclusive roll-out algorithm is also applied. This is achieved by a greedy algorithm implementation. Overall, this helps to solve the multienergy scheduling problem that occurs due to variability in renewable energy generation. The downside of the above methods is that the computational time complexity is not taken into consideration. But to ensure reliable, efficient, and sustainable operation, the emphasis is laid on the inclusion of electric vehicles, mitigation of environmental emission, costs incurred for communication purposes due to decentralized operations of the microgrid, etc.

4.3.3 *Model Predictive Control-Based Solution*

For ensuring stable operation of a microgrid in islanded mode, a nonlinear model predictive control is formulated in Luo (2017). This model uses data related to the active power generated from renewable energy systems, forecasted load demand, and state of charge (SoC) of the battery to train an artificial neural network that identifies the forthcoming variations in the active power to initiate load shedding for non-important loads. A two-stage coordinated control approach is proposed for a grid-connected microgrid in Prodan et al. (2015). The primary stage deals with economic dispatch, whereas the secondary is a real-time adjusting one. In the Economic dispatch stage (EDS), the control model makes use of the piecewise linear thermal and electric efficiency curves for operation schedules based on the values of the energy forecast. Whereas the real-time adjusting stage makes use of real-time energy production values to deal with the power variations. A soft constrained model predictive control and mixed-integer programming-based framework are introduced in Sawle and Gupta (2017) to minimize operational costs of microgrids such as expenses due to energy purchase from the utility grid, lifecycle of batteries, and imbalance occurring in the demand load. The framework allows the formulation of a fault-tolerant EMS, taking into consideration the variations in the renewable energy produced, generator faults, load demand, and the market price of energy.

4.3.4 Metaheuristic Approaches

4.3.4.1 Swarm Optimization Method

Particle Swarm Optimization (PSO) based algorithm is employed for an EMS in a microgrid with multiple renewable energy generation units and energy storage systems in Moghaddam (2011). The suggested approach helps in minimizing the various operational and energy costs of the grid by considering optimum values of control variables. The EMS is modeled using an estimate or probabilistic method owing to uncertainty in the energy produced through renewable sources like wind and solar, energy market prices, and load demand. The limitations of this optimization technique are that the costs of environmental emissions and the demand response are not considered.

An Adaptive Modified Particle Swarm Optimization (AMPSO) that is a combination of Fuzzy self-adaptive structure and Chaotic local search mechanism-based PSO is presented in Abedini et al. (2016). A multiobjective nonlinear constraint optimization problem is formulated for the optimized operation of a microgrid consisting of several energy generations and storage units. This approach helps in the reduction of operation and emission costs. A Gaussian Mutation algorithm for a novel guaranteed convergence PSO is developed and implemented in Sawle and Gupta (2015). This gives an optimal strategy for the management of microgrids in the islanded mode of operation and also the optimum capacity of the renewable energy generation systems in it. It is implemented on 69 and 94 bus systems to reduce the fuel and capital costs. The shortcoming of this method is that the environmental emissions and power losses in the microgrid system are not considered.

The authors in Golshannavaz et al. (2016) propose a dual-stage energy management strategy for the optimized operation of a microgrid. In the primary stage, energy scheduling is done according to the variations in a generation. Stage two involves expressing the generation intermittencies as perturbations in nodal power injections. Sensitivity analysis is used for the estimation of the corresponding optimal spinning reserves. The energy of microgrid purchase from the utility grid, remaining demand response capacity, and other conventional sources are utilized by the real spinning reserves for the difference between the actual and optimal values. Hence, a stochastic weight trade-off PSO and affine arithmetic-based power flow is suggested for minimizing the costs incurred due to environmental emissions, power trade with the utility grid, fuel, and load shedding. Also, the mismatches in the active and reactive power, deviation in the voltage are reduced.

4.3.4.2 Genetic Algorithm Method

Microgrids face challenges related to scheduling of power generation and load demand due to intermittencies in non-conventional energy sources. An efficient, high-performance EMS is formulated in Askarzadeh (2018) incorporating automatically controlled switches. The system uncertainties are modeled by the generation of various scenarios as well as the usage of the reduction method. An optimized network topology is also suggested for every power scheduling interval. A genetic algorithm is proposed for the nonlinear mixed-integer problem-based optimization, which aims at optimal scheduling of active, reactive, and reserved power. Energy storage systems and power loss-related expenditures are not considered. Communication of the energy management system of a microgrid with distributed generation units is crucial for optimum energy allocation. A memory-based genetic algorithm is formulated in Chen (2011) for the same. It also helps in the reduction of costs related to the production of energy. This algorithm proves to be more efficient than particle swarm optimization with constriction and inertia factors. The authors in Vivek Tamrakar et al. (2015) have formulated an economical energy storage system for a microgrid in the grid-connected mode of operation. A matrix real coded genetic algorithm-based EMS helps in the minimization of operational costs through the inclusion of storage and generation bids as well as profits obtained from energy trade.

4.3.5 Agent-Based Approach

The distributed generation units, energy storage systems, consumers, and the microgrid are all said to be agents in the agent-based approach. For a microgrid in the islanded mode of operation and EMS, as well as a scheme for load shedding, is modeled in Karavas (2015) for optimal energy supply to satisfy load demand via effective coordination among agents. This is based on the forecasted values of energy generated and load powers as well as the multiagent system concept. The solar energy system, wind turbine system, load, fuel cells, and batteries are the agents. An autoregressive moving average method-based model is used for the prediction of solar irradiance, wind speed, temperature, and loads. If the energy generation and storage systems are not able to meet the load demand, the multi-agent system algorithm schedules the load shedding accordingly. To mitigate fluctuations in the voltage and the notches in the A.C. bus, STATACOM based compensation is employed. A decentralized EMS based on fuzzy cognitive maps-based multiagent system algorithm is suggested in Shi et al. (2015) for a multigeneration autonomous microgrid. A decentralized EMS allows individual control of different units of a microgrid and hence proves to be a much better alternative in case of partial failure in grid operations. The aim of this approach is the minimization of operational, penalty, and energy storage costs wherein the battery, electrolyzer, fuel cell, and renewable energy resources are all agents.

4.3.6 Other Solutions

An EMS for a microgrid in the grid-connected mode of operation with decentralized supervisory control is proposed in Mohamed and Koivo (2010) since a decentralized approach proves to be more efficient in computational time complexity at the central control of the microgrid as well as is more economical. A predictor–corrector proximal multiplier algorithm is applied for the optimization of the objective function that includes generator operation cost, load shedding costs, power losses, and costs due to trade of energy with the utility grid. Therefore, the suggested EMS is such that the central control of the microgrid and the local controllers together ensure an optimal power flow. A mesh adaptive direct search algorithm is introduced in Leitão et al. (2020) for the minimization of cost functions that consider expenses incurred due to the emission of greenhouse gases like CO_2, SO_2, and NOx as well as operation and maintenance.

4.4 Energy Management Analysis in Smart Grid

4.4.1 Introduction

In today's generation, transmission, and distribution networks, rapid economic growth is causing tremendous stress, as they are unable to keep pace with rising demand. The construction and integration of a large number of electrical power generation systems with increased ability to cope with increasing demand have a negative environmental impact, so successful energy management is imperative. In response to this anticipated energy crisis and management, Smart Grid could be the solution. For handling extensive and complex power systems, traditional instrumentation has proven inadequate in order to improve overall system performance and reliability, the electric grid. Most of the latest infrastructure used by the grid is obsolete and, in many ways, inefficient; there are two key aspects of Smart Grid, the smart grid (generation, transmission, and distribution system) on the one hand and programmable smart devices on the other.

4.4.2 Energy Management Analysis

Energy efficiency is a major concern for activities in the field of sustainable growth since rising energy use typically means increasing CO_2 emissions and a long-term effect on global warming (Leitão et al. 2020). Over the last few years, energy demand has been gradually growing, partially due to the advent of new electrical applications, such as new transportation systems and technologies, which require increased investment in the energy-producing sector. In addition, because of high power demand, the electricity distribution network may be under stress at certain particular times (Bohre et al. 2021). A variety of options for efficient energy consumption can be sought in order to cope with increasing electricity demand. Indeed, energy management includes all steps that could affect energy demand, such as measures to suppress inefficient energy consumption and measures to minimize energy consumption on a large or medium scale (Mohsin 2021).

Main Goals of Energy Management System:

- The study should take into account standby usage, as well as the fact that it eliminates demand peaks, decreases energy losses, to understand the environmental benefits of energy management through linked household appliances with consistent figures.
- To measure the reduction of CO_2 emissions by improved consumption and control of loads.
- To conduct a cost–benefit analysis of the energy management system of a building.

The working method for the energy management system is presented in Fig. 4.4

4.4.3 Importance of Energy Management System

Globally, where power request is mounting, power production should also rise to come across the requirements of users and the development of their normal lives (Said et al. 2020). Moreover, since the amount of consumers is increasing, and due to irregularity in the energy load, the request for electricity will cause difficulties for utility providers as well as the system participants. High load peaks are extremely likely to arise in numerous sets and can negotiate the operations of the device (Thirunavukkarasu and Sawle 2020). To address this subject, the utilities, as well as the system participants, have two possibilities:

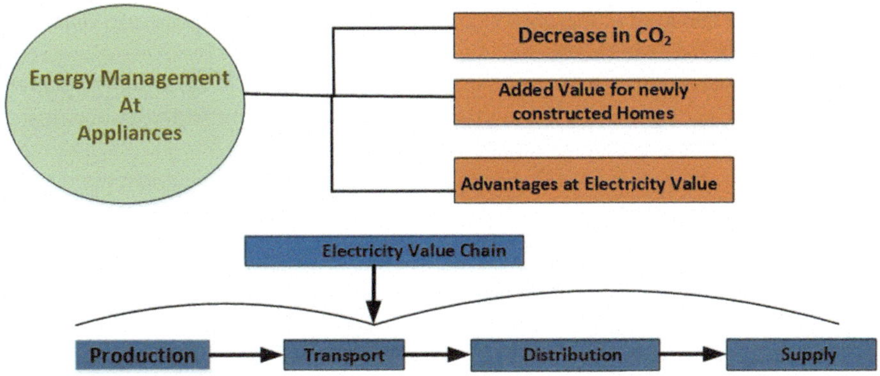

Fig. 4.4 Working method for energy management system

- It is increasing the magnitude as well as dimensions of the system that is expensive and needs time to carry out.
- Utilizing the E.M. so that to decrease the risk of high peak load for the period of peak times.

The second way out is more realistic; Moreover, it requires more complex computations and approaches to be accomplished in handling E.M. E.M. is necessary for a smart grid for several reasons such as:

- It is automatic, and it omits the involvement of direct interference from users.
- It provides precise outcomes and estimates.
- It supports the utilities for better optimization of its production units as well as minimizes the production costs.
- It supports the system participants in minimizing the power losses on the system and lines. It may reduce the indirect distribution of electricity cost drastically
- It benefits the consumers to achieve their own load demand and as well as minimize their charges.

4.4.4 Energy Management Application

It is possible to split E.M. into two main categories. The first is from the perspective of the energy producer, while the second is from the perspective of the electricity user (Kumar et al. 2020).

- Suppliers of electricity can use energy management to monitor their generating units effectively. For example, some generators may be turned on by the electricity supplier, which may have the lowest operating cost, in order to meet some customer power demands, using energy management. While Diesel generators with maximum operating costs are deserted. It offers extra peak load for definite peak hours. The goal of the power company is to reduce the operating costs of its generating systems in this way.

- The transmission and distribution system operators can use E.M. to control current flow that reduces power losses on the system. It effectively increases the degree of infiltration of non-conventional energy sources.
- Energy management is used by end consumers (such as households, residential, and industrial buildings, warehouses, schools, etc., to reduce their electric current bills and plan their load demand effectively.

4.4.5 *Tools Used in Energy Management System*

Historically, the management of production sectors and electrical machines was labor-intensive, or minor management systems were utilized. With the advancement of computational systems as well as smart algorithms into the grid, it is becoming stress-free to manage loads in modern times (Leonori 2020). It is possible to quote nearly some of the popularly utilized systems as follows:

- PLC (Programmable Logic Controller)
- SCADA (Supervisory Control and Data Acquisition)
- EMS (Energy Management System)
- BMS (Building Management System).

4.4.6 *Current Situation and Barriers*

The method requires time and effort to switch from one technology to another. The introduction of energy conservation is in progress. The introduction of energy conservation is in progress (Azeroual 2020).

- The introduction of energy conservation systems has high costs. Over the long term, the rate of return is low.
- The price of electricity must be time-variable. For utility grids as well as the power traders, converting from conventional tariff to a newer tariff system is not easy.
- The present network's architecture, which could charge the system operator a lot of money, should be upgraded.
- Two-way power flow is in the trial process. It deals with an interruption in perfect resource administration.

In defining how firm, the population performs the main part (Sola and Mota 2020).

- There are high costs for implementing energy management programs. The rate of return is, however, minimum in the long term.
- It is not easy to move from a conventional tariff to a newer one for electric utilities and power retailers.

- The systems architecture must be advanced, which could charge the system operator a lot of money.
- The population's understanding plays a major part in deciding how quick employment can be.
- The main explanation for the transition from the conventional framework to a smarter one could be the greenhouse effect and global warming. Nevertheless, there is an event that does not profit from the change towards green energy as well as a healthy climate.

4.5 The Architecture of Energy Management System in Smart Grid

4.5.1 History of Energy Management System

Construction of the EMS, which was called a management unit, and further it is known as an Energy management unit in the course of 1973, started in the 1960s (Razmjoo 2021). It was called as Energy Management System (SCADA) as soon as advanced computation-based SCADA system in the primary 1991 and ultimately progressed in current time called as Energy Management System that involves numerous controlling methods such as load control, demand-side management, and distribution management system. EMS's operation is to ideally assign diverse power sources to users, with the integration of renewable energy sources, without sacrificing the efficiency, safety, and security of the network. An Energy Management System can observe, track, enhance, and manage client, delivery, transmission, and generation facilities. It also works in the current time for Supervisory Control and Data Acquisition, track power forecast, and accounting applications, and management of transmission security (Oehler and Jimenez 2017).

4.5.2 Objectives of Energy Management System

An EMS can have a single technical, economic, technoeconomic, environmental, and social-economic goal or a set of objectives (Byrne et al. 2018). The technical goals of EMS include power efficiency, device performance, and degradation of the transformer, taking into account improved system performance, improved energy efficiency, longer lifetime, as well as reduced maintenance cost (Xiong 2020). Economic goals refer to the maximization of the net operating expense of energy prices for the customer gain of aggregators and parking lot managers, etc. Energy Management System enhancement may have the finest economic result, although without taking into account technical limitations, the network might be subject to blackout, or else devices running under a distributed system.

4.5.3 Construction of Energy Management System

The effectiveness of the Energy Management System determines on the control planning of the Energy Management System, widely used designs are centralized, decentralized (distributed), and hierarchical EMS with the adopted solution approach. Below are the three control architectures of EMS discussed.

Centralized Energy management system:

One central controller with a high-performance computing system and a dedicated, secure communication network to manage the use of resources describes the unified Energy Management System architecture. The prime controller might be a utility company or aggregator that gathers information from all nodes for efficient operation, such as DER power generation, load/consumer energy consumption trends, meteorological data, other appropriate market operator information, etc., to enforce optimization programs to achieve its objectives (Lü 2020). This central control network delivers optimal productivity, but it also has its fair share of drawbacks.

Decentralized Energy management system:

Particular nodes consuming independent control capabilities and point-to-point communication with other nodes, the decentralized Energy Management System construction is characterized by a distributed processing system. The decentralized EMS architecture, therefore, overcomes the shortcomings of the centralized architecture by improving expandability, enabling superior organizational flexibility, and preventing single-point failure (Espín-Sarzosa et al. 2020).

Hierarchical Energy management system:

Based on decentralized EMS, few researchers propose and test the hierarchical EMS architecture, particularly for a microgrid community network where specific microgrids are interconnected to make a microgrid community network. The structure is allocated into many levels of control with diverse control goals at each stage in the hierarchical architecture. Two-level or three-level systems are usually suggested. The flow of information is between adjacent levels only, and no information is shared at the same level between units. Most hierarchical Energy Management System includes the level of supervisory control, the level of optimization control, and the level of execution control.

EMS is implemented in the optimization control layer, which, if modeled with suitable uncertainty management techniques, provides cost-effective and efficient device operation. As coordination between adjacent levels is needed for this architecture, if there is a communication fault in the higher level, there will be no transmission of data as well as energy. The unified architecture, therefore, enables the economic application as well as easy maintenance of Energy Management System, which provides ideal results globally; moreover, there are few problems with it: high technical burden, high communication costs, data privacy, and reliability. For microgrids, distributed processing decentralized architecture provides higher efficiency, low computational burden but does not guarantee optimal global performance. Hierarchical architecture is a suitable choice for MGC/MMG, offering different levels of power. The difficulty of implementation, however, is high. The robustness of the EMS can be described as the capacity to deal with uncertainty in the presence of uncertain loads, sources, and uncertain electricity prices in the context of the EMS for smart grids, in order to provide optimal performance. Precise modeling of uncertainty is the primary criterion for achieving robustness in the Energy Management System (Weinand et al. 2020; Sawle et al. 2017; Sawle and Gupta 2014; Tamrakar et al. 2015).

A significant cost factor in real-time applications is the Energy Management System. Enhancing energy effectiveness is a significant way of reducing prices as well as increasing stable earnings, especially in periods of high volatility in energy prices. In the current method, autotransformers are currently used as an energy storage system in industries. Industries in the manufacturing units, in particular, are a matter of major concern, as Energy Conservation's minimal charges mean more sustainable commodity prices on the world market. An autotransformer preserving the frequency is the hardest task in the current system since this power source failure also causes the entire process to shut down. Seasonal failure is the biggest downside of autotransformers. So even the power resource often needs to be changed at the time of expansion. Due to the direct type system. So they often get an individual supply for it or often run through generators. The main purpose is to minimize the completion of the autotransformer that often fails.

4.6 Tools Utilized in Energy Management Systems in Smart Grid

As we have seen in Sect. 4.4, about the various tools utilized in energy management systems in smart grid. In this section, we will be discussing in detail the tools utilized in energy management analysis in smart grids.

4.6.1 Introduction

A Programmable Logic Controller is a computer used to control devices and processes in a specialized way and originally designed to substitute logic boards for relays, i.e., actuation of the sequence device and coordinating activities. Accepts input from a switch series. Sends devices or relays output (Gholinejad 2020). To store instructions and execute specific functions that include on/off the power, timing, counting, sequencing, arithmetic, and data handling, it utilizes a programmable memory. PLC is an arrangement of solid-state elements designed to provide results and allow rational and sequential choices. PLCs are used to manage and operate equipment and machinery for manufacturing processes. Most of the hard wiring associated with traditional relay control circuits has been replaced by PLC. The following parts can be separated into a standard PLC.

- Central Processing Unit (CPU)
- The Input/output (I/O) section
- The Power supply and
- The Programming device.

In the field of automation, PLCs play a key role, being part of a larger SCADA scheme. A PLC may be configured in accordance with the process' operational requirements. In the manufacturing sector, because of the shift in the design of production, there would be a need for reprogramming. PLC-based control systems were introduced to address this challenge.

4.6.2 Scada

These are process control systems, which are used to track, capture, as well as analyze current time ecological data (Ajeya and Vincent 2020). Based on a predetermined collection of conditions, such as traffic control or power grid management, process control system are built to automate electronic systems. Some process control systems are made up of many Remote Terminal Units as well as PLCs linked to many sensors that transmit information for analysis to a master collection system. SCADA systems with the following elements have been defined as follows (Pliatsios et al. 2020):

- Operating equipment: Pumps, valves, conveyors, and breakers of substations that can be operated by actuators or relays being energized.
- Local processors: Connect with the tools and operating equipment of the site. This includes the Programmable Logic Controller (PLC), Intelligent Electronic System (IED), Remote Terminal Unit (RTU), and Process Automation Controller. Dozens of inputs from instruments and outputs to operating equipment can be responsible for a single local processor.
- Tools: in the field or in a facility that senses conditions such as pH, temperature, pressure, level of power, and rate of flow.
- Short-range communications: between processors, tools, and running equipment locally. Using electrical characteristics such as voltage and current or other developed industrial communications protocols, these relatively short cables or wireless links carry analog and discrete signals.
- Long-range communications: between host computers and local processors. Using methods such as leased phone lines, radio, microwave, frame relay, and cellular packet data, this contact usually covers miles.
- Host computers: they serve as a central monitoring and control point. The host machine is where a human operator, as well as alarms, obtained, data checked, and exercise control can control the operation (Upadhyay and Sampalli 2020).

4.6.3 *Building Energy Management System (BEMS)*

To reduce total energy usage, BEMS should regulate and track building energy needs. HVAC, lighting, and charging loads are important peaks considered in Building Energy Management System forming (Maldonado-Correa 2020). For commercial as well as residential buildings, Building Energy Management System could be introduced. In Ożadowicz and Grela, a building automation and control system (BACS) is suggested to reduce energy intake, using the current time as well as event-driven demand measurement mechanisms (Mariano-Hernández 2021). For the functional application of HEMS and BEMS, the following points need to be considered (Nizami 2020).

- Computerized time plays a crucial role, as well, as robust controllers need to reduce it.
- A significant issue that needs to be addressed is the protection of the user's data.
- Local implementation lowers contact and connectivity costs.
- The customer's comfort level should be a priority in order for any end-user to engage in such programs.

4.7 Conclusion

Energy Management Systems guarantee that dispersed energy sources are used optimally in order to achieve the affordable, sustainable, and dependable operation of microgrids and smart grids. In this chapter, we have discussed in detail about the concepts of microgrid and smart grid. We have also discussed several optimization techniques that have been described thoroughly to deal with the problems faced by them and for the optimization of objective functions. We have also tried to give detailed solutions for optimization framework-based problems encountered while formulating an energy management system for a microgrid. This chapter also addresses about the importance of energy management system in smart grids and tools utilized in energy management systems in smart grids.

4.8 Future Scope

In the recent decade, there has been a remarkable growth in research and development on smart grids as well as microgrids. The main research works should be evaluated for future study:

(a) The development of a cost-efficient smart grid and microgrid network with efficient and robust communication.
(b) Rapid as well as accurate simulation of uncertainties.

(c) Real-time, efficient, and cost-effective deployment of EMS through technology.

References

Abedini M, Moradi MH, Mahdi Hosseinian S (2016) Optimal management of microgrids including renewable energy sources using GPSO-GM algorithm. Renew Energy 90:430–439. https://doi.org/10.1016/j.renene.2016.01.014

Ajeya B, Vincent S (2020) Integration of contactless power measuring instruments to PLC and SCADA through industrial wireless sensor network for EMS. Optical and wireless technologies. Springer, Singapore, pp 279–292

Armaroli N, Balzani V (2007) The future of energy supply: challenges and opportunities. Angew Chem Int Ed 46(1–2):52–66. https://doi.org/10.1002/anie.200602373

Askarzadeh A (2018) A memory-based genetic algorithm for optimization of power generation in a microgrid. IEEE Trans Sustain Energy 9(3):1081–1089. https://doi.org/10.1109/TSTE.2017.2765483

Azeroual M et al (2020) Simulation tools for a smart grid and energy management for microgrid with wind power using multi-agent system. Wind Eng 44(6):661–672. https://doi.org/10.1177/0309524X19862755

Birol F (2006) World energy prospects and challenges. CESifo forum, vol 7, no 2. München: ifo Institut für Wirtschaftsforschung an der Universität München

Bohre AK, Acharjee P, Sawle Y (2021) Analysis of grid connected hybrid micro-grid with different utility tariffs. In: 2021 1st international conference on power electronics and energy (ICPEE), Bhubaneswar, India, pp 1–6. https://doi.org/10.1109/ICPEE50452.2021.9358603

Brown MA, Zhou S (2013) Smart-grid policies: an international review. Wiley Interdiscip Rev Energy Environ 2(2):121–139. https://doi.org/10.1002/wene.53

Byrne RH, Nguyen TA, Copp DA, Chalamala BR, Gyuk I (2018) Energy management and optimization methods for grid energy storage systems. IEEE Access 6:13231–13260. https://doi.org/10.1109/ACCESS.2017.2741578

Carvalho M, Perez C, Granados A (2012) An adaptive multi-agent-based approach to smart grids control and optimization. Energy Syst 3(1):61–76. https://doi.org/10.1007/s12667-012-0054-0

Cau G et al (2014) Energy management strategy based on short-term generation scheduling for a renewable microgrid using a hydrogen storage system. Energy Convers Manag 87:820–831. https://doi.org/10.1016/j.enconman.2014.07.078

Chaouachi A, Kamel RM, Andoulsi R, Nagasaka K (2013) Multiobjective intelligent energy management for a microgrid. IEEE Trans Industr Electron 60(4):1688–1699. https://doi.org/10.1109/TIE.2012.2188873

Chen C et al (2011) Smart energy management system for optimal microgrid economic operation. IET Renew Power Gener 5(3):258–267. https://doi.org/10.1049/iet-rpg.2010.0052

Davis III, Thompson E et al (2007) Cognitive-behavioral treatment for specific phobias with a child demonstrating severe problem behavior and developmental delays. Res Dev Disabil 28(6):546–558. https://doi.org/10.1016/j.ridd.2006.07.003

De Santis E, Rizzi A, Sadeghian A (2017) Hierarchical genetic optimization of a fuzzy logic system for energy flows management in microgrids. Appl Soft Comput 60:135–149. https://doi.org/10.1016/j.asoc.2017.05.059.

Duan L, Zhang R (2013) Dynamic contract to regulate energy management in microgrids. In: 2013 IEEE international conference on smart grid communications (SmartGridComm), pp 660–665. https://doi.org/10.1109/SmartGridComm.2013.6688034

El-Hawary ME (2014) The smart grid—state-of-the-art and future trends. Electr Power Compon Syst 42(3–4):239–250. https://doi.org/10.1080/15325008.2013.868558

Espín-Sarzosa D, Palma-Behnke R, Núñez-Mata O (2020) Energy management systems for microgrids: main existing trends in centralized control architectures. Energies 13(3):547. https://doi.org/10.3390/en13030547

EUR 22040 (2006) European technology platform smart grids: vision and strategy for Europe's electricity networks of the future. Publication of European Commission. ISBN 92-79-01414-5

Gamez Urias ME, Sanchez EN, Ricalde LJ (2015) Electrical microgrid optimization via a new recurrent neural network. IEEE Syst J 9(3):945–953. https://doi.org/10.1109/JSYST.2014.2305494

Gholinejad HR et al (2020) A hierarchical energy management system for multiple home energy hubs in neighbourhood grids. J Build Eng 28:101028. https://doi.org/10.1016/j.jobe.2019.101028

Golshannavaz S, Afsharnia S, Siano P (2016) A comprehensive stochastic energy management system in reconfigurable microgrids. Int J Energy Res 40(11):1518–1531. https://doi.org/10.1002/er.3536

Gungor VC et al (2011) Smart grid technologies: communication technologies and standards. IEEE Trans Industr Inf 7(4):529–539. https://doi.org/10.1109/TII.2011.2166794

Hamilton B, Summy M (2011) Benefits of the smart grid [In My View]. IEEE Power Energy Mag 9(1):104–102. https://doi.org/10.1109/MPE.2010.939468

Heffner G (2011) Smart grid-smart customer policy needs. Workshop report for the IEA energy efficiency working party, April, Paris

Hirsch A, Parag Y, Guerrero J (2018) Microgrids: a review of technologies, key drivers, and outstanding issues. Renew Sustain Energy Rev 90. https://doi.org/10.1016/j.rser.2018.03.040

Hooshmand A, Asghari B, Sharma R (2013) A novel cost-aware multi-objective energy management method for microgrids. In: 2013 IEEE PES innovative smart grid technologies conference (ISGT), pp 1–6. https://doi.org/10.1109/ISGT.2013.6497882

IEA (2015) International Energy Agency. Key world energy statistics

Joskow PL (2012) Creating a smarter US electricity grid. J Econ Perspect 26(1):29–48. https://doi.org/10.1257/jep.26.1.29

Joskow PL, Wolfram CD (2012) Dynamic pricing of electricity. Am Econ Rev 102(3):381–385

Karavas C-S et al (2015) A multi-agent decentralized energy management system based on distributed intelligence for the design and control of autonomous polygeneration microgrids. Energy Convers Manag 103:166–179. https://doi.org/10.1016/j.enconman.2015.06.021

Kumar PS, Chandrasena RPS, Ramu V, Srinivas GN, Babu KVSM (2020) Energy management system for small scale hybrid wind solar battery based microgrid. IEEE Access 8:8336–8345. https://doi.org/10.1109/ACCESS.2020.2964052

Kyriakarakos G et al (2012) A fuzzy logic energy management system for polygeneration microgrids. Renew Energy 41:315–327. https://doi.org/10.1016/j.renene.2011.11.019

Lasseter RH, Paigi P (2004) Microgrid: a conceptual solution. In: 2004 IEEE 35th annual power electronics specialists conference (IEEE Cat. No.04CH37551), vol 6, pp 4285–4290. https://doi.org/10.1109/PESC.2004.1354758

Leitão J, Gil P, Ribeiro B, Cardoso A (2020) A survey on home energy management. IEEE Access 8:5699–5722. https://doi.org/10.1109/ACCESS.2019.2963502

Leonori S et al (2020) Microgrid energy management systems design by computational intelligence techniques. Appl Energy 277:115524. https://doi.org/10.1016/j.apenergy.2020.115524

Lü X et al (2020) Energy management of hybrid electric vehicles: a review of energy optimization of fuel cell hybrid power system based on genetic algorithm. Energy Convers Manag 205:112474. https://doi.org/10.1016/j.enconman.2020.112474

Luo Z et al (2017) A two-stage optimization and control for CCHP microgrid energy management. Appl Therm Eng 125:513–522. https://doi.org/10.1016/j.applthermaleng.2017.05.188

Ma L, Liu N, Zhang J, Tushar W, Yuen C (2016) Energy management for joint operation of CHP and P.V. Prosumers inside a grid-connected microgrid: a game theoretic approach. IEEE Trans Industr Inf 12(5):1930–1942. https://doi.org/10.1109/TII.2016.2578184

Ma K, Hu G, Spanos CJ (2013) Energy consumption scheduling in smart grid: a non-cooperative game approach. In: 2013 9th Asian control conference (ASCC), pp 1–6. https://doi.org/10.1109/ASCC.2013.6606274

Maldonado-Correa J et al (2020) Using SCADA data for wind turbine condition monitoring: a systematic literature review. Energies 13(12):3132. https://doi.org/10.3390/en13123132

Mariano-Hernández D et al (2021) A review of strategies for building energy management system: model predictive control, demand side management, optimization, and fault detect & diagnosis. J Build Eng 33:101692. https://doi.org/10.1016/j.jobe.2020.101692

Meliopoulos APS (2002) Challenges in simulation and design of/spl mu/grids. In: 2002 IEEE power engineering society winter meeting. In: Conference proceedings (Cat. No. 02CH37309), vol 1, pp 309–314. https://doi.org/10.1109/PESW.2002.985004

Moghaddam AA et al (2011) Multi-objective operation management of a renewable MG (micro-grid) with back-up micro-turbine/fuel cell/battery hybrid power source. Energy 36(11):6490–6507. https://doi.org/10.1016/j.energy.2011.09.017

Mohamed FA, Koivo HN (2010) System modelling and online optimal management of microgrid using mesh adaptive direct search. Int J Electr Power Energy Syst 32(5):398–407

Mohamed FA, Koivo HN (2011) Multiobjective optimization using modified game theory for online management of microgrid. Eur Trans Electr Power 21(1):839–854. https://doi.org/10.1002/etep.480

Mohsin M et al (2021) Assessing the impact of transition from non-renewable to renewable energy consumption on economic growth-environmental nexus from developing Asian economies. J Environ Manag 284:111999. https://doi.org/10.1016/j.jenvman.2021.111999

Nizami MSH et al (2020) A residential energy management system with bi-level optimization-based bidding strategy for day-ahead bi-directional electricity trading. Appl Energy 261:114322. https://doi.org/10.1016/j.apenergy.2019.114322

Oehler D, Jimenez M (2017) History of energy management at Judson ISD

Olivares DE et al (2014) Trends in microgrid control. IEEE Trans Smart Grid 5(4):1905–1919. https://doi.org/10.1109/TSG.2013.2295514

Ornelas-Tellez F, Jesus Rico-Melgoza J (2013) Optimal tracking control for energy management systems in microgrids. In: 2013 IEEE 56th international Midwest symposium on circuits and systems (MWSCAS), 2013, pp 489–492. https://doi.org/10.1109/MWSCAS.2013.6674692

Peng FZ, Li YW, Tolbert LM (2009) Control and protection of power electronics interfaced distributed generation systems in a customer-driven microgrid. In: 2009 IEEE power & energy society general meeting, pp 1–8. https://doi.org/10.1109/PES.2009.5275191

Pliatsios D, Sarigiannidis P, Lagkas T, Sarigiannidis AG (2020) A survey on SCADA systems: secure protocols, incidents, threats and tactics. IEEE Commun Surv Tutor 22(3):1942–1976. https://doi.org/10.1109/COMST.2020.2987688

Pratt RG et al (2010) The smart grid: an estimation of the energy and CO2 benefits. No. PNNL-19112 Rev 1. Pacific Northwest National Lab. (PNNL), Richland, WA (United States)

Prodan I, Zio E, Stoican F (2015) Fault tolerant predictive control design for reliable micro-grid energy management under uncertainties. Energy 91:20–34. https://doi.org/10.1016/j.energy.2015.08.009

Razmjoo A et al (2021) Effective policies to overcome barriers in the development of smart cities. Energy Res Soc Sci 79:102175. https://doi.org/10.1016/j.erss.2021.102175

Rezaei N, Kalantar M (2015) Stochastic frequency-security constrained energy and reserve management of an inverter interfaced islanded microgrid considering demand response programs. Int J Electr Power Energy Syst 69:273–286. https://doi.org/10.1016/j.ijepes.2015.01.023

Said O, Al-Makhadmeh Z, Tolba A (2020) EMS: an energy management scheme for green IoT environments. IEEE Access 8:44983–44998. https://doi.org/10.1109/ACCESS.2020.2976641

Sawle Y, Gupta SC (2015) A novel system optimization of a grid-independent hybrid renewable energy system for telecom base station. Int J Soft Comput Math Control (IJSCMC) 4(2):49–57

Sawle Y, Gupta SC, Bohre AK (2016) PV-wind hybrid system: a review with case study. Cogent Eng J 3(1):1189305

Sawle Y, Gupta SC, Bohre AK (2017) Review of hybrid renewable energy systems with comparative analysis of off-grid hybrid system. Renew Sustain Energy Rev 81:2217–2235

Sawle Y, Gupta SC, Bohre AK (2018) Socio-techno-economic design of hybrid renewable energy system using optimization techniques. Renew Energy 119:459–472

Sawle Y, Jain S, Babu S, Nair AR, Khan B (2021) Prefeasibility economic and sensitivity assessment of hybrid renewable energy system. IEEE Access 9:28260–28271. https://doi.org/10.1109/ACCESS.2021.3058517

Sawle Y, Gupta SC (2014) Optimal sizing of photovoltaic/wind hybrid energy system for rural electrification. In: 2014 6th IEEE power India international conference (PIICON), pp 1–4. IEEE

Sawle Y, Gupta SC (2015) Optimal sizing of a PV/biomass hybrid system with battery, storage for rice mill electrification to an off-grid remote area. Ciencia e Tecnica vitivinicola J 30(3):173–194

Sawle Y, Gupta SC (2017) Optimization of a standalone hybrid renewable energy system for telecom base station. Progress Petrochem Sci Crimson J 2(5):1–10. PPS.000546.2018. https://doi.org/10.31031/PPS.2018.02.000546. ISSN 2637-8035

Sawle Y, Gupta SC, Bohre AK (2017) Optimal sizing of standalone PV/Wind/Biomass hybrid energy system using GA and PSO optimization technique. In: 1st international conference on power engineering, computing and control, PECCON-2017, 2–4 Mar 2017, VIT University, Chennai Campus, Elsevier, Energy Procedia 00 (2017), vol 117, pp 690–698 (Scopus Indexed)

Sawle Y, Gupta SC, Bohre AK (2018) A novel methodology for scrutiny of off-grid hybrid renewable system. Int J Energy Res 42:570–586. Wiley

Sawle Y, Gupta SC, Bohre AK (2018) Techno-economic scrutiny of HRES through GA and PSO technique. Int J Renew Energy Technol 9(1/2):84–107. 69525. Inderscience

Shi W, Xie X, Chu C, Gadh R (2015) Distributed optimal energy management in microgrids. IEEE Trans Smart Grid 6(3):1137–1146. https://doi.org/10.1109/TSG.2014.2373150

Shortfall, managing an electricity (2010) The Energy Sector Management Assistance Program (ESMAP) is a global knowledge and technical assistance program administered by the World Bank and assists low-and middle-income countries to increase know-how and institutional capability to achieve environmentally sustainable energy solutions for poverty reduction and economic growth

Sola AVH, Mota CMM (2020) Influencing factors on energy management in industries. J Clean Prod 248:119263. https://doi.org/10.1016/j.jclepro.2019.119263

Sukumar S et al (2017) Mix-mode energy management strategy and battery sizing for economic operation of grid-tied microgrid. Energy 118:1322–1333. https://doi.org/10.1016/j.energy.2016.11.018

Tamrakar V, Gupta SC, Sawle Y (2015) Study of characteristics of single and double diode electrical equivalent circuit models of solar PV module. In: 2015 international conference on energy systems and applications, Pune, 2015, pp 312–317

Tenfen D, Finardi EC (2015) A mixed integer linear programming model for the energy management problem of microgrids. Electric Power Syst Res 122:19–28. https://doi.org/10.1016/j.epsr.2014.12.019

Thirunavukkarasu M, Sawle Y (2020) Design, analysis and optimal sizing of standalone PV/diesel/battery hybrid energy system using HOMER. IOP Conf Ser Mater Sci Eng 937(1). IOP Publishing

Thirunavukkarasu M, Sawle Y (2021) A Comparative Study of the Optimal Sizing and Management of Off-Grid Solar/Wind/Diesel and Battery Energy Systems for Remote Areas. Front. Energy Res 9:752043

Tuballa ML, Abundo M (2016) A review of the development of smart grid technologies. Renew Sustain Energy Rev 59:710–725. https://doi.org/10.1016/j.rser.2016.01.011

Upadhyay, Darshana, and Srinivas Sampalli. "SCADA (Supervisory Control and Data Acquisition) systems: Vulnerability assessment and security recommendations." Computers & Security 89 (2020): 101666, doi: https://doi.org/10.1016/j.cose.2019.101666

Urmee T, Harries D, Schlapfer A (2009) Issues related to rural electrification using renewable energy in developing countries of Asia and Pacific. Renew Energy 34(2):354–357. https://doi.org/10.1016/j.renene.2008.05.004

Vergara PP et al (2017) Security-constrained optimal energy management system for three-phase residential microgrids. Electric Power Syst Res 146:371–382. https://doi.org/10.1016/j.epsr.2017.02.012

Vivek Tamrakar S, Gupta C, Sawle Y (2015) Single-diode and two-diode PV cell modeling using Matlab for studying characteristics of solar cell under varying conditions. Electr Comput Eng Int J (ECIJ) 4(2):67–77. ISSN 2201-5957

Weinand JM, Scheller F, McKenna R (2020) Reviewing energy system modelling of decentralized energy autonomy. Energy 203:117817. https://doi.org/10.1016/j.energy.2020.117817

Xiong L et al (2020) Multi-agent based multi objective renewable energy management for diversified community power consumers. Appl Energy 259:114140. https://doi.org/10.1016/j.apenergy.2019.114140

Yu X, Cecati C, Dillon T, Simões MG (2011) The new frontier of smart grids. IEEE Ind Electron Mag 5(3):49–63. https://doi.org/10.1109/MIE.2011.942176

Zhao B, Shi Y, Dong X, Luan W, Bornemann J (2014) Short-term operation scheduling in renewable-powered microgrids: a duality-based approach. IEEE Trans Sustain Energy 5(1):209–217. https://doi.org/10.1109/TSTE.2013.2279837

Zia MF, Elbouchikhi E, Benbouzid M (2018) Microgrids energy management systems: a critical review on methods, solutions, and prospects. Appl Energy 222:1033–1055. https://doi.org/10.1016/j.apenergy.2018.04.103

Chapter 5
A Comprehensive Review on the Advancement of Biogas Production Using Leftover Food and Kitchen Waste

Anup Kumar Rajak, Amit Kumar, Devendra Deshmukh, Rajkumar Singh, and Shalendra Kumar

Abstract This review aims to get detailed information about biological reactions in biogas production and the Pressure Swing Adsorption (PSA) approach for biogas upgrading system based on biogas system installed at Rewa Engineering College, Rewa, Madhya Pradesh. The first part of this paper is the pretreatment of microbes, fungal reactions, enzymatic reactions, and metabolic engineering methods. The second part of this paper presents the up-gradation of biogas and their reaction with the PSA technique. The impacts of advancement of biogas production and their potential in advance improving the biogas industry are widely scrutinized. Methane (CH_4) (50–65%) in biogas obtained from biogas digester also consists of ammonia (NH_3), hydrogen (H_2), hydrogen sulfide (H_2S) (1–2%), nitrogen (N_2) and oxygen (O_2) (1–2%), and carbon dioxide (CO_2) (25–40%).

A. K. Rajak (✉)
Department of Mechanical Engineering, Rewa Engineering College, Rewa 486001, Madhya Pradesh, India
e-mail: anuprajak14@gmail.com

A. Kumar
Department of Electrical Engineering, National Institute of Technology, Kurukshetra, Haryana, India

D. Deshmukh
Department of Mechanical Engineering, Indian Institute of Technology, Indore Madhya Pradesh,, India

R. Singh
Department of Mechanical Engineering, Rewa Engineering College, Rewa 486001, Madhya Pradesh, India

S. Kumar
Department of Mechanical Engineering, National Institute of Technology, Jamshedpur, Jharkhand, India

A. K. Bohre et al. (eds.), *Planning of Hybrid Renewable Energy Systems, Electric Vehicles and Microgrid*, Energy Systems in Electrical Engineering,
https://doi.org/10.1007/978-981-19-0979-5_5

Graphical Abstract

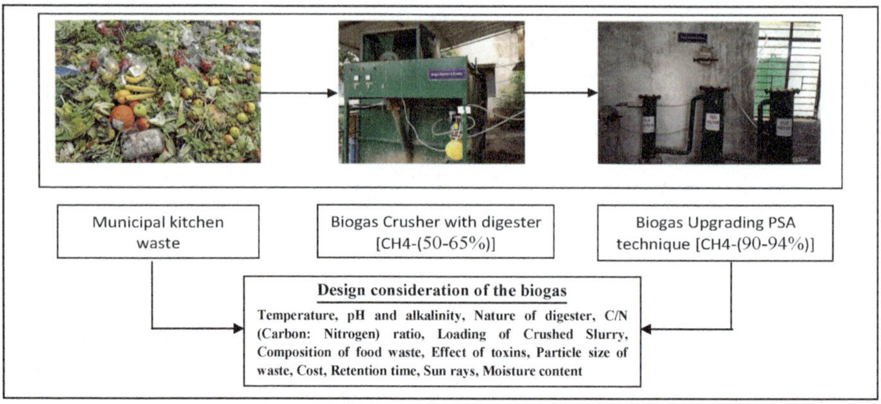

Keywords Biogas anaerobic digestion · Kitchen waste biogas digester design · Biogas chemical reactions · Design consideration

Nomenclature

AD	Anaerobic Digestion
C/N	Carbon: Nitrogen
VFA	Volatile Fatty Acids
PSA	Pressure Swing Adsorption
CH_4	Methane
N_2	Nitrogen
O_2	oxygen
NH_3	Ammonia
H_2S	Hydrogen sulphide
CO_2	Carbon dioxide

5.1 Introduction

Biogas is a sustainable energy source which is the most challenging issue worldwide. Nowadays, lots of renewable and nonrenewable energy sources have been used to fulfil energy demands (Hussain et al. 2019). Biogas is generated mainly from cow dung, buffalo, horse, and other wastes in those days. Kitchen waste can also be utilized to produce biogas due to its high nutritive value (Kumar and Majid 2020). As per the literature, the production ability of biogas using the kitchen is an alternative source with co-digestion of cow manure (Kumar and Tewary 2021). Solid, liquid, or other types of municipality waste are deposited over an open land surface which causes air pollution, human health problems, water pollution, and a totally unbalancing ecosystem (Rawoof et al. 2021).

There are two types of energy sources spread worldwide: controlled energy sources and uncontrolled energy sources. Controlled energy sources can increase productivity, and on the other hand, uncontrolled energy sources can cause problems. By using effective methods or procedures, these uncontrolled sources can be utilized to improve human life. Municipality waste management is also one of the same types of approach. Biogas/Bio-methane technique is used for successful and controlled management and generates methane as fuel and compost caused by leftover food, kitchen waste, and bio-degradable waste (Aravind et al. 2020; Khalid et al. 2021).

The paper reviewed the following constraints:

- Constraint involved in AD from 250 kg municipal solid waste for the production of biogas.
- The design considerations like temperature, waste particle size, hydrogen ion concentration, nature of digester, C/N ratio, organic loading rate, the composition of waste, retention time, moisture content, and cost are described and also discussed as cost-effective, environment-friendly, and optimally designed digester of 250 kg kitchen/municipal solids waste.
- AD processes are straightly connected to all aspects to improve methane yields, such as chemical, operation, and microbiological aspects. These literature studies have demonstrated that the energy potential of AD is to recover at a high priority level.
- There are quite a few other issues that have been addressed in previous research in AD systems, such as C/N ratio, pH, particle size, temperature, alkalinity loading rate, and retention time.

5.2 Literature Survey

Biogas is an eco-friendly heating energy source and also a combustible mixture of gases. It consists of methane (CH_4), carbon dioxide (CO_2), and is also produced from bacterial decomposition of organic compounds with AD. The biogas production from food waste is a complex method in which different type of bacteria are involved (Kadier et al. 2019). Table 5.1 shows the percentage composite of CH4 which usually ranges from 50 to 65% produced while that of CO_2 25–40 and other gases from AD.

Using AD to produce sustainable energy in tropical regions by converting biomass waste into bio-energy products is an opportunity. Energy demand has been increased and also there is a shortage of fossil fuels. Therefore, the awareness of people worldwide has moved toward the biogas energy source (Khalid et al. 2021; Kadier et al. 2019).

Table 5.1 Composition of landfill gas and biogas (Voelklein et al. 2019; Bonk et al. 2019)

S. no	Content	Landfill gas	Biogas percentage (%)
1	Methane, CH_4	35–65	50–65
2	Carbon dioxide, CO_2	15–40	25–40
3	Carbon monoxide, CO	–	1–5
4	Nitrogen, N_2	15	0–3
5	Hydrogen sulfide, H_2S	5 ppm	100 ppm
6	Oxygen, O_2	0–5	0–2
7	Ammonia, NH_3	0–5	0–2
8	Total chlorine as CI	20–200 Mg/Nm^3	0–5 Mg/Nm^3

5.3 Methodology

The breakdown of feedstock without oxygen is encouraged by the amalgamation of bacteria at each phase of the digestion process, prompting the decomposition of feedstock and incorporating CH_4 gas as the principal component with a mixture of gases (Ali et al. 2019; Kasirajan and Maupin-Furlow 2020). The chemical reaction sequences in these steps are also described (Dar et al. 2021).

The entire biogas generation from compound or straightforward organic matters can be divided into four chemical reactions.

5.3.1 *Hydrolysis Reaction*

In hydrolysis reaction, acid, base, and water can be used to step up the reaction in the presence of enzymes. In this reaction, starch, cellulose, and simple sugars can be broken down by water and enzymes. The enzymes are exoenzymes of cellulose, proteins, etc. from a number of bacteria, fungi, and protozoa in AD see Eq. (5.1) (Zhurka et al. 2020; Piotrowska-Długosz 2020).

$$(C_6H_{10}O_5)n + nH_2O \rightarrow nC_6H_{12}O_6 + nH_2 \tag{5.1}$$

The cellulose conversion in Eq. (5.1) involves the breaking of the β-1,4-glycosidic linkage. In connection with this stage of AD, insoluble cellulose consists of organic compounds that are converted. The dissociation of water from protons (H+) and hydroxide ions (OH−) results in the formation of homogeneous and/or heterogeneous acid catalysis with the presence of species. Bacteria cells can be used to break chemical bonds to form soluble compounds. Some are insoluble in organic H_2O, and microorganisms in AD are responsible for the formation of soluble compounds.

5.3.2 Acidogenesis Reaction

During acidogenesis, the molecules are further broken down into volatile fatty acids, NH_3, H_2S, and H_2 by bacteria. In other words, soluble monomers are transformed into small organic compounds such as alcohol (ethanol, methanol), ketones (glycerol, acetone), and volatile acids (butyric, formic, lactic, propionic, succinic acids) in this process (Auma 2020). Equations (5.2) – (5.4) show the acidogenic stage:

$$C_6H_{12}O_6 \rightarrow 2CH_3CH_2COOH + 2H_2O \tag{5.2}$$

$$C_6H_{12}O_6 \rightarrow 2CH_3CH_2OH + 2CO_2 \tag{5.3}$$

$$C_6H_{12}O_6 \rightarrow 2CH_3COOH \tag{5.4}$$

These organisms are able to live under both aerobic and anaerobic conditions with coli, Desulfomonas, Escherichia, Micrococcus, Peptococcus, and Streptococcus in isolated species. However, the major determinants of bacteria that predominate the properties of the material used as feedstock.

5.3.3 Acetogenesis Reaction

In this process, acetogenesis bacteria reacts, and products consist of acetic acid, CO_2, and H_2. Equations (5.5) – (5.8) represent the reactions related to the acetogenesis stage:

$$C_6H_{12}O_6 + 2H_2O \rightarrow 2CH_3COOH + 4H_2 + CO_2 \tag{5.5}$$

$$CH_3CH_2COO^- + 3H_2O \rightarrow CH_3COO^- + H^+ + HCO_3{}^- + 3H_2 \tag{5.6}$$

$$CH_3CH_2OH + 2H_2O \rightarrow CH_3COO^- + H^+ + 2H_2 \tag{5.7}$$

$$HCO_3 + 4H_2 + H^+ \rightarrow CH_3COO^- + 4H_2O \tag{5.8}$$

Several bacteria contribute to acetogenesis, such Syntrophomis wolfci, Syntrophobacter wolinii, butyrate decomposer, propionate decomposer, Peptococcus anaerobes, Actinomyces, Clostridium spp. and lactobacillus are acid formers. In the previous step, volatile fatty acids are produced, which are further broken down to the formation of CH_3COOH, CO_2, and H_2 in the acetogenesis step by binding to hydrogen-producing acetogenic microbes. Some amount of H_2O from the previous steps serves as an electron source to aid in the transformation of the VFA (Anthony et al. 2019).

5.3.4 Methanogenesis Reaction

The last phase is the methanogenesis reaction of AD. These intermediate products formed in the other phases converted into the key product methane CH_4 (Logroño et al. 2020). The reaction Eqs. (5.9) – (5.14) represent the condition that takes place in the methanogenic stage:

$$2CH_3CH_2OH + CO_2 \rightarrow 2CH_3COOH + CH_4 \tag{5.9}$$

$$CH_3COOH \rightarrow CH_4 + CO_2 \tag{5.10}$$

$$CO_2 + 4H_2 \rightarrow CH_4 + 2H_2O \tag{5.11}$$

$$CH_3OH \rightarrow CH_4 + H_2O \tag{5.12}$$

$$CH_3COO^- + SO_4^{2-} + H^+ \rightarrow HCO_3 + H_2S \tag{5.13}$$

$$CH_3COO^- + NO^- + H_2O + H^+ \rightarrow 2HCO_3 + NH_4^+ \tag{5.14}$$

The conversion of CH_3CH_2OH into CH_3COOH is represented in Eq. (5.9) and is further converted into CH_4 and CO_2 as represented in Eq. (5.10). The formed CO_2 reacts with H_2 and is converted into CH_4 gas as represented in Eq. (5.11). HCO_3, H_2S, and NH_4^+ are formed in Eqs. (5.13) and (5.14) (Voelklein et al. 2019; Bonk et al. 2019).

Based on these four steps, a biogas digester combined with a crusher has been developed to study the production of biogas. The CH_4 content in biogas streams exceeds 50% and CH_4 emissions to the environment result in a resourceful hydrocarbon waste that also has a greenhouse warming potential 23 times greater than CO_2. Therefore, adequate consumption and collection of the CH_4 limited in biogas ward off emissions of CH_4 to the environment.

5.4 Design Consideration of the Bio-Gasification

The generation of gas is affected by various factors. Some of the environmental factors also influence bio-gasification (Schulzke 2019; Nsair et al. 2020; Igoni et al. 2008). Some of them are as follows.

5.4.1 Temperature

The optimum temperature range for biogas production is around 25–40 °C and can be achieved without additional heat. Additional heat input is required to raise the temperature to 50–60 °C for additional biogas production. At present, it is observed that this temperature range is normal for biogas production (Igoni et al. 2008; Hamzah et al. 2019).

5.4.2 PH and Alkalinity

Due to excess loading and the presence of toxic materials (acidic), the pH value is reduced below 6.5 and a decrease in the production of biogas is caused. The process is delayed in an inactive state. When the pH value is very low, the loading to the digester is stopped and the recommended time to recover the pH for this temperature range is acquired (Nsair et al. 2019). In general, in AD, pH = 7 is optimum, which is close to neutral.

5.4.3 Nature of Digester

Nowadays, the production of gas from household waste is insignificant since current digesters are not proficient for small-scale uses. Therefore, the nature of feedstock also needs research. Low-cost community-based digesters, low-tech natural digesters, or modern digesters may be used. Many researchers used digesters like batch systems, high solid, anaerobic sequencing batch reactors, or continuous one or two-stage systems. However, based on the nature of solids, excess digesters are available.

5.4.4 Nutrient Concentration

An ideal carbon: nitrogen ratio of 25:1 is to be maintained for efficient plant operation in the digester [48]. If the C/N ratio is too high, in that case, biogas production can be improved to N_2 and vice versa. If nitrogen is too high, it inhibits methanogenic activity. Carbon provides energy to maintain microorganisms, while N_2 helps build their cells. This can be increased by adding carbon. In AD, not only does it convert plant material into CH_4 gas, but it also releases plant nutrients such as potassium (K), nitrogen (N), and phosphorus (p). It is also converted into compost which can be useful for plants. 22 and 25 C/N ratio is best for AD of food waste (Bougrier et al. 2018; Yuan et al. 2019).

5.4.5 Loading of Crushed Slurry

The loading rate of crushed slurry is the biological translation of a reactor. It determines the amount of volatile solids that are possible as inputs to the digester. The feed rate is given by the mass of total solids (m-kg) fed per day, divided by the ratio of total solids (Ts) in the mixed slurry (Babaei and Shayegan 2020). Hydraulic loading is also an important factor for digester volume. A high organic loading rate may cause a rise in fatty acids and result in low biogas yield. This will lead to the mass death of methanogenic bacteria, decrease in pH, and propagation of acidogenesis.

5.4.6 Composition of Food Waste

The composition of solid waste (fruits and vegetables) is an essential consideration to predict efficient design. Residential, commercial, and institutional establishments such as cafeterias or canteens are active sources of food waste. When an excess of protein is followed by carbohydrates and cellulose and results, methanization is accelerated. In line with previous literature, legumes and milk powders show wide variations among compound constituents. In general, legumes show a high content of milk and carbohydrates and confirm a high lipid content (Sarker et al. 2019).

5.4.7 Effect of Toxins

The presence of toxic substances in AD, such as chlorinated hydrocarbons (such as chloroform) and other organic solvents, is mostly toxic to digestion. This is the important factor that prevents the production of gas. The toxins are difficult to remove if the digester is badly poisoned. In this case, before loading the fresh solution, the digester should be emptied and cleaned with plenty of water (Nsair et al. 2020; Annibaldi et al. 2019).

5.4.8 Retention Time

The time during of feedstock remain in the reactor is known as the retention time. 15–30 days for mesophilic digester and 12–24 days for thermophilic are optimal retention times for complete biological conversion. It depends upon the intended use of digested material, type of substrate, and environmental conditions. For reducing the instability of the digester, parameters like temperature, hydraulic retention time, and organic loading rate must be monitored.

5.4.9 Particle Size of Waste

The particle size of waste id directly affects the breakdown in AD. Particle size can be reduced by grinding, crushing, and shredding. Hence, these methods increase the surface area for microbe's action and eventually recover the efficiency of the digester. Mostly, mechanical grinding, thermal, chemical, microwave, ultrasound methods are used for the disintegration of waste (Gollakota and Meher 1988).

5.4.10 Cost

Construction and maintenance costs, capital and operating costs, substrate receiving, and waste processing are essential considerations in the selection of the type and size of the digester (Babaei and Shayegan 2020; Sarker et al. 2019).

5.4.11 Sun Rays

One more factor of sun rays is the most important to maintain the biogas depending on the solar intensity of the atmosphere.

5.4.12 Moisture Content

An anaerobic digestion process with 70–80% moisture is carried out at different levels. The bioreactor operates at 70% moisture content and produces more methane than bioreactors operating at 80% moisture content. Hence, high moisture contents usually facilitate AD. However, keeping the availability of the same water level is very difficult throughout the digestion cycle (Gollakota and Meher 1988).

5.5 Operational Performance Data

The main factor affecting biogas production are pH, chemical oxygen demand, volatile solid, total solid, time, temperature, hydraulic retention time. Table 5.1, 5.2 show a literature review on biogas production from municipal solid waste from 2010 to 2021.

5.6 Kitchen Waste-based Biogas Plant Design

5.6.1 Digester

The digester is to undergo the fermentation of the food waste which is available in the hostel mess and inside the premises of the college. Figure 5.1 shows the biogas digester, which is installed at Rewa Engineering College, Rewa. The digester is designed to have the daily food waste with 250 kg maximum per day. The retention time is 30 days and the waster added is 100ltr max. The digester has 1 inlet, which is accompanied by the crusher (2hp motor) to crush the waste to 5 to 8 mm size. The outlet is attached with the 4″ diameter pipe to carry the manure.

5.6.2 Floater

The floater is used to collect the biogas generated by the fermentation of the food inside the digester. The floater is also called the floating dome of the digester. The biogas generated/produced due to anaerobic fermentation of the food waste is collected in it. The floating dome, as the gas is produced, is lifted automatically, and as the gas is used, the dome comes to its original form (means touches the digester). For the proper function of the floater, always check the liquid in the water jacket (oil/water can be used in the jacket). Check the size of the floater in the manual. The floater is attached with the flexible outlet pipe to suck the biogas and to transfer it to the filter at the pressure of 5 bar, with the help of a small booster pump (Fig. 5.2).

Table 5.2 A literature review on biogas production from municipal waste (Gollakota and Meher 1988; Izhar et al. 2021; Masih et al. 2019; Suelen et al. 2017; Akpan et al. 2015; Sathish and Vivekanandan 2015; Joel et al. 2015; Dasgupta and Mondal 2012; Biswas et al. 2010)

Sources	Waste material	pH	C/N ratio	Temp	Total solid	Volatile solid	Volatile fatty acids	Retention time	Methane yield
Izhar et al. (2021)	Kitchen waste: University canteen Fruit and vegetable waste	5.08 to 5.28	13.98–17.21	42	22.17 7.94	17.87 6.74	–	80 days	0.642 m3 CH_4/kg VS added
Masih et al. (2019)	Cattle dung using with jaggery	5.7 to 7.3	46.4	12 to 24	12.5% to 10.7%	84.6% to 80.9%	1189 to 1110 mg/kg	Nine week	40 l/kg
Suelen et al. (2017)	Fruit and vegetable waste	8– 8.2	34.7	35	–	54.6%	3526.6	20 days	396.6 Nml/gVS
Akpan et al. (2015)	Biodegradable component of MSW in Nsukka metropolis	6.7 to 7.1	–	28–37	–	–	–	28 days	84,277.17m^3
Sathish and Vivekanandan (2015)	Using industrial waste (press mud)	7.1	Less than 18	30–35	–	–	–	45 days	0.68 m3
Abishek et al. (2015)	Using fungi culture with methanogens	6.7	–	30–37	75.55%	93.36%	–	60 days	59.3%
Dasgupta and Mondal (2012)	Bioenergy conversion of Varanasi's	8.9	25–30	–	30%	80%	–	15 days	0.40 Nm3/kgVS
Yilmaz and Demirer (2011)	Cattle manure: Effect of phase separation	6.5 to 7.5	–	35	–	35%	1700 and 1300 mg / l	57 days	313 and 221 ml CH_4/g.VS
Biswas et al. (2010)	Growing up biogas yield of compost by waste	–	–	180	20%	14.8%	–	10 min	136%

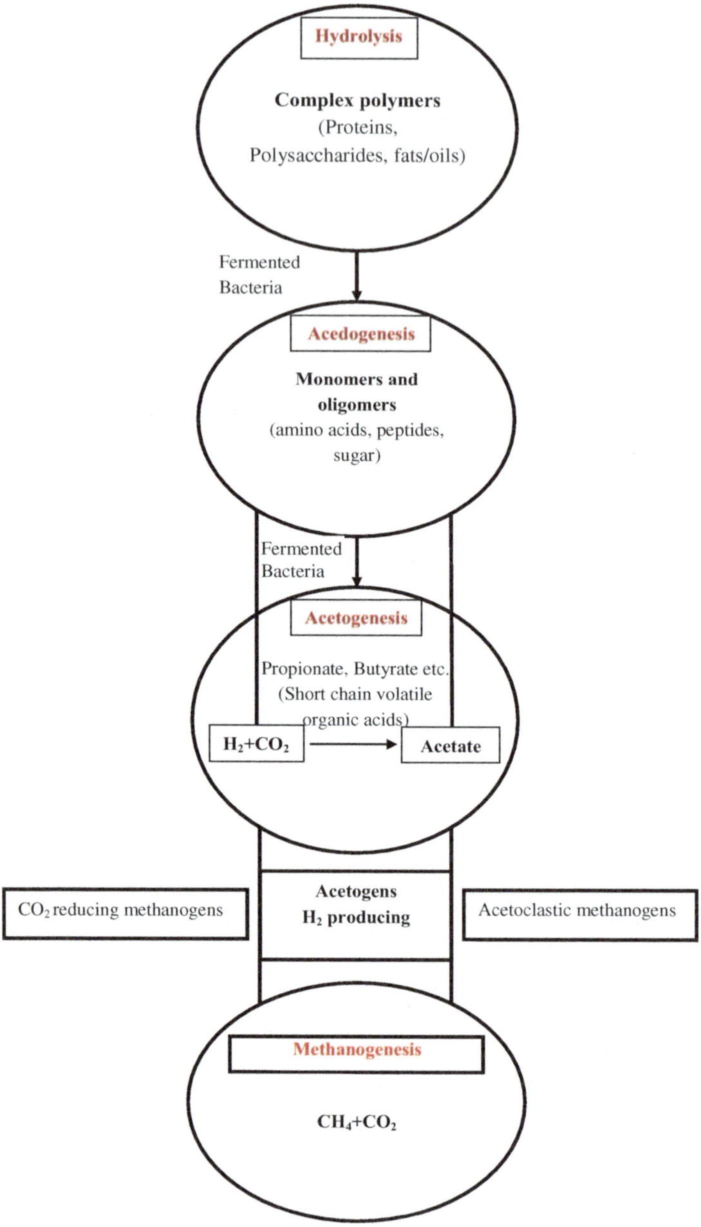

Fig. 5.1 Anaerobic digestion process

Fig. 5.2 250 kg Kitchen waste biogas digester installed at Rewa Engineering College, Rewa, Madhya Pradesh

5.6.3 Filter Unit

The filters use to filter the CO_2, H_2O, H_2S from the raw biogas and make it effective to 90 to 94% methane (separation of other gases and retaining the CH_4). The technique used for filtration is PSA.

Three different cylindrical vessels ate used with adsorption material filled inside them to do the surface adsorption and then released from their surfaces after the period as mentioned. In this process, biogas is compressed into 4–10 bar pressure. Compressed raw biogas is fed to multiple vessels filled with adsorbent materials. In this process activated carbon, zeolites, and other materials (titanosilicates) are engaged as an adsorbent. These adsorbents have a high surface area due to the porosity and filtered CO_2, H_2O, and H_2S from the biogas in contact with these adsorbent materials.

The adsorbents become saturated with CO_2 after a certain time, and the column needs to be regenerated by reducing the pressure. Normally, the regeneration of biogas is pressurized at vacuum. The adsorbed material is toxic by H_2S because it adsorbs H_2S irreversibly. To overcome this issue, the PSA process also includes an initial H_2S removal step. Another adsorption vessel removes the moisture of this biogas. Figure 5.3 shows biogas filtered vessels in which pure methane is recovered at the top of the final column with very little pressure drop.

Fig. 5.3 Biogas upgrading system installed at Rewa Engineering College, Rewa, Madhya Pradesh

To simulate a continuous process, multi-column arrays are used. Storage tanks with a single column can also be used for smaller size applications. They are connected with biogas meter to read raw biogas. The most important property of pressure swing adsorption technology is that it can be adapted to biogas purification systems in any location in the world as it is available at hot or cold sources.

5.6.4 Biogas Analyzer

Three sensors-based problem systems are used to analyze the flowing gases at the pressure of 0.5 l/m. The analyzer is having an input (for biogas) of 4 mm with a soft, flexible pipe. The analyzer runs on the direct supply and also has the 6v battery inside the apparatus. The display shows the data for CH_4, CO_2, and H_2S. The reading can also be stored in the apparatus and also be transferred to the computer with the cable provided. Size of the apparatus is 9" × 4" × 7", weight around 250gm.

5.7 Procedure

Check all the connections of the pipes and nozzles are airtight (prevent the leakages of the gas. The digester is to be provided with a 5A power supply and connected to the panel of the plant, and then the water jacket is to be filled with oil + water (400–500 L). The crusher is ready for feeding.

The first feed of the plant is to be done 1.5–2 tons with the cow dung so that it will help the fast production of the biogas. Now, the daily feeding of kitchen waste (250 kg food waste + water 100 litter) is to be started. After 15 to 20 days, the floater on the top starts inflating with the gas produced in the digester, but it is not sure that it is enriched in methane or not or in other words, the gas is burnable or not (so we should check it by connecting with the burner if the gas is burning. If the gas is burnable, we start with the filtration of the gas. The raw biogas from the digester is forced to enter the filtration chambers at 5 to 6 bar by switching the booster pump. The gas entered in the filter is passed from the flow meter to get the reading (to calculate the volume of the biogas). Calibrate the gas analyzer and use it to take the reading of the filter gas from filters chambers from the sample valves. Three gas readings can be taken from the analyzer CH_4, CO_2, H_2S gas.

5.8 Conclusion

This review paper is based on a 250 kg kitchen waste biogas digester system, which is installed at Rewa Engineering College, Rewa. Some salient features are immerged with this reviewed study that biogas is a better alternative of energy source, produced from the kitchen waste, cow dung, or other wastes. AD processes convert biomass energy into recycling organic waste and reduce harmful effects on the environment. Biogas digesters work in two ways: one is to reduce waste and the other is to generate valuable energy.

According to this review, more sophisticated research is needed for the following:

- Information collection and, based on experiments, design environmentally friendly viable digesters for municipal waste.
- These considerations affect feed rate, feedstock moisture content, fluid flow patterns (such as unsteady and stratified fluid flows), and co-digestion of different feedstocks. However, those highlighted ideas have been widely reported and the advances made in the optimization of AD technology have been confirmed.

In terms of the various potential applications of AD, future work will be devoted to achieving full optimization of the system. Shown is the two-stage biodegradation of food waste system with good capacity and efficiency. Therefore, additional attention should be paid to the development of such digester systems for municipal waste. However, the single-phase should not be overlooked as it is effective.

Acknowledgements The study is based on the Research project entitled "Modeling and optimisation of biogas production for advanced fuel, heat and power production" that has been approved under "TEQIP Collaborative Research Scheme." The authors would like to thank Dr. B.K. Agrawal, Dr. S.D. Sharma, Dr. R P Tiwari, and Dr. R K Jain for their support and are also very grateful to Rewa Engineering College, Rewa, for the academic and technical support.

Funding This research has been financed by the National Project Implementation Unit (NPIU) and governed by AICTE India, under the project of Collaborative Research Scheme- (CRS ID: CRS-1–5,728,212,545).

References

Akpan PU, Omeife MA, Onyishi HO, Okoye OC (2015) Biogas production from biodegradable component of municipal solid waste in Nsukka Metropolis. International Conference on Electric Power Engineering (ICEPENG 2015), OCTOBER 14–16, 14–16. https://doi.org/10.13140/RG.2.1.4310.3448

Ali AM, Atefeh S, Akbari NK, Tala B, Antonio V (2019) Halophiles and their vast potential in biofuel production. Front Microbiol 10:1895. https://doi.org/10.3389/fmicb.2019.01895

Annibaldi V, Cucchiella F, Gastaldi M, Rotilio M, Stornelli V (2019) Sustainability of biogas based projects: technical and economic analysis. International Conference on Green Energy and Environment Engineering. In Proceedings of the E3S Web of Conferences, Kitahiroshima, Japan, vol 93, 03001. https://doi.org/10.1051/e3sconf/20199303001

Anthony A, Ali M, Muhammad N, Karin G (2019) A review of the chemistry of anaerobic digestion: methods of accelerating and optimizing process efficiency. Processes 7(8):2227–9717. https://doi.org/10.3390/pr7080504

Aravind S, Kumar PS, Kumar NS, Siddarth N (2020) Conversion of green algal biomass into bioenergy by pyrolysis—a review. Environ Chem Lett 18:829–849. https://doi.org/10.1007/s10311-020-00990-2

Auma EO (2020) Anaerobic co-digestion of water hyacinth (Eichhornia crassipes) with ruminal slaughterhouse waste under mesophilic conditions. Thesis, University of Nairobi. http://erepository.uonbi.ac.ke/handle/11295/154487.

Babaei A, Shayegan J (2020) Effects of temperature and mixing modes on the performance of municipal solid waste anaerobic slurry digester. J Environ Health Sci Eng 17:1077–1084. https://link.springer.com/article/10.1007%2Fs40201-019-00422-6.

Biswas R, Uellendahl H, Ahring BK (2010) Top of formIncreasing the biogas yield of manure by wet explosion of the digested fiber fraction bottom of form. 12th World Congress on Anaerobic Digestion, Guadalajara, Mexic

Bonk F, Popp D, Weinrich S, Sträuber H, Becker D, Kleinsteuber S, Harms H, Centler F (2019) Determination of microbial maintenance in acetogenesis and methanogenesis by experimental and modeling techniques. Front Microbiol 10:1–13. https://doi.org/10.3389/fmicb.2019.00166

Bougrier C, Dognin D, Laroche C, Gonzalez V, Benali-Raclot D, Rivero JAC (2018) Anaerobic digestion of Brewery Spent Grains: trace elements addition requirement. Biores Technol 247:1193–1196. https://doi.org/10.1016/j.biortech.2017.08.211

Dar RA, Parmar M, Dar EA, Sani RK, Phutela UG (2021) Biomethanation of agricultural residues: potential, limitations and possible solutions. Renew Sustain Energy Rev 135, 110217, ISSN 1364–0321. https://doi.org/10.1016/j.rser.2020.110217.

Dasgupta BV, Mondal MK (2012) Bio energy conversion of organic fraction of varanasi's municipal solid waste. Energy Procedia 14:1931–1938. https://doi.org/10.1016/j.egypro.2011.12.1190

Gollakota K, Meher K (1988) Effect of particle size, temperature, loading rate and stirring on biogas production from castor cake (oil expelled). Boil. Wastes 24:243–249. https://doi.org/10.1016/0269-7483(88)90109-7

Hamzah MAF, Jahim JM, Abdul PM, Asis AJ (2019) Investigation of temperature effect on start-up operation from anaerobic digestion of acidified palm oil mill effluent. Energies 12:2473. https://doi.org/10.3390/en12132473

Hussain A, Sarangi GK, Pandit A, Ishaq S, Mamnun N, Ahmad B, Jamil MK (2019) Hydropower development in the Hindu Kush Himalayan region: issues, policies and opportunities. Renew Sustain Energy Rev 107:446–461. https://doi.org/10.1016/j.rser.2019.03.010
Igoni H, Ayotamuno MJ, Eze CL, Ogaji SOT, Probert SD (2008) Designs of anaerobic digesters for producing biogas from municipal solid-waste. Appl Energy 85(6):430–438. https://doi.org/10.1016/j.apenergy.2007.07.013
Izhar TNT, Zakarya IA, Zaaba SK, Yusof AHM, Shahril NM (2021) A review of food waste characterization and treatment in anaerobic digestion. International Conference on Civil and Environmental Engineering, IOP Conf. Series: Earth and Environmental Science 646, 012004. https://doi.org/10.1088/1755-1315/646/1/012004
Joel JA, Murali G, Ravishankar M, Sibichakravarthy M, Sundhirasekar A (2015) Performance analysis of anaerobic digestion to extract biogas from kitchen waste. Int J Sci Eng Res vol 6(3):703
Kadier A, Kalil MS, Rai PK, Kumar SS, Abdeshahian P, Sivagurunathan P, Hasan HA, Hamid AA, Mohamed A (2019) A promising and green approach for bioenergy and biochemical production from waste resources. Microbial Electrolysis Cells (MECs). https://doi.org/10.1002/9781119611103.ch12
Kasirajan L, Maupin-Furlow JA (2020) Halophilic archaea and their potential to generate renewable fuels and chemicals. Biotechnol Bioeng. https://doi.org/10.1002/bit.27639
Khalid ZB, Siddique MNI, Nayeem A, Adyel TM, Ismail SB, Ibrahim MZ (2021) Biochar application as sustainable precursors for enhanced anaerobic digestion: a systematic review. J Environ Chem Eng 9, 4, 105489, ISSN 2213–3437. https://doi.org/10.1016/j.jece.2021.105489.
Kumar JCR, Majid MA (2020) Renewable energy for sustainable development in India: current status, future prospects, challenges, employment, and investment opportunities. Energy, Sustainability and Society 10:2. https://doi.org/10.1186/s13705-019-0232-1
Kumar D, Tewary T (2021) Techno-economic assessment and optimization of a standalone residential hybrid energy system for sustainable energy utilization. Int J Energy Res. https://doi.org/10.1002/er.6389
Logroño W, Popp D, Kleinsteuber S, Sträuber H, Harms H, Nikolausz M (2020) Microbial resource management for ex situ biomethanation of hydrogen at alkaline pH. Microorganisms 8:4. https://doi.org/10.3390/microorganisms8040614
Masih JC, Mehta S, Malik K, Sindhu S, Anand RC (2019) Enhancement of biogas production from cattle dung using additive. J Entomol Zool Stud 7(1):1625–1630
Nsair A, Cinar SÖ, Abu-Qdais H, Kuchta K (2019) Optimizing the performance of a large-scale biogas plant by controlling stirring process: a case study. Energy Convers Manage 198. https://doi.org/10.1016/j.enconman.2019.111931
Nsair A, Onen Cinar S, Alassali A, Abu Qdais H, Kuchta K (2020) Operational parameters of biogas plants: a review and evaluation study. Energies 13(15):3761. https://doi.org/10.3390/en13153761
Piotrowska-Długosz, A (2020). Significance of the enzymes associated with soil C and N transformation. Carbon and Nitrogen Cycling in Soil. Springer, Singapore. https://doi.org/10.1007/978-981-13-7264-3_12
Rawoof SAAP., Kumar S, Subramanian DNVS (2021) Sequential production of hydrogen and methane by anaerobic digestion of organic wastes: a review. Environmental Chemistry Letters vol 19, pp 1043–1063
Sarker S, Lamb JJ, Hjelme DR, Lien KM (2019) A review of the role of critical parameters in the design and operation of biogas production plants. Appl Sci 9:1915. https://doi.org/10.3390/app9091915
Sathish S, Vivekanandan S (2015) Experimental investigation on biogas production using industrial waste (press mud) to generate renewable energy. Int J Innovative Res Sci Eng Technol 4(2):388–392
Schulzke T (2019) Biomass gasification: conversion of forest residues into heat, electricity and base chemicals. Chem Pap 73:1833–1852. https://doi.org/10.1007/s11696-019-00801-1

Suelen P, Kramer LE, Gomes LP, Miranda LAS (2017) Biogas production from co-digestion of organic fraction of municipal solid waste and fruit and vegetable waste. Bioresour Technol 228. https://doi.org/10.1016/j.biortech.2017.01.003.

Voelklein MA, Rusmanis D, Murphy JD (2019) Biological methanation: strategies for in-situ and ex-situ upgrading in anaerobic digestion. Applied Energy 235:1061–1071. https://doi.org/10.1016/j.apenergy.2018.11.006.

Yuan Y, Bian A, Zhang L, Chen T, Pan M, He L, Wang A, Ding C (2019) Combined process for efficient biomethane production from corn straw and cattle manure: optimizing C/N ratio of mixed hydrolysates. BioResources 14:1347–1363

Zhurka M, Spyridonidis A, Vasiliadou IA, Stamatelatou K (2020) Biogas production from sunflower head and stalk residues: effect of alkaline pretreatment molecules 25(1):164. https://doi.org/10.3390/molecules25010164

Chapter 6
Design and Analysis of Renewable-Energy-Fed UPQC for Power Quality Improvement

Miska Prasad, Yogesh Kumar Nayak, Rajesh Ranjan Shukla, Rajagopal Peesapati, and Sudhansu Mehera

Abstract Z-source inverter (ZSI) is a new topology in power converter, especially in DC–AC converter at a very interesting power level. For instance, it only uses a single-stage power converter with the ability of buck–boost characteristic operations. This work introduces a combination of a solar system with unified power quality conditioner (UPQC) based Z-source inverter for reducing the voltage swell and harmonics under the sudden addition of a balanced three-phase nonlinear load. This article additionally proposed another mixture MPPT system, which is the combination of perturbation and observation (P&O) and incremental conductance (InC) methodologies. The modeling and simulation of the proposed UPQC with ZSI has been executed in MATLAB/Simulink for relief of voltage swell and harmonics and the obtained results are contrasted and UPQC with VSI and CSI.

Keywords ZSI · UPQC · Sag · Swell · Solar Energy

6.1 Introduction

These days voltage quality unsettling influences, in particular, voltage droops, and swells represent a genuine danger to the clients (Venkatesh et al. 2011; Alam et al.

M. Prasad (✉)
EEE Department, ACE Engineering College, Hyderabad, India
e-mail: pmiska25@gmail.com

Y. K. Nayak · R. R. Shukla
Electrical Engineering Department, Government College of Engineering, Keonjhar, Odisha, India
e-mail: yogeshnayak_fee@gcekjr.ac.in

R. Peesapati
EEE Department, Raghu Engineering College, Vizag, India
e-mail: rajgopal.peesapati@raghuenggcollege.in

S. Mehera
Department of Mechanical Engineering, Government College of Engineering, Keonjhar, Odisha, India
e-mail: sudhansumeher_fme@gcekjr.ac.in

A. K. Bohre et al. (eds.), *Planning of Hybrid Renewable Energy Systems, Electric Vehicles and Microgrid*, Energy Systems in Electrical Engineering,
https://doi.org/10.1007/978-981-19-0979-5_6

2015). Earlier conventional strategies are proposed to mitigate power quality issues. Yet, of late, to improve the voltage quality, Custom Power Device (CPD) gadgets are utilized. UPQC is the arrangement of shunt associated CPD gadget that is used to limit the voltage hang, swell, and sounds in a force dispersion framework (Sundarabalan and Selvi 2015; Sadigh and Smedley 2016; Rahman et al. 2015; Yunfei et al. 2016). Apart from this, the issue of shoot-through or cross-conduction may obliterate the entire IGBT exchanging gadgets. The CSI works as a step-up inverter and the open circuit across the inductor is the main problem. But UPQC with ZSI works as a step-down and step-up inverter (Vodapalli et al. 2015; Reisi et al. 2013; Balamurugan et al. 2013). The voltage infusion ability has both buck and boost capability (Hanif et al. 2011; Zope and Somkuwar 2012; Pilehvar et al. 2015). Because of the nearness of this one-of-a kind character, it licenses inverters to be worked in the shoot-through mode (Tang et al. 2011; Saravanan et al. 2015; Tajuddin et al. 2015a). Solar energy is one of the most reliable energy sources for renewable energy power generation. It changes the energy from sunlight and converts it using power converters process in order to deliver the generated power to an existing electrical network. As has been known, electrical power consumption increases almost 3.5% every year (Wu et al. 2017). It shows that solar power generation is a very promising electrical generation, especially in a country that receives high sunlight penetration during the day. However, the conventional PV power converters circuit requires two-stage converters (Tajuddin et al. 2015b); the first is to boost the PV voltage to the desired level and then convert the DC input back into AC before it can be fed to the existing electrical grid. Though this configuration is proven to be performing well, it creates lower efficiency, lower reliability, larger in size, more circuitry converters, and produces high power losses as explained in Kannan et al. (2014), Ali (2018) at the output of the converter. As a solution, a Z-source inverter (ZSI) which is based on a passive circuitry process is proposed to overcome the drawbacks of a conventional PV-inverter connection. Currently, ZSI has become an emerging topology in power converter especially in DC–AC converter at a very interesting power level. For instance, it only uses a single-stage power converter with the ability of buck–boost characteristic operations with added features of low ripple input current and high value of voltage gain. These advantages make use of this converter for photovoltaic power generation with high tracking efficiency and achieve better performance (Carrasco and Mancilla-David 2016). Since ZSI is considered a new type of inverter, a lot of research on this area has increased. The focus of the research is to improve the control algorithm, reduce switching schemes, or add a new topology or others (Kannan et al. 2014). However, all of these techniques could not work if the current maximum power point tracking (MPPT) has not been modified to extract more power from the PV for the grid supply (Ahmed and Salam 2016). Consequently, numerous MPPT techniques are reported in the literature (Kandemir et al. 2017; Fathabadi 2016; Alik and Jusoh 2018) and it has been discussed that an emphasis on an improved MPPT can be applied in a Z-source inverter. MPPT approaches can be classified into two categories, which are the conventional and soft computing approaches (Ahmed and Salam 2015). At the moment, the conventional MPPT algorithms such as perturb and observe (P&O) (Umarani and Seyezhai 2016; Tey and Mekhilef 2014), incremental conductance

(InC) (Farhat et al. 2015), hill climbing (HC), etc. have been used intensively in tracking power generated from a solar panel. Several researchers had utilized the Artificial Intelligence techniques such as fuzzy logic control (FLC) (Gheibi et al. 2016), adaptive fuzzy controller (Messalti and Harrag 2015), neural network (NN) (Kermadi and Berkouk 2017) to operate in soft MPPT methods or others names called as soft computing approach. All of them are proven to be very effective in dealing with nonlinear characteristics of solar cells based on the PV I–V curve characteristics.

6.2 PV UPQC with VSI, CSI, and ZSI

Solar photovoltaic UPQC with VSI, CSI, and ZSI is highlighted in Figs. 6.1, 6.2 and 6.3. The DC-connection of both dynamic force channels is associated with a typical DC-interface capacitor for a situation of VSI-UPQC, regular DC-interface inductance for a situation of CSI-UPQC, and both if there should arise an occurrence of ZSI-UPQC. The arrangement gadget of the UPQCs, otherwise called an arrangement dynamic force channel, keeps the customer load end voltages inhumane toward the inventory voltage quality issues like hang/swell, floods, vacillations, and unbalance. The shunt gadget of the UPQCs, otherwise called a shunt dynamic force channel, gives responsive force remuneration, load adjusting, and disposal of sounds (Fig. 6.2).

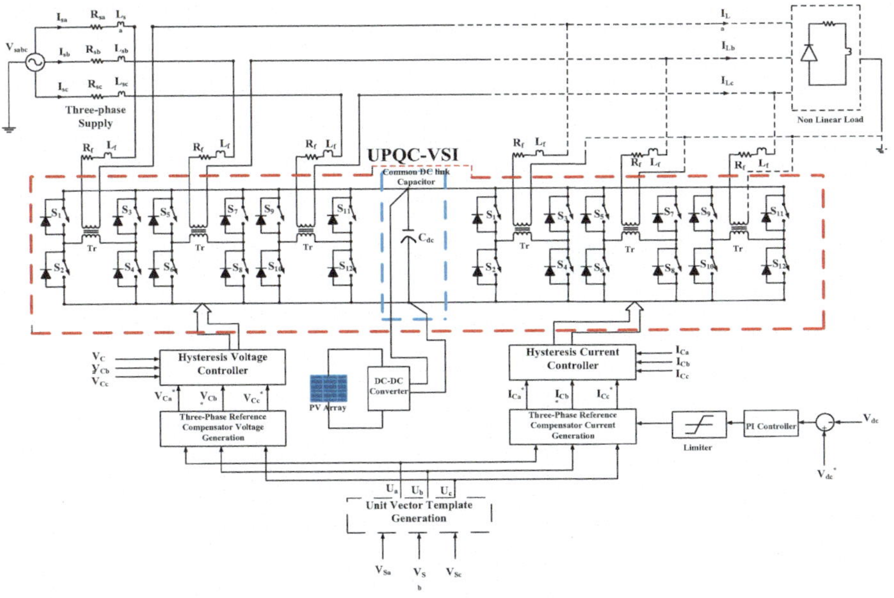

Fig. 6.1 UPQC with VSI

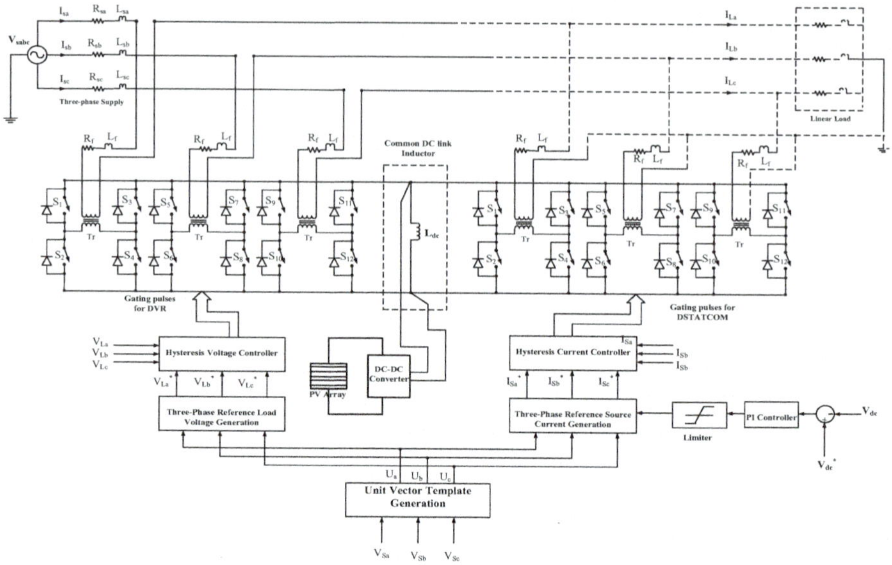

Fig. 6.2 UPQC with CSI

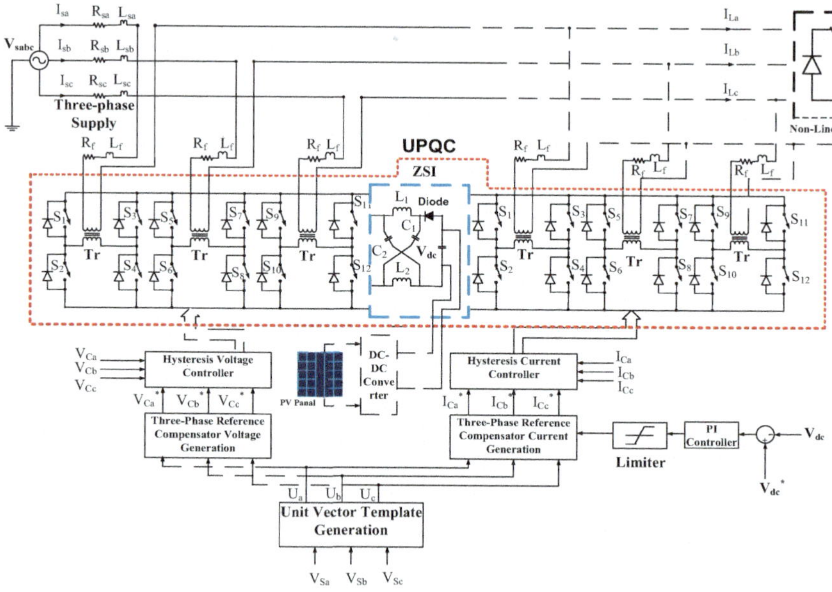

Fig. 6.3 UPQC with ZSI

6.3 Working States of ZSI

The action of ZSI can be finished after three modes such as active mode, zero-state mode, and shoot-through (ST) mode as shown in Fig. 6.4(a–c). The parameters in Figs. 6.1–6.3 are highlighted in Table 6.1.

Let us consider impedance network elements have a similar value ($L_1 = L_2 = L$ and $C_1 = C_1 = C$).

Hence, the voltage relation of ZSI is (Hanif et al. 2011; Pilehvar et al. 2015)

$$\left.\begin{aligned} V_{L1} &= V_{L2} = V_L \\ V_{C1} &= V_{C2} = V_C \end{aligned}\right\} \tag{6.1}$$

where t_0 is the range of shoot-through mode (Hanif et al. 2011; Pilehvar et al. 2015).

$$\left.\begin{aligned} V_L &= V_C \\ V_{diode} &= 2V_C \\ V_{dc-1} &= 0\,(because \text{ shoot-through}) \end{aligned}\right\} \tag{6.2}$$

where t_1 is the range of normal and zero-state mode (Hanif et al. 2011; Pilehvar et al. 2015).

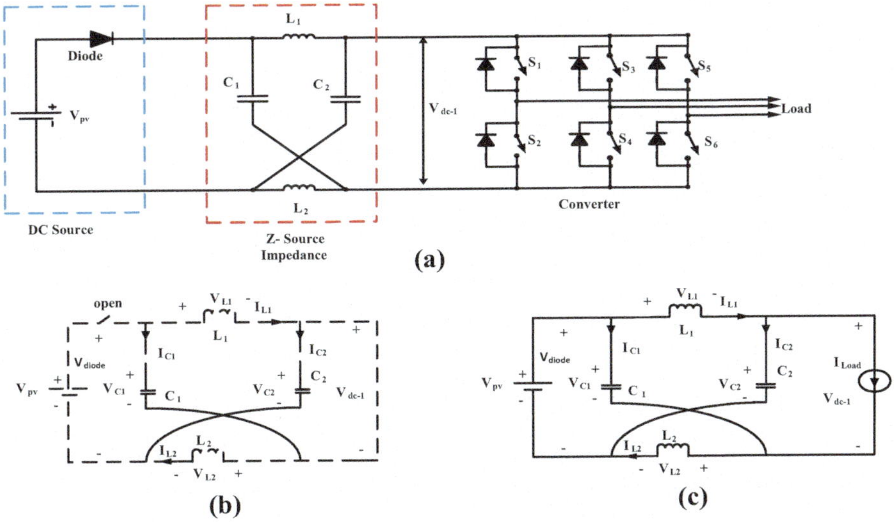

Fig. 6.4 **a** ZSI, **b** ST, and **c** active and zero mode

Table 6.1 Value of elements in Figs. 6.1–6.3

Parameter	Value
Solar photovoltaic output voltage (V_{PV})	35.5 V
ZSI Inductance (L1 = L2)	0.6 mH
ZSI Capacitor (C1 = C2)	0.05 μF
Peak DC-link voltage Vdc-1 (during shoot-through mode)	0 V
Vdc-1 (during active and null mode)	300 V
Voltage across capacitor (V_C)	167.75 V
Voltage across inductor (V_L)	−132.25 V
Modulation index (M)	0.5
Shoot-through duty ratio (D)	0.49
Shoot-through time per cycle (t_0)	9.83 μs
Non-shoot-through time per cycle (t_1)	12.8 μs
Total Switching Period (t)	0.02 ms
Average current through inductor($\bar{I}_{L,Avg}$)	4.28 A
Maximum inductor current ($I_{L,Max}$)	5.56 A
Minimum inductor current ($I_{L,Min}$)	2.99 A
Switching frequency (fs)	50 kHz

$$\left.\begin{array}{l} V_L \neq V_C \\ V_{diode} = V_{PV} = V_L + V_C \\ V_L = V_{PV} - V_C = V_C - V_{dc-1} \\ V_{dc-1} = V_C - V_L = 2V_C - V_{PV} \end{array}\right\} \tag{6.3}$$

where V_{pv} is solar photovoltaic output voltage.

At switching period t, the mean voltage of the inductor during steady state is zero (V_{LAvg}) (Hanif et al. 2011; Pilehvar et al. 2015).

$$V_{LAvg} = V_C * t_0 + (V_{PV} - V_C) * t_1 = 0$$

$$V_C = V_{PV}\left(\frac{t_1}{t_1 - t_0}\right) \tag{6.4}$$

Normal DC-connect voltage of inverter (Vdc-1).
$V_{dc-1} = t_0 * 0 + t_1(2V_C - V_{PV})$ using Eq. (6.4)

$$V_{dc-1} = \left(\frac{t_1}{t_1 - t_0}\right)V_{PV} = V_C \tag{6.5}$$

Maximum DC-link voltage

$$\hat{V}_{dc-1} = (V_C - V_L) = V_C - (V_{PV} - V_C)$$

$$\hat{V}_{dc-1} = (2V_C - V_{PV}) \tag{6.6}$$

$$\hat{V}_{dc-1} = BV_{PV} \tag{6.7}$$

$$B = \left(\frac{t}{t_1 - t_0}\right) = \frac{1}{1 - \left(\frac{2t_0}{t}\right)} \geq 1\,,\ t = t_0 + t_1 \tag{6.8}$$

The converter output voltage can be calculated as (Hanif et al. 2011; Pilehvar et al. 2015)

$$\hat{V_{AC}} = \frac{M\,\hat{V}_{dc-1}}{2} = \left(\frac{MBV_{PV}}{2}\right) \tag{6.9}$$

Voltage across the capacitor (Hanif et al. 2011; Pilehvar et al. 2015).

From Eq. (6.4), we have

$$V_C = \left(\frac{t_1}{t_1 - t_0}\right)V_{PV}$$
$$V_C = \left(\frac{t_1}{t}\right)\left(\frac{t}{t_1 - t_0}\right)V_{PV} \tag{6.10}$$
$$V_C = \left(\frac{1 - \left({t_0}/{t}\right)}{1 - \left({2t_0}/{t}\right)}\right)V_{PV}$$

$$V_C = \frac{B+1}{2B} * B * V_{PV} = \left(\frac{B+1}{2}\right)V_{PV} \tag{6.11}$$

6.4 Modulation Algorithm With Timing Diagram of the Proposed Z-Source Inverter

Figure 6.5 shows the structure and Table 6.2 lists the 83 switching states of the proposed Z-Source inverter, which includes 40 active states, 2 null states, and 41 shoot-through states.

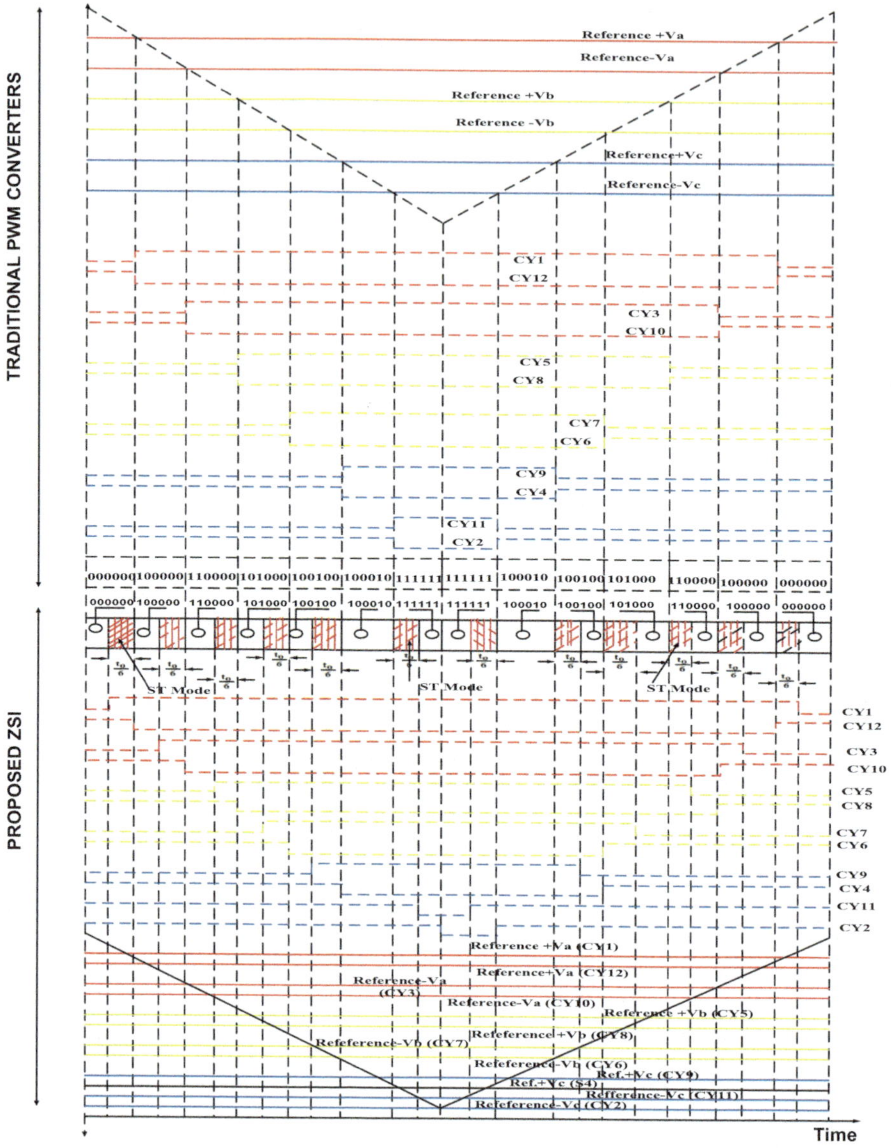

Fig 6.5 Z-Source Inverter Modulation

6.5 Proposed Hybrid Technique

The most regularly utilized MPPT regulators are P&O (Umarani and Seyezhai 2016; Tey and Mekhilef 2014) and the InC (Farhat et al. 2015) MPPT regulator. These regulators likewise have not many downsides which now and again could be an issue

Table 6.2 Switching States of the proposed ZSI (!CY represents complement of SX, where Y = 1, 3, 5, 7, 9, or 11)

State{100,000}(finite)	C1	C12	C3	C10	C5	C8	C7	C6	C9	C4	C11	C2
State{110,000}(finite)	1	0	0	1	0	1	0	1	0	1	0	1
State{101,000}(finite)	1	0	1	0	0	1	0	1	0	1	0	1
State{100,100}(finite)	1	0	0	1	1	0	0	1	0	1	0	1
State{100,010}(finite)	1	0	0	1	0	1	0	1	1	0	0	1
State{100,001}(finite)	1	0	0	1	0	1	0	1	0	1	1	0
State{010,000} finite	0	1	1	0	0	1	0	1	0	1	0	1
State{011,000} finite	0	1	1	0	1	0	0	1	0	1	0	1
State{010,100} finite	0	1	1	0	0	1	1	0	0	1	0	1
State {010,010} finite	0	1	1	0	0	1	0	1	1	0	0	1
State {010,001} finite	0	1	1	0	0	1	0	1	0	1	1	0
State {001,000} finite	0	1	0	1	1	0	0	1	0	1	0	1
State {001,100} finite	0	1	0	1	1	0	1	0	0	1	0	1
State {001,010} finite	0	1	0	1	1	0	0	1	1	0	0	1
State {001,001} finite	0	1	0	1	1	0	0	1	0	1	1	0
State {000,100} finite	0	1	0	1	0	1	1	0	0	1	0	1
State {000,110} finite	0	1	0	1	0	1	1	0	1	0	0	1
State {000,101} finite	0	1	0	1	0	1	1	0	0	1	1	0
State {000,010} finite	0	1	0	1	0	1	0	1	1	0	0	1
State {000,011} finite	0	1	0	1	0	1	0	1	1	0	1	0
State {111,000} finite	1	0	1	0	1	0	0	1	0	1	0	1
State {110,100} finite	1	0	1	0	0	1	1	0	0	1	0	1
State {110,010} finite	1	0	1	0	0	1	0	1	1	0	0	1
State {110,001} finite	1	0	1	0	0	1	0	1	0	1	1	0
State {011,100} finite	0	1	1	0	1	0	1	0	0	1	0	1
State {011,010} finite	0	1	1	0	1	0	0	1	1	0	0	1
State {011,001} finite	0	1	1	0	1	0	0	1	0	1	1	0
State {001,110} finite	0	1	0	1	1	0	1	0	1	0	0	1
State {001,101} finite	0	1	0	1	1	0	1	0	0	1	1	0
State {000,111} finite	0	1	0	1	0	1	1	0	1	0	1	0
State {100,011} finite	1	0	0	1	0	1	0	1	1	0	1	0
State {111,100} finite	1	0	1	0	1	0	1	0	0	1	0	1
State {111,010} finite	1	0	1	0	1	0	0	1	1	0	0	1
State {111,001} finite	1	0	1	0	1	0	0	1	0	1	1	0
State {011,110} finite	0	1	1	0	1	0	1	0	1	0	0	1
State {011,101} finite	0	1	1	0	1	0	1	1	0	1	1	0

(continued)

Table 6.2 (continued)

State{100,000}(finite)	C1	C12	C3	C10	C5	C8	C7	C6	C9	C4	C11	C2
State {001,111} finite	0	1	0	1	1	0	1	0	1	0	1	0
State {100,111} finite	1	0	0	1	0	1	1	0	1	0	1	0
State {111,110} finite	1	0	1	0	1	0	1	0	1	0	0	1
State {011,111} finite	0	1	1	0	1	0	1	0	1	0	1	0
Null {000,000} (0 V)	0	1	0	1	0	1	0	1	0	1	0	1
Null{111,111} (0 V)	1	0	1	0	1	0	1	0	1	0	1	0
Shoot-through F1 (0 V)	1	1	C3	!C3	C5	!C5	C7	!C7	C9	!C9	C11	!C11
Shoot-through F2 (0 V)	C1	!C1	1	1	C5	!C5	C7	!C7	C9	!C9	C11	!C11
Shoot-through F3 (0 V)	C1	!C1	C3	!C3	1	1	C7	!C7	C9	!CS9	C11	!C11
Shoot-through F4 (0 V)	C1	!C1	C3	!C3	C5	!C5	1	1	C9	!C9	C11	!C11
Shoot-through F5 (0 V)	C1	!C1	C3	!C3	C5	!C5	C7	!C7	1	1	C11	!C11
Shoot-through F6 (0 V)	C1	!C1	C3	!C3	C5	!C5	C7	!C7	C9	!C9	1	1
Shoot-through F7 (0 V)	1	1	1	1	C5	!C5	C7	!C7	C9	!C9	C11	!C11
Shoot-through F8 (0 V)	1	1	C3	!C3	1	1	C7	!C7	C9	!C9	C11	!C11
Shoot-through F9 (0 V)	1	1	C3	!C3	C5	!C5	1	1	C9	!C9	C11	!C11
Shoot-through F10(0 V)	1	1	C3	!C3	C5	!C5	C7	!C7	1	1	C11	!C11
Shoot-through F11(0 V)	1	1	C3	!C3	C5	!C5	C7	!C7	C9	!C9	1	1
Shoot-through F12(0 V)	C1	!C1	1	1	1	1	C7	!C7	C9	!C9	C11	!C11
Shoot-through F13(0 V)	C1	!C1	1	1	C5	!C5	1	1	C9	!C9	C11	!C11
Shoot-throughF14 (0 V)	C1	!C1	1	1	C5	!C5	C7	!C7	1	1	C11	!C11
Shoot-through F15(0 V)	C1	!C1	1	1	C5	!C5	C7	!C7	C9	!C9	1	1
Shoot-through F16(0 V)	C1	!C1	C3	!C3	1	1	1	1	C9	!C9	C11	!C11
Shoot-through F17(0 V)	C1	!C1	C3	!C3	1	1	C7	!C7	1	1	C11	!C11
Shoot-through F18(0 V)	C1	!C1	C3	!C3	1	1	C7	!C7	C9	!C9	1	1
Shoot-through F19(0 V)	C1	!C1	C3	!C3	C5	!C5	1	1	1	1	C11	!C11
Shoot-through F20(0 V)	C1	!C1	C3	!C3	C5	!C5	1	1	C9	!C9	1	1
Shoot-through F21(0 V)	C1	!C1	C3	!C3	C5	!C5	C7	!C7	1	1	1	1
Shoot-through F22(0 V)	1	1	1	1	1	1	C7	!C7	C9	!C9	C11	!C11
Shoot-through F23(0 V)	1	1	1	1	C5	!C5	1	1	C9	!C9	C11	!C11
Shoot-through F24(0 V)	1	1	1	1	C5	!C5	C7	!C7	1	1	C11	!C11
Shoot-through F25(0 V)	1	1	1	1	C5	!C5	C7	!C7	C9	!C9	1	1
Shoot-through F26(0 V)	C1	!C1	1	1	1	1	1	1	C9	!C9	C11	!C11
Shoot-through F27(0 V)	C1	!C1	1	1	1	1	C7	!C7	1	1	C11	!C11
Shoot-through F28(0 V)	C1	!C1	1	1	1	1	C7	!C7	C9	!C9	1	1
Shoot-through F29(0 V)	C1	!C1	C3	!C3	1	1	1	1	1	1	C11	!C11

(continued)

Table 6.2 (continued)

State{100,000}(finite)	C1	C12	C3	C10	C5	C8	C7	C6	C9	C4	C11	C2
Shoot-through F30(0 V)	C1	!C1	C3	!C3	1	1	1	1	C9	!C9	1	1
Shoot-through F31(0 V)	C1	!C1	C3	!C3	C5	!C5	1	1	1	1	1	1
Shoot-through F32(0 V)	1	1	1	1	1	1	1	1	C9	!C9	C11	!C11
Shoot-through F33(0 V)	1	1	1	1	1	1	C7	!C7	1	1	C11	!C11
Shoot-through F34(0 V)	1	1	1	1	1	1	C7	!C7	C9	!C9	1	1
Shoot-through F35(0 V)	C1	!C1	1	1	1	1	1	1	1	1	C11	!C11
Shoot-through F36(0 V)	C1	!C1	1	1	1	1	1	1	C9	!C9	1	1
Shoot-through F37(0 V)	C1	!C1	C3	!C3	1	1	1	1	1	1	1	1
Shoot-through F38(0 V)	1	1	1	1	1	1	1	1	1	1	C11	!C11
Shoot-through F39(0 V)	1	1	1	1	1	1	1	1	C9	!C9	1	1
Shoot-through F40(0 V)	C1	!C1	1	1	1	1	1	1	1	1	1	1
Shoot-through F41(0 V)	1	1	1	1	1	1	1	1	1	1	1	1

while following the force from PV modules. Henceforth, another crossbreed MPPT regulator is proposed in this paper that disposes of the defeats of both P&O and INC MPPT regulators. The proposed MPPT regulator likewise expects to join the advantages of both the existing MPPT regulators. The cross-breed MPPT method is proposed to dispose of the constraints of both P&O and InC procedures as portrayed in Fig. 6.6. The principal focus of this proposed method is to gain the merge benefits of bother and perception and steady conductance techniques. It estimates voltage and current from the sun-oriented PV board and computes yield power. Subsequent to ascertaining the yield power, the proposed procedure stores power, estimated voltage, and estimated current in a brief memory for examination. Presently, it contrasts the force and the past estimation of force by finding the adjustment in power. The proposed computation checks whether $\Delta I/\Delta V$ is more noticeable than, not actually, or identical to – I/V and makes its decision whether to augmentation or reduction the terminal voltage.

Fig. 6.6 Hybrid strategy

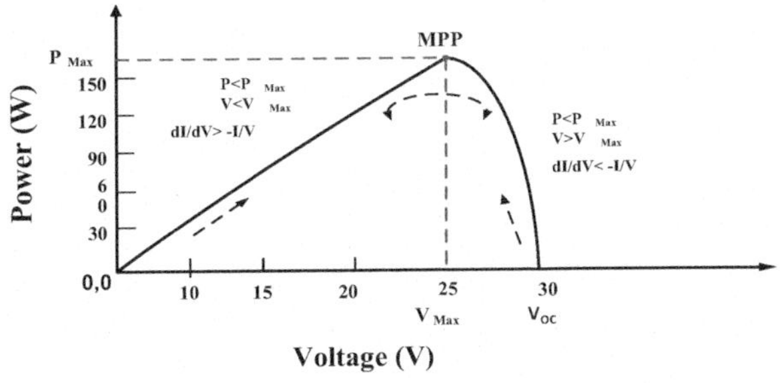

6.6 Results and Discussion

To show the usefulness of the PV dealt with VSI-, CSI-, and ZSI-based UPQCs with its connected UVT control strategy and the power circuit given in Figs. 6.1–6.3 have been set up with MATLAB/Simulink programming. The PV exhibit with chopper gives a more prominent yield voltage. Figure 6.7 shows the PV cluster voltage without and with a chopper. Figure 6.8a–c highlights that the force yield of the proposed half and half strategy is more viable in separating the most extreme force point (MPP = 152 W) from a sun-based PV framework contrasted with the greatest force point (MPP = 151 W) for a situation of P&O procedure and greatest force point (MPP = 152.58 W) for a situation of gradual conductance technique. The primary goal of this part is to assess the presentation of the PV took care of UPQC with ZSI in examination with that of PV took care of UPQC with VSI and UPQC with CSI for the easing of force quality occasions, for example, voltage swells and sounds in supply current just as burden voltage.

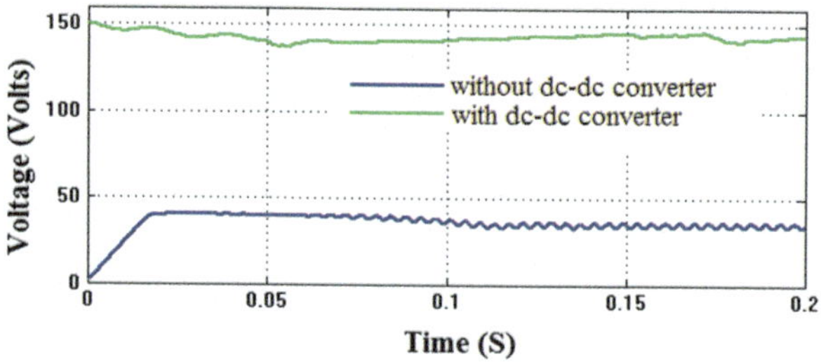

Fig. 6.7 Solar system output voltage with and without a chopper

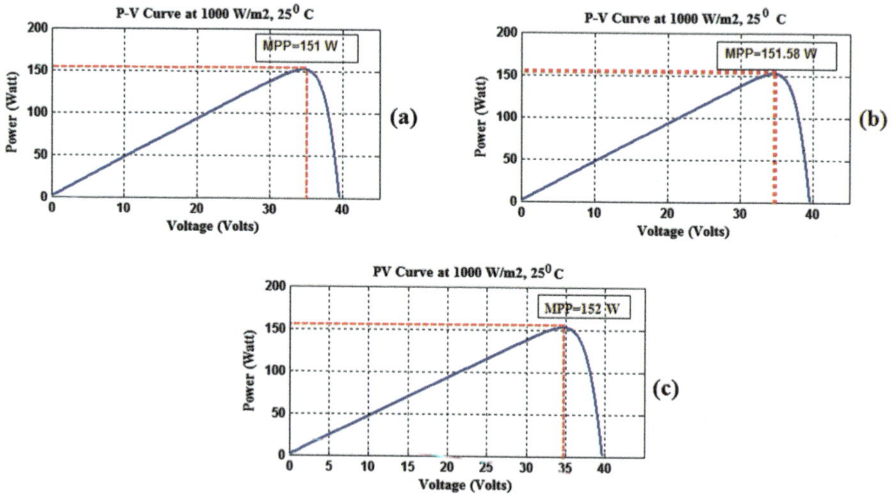

Fig. 6.8 **a** P&O, **b** InC, and **c** the proposed hybrid scheme

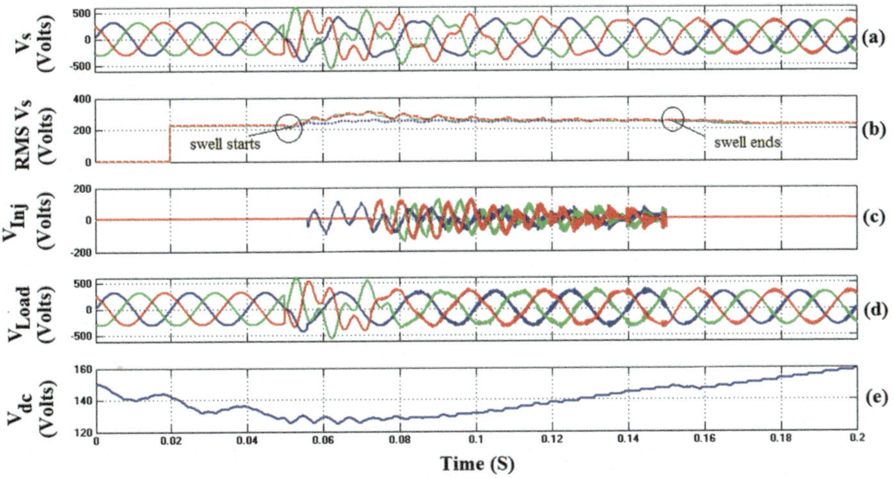

Fig. 6.9 Swell alleviation by UPQC with VSI

6.6.1 *Swell Alleviation by UPQC with VSI*

The presentation of UPQC with VSI is featured in Fig. 6.9a–e for limiting the voltage enlarge. A voltage swell of size 20% is noted as displayed in Fig. 6.9a–b. To limit this issue, sun-powered photovoltaic-based UPQC with VSI comes right into it and infuses missing voltage as displayed in Fig. 6.9c. The infused voltage is added to supply voltage and, consequently, the load voltage is liberated from voltage expansion as displayed in Fig. 6.9d. The conduct of DC-connect voltage during voltage expansion occasion uncovered in Fig. 6.9e.

6.6.2 *Swell Alleviation by UPQC with CSI*

A swell of size 20% saw as displayed in Fig. 6.10a–b. To restrict this voltage enlargement, sun-based photovoltaic-based UPQC with CSI is related and produces the required measure voltage as shown in Fig. 6.10c. To restrict this issue, the solar system connected UPQC with CSI comes right into it and injects the required voltage as featured in Fig. 6.10c. The infused voltage is added to supply voltage and, consequently, the load voltage is liberated from voltage enlarge as displayed in Fig. 6.10d. The conduct of inductor current during voltage growth occasion is uncovered in Fig. 6.10e.

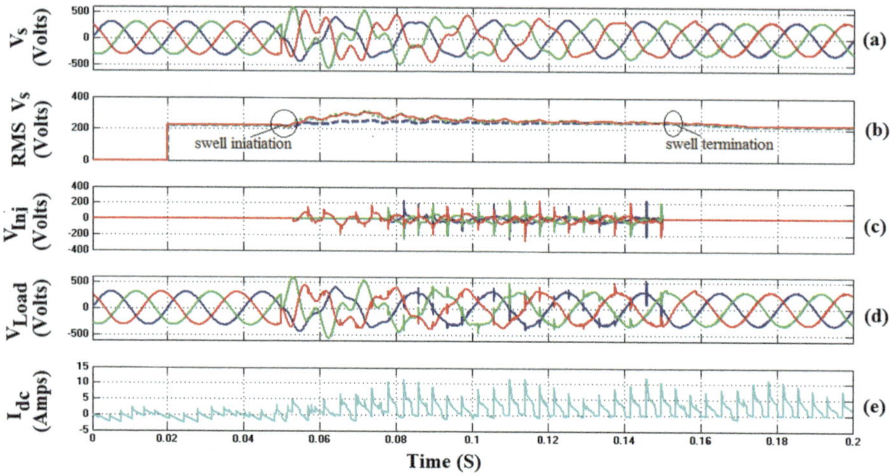

Fig. 6.10 Swell alleviation by UPQC with CSI

6.6.3 Swell Alleviation by UPQC with ZSI

The strength of UPQC with ZSI is featured in Fig. 6.11a–f for alleviation of voltage enlargement. Without UPQC, a voltage swell of size 20% is noted as displayed in Fig. 6.11a–b. To decrease the impact of voltage enlarge, the solar system connected UPQC with ZSI comes right into it and infuses the required voltage as featured in Fig. 6.11c. The UPQC infused voltage is added to source voltage and, consequently, the load voltage is liberated from voltage expansion as displayed in Fig. 6.11d. The

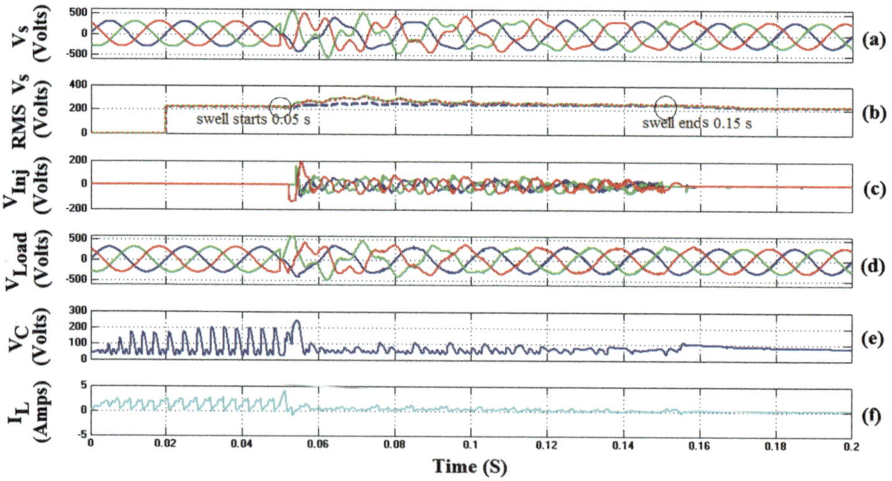

Fig. 6.11 Response of UPQC with ZSI

conduct of DC-interface voltage and inductor current during voltage growth occasion uncovered in Fig. 6.11e and f.

Changed and uncompensated burden voltages under voltage swell conditions are depicted in Fig. 6.12. From the outset, when UPQC with VSI, CSI, and ZSI is not related to the structure, the system encounters a voltage swell of a degree of 20% (60 V) of the load voltage. During voltage grow event UPQC with VSI implants the voltage of 120 V, UPQC with CSI injects the voltage of 200 V and UPQC with ZSI imbues the voltage of 75 V as exhibited in Fig. 6.12a. The PV connected UPQC with ZSI shows the preferable display over make the reasonable proportion of imbued voltage appeared differently in relation to UPQC with VSI and UPQC with CSI. Appropriately, load voltage gets sinusoidal as exhibited in Fig. 6.12b.

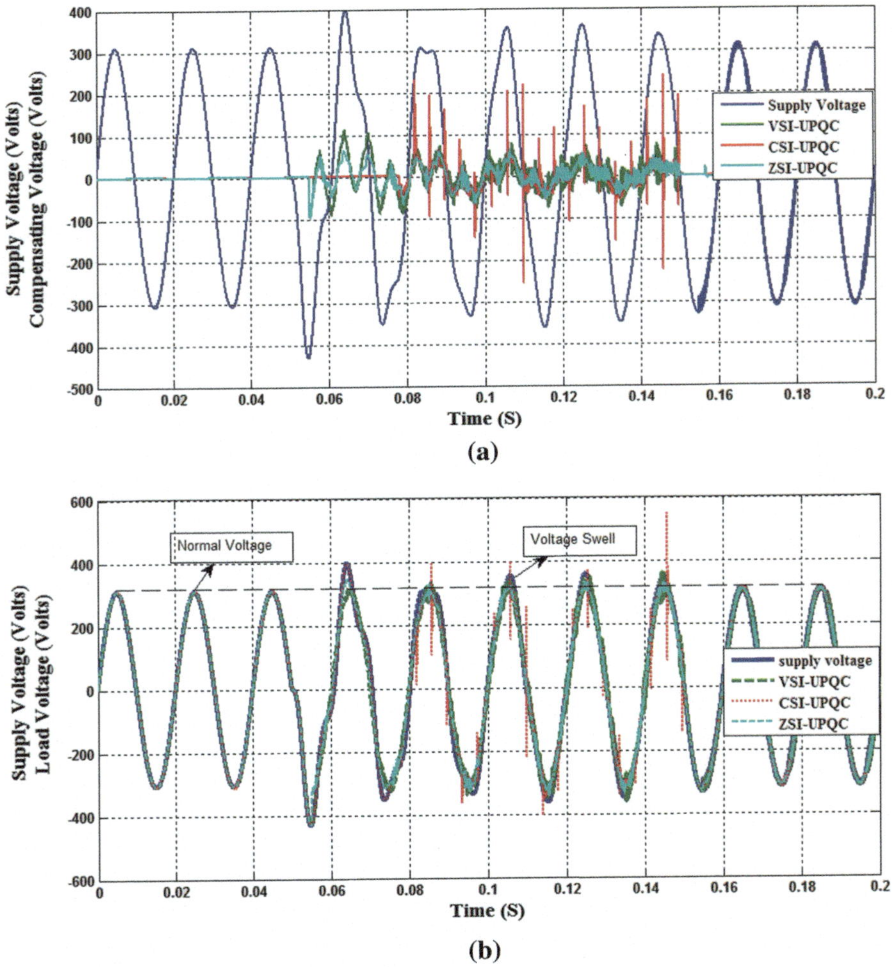

Fig. 6.12 **a** Injected voltages and **b** load voltages

Table 6.3 Supply current harmonics THD_i of VSI, CSI, and ZSI UPQCs

Without UPQCs	UPQC-VSI		UPQC-CSI		UPQC-ZSI	
	With UPQC-VSI	Improvement in THDi (%)	With UPQC-CSI	Improvement in THDi (%)	With UPQC-ZSI	Improvement in THDi (%)
7.28	1.67	77.06	1.66	77.20	1.60	78.02

Table 6.4 Load voltage harmonics THD_S of VSI, CSI, and ZSI UPQCs

Without UPQCs	UPQC-VSI		UPQC-CSI		UPQC-ZSI	
	With UPQC-VSI	Improvement in THDs (%)	With UPQC-CSI	Improvement in THDs (%)	With UPQC-ZSI	Improvement in THDs (%)
18.62	2.31	87.60	2.48	86.68	1.48	92.05

The exhibition of supply current as well as load harmonics filtering of PV took care of UPQCs during load exchanging condition as shown in Tables 6.3–6.4. During load exchanging, the source current THDi before pay is discovered to be 7.97%, while THDi of the source current after remuneration is discovered to be 0.72% in the presence of UPQC-VSI with a 90.96% decrease in THDi, 0.83% if there should arise an occurrence of UPQC-CSI with an 89.58% decrease in THDi and 0.72% in the presence of ZSI-UPQC with a 90.96% decrease in THDi individually as demonstrated in Table 6.3. In this way, VSI-UPQC and CSI-UPQC are decreasing a lesser measure of THDi in contrast with UPQC-ZSI.

Table 6.4 summed up the THDv correlation of PV connected UPQC with VSI, CSI, and ZSI during load exchanging. From table 6.4, it is featured that THDv of load voltage before remuneration is recorded to be 18.62%, and the THDv of load voltage after pay is discovered to be 2.31% if there should arise an occurrence of UPQC with VSI, 2.48% in presence of UPQC with CSI and 1.48% in presence of ZSI-UPQC. Along these lines, an 87.60% decrease in THDv has been accomplished utilizing UPQC with ZSI contrasted with an 86.68% decrease in THDv of UPQC-VSI and 92.05% decrease in THDv of UPQC-CSI.

6.7 Conclusion

The proposed methodology shows the preferred introduction over produce most limit power yield appeared differently in relation to P&O and InC methodologies. The PV dealt with UPQCs are liable for the fast and exact making of implanting voltage and injects it into the structure for compensation of supply voltage agitating impacts, for instance, voltage swells. The obtained reproduction results show that UPQC with ZSI implants the reasonable proportion of three-stage mixing voltage stood out from UPQC with VSI and CSI. The reenactment results moreover exhibit that the proposed

UPQC-ZSI shows a preferable capacity over annihilating the source current and load voltage harmonic stood out from UPQC-VSI and UPQC-CSI.

References

Ahmed J, Salam Z (2015) An improved perturb and observe (P & O) maximum power point tracking (MPPT) algorithm for higher efficiency. Appl Energy 150:97–108

Ahmed J, Salam Z (2016) A modified P&O maximum power point tracking method with reduced steady state oscillation and improved tracking efficiency. IEEE Trans Sustain Energ 7:1506–1515

Alam MR, Muttaqi KM, Bouzerdoum A (2015) Characterizing voltage sags and swells using three-phase voltage Ellipse parameters. IEEE Trans Indus Appl 51:2780–2790. https://doi.org/10.1109/TIA.2015.2397176

Ali US (2018) Impedance source converter for photovoltaic stand-alone system with vanadium redox flow battery storage. Mater Today Proc 5:241–247

Alik R, Jusoh A (2018) An enhanced P&O checking algorithm MPPT for high tracking efficiency of partially shaded PV module. Sol Energy 163:570–580

Balamurugan M, Sivakumaran TS, Devi MA (2013) Voltage sag/swell compensation using Z-source inverter DVR based on FUZZY Controller. IEEE International Conference on Emerging Trends in Computing, Communication and Nanotechnology (ICECCN), 648–653

Carrasco M, Mancilla-David F (2016) Maximum power point tracking algorithms for single-stage photovoltaic power plants under time-varying reactive power injection. Sol Energy 132:321–331

Farhat M, Barambones O, Sbita L (2015) Efficiency optimization of a DSP-based standalone PV system using a stable single input fuzzy logic controller. Renew Sustain Energy Rev 49:907–920. https://doi.org/10.1016/j.rser.2015.04.123

Fathabadi H (2016) Novel fast dynamic MPPT (maximum power point tracking) technique with the capability of very high accurate power tracking. Energy 94:466–475

Gheibi· A, Mohammadi· SMA, Farsangi MM (2016) A proposed maximum power point tracking by using adaptive fuzzy logic engineering and electrical engineering 23:1272–1281

Hanif M, Basu M, Gaughan K (2011) Understanding the operation of a Z-source inverter for photovoltaic application with a design example. IET Power Electron 4:278–287

Kandemir E, Cetin NS, Borekci S (2017) A comprehensive overview of maximum power extraction methods for PV systems. Renew Sustain Energy Rev 78:93–112

Kannan SA, Rakesh R, Amal MR, Kamala DV (2014) Performance analysis of PV single phase Z-source inverter. Int J Innovative Res Electr Electron. Instrum Control Eng 2:1069–1075

Kermadi M, Berkouk EM (2017) Artificial intelligence-based maximum power point tracking controllers for photovoltaic systems: comparative study. Renew Sustain Energy Rev 69:369–386. https://doi.org/10.1016/j.rser.2016.11.125

Messalti S, Harrag AG (2015) A new neural networks MPPT controller for PV systems. IEEE International Conference on Renewable. Energy Congress, 1–6, https://doi.org/10.1109/IREC.2015.7110907

Pilehvar MS, Mardaneh RM, A (2015) An analysis on the main formulas of Z-source inverter. Scientia Iranica Transactions d: Computer Science & Engineering and Electrical Engineering 22(3):1077–1084

Rahman SA, Janakiraman PA, Somasundaram P (2015) Voltage sag and swell mitigation based on modulated carrier PWM. Electr Power Energy Syst 66:78–85. https://doi.org/10.1016/j.ijepes.2014.09.017

Reisi AR, Moradi MH, Showkati H (2013) Combined photovoltaic and unified power quality controller to improve power quality. Sol Energy 88:154–162

Sadigh AK, Smedley KM (2016) Fast and precise voltage sag detection method for dynamic voltage restorer (DVR) application. Electric Power Systems Research 130:192–207. https://doi.org/10.1016/j.epsr.2015.08.002

Saravanan V, Ramanujam R, Arumugam M (2015) Novel improved three level Z-source inverter. Aust J Electr Electron Eng 12:133–140

Sundarabalan CK, Selvi K (2015) Compensation of voltage disturbances using PEMFC supported dynamic voltage restorer. Electr Power Energy Syst 71:77–92. https://doi.org/10.1016/j.ijepes.2015.02.032

Tajuddin MFN, Arif MS, Ayob SM, Salam Z (2015a) Perturbative methods for maximum power point tracking (MPPT) of photovoltaic (PV) systems: a review. Int J Energy Res 39(9):1153–1178

Tajuddin MFN, Ismail B, Azmi A, Ayob SM, Salam Z (2015b) Single phase Z-source inverter with differential evolution (DE) based maximum power point tracker. TELKOMNIKA Indonesian J Electr Eng 14:80–89

Tang Y, Xie S, Zhang C (2011) An improved Z-source inverter. IEEE Trans Power Electron 26:3865–3868

Tey KS, Mekhilef S (2014) Modified incremental conductance MPPT algorithm to mitigate inaccurate responses under fast-changing solar irradiation level. Solar Energy 101:333–342. https://doi.org/10.1016/j.solener.2014.01.003

Umarani D, Seyezhai R (2016) Modeling and control of quasi z-source cascaded h-bridge multilevel inverter for grid connected photovoltaic systems. Energy Procedia 90:250–259

Venkatesh C, Siva Sarma DVSS, Sydulu M (2011) Mitigation of voltage sag/swell using peak detector based pulse width modulation switched autotransformer. Electric Power Compon Syst 39:1117–1133. https://doi.org/10.1080/15325008.2011.559188

Vodapalli P, Reddy TRS, Kalyani ST (2015) A new unified power quality conditioner for grid integration of PV system and power quality improvement feature distribution system. IEEE International Conference on Electrical, Electronics, Signals, Communication and Optimization EESCO

Wu JC, Jou HL, Tsai JH (2017) A flexible grid interface for a PV power system. Sol Energy 144:540–547

Yunfei XU, Xiao X, Sun T, Long Y (2016) Voltage swell compensation strategy for unified power quality conditioner with simultaneous reactive power injection. Int J mod Power Syst Clean Energy 4(1):113–122

Zope PH, Somkuwar A (2012) Design and Simualtion of single-phase Z-source inverter for utility interface. Int J Electr Eng Technol 1:127–143

Chapter 7
Energy Storage Technologies; Recent Advances, Challenges, and Prospectives

Ababay Ketema Worku, Delele Worku Ayele, Nigus Gabbiye Habtu, Bimrew Tamrat Admasu, Getu Alemayehu, Biniyam Zemene Taye, and Temesgen Atnafu Yemata

Abstract Fossil fuels are the origins of conventional energy production, which has been progressively transformed into modern innovative technologies with an emphasis on renewable sources such as wind, solar, and hydrothermal. Recently, the challenges concerning the environment and energy, the growth of clean and renewable energy-storage devices have drawn much attention. Renewable energy sources are the primary choice, which addresses some critical energy issues like energy security and climate change. But, renewable energy sources have interrupted and irregular supplies that should be stored in efficient, safe, efficient, reliable, affordable, and clean ways. Hence, energy storage is a critical issue to advance the innovation of energy storage for a sustainable prospect. Thus, there are various kinds of energy storage technologies such as chemical, electromagnetic, thermal, electrical, electrochemical, etc. The benefits of energy storage have been highlighted first. The classification of energy storage technologies and their progress has been discussed in this chapter in detail. Then metal–air batteries, supercapacitors, compressed air, flywheel, thermal energy, superconducting magnetic, pumped hydro, and hybrid energy storage devices are critically discussed. Finally, the recent progress, problems, and future prospects of energy storage systems have been forwarded. The chapter is vital for scholars and

A. K. Worku (✉) · G. Alemayehu · B. Z. Taye
Bahir Dar Energy Center, Bahir Dar Institute of Technology, Bahir Dar University, P.O. Box 26, Bahir, Ethiopia
e-mail: Ababay.Ketema@bdu.edu.et

D. W. Ayele
Department of Chemistry, Bahir Dar University, P.O. Box 79, Bahir Dar, Ethiopia

N. G. Habtu · T. A. Yemata
Department of Chemical Engineering, Bahir Dar Institute of Technology, Bahir Dar University, P.O. Box 26, Bahir Dar, Ethiopia

B. T. Admasu
Department of Mechanical Engineering, Bahir Dar Institute of Technology, Bahir Dar University, P.O. Box 26, Bahir Dar, Ethiopia

B. Z. Taye
Faculty of Electrical and Computer Engineering, Bahir Dar Institute of Technology, Bahir Dar University, P.O.Box 26, Bahir Dar, Ethiopia

A. K. Bohre et al. (eds.), *Planning of Hybrid Renewable Energy Systems, Electric Vehicles and Microgrid*, Energy Systems in Electrical Engineering,
https://doi.org/10.1007/978-981-19-0979-5_7

scientists, which provides brief background knowledge on basic principles of energy storage systems.

Keywords Renewables · Metal–air batteries · Energy storage · Thermal energy · Pumped hydro storage · Superconducting magnetic

Nomenclature

ESSs	energy storage systems
EMES	electromagnetic energy storage
CESTs	chemical energy storage technologies
EMES	electromagnetic energy storage
PHES	pumped hydroelectric energy storage
MESTs	mechanical energy storage technologies
FEST	flywheel Energy Storage technology
TES	thermal energy storage
FC	fuel Cells
SHS	sensible heat storage
LHS	latent heat storage
ECES	electrochemical energy storage
NiCd	nickel-cadmium batteries
FBs	flow batteries
LIBs	lithium-ion batteries
MABs	metal-Air Batteries
HESSs	hybrid energy storage systems

7.1 Introduction

Recently, the world population is increased in an amazing manner, which leads to the growth of global energy demand. Thus, this demand has been maintained using fossil fuels as a source of energy (Sadeghi et al. 2021). However, their inadequate assets, climate change issues, and energy security issues have been forced to focus on alternative energy technologies. Renewable energy sources have great advantages related to environmental effects and energy security, which is not a constant supply of energy (Zhao and Guo 2021). Hence, renewables need to be stored in safe, eco-friendly, effective, and reliable ways for later use. Energy storage systems (ESSs) can be divided according to different principles (Komala et al. 2021). They can be divided as chemical, electromagnetic, thermal, mechanical, and electrochemical, associated with the kind of stored energy. Energy in the form of potential or kinetic can be stored in mechanical ESSs (Cheng et al. 2021). Betties gained special attrition for ESSs because this electrochemical energy storage was studied highly. Moreover, chemical energy storage such as ammonia, methane, and hydrogen are frequently studied technologies (Hu et al. 2021). Additionally, latent or sensible heat storage is a type of thermal ESSs. Electromagnetic energy storage is an emerging technology, which needs special attrition. The purpose of this chapter is to deliver a detailed discussion on energy storage technologies, which is used as a reference for different scholars and industries involved in the area. However, there are a limited number of reviews on energy storage technologies and their application (Wang et al. 2021). Hence, in this chapter, we discussed the recent advancements in basic energy storage tools such as electromagnetic, electrochemical, thermal, mechanical, and chemical, energy storage devices (Nguyen et al. 2014). Finally, challenges and prospectives are discussed to identify the gaps and to forward import directions for the enhancement of energy storage technologies.

7.2 Benefits of Storing Energy

ESSs can be classified based on different systems such as (Pickard 2012).

- Chemical,
- Electrochemical,
- Electromagnetic,
- Thermal, and
- Mechanical.

Thus, each system has its own characteristics and efficiency. The criteria for choosing suitable ESSs are shown in Fig. 7.1. The criteria for selecting ESSs, such as storage cost, adaptability, environmental impact, capacities, and efficiency, can be used in the selection process.

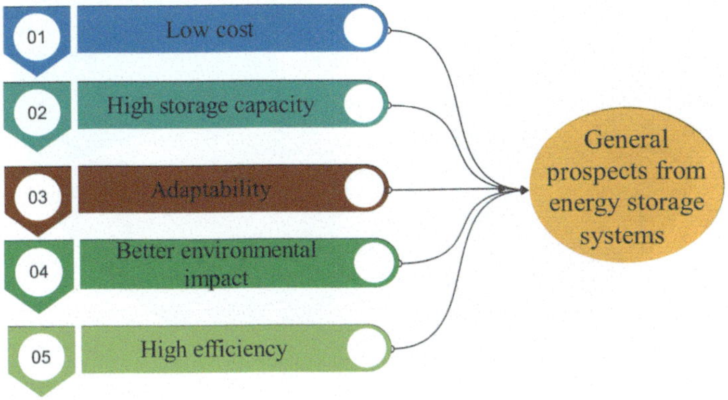

Fig. 7.1 Schematic illustration of criteria for selecting appropriate energy storage systems

In fact, ESSs have many characteristics, and each energy storage system has different expectations, depending on the requirements of the end-user. Though, when demand and supply do not balance each other, all will be used together to provide clean, affordable, efficient, safe, and reliable energy. Recently prepared various ESSs have similar features with traditional energy generation technologies. Additionally, they produce some different properties, which make the energy method further complex (Papaefthymiou et al. 2010). Various energy production technologies from hydroelectric power plants, the energy produced by storage systems are restricted, which means in an energy storage system, the peak power production can be kept for a certain period of time, associated with the energy previously stored in the system. Moreover, furthermore to limited power generation capacity, most energy storage systems also have cycle limits. Though, in addition to the problems, ESSs still have significant advantages and can meet energy needs without or with limited supply. In addition to these main benefits, they also have technological, environmental, and economic merits, which make them an essential foundation in energy technology. Figure 7.2 displays selected vital conditions to be provided in the designing process of ESSs. For instance, if the system contains a higher discharging and charging power rating, however a lower capacity, it may be utilized for fast and short-lived emergencies, mobile power supplies, etc. It is a good choice, but it is not appropriate for periodic energy storage. Moreover, systems with lower capital costs and higher operating costs will be more suitable for short-term storage such as emergency and peak

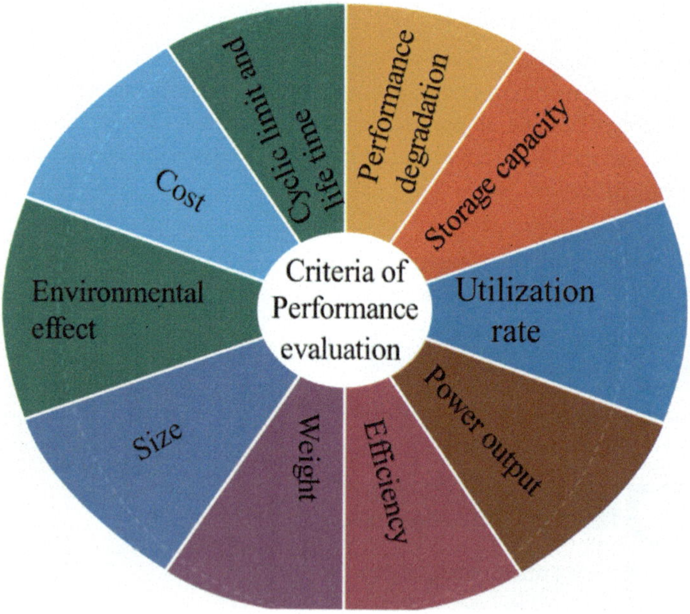

Fig. 7.2 Schematic illustration of criteria used to estimate the performance of ESSs

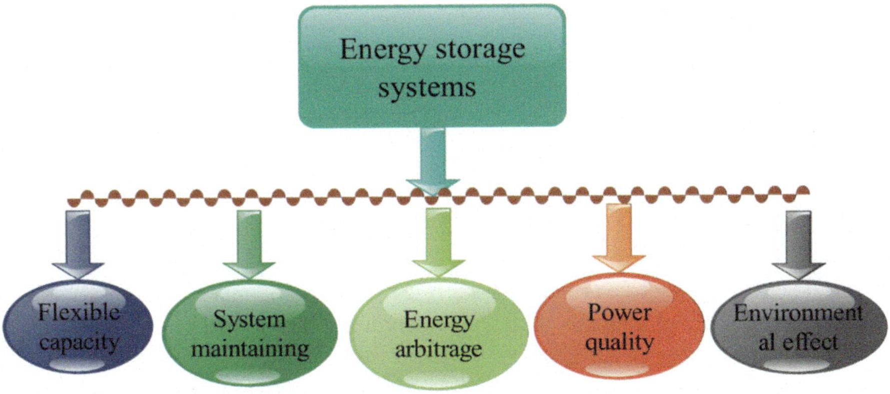

Fig. 7.3 Schematic illustration of benefits of ESSs

demand needs. Oppositely, technologies with low operating costs and high capital are further appropriate for long-term energy requirements like periodic storage.

Depending on the location of the facility, utility rates and power load, and other factors, energy storage can be the best means for facilities to cut electricity bills. The price of ESSs is declining, and the figure of customer-defined ESSs that has been installed is rapidly increasing. Moreover, to increase the use of renewable energy for power generation, improved energy storage technology also has the following advantages (Fig. 7.3) (Liu et al. 2010):

- **Environmental issues**: Energy storage has different environmental advantages, which make it an important technology to achieving sustainable development goals. Moreover, the widespread use of clean electricity can reduce carbon dioxide emissions (Faunce et al. 2013).
- **Cost reduction**: Different industrial and commercial systems need to be charged according to their energy costs. Solar photovoltaic power generation can decrease total power consumption, but these merits do not permanently coincide with the peak usage hours of buildings (Luo et al. 2015).
- **Maximize usage time**: ESSs can transform power consumption from expensive periods when demand is high to low-cost power periods when demand is low. If the electricity price structure changes over time, and the peak demand period shifts to the evening when there is no light, this can reduce the risk of reducing the value of on-site solar energy. This also enables facilities to take full advantage of time-of-use pricing and reduces the risk of electricity price structure changing the cost of electricity.
- **Emergency backup**: Power backup is a significant part of a resilient plan. Moreover, by using this infrastructure on a daily basis to reduce demand costs, its reliability and availability can be improved during shutdowns compared to independent battery systems and diesel generators used only during shutdowns (Tewari and Mohan 2013).

- **Economy**: Increase the economic value of wind energy and solar energy (Pearre and Swan 2015).
- **Work**: Creates work in transportation, engineering, construction, financial, and manufacturing departments (Heymans et al. 2014).

7.3 Energy Storage Technologies

In this section, a brief overview of chemical, electromagnetic, electrochemical, mechanical, and thermal ESSs with their technical status will be presented. Thus, ESSs can store energy in different systems for future utilization (Zhao et al. 2015). The prospect of energy storage is to be able to preserve the energy content of energy storage in the charging and discharging times with negligible loss. Hence, the selected technologies primarily change electrical energy into various forms during the charging process for efficient storage (Kirubakaran et al. 2009). The most widely used storage technologies can be categorized according to the kind of energy stored, as shown in Fig. 7.4. Moreover, there are various types of technologies such as end-use applications, which are used as ESSs. The typical example is the adjustment of energy consumption peak and time demand. Other examples include utility control of electric water heaters, pre-cooling, adjustment of municipal water time, etc. to reduce cooling requirements during the day (Whittingham 2008).

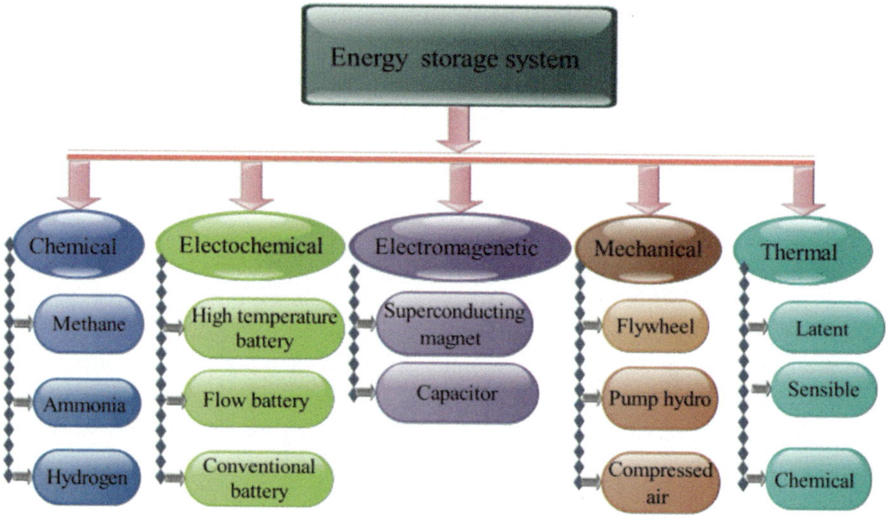

Fig. 7.4 Schematic illustration of the classification of selected ESSs

7.3.1 *Chemical Energy Storage Technologies (CESTs)*

In CESTs, energy can be stored using various materials in the form of chemical energy. It can be categorized as follows:

- Biofuels,
- Hydrogen storage.

7.3.1.1 Hydrogen Storage

Hydrogen is a type of energy that can be transported and stored. Moreover, hydrogen gas has expensive storage, low energy density, and non-toxicity with combustion product of H_2O. Hydrogen can be fabricated via several methods such as electrolysis, natural gas, coal, and oil. It can be stored in various forms such as in metal-hydride, liquid, and gaseous forms. Thus, hydrogen storage in the form of metal-hydride and gas are very mature systems for hydrogen storage. However, the boiling point of hydrogen is 20 K, which is a challenge of hydrogen storage in the form of liquid. Hydrogen storage in the form of gaseous is associated with mechanical stability and material permeability in extreme pressure.

7.3.1.2 Biofuels

Biofuels are formed via biological methods instead of geological methods. Biomass is an organic substance acquired from the residues of animal and plant manure. Hence, biomass is employed to generate the biogas that can be utilized for local application or can be transformed into electricity via a generator. There are various types of biofuels such as solid biomass, biodiesel, gasoline, biofuel syngas, bio-alcohols, biogas, bio ether, green diesel, ethanol, and vegetable oils.

7.3.2 *Electromagnetic Energy Storage (EMES)*

In superconductors, the flow of direct current produces energy, which can be stored in the form of a magnetic field. Electricity storing in the form of electrical energy is a challenging activity because of different causes such as low efficiencies and high system losses. Hence, electrical energy might be changed to different types of energy for storage purposes in an affordable, safe, environmentally benign, and reliable way. In EMES, electrical or different type of energy is changed to electromagnetic energy using different devices such as superconducting electromagnets and capacitors. The two electrical conductors of a capacitor are separated using a dielectric material. Charge can be accumulated on the side of the applied current, while current is applied to the conductor. Thus, the conductor plates can be stored energy in the form of an

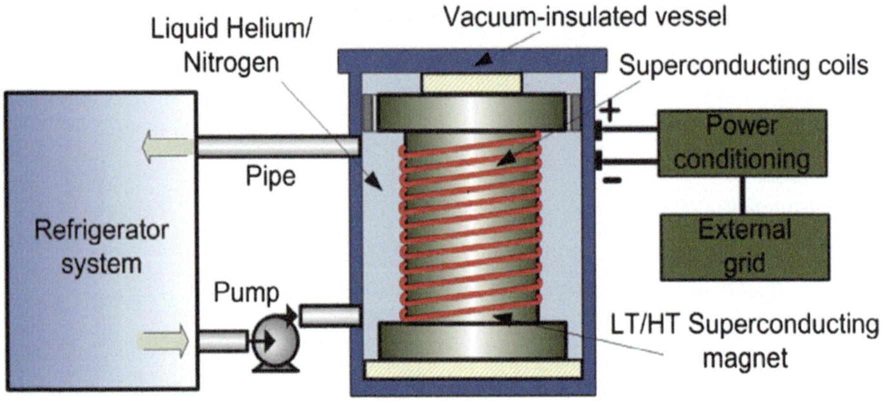

Fig. 7.5 Schematic diagram of electromagnetic energy storage technology. Reproduced with permission (Luo et al. 2015) Copyright (2015), Elsevier

electric field. Capacitors with higher energy density are called supercapacitors. For the generation of a magnetic field, superconducting magnetic energy storage is used via a cryogenically cooled superconducting coil. Hence, such types of technologies are appropriate for high-power requests when storing fluctuating and intermittent energy sources. EMES have various merits such as sensitivity to battery voltage imbalance maximum voltage threshold, and battery interdependence, as well as safety issues, such as explosion, chemical, fire, and hazards. Figure 7.5 displays the diagram of electromagnetic energy storage technology.

7.3.3 Mechanical Energy Storage Technologies (MESTs)

In MESTs, excess energy is changed into potential or kinetic energy for future utilization. There are various types of MESTs used as energy storage the typical examples are listed as follows:

- Flywheel,
- Compressed air storage, and
- Pumped storage.

7.3.3.1 Pumped Hydroelectric Energy Storage (PHES)

PHES is the best and most advanced technology utilized for energy storage. Presently, approximately 129 GW of pumped storage capacity has been installed worldwide. The basic working mechanism of pumped storage can be categorized into two steps. Primarily, electricity is applied to pump water from the lower reservoir to the upper reservoir. Then, water flows from the higher reservoir to the lower reservoir, the input energy is recovered through the turbine (Figueiredo and Flynn 2006). From this, we can conclude that pumped storage has very similar working principles with a hydroelectric power plant. According to reports, the total proficiency of the pumped storage system is between 70 and 85%, which depends on construction, size, service life, condition, and location status. The principal merits of pumped storage are its flexibility, which can be utilized as energy storage several times. The response time of the pumped storage system is also very short (a few seconds to a few minutes). The other merits of pumped storage are long service life, low operating cost, lack of circulating energy consumption, and low maintenance cost. However, the pumping system has very special location conditions. Furthermore, pumped storage usually needs high asset costs. However, pumped storage has been regarded as an efficient solution that can be utilized to balance the load of the power system and reduce peak energy demand. The PHES devices store energy in the form of potential energy, which is pumped from lower reservoirs to higher reservoirs (Fig. 7.6). In such type of technology, low-cost electricity (power during off-peak hours) is applied to run pumps to lift water from the lower reservoir to the upper reservoir. At the time of high power demand, the stored water is released via the hydraulic turbine to generate electrical energy. When necessary, the reversible generator assembly acts as a turbine. Recently, PHES systems have solar photovoltaic and wind power generation systems that can transfer water from lower reservoirs to upper reservoirs. This technology is currently used as a low-cost way of storing huge amounts of electrical energy; however, suitable geographic location and capital cost are crucial decisive issues. The preparations of most PHES power plants are extremely associated with site properties. If the topography and geological conditions of the area are favorable, it is said that there is sufficient water available for the development of PHES plants

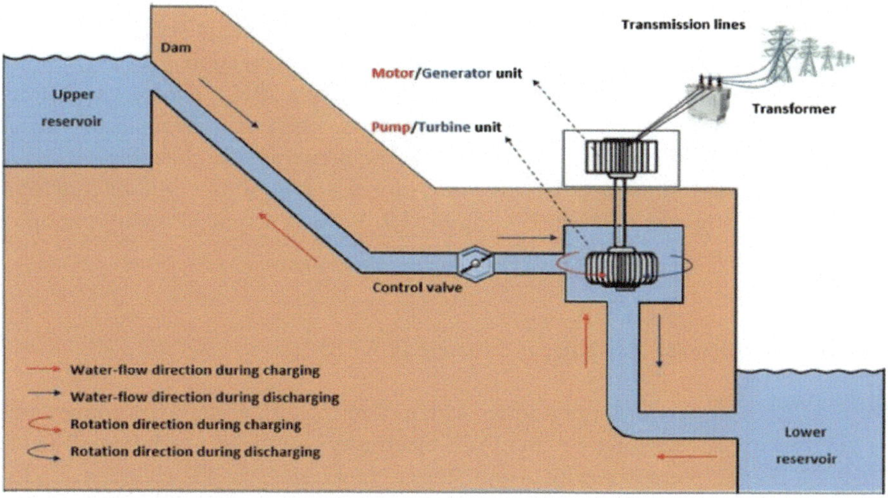

Fig. 7.6 Schematic diagram of pumped hydro storage plant. Reproduced with permission (Shaqsi et al. 2020) Copyright (2020), Elsevier

in the area (Zuo et al. 2015). Generally, PHES is considered to be the most effective technology to enhance the penetration rate of renewable energy in the power system, especially in small autonomous island power grids. Pumped storage technology and wind power, which are called hybrid power plants establish a feasible and realistic way for achieving high penetration rates of renewable energy, given that their elements are appropriately sized. At present, the pumped storage solution provides the most important commercial means for large-scale grid energy storage and increases the daily power generation capacity of the power generation technology (Beaudin et al. 2010).

7.3.3.2 Flywheel Energy Storage Technology (FEST)

In the flywheel, charging and discharging are performed by accelerating the inertial mass (rotor). The rotor is the main component of the flywheel. The capacity and energy density of the FEST are mainly influenced by the properties of the rotor, containing maximum speed, and inertia (Amodeo et al. 2009). The flywheel is suitable for various power applications in between 100 kW and 2 MW. The discharge time of the flywheel ranges from 5 s to 15 min. The capacity of the flywheel is approximately 0.5 to 1 kWh. The overall efficiency of the flywheel is about 70–80%, and its rated standby power consumption is about 1–2%. The merits of the 30 flywheels are its fast discharge and fast response time. Therefore, they are supposed for uninterruptible power supply and high-quality power supply utilization. However, the flywheel assembly is easy to wear during continuous operation, so the service life is short (about 100,000 charge and discharge cycles) (Amiryar and Pullen 2017). Therefore, reducing and ultimately eliminating the friction of all components is the main challenge facing the flywheel (Connolly et al. 2010). Though, because of the improvement of materials such as power electronics, magnetic bearings, and the development of high-speed motors, FEST has been recognized as a reliable choice for energy storage utilization (Sebastián and Peña Alzola 2012). The flywheel stores energy according to the principle of rotating mass. FEST is a mechanical storage technology that simulates the storage of electrical energy via changing electrical energy to mechanical energy. The flywheel stored energy in the type of rotational kinetic energy (Suzuki et al. 2005). Figure 7.7 displays a typical configuration of flywheel technology.

7.3.3.3 Compressed Air Energy Storage (CAES)

The CAES is a means of energy storage, which stored electrical energy as compressed air via a compressor. Moreover, in CAES electricity is utilized to compress the air, which stores the pressurized air using storage tanks such as gas chamber, underground mine, expired wells, and underground salt caverns at the energy storage time (Fig. 7.8). Hence, the mechanical level of CAES is determined by its charge life. At the time of charging, the compressor applies off-peak electric energy, which uses solar

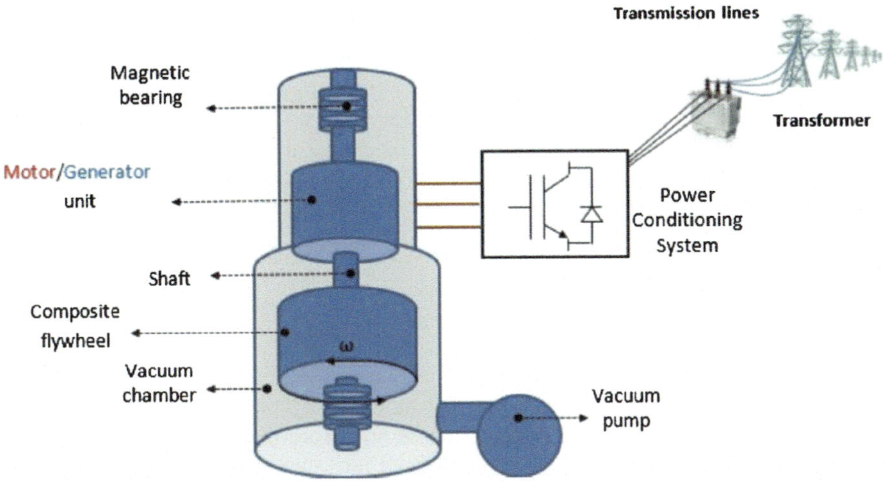

Fig. 7.7 Schematic diagram of the configuration of a flywheel technology. Reproduced with permission (Shaqsi et al. 2020) Copyright (2020), Elsevier

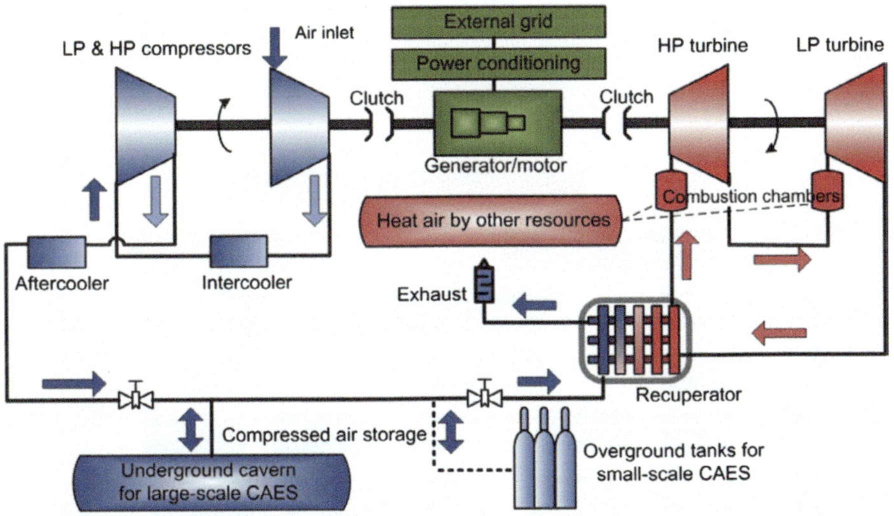

Fig. 7.8 Schematic illustration of CAES plant source. Reproduced with permission (Luo et al. 2015) Copyright (2015), Elsevier

and wind power to compress ambient air. The key variation between different CAES structures is associated with thermal engineering. However, CAES technologies can store energy for a long period of time associated with batteries. The main challenge of CAES design in large-scale application is laid in the management of thermal energy. In CAES, the compression produces unwanted temperature, which damages and decreases the operational efficiency of the technology (Ries and Neumueller 2001). Thus, the efficiency of CAES systems is between 40 and 75%, and the start-up time is around 5–15 min (Yang et al. 2011). In addition to pumped storage, flywheel, and compressed air storage, there are also different types of new mechanical energy technology under development. For instance, mechanical energy storage technology is based on the slope of a tram carrying rocks or sand in an electric car equipped with a motor-generator (Chen et al. 2009).

7.3.4 Thermal Energy Storage (TES)

TES is a means of thermal energy storage using heating (cooling) a condition, which is used for later application (Sharma et al. 2009). Moreover, in TES system, electrical energy (other types of energy) is changed into thermal energy in a cold state (for instance, coolers and ice storage) or in a hot state (for instance, solar thermal collectors). The typical example of high-temperature TES is a concentrated solar power plant, where the stored heat is utilized at cloudy and night time while solar energy does not exist (Fasano et al. 2015). The TES technologies can be classified into three kinds (Fig. 7.9) as follows:

- Latent heat,
- Chemical reaction storage, and

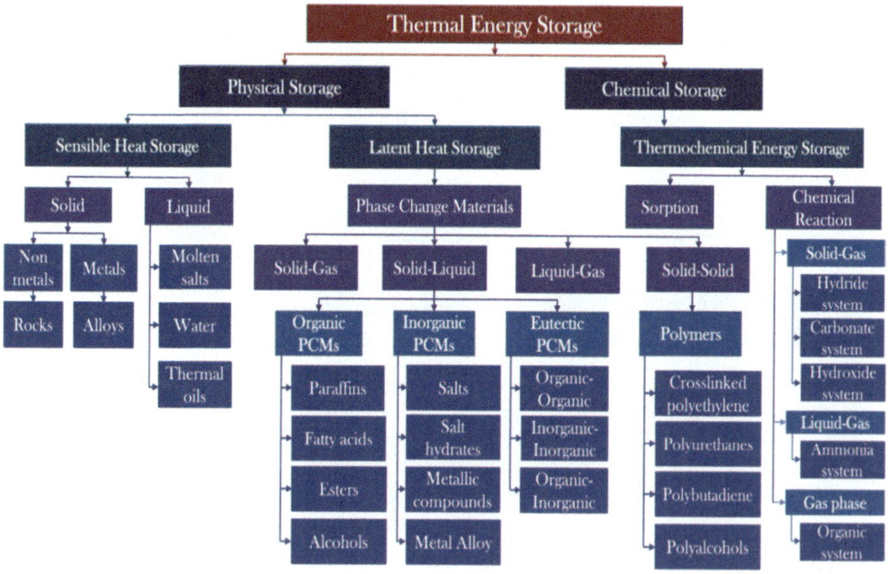

Fig. 7.9 Schematic illustration of TES materials. Reproduced with permission (Nazir et al. 2019) Copyright (2019), Elsevier

- Sensible heat.

The division of TES technologies materials is shown in Fig. 7.9. The overall efficiency of the TES system based on molten salt is relatively low, about 25% to 35%. However, their investment costs are also relatively low, which makes them ideal for TES technologies (Pintaldi et al. 2015).

7.3.4.1 Sensible Heat Storage (SHS)

The SHS is prepared via storing heat energy in different materials according to the change of temperature and heat capacity of the material throughout the charging and discharging route. The key merit of SHS is that discharging process and charging process is totally reversible as well as unlimited life cycles. Thus, the heat energy can be stored in different mediums such as solid, dual, and liquid. For SHS, we can use energy input either solar energy or electricity (Asjid et al. 2021; Velasco-Fernández et al. 2015). The figure displays that water is circulated in the loop easily because of the thermosiphon effect.

7.3.4.2 Latent Heat Storage (LHS)

LHS is associated with the amount of heat absorbed or released in the phase conversion of different materials (Rashidi et al. 2021). It depends on the phase conversion of the medium, for example, the phase change of solid to liquid using latent heat for energy storage. LHS materials typically include the following:

- **Organic materials**. These materials are used in buildings for cooling and heating due to their melting point, which lies in the range of 20–32 °C. Moreover, these groups of materials are flammable, noncorrosive, nontoxic, and chemically stable.
- **Inorganic materials**. This group includes materials such as metallic salt compositions and alloys, which are utilized in solar thermal applications. They have different characteristics such as cooling down quickly, good thermal conductivity, corrosive property, and high heat capacity (Alvi et al. 2021).
- **Eutectic mixtures**. This includes materials such as organic–inorganic, inorganic–inorganic, and organic–organic. They can be utilized in different buildings (Fatih Demirbas 2006). The schematic illustration of LHS is exhibited in Fig. 7.10.

7.3.5 Electrochemical Energy Storage (ECES)

In ECES technology, electrical energy is changed into chemical energy and stored for later use, which is changed back to electricity when the energy is needed (Dunn et al. 2011).

Five kinds of ECES technologies are discussed in this section:

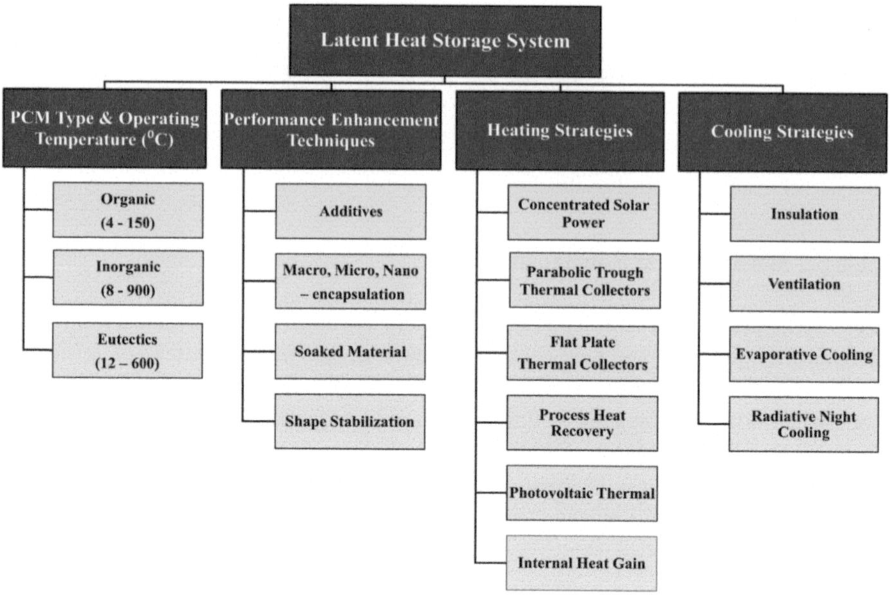

Fig. 7.10 Schematic illustration SHS in the solar water heater. Reproduced with permission (Nazir et al. 2019) Copyright (2019), Elsevier

- High-temperature batteries,
- Conventional batteries,
- Metal–air batteries,
- Flow batteries, and
- Fuel cells.

However, batteries have high maturity and cost-effective technologies yet there are a number of challenges for these systems (Singh et al. 2021). The main problems for this technology are the distortion of components and electrolyte degradation associated with electrochemical reactions that decrease the performance of the battery (Ibrahim et al. 2008).

7.3.5.1 Conventional Batteries

These electrochemical cells are composed of an anode, separator, electrolyte, and cathode. In the charging process, the electrolyte is ionized, and in the discharge process, a redox reaction occurs to recover the chemical energy stored in the ions (Kondoh et al. 2000). The kind of electrolyte utilized determines the kind of battery such as lithium–ion, nickel–cadmium, and lead–acid. Lead–acid batteries are cost-effective with the highest technological maturity in conventional battery technology. Though, lead–acid batteries (LABs) show some drawbacks such as low power, needs high maintenance, low specific energy, short life cycle, and toxicity. When the end-user needs high power quality, lead–acid batteries are the most desirable. Recently, LABs efficiency reached between 75 and 85%. Lead–acid batteries' lifespan is affected by different factors such as the quality of component materials, which is expected to serve 3 - 10 years (Thounthong et al. 2009). An LAB is composed of electrolyte (dilute aqueous sulfuric acid), lead (the negative active material), highly porous PbO_2 (positive active material), and separator (Fig. 7.11). Moreover, a lead alloy grid is used as a current collector and provides mechanical support (Divya and Østergaard 2009).

Nickel–cadmium batteries (NiCd): These types of batteries have a longer service life, about 10 to 15 years. Though, it has problems related to toxicity, which is because of cadmium. Moreover, the overall efficiency of NiCd batteries is slightly lower than that of LABs (60–70%), and the price is higher (Zhu et al. 2013). Despite these shortcomings, NiCd batteries have reached technological maturity and can be used commercially. Recently, the total output power of NiCd batteries installed worldwide is approximately 27 MW (Lacerda et al. 2009).

Advantages

- Compared to new technologies, NiCd is relatively cheap;
- NiCd shows better specific energy, as compared to LABs;
- NiCd batteries are table than other battery technologies;
- NiCd batteries can be assembled easily;
- NiCd has an advanced performance cycle life.

Disadvantages

- Cadmium is heavy metal with a toxic property, which causes diseases and thus needs to be recycled instead of thrown away;
- Compared with different batteries such as Li-ion, it has low energy density;
- It shows high self-discharge level.

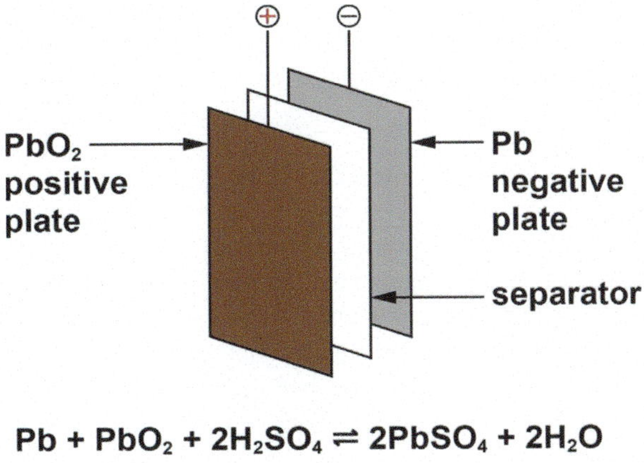

Fig. 7.11 Schematic diagram of components and working principle of a lead–acid battery. Reproduced with permission (May et al. 2018) Copyright (2018), Elsevier

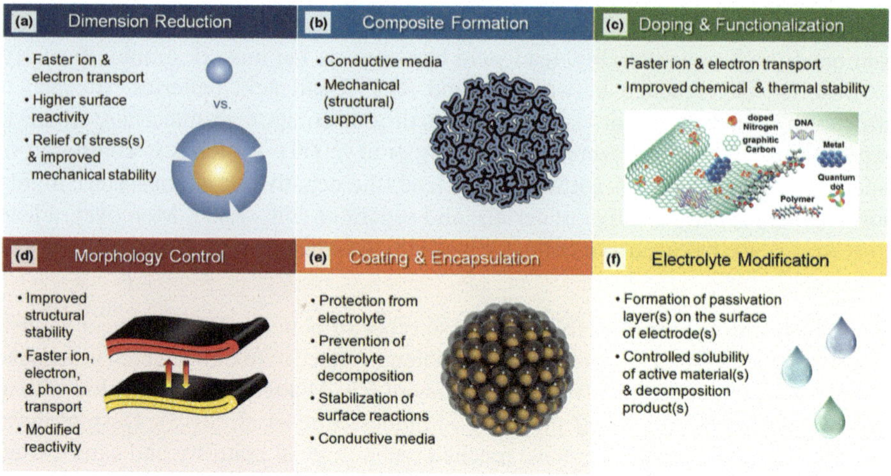

Fig. 7.12 Schematic illustration of approaches for performance improvement and their rationale. Reproduced with permission (Nitta et al. 2015) Copyright (2015), Elsevier

Lithium-ion batteries (LIBs): This battery system has longer service life, low standby loss, and high energy density that make the technology a promising choice and popular for different applications. Moreover, with the increasing acceptance of electric vehicles, LIBs have received more and more attention. Despite the cost challenges of LIBs (especially for larger applications), it has proven to be a huge benefit to connect them to the grid to support utilities. Furthermore, LIBs are more efficient (about 85–95%) and have a service life of about 10–15 years. The cost of LIBs is higher than that of NiCd and lead batteries because they are still somewhat new. Though, as this battery technology matures, their cost is expected to decrease (Hadjipaschalis et al. 2009). Li-ion batteries are the appropriate source of different portable electrochemical energy storage, which needs to enhance their performance and cost (Alvi et al. 2021). With the purpose of enhancing the electrode Li-ion batteries, different approaches have been designed (Wen et al. 2021). These approaches are presented in Fig. 7.12 and they are the same regardless of operating mechanism, crystal structure, and material type.

7.3.5.2 High-Temperature Batteries

These types of batteries show similar working mechanisms with conventional batteries, which are also named molten salt batteries. The difference between high-temperature batteries and conventional batteries is that high-temperature batteries contain solid electrolytes and operate at high temperatures. The most common types of high-temperature batteries utilized currently are sodium–nickel chloride and sodium–sulfur (NaS) batteries. The operating temperature of an NaS battery is approximately 300 °C to 360 °C. Associated with nickel chloride batteries, they have the advantage of longer energy storage time. Though, they are in the early stages of commercialization, they have not been admitted for large-scale grid application. The overall efficiency of NaS batteries is comparatively high, which is between 70 and 90%. Compared with nickel chloride batteries, NaS batteries have less impact on the environment. Sodium–sulfur batteries still face some challenges, namely toxicity, safety hazards caused by the high operating temperature and explosiveness of sodium, and high costs. The increase in material science and technological advancement of these batteries is expected to solve these challenges. Sodium–nickel chloride batteries operate at a temperature of 270 °C, which is slightly lower than that of sodium–sulfur batteries. The total efficiency of the sodium chloride nickel battery is about 85% to 90%, and the response time is relatively fast. Thus, corrosion at higher temperature, the service life of these batteries is not good; however, with recent developments in materials science, cost, life and other problems associated with these batteries can be solved (Baharoon et al. 2015).

7.3.5.3 Flow Batteries (FBs)

The electrochemical reactions that occur in FBs are very similar to high-temperature and conventional batteries. The difference between FBs is that in FBs, the electrolyte is retained in a container outside the reaction cell and is continuously pushed out of the reactor and into the reactor (Nguyen and Savinell 2010). FBs have various advantages such as easy control and monitoring of electrolyte concentration, easier electrolyte replacement, and continuous operation. However, these types of batteries have drawbacks such as high cost and high maintenance service, because of the extra technology requests to maintain electrolyte out/inflow. Moreover, FBs shows lower efficiency because of the energy demands of their auxiliary tools. Hence, in FBs, size has a great impact related to other batteries due to the space required additional equipment (Bueno and Carta 2006). FBs are divided into two categories: hybrid batteries and redox batteries. In hybrid FBs, one of the electroactive components deposited on the electrode surface is the fuel cell electrode and the other is the battery electrode (Weber et al. 2011). The hybrid FBs' storage capacity is linked with the size of the battery electrode. Redox FBs are composed of electroactive parts that are dissolved in the electrolyte. The redox FBs' storage capacity is associated with the capacity of the electrolyte as well as its power capacity is associated with the area of the electrode. The typical examples of redox FBs are vanadium and zinc bromide. Vanadium redox batteries can be utilized in various utilization such as peak demand management and renewable power storage (Leung et al. 2012). Currently, commercial-level vanadium redox batteries show 5 kW power capacities. Although zinc bromide batteries are in the early stages of advancement, they are affordable, have promising storage and high energy density technology. The zinc bromide battery has an important problem: due to the uneven accumulation of zinc on the electrode, it must be completely discharged every 5 to 10 cycles. Moreover, the corrosive issue of bromine is another problem (Leung et al. 2012).

7.3.5.4 Metal–Air Batteries (MABs)

MABs are composed of four components: an air cathode, an electrolyte, a metal anode, and separator, as shown in Fig. 7.13 (Clark et al. 2018; Worku et al. 2021a). The separator is an insulator that only allows ion conversion. During the discharge process, the metal is dissolved in the electrolyte, the metal anode undergoes an oxidation reaction, and the air cathode initiates an oxygen reduction reaction (ORR). Because of the open battery structure, MABs utilize air as the reactant, which has a higher specific capacity (Zhang et al. 2014). Despite their high energy density, these huge problems must be addressed before these systems can be put into practical use (Sharma and Bhatti 2010). In the next section among different metal–air batteries, two potential battery technologies are presented.

Li–Air Batteries (LABs): LABs are the most preferable battery technology, which can be used for electric vehicles because of their high energy density. Based on the type of electrolyte utilized, LABs can be categorized as follows (Fig. 7.14):

- Solid-state LAB,
- Aqueous LAB,
- Aprotic LAB, and
- Mixed aqueous/aprotic LAB.

Hence, in the above configuration, all battery systems have oxygen gas as cathode and lithium metal as anodic materials. Though, the configuration is similar, they possess different reaction mechanisms, which depend on the electrolyte applied.

Zinc–air Batteries (ZABs): ZABs are made up of metal electrolyte, separator, cathode, and anode. Figure 7.15 shows the schematic diagram of typical rechargeable ZABs. The air electrode is composed of a catalytically active layer and a gas diffusion layer, which are the essential parts in the charging and discharging process

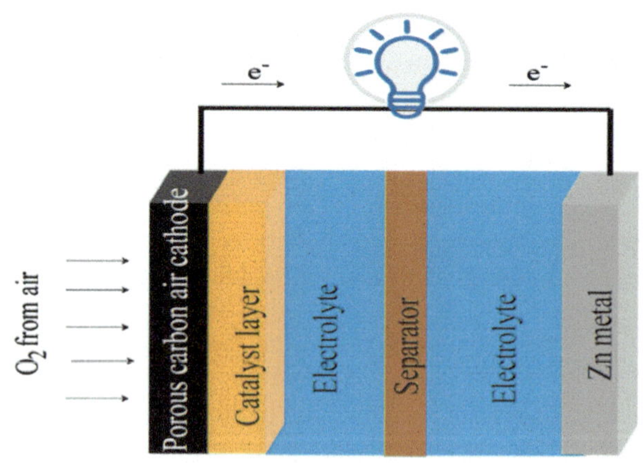

Fig. 7.13 Schematic illustration of MABs

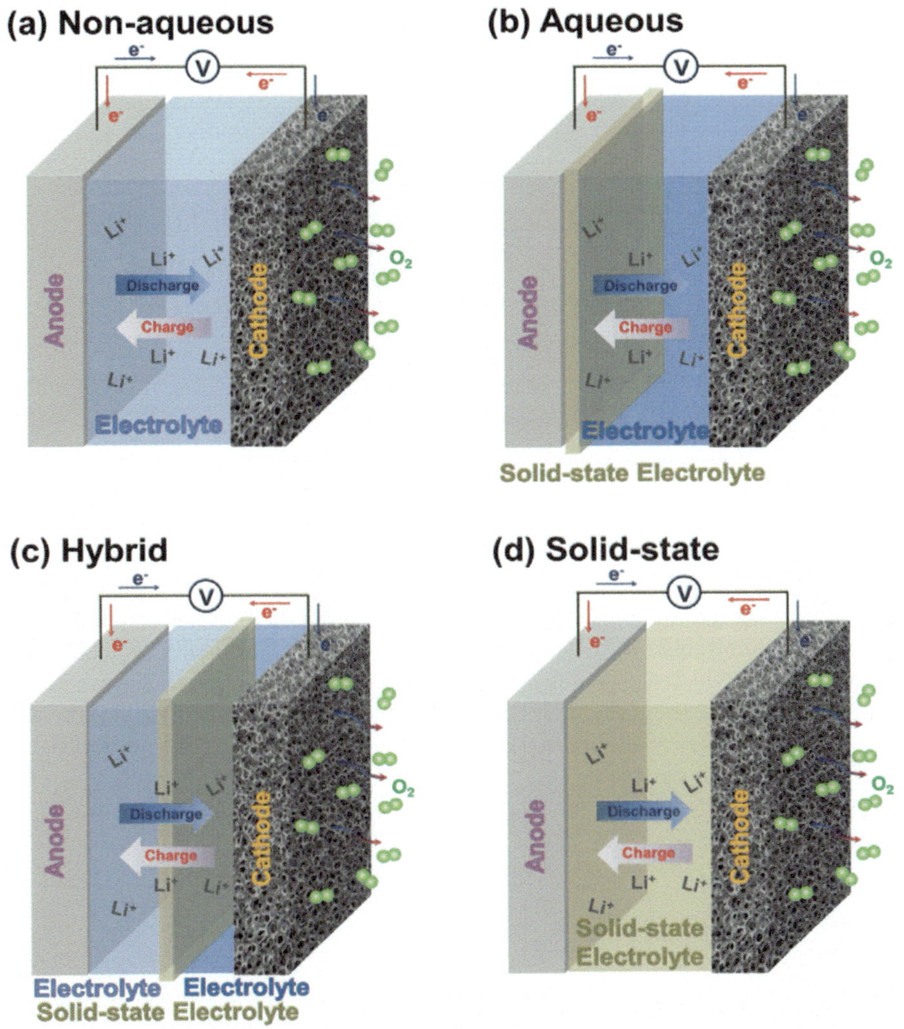

Fig. 7.14 Schematic illustration of Li–air batteries. Reproduced with permission (Tan et al. 2017) Copyright (2017), Elsevier

(Cho et al. 2015). Atmospheric oxygen in the gas phase is used as an active material for ZABs at the cathode part. ZABs are classified into two types as follows:

- Primary,
- Rechargeable (secondary).

Primary-based ZABs are stable with long storage life. Sealed primary ZABs show 2% capacity loss after a year of storage life. Thus, ZABs are found in different voltages and sizes. Like other primary batteries, primary ZABs can be connected

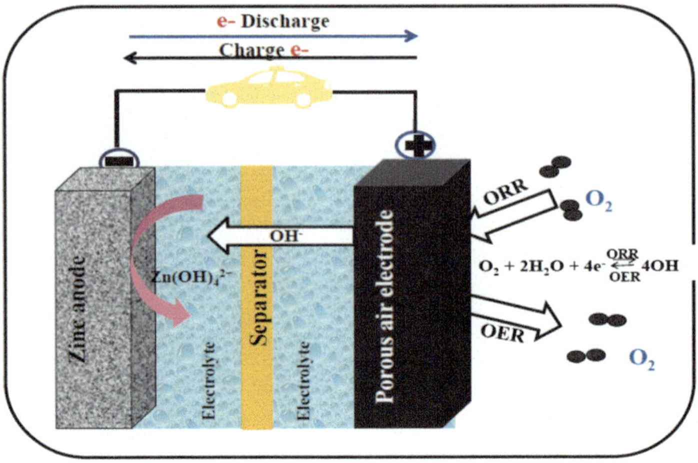

Fig. 7.15 Schematic illustration of rechargeable zinc–air battery

in series to fabricate higher voltage a battery (Harting et al. 2012). Primary ZABs are used in buoys railroad, and hearing aid applications. In mechanically recharged ZABs, Zn is consumed in the charge–discharge process and finally replaced by a fresh electrode, which can be utilized in grid storage. Hence, because of the different advantages, most researchers are focusing on electrically rechargeable ZABs (Worku et al. 2021b). The key challenge in the anode part is low utilization efficiency because of dendritic growth, corrosion, and passivation. Moreover, the main challenge of the air electrode is high overpotential and the sluggish property of oxygen reactions (Xu et al. 2018).

7.3.5.5 Fuel Cells (FC)

A fuel cell (FC) generates electricity by using oxygen and hydrogen through an electrochemical reactor. Recently, different types of FC devices have been fabricated, which have similar working principles. It uses pure hydrogen gas as a fuel; however, methanol, methane, natural gas, and other hydrocarbons can be utilized combined with oxygen. FC can be divided by the type of electrolyte they used. This division limits the type of electrochemical reactions that occur in the cell, the fuel required, cell operation the temperature range, the type of catalysts utilized, and other issues. Recently, there are various kinds of FC under development every one with its own potential applications, limitations, and advantages. The classifications of FC based on the electrolyte used are listed below:

- Solid oxide FC,
- Alkaline FC,
- Polymer electrolyte membrane FC,

- Reversible FC,
- Molten carbonate FC,
- Phosphoric acid FC, and
- Direct methanol FC.

7.4 Hybrid Energy Storage Systems (HESSs)

The energy storage technologies are built in a grid by integrating multiple devices, the system is termed as a HESSs (Bocklisch 2016). As a result, the merits of each system in an integrated device face difficult conditions can add up to meet specific needs, and improve technology performance (Komala et al. 2021). The main goal is to deal with the real-time harsh working environment that a single system cannot accomplish (Mandelli et al. 2015). Thc HEESS also helps to increase many ideal technologies such as power level, cost operating temperature, life cycle, discharge rate, and energy density. By combining the unique advantages of a single device or system, cycle efficiency can be improved. Generally, in the HESS system, a slow response system and a fast response system are mixed together to achieve higher and higher characteristics (Blechinger et al. 2014). The role and use of HESSs have been proven in various fields. In the field of electric transportation, the hybrid power of batteries and supercapacitors has been proven effective when used in electric vehicles (Henson 2008). In wind energy systems, the most practical method is to use battery supercapacitors to achieve energy smoothing and grid integration. Proposals to use battery–supercapacitor or fuel cell–battery hybrid power to support photovoltaic power plants have been widely proposed (Bocklisch 2016). The fuel cell–battery combination can well support the hybrid wind-PV renewable energy system (Álvaro et al. 2019). Therefore, the use of HESSs can be regarded as an ideal solution for different utilization in the future. However, to prove the feasibility and functionality of HESSs, further research and development must be carried out. Various storage technologies have been combined for different applications as shown in Fig. 7.16 Most commonly used in renewable energy sources can be classified as fuel cell /flywheel HESSs, supercapacitor/battery, fuel cell/supercapacitor, battery/flywheel, battery/CAES, SMES/battery, and fuel cell /battery (Samweber et al. 2015).

7.5 Challenges and Prospects of Energy Storage Technologies

The development and innovation of energy storage technologies have faced many challenges. For the commercialization, widespread dissemination, and long-term adaptation of the latest inventions in this field, these challenges must also be met. When ESSs are used and the storage system is in operation to store excess generated

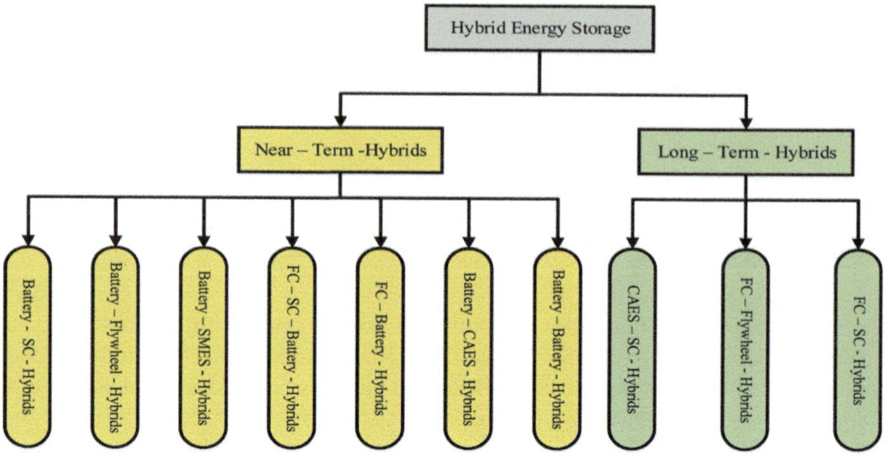

Fig. 7.16 Schematic illustration of different combination methods for hybrid energy storage technologies. Reproduced with permission (Hajiaghasi et al. 2019) Copyright (2019), Elsevier

energy, the world faces some constraints and challenges. According to reports, all equipment and systems have not released 100% of the stored energy for later use, which means that waste will definitely occur during storage and release. The implementation, operation, and replacement of energy storage technologies also require a large amount of capital. Certain energy storage devices may cause environmental impact, which starts from the extraction of materials used for manufacturing and continues until the end of their useful life until disposal. Therefore, research is needed to develop equipment that is not only more efficient, but must also be cost-effective and must have minimal environmental issues, particularly for the disposal of utilized equipment after the life cycle is completed. The current energy production mainly relies on fossil fuel power generation, which is not only costly but also impossible to update, so it cannot be maintained indefinitely. Furthermore, the electricity fabrication of fossil fuel power plants is bound to be related to carbon dioxide emissions, which can cause serious environmental pollution. In order to provide an effective power supply, optimal management of ESSs is a problem in modern power grids. Therefore, there is a big voice that can gradually reduce the dependence on the use of oil and natural gas for electricity production.

7.6 Conclusion

Energy storage systems have different merits, disadvantages, functions, and system maturity. Hence, the purpose of this chapter is to overview the advancement of key energy storage technologies, such as chemical, electromagnetic, thermal, electrical, and electrochemical energy storage systems. Self-discharge rate, specific power, environmental impact efficiency, power density, lifetime, power capital cost, specific energy, energy capital cost, and energy density are performance indicators for evaluating the state of ESSs. When we consider all selected economic, energy, environmental, and technical criteria at the same time, the findings of this chapter indicate the following:

- From MESSs, the average performance of pumped storage systems ranks the highest.
- From electrochemical energy storage technologies, high-temperature batteries showed the highest performance.
- From CESSs, ammonia shows the highest performance level.
- The TESSs based on molten salt have the highest performance level.
- From electromagnetic energy storage technologies, superconducting magnets showed an excellent performance level.

Hence, from electromagnetic electrochemical, thermal, chemical, and mechanical energy storage technologies, chemical energy storage technology showed the highest performance, whereas, electrochemical energy storage technology showed the lowest performance levels.

References

Alvi JZ, Feng Y, Wang Q et al (2021) Effect of working fluids on the performance of phase change material storage based direct vapor generation solar organic Rankine cycle system. Energy Rep 7:348–361. https://doi.org/10.1016/j.egyr.2020.12.040

Álvaro D, Arranz R, Aguado JA (2019) Sizing and operation of hybrid energy storage systems to perform ramp-rate control in PV power plants. Int J Electr Power Energy Syst 107:589–596. https://doi.org/10.1016/j.ijepes.2018.12.009

Amiryar ME, Pullen KR (2017) A review of flywheel energy storage system technologies and their applications. Appl Sci 7. https://doi.org/10.3390/app7030286

Amodeo SJ, Chiacchiarini HG, Solsona JA, Busada CA (2009) High-performance sensorless nonlinear power control of a flywheel energy storage system. Energy Convers Manag 50:1722–1729. https://doi.org/10.1016/j.enconman.2009.03.024

Asjid M, Ali M, Waqas A et al (2021) Thermal performance evaluation of circular-stadium double pipe thermal energy storage systems. J Energy Storage 36. https://doi.org/10.1016/j.est.2021.102403

Baharoon DA, Rahman HA, Omar WZW, Fadhl SO (2015) Historical development of concentrating solar power technologies to generate clean electricity efficiently–a review. Renew Sustain Energy Rev 41:996–1027

Beaudin M, Zareipour H, Schellenberglabe A, Rosehart W (2010) Energy storage for mitigating the variability of renewable electricity sources: an updated review. Energy Sustain Dev 14:302–314

Blechinger P, Seguin R, Cader C, et al (2014) Assessment of the global potential for renewable energy storage systems on small islands. In: Energy Procedia. Elsevier Ltd, pp 294–300

Bocklisch T (2016) Hybrid energy storage approach for renewable energy applications. J Energy Storage 8:311–319. https://doi.org/10.1016/j.est.2016.01.004

Bueno C, Carta JA (2006) Wind powered pumped hydro storage systems, a means of increasing the penetration of renewable energy in the Canary Islands. Renew Sustain Energy Rev 10:312–340

Chen H, Cong TN, Yang W et al (2009) Progress in electrical energy storage system: a critical review. Prog Nat Sci 19:291–312
Cheng L, Zhai Q, Chen S et al (2021) Component-tunable hierarchical flower-shaped bimetallic zinc-cobalt selenides for high-performance hybrid supercapacitor. J Energy Storage 36. https://doi.org/10.1016/j.est.2021.102374
Cho J, Jeong S, Kim Y (2015) Commercial and research battery technologies for electrical energy storage applications. Prog Energy Combust Sci 48:84–101. https://doi.org/10.1016/j.pecs.2015.01.002
Clark S, Latz A, Horstmann B (2018) A review of model-based design tools for metal-air batteries. Batteries 4
Connolly D, Lund H, Mathiesen BV, Leahy M (2010) A review of computer tools for analysing the integration of renewable energy into various energy systems. Appl Energy 87:1059–1082
Divya KC, Østergaard J (2009) Battery energy storage technology for power systems-an overview. Electr. Power Syst. Res. 79:511–520
Dunn B, Kamath H, Tarascon JM (2011) Electrical energy storage for the grid: a battery of choices. Science (80-) 334:928–935
Fasano M, Bozorg Bigdeli M, Vaziri Sereshk MR et al (2015) Thermal transmittance of carbon nanotube networks: guidelines for novel thermal storage systems and polymeric material of thermal interest. Renew Sustain Energy Rev 41:1028–1036
Fatih Demirbas M (2006) Thermal energy storage and phase change materials: an overview. Energy sources. Part B Econ. Plan. Policy 1:85–95
Faunce TA, Lubitz W, Rutherford AW et al (2013) Energy and environment policy case for a global project on artificial photosynthesis. Energy Environ Sci 6:695–698
Figueiredo FC, Flynn PC (2006) Using diurnal power price to configure pumped storage. IEEE Trans Energy Convers 21:804–809. https://doi.org/10.1109/TEC.2006.877373
Hadjipaschalis I, Poullikkas A, Efthimiou V (2009) Overview of current and future energy storage technologies for electric power applications. Renew Sustain Energy Rev 13:1513–1522
Hajiaghasi S, Salemnia A, Hamzeh M (2019) Hybrid energy storage system for microgrids applications: a review. J Energy Storage 21:543–570. https://doi.org/10.1016/J.EST.2018.12.017
Harting BK, Kunz U, Turek T (2012) Zinc-air batteries: prospects and challenges for future improvement. Z Phys Chem 226:151–166. https://doi.org/10.1524/zpch.2012.0152
Henson W (2008) Optimal battery/ultracapacitor storage combination. J Power Sources 179:417–423. https://doi.org/10.1016/j.jpowsour.2007.12.083
Heymans C, Walker SB, Young SB, Fowler M (2014) Economic analysis of second use electric vehicle batteries for residential energy storage and load-levelling. Energy Policy 71:22–30. https://doi.org/10.1016/j.enpol.2014.04.016
Hu L, Li X, Ding L et al. (2021) Flexible textiles with polypyrrole deposited phase change microcapsules for efficient photothermal energy conversion and storage. Sol Energy Mater Sol Cells 224. https://doi.org/10.1016/j.solmat.2021.110985
Ibrahim H, Ilinca A, Perron J (2008) Energy storage systems-characteristics and comparisons. Renew Sustain Energy Rev 12:1221–1250
Kirubakaran A, Jain S, Nema RK (2009) A review on fuel cell technologies and power electronic interface. Renew Sustain Energy Rev 13:2430–2440
Komala K, Kumar KP, Cherukuri SHC (2021) Storage and non-storage methods of power balancing to counter uncertainty in hybrid microgrids—a review. J Energy Storage 36
Kondoh J, Ishii I, Yamaguchi H et al (2000) Electrical energy storage systems for energy networks. Energy Convers Manag 41:1863–1874. https://doi.org/10.1016/S0196-8904(00)00028-5
Lacerda VG, Mageste AB, Santos IJB et al. (2009) Separation of Cd and Ni from Ni-Cd batteries by an environmentally safe methodology employing aqueous two-phase systems. J Power Sources 193:908–913. https://doi.org/10.1016/j.jpowsour.2009.05.004
Leung P, Li X, Ponce De León C et al (2012) Progress in redox flow batteries, remaining challenges and their applications in energy storage. RSC Adv 2:10125–10156. https://doi.org/10.1039/c2ra21342g

Liu C, Li F, Lai-Peng M, Cheng HM (2010) Advanced materials for energy storage. Adv. Mater. 22
Luo X, Wang J, Dooner M, Clarke J (2015) Overview of current development in electrical energy storage technologies and the application potential in power system operation. Appl Energy 137:511–536. https://doi.org/10.1016/j.apenergy.2014.09.081
Mandelli S, Molinas M, Park E, et al (2015) The role of storage in emerging country scenarios. In: Energy Procedia. Elsevier Ltd, pp 112–123
May GJ, Davidson A, Monahov B (2018) Lead batteries for utility energy storage: a review. J Energy Storage 15:145–157. https://doi.org/10.1016/J.EST.2017.11.008
Nazir H, Batool M, Bolivar Osorio FJ et al (2019) Recent developments in phase change materials for energy storage applications: a review. Int J Heat Mass Transf 129:491–523. https://doi.org/10.1016/j.ijheatmasstransfer.2018.09.126
Nguyen T, Savinell RF (2010) Flow batteries. Electrochem Soc Interface 19:54–56. https://doi.org/10.1149/2.F06103if
Nguyen TH, Fraiwan A, Choi S (2014) Paper-based batteries: a review. Biosens Bioelectron 54:640–649. https://doi.org/10.1016/j.bios.2013.11.007
Nitta N, Wu F, Lee JT, Yushin G (2015) Li-ion battery materials: present and future. Mater Today 18:252–264
Papaefthymiou SV, Karamanou EG, Papathanassiou SA, Papadopoulos MP (2010) A wind-hydro-pumped storage station leading to high RES penetration in the autonomous island system of Ikaria. IEEE Trans Sustain Energy 1:163–172. https://doi.org/10.1109/TSTE.2010.2059053
Pearre NS, Swan LG (2015) Technoeconomic feasibility of grid storage: mapping electrical services and energy storage technologies. Appl Energy 137:501–510. https://doi.org/10.1016/j.apenergy.2014.04.050
Pickard WF (2012) The history, present state, and future prospects of underground pumped hydro for massive energy storage. In Proceedings of the IEEE. Institute of Electrical and Electronics Engineers Inc., pp 473–483
Pintaldi S, Perfumo C, Sethuvenkatraman S et al (2015) A review of thermal energy storage technologies and control approaches for solar cooling. Renew Sustain Energy Rev 41:975–995
Rashidi S, Esfahani JA, Hosseinirad E (2021) Assessment of solar chimney combined with phase change materials. J Taiwan Inst Chem Eng. https://doi.org/10.1016/j.jtice.2021.03.001
Ries G, Neumueller HW (2001) Comparison of energy storage in flywheels and SMES. Phys C Supercond Its Appl 357–360:1306–1310. https://doi.org/10.1016/S0921-4534(01)00484-1
Sadeghi S, Jahangir H, Vatandoust B et al (2021) Optimal bidding strategy of a virtual power plant in day-ahead energy and frequency regulation markets: a deep learning-based approach. Int J Electr Power Energy Syst 127. https://doi.org/10.1016/j.ijepes.2020.106646
Samweber F, Fischhaber S, Nobis P (2015) Electric mobility as a functional energy storage in comparison to on-site storage systems for grid integration. In: Energy Procedia. Elsevier Ltd, pp 94–102
Sebastián R, Peña Alzola R (2012) Flywheel energy storage systems: review and simulation for an isolated wind power system. Renew Sustain Energy Rev 16:6803–6813
Shaqsi AL, AZ, Sopian K, Al-Hinai A, (2020) Review of energy storage services, applications, limitations, and benefits. Energy Rep 6:288–306. https://doi.org/10.1016/j.egyr.2020.07.028
Sharma A, Tyagi VV, Chen CR, Buddhi D (2009) Review on thermal energy storage with phase change materials and applications. Renew Sustain Energy Rev 13:318–345. https://doi.org/10.1016/j.rser.2007.10.005
Sharma P, Bhatti TS (2010) A review on electrochemical double-layer capacitors. Energy Convers Manag 51:2901–2912. https://doi.org/10.1016/j.enconman.2010.06.031
Singh VP, Kumar M, Srivastava RS, Vaish R (2021) Thermoelectric energy harvesting using cement-based composites: a review. Mater Today Energy 21. https://doi.org/10.1016/j.mtener.2021.100714
Suzuki Y, Koyanagi A, Kobayashi M, Shimada R (2005) Novel applications of the flywheel energy storage system. Energy. Elsevier Ltd, pp 2128–2143

Tan P, Jiang HR, Zhu XB et al (2017) Advances and challenges in lithium-air batteries. Appl Energy 204:780–806. https://doi.org/10.1016/j.apenergy.2017.07.054

Tewari S, Mohan N (2013) Value of NAS energy storage toward integrating wind: results from the wind to battery project. IEEE Trans Power Syst 28:532–541. https://doi.org/10.1109/TPWRS.2012.2205278

Thounthong P, Chunkag V, Sethakul P et al (2009) Comparative study of fuel-cell vehicle hybridization with battery or supercapacitor storage device. IEEE Trans Veh Technol 58:3892–3904. https://doi.org/10.1109/TVT.2009.2028571

Velasco-Fernández R, Ramos-Martín J, Giampietro M (2015) The energy metabolism of China and India between 1971 and 2010: studying the bifurcation. Renew Sustain Energy Rev 41:1052–1066

Wang H, Cao M, Huang R et al (2021) Preparation of BaTiO3@NiO core-shell nanoparticles with antiferroelectric-like characteristic and high energy storage capability. J Eur Ceram Soc 41:4129–4137. https://doi.org/10.1016/j.jeurceramsoc.2021.02.042

Weber AZ, Mench MM, Meyers JP et al (2011) Redox flow batteries: a review. J Appl Electrochem 41:1137–1164

Wen T, Luo Y, Wang M, She X (2021) Comparative study on the liquid desiccant dehumidification performance of lithium chloride and potassium formate. Renew Energy 167:841–852. https://doi.org/10.1016/j.renene.2020.11.157

Whittingham MS (2008) Materials challenges facing electrical energy storage. MRS Bull 33:411–419. https://doi.org/10.1557/mrs2008.82

Worku AK, Ayele DW, Habtu NG et al (2021a) Recent progress in MnO2-based oxygen electrocatalysts for rechargeable zinc-air batteries. Mater Today Sustain 13. https://doi.org/10.1016/j.mtsust.2021.100072

Worku AK, Ayele DW, Habtu NG (2021b) Recent advances and future perspectives in engineering of bifunctional electrocatalysts for rechargeable zinc–air batteries. Mater Today Adv 9:100116. https://doi.org/10.1016/j.mtadv.2020.100116

Xu K, Loh A, Wang B, Li X (2018) Enhancement of oxygen transfer by design nickel foam electrode for zinc-air battery. J Electrochem Soc 165:A809–A818. https://doi.org/10.1149/2.0361805jes

Yang Z, Zhang J, Kintner-Meyer MCW et al (2011) Electrochemical energy storage for green grid. Chem Rev 111:3577–3613

Zhang T, Tao Z, Chen J (2014) Materials Horizons Magnesium–air batteries : from principle to application. Mater Horiz 1:196–206. https://doi.org/10.1039/c3mh00059a

Zhao H, Wu Q, Hu S et al (2015) Review of energy storage system for wind power integration support. Appl Energy 137:545–553. https://doi.org/10.1016/j.apenergy.2014.04.103

Zhao H, Guo W (2021) Coordinated control method of multiple hybrid energy storage systems based on distributed event-triggered mechanism. Int J Electr Power Energy Syst 127. https://doi.org/10.1016/j.ijepes.2020.106637

Zhu WH, Zhu Y, Davis Z, Tatarchuk BJ (2013) Energy efficiency and capacity retention of Ni-MH batteries for storage applications. Appl Energy 106:307–313. https://doi.org/10.1016/j.apenergy.2012.12.025

Zuo Z, Liu S, Sun Y, Wu Y (2015) Pressure fluctuations in the vaneless space of high-head pump-turbines—a review. Renew Sustain Energy Rev 41:965–974

Chapter 8
Hydrogen Production from Renewable Energy Sources, Storage, and Conversion into Electrical Energy

El Manaa Barhoumi, Paul C. Okonkwo, Slah Farhani, Ikram Ben Belgacem, and Faouzi Bacha

Abstract This chapter discusses the electrolysis process used to produce green hydrogen from renewable energy sources and the conversion of hydrogen into electrical energy by using fuel cells. Hydrogen can be produced from renewable energy sources, stored, and used whenever electrical energy is required by the loads. The process of electrolysis is the use of electrical energy and water to produce hydrogen. The different electrolyzers: solid oxide, alkaline, and proton exchange membrane have different characteristics and efficiencies. The cost of hydrogen production depends on the worth of renewable energy systems and hydrogen production equipment. On the other hand, the overall efficiency of hydrogen production depends on the renewable energy system efficiency. So, the optimization of the renewable energy system and the selection of adequate sites are characterized by their high renewable energy potential allow maximizing the effectiveness of hydrogen production. The design of a photovoltaic system to generate the electrical energy required to produce 100 kg of hydrogen per day highlights the potential future of green hydrogen produced from solar energy using photovoltaic systems. This hydrogen gas power station requires the installation of 2662.2 kWp of the PV system to produce 13,311

E. M. Barhoumi (✉)
Department of Electrical and Computer Engineering, College of Engineering, Dhofar University, Salalah, Oman
e-mail: ebarhoumi@du.edu.om

P. C. Okonkwo
Department of Mechanical and Mechatronics Engineering, College of Engineering, Dhofar University, Salalah, Oman
e-mail: pokonkwo@du.edu.om

S. Farhani
Laboratory of Computer for Industrial Systems, INSAT, University of Carthage, Carthage, Tunisia

I. B. Belgacem
Laboratoire de Génie Mécanique, Ecole Nationale d'Ingénieurs de Monastir, Université de Monastir, Monastir, Tunisie

F. Bacha
Department of Electrical Engineering, University of Tunis, ENSIT, Tunis, Tunisia
e-mail: faouzi.bacha@esstt.rnu.tn

A. K. Bohre et al. (eds.), *Planning of Hybrid Renewable Energy Systems, Electric Vehicles and Microgrid*, Energy Systems in Electrical Engineering,
https://doi.org/10.1007/978-981-19-0979-5_8

kWh of electrical energy per day to run four proton exchange membrane electrolyzers during 5 h per day. The produced hydrogen can be used to charge fuel cell vehicles, generate electricity for buildings during the night, or be transported to be consumed in any other industrial applications.

Keywords Electrolyzers · Fuel cells · Renewable energy · Photovoltaic-Hydrogen Refueling Station

Nomenclature

PV	photovoltaic
HGS	hydrogen gas station
PEM	proton exchange membrane
SOE	solid oxide electrolyzer
FCV	fuel cell vehicle
H2	hydrogen
V	volt
Wp	watt peak
Vohm	voltage of ohmic losses
O2	dioxygen
H_2O	water
OH-	hydroxide
η	Efficiency
$P_{DC\text{-}DC\ converter}$	power of one DC-DC converter
$P_{PV\text{-}System}$	power of the PV system
$N_{pv\ panels}$	number of PV panels
V_{act}	voltage of activation losses
V_{conc}	voltage of concentration losses

8.1 Introduction

The population increase, the urbanization, and industrialization development lead to an increase in electricity consumption (Yoo and Lee 2010). The excess of fossil fuels exploitation to produce electricity results in the pollution of the environment and the decrease of fuel reserve (Razmjoo et al. 2021). Renewable energy sources represent an alternative solution to produce electrical energy from clean and renewable sources (Boran et al. 2012). Thereafter, solar, wind, geothermal, and other renewable sources are exploited to produce electrical energy. However, the electrical energy produced from renewable sources is dependent on weather conditions and cannot be controlled according to load consumption (Barhoumi et al. 2019; Barhoumi et al. 2021). Therefore, the storage of the surplus of produced electrical energy from renewable sources consists of an adequate solution to maximize energy efficiency (Marano et al. 2012). (Abdin and Mérida 2019). However, the lifetime and the cost of batteries discourage the implementation of this solution (Duggal and Venkatesh 2015; Hammad and Ebaid 2015). To avoid the dependence on batteries for the storage of electrical energy, the electrical energy is converted into hydrogen to be stored in gas or liquid state (Boulmrharj et al. 2020; Gondal et al. 2018). Thereafter, hydrogen is converted into electrical energy to be used when loads required electrical energy. So, hydrogen represents an effective alternative to store electrical energy. Hydrogen energy is leading to a clear change of renewable energy exploitation. Indeed, the surplus of electrical energy is used to produce hydrogen. Hydrogen can be stored in huge quantities, transported to different locations, and used to produce electrical energy (Christensen 2020; HassanzadehFard et al. 2020). The electrolysis technologies and the fuel cells are key solutions in the conversion of energy to gas and gas to energy (Chi and Yu 2018).

This chapter discusses the electrolysis technologies used to produce hydrogen from renewable energy sources, the hydrogen industrial applications, and the conversion of hydrogen into electrical energy. Then, this chapter is organized as follows. Section 8.2 presents the electrolysis process and a comparison of three types of electrolyzers used in hydrogen production. In Sect. 8.3, the production of hydrogen from renewable energy sources is presented and discussed. The requirements of hydrogen storage are presented in Sect. 8.4. The principle of operation, modeling, and applications of the fuel cell is presented in Sect. 8.5. In Sect. 8.6, the design of the photovoltaic hydrogen station is presented and analyzed. The conclusion is given in Sect. 8.7.

8.2 Electrolysis

The processes used to produce hydrogen from renewable sources are summarized in Fig. 8.1 (Shiva Kumar and Himabindu 2019). These processes have different efficiencies and different costs. The hydrogen is called "green" when is produced from renewable energy sources (Gondal et al. 2018). In the process of water splitting, the hydrogen can be produced by photolysis, thermolysis, or electrolysis method. In electrolysis, hydrogen is produced using PEM, solid oxide, or alkaline electrolyzers. The process using electrical current to produce hydrogen is called electrolysis (Yan et al. 2020). This process is based on water-splitting using electrical current to produce hydrogen and oxygen. The different components required for the design of an electrolyzer are the cathode, the anode, the membrane, and the electrolyte (Shiva Kumar and Himabindu 2019). The anode of the electrolyzer is connected to the positive terminal of the DC power source. The negative terminal of the power source is connected to the cathode. The membrane ensures electrical separation

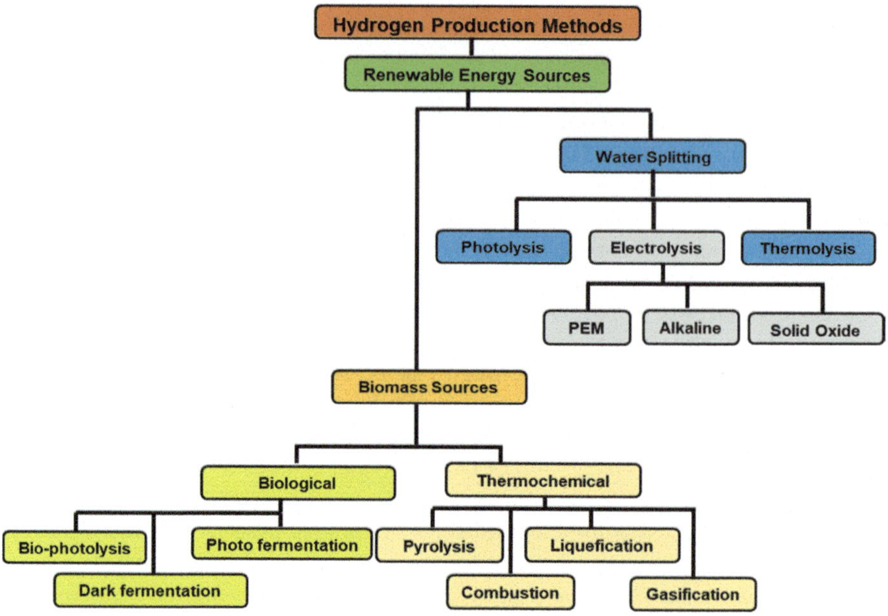

Fig. 8.1 Processes of hydrogen production (Nikolaidis and Poullikkas 2017)

between the anode and the cathode circuits. At the cathode, hydrogen gas is generated. The oxygen is generated at the cathode. The membrane requirements are ion permeability, gas-tight, chemical stability, electrically insulating, and mechanical robustness (Okonkwo et al. 2021).

The electrolysis is described by the following chemical reaction expression (Chi and Yu 2018):

$$H_2O_{(\text{liquid})} \rightarrow H_{2(\text{gaz})} + \frac{1}{2} O_{2(\text{gaz})} \tag{8.1}$$

8.2.1 PEM Electrolyzer

The PEM electrolyzer is mainly composed of anode, cathode, and membrane. In the PEM electrolyzer, the electrodes are made of platinum and the membrane is made of Nafion (Shiva Kumar and Himabindu 2019). The membrane allows separating the electrodes and the gases produced at the cathode and the anode. The flow of the current through the water leads to the apparition of hydrogen and oxygen. At the cathode, the reduction reaction takes place (Abdin et al. 2015):

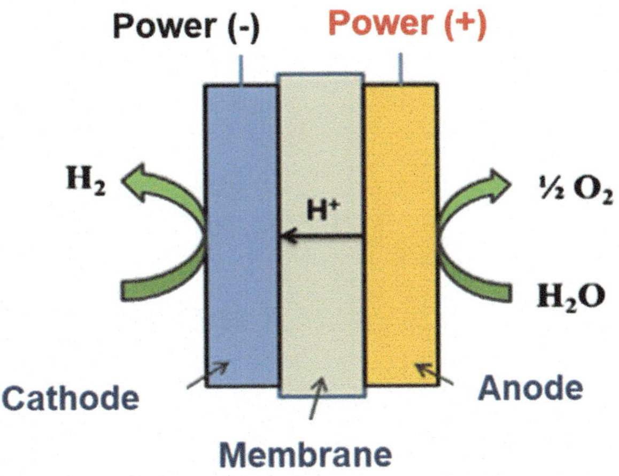

Fig. 8.2 Schematic diagram of the PEM electrolysis process (Barbir May 2005)

$$2H^+ + 2e^- \rightarrow H_{2(gaz)} \tag{8.2}$$

The cathode attracts the cations, and the anode attracts anions. At the anode, the oxidation reaction takes place.

$$H_2O_{(liquid)} \rightarrow 2H^+_{(gaz)} + \frac{1}{2}O_{2(gaz)} + 2e^- \tag{8.3}$$

The overall reaction is expressed as

$$H_2O_{(liquid)} \rightarrow H_{2(gaz)} + \frac{1}{2}O_{2(gaz)} \tag{8.4}$$

The different reactions of water splitting using the PEM electrolyzer are summarized in Fig. 8.2.

The EPM electrolysis is used and implemented in many countries to produce hydrogen from water despite it having the lowest efficiency compared to other electrolyzers.

8.2.2 *Solid Oxide Electrolyzer (SOE)*

The SOE consists of multiple SOE stacks, a fan for circulating oxygen, and a separator to separate produced water from the water (Wang et al. 2019). The SOEs are usually operating in exothermic condition (Nieminen et al. 2010). For lower potential operation, heat is required to start up. This heat and electricity can be generated from solar energy. The electrolysis takes place at high temperatures (400–500 °C). The SOE allows to produce pure oxygen which is considered a valuable product. The

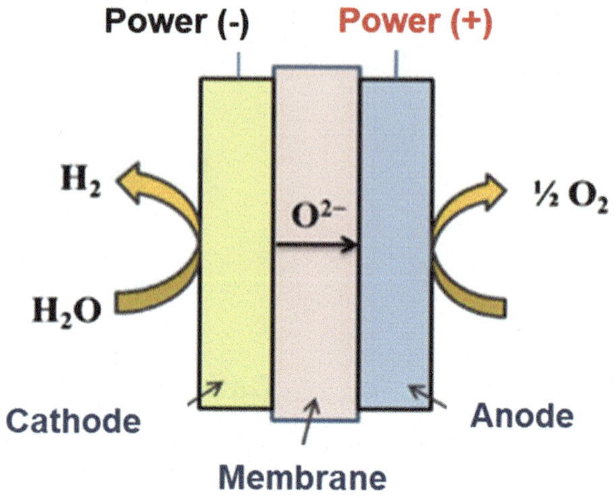

Fig. 8.3 Schematic diagram of the water electrolysis using the SOE

overall efficiency of the SOE is higher than the PEM electrolyzer (Nieminen et al. 2010). The elementary stack of the SOE is depicted in Fig. 8.3.

The chemical reactions at the anode and cathode sides, respectively, are expressed as follows (Ni et al. 2008):

$$O^{2-} \rightarrow 2e^- + \frac{1}{2}O_2 \tag{8.5}$$

$$H_2O + 2e^- \rightarrow H_2 + O^{2-} \tag{8.6}$$

The overall reaction is

$$H_2O_{(liquid)} \rightarrow H_{2(gaz)} + \frac{1}{2}O_{2(gaz)} \tag{8.7}$$

8.2.3 Alkaline Electrolyzer

The electrodes of the alkaline electrolyzer are merged in the aqueous electrolyte of approximately 30% Sodium hydroxide or potassium hydroxide (David et al. 2019; Chi and Yu 2018). The cathode is made of nickel and the anode is made of nickel or copper. At the cathode, the water is decomposed into hydrogen and anions OH-. At the anode, the oxidation of OH- taking place at the anode generates water and oxygen. The principle of operation of the alkaline electrolyzer is given in Fig. 8.4.

The chemical reactions at anode and cathode sides are expressed by David et al. (2019)

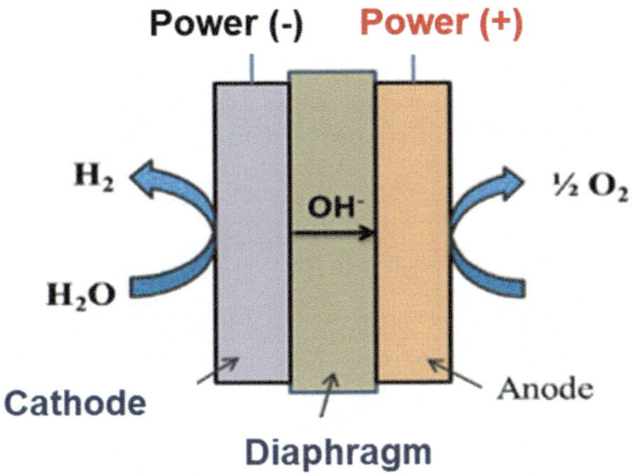

Fig. 8.4 Schematic diagram of the water electrolysis using the alkaline electrolyzer

$$2OH^- \rightarrow 2e^- + \frac{1}{2}O_2 + H_2O \tag{8.8}$$

$$2H_2O + 2e^- \rightarrow H_2 + 2OH^- \tag{8.9}$$

The overall reaction is

$$H_2O_{(liquid)} \rightarrow H_{2(gaz)} + \frac{1}{2}O_{2(gaz)} \tag{8.10}$$

8.2.4 Comparison of Electrolyzers

The efficiency of an electrolyzer is calculated by Nieminen et al. (2010)

$$\eta = \frac{P_{out}}{P_{input}} \tag{8.11}$$

where P_{out} is the output power calculated in kWh equivalent to the produced hydrogen P_{input} is the electrical energy required to run the electrolyzer to produce the hydrogen. In other words, the efficiency is equivalent to the energy of produced hydrogen divided by the amount of absorbed electricity. 1 kg of hydrogen is equivalent to 33.4 kWh of electrical energy. This value is called the higher heating value of hydrogen.

Example: An electrolyzer is absorbing 50 kWh of electrical energy to produce 0.5 kg of hydrogen. Then, the efficiency of the electrolyzer is.

Table 8.1 Comparison of electrolyzers' characteristics

Electrolyzer characteristics	PEM	SOE	Alkaline
Material of anode	Pt black, Ir, Ru, Rh	Lanthanum strontium manganate (LSM)	Ni or Cu
Material of cathode	Pt black, Ir, Ru, Rh	Ni-doped YSZ	Ni
Electrolyte	Solid polymer	Solid oxide or ceramic material	Potassium hydroxide (KOH) or sodium hydroxide (NaOH)
Efficiency	57–69%	80%	90%
Working temperature	50–100 °C	500–950 °C	80 °C

$$\eta = \frac{0.5 \times HeatValue(H2)}{50 \times HeatValueof 1kWh} = 33.4\% \tag{8.12}$$

The different performances and main characteristics of electrolyzers used to produce hydrogen are summarized in Table 8.1.

8.3 Renewable Energy Sources for Hydrogen

Even though hydrogen is produced from renewable sources using different techniques, electrolysis still considered the main process used to produce hydrogen (Barbir 2005). When the electrical current used in electrolysis is produced from classic power generating stations, e.g., thermal power station and gas power station, the process is considered a pollutant and non-green (Chi and Yu 2018). Indeed, hydrogen is produced from pollutant sources. However, electrolysis can use electrical current produced from renewable energy sources where no dioxide of carbon is produced (Yan et al. 2020). The green hydrogen is produced from carbon-free and environmentally friendly sources. Recently, governmental authorities in many countries showed interest to convert renewable energy sources into hydrogen. The process of the conversion of wind energy and solar energy into hydrogen through the electrolysis is presented in Fig. 8.5. The hydrogen produced from renewable sources by electrolysis is used on-site for refueling FCVs or transported and used in other places. Therefore, the hydrogen is considered an effective method for the storage of electrical energy produced from renewable energy sources.

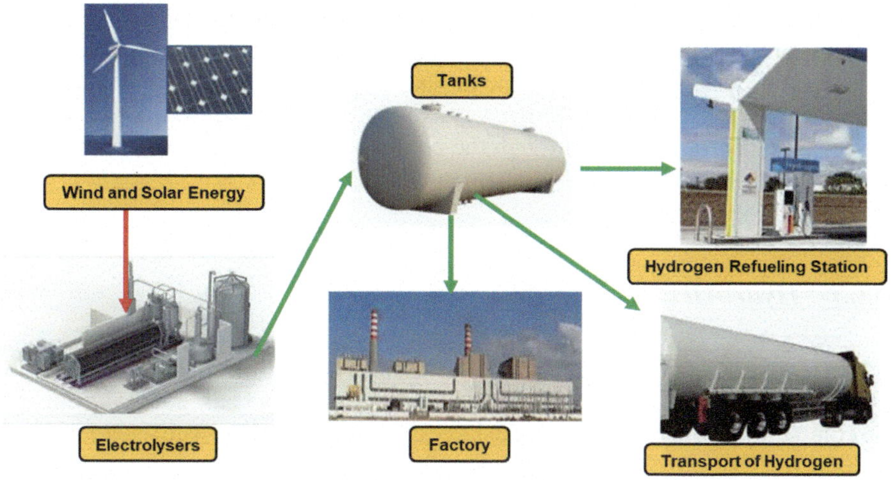

Fig. 8.5 Conversion of electrical energy produced from renewable energy sources into hydrogen

8.3.1 *Solar Energy*

Solar energy has been used to produce electrical energy to run different industrial processes (Hammad and Ebaid 2015; Albadi et al. 2019; Shaner et al. 2016). Nowadays, large-scale PV plants is installed around the world to produce electrical energy from solar energy using large-scale PV systems (El Ouderni et al. 2013). To control the flow of produced electrical energy into the power grid, the surplus of electrical energy that cannot be fed to the power grid is converted into hydrogen (Shaner et al. 2016). The hydrogen will be used later to produce electrical energy when required. A photovoltaic system feeding an electrolyzer to provide 58,400 kg of hydrogen in London city on a yearly basis required the installation of 2.98 MWp as DC capacity of PV system (Abdin and Mérida 2019). The results show that 47.82 kWh of electrical energy is required to produce 1 kg of hydrogen. The economic analysis shows that 1 kg of hydrogen costs between 3 and 8 Euro according to the price of the PV systems, civil works, compressors, and storage tanks.

8.3.2 *Wind Energy*

Recently, several plants are used to produce hydrogen from wind power (Xiao et al. 2020). The basic idea consists of the production of electricity from wind energy using wind turbines, electric generators, and power converters. In the second stage, electricity produced is used to produce hydrogen (Schnuelle et al. 2020). The method has a lot of advantages. Indeed, hydrogen is produced from clean and cost-free sources. The hydrogen is then stored in special tanks to be used to produce electricity by the fuel cell. Xcel Energy in cooperation with NREL's installed the wind turbines to produce the hydrogen (Givler and Lilienthal 2005). The project, Wind2H2, uses two wind turbines 1of 00-kW and one turbine of 10-kW. The project uses two PEM and one alkaline electrolyzer. The hydrogen produced during the electrolysis is stored in special tanks for later use through the conversion back to electrical power to be fed into the utility grid ("Wind-to-Hydrogen Project" 2021).

8.3.3 Hydroelectric Energy

Hydrogen can be produced from hydroelectric electric energy by converting the hydro energy into electrical power to fed electrolyzers (Edwards et al. 2021). In Canada, the company Air Liquide has built the largest PEM electrolyzer with a total capacity of 20 MW.

8.3.4 Geothermal Energy

Geothermal energy consists of an essential renewable source of electrical energy around the world. So, the production of green hydrogen using geothermal energy represents another effective alternative to green hydrogen (Balta et al. 2010).

8.4 Storage of Hydrogen

The electrical energy to hydrogen energy to electrical energy process can be summarized in three stages: hydrogen production by electrolysis, storage in special tanks, and electricity generation by fuel cells. The hydrogen is converted later into electrical energy to feed fuel cells and produce electrical energy. The storage stage of hydrogen represents a delicate step due to the safety requirements and exigencies. The hydrogen gas storage process is described in Fig. 8.6. The heat exchanger is shown in Fig. 8.6. allowing to minimize the compression work during the compression and intercooling, which allows the increasing of the efficiency of the storage process. Hydrogen can be stored in gas or liquid state. The storage of hydrogen in the gas state shall be done in tanks that can resist high pressure (350–700 Bar). The boiling temperature of hydrogen is -252.8 °C at one-atmosphere pressure. Therefore, cryogenic temperatures are required to perform the storage of hydrogen in the liquid state.

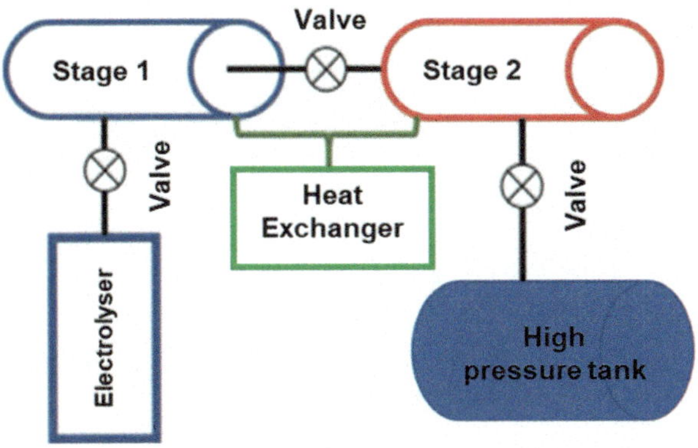

Fig. 8.6 Storage of hydrogen

8.5 The Generation of Electrical Energy from Hydrogen

8.5.1 The Fuel Cells

The Alkaline and the PEM fuel cells are the main fuel cells used in the production of electrical energy from hydrogen. These two types operate at temperatures below 100 °C. Other fuel cells are characterized by electrolytes requiring higher temperatures to become ion-conducting. The two electrodes where the chemical reactions take place are indispensable. These two electrodes emerge in a solution called electrolyte. To produce electrical energy, two chemical elements are required which are hydrogen and oxygen. The hydrogen (H_2) reaches the fuel cell at the anode. Then, a chemical reaction takes place to strip the hydrogen molecules of their electrons. Hence, the atoms become ionized to form the ion H+ . The electrons produced at the anode from the chemical reaction will flow through wires to constitute the electrical current. On the other side, the oxygen reaches the cathode. The oxygen is usually coming from the air. The oxygen attracts the electrons that flow through the wires connecting the anode and the cathode. The voltage produced in a single fuel cell from a chemical reaction is only about 0.7 V. Single cells are usually stacked in series to obtain a higher voltage at the terminal of the fuel cell and make the fuel cell used in residential applications, fuel cell cars, and any other equipment requiring electrical power.

The chemical reactions happening in a single fuel cell stack are as described by the following expressions. At the anode side, an oxidation reaction described by the following equation will take place.

$$H_2 \rightarrow 2H^+ + 2e^- \tag{8.13}$$

At the cathode side, a chemical reaction called a reduction reaction will take place.

$$\frac{1}{2}O_2 + 2H^+ + 2e^- \rightarrow H_2O \tag{8.14}$$

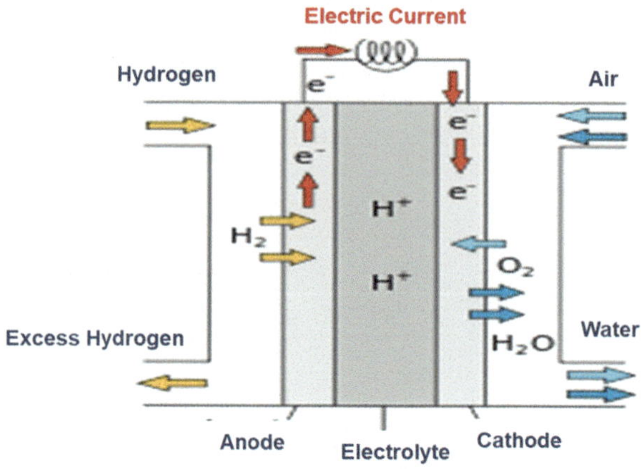

Fig. 8.7 Principle of working of the fuel cell

The net reaction, called the oxidation–reduction or redox reaction, is described by

$$\frac{1}{2}O_2 + 2H^+ \rightarrow H_2O \tag{8.15}$$

Figure 8.7 shows the main components and the fuel cell principle of working.

8.5.2 Comparison of Fuel Cells and Batteries

Fuel cells are distinct from batteries in dependence on integrating elements required to produce electricity. In the battery, there is no external fuel source. Indeed, the battery should be charged and after that connect to the load until a defined discharging point. The fuel cells are not charged or discharged. Indeed, the hydrogen and oxygen required to produce electrical energy are provided from external tanks. This property gives the fuel cell more effectiveness in terms of working hours and avoiding the charging time. Moreover, fuel cells can provide electricity when hydrogen is available, only hydrogen tanks need to be charged. In conventional batteries, the battery components are the basis of energy generation, where the reaction of chemical components of the battery produces electricity. This process continues until the end of the battery reactant chemicals, while the fuel cells operate on a continuous basis as the hydrogen and the oxygen are provided from external sources. The principles of operation made a big difference in terms of lifetime and cost. Therefore, fuel cells have more lifetime and high cost due to their high reliability and efficiency. Moreover, the fuel cells have high efficiency and showed better performances in transportation applications.

8.5.3 Modeling of the Fuel Cell

During the phase of generating current, the voltage of the fuel cell Vcell is less than the voltage produced in open-circuit Vnern (Barhoumi et al. 2020):

$$V_{Cell} = E_0 - \Delta V_{act} - \Delta V_{ohm} - \Delta V_{con} \tag{8.16}$$

The voltage drop is due to the Ohmic losses, activation, and concentration losses. The activation losses are described by the activation overvoltage, Vact:

$$\frac{d\Delta V_{act}}{dt} = \frac{\Delta V_{act} I}{\eta_{act}\, CdI} + \frac{I}{CdI} \tag{8.17}$$

The cathodic activation drop, ηact, is given by the following equation:

$$\eta_{act} = -(\xi_1 + \xi_2\, T + \xi_3 T \ln C_{O2} + \xi_4 T \ln I) \tag{8.18}$$

where:

I: is the current produced by the fuel cell,

$$\xi_1 = -0.948$$
$$\xi_2 = 0.00286 + 0.0002\ \ln(A_{cell}) + 4.3\ \ 10^{-5} \ln(C_{H2})$$
$$\xi_3 = 7.6\ \ 10^{-3}$$
$$\xi_4 = -1.93\ \ 10^{-4}$$

The Ohmic voltage is given by

$$\Delta V_{ohm} = iR_{ohm} = i(R_{mem} + R_e) \tag{8.19}$$

where: $R_{ohm} = \frac{t_m}{\sigma_m}$.

The membrane conductivity is given by Yoo and Lee (2010)

$$\sigma_m = b_1 \exp\left[b_2\left(\frac{1}{303} - \frac{1}{T_{fc}}\right)\right] \tag{8.20}$$

The concentration voltage is expressed by

$$\Delta V_{act} = \frac{RT}{nF} \ln\left(1 - \frac{I_{fc}}{I_{lim}}\right) \tag{8.21}$$

This voltage drop is due to the concentration of reactants consumed during the chemical e reaction.

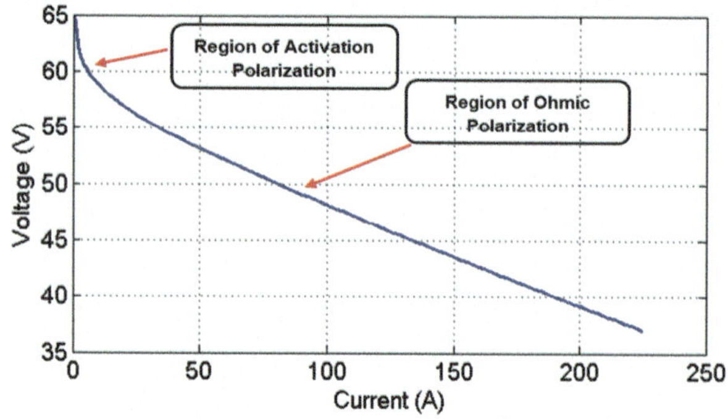

Fig. 8.8 Voltage–current (V–I) characteristic of the fuel cell

I_{lim} is the maximum current density of the cell. The practical characteristic of voltage–current (V–I) of the fuel cell type PEM 1.6 kW is presented in Fig. 8.8.

8.5.4 Characteristics of the PEM Fuel Cell

The output voltage and power of a PEMFC depend on the membrane thickness. To determine the effect of the thickness variation on the output power and the output voltage, a simulation using MATLAB/Simulink was carried out. All parameters are fixed and only, the thickness was varied from 10 to 200 mm. The results presented in (Fig. 8.9) show the variation of the output voltage versus the membrane thickness when the electrical current delivered by the PEMFC is more than 20A. Indeed, for a current equal to 30 A, the voltage decreases from 60 to 30 V when the thickness varies from 10 to 200 mm. This study shows the important effect of the membrane thickness on the output voltage. Hence, a judicious choice for this parameter will be selected depending on the required performances of the PEMFC.

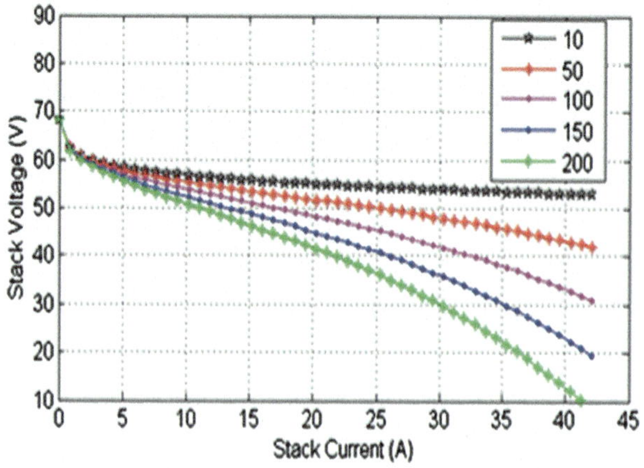

Fig. 8.9 Output voltage versus the current for different values of the membrane thickness

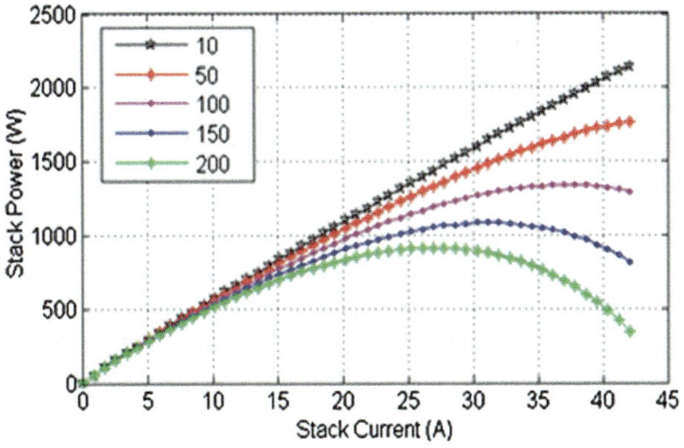

Fig. 8.10 Output power versus the current for different values of the membrane thickness

Figure (8.10) shows the variation of the output power versus the stack current for different membrane thicknesses. The presented results show a maximum output power corresponding to a minimum membrane thickness and maximum stack current.

Figure (8.11) shows the variation of the output voltage versus the hydrogen concentration for different levels of electrical current. The variation of the hydrogen does not affect considerably the output power and voltage. The voltage increases with the same rate when the current increases with the same rate for all levels of the hydrogen concentration.

The variation of the power generated by the fuel cell versus the variation of the temperature is given in Fig. 8.12. The power increases when the temperature increases, especially at high values of generated current. The power reaches a maximum value for a temperature 353 K and a current generated equal to 35A. Hence, it is better to operate at a selected temperature level and optimum current to generate optimum power from the fuel cell.

A fuel cell is an electrochemical machine used for the generation of electrical energy according to a chemical reaction between hydrogen and oxygen. The output

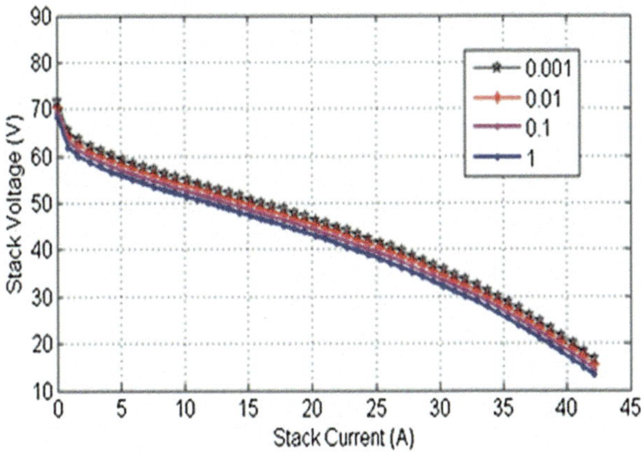

Fig. 8.11 Variation of the output voltage versus the current for different values of hydrogen concentration

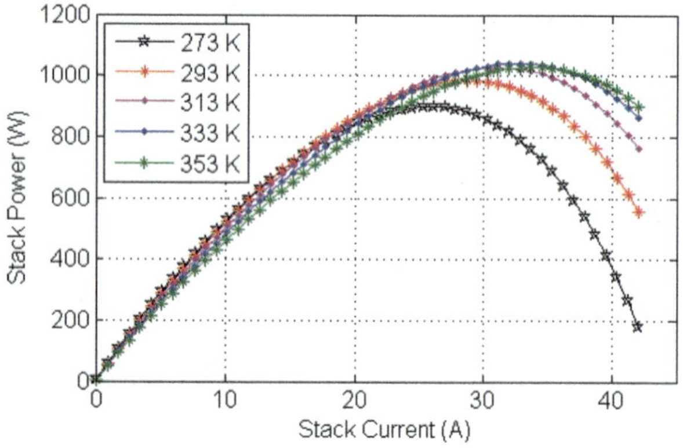

Fig. 8.12 Output power versus the current for different values of the temperature

power generated depends on the geometrical parameters and the pressure and concentration of the hydrogen and the temperature variation. Hence, the output power is variable. To maintain the power and the voltage at a given value, it is usually required to associate the fuel cell stack with a power converter to control the power flow.

8.5.5 Applications of Fuel Cell

Fuel cells were used in the beginning in space applications and then began to evolve and spread more in military circles, where they had been relying on fuel cells as a source of reserves because of their high reliability. Currently, all auto companies developed new hybrid products that use fuel cells in addition to classic engines used to run vehicles. Fuel cells began to spread in buses, trains, planes, and scooters. Fuel cells are also used in small applications such as telephones, cellular phones, and laptops. In the near future, the fuel cells will be used to feed hospitals, banks, and ATMs. The fuel cells will be installed in the waste treatment plants to convert methane gas into electricity. Multiple fuel cell applications do not end with the technical development of the areas of increased use (Barhoumi et al. 2018).

Many auto manufacturers are working to develop new hybrid or fuel cell vehicles (FCV), where fuel cells are used to provide electrical power to electrical motors used in these cars. In addition, many companies are installing commercial hydrogen stations for recharging vehicles with hydrogen gas. All vehicles that run on fuel cells, (General Motors) (Toyota) Daimler-Chrysler)), which consists of about 40% of the total investments in the field of FCV in the world, they are supporting the development of research on fuel cell and hydrogen production, in order to make the hydrogen the best eco-friendly choice. Figure (8.13) shows the main elements used in the FCV. The vehicle contains a fuel tank for the storage of hydrogen, batteries used as a buck up source of electrical energy, and power converters to control the flow of the power from the fuel cell and batteries to the electrical motor.

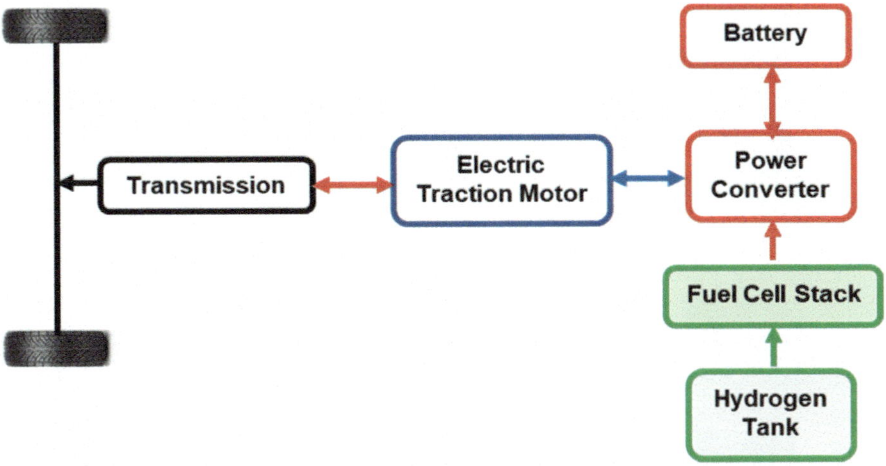

Fig. 8.13 Hydrogen fuel cell car

Fig. 8.14 Portable power applications of fuel cells

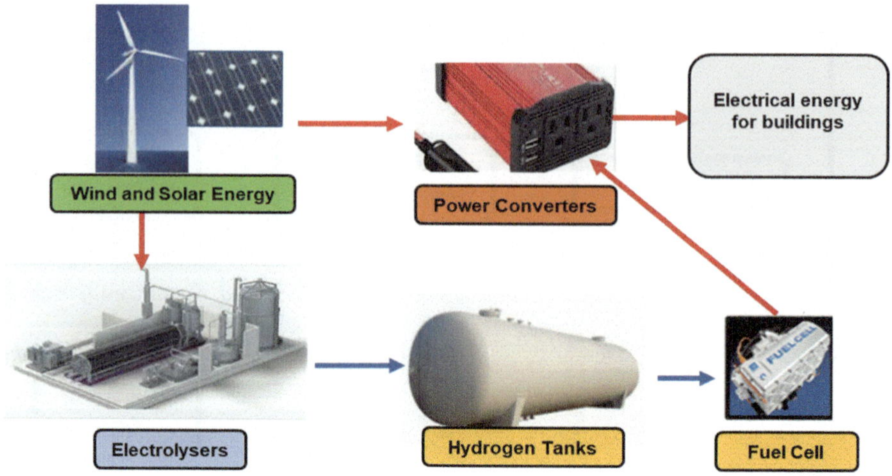

Fig. 8.15 Fixed applications of the fuel cells

Nowadays, fuel cells are used in military planes because of their low noise and compact size. One of the largest companies that developed this technique is the Boeing Company. The fuel cell can provide electrical power to the fed electric power grid in fault cases and emergency situations. Indeed, the fuel cells are more efficient than batteries working longer and lighter. Hence, they are used in a lot of industrial applications (Fig. 8.14).

More than 2500 applications of the fuel cell are developed around the world, in hospitals, hotels, offices, and schools. Fuel cell stations are connected to the public electricity grid to secure the support of the network or independent power stations in areas that are difficult to connect with the network.

Fuel cell systems operate efficiently 40%, without noise, without polluting the air. When using fuel cells in a co-generation power system where taking advantage of the thermal energy produced can increase the efficiency to 85%. Figure (8.15) shows the principle of fixed application of the fuel cell system where the fuel cell is used to produce the electricity required to run the different equipment and lighting in this building. Another advantage of this application is to convert the surplus of electrical energy produced from renewable sources into hydrogen. Hydrogen is used to produce electrical energy during cloudy days and nights.

The fuel cells are used in telecommunications applications to generate electrical energy required by the communications equipment in case of emergency, as shown in Fig. 8.16. With the increased use of computers, the Internet, and telecommunications networks emerged the need for a more reliable source of electrical power. As a source of electrical energy, fuel cells have shown high reliability.

Overall, the fuel cell consists of a reliable source of electrical power for telecommunications systems. They are silent and environmentally friendly and can be designed to be robust bear any ambient conditions. They are currently being used

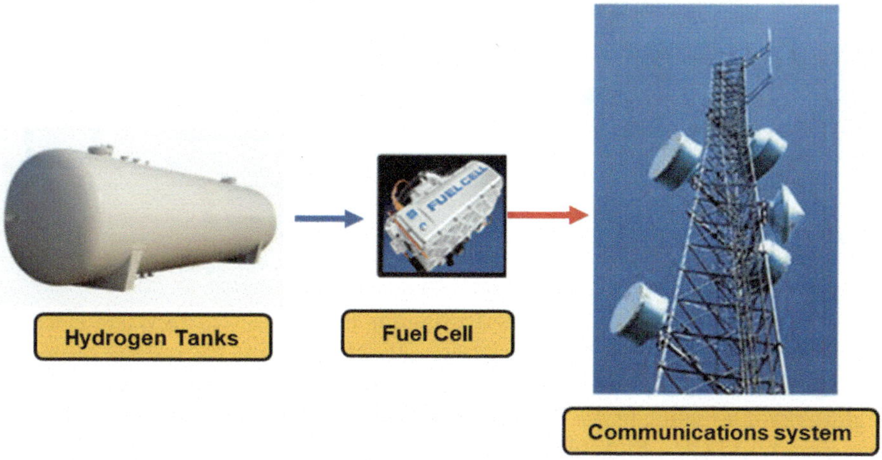

Fig. 8.16 Fuel cell application in the telecommunication system

mainly as a source of energy or support in the communication towers cell towers as the source, point of conversion telecom switch nodes.

The fuel cells are currently installed in waste and water treatment plants all over the world. This technique has proven its ability to reduce emissions where using methane gas produced from these sites as fuel. As it has been installed in several factories for the production of beer, where alcohol is passed untreated breaks organic compounds, and generates methane, which is rich in hydrogen.

8.6 Design of a Photovoltaic-Hydrogen Gas Station

8.6.1 Solar to Hydrogen

The conversion of solar energy into hydrogen energy is possible through the production of electrical energy using PV systems and the production of hydrogen by electrolysis process. The exponential growth of the price of car fuels has pushed researchers and engineers to look at cheap sources of fuels. On the other hand, the production of hydrogen from renewable energy sources leads to a decline in the cost of hydrogen production. In many countries around the world, e.g., Europe and United States of America, renewable energy has been used to produce green hydrogen. In Tunisia, the government sets a plan to produce electrical energy from renewable energy sources. In this context, the implementation of many small-scale and medium-scale PV stations is ongoing. The FCVs consist of a key solution for the transport problems by reducing the cost of fuel prices. The efficiency of such projects depends mainly on the PV potential and the cost of PV panels, MPPT converters, electrolyzers, and storage

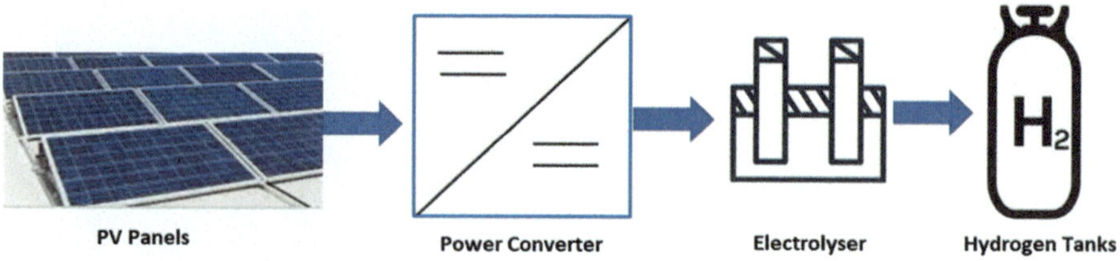

Fig. 8.17 Hydrogen production using PV system

tanks. According to the PV potential distribution in Tunisia, 4 to 5 kWh of electrical energy can be produced daily by installing only 1kWp of the PV system. Moreover, hydrogen production stations using PV systems can be implemented near the main highways for refueling FCV. The main idea is to produce green hydrogen from solar energy to feed FCVs. In the next section, the sizing of the PV system and hydrogen gas station to produce a quantity of hydrogen of 100 kg daily is discussed in detail.

8.6.2 Sizing of Hydrogen Gas Station

The sizing of different components is performed for one hydrogen gas station producing 100 kg of hydrogen daily. The hydrogen PV power station requires the PV system, the power converters, the electrolyzers, and the storage tanks. The process of hydrogen production from solar energy using PV panels is depicted in Fig. 8.17.

- The electrolyzer

A 400 kW PEM electrolyzer with a capacity of production of 5 kg of hydrogen/hour for each electrolyzer is proposed for this example. As the solar energy is approximately available only for 5 h per day, then the electrolyzer will operate only 5 h per day. Therefore, one electrolyzer produces about 25 kg of hydrogen per day. The total number of required electrolyzers is calculated as follows:

$$\mathrm{N}_{\mathrm{Electrolysers}} = \frac{\text{Total required hydrogen}}{\text{Daily capacity of one electrolyser}} = \frac{100}{25} = 4 \tag{8.22}$$

Then, 4 electrolyzers are required to produce 100 kg of hydrogen per day.

With an efficiency of 50%, an electrolyzer requires 80 kWh of electrical energy to produce 1 kg of H2. So, to produce 5 kg of H2 per hour, one electrolyzer requires 400 kWh. The 4 electrolyzers consume 1600 kWh of electrical energy per hour.

- The power converter

The power converters required are DC–DC converters controlled with maximum power point tracking system (MPPT). The DC–DC converters allow to optimize the flow of the power from the PV system to the electrolyzers. The rated power of one electrolyzer is 400 kW. Considering a coefficient of 0.25 to model the losses in the DC–DC converter, then the rated power of one DC–DC converter feeding one electrolyzer is calculated using the following expression:

$$P_{DC-DCconverter} = P_{Electrolyser} \times 1.25 = 400 \times 1.25 = 500\text{kW} \quad (8.23)$$

Each electrolyzer is connected in series with one DC–DC converter. So, 4 (500 kW) DC–DC converters are required. The total input power required for the 4 electrolyzers is 2000 kW.

- The PV panels

The PV panels convert solar energy into electrical energy. Then, the PV panels represent the power source, and they should provide the required power to the load. A coefficient of 0.25 is added to the power of the DC–DC converter for modeling the cable losses, then the rated power of the PV system is calculated as follows:

$$P_{PV-system} = P_{DC-DCconverter} \times N_{DC-DCconverter} \times 1.25 = 500 \times 4 \times 1.25 = 2500\text{kW} \quad (8.24)$$

As many other equipment are required for the hydrogen gas station, like dispensers, compressors, and others, a total capacity of 2662.2 kW of the PV system is proposed to provide electrical energy to all equipment in the hydrogen station.

The technical characteristics of the selected PV panels are given in Table 8.2.

Based on the data given in Table 8.2, the number of required panels is calculated as follows:

$$N_{PV\,panels} = \frac{P_{PV-system}}{P_{Panel}} = \frac{2662,200}{435} = 6120\,\text{panels} \quad (8.25)$$

Then, 6120 panels (435 Wp each) provide the electrical power to the hydrogen gas station. The connection of PV panels is performed according to the requirements of the DC input voltage of the DC–DC converter. In case of the maximum voltage of the DC–DC converter is 1250 V, the maximum number of panels per string is calculated as follows:

Table 8.2 Technical characteristics of PV panels (Canadian_Solar-Datasheet 2021)

Technical characteristics	Specifications
Rated power	435 W
Open circuit voltage	48.6 V
Short circuit current	11.35 A
MPP voltage	40.1 V
MPP current	10.85 A

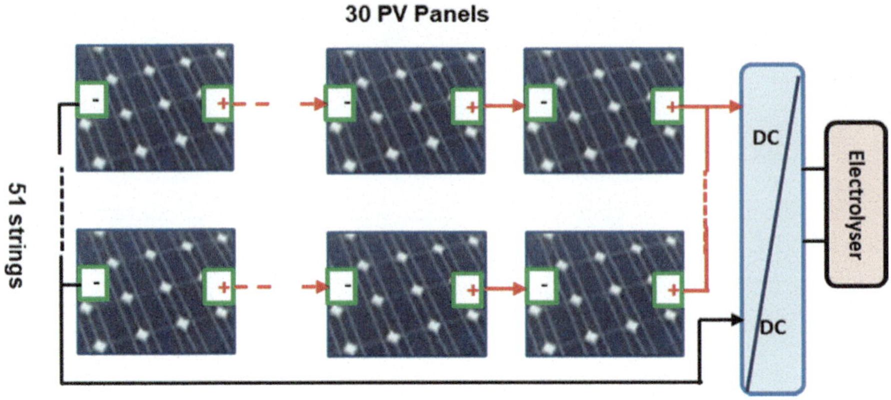

Fig. 8.18 Connection of PV panels to the DC–DC converter

$$N_{\text{PV panels per string}} = \frac{\text{Maximum Voltage of DC} - \text{DC converter}}{\text{Mpp Voltage of PV panel}} = \frac{1250}{40.1} = 31.17 \text{ pa}\quad (8.26)$$

Then, the proposed number of panels connected in series to form one string is 30 panels.

The number of strings connected to each converter depends on the maximum input current of the converter. The maximum input current of the DC–DC converter is 600 A, then, the maximum number of strings is given by Eq. (8.27).

$$N_{\text{strings per converter}} = \frac{\text{Max input current of the converter}}{\text{Mpp current of PV panel}} = \frac{600}{10.85} = 55.29 \text{ strings} \quad (8.27)$$

According to Eq. (8.27), 55 strings are the maximum number of strings that can be connected to each power converter.

The total number of PV panels is distributed between the four converters. Then, 1530 panels are connected to each converter. So, for each converter, the PV panels will be distributed into 51 strings connected in parallel, each string is formed by 30 PV panels connected in series. The connection of PV strings to DC–DC converters and electrolyzers is clarified in Fig. 8.18. The hydrogen station formed by 4 electrolyzers, 4 DC–DC converters, and PV systems is given in Fig. 8.19.

8.7 Conclusion

This chapter has discussed the life cycle of hydrogen. Indeed, the electricity generated from renewable energy sources is converted into hydrogen through the electrolysis

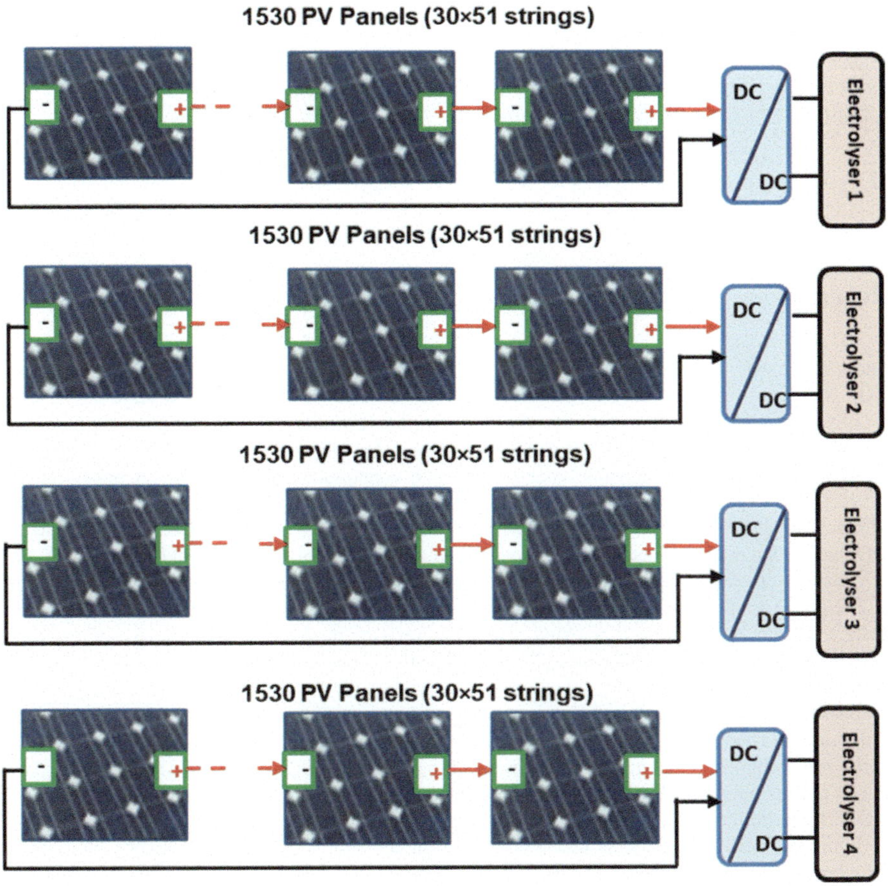

Fig. 8.19 Connection of all PV panels to the DC–DC converters and electrolyzers

process. The hydrogen is converted into electrical energy using PEMFC. The main electrolyzers used to produce hydrogen are the PEM, the solid oxide, and the alkaline electrolyzers. Among these three electrolyzers, the alkaline electrolyzer has the highest efficiency.

The hydrogen produced from renewable energy sources represents an effective solution for the storage of the surplus of electrical energy produced from renewable energy sources. Indeed, the hydrogen is converted into electrical energy using fuels cells. The applications of hydrogen are continuously increasing to cover a wide range of industrial applications, mobile applications, and transportation sectors.

The principle of operation, the characteristics of the PEMFC were presented and analyzed in this chapter. The PEMFC power and output voltage depend on different parameters, especially the operation temperature, the hydrogen concentration, and the membrane thickness.

To shed the light on the principle of sizing of the PV systems for hydrogen production, the design of a hydrogen gas station was presented and discussed in this chapter. The obtained results showed that a standalone PV system with a total capacity of 2662.2 Wp is required to provide electricity to the hydrogen station. The PV system produces about 13.311 MWh of electrical energy per day. The quantity of hydrogen produced per day is 50 kg.

References

Abdin Z, Webb CJ, MacA E, Gray (2015) Modelling and simulation of a proton exchange membrane (PEM) electrolyser cell. Int J Hydrog Energy 40(39):13243–13257, Oct. https://doi.org/10.1016/j.ijhydene.2015.07.129

Abdin Z, Mérida W (2019) Hybrid energy systems for off-grid power supply and hydrogen production based on renewable energy: a techno-economic analysis. Energy Convers Manag 196:1068–1079, Sep. https://doi.org/10.1016/j.enconman.2019.06.068

Albadi MH, Al-Hinai AS, Maharbi MJA, Hosni AMA, Hajri MAA (2019) Economic dispatch of Oman's main interconnected system in presence of 500MW solar PV plant in Ibri. In: 2019 IEEE Jordan International Joint Conference on Electrical Engineering and Information Technology (JEEIT), Apr., pp. 204–208. https://doi.org/10.1109/JEEIT.2019.8717418.

Balta MT, Dincer I, Hepbasli A (May 2010) Potential methods for geothermal-based hydrogen production. Int J Hydrog Energy 35(10):4949–4961. https://doi.org/10.1016/j.ijhydene.2009.09.040

Barbir F (2005) PEM electrolysis for production of hydrogen from renewable energy sources. Sol Energy 78(5):661–669, May. https://doi.org/10.1016/j.solener.2004.09.003

Barhoumi E, Ben Belgacem I, Khiareddine A, Zghaibeh M, Tlili I (2018) A neural network-based four phases interleaved boost converter for fuel cell system applications. Energies 11(12):3423, Dec. https://doi.org/10.3390/en11123423

Barhoumi EM et al (2019) Renewable energy resources and workforce case study Saudi Arabia: review and recommendations. J Therm Anal Calorim, Dec. https://doi.org/10.1007/s10973-019-09189-2

Barhoumi EM, Chukwuleke Okonkwo P, Belgacem IB, Zghaibeh M (2020) MPPT control of an interleaved boost converter for a polymer electrolyte membrane fuel cell applications. In: 2020 International Conference on Electrical and Information Technologies (ICEIT), Mar, pp. 1–5. https://doi.org/10.1109/ICEIT48248.2020.9113228

Barhoumi EM, Farhani S, Okonkwo PC, Zghaibeh M, Bacha F (2021) Techno-economic sizing of renewable energy power system case study Dhofar Region-Oman. Int J Green Energy 1–10, Feb. https://doi.org/10.1080/15435075.2021.1881899

Boran FE, Boran K, Menlik T (2012) The evaluation of renewable energy technologies for electricity generation in Turkey using intuitionistic fuzzy TOPSIS. Energy Sources Part B Econ. Plan Policy 7(1):81–90, Jan. https://doi.org/10.1080/15567240903047483

Boulmrharj S, Khaidar M, Bakhouya M, Ouladsine R, Siniti M, Zine-dine K (2020) Performance assessment of a hybrid system with hydrogen storage and fuel cell for cogeneration in buildings. Sustainability 12(12):4832, Jun. https://doi.org/10.3390/su12124832

Canadian_Solar-Datasheet-BiHiKu_CS3W-PB-AG_High-Efficiency_1000V1500V_EN-2.pdf (2021) Accessed 21 Apr, 2021. https://www.canadiansolar.com/wp-content/uploads/2019/12/Canadian_Solar-Datasheet-BiHiKu_CS3W-PB-AG_High-Efficiency_1000V1500V_EN-2.pdf

Chi J, Yu H (2018) Water electrolysis based on renewable energy for hydrogen production. Chin J Catal 39(3):390–394, Mar. https://doi.org/10.1016/S1872-2067(17)62949-8

Christensen, A (2020) Assessment of hydrogen production costs from electrolysis: United States and Europe, p. 73

David M, Ocampo-Martínez C, Sánchez-Peña R (2019) Advances in alkaline water electrolyzers: a review. J. Energy Storage 23:392–403, Jun. https://doi.org/10.1016/j.est.2019.03.001

Duggal I, Venkatesh B (2015) Short-term scheduling of thermal generators and battery storage with depth of discharge-based cost model. IEEE Trans Power Syst 30(4):2110–2118, Jul. https://doi.org/10.1109/TPWRS.2014.2352333

Edwards RL, Font-Palma C, Howe J (2021) The status of hydrogen technologies in the UK: a multi-disciplinary review. Sustain. Energy Technol. Assess. 43, Feb. https://doi.org/10.1016/j.seta.2020.100901

El Ouderni AR, Maatallah T, El Alimi S, Ben Nassrallah S (2013) Experimental assessment of the solar energy potential in the gulf of Tunis, Tunisia. Renew. Sustain. Energy Rev., vol. 20, pp. 155–168, Apr. https://doi.org/10.1016/j.rser.2012.11.016

Givler T, Lilienthal P (2005) Using HOMER software, NREL's micropower optimization model, to explore the role of gen-sets in small solar power systems; case study: Sri Lanka. National Renewable Energy Lab., Golden, CO (US), NREL/TP-710–36774, May. https://doi.org/10.2172/15016073

Gondal IA, Masood SA, Khan R (2018) Green hydrogen production potential for developing a hydrogen economy in Pakistan. Int J Hydrog Energy 43(12):6011–6039, Mar. https://doi.org/10.1016/j.ijhydene.2018.01.113

Hammad M, Ebaid MSY (2015) Comparative economic viability and environmental impact of PV, diesel and grid systems for large underground water pumping application (55 wells) in Jordan. Renew. Wind Water Sol. 2(1):12, Jul. https://doi.org/10.1186/s40807-015-0012-2

HassanzadehFard H, Tooryan F, Collins ER, Jin S, Ramezani B (2020) Design and optimum energy management of a hybrid renewable energy system based on efficient various hydrogen production. Int J Hydrog Energy 45(55):30113–30128, Nov. https://doi.org/10.1016/j.ijhydene.2020.08.040

Marano V, Rizzo G, Tiano FA (2012) Application of dynamic programming to the optimal management of a hybrid power plant with wind turbines, photovoltaic panels and compressed air energy storage. Appl Energy 97:849–859, Sep. https://doi.org/10.1016/j.apenergy.2011.12.086

Ni M, Leung MKH, Leung DYC (2008) Technological development of hydrogen production by solid oxide electrolyzer cell (SOEC). Int J Hydrog Energy 33(9):2337–2354, May. https://doi.org/10.1016/j.ijhydene.2008.02.048

Nieminen J, Dincer I, Naterer G (2010) Comparative performance analysis of PEM and solid oxide steam electrolysers. Int J Hydrog Energy 35(20):10842–10850, Oct. https://doi.org/10.1016/j.ijhydene.2010.06.005

Nikolaidis P, Poullikkas A (2017) A comparative overview of hydrogen production processes. Renew Sustain Energy Rev 67:597–611, Jan. https://doi.org/10.1016/j.rser.2016.09.044

Okonkwo PC et al. (2021) Platinum degradation mechanisms in proton exchange membrane fuel cell (PEMFC) system: a review. Int J Hydrog Energy 46(29):15850–15865, Apr. https://doi.org/10.1016/j.ijhydene.2021.02.078

Razmjoo A, Gakenia Kaigutha L, Vaziri Rad MA, Marzband M, Davarpanah A, Denai M (2021) A technical analysis investigating energy sustainability utilizing reliable renewable energy sources to reduce CO2 emissions in a high potential area. Renew. Energy, vol. 164, pp. 46–57, Feb. https://doi.org/10.1016/j.renene.2020.09.042

Schnuelle C, Wassermann T, Fuhrlaender D, Zondervan E (2020) Dynamic hydrogen production from PV & wind direct electricity supply – Modeling and techno-economic assessment. Int J Hydrog Energy 45(55):29938–29952, Nov. https://doi.org/10.1016/j.ijhydene.2020.08.044

Shaner MR, Atwater HA, Lewis NS, McFarland EW (2016) A comparative technoeconomic analysis of renewable hydrogen production using solar energy. Energy Environ Sci 9(7):2354–2371. https://doi.org/10.1039/C5EE02573G

Shiva Kumar S, Himabindu V (2019) Hydrogen production by PEM water electrolysis–a review. Mater. Sci. Energy Technol, vol. 2, no. 3, pp. 442–454, Dec. https://doi.org/10.1016/j.mset.2019.03.002.

Wang L et al. (2019) Power-to-fuels via solid-oxide electrolyzer: operating window and techno-economics. Renew Sustain Energy Rev 110:174–187. https://doi.org/10.1016/j.rser.2019.04.071
Wind-to-Hydrogen Project (2021) https://www.nrel.gov/hydrogen/wind-to-hydrogen.html, 20 Apr 2021
Xiao P, Hu W, Xu X, Liu W, Huang Q, Chen Z (2020) Optimal operation of a wind-electrolytic hydrogen storage system in the electricity/hydrogen markets. Int J Hydrog Energy 45(46):24412–24423, Sep. https://doi.org/10.1016/j.ijhydene.2020.06.302
Yan Z, Hitt JL, Turner JA, Mallouk TE (2020) Renewable electricity storage using electrolysis, p. 6
Yoo S-H, Lee J-S (2010) Electricity consumption and economic growth: a cross-country analysis. Energy Policy 38(1):622–625, Jan. https://doi.org/10.1016/j.enpol.2009.05.076

Chapter 9
Planning and Impact of Electric Vehicle Charging Stations in Distribution System Using Optimization Techniques

Ramendra Kumar Rai, Aashish Kumar Bohre, Pradyumn Chaturvedi, Mohan Lal Kolhe, and Sri Niwas Singh

Abstract Global warming concerns have led developing as well as developed countries to promote the use of clean & green energy more than ever. Electric vehicle seems to be one of the most adaptable option to counter the concern. But to tackle environmental issues with electric vehicles, there are other factors that need to be addressed and understood. This chapter will showcase the impact of electric vehicle penetration on the power distribution grid and how the impact can be minimized using an optimization technique. The effects of introducing electric vehicles on the voltage profile, cost of generation, active and reactive power losses are discussed in view of emerging markets like India. A 69–bus system is considered for this study and different loads are applied to each branch of the system. A gradual increase in the load is provided and using AC- Optimal Power Flow (OPF) method, behaviour and impacts of different parameters is calculated and compared graphically. Finally, the Artificial Bee Colony (ABC) optimization technique has been employed to find the optimal location for EV charging stations Graphical comparisons are also done between base cases, scenario cases and systems with EV charging stations to reach a conclusion.

R. K. Rai (✉) · A. K. Bohre
National Institute of Technology, Durgapur, India
e-mail: ramendra.ee@gmail.com

A. K. Bohre
e-mail: aashishkumar.bohre@ee.nitdgp.ac.in

P. Chaturvedi
Visvesvaraya National Institute of Technology (VNIT), Nagpur, India
e-mail: pradyumn.c@eee.vnit.ac.in

M. L. Kolhe
University of Agder, Norway, India
e-mail: mohan.l.kolhe@uia.no

S. N. Singh
Indian Institute of Technology (IIT), Kanpur, India

A. K. Bohre et al. (eds.), *Planning of Hybrid Renewable Energy Systems, Electric Vehicles and Microgrid*, Energy Systems in Electrical Engineering,
https://doi.org/10.1007/978-981-19-0979-5_9

Keywords Electric Vehicle (EV) · Optimal power flow · Distribution system · EV load penetration · Artificial Bee Colony (ABC)

Nomenclature

ABC	Artificial Bee Colony
DG	Distributed Generation
ENS	Energy Not Supplied
ICE	Internal Combustion Engine
OPF	Optimal Power Flow
THD	Total Harmonic Distortion
QGA	Quantum Genetic Algorithm
BSS	Battery Swap Station
EV	Electric Vehicle
GHG	Green House Gases
EVCS	Electric Vehicle Charing Station
BPSO	Butterfly Particle Swarm Optimization
LMP	Locational Price Margin
QPmR	Quasi Polynomial mapping based Rootfinder

9.1 Introduction

With the increasing concern of global warming around the world, Governments are promoting the use of clean & green energy. Electric vehicles seem to be a viable option in the transportation sector to support the cause. Electric vehicles have major advantages over Internal Combustion Engine (ICE) vehicles, such as reduced noise from engine and zero to none greenhouse gas (GHG) release. Column-chart shown below in Fig. 9.1, highlights the carbon emission data (2020) of different countries around the world (www.ucsusa.org/resources/each-countrys-share-co2-emissions). India accounts for ~ 7% of the world's total emission. These stats call for stringent policies around carbon emission with a strong base for implementation as well. A number of lucrative policies are being proposed and enacted regarding the induction of electric vehicles, which has increased participation from private entities and major organizations from around the world. But, a huge induction of electric vehicles in power distribution networks can have effects that are adversative in terms of - increment in energy losses, voltage drops, power quality issues, non-desired load peaks, overload on grid components, reliability indices deterioration and load factor reduction (Deb et al. 2018). Various studies have been done around interfacing of electric vehicles with power distribution networks. The IEEE Xplore database, which is a premium database for worldwide scientific research, consists of strongly addressed

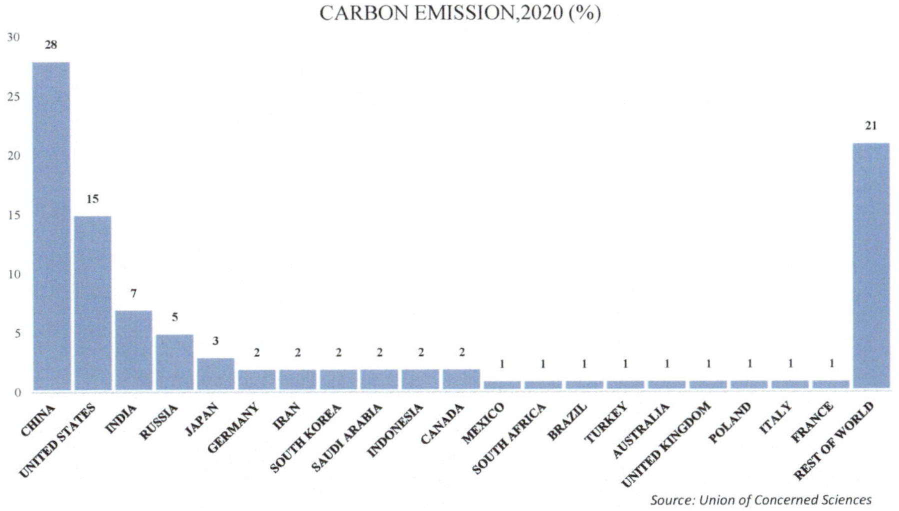

Fig. 9.1 Net carbon emission of different countries, 2020

related topics of more than fifteen years (Salihi 1973). Few literature reviews highlight the use of electric vehicles as a distribution system services provider, broadly categorized in active/reactive power provision and source integration of renewable energy aid, while few of them point out the limitations in control tactics of various methods employed in order to boost the study of new aspects associated with the existing necessities of smart grids. In the literature review (Heydt 1983), a thorough review to assess and alleviate the effects of charging electric vehicles on residential distribution systems has been done.

Adverse effects of electric vehicles on power distribution networks are the most speculated topics by researchers in the last half-decade. These effects include but are not limited to, overload on power grid gears/components, voltage drops, non-desired load peaks, rise in energy losses, reliability indices deterioration, power quality problems and reduction in load factors (Arias-Londoño et al. 2020). Various studies shows the analysis of real-time impacts on power grids due to insertion of electric vehicles as load.

Electric vehicle components are considered non-linear and sources of harmonic signals (Karmaker et al. 2019). These signals affect power quality. Reliability and harmonics levels have also been discussed in terms of Total Harmonic Distortion (THD) in several research papers. In this sense, (Bohre et al. 2015) has presented work on optimal location and sizing of distributed generation (DG) using optimal power flow (OPF) and Butterfly-particle swarm optimization (PSO) methods. Various optimization techniques have been used to optimize the output of similar case studies (Kumar et al. 2020). In (Singh and Bohre 2020), real and reactive power losses, voltage and economic aspects have been considered in order to establish an efficient charging location using PSO as an optimization technique. Huang et al. (2018) has

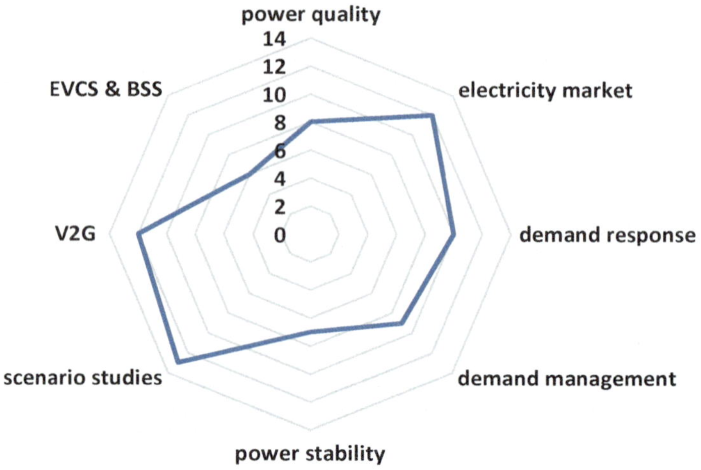

Fig. 9.2 Different aspects of electric vehicle impacts

discussed the Electric Vehicle Charging Station (EVCS) planning (sizing and siting) from the economic benefit point of view using Quantum genetic Algorithm (QGA). In (Naveed et al. 2020), a graphical method describing various stability conditions is performed and impact of electric vehicles, qualitatively, on stability margin is analyzed using Quasi Polynomial mapping based Rootfinder (QPmR) algorithm. In paper (Bastida-Molina et al. 2021), temporal valleys-based technique is used to avoid peak hour demand for the charging of EVs in Spain in order to reduce the load on the grid. Different aspects of study in Arias-Londoño et al. (2020) can be broadly put in the categories as shown in Fig. 9.2 along with the number of research studies considered in each aspect.

9.2 Methodologies

To assess the impact of Electric vehicles on various grid parameters, AC-OPF (Optimal Power Flow) method has been used. Data points related to Real power (sending & receiving end), Reactive power (sending & receiving end), Generated power, Voltage at sending and receiving end, Bus voltage angle, Real & Reactive power losses, etc. are obtained using AC-OPF.

After the real and reactive power at both ends is obtained, nodal cost estimation is performed using a mathematical model shown and discussed below.

Due to failures in transmission lines and other factors, the volume of electricity generated is not transferred totally to the consumer, this lost energy is termed as Energy Not Supplied (ENS). Now, Energy Not Supplied (ENS) and reliability of the system are calculated to measure the efficacy of the distribution network.

At last, ABC optimization technique has been employed to find the optimal location for EV charging stations. This will improve the voltage profile and minimize the overall losses of the system as can be seen from the comparison tables below in the results and discussion section.

9.2.1 AC- optimal Power Flow

In order to minimize the generation cost the optimal power flow (OPF) method is used, where a total generation cost is defined as a 2nd order polynomial generation cost function (Bohre et al. 2015; Zimmerman and Murillo-Sanchez 2011). On the basis of active and reactive power generation cost, the cost function may be defined as:

$$\mathrm{F}(P_g) = \mathrm{m} + \mathrm{n} \times P_g + \mathrm{r} \times P_g^2 \tag{9.1}$$

$$\mathrm{F}(Q_g) = \mathrm{m}' + \mathrm{n}' \times Q_g + \mathrm{r}' \times Q_g^2 \tag{9.2}$$

$$F_f = \min \sum_{k=1}^{n} (\mathrm{F}(P_{gk}) + \mathrm{F}(Q_{gk}) \tag{9.3}$$

Power balancing equation at i^{th} bus (DG) without distributed generation is:

$$\sum P_{Di} + \sum P_{li} - \sum_{k=1}^{n} |V_i|V_k||Y_{ik}|\cos(\theta_{ik} + \delta_k - \delta_i) \tag{9.4}$$

$$\sum Q_{Di} + \sum Q_{li} - \sum_{k=1}^{n} |V_i||V_k||Y_{ik}|\cos(\theta_{ik} + \delta_k - \delta_i) \tag{9.5}$$

Constraints related to with-DG is given by:

$$\sum P_{Di} + \sum P_{li} - P_{DGi} \sum_{k=1}^{n} |V_i||V_k||Y_{ik}|\cos(\theta_{ik} + \delta_k - \delta_i) \tag{9.6}$$

$$\sum Q_{Di} + \sum Q_{li} - Q_{DGi} - \sum_{k=1}^{n} |V_i||V_k||Y_{ik}|\cos(\theta_{ik} + \delta_k - \delta_i) \tag{9.7}$$

Here,
$\delta_{min} \leq \delta_i \leq \delta_{max}$; $V_{min} \leq V_i \leq V_{max}$; $P_{min} \leq P_i \leq P_{max}$; $Q_{min} \leq Q_i \leq Q_{max}$

$P_{DG\min} \leq P_{DGi} \leq P_{DG\max}; \; Q_{DG\min} \leq Q_{DGi} \leq Q_{DG\max}$
where,
P_i = Real power flow at ith bus
Q_i = Reactive power flow at ith bus
P_{Di} = Real power demand at ith bus
Q_{Di} = Reactive power demand at ith bus
V_i = Voltage magnitude at ith bus
V_k = Voltage magnitude at kth bus
P_{DGi} = Real power flow at ith bus
P_i = Real power flow at ith bus
δ_i, δ_k = bus voltage angle at ith & kth bus
Y_{ik} = $ikth$ element in bus admittance matrix
θ_{ik} = $ikth$ element's angle in bus admittance matrix
n = Total number of buses.

9.2.2 Real & Reactive Power (Combined) for Nodal Cost Estimation

The reactive to real power demand will be constant, considering i^{th} dispatchable load to be modelled as constant power factor. The uniform nodal cost estimation is represented based on per MW or per MVAr (Zimmerman and Murillo-Sanchez 2011). Let's suppose load is at bus i and cost of real and reactive power are $\lambda_{Pi} and \lambda_{Qi}$, respectively. Now the combined power ψ can be written as:

$$\Psi = \lambda_{Pi} P_{Di} + \lambda_{Qi} Q_{Di} \tag{9.8}$$

$$\Psi = \lambda_{Pi} P_{Di} + \lambda_{Qi} \frac{Q_{Di}}{P_{Di}} P_{Di} \tag{9.9}$$

$$\Psi = (\lambda_{Pi} + \lambda_{Qi} k_i) P_{Di} \tag{9.10}$$

where, $k_i = \frac{Q_{Di}}{P_{Di}}$ = constant.

9.2.3 Energy not Supplied (ENS) & Reliability

Energy Not Supplied (ENS) is the volume of energy lost due to reasons like faults or failures in the network while supplying electricity to customers. It is a key factor in the calculation of Reliability. Energy Not Supplied (ENS) can be calculated as:

$$ENS = Total\,Generation\,(S_G) - Total\,Load\,Demand\,of\,the\,system \tag{9.11}$$

After the energy not supplied (ENS) is calculated, Reliability of the 69–Bus radial distribution system can be obtained using the formula:

$$Reliability = 1 - \frac{ENS}{Total\,Load\,Demand\,(S_D)} \tag{9.12}$$

9.2.4 *Artificial Bee Colony (ABC) Optimization Technique*

Artificial Bee Colony optimization technique is implemented to choose the optimal location of EV charging stations in our 69–bus radial distribution system. The ABC method employs four phases, which are discussed below in detail. Initialization is done and input parameters are decided. An objective function is made considering loss index function as (f1) and voltage penalty function as (f2). W1 and w2 are weighing factors with values 0.6 and 0.4, respectively.

$$Objective\,Function = (w1 * f1) + (w2 * f2) \tag{9.13}$$

The ABC method aims to minimize the objective function and obtain the optimal location for EV charging stations. Our main purpose for using the optimization is to improve the voltage profile and minimize the system overall losses.

9.3 Results and Discussion

The algorithm used in this study is implemented on 69–Bus radial distribution systems. Base MVA throughout the study is 10 MVA. Methodology proposed in this paper is processed in MATLAB (2018a)/Matpower 7.0 tool with system specifications being Windows 10, Intel Core i5, 1.8Ghz, 4 GB RAM.

9.3.1 *Analysis of 69–Bus Radial Distribution System—Based on Load Growth Method*

This radial system has a base load of 3.8 MW and 2.69 MVAr real and reactive power loads, respectively. With base load, real and reactive power losses of 69–Bus systems are 180 kW and 80 kVAr, respectively. All the data related to branch and load of 69–Bus system is given in (Acharya et al. 2006, Murthy and Kumar 2013;

Table 9.1 Comparative analysis of 69–Bus radial distribution system parameters

Case	Overall load (kVA)	Overall losses (kVA)	Overall generation (MVA)	Generation cost ($/hr)	Reliability (%)
Base case	4655.7	196.97	4.86	79.65	94.88
SCENARIO 1	5126.99	243.1	5.38	88.07	94.28
SCENARIO 2	5593.08	293.66	5.88	96.65	93.67
SCENARIO 3	6059.17	349.96	6.40	105.23	93.04
SCENARIO 4	6525.26	412.4	6.90	113.97	92.39

Table 9.2 Comparative analysis of voltage profile and real and reactive power losses

Case	Maximum voltage (p.u.), Bus number	Minimum voltage (p.u.), Bus number	Real power loss (kW)	Reactive power loss (kVAr)
Base case	1.1, 1	1.019, 65	180	80
SCENARIO 1	1.1,1	1.010,65	221.3	100.6
SCENARIO 2	1.1,1	1.001,65	267.4	121.4
SCENARIO 3	1.1,1	0.992, 65	318.7	144.6
SCENARIO 4	1.1,1	0.982,65	375.6	170.3

Mishra 2015; Gopiya Naik et al. 2015). Mixed load models have been considered in 69–Bus radial distribution system. The results for this system with mixed load scenarios are given in Table 9.1. And further, changes in voltage profile and real & reactive power losses in each case is shown below in Table 9.2.

Four cases have been analyzed with respect to the base load scenario. The active (kW) and reactive power (kVAr) losses, voltage profile (p.u.), generation cost ($/hr), energy not supplied (ENS) and reliability percentage have been analyzed in all the four cases by tabular method as well as graphically. The energy not supplied (ENS) and reliability percentage for base load scenario is found to be 194.6 kVA and 94.88%, respectively.

All the four cases under the study are as defined below:

SCENARIO 1 – 10% Penetration of EV load with base system load.

SCENARIO 2 – 20% of base load increment.

SCENARIO 3 – 30% of base load increment.

SCENARIO 4 – 40% of base load increment.

Load is varied with respect to base load of the 69–Bus system. This load variation on different buses is shown in Fig. 9.3. Maximum load is observed on bus 56, which is 1681.27 kVA in base case and 2139.79 kVA in case of 40% increment in load as electric vehicle load. As the electric vehicle load is increased with respect to base load, voltage profile is changed. This variation in voltage profile is shown in Fig. 9.4.

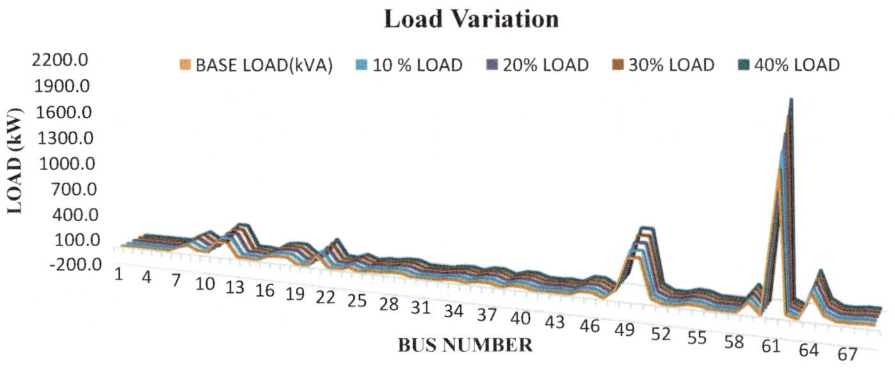

Fig. 9.3 Branch wise load variation profile

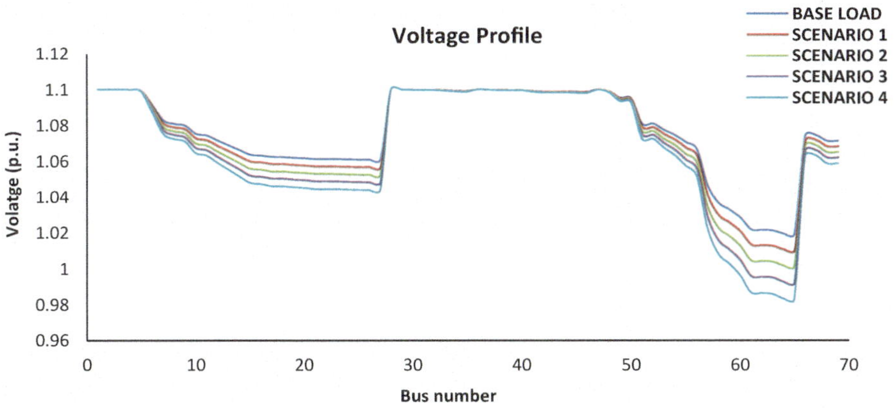

Fig. 9.4 Voltage profile for different load scenario

Major variation in voltage profile can be seen on bus numbers from 10 to 27 and 56 to 66. Maximum voltage is observed on bus 1, i.e. 1.1 p.u.in each case and minimum voltage is obtained at bus 65 which is 0.982 p.u. in case of scenario 4. Losses attached with each bus have increased as the electric vehicle load is increased. These losses are shown bus wise in Fig. 9.5. Maximum loss is observed on bus 56 which is 41.774 kVA in case of base load and 87.880 kVA in case of scenario 4. The economic aspect of the study shows an increase of 43% in electricity production cost from base case to scenario 4 where 40% load is imparted as electric vehicle load. The cost of generation for base case is found to be 79.65 \$/hr and 113.97 \$/hr for scenario 4. Reliability of the system is also decreased as the load is increased. Base case reliability is found to be 94.88% whereas scenario 4 case reliability is reduced by 2.69% to 92.39%. Results in Fig. 9.6 show the graphical comparison of load increment, losses induced, cost of electricity generation and reliability of the system of all the cases that are under study in this paper. So, we can see that a gradual increase in load results in

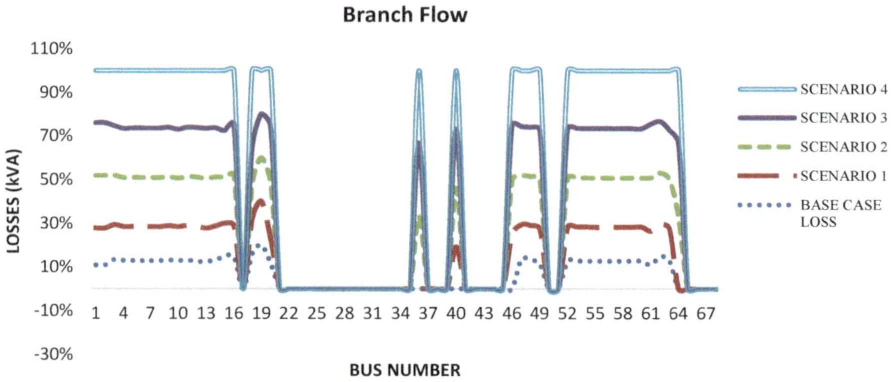

Fig. 9.5 Branch flow with different load scenario

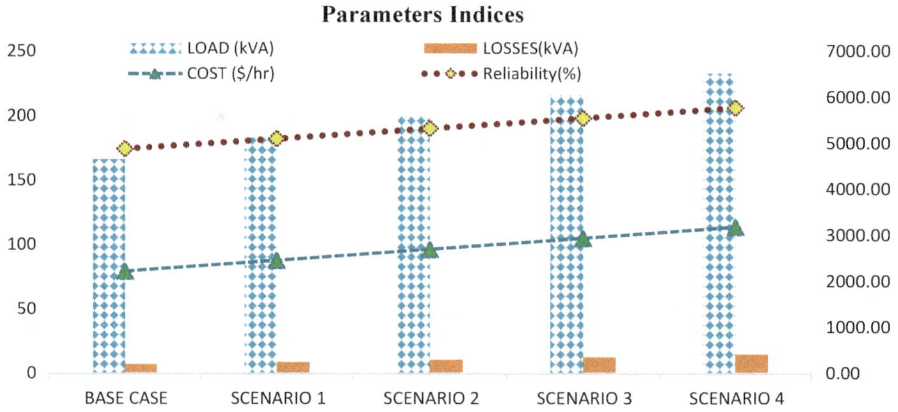

Fig. 9.6 Post load increment analysis of parameter indices

increasing overall line losses, overall generation cost and decreases the reliability of the system. These impacts are later minimized by finding the optimal location of EV charging stations using an optimization technique.

9.3.2 Analysis of 69–Bus Radial Distribution System with EV Charging Station Using ABC Optimization Technique

In our study, Artificial Bee Colony (ABC) optimization technique is used to find the optimal location of the EV charging stations on different feeder line of our study system of 69–bus radial distribution system.

Assumptions:

An EV charging station of size 0.93 MW is considered during our study, which is capable of charging around 10 T Model S electric vehicle at a time which have a battery capacity of 85 kWh with charge efficiency of 92%.

The Artificial Bee Colony (ABC) algorithm is a swarm-based meta-heuristic algorithm that was introduced by Karaboga in 2005 for optimizing numerical problems. It consists of four phases-

(a) Initialization phase
(b) Employed phase
(c) Onlooker phase
(d) Scout phase.

Flow chart shown below in Fig. 9.10, shows the different phases and steps employed in ABC algorithm technique. Using this technique, optimal locations for the EV charging stations have been found, which in turn have improved the voltage profile as can be seen in Fig. 9.7. and minimized the line losses even after the load is increased. Table 9.3 shows the comparative analysis of base case, scenario 4 case and system with EV charging station after optimization. It can be seen that reliability has increased to 97.45% after the optimization process. Overall losses have been decreased to 199.4 kVA despite an increase in overall load from scenario 4. Table 9.4 compares the real & reactive power losses before and after the optimization is done. Figure 9.8 and Fig. 9.9 show the comparison between base case real and reactive power losses and line losses with EV charging stations. A slight increase in losses can be observed at certain busses of the 69–bus distribution system. Cost of generation is also increased from 79.65 $/hr to 154.06 $/hr after the EV charging stations are deployed at different feeder lines of the radial distribution network.

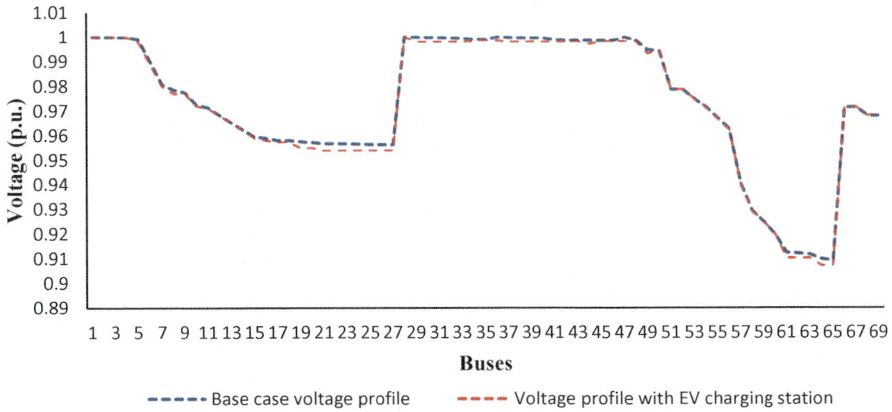

Fig. 9.7 Voltage profile comparison

Table 9.3 System parameters with optimal planning of EV charging station using ABC method

Case	Overall load (kVA)	Overall losses (kVA)	Overall generation (MVA)	Generation cost ($/hr)	Reliability (%)
Base case	4660.9	198	4.88	79.65	95.29
SCENARIO 4	6525.26	412.4	6.90	113.97	92.39
System with EV charging station	7986.6	199.4	8.19	154.06	97.45

Table 9.4 Comparative analysis of voltage profile, real and reactive power losses with optimal planning of EV charging station using ABC method

Case	Maximum voltage (p.u.), Bus number	Minimum voltage (p.u.), Bus number	Real power loss (kW)	Reactive power loss (kVAr)
Base case	1.1, 1	1.019, 65	180	80
SCENARIO 4	1.1,1	0.982,65	375.6	170.3
System with EV charging station	1.1, 1	1.019, 65	181	83.8

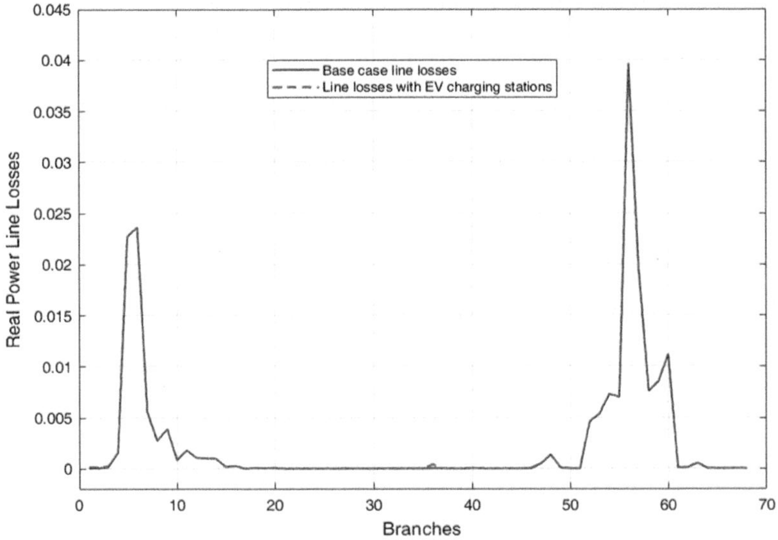

Fig. 9.8 Real power loss comparison graph

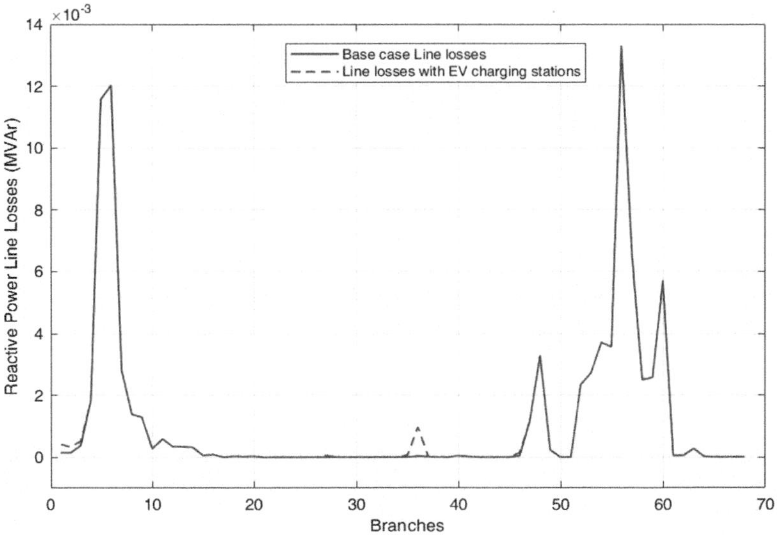

Fig. 9.9 Reactive power loss comparison graph

9.4 Conclusion

Induction of electric vehicles in emerging market, like India, is likely to receive positive feedback. With government aiming to build the required infrastructure by 2025, it is important to understand the power grid performance parameters before a huge electric vehicle induction hits the market. EVs will have a huge impact on distribution as well as transmission system performance. This study is done to present an overview of how different power grid indices like voltage profile, losses associated with different buses of 69–Bus system, cost of generation and reliability vary when the electric vehicle is inducted as load into the system. A comparative analysis is also done between the power grid parameters, which shows the increment in losses as well as in cost of generation and decrement in the reliability of the overall system when the load is gradually increased. In the base case scenario, maximum voltage is found at bus 1 which is 1.1 p.u. and minimum voltage is found on bus 65 which is 1.019 p.u. After an increase of 10% electric vehicle load, real and reactive power losses increased by 22.7% and 25%, respectively, to 221 kW and 100 kVAr from the base case losses which were 180 kW and 80 kVAr. A tabular comparison has been done for the maximum and minimum values of different parameters in Table 9.2.

An optimization technique (ABC algorithm) is used to find the optimal location of EV charging stations on different feeder lines of the 69–bus distribution network and to improve the voltage profile as well as reduce the real and reactive power losses of the system. After the EV chargers are placed at optimal locations, overall generation is increased to 8.19 MVA, losses are decreased as compared to scenario 4 case study and reliability of the system is increased to 97.45% (Table 9.3).

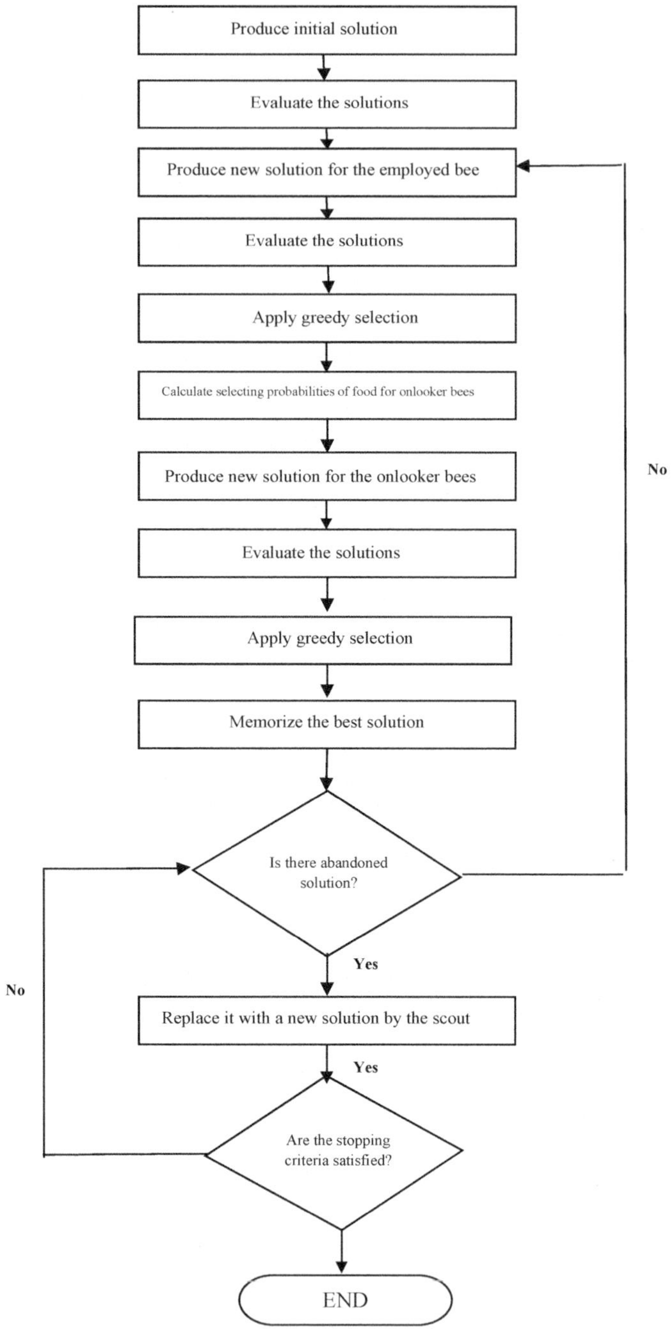

Fig. 9.10 Flow chart for ABC technique

References

www.ucsusa.org/resources/each-countrys-share-co2-emissions

Acharya N, Mahat P, Mithulananthan N (2006) An analytical approach for DG allocation in primary distribution network. Int J Electr Power Energy Syst 28(10):669–678

Arias-Londoño A, Montoya OD, Grisales-Noreña LF (2020) A chronological literature review of electric vehicle interactions with power distribution systems. Energies 13:3016. https://doi.org/10.3390/en13113016

Bastida-Molina P, Hurtado-Pérez E, Pérez-Navarro Á et al (2021) Light electric vehicle charging strategy for low impact on the grid. Environ Sci Pollut Res 28:18790–18806. https://doi.org/10.1007/s11356-020-08901-2

Bohre A, Agnihotri G, Dubey M (2015) The optimal planning of distributed generation using OPF and Butterfly-PSO (BF-PSO) technique. 15:127–138

Deb S, Tammi K, Kalita K, Mahanta P (2018) Impact of electric vehicle charging station load on distribution network. Energies 11:178. https://doi.org/10.3390/en11010178

Gopiya Naik SN, Khatod DK, Sharma MP (2015) Analytical approach for optimal siting and sizing of distributed generation in radial distribution networks. IET Gener Transm Distrib 9(3):209–220. https://doi.org/10.1049/iet-gtd.2014.0603

Heydt GT (1983) The impact of electric vehicle deployment on load management strategies. IEEE Trans Power Appar Syst 5:1253–1259

Huang X, Chen J, Yang H, Cao Y, Guan W, Huang B (2018) Economic planning approach for electric vehicle charging stations integrating traffic and power grid constraints. IET Gener Transm Distrib 12:3925–3934

Karmaker AK, Roy S, Ahmed MR (2019) Analysis of the impact of electric vehicle charging station on power quality issues. 2019 International Conference on Electrical, Computer and Communication Engineering (ECCE), pp. 1–6. https://doi.org/10.1109/ECACE.2019.8679164

Kumar B, Saw BK, Bohre AK (2020) Optimal distribution network reconfiguration to improve the system performances using PSO with multiple-objectives. International Conference on Computational Intelligence for Smart Power System and Sustainable Energy (CISPSSE), Keonjhar, India, pp. 1–6. https://doi.org/10.1109/CISPSSE49931.2020.9212262

Mishra, M (2015) Optimal placement of DG for loss reduction considering DG models. IEEE Int. Conf. on Electrical, Computer and Communication Technologies (ICECCT), 5–7 March 2015, pp 1–6. https://doi.org/10.1109/ICECCT.2015.7225996

Murthy VVSN, Kumar A (2013) Comparison of optimal DG allocation methods in radial distribution systems based on sensitivity approaches. Int J Electr Power Energy Syst 53:450–467

Naveed A, Sönmez Ş, Ayasun S (2020) Impact of electric vehicle aggregator with communication time delay on stability regions and stability delay margins in load frequency control system. In J Mod Power Syst Clean Energy. https://doi.org/10.35833/MPCE.2019.000244

Salihi JT (1973) Energy requirements for electric cars and their impact on electric power generation and distribution systems. IEEE Trans. Industry Application 5:516–532

Singh DK, Bohre AK (2020) Planning of EV fast charging station including DG in distribution system using optimization technique. IEEE International Conference on Power Electronics, Drives and Energy Systems (PEDES), Jaipur, India, pp. 1–6. https://doi.org/10.1109/PEDES49360.2020.9379351

Zimmerman RD, Murillo-Sanchez CE: Matpower4.1 (2011) December. http://www.pserc.cornell.edu//matpower/

Chapter 10
Planning of Electric Vehicle Charging Station with Integration of Renewables in Distribution Network

Arnab Pal, Aniruddha Bhattacharya, and Ajoy Kumar Chakraborty

Abstract Electric vehicle (EV) plays an important role to escape from environmental issues like increasing air pollution and limited stock of fossil fuel. EVs run on an electric motor, which is powered by its onboard rechargeable battery. EVs need a charging station to fulfil the energy demand of the battery. Recently, the penetration of EVs is increasing drastically in the modern power system. Therefore, the optimal planning of the electric vehicle charging stations (EVCSs) is necessary. The charging stations should maintain the power system parameters as well as it should provide services to the maximum EV users, which is a challenging job to the system planning engineers. However, the environmental benefits of EVs cannot be achieved if renewable energy resources are not incorporated into the system. The renewable-based distributed generation (DG) also helps to reduce the losses and improve the voltage profile of the system. Therefore, the distribution system with the charging stations must be renewably supported. However, the flow of EVs and generation from renewables are uncertain which is a matter of concern. This chapter describes the available planning for EVCSs in different distribution networks along with their methodologies. The results are also analyzed with their merits and demerits. The results of differenttypes of literature review show that the optimal allocation of charging stations helps to reduce the power loss of the distribution system as well asthe installation cost. Optimal allocation of DG contributes to additional power loss minimization. Moreover, technical information and specifications about EV and its charging infrastructure are presented in this chapter. The uncertain attributes associated with EV and DG are described. Also, the steps of different established methods to handle uncertainties are explained in this chapter.

A. Pal (✉) · A. K. Chakraborty
Department of Electrical Engineering, National Institute of Technology Agartala, Agartala, India
e-mail: arnabee.sch@nita.ac.in

A. K. Chakraborty
e-mail: akcall58.ee@nita.ac.in

A. Bhattacharya
Department of Electrical Engineering, National Institute of Technology Durgapur, Durgapur, India
e-mail: aniruddha.bhattacharya@ee.nitdgp.ac.in

A. K. Bohre et al. (eds.), *Planning of Hybrid Renewable Energy Systems, Electric Vehicles and Microgrid*, Energy Systems in Electrical Engineering,
https://doi.org/10.1007/978-981-19-0979-5_10

Keywords Distribution network · Distributed generation · Electric vehicle · Electric vehicle charging station · Optimal planning · Uncertainty

Abbreviations

EV	Electric vehicle
EVCS	Electric vehicle charging station
DG	Distributed generation
PEM	Point estimation method
IC	Internal combustion
EM	Electric motor
HEV	Hybrid electric vehicle
PHEV	Plug-in hybrid electric vehicle
BEV	Battery electric vehicle
BSS	Battery swapping station
G2V	Grid to vehicle
V2G	Vehicle to grid
SOC	State of charge
STD	Subsequent trip distance
AER	All-electric range
MCS	Monte-carlo simulation
RES	Renewable energy source
BES	Battery energy storage
PV	Photovoltaic
MT	Micro turbine

10.1 Introduction

10.1.1 Electric Vehicle

Vehicles with an internal combustion engine (ICE) run with fossil fuels and contribute to massive air and sound pollution. The electric vehicle is the solution for pollution less transportation system. EV also helps to stop the burden on fossils fuel consumption which is already at a diminished level. EV is based on an electric motor (EM) which is driven by a battery pack via a power converter. A three-phase induction motor and lithium-ion battery are mostly used in EV (Fathabadi 2019). The all-electric range (AER) is the distance that can be covered by an EV using only battery power with its full charge. The recent technologies have enhanced the EV driving range by more than 400 km in one charge (Al-Adsani et al. 2020). In broad, there are

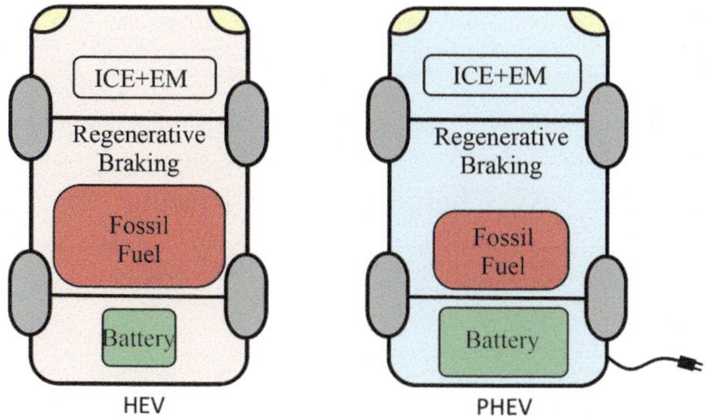

Fig. 10.1 Types of EV

three types of EVs, which are shown in Fig. 10.1: (i) hybrid electric vehicle (HEV), (ii) plug-in hybrid electric vehicle (PHEV), (iii) battery electric vehicle (BEV).

(i) *HEV*: This type of EV is having both an IC engine and electric motor, but the battery is not facilitated to charge from an external power supply. The battery takes charge from the braking system of the vehicle, called regenerative braking. The electric motor gives support to the IC engine to drive the vehicle as per requirement, that's why it is called "hybrid". The battery capacity of HEV is around 6 kWh (Iclodean et al. 2017).

    Example: Honda Civic Hybrid, Toyota Prius Hybrid, Toyota Camry Hybrid, etc.

(ii) *PHEV*: This is also a hybrid vehicle with an IC engine and electric motor, like HEV. Additionally, the battery of PHEV can be charged from regenerative braking as well as from external power, that's why it is called "plug-in". The battery capacity of BEV is around 100 kWh (Iclodean et al. 2017).

    Example: Chevy Volt, Hyundai Sonata, Toyota Prius, etc.

(iii) *BEV*: Only electric motor and battery are present in this type of vehicle. The battery is charged by the external EV charger. BEV also may be supported with regenerative braking. As there is no IC engine, therefore, the motor rating and battery capacity are quite higher than other types of vehicles. The battery capacity of BEV is around 5 kWh (Iclodean et al. 2017).

    Example: BMW i3, Nissan LEAF, Tesla Model S, etc.

10.1.2 EV Charging

The charging station is mandatory to charge the EV, as nowadays most of the vehicles are PHEV and BEV. EVCSs are connected to the nodes in the distribution network. There are few standards for the connector of EV charger, such as CHAdeMO, Society of Automotive Engineers (SAE), International Electromechanical Commission (IEC) and Combined Charging System (CCS) (Trentadue et al. 2018). Furthermore, the EV

Fig. 10.2 Classification of EV charging level

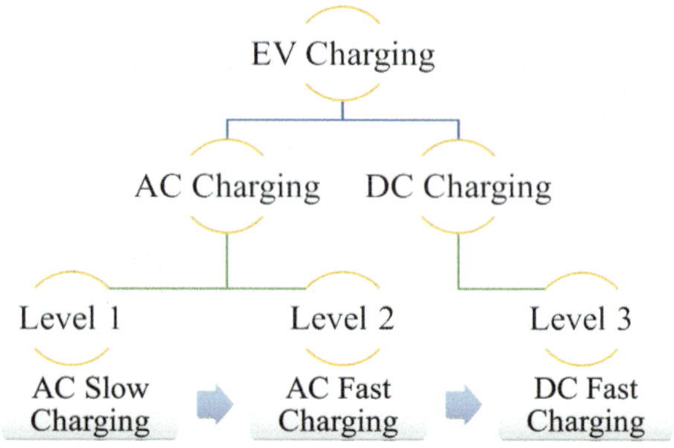

charging facilities can be categorized as per the charging rate, Fig. 10.2 present the different levels of EV charging. CHAdeMO is the most popular among those for DC fast charging (Ahmadian et al. 2020).

However, the health of the batteries should be kept in mind, therefore, the SOC limits of EV's battery are maintained for all the scenarios. Typically, 20% is the lower level of the SOC and 90% is the upper level, which is maintained (Pal et al. 2021). Charging efficiency and discharging efficiency are taken as 95 and 90%, respectively, for the simulation work with EV (Pal et al. 2021). Rating and uses of different charging levels according to SAE are as follows (Ahmadian et al. 2020):

Level 1: Single-phase AC 120 V, current 12–16 Amp, power 1.44–1.92 kW. This charging level is used for home charging and parking lot charging.
Level 2: Single/three-phase AC 240 V, current up to 80 Amp, power up to 19.2 kW. This charging level is used for the public charging station as well as home charging.
Level 3: DC 200–450 V, current up to 80 Amp, power up to 36 kW. This is particularly applicable only for the public charging station. The DC fast charging of CHAdeMO is with 50–500 V, current up to 125 Amp and output power up to 50 kW (Aziz and Oda 2017).

During charging, EV takes power from the grid, which is termed as "grid to vehicle". Whereas, "vehicle to grid" technology provides the facility to feed power from EV to power system grid. V2G is beneficial for EV users and service providers along with the grid (Ma et al. 2012). EV owners can trade the power with the grid according to the real-time prices. The service providers can gain revenue from the dividend. The grid can mitigate the peak power demand from EVs and able to improve the node voltages when it is required. The rate of discharging during V2G is 2–4 kW (Pal et al. 2021). The drawback of the V2G is the battery degradation due to additional cycles of charging and discharging.

There is another solution for charging the EV batteries, i.e. battery-swapping station (BSS). In BSS, a charged battery is exchanged with a nearly empty battery of EV. The BSS is not much convenient because more batteries are required and annoyed users about nonidentical batteries' health.

10.1.3 Distributed Generation

Distributed generation is a small-scale generation typically renewable-based, connected to the nodes of the distribution network. As DGs are installed near to the loads, therefore, it can supply the power locally to the load and helps to reduce the power losses as well as to improve the voltage profile of the distribution system (Pal et al. 2020). As the main motto of EVs is to reduce emissions, therefore, the conventional thermal power plant should not be utilized to charge the EVs. Incorporating renewable-based DG such as solar photovoltaic (PV), wind generators, will help to reduce pollution to some extent. However, the sizes and locations of DG in the distribution network are most significant to minimize maximum possible loss along with improvement of the voltages. Moreover, unplanned DG allocation may lead to undervoltage or overvoltage. The locations and demand of EVCSs can influence the allocation of DGs. Hence, simultaneous solution or two-layered optimization for EVCSs and DGs allocation is required to make optimal renewable supported distribution system including EVCSs. Many literature consider that DGs are connected with the EVCS. However, in case of renewable-based DG, uncertainties are a big issue to design the system optimally. Also, RESs are non-dispatchable generations that may require BES to store the surplus power and to utilize it as per requirement.

10.1.4 Uncertainty

Uncertainties are the critical concern to achieve realistic solutions for the research problem with EV and PV along with existing loads. The driving pattern of EVs and solar irradiance are the uncertain parameters for the EVCSs and DG power outputs. The uncertain parameters related to EV are:

(i) *Arrival Time (hh:mm)*: The time when an EV arrives at a charging station, is termed as the arrival time of that particular EV.

(ii) *Arrival SOC (%)*: The amount of SOC remaining of an EV at the moment of arrival at EVCS is called the arrival SOC of that EV.

(iii) *First Trip Distance (km)*: The distance covered by the EV before coming to the charging station.

(iv) *Subsequent Trip Distance (km)*: Expected distance will be travelled by the EV after getting charged from the EVCS.

(v) *Departure SOC (%)*: It is the desired SOC level by the EV, in other words, SOC of the EV while leaving the EVCS after charging. Departure SOC depends on the STD.

(vi) *Energy Requirement (kWh)*: It is the energy demand by the EV, which will be fulfilled from the EVCS. Energy demand is calculated from SOC at arrival, SOC at departure and battery capacity of the EV.

(vii) *Departure Time (hh:mm)*: The time when EV departs the EVCS after getting its desired charge. It depends on the energy requirement by the EV and the rate of charging/discharging.

The uncertain parameters related to PV power output are *i) Solar Irradiance, ii) Temperature.*

There are many established uncertainty handling methods that are able to take the uncertainties into consideration and provide accurate solutions. K-mean algorithm, Monte Carlo simulation and 2 m point estimation method are widely used stochastic tools to deal with uncertainties.

10.1.5 Highlights of the Chapter

This chapter presents the planning procedure of EVCSs with DGs and analyzed the results of different planning of EVCSs. To formulate the problem for EVCS allocation, some basic knowledge is necessary, which are given in the introduction section as:

- Basic information of electric vehicle and its types with specifications.
- Technical information about EV charging facility with its specifications and different levels.
- Benefits of distributed generation.
- Uncertainties associated with EVs and PVs.

The mathematical formulations, methodologies and results from different literature are shown and discussed in this chapter as:

- Calculation of different parameters of EV and load at EVCS.
- Objective functions to allocate EVCS and DG.
- Constraints related to distribution network, EV, EVCS and DG.
- Flow chart to solve allocation problem using optimization technique.
- Different methods to handle uncertainties.
- Result analyses of EVCS and DG allocation in different distribution networks for various objectives.

10.2 Related Work

There are many works related to EVCS allocation. Some of them provide realistic solutions considering different practical parameters. In Pal et al. (2019), fast-charging stations are placed optimally in IEEE 33 distribution system. The distribution network is divided into three zones to allocate EVCS. The power loss is minimized using grey wolf optimizer and results are confirmed using whale optimization algorithm. In Cui et al. (2019), ECVSs are located in the distribution network in view of practical factors associated with the urban area. In Deb et al. (2019), EVCSs are placed in Guwahati highway where multi-objective problem is formulated and solved using soft-computing techniques. In Dong et al. (2019), EVCS allocation is executed in London city. The EVCSs are allocated in the roads of the U.S. in He et al. (2019). Various optimization techniques are applied in Zhang et al. (2019) to obtain the locations of EVCS by taking care of area-wise charging demand. In Kong et al. (2019) and Guo and Zhao (2015), the optimal locations of EVCS are determined in Beijing. Conditions-based place selection for EVCS is performed in Hosseini and Sarder (2019). Again, (He et al. 2018) is also a distinguished investigation where EVCSs are planned on different roads and maps of cities.

In Awasthi et al. (2017), a hybrid optimization is used to allocate EVCSs in the Allahabad distribution network by minimizing installation cost and maximizing power quality. In Mehta et al. (2019), EVCSs are placed in a 33 bus distribution network by two-layer optimization. In Shojaabadi et al. (2016), EVCSs are planned considering the demand-side response. EVCSs are installed at optimal places with optimal sizes in Aljaidi et al. (2019). Charging station set-up is found out optimally in Davidov and Pantoš (2019) by various cost minimizations. In Yi et al. (2019), optimal locations and sizes are found out for EVCS. The costs related to EVCS instalment and operation are minimized in Liu and Bie (2019) and (Andrade et al. 2020) by allocating at optimal places. In Zhang et al. (2019), road networks and distribution networks are overlaid to find out the appropriate locations of the EVCS using soft-computing techniques. In Zhang et al. (2019), optimal locations of EVCS are determined in Beijing city to minimize power loss of the electrical network and EVCS cost. Optimal charging/discharging scheduling is achieved along with optimal EVCS locations in Hadian et al. (2020). In Pal et al. (2021), EVCS and DG are allocated simultaneously in the radial distribution network. The problem is solved using two-layer optimization. The uncertainties associated with EVs and DGs are taken care of. The results show that DG helps to increase voltage profile and to reduce the energy loss of the distribution system. It can be observed that optimal locations of EVCS reduce the energy loss. Also, locations influence the total land cost for the EVCSs.

In Pal et al. (2021), EVCSs are placed optimally in 33 radial distribution network which is merged with a road network. Multi-objective problem is formulated where energy loss, voltage deviation, last cost are minimized and weightage of the location is maximized. The results show that allocated EVCSs are distributed in the road network to serve the users suitably. In Shaaban et al. (2019) and Atat et al. (2020), EVCSs, as well as DGs, are located optimally. In Pal et al. (2021), a superimposed network is used to allocate EVCS which is taken as a public fast-charging station. Renewable-based DGs (PV) are also allocated optimally to minimize more loss of energy and to enhance the voltages of the distribution network. BESs are incorporated with PV to manage the variable renewable generation and variable load demand. BESs are also optimally scheduled. Moreover, EVs are assigned at apt EVCS considering their routes and traffic congestions. The BESs are also installed with EVCS in Dai et al. (2019), Yan et al. (2017), Gong et al. (2019), Kasturi et al. (2020). Allocation of EVCS and BSS is performed on a 15-bus and 43-bus network in Zheng et al. (2014).

In Moradijoz et al. (2013), parking lots are placed in a distribution network. Again in Ahmadi et al. (2021), parking lots are optimally planned in IEEE 69 network with the appropriate charging strategies of the EVs to reduce the total costs associated with a parking lot. Also, in Liu et al. (2020)–(Ponnam and Swarnasri 2020), EVCSs are planned optimally in the distribution system.

The EV movement and charging requirement are uncertain parameters for this type of work. In Sokorai et al. (2018), the uncertainty of daily usages of EVCSs is modelled using the Markov chain's tool. In Yi et al. (2019), uncertainties are handled using an artificial immune algorithm to find out the EVCS locations. K-means clustering method is applied in Shaaban et al. (2019) to look into uncertainties associated with DG and EV. MonteCarlo simulation-based algorithm is established in Andrade et al. (2020) to get the uncertain EV driving pattern for the allocation of EVCS. MCS is applied in Shojaabadi et al. (2016); Shojaabadi et al. 2016) to tackle the uncertainties related to EV. Still, these methods have drawbacks such as computational time, mathematical complexity, hard to deal with huge uncertain variables (Saha et al. 2019). 2 m PEM has many advantages over other techniques, which are used in Pal et al. (2021), (2021), (2021) to deal with uncertainties for EVCS allocation.

10.3 Problem Formulation

10.3.1 Mathematical Calculation of EV Parameters

The load at any charging station depends on the number of vehicles that are in charging/discharging mode and the rate of charging/discharging. The calculation of total load $\left(EvL_S^{(t)}\right)$ on Sth EVCS due to EVs at tth time is calculated as follows (Pal et al. 2021):

$$EvL_S^{(t)} = \left(nv_{G2V}^{(t)} \times C_r\right) - \left(nv_{V2G}^{(t)} \times D_r\right) \tag{10.1}$$

where C_r and D_r charging and discharging rate of G2V and V2G, $nv_{G2V}^{(t)}$ and $nv_{V2G}^{(t)}$ are number of vehicles in G2V mode and V2G mode, respectively, at tth time. The energy demand $\left(Req_v^{eng}\right)$ by the vth vehicle is determined as

$$Req_{vc}^{eng} = \begin{cases} (Req_v^{soc} \times Bc_v)/\eta_{ch}, & if\ V2G\ mode \\ (Req_v^{soc} \times Bc_v \times \eta_{dis}, & if\ V2G\ mode \end{cases} \tag{10.2}$$

where Req_v^{soc}, Req_v^{eng} and Bc_v indicate required SOC in %, required energy in kWh and battery capacity in kWh of vth vehicle, respectively. η_{ch} and η_{dis} are charging and discharging efficiency of EV, respectively. The required SOC $\left(Req_v^{soc}\right)$ and required time $\left(Req_v^{time}\right)$ to charge the vth vehicle is calculated as

$$Req_v^{soc} = D_v^{soc} - A_v^{soc} \tag{10.3}$$

$$Req_v^{time} = \begin{cases} Req_v^{eng}/C_r, if G2V \\ \left|Req_v^{eng}\right|/D_r, if V2G \end{cases} \tag{10.4}$$

Req_v^{soc} is positive when departure SOC $\left(D_v^{soc}\right)$ is more than arrival SOC $\left(A_v^{soc}\right)$, that means the EV is connected in G2V/charging mode. If the value is negative, the state of EV is V2G/discharging. The arrival SOC $\left(A_v^{soc}\right)$ can be found by

$$A_v^{soc} = 1 - (ftd_v/AER_v) \tag{10.5}$$

where ftd_v is the first trip distance which is travelled by the EV to come to the EVCS and AER_v is the all- electric range of the vth EV. The departure SOC $\left(D_v^{soc}\right)$ and departure time $\left(D_v^{time}\right)$ of vth vehicle is calculated as (Pal et al. 2021)

$$D_v^{soc} = (STD_v/AER_v) + 0.20 \tag{10.6}$$

$$D_v^{time} = A_v^{time} + Req_v^{time} \tag{10.7}$$

where STD_v is the subsequent trip distance of vth vehicle. 20% is the minimum SOC level that should be maintained, to be specific, the departure SOC should not be less than 20%. The STD is obtained by (8), where td_v is the total distance is covered by the EV.

$$STD_v = td_v - ftd_v \tag{10.8}$$

In the above mentioned calculations, arrival time, first trip distance, total travel distance, AER and the battery capacity are the input parameters.

10.3.2 EVCS and DG Allocation

To allocate the EVCSs and DGs in the distribution system, the total load due to EVCS and generation from DG need to be calculated. If Sth EVCS and DG are connected at buth node in the distribution network, the total load (P_{bu}) on that node at tth hour can be expressed as

$$P_{bu}^{(t)} = ExL_{bu}^{(t)} + EvL_S^{(t)} - DG_S^{(t)} \tag{10.9}$$

where $ExL_{bu}^{(t)}$ is the existing load at buth node. $EvL_{S}^{(t)}$ is load due to EVCS at tth time period which is obtained by (1). $DG_{S}^{(t)}$ is the power generated from DG at time t. The EVCS load and DG generation would be zero if they are not connected at bus bu.

10.3.2.1 Objectives for the Allocation Problem

Different objectives are considered in various research works to allocate EVCS in the distribution network as well as traffic network. Maximization/minimization problem depends on the type of objective functions, such as electrical loss and instalment cost due to EVCSs being minimized and service to the users being maximized.

- The total cost of EVCSs is minimized in Liu et al. (2013) considering the costs for operation and maintenance, investment and power loss to place the EVCSs at optimal locations. The mathematical objective function for the planning of EVCSs in the distribution network is expressed in Liu et al. (2013) as follows:

$$\min(F) = \sum_{t=1}^{T} \frac{1}{(1+\eta)^t} \left[\sum_{i=1}^{N_{EVCS}} C_{EVCSi}^{I}(t) + C_{EVCSi}^{O}(t) + C_{EVCSi}^{M}(t) + C_{PS}^{L}(t) \right] \tag{10.10}$$

  where $C_{EVCSi}^{I}(t)$, $C_{EVCSi}^{O}(t)$ and $C_{EVCSi}^{M}(t)$ are the instalment cost, operation cost and maintenance cost of ith EVCS at tth time period. $C_{PS}^{L}(t)$ is the cost of the power loss in the system at tth time. η is the discount rate for reverse calculation of upcoming price, N_{EVCS} is the number of EVCS is being allocated and T is the total project period.
- For allocation of EVCSs. three objectives are considered in Wang et al. (2013) to cover maximum EV users by the EVCSs with keeping power losses and voltage deviations minimum. The multi-objective problem is converted to a single objective function by multiplying weighting factors ($\alpha_1, \alpha_2, \alpha_3$) as below.

$$\min(F) = \min(\alpha_1 F_1 + \alpha_2 F_2 + \alpha_3 F_3) \tag{10.11}$$

  where (11) is a minimization problem and F_1, F_2 and F_3 are the normalized values of three objective functions. The objective functions as follows:

$$\max(F_1) = \sum_{q \in Q} f_q y_q \tag{10.12}$$

$$\min(F_2) = P_{Loss}(U_i P_{Si}, U_i Q_{Si}) \tag{10.13}$$

$$\min(F_3) = \sum_{i=1}^{N_d} \frac{|V_i - V_0|}{V_0} \tag{10.14}$$

where F_1 is to serve the maximum number of EV by the EVCSs. F_2 and F_3 are to minimize power loss and voltage deviation of the distribution network, respectively. f_q is the EV flow on qth road, Q is the total number of roads and y_q is binary variable for the response of all EVs, served or not. Binary U_i denotes EVCS is connected at ith bus or not,P_{Si} and Q_{Si} are the real and reactive power drawn by the EVCS at ith bus.V_0 and V_i are the rated voltage and the voltage at bus i, respectively. N_d a total number of buses present in the system.

- The EVCSs are allocated in Awasthi et al. (2017) taking multi-objective problem (15) consisting of minimization of active and reactive power loss along with EVCS cost and maximization of voltage improvement index.

$$\min(F) = \min\left\{\frac{1}{w_1 VPII_i + w_2 PLRI_i + w_3 QLPI_i}\right\}(i \in NC) \tag{10.15}$$

where w_1, w_2 and w_3 are the weighting coefficients to make a single objective problem to multi-objective and NC is the total number of charging stations to be allocated. IC_i is the initial cost for ith EVCS and X_i is binary variable is 1 if the EVCS is connected at ith bus, otherwise 0. $VPII_i$, $PLRI_i$ and $QLPI_i$ are the indexes at ith bus for improvement of voltage problem, reduction of active power loss and reduction of reactive power loss, respectively, as shown below.

$$VPII_i = \frac{1}{\alpha + \max_{i=1}^{n}\left(\left|1 - V_{(CS_i)}\right|\right)} \tag{10.16}$$

$$PLRI_i = \frac{P_{L(base)} - P_{L(CS_i)}}{P_{L(base)}} \tag{10.17}$$

$$QLPI_i = \frac{Q_{L(base)} - Q_{L(CS_i)}}{Q_{L(base)}} \tag{10.18}$$

$$IC_i = C_{in} + 25C_{land}.N_i + C_p.C_{dev}.(N_i - 1) \tag{10.19}$$

where $V_{(CS_i)}$ is the voltage of bus i once the EVCSs are connected and α is a scalar variable. $P_{L(base)}$ and $Q_{L(base)}$ are the active and reactive power loss respectively at ith bus without any EVCS and $P_{L(CS_i)}$ and $Q_{L(CS_i)}$ are the active and reactive power loss respectively at ith bus after placing the EVCS. C_{in}, C_{land}, C_p and C_{dev} are investment cost, land cost, rated power of connector and development cost of connector, respectively, for charging station.

- In Pal et al. (2021), multi-objective function is considered for EVCS allocation, which is transformed to a single objective as a minimization problem by applying weighting coefficients ($\omega_1, \omega_2, \omega_3, \omega_4$) as follows:

$$\min(F) = \min(\omega_1 f_1 + \omega_2 f_2 + \omega_3 f_3 - \omega_4 f_4) \tag{10.20}$$

where f_1, f_2, f_3, f_4 are the four objective functions, which are expressed below

$$\min(f_1) = \sum_{t=1}^{24} \sum_{br=1}^{N_{br}} \left(I_{br}^{(t)}\right)^2 \times R_{br} \tag{10.21}$$

$$\min(f_2) = \sum_{t=1}^{24} \sum_{bu=1}^{N_{bu}} \left|1 - V_{bu}^{(t)}\right| \tag{10.22}$$

$$\max(f_3) = \sum_{S=1}^{N_S} w_S \tag{10.23}$$

$$\min(f_4) = \sum_{S=1}^{N_S} c_S \tag{10.24}$$

where f_1 is energy loss and f_2 is voltage deviation of the distribution system, f_3 is weightage of the locations of EVCS and f_4 is the land cost for installing EVCSs. $I_{br}^{(t)}$ is current flowing through branch br at time t, R_{br} is the resistance of brth branch. $V_{bu}^{(t)}$ is the voltage of bus bu at time t. w_S and c_S are the weightage and land cost of the Sth EVCS's location, respectively. N_{br}, N_{bu} and N_S are the total number of branches, number of buses and number of EVCS, respectively.

- In Mozafar et al. (2017), minimization of a multi-objective problem is formulated involving four objectives, i.e. power loss, voltage fluctuation, EV charging cost and battery degradation cost.

$$\min(F) = \min\left(\tau f_1 + \beta f_2 + \gamma f_3 + \alpha f_4\right) \tag{10.25}$$

where τ, β, γ and α are the weighting factors, set as 0.4, 0.3, 0.2 and 0.1, respectively. f_1, f_2 are the objective functions for power loss and voltage fluctuation for 24 h, which can be calculated as (21, 22), respectively. $f_3$is the objective function for the cost of charging (G2V) including benefits from discharging (V2G) and f_4 is for downgrading costs of battery, as follows:

$$f_3 = \sum_{t=1}^{24} \left(P_{sub,t} \times \pi^{TOU}\right) + f_{ch} - f_{dc} \tag{10.26}$$

$$f_4 = \frac{C_b.BCAP + C_r}{L_c.BCAP.DOD} P_{dc}^{PHEV} \tag{10.27}$$

where $P_{sub,t}$ is power from the substation at time t. The dynamic electricity price is $\pi^{TOU}.f_{ch}, f_{dc}$ are the cost for charging and profit from discharging (V2G). C_b, C_r and L_c are the battery cost, cost for battery replacement and lifecycle of the battery, respectively, $BCAP\text{and}DOD$ is the battery capacity and depth of discharge, respectively. P_{dc}^{PHEV} is the energy discharged by the PHEV during V2G.

- Single objective function is taken in Alhazmi et al. (2017) to maximize the coverage of the charging station to the EV users by allocating EVCS at proper places. The maximization problem is presented in Alhazmi et al. (2017) as

$$\max(F) = \max\sum_{i=1}^{N_T} TD_i w_i \tag{10.28}$$

where TD_i is the demand by the EV transportation at ith location and w_i is 1 if demand is fulfilled by the EVCS, otherwise $0.N_T$ is the total nodes in the transportation network.

- In Kandil et al. (2018), charging stations for PEVs are assigned along with renewable-based DG and battery energy storage. The yearly cost of energy is minimized to find out the appropriate locations, as follows:

$$\min(F) = \min_{\Omega_1,\Omega_2}\sum_s Pro_{(s)}\times \left(\sum_i \left[C_{RE(i)} - R_{RE(i,s)} + C_{ES(i)} + C_{EV(i)}\right] + \left[C_{loss(s)} + C_{cons(s)}\right]\right) \tag{10.29}$$

where Ω_1, Ω_2 are the installation and operation variables, respectively,$Pro_{(s)}$ is the probability of sth scenario. $C_{RE(i)}, C_{ES(i)}$ and $C_{EV(i)}$ are the capital investment and running costs for renewable energy, energy storage and EV charging station, respectively, connected at ith bus. $R_{RE(i,s)}$ is the profit from renewable power generation. $C_{loss(s)}, C_{cons(s)}$ are the energy loss cost and energy consumption cost by the loads, respectively.

- DGs and EVCSs are allocated in Liu et al. (2020) using two-level optimization, where both levels are maximization problems. The objective function (30) of the upper level is consisting annual return $\left(C_{pro}\right)$, power loss cost (C_{loss}) and environmental benefit (C_{env}).

$$\max\left(F_{upper}\right) = C_{pro} - C_{loss} + C_{env} \tag{10.30}$$

The objective function in lower-level optimization is

$$\max(F_{lower}) = C_{l,S} + C_{l,B} - C_{l,OM} - C_{l,loss} + C_{l,env} \tag{10.31}$$

where l is the probabilistic scenario, $C_{l,S}$ denotes profit from the utility,$C_{l,B}$ represents government support for renewable generation, $C_{l,OM}$ is the operation and maintenance cost of EVCS and DG, $C_{l,loss}$, $C_{l,env}$ are the power loss and environmental benefit, respectively.

- Authors in Luo et al. (2020) propose a cost minimization objective considering DG and EVCS for optimal allocation. The total annual cost in the objective function involves the cost of investment $\left(C^{I}\right)$, operation and maintenance cost $(C^{O\&M})$, fuel cost with carbon emission $(C^{F\&E})$, electricity purchase cost from utility $\left(C^{P}\right)$, power loss cost (C^{NL}), energy loss for charging/discharging of EV (C^{CL}) and battery degradation cost (C^{B}), as

$$\min(F) = \min\left(C^{I} + C^{O\&M} + C^{F\&E} + C^{P} + C^{NL} + C^{CL} + C^{B}\right). \quad (10.32)$$

10.3.2.2 Constraints for the Allocation

Different constraints are considered related to EV, charging station, traffic network and distribution system for EVCS and DG allocation. Some required constraints are presented below.

- **Constraint Related to Distribution Network**

- *Constraint for power flow balance*: This is a compulsory condition that need to be satisfied for any distribution network at any point of time. The total power supplied from substation and DGs should be equal to total power loss and total load including EVCSs, can be written as:

$$P_{sub} + P_{DG} - P_{loss} - P_{load} = 0 \quad (10.33)$$

where P_{sub}, P_{DG}, P_{loss} and P_{load} are the power from substation, power from DG, power losses in the system and load, respectively.

- *Constraint for node voltages*: To provide quality power to the users, the voltage of the nodes of the distribution network should be maintained within permissible limits as (34) where V^{min} and V^{max} are the lower and upper limit of the bus voltage. respectively.

$$V^{\min} < V_{bu} < V^{\max} \quad (10.34)$$

- *Constraint for line currents*: This constraint is for a limit of the current flow through each branch because of the conductors' maximum current carrying capacity (I^{max}) as

$$|I_{br}| < I^{max} \quad (10.35)$$

- Constraint Related to EVCS

- *The constraint for the number of ports*: Number of maximum and minimum ports in an EVCS can be restricted according to EV demand and power supply capacity of distribution node as (Awasthi et al. 2017)

$$N_i^{min} \leq N_i \leq N_i^{max}, i = 1, 2, \ldots, N_{EVCS} \tag{10.36}$$

where N_i is the number of the port at ith charging station, N_i^{min} and N_i^{max} are the minimum and maximum limit, respectively. N_{EVCS} is the number of EVCS in the system.

- *Constraint for EVCS power*: The total power drawn by the EVCSs in a distribution system should be limited by permissible level because of the supply capacity of the distribution system as (Liu et al. 2013)

$$\sum_{i=1}^{N_{EVCS}} P_{EVCSi} \leq P_{EVCS}^{\max} \tag{10.37}$$

where P_{EVCSi} power is drawn by the ith EVCS and $P_{EVCS}^{max}$is the maximum permissible limit.

- Constraint Related to EV

- *Constraint for EV's SOC*: The SOC level of the EV has to be within limits (38) because overcharge or overdischarge impacts the battery health (Pal et al. 2021).

$$SOC^{min} \leq SOC \leq SOC^{max} \tag{10.38}$$

where SOC^{min} and SOC^{max} are the lower and upper limit of the SOC which should be maintained in the EV battery.

- Constraint Related to DG

- *Constraint for DG power*: The DG output power should not exceed the maximum limit because it may cause over voltage, high short circuit current and reverse power flow (Pal et al. 2020). The constraint is shown below, where P_{DGi} is the power generated from ith DG and N_{DG} is the total number of DG presence in the system.P_{DG}^{max} is the maximum allowable power which can be penetrated from the DGs.

$$\sum_{i=1}^{N_{DG}} P_{DGi} \leq P_{DG}^{\max} \tag{10.39}$$

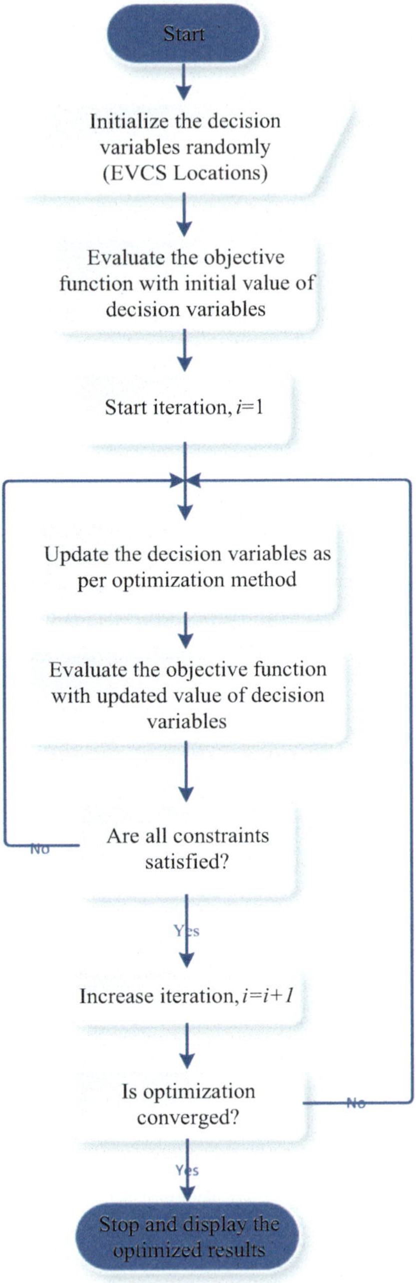

Fig. 10.3 Flowchart of the generalized algorithm of EVCS allocation using optimization technique

10.3.2.3 Generalized Algorithm of EVCS Allocation

A simple algorithm is illustrated in Fig. 10.3 to allocate the EVCSs using the optimization technique. First, the locations are generated randomly in the initialization stage. Then the locations are updated using an optimization algorithm to minimize/maximize the objective function. If all the constraints are satisfied then only the solution is acceptable. The updating procedure is stopped when the convergence criteria are reached.

10.4 Uncertainty Modelling

In the allocation problem of EVCS and DG, the EV flow and PV power output are not deterministic variables. Researchers have to take care of the uncertain variables to find out the optimal locations, otherwise, the solutions would not be applicable in a practical scenario. The Monte Carlo simulation, K-mean algorithm and 2 m point estimation method are the established method to handle the uncertain variables.

- *K-mean algorithm*: In Liu et al. (2020), K-mean + + algorithm is employed for uncertain variables such as PV, wind power output and load demand. In this technique, the number of uncertain scenarios is reduced by the clustering process using historical data. The procedure is given below as (Liu et al. 2020):

(i) The historical original data of uncertain variables (PV output, EV charging demand) are normalized.
(Ii) Total K number of clustering centres $(\xi_1, \xi_2, \xi_3, \ldots, \xi_K)$ are created as follows:

(a) A random set is selected from the original 365 number of sets $(N_i, i = 1, 2, 3, \ldots, 365)$ for the 1st cluster (ξ_1).
(b) The shortest distance is calculated between N_i and ξ_1, i.e. $d(N_i, \xi_1)$. A summation of all distances (d) is needed.
(c) A random distance is generated which is less than $\text{sum}(d)$ and it is updated to $d(N_i, \xi_1)$. N_i will be the new centre of the cluster if the random distance is less than 0.
(d) Repeat step b and c to select K number of centres $(\xi_1, \xi_2, \xi_3, \ldots, \xi_K)$.
(e) The distances between remaining original sets to centres of the clustering. The original sets are allocated to the nearest cluster.
(f) The centres of K number of cluster are found out and new generations' centres are updated.
(g) Repeat step c and d till clustering are unchanged.

- *Monte Carlo simulation*: The authors in Kandil et al. (2018) have applied a Markov chain-based MCS to capture uncertain arrival time and departure time of the EVs. The MCS model is used in Alhazmi et al. (2017) to tackle the uncertainties of travel distance by the EVs and energy remain in EVs' batteries. The simple steps are:

(i) Generate random numbers of the uncertain variables from their distribution function.
(ii) Make numerous sets with the random numbers as different scenarios.
(iii) Solve the problem for each scenario with a fixed deterministic set.
(iv) The average value of the results of all scenarios is taken.

- *2 m point estimation method*: This method is utilized in Pal et al. (2021) to handle the uncertainties of arrival time, arrival SOC and travelling distance of the EVs. In this method, dual scenarios are created for each uncertain variable using a distribution function based on past data. The major steps are as follows:

(i) Generate two random values of each uncertain variable using distribution function and central moment.
(ii) Make two sets with a random value and mean value for a fixed deterministic set.
(iii) Repeat step 1 and 2 for all the uncertain variables. The total number of the set will be double the number of uncertain variables.
(iv) Solve the problem for each set with a fixed deterministic set.
(v) Determine the mean value of the outcomes from all sets.

10.5 Result and Discussion

- In Pal et al. (2019), fast-charging stations are allocated in IEEE 33 radial distribution network. The whole network is divided into separate zones, where each EVCS is placed at each zone optimally as Fig. 10.4. In this work, the power loss of the distribution network is minimized. The locations are 2, 21 and 30, which are scattered in the area. The power loss after optimal allocation of EVCS is

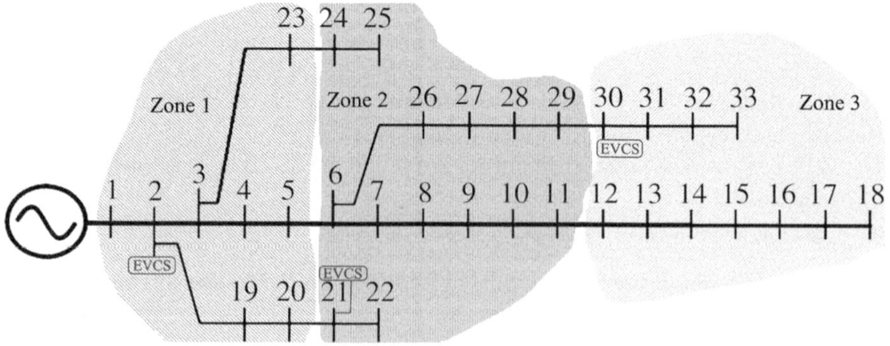

Fig. 10.4 Allocation of fast-charging station in IEEE 33 node system (Pal et al. 2019)

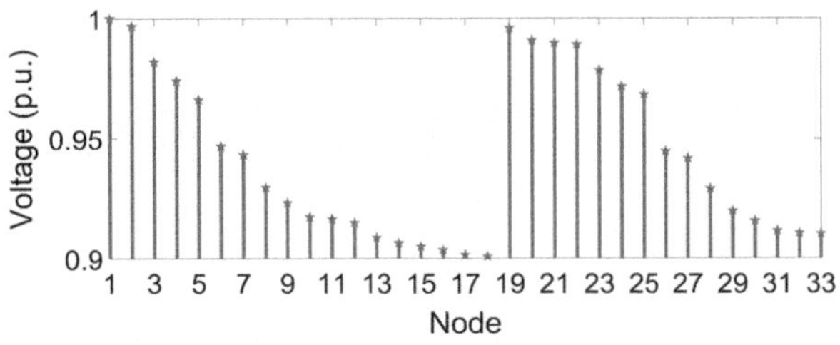

Fig. 10.5 Voltage profile of IEEE 33 node system after optimal allocation of EVCSs (Pal et al. 2019)

Table 10.1 Optimal allocation of EVCSs in IEEE 123 distribution system (Liu et al. 2013)

Type	Location	EVCS size (kVA)	Total cost ($)
1	1	500	1,313,917.16
3	31	80	
3	39	80	
3	87	50	
2	107	200	

234.6 kW for 30 EVs. The voltage profile of the distribution network is shown in Fig. 10.5 after zone-wise allocation of EVCSs. It is seen that the voltage of each node maintains permissible 10% limits.

- EVCSs are placed optimally on IEEE 123 distribution network in Liu et al. (2013). The optimization problem is solved using the cross-entropy technique. Three types of EVCSs are considered based on their capability to serve the area, where type-1 is capable to serve 1700–2000 m^2 area, type-2 is capable to serve 1000 m^2 area and type-3 is capable to serve 50–100 m^2 area. Five charging stations are allocated at optimal locations based on cost minimization (10) with the decision of the type of EVCSs as mentioned in Table 10.1, where the total cost is $1,313,917.16 including investment cost, operation and maintenance cost of EVCSs, and power loss cost of the distribution system. The capacity of the respective EVCSs is also mentioned in Table 10.1.

- In Wang et al. (2013), the EVCSs are allocated in an overlapped network with a 33 bus distribution system and a 25 node traffic network. A modified primal–dual interior point method is used to solve the problem. Four charging stations are taken for 400 EVs in the study area to serve maximum EVs with minimum power loss and minimum voltage deviation (11). The four cases are performed and the optimal locations are mentioned in Table 10.2 (Wang et al. 2013).

The four cases are:

i Allocation in only distribution network for power loss and voltage deviation minimization.
ii Allocation in only traffic network to serve the maximum EVs.
iii Allocation considering both the network overlapped with all the objectives.

Table 10.2 Optimal allocation of EVCS in a 33 distribution system (Wang et al. 2013)

Case	Location	EVCS size (MW)	Power loss (MW)	Voltage deviation (%)	EV served (%)
i	12, 13, 14, 16	0.1, 0.2, 0.4, 0.1	0.1953	2.493	24.93
ii	8, 14, 18, 23	0.2, 0.1, 0.2, 0.3	0.2575	3.251	45.83
iii	14, 15, 18, 23	0.1, 0.1, 0.4, 0.2	0.2068	2.719	32.25
iv	2, 19, 22, 20	0.1, 0.1, 0.2, 0.4	0.1926	2.491	53.27

iv Case-iii without considering the EVs' battery limits.

It is seen in Table 10.2 that case-ii creates more power loss and voltage deviation than case-i, but case-ii successfully served 45.83% EVs because case-ii only cares about traffic network and maximum services. Case-i serves less EV because it only considers electrical parameters of the distribution system. Case-iii is a balanced solution among all, because it considers both the networks will all the objectives. Case-iv achieves minimum power loss and voltage deviation with maximum EV served, which indicates large battery capacity of EV can help distribution network as well as traffic network. In case-iv, the power loss is 206.8 kW, voltage deviation is 2.49% and 53.27% EVs are served.

- Authors in Awasthi et al. (2017) consider the distribution system of Allahabad town to allocate the EVCSs. Six sub-feeders (1, 2, 3, 4, 5 and 6) are there with different number of node. Three nodes are in sub-feeder 1, denoted as 1.1, 1.2 and 1.3, in this way, there are 6 nodes in feeder 2, 3 nodes in feeder 3, 5 nodes in feeder 4, 8 nodes in feeder 5 and 5 nodes in feeder 6. Total 14 EVCSs are placed in Allahabad city. A hybrid optimization technique with particle swarm optimization and genetic algorithm is employed to solve this problem. The optimized location of EVCSs, number of ports in respective EVCSs, required land for EVCSs and cost of the EVCSs are shown in Table 10.3.

- In Mozafar et al. (2017), two EVCSs are allocated along with two renewable energy resources considering uncertainties and battery degradation. IEEE 33 node test system is taken for investigation of the EVCS allocation problem in presence of PV/wind generation. A combination of particle swarm optimization and genetic algorithm is used for optimization to find out the locations and sizes of the EVCS and RES which are presented in Table 10.4 with the values of objective functions in (25). It is seen that nodes 23 and 6 are the optimal locations for EVCS and nodes 13, 23 are optimal locations for DG.

- A 23 node radial distribution system is considered along with a 20 node road network in Alhazmi et al. (2017). The EVCSs are allocated to cover the maximum area of a city. The uncertain parameters are handled using Monte Carlo simulation.

Table 10.3 Optimal allocation of EVCS in the Allahabad distribution system (Awasthi et al. 2017)

Location	No. of port	Required land m^2	Cost ($) $\times 10^3$	Active power loss (kW)	Reactive power loss (kVar)
2.4	29	725	100.42	18,532	73,989
3.1	29	725	99.21		
4.3	41	1025	115.12		
4.5	31	775	106.46		
5.1	28	700	97.58		
5.2	28	700	98.46		
5.3	28	700	100.03		
5.4	28	700	97.00		
5.5	28	700	97.87		
5.6	28	700	98.60		
5.7	28	700	99.62		
5.8	28	700	101.08		
6.1	29	725	100.11		

Table 10.4 Best solution of locations and sizes of EVCS and RES in IEEE 33 distribution system (Mozafar et al. 2017)

EVCS location	EVCS size (MW)	RES location	RES size (MW)	(f_1) Power loss (MW)	(f_2) Voltage fluctuation	(f_3) Cost of G2V-V2G	(f_4) Battery downgrading costs
23	1.0512	13	0.6939	2.108906	24.1073 pu	$8,221,640	$147,189
6	0.9035	30	0.7126				

The results are illustrated in Table 10.5 considering the service range of each EVCS is 25 km. For the optimal locations, the total installation cost is 4 M$.

- In Pal et al. (2021), EVCSs and DGs are allocated in a 33 node radial distribution system which is merged with a road network. Two-layered optimization problem is formulated, where the first layer is to place the EVCSs optimally and the second layer is for allocation of DGs. In this work, the sizes of EVCS and the sizes of DG are not optimized. Energy loss, voltage deviation, weightage and cost of the locations are taken as objectives. Table 10.6 presents the locations of EVCS and DG. Nodes 10, 23 and 28 are the optimal locations for EVCS. Nodes 16, 17 and 18

Table 10.5 Optimal locations of EVCS in traffic + distribution system (Alhazmi et al. 2017)

Location in traffic network	Construction cost
6, 7, 11, 17, 20	4 Million $

Table 10.6 Optimal allocation of EVCS with DG in a 33 bus system (Pal et al. 2021)

EVCS location	DG location	Energy loss
28	16	6111.5 kWh
23	17	
10	18	

are the optimal locations for DG. For this solution, the energy loss is minimized to 6111.5 kWh for 100% penetration of EV.

Figure 10.6 illustrates the energy loss for different penetration levels of EV with and without DG. It can be seen that energy losses are increasing with EV penetration. However, the energy loss is less with DG for any penetration level.

Figure 10.7 depicts the differences of voltages with DG and without DG of nodes 11–18. It can be seen that node voltages are increased after the incorporation of DGs. However, due to non-optimal sizes of DG, the voltage improvement is not much significant.

- Authors in Kandil et al. (2018) allocate EVCSs, RESs and BESs simultaneously in a 38 node distribution network considering the uncertainties using Monte Carlo

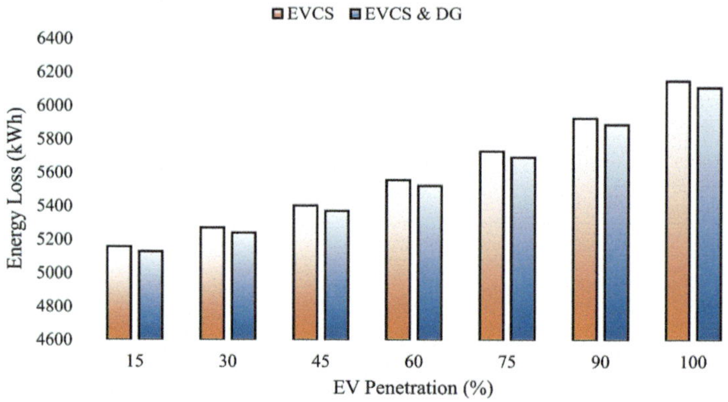

Fig. 10.6 Comparison between with and without DG for various EV penetration levels (Pal et al. 2021)

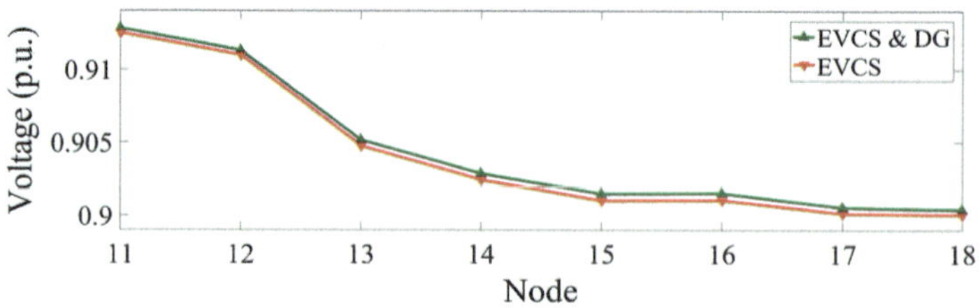

Fig. 10.7 Voltage profiles with and without DG (Pal et al. 2021)

simulation. PV is considered as a renewable energy source. The cost of energy is minimized to find out the optimal solution. Three cases are shown in Table 10.7, where case-1 is locations and sizes of PV and BES, case-2 is locations and sizes of EVCS and BES and case-3 is with PV, EVCS and BES. For the base case, the energy loss and cost of energy are 427 MWh and 0.0702 $/kWh, respectively.

It can be seen in Table 10.7 that for case-1, the cost of energy is reduced to 0.0169 $/kWh from the base case due to the optimal allocation of RESs and BESs. But in case-2, the loss is increased because there is no renewable resource as a DG, however, the cost of energy is less due to BESs. The simultaneous allocation of EVCS, PV and BES is attempted in case-3, and loss is decreased with a lower cost of energy. However, the case-3 results show that the BESs are not desired due to high cost. EVs can be charged when PV power output is available, to avoid the BES. Therefore, BESs are not needed to manage renewable energy according to load requirements with additional investment costs.

- DGs and EVCSs are allocated in Liu et al. (2020) considering the uncertainties using K-mean ++ algorithm. Each DG's size is 10 kW. Various types of EV are considered, such as bus, taxi and car. The allocation problem is executed for the distribution system of IEEE 33, PG&E-69 and a local 30 node network. An enhanced harmonic particle swarm optimization is used to solve the allocation problem. The case studies are performed in Liu et al. (2020) as:

(i) Coordinated allocation of DGs and EVCSs using an active management strategy.
(ii) Coordinated allocation of DGs and EVCSs without using an active management strategy.
(iii) Uncoordinated allocation of DGs and EVCSs.

The results are presented in Tables 10.8, 10.9 and 10.10 for IEEE 33, PG&E-69 and local 30 node network distribution system, respectively.

It is seen in the above three tables that in case-iii, the cost of the power loss is highest and annual return and environmental benefit are lowest for all the distribution systems due to uncoordinated allocation. Case-i provides better results than case-ii because of the adoption of active management techniques with coordinated allocation.

- Also, in Luo et al. (2020), coordinated allocation of DG and EVCS is achieved. V2G technology is considered along with placement problems. A practical distribution network of China with 31 nodes is chosen to allocate EVCS, PV and MT. Seasonal variations in EV charging demand are considered in this work. The annual cost is minimized to allocate the EVCS. The locations and sizes for the three cases are illustrated in Table 10.11 and the value of the objectives in (32) are shown in Table 10.12. Three cases are carried out as (i) without V2G, (ii) unidirectional V2G, (iii) bidirectional V2G (Luo et al. 2020).

Table 10.12 shows that case-i is the poorest solution among all because it involves only charging not V2G. Case-iii provides better results compared to case-ii because

Table 10.7 Allocation of EVCSs, RESs and BESs in a 38 distribution system (Kandil et al. 2018)

Case	EVCS location	No. of port	RES location	RES size (kW)	BES location	BES size (kWh)	Energy loss (MWh)	Loss reduction (%)	Cost of energy ($/kWh)
1	–	–	4, 5, 6, 7, 10, 11, 12, 13, 16, 17, 18, 20, 23, 24, 25, 26, 29, 30, 31, 32, 38	90, 20, 215, 25, 75, 80, 165, 160, 125, 120, 130, 50, 25, 565, 505, 55, 125, 505, 60, 50, 30	38	3500	459	7.49	0.0169
2	29, 33, 34, 38	98, 10, 76, 100	–	–	35	5000	1042	144.03	0.0405
3	29, 34, 35, 38	60, 100, 28, 100	6, 7, 8, 9, 10, 11, 12, 13, 14, 15, 16, 17, 18, 22, 24, 25, 26, 27, 28, 29, 30, 31, 32, 33, 34, 38	115, 20, 100, 15, 60, 80, 80, 120, 200, 100, 105, 100, 150, 25, 130, 225, 30, 15, 30, 610, 560, 185, 390, 85, 395, 75	Nil	NA	369	–13.58	0.0096

Table 10.8 EVCS and DG allocation in IEEE 33 node distribution system (Liu et al. 2020)

Case	EVCS location	EVCS size (kW)	DG location	Number of DG	Annual return/10000 RMB	Power loss cost/10000 RMB	Environmental benefit/10000 RMB
i	20, 4, 8, 14, 29	237, 194, 223, 309, 203	13, 23, 31, 7, 21, 28	8, 10, 10, 7, 2, 6	709.11	9.16	83.72
ii	20, 4, 8, 14, 29	252, 197, 186, 173, 208	13, 23, 31, 7	10, 5, 8, 7	643.69	18.59	72.33
iii	20, 4, 8, 14, 29	185, 186, 185, 184, 185	13, 23, 31, 21, 28	9, 4, 10, 3, 1	625.81	19.53	69.71

Table 10.9 EVCS and DG allocation in PG&E-69 node distribution system (Liu et al. 2020)

Case	EVCS location	EVCS Size (kW)	DG location	Number of DG	Annual return/10000 RMB	Power loss cost/10000 RMB	Environmental benefit/10000 RMB
i	14, 32, 40, 45, 61	186, 273,190, 266, 339	10, 33, 38, 21, 50, 66	9, 2, 10, 3, 10, 10	687.17	2.73	77.34
ii	14, 32, 40, 45, 61	185, 294, 238, 224, 214	10, 33, 38, 21, 50, 66	2, 5, 10, 4, 10, 10	661.04	4.57	71.53
iii	14, 32, 40, 45, 61	190, 181, 193, 188, 191	10, 33, 38, 21, 50, 66	5, 2, 10, 2, 9, 9	641.33	4.60	67.90

case-iii can operate bidirectional V2G, which means it can charge the EV as well as discharge when necessary. On the other hand, the charging port can't perform V2G and vice versa.

- Authors in Pal et al. (2021), use a 33 node distribution system along with the road network with three types of road to allocate the EVCSs. Energy loss is minimized in terms of distribution network and land cost is minimized in terms of the road network. The optimization issue is resolved by the harris hawks optimization technique. The uncertainties related to EVs are handled by 2 m PEM. The EVCSs are allocated in a distributed manner in the area for better service to the users. The optimized locations of EVCSs are nodes 10, 23 and 28, shown in Fig. 10.8. For

Table 10.10 EVCS and DG allocation in local 30 node distribution system (Liu et al. 2020)

Case	EVCS location	EVCS Size (kW)	DG location	Number of DG	Annual return/10000 RMB	Power loss cost/10000 RMB	Environmental benefit/10000 RMB
i	5, 8, 16, 22, 25	470, 494, 562, 372, 660	8, 9, 10, 11, 12, 13, 14, 23, 24, 25, 26, 27, 28, 29	13, 12, 15, 15, 10, 11, 15, 8, 12, 11, 11, 14, 15, 6	2981.13	27.75	299.66
ii	5, 8, 16, 22, 25	454, 537, 316, 482, 203	8, 9, 10, 11, 12, 13, 14, 23, 24, 25, 27, 28, 29	15, 15, 15, 12, 7, 13, 15, 8, 12, 1, 7, 12, 3	2873.41	44.65	282.30
iii	5, 8, 16, 22, 25	375, 369, 371, 371	8, 9, 10, 11, 12, 13, 14, 23, 24, 25, 26, 27, 28, 29	14, 14, 14, 7, 13, 14, 14, 2, 6, 7, 3, 8, 3, 7	2809.21	46.48	271.77

Table 10.11 EVCS and DG allocation in PG&E-69 node distribution system (Luo et al. 2020)

Case	PV location	PV size (kVA)	MT location	MT size (kVA)	EVCS location	No. of port
i	2, 3, 6, 20, 21, 23, 25, 29	4720, 1140, 480, 540, 690, 4390, 450, 160	5, 7, 16, 17	550, 180, 150, 1710	2, 12, 17, 20, 26, 28, 29	23, 26, 17, 16, 8, 12, 5
ii	2, 3, 6, 20, 21, 23, 25, 29	5530, 690, 740, 560, 620, 4540, 540, 170	5, 7, 17	330, 70, 1060	2, 12, 17, 20, 26, 28, 29	11, 17, 11, 9, 5, 7, 3
iii	2, 3, 6, 20, 21, 23, 25, 29	2030, 1180, 480, 3930, 710, 4760, 430, 60	5, 17	40, 20	2, 12, 17, 20, 26, 28, 29	21, 26, 13, 13, 5, 7, 3

optimal allocation of fast-charging stations, the daily energy loss of the distribution network is 6151.58 kWh and the required land cost is $143,019.8.

- In Pal et al. (2021), EVCS, DG and BES are installed in the distribution system. The test network is considered as Fig. 10.8, which is a superimposed network. In this work, EVCSs and DGs are allocated at optimal nodes. The quantity of

Table 10.12 Numerical value of the different objectives (Luo et al. 2020)

Case	Investment cost (×106 $)	Operation and maintenance cost (×104 $)	Fuel and emission cost (× 105 $)	Electricity cost (×105 $)	Power loss cost (×103 $)	Energy loss cost (×104 $)	Battery degradation cost (×105 $)
i	1.1347	9.9184	1.4839	1.8244	1.1929	4.3420	1.6283
ii	1.0751	8.1254	0.5845	1.1151	1.2541	4.3421	1.6283
iii	0.9814	9.1477	0.0225	1.1453	1.9992	5.1447	1.9293

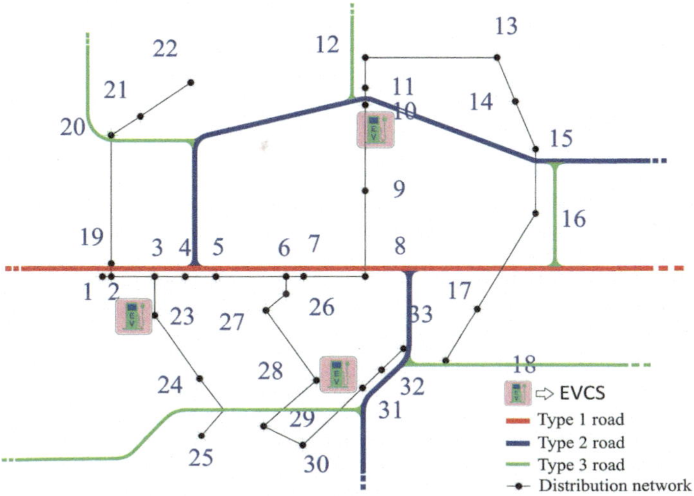

Fig. 10.8 Allocation of the EVCSs in a 33 node distribution network with road network (Pal et al. 2021)

charging ports in the EVCSs and the size of the DGs are also optimized. DGs are placed to decrease the loss of energy of the distribution system. To deal with the variable renewable generation and variable EV demand, BESs are optimally scheduled. Moreover, EVs are assigned to apt EVCS considering the shortest route and traffic congestions. The shortest route is found out using Dijkstra's algorithm and EVs are assigned using Integer Linear Programming. This helps EVs to reach EVCS by consuming less battery energy and the arrival SOC of the EVs are increased.

Table 10.13 presents the optimal locations and sizes of EVCS and DG. Figure 10.9 shows the 24 h optimal charging/discharging scheduling of three BESs. EVCS is considered as a public fast-charging station. The optimal locations of EVCS are 2, 16 and 31 with sizes of 198 kW, 154 kW and 176 kW, respectively. The DG locations are nodes 9, 7 and 8 with sizes of 2.08, 1.48 and 1.35 kW. The energy loss for the solution is 2623.72 kWh. However, when the installation, operation and maintenance

Table 10.13 Optimal allocation of EVCS and DG in presence of BES (Pal et al. 2021)

EVCS locations	EVCS sizes (kW)	Number of ports	DG locations	DG sizes (kW)	Energy loss (kWh)
2, 16, 31	198, 154, 176	9, 7, 8	6, 14, 31	2.08, 1.48, 1.35	2623.72

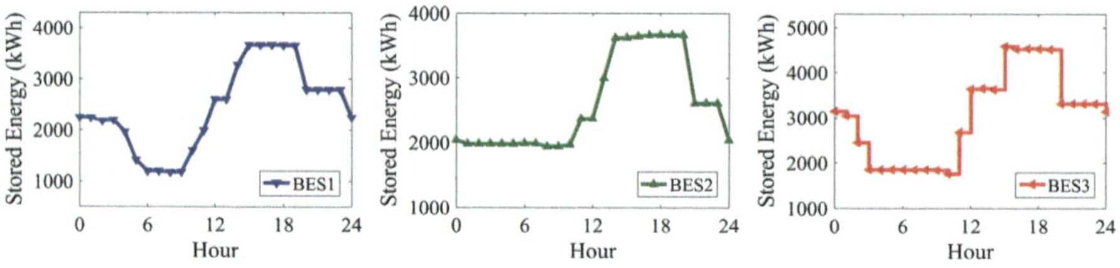

Fig. 10.9 Optimal scheduling of BESs (Pal et al. 2021)

costs of EVCS, DG and BES are taken, the solution is changed. The EVCS locations are 3, 14 and 32 considering the costs.

The voltage profiles for only EVCS allocation, EVCS with DG allocation and EVCS, DG allocation with BES scheduling, are shown in Fig. 10.10. It can be observed that optimal locations and sizes of DG can improve voltage significantly. BES and its optimal scheduling offer more voltage improvement.

- In Hadian et al. (2020), two types of EVCS are allocated in IEEE 69 radial distribution network. Types of EVCS are (i) administrative and (ii) residential. The optimization problem is solved by particle swarm optimization and MCS is applied for uncertainties. Power loss, voltage deviation and reliability of the electrical network are taken as objective. In this work, three charging strategies are presented, i.e. (i) uncontrolled, (ii) controlled and (iii) smart strategy. Weekdays and weekends are taken care of in this study. The optimal locations of EVCS

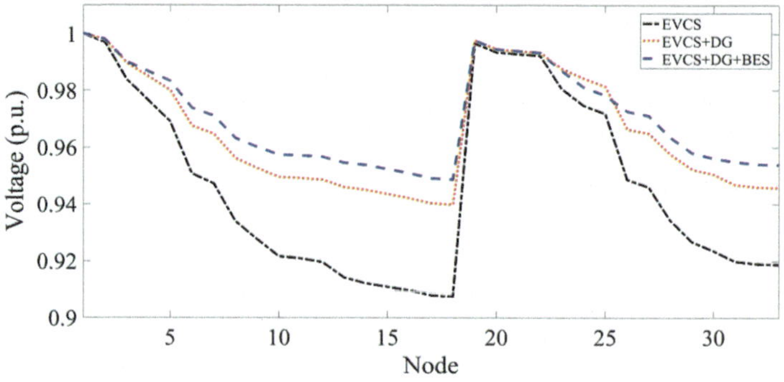

Fig. 10.10 Voltage profile assessment (Pal et al. 2021)

Table 10.14 Charging strategies and respective EVCS locations (Hadian et al. 2020)

Type of EVCS	Administrative EVCS locations			Residential EVCS locations		
charging strategy	Uncontrolled	Controlled	Smart strategy	Uncontrolled	Controlled	Smart strategy
EVCS location	7	16	20	32	37	39
No of EVs	103	115	123	117	120	125

in the radial distribution network are shown in Table 10.14 for three types of charging strategies. It is noticed that the optimal place is changed with the charging strategies.

10.6 Conclusion

This chapter provides technical specifications of EV and EVCS which are necessary for EVCS related work. Mathematical calculations of different parameters of EV and EVCS are shown, such as SOC requirement of EVs, load at EVCS due to EV charging. However, the allocation of electric vehicle charging stations is a truly challenging and practical issue, which is analyzed in this chapter. The electrical network and road network, both are influencing factors for the locations of the EVCS. Many researches consider the distribution network and the road network simultaneously. However, location-wise EV flow, SOC requirement is unpredictable, which need to be fulfilled by the EVCSs at proper locations with appropriate sizes. Therefore, the uncertainty handling tool is required to incorporate to achieve practical applicable solutions. However, using EVs is not the solution to the environmental worries until and unless the power to charge the EVs is not generated from renewable resources. Therefore, renewable-based DG should be essential to instal in the distribution system. DG also helps to reduce power losses and improve the voltage profile. The DGs also have to be optimally allocated in the network to get the maximum benefit from it. The results show that the simultaneous allocation of EVCS and DG is having a better impact. Some researchers allocate the BES along with EVCS and DG, which helps to manage the dynamic power demand by the EVs because BES can store the renewable generated power which can be utilized as per necessity. The allocation of EVCS considering distribution network, as well as road network, provides a realistic solution in terms of EV users and grid. Moreover, selecting proper EVCS for each EV with its suitable route is vital in terms of SOC requirements.

References

Ahmadi M, Hosseini SH, Farsadi M (2021) Optimal allocation of electric vehicles parking lots and optimal charging and discharging scheduling using hybrid metaheuristic algorithms. J Electr Eng Technol 16(2):759–770. https://doi.org/10.1007/s42835-020-00634-z

Ahmadian A, Mohammadi-Ivatloo B, Elkamel A (2020)A review on plug-in electric vehicles: introduction, current status, and load modeling techniques. J Mod Power Syst Clean Energy 8(3):412–425. https://doi.org/10.35833/MPCE.2018.000802

Al-Adsani AS, Jarushi AM, Beik O (2020) ICE/HPM generator range extender for a series hybrid EV powertrain. IET Elect Syst Transp 10(1):96–104. https://doi.org/10.1049/iet-est.2018.5097

Alhazmi YA, Mostafa HA, Salama MMA (2017) Optimal allocation for electric vehicle charging stations using Trip Success Ratio. Int J Electr Power Energy Syst 91:101–116. https://doi.org/10.1016/j.ijepes.2017.03.009

Aljaidi M, Aslam N, Kaiwartya O (2019) Optimal placement and capacity of electric vehicle charging stations in urban areas: survey and open challenges: in 2019 IEEE Jordan international joint conference on electrical engineering and information technology (JEEIT), Amman, Jordan, pp 238–243. https://doi.org/10.1109/JEEIT.2019.8717412

Andrade J, Ochoa LF, Freitas W (2020) Regional-scale allocation of fast charging stations: travel times and distribution system reinforcements. IET Gener Transm Distrib 14(19):4225–4233. https://doi.org/10.1049/iet-gtd.2019.1786

Atat R, Ismail M, Serpedin E, Overbye T (2020) Dynamic joint allocation of EV charging stations and DGs in spatio-temporal expanding grids. IEEE Access 8:7280–7294. https://doi.org/10.1109/ACCESS.2019.2963860

Awasthi A, Venkitusamy K, Padmanaban S, Selvamuthukumaran R, Blaabjerg F, Singh AK (2017) Optimal planning of electric vehicle charging station at the distribution system using hybrid optimization algorithm. Energy 133:70–78. https://doi.org/10.1016/j.energy.2017.05.094

Aziz M, Oda T (2017) Simultaneous quick-charging system for electric vehicle. Energy Procedia 142:1811–1816. https://doi.org/10.1016/j.egypro.2017.12.568

. Cui Q, Weng Y, Tan C-W (2019) Electric vehicle charging station placement method for urban areas. IEEE Trans Smart Grid:1–1. https://doi.org/10.1109/TSG.2019.2907262

Dai Q, Liu J, Wei Q (2019) Optimal photovoltaic/battery energy storage/electric vehicle charging station design based on multi-agent particle swarm optimization algorithm. Sustainability 11(7):1973. https://doi.org/10.3390/su11071973

Davidov S, Pantoš M (2019) Optimization model for charging infrastructure planning with electric power system reliability check. Energy 166:886–894. https://doi.org/10.1016/j.energy.2018.10.150

Deb S, Tammi K, Kalita K, Mahanta P (2019) Charging station placement for electric vehicles: a case study of Guwahati City, India. IEEE Access 7:100270–100282. https://doi.org/10.1109/ACCESS.2019.2931055

Dong G, Ma J, Wei R, Haycox J (2019) Electric vehicle charging point placement optimisation by exploiting spatial statistics and maximal coverage location models. Transp Res Part d: Transp Environ 67:77–88. https://doi.org/10.1016/j.trd.2018.11.005

Fathabadi H (2019) Combining a proton exchange membrane fuel cell (PEMFC) stack with a Li-ion battery to supply the power needs of a hybrid electric vehicle. Renew Energy 130:714–724. https://doi.org/10.1016/j.renene.2018.06.104

Gong D, Tang M, Buchmeister B, Zhang H (2019) Solving location problem for electric vehicle charging stations—a sharing charging model. IEEE Access 7:138391–138402. https://doi.org/10.1109/ACCESS.2019.2943079

Guo S, Zhao H (2015) Optimal site selection of electric vehicle charging station by using fuzzy TOPSIS based on sustainability perspective. Appl Energy 158:390–402. https://doi.org/10.1016/j.apenergy.2015.08.082

Hadian E, Akbari H, Farzinfar M, Saeed S (2020) Optimal allocation of electric vehicle charging stations with adopted smart charging/discharging schedule. IEEE Access 8:196908–196919. https://doi.org/10.1109/ACCESS.2020.3033662

He J, Yang H, Tang T-Q, Huang H-J (2018) An optimal charging station location model with the consideration of electric vehicle's driving range. Transp Res Part C: Emerg Technol 86:641–654. https://doi.org/10.1016/j.trc.2017.11.026

He Y, Kockelman KM, Perrine KA (2019) Optimal locations of U.S. fast charging stations for long-distance trip completion by battery electric vehicles. J Clean Prod 214:452–461. https://doi.org/10.1016/j.jclepro.2018.12.188

Hosseini S, Sarder M (2019) Development of a Bayesian network model for optimal site selection of electric vehicle charging station. Int J Electr Power Energy Syst 105:110–122. https://doi.org/10.1016/j.ijepes.2018.08.011

Iclodean C, Varga B, Burnete N, Cimerdean D, Jurchiş B (2017) Comparison of different battery types for electric vehicles. IOP Conf Ser: Mater Sci Eng 252:012058. https://doi.org/10.1088/1757-899X/252/1/012058

Kandil SM, Farag HEZ, Shaaban MF, El-Sharafy MZ (2018) A combined resource allocation framework for PEVs charging stations, renewable energy resources and distributed energy storage systems. Energy 143:961–972. https://doi.org/10.1016/j.energy.2017.11.005

Kasturi K, Nayak M, Nayak C (2020) PV/BESS to support electric vehicle charging station integration in a capacity constrained power distribution grid using MCTLBO. Sci Iran 0(0):0–0. https://doi.org/10.24200/sci.2020.5128.1112

Kong W, Luo Y, Feng G, Li K, Peng H (2019) Optimal location planning method of fast charging station for electric vehicles considering operators, drivers, vehicles, traffic flow and power grid. Energy 186:115826. https://doi.org/10.1016/j.energy.2019.07.156

Liu X, Bie Z (2019) Optimal allocation planning for public EV charging station considering AC and DC integrated chargers. Energy Procedia 159:382–387. https://doi.org/10.1016/j.egypro.2018.12.072

Liu Z, Wen F, Ledwich G (2013) Optimal planning of electric-vehicle charging stations in distribution systems. IEEE Trans Power Deliv 28(1):102–110. https://doi.org/10.1109/TPWRD.2012.2223489

Liu L, Zhang Y, Da C, Huang Z, Wang M (2020) Optimal allocation of distributed generation and electric vehicle charging stations based on intelligent algorithm and bi-level programming. Int Trans Electr Energ Syst. https://doi.org/10.1002/2050-7038.12366

Liu L, Zhang Y, Da C, Huang Z, Wang M (2020) Optimal allocation of distributed generation and electric vehicle charging stations based on intelligent algorithm and bi-level programming. Int Trans Electr Energ Syst 30(6). https://doi.org/10.1002/2050-7038.12366

Luo L, Wu Z, Gu W, Huang H, Gao S, Han J (2020) Coordinated allocation of distributed generation resources and electric vehicle charging stations in distribution systems with vehicle-to-grid interaction. Energy 192:116631. https://doi.org/10.1016/j.energy.2019.116631

Ma Y, Houghton T, Cruden A, Infield D (2012) Modeling the benefits of vehicle-to-grid technology to a power system. IEEE Trans Power Syst 27(2):1012–1020. https://doi.org/10.1109/TPWRS.2011.2178043

Mehta R, Verma P, Srinivasan D, Yang J (2019) Double-layered intelligent energy management for optimal integration of plug-in electric vehicles into distribution systems. Appl Energy 233–234:146–155. https://doi.org/10.1016/j.apenergy.2018.10.008

Moradijoz M, Parsa Moghaddam M, Haghifam MR, Alishahi E (2013) A multi-objective optimization problem for allocating parking lots in a distribution network. Int J Electr Power Energy Syst 46:115–122. https://doi.org/10.1016/j.ijepes.2012.10.041

Mozafar MR, Moradi MH, Amini MH (2017) A simultaneous approach for optimal allocation of renewable energy sources and electric vehicle charging stations in smart grids based on improved GA-PSO algorithm. Sustain Cities Soc 32:627–637. https://doi.org/10.1016/j.scs.2017.05.007

Pal A, Bhattacharya A, Chakraborty AK (2021) Allocation of electric vehicle charging station considering uncertainties. Sustain Energy Grids Netw 25:100422. https://doi.org/10.1016/j.segan.2020.100422

Pal A, Bhattacharya A, Chakraborty AK (2021) Placement of public fast-charging station and solar distributed generation with battery energy storage in distribution network considering uncertainties and traffic congestion. J Energy Storage 41:102939. https://doi.org/10.1016/j.est.2021.102939

Pal A, Bhattacharya A, Chakraborty AK (2019) Allocation of ev fast charging station with v2g facility in distribution network. In: 2019 8th international conference on power systems (ICPS), Jaipur, India, pp 1–6. https://doi.org/10.1109/ICPS48983.2019.9067574

Pal A., Chakraborty AK, Bhowmik AR (2020) Optimal placement and sizing of dg considering power and energy loss minimization in distribution system. Ijeei 12(3):624–653. https://doi.org/10.15676/ijeei.2020.12.3.12

Pal A, Bhattacharya A, Chakraborty AK (2021) Placement of electric vehicle charging station and solar dg in distribution system considering uncertainties. Scientia Iranica, vol. Articles in Press. https://doi.org/10.24200/SCI.2021.56782.4908

Ponnam VKB, Swarnasri K (2020) Multi-objective optimal allocation of electric vehicle charging stations and distributed generators in radial distribution systems using metaheuristic optimization algorithms. Eng Technol Appl Sci Res 10(3):5837–5844. https://doi.org/10.48084/etasr.3517

Saha A, Bhattacharya A, Das P, Chakraborty AK (2019) A novel approach towards uncertainty modeling in multiobjective optimal power flow with renewable integration. Int Trans Electr Energ Syst 29(12). https://doi.org/10.1002/2050-7038.12136

Shaaban MF, Mohamed S, Ismail M, Qaraqe KA, Serpedin E (2019) Joint planning of smart EV charging stations and DGs in eco-friendly remote hybrid microgrids. IEEE Trans Smart Grid 10(5):5819–5830. https://doi.org/10.1109/TSG.2019.2891900

Shojaabadi S, Abapour S, Abapour M, Nahavandi A (2016) Simultaneous planning of plug-in hybrid electric vehicle charging stations and wind power generation in distribution networks considering uncertainties. Renew Energy 99:237–252. https://doi.org/10.1016/j.renene.2016.06.032

Shojaabadi S, Abapour S, Abapour M, Nahavandi A (2016) Optimal planning of plug-in hybrid electric vehicle charging station in distribution network considering demand response programs and uncertainties. IET Gener Transm Distrib 10(13):3330–3340. https://doi.org/10.1049/iet-gtd.2016.0312

Sokorai P, Fleischhacker A, Lettner G, Auer H (2018) Stochastic modeling of the charging behavior of electromobility. WEVJ 9(3):44. https://doi.org/10.3390/wevj9030044

Trentadue G, Lucas A, Otura M, Pliakostathis K, Zanni M, Scholz H (2018) Evaluation of fast charging efficiency under extreme temperatures. Energies 11(8):1937. https://doi.org/10.3390/en11081937

Wang G, Xu Z, Wen F, Wong KP (2013) Traffic-constrained multiobjective planning of electric-vehicle charging stations. IEEE Trans Power Deliv 28(4):2363–2372. https://doi.org/10.1109/TPWRD.2013.2269142

Yan Q, Zhang B, Kezunovic M (2019) Optimized operational cost reduction for an EV charging station integrated with battery energy storage and PV generation. IEEE Trans. Smart Grid 10(2):2096–2106. https://doi.org/10.1109/TSG.2017.2788440

Yi T, Cheng X, Zheng H, Liu J (2019) Research on location and capacity optimization method for electric vehicle charging stations considering user's comprehensive satisfaction. Energies 12(10):1915. https://doi.org/10.3390/en12101915

Zhang H, Tang L, Yang C, Lan S (2019) Locating electric vehicle charging stations with service capacity using the improved whale optimization algorithm. Adv Eng Inform 41:100901. https://doi.org/10.1016/j.aei.2019.02.006

Zhang Y, Zhang Q, Farnoosh A, Chen S, Li Y (2019) GIS-based multi-objective particle swarm optimization of charging stations for electric vehicles. Energy 169:844–853. https://doi.org/10.1016/j.energy.2018.12.062

Zheng Y, Dong ZY, Xu Y, Meng K, Zhao JH, Qiu J (2014) Electric vehicle battery charging/swap stations in distribution systems: comparison study and optimal planning. IEEE Trans Power Syst 29(1):221–229. https://doi.org/10.1109/TPWRS.2013.2278852

Chapter 11
Techno-Economic Analysis of Hybrid Renewable Energy Systems—A Review with Case Study

Arnab Ari, Aashish Kumar Bohre, Pradyumn Chaturvedi, Mohan Lal Kolhe, and Sri Niwas Singh

Abstract Although renewable sources of energy provide multiple benefits, their intermittent nature makes it difficult for application as individual sources of energy. A hybrid renewable energy system integrates different non-renewable and renewable sources along with storage systems to overcome this drawback. This work aims to shed light on the various techno-economic aspects of HRES discussed in recent papers. Also the different components and their mathematical modelling have been highlighted. Diverse range of optimizing techniques used for sizing and optimization of cost, reliability and environmental parameters has been discussed. A brief look at the applications studied in recent papers is taken and future prospects have been discussed. Additionally, In this work, the case study of rural area of West Bengal is presented. The study location is Digha village (21° 37.6'N, 87° 30.4'E), West Bengal, India. The techno-economic analysis using HOMER software of hybrid renewable energy system is presented for possible best optimized solution based on Reliability, Net present costs and Levelized Cost of Energy.

Keywords HRE systems · Integrated renewable energy systems · Technical and economic analysis · Hybrid renewable energy systems

---

A. Ari (✉) · A. K. Bohre
National Institute of Technology Durgapur, Durgapur, West Bengal, India
e-mail: aa18u10065@btech.nitdgp.ac.in

A. K. Bohre
e-mail: aashishkumar.bohre@ee.nitdgp.ac.in

P. Chaturvedi
Visvesvaraya National Institute of Technology (VNIT) Nagpur, Nagpur, Maharashtra, India
e-mail: pradyumn.c@eee.vnit.ac.in

M. L. Kolhe
University of Agder, Kristiansand, Norway
e-mail: mohan.l.kolhe@uia.no

S. N. Singh
Indian Institute of Technology (IIT) Kanpur, Kalyanpur, Kanpur, India

A. K. Bohre et al. (eds.), *Planning of Hybrid Renewable Energy Systems, Electric Vehicles and Microgrid*, Energy Systems in Electrical Engineering,
https://doi.org/10.1007/978-981-19-0979-5_11

Nomenclature

HRES	Hybrid renewable energy system
WECS	Wind energy conversion system
PV	Photovoltaic
DG	Diesel generator
FC	Fuel-cell
BG	Biogas
WT	Wind turbine
FEE	Final excess energy
BESS	Battery energy storage system
PHS	Pumped hydro storage
NPC	Net present cost
COE	Cost of energy
ASC	Annual systems cost
LCC	Life cycle cost
LCOE	Levelized cost of energy
LPSP	Loss of power supply probability
LLP	Loss of load probability
UL/UEL	Unmet electric load
RF/REF	Renewable energy fraction
GHGE	Greenhouse gas emission
PSO	Particle swarm optimization
BSO	Brain storm optimization
GA	Genetic algorithm
ANN	Artificial neural network
WCA	Water cycle algorithm
SSO	Social spider optimization
MFO	Moth-Flame optimization
GWO	Grey wolf optimizer
EO	Equilibrium optimizer
HHO	Harris hawks optimizer
AEFA	Artficial electric field algorithm
STOA	Sooty tern optimization algorithm

11.1 Introduction

As the conventional sources of energy, having limited supply, continue to deplete and growth of industry and economy elevates the demand for energy; alternative resources of energy resource has become a necessity. Also, use of fossil fuels has contributed to climate change and global warming over the past decades and it has become essential to look towards environment-friendly energy sources. Renewable

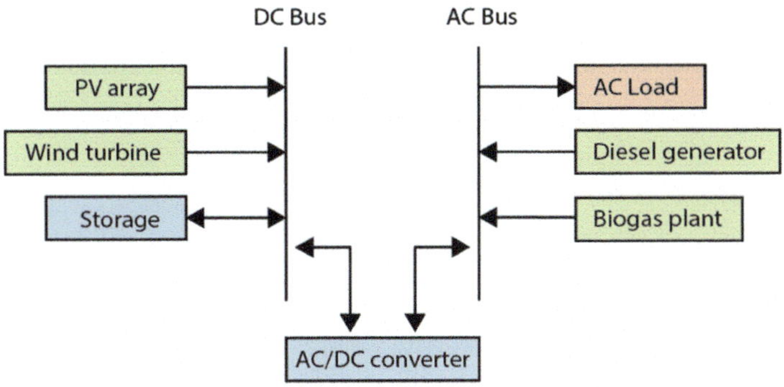

Fig. 11.1 Block diagram of a general HRE system

sources like solar energy, wind energy, biomass are seen as promising alternatives to conventional sources (Sawle et al. 2021). However, they are not without drawbacks. In solar energy systems, generation drastically falls on cloudy days and at night; windmill output is low at low speeds and prone to breakdown at high speeds; biomass plant performance drops at low temperature and so on. Compared to the conventional sources, their supply is irregular, uncertain and has low energy densities, making them less reliable. A hybrid renewable energy system attempts to overcome these drawbacks by integrating more than one power generation system into a single unit.

It usually combines renewable and conventional sources along with storage units to provide greener and more reliable energy. Different storage systems can be used to ensure an uninterrupted power supply when the power obtained from the sources is not enough to feed the load. Recent studies focus on the integration of multiple renewable energy systems such as solar, wind, biomass, micro-hydro and tidal and their various permutations and topologies, depending on the availability and feasibility (Oladigbolu et al. 2020). Many researches on hybrid wind and PV systems have been carried out in terms of HRES modelling, size optimization, reliability analysis, environmental and economic assessment. The application of biomass and micro-hydro systems has also been studied. This review aims to highlight and analyse advances in the sector at present times. The block diagram of a general HRE System is shown in Fig. 11.1.

11.2 Components and Modelling of HRES

The energy generation, control and storage elements used in HRES vary from application to application depending mainly upon availability, demand and affordability. After the selection of components, firstly, mathematical modelling for each component needs to be done, for further optimization (as discussed in the next section). HRES are either grid-connected or off-grid. While grid-connected systems have the advantage of selling surplus power and drawing power when demand is high (Taha et al. 2018), off-grid systems are suitable for isolated remote locations (Rezk et al. 2020). The components and their general modelling approach are described below.

11.2.1 Solar Photovoltaic Module (PV)

Solar power's potential is enormous and abundant and is thus used in most HRE systems. However, time duration and quantity of solar irradiation received varies globally, so every location is not equally potent. Solar energy has an intermittent nature, thus usually used in conjunction with power storage systems (Tsai et al. 2020). The general expression for output power of the solar PV i.e. $P_{pv}(t)$ in terms of the atmospheric temperature and solar irradiation is (Li et al. 2019):

$$P_{pv}(t) = Y_{pv} \times D_{pv}\left(\frac{R_t}{R_{tSCT}}\right)\left[1 + \partial_p(T_c - T_{cSCT})\right] \tag{11.1}$$

where based on standard testing conditions, the rated power capacity of the solar module is denoted by Y_{pv} (kW), D_{pv} denotes the derating factor of solar module in percentage, R_t denotes the incident solar radiation in kW/m^2, R_{tSCT} is the radiation (kW/m^2), ∂_p is the power temperature coefficient (%/°C), T_c is the PV cell temperature (°C), T_{cSCT} is the solar module temperature at standard testing conditions. The variables are obtained from historic Meteorological database of the region of implementation. The useful incident solar radiation depends on PV module tilt angle, latitude of the region, etc., so an attenuated value is usually used for more practical implementation (Singh et al. 2018).

11.2.2 Wind Energy Conversion System (WECS)

One of the fastest-growing and most promising sources of energy is WECS, because of its wide availability and improvement in technological aspects. Wind turbines (WT) are often used with PV systems to mitigate their seasonal intermittent nature (Elkadeem et al. 2020). The performance of the wind turbine is dependent upon the weather conditions, its installation height and its components' efficiency. Its output power is given by Murugaperumal and Raj (2019):

$$P_w(t) = 0.5\eta_w\eta_g\rho_a C_p A v_r^3 \tag{11.2}$$

where $P_w(t)$ denotes the output power of turbine at some time instant t, η_w is the turbine's efficiency, η_g is the generator's efficiency, ρ_a denotes air density, C_p is the turbine's power coefficient and the swept area is denoted by A and v_r denotes the wind speed corresponding to the turbine-hub height. Maximum Power Point

Tracking (MPPT) device is necessary for both photo-voltaic and wind power systems to derive maximum possible power from the solar irradiation and dynamic wind speed conditions. Kumar et al. (2017) use RBFN-based intelligent MPPT controller aiming for reduction in cost and complexity of MPPT systems.

11.2.3 Diesel Generator (DG)

Diesel run generator is primarily used in HRE System as an emergency back power supply, to improve the system reliability. Since it is a non-renewable source and a source of GHG, optimization is done to reduce the usage of DG by optimal scheduling. Sawle et al. (2018) discussed the advantages of using optimal dispatch strategies with Variable speed diesel generator (VSDG). Rezk et al. (2020) uses optimization and energy control dispatch strategies to reduce emissions and fuel consumption. The diesel fuel consumption is largely determined by power output of DG, and it is calculated as follows (Habib et al. 2019):

$$Fuel_{c,DG} = a.P_{DG,n} + b.P_{DG,g} \tag{11.3}$$

where $Fuel_{c,DG}$ is the fuel consumption rate (L/h), the nominal & generated power of the DG are denoted by $P_{DG,n}$ and $P_{DG,g}$ (kW), while diesel fuel consumption curve coefficients are a and b.

11.2.4 Fuel Cell (FC) and Electrolyser

FC processes the chemical energy obtained by combining hydrogen (H_2) gas and oxygen (from the air) and transforms it into electrical energy without combustion. Excess energy is diverted to electrolyser, which breaks water to produce hydrogen. This hydrogen is then stored in compressed hydrogen storage tanks as power backup when there is energy deficit (Ani 2021). Khalilnejad et al. (2018) discusses the optimization of solar/wind hybrid systems with electrolyser for maximum production of hydrogen. The hydrogen production by electrolyser can be calculated as (Gabbar et al. 2020):

$$H_{pr}(t) = \frac{\eta_{ele} \times E_{ele}(t) \times D_t}{HV_H} \tag{11.4}$$

where η_{ele} is the electrolyser efficiency, $E_{ele}(t)$ denotes electrolyser input energy (kWh), D_t denotes hydrogen density, HV_H is the heating value of H_2 under standard testing conditions (3.4 kWh/m^3).

Eriksson and Gray (2017) reviews integration of fuel-cell in HRE systems, its optimization and analyses its various aspects and challenges. The fuel-cell energy produced can be obtained by Gabbar et al. (2020).

$$E_{FC}(t) = \frac{[H_t(t-1) - H_t] \times \eta_{FC} \times HV_H}{D_H} \tag{11.5}$$

where η_{FC} is the FC efficiency, and H_t denotes the quantity of hydrogen stored (in kg) at time t.

11.2.5 Micro Hydro Power Plant

Small-scale hydro power plants generate energy from flow of water in streams. A typical micro-hydropower system produces up to 100 kW by using turbine-alternator systems (Wang et al. 2019). They are efficient, reliable with no need for reservoir. Tiomo and Wamkeue (2019) discusses the modelling of micro-hydro power generation system for small scale applications. The output of hydropower generator can be estimated using the following equations (Gabbar et al. 2020; Oladigbolu et al. 2020):

$$H_{eff} = H_a(1 - H_{loss}) \tag{11.6}$$

$$P_h(t) = H_{eff} \times \rho_w \times g \times Q_t(t) \times \eta_t \tag{11.7}$$

where H_{eff}, H_{loss}, H_a, ρ_w, g, $Q_t(t)$ and η_t represents effective water head (in m), head loss of pipe, available water head, density of water, gravitational constant, water flow rate and turbine efficiency respectively.

11.2.6 Biogas Plant

Biomass is organic matter used as fuel or for the production of Biogas which is a secondary fuel. Biomass is produced in form of wastes from industrial sludge that are biodegradable, crop residues based agricultural waste, cattle dung, food waste and other forest bio-energy resources. Choice of biomass depends on the availability and calorific value. Sawle et al. (2018) demonstrates the use of rice husk as a source of biomass energy. Sawle et al. (2021) uses rice straw, cow manure and other wastages along with rice husk in this case. Das et al. (2019) considered the use of biogas generator. The power produced from the bio-gas plant is given by:

$$E_{BG}(t) = \eta_{BG} \times F_{BG}(t) \times LHV_{BG} \tag{11.8}$$

where η_{BG} is the electrical efficiency, $F_{BG}(t)$ is the consumption of biogas in m^3/hour, and LHV_{BG} is the lower heating value of biogas in kWh/m^3.

11.2.7 Storage

Energy storage devices are a necessary component of a HRE System. It acts as a backup and improves the reliability by mitigating the effects of intermittent behaviour of some renewable energy sources (Rinaldi et al. 2021). They are utilized for providing flexibility, smoothening fluctuations, shaving off peak and increasing the response speed when needed. Storage of energy can be classified as mechanical, chemical or electrical in terms of forms of energy storage and in terms of duration of storage as short term, medium term or long term.

Battery energy storage system (BESS) is the most used method of energy storage (Murty and Kumar 2020). Mainly Lead-acid type (Rezk et al. 2020) and Lithium-ion (Falk et al. 2020) type storage is preferred. Pumped hydro storage (PHS) (Diab et al. 2020) and Hydrogen storage (Xu et al. 2020) are two other popular methods for long term energy storage purposes. Nguyen XP (Nguyen and Hoang 2020) presents a structure of Flywheel Energy Storage System to be used in microgrids as a regulation element. Romanos et al. (2019) discusses use of thermal energy storage tanks for isolated power systems. Akram et al. (2017) used super-capacitor along with BESS to handle highly fluctuating loads. They make use of complementary properties of SC and BESS and provide large economic energy supply and fast dynamic response at the same time effectively. Their paper demonstrates the use of hybrid energy storage system to enhance the performance and ensure longer discharge life cycle.

11.2.8 Power Converter

In a typical HRES application, both DC and AC sources and loads are utilized. PV and storage systems are connected to DC bus whereas load, generators and wind energy converters primarily connect to AC bus. These buses are linked with AC/DC power converters. The size and capacity of the converter depend on the system ratings and the power conversion requirements. The power converter's installed capacity size is considered to be higher than electric power transferred in either direction as shown (Elkadeem et al. 2020):

$$\eta_{DC/AC}^{Con} \times [P_{BAT}(t) + P_{PV}(t) + P_{FC}(t)] \leq N_{conv} \tag{11.9}$$

$$\eta_{AC/DC}^{Con} \times [P_{wt}(t) + P_{DiG}(t)] \leq N_{conv} \tag{11.10}$$

where, $\eta_{DC/AC}^{Con}$ and $\eta_{AC/DC}^{Con}$ are the AC to DC power conversion efficiency and vice versa. N_{conv} denotes the converter capacity in kW and P(t) denotes the power supply of each component.

11.3 Sizing and Performance Optimizations

Optimal sizing and control of HRES are necessary for optimal resource management. If the HRES is over-sized, there will be an increase in net present cost (NPC) and the project may not be suitable from the economic point of view. Alternatively, if the HRE system is undersized, it may not be able to deliver the required load in peak load, thus the system's reliability will be compromised. Hence, optimization techniques must be used to make a trade-off, to improve one or more performance metrics discussed in the next section. Optimization aims to fulfil the necessary objectives by balancing various parameters within the constraints defined. Since sizing and cost optimization is an important and necessary step in the design of HRES, several methods have been explored over the years in different configurations to find a quick, accurate solution. These optimization methods can be classified as follows.

11.3.1 Classical Optimization Techniques

Classical optimization methods make use of differential calculus for finding optimal solutions to continuous objective functions. It can take inequality constraints into consideration, however; non-differentiable functions cannot be optimized using these methods. Dahlan et al. (2021) utilizes linear programming model to optimize HRES and study its economic and social impacts. Taha et al. (2018) used mixed-integer linear (MIL) programming using model predictive control for optimizing total operational cost and minimizing pollutant emissions. Lee and Kum (2019) shows application of optimal control based on dynamic programming and compares its performance to simple and advanced rule-based (RB) control respectively.

11.3.2 Meta-Heuristic Optimization Techniques

These are computational procedures that iteratively try to reach the optimal solutions by following a set of rules. Their primary objective is to provide a reasonably accurate solution within limited time. Some of the popular heuristic optimization techniques include Particle swarm optimization (PSO), Genetic algorithm (GA), Ant colony (AC) algorithm, Artificial bee colony (ABC) etc. PSO is inspired by swarm behaviour of social organisms and is often used as a meta-heuristic tool for many analyses. Gabbar et al. (2020) implements Particle Swarm Algorithm in the MATLAB environment to minimize net present cost. Liaquat et al. (2020) analyses the APSO and firefly algorithm for hybrid system scheduling. Kharrich et al. (2021) applies Equilibrium Optimizer (EO) for optimal performance of Solar/Wind/DG/Battery based microgrid and compares it to popular optimization techniques like Artificial Electric Field Algorithm (AEFA), Harris Hawks Optimizer (HHO), Sooty Tern Optimization Algorithm (STOA) and Grey Wolf Optimizer (GWO). It is found that EO provides better solution to sizing as compared to others comparing Net Present Cost. Chen et al. (2019) implements HRES using Improved Brain Storm Optimization for minimising ACS, LPSP and total fuel emissions. Sawle et al. (2018) used PSO, BFPSO, GA and TLBO optimization techniques are used for optimization for six different configurations. As per analysis, the TLBO outperforms BFPSO, PSO and GA. Diab et al. (2020) compares Water Cycle Algorithm (WCA), Whale Optimization Algorithm (WOA), Grey Wolf optimizer (GWO) and Salp Swarm Algorithm (SSA), and finds that over other techniques, WOA offers the best performance.

11.3.3 Hybrid Optimization Techniques

In order to utilize the positives of multiple techniques, hybrid techniques are used by combining classical and heuristic or two or more heuristic techniques. Geleta and Manshahia (2021) implements "hybrid of grey wolf optimization and genetic algorithm" (HGWOGA) in MATLAB environment. Compared to GA, GWO, artificial bee colonies (ABC) and PSO, this method offers superior results. For multi-objective constraint models, Chen et al. (2021) presented a method, by integration of relaxed constraint handling technique into NSGAII (Non-dominated sorting algorithm II). The method they derived is thus denoted as ϵ-CNSGAII. In this study, this method is found to be more effective than NSGAII.

11.3.4 Software-Based Optimization Techniques

Many softwares have been developed to provide a user-friendly interface and facilitate faster optimization such as HOMER, Lingo, iHOGA, etc. HOMER is one of the most widely used commercially available HRE Systems design software programs. HOMER stands for Hybrid Optimization Model for Electric Renewables. Elkadeem et al. (2020) utilizes HOMER pro to minimize the total LCC and simulation and optimization analysis of technical–economic parameters is performed to find the optimum size of HRE-based micro-gird. Wang et al. (2019) converts multi-objective model into an equivalent single-objective model utilising ε-constraint method and solves the model using LINGO 13.0 software. Software like MATLAB/Simulink (Gabbar et al. 2020) and TRNSYS (Anoune et al. 2019) are utilized to design models and simulate but do not directly assist in optimization.

11.4 Performance Metrics Analysis

Optimization is done based on various technical, economical and environmental parameters. Objective function is generally formulated taking these parameters into account. Optimum results thus obtained are used as performance metrics to judge the economic viability, reliability and effects on the environment. These metrics can be broadly classified as follows:

1. Cost indices—To measure the economic viability of the HRE system
2. Reliability indices—To determine the usefulness and reliability of the HRE system
3. Environmental indices—To provide a quantitative measure of effects on the environment.

The different HRES components & optimizing techniques and the Objective, Methodology and Outcome Summary on HRES are tabulated in Tables 11.1 and 11.2, respectively.

11.4.1 Cost Indices

Cost analysis and optimization are central to HRES design because the components used tend to be costly. Hence various parameters are used to determine the economic feasibility of the study.

11.4.1.1 Net Present Cost (NPC)

NPC of any piece of equipment is the sum of the present-day value of total capital cost, cost of replacement, operating and maintenance cost and cost of fuel. Oladigbolu et al. (2020) compares different configurations in terms of their total NPC to find the scheme that is most economic amongst the studied schemes. Kaluthanthrige et al. (2019) uses HOMER to optimize NPC to obtain a rough estimate of the cost of operating the system over its life cycle.

11.4.1.2 Cost of Energy (COE)

Energy costs associated with generating one-kilowatt-hour are defined as COE. COE facilitates the comparison of different designs and configuration, that have a vast difference in cost of installation, capacities, return and lifetime. Elkadeem et al. (2020) used different configuration plans to achieve low levels of COE by maximizing utilization of renewable energy system.

Table 11.1 HRES components and optimizing techniques

S. No.	Grid type	Sources	Storage	Optimization technique
1	Off-grid & Grid connected	PV, Wind, Non-intermittent source	Battery	Reformed Electric System Cascade Analysis (RESCA)
2	Off-grid	PV, Wind energy, Biomass, Diesel	Battery	Particle swarm optimization (PSO), Teaching–Learning-Based Optimization (TLBO), Genetic algorithm (GA), Butterfly PSO (BFPSO) and
3	Off-grid	Wind, PV	Battery	Reformed Electric System Cascade Analysis (RESCA)
4	Grid connected	Diesel, Wind, Solar	Battery	Mixed Integer Linear Programming
5	Off-grid	Diesel, PV	NaS Battery	Mixed Integer Linear Programming
6	Off-grid	Wind, Solar	Battery	HOMER
7	Off-grid	Wind, Biomass, Solar	Battery	HOMER
8	Off-grid	Biogas, PV	Battery, Pumped hydro energy storage (PHES)	Moth-Flame Optimization (MFO), Water Cycle Algorithm (WCA)
9	Off-grid	Diesel, Wind	Battery	HOMER
10	Off-grid	PV, Micro-Hydro	Pumped hydro storage (PHS)	Modified Multi-Objective PSO
11	Off-grid	Diesel, Wind, Solar	Battery	Brain Storm optimization (BSO)
12	Off-grid	Diesel, Wind, Solar	Battery	HOMER
13	Off-grid	Diesel, Solar	Battery	HOMER
14	Off-grid	PV, Wind	Battery	Opposition based Social Spider Optimization (OSSO)
15	Off-grid	PV, Wind	Battery	Dynamic simulation
16	Off-grid	Diesel, Wind, Solar	Battery	Dynamic programming (DP)
17	Off-grid	Diesel, Wind, Solar, Fuel cell	Battery	HOMER pro
18	Off-grid	Diesel, Wind, Solar	Battery	HOMER
19	Off-grid	Diesel, Solar	Battery	HOMER
20	Off-grid	PV, Wind, Hydro, Nuclear, Fuel-cell	Hydrogen Storage, Battery	Particle swarm optimization (PSO)

(continued)

Table 11.1 (continued)

S. No.	Grid type	Sources	Storage	Optimization technique
21	Off-grid	PV, Wind, Diesel, Micro-Hydro	Battery	HOMER
22	Off-grid	PV, Diesel	Battery	HOMER
23	Off-grid	PV, Wind, Diesel	Battery	Nomadic People Optimizer (NPO)
24	Off-grid	PV, Wind, DG	Battery	Social Spider Optimizer (SSO)
25	Grid-connected	PV, Wind	PHS	Water Cycle Algorithm (WCA), Whale Optimization Algorithm (WOA), Grey Wolf optimizer (GWO) and Salp Swarm Algorithm (SSA)
26	Off-grid	Fuel cell, Wind, Solar	Hydrogen Storage	Modified Non-dominated Sorting GA (NSGA-I)
27	Grid-connected	Diesel, Wind, Solar, Biomass, Hydro	Battery	HOMER
28	Off-grid	PV, Wind, Diesel	Battery	Equilibrium Optimizer (EO)
29	Off-grid	PV, Wind, DG	Battery	HOMER
30	Grid-connected	Wind, Solar	Battery	Particle Swarm Optimization (PSO)
31	Off-grid	Diesel, Wind, Solar	Battery	HOMER
32	Off-grid	PV, WT	Battery	Improved Crow Search Algorithm
33	Off-grid	Diesel, Wind, Solar	Battery	ϵ constraint Non-dominated sorting algorithm II (ϵ-CNSGAI)
34	Off-grid	Diesel, Solar	Battery	Linear Programming Model
35	Off-grid	Wind, Solar	NIL	HOMER

11.4.1.3 Annual System Cost (ASC)

ACS consists of various components of system-related costs incurred yearly like total capital cost, maintenance cost and replacement cost. Singh et al. (2018) proposed use of ASC as a secondary constraint for HRE system optimization to get the idea about initial investment required for such system. ACS is also used for comparison of different configurations.

Table 11.2 Objective, methodology and outcome summary on HRES

S. No.	Author	Objective	Methodology	Outcome
1	Singh et al. (2018)	To present the application of RESCA (Reformed electric-system cascade analysis) for optimization of grid connected and isolated HRES	Four systems configurations with Photo-voltaic systems, Wind energy conversion systems, grid and non-intermittent source are simulated in MATLAB environment, programmed with RESCA optimization method for a period of one month	RESCA optimization method shows improved versatility and robustness while improving upon the drawbacks of other optimization techniques
2	Sawle et al. (2018)	To design HRES, comparing different configurations and optimization techniques, optimizing based on technical, economical and social parameters	Design of HRES by minimising a multi-objective function based on socio-techno-economic parameters. TLBO, GA, PSO and BFPSO optimization techniques are used for optimization for six different configurations and compared	It is found that TLBO is better than PSO, BFPSO and GA in terms of performance according to the analysis of results for six cases. Also configuration with Biomass, Solar, Diesel & Battery was found to be the most efficient solution
3	Singh and Bansal. (2018)	To implement Reformed Electric System Cascade Analysis (RESCA) technique to optimize an autonomous HRES consisting of PV, WECS and BESS	HRES comprising of PV-Wind-BESS is optimized with FEE, NPC, COE, EGR and REF constraints using RESCA technique. Results are compared with HOMER stimulation	RESCA has shown versatility in handling multi-criteria & multi-source optimization. The results are comparable to HOMER, validating the method. Various other advantages are highlighted

(continued)

Table 11.2 (continued)

S. No.	Author	Objective	Methodology	Outcome
4	Taha et al. (2018)	To design an online energy management system based on MPC, of the HRS using 2 level optimization: rolling horizon & feedback correction to compensate for the uncertainties in the energy system and minimize operation cost	EMS is designed with two levels of optimization algorithm: Rolling optimization & feedback intra-sample correction methods. Model Predictive Control (MPC) approach is used in rolling optimization. Mixed-integer linear (MIL) programming framework was used to formulate the optimization problem. MATLAB with YALMIP toolbox is used to implement EMS for optimization	The result indicates reduction in the cost of system operations when the EMS developed is applied in a deregulated electricity market. But in flat-shaped pool price, the total system cost was marginally increased over the traditional one
5	Dolatabadi and Mohammadi-Ivatloo (2018)	To analyse a risk-based two stage stochastic method in order to investigate the optimal sizing of the hybrid PV/diesel/storage merchant marine vessel HRES	A general hybrid energy-powered ship model with the DG, ESS and solar energy system has been considered. Monte Carlo simulations are used to account for solar radiation intermittency in the optimization problem. CVaR (conditional value-at-risk) methodology is used to identify the potential risk of the sizing problem	Based on the results, the proposed framework to study the impact of solar insolation uncertainties and potential risks on the system merchant ships' sizing was found to be both efficient and applicable

(continued)

Table 11.2 (continued)

S. No.	Author	Objective	Methodology	Outcome
6	Mohammadi et al. (2018)	To study the Optimal planning of RES for residential housing, considering multiple reliability and economic criteria	The demand and resources are assessed to determine the possible configurations. HOMER software is used for modelling the system and optimal planning of HRE System considering impact on reserve system. The best HRES combination is selected based on a trade-off between economic and reliability criteria	A summary of the characteristics of the final optimal solution has been provided for this case study. Low cost reliable system is found after optimization
7	Murugaperumal and Raj. (2019)	To find a cost-effective solution for electrifying isolated rural areas with hybrid energy that is technically feasible and economically viable	ANN is used to anticipate future loads from existing data and growth rate is obtained based on Levenberg Marquardt (LM) and feed-forward back propagation (BP) network type of training functions. Then HOMER is used for sizing optimization and the model is simulated in MATLAB environment	ANN predictions are found better than previously used regression trees and linear regression method. A cost-effective solution for rural development is provided by this method, along with accurate load predictions

(continued)

Table 11.2 (continued)

S. No.	Author	Objective	Methodology	Outcome
8	Das et al. (2019)	To optimize HRES using WCA (Water Cycle Algorithm) and MFO (Moth-Flame Optimization) meta-heuristic techniques for powering an Indian radio transmission station	HRES is modelled with PV-BG-PHES-Battery combination and optimized using WCA and MFO, and GA is used as benchmark to compare their techno-economic performances of the algorithms	In terms of techno-economic design of the HRES, WCA algorithm is found to outperform MFO algorithm. It also outperforms other algorithms in terms of LCOE. It also reveals that HRE system based on isolated Solar-Biogas-PHES-Battery, is effective for a radio transmission station in a standalone manner
9	Habib et al. (2019)	To design techno-economic based HRES simulation and optimize Wind-DG-battery-converter configurations and PMS for load balance using Model predictive control (MPC) approach	Using HOMER software, different HRES configurations are modelled, examined and compared in terms of economical, technical and environmental parameters and optimal sizing is analysed. Then suitable PMS is implemented in MATLAB/Simulink. The parameters for each configuration are compared	The optimized PV/Wind/DG/battery configuration with power management and controllers, adequately supply load with promising results under different conditions while carbon emission and fuel consumption are minimized
10	Li et al. (2019)	To optimize HRE systems integrated with Solar, Micro-Hydro & PHS considering similarity of load/generation and evaluate the model's performance	Solar-MHP-PHS hybrid system is modelled and studied by use of MOPSO based on similarity of generation or load and economy from single objective and double objective perspectives. The economic revenue (ER) and similarity value (SIM) are analyzed for their sensitivity	Here two single-objective optimization processes are used to verify trade-off characteristics between similarity of load /generation and the economic analysis of Pumped-hydro storage is done. The optimal operating strategy is determined and no financial penalties were found

(continued)

Table 11.2 (continued)

S. No.	Author	Objective	Methodology	Outcome
11	Chen et al. (2019)	To optimize HRES using an improved BSO (Brain Storm Algorithm) algorithm for minimising ACS, LPSP and total fuel emissions	BSO algorithm is worked to optimize the ACS, LPSP and the total fuel emissions. Simulation of a time series for one year is carried out and different solutions are compared	The BSO algorithm is found to have a faster convergence rate, and thus optimal values of Pareto solution are found quicker compared to other similar algorithms
12	Razmjoo and Davarpanah (2019)	To analyse Solar-DG-Wind based HRE System from the technical and economic point of view, in Semirom city of Iran and study its feasibility and sustainability	HRES modelled and optimization on meteorological data of Semirom city is done using HOMER software. Economic and environmental parameters are calculated and analysed	According to the calculated analyses, the potential of solar-wind energy resources of the city is adequate to construct an HRE plant and such construction, is found to be an economic investment
13	Kaluthanthrige et al. (2019)	To identify the optimal power system configuration for an average isolated community in Northern Canada integrating Solar, DG and battery as possible energy options	System is modelled and optimized in HOMER for Net Present Cost (NPC) in order to denote the LCC for one year. Then economic, reliability and environment assessment is done	Results demonstrate the suitability of Solar-DG-Battery configuration at high degree of renewable penetration and reduction in fuel usage and environment protection compared to DG system

(continued)

Table 11.2 (continued)

S. No.	Author	Objective	Methodology	Outcome
14	Sandeep and Nandihalli (2020)	To study the use of opposition-based social spider optimization (OSSO) for design of HRE systems	PV-battery-wind energy system is designed. Optimization is done in MATLAB environment for finding optimal quantities of solar-wind components & amount of battery power backup necessary in order to serve the demand economically. AGA-Cauchy, AFSO, PSO & SSO swarm intelligence algorithms were used for optimal size determination & compared to OSSO	The results of the study reveal that OSSO is a proficient swarm intelligence algorithm compared to other techniques and is efficient in providing solutions
15	Anoune et al. (2019)	To design and size HS based of solar and wind energy system in order to fulfil load demand of industrial laboratory prototype and investigate its techno-economic-environmental feasibility	A HRE System design based on TRNSYS and Matlab is studied in order to evaluate and implement weather conditions. A dynamic simulation of energy output performance for different configurations of HRES is analyzed for duration of 1 year based on load profile. Techno-economic-environmental assessment is done for the industrial laboratory prototype	It is found that to solve the sizing problem optimally, each selected location has specific configuration requirements. The alternative is found to be more affordable than conventional sources of energy

(continued)

Table 11.2 (continued)

S. No.	Author	Objective	Methodology	Outcome
16	Lee and Kum (2019)	To explore the complete design and configuration of HRE system by dynamic programming (DP) method based optimal control	Optimal design is determined on the basis of optimal control trajectory using dynamic programming, considering four economic parameters. Influence of sizing of each component is studied. DP and rule-based (RB) control are compared to investigate the impact of energy dispatch strategy	It is found that the system's overall economic performance is mainly determined by the sizes of PV & WECS and battery system as opposed to the DG-size. Also, increasing PV size in proportion to BESS is necessary, to reduce net operating cost
17	Elkadeem et al. (2020)	To plan and access HRE based microgrids with techno-environmental-economic optimization with case study in Egypt	Different configuration scenarios are modelled using HOMER pro and compared based on NPC saving, COE saving and GHGE	Amongst all feasible configurations simulated, the Solar/ Wind/ FC/ Battery/ CONV HRE-microgrid was found to be superior for fulfilment of the energy demand. The simulation shows substantial decrease in NPC and COE and GHGE
18	Murty & Kumar. (2020)	To evaluate the techno-economic benefits of standalone microgrids with hybrid renewable energy sources and battery storage and compare different configurations	HRES based on PV/WT/DG/BESS are simulated and optimized in HOMER software with NPC, COE & GHGE as constraints. Performance of eight different configurations is compared. Impact of demand response is also demonstrated	PV + BESS is found to be the most economically viable configuration Compared to the conventional grid power distribution, the cost-of-energy in case of microgrid is greater. However, GFG emissions were reduced by 68% compared to conventional energy sources

(continued)

Table 11.2 (continued)

S. No.	Author	Objective	Methodology	Outcome
19	Tsai et al. (2020)	To study the techno-economic analysis of isolated DG system and isolated Solar/DG/energy storage configurations for application in the Pratas island, Taiwan	Different configuration scenarios are modelled using HOMER and compared based on NPC, COE, fuel saving and the CO2 emissions are studied	It is found that the cost-of-energy of Solar/DG/storage hybrid is higher than that of isolated DG system. A reduction of 31.63% is achieved in the annual total carbon dioxide emissions, which is of great benefit to the environment
20	Gabbar et al. (2020)	To evaluate nuclear-renewable hybridization methods with different coupling techniques and identify the most effective one	Three different methods of hybridization -Direct coupling, multiple product-based coupling and single resource multiple product-based coupling and multiple resources are evaluated in MATLAB environment using PSO optimization. Their NPC, reliability and sensitivity are compared to find the best method	All the systems are found to be reliable however; "Multiple Resources and Multiple Products-based nuclear-renewable HRES" was the most efficient hybrid system in practical scenario, in terms of load fulfilment, dynamic performance and profitability
21	Oladigbolu et al. (2020)	To perform comparative analysis of HRES and determine its feasibility for isolated rural electrification in a rural area	HOMER was used to simulate four configurations and NPC, COE, operation cost, excess electricity, renewable fraction, fuel consumption of generator, load fulfilment and CO_2 emission savings were determined for each case and compared	Hybrid hydro/solar/diesel /wind /battery system was found to be the most suitable option for providing power to remote village area. This system is both environment friendly and economically feasible

(continued)

Table 11.2 (continued)

S. No.	Author	Objective	Methodology	Outcome
22	Rezk et al. (2020)	To analyse feasibility, technical-economical parameters and management of power for isolated hybrid PV/ Battery/DG-based HRE System	A real case study based on Minya city, Egypt is considered. HOMER software is used for optimal sizing and to obtain a strategy for energy management. Load following control strategy is used. During the analysis of the proposed system, grid extension and isolated DG system were compared	It is found that all configurations of hybrid Solar/DG/Battery are more economically feasible in comparison of with grid connection for RO-500 unit however grid connection is more economical for lower capacity RO-250 unit
23	Mohammed et al. (2020)	To develop HES for residential complex in Iraq using NPO (Nomadic People Optimizer)	Different Solar /wind turbine/diesel generator and battery configurations are designed, optimized using Nomadic People Optimizer (NPO) and their performances are compared	The solar/wind/large DG/battery system when the batteries are charged by the DG, was found to be most effective in terms of LCC^T, COE, Total Dump Energy and total CO_2 emissions
24	Fathy et al. (2020)	To find the design variables including number of wind turbines, number of solar panels and number of battery autonomy days in order to minimize cost of energy of a HRE system based integrated microgrid	Hybrid Solar/Wind/Battery/DG integrated microgrid is designed and Social Spider optimizer (SSO) is utilized for determination of the optimal size. A proposed energy management strategy (EMS) manages the distribution of energy between different components. Sensitivity analysis with different configurations is obtained	Results display that the approach proposed in the paper provides optimal configuration of HRE System in comparison with MVO, GWO, HHO, ALO & WOA, also proposed system is superior in reliability and cost-effectiveness

(continued)

Table 11.2 (continued)

S. No.	Author	Objective	Methodology	Outcome
25	Diab et al. (2020)	To study the HRE system involving grid-connected Solar/wind/PSH system for Ataka region, Egypt; using different optimising algorithms	MATLAB has been used to test the different optimization algorithms. Detailed comparative analysis, including parameters of HRE-system at 10 iterations and at the best scenario between WCA, WOA, GWO and SSA, is presented	Results prove that in comparison with other algorithms, WOA shows the most promise. WOA produces most accurate COE and superior in convergence speed. Additionally, the design is acceptable with respect to COE and LPSP
26	Xu et al. (2020)	To investigate the optimal configuration of an isolated wind/solar/hydrogen system using modified NSGA-II & CRITIC-TOPSIS	Decision framework involving dual stages is applied. A reinforcement learning method is used to find pareto solutions based on (NSGA)-II. Using the CRITIC method, importance is given to the three objectives, while the TOPSIS method is employed to determine the best result amongst the Pareto solutions. Applications are analysed for verification	Load demand changes are found to affect LCOE values more so than any other factor. RL-based NSGA-II is found to be superior to that of conventional NSGA-II for obtaining Pareto front. This dual-stage decision framework proves to be useful for the development of HRES
27	Sawle et al. (2021)	To perform financial and technical evaluation and estimation of renewable factor in order to determine the amount of harmful emissions, and analysing the sensitivity of HRES	Analysis of four different configurations RES is done using HOMER pro. The size cost and %RF of the configurations are compared and sensitivity is studied	It is found that Solar-Wind-MH-CT-BT-DG-BG is superior in terms of COE, NPC and cost of operation. It is also more environment friendly and reliable

(continued)

Table 11.2 (continued)

S. No.	Author	Objective	Methodology	Outcome
28	Kharrich et al. (2021)	To design a Solar/Wind/DG/ battery system based microgrid, with Equilibrium Optimizer for Optimal Design with analysis and comparison	EO approach is used for optimal design of HRE-based microgrid. The objective function is based on NPC and LCOE. It is compared with recent algorithms like HHO, AEFA, GWO and STO. Two stages are involved in the sensitivity analysis for checking variations in different parameters	Results show that performance is best with EO where it is quicker than AEFA, HHO, STOA and GWO. It also shows that EO is more economical compared with other algorithms
29	Shafiullah (2021)	To develop a Hybrid REPS considering COE, NPC and RF and determine the feasibility of different renewable energy systems that can fulfil the present and projected future load demand of rural region of Chipendeke, Zimbabwe	A model is developed to estimate possible power demand of the Chipendeke community. Using HOMER, development of models is done and optimization. To find an optimum model, sensitivity analyses of the models are done for different input parameters	Among all cases, isolated hybrid Micro-hydro/Solar/Diesel/Battery storage system was found to be cost-effective. This model is found suitable for application for socio-economic growth
30	Martinez-Rico et al. (2021)	To perform a production scheduling optimization for the day-ahead market is conducted in an HRES plant with a BESS using Particle Swarm Optimization	Multi-objective cost function optimizes both the net profitability of the plant and the loss of value of the battery, which are calculated with the number of cycles and the change of market price. The global performance and profit of the plant are optimized by implementing a PSO algorithm. The data is simulated for a complete year	It is found that global performance and profit of the plant are improved by implementing a PSO algorithm. However, the profitability is low by performing energy arbitrage uniquely in the day-ahead market

(continued)

Table 11.2 (continued)

S. No.	Author	Objective	Methodology	Outcome
31	Rinaldi et al. (2021)	To analyse techno-economic aspects of hybrid systems (Solar–wind–DG) to electrify Peru's off-grid communities and determine optimal configuration for the system	Different locations are selected for seven scenarios simulations. A system's optimal size is determined by the use of the HOMER software. Different configurations are then compared using economic indices, renewable fraction and emissions	The results show that of different investigated configurations, solar–wind–DG hybrid system is the most cost-effective configuration, despite of having high initial cost. CO2 emissions are significantly reduced
32	Moghaddam et al. (2021)	To design a hybrid isolated solar/wind/battery energy system to reduce the overall expenses of a hybrid system and enhance reliability for the city of Zanjan, Iran	Hybrid Solar/WT/Battery is modelled and improved crow search algorithm is utilized for optimization on the basis of mutation and crossover operators of the GA. Monte Carlo simulation is used to include uncertainty for optimal design. The cost and reliability parameters are compared	It is found that compared to conventional CSA and PSO methods; the improved crow search algorithm provides superior results. Although the system cost increases, the reliability is considerably improved
33	Chen et al. (2021)	To develop constraint multi-objective a modelling approach to optimize HRE System sizing in cases of high-load demand of intermittent nature	Optimum environmental, economical and reliability criteria are considered simultaneously in an optimization model for high loads having intermittent nature. ϵ-CNSGAII is presented to solve the model	It shows that the designed HRE System is reliable and delivers the load satisfactorily. ϵ-CNSGAII based optimization is found to be effective sizing strategy
34	Dahlan et al. (2021)	To access the social, technological and economic impact of HRES using linear programming model	Scenarios of HRE System with and without battery storage are considered. A linear programming model is used to optimize economic and reliability parameters. Sensitivity analysis is done with variation in load	It is found that pairing diesel generator, PV and battery to generate electricity is a cost-effective solution to successfully meet the load demand. It also shows the importance of battery to reduce losses as well as to store excess energy

Table 11.2 (continued)

S. No.	Author	Objective	Methodology	Outcome
35	Nebey (2021)	To design an optimal HRES in order to supply Ethiopian rural areas using MATLAB and HOMER	HRES is designed and optimized for NPC by using HOMER software. Fuzzy logic-based controller is designed in order to regulate the load distribution. Different supply and demand combinations are used to study performance	It is found that the HRE System along with the controller is able to meet the demand without the need of battery; thereby cost of the system is diminished. Wind energy was found to be the biggest contributor in the region studied

11.4.1.4 Life Cycle Cost (LCC)

It is defined as the summation of rehashing cost and one-time costs for lifetime duration. Khan et al. (2017) uses LCC consisting of installation cost, operation cost and initial capital cost over the lifespan of the system to optimize the system using HOMER software.

11.4.1.5 Levelized Cost of Energy (LCOE)

Levelized cost of energy is described as the cost of a unit of energy. Kharrich et al. (2021) analyses the impact of solar radiation, speed of wind, interest rate and DG fuel on LCOE. It is used as a parameter in the objective function for cost minimization. Das uses LCOE for system reliability analysis (Singh and Bansal 2018).

11.4.2 Reliability Indices

Since renewable resources are intermittent in nature, reliability analysis is necessary to determine whether the system is capable of fulfilment of load requirements. Various parameters are used to determine the reliability, such as:

11.4.2.1 Loss in Power Supply Probability (LPSP)

In case, the system generation is less than the demand of power; a power supply loss is experienced by the system. It is important to keep LPSP value below a margin to ensure system reliability. Gabbar et al. (2020) implements LPSP as a constraint in PSO optimization of the HRES design. Kharrich et al. (2021) studied the effects of variation in size on loss of power supply probability (LPSP).

11.4.2.2 Loss of Load Probability (LLP/LOLP)

It is the ratio of duration of power failure and total working time of the HRE System. Das performs a sensitivity analysis on the basis of LLP type reliability criteria for investigation of the feasibility of the HRES design (Singh and Bansal 2018).

11.4.2.3 Expected Demand Not Served

It is defined as the product of total load lost and probability of capacity outrage. Dahlan et al. (2021) uses capacity outage probability tables to determine the expected demand not served of the system to examine the socio-economic effects of HRES.

11.4.2.4 Unmet Electric Load (UL/UEL)

It is defined as the load that the HRE system is unable to serve in a given time period. It is necessary to measure UL when demand is more than the supply. Kaluthanthrige et al. (2019) uses UL as a reliability parameter in order to find out how much load was not served due to insufficiency in generation. Mohammadi et al. (2018) uses UL for selecting the most efficient combination of HRES in the case study considered.

11.4.2.5 Surplus Energy Fraction

This parameter is specific to planning of nuclear integrated HRES. For an efficient energy system, a higher percent of surplus energy fraction is unacceptable, so it needs to be minimized. Gabbar et al. (2020) uses this parameter as a constraint for optimization of Nuclear-HRE System by Particle Swarm Optimization method.

11.4.3 Environmental Indices

Reduction of greenhouse gasses and carbon emissions has been one of the focuses of recent HRES designs. Sometimes conventional sources are used as backup to intermittent ones. Therefore effect of HRES on environment is studied and optimized in its design using some indices. *Renewable energy fraction* (*REF/RF*) is widely used as a measure of sustainability of a system. REF is the indicator that shows the share of the load supplied by the renewable components of a HRE system; it can be mathematically denoted as (Singh et al. 2018):

$$REF = 1 - \frac{\sum_{t=0}^{t=T} E_{NRC}(t)}{\sum_{t=0}^{t=T} L(t)} \tag{11.11}$$

where energy generated from conventional resources is referred to as $E_{NRC}(t)$, the load component is given by $L(t)$ and analysis period is Oladigbolu et al. (2020) compares different configurations of HRE systems to minimize REF using HOMER. Rezk et al. (2020) uses renewable fraction as one of the main criteria for optimization and determined the environmental effects based on it. Troullaki et al. (2019) examines the environmental impact of isolated HRE system during its life-cycle. The locally manufactured system under consideration has proven not only to be affordable, but also to be sustainable.

11.5 Applications and Future Scope

Scope of Hybrid Renewable Energy Systems is vast and evidently it is expanding to new applications. Some of these applications have been highlighted below:

Khan et al. (2017) discusses the application of HRES in telecommunication tower industry in the cities of Punjab in India with an aim of finding the most economic configuration by decreasing dependence on fossil fuels and limiting carbon emission. For isolated areas where either grid-connection is either not possible or becomes non-economic, renewable resources of the area can be utilized to generate required energy. Tsai et al. (2020) investigate the possibility of off-grid HRES for the Pratas Island in Taiwan. They aim to reduce the fossil fuel usage cost and its impact on the environment. Oladigbolu et al. (2020) plans decentralized energy systems in isolated village areas with lack of access to power supply, which can provide better living conditions for residents in these areas. Dolatabadi and Mohammadi-Ivatloo (2018) discusses the integration of Photo-voltaic modules and battery storage with diesel generators of a merchant marine vessel. Rezk et al. (2020) discusses the evaluation and energy management of hybrid solar-battery-diesel system for pumping and desalination of water at remote areas. Dahlan et al. (2021) aims to use alternate energy resources to power the rural off-grid regions. Rzetelska and Combrinck (2021) discussed the use of an integrated HRE system to alleviate the problem of fuel poverty as a means of reducing heating system carbon emissions. Ssenyimba et al. (2020) studied the application of wind and solar-based HRE Systems for small-scale irrigation purposes.

Some areas still require further investigation has the scope of improvement in future studies as highlighted below:

1. The manufacturing cost of renewable energy systems is still high and this leads to an increase in initial investment cost and thus the payback time. Work needs to be done to improve the manufacturing process by the use of innovative technologies. It will make these systems more accessible and incentivize the industry to implement such systems.
2. Although cost of energy analysis of isolated regions has been carried out in many studies, the affordability of the power generated from the system requires more attention to determine its real utility and adoption by local residents.
3. In the applications using diesel generators as a backup supply, the use of locally produced bio-diesel should be considered as a sustainable alternative to fossil fuels. Integration of hybrid renewable resources with hybrid storage systems using micro-grid for optimum utilization is necessary for large scale applications.
4. Most HRES systems developed present use on battery storage systems. The use of integrated hybrid storage system with an appropriate control strategy should be introduced to increase the overall lifespan of the system while minimising the life cycle cost.

11.6 Case Study

The case study of Digha village of West Bengal is considered here including residential and public and commercial loads. The factors that come into consideration for the evaluation of the total loads are the no. of residential houses, types, number and rating of appliances, hours operated and individual watt-hour. From the calculation is has been generated that the high class, middle class and low class houses' loads are 646, 595 and 94.5 kWh/day respectively. The Daily Load Profile of Digha village and the geographical study location are shown in Figs. 11.2 and 11.3.

The average electric energy consumption of study area is about 2281.9 kWh/day on a daily basis. The monthly average electric energy consumption is 69.4 MWh/month, evaluating to about 832.6 MWh annually. As the main objective of the proposed hybrid renewable energy system is to lower the Net Present Cost (NPC) and the Levelized Cost of Energy (COE), the proposed system architecture is the Grid utility with just the Solar PV arrays in place. The projected annual savings on the system is ₹26,07,724. The system capital cost is about ₹2,25,16,810 and over the project lifetime of 25 years, the savings are estimated as ₹6,51,93,100. The Internal

Fig. 11.2 Daily load profile of Digha village, WB

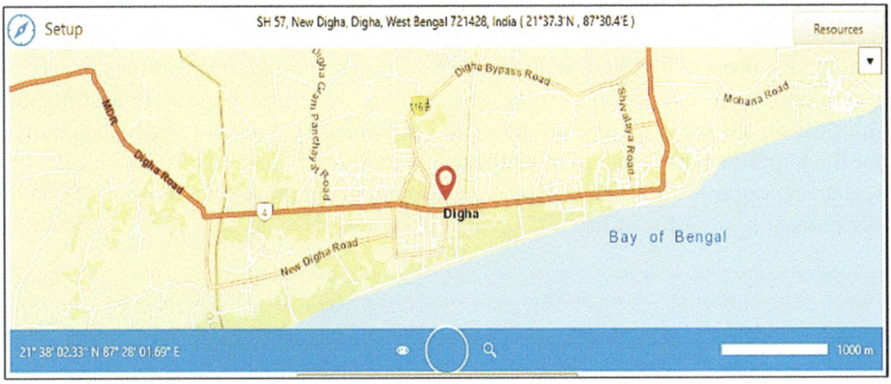

Fig. 11.3 Case study location in Digha, WB

Table 11.3 Annual electric bill for the current system architecture with grid utility

Month	Energy purchased (kWh)	Energy sold (kWh)	Net Energy purchased (kWh)	Peak load (kW)	Energy charge	Demand charge	Fixed charge	Total
January	50,756	0	50,756	353	₹3,55,295	₹5,296	₹15.00	₹3,60,606
February	45,328	0	45,328	385	₹3,17,297	₹5,768	₹15.00	₹3,23,080
March	54,137	0	54,137	390	₹3,78,957	₹5,856	₹15.00	₹3,84,828
April	50,610	0	50,610	354	₹3,54,268	₹5,308	₹15.00	₹3,59,591
May	50,626	0	50,626	338	₹3,54,383	₹5,068	₹15.00	₹3,59,466
June	50,736	0	50,736	344	₹3,55,149	₹5,162	₹15.00	₹3,60,326
July	51,090	0	51,090	343	₹3,57,632	₹5,147	₹15.00	₹3,62,794
August	54,071	0	54,071	367	₹3,78,497	₹5,504	₹15.00	₹3,84,016
September	51,061	0	51,061	333	₹3,57,426	₹4,994	₹15.00	₹3,62,435
October	51,716	0	51,716	357	₹3,62,009	₹5,350	₹15.00	₹3,67,374
November	49,599	0	49,599	319	₹3,47,191	₹4,790	₹15.00	₹3,51,996
December	51,595	0	51,595	318	₹3,61,163	₹4,772	₹15.00	₹3,65,950
Annual	6,11,324	0	6,11,324	390	₹4.28 M	₹63,015	₹180.00	₹4.34 M

Rate of Return (IRR) and the payback time are calculated to be 8.46% and 9.8 years respectively. Generic flat plate type PV arrays of 500 kW size are recommended to be installed in place having price of ₹37.00/watt. The installation cost is calculated to be ₹1,85,00,000 with an annual expense of about ₹2,50,000/yr (Tables 11.3 and 11.4).

The Detailed overview of the Utility Bill, Projected annual savings on the Utility Bill, Cumulative cash flow over project lifetime for proposed and different possible configurations are illustrated in Figs. 11.4, 11.5 and 11.6, respectively. Also, the Monthly Utility Bill savings and Time Series charts without EV charging conditions for different renewable sources and loads are demonstrated in Figs. 11.7 and 11.8, respectively. The analysis clearly indicates that the more economic solution is obtained for proposed system with solar PV. The Annual Electric Bill for the current system architecture with grid utility only and for the proposed system architecture with grid utility and solar PV are tabulated in Tables 11.3 and 11.4 which specifies that the superior performance of proposed system over current system for all technical and economic parameters. The comparision of parameters for proposed system over current system for utility bill savings is shown in Table 11.5:

Table 11.4 Annual electric bill for the proposed system architecture with grid utility and solar PV

Month	Energy purchased (kWh)	Energy sold (kWh)	Net energy purchased (kWh)	Peak load (kW)	Energy charge	Demand charge	Fixed charge	Total
January	19,243	20,754	1,511	353	₹1,34,700	₹3,743	₹15.00	₹1,38,457
February	13,902	19,069	5,167	385	₹97,314	₹3,369	₹15.00	₹1,00,698
March	17,675	21,405	3,730	390	₹1,23,726	₹3,462	₹15.00	₹1,27,203
April	16,038	22,452	6,415	354	₹1,12,266	₹4,069	₹15.00	₹1,16,349
May	17,185	23,027	5,842	338	₹1,20,297	₹3,407	₹15.00	₹1,23,719
June	22,415	14,712	7,703	344	₹1,56,906	₹4,284	₹15.00	₹1,61,205
July	24,108	13,131	10,977	343	₹1,68,759	₹3,748	₹15.00	₹1,72,521
August	26,466	12,945	13,520	367	₹1,85,260	₹5,119	₹15.00	₹1,90,394
September	24,421	14,971	9,450	333	₹1,70,945	₹4,302	₹15.00	₹1,75,262
October	20,749	19,547	1,202	357	₹1,45,245	₹3,502	₹15.00	₹1,48,762
November	19,896	19,055	841	319	₹1,39,273	₹3,792	₹15.00	₹1,43,080
December	19,033	20,163	1,129	318	₹1,33,232	₹3,839	₹15.00	₹1,37,087
Annual	2,41,132	2,21,231	19,901	390	₹1.69 M	₹46,635	₹180.00	₹1.73 M

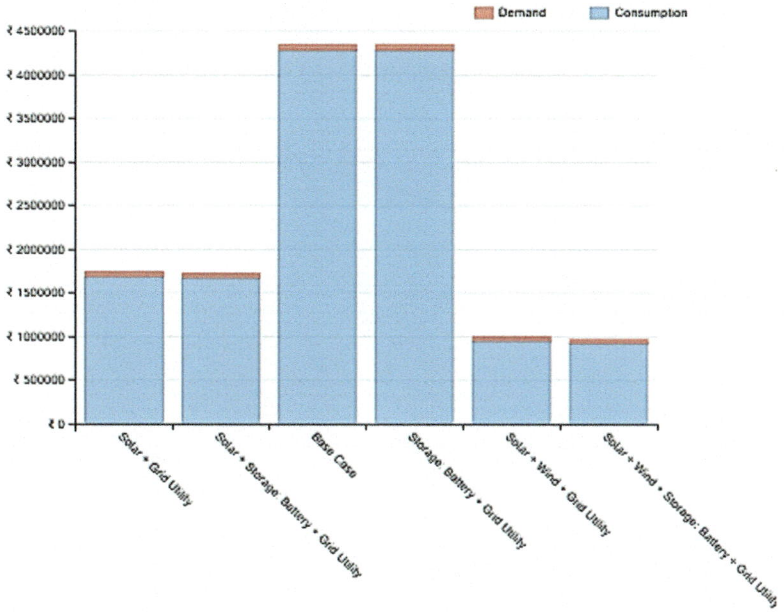

Fig. 11.4 Detailed overview of the utility bill

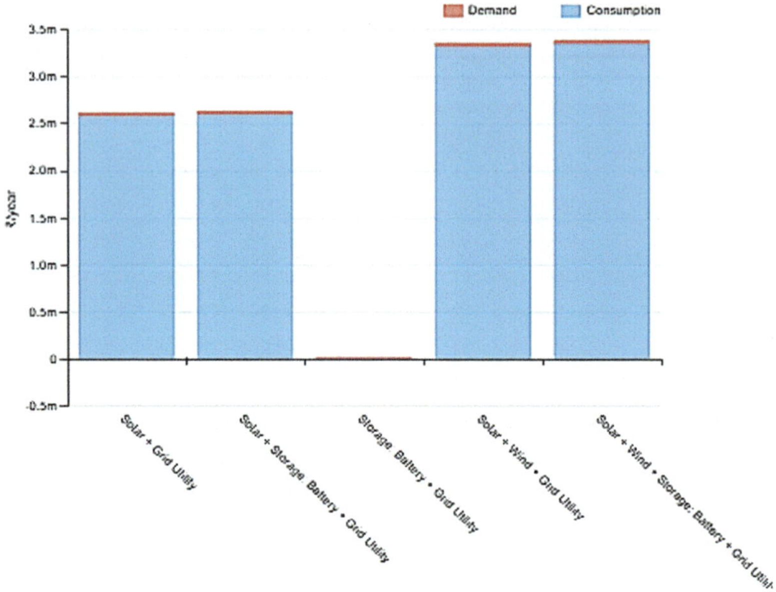

Fig. 11.5 Projected annual savings on the utility bill

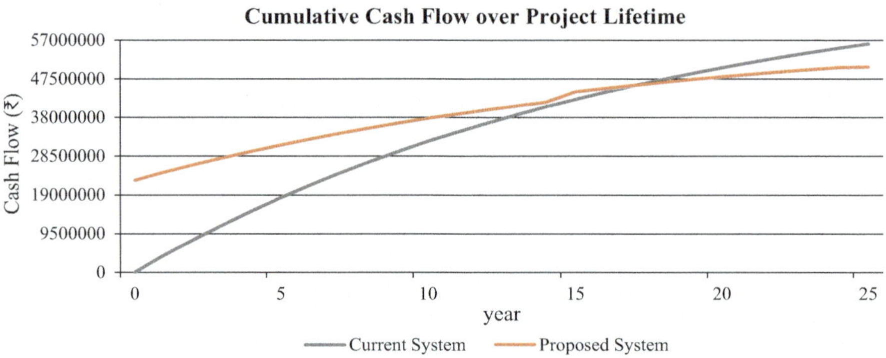

Fig. 11.6 Cumulative cash flow over project lifetime

11.7 Conclusion

Recent studies are found to be focused on optimization procedures and improvement of various performance parameters like NPC, COE & LPSP. The main focus is on the development of isolated systems to improve the reach of power supply in secluded areas and help promote economic activities in those places. However, a few studies also focus on the benefits of connecting to the grid. The review summarizes different configurations of HRE-systems, their modelling as well as optimization methods

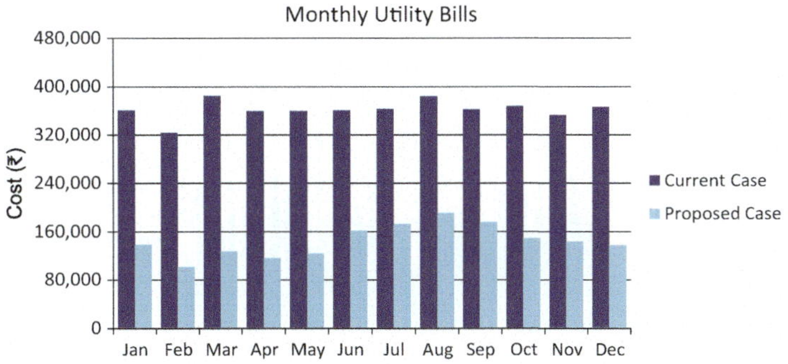

Fig. 11.7 Monthly utility bill savings

that have been studied in the past few years. The hybridization of renewable energy sources is found to be a realistic way of harnessing renewable energy resources by successfully mitigating its intermittent nature and providing a reliable supply of power. HRES has shown promising results in the reduction of carbon footprints in the energy generation sector and moving towards a sustainable future. This review highlights various aspects of HRES from a techno-economic point of view and throws light on the recent advances in the sector. It is clear that HRES will play crucial role in the development of future energy systems.

The optimization analysis is divided into two different cases: namely, base case for Grid only and proposed system including grid utility and solar PV. Here, the base case and proposed system cases are compared based on the evaluated best case for the designed system. In the first case system, i.e. base case has a NPC of ₹56.1 M, including an annual demand charge of ₹63,015.00 and annual energy charge of ₹4.34 M. With the proposed hybrid system, the NPC is reduced to ₹50.4 M, with annual demand charges and annual energy charges of ₹46,635 and ₹1.69 M, respectively. Therefore, the system economy is enhanced with proposed system including renewable as compared to base case system. In similar way the analysis of reliability and renewable fraction can be given for proposed over base case system. Hence, the above-mentioned analysis shows that hybrid renewable systems are more techno-economic solution for proposed case study.

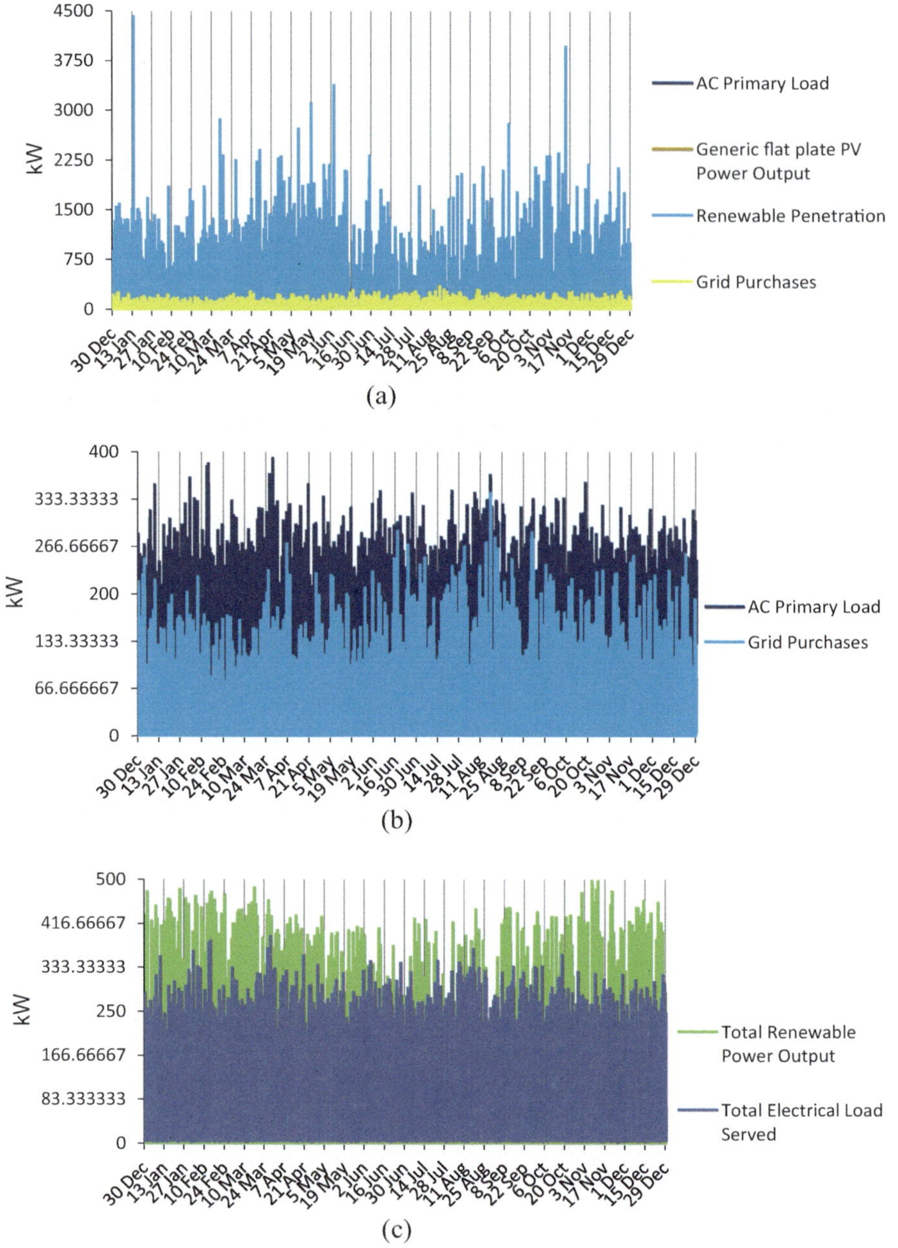

Fig. 11.8 Time series charts without EV charging conditions for different renewable sources and loads: **a** load, power output and renewable penetration **b** primary load output and grid purchases **c** renewable power output and total load served **d** inverter power output and total demand rate

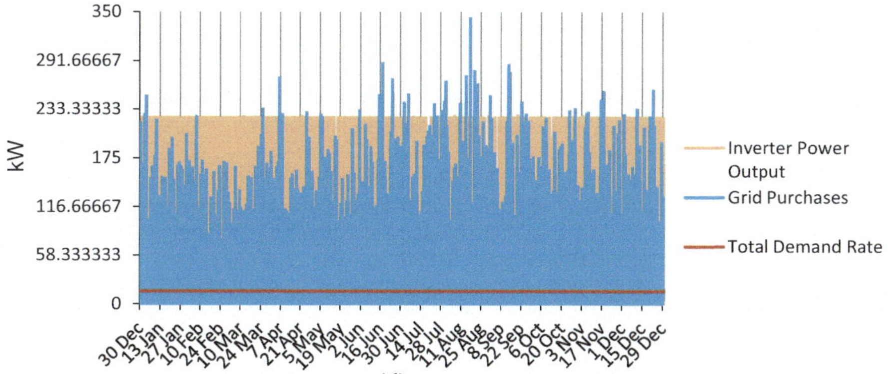

(d)

Fig. 11.8 (continued)

Table 11.5 Categorized utility bill savings in the proposed system v/s Bsae (current) system

Cases	COE	Consumption charge	Demand charge	Total
Base case with grid only	₹7.10	₹4.28 M	₹63,015	₹4.34 M
Proposed case with Solar system	₹4.68	₹1.69 M	₹46,635	₹1.73 M
Annual savings	–	₹2.59 M	₹16,380	₹2.61 M

References

Akram U, Khalid M, Shafiq S (2017) An innovative hybrid wind-solar and battery-supercapacitor microgrid system—development and optimization. IEEE Access 5:25897–25912. https://doi.org/10.1109/ACCESS.2017.2767618

Ani VA (2021) Development of an intelligent power management system for solar PV-wind-battery-fuel-cell integrated system. Front Energy Res 9:31. https://doi.org/10.3389/fenrg.2021.613958

Anoune K, Bouya M, Laknizi A, Abdellah AB, Astito A, Ghazouani M (2019) Techno-economic investigation of solar-wind potential to power an industrial prototype using a hybrid renewable energy system. SN Appl Sci 1(11):1–12. https://doi.org/10.1007/s42452-019-1358-7

Chen XR, Li JQ, Han Y, Niu B, Liu L, Zhang B (2019) An improved brain storm optimization for a hybrid renewable energy system. IEEE Access 7:49513–49526. https://doi.org/10.1109/ACCESS.2019.2908227

Chen Y, Wang R, Ming M, Cheng S, Bao Y, Zhang W, Zhang C (2021) Constraint multi-objective optimal design of hybrid renewable energy system considering load characteristics. Complex IntellSyst 1–15. https://doi.org/10.1007/s40747-021-00363-4

Dahlan NY, Abdullah NA, Razali NS, Ramachandaramurthy VK, Alrazi B, (2021) Assessing socio and techno-economic impacts of hybrid renewable energy system with energy storage for a rural development in malaysia. Int J Renew Energy Res (IJRER) 11(2):673–690. https://www.ijrer.ijrer.org/index.php/ijrer/article/view/11781

Das M, Singh MAK, Biswas A (2019) Techno-economic optimization of an off-grid hybrid renewable energy system using meta-heuristic optimization approaches–case of a radio transmitter station in India. Energy Convers Manag 185:339–352. https://doi.org/10.1016/j.enconman.2019.01.107

Diab AAZ, Sultan HM, Kuznetsov ON (2020) Optimal sizing of hybrid solar/wind/hydroelectric pumped storage energy system in Egypt based on different meta-heuristic techniques. Environ Sci Pollut Res 27(26):32318–32340. https://doi.org/10.1007/s11356-019-06566-0

Dolatabadi A, Mohammadi-Ivatloo B (2018) Stochastic risk-constrained optimal sizing for hybrid power system of merchant marine vessels. IEEE Trans Industr Inf 14(12):5509–5517. https://doi.org/10.1109/TII.2018.2824811

Elkadeem MR, Wang S, Azmy AM, Atiya EG, Ullah Z, Sharshir, SW (2020) A systematic decision-making approach for planning and assessment of hybrid renewable energy-based microgrid with techno-economic optimization: a case study on an urban community in Egypt. Sustain Cities Soc 54:102013. https://doi.org/10.1016/j.scs.2019.102013

Eriksson ELV, Gray EM (2017) Optimization and integration of hybrid renewable energy hydrogen fuel cell energy systems–a critical review. Appl Energy 202:348–364. https://doi.org/10.1016/j.apenergy.2017.03.132

Falk J, Nedjalkov A, Angelmahr M, Schade W (2020) Applying Lithium-Ion second life batteries for off-grid solar powered system—a socio-economic case study for rural development. Zeitschrift Für Energiewirtschaft 44(1):47–60. https://doi.org/10.1007/s12398-020-00273-x

Fathy A, Kaaniche K, Alanazi TM (2020) Recent approach based social spider optimizer for optimal sizing of hybrid PV/wind/battery/diesel integrated microgrid in Aljouf region. IEEE Access 8:57630–57645. https://doi.org/10.1109/ACCESS.2020.2982805

Gabbar HA, Abdussami MR, Adham MI (2020) Optimal planning of nuclear-renewable micro-hybrid energy system by particle swarm optimization. IEEE Access 8:181049–181073. https://doi.org/10.1109/ACCESS.2020.3027524

Geleta DK, Manshahia MS (2021) A hybrid of grey wolf optimization and genetic algorithm for optimization of hybrid wind and solar renewable energy system. J Oper Res Soc China 1–14. https://doi.org/10.1007/s40305-021-00341-0

Habib HUR, Wang S, Elkadeem MR, Elmorshedy MF (2019) Design optimization and model predictive control of a standalone hybrid renewable energy system: a case study on a small residential load in Pakistan. IEEE Access 7:117369–117390. https://doi.org/10.1109/ACCESS.2019.2936789

Kaluthanthrige R, Rajapakse AD, Lamothe C, Mosallat F, (2019) Optimal sizing and performance evaluation of a hybrid renewable energy system for an off-grid power system in northern Canada. Technol Econ Smart Grids Sustain Energy 4(1):4. https://link.springer.com/article/https://doi.org/10.1007/s40866-019-0061-5

Khalilnejad A, Sundararajan A, Sarwat AI (2018) Optimal design of hybrid wind/photovoltaic electrolyzer for maximum hydrogen production using imperialist competitive algorithm. J Modern Power Syst Clean Energy 6(1):40–49. https://doi.org/10.1007/s40565-017-0293-0

Khan MJ, Yadav AK, Mathew L (2017) Techno economic feasibility analysis of different combinations of PV-Wind-Diesel-Battery hybrid system for telecommunication applications in different cities of Punjab, India. Renew Sustain Energy Rev 76:577–607. https://doi.org/10.1016/j.rser.2017.03.076

Kharrich M, Kamel S, Abdeen M, Mohammed OH, Akherraz M, Khurshaid T, Rhee SB (2021) Developed approach based on equilibrium optimizer for optimal design of hybrid pv/wind/diesel/battery microgrid in Dakhla, Morocco. IEEE Access 9:13655–13670. https://doi.org/10.1109/ACCESS.2021.3051573

Kumar K, Babu NR, Prabhu KR (2017) Design and analysis of RBFN-based single MPPT controller for hybrid solar and wind energy system. IEEE Access 5:15308–15317. https://doi.org/10.1109/ACCESS.2017.2733555

Lee K, Kum D (2019) Complete design space exploration of isolated hybrid renewable energy system via dynamic programming. Energy Convers Manag 196:920–934. https://doi.org/10.1016/j.enconman.2019.05.078

Li T, Hu W, Xu X, Huang Q, Chen G, Han X, Chen Z (2019) Optimized operation of hybrid system integrated with MHP, PV and PHS considering generation/load similarity. IEEE Access 7:107793–107804. https://doi.org/10.1109/ACCESS.2019.2932851

Liaquat S, Fakhar MS, Kashif SAR, Rasool A, Saleem O, Padmanaban S (2020) Performance analysis of APSO and firefly algorithm for short term optimal scheduling of multi-generation hybrid energy system. IEEE Access 8:177549–177569. https://doi.org/10.1109/ACCESS.2020.3027436

Martinez-Rico J, Zulueta E, de Argandoña IR, Fernandez-Gamiz U, Armendia M (2021) Multi-objective optimization of production scheduling using particle swarm optimization algorithm for hybrid renewable power plants with battery energy storage system. J Mod Power Syst Clean Energy 9(2):285–294. https://doi.org/10.35833/MPCE.2019.000021

Moghaddam S, Bigdeli M, Moradlou M (2021) Optimal design of an off-grid hybrid renewable energy system considering generation and load uncertainty: the case of Zanjan city, Iran. SN Appl Sci 3(8):1–15. https://doi.org/10.1007/s42452-021-04718-x

Mohammadi M, Ghasempour R, Astaraei FR, Ahmadi E, Aligholian A, Toopshekan A (2018) Optimal planning of renewable energy resource for a residential house considering economic and reliability criteria. Int J Electr Power Energy Syst 96:261–273. https://doi.org/10.1016/j.ijepes.2017.10.017

Mohammed AQ, Al- KA, Hannun RM (2020) Optimal combination and sizing of a stand-alone hybrid energy system using a nomadic people optimizer. IEEE Access 8:200518–200540. https://doi.org/10.1109/ACCESS.2020.3034554

Murty VV, Kumar A, (2020) Optimal energy management and techno-economic analysis in microgrid with hybrid renewable energy sources. J Mod Power Syst Clean Energy 8(5):929–940. https://doi.org/10.35833/MPCE.2020.000273

Murugaperumal K, Raj PADV (2019) Feasibility design and techno-economic analysis of hybrid renewable energy system for rural electrification. Sol Energy 188:1068–1083. https://doi.org/10.1016/j.solener.2019.07.008

Nebey AH (2021) Design of optimal hybrid power system to provide reliable supply to rural areas of Ethiopia using MATLAB and Homer. RenewS Wind Water Sol 8(1):1–7. https://doi.org/10.1186/s40807-021-00067-w

Nguyen XP, Hoang AT (2020) The flywheel energy storage system: An effective solution to accumulate renewable energy. In 2020 6th International conference on advanced computing and communication systems (ICACCS). IEEE. 1322–1328. https://doi.org/10.1109/ICACCS48705.2020.9074469

Oladigbolu JO, Ramli MA, Al- YA (2020) Feasibility study and comparative analysis of hybrid renewable power system for off-grid rural electrification in a typical remote village located in Nigeria. IEEE Access 8:171643–171663. https://doi.org/10.1109/ACCESS.2020.3024676

Razmjoo AA, Davarpanah A (2019) The role of renewable energy to achieve energy sustainability in Iran. An economic and technical analysis of the hybrid power system. Technol Econ Smart Grids Sustain Energy 4(1):1–11. https://link.springer.com/article/https://doi.org/10.1007/s40866-019-0063-3

Rezk H, Al- M, Hassan YB, Ziedan HA (2020) Optimization and energy management of hybrid photovoltaic-diesel-battery system to pump and desalinate water at isolated regions. IEEE Access 8:102512–102529. https://doi.org/10.1109/ACCESS.2020.2998720

Rinaldi F, Moghaddampoor F, Najafi B, Marchesi R (2021) Economic feasibility analysis and optimization of hybrid renewable energy systems for rural electrification in Peru. Clean Technol Environ Policy 23(3):731–748. https://doi.org/10.1007/s10098-020-01906-y

Romanos P, Voumvoulakis E, Markides CN, Hatziargyriou N (2019) Thermal energy storage contribution to the economic dispatch of island power systems. CSEE J Power Energy Syst 6(1):100–110. https://doi.org/10.17775/CSEEJPES.2019.00610

Rzetelska D, Combrinck M (2021). Fuel poverty and low carbon emissions: a comparative study of the feasibility of the hybrid renewable energy systems incorporating combined heat and power technology. Front Energy 1–21. https://doi.org/10.1007/s11708-021-0748-x

Sandeep SR, Nandihalli R (2020) Optimal sizing in hybrid renewable energy system with the aid of opposition based social spider optimization. J Elect Eng Technol 15(1):433–440. https://doi.org/10.1007/s42835-019-00184-z

Sawle Y, Gupta SC, Bohre AK (2018) Socio-techno-economic design of hybrid renewable energy system using optimization techniques. Renewable Energy 119:459–472. https://doi.org/10.1016/j.renene.2017.11.058

Sawle Y, Jain S, Babu S, Nair AR, Khan B (2021) Prefeasibility economic and sensitivity assessment of hybrid renewable energy system. IEEE Access 9:28260–28271. https://doi.org/10.1109/ACCESS.2021.3058517

Shafiullah GM, Masola T, Samu R, Elavarasan RM, Begum S, Subramaniam U, Romlie MF, Chowdhury M, Arif MT (2021) Prospects of hybrid renewable energy-based power system: a case study, post analysis of Chipendeke Micro-Hydro, Zimbabwe. IEEE Access 9:73433–73452. https://doi.org/10.1109/ACCESS.2021.3078713

Singh R, Bansal RC (2018) Optimization of an autonomous hybrid renewable energy system using reformed electric system cascade analysis. IEEE Trans Industr Inf 15(1):399–409. https://doi.org/10.1109/TII.2018.2867626

Singh R, Bansal RC, Singh AR, Naidoo R (2018) Multi-objective optimization of hybrid renewable energy system using reformed electric system cascade analysis for islanding and grid connected modes of operation. IEEE Access 6:47332–47354. https://doi.org/10.1109/ACCESS.2018.2867276

Ssenyimba S, Kiggundu N, Banadda N (2020) Designing a solar and wind hybrid system for small-scale irrigation: a case study for Kalangala district in Uganda. Energy Sustain Soc 10(1):1–18. https://doi.org/10.1186/s13705-020-0240-1

Taha MS, Abdeltawab HH, Mohamed YARI (2018) An online energy management system for a grid-connected hybrid energy source. IEEE J Emerg Select Topics Power Electron 6(4):2015–2030. https://doi.org/10.1109/JESTPE.2018.2828803

Tiomo D, Wamkeue R (2019) May. Dynamic modeling and analysis of a micro-hydro power plant for microgrid applications. In 2019 IEEE Canadian conference of electrical and computer engineering (CCECE). IEEE, pp. 1–6. https://doi.org/10.1109/CCECE.2019.8861875

Troullaki A, Latoufis K, Marques P, Freire F, Hatziargyriou N (2019) Life cycle assessment of locally manufactured small wind turbines and pico-hydro plants. In 2019 international conference on smart energy systems and technologies (SEST). IEEE, pp 1–6. https://doi.org/10.1109/SEST.2019.8849074

Tsai CT, Beza TM, Molla EM, Kuo CC (2020) Analysis and sizing of mini-grid hybrid renewable energy system for islands. IEEE Access 8:70013–70029. https://doi.org/10.1109/ACCESS.2020.2983172

Wang F, Xie Y, Xu J (2019) Reliable-economical equilibrium based short-term scheduling towards hybrid hydro-photovoltaic generation systems: case study from China. Appl Energy 253:113559. https://doi.org/10.1016/j.apenergy.2019.113559

Xu C, Ke Y, Li Y, Chu H, Wu Y (2020) Data-driven configuration optimization of an off-grid wind/PV/hydrogen system based on modified NSGA-II and CRITIC-TOPSIS. Energy Convers Manag 215:112892. https://doi.org/10.1016/j.enconman.2020.112892

Chapter 12
Modeling and Control of PV Systems for Maximum Power Point Tracking and Its Performance Analysis Using Advanced Techniques

Md Tuhin Rana and Partha Sarathee Bhowmik

Abstract Maximum power point tracking (MPPT) is a necessary and primary concern in modern photovoltaic (PV) energy systems. The nonlinear nature of the output characteristics of PV systems causes it to supply maximum amount of power at a particular point of operation which is known as the maximum power point (MPP). For optimal utilization of the PV modules effective tracking of this particular operating point is necessary for most of the PV energy systems. Over the years, many different approaches have been proposed for maximum power point tracking in PV energy systems. But most of these literatures do not draw a complete picture of the design, control and operation process of the whole system involved in MPPT. To alleviate such difficulties, this chapter discusses the MPPT system in an exhaustive manner using a novel integrated model of the system for designing robust and effective controllers for MPPT. This chapter considers a system where a PV module is connected to a DC-DC converter system for demonstrating the process. Firstly, a novel *small-signal model* of the system where a PV module is connected to any of the three basic DC to DC converters is obtained by utilizing the *perturbation and linearization* method on the available large signal models. Then, using these small-signal models, generalized transfer function models of the systems are obtained by application of simple circuit theory. The intrinsic nonlinearities of the system and other inherent factors that create difficulty in controller design are also pointed out. As an effective solution to this problem of nonlinearity, the design and implementation process of a gain-scheduled PID controller for controlling such highly nonlinear systems is presented. Moreover, the design and implementation of an MPPT control loop around the voltage control loop are described using some advanced computation-based algorithms, namely, Particle Swarm Optimization (PSO), Differential Evolution Algorithm (DEA) and Binary Coded Genetic Algorithm (BCGA)

Md T. Rana (✉) · P. S. Bhowmik
National Institute of Technology Durgapur, Durgapur, West Bengal, India
e-mail: tuhinrana3104@gmail.com

P. S. Bhowmik
e-mail: psbhowmik@ee.nitdgp.ac.in

A. K. Bohre et al. (eds.), *Planning of Hybrid Renewable Energy Systems, Electric Vehicles and Microgrid*, Energy Systems in Electrical Engineering,
https://doi.org/10.1007/978-981-19-0979-5_12

for MPPT. In addition to that, a comparative analysis of the three metaheuristic algorithms implemented here is also presented in this chapter. All the discussed theoretical aspects were validated through simulation in MATLAB/SIMULUNK implementing various real-world scenarios such as partial shading, load disturbance, etc. The novel modeling approach, the systematic control system design approach, the MPPT algorithm design methods and implementation and their comparative analysis can be of great usefulness to a designer.

Keywords Binary coded genetic algorithm (BCGA) · Differential evolution algorithm (DEA) · Gain scheduling · Maximum power point tracking (MPPT) · Particle swarm optimization (PSO) · Photovoltaic (PV) system · Proportional integral derivative controller (PID) · Small-signal modeling

Nomenclature

ANN	Artificial neural network
SA	Simulated annealing
MPPT	Maximum power point tracking
DEA	Differential evolution algorithm
PSO	Particle swarm optimization
FL	Fuzzy logic
PV	Photovoltaic
FLC	Fuzzy logic controller
TLBO	Teaching and learning based optimization
BCGA	Binary coded genetic algorithm
GA	Genetic algorithm
DC	Direct current
AC	Alternating current
PID	Proportional integral derivative
InC	Incremental conductance
P&O	Perturb and observe
LHS	Left hand side
RHS	Right hand side
PWM	Pulse width modulation
LUT	Look-up table
CCM	Continuous conduction mode
PI	Proportional integral
KCL	Kirchhoff's current law
KVL	Kirchhoff's voltage law

12.1 Introduction

In general, a photovoltaic energy conversion system comprises of photovoltaic modules, power electronic converters, controllers, and loads. Figure 12.1 depicts the structure of a typical PV energy conversion system. Such PV systems are classified into different categories depending on the type of loads connected to the system, the number of power electronic conversion stages present in the system, grid connection, etc.

On the basis of interaction with the grid, photovoltaic energy systems are categorized into three categories, namely, grid-tied system, off-grid system and grid-interactive hybrid system. In a typical grid-tied PV energy system the power generated by the PV modules is fed directly into the utility grid with the help of some power electronic interfaces (Malinowski et al. 2017), whereas, in case of an off-grid system the loads are fed from the PV modules which are assisted by energy storage systems/batteries (Malinowski et al. 2017). The structure of a grid-interactive hybrid PV-battery system is a hybrid combination of off-grid and grid-tied PV systems (Khezri et al. 2020). Such systems are made of a typical energy storage systems along with the PV modules and these can be operated in either of the grid-tied of off-grid mode of operation.

The majority of grid-tied and off-grid PV systems are either single-stage (Guo et al. 2020) conversion system where the inverter itself tracks the MPP or double-stage conversion system (Fahad et al. 2019) where a DC-DC converter tracks the MPP and the DC-AC inverter pumps the energy into the grid or load from the DC-DC converter. Even though single-stage conversion is more efficient it is hard to apply in low or medium voltage PV systems (Guo et al. 2020). In this chapter the modeling, control and MPPT in a typical PV system with a DC-DC converter in the first stage, as shown in Fig. 12.2 is discussed. In case of such a system, the representation of the inverter input port as a resistor drastically simplifies the analysis of the system without any significant loss of actual system characteristics (Li et al. 2019).

Even though, the topology, operation and control system structures are different for the different types of systems mentioned above, the design and implementation of MPPT in those systems are somewhat similar. Many different algorithms and

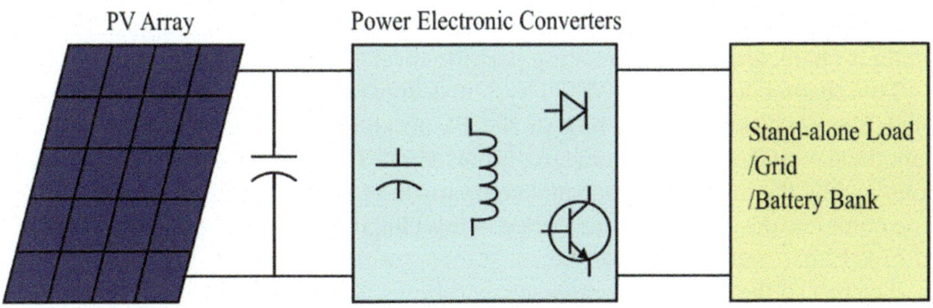

Fig. 12.1 General representation of photovoltaic energy systems

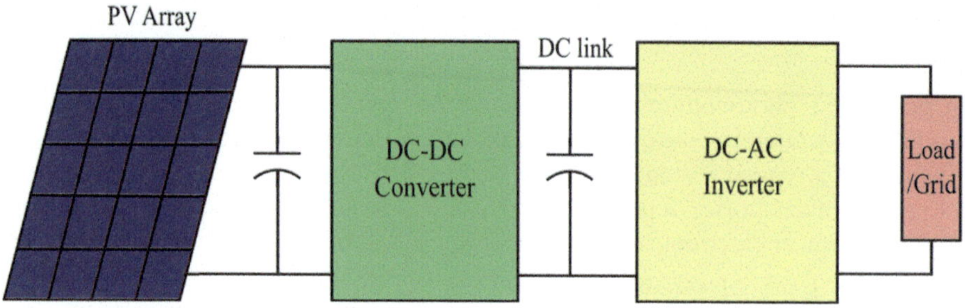

Fig. 12.2 Schematic of double-stage PV system

methods have been suggested throughout the years for both single-stage and double-stage conversion of solar energy. There are some conventional techniques, such as, "*perturb and observe (P&O) technique*" (Haque 2014), "*incremental conductance technique (InC)*" (Shang et al. 2020), "*open circuit voltage method*" (Ko et al. 2020), "*short circuit current method*" (Ko et al. 2020), etc. Even though, these conventional methods are relatively easy to implement, most of the time these methods fail to track the "*global MPP*" under "*partial shading conditions.*" To solve the problem of stagnation of the classical methods many new techniques have been proposed. In recent years, with the advent of fast computational systems the MPPT problem is being considered as an optimization problem. Consequently, many optimization algorithms have been suggested in literature. Implementation of"*particle swarm optimization (PSO)*" in MPPT is discussed by Li et al. (2019). References (Fathy et al. 2018; Lyden et al. 2020; Rezk et al. 2019) and (Allahabadi et al. 2021) shows the application of "*teaching and learning based optimization (TLBO),*" "*simulated annealing (SA),*" "*fuzzy logic control (FLC),*" and "*artificial neural network (ANN)*" respectively. In terms of speed and tracking of global MPP performance of these methods are satisfactory. But these improvements in tracking performance and efficiency come at a cost of increased complexity, requirement of more costly components and requirement of more numbers of sensors. The target of this chapter is to discuss the various aspects of the advanced computation-based MPPT system. It discusses different aspects of modeling and control system design for MPPT in PV systems. It also analyses some popular advanced computation-based metaheuristic algorithm-based MPPT systems and presents a comparative analysis of the presented algorithms based on their characteristics and tracking performance.

This chapter describes the MPPT system design procedure in a systematic manner. First, the small signal modeling of the PV module and its combination with the canonical model of DC-DC converter is presented. Then the controller design process followed by the implementation and comparison of MPPT algorithms are discussed. The complete design process described in this chapter is summarized in the flowchart of Fig. 12.3.

The rest of the chapter is organized as follows. Section 2 derives the small signal modeling of a PV module based on perturb and observe technique which is then

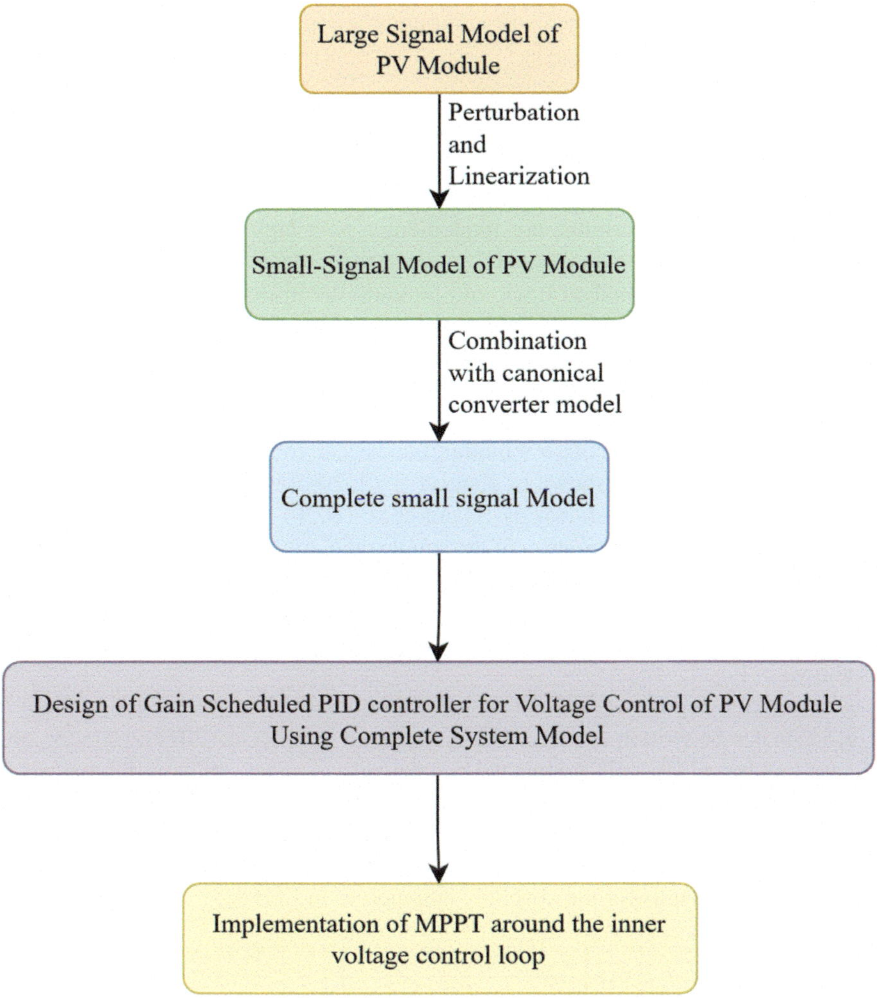

Fig. 12.3 Flowchart of the described design Process

combined with a canonical model of the DC-DC converters in Sect. 3. A gain scheduled PID controller is then designed in Sect. 4 depending on the system model. After that, the simulation results for different advanced algorithm-based MPPT algorithms are presented in Sect. 5. Section 6 presents comparative analysis of these algorithms when applied for MPPT which is followed by the evaluation of system performance subjected to load change in Sect. 7. Finally, the chapter is summarized in Sect. 8. The key contributions of this chapter are:

- Novel Small-signal model of PV module and its combination with existing canonical model of DC-DC converters.
- Design of gain scheduled PID controller using the derived system models.

- Implementation of advanced soft-computing algorithm-based MPPT algorithms and their comparative analysis.

12.2 Small-Signal Model of Photovoltaic Module

Majority of MPPT algorithms are implemented as a high level control algorithm which controls the command signal for an internal control loop. The inner control loop may be implemented to track and regulate the system voltage or current in accordance with the command signal generated by the MPPT algorithm. In order to properly design and implement this necessary control system one requires an accurate small-signal model representation of the actual system at the concerned operating points.

The small-signal model of a PV module can be analytically derived from its large-signal nonlinear model as shown by Rana et al. (2020). In this chapter a single-diode model of the PV cell, as shown by Shang et al. (2020) is used as the large-signal model to be utilized for derivation of the small signal model. Generally, several such PV cells are interconnected in both series–parallel configuration with each other to construct a PV module. Li et al. (2019) stated that the equivalent circuit schematic of such PV module can be drawn as shown in Fig. 12.4a having terminal characteristics as shown in Fig. 12.4b.

The relationship between the terminal current and voltage of the PV module in Fig. 12.4a can be written as in Eq. (12.1) as shown by Li et al. (2019). Here, N_s and N_p are number of PV cells connected respectively in series and parallel manner to construct the module, R_{sh} and R_s are equivalent shunt and series resistances of each cell, η is the quality factor of the material, q denotes the value of the charge of a single electron, I_{rs} is the diode reverse saturation current, K represents Boltzmann's constant, and T stands for the absolute temperature in Kelvin.

$$I = N_p I_{ph} - N_p I_{rs}\left[\exp\left\{\frac{q(V/N_s + IR_s/N_p)}{\eta KT}\right\}\right] - \left[\frac{N_p V/N_s + IR_s}{R_{sh}}\right] \quad (12.1)$$

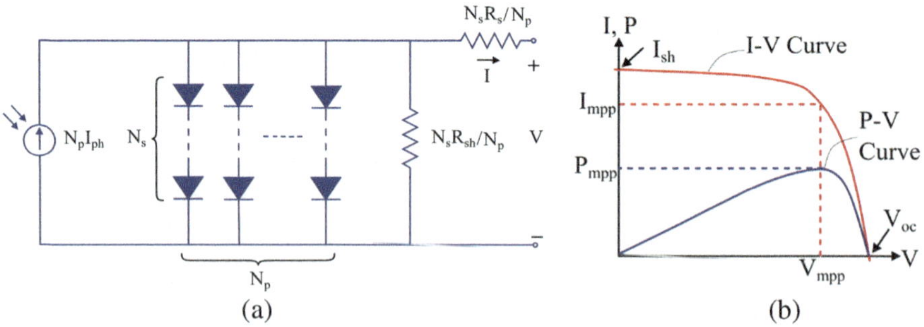

Fig. 12.4 **a** Equivalent circuit representation of a PV module. **b** Characteristics of a PV module

"The nonlinear large-signal model represented by (1) can be linearized by using analytical methods to derive a small-signal model at a particular operating point of the system. In this chapter the "*perturbation and linearization*" technique is used to linearize the system around a quiescent operating point and obtain the small-signal model" (Rana et al. 2020). Equation (12.2) is obtained when small perturbations in the PV terminal voltage and current, ($\widehat{v}_{pv}$, $\widehat{i}_{pv}$) are applied around the operating point, (V_{pv}, I_{pv}) in Eq. (12.1).

$$I_{pv} + \widehat{i}_{pv} = N_p I_{ph} - N_p I_{rs}\left[\exp\left\{\frac{q(V_{pv}/N_s + I_{pv}R_s/N_p) + q(\widehat{v}_{pv}/N_s + \widehat{i}_{pv}R_s/N_p)}{\eta KT}\right\}\right] - \left[\frac{(N_p V_{pv}/N_s + I_{pv}R_s) + (N_p \widehat{v}_{pv}/N_s + \widehat{i}_{pv}R_s)}{R_{sh}}\right] \tag{12.2}$$

After expanding the exponential term in (2) using Taylor series expansion with respect to the small perturbation terms (3) is obtained.

$$\begin{aligned} I_{pv} + \widehat{i}_{pv} = {} & N_p I_{ph} - \left[\frac{(N_p V_{pv}/N_s + I_{pv}R_s)}{R_{sh}}\right] - \left[\frac{(N_p \widehat{v}_{pv}/N_s + \widehat{i}_{pv}R_s)}{R_{sh}}\right] \\ & - N_p I_{rs}\left[\exp\left\{\frac{q(V_{pv}/N_s + I_{pv}R_s/N_p)}{\eta KT}\right\}\right. \\ & \left\{1 + \frac{q(\widehat{v}_{pv}/N_s + \widehat{i}_{pv}R_s/N_p)}{\eta KT} + \frac{1}{2!}\frac{q^2(\widehat{v}_{pv}/N_s + \widehat{i}_{pv}R_s/N_p)^2}{(\eta KT)^2}\right. \\ & \left.\left. + \frac{1}{3!}\frac{q^3(\widehat{v}_{pv}/N_s + \widehat{i}_{pv}R_s/N_p)^3}{(\eta KT)^3}\ldots\right\} - 1\right] \end{aligned} \tag{12.3}$$

Since, the perturbations are assumed to be very small, the higher power of these terms would result in even smaller terms which have very little effect on the system. So, a first order approximation may be used to neglect the terms containing higher powers of the perturbations. After the approximation and some rearrangements (3) results in (4).

$$
\begin{aligned}
I_{pv} + \underset{pv}{\widehat{i}} &= N_p I_{ph} - N_p I_{rs}\left[\exp\left\{\frac{q(V_{pv}/N_s + I_{pv}R_s/N_p)}{\eta KT}\right\} - 1\right] \\
&- \left[\frac{(N_p V_{pv}/N_s + I_{pv}R_s)}{R_{sh}}\right] \\
&- N_p I_{rs}\left[\exp\left\{\frac{q(V_{pv}/N_s + I_{pv}R_s/N_p)}{\eta KT}\right\}\left\{\frac{q(\underset{pv}{\widehat{v}}/N_s + \underset{pv}{\widehat{i}}R_s/N_p)}{\eta KT}\right\}\right] \\
&- \left[\frac{(N_p \underset{pv}{\widehat{v}}/N_s + \underset{pv}{\widehat{i}}R_s)}{R_{sh}}\right]
\end{aligned}
\tag{12.4}
$$

It can be observed that the first term on LHS and the first three terms on the RHS of (4) represent the DC relationship between PV current and voltage, whereas, the other terms in the equation represent the small-signal relationship. After equating the small-signal terms on both sides of (4) one can derive (5).

$$
\underset{pv}{\widehat{i}} = -\alpha(\underset{pv}{\widehat{v}}/N_s + \underset{pv}{\widehat{i}}R_s/N_p) - (N_p \underset{pv}{\widehat{v}}/N_s R_{sh}) - (\underset{pv}{\widehat{i}}R_s/R_{sh}) \tag{12.5}
$$

$$
\text{where, } \alpha = \left[qN_p I_{rs}\exp\left\{\frac{q(V_{pv}/N_s + I_{pv}R_s/N_p)}{\eta KT}\right\}\right]/\eta KT \tag{12.6}
$$

Some rearrangements of (5) result in the required linear model representation of the PV module as shown in (7). Here, σ is the small-signal conductance of the PV module, mathematically its value is equal to the negative value of the slope of the I-V curve of the module.

$$
\underset{pv}{\widehat{i}} = -\sigma.\underset{pv}{\widehat{v}} \tag{12.7}
$$

$$
\text{where, } \sigma = \frac{(\alpha/N_s) + (N_p/N_s R_{sh})}{1 + (\alpha R_s/N_p) + (R_s/R_{sh})} \tag{12.8}
$$

Using the relationship in (7) the small-signal model of the PV module can be interpreted as a variable resistor as in Fig. 12.5a whose value depends on the operating point of the system. If the relative directional representation between the voltage and current is reversed then the model can be drawn with a resistor having positive value as in Fig. 12.5b.

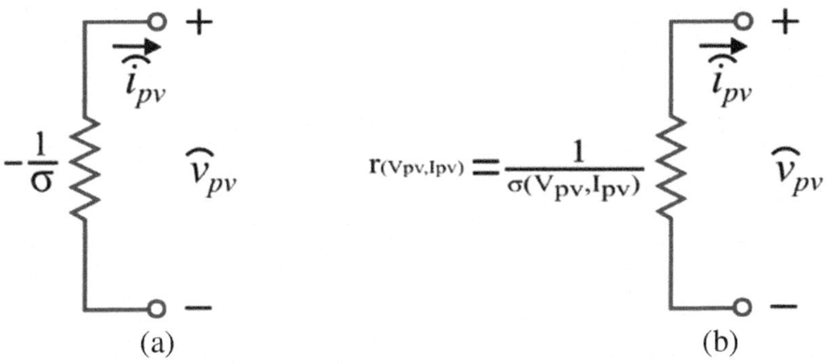

Fig. 12.5 Small signal model of PV module **a** equivalent model from Eq. (12.7). **b** With reversed current direction

12.3 The Integrated Small-Signal Model

In a typical double-stage PV system the MPPT operation is performed in the DC-DC conversion stage. Different types of converters can be used to serve this purpose. Figure 12.6 shows the circuit configuration of such systems with the three well-known basic DC to DC converters, namely, Buck, Boost and Buck-boost converter. Here, the load, which might be an inverter or any other kind of load is modelled as a resistor R_O.

The liner small-signal model of PV system derived in the previous section can be used for system modeling purpose. In this chapter, the linear resistive model is integrated with a general *canonical model* of the DC-DC converters operating in *continuous conduction mode* (CCM) to obtain the complete small-signal model representation of the system. "The *canonical model* of DC-DC converter is a generalized representation of the converters where the basic structure of the equivalent circuit remains same for all the converters" (Cuk 1977). The schematic of the system can be seen in Fig. 12.7, where, the small-signal resistance of the PV module is represented by r.

The expression of the values of the different passive circuit components in the canonical converter model is different for the different converters. These expressions for the aforementioned DC-DC converters used in this chapter are given in Cuk (1977), a modified version of which are reproduced in Table 12.1.

With the help of Fig. 12.7 and Table 12.1, one can obtain various transfer functions of the system. In this chapter, the voltage across the terminals of the PV module is taken as the parameter to be controlled for MPPT applications. Which necessitates the derivation of the control to output transfer function, where the duty cycle of the PWM converter switch is the control signal and the voltage across the terminals of the PV module is considered as the output signal.

Applying conventional circuit theories, such as KVL and KCL, the necessary transfer function can be derived in the form shown in (9).

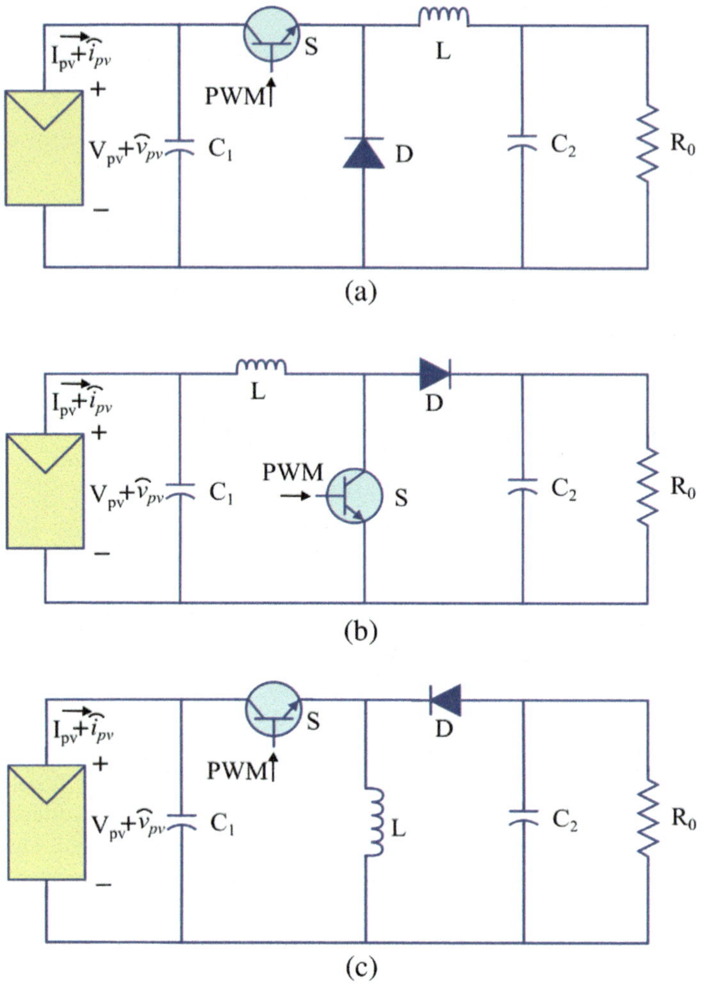

Fig. 12.6 DC-DC conversion stage of PV systems incorporating (**a**) Buck converter (**b**) Boost converter (**c**) Buck-boost converter

$$G_{vd}(s) = \frac{\widehat{v}_{pv}(s)}{\widehat{d}(s)} = \frac{b_2 s^2 + b_1 s + b_0}{a_3 s^3 + a_2 s^2 + a_1 s + a_0} \tag{12.9}$$

$$\text{Here, } a_0 = \frac{R_0}{M^2} + r \tag{12.10}$$

$$a_1 = \frac{L_e}{M^2} + \frac{r C_1 R_0}{M^2} + r C_2 R_0 \tag{12.11}$$

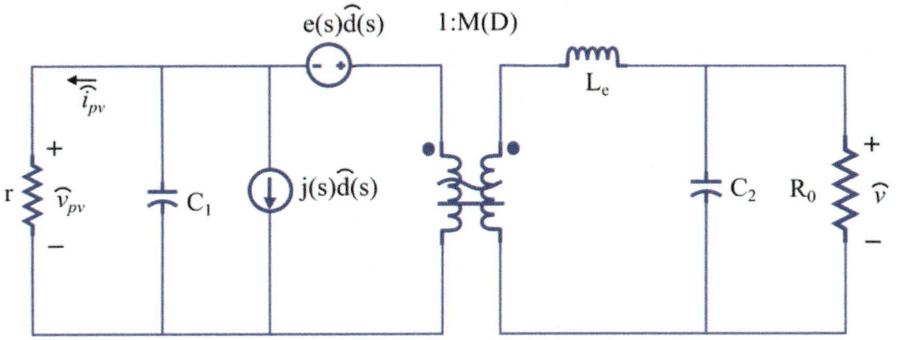

Fig. 12.7 Complete Linear Small-Signal Model Representation of the System

Table. 12.1 Parameters of canonical model for basic DC-DC Converters

Converter type	M(D)	e(s)	j(s)	L_e
Buck	D	$\frac{V_{pv}}{D}$	$\frac{DV_{pv}}{R_0}$	L
Boost	$\frac{1}{(1-D)}$	$\frac{V_{pv}}{(1-D)}\left(1-\frac{sL}{(1-D)^2R_0}\right)$	$\frac{V_{pv}}{(1-D)^3R_0}$	$\frac{L}{(1-D)^2}$
Buck-boost	$-\frac{D}{(1-D)}$	$\frac{V_{pv}}{D(1-D)}\left(1-\frac{sDL}{(1-D)^2R_0}\right)$	$\frac{DV_{pv}}{(1-D)^3R_0}$	$\frac{L}{(1-D)^2}$

$$a_2 = \frac{rC_1L_e}{M^2} + \frac{L_eC_2R_0}{M^2} \tag{12.12}$$

$$a_3 = \frac{rL_eC_1C_2R_0}{M^2} \tag{12.13}$$

$$b_0 = -r\left\{e(s) + \frac{j(s)R_0}{M^2}\right\} \tag{12.14}$$

$$b_1 = -r\left\{e(s)R_0C_2 + \frac{j(s)L_e}{M^2}\right\} \tag{12.15}$$

$$b_2 = -\frac{j(s)rR_0L_eC_2}{M^2} \tag{12.16}$$

The expression of the different elements of the polynomials in numerator and denominator of (9) for different DC-DC converters can be derived using the parameter values from Table 12.1. The derived coefficients are tabulated in Table 12.2. It needs to be mentioned that, since some of the canonical model parameters are complex frequency dependent, the expression of the different coefficients need to be reorganized to obtain the final form in Table 12.2.

Table 12.2 Coefficients of transfer function polynomials

Coefficient	Buck	Boost	Buck-boost
a_0	$\frac{R_O}{D^2}+r$	$R_O(1-D)^2+r$	$\frac{R_O(1-D)^2}{D^2}+r$
a_1	$\frac{L}{D^2}+rR_O\left(\frac{C_1}{D^2}+C_2\right)$	$L+rR_O\{C_1(1-D)^2+C_2\}$	$\frac{L}{D^2}+rR_O\left\{\frac{C_1(1-D)^2}{D^2}+C_2\right\}$
a_2	$\frac{L}{D^2}(rC_1+R_OC_2)$	$L(rC_1+R_OC_2)$	$\frac{L}{D^2}(rC_1+R_OC_2)$
a_3	$\frac{LR_OC_1C_2}{D^2}$	$LR_OC_1C_2$	$\frac{LR_OC_1C_2}{D^2}$
b_0	$-2\frac{rV_{pv}}{D}$	$-2\frac{rV_{pv}}{(1-D)}$	$-2\frac{rV_{pv}}{D(1-D)}$
b_1	$-\frac{rV_{pv}}{D}\left(R_OC_2+\frac{L}{R_O}\right)$	$-\frac{rV_{pv}}{(1-D)}(R_OC_2)$	$-\frac{rV_{pv}}{D(1-D)}\left(R_OC_2+\frac{L}{R_O(1-D)}\right)$
b_2	$-\frac{rV_{pv}LC_2}{D}$	0	$-\frac{rV_{pv}}{D(1-D)}\frac{LC_2}{(1-D)}$

The transfer function model in (9) along with Table 12.2 can be used for analysing the system behaviour and design a suitable controller to control the terminal voltage of the PV modules.

12.4 System Analysis and Controller Design

The transfer functions that have been derived so far can be used to analyse the system before designing the control system for the system. In this chapter, a typical PV system with a Boost converter as in Fig. 12.6b having parameter values as depicted in Table 12.3 is considered for the system analysis purpose. The bode plots of the

Table 12.3 Parameter values of PV module and converter

Parameter name	Symbol	Value
Open circuit voltage of each PV module	V_{OC}	21.75 V
Short circuit current of each PV module	I_{SC}	6.36 A
Each module voltage at MPP	V_{MPP}	17.83 V
Each module current at MPP	I_{MPP}	5.915 A
Series resistance of each cell	R_s	0.001 Ω
Parallel resistance of each cell	R_{sh}	5 k Ω
Number of cells connected in series	N_s	36
Number of cells connected in parallel	N_p	1
Reverse saturation current of diode	I_{rs}	1 μA
Load resistance	R_O	100 Ω
Inductor	L	10 mH
Input capacitor	C_1	100 μF
Output capacitor	C_2	100 μF

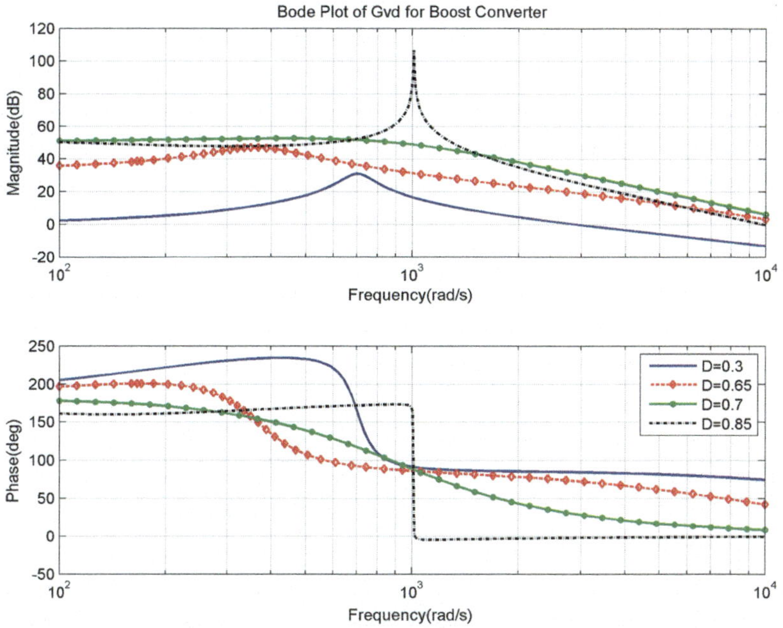

Fig. 12.8 Bode plots of the transfer functions for boost converter at different operating points

transfer functions at different operating points, i.e. at different values of converter's duty cycle are illustrated in Fig. 12.8.

From Fig. 12.8 one can interpret that as the operating point varies, the dynamic response of the open-loop system also change. The frequency responses presented here are similar to a typical three pole system response with a high frequency pole at relatively lower duty cycles. It can be observed that the bandwidth of the open loop system increases with the PWM duty cycle up to a certain limit then it decreases again, and the phase margin of the system keeps decreasing as the duty cycle increases. Moreover, the natural damping of the system is not same at all the operating points; the system is well damped around the MPP which happens to occur at 70% duty cycle but the damping is not good enough for relatively smaller and larger duty cycle than this. During the design process of a controller for the system one should consider these key features of the frequency response of the system.

Even though the dynamic response of the system changes with operating point, it is still possible to implement conventional control approaches, such as PI or PID control for the system with the help of "*quasi-static approximation*" (Kocher and Steigerwald 1982). But in such cases, the performance of the system will be suboptimal because of the non-adaptability of the controller. But the overall response of the MPPT system depends largely on the performance of the control system, which is why the control system has to be robust and optimal. Consequently, different kinds of adaptive control approaches are proposed in literature. Such as *sliding mode control* (Meng et al. 2018), *fuzzy logic control* (Chamanpira et al. 2019), etc. In this chapter, a *gain scheduled PID controller* (Shao et al. 2019) is considered for controlling the system and ensuring satisfactory response. The structure of the inner control system is depicted in Fig. 12.9. The gain scheduling could be done on the basis of different system variable. The PV module voltage varies linearly with irradiance but current varies logarithmically with the irradiance the variation in the voltage is smaller than current. That is why the voltage of the PV module is taken as the variable depending on which the gain scheduling is implemented. Using the transfer functions of the system the PID controllers were tuned at the different operating points and using these tuning data the look-up tables (LUTs) were implemented. The differential gain was filtered using a first order filter and a clamping algorithm was implemented to avoid windup problem of the integrator.

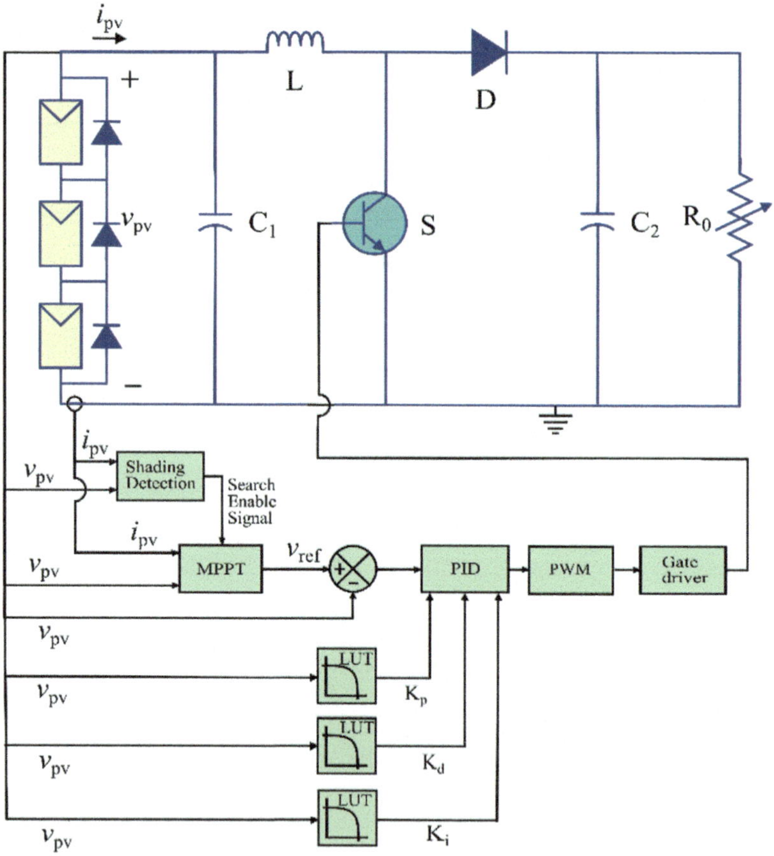

Fig. 12.9 Detailed schematic with inner control loop of the system with gain scheduled PID control

12.5 Advanced Soft-Computing Algorithms for MPPT

Soft computing algorithms are generally algorithms that are based on some form of artificial intelligence or are inspired by some natural phenomena. In recent days, these algorithms have found an overwhelmingly rising application in the field of MPPT in various systems. Some of these instances include application of "*artificial neural network (ANN)*" (Allahabadi et al. 2021), "*simulated annealing (SA)*" (Lyden et al. 2020), "*teaching and learning based optimization (TLBO)*" (Fathy et al. 2018), etc. In this chapter the application and performance analysis of "*particle swarm optimization (PSO)*" based MPPT (Rana et al. 2020), "*differential evolution algorithm (DEA)*" (Rana et al. 2020) and "*binary coded genetic algorithm (BCGA)*" (Nagarani and Nesamony 2019) is discussed. This chapter also compares these algorithms based on their tracking performance and convergence characteristics.

The system performance for the different aforementioned MPPT algorithms is examined by simulating them in Matlab/Simulink. The schematic of the simulations model is shown in Fig. 12.9. As discussed earlier the MPPT algorithm is implemented as a higher level control algorithm which generates the reference or command signal for the internal control loop. For a system with structure like this, the designer has to ensure that the internal control loop is significantly faster than the external control loop. For demonstration purpose the outer control or the MPPT control algorithms are implemented in such a way that the fastest change that they can make to the command signal has an interval of at least 15 ms between them whereas, the settling time of the inner control loop is maintained at around 5 ms. The simulations are performed for two different shading conditions with the power-voltage curve as shown in Fig. 12.10. It can be inferred that one of the conditions are uniform irradiation condition which is applied to the system for the first half of the simulations while the other one is a typical partial shading condition which is applied to the system during the second half of the simulations. The irradiance condition on the different PV modules and the power and voltage at the MPP for those conditions are given in Table 12.4. Using this format of simulation the different algorithms are evaluated.

12.5.1 Particle Swarm Optimization (PSO) for MPPT

The "*particle swarm optimization algorithm (PSO)*," first introduced by Kennedy and Eberhart (Kennedy and Eberhart 1995), is inspired by two natural phenomena known as "*Bird-Flocking* and *Fish-Schooling*" (Khan et al. 2021), where each particle use the collective intelligence and experience of all the agents or particles to move towards an optimal point in the search space. The application of PSO for MPPT in PV system is described in Li et al. Jan. (2019). The flowchart of MPPT algorithm based on PSO is presented in Fig. 12.11. The mathematical process for performing the various steps of the PSO algorithm is summarized in (17), (18) and (19) (Li et al. 2019). Here, p_i^n and v_i^n are the position and the velocity of the *ith* particle in

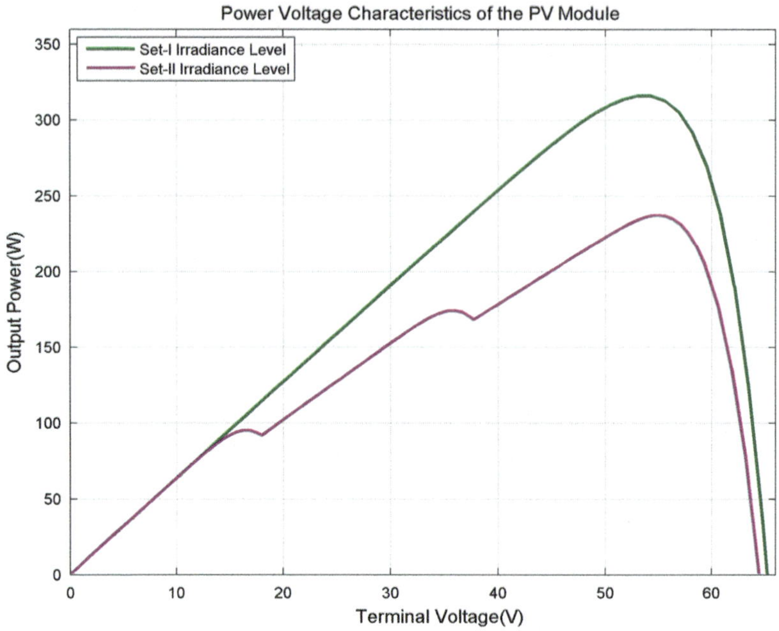

Fig. 12.10 Output power versus output voltage curve of PV array for different shading conditions

Table 12.4 Parameters of a single PV module

Set of irradiance	Module-1	Module-2	Module-3	MPP voltage (V_{mpp})	MPP power (P_{mpp})
Set-I	1000 W/m^2	1000 W/m^2	1000 W/m^2	53.64 V	316.4 W
Set-II	1000 W/m^2	800 W/m^2	700 W/m^2	54.78 V	236.9 W

nth iteration respectively, C_1 and C_2 are tuning parameters, r_1 and r_2 are random numbers between 0 and 1, *w* is the *dynamic inertia weight* factor with maximum and minimum values w_{max} and w_{min}, respectively, *m* is the number of maximum iteration, and P_{best} and G_{best} are *personal best* and *global best* solution. It is worth mentioning that the weight factor *w* reduces gradually as the search process progresses; this is implemented to let the particles explore the search space more at the beginning and make them converge and move less near the ending of the search process.

$$w = w_{\max} - [\{(w_{\max} - w_{\min})n\}/m] \tag{12.17}$$

$$v_i^{n+1} = wv_i^n + C_1r_1(P_{best_i} - p_i^n) + C_2r_2(G_{best_i} - p_i^n) \tag{12.18}$$

$$p_i^{n+1} = p_i^n + v_i^{n+1} \tag{12.19}$$

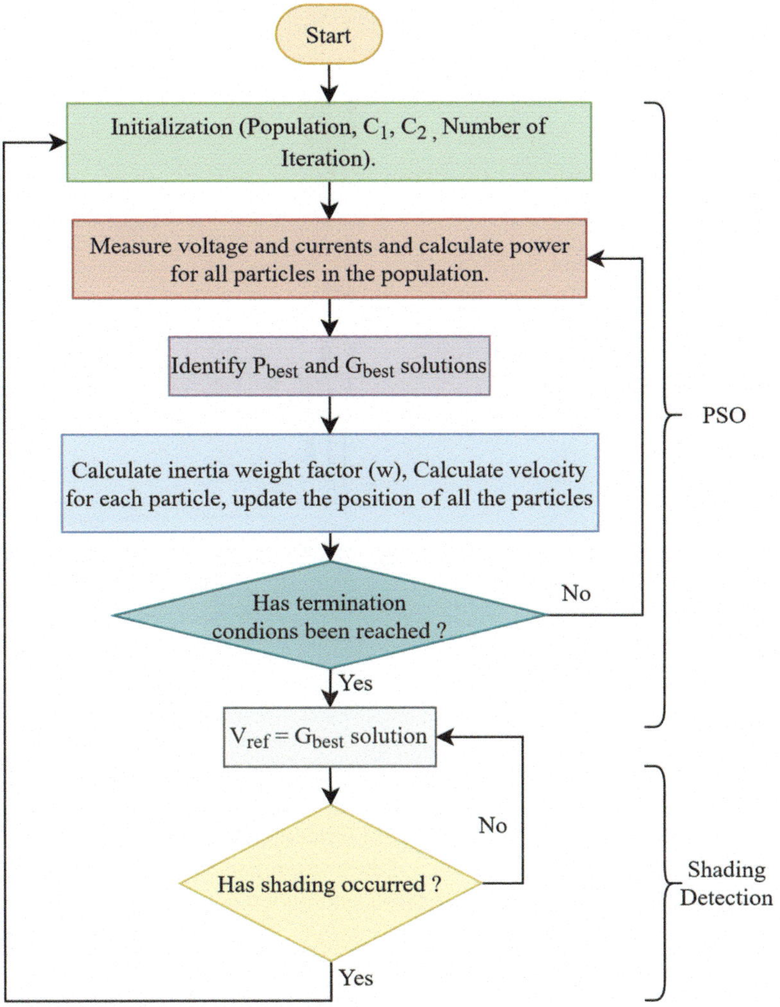

Fig. 12.11 Flowchart of PSO for MPPT

The shading detection algorithm as shown in Fig. 12.11 is a simple algorithm for *PSC detection* (Wellawatta and Choi 2018). It keeps track of the power available at a certain set point and whenever the power changes more than a certain percentage of the previous power it triggers a new search for MPP. The *Maximum number of iteration that are allowed* is used as the stopping criteria for the *PSO algorithm.*

The variation in power and voltage levels of the PV array with time which are obtained from the simulation model with the simulation setup as discussed earlier is illustrated in Fig. 12.12. It can be observed that the MPPT algorithm based on PSO tracks the MPP with fair accuracy. The system parameters converge to the MPP with gradually reducing oscillations. Since, the MPPT based on PSO algorithm is bursty

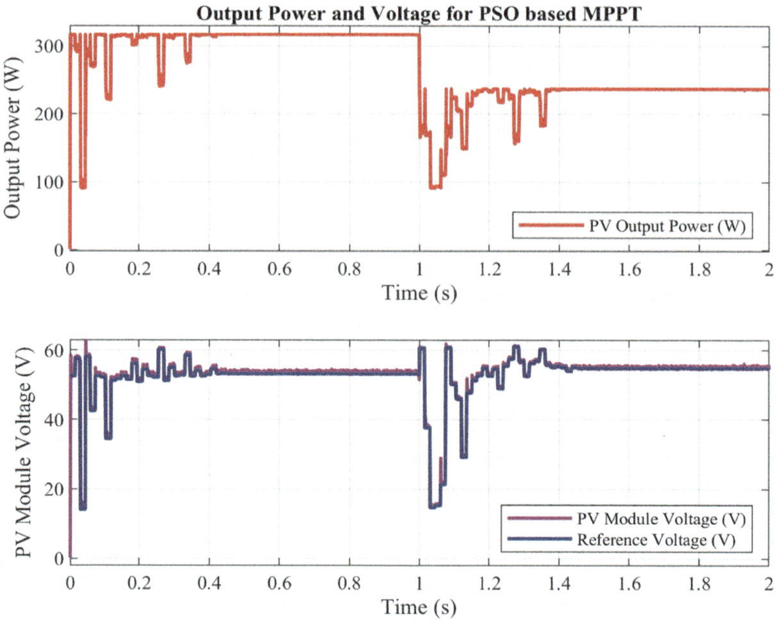

Fig. 12.12 Power and voltage of PV array for MPPT based on PSO

in nature, i.e. the search process is carried out in intervals or only when necessary, the oscillations after a complete search process are very less or negligible. Moreover, the performance of the inner control loop with the gain scheduled PID controller can also be said to be satisfactory. The gain scheduled controller makes the system voltage to settle down at the reference voltage commanded by the PSO algorithm within 5 ms with no or very less overshoots and undershoots.

12.5.2 Differential Evolution Algorithm (DEA) for MPPT

"*Differential evolution algorithm* (*DEA*)" was invented by Storn and Price (1997). *DEA* is a nature-inspired *evolutionary* algorithm where each element in the population (chromosome) undergoes a process called mutation followed by a process called crossover to obtain the optimal solution in the search space. Figure 12.13 captures the flowchart of the MPPT based on DEA algorithm implemented in this study (Zhang and Sui 2020). In this case, also the partial shading detection algorithm is similar to the algorithm in case of the PSO-based MPPT mentioned in the previous discussion. Equations (12.20)–(12.24) represents the mathematical process of mutation, crossover, population update and parameter update which are undergone in course of the run of this algorithm. Instead of a static parameter-based DE algorithm a dynamic parameter-based algorithm as shown by Brest et al. (2006) is implemented

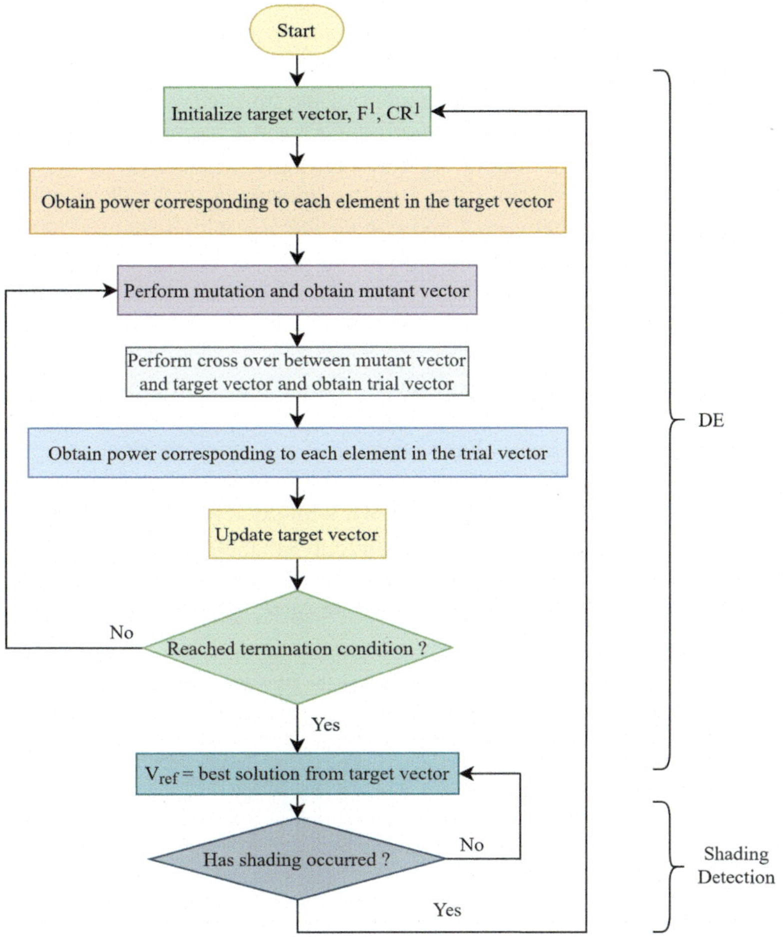

Fig. 12.13 Flowchart of DEA for MPPT

here for better tracking performance and smoother convergence.

$$v_i^{n+1} = x_{r1}^n + F^{n+1}(x_{r2}^n - x_{r3}^n) \tag{12.20}$$

$$u_i^{n+1} = \begin{cases} v_i^{n+1}, \text{ if } r \leq CR^{n+1} \text{ or } i = I_{rand} \\ x_i^n, \text{ if } r > CR^{n+1} \text{ and } i \neq I_{rand} \end{cases} \tag{12.21}$$

$$x_i^{n+1} = \begin{cases} u_i^{n+1}, \text{ if } f(u_i^{n+1}) \geq f(x_i^n) \\ x_i^n, \text{ Otherwise} \end{cases} \tag{12.22}$$

$$CR^{n+1} = CR^n - [(CR^1 - CR_{\min})/m] \tag{12.23}$$

$$F^{n+1} = F^n - [(F_{\max} - F^1)/m] \quad (12.24)$$

Here, v_i^n, u_i^n and x_i^n are the *ith* element of the mutated-vector, trial-vector and target-vector respectively in the *nth* iteration, CR^n and F^n are the crossover rate and the scaling factor in *nth* iteration, the variables I_{rand}, *r1*, *r2* and *r3* are randomly generated integers. I_{rand} ranges from 1 to the *dimension* of the elements, and the others are with values in between 1 and *population size* which are used to select the elements from the population to undergo mutation, *r* is also a randomly generated number that ranges from 0 to 1, The *fitness value* of the element *x* is represented by *f*(*x*). CR_{min} is the lowest limit of *crossover rate* and $F_{\max}$ is upper limit of the *scaling factor*.

The variation in power and voltage levels of the PV array with time for the MPPT based on DEA algorithm is visualized in Fig. 12.14. It can be perceived that the oscillations in the power and voltage waveforms are much more chaotic than that PSO-based MPPT algorithm. But as the search progress, the oscillation gradually reduces and finally, the MPP is tracked down with good accuracy. Apart from that, it can also be observed that the inner control loop's performance is also satisfactory. It makes the voltage of the PV module track the set point with fast response (settles within 5 ms) and with less number of overshoots and undershoots.

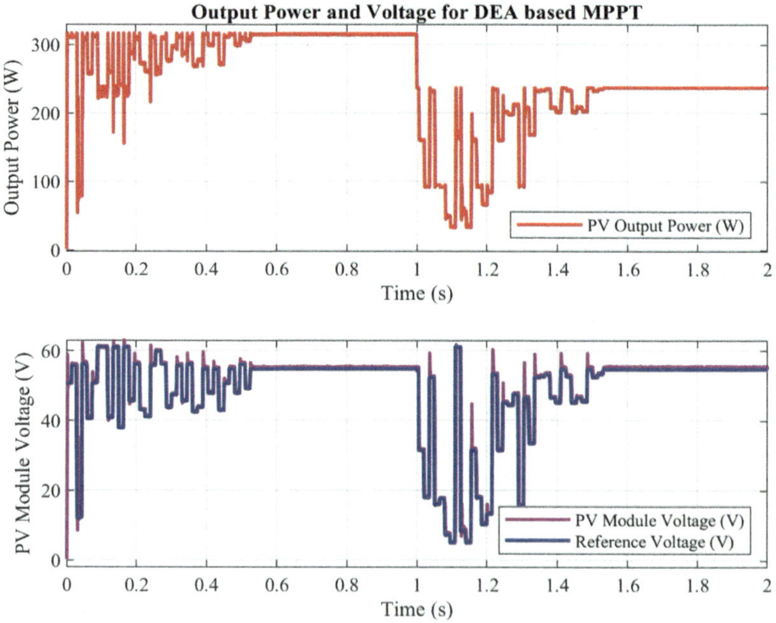

Fig. 12.14 Power and voltage of PV array for MPPT based on DEA

12.5.3 Binary Coded Genetic Algorithm (BCGA) for MPPT

"The *binary coded genetic algorithm (BCGA)* also known as *genetic algorithm (GA)* is a metaheuristic algorithm that employs the principle of natural evolution by employing a systematic process of selection, crossover, mutation, etc., to find the best-fit solution of a problem in the search space" (Chaturvedi 2008). Reference (Garud et al. 2021) presents the implementation of GA for modeling PV system. The flowchart of the BCGA based MPPT algorithm implemented in this study is presented in Fig. 12.15. Similar to the previous cases, in this case also the PSC

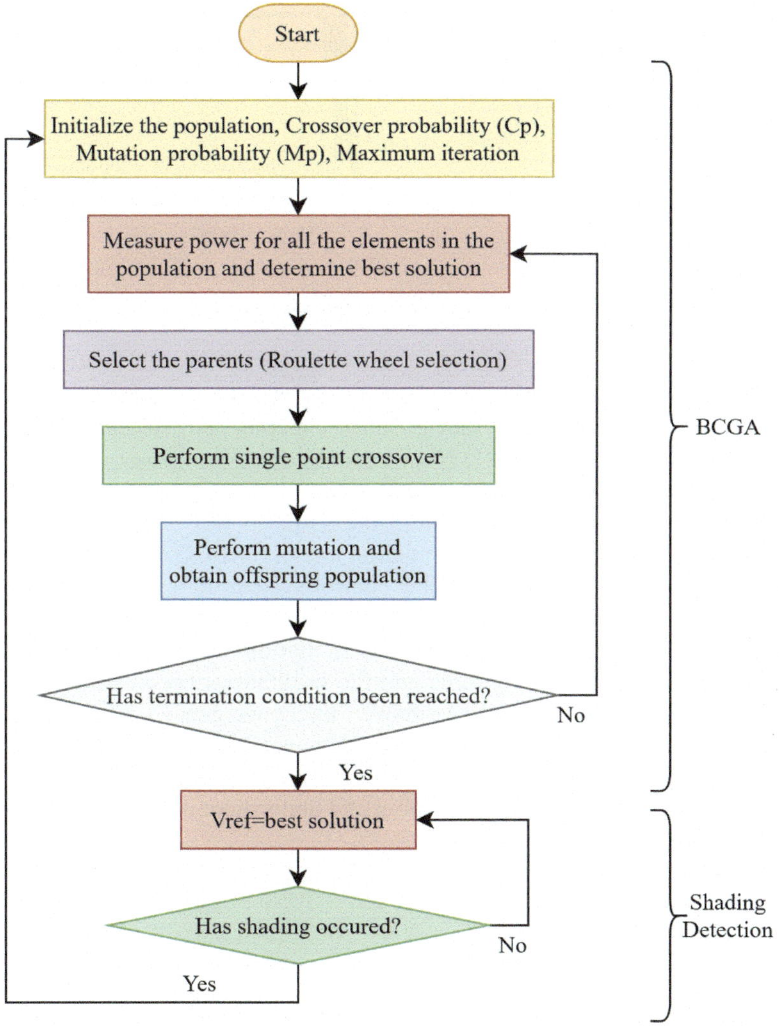

Fig. 12.15 Flowchart of BCGA for MPPT

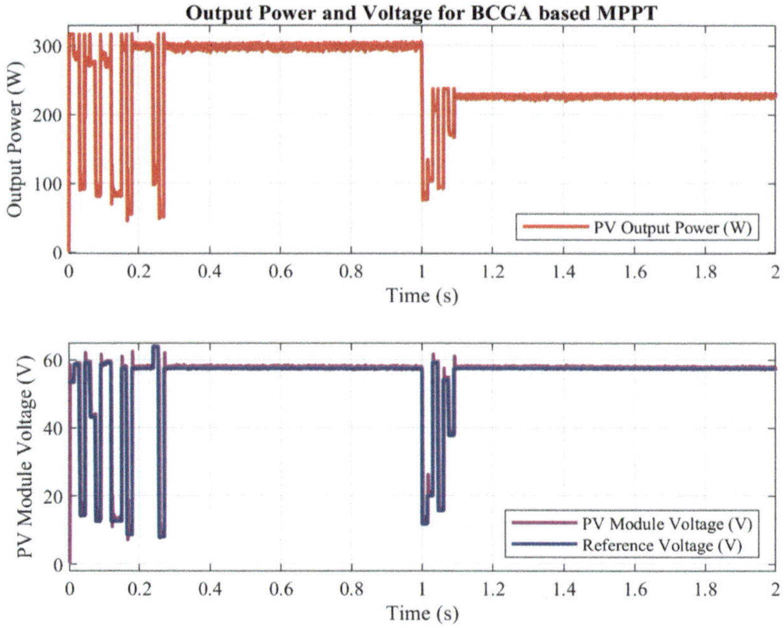

Fig. 12.16 Power and voltage of PV array for MPPT based on BCGA

detection algorithm is a simple power monitoring-based algorithm. For the purpose of selection of the parents before they go into the mating pool, *Roulette Wheel* selection technique (Katoch et al. 2021) is used and for cross-over, *single point crossover* is implemented (Chaturvedi 2008).

Figure 12.16 captures the variation of the output power and the output voltage of the PV array obtained from the simulation model for BCGA based MPPT. It can be noticed that the BCGA based MPPT control system tracks the global MPP for both uniformly irradiated as well as partially shaded condition with fair degree of accuracy. In this case, it can be seen that the oscillations during the search are lesser compared to the previous algorithms and the oscillations become negligible after the convergence of the algorithm. Moreover, the convergence time is significantly lesser than the other algorithms in this case. The inner gain scheduled controller also performs satisfactorily. The setpoint tracking is performed with very good accuracy, fast response and less number of overshoot and undershoot in the voltage waveform of the PV module.

12.6 Comparative Analysis

The MPPT algorithms described in the previous section are all based on some meta-heuristic algorithm. But at this point of discussion, it is still somewhat unclear which one of them performs better than the others or what are the criteria that sets them apart from each other. The following comparative study ponders on some of these questions. The comparison of the different algorithms are done on the basis of their convergence characteristic (Lolla et al. 2021) and tracking performance.

12.6.1 Convergence Characteristics

The convergence characteristic of an algorithm reveals many useful information about the performance of the algorithm. The characteristic typically shows how the error value changes with respect to the progression of the algorithm. In this chapter, the root mean squared error is plotted with respect to the iteration number of the algorithm. The formula for calculating the *root mean squared (RMSE)* error is shown in Eq. (12.25). Here, x_{mpp} represents the MPP voltage (V_{mpp}) and power (P_{mpp}), x_i is the *ith* element in the population in a particular iteration, and N is the aggregate number of elements in that set of population.

$$RMSE = \sqrt{\left[\left\{\sum_{i=1}^{N} (x_{mpp} - x_i)^2\right\} / N\right]} \tag{12.25}$$

Figure 12.17 portrays the convergence characteristics of the power and voltage obtained for the both uniform and partial shading case studies for the three algorithms used for MPPT in this chapter. From Fig. 12.17a and b, it can be observed that the PSO and DEA converges to the MPP with a monotonically decreasing error characteristic, whereas, the BCGA tends to diverge at the beginning but converges hastily at the ending part of the search. Fig. 12.17c and d show similar kind of behaviour for the DEA. Since, the DEA and the BCGA algorithms are inspired by the natural evolution process where the fate of the future population depends greatly on their predecessor, the performance of these algorithms is greatly dependent on the fitness and quality of the initial population. As a result, when the randomly generated initial population is of poor quality the performance of the algorithms degrades. So, for these two algorithms the initial population has to be generated carefully. On the other hand, the PSO algorithm is not that much dependent on the initial population. Its performance depends on the velocity of the particles, and diversity of the population. This is why, the PSO algorithms converge to the MPP in both the cases with a smoother characteristics.

Apart from that, it should be noted that the BCGA shows a snappy characteristics, i.e. the algorithm has a tendency to converge faster to the best solution found in a particular iteration in the search space. Which results in less exploration in the search space. In some cases, this might cause the algorithm to obtain a suboptimal solution (local MPP) and miss the optimal solution (global MPP).

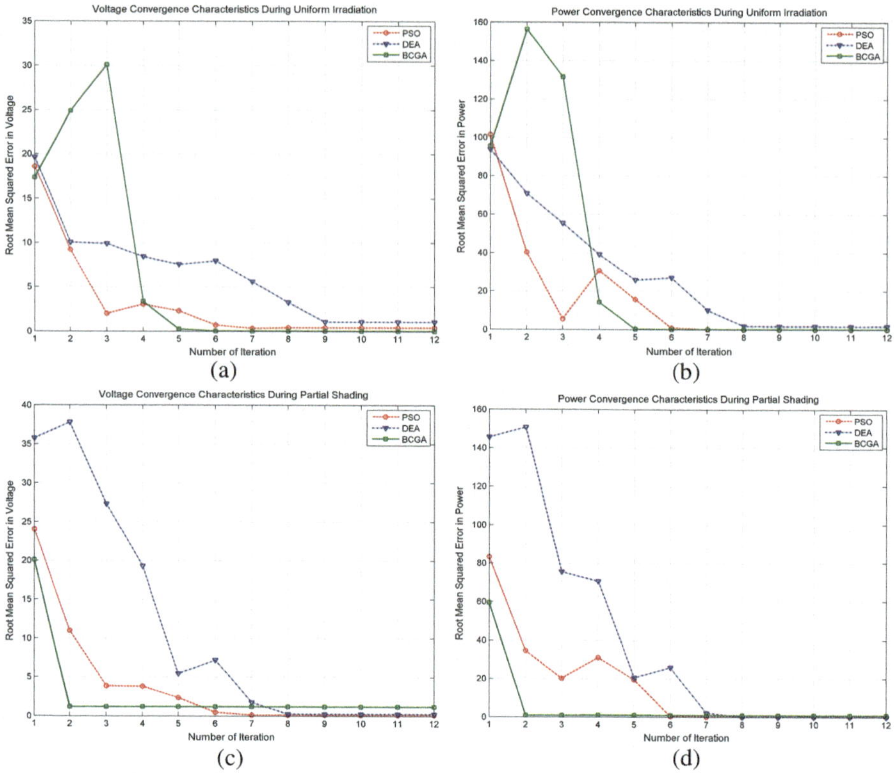

Fig. 12.17 Convergence characteristics of the algorithms. **a** Voltage convergence for uniform irradiation **b** Power convergence for uniform irradiation **c** Voltage convergence for PSC **d** Power convergence for PSC

12.6.2 Tracking Efficiency

It was previously discussed that the oscillations during the search process are largest for the DEA algorithm, least for the BCGA and moderate for the PSO algorithm-based MPPT. It was also shown in the previous section that the BCGA converges fastest and the DEA converges slowest. So, assuming that the runtime for all the algorithms are same and they all successfully track the global MPP, the energy lost during the search process will be most for the DEA, least for the BCGA, and moderate for the PSO based MPPT. This means that the BCGA will be most efficient and then there will be PSO and DEA in subsequent order.

But it has to be noted that all the algorithms do not track the exact same operating point. Consequently, their tracking efficiency changes. To obtain their steady state efficiency the power loss during the search process are neglected and the tracking efficiency is calculated on the basis of the expression in (26) as given by Islam et al. (2018). The efficiencies calculated for the case studies presented in this chapter are

Table 12.5 Tracking efficiency of the MPPT algorithms

Shading condition	Efficiency of PSO based MPPT (η_{PSO}) (%)	Efficiency of DEA based MPPT (η_{DEA}) (%)	Efficiency of BCGA based MPPT (η_{BCGA}) (%)
Uniform irradiation	99.97	99.75	99.90
Partial shading condition	99.99	99.99	99.98

shown in Table 12.5. It can be seen that the efficiencies are more or less same. So, it can be said that only the performance during search process sets them apart.

$$\eta = \frac{Power\ \text{Tarcked}}{P_{mpp}} \times 100\% \tag{12.26}$$

12.7 Effect of Load Variation

In a PV system, the performance of the whole control system depends largely on the inner control loop, i.e. the gain scheduled PID controller in this chapter. The controller must be immune to the various disturbances that are present in the system. Otherwise, the system will have to frequently search for the MPP whenever a disturbance occurs. One of the major disturbances that occur in such systems is load disturbance. The performance of the system for changing load conditions is shown in Fig. 12.18. The load change is realized as a step change in the load resistance (R_O) at 0.2 s and ramp change between 0.6 s and 1 s of simulation. It can be observed that the system performs very satisfactorily. The voltage regulation is very accurate and the system successfully maintains the MPP rejecting the effects of these disturbances.

12.8 Summary

Advanced computation-based metaheuristic algorithms are increasingly being applied in the case of MPPT applications in PV systems. These algorithms ensure better performance under both uniform irradiation and partial shading condition. To design an effective MPPT system implementing these advanced algorithms involves complex tasks such as proper system modeling, robust controller design. etc. This chapter discussed some of these aspects in detail. An analytical method-based generalized modeling approach for PV system with DC to DC converters that operate in CCM is presented and then based on this model the controller design process is presented. The performance of the system is shown to be satisfactory when different

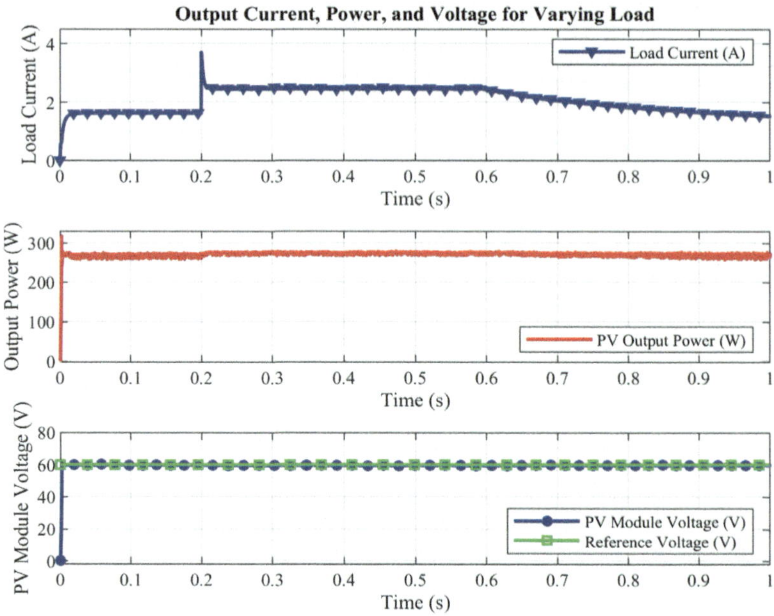

Fig. 12.18 System performance under load disturbances

soft-computing algorithms. The comparative analysis revealed the strength and weaknesses of the algorithms. It was shown that the soft-computing algorithm-based MPPT system tracks the MPP with good accuracy and fast response, but it has to be noted that since these algorithms are not gradient-based continuous running algorithm but rather bursty in nature, they may not be the most effective choice to track MPP under slowly varying irradiance condition. Especially, when the rate of change of irradiance is close to the convergence time of the algorithms. To solve such problems, hybrid algorithms can be used, where the metaheuristic algorithm first finds the MPP and then a gradient-based traditional MPP algorithm is used to maintain the MPP.

So, the key takeaways and inferences from this chapter are:

- The small-signal model of the PV system with DC-DC converters are not the same at all operating points. Rather, they vary with the operating points. But quasi-static approximation can be used to design the controllers at different strategic operating points and they can be combined to control the system properly. A gain-scheduled PID controller is such a controller.
- The soft-computing algorithms perform satisfactorily under fast changing shading conditions but they reduce the system efficiency when subjected to slowly varying shading condition.
- Even though, the tracking efficiency of all the MPPT algorithms presented in this chapter are more or less similar, they have different convergence characteristics

and different computational complexity. As a result, they should be chosen based on the type of application.

References

Allahabadi S, Iman-Eini H, Farhangi S (2021) Fast Artificial neural network based method for estimation of the global maximum power point in photovoltaic systems. IEEE Trans Ind Electron 2021. https://doi.org/10.1109/TIE.2021.3094463

Brest J, Greiner S, Boskovic B, Mernik M, Zumer V (2006) Self-adapting control parameters in differential evolution: a comparative study on numerical benchmark problems. IEEE Trans Evol Comput 10(6):646–657. https://doi.org/10.1109/TEVC.2006.872133

Chaturvedi DK (2008) Genetic algorithms. In: Soft computing. Studies in computational intelligence, vol 103. Springer, Berlin, Heidelberg. https://doi.org/10.1007/978-3-540-77481-5

Chamanpira M, Ghahremani M, Dadfar S, Khaksar M, Rezvani A, Wakil K (2019) A novel MPPT technique to increase accuracy in photovoltaic systems under variable atmospheric conditions using Fuzzy Gain scheduling. Energy Sources, Part A: Recov Utiliz Environ Effects. https://doi.org/10.1080/15567036.2019.1676325

Cuk S (1997) Modelling, analysis, and design of switching converters. Dissertation (Ph.D.), California Institute of Technology. https://doi.org/10.7907/SNGW-0660

Fahad S, Ullah N, Mahdi AJ, Ibeas A, Goudarzi A (2019) An advanced two-stage grid connected PV system: a fractional-order controller. In: International journal of renewable energy research, vol 9, no. 1. https://arxiv.org/abs/2004.14106

Fathy A, Ziedan I, Amer D (2018) Improved teaching–learning-based optimization algorithm-based maximum power point trackers for photovoltaic system. Electr Eng 100:1773–1784. https://doi.org/10.1007/s00202-017-0654-8

Garud KS, Jayaraj S, Lee MY (2021) A review on modeling of solar photovoltaic systems using artificial neural networks, fuzzy logic, genetic algorithm and hybrid models. Int J Energy Res 45:6–35. https://doi.org/10.1002/er.5608

Guo B et al (2020) Optimization design and control of single-stage single-phase PV inverters for MPPT improvement. IEEE Trans Power Electron 35(12):13000–13016. https://doi.org/10.1109/TPEL.2020.2990923

Haque A (2014) Maximum power point tracking (MPPT) scheme for solar photovoltaic system. Energy Technol Policy 1(1):115–122. https://doi.org/10.1080/23317000.2014.979379

Islam H, Mekhilef S, Shah NBM, Soon TK, Seyedmahmousian M, Horan B, Stojcevski A (2018) Performance evaluation of maximum power point tracking approaches and photovoltaic systems. Energies 11(365):2018. https://doi.org/10.3390/en11020365

Katoch S, Chauhan SS, Kumar V (2021) A review on genetic algorithm: past, present, and future. Multimed Tools Appl 80:8091–8126. https://doi.org/10.1007/s11042-020-10139-6

Khan RA, Yang S, Fahad S, Khan SU, Kalimullah (2021) A modified particle swarm optimization with a smart particle for inverse problems in electromagnetic devices. IEEE Access 9:99932–99943. https://doi.org/10.1109/ACCESS.2021.3095403

Khezri R, Mahmoudi A, Haque MH (2020) Optimal capacity of solar PV and battery storage for australian grid-connected households. IEEE Trans Ind Appl 56(5):5319–5329. https://doi.org/10.1109/TIA.2020.2998668

Kennedy J, Eberhart R (1995) Particle swarm optimization. In: Proceedings of ICNN'95—international conference on neural networks, Perth, WA, Australia, vol. 4, pp. 1942–1948. https://doi.org/10.1109/ICNN.1995.488968

Kocher MJ, Steigerwald RL (1982) An AC to DC converter with high quality input waveforms. IEEE Power Elect Spec Conf 1982:63–75. https://doi.org/10.1109/PESC.1982.7072396

Ko J-S, Huh J-H, Kim J-C (2020) Overview of maximum power point tracking methods for PV system in micro grid. Electronics 9(5):816. https://doi.org/10.3390/electronics9050816

Li H, Yang D, Su W, Lü J, Yu X (2019) An overall distribution particle swarm optimization mppt algorithm for photovoltaic system under partial shading. IEEE Trans Industr Electron 66(1):265–275. https://doi.org/10.1109/TIE.2018.2829668

Lolla PR, Rangu SK, Dhenuvakonda KR, Singh AR (2021) A comprehensive review of soft computing algorithms for optimal generation scheduling. Int J Energy Res 45:1170–1189. https://doi.org/10.1002/er.5759

Lyden S, Galligan H, Haque ME (2020) A hybrid simulated annealing and perturb and observe maximum power point tracking method. IEEE Syst J. https://doi.org/10.1109/JSYST.2020.3021379

Malinowski M, Leon JI, Abu-Rub H (2017) Solar photovoltaic and thermal energy systems: current technology and future trends. Proc IEEE 105(11):2132–2146. https://doi.org/10.1109/JPROC.2017.2690343

Meng Z, Shao W, Tang J, Zhou H (2018) Sliding-mode control based on index control law for MPPT in photovoltaic systems. CES Trans Elect Mach Syst 2(3):303–311. https://doi.org/10.30941/CESTEMS.2018.00038

Nagarani B, Nesamony J (2019) Performance enhancement of photovoltaic system using genetic algorithm- based maximum power point tracking. In: Turkish journal of electrical engineering & computer sciences. https://doi.org/10.3906/elk-1801-189

Rana MT, Bhowmik PS (2020) An integrated small-signal model of photovoltaic system for MPPT applications. In: 2020 21st national power systems conference (NPSC), Gandhinagar, India, pp 1–6. https://doi.org/10.1109/NPSC49263.2020.9331858

Rezk H, Aly M, Al-Dhaifallah M, Shoyama M (2019) Design and hardware implementation of new adaptive fuzzy logic-based MPPT control method for photovoltaic applications. IEEE Access 7:106427–106438. https://doi.org/10.1109/ACCESS.2019.2932694

Shang L, Guo H, Zhu W (2020) An improved MPPT control strategy based on incremental conductance algorithm. Prot Control Mod Power Syst 5(14). https://doi.org/10.1186/s41601-020-00161-z

Shao P, Wu J, Wu C, Ma S (2019) Model and robust gain-scheduled PID control of a bio-inspired morphing UAV based on LPV method. Asian J Control. 21:1681–1705. https://doi.org/10.1002/asjc.2187

Storn R, Price K (1997) Differential evolution—a simple and efficient heuristic for global optimization over continuous spaces. J Global Optim 11:341–359. https://doi.org/10.1023/A:1008202821328

Wellawatta TR, Choi S (2018) A novel partial shading detection algorithm utilizing power level monitoring of photovoltaic panels. In: 2018 international power electronics conference (IPEC-Niigata 2018-ECCE Asia), Niigata, pp 1409–1413. https://doi.org/10.23919/IPEC.2018.8507897

Zhang P, Sui H (2020) Maximum power point tracking technology of photovoltaic array under partial shading based on adaptive improved differential evolution algorithm. Energies 13(5):1254. https://doi.org/10.3390/en13051254

Chapter 13
Design of Vertical Axis Wind Turbine in Recent Years—A Short Review

Vijayakumar Mathaiyan, Vijayanandh Raja, S. Srinivasamoorthy, Dong Wong Jung, M. Senthilkumar, and Sivaranjani Sivalingam

Abstract Though the available models cannot produce the efficiency or power as Horizontal Axis Wind Turbine (HAWT), the Vertical Axis Wind Turbine (VAWT) design in recent works was reviewed for its aesthetic value and efficiency. This review will be a useful guide to modify available design for any intended purpose or provide a futuristic design which can be efficient in power generation and be an ornamental device. Besides these, the overview of recent researches in the field of wind turbine technology is covered in this book chapter. The work provides the guide to design VAWT with the information about the implementation of farm, reduction of noise, and computational techniques used in recent researches. The review of this kind always has greater importance because of the up to date information about the ongoing researches.

Keywords VAWT · Aerodynamic design · Computational analysis · Experiment method · Renewable energy

Nomenclature

VAWT	Vertical Axis Wind Turbine
VIVWT	Vortex Induced Vertical Axis Wind Turbine
BWT	Butterfly Wind Turbine
QMS	Relative Quadruple-multiple Stream Turbine Model
DFBI	Dynamic Fluid Body Interaction Method
CU	Counter Up
ANN	Artificial Neural Network

V. Mathaiyan · S. Srinivasamoorthy · D. W. Jung (✉) · S. Sivalingam
Department of Mechanical Engineering, Jeju National University, Jeju City, South Korea
e-mail: jungdw77@gmail.com; jungdw77@naver.com

V. Raja · M. Senthilkumar
Kumaraguru College of Technology, Coimbatore, India
e-mail: senthilkumar.m.aeu@kct.ac.in

A. K. Bohre et al. (eds.), *Planning of Hybrid Renewable Energy Systems, Electric Vehicles and Microgrid*, Energy Systems in Electrical Engineering,
https://doi.org/10.1007/978-981-19-0979-5_13

ALM	Actuator Line Model
CR-VAWT	Co-axial Contra-rotating Vertical Axis Wind Turbine
UDF	User Defined Function
SHM	Structural Health Monitoring
HAWT	Horizontal Axis Wind Turbine
TTWT	Tornado Type Wind Turbine
OCS	Overspeed Control System
TSR	Tip Speed Ratio
CD	Counter Down
KEPC	Korea Electric Power Corporation
POD	Proper Orthogonal Decomposition
C-VAWT	Conventional Vertical Axis Wind Turbine
FSI	Fluid Structure Interaction

13.1 Introduction

The solar energy from nuclear fusion creates wind energy. Wind moves because of the Atmospheric pressure gradient created by uneven solar heating, Coriolis effect of earth rotation, and geographical conditions (Tong 2010). The concept of harnessing the energy from wind is one of the most promising technologies to replace the available methodology which pollutes the earth. Recent climatic changes and other adverse effects on the health of human and other living creatures explicitly or implicitly. With the increasing population and power requirement, the peoples were not only concerned about the power generation but also the way it looks. We believe that if wind turbines were decorative enough, it will be easy for the employment of the device and reduce pollution. It is notable that noise from the device should also be as minimal as possible without affecting the peace. Even the governments are encouraging the implementation of renewable energy sources (Weiss and Tsuchida 2017). Blade thickness and blade loads in the rotor is the major source of aerodynamic noise. Unsteady flow of VAWT reduces the generated noise and the maximum power coefficients reaches when $\lambda = 0.5$ (Dumitrache et al. 2015). The noise from the wind turbine is a major problem and the study about effects to human is done by Amelia et al. Even the tonal components at low frequencies create the discomfort, and annoyed people in % is higher for the wind turbines than the other sources of noise pollution (Trematerra et al. 2017).

Vertical Axis Wind Turbines (VAWTs) are unresponsive to the wind direction due to its omni-directionality. Less moving parts of a simple mechanical design debarred the yawing equipment and outcome is the pitching mechanism. The total energy cost consumption for the operation and maintenance, stable structure, and scalability are the advantages of VAWT over HAWT. Difference in blade's circulation along the rotational axis of turbine is the energy conversion in VAWT and only by tip vortices in HAWT (Victor et al. 2017). Though the power generations from HAWT are better than in VAWT, it is notable that HAWT has robust unalterable design. VAWT is of primary interest only because its design can be volatile. In this short review, we have provided the VAWT design calculation to meet the power requirement while providing a view on the available attractive designs. Some of the commonly available VAWT are Darrieus, Savonius, H-Darrieus-Rotor, helical Darrieus turbine, etc. The authors believe that the helical Darrieus turbine has an attractive design, so its design is reviewed. Note that the VAWT is much safer than the HAWT, if there is any catastrophe.

This review paper covers the major design factors and parameters to be considered. We have provided commonly followed design process including the Aerodynamic design of blade and the effects of Low Reynolds Number. It is notable that the efficient design of blade provides good result in generating power. Lithium batteries can be used as power storage units in renewable energy technologies (Meng and Li 2019). Blade element momentum theory, BEM theory application, Navier–Stokes solutions, and Hybrid methods are the available aerodynamic modelling of wind turbines (Das and Amano 2014). The outline of the entire work is shown in the flow chart below. The chapter covers the commonly available VAWT and introduces the design methodology. The farm design and the numerical results with a technique to validate the results are elaborated to an extent.

13.2 Commonly Available VAWT

Most of the VAWT usually has blade arrangement to create a mechanical energy. Power generation system is directly engaged with the arrangements to generate electricity. Some of the VAWT arrangements are discussed below.

13.2.1 Vortex Induced Vibrations Wind

Williamson et al. review work on the vortexes around the circular cylinder provides more information about the possible vortexes from Vortex Induced Vibrations Wind Turbine (Williamson and Govardhan 2004). Typical arrangements in the VIVWT from Sassi et al. (2020) are illustrated in Fig. 13.1. The arrangement is simple and does not have any sharp edges which may create disaster when the structure collapses.

The mast, electricity generation system, and support system are the components in this bladeless type wind turbines. Phenomena of vorticity create the mechanical movement of Mast. The mechanical energy is converted into electrical energy using the electricity generation system. Stator is placed onto the supporting system which keeps it stable on the ground and the rotor is attached to the mast. The arrangement is proved to be efficient than the other available wind turbines.

Villarreal patent work yields more information about this turbine (Villarreal 2016). From the illustration, it is understood that the power generation depends on the mast.

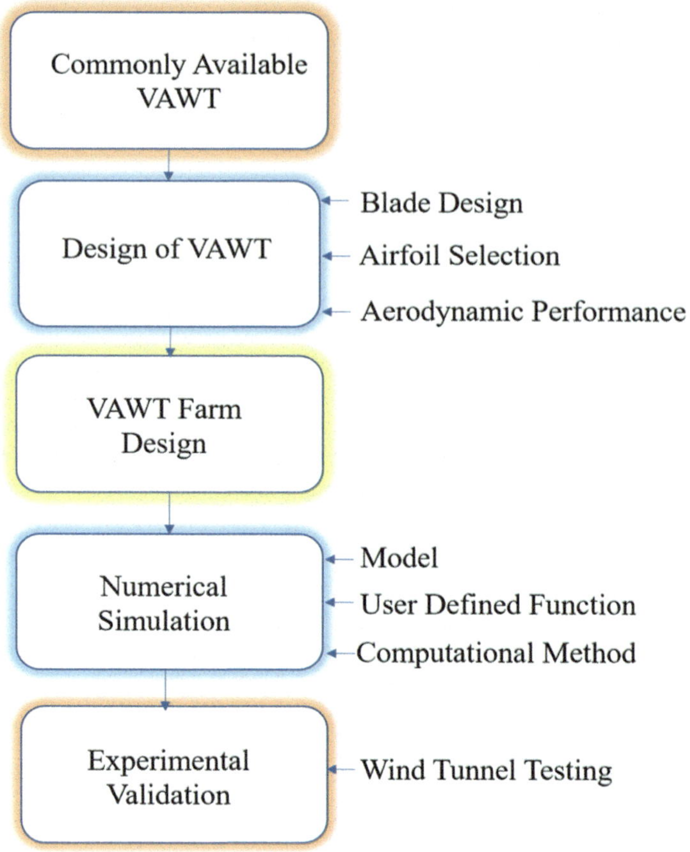

Fig. 13.1 Flowchart for contents in the chapter

The design of the mast with different cross-section is the ongoing research work. The cross-section can even be the symmetrical airfoils other than the circular and ellipse sections.

13.2.2 Tornado Type Wind Turbine

Tornado Type Wind Turbine (TTWT) uses the phenomena of tornado to generate electricity. The circular and spiral models are available in TTWT. The circular model has the circular wind and tornado generation chamber. From Hsu (1984), the setup of the turbine has intake chamber, wind chamber, turbine arrangement, and vent are shown in Fig. 13.2. The air flows through the intake chamber and the wind chamber

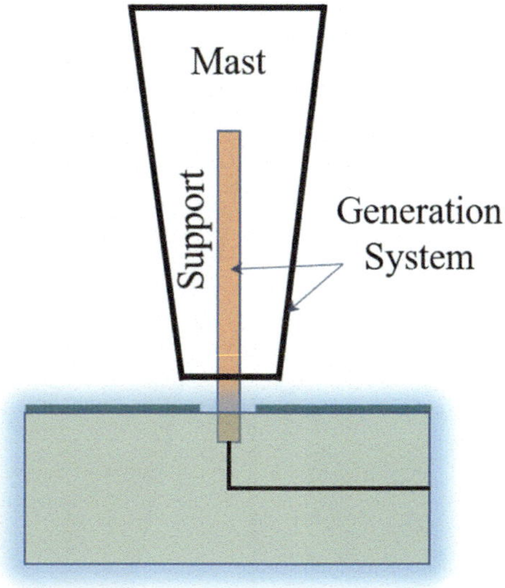

Fig. 13.2 Typical arrangement of VIVWT

arrangement creates the tornado or vortex which then hits the turbine. The mechanical energy of the turbine is converted to electrical energy.

Yen provided the vortex or tornado as three rotating flow regimes (Yen et al. 1978). The viscous effect dominated first flow regime has the tangential velocity increases with radial distance. In the second flow regime, the flow is laminar with almost negligible viscous effects. The circulation for the laminar second flow regime is given by $\Gamma = V\ \mathrm{Re}$, where Re is the Reynolds number, and V is the velocity. Third flow regime has the turbulence with eddies in it. Since the reverse flow or ground effect is not desirable, the vent designs are also an important design parameter. It is notable that the pathway of air in wind chamber is affected by the artificial tornado generation systems.

Most of the research work suggests that the optimization of geometry for tornado strength decides the power generation capability of TTWT. Tower to turbine diameter ratio is a key parameter in the TTWT design. Coefficient of power is considered as the performance indicator (Fig. 13.3).

13.2.3 Darrieus Type Wind Turbine

Tjiu et al. reviewed the Darrieus Vertical Axis Turbine and classified it into curved blades and straight blades (Tjiu et al. 2015). Guy-wired Phi rotor, Variable-Geometry, Variable-Pitch, Delta rotor, Diamond rotor, and V/Y rotor are some of the turbines discussed in the review. The small-sized turbines are researched to reduce the cost of energy, which resulted in an armless VAWT with looped blades called Butterfly Wind Turbine (BWT). In other words, Butterfly type wind turbine is a type of Darrieus

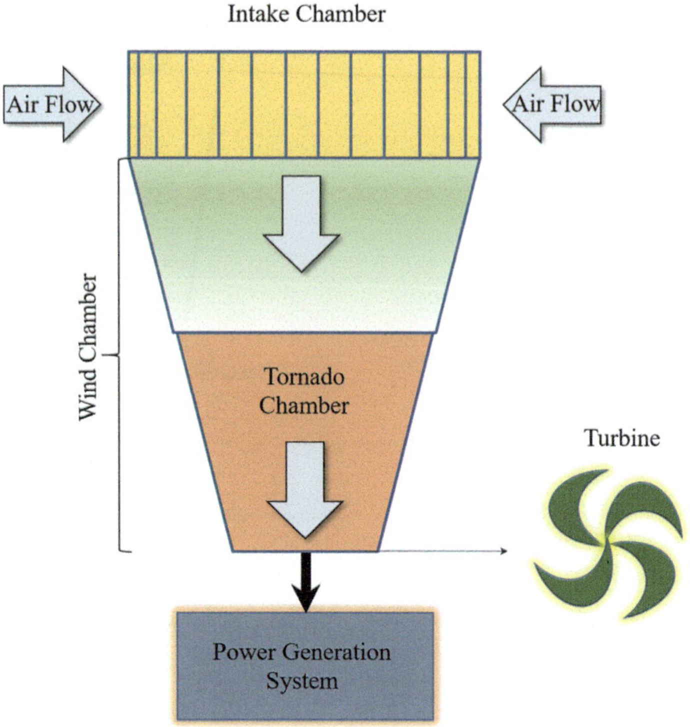

Fig. 13.3 Illustration of TTWT

Type wind turbine with looped blades. Low aerodynamic resistance and blade tip loss are possible in BWT, because of its armless rotor construction and an additional feature. Over speed control system (OCS) is used to control the speed of the BWT from centrifugal force acting on the rotor. OCS is recommended as a prerequisite to reduce the cost of small VAWT and increase safety. The blade rotates about the blade axis which is connected to OCS or generator. OCS controls electric power value around 3.2 kW, even when functioned at 12 m/s or more (Hara et al. 2017, 2018).

Figure 13.4 shows the typical VAWT. When the wind blows, the force of the wind acts on the blade. The blade movement creates the centrifugal force and makes the rotor rotate. In the power generation system, the rotor movement cuts the magnetic field in the stator resulting in power generation.

13.3 Design of VAWT

Designing process is a series of steps that involves the design of VAWT with the increased cross-sectional area, selection of materials, airfoil selection, aerodynamic

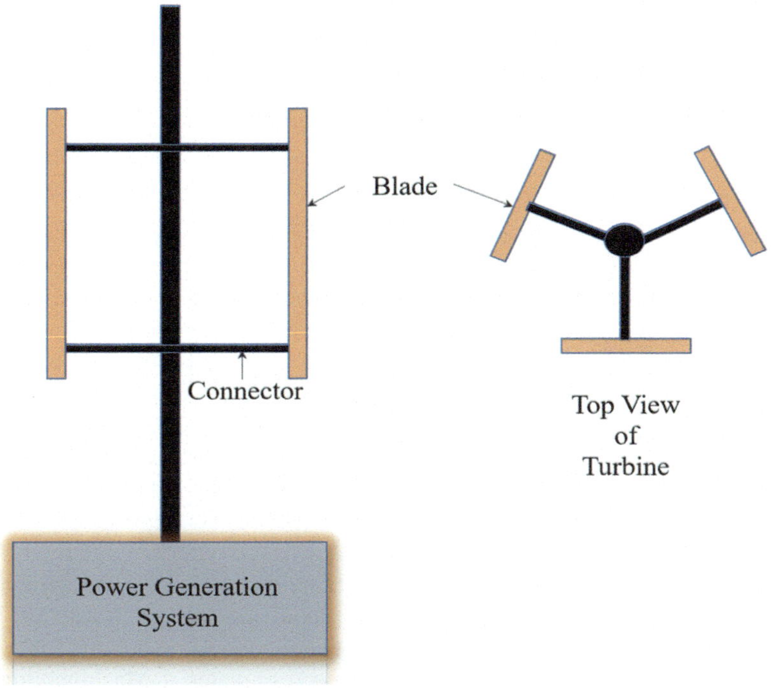

Fig. 13.4 Illustration of Darrieus Type wind turbine

design of blade, number of blades and supporting arrangements, prototype design and computational analysis, etc., Efforts were made in this work to cover the overall design process of VAWT.

13.3.1 Design Requirements

The major design requirement for the VAWT is maximum power generation, aesthetic design, and noise-free operation. From Whittlesey (2017), the power *P* productivity equation for any type of VAWT is given by

$$P = c_p \frac{1}{2} \rho v^3 A \tag{13.1}$$

where c_p is the power coefficient, ρ is the density of air, v is the velocity of wind, and *A* is the cross-sectional of the VAWT. Equation (13.1) shows that in order to increase the power generation using design, the cross-sectional area of VAWT should be increased. Cross-sectional area is the radius of the turbine multiplied with the length

Table 13.1 Shows the VAWT and the peak efficiency from Schubel et al.

VAWT type	Peak efficiency (%)
Savonius rotor	16
Cup	8
Darrieus	40

of blades from Whittlesey et al. The material employed in making and appropriate placement of the VAWT is of major importance.

Arms of a vertical axis wind turbine effects the pressure drag. Blade drag based on pressure is high for the rectangular arm when compared with the aerofoil and cylindrical arms. Whereas the pressure-based total drag coefficients are high for cylindrical arms and air foil has low pressure-based total drag coefficients (Hara et al. 2019). The poor self-starting performance of the lift-driven VAWTs can be improved by the Double-blade VAWTs. Quadruple multiple stream tube model (QMS) is used to simulate the performance of DB-VAWT. With a correction factor added to QMS model, the values are in good compliance with CFD results (Hara et al. 2014).

From Schubel and Crossley (2012), relative wind velocity is given by difference between wind and blade velocity. Minor losses are from the tip, blade aerodynamics from pre and post wind impact, and driving mechanism. Some of the available VAWT and their predicted efficiency from Schubel et al. are tabulated below (Table 13.1).

One should keep the factors such as the property of material, Machinability, resistance to the atmospheric effects, safety, availability, cost of material, and reliability (Babu et al. 2006). Commonly used materials in making wind turbines are carbon fibers, Aramid fibers, glass fibers, aluminum, stainless steel, and wood. Among all the commonly available materials, Maskepatil et al. (2014) and Babu et al. (2006) claimed that an opted choice for wind turbine is carbon fiber. They have considered the factors such as density, mechanical strength, resistance to environmental changes, costs, and availability to determine the most suitable material.

Huang (2009) reviewed the fabrication and properties of Carbon fibers in detail. Carbon fibers are the fibers with at least 92wt% carbon and it usually has high tensile properties, low densities, conductivity, and stability to the thermal and chemical stress, and resistance to creep. Manufacturing process involves the controlled pyrolysis of stabilized precursor fibers. Oxidation process of the precursor fibers at around 200–400 °C in air stabilizes the material. Then the resultant material undergoes a process called carbonization. In this process, the non-carbon elements like hydrogen, oxygen, nitrogen, etc., are removed from the stabilized material by treating it around 1000 °C in an inert atmosphere. Carbonized fibers are then process of graphitization under a temperature of about 3000 °C. Young's modulus of the carbon fiber depends on the carbon content in carbon fibers. Carbon fibers are classified based on the Young's modulus and temperature at which it was treated. The atomic structure of carbon fiber and graphite are almost similar. Because of carbon fibers property, it finds application in the field of aerospace, automotive industries, and so on.

13.3.2 Blade Design

Schaffarczyk (2014) mentioned that wind turbine blade involves the aerodynamics, structural design and manufacturing. Aerodynamics involves the fluid physics on the surface of blade and the structure design involves reduction of materials in the blade without compromising the desired property. Structural integrity of the design is utmost necessity since the turbines are expected to work for decades. The Manufacturing includes the process of machining the blade according to the requirements with minimum cost. Besides the aerodynamic design of the blade, there are other removable accessories like vortex generators and Gurney flaps available to enhance the wing aerodynamics. Kariamian and Abdolahifar (2019) research claimed that the proposed 3part blade design yields better performance than the helical blade with increased tip speed ratio (TSR). Increasing the area of the blades or fins is considered as one of the options to improve aerodynamic performance in the study of Utomo et al. (2018). The size of wind turbines is increased for more energy in recent days. The optimization of root area of the wind turbine blade is carried using genetic algorithm and it is suitable for complex optimization (Herrmann et al. 2019).

13.3.3 Aerodynamic Design

Airfoil selection can be made based on the Drag polar or lift drag bucket. Symmetrical airfoil like NACA-0018 is an application for the designed wind turbine because it has lower maximum lift coefficient than the asymmetrical airfoil (Shahariar and Hasan 2014). Castelli et al. researched the blade inclination in Darrieus type wind turbine using NACA 0021 (Castelli et al. 2012). Selig et al. provided high lift airfoil at low Reynolds number named S1223 (Selig and Guglielmo 1997). This high performance is attained using the concave pressure recovery and aft loading which are the favorable effects. The results show that maximum coefficient of lift $C_{l,\, max} = 2.2$ at a Re $= 2 \times 10^5$. Even the airfoil coordinates are presented in their work. Madasamy et al. showed that the Unsymmetrical airfoils can be used in Horizontal Type Wind Turbine. These airfoils can also be used for VAWT (Madasamy et al. 2021) (Figs. 13.5, 13.6 and Graph 13.1).

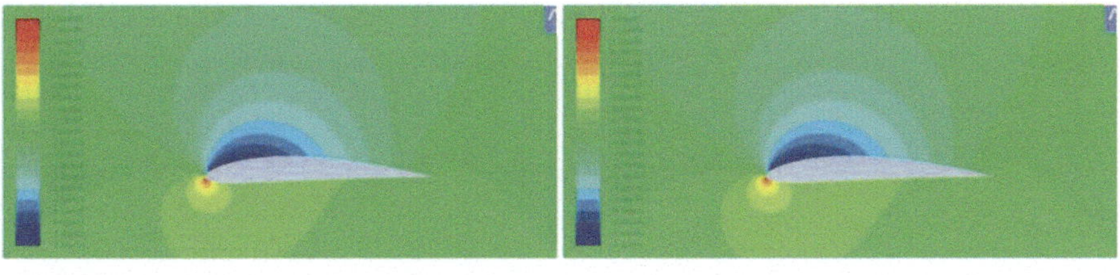

Fig. 13.5 Pressure contour over NACA 4412 at $\alpha = 4°$ and $\alpha = 6°$ (Madasamy et al. 2021)

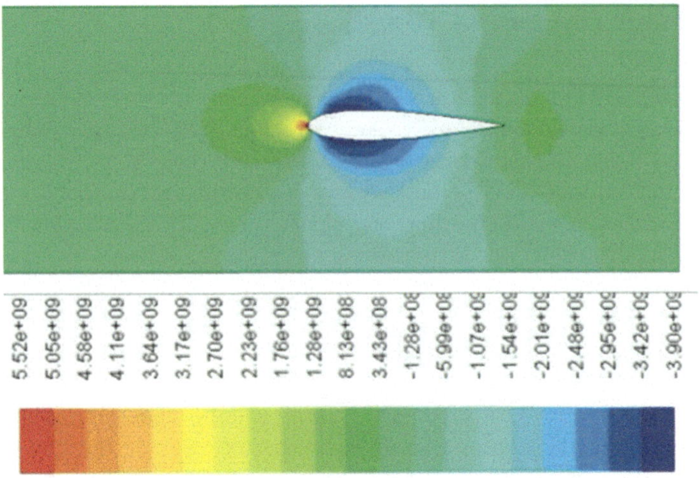

Fig. 13.6 Pressure Contour for NACA 0018 at velocity of 5 m/s (Madasamy et al. 2021)

Graph 13.1 Graph shows the aerodynamic performance of NACA airfoil 4412 (Madasamy et al. 2021)

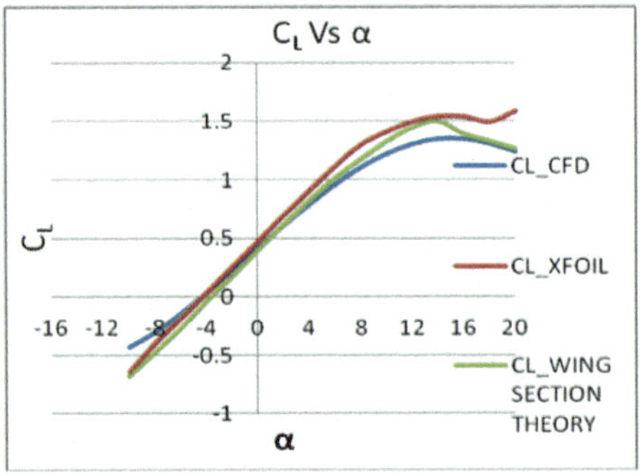

NACA 4412 C_L Vs Angle of attack

The characteristics of aerofoil determine the power generation of a wind turbine. The flow analyses of aerofoil are used to predict the lift to drag ratio and the sound pressure level. The genetic algorithm and Fuzzy Bellman-Zadeh decision making method are the optimization techniques that provided the increase in lift to drag ratio and decrease in sound pressure level (Sanaye et al. 2014). Aerofoil optimization depends on the tip speed ratio and the optimization can wither to better aerodynamic performance.

Table 13.2 Design data of VAWT from the literature review (Kariamian and Abdolahifar 2019; Elkhoury et al. 2015)

Turbine type	Turbine radius (m)	Turbine height (m)	Blade Chord Length (m)	Number of blades	Airfoil selection
3 Part Blade (Kariamian and Abdolahifar 2019)	0.99	1.15	0.3	3	NACA 0021
Straight Blade (Elkhoury et al. 2015)	0.4	0.8	0.2	3	NACA 0021

13.4 Methods to Enhance Lift at Low Reynolds Number

Other than designing the airfoil for the desired performance, there are other techniques like Morphing airfoils and additional arrangements such as slats and flaps to attain high performance. The arrangements were divided into trailing edge and leading-edge high lift mechanisms. The classification is based on the location where the arrangements were employed in the blade. Fowler flaps were used in the study of Zaccai et al. The research detailed the mechanism with useful results (Zaccai et al. 2017) (Table 13.2).

Some of the mechanisms used in the study include dropped-hinge, four-bar, link-track, and hooked-track. The leading-edge flaps and the surface-mounted trip can increase lift at low Reynolds number. Employing the leading-edge flaps for the low Reynolds number vehicle is suggested from the work of Jones et al. (2008). Torres et al. study showed that the low aspect ratio also leads to high lift coefficient (Torres and Mueller 2004).

13.5 Power Transmission Units and Generators

Drive train can be implemented for a smooth power transmission. The arrangement for the VAWT is simple with two intermeshing gear arrangements (Govind 2017). The ratios such as gear ratio, transmission ratio and contact ratio should be considered while designing the gear drive for effective power transmission (Jelaska 2012). From Hillier's, the gear ratio for any system can be given by Hillier (2012)

$$\text{Gear Ratio} = \frac{\text{Number of teeth on driven wheel}}{\text{Number of teeth on driving wheel}}$$

It is seen that most of the VAWT transmission unit has only fixed gear arrangement. In four wheelers, the number of gear arrangement is variable. Manual or automatic gear transmission is employed to change gears based on the increase in speed. If the automatic gear transmission is utilized, we can operate VAWT in different running condition. When Involute Contact ratio is 2 or 3, the design is good to control the noise generated from the spur gears (Kahraman and Blankenship 1999). If the size of the wind turbine is large, the lubrication of the gear drive should be considered.

13.6 VAWT Farms

Wind turbines in large arrays reduce the efficiency of the turbines. Large-eddy simulation is the model used to study wind-turbine array boundary layers (Calaf et al. 2010). Study with Dynamic Fluid Body Interaction method (DFBI) of two wind turbines in close proximity in 2D showed that the Counter Down layout yields more power than Counter Up layout (Hara 2021). Hezaveh et al.'s parametric study on the Vertical Axis Wind Turbine farms shows that the triangular arrangements are desirable (Hezaveh et al. 2018). The results showed that the triangular design makes use of the flow acceleration and has minimum effects due to wake. The parametric study from Hezaveh et al. for the VAWT farm is shown in the fig. Note that the greyed sport in fig is a VAWT. It is concluded that the normalized coefficient of pressure value is high for the cluster staggered (60°) and low for the aligned. Staggered has the C_P^* higher than aligned and lower than the cluster staggered (0°). C_P^* is the direct measure of the amount of power generated from the wind farms.

C_P^* is introduced in their work and the normalized relation of the quantity per unit power input is given by

$$C_P^* = \frac{2P}{\rho A u^2 U} \tag{13.2}$$

$$u^2 = \Delta P_{\text{drop}} \frac{L_x}{L_z} \tag{13.3}$$

where P is the average power per VAWT, ρ is the density of the air, u is the root mean square of the total mean drag, ΔP_{drop} is the total pressure drop in the VAWT farm, U is the averaged streamwise velocity component, A is the farm area, L_x is the length of the farm and L_z is the height of VAWT (Figs. 13.7 and 13.8).

The wind power forecasting is replaced with a technique called Advance Neural Networks (ANN) (Hezaveh et al. 2018).

$$x_i(t+1) = a_i x_i(t) + b_i \boldsymbol{\Sigma} w_{i,j}(t) y_j(t) \tag{13.4}$$

where a and b are the constants, x is the state of the respective neuron, w is the synaptic weight for the jth input of ith neuron, y is an external input or state of neuron inputted from the sigmoidal function. The wind park power profile, an important factor used to measure the power performance, is the ratio of the power from total number of wind turbines at a certain time to the power from the number of wind turbines operated at the time. The prediction of this quantity is difficult because of the external factors and

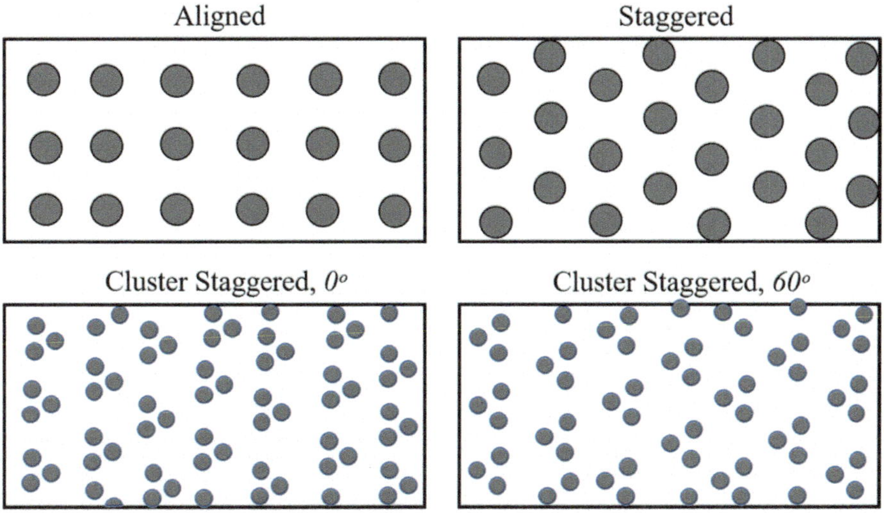

Fig. 13.7 The parametric study arrangement of the VAWT wind farm from Hezaveh et al.

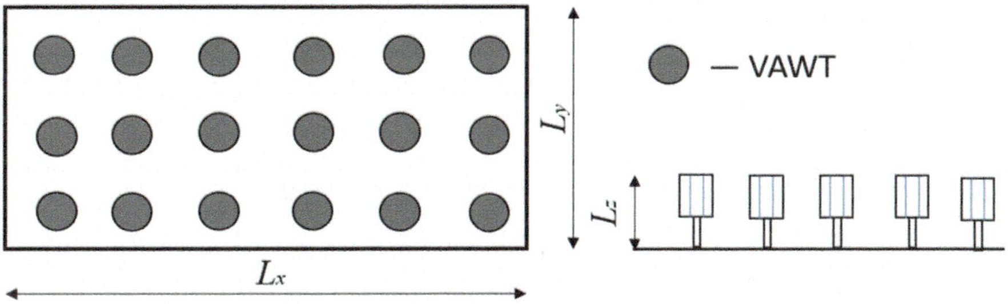

Fig. 13.8 From Reference Hezaveh et al. (2018) to showing the dimensions of the VAWT farm

the measurement was made based on the data acquired from the farm or park. The test is carried with the three-basic configuration viz the 12 output from a network for every time step, 1 output network with 12-time steps forecasted iteratively, error optimized network of the 2nd configuration. Though there are HAWT farms widely used in the world, the VAWT farms should be considered for its efficiency. Some of the areas which have HAWT farm in India generate considerable amount of power. The top 5 states contributing to the power generation with its capacity are tabulated below (https://www.mnre.gov.in/wind/current-status/). Korea Electric Power Corporation (KEPC) is doing notable works to generate green energy in South Korea. Countries like Japan are prone to natural disasters, it is important to keep these external factors in the design for a reliable structure (Table 13.3).

Table 13.3 Shows the top contributors of wind energy in India

State	Capacity in MW
Gujarat	84,431
Rajasthan	18,770
Maharashtra	45,394
Tamil Nadu	33,799
Madhya Pradesh	10,483

13.7 Numerical Simulation for the VAWT Design

Commercially software packages like Ansys, Hyper works, Abaqus, etc., are used to visualize the fluid dynamics of the VAWT. The visualization provides information about the changes which may be required for better aerodynamic efficiency. Though the theoretical calculation equips us with data, this commercial software gives better insight required to proceed further in the designing process. The pisoFoam Turbine solver is the apt choice for the analysis when the horizontal gradient pressure and Coriolis force in momentum equation are negligible. ALM (Actuator Line model) captures the flow simulation around the blades in a wind farm of 12 wind turbines (Strijhak et al. 2019). Usage of Blade element theory in the aerodynamic load of an offshore floating wind turbine may be not accurate because of the complicated phenomena such as ground effects, variations in wind shear, and unsteady yaw and aerodynamics. Coding for Unsteady vortex lattice method is integrated with the FAST AeroDyn software and was validated. This method had an acceptable level of accuracy and performance in unsteady yaw condition (Jeon et al. 2014).

Ansys Fluent covers wide range of fluid problems. From the literature review (Fluent Manual 2009), the best model to analyze the wind turbine is the Shear Stress Transport (SST) kinetic energy k and dissipation ω model. Menter et al. work shows that this model can be used for the prediction of adverse pressure gradient flows (Menter 1994).

$$\frac{\partial(\rho k)}{\partial t} + \frac{\partial(\rho k u_i)}{\partial x_i} = \frac{\partial}{\partial x_j}\left(\Gamma_k \frac{\partial k}{\partial x_j}\right) + G_k - Y_k + S_k \quad (13.5)$$

$$\frac{\partial(\rho \omega)}{\partial t} + \frac{\partial(\rho \omega u_i)}{\partial x_i} = \frac{\partial}{\partial x_j}\left(\Gamma_\omega \frac{\partial \omega}{\partial x_j}\right) + G_\omega - Y_\omega + S_\omega \quad (13.6)$$

where G_k is the generation of turbulence kinetic energy due to mean velocity gradient, G_ω is the generation of specific rate of dissipation, S is the source terms, Y is the dissipation due to turbulence, u is the velocity term, Γ is the effective diffusivities.

$$\Gamma_k = \mu + \frac{\mu_t}{\sigma_k} \quad (13.7)$$

$$\Gamma_\omega = \mu + \frac{\mu_t}{\sigma_\omega} \quad (13.8)$$

Turbulent viscosity μ_t is calculated using the following equation.

$$\mu_t = \frac{\rho k}{\omega} \frac{1}{\max\left[\frac{1}{\alpha^*}, \frac{SF_2}{a_{1\omega}}\right]} \quad (13.9)$$

where S is the Strain Rate Magnitude.

$$\sigma_k = \frac{1}{F_1/\sigma_{k,1} + (1 - F_1)/\sigma_{k,2}} \quad (13.10)$$

$$\sigma_\omega = \frac{1}{F_1/\sigma_{\omega,1} + (1 - F_1)/\sigma_{\omega,2}} \quad (13.11)$$

The blending functions F_1 and F_2 is provided in the reference for a distance y to the next surface as below.

$$F_1 = \tanh\left(\phi_1^4\right) \quad (13.12)$$

$$\phi_1 = \min\left[\max\left(\frac{\sqrt{k}}{0.09\omega y}, \frac{500\mu}{\rho y^2 \omega}\right), \frac{4\rho k}{\sigma_{\omega,2} D_\omega^+ y^2}\right] \quad (13.13)$$

$$D_\omega^+ = \max\left[2\rho \frac{1}{\sigma_{\omega,2}} \frac{1}{\omega} \frac{\partial k}{\partial x_j} \frac{\partial \omega}{\partial x_j}, 10^{-10}\right] \quad (13.14)$$

$$F_2 = \tanh\left(\phi_2^4\right) \quad (13.15)$$

$$\phi_2 = \left[\max\left(\frac{2\sqrt{k}}{0.09\omega y}, \frac{500\mu}{\rho y^2 \omega}\right)\right] \quad (13.16)$$

where D_ω^+ is the cross-diffusion term.

The production of turbulence kinetic energy is $G_k^* = \min(G_k, 10\rho\beta^* k\omega)$

$$G_\omega = \frac{\alpha}{v_t} G_k^* \quad (13.17)$$

The constant $a_\infty = F_1 a_{\infty,1} + (1 - F_1) a_{\infty,2}$

$$a_{\infty,1} = \frac{\beta_{i,1}}{\beta_\infty^*} - \frac{k^2}{\sigma_{\omega,1}\sqrt{\beta_\infty^*}} \quad (13.18)$$

$$a_{\infty,1} = \frac{\beta_{i,2}}{\beta_\infty^*} - \frac{k^2}{\sigma_{w,2}\sqrt{\beta_\infty^*}} \quad (13.19)$$

Dissipation of ω,

$$Y_k = \rho \beta^* k \omega \tag{13.20}$$

$$Y_k = \rho \beta^* \omega^2 \tag{13.21}$$

The value of β_i is calculated using the following relation.

$$\beta_i = F_1 \beta_{i,1} + (1 - F_1)\beta_{i,2} \tag{13.22}$$

Cross diffusion modification D_ω,

$$D_\omega = 2(1 - F_1)\rho\sigma_{\omega,2}\frac{1}{\omega}\frac{\partial k}{\partial x_j}\frac{\partial \omega}{\partial x_j} \tag{13.23}$$

Model constants value is given below,

$$\sigma_{k,1} = 1.176, \sigma_{\omega,1} = 2.0, \sigma_{k,2} = 1.0, \sigma_{\omega,2} = 1.168$$
$$a_1 = 0.31, \beta_{i,1} = 0.075, \beta_{i,2} = 0.0828$$

The other constants in the above equations remain the same for the model.

User Defined Functions (UDF) in the Fluent allows us to analyze the rotatory component using the C program. The function usually requires the boundary condition which must be imposed on to the desired moving components. Figure 13.9 shows the computational domain used for such analysis. Using UDF in Fluent, the author imposed the angular velocity to the airfoils and analysis was carried on. When the analysis is made with moving object, in the computational domain, dynamic mesh will be used. After running the computational fluid dynamics, Ansys allows us to

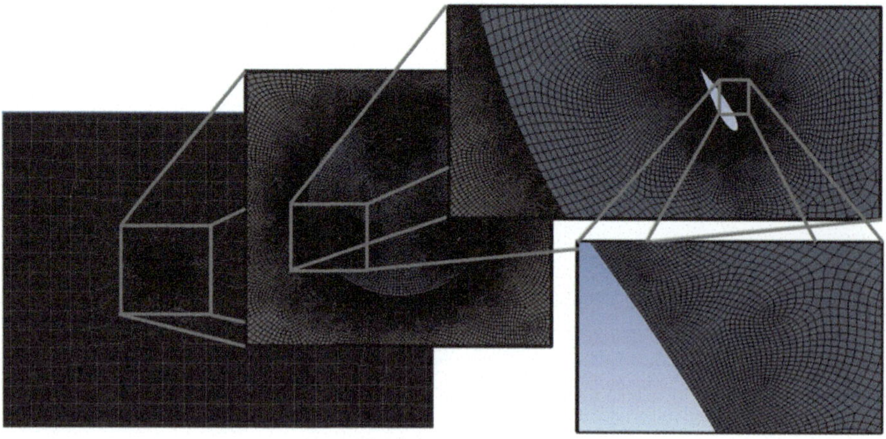

Fig. 13.9 Computational domain for the study from Madasamy et al. (2021)

import the values to structural analysis. The technique is called as Fluid–Structure Interaction (FSI).

The free rotation of the wind turbine blades completely depends upon the rigid support and its behavior. Owing to this reason only lightweight cum high strength-based materials have been proposed and imposed on the construction of modern wind turbines. The ultimate motto of the wind turbine is to extract a high amount of wind energy from the free stream velocity of the fluid through aerodynamic rotors. With the help of high aerodynamic efficient design profiles of the rotors, the induced aerodynamic forces and thereby rotations of the wind turbine rotors are been increased at an acceptable rate. By means of the implementation of lightweight cum high strength materials in wind turbines, the induced aerodynamic forces are successfully converted into real-time movement of the rotors. Therefore, the investigation about the suitable lightweight material for the use of real-time applications and its fatigue life estimations are much needed in the wing energy sector. The working environment of the wind turbine is very complicated, which is the cumulative impact of the aero and rotodynamic effect, so a normal approach does not have the capacity to capture the external aerodynamic load for structural simulation. Thus, an advanced approach can have the capability to solve aforesaid critical environment-based problems, which is Fluid–Structure Interaction [FSI] based Coupling computation. In FSI, the very broad two different environments have been coupled effectively in order to attain the structural outcomes of the given wind turbine rotors and thereby provide a strong conclusion on the above-mentioned rotors based on the results such as equivalent stress, fatigue life, strain energy, etc. Through the help of an advanced coupling scheme, the complete aerodynamic load has been transferred from fluid environment to structural environment, which enhances the reliability of acceptable rate of the structural outcomes. Based on the way of coupling scheme, two major classifications are available in the advanced computations, which are one way coupling and two-way coupling. One way coupling facility-based FSI is mostly available in all kinds of FEA and CFD solvers whereas the two-way coupling facility-based FSI is suitable and adaptable for few advanced computational solvers. Because, the two-way coupling needs to be computed by means of a separate cum unique tool, which is named as system coupling. Comparatively, two-way coupling approach is the best one to provide trustworthiness structural outcomes on wind turbines. Two-way coupling approach is based on parallel computation so unless otherwise good boundary conditions, the attainment of results is very tough. On the other hand, good boundary conditions are only having the capacity for the production of actual results thus two-way coupling is more preferable to implement in wind turbine problems.

The input inlet velocity of Vi is given as the inlet boundary condition and boundary chosen for outlet is Pressure Far field. A constant angular velocity ω is imposed on the blade arrangements. The number of elements used in the computational domain is 256014 and the time step of 0.0016 s is simulated with the 10^4—time steps. The converged solution had the input of 7 ± 1 m/s and the blades are set to rotate in a constant angular velocity of 10 rad/s. The velocity contour, obtained by solving the above computational grid, shows that there is a considerable pressure variation on the blades because of the inlet velocity provided. This shows that there will be a notable change in the rotational speed of the turbine blades resulting in generating more power. The input velocity is chosen as 7 m/s as it is the average wind speed in the work.

The computational domain is the 3 turbine blade arrangements in 2D excluding the mounting equipment. The result from the preliminary research in 2D is in a good compliance with the results from other research articles. Because of the computational cost, the author is performed the research on computational domain with UDF. UDF imposes the angular velocity on the blade using the following equation.

$$\omega = \frac{V\lambda}{R} \tag{13.24}$$

where V is the freestream velocity, λ is the tip speed ratio of the blade and R is the radius of the blade. In Fluent software package, the author is planning use the DEFINE_ZONE_MOTION or DEFINE_TRANSIENT_PROFILE UDF. Both the UDFs allow the rotational rate of the domain with the angular velocity and time t. UDF also allows us to input unsteady pressure which is the real-time case in for the wind turbine. There is always unsteady airflow flow in the atmosphere of earth. Though there are plenty of sophisticated commercial packages available to do the computational analysis, the care should be made in validating the results. Common errors from the mesh quality, dirty geometry, model used for analysis, etc., The experimental and or analytical validation is always recommended even if the individual is convinced with computational results. An example of UDF coding for unsteady pressure input is shown below. The information such as the profile definition and boundary to face is provided in the program (Fig. 13.10).

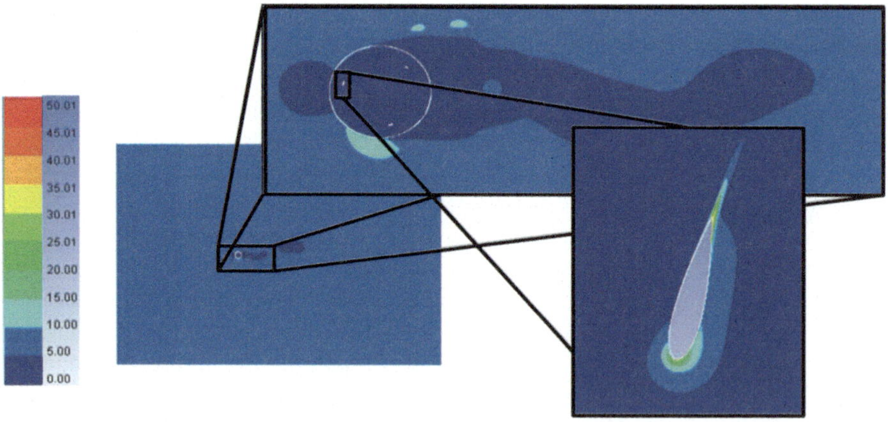

Fig. 13.10 Velocity (V) contour of the turbine blades rotating at the angular velocity of 10 rad/s where input velocity is 7 ± 1 m/s from the reference Madasamy et al. (2021)

```
*************************************************************************** ********
                              UDF CODING
          (c programming language integrated with Fluent)
*************************************************************************** ********
#include "udf.h"
 DEFINE_PROFILE (Unsteady_Pressure, thread, position)
  {
    face_t f;
    real t = CURRENT_TIME;
    begin_f_loop(f, thread)
     {
       F_PROFILE (f, thread, position) = 101325.0 + 7.0*sin(10.0*t);
     }
    end_f_loop(f, thread)
  }
***************************************************************************
```

13.8 Computation Method

13.8.1 Pressure-Based Segregated Solver

The solution algorithm employed in pressure-based segregated solver solves the governing equations, sequentially. Solution loop is carried out iteratively to attain a converged numerical solution. Nonlinear and coupled governing equations are solved one after another. The solution procedure is given below.

13.8.1.1 Solution Procedure

When a two-step initialization sequence is executed outside the solution iteration loop, solution process commences. The inputs from user interface initialize equations after the sequence begun. PROFILE and INIT UDFs (User Defined Function) are called consecutively. Initialization overwrites the previously set initialized values. Execution of PROFILE UDFs and Adjust UDFs are carried out when the loop begins. The governing equations are solved segregatively and properties are updated. Depending upon the solution convergence, the continuation of the loop is determined (Jeon et al. 2014) (Fig. 13.11).

PISO has quicker convergence than the SIMPLE Scheme but poor accuracy of the results (Barnes et al. 2021). Niculescu et al. research showed that all the existing turbulence model is not suitable for the Turby Darrieus Wind Turbine and suggested that Proper Orthogonal Decomposition (POD) can cover the physics (Niculescu et al. 2017). The POD can explain the different projection in the decomposition process and the analysis is carried to the finite dimensional domain by restricting the approach to the separation of variables in space and time (Weiss 2019). POD data analysis often extracts the modal shapes or its basic functions. The eigen values are solved for the infinite-dimensional POD. The method is sensitive to coordinate changes and unable to distinguish the featureless points (Chatterjee 2000).

ICE3D is used to study the effect of icing on the VAWT performance. Blade moment coefficient difference between the clean and iced blades is negative. The positive blade moment is reduced to 60% in rime ice condition (Manatbayev et al. 2021). Asymmetric-Primary Axis-Flux Hybrid Generator reduces the ripple in its axis with magnetic levitation force and the work is presented with results from 3D Finite Element Method (Jing et al. 2021). Co-axial Contra-rotating Vertical Axis Wind Turbine (CR-VAWT) performance increases with an increase in speed when compared with Conventional Vertical Axis Wind Turbine (C-VAWT) (Poguluri et al. 2021). 3D FEA is preferable to the PreComp structural analysis tool and SHELL281 is utilized for the study (Hand et al. 2021).

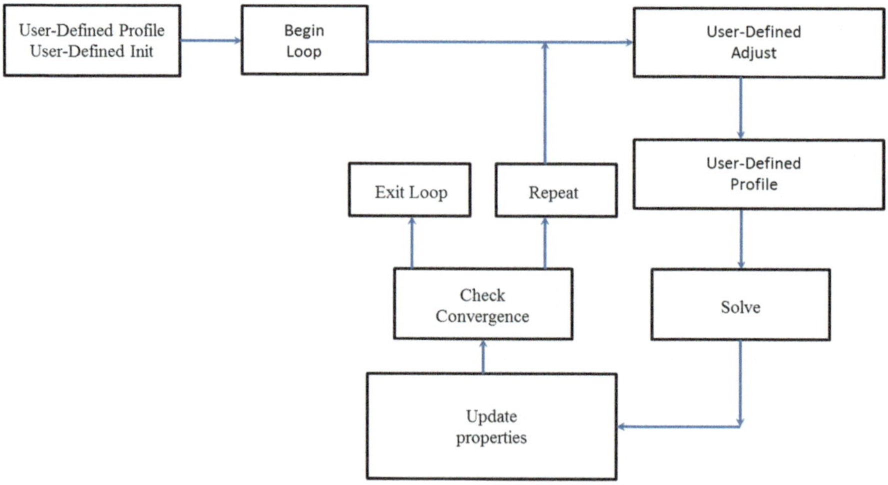

Fig. 13.11 Flow chart of solution procedure for SIMPLE scheme (Fluent Manual 2009)

13.9 Experimental Methods for Validation

Smoke visualization, hotwire measurement, and thrust force measurement are the techniques used to experimentally study the VAWT in skewed flow condition. For the flow visualization, the smoke is injected in the upwind of rotor and stroboscopic light were used to analyze the skewed flow physics. Hotwire is employed to find the velocity. Thrust is measured by balancing the VAWT model and for the skewed model, the measurements were based on the velocity in and perpendicular to the flow (Ferreira et al. 2006) (Fig. 13.12).

In the typical wind turbine, the inlet, wind tunnel, and outlet are the components. The wind will be forced out from the inlet using a fan in the outlet. VAWT to be tested is placed inside the wind tunnel for possible flow visualization or any other measurements to be done. The arrangements inside the wind tunnel usually have the contraction, test, and diffusion part. There are various types of wind tunnel based on the speed of fluid flow. If the flow is steady and incompressible, neglecting the viscosity, the speed of the fluid V is given by,

$$V = \sqrt{\frac{2\Delta p}{\rho}} \tag{13.25}$$

where Δp is the pressure difference total and static pressure, ρ is the density of fluid. It is considered that creating prototype and testing in wind tunnel is expensive and time-consuming. In modern world, the computational analysis is used to replace the experimental techniques. While testing the VAWT of small diameter of 1 m and chord length of 75 mm, there was an issue in starting torque. The acceleration to

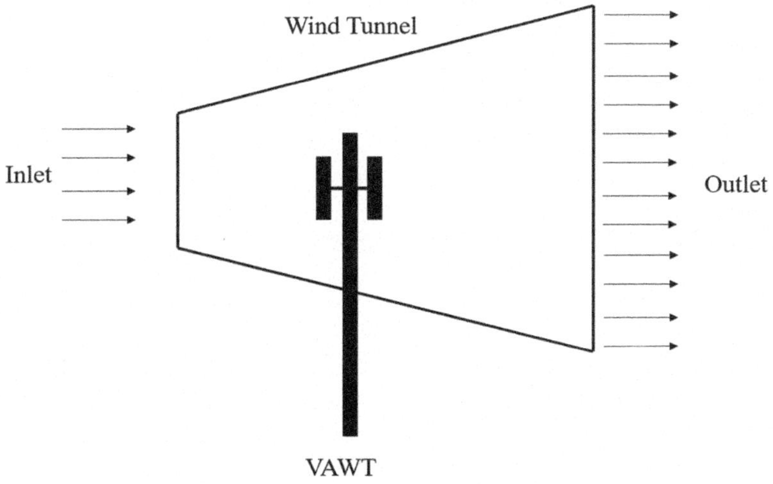

Fig. 13.12 Shows the typical wind tunnel arrangement for the validation of results

nominal speed is not met, hence the system is initially operated with the engine in the wind tunnel (Pfeffermann et al. 2019).

13.10 Installation of VAWT

The VAWT can be installed in urban and sub-urban areas where there are steady winds. Highways is another source of steady winds and the placement can also be on tall building and skyscrapers. The main rotor shaft is installed vertically in VAWT (Mahale et al. 2015). Installation of VAWT can be done by keeping the following points in mind (Chong et al. 2013).

a. Wind speed profile in the location.
b. Effects of the VAWT installation to the environment.
c. Areas like Cooling tower and ventilation can also be the place for installation.
d. Power generation capability of the device.

Structure Health Monitor (SHM) is a process of monitoring the health of engineering structure for the entire life cycle. Usage of biocomposites can reduce the dependency on synthetic fiber composite without compromising the structural requirement. Materials like piezoelectric is a smart material and can perform operations like sensing, actuating, and harvesting energy (Hamdan et al. 2014). The turbine can be protected from atmosphere by choosing appropriate material and coating. Clysar provides good adhesion to the blade and thermo-retraction property (Mauri et al. 2014).

13.11 Advantages and Limitations of VAWT Over HAWT

Besides power generation and free from pollution, some of the advantages and limitations from the literature review are given below.

a. Sensitivity of the VAWT to the turbulence and fluctuation is less (Tasmeen et al. 2020).
b. Low Efficiency when compared with HAWT (Tasmeen et al. 2020).
c. VAWT can have aesthetic design and multi-stages.
d. Low installation and maintenance cost when compared to HAWT (El-Thalji et al. 2012).
e. Low life cycle cost when compared to HAWT (El-Thalji et al. 2012).
f. Flickering shadow from the VAWT is considered as a disturbance (Mauri et al. 2014).
g. Less requirements for the setup of VAWT.
h. VAWT can be bladeless and less dangerous than HAWT.
i. VAWT can be produced in various size ranges.

j. Less space consumption.
k. VAWT can be operated by the wind around 360° (Mahale et al. 2015).
l. VAWT creates less noise (Mahale et al. 2015).
m. No visual Impact (Chong et al. 2013).
n. VAWT can also be used to change the wind flow.
o. VAWT can be placed in offshore as well as onshore.

13.12 Conclusion

The wind turbines can produce sustainable energy source free from any sort of pollution because of fossil fuels. VAWT can be made in various shapes and sizes. It can offer wide range of aesthetic designs with less Terrestrial land consumption. Less dependency on the land is a major advantage when considering the VAWT. VAWT can be less noisy and can be employed in any place or vehicles. They are less dangerous structures compared to the HAWT. Being the advantages over the HAWT, the book chapter concentrated on the design and analysis with validation. The author would like to emphasize on the importance of VAWT over HAWT and expecting the reader to use design inputs from various articles for the optimal design. We have highlighted that the asymmetrical airfoils can also be used for design of VAWT. Tip Speed ratio determines the dynamic stall of the airfoil and optimization techniques can be employed for better result (Tirandaz et al. 2021). Detailed derivation of the model SST k-ω and computational method presented in the chapter helps the reader to understand the process in numerical simulations.

Acknowledgements The authors would like to thank Professor Hara Yutaka, Department of Mechanical Engineering, Tottori University for his guidance in 2D analysis from Madasamy et al. (2021).

References

Babu SK, Subba Raju NV et al (2006) The material selection for typical wind turbine blades a MADM approach and analysis of blades. MCDM, pp 1–12

Barnes A et al (2021) Towards a standard approach for future vertical axis wind turbine aerodynamics research and development. Renew Sustain Energy Rev 148:1–20

Calaf M et al (2010) Large Eddy simulation study of fully developed wind-turbine array boundary layers. Phys Fluids 015110-1–17

Castelli MR et al (2012) Effect of blade inclination angle on a Darrieus wind turbine. J Turbomach ASME 031016-1–10

Chatterjee A (2000) An introduction to the proper orthogonal decomposition. Curr Sci 78(7):10, 808–817

Chong WT et al (2013) Early development of an energy recovery wind turbine generator for exhaust air system. Appl Energy 112:568–575

Das PMM, Amano RS (2014) Basic theory for wind turbine blade aerodynamics. In: Aerodynamics of wind turbines, vol 81. WIT Press, pp 11–23

Dumitrache A et al (2015) Influence of unsteady flow on the aerodynamics of vertical axis wind turbines. In: AIP conference proceedings, pp 500007-1–4
Elkhoury M, Kiwata T, Aoun E (2015) Experimental and numerical investigation of a three-dimensional vertical-axis wind turbine with variable-pitch. J Wind Eng Ind Aerodyn 111–123
El-Thalji I et al (2012) Scalable and customer-oriented life cycle costing model: a case study of and innovative vertical axis wind turbine concept (Case-VAWT). In: The twenty-second international offshore and polar engineering conference, ISOPE, Greece, pp 1–3
Fluent Manual (2009) Ansys 12, Ansys Inc, PA
Ferreira C et al (2006) Wind tunnel hotwire measurements, flow visualization and thrust measurement of a VAWT in skew. In: 44th AIAA aerospace sciences meeting and exhibit, pp 1–12
Govind B (2017) Increasing the operational capability of a horizontal axis wind turbine by its integration with a vertical axis wind turbine. Appl Energy 479–494
Hamdan A et al (2014) A review on the micro energy harvester in structural health monitoring (SHM) of biocomposite material for vertical axis wind turbine (VAWT) system: a Malaysia perspective. Renew Sustain Energy Rev 35:23–30
Hand B et al (2021) Structural analysis of an offshore vertical axis wind turbine composite blade experiencing an extreme wind load. Marine Struct 75:102858, 1–13
Hara Y et al (2017) Numerical simulation on fluid forces and structure of triangular-blade butterfly wind turbine, ICJWSF, October, Ohio, USA
Hara Y et al (2021) Numerical analysis of the dynamic interaction between two closely spaced vertical-axis wind turbines. Energies, MDPI 14:2286, 1–22
Hara Y, Horita N et al (2019) Numerical analysis of effects of arms with different cross-sections on straight-bladed vertical axis wind turbine. Energies, MDPI Publication 1–24
Hara Y et al (2014) Predicting double-blade vertical axis wind turbine performance by a quadruple-multiple stream tube model. Int J Fluid Mach Syst
Hara Y et al (2018) Development of a butterfly wind turbine with mechanical over-speed control system, vol 2, MDPI Publishing, pp 1–25
Herrmann J et al (2019) Multi-objective optimization of a thick blade root airfoil to improve the energy production of large wind turbines. J Renew Sustain Energy 043304-1–25
Hezaveh SH et al (2018) Increasing the power production of vertical-axis wind-turbine farms using synergistic clustering. In: Boundary-layer meteorology, Springer, pp 1–22
Hillier VAW (2012) Hillier's fundamentals of moto vehicle technology, 6th edn. Nelson Thornes Ltd. Publication
Hsu C-T (1984) Tornado type wind turbines, US Patent, pp 1–4
https://www.mnre.gov.in/wind/current-status/
Huang X (2009) Fabrication and properties of carbon fibers. Materials. MDPI Publishing 2369–2403
Jelaska D (2012) Gears and gear drive. Wiley
Jeon M et al (2014) Unsteady vortex lattice method coupled with a linear aeroelastic model for horizontal axis wind turbine. J Renew Sustain Energy 042006-1–7
Jing L et al (2021) An asymmetric-primary axis-flux hybrid-excitation generator for the vertical axis wind turbine, special section on advanced energy conversion systems based on multi-port electrical machines. IEEE Access 9:92318–92325
Jones AR et al (2008) Low Reynolds number aerodynamics of leading-edge flaps. J Aircraft 45(1):342–345
Kahraman A, Blankenship GW (1999) Effect of involute contact ratio on spur gear dynamics. Trans ASME 112–118
Kariamian SMH, Abdolahifar A (2019) Performance investigation of a new Darrieus vertical axis wind turbine. Energy 1–48
Madasamy SK et al (2021) Investigation of NACA airfoils for aerodynamic performance of wind turbines, pp 1–12 [Yet to be published]
Mahale P et al (2015) Vertical axis wind turbine: a lucid solution for global small-scale energy crisis. J Acad Ind Res, Research Article 3(8):393–396

Manatbayev R et al (2021) Numerical simulation on static vertical axis turbine blade icing. Renew Energy 170:997–1007
Maskepatil LP, Gandigude AU, Kale SA (2014) Selection of material for wind turbine blade by analytic hierarchy process (AHP) method. Appl Mech Mater. Trans Tech Publications, 145–150
Mauri M et al (2014) Design and realisation of a high-performance active pitch-controlled H-Darrieus VAWT for urban installations. In: IEEE, 3rd renewable power generation conference, pp 1–6
Meng H, Li Y-F (2019) A review on prognostics on health management (PHM) methods of lithium-ion batteries. Renew Sustain Energy Rev 1–12
Menter FR (1994) Two-Equation Eddy-viscosity turbulence models for engineering applications. AIAA J 32(8):1598–1605
Niculescu ML et al (2017) CFD analysis of a Darrieus wind turbine. In: AIP conference proceedings, 1863, pp 420003-1–4
Pfeffermann DL et al (2019) Installation and design of a new wind tunnel for measurement of vertical axis wind turbines. Arch Chem Chem Eng 1(2):1–10
Poguluri SK et al (2021) An investigation on the aerodynamic performance of a co-axial contra-rotating vertical-axis wind turbine. Energy 219:119547, 1–14
Sanaye S et al (2014) Multi-objective optimization of airfoil shape for efficiency improvement and noise reduction in small wind turbines. J Renew Sustain Energy 053105-1–21
Sassi P, Freiria J et al (2020) Simulation of vorticity wind turbines. Heliyon 6:1–12
Schaffarczyk AP (2014) Introduction to wind turbine. In: Green energy and technology, Springer Publications
Schubel PJ, Crossley RJ (2012) Wind turbine blade design, energies. MDPI J 5:3425–3449
Selig MS, Guglielmo JJ (1997) High-lift low Reynolds number airfoil design. J Aircraft 34(1):72–79
Shahariar HGM, Hasan MR (2014) Design and construction of a vertical axis wind turbine. In: The 9th international forum on strategic technology (IFOST), pp 326–329
Strijhak SV et al (2019) Simulation of turbulent wakes in model wind farm with arbitrary location for wind turbines. J Phys Conf Ser 1382:012043-1–7
Tasmeen Z et al (2020) An analytical review on the evaluation of wind resources and wind turbine for urban application: prospect and challenges. Develop Built Environ 4:100033, 1–15
Tirandaz MR et al (2021) Effect of airfoil shape on power performance of vertical axis wind turbines in dynamic stall: symmetric airfoils. Renew Energy 173:422–441
Tjiu W et al (2015) Darrieus vertical axis wind turbine for power generation I: assessment of Darrieus VAWT configurations. Renew Energy 75:50–67
Torres GE, Mueller TJ (2004) Low-aspect-ratio wing aerodynamics at low Reynolds numbers. AIAA J 42(5):865–873
Trematerra A et al (2017) Wind turbines acoustic measurements. In: AIP conference proceedings, pp 020001-1–7
Utomo IS et al (2018) Experimental studies of Savonius wind turbines with variations sizes and Fin numbers towards performance. In: AIP conference proceedings, pp 030041-1–7
Victor M et al (2017) Wake flow simulation of a vertical axis wind turbine under the influence of wind shear. J Phys Conf Ser 012031-1–10
Villarreal DJY (2016) Vortex resonance wind turbine, US Patents, pp 1–9
Tong W (2010) Wind power generation and wind turbine design. WIT Press
Weiss J (2019) A tutorial on the proper orthogonal decomposition. In: AIAA Conference Publication, pp 1–21
Weiss J, Tsuchida BT (2017) Integration into national grids. In: Wind energy engineering, a handbook for onshore and offshore wind turbines, pp 419–436
Whittlesey R (2017) Vertical axis wind turbines: farm and turbine design. In: A handbook for onshore and offshore wind turbines, wind energy engineering, pp 185–202

Williamson CHK, Govardhan R (2004) Vortex induced vibrations. Ann Rev Fluid Mech 36(1):413–455
Yen JT et al (1978) Tornado-type wind turbine, US Patent, pp 1–8
Zaccai D, Bertels F, Vos R (2017) Design methodology for trailing-edge high lift mechanisms. CEAS Aeronaut J 521–534

Chapter 14
Theoretical Modelling, Analysis and Energy Yield Prediction for Horizontal Axis Wind Turbine Rotors

Vasishta Bhargava Nukala, Rahul Samala, Satya Prasad Maddula, Swamy Naidu Neigapula Venkata, and Chinmaya Prasad Padhy

Abstract In this work, aerodynamic modelling of HAWT rotors and energy yield prediction using power curve are studied. The influence of wake losses and rotor solidity on the performance of HAWT is investigated using blade element momentum theory (BEMT). Empirical corrections to a rotor thrust and induced velocities proposed by Glauert, Buhl and Wilson–Walker were evaluated for various Prandtl tip loss factors and validated with experiment thrust data. For axial induction factor less than 0.5, the maximum thrust coefficient obtained using BEMT varied between 0 and 1. When the axial induction factor exceeded 0.5, the Glauert thrust correction factor showed discontinuity for which standard BEMT is invalid. Power coefficient was evaluated for various rotor lift to drag ratios with a rotor diameter of 36 m. The results demonstrated a 17% increase in power coefficient when lift to drag ratios is increased up to three times for tip speed ratios less than 7. A maximum change of ~50% in the power coefficient is obtained when the blade count is increased from one to three. Power curves for 2 MW machine with a diameter of 95 m at two air densities of 1.225 and 1.115 kg/m^3 showed a maximum power error of 29% at cut in wind speed. Time series of 10-min averaged data showed that nominal power is obtained when the blade pitch angle position varied between 0° and 5°. Bin analysis also revealed that maximum value for power coefficient obtained is 0.43, while thrust coefficient of 0.94 at wind speed of 7.6 m/s. Energy yield for a 2 MW turbine is also predicted based on two parameter Weibull distribution for different scale and shape factors.

V. B. Nukala (✉)
Sreyas Institute of Engineering and Technology, Hyderabad, India
e-mail: vasishtab@gmail.com

R. Samala
Indian Institute of Technology Madras, Chennai, India

S. P. Maddula · C. P. Padhy
GITAM (Deemed to be) University, Hyderabad, India
e-mail: cpadhy@gitam.edu

S. N. N. Venkata
National Institute of Technology, Raipur, Raipur, India

A. K. Bohre et al. (eds.), *Planning of Hybrid Renewable Energy Systems, Electric Vehicles and Microgrid*, Energy Systems in Electrical Engineering,
https://doi.org/10.1007/978-981-19-0979-5_14

Keywords Solidity · Airfoil · Rotor blade · Wake · Tip · Wind speed · HAWT · Power coefficient · Thrust coefficient · Power curve

Nomenclature

BEM	Blade element momentum
TSR	Tip speed ratio
L/D	Lift to drag
RPM	Rotations per minute
MSRD	Measured
STRD.	Standard
WPD	Wind power density
O&M	Operations & maintenance
GW	Gigawatt
MW	Megawatt
LCOE	Levelized cost of energy
RANS	Reynolds Averaged Navier Stokes
HAWT	Horizontal axis wind turbine
NACA	National advisory committee for aeronautics
IEC	International electronical commission
TI	Turbulence intensity
MCP	Measure, correlate, predict
BEMT	Blade element momentum theory
DFIG	Doubly fed induction generator
WS	Wind speed
CFD	Computational fluid dynamics
NREL	National renewable energy laboratory

14.1 Introduction

For growing energy demands of future generations, wind turbines are one of the cleanest and cost-effective source of energy supply among all renewable energy technologies. The cumulative wind power capacity installed globally has increased by more than 100% over the last two decades. Countries like US, Germany, Spain and Denmark have the largest number of cumulative wind power installations totalling nearly 200 GW as of year 2018. Further, in China alone the cumulative wind power capacity installed as of year 2019 has reached 200 GW with average size of installed turbine capacity as 1 MW for onshore applications and 3 MW for offshore wind sites. In India, cumulative installed wind power capacity has reached 35 GW with its largest wind park located in the state of Tamil Nadu. Over the past two decades improved conversion efficiencies from wind turbines have occurred due to innovations in turbine design, manufacturing technologies as well as operation and maintenance procedures. The annual energy production from a wind turbine or a wind farm array is highly improved due to efficient operation and maintenance practices which contributed to low levelized cost of energy (LCOE) from wind power plants (Bhadra et al. 2010; Bossanyi et al. 2005; Emblemsvag 2020; Goudarzi and Zhu 2013). Latest technology in land-based wind power plants consist of rotors installed on tall towers located 50 to 100 m above ground level, while offshore wind towers are floating structures mounted to a depth of 100 m below sea level. At industrial scale operation, there are two orientations for horizontal axis wind turbine rotors, viz. upwind and downwind. In the upwind orientation, rotor is assembled ahead of the tower structure facing the wind direction. In uniform inflow, the risk of blade striking the tower is low in upwind orientation due to positive tilt angle of rotor in the lift producing blades and experiences no tower shadow. However, in downwind configuration, the rotor assembly is placed behind the tower structure. Due to the presence of tower, aerodynamic wake is formed which causes a velocity deficit and an equivalent to velocity potential. Also, when the rotor is inclined downwind of the uniform free stream, the unsteady flow over the blades is responsible for production of aerodynamic noise. This flow also causes increased fatigue loads on structural components of turbine. Although rotor loading influences the turbine efficiency, increase in tip losses contribute to high aerodynamic power loss and is considered an important subject of study by many researchers in wind energy.

Horizontal axis wind turbines are also classified according to pitch and stall control, variable and constant speed types. Machines that utilize the stall phenomenon over the blades to control the power output are considered cost effective than pitch regulated machines. In contrast, pitch operated machines use the blade pitch mechanism powered by induction motors to rotate the blades and orient them into direction of wind. As of year 2019, nearly 70% of turbines installed in the world constitute fixed speed stall-controlled category. The electricity from a wind power plant is regulated using a suitable blade inclination sometimes referred to as pitch angle as well as angular speed of rotor. The blades of turbine rotor are subjected to unsteady aerodynamic and cyclical gravitational force acting on components in nacelle and tower structure due to periodic rotations of blade. For a typical wind turbine, efficiency can be described using its power curve (Snel and Schepers 1995; Vermeer et al. 2003; Directory of Indian Wind Power 2012). The efficiency of turbine varies with thrust (C_T), torque (C_t) and power (C_p) coefficients as well as due to the rotor solidity, blade length, rotational speed and controller algorithms (Jiang et al. 2012; Bhargava et al. 2020; Krishna et al. 2018). Tip loss is considered important when turbine operates in a wind farm array since wakes produced by rotating blade cause a velocity deficit for neighbouring turbine and affect the power production significantly. The Prandtl tip loss factor provides a detailed understanding of dynamics of the wake produced by rotating turbine blade which includes blade count as well as the velocity deficit downstream of turbine. The C_p also known as mechanical power coefficient is evaluated using a coupled form of 1D momentum and blade element theories. Most common theories, viz. generalized momentum theory, blade element and lifting surface or actuator disc theory that were developed in the past and extension of those theories by using corrections show a profound improvement of thrust and torque prediction on the turbines and propellers.

Shen et al. (2005) analyzed tip loss corrections based on Prandtl (1935) and Glauert (1948) propositions that wake expansion behind turbine rotor is proportional to bound circulation along the blade span. It is known that bound circulation has finite strength and is dependent on lift force acting on blade, density of fluid and free stream velocity. The force coefficients and blade loads are represented by body forces acting on rotor in 3D flow field. They found an approximation of bound circulation as zero towards the tip region led to erroneous results in estimation of induced velocity in axial and tangential directions even for low solidity rotors and at low tip speed ratios. Hence, improved tip corrections to predict the flow field near the blade tip were proposed by Shen et al. (2005) and Mikkelsen et al. (2009) to study the rotor blade loading using combined RANS solver and actuator line solution method (Mikkelsen et al. 2009). This method eliminated Glauert thrust correction for 2D and 3D flow fields since it was able to resolve the turbulent eddies or vortices at the tip region completely, and hence predicted mechanical loads on rotor plane and wake effects on downstream turbines accurately. Similarly, Wilson and Lissamon also proposed empirical thrust coefficient corrections based on the Prandtl tip loss model which assumes the orthogonal nature of the axial induced and relative velocity (Sherry et al. 2013; Shen et al. 2005).

De Vries (1979) showed that orthogonal condition between the axial induced velocity and relative velocity at individual blade element do not yield physically possible flows behind the turbine rotor (Vries 1979). He suggested higher order tip loss corrections which are more realistic and were based on the axial and tangential interference factors similar to that proposed by Prandtl (1935). Similarly, Shen et al. 2005 new tip loss model has also shown that correction of airfoil data towards the tip region is necessary due to the 3D rotational effects of rotor. Improved force coefficients were derived using exponential function of tip speed ratio and blade count. This formulation was applicable to rotors consisting of symmetric and cambered airfoils. Shen et al. 2005 tip correction model was applied on NREL Phase VI experiment rotor blades with S809 airfoil and Swedish WG 500 rotor with blade radius of 5.03 and 2.67 m. It must be noted that orthogonal condition between axial induced velocity and relative velocity was not satisfied when the boundary layer or pressure drag is dominant or included in blade element momentum (BEM) computations. The results from Shen's tip correction model demonstrated better aerodynamic loading in tip region of blade for wide range of tip speed ratios.

Javaherchi et al. (2014) investigated the computational wake modelling around horizontal axis wind turbine (HAWT) rotors using a combination of classical BEM and actuator disc theory based on the fixed and rotating reference frame methods. Although experiment studies and classical BEM theories provide realistic behaviour of wake dynamics to assess the performance of wind turbines, implementation of high-order turbulence models of computation fluid dynamics (CFD) were found more accurate and complemented the experiment data to establish the fidelity of results. Results of CFD simulation tools such as 3D Reynolds averaged Navier Stokes

(RANS) model for axisymmetric flows for modelling of wind turbine wake dynamics agreed well for wide range of tip speed ratios in contrast to BEM and actuator disc theories.

Bontempo and Manna (2019) investigated on errors in axial momentum theory applied to uniform loading of rotors relevant for turbofans' axial flow propellers without wake rotation. In this prediction of pressure and velocity field characteristics, panel methods, lifting line and lifting surface theories have been used. Further 1D axial momentum theory proposed by Glauert (1948) was applied to marine or aerodynamic propellers to examine the tip singularities to investigate unrealistic flows that occur during the wake expansion of rotor (Glauert 1948; Bontempo and Manna 2019; Javaherchi et al. 2014). For steady inviscid incompressible flow assumptions, it is known that generalized momentum theory does not take account of wake rotation; Bontempo noticed that for uniform loaded propellers the axial flow velocity at the tip produced significant errors and hence used combined axial momentum and CFD turbulence model for assessing the flow field near tip region of propeller blades. He also studied the vortex ring method proposed by Oye and improved the BEM model by superposition of ring vortices to predict the axial flow velocity at tip more accurately.

It must be noted that free ring vortex model is based on Kutta–Jukouwski theorem, which is essentially related to bound circulation that is responsible for producing lift forces on annular blade sections (Djojodihardjo et al. 2013). For uniformly loaded propellers, unsteady lift forces produced vortex flux behind the trailing edge of rotor that are assumed to originate from blade root or hub section. In contrast to trailing edge vortex, bound vortex is confined to annular blade section till the pressure gradient within boundary layer cause centrifugal pumping force action in radial direction that was enough to create the helicoidal wake structure downstream of the rotor. Further, he verified that induced velocity also varied predominantly in radial direction but remained constant when the axial thrust coefficient is less than 1.

In another study by Ouikke and Arbaoui (2020), corrections were applied to conventional axisymmetric 1D momentum theory using improved BEM model based on rotor-induced velocities and angle of attack. Experiments on NREL Phase VI rotor were performed to predict the blade loading near the outboard region which takes account of centrifugal and Coriolis forces along radial flow directions. Closed form system of equations was derived, and computer simulations were performed for low solidity wind turbine rotors at moderate tip speed ratios. The new BEM model used iterative methods which also included improved stall delay and 3D rotational augmentation effects in far wake based on the Kutta–Jukouwski theorem. As opposed to classical BEM, this iterative method incurred low computational cost per time step as well as reduced the uncertainty in predicting the aerodynamic forces and induced velocities at tip region of blade (Ouakki and Arbaoui 2020; Sebastian et al. 2019).

This paper is outlined as follows. In Sect. 14.2, a broad classification of horizontal axis wind turbine based on turbine parameters is presented along with historic overview of windmill development. In Sect. 14.3, we describe the geometric model and fundamental aerodynamic principle of HAWT rotors. In Sect. 14.4, aerodynamic force modelling on wind turbine blades is described using blade element momentum theory. In Sect. 14.5 a pseudocode which can be used to derive the power curve using manufacturer supplied power data is presented. In Sect. 14.6 a mathematical relation between scale and shape factor is derived for predicting annual energy yield at a given location. In Sect. 14.7, results are presented highlighting effects of rotor solidity and tip loss factors on aerodynamic performance based on BEM theory. The combined tip and hub losses for HAWT rotors are calculated for different rotor solidities. Thrust coefficient corrections proposed by Glauert, Buhl, and Wilson–Walker were also computed and validated with the experiment data. The energy yield prediction based on machine power curves is done using two-parameter Weibull distribution. The energy yield is estimated for two different air densities, and reasons for difference in energy yields for varying scale and shape factors is discussed. Time history of 10-min averaged data for electric power, main shaft torque and wind speed are illustrated for pitch-controlled turbine. Finally, in Sect. 14.8 conclusions are presented based on the methods studied.

14.2 Classification of Wind Turbines

Figure 14.1a depicts a chart which shows a broad classification of wind turbine types based on several technology parameters. They include rotation axis, number of blades, rotor orientation, generator technology, speed regulation, power converter, foundation, power regulation, drive train technology and tower structures. Figure 14.1b shows the various types of windmills that have been used over the centuries. Modern wind turbines consist of several components and the blades are critical components that undergo millions of revolutions in its lifetime and experience highly unsteady aerodynamic forces during operation which may often lead to structural component failures of turbine.

14.3 Geometric Model and Aerodynamic Principle of Wind Turbine Rotor

The efficiency of a turbine is dependent upon the geometry of airfoil profiles used for the blade. An airfoil has typically the following important geometrical parameters which affect the efficiency of a turbine blade, viz. chord, thickness, camber and solidity. These properties influence the lift coefficient across the blade span which is responsible for producing the torque of a turbine (Emblemsvag 2020; Directory of Indian Wind Power 2012). Also, at a wind velocity, the force coefficients such as lift and drag on a blade influence the power output from turbine. The solidity is related to the blade element chord at a given span-wise station and expressed as area of plan to swept area of blade section.

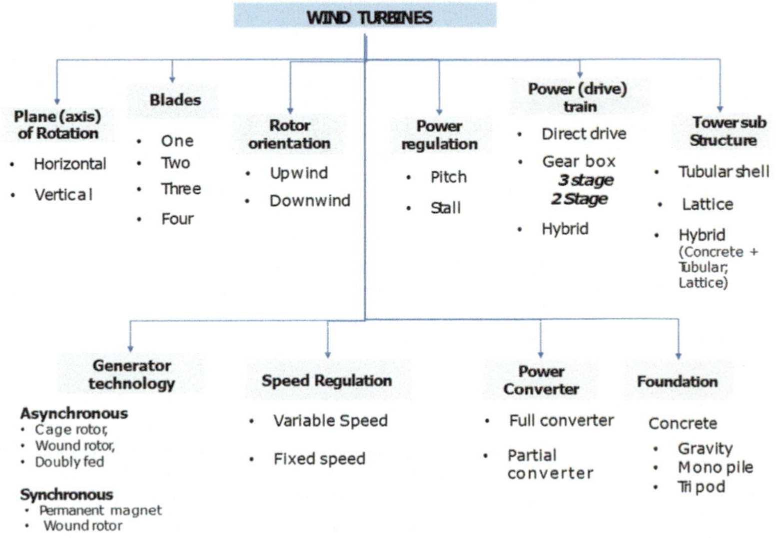

(a)

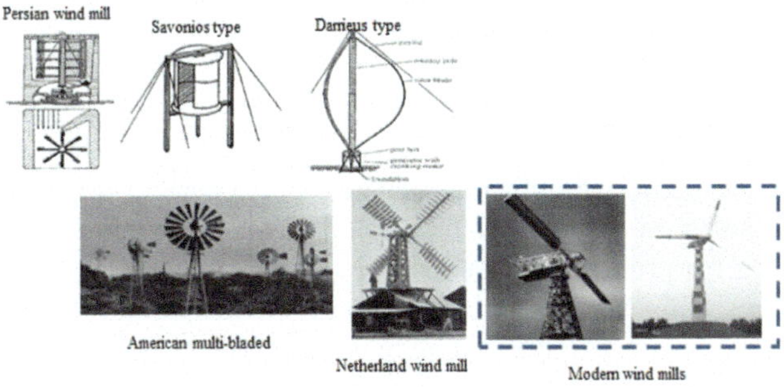

(b)

Fig. 14.1 Chart of wind turbine classification based on **a** several turbine technologies, **b** historic snapshots of windmill development over ancient, medieval and modern times (Gasch and Twele 2012)

Figure 14.2a shows the top view of blade element and forces acting on it. From the rotation plane it can be seen that lift (L) and drag (D) forces act along mutually perpendicular directions. Similarly, the normal (F_n) and tangential (F_t) forces are orthogonal to each other. The normal force is oriented out of rotor plane and inclined to the apparent wind speed. The tangential force acts in the rotor plane obtained after resolving lift and drag forces on the blade element and responsible for producing torque in horizontal axis wind turbine rotors. For airfoils with finite trailing edge thickness, the angle of incidence α is angle between the chord line and relative inflow velocity in the direction of fluid motion. The thickness of such an airfoil

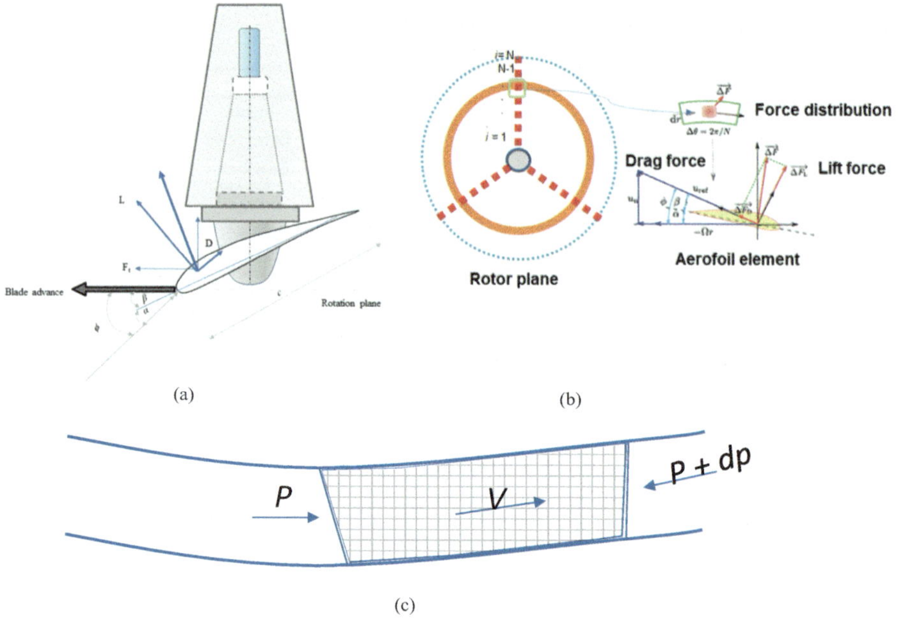

Fig. 14.2 Blade element and forces acting on airfoil of a wind turbine blade **a** top view (adapted from El-Okda (2015)), **b** front view, **c** illustration of stream tube and differential pressure acting on control volume of a fluid element

can be obtained by taking straight line distance between the upper (suction) and lower (pressure) surfaces. Figure 14.2b shows the blade element and aerodynamic forces responsible for producing lift and drag on a rotating airfoil. As mentioned, the lift and drag resulting forces act along mutually perpendicular directions and are inclined at angle of ϕ with respect to apparent wind speed. To considerable extent, aerodynamic power extraction from the wind is attributed to geometric properties shown in Table 14.1. Near blade root, the chord and twist angle for both blades are high and decrease towards the outboard region. The higher twist and chord provide a necessary starting torque required for the turbine. In addition, structural properties such as flexure and torsional stiffness of blade and tower components determine dynamic stability, using natural frequencies and mode shapes of turbine. Table 14.1 shows the geometric properties of blade lengths of 17 and 47 m found common in wind turbine industry. In the present work, BEM analysis has been performed on turbine with blade length of 17 m, while the power curve and energy yield analysis were done for turbine with blade length of 47 m.

Figure 14.2c shows the fluid flow as a stream tube relating pressure and velocity in a finite cross-sectional area. The shaded area represents the control volume for flow field within a stream tube. According to Bernoulli's equation for steady incompressible flows, the pressure differential or pressure head along the streamlines defining the control volume must remain constant (Glauert 1948). This pressure differential for

Table 14.1 Geometric properties of rotor blades of lengths 17 and 47 m

Blade length: 17 m				Blade length: 47 m			
Sl. No.	r/R (–)	Chord, (m)	Twist (deg)	Sl. No.	r/R (–)	Chord (m)	Twist (deg)
1	0.2	1.085	15	1	0.2	3.99	11.1
2	0.25	1.045	12.1	2	0.25	3.9014	10.7803
3	0.3	1.005	9.5	3	0.3	3.6762	9.9358
4	0.35	0.965	7.6	4	0.35	3.454	9.038
5	0.4	0.925	6.1	5	0.4	3.235	8.0833
6	0.45	0.885	4.9	6	0.45	2.9602	7.1666
7	0.5	0.845	3.9	7	0.5	2.6618	6.2659
8	0.55	0.805	3.1	8	0.55	2.4991	5.4173
9	0.6	0.765	2.4	9	0.6	2.348	4.5814
10	0.65	0.725	1.9	10	0.65	2.2016	3.7932
11	0.7	0.685	1.5	11	0.7	2.0419	3.0478
12	0.75	0.645	1.2	12	0.75	1.86	2.3739
13	0.8	0.605	0.9	13	0.8	1.75	1.7
14	0.85	0.565	0.6	14	0.85	1.5601	1.0853
15	0.9	0.525	0.4	15	0.9	1.3688	0.5101
16	0.95	0.485	0.2	16	0.95	1.171	0.1306
17	0.99	0.445	0	17	0.99	0.715	3.0625

a given control volume is a main factor responsible for lift generation on the blades. Figure 14.3a depicts the geometric model of wind turbine rotor with an annular blade section width, *dr*, located at a distance of *r* from the hub center. An increase in effective blade length in a rotor, as well as chord length, tends to increase the rotor solidity. Figure 14.3b demonstrates the rotor disk with finite solidity. Figure 14.3b also shows that an increase of blade count from two to three changes the rotor solidity by factor of 1.7. The blue color line shows the solidity factor for three blades and red line for two blades. Evidently, the solidity factor continues to increase near hub where profiles are thick and wide. However, it reduces outboard of span where the chord length becomes appreciably low compared to maximum chord of the blade. Like a wind turbine blade, an aircraft wing also exhibits maximum lift generation along its span for a given incidence angle. The tip vortices that are observed on aircraft wings have finite aspect ratios. As the aspect ratio is increased the maximum lift coefficient on wing is also increased. The formation of tip vortices produces a downwash on the wing that varies with flow angle of attack as well as Reynolds number. Downwash tends to reduce effective lift generated on aircraft wing due to high turbulence and is considered disadvantageous. Therefore, modern high speed aircraft wing tips are bent at an angle inclined to vertical plane. To counter reduced lift, active flow control methods such as suction and blowing are created to produce sufficient momentum and prevent the boundary layer flow separation at moderate to

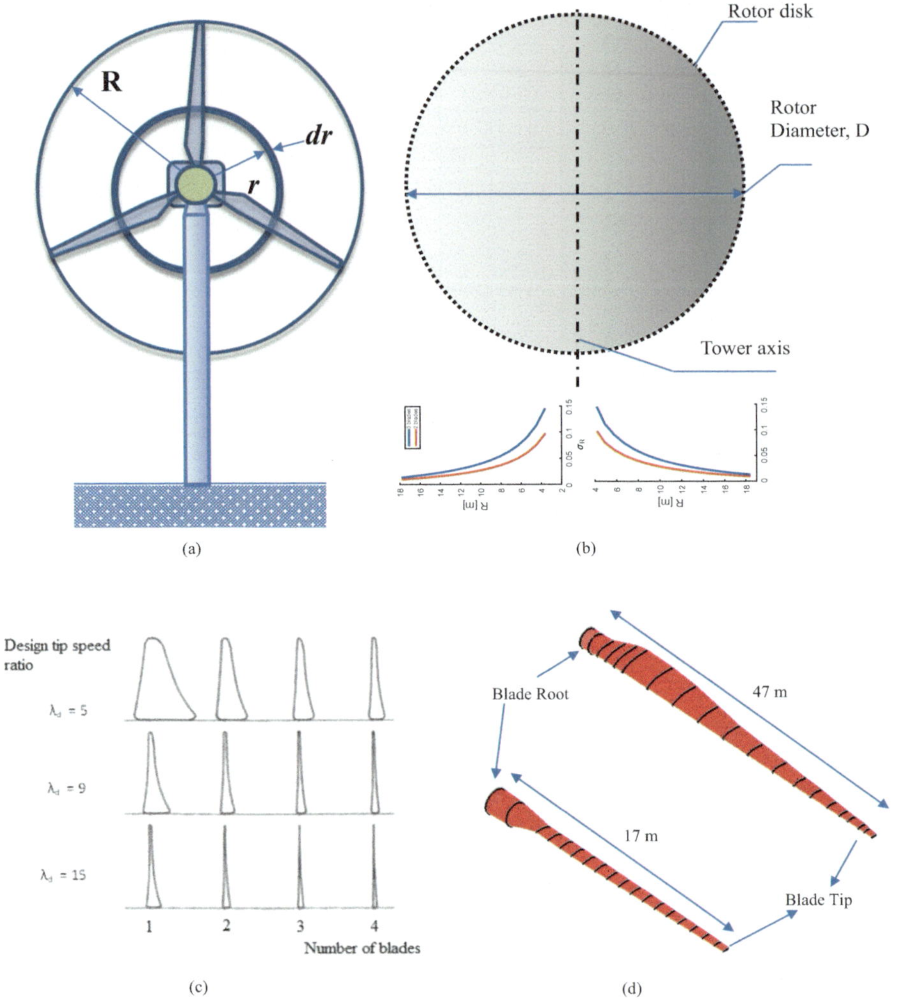

Fig. 14.3 **a** Illustration of annular blade section for three-bladed wind turbine rotor (Bhargava et al. 2020). **b** Comparison of rotor solidity from root to the tip of blade for three- and two-bladed wind turbine rotors. **c** Standard blade plan forms for various design values of λ and blade count (Jordan, et al. 2019). **d** Isometric view of the 17 and 47 m blades used in current study, developed in NuMAD software (https://energy.sandia.gov/programs/renewable-energy/wind-power/rotor-innovation/numerical-manufacturing-and-design-tool-numad/)

high flow angles oriented to wing (Genc et al. 2016). However, commercial use of such techniques has not been used in wind turbine industry till date to improve the aerodynamic efficiency of rotor. For wind turbines, at a given free stream velocity, the axial induction factors at the rotor disc continuously vary according to the inflow angles and the lift and drag forces on the blade. This unsteady nature of lift and drag forces are resolved in axial and tangential directions of rotor plane to produce thrust

Table 14.2 Key turbine parameters

Parameter	Value/Description	
Wind zone	IIA/IIIB	IIIB
Top height	80 m	50 m
Nominal power	2100 kW	500 kW
Diameter of rotor	95 m	36 m
Blade count	3	3
Cone angle	1	0
Tilt angle	3	2
Pitch imbalance	–	0.1°
Mass imbalance	–	–
Airfoils	NACA 63-xxx; FFA-W3-301	NACA 0012, NACA 4412, NACA632xx series
Power regulation	Stall control	Pitch control

and torque coefficients. In general, rotor solidity is given by Eq. (14.1)

$$\sigma = \frac{Nc}{2\pi \frac{r}{R}} \tag{14.1}$$

where N is number of blades; c is local blade chord; r/R is the normalized blade span.

Table 14.2 shows the turbine design parameters along with type of NACA airfoil family used for rotor blades. Aerodynamic data for airfoils are based on wind tunnel tests conducted on 2D sections. Conventional turbine designers relied on airfoil data that were used in aircraft industry which were typically 4-, 5- and 6-digit series, e.g. NACA 230xx, NACA 44xx, NACA 63-2xx. Due to rapid technology changes and adoption of new technologies from aircraft industry, the aerodynamics of blade design used devices such as winglets, flaps and vortex generators has resulted in the continuous increase of maximum lift coefficient and hence improved energy yield from a wind turbine.

Figure 14.3c shows that for a tip speed ratio range, increasing thickness to chord ratio and chord length of the blade reduces the design tip speed ratio. It also renders that for moderate thickness to slender blades the design tip speed ratio is higher and typically range between 9 and 15. According to Ragheb and Ragheb (2011), the theoretically optimum tip speed ratio is inversely proportional to number of blades in turbine and is given by $\lambda = 2\pi/B$ where B is blade count. Higher values of tip speed ratio (TSR) show that blades rotate faster in the wind field and convert more kinetic energy to mechanical energy. However, it must be noted that too low values of TSR explains that blades tend to escape the free wind and hence increase the uncertainty to extract aerodynamic power. Further explanation about the design tip speed ratio is demonstrated in Sect. 14.7.2. Figure 14.3d shows the comparison of the plan form of 17 m and 47 m blade length developed using NuMAD

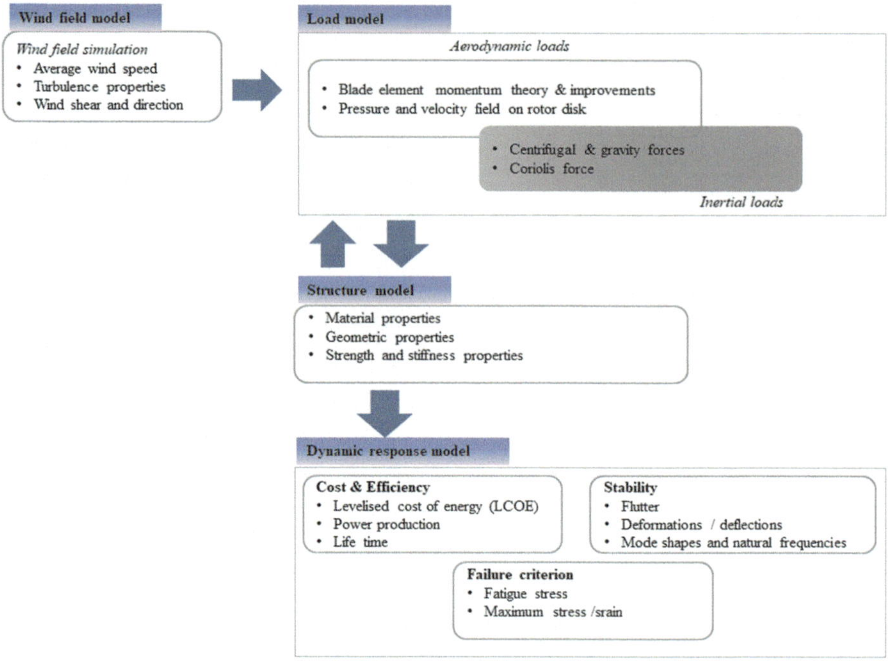

Fig. 14.4 Schematic showing interaction between structural and aerodynamic processes for a HAWT (adapted from Andrew and Flay 1999)

software (https://energy.sandia.gov/programs/renewable-energy/wind-power/rotor-innovation/numerical-manufacturing-and-design-tool-numad/).

Figure 14.4 depicts the aero-elastic response model for a wind turbine system. Modern wind turbine blades are built using composite materials typically made of glass fiber reinforced plastic (GFRP). They are designed to deliver improved mechanical properties such as strength and stiffness at a low weight and cost than conventional materials. The structural model of an aero-elastic system achieves these properties by using sandwich laminates which can be oriented in symmetric or anisotropic manner. By doing so, the geometry of blade is coupled to material characteristics to attain desired bending and torsion stiffness enhancements at various loading modes. It can be said that the response characteristics of an aero-elastic system utilizes the finite element method (FEM) approach and depend on the wind field, loading and structural models. The loading model shown in Fig. 14.4 includes the forces and moments due to aerodynamic, inertial, gravity, centrifugal and operational modes in the chord and span directions of blade.

14.4 Blade Element Momentum (BEM) Theory

The earliest known theoretical model for horizontal axis wind turbines was based on the actuator disk theory conceptualized by Rankine (1865) and Froude (1885) and was originally intended for propellers and helicopters. Actuator disk theory considered wind turbine rotor as stationary solid disk that could extract translational kinetic energy from wind to rotational energy. During this process, a change in momentum occurs and energy extraction is made possible through mechanical devices. Later, Lanchester and Betz proved that maximum possible power that can be extracted from a wind turbine as 59.3%. He essentially utilized the Bernoulli's principle of conservation of energy by assuming that the kinetic energy in wind and pressure drop across the turbine can be converted to useful mechanical work. Prandtl (1918) and Glauert (1935) later redesigned it with significant improvements to the assumptions of basic actuator disk theory. It is renamed as actuator line model which consists of imaginary streamlines in form of tubes that extended upstream and downstream of the rotor disk. The force distribution on lifting surface such as wind turbine blade can be governed by actuator line model. This model was modified by introducing the rotation of wake behind the rotor to find aerodynamic power losses in a reliable manner. This led to the foundation of BEM theory which is discussed in next section.

Lifting line theory was first proposed by Prandtl and Lanchester during World War I and based on assumption that flow around wings of finite span was approximated by discrete summation of flow around the thin airfoil elements of wing. Also, the wing is assumed as flat plate and the nature of flow around each airfoil element is predominantly 2D. It considered the 3D flow behaviour taking account of trailing edge vortex on wingspan which is depicted as helical wake that expands downstream producing downwash on the wing. For uniform flows, the downwash is caused due to bound circulation which varies uniformly along the wingspan. The strength of such bound circulation at arbitrary distance in radial direction can be determined using Biot–Savart law. An important drawback of this theory was that it does not account for thickness distribution, dihedral angle and in-plane sweep angle along span but considers twist and taper. It can be noted that downwash is produced due to tangential-induced velocity along the wingspan causing a trailing edge vortex which expands far downstream of wing. This theory also considers normal induced velocities along wingspan to predict the strength of downwash computed by means of Biot–Savart law. The lift produced is proportional to strength of circulation which is essential to predict the inflow angles and angle of attack. At higher or near stall AOA, higher lift forces and thrust on the wingspan or a rotor blade are expected (Snel and Schepers 1995).

14.4.1 BEM Theory Assumptions and Formulation

From classical BEMT, velocity field on each airfoil having a finite width dr is different from remaining airfoils (Bhadra et al. 2010; Emblemsvag 2020). The flow is fundamentally free of any shear forces and constant fluid density along the rotor plane which means no wake trails of fluid rotation are present. Further, wake losses due to tip are not modelled due to rotor being assumed as solid disk (Bhadra et al. 2010; Sherry et al. 2013). No yawing is accounted in BEMT. As mentioned earlier, the limiting value for aerodynamic power extracted by a horizontal axis turbine is 16/27 = ~59.3% which according to Betz is impossible to achieve practically. This limit occurs when the axial induction factor at rotor disk reaches a value of 0.33 (Hansen 2010). In general, the C_p for a turbine containing finite number of blades can be approximated using velocity triangles method. C_p and C_T for machine can be evaluated using below steps for a given L/D ratio

- At every blade section, the inflow angle is computed using Eq. (14.2). The inflow angle at any blade section is function of λ.
- From the inflow angle, the axial induction factor is computed at individual airfoils given by Eq. (14.3.1).
- For a given tip speed ratio, airfoil efficiency and tip loss factor for each blade are integrated to evaluate peak aerodynamic efficiency of rotor.
- For L/D values of 40, 80 and 120, the C_p is given using Eqs. (14.2) and (14.3.1, 14.3.2).

$$\emptyset = \tan^{-1}\left[\frac{1}{\left(\frac{r}{R}\right)\lambda}\right] \tag{14.2}$$

$$a = 4\lambda\mu^2\left[\frac{\sin\left[\frac{2}{3}\emptyset\right]^3}{\sin[\emptyset]^2}\right]d\mu \tag{14.3.1}$$

$$\eta = 1 - \frac{\lambda}{C_l/C_d} \tag{14.3.2}$$

- Prandtl tip correction is expressed as function of blade count, the increase in tip velocity with respect to free stream and inflow angle of blade. The overall turbine efficiency can therefore be evaluated by summing airfoil's efficiency wake loss factor due to tip and wind velocity deficit factors as well as lift to drag ratio along the blade span.
- The ratio of aerodynamic efficiency to the maximum possible efficiency for the assumed lift to drag ratios includes the tip loss factor (Bossanyi et al. 2005; Snel and Schepers 1995; Vermeer et al. 2003).

More detail steps on how to derive power coefficient using velocity triangles are explained later in this section. For smaller turbine conservative ratios of L/D are found to vary commonly in the range 40–60. In actual cases, this will vary up to L/D

= 120, for large megawatt scale wind turbines at high Reynolds number. A large value of L/D ratios means higher C_p for a turbine and implies the exceedance of Betz limit. The Prandtl tip and hub loss factor can be expressed mathematically from Eqs. (14.4) to (14.6)

$$F(\mu) = ft(\mu) \cdot fr(\mu) \tag{14.4}$$

where F is total wake velocity deficit, $f_r(\mu)$ is root wake velocity deficit, $f_t(\mu)$ is tip wake loss factor, μ is the normalized blade radius, r/R, μ_R is normalized blade root radius, N is number of blades, a is axial induction factor at the rotor disc and λ is tip speed ratio

$$f_r(\mu) = \frac{2}{\pi}\cos^{-1}\left[e^{-\left(\frac{N}{2}\right)(\mu-\mu_R)/\mu\sqrt{1+(\lambda\mu)^2/(1-a)^2}}\right] \tag{14.5}$$

$$f_t(\mu) = \frac{2}{\pi}\cos^{-1}\left[e^{-\left(\frac{N}{2}\right)(1-\mu)/\mu\sqrt{1+(\lambda\mu)^2/(1-a)^2}}\right] \tag{14.6}$$

The rotation of blade causes the flow movement in axial, radial and tangential directions. In this process, angular momentum of rotor is not constant and dynamic stall phenomenon occurs at an AOA beyond the critical angle, typically about 16°. For chord-wise direction of blade, near stall angles of attack, flow becomes more turbulent and causes local circulation to expand to the tip of the blade. Similar phenomenon is also observed near center of rotational axis of machine. These vortices cause the aerodynamic power loss at high tip speed ratios which are undesirable. In the current study, only the wake loss from tip is examined (Bossanyi et al. 2005). Wake loss from tip is high for turbines operating in large wind farms due to interaction of wake from neighbouring turbines which causes a velocity reduction and affects the overall power output from a wind farm.

In case of wind turbine rotors, the maximum C_p and C_T during yaw is expressed in terms of yaw angle and given from Eqs. (14.7) to (14.10)

$$C_p = 4a(\cos\gamma - a)2 \tag{14.7}$$

$$a = \frac{\cos\gamma}{3} \tag{14.8}$$

$$C_{p,\max} = \frac{16}{27}(\cos\gamma)^3 \tag{14.9}$$

$$C_T = 4a\sqrt{1 - a(2\cos\gamma - a)} \tag{14.10}$$

γ is the yaw inclination, and a is the velocity deficit factor. In the present study, the maximum power coefficient is evaluated using axial momentum theory which predicts the C_p value better than Glauert's correction to momentum theory. For steady incompressible aerodynamic flows, actuator disc and BEM theory can be evaluated at all values of axial induction factor. In practice, however, results from BEM theory are not accurate when the axial induction factor exceeds 0.33 but is less than 0.5 (Strangfeld et al. 2015; Hansen 2010). When the value of a crosses 0.5, the BEMT produces invalid or inaccurate loading results for a turbine. Further, Mansberger (2016) derived the blade element theory based on thermodynamic model which states that kinetic energy in the flow field is converted to internal or thermal energy and vice-versa within the control volume (Mansberger 2016). It predicts higher values for C_p for all values of axial induction factor, a compared to Betz limit derived from classic BEMT and claims maximum C_p to be 65.1% when a is 0.457 thus disproving the Betz limit. Also, it assumes that diameter of near wake expands causing an increase in velocity deficit while the diameter of far wake contracts leading to reduced velocity deficit. Although velocity deficit has been proven and validated using laser doppler velocity measurement data the details of this thermodynamic model are beyond the scope of present study.

As mentioned earlier velocity deficit that occurs across the rotor disk can be expressed by Eq. (14.11)

$$a = \frac{U_0 - U_d}{U_0} \tag{14.11}$$

This factor varies on the extent of pressure drop and velocity reduction of free stream air flow downstream of rotor disk. It can be said that axial induction factor a shows large variation when the tip speed ratio of rotor is high. Furthermore, a large velocity deficit produced far downstream of rotor will induce a higher thrust coefficient. This incremental change in thrust coefficient far downstream is an important parameter in the design of wind farms for a given site. However, the resultant velocity seen by blade section located at radius r varies with both axial and angular induction factors and is expressed by Eq. (14.12)

$$W = U_r = \sqrt{\{(U_0(1-a))\}^2 + \{r\Omega(1+a')\}^2} \tag{14.12}$$

It must be noted that axial induction factor according to Eq. (14.3.1) can be rearranged in terms of inflow angle, ϕ and local blade solidity, σ_r, aerodynamic force coefficients, C_l and C_d to yield from Eqs. (14.13) to (14.15), respectively.

$$\emptyset_r = \tan^{-1}\left(\frac{1-a}{(1+a')\lambda_r}\right) \tag{14.13}$$

$$\frac{a}{(1-a)} = \frac{\sigma_r[C_l\cos\emptyset + C_d\sin\emptyset]}{4\sin^2\emptyset} \tag{14.14}$$

$$\frac{a'}{(1+a')} = \frac{\sigma_r[C_l\sin\emptyset - C_d\cos\emptyset]}{4\sin\emptyset\cos\emptyset} \tag{14.15}$$

The change in the lift and drag force vectors on a local chord of an airfoil is given by Eqs. (14.16) and (14.17), while the resolved force components in axial and tangential directions of the rotor plane are given from Eqs. (14.18) to (14.20)

$$dF_L = 0.5\rho U_r^2 c_r C_l dr \tag{14.16}$$

$$dF_D = 0.5\rho U_r^2 c_r C_d dr \tag{14.17}$$

$$dF_N = dF_l \cos\emptyset + dF_d \sin\emptyset \tag{14.18}$$

$$dF_T = dF_l \sin\emptyset - dF_d \cos\emptyset \tag{14.19}$$

$$dM = dT = [dF_l \sin\emptyset - dF_d \cos\emptyset] \cdot r \tag{14.20}$$

During the rotation, a centrifugal pumping action of the airstream tends to occur in radial direction of the rotor disk and is responsible for delaying stall on the blades. The air flow near root or hub rotates in rotation plane and moves towards the tip of blades and flow eventually forms a bound vortex along blade span and tip vortex at tip of blade which is responsible for local lift generation. These vortices are shed in a cyclical manner in the form of a helix which extends downstream of turbine rotor. The differential thrust and torque equations are expressed in terms of total wake loss or deficit factor F and are given by Eqs. (14.21) and (14.22)

$$dA_{\text{Thrust}} = 4\pi\rho a(1-a)U_0^2 \cdot rdr\,\mathrm{F} \tag{14.21}$$

$$dA_{\text{Torque}} = 4\pi\rho r\omega a'(1-a)U_0 \cdot r^2 dr\,\mathrm{F} \tag{14.22}$$

The torque, power and thrust coefficients can be expressed from Eqs. (14.23) to (14.25)

$$C_t = \frac{T}{0.5\rho\pi R^3 V^2} \tag{14.23}$$

$$C_p = \frac{P}{0.5\rho\pi R^2 V^3} \tag{14.24}$$

$$C_T = \frac{F}{0.5\rho\pi R^2 V^2} \tag{14.25}$$

where T is mechanical torque, N-m produced by the main shaft of turbine. R is blade radius, m, V is uniform speed measured in m/s, P is aerodynamic power, kW, ρ is air density in kg/m^3, F is the resultant normal force acting on rotor. A simple algorithm to implement BEM for HAWT rotors consist of following steps.

1. Begin with a and a' as zeros.
2. Compute the blade flow angle relative to apparent velocity field, ϕ.
3. Determine chord angle relative to apparent velocity field, α.
4. Import the C_l (α) and C_d (α) table data.
5. Calculate the force coefficients in axial and tangential directions relative to free stream.
6. Determine the velocity deficit factors in normal and along rotor plane, a and a'.
7. Fix a condition to a and a' for solution stability; otherwise repeat process from step 2.
8. Apply the tip wake loss factors.

14.4.2 Wind Shear Approximation

Wind shear is a measure of gradient in wind speed and it can vary in horizontal and vertical direction in atmosphere. Wind speed in atmosphere can be considered as stochastic and is described using parameters such as turbulence intensity and roughness height. Further mean wind speeds are found to increase with height above ground. For a given geographic location, it is function of orography factors like obstacles, e.g. buildings, trees, vegetation and expressed by reference surface roughness, z_0. The turbulence characteristics of wind field seen by a wind turbine rotor can thus be conveniently expressed in terms of roughness height, z, above the ground and wind shear exponent, ε. For a turbine operating at a windy site, the wind shear approximation can be done either by power law or logarithmic law given by Eq. (14.26) and Eq. (14.27), respectively. For BEM studies, a wind shear exponent of 0.1 for power law in Eq. (14.26) has been assumed which represents the plain or open grass land with few dispersed vegetation.

$$\frac{U(z)}{U_{ref}} = \left(\frac{z}{z_{ref}}\right)^{\varepsilon} \tag{14.26}$$

$$\frac{U(z)}{u_*} = \frac{1}{k}\ln\left(\frac{z}{z_0}\right) \tag{14.27}$$

U is mean wind speed in m/s; u_* is the friction velocity; k is the von-Karman constant, 0.4; z is the height above ground in m; z_0 is the reference surface roughness length in m.

14.4.3 Limitations and Improvements to Classical BEM Theory

As mentioned in the earlier section, BEM is based on 1D linear momentum theory in which a turbine rotor is modelled as uniform disk consisting of finite number of blades with discrete blade elements. It has several limitations since it cannot model complex flows around the rotor and its accuracy varies during stalled and unsteady flow conditions during yaw operation. Also, there is no rotation and expansion of wake downstream of turbine rotor. More advanced aerodynamic models for load modelling are dependent on the wake structure behind the rotor such as the prescribed wake or free wake vortex acceleration potential methods which have been successfully implemented for propeller and helicopter aerodynamics and adapted to HAWT rotors. Simulating 3D non-linear wake structures requires advanced wind turbine design codes based on CFD techniques that are computationally intensive and prohibitive in nature. In the recent past, modern wind turbine design codes have been modified and aimed to improve accuracy by implementing corrections to the standard BEM code by integrating engineering models that take account two important factors, viz. dynamic stall and dynamic wake. In each case, 2D static airfoil profile data is corrected for 3D effects that include tip and hub loss as well as stall delay along the blade span to increase section lift coefficient or limit the blade loading. An empirical stall delay model proposed by Corrigan and Schillings (1994) given by Eqs. (14.31) and (14.32) in Corrigan and Schillings (1994) takes account of vortex wake rotation effects observed at high AOA. It utilizes the trailing edge angle, local radius, local chord length of blade segment and velocity gradient parameter to determine the change in AOA. This model was applied by Tangler and Selig (1997), Xu and Sankar (2002) for determining increment C_L values (Tangler and Selig 1997; Xu and Sankar 2002). Empirical relations often give a useful estimate for verifying the BEM results. A simpler approximation to C_p according to Manwell et al. (2009) is given by Eq. (14.28)

$$C_p = \frac{16}{27}\left[\frac{\lambda B^{0.67}}{1.48 + \left(B^{0.67} - 0.04\right)\lambda + 0.0025\lambda^2} - \left(\frac{1.92\lambda^2 B}{1 + 2\lambda B}\right)\frac{D}{L}\right] \tag{14.28}$$

Another essential approach is the inverse BEM method in which angle of attack (AOA) and inflow angle values are obtained using the blade loading functions. Transfer functions are utilized in the process to determine the inflow angle pattern for different flow conditions; however, such predictions are often difficult to verify limited extensive experimental data and hence a major constraint in BEM theory validation.

14.4.4 Turbulence Intensity and Empirical C_p Model

Turbulence intensity (TI) is an indicator of *randomness* or *gustiness* in wind. It is the ratio of standard deviation of longitudinal value of wind speed component to its magnitude of free stream velocity and is expressed using Eq. (14.29). For wind data obtained at a given location, it is an important parameter which is used for predicting the fatigue life of wind turbine components. Typically estimates for turbulence intensity parameter vary with long-term data and also according to orography properties of a geographic site. According to Germanischer Lloyd wind turbine certification standards, there are broadly two turbulence intensity classes which are termed as A and B and are equal to 18% and 16%, respectively (Germanisher 2007).

$$TI = \frac{\sigma}{\overline{V}} \tag{14.29}$$

As mentioned in Sect. 14.3, the power regulation for a pitch-controlled turbine changes due to blade angle. The C_p is function of blade pitch angle β and TSR λ. A model proposed by Heier (1998) for double fed induction generator (DFIG) has been analyzed in the present study. In the present study, the power coefficient C_o is given according to empirical exponential formulation using Eqs. (14.30) and (14.31). It can be seen that Eq. (14.30) is dependent on coefficients c_1, c_2, c_3, c_4, c_5 and c_6, blade pitch angle and tip speed ratio of turbine. In the present case, the values for those coefficients have been tested for turbine with 95 m rotor diameter in such a way that power coefficient values remained within 1% of the manufacturer obtained power coefficient.

$$C_p(\lambda, \beta) = c_1\left(\frac{c_2}{\lambda_i} - c_3\beta - c_4\right)e^{-\frac{c_5}{\lambda_i}} + c_6\lambda \tag{14.30}$$

$$\frac{1}{\lambda_i} = \frac{1}{\lambda + 0.08\beta} - \frac{0.035}{\beta^3 + 1} \tag{14.31}$$

14.5 Power Curve Determination

Accurate modelling of power curve is essential to forecast the power output from a single or group of turbines. In the power curve analysis, averaged wind speed and electric power are often essential to estimate energy. Data reduction and corrections are necessary with respect to cut-in and cut-out wind speeds by using the method of bins. A cut-in wind speed is a minimum value at which turbine starts delivering power to grid.

Rated wind speed is a value when the generator produces nominal power. Both cut-in and rated wind speeds are fixed values while a cut-out wind speed is not a

Table 14.3 Pseudo code: wind turbine power performance

```
STEP 1: Initialize data

PowerData :            i ←1 to N      // N is data length or samples
ThrustData :           i ← 1 to N
TorqueData :           i←1 to N
WindSpeedData :        i ←1 to N

STEP 2: Assign counters

ThresholdPowerData : = 1 to 25     //set threshold values of electric power corresponding to wind speed
CountWS := 1 to 25                 // wind speed bin
 k: = 1                            // set counter for bin average
 i: = 1                            // set counter for data increment

STEP 3: Algorithm
WHILE k <= 25, Do
        FOR i ← 1 to N Do
            1.  Check for wind speed bin limits
                IF WindSpeedData_k >= CountWS_k and WindSpeedData_k <= CountWS_k Do
                        PData_i,k := PowerData_i,k                // extract electric power data
values
                        ThrustData_i,k := ThrustData_i,k          // extract thrust data values
                        TorqueData_i,k := TorqueData_i,k          // extract torque data values
                END IF
            2.  If invalid data is found in extracted bins, replace with empty character
                Else Go to 3
            3.  i ← i+1
        END FOR
P_m_k := Do the average of (PowerData_i,k)
Thrust_m_k := Do the average of (ThrustData_i,k)
Torque_m_k := Do the average of (TorqueData_i,k)

IF P_m_k < twice ThresholdPowerData_k || P_m_k >= twice ThresholdPowerData_k Do
  P_m_k := P_m_k ± Standard deviation of PowerData_i,k //set standard deviation according to bin count
END IF
        k ← k +1
END WHILE

STEP 4: Display outputs

PRINT CountWS, P_m
PRINT CountWS, Thrust_m
PRINT CountWS, Torque_m

STOP
```

constant. Power hysteresis algorithms are usually implemented by turbine operators for above rated wind speeds in order to reduce loads and extend lifetime of turbine components. For certification of power curve, IEC 61400-12-1 is followed for power performance measurements in case of single turbine. The standards tell about the measurement method for actual power curve determined using time varying values for wind velocity and power at a test site. The typical bin width is taken as 1 m/s for below nominal wind velocity and 2 m/s when wind velocity is above nominal values.

Table 14.3 shows the pseudo code implemented for power curve analysis in the present study. The power curves vary with size of turbine and its manufacturer specifications. Between minimum threshold and nominal values, the curve varies almost linearly in response to the wind speed for both pitch control and stall-controlled turbines. For above rated wind speeds, the power output remains constant for pitch-controlled turbines while it reduces in case of stall-controlled turbine. For any case, bin averaging procedures are applied to measured data. It can be noted that bin averaging of data is necessary for each performance parameter in which data range for each bin needs to be sorted, summed and divided with the total data samples per bin. For some wind speed bins around cut-in values, the bin values for power data appear to be negative. The negative values indicate that generators consume reactive power from grid.

Operational data is acquired through a digital data acquisition system and monitored for accuracy of measured signals which can impact the energy production for given period. The system must ensure that error free data is recorded. This can be achieved by checking the accuracy and linearity of sensors, transducers, and preamplifiers. The minimum sampling rate for measurement is 0.5 Hz according to IEC 61400-12-1 standards. Erroneous data can be measured by anemometer, and when the anemometer is present in the wake can thus be eliminated by following standards and procedures. It can be noted that sensors used to measure electric power data must ensure $\pm 0.5\%$ accuracy of the rated power of wind turbine required for power curve verification (Power performance testing 1990).

Commercial wind farm design tools such as WAsP and Wind Farmer are software which can model the complex flow conditions at given site. They use quantitative methods for forecasting long-term wind speed and consists of measure, correlate and predict (MCP) procedures. MCP is a method to predict long-term wind conditions, i.e. speed and direction at a target site by correlating it with the short-term concurrent or time series measured wind data from a nearby reference mast. To avoid energy losses due to wind farm array design, factors such as inter-turbine spacing, closeness of wind farm electrical infrastructure and orography variables are important. Further, soiling losses that occur due to change in surface roughness of blade can be minimized when best practices in operation and maintenance are followed. Knowing the accuracy of power output from a turbine at a given wind speed provides useful information in the design process of turbine or its components. It is also useful for wind turbine operators and power grid companies to forecast the energy prices and power trading in electricity markets. A flow chart for an iterative numerical prediction algorithm used in data analysis is shown in Fig. 14.5. It must be noted that these factors affect the gross energy output from a wind farm. The data required for energy yield assessment is quantified using power and thrust curves and depends on measured wind velocity data. The quality of measured wind speed data at a site also depends on the distance of met mast to a neighboring turbine. Also, the boom orientation on mast is critical to measuring the wind velocity accurately. The number of velocity sensors mounted on mast and hysteresis of wind vane error also contributes to uncertainty in power productions from a wind farm (Astolfi et al. 2020). Direction measurement must be accurate within $\pm 3^{\circ}$. According to Garrad Hassan sensor errors contribute to nearly

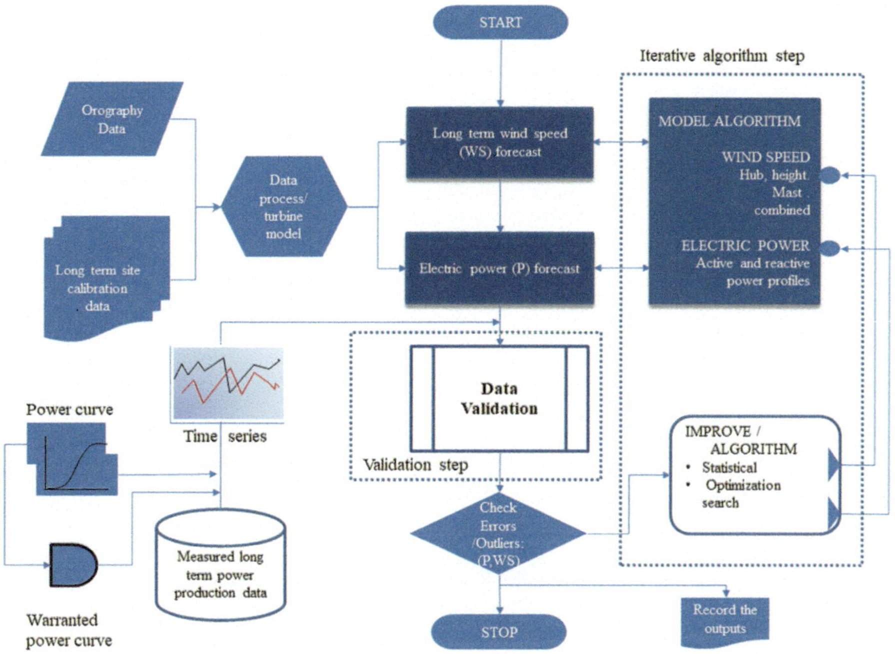

Fig. 14.5 Schematic for iterative numerical prediction algorithm used in wind farm data analysis

2% reduction in energy production. Also, the recommended distance for placement of test anemometers is between two and six rotor diameters and at a height equal to hub height of wind turbine. Further, another factor which affects energy yield from a turbine is surface roughness condition of blade during operation which varies due to accumulation of dirt, ice, rime or insects. This leads to blade degradation and accounts between 3 and 5% reduction in annual energy yield from a wind turbine.

It must be noted that as average distance of met mast in a site increases, the uncertainty of wind speed measurement also increases. According to IEC 61400-12-1, the gross errors in the annual energy yield would double for every 1% increase in wind speed measurement error. Overall, about 10–15% reduction of gross energy yield from wind park can be possible due to the above factors considered (Wind energy, The facts, Part-I; Technology 2014).

14.6 Mathematical Relation Between Scale and Shape Factor

In this section, the relationship between the shape and scale factor using Weibull distribution is derived using Eqs. (14.32) to (14.38)

$$f(V) = k \cdot \frac{V^{k-1}}{c^k} \cdot e^{-\left(\frac{V}{c}\right)^k} \tag{14.32}$$

$$\overline{V} = \int_0^\infty V f(V) dV \tag{14.33}$$

$$\text{Let} \left(\frac{V}{c}\right)^k = y \tag{14.34.1}$$

$$V = c y^{\frac{1}{k}} \tag{14.34.2}$$

$$dV = \frac{c}{k} y^{\frac{1}{k}-1} dy \tag{14.34.3}$$

Substituting Eqs. (14.32) and (14.34.1, 14.34.2, 14.34.3) in Eq. (14.33) we get

$$\overline{V} = c \int_0^\infty e^{-y} y^{\frac{1}{k}} dy \tag{14.35}$$

After rearrangement Eq. (14.35) reduces to Eq. (14.36) and using identity of Eq. (14.37.1) we obtain Eq. (14.37.2)

$$\overline{V} = c \int_0^\infty e^{-y} y^{1+\frac{1}{k}-1} dy \tag{14.36}$$

$$\int_0^\infty e^{-y} y^{1+\frac{1}{k}-1} dy = \Gamma\left(1 + \frac{1}{k}\right) \tag{14.37.1}$$

$$\overline{V} = c\Gamma\left(1 + \frac{1}{k}\right) \tag{14.37.2}$$

$$c = \frac{\overline{V}}{\Gamma\left(1 + \frac{1}{k}\right)} \tag{14.38}$$

where k and c are shape and scale parameters, respectively, and $\overline{V}$ is wind velocity. $f(V)$ is probability density function. One can obtain scale factor using gamma function when $\overline{V}$ annual mean wind speed at site is known. (Bhadra et al. 2010)

Equation (14.38) is used to estimate the energy for a given period based on scale and shape factors. From Eq. (14.38) one can estimate the wind power density (WPD) at a site using the following relation

$$WPD = \frac{1}{2}\rho c^3 \Gamma\left(1 + \frac{3}{k}\right) \quad (14.39)$$

Equation (14.39) is obtained by substituting the value of $\overline{V}$ in Eqs. (14.38) into (14.24). One can note that wind power density is a function of scale factor and shape factor which varies according to site location (Saxena and Rao 2016). More detailed analysis of energy yield assessment is presented in Sect. 14.7.3.

14.7 Results and Discussion

14.7.1 Turbine Efficiency and Dynamics

As mentioned in Sect. 14.4, the performance of a HAWT is dependent on aerodynamic lift and drag force coefficients which to a large extent depend on geometric and structural properties of blade. The C_p of turbine is computed for variable range of lift to drag coefficient ratios. For a turbine operating in wind farm array, the expanding helical wake behind the rotor affects the net power production and thus efficiency. The highest efficiency is obtained when wakes are spaced equidistant behind the rotor. Figure 14.6 depicts the influence of force coefficient ratio (L/D) on efficiency of wind turbine rotor element. The value of this ratio changes with the type of airfoils, operating tip speed ratio of rotor and surface roughness profiles of blade. Typically, thin airfoils such as NACA 4 digit and 6 series are slender wing elements present in outer annular ring while thick airfoils are present in inner annular ring (Ladson and Brooks 1996). The maximum blade lift usually varies in a range between 30 and 80% of blade span. From Fig. 14.6 it can be observed that as L/D ratios are increased the C_p values also tend to increase steadily for moderate to high values of λ between 4 and 10. For tip speed ratio (TSR) of 8, a step change in L/D from 40 to 120 showed

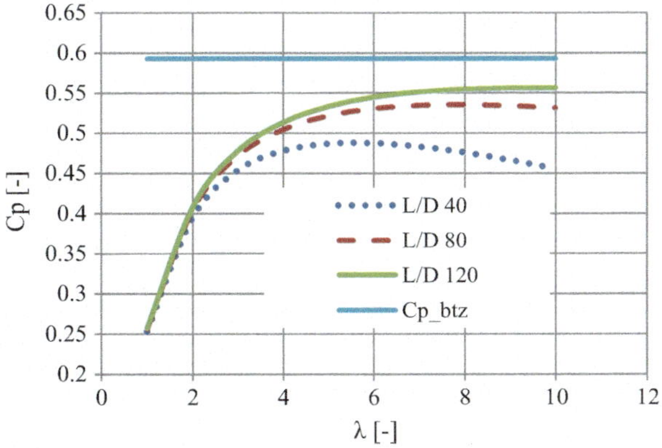

Fig. 14.6 Computed mechanical power coefficient for different L/D ratios of wind turbine rotor element

a 17% increase in C_p value. This increasing trend is seen for TSR values between 6 and 10. The horizontal straight line shows the maximum power coefficient obtained considering no wake rotation and possible only for idealized wind turbine.

However, from Fig. 14.7 for a given L/D ratio, significant change in C_p value is found for tip speed ratios less than 6 at different rotor solidity values. It corresponds to ~50% increase in C_p when three blades are present in turbine than when only one blade is used in rotor. During the Cp calculation, tip losses which account for aerodynamic power loss have been considered so that computed values can be compared readily with values obtained using measured data. Figure 14.8 shows the angle of attack (AOA) for varying tip speed ratios. It can be seen that for low tip speed ratios, the AOA are found as high as 50° towards the blade root region. As the tip speed ratio is increased, the AOA is reduced to as low as 5° and affect the maximum lift coefficient on the blade. In comparison to wind turbine rotors, commercial aircraft wings operate at low incidence angles at remarkably high Reynolds number.

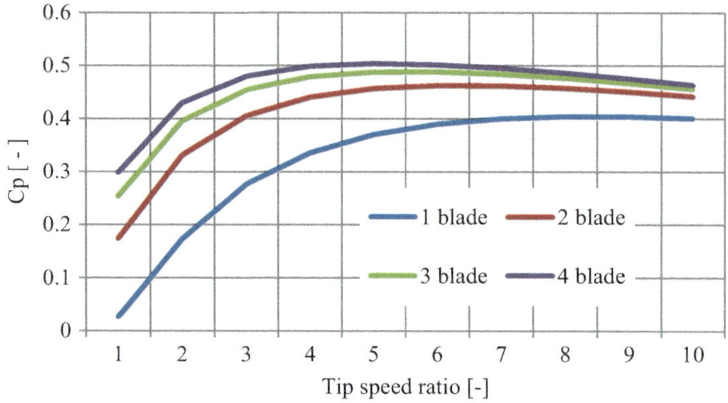

Fig. 14.7 Mechanical C_p computed for a turbine consisting of 1–4 blades with a L/D = 40 and considering tip loss correction, but no drag

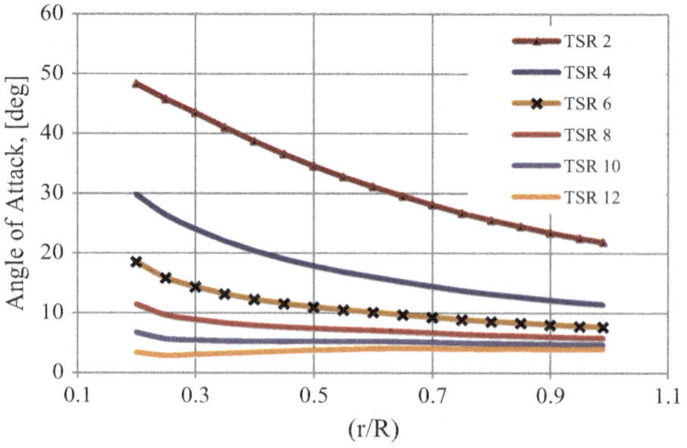

Fig. 14.8 Angle of attack distribution for a blade length of 17 m at varying tip speed ratios

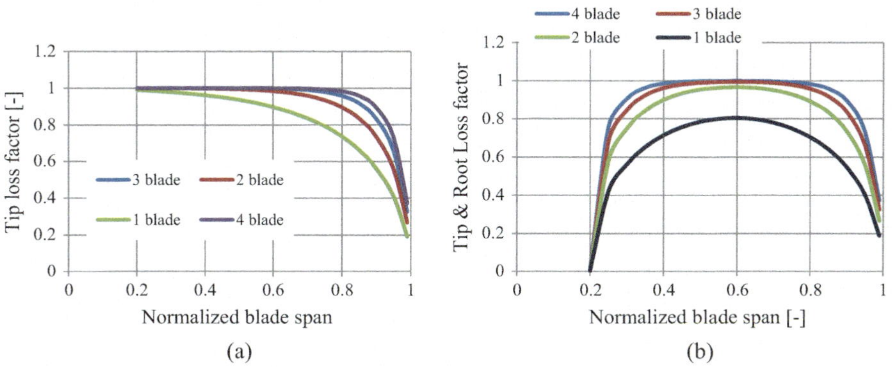

Fig. 14.9 **a** Prandtl tip vortex loss correction factor for varying number of blades, **b** combined with hub vortex loss

It can be noted that vortices are formed at tip and root regions that develops as cyclical wake screw and propagate leeward side of the rotor (Snel and Schepers 1995). Also, the magnitude of axial-induced velocity varies with the wake distance downstream and the thrust coefficient. When the wake expands downstream of the rotor, flow becomes more turbulent, and a higher velocity deficit is experienced by downstream turbine. As a result, higher thrust coefficient far downstream in the wake is produced and causes increase in fatigue loads on blade.

Figure 14.9a shows the Prandtl tip loss correction factor evaluated for one, two, three and four blades of a turbine. It is apparent that tip losses are higher towards the inboard region of blade where the thickness to chord ratio for airfoils is high but reduce eventually to a finite non-zero value. This trend is true for all horizontal axis turbines irrespective of number of blades. The total wake loss factor is depicted in Fig. 14.9b. Both wake and hub velocity deficits are computed at 20% of total blade length. The maximum wake loss was observed between 40 and 80% of length of blade for incremental blade count. It must be noted that tip vortices increase with varying aspect ratio of blade. As the aspect ratio is reduced to less than one, the maximum lift coefficient of a given blade section would increase. However, for low to moderate Reynolds number flows, vortices can be formed quickly due to flow separation which occurs at the leading edge of blade (Genc et al. 2016). The axial induction factor variation along the blade span is shown for the cases local to blade and azimuthal directions which seem to diverge at the tip as depicted in Fig. 14.10.

As discussed in Sect. 14.3, the axial induction factor a at rotor disk is indicator of velocity deficit used to test BEMT. This value also varies according to the blade azimuth orientation. Figure 14.11 shows the computed values for a at various span positions of blade for azimuth angles between 0° and 360°. It can be seen for span position of 0.99 R, a value fluctuates between 0.14 and 0.65. In the present study, the reference blade azimuth angles for three blades are assumed to be located at 90°, 210° and 330° with respect to the tower axis (Chaudhry et al. 2014; Arramach et al. 2017).

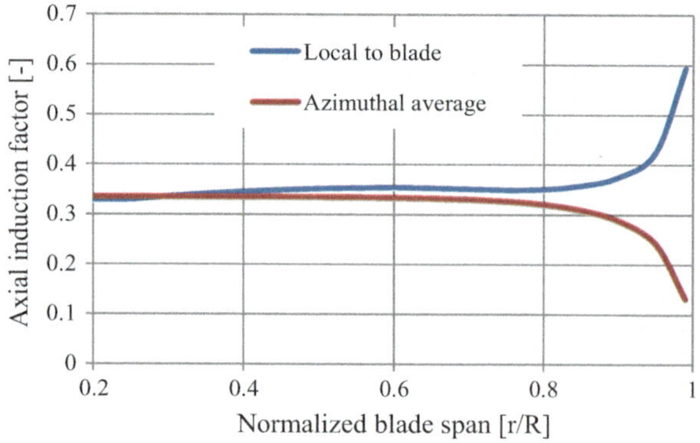

Fig. 14.10 Distribution of axial induction factor values local to blade and in rotor azimuth averaged conditions for a blade length of 17 m

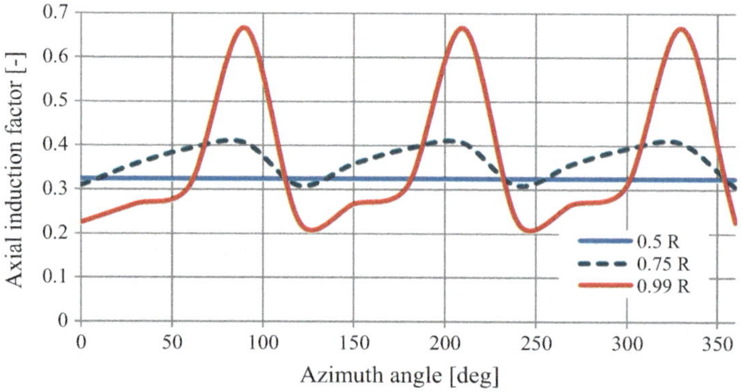

Fig. 14.11 Computed axial induction factors at 0.5R, 0.75R and 0.99R locations along blade span of length 17 m for various rotor blade azimuth angles

However, it can also be seen that for span positions of 0.75R, the maximum values of axial induction factor reached 0.43 only. Similarly, for span position of 0.5R the axial induction factor is constant at 0.5. For outboard span positions, the value of a fluctuates rapidly when the blades pass the tower structure and produce higher velocity deficit in wake. It can be said that for small size wind turbines with high twist distribution, the angle of attack and Reynolds number dependency affect the boundary layer flow behaviour. At high values of AOA, the span-wise flow become turbulent and degenerates towards the tip region of blade forming vortices. The Reynolds number for blade lengths of order of 17 m vary from low to moderate values, viz. 3×10^5 to 8×10^5. In contrast for large size turbines, with rotor diameters above 80 m equipped with high thickness to chord ratios, Reynolds numbers up to 3 million are possible.

Figure 14.12 shows that Glauert's momentum theory predicts higher power coefficient values up to 50% even for moderate to high yaw angles. For such angles, wake expansion and rotation effect behind the rotor increases faster (Bhadra et al. 2010;

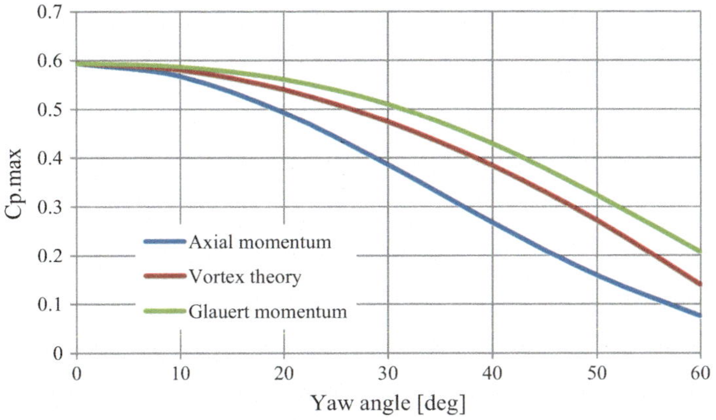

Fig. 14.12 Comparison of maximum mechanical power coefficient using axial momentum, vortex cylinder and Glauert theories for varying yaw angles

Hansen 2010). This approach is less conservative in nature as the wake expansion behind the rotor introduces significant increase in the axial thrust coefficient on the rotor. Also, it can be observed that axial momentum theory predicts 10 to 15% lower maximum power coefficient compared to vortex cylinder theory and Glauert theories. The C_{pmax} predicted according to axial momentum, Glauert's momentum and vortex cylinder model of yawed actuator disc is according to Eq. (3.90), Eq. (3.106) and Eq. (112) given in Bossanyi et al. (2005). As mentioned in Sect. 14.3, the magnitude of aerodynamic forces varies along the blade span which can be resolved in normal and tangential directions to rotor plane. The normal force per unit length at a given blade section increases towards the tip and contributes to thrust while tangential force per unit length contributes to torque. Typically, the tangential force per unit length is approximately 2–3 times the normal force per unit length (Bhargava et al. 2020; Krishna et al. 2018). In Fig. 14.13, a comparison of thrust coefficient evaluated according to Eq. (14.10) for varying yaw angles of turbine is depicted. As the yaw angle is increased, the axial force on blade decreases. For yawed rotor the wake becomes turbulent for higher yaw angles. Notice that thrust coefficient is reduced for axial induction factors beyond 0.5 where the momentum theory becomes invalid. One reason is that for large wake expansions, the axial-induced velocity reaches a limiting value relative to wind direction flow. For un-yawed rotor the change in wind velocity in wake of turbine is found to be half its initial value (Bossanyi et al. 2005). From Fig. 14.14a thrust coefficient obtained from BEM theory is valid until C_T reached a maximum value of 1 at axial induction factor of 0.5. Beyond this value the thrust coefficient predicted by BEM theory is invalid and starts reducing even when the axial induction factor a increases and reaches zero when the value of a is 1. For axial induction factor greater than 0.5, the wake expansion behind the rotor occurs rapidly and becomes highly turbulent for which predicted thrust coefficient is not accurate. This behavior is not observed in case of Glauert thrust coefficient correction, which increases up to value of 2. This shows that Glauerts proposition is

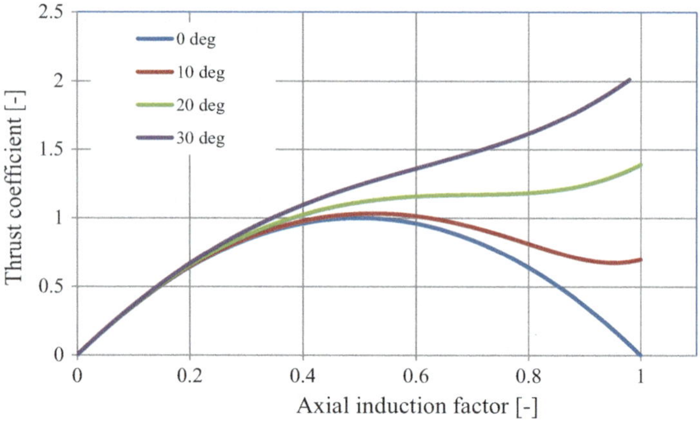

Fig. 14.13 Comparison of thrust coefficients for zero and non-zero yaw angles at different axial induction factors using BEM theory

less conservative in the prediction of thrust coefficient compared to vortex cylinder and axial momentum theories. To overcome these drawbacks, corrections for axial induction factor and thrust coefficient were proposed by Spera et al. (1979), Marshal and Buhl (2005), and Aagaard Madsen et al. (2012). The axial induction factor and thrust coefficient corrections are expressed using Eqs. (40) to (46) given in El-Okda (2015).

Figure 14.14a also compares the thrust coefficient values obtained from experiment (Lock et al. 1925) with the Buhls and Wilson–Walker correction factors. It must be noted that correction factors for thrust coefficient are evaluated when the Prandtl tip loss factor for rotor blades $F = 1$. Similarly, Fig. 14.14b, d represent the comparison of BEM predicted thrust coefficient with the correction factors proposed by Glauert, Buhl and Wilson–Walker and validated them using the experiment's obtained values. As the tip loss factor, F reduced the maximum thrust coefficient predicted by BEM theory which varied from 1 to 0.5. It can be seen that peak values for Glauert and Wilson–Walker correction factors also reduced for axial induction factors greater than 0.5. Also, the experiment data obtained by Lock et al. (1925) agreed well with thrust coefficient for axial induction factor more than 0.5 (Lock et al. 1925; Marshal and Buhl 2005). It can be noticed that Glauert's correction factor shows a discontinuity for values of $a > 0.4$ while Wilson correction factor is a tangent line to the BEMT curve and show discontinuity for $a > 0.2$. In contrast, Buhl correction factor is a continuous curve throughout the whole range of axial induction factor between 0 and 1. The discontinuity indicates aerodynamic instability when the BEMT produces invalid results.

Figure 14.15 shows that maximum power extraction ability increases towards the outboard of blade based on Eqs. (3.72) and (3.73) given in Bossanyi et al. (2005). Considering the Prandtl tip loss on the blade, the aerodynamic power extracted reduced to zero towards the tip of blade. In contrast, without tip loss correction,

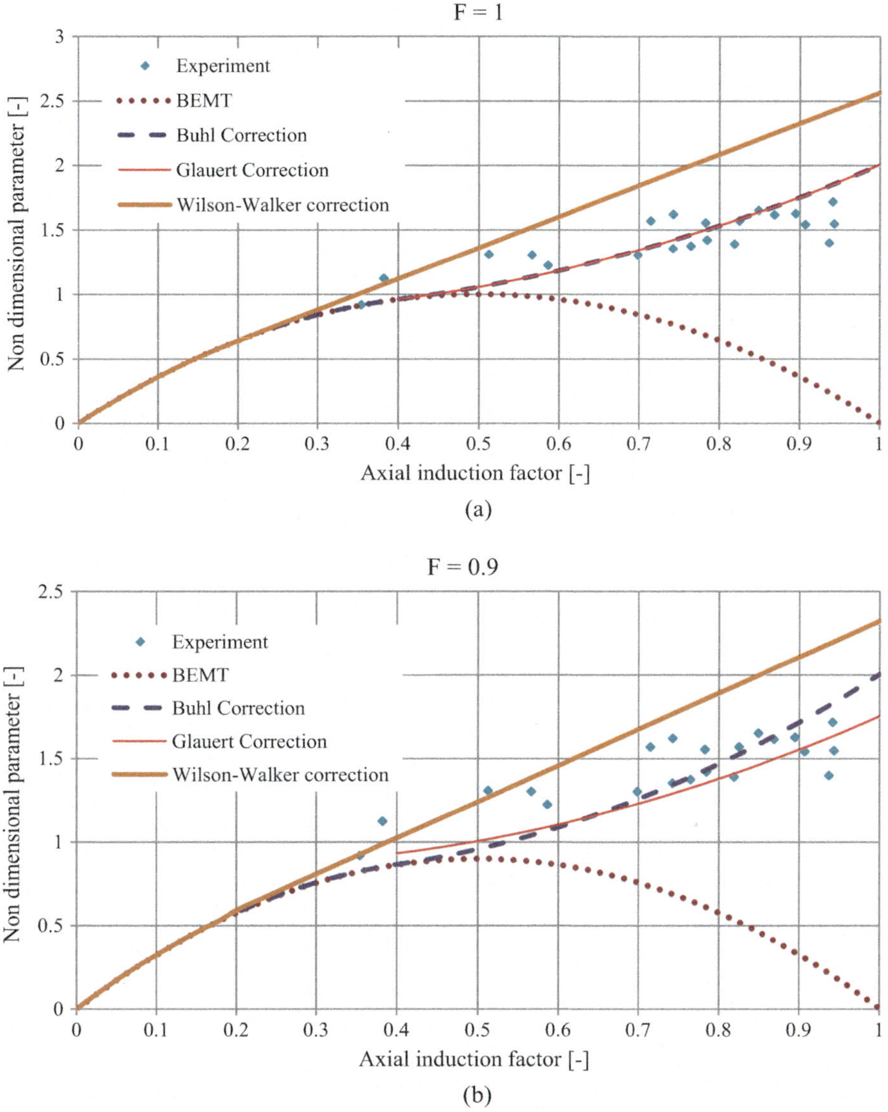

Fig. 14.14 Computed thrust coefficient using BEMT and Glauert, Buhl and Wilson–Walker correction factors and validated with experiment data for **a** $F = 1$, **b** for $F = 0.9$, **c** for $F = 0.7$, **d** for $F = 0.5$

the energetic flow continues to propagate if flow remains attached below the stall angle of attack; therefore, power extraction ability continues to increase.

Figure 14.16a, b show the aerodynamic forces per unit span length on the 47 and 17 m blades in normal (F_n) and tangential (F_t) directions of the rotor plane. The aerodynamic forces have been computed using BEMT including Prandtl tip

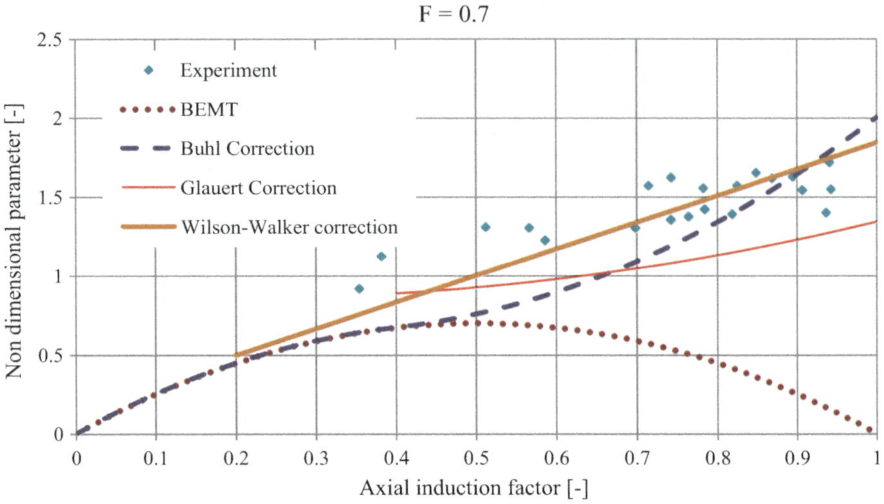

(c)

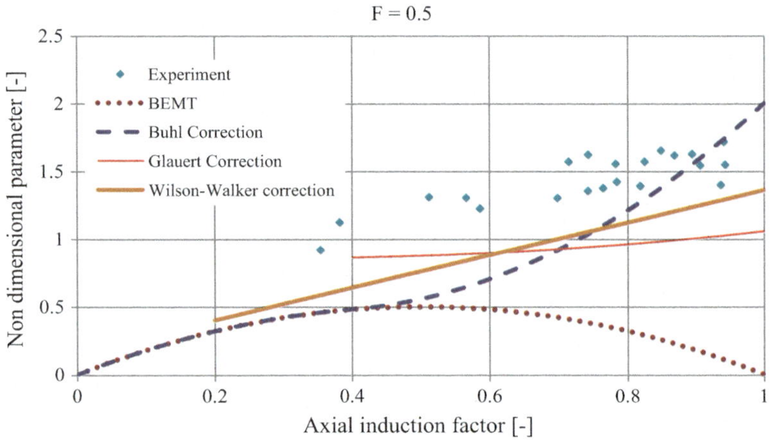

(d)

Fig. 14.14 (continued)

correction factor at wind speed of 12 m/s and 0° yaw. The forces can be seen to increase linearly towards the outboard region of blade but reduce near the tip. For 47 m blade, the maximum axial and tangential forces of 2889 N/m and 536 N/m are found while for 17 m blade it is 862 N/m and 171 N/m, respectively. This shows that BEMT predicts non-linear aerodynamic forces accurately for wide range of hub height wind speeds. Further, the magnitude of axial force on the rotor is nearly six times higher than the tangential force. In comparison to 47 m blade, force on the 17 m blade length varied by more than three times. This implies that loading on rotor is also increased significantly around rated wind speeds. Also, for pitch-controlled

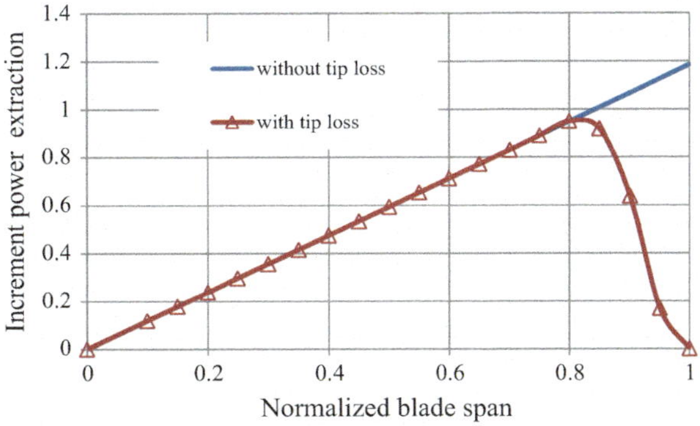

Fig. 14.15 Incremental power extracted with and without tip loss factors along blade span having a length of 17 m

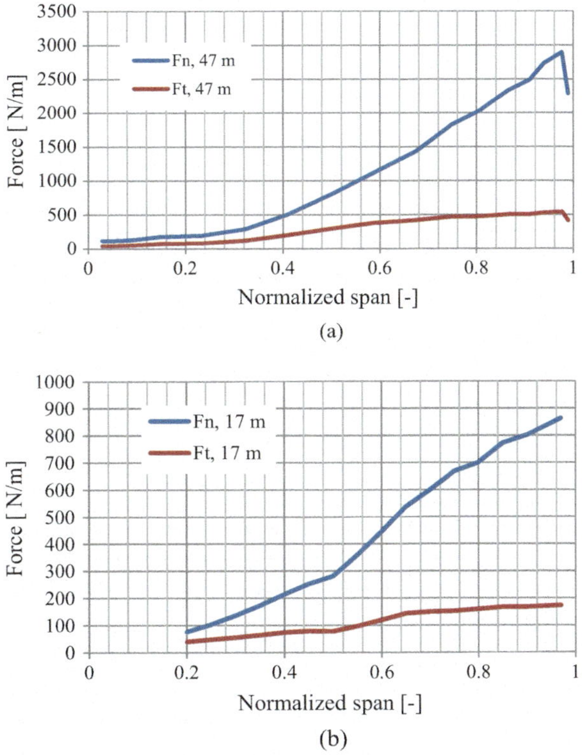

Fig. 14.16 BEM computed normal (F_n) and tangential (F_t) aerodynamic forces per unit span at wind speed of 12 m/s and 0° yaw using BEMT for **a** 47 m blade. **b** 17 m blade

turbines at near and above rated wind speed operation, loading on rotor in the axial direction increases heavily due to continuous blade pitching action to optimize the aerodynamic power of the turbine. Specifically, the flap-wise loads on blade vary dramatically compared to edge-wise load in response to the unsteady wind speeds.

14.7.2 Power Curve Assessment and Efficiency

The annual energy production from a turbine operating in wind farm can be predicted using long-term mean wind speed. At a given site, power curve modelling is an essential tool to estimate energy production from a wind turbine. Stochastic and deterministic methods can be used to develop power curve models to determine a relation between non-linear power outputs and wind speed at a site. Advanced power curve models utilize higher order polynomial models involving power coefficients (Teyabeen et al. 2017). The power output varies with wind bounded by regions known as cut-in wind speed at which the power is supplied to grid usually when the wind velocity is in the range of 2.5 and 5 m/s. The rated wind speed is expected when the machine delivers its nominal or nameplate capacity of electric power to grid, and finally cut-out wind speed when the generator is disconnected from the grid due to capacity constraints. Since instantaneous wind speed at a site is highly stochastic, it can be approximated using probabilistic function for accurate prediction of energy yield. Also, analytical methods such as measure correlate and predict (MCP) techniques are fairly accurate for estimating long-term wind speed data that trends at potential site. Based on IEC 61400-12-1 regulations for power curve measurements wind speed, direction and electrical power quantities are essential. Further measurements require air density, pressure, temperature and turbulence intensity corrections according to the bin averaging procedures. To evaluate turbulence intensity, standard deviation of longitudinal wind speed data is necessary. Using the short-term data for wind speed and direction, the turbulence intensity cannot be known accurately because binning method does not take account of data for the maximum hub height averaged wind speed over long term. Table 14.4 indicates the comparison of power curves for the machine at two different site air densities 1.115 and 1.225 kg/m^3 along with turbulence intensity. The TI values decay for above rated wind speeds but remain strongest between cut-in and rated wind speed operation. The standard deviation of wind speed did not vary appreciably between lower threshold and nominal wind velocities; however, for above nominal wind velocities, the deviation can be seen higher due to fewer 10 min averaged data present in each bin. It can be noted that a deviation between measured and manufacturer power data is often due to change in atmospheric variables, e.g. density of air. Therefore, for power curve performance air density correction is done based on IEC 61400-12 procedures mentioned in Bossanyi et al. (2005).

From Fig. 14.17a, the bin averaged power curves at two different air densities 1.225 and 1.115 kg/m^3 are illustrated along with measured power data between cut-in and rated wind speeds. It can be seen that data points that are scattered near cut-in wind speed are non-linear and show an offset with respect to both power curves. The power curve in red color is obtained at site air density of 1.115 kg/m^3. The manufacturer power curve is shown in green color line and taken from technical data specifications of model turbine. The comparison shows that bin averaged power curves fit the power data well in upper quartile range. It also indicates that distortion due to binning and occurrence of power errors are likely when power curves are derived from 10-min

Table 14.4 C_p obtained for 95 m rotor using wind speed frequency distribution

	Power coefficient (–)		Wind speed		Turbulence Intensity
WS, m/s	Standard	Measured	Frequency count	Stdev	%
1	0.000	−0.298	7	0.151	15.12
2	0.000	−0.019	47	0.261	13.04
3	0.000	0.099	182	0.269	8.95
4	0.174	0.164	355	0.287	7.18
5	0.351	0.332	363	0.293	5.86
6	0.409	0.386	359	0.296	4.93
7	0.424	0.405	476	0.289	4.12
8	0.430	0.411	579	0.286	3.57
9	0.423	0.404	687	0.289	3.21
10	0.402	0.389	496	0.294	2.94
11	0.361	0.353	454	0.301	2.74
12	0.311	0.301	261	0.286	2.38
13	0.251	0.245	108	0.263	2.03
14	0.201	0.196	21	0.311	2.22
15	0.163	0.160	10	0.300	2.00
16	0.135	0.132	3	0.321	2.01

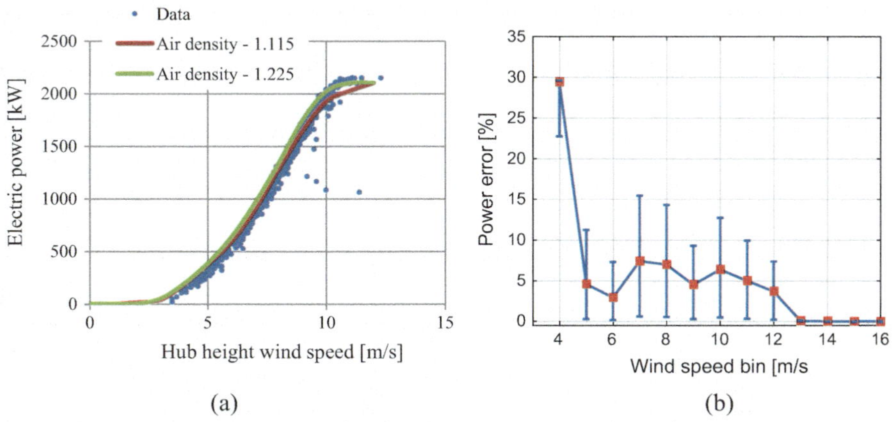

Fig. 14.17 **a** Power curve of manufacturer at air density of 1.225 kg/m^3 and observed data at 1.115 kg/m^3. **b** Graph of electric power error in % at air densities of 1.225 and 1.115 kg/m^3

averaged data. As the rated wind speed is approached, the linearity trend observed for data is lower compared to bin averaged power curves. It can also be observed that few outliers for power data are found at near rated wind speeds ~10 m/s. This can be due to distortion of bin data or from power curtailment operation imposed by controller to reduce the mechanical loads on the blade. Figure 14.17b illustrates the electric power error between cut-in and rated wind speeds. Maximum power error of 29% is seen near cut-in wind speed with a low uncertainty. However, power error reduced to 5% at 5 m/s wind speed bin and increased to 7% at the rated wind speed. The blue color bar shows the uncertainty in the power error which is highest for 7 and 8 m/s wind speed bins. The present bin analysis is based on the 10 min bin averaged data which does not capture the dynamics of power curve accurately. So according to Milan et al. (2014) to overcome this limitation, combined stochastic and mathematical modelling of power curve at sampling frequency of 1 Hz would be useful for accurate power predictions (Milan et al. 2014). The annual energy production from wind turbine can be monitored with the help of central data control and acquisition software to log the operational data continuously. Following are main turbine operating conditions.

- Machine is started using sequential controller steps.
- Machine is connected to grid to be able to supply power.
- Machine is stopped whenever external conditions and uncertain events occur. For example 50-year gust, loss of grid.
- Grid curtailment mode according to network grid conditions.
- Rotor freewheeling when there is no wind enough for power production to grid.

Figure 14.18 shows the blade pitch angle between minimum and nominal wind velocities. It can be seen that rotor speed increase steadily soon after cut-in wind speed and remains constant up to rated wind speed. During this period, the blade pitch angle varied between 0° and 5° where aerodynamic power extraction is maximized by controller algorithm (Chaudhry et al. 2014). So, for below rated wind speeds one

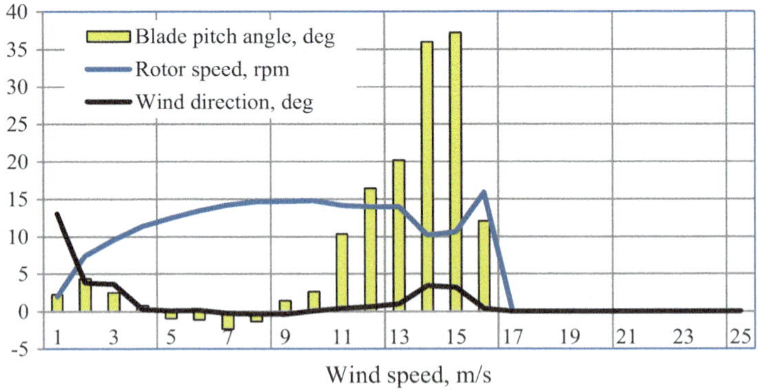

Fig. 14.18 Bin averaged values for blade pitch angle, rotor speed and wind direction at different wind speeds for pitch-controlled turbine

must note that optimum turbine performance is regulated continuously to track the wind. Also, for below rated wind speed the tip speed is found as 79.08 m/s when TSR varied between 4 and 8. In case of turbines with fixed speed, this value reaches set point for specific wind velocity; however, for machines which can produce power at different rotor speeds, it is varied in such a way that $C_{p\max}$ is obtained. A feedback control loop ensures to optimize the C_p even for low wind regimes. Also, for a turbine in which power is regulated by pitch movement of the blade with its leading edge into wind, a phenomenon known as *pitch to feather* occurs for above rated wind speed operation. Pitch to feather action reduces the unsteady aerodynamic lift on blades which lowers the fatigue loads on rotor. In contrast, for stall-controlled turbines operating at above rated wind speeds, the blade experiences dynamic stall on its suction side for high angles of attack. This phenomenon also occurs in a non-linear manner which tends to increase fatigue loads on turbine components. So, the controller regulates power output from turbine by turning the trailing edge into wind and this process otherwise known as *pitch to stall.* Since bin averaging is applied to the 10-min measured data set for each wind speed between cut-in and cut-out, the power errors for individual bin are corrected by applying standard deviation to the average values. A maximum power value of 2141 kW can be found for above rated wind speed bins. The standard deviation for the bin averaged values of wind speed and electric power increased due to the low data samples used for averaging.

Figure 14.19 shows the 10 min bin values for thrust coefficient for below rated wind speed operation or for high tip speed ratio. It varied between 0.9 and 0.98 with fluctuations seen close to cut-in wind speed. This value is turbine specific which cannot be compared with others operating in a wind farm array. Similarly, for power coefficient it varies between 0.32 and 0.44, respectively.

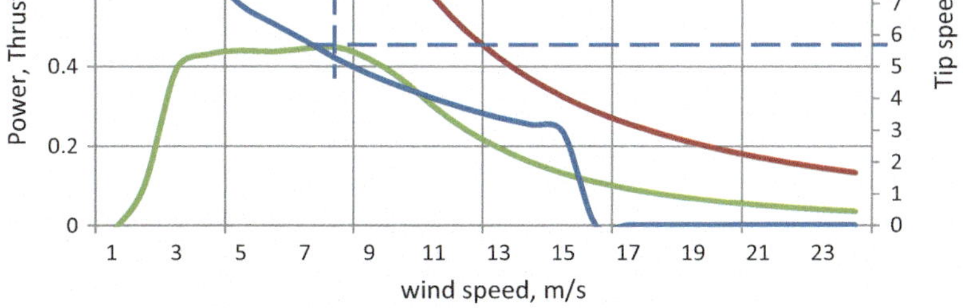

Fig. 14.19 Thrust and power coefficients for 95 m turbine rotor and tip speed ratio between cut-in and cut-out wind speeds

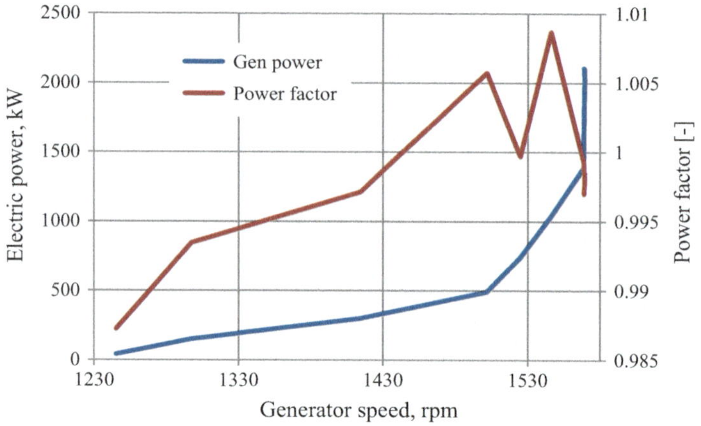

Fig. 14.20 Comparison of generated power and power factor versus speed characteristic for variable pitch control turbine

To know wind speed and directions, an anemometer and wind vane installed on a met mast at height equal to hub height located few rotor diameters away from turbine are required. It can be noted that readings from sensors located on the nacelle may not be accurate. Therefore, corrections have been applied to power curve as described in Sect. 14.5. The power produced by any turbine can be controlled using the measured rotor speed and blade pitch angle sensors located in hub controller. This signal is sent to main controller via a feedback loop to compare and track the set-point values of rotor speed and blade pitch angle. To measure the generator, speed proximity inductive sensor is used. The sensor data is continuously monitored and can be used to determine the operating state of turbine. It can be observed from Fig. 14.20 that rated power for generator is ~1569 rpm and the power factor value is ~1. For below rated wind speed operation, the generator speeds vary between 1250 and 1500 rpm, and ramp-up is observed. When the rotor speed is operating near rated values, the generator speed remains nearly constant after which the generator torque output is kept constant by controller once it reaches the rated value. When approaching synchronous speeds of the generator a non-linear behaviour in power production can be seen. It can also be seen that power factor also known as cosϕ fluctuates near rated generator speed, exact reason of which could be attributed to grid stability and turbine control procedures near rated wind speed (Bhadra et al. 2010).

Figure 14.21 shows a comparison of power coefficient, obtained using bin averaging method at various λ values and for blade pitch of 0°. The graph showing the power coefficient value was obtained using the empirical relation using Eqs. (14.30) and (14.31) with following values for constants, $c_1 = 0.4176$, $c_2 = 135$, $c_3 = 2.1$, $c_4 = 9.3$, $c_5 = 15.7$ and $c_6 = 0.0088$, respectively. For TSR less than 7, C_p predicted using bin method and model agreed within 1% demonstrating that for low blade pitch angles between 0° and 5° constants in Eqs. (14.30) and (14.31) produce expected

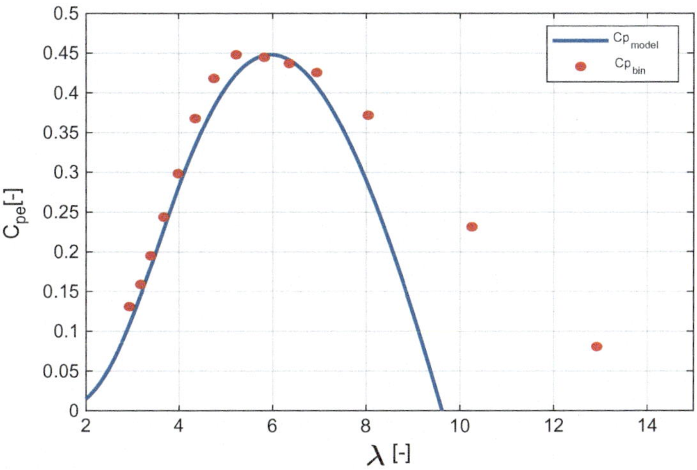

Fig. 14.21 Computed power coefficient for various tip speed ratios using bin averaged data and model output at 0° blade pitch angle

power output with a high fidelity. On the other hand, for TSR values greater than 7, divergence for C_p values can be observed clearly. Apparently, a difference of ~25% can be seen at high tip speed ratios. One reason for this difference is attached to quantity of data used in the binning procedures. It can be noted that design tip speed ratio predicted by model agreed well within 1% of value obtained from measured data.

The time-series history for low-speed shaft torque, power and wind velocity are shown in Fig. 14.22. It can be observed from data that turbine experienced high wind conditions for about 1.5 days. However, power and low speed shaft moment

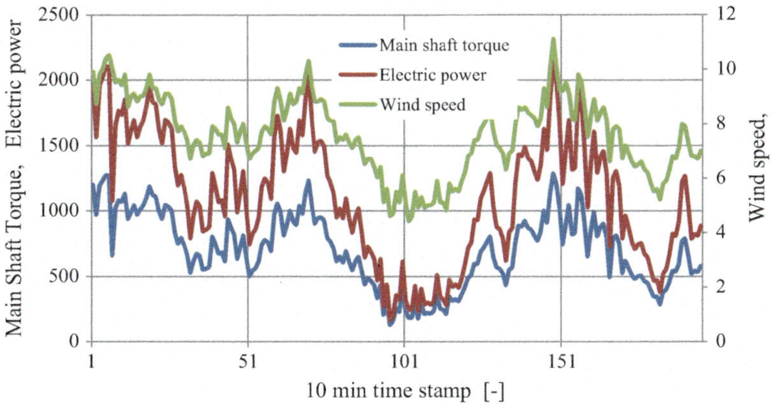

Fig. 14.22 Time-series history of low-speed shaft torque, kN-m, electric power, kW and wind velocity, m/s for period of 34 h

are 201 kW and 124.3 kN-m, between time stamp values of 90 and 110, as result of direction change of wind. At any given location, direction of wind velocity can be obtained using a wind vane. The nacelle controller computes the directional turns of nacelle and prevents the turbine from failure. In some turbines, the transformer room is integrated into nacelle to reduce the cabling costs and to prevent them from breaking during yaw. For wind speeds close to cut-in values, the low-speed shaft torque and power fluctuate. For the sampled data, it was found to be 160 kW and 124.3 kN-m. However, near rated wind speeds between 11 m/s and 12 m/s the generator produces ~2100 kW and ~1285 kN-m. For measured data, total axial rotor force of 323.6 kN, turning moment of 1289 kN-m and power of 2140.1 kW for the turbine are apparent.

14.7.3 Energy Yield Prediction

In this section, the energy yield prediction from manufacturer and bin averaged power curves are demonstrated (Directory of Indian Wind Power 2012). From the manufacturer power curve table, the minimum wind velocity for the turbine is 3.5 m/s while nominal speed is 11 m/s. These values are recorded at air density of 1.225 kg/m^3 and at standard temperature and pressure conditions. But the minimum and maximum wind velocities vary according to size of turbine. As the rotor diameter becomes larger, the power extraction ability of rotor increases. To predict energy yield from a machine, the long-term wind speed modelling is essential. For meteorology and wind turbine applications a Weibull probability density function is most suitable for analysis of long-term wind speed frequency distribution at a place (Sohoni et al. 2016). Fundamentally it consists of two random variables, viz. shape and scale factors for predicting the wind speed frequency. The shape factor considers geographic features of terrain, while the scale factor indicates the long-term wind speed for a location. The likelihood of velocity occurrence and corresponding energy production data for a given period shows the likelihood of energy generation from a machine. Turbine designers and wind resource analysts use such information for sizing and design of turbine components as well as in site prospecting. Also, the net power from wind farm is calculated using measured data for any period based on active and reactive values of power. It has been found to be 0.77 and 0.067 GWh and resultant energy for duration of ~742 h is found to be 0.701 GWh. Furthermore, the difference in energy yield can be accounted to several factors as mentioned in Sect. 14.5. According to Ehrmann et al. (2017), blade surface roughness affects the wind turbine performance by reducing the lift to drag ratio of rotor blades. For surface roughness height ranging from 140 μm to 1 mm on blade, an estimated 2–3% loss in annual energy production is possible due to 40% reduction of lift to drag ratio. The Weibull probability density function for various shape factors at a scale factor of 8.5 m/s is shown in Fig. 14.22.

One can observe that as annual average wind velocity increases, the likelihood of energy capture increases at a location. Further, from the occurrence of bin averaged air velocity are higher for annual mean wind speeds of 8.5 m/s and 7.5 m/s, respectively.

Figures 14.23 and 14.24 show the influence of shape factor and scale factor on likelihood of occurrence of a given wind speed at a given location. Figure 14.23 shows that the probability density function is skewed towards left and has amplitude that range between 0.08 to 0.1 for wind speeds lower than 7 m/s. The monthly shape factor range between 1.25 and 2.0. However, as the shape factor is increased beyond 2 the wind speed likelihood increased in amplitude for wind speeds between 7 and 10 m/s. On the other hand, from Fig. 14.24, as scale factor is reduced from 8.5 to 7.5 m/s the amplitude of density function varied from 0.114 to 0.1 without appreciable change. One can notice the tail of distribution in both Figs. 14.23 and 14.24 is similar and approach same limiting values with amplitudes that range between 0 and 0.02 for higher values of wind speeds. However, in Fig. 14.24 the tails appear to be fatter as the shape factor is reduced. This indicates that for a given scale factor, the likelihood of energy extraction at such wind speed increased by ~1%. Figure 14.25

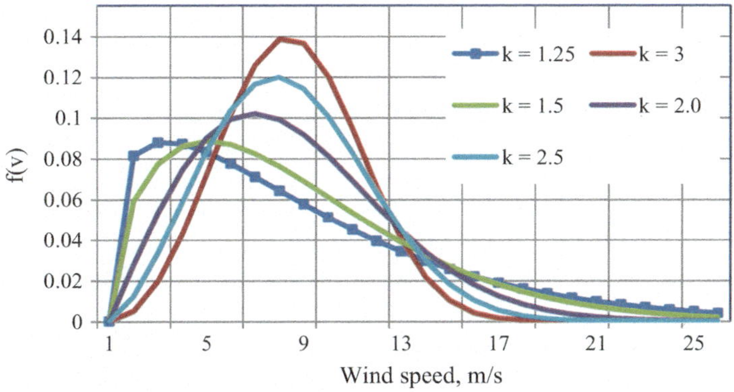

Fig. 14.23 Weibull wind speed probability density function at different shape factors for given scale factor of 8.5 m/s

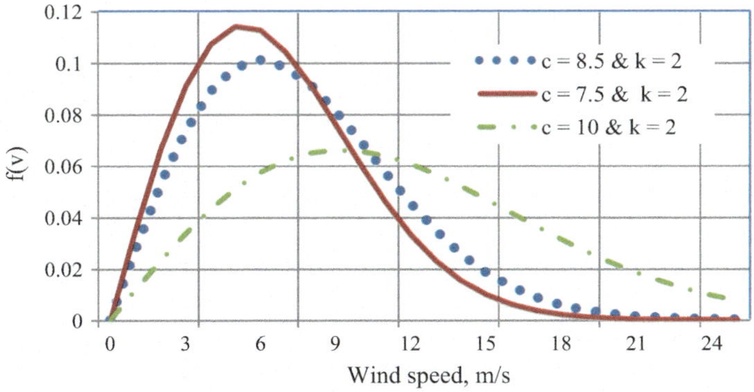

Fig. 14.24 Weibull wind speed probability density function at different scale factors at constant shape factor of 2

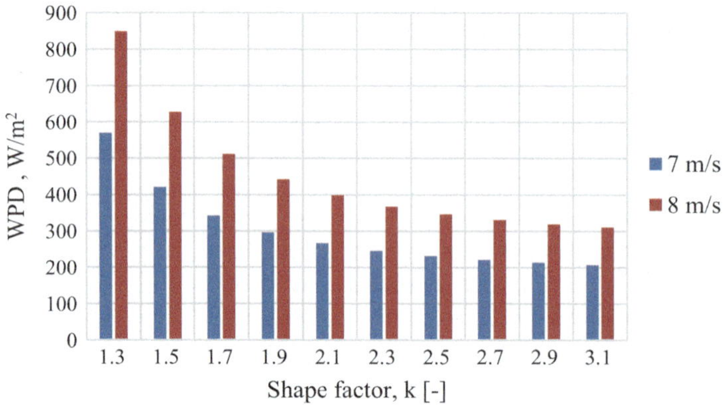

Fig. 14.25 Graph of annual wind power density for different annual scale and shape factors obtained using two-parameter Weibull distribution

shows the annual wind power density (WPD) computed for different shape and scale factors using Eq. (14.39) presented in Sect. 14.6. Wind power density is a quantity which shows the kinetic energy available in the wind at any specified location and is usually calculated at specific height above ground. On the other hand, specific power indicates amount of aerodynamic power that can be extracted by turbine over a finite swept area of rotor. A geographic area can be categorized according to wind power density and are known as wind zones. According to Directory of Indian Wind Power (2012), there are seven wind zones with distinct mean wind speed values which range from 6 to 10 m/s. So, wind power density at a site could be estimated from Weibull distribution parameters, viz. scale and shape factor as discussed earlier in this section. Practical values for wind power density are measured at specific height above the ground level. Typically met mast installed at 100 m or 50 m above ground level are used as reference for measuring wind power density values. It can be noted that for annual scale factor of 7 m/s the WPD varied from 568 to 201 W/m^2 as the annual shape factor is increased from 1.3 to 3.1. This shows that high values of shape factor showed a negative effect on wind power density which implies lower energy available in wind for power production. However, when the annual scale factor is increased from 7 m/s to 8 m/s the annual wind power density is increased by ~ 49%. It must be noted that computed values are based on the theoretical estimates of two parameter Weibull distribution. Hence, Weibull distribution is a useful tool in site prospecting and wind resource assessment.

Table 14.5 shows the comparison of monthly Weibull energy GWh, estimated using manufacturer and measured power curves for different scale and shape factors. The analysis used the excel WEIBULL function which accepts two input parameters to estimate the approximate number of hours for bin averaged wind speeds. Theoretical hours data and power data for the corresponding bin are obtained from standard power curve to forecast energy output. Weibull energy variation was found to be ~33% when shape factor changed from 1.3 to 2.1 while a ~13% when scale

Table 14.5 Weibull energy yield comparison using manufacturer and measured power curves of 95 m rotor at air densities 1.115 and 1.225 kg/m^3

Power curve type	Weibull hours	Scale factor	Shape factor	Energy (GWh)
				Weibull
MSRD	712.1	9.5	2.1	0.652
STRD	712.1	9.5	2.1	0.725
MSRD	641.3	9.2	1.3	0.466
STRD	642.1	9.2	1.3	0.510
MSRD	624.4	8.4	2.1	0.567
STRD	624.3	8.4	2.1	0.634
MSRD	416.1	7.77	1.1	0.379
STRD	416.2	7.77	1.1	0.416

factor changed from 7.7 to 9.5 m/s. The predicted values for energy yield are found when the turbine is not under any major shutdown events. In practice O&M includes downtime factors which account up to 2–4% reduction in the annual yield production.

From Table 14.6 the annual energy yield for turbine showed an increase of 16% when the annual scale factor is varied from 7.5 to 8.5 m/s. For rotor diameter of 36 m, the expected annual production is 0.835 GWh when the annual scale factor is 7.5 m/s and 0.977 GWh when the scale factor is increased to 8.5 m/s. The estimated Weibull energy is obtained using standard power curves for the turbines measured at 1.225 kg/m^3. When the rotor diameter is increased by 3 times, the expected annual energy output from the turbines increase by ~6 times. A large rotor diameter implies higher amount of aerodynamic power extracted per swept area of turbine. However, one must note that power extraction ability depends on the operating state and power curve characteristics of turbine as discussed in Sect. 14.7.2. Hence, performance and energy yield prediction of wind turbines can be done effectively through simulation of high-fidelity theoretical aerodynamic models, as well as application of advanced empirical or analytical methods.

Table 14.6 Weibull annual energy yield computed for turbines with 95 and 36 m rotor diameter at scale factors of 8.5, 7.5 m/s and shape factor of 1.4

Rotor diameter (m)	Power curve type	Annual Weibull hours	Shape factor	Scale factor	Energy (GWh)
					Weibull
95	Standard	8545.2	1.4	7.5	6.31
36	Standard	8547.1	1.4	7.5	0.835
95	Standard	8525.4	1.4	8.5	7.34
36	Standard	8521.3	1.4	8.5	0.977

14.8 Conclusions

- In this paper, we studied blade element momentum theory and its improvements to find operating characteristics of HAWT rotors. Various design factors such as tip speed ratio, rotor diameter, blade count and pitch angle, L/D ratios were investigated using sampled operational data. Power and thrust coefficients are key parameters to measure turbine efficiency. Bin data analysis showed that peak thrust and power coefficient values were found to be 0.945 and 0.44 for wind turbine with 95 m rotor diameter. BEM analysis revealed that turbines with small rotor diameters exhibited aerodynamic forces which were three to four times lower compared to megawatt scale turbines.
- Induction variable along rotor axis and thrust corrections proposed by Glauert, Buhl and Wilson–Walker showed good agreements with those predicted from BEMT for values less than 0.5; however, thrust coefficient predicted using conventional BEMT agreed well with experiment data when the induction variable is greater than 0.5. For all correction models, the maximum values for thrust coefficients reduced as the combined tip and hub loss factors are reduced from 1 to 0.5.
- With increase in blade count, the wake losses from tip of rotor were found to be lower. The maximum aerodynamic power from a blade becomes higher towards the outboard of blade. When the lift to drag ratio is changed from 40 to 120, it resulted in 17% improvement of aerodynamic efficiency of turbine. Hence, it can be concluded that increasing the rotor solidity and lift to drag ratios will increase the aerodynamic power extraction ability.
- A method for performing data binning to analyze SCADA operational data using a pseudocode has been presented. The code was designed to extract power, torque and thrust coefficients for given wind speed data. Results of binning procedure showed that maximum power error of 28% is found at cut-in wind speed due to power outliers present in bin data. An empirical relation for predicting C_p using TSR and blade pitch angle was proposed to verify the accuracy of the predictions. The model predictions were validated using the bin averaged power coefficient data. The results for power coefficient agreed within 1% of the measured data for tip speed ratios less than 7 but found to disagree between 7 and 10.
- Long-term energy yield using two-parameter Weibull distribution requires different shape and scale factors, and bin-wise frequency distribution of wind velocity. In contrast to fixed speed stall-controlled machines, the maximum power production is not constant for variable speed machines between minimum and nominal wind velocities for which the tip speed ratios varied between 5 and 7.

References

Aagaard Madsen H, Riziotis V, Zahle F, Hansen MOL, Snel H, Grasso F, Larsen TJ, Politis E, Rasmussen F (2012) Blade element momentum modelling of inflow with shear in comparison with advanced model results. Wind Energy 15(1):63–81
Andrew TL, Flay RGJ (1999) Compliant blades for wind turbines. IPENZ Trans 26:7–12. https://doi.org/10.1260/0309524001495369
Arramach J, Boutammachte N, Bouatem A, Al Mers A (2017) Prediction of the wind turbine performance by using a modified BEM theory with an advanced brake state model. Energy Procedia 118:149–157. https://doi.org/10.1016/j.egypro.2017.07.033
Astolfi D, Castellani F, Bechetti M, Lombardi A, Terzi L (2020) Wind turbine systematic yaw error: operation data, analysis techniques for detecting it and assessing its performance impact. Energies 13:2351. https://doi.org/10.3390/en13092351
Bhadra SN, Kastha D, Banerjee S (2010) Wind electrical systems. Oxford Education, New Delhi
Bhargava V, Kasuba S, Maddula SP, Donepudi J, Akhtar MDK, Padhy C, Chinta HP, Sekhar CS, Dwivedi YD (2020) A case study of wind turbine loads and performance using steady state analysis of BEM. Int J Sustain Energy 40:22–40. https://doi.org/10.1080/14786451.2020.1787411
Bontempo R, Manna M (2019) Verification of the axial momentum theory for propellers with a uniform load distribution. Int J Turbomachinery Propulsion Power 4:8. https://doi.org/10.3390/ijtpp4020008
Bossanyi E, Sharpe D, Jenkins N, Burton T (2005) Wind energy handbook. Wiley, London
Chaudhry U, Mondal P, Tripathy P, Nayak SK, Saha UK (2014) Modeling and optimal design of small HAWT blades for analyzing the starting torque behavior. In: 18th National power systems conference, Guwahati, India, pp 1–6. https://doi.org/10.1109/NPSC.2014.7103886
Corrigan JJ, Schillings JJ (1994) Empirical model for stall delay due to rotation. In: American Helicopter Society aeromechanics specialists conference, San Francisco CA
De Vries O (1979) Fluid dynamic aspects of wind energy conversion. AGARD report, AG-243, Chapter 4
Directory of Indian Wind Power. Consolidated Energy Consultants, Bhopal. India (2012)
Djojodihardjo H, Abdul Hamid MF, Jaafar AA, Basri S, Romli FI, Mustapha F, Mohd Rafie AS, Abdul Majid DLA (2013) Computational study on the aerodynamic performance of wind turbine airfoil fitted with Coanda Jet. J Renew Energy 2013. https://doi.org/10.1155/2013/839319
Ehrmann RS, Wilcox B, White EB, Maniaci DC (2017) Effect of surface roughness on wind turbine performance. Sandia report, SAND2017-10669. https://doi.org/10.2172/1596202
El-Okda MY (2015) Design methods of horizontal axis wind turbine rotor blades. Int J Ind Electron Drives 2:135–150
Emblemsvag J (2020) On the levelised cost of energy of wind farms. Int J Sustain Energy 39:700–718. https://doi.org/10.1080/14786451.2020.1753742
Froude (1885). https://en.wikipedia.org/wiki/Momentum_theory#:~:text=In%20fluid%20dynamics%2C%20momentum%20theory,Robert%20Edmund%20Froude%20(1889)
Gasch R, Twele J (2012) Wind power plants, fundamentals, design, construction and operation, 2nd edn. Berlin
Genc SM, Ozkan G, Acikel HH, Kiris MS, Yildiz R (2016) Effect of tip vortices on flow over NACA4412 aerofoil with different aspect ratios. https://doi.org/10.1051/epjconf/201611402027
Germanisher L (2007) Guideline for certification of wind turbines, Edition 2003 with supplement 2004, Reprints Hamburg
Glauert H (1948) The elements of airfoil and airscrew theory, 2nd edn. Cambridge University Press
Goudarzi N, Zhu WD (2013) A review on the development of wind turbine generators across the world. Int J Dyn Control 1:192–202. https://doi.org/10.1007/s40435-013-0016-y
Hansen MOL (2010) Aerodynamics of wind turbines. James & James, London, UK
Heier S (1998) Grid integration of wind energy conversion systems. Wiley, Kassel, Germany

Javaherchi T, Antheaume S, Aliseda A (2014) Hierarchical methodology for the numerical simulation of the flow field around and in the wake of horizontal axis wind turbines: rotating reference frame, blade element method and actuator disk model. J Wind Eng 38:181–201
Jiang H, Cheng Z, Zhao Y (2012) Torque limit of horizontal axis wind turbine. In: Proceedings of international conference mechanical engineering & materials sciences. MEMS. https://doi.org/10.2991/mems.2012.114
Jordan D et al (2019) Wind turbine technical report
Krishna JM, Bhargava V, Donepudi J (2018) BEM prediction of wind turbine operation and performance. Int J Renew Energy Res 8:1962–1973
Ladson CL, Brooks CW (1996) Computer program to obtain ordinates for NACA airfoil. Langley. Research Center, document ID: 19970008124, Hampton, Virginia
Lock CNH, Bateman H, Townsend HCH (1925) An extension of the vortex theory of airscrews with applications to airscrews of small pitch including experimental results No. 1014. Aeronautical research committee reports and memoranda, London, Her Majesty Stationery Office
Mansberger L (2016) Thermodynamic wind turbine model addendum. Mansberger Aircraft Inc., Fortworth, Texas, USA
Manwell JF, McGowan JG, Rogers AL (2009) Wind energy explained, theory, design and application, 2nd edn, Wiley, London, UK
Marshal L, Buhl Jr (2005) A new empirical relationship between thrust coefficient and induction factor for the turbulent wind mill state, NREL Technical report, NREL/TP-500-36834
Mikkelsen R, Sorensen JN, Ivanell S, Henningson D (2009) Analysis of numerically generated wake structures. Wind Energy 12:63–80. https://doi.org/10.1002/we.285
Milan P, Wachter M, Peinke J (2014) Stochastic modelling and performance monitoring of wind farm power production. J Renew Sustain Energy. https://doi.org/10.1063/1.4880235
Numerical Manufacturing and Design Tool, NuMAD. https://energy.sandia.gov/programs/renewable-energy/wind-power/rotor-innovation/numerical-manufacturing-and-design-tool-numad/
Ouakki Y, Arbaoui A (2020) Accurate loads and velocities on low solidity wind turbines using an improved blade element momentum model. Wind Energy Sci. https://doi.org/10.5194/wes-2020-43
Power performance testing, 2nd edn, Expert group study on recommended practices for wind turbine testing and evaluation (1990)
Prandtl (1935). https://en.wikipedia.org/wiki/Blade_element_theory
Ragheb M, Ragheb AM (2011) Wind turbine theory—the Betz equation and optimal rotor tip speed ratio. In: Fundamental and advanced topics in wind power. IntechOpen
Saxena BK, Rao KVS (2016) Estimation of wind power density at a wind farm site located in western Rajasthan region of India. In: International conference on emerging trends in engineering, science and technology, ICETEST
Sebastian PB, Francesco P, Joseph S, David M, Alessandro B, Oliver CP (2019) Is the blade element momentum theory overestimating wind turbine loads?—a comparison with a lifting line free vortex wake method. Wind Energy Sci 5:721–743. https://doi.org/10.5194/wes-5-721-2020
Shen WJ, Mikkelsen R, Sorensen JN, Bak C (2005) Tip loss corrections for wind turbine computations. J Wind Energy 8:457–475. https://doi.org/10.1002/we.153
Sherry M, Sheridan J, Jacono DL (2013) Characterization of a horizontal axis wind turbine's tip and root vortices. Exp Fluids 54:1417. https://doi.org/10.1007/s00348-012-1417-y
Snel H, Schepers JG (1995) Joint investigation of dynamic inflow effects and implementation of an engineering method. Energy Research Centre Netherlands, ECN–C-94-107
Sohoni V, Gupta SC, Nema RK (2016) A critical review on wind turbine power curve modelling techniques and their applications in wind-based energy systems. J Energy 2016:1–18. https://doi.org/10.1155/2016/8519785
Spera DA, Viterna LA, Richards TR, Neustadter HE (1979) Preliminary analysis of performance and loads data from the 2-Megawatt MOD-1 wind turbine generator. In: 4th Biennial conference and workshop on wind energy conversion systems. Department of energy, Washington D.C, U.S

Strangfeld C, Rumsey CL, Muller Vahl H, Greenblatt D, Nayeri CN, Paschereit CO (2015) Unsteady thick airfoil aerodynamics: experiments, computation and theory. In: 45th AIAA fluid dynamics conference, Dallas, Texas, USA. https://doi.org/10.2514/6.2015-3071
Tangler JL, Selig MS (1997) An evaluation of an empirical model for stall delay due to rotation for HAWTs. In: Proceedings windpower, Austin TX, pp 87–96
Teyabeen AA, Akkari FR, Jwaid AE (2017) Power curve modelling for wind turbines. In: 19th international conference on modelling and simulation
Vermeer LJ, Sorensen JN, Crespo A (2003) Wind turbine wake aerodynamics. Prog Aerosp Sci 39:467–510. https://doi.org/10.1016/S0376-0421(03)00078-2
Wind energy, The facts, Part-I; Technology. Garrad Hassan and Partners Ltd, Bristol, UK (2014)
Xu G, Sankar LN (2002) Application of a viscous flow methodology to the NREL phase-IV rotor. In: ASME wind energy symposium, Reno, AIAA-2002-0030, pp 83–93

Chapter 15
Energy Sources for Electric Vehicles

Irfan Ahmed, Aashish Kumar Bohre, Tushar Kanti Bera, and Aniruddha Bhattacharya

Abstract The tremendous increase in pollution levels caused by automobiles energized through fossil fuels as well as the eventual depletion of these fuels has led to an increase in the interest for electric and hybrid electric vehicles. Electric vehicles (EVs) provide cleaner means of transportation with the pollution being limited to the locations of electric power generating plants. They are the vehicles of the future, without any doubt. There has been a tremendous amount of research in EV technology in recent times, and a lot of research has reached mature levels. The only impediment to the complete commercial use of EVs is the energy sources required to power them. In this chapter, we take a look at the conventional sources of energy for EVs, their current status and developments, as well as technologies to look out for in the future. The focus is on the basic working principles of these sources without delving too much into their chemical reactions.

Keywords Battery energy sources · Fuel cells · Ultracapacitors · Ultrahigh-speed flywheels

Nomenclature

EV	Electric Vehicle
NiCd	Nickel–cadmium

I. Ahmed (✉) · A. K. Bohre · T. K. Bera · A. Bhattacharya
National Institute of Technology Durgapur, Durgapur, India
e-mail: ahmed.irfan@ee.nitdgp.ac.in

A. K. Bohre
e-mail: aashishkumar.bohre@ee.nitdgp.ac.in

T. K. Bera
e-mail: tusharkanti.bera@ee.nitdgp.ac.in

A. Bhattacharya
e-mail: aniruddha.bhattacharya@ee.nitdgp.ac.in

A. K. Bohre et al. (eds.), *Planning of Hybrid Renewable Energy Systems, Electric Vehicles and Microgrid*, Energy Systems in Electrical Engineering,
https://doi.org/10.1007/978-981-19-0979-5_15

Li-ion	Lithium-ion
NaS	Sodium–sulfur
SoC	State of Charge
DoD	Depth of Discharge
PEM	Proton Exchange Membrane
PAFC	Phosphoric Acid Fuel Cell
SOFC	Solid Oxide Fuel Cell
HEV	Hybrid Electric Vehicle
NiMH	Nickel-metal-hydride
Li-poly	Lithium-polymer
Zn-Air	Zinc-air
SoD	State of Discharge
AFC	Alkaline Fuel Cell
DMFC	Direct Methanol Fuel Cell
MCFC	Molten Carbonate Fuel Cell

15.1 Introduction

One of the prominent features of an electric vehicle (EV) is that it has a portable energy source, which may be chemical or electromechanical in nature. Portable source implies that the energy source must be able to move with the vehicle. There are some exceptions to this rule, such as in the case of electric trains and electric trams. However, in general, the energy source must be portable and implanted inside the vehicle. Examples of chemical sources include batteries and fuel cells; while those of electromechanical sources include flywheels or the motor itself acting as a generator during regenerative braking.

Out of the number of sources that are available, only the battery energy source technology is mature enough to find commercial applications in EVs. The other sources such as fuel cells, ultracapacitors and flywheels, while not yet in commercial use, exhibit potential for the future. This chapter attempts to provide an overview of all these energy sources that can be used as portable sources for EVs, their basic working and current status.

15.2 Battery

The chemical battery is the only energy storage technology mature enough to be used commercially in EV applications. The batteries used for EVs are usually rechargeable (secondary). Earlier lead-acid batteries were mostly employed, but with improvements in technology, the lithium-ion battery is now more commonly used; although this may change in the future as well on account of ongoing research (Iclodean et al.

2017). The different battery technologies do share the same basic working principle and this is where we shall start with.

15.2.1 Battery Basics

Electric as well as hybrid electric vehicles use high-voltage battery packs. Each battery pack consists of individual modules and cells connected in series and/or parallel. Series connection is required to increase the battery pack voltage rating, whereas parallel connection is required to increase the current handling capacity of the pack. A cell is the smallest packaged form that a battery can take. Generally, it is of the order of 1 to 6 V. A single cell is not sufficient for our requirement, so a number of cells are connected in series and/or parallel to form a module. A number of modules connected together form a battery pack.

A battery is a device that stores energy in chemical form and converts it into electrical energy when required. Chemical reactions take place in a battery and involve the flow of electrons from one material to another. These materials are called electrodes. The electrons will flow through an external circuit. This flow of electrons can be utilized to do some useful work. In other words, the flow of electrons is providing an electric current that can be used to do useful work. To balance the electric charge, there has to be a flow of ions also. The electrons flow through conducting wires (external circuit), whereas charged ions flow in the opposite direction inside the battery through an electrolyte. Different combinations of electrodes and electrolytes produce different chemical reactions that determine the working of each battery type, the amount of specific energy that can be stored in the battery, as well as its voltage rating.

The first battery in the world was produced by Alessandro Volta, an Italian physicist, chemist and one of the pioneers of electricity (Britannica 2021). One of Volta's friends, Luigi Galvani, performed an experiment with a frog in which he had tied the frog's legs to a metal. When Galvani touched the frog's legs with an electrode, the legs twitched. This led Galvani to conclude that there was electricity inside the frog. However, Volta did not believe this. He repeated Galvani's experiment, but instead of using the frog, he used metals as electrodes separated by an electrolyte (brine-soaked paper), and proved that the electricity was coming from the electrodes and not from the frog. This is how the first battery was produced. Till today, the same concept that Volta introduced is being used to produce the batteries. The materials have changed, battery sizes and battery technology have improved, but the basic principle still remains the same.

In summary, the essential components of each battery cell are two electrodes (one positive and one negative) and an electrolyte in contact with both the electrodes. Current will flow between the two electrodes only when there is an external connection between the two electrodes; otherwise, ideally, no current will flow between the two electrodes. The functions of the electrolyte are to provide a path for the flow of electric ions and to provide chemical reactions between the electrodes and the electrolyte which result in the production and consumption of electrons and hence to an electric current. The electrolyte should be such that it should allow the flow of ions but not the flow of electrons (Hussain 2003; Chan and Chau 2001; Mi et al. 2011). The essential battery components are described in further detail below.

15.2.1.1 Electrodes

To produce a flow of electrons, we need to have a source material where the electrons are produced, and a destination material where the electrons are consumed (How a battery works, nd). These materials are called the cell's electrodes. The electrons are produced through chemical reactions at the electrode called the anode or negative electrode, and are again consumed through chemical reactions at the cathode or positive electrode. The battery terminology uses the anode notation for the negative electrode or the electrode at which electrons are produced, whereas the positive electrode or the electrode at which electrons are consumed is called the cathode. The electrodes are generally different types of metals or other chemical compounds. The anode material should be such that a chemical reaction between the anode and the electrolyte should produce electrons. These electrons get accumulated at the anode and should not be able to flow to the cathode unless there is an electrical connection between the anode and the cathode. On the other hand, the cathode material should be such that a chemical reaction between the cathode and the electrolyte must be able to absorb/accept electrons received from the anode. The electrons flow should only be through an external connection between the anode and the cathode; no flow of electrons should be possible through the electrolyte.

The technical chemical term for a reaction that involves an exchange of electrons is a reduction–oxidation reaction, more commonly called a redox reaction. So electricity is generated in a chemical cell because of a redox reaction. This redox reaction can be separated into two half-reactions. In the case of an electrochemical cell, one half-reaction occurs at the anode and the other half-reaction occurs at the cathode. Reduction is the gain of electrons, and is what occurs at the cathode. At the cathode, electrons are being accepted because of a chemical reaction. The cathode is gaining electrons, and this process of gaining of electrons is called as reduction; we say that the cathode is reduced during the reaction. Oxidation is the loss of electrons. We say that the anode is oxidized; the anode is giving up electrons. The electrode materials should be such that they should interact chemically with the electrolyte to facilitate the redox reaction.

15.2.1.2 Electrolyte

The earlier electrolytes used were predominantly liquids, but now we also have gels and solid substances being used as electrolytes (Li et al. 2016). The only condition is that they must be able to allow the movement of charged ions, while not allowing the flow of electrons through them. In other words, the electrolyte should be such that it has high and selective conductivity for ions that take part in the electrode chemical reactions, but it should be a non-conductor for electrons. They should also be able to react with the materials of the anode and cathode in order to aid in the redox reaction. Ideally, the electrolyte will not allow the flow of electrons, but in practice some electrons will be able to flow through the electrolyte. That is why, even when we are not using the battery, it will self-discharge. The amount of self-discharge depends on the type of battery being used. Some batteries discharge up to 10–20% in a month, whereas some discharge by only 2–4% in a month.

Electrons have a negative charge, and we are sending electrons from the anode to the cathode; so we are sending negative charge from the anode to the cathode. To maintain charge balance, there has to be a flow of positive charge also in the same direction. This positive charge flow will be in the form of positive ions. These positive ions will flow from the anode to the cathode through the electrolyte. So the electrolyte provides a medium through which charge balancing positive ions can flow through.

At the anode, the electrode material will be involved in a chemical reaction with the electrolyte to produce electrons (oxidation); whereas at the cathode, another chemical reaction will take place between the material of the cathode and the electrolyte due to which electrons will be absorbed by the cathode (reduction). The electrons flow through the external wire, while the electrolyte provides a pathway for the transfer of positively charged ions to balance the negative flow. This flow of positively charged ions is just as important as the flow of electrons through the external circuit. The charge balancing role performed by the electrolyte is necessary to keep the entire process running.

The electrolyte can be acidic or alkaline, depending on the type of battery. Traditional batteries such as lead-acid and nickel–cadmium use liquid electrolytes. Advanced batteries currently under development such as nickel-metal-hydride and lithium-ion batteries use an electrolyte that is gel, paste or resin. Lithium-polymer batteries use a solid electrolyte (Hussain 2003).

15.2.1.3 Separator

The separator is an electrically insulating layer of material that physically separates electrodes of opposite polarity. Effective separators must possess high porosity so that they will have low electrical resistance, low pore diameter in order to achieve good separation and, finally, resistance to oxidation and stability in highly acid conditions (Vincet and Scrosati 1998). Separators must be permeable to ions; they must allow the flow of ions but not the flow of electrons. Present-day separators are made from synthetic polymers or glass filters.

15.2.2 Types of Batteries

The basic working principle of all batteries is the same. The only things that change are the materials that are being used for the electrodes and the electrolyte and the associated chemical reactions. Different types of batteries are in commercial production, each suited to its own specific application.

There are two basic types of batteries: primary batteries and secondary batteries (Hussain 2003). Primary batteries are those which cannot be recharged. They are designed for a single discharge. Examples are lithium batteries used in watches, calculators, cameras, etc. and manganese dioxide batteries used to power toys, radios, torches, etc. Secondary batteries are those which can be recharged by flowing current in a direction opposite to that during discharge. The chemical reactions are the reverse of those during discharge. Secondary batteries can be reused after charging. However, after every successive charge and discharge, the condition of the battery will go on deteriorating. There is a limit to how many times/cycles the battery can be charged and discharged. All the batteries used for electric vehicles are secondary batteries.

The major types of rechargeable batteries considered for EV and HEV applications are as follows (Hussain 2003):

- Lead-acid (Pb-acid)
- Nickel–cadmium (NiCd)
- Nickel-metal-hydride (NiMH)
- Lithium-ion (Li-ion)
- Lithium-polymer (Li-poly)
- Sodium–sulfur (NaS)
- Zinc-air (Zn-Air)
- Sodium nickel chloride.

Out of these, only the lead-acid, lithium-ion, lithium-polymer and nickel-metal-hydride have been commercially used for EV applications. The lead-acid battery was actually discovered around 1860. The basic process is still the same; there are improvements, but the improvements are minor. It is the earliest type of rechargeable battery that was created. It has a low energy-to-weight ratio, also a low energy-to-volume ratio. To have a high amount of energy, the weight as well as the volume of the battery will be quite high. It does, however, have a relatively large power-to-weight ratio as compared to other batteries. It can provide short bursts of high power. That is the reason why the lead-acid batteries are very popular to provide the high current required by starter motors. The batteries used in earlier cars to provide the high current to start the motor were lead-acid batteries. They were perfectly suited for that purpose on account of their high power-to-weight ratio. However, their energy-to-weight ratio is poor, as is their energy-to-volume ratio. When EVs started becoming popular, around the late 1990s, the lead-acid battery was a sufficiently mature technology. They were also easily available and had a low cost. That is why the earlier EVs used lead-acid batteries. However, they constituted around 25–50% of the final vehicle mass. Also, the efficiency of the lead-acid battery is 70–75%, and decreases at lower temperatures. The nickel-metal-hydride battery was the next to be considered for commercial EV applications. It was also a relatively mature technology. The NiMH battery is even less efficient than the lead-acid battery; its efficiency is around 60 to 70%. It does, however, have a high specific energy of 30–80 Wh/kg, as compared to 30–50 Wh/kg for the lead-acid battery. Note that these efficiencies are ideal or theoretical. They depend on the temperature and will continuously decrease as the battery continues to be used. The NiMH batteries have exceptionally long lives; typically, 160,000 km and over a decade of service for the first-generation Toyota RAV4 EVs. (The second-generation Toyota RAV4 EVs used the Li-ion battery instead.) The drawbacks of the NiMH battery were its poor efficiency, its high self-discharge and poor performance in cold temperatures. Due to these reasons, the NiMH battery was superseded by the Li-ion and Li-poly batteries. Both of them are similar batteries. They were initially developed and commercialized for use in laptops and consumer electronics. Their high energy density and long cycle life have now made them the most popular choice for EV applications.

The nickel–cadmium battery, which used nickel-oxide-hydroxide and metallic cadmium as electrodes, had good cycle life and performance at low temperatures as well as the ability to deliver their full rated capacity at high discharge rates. However, it was costly than the lead-acid battery and had a high self-discharge rate; and was supplanted by the NiMH battery. Disposal of toxic metal cadmium was also an issue. Perhaps, the best battery to be considered was the sodium–sulfur (NaS) battery, also known as the molten salt battery. It was constructed from liquid sodium and sulphur. It had high energy density, high charge and discharge efficiency, long cycle life and was made from inexpensive materials. It was, however, difficult to manufacture and was hazardous. Thus, safety and manufacturing difficulties led to the abandonment of the technology.

15.2.2.1 Lead-Acid Battery

The lead-acid battery was the most popular choice for EVs for a considerable period of time. At the time electric vehicles started gaining some traction, the lead-acid battery was the most developed battery technology. Also, it was high-powered, inexpensive, safe and reliable. Hence, the lead-acid battery, which was earlier used for starting the car, started being used for driving the EVs also. Recycling structure was also well established. It had three significant features: low cost, easy availability of raw materials and ease of manufacture. However, its specific energy was low, its cold temperature performance was poor and it had a short calendar and cycle life. In fact, the battery life was less than the car life; around every three years the battery had to be replaced.

In the lead-acid battery, the negative electrode (anode) is made of spongy or porous lead (Pb), whereas the positive electrode (cathode) is made from lead oxide (PbO_2). The electrolyte used is sulphuric acid (H_2SO_4).

During cell discharge, at the anode, the lead will interact with the sulphuric acid to produce lead sulphate ($PbSO_4$) and two electrons. These electrons flow through the external conductor to reach the cathode. At the cathode, the lead oxide will absorb the two electrons in a chemical reaction with the sulphuric acid to produce lead sulphate and water.

During battery charging, the lead sulphate is once again broken into lead and lead oxide. However, some amount of lead sulphate, in the form of a fine white powder, keeps on depositing on the electrodes. On account of this, the internal resistance of the cell goes on increasing with time. Also, as H_2SO_4 is consumed, the electrolyte conductivity goes on decreasing. The lead sulphate deposited on the electrodes in a dense fine grain form leads to sulfatation. This reduces cell capacity significantly. The lead sulphate is not soluble in water, and keeps on depositing on the plates as well as electrodes. As a result, the battery capacity as well as efficiency keeps on decreasing.

The chemical reactions during charging of the lead-acid battery are opposite to those during discharging. The lead sulphate will be converted back to the reactant states of lead and lead oxide. For this, a current has to flow into the positive electrode through an external source; the electrons have to flow in the opposite direction from the positive electrode to the negative electrode. The external source delivers electrical energy to the cell, where it gets converted into chemical energy and can be stored. At the positive electrode, the lead sulphate consumes the water produced at the time of discharging and produces lead oxide, $4H^+$ and SO_4^{2-} ions and two electrons. At the negative electrode, the lead sulphate consumes the two electrons to produce lead and SO_4^{2-} ions.

The lead-acid battery continues to be popular for a lot of applications. However, it is not suited for driving the electric vehicle over long distances. Specifically, its low specific energy is a major hindrance, causing a considerable increase in the volume and mass of the battery for driving the EV. Other better suited battery technologies have now been developed, as a result of which, the lead-acid battery is no longer being used for driving long range electric vehicles.

15.2.2.2 Nickel–Cadmium Battery

The nickel–cadmium (NiCd) battery is an alkaline battery; the electrolyte used in the battery is alkaline. In this case, the electrical energy is derived from the chemical reaction of a metal (cadmium) with oxygen in an alkaline electrolyte medium. One major disadvantage of the NiCd battery was its low specific energy due to the addition of the weight of the metal.

In these batteries, the positive electrode is nickel oxide, while the negative electrode is made from metallic cadmium. The alkaline electrolyte is potassium hydroxide (KOH). The practical cell voltage obtainable from NiCd batteries is 1.2–1.3 V. The specific energy is 30–50 Wh/kg, similar to the lead-acid batteries.

The advantages of the nickel–cadmium batteries were superior low-temperature performance as compared to lead-acid batteries, flat discharge voltage characteristic, long life, excellent reliability and lower maintenance requirements. However, they had a number of limitations also. Their cost was high because of the metal involved, as cadmium was highly toxic. In fact, after 2004, the NiCd battery was not allowed in Europe, specifically for home applications. Their power output was also low, and was insufficient for EV applications. The popularity of the nickel-metal-hydride battery, which is a much improved alkaline battery, made the NiCd battery redundant for EV applications.

15.2.2.3 Nickel-Metal-Hydride Battery

The nickel-metal-hydride (NiMH) battery is the successor to the nickel–hydrogen battery. A number of hybrid electric vehicles have used the NiMH battery for providing electric power for traction purposes. Its operating voltage is almost the same as the NiCd battery, with a flat discharge characteristic. The specific energy is 60–80 Wh/kg, which is higher than the lead-acid and NiCd batteries. The specific power can be as high as 250 W/kg.

The NiMH battery also uses nickel oxide as the material for the positive electrode, but the negative electrode is a metal hydride. When certain metals are exposed to hydrogen, they have the ability to absorb that hydrogen. They also have the ability to release that hydrogen when required. Such metals, which can absorb hydrogen, are called as metal hydrides. Fine particles of certain metallic alloys, when exposed to hydrogen at certain pressures and temperatures, absorb large amounts of hydrogen to form metal hydride compounds. Metal hydrides can absorb and release hydrogen many times without getting deteriorated.

The advantages of the NiMH battery are a much longer life cycle than lead-acid batteries, as well as being safe and abuse tolerant. Abusive conditions for a battery include short circuiting a battery, overcharging a battery, physically deforming a battery. The NiMH is abuse tolerant. Specifically, the exceptionally long-life cycle of the NiMH battery is a significant advantage. However, its cost is relatively higher, has a higher self-discharge rate as compared to the NiCd battery and has poor charge acceptance capability at elevated temperatures and low cell efficiency. Although the NiMH battery was used to drive EVs and Hybrid EVs (HEVs) earlier, they are no longer preferred on account of the improvements in the lithium-ion battery technology.

15.2.2.4 Lithium-Ion Battery

The lithium-ion (Li-ion) battery is one of the most promising batteries for EV applications. Right now, the Li-ion battery is the battery of the future that will dominate the EV and HEV applications for decades. Of course, it is possible that some other battery technology may come along and replace the Li-ion battery, but right now, the Li-ion battery is the most promising battery for the future. The reason is that lithium metal has a high electrochemical reduction potential (3.045 V) and lowest atomic mass (6.94). Reduction potential is related to the oxidation potential, or the redox potential. The tendency of a chemical to acquire/release electrons from an electrode and thereby be reduced or oxidized is called the reduction potential. Lithium has the potential to acquire large number of electrons. It is possible to generate up to 3.045 V, while still having the lowest atomic mass. This shows promise for a battery of 3 V cell when combined with a suitable positive electrode. However, lithium metal is highly reactive with moisture, which limited their use earlier. It was only the discovery in the late 1970s that lithium can be intercalated (absorbed) into the crystal lattice of cobalt or nickel to form lithium cobalt oxide ($LiCoO_2$) or lithium oxide nickel ($LiNiO_2$) that paved the way toward the development of lithium-ion batteries.

The Li-ion battery uses lithium intercalated (absorbed) carbons (Li_xC) in the form of graphite or coke as the negative electrode. The positive electrode is made up of lithium metallic oxides. The most satisfactory lithium metallic oxide is the cobalt oxide, but it is very expensive. Another lithium metallic oxide is the lithium oxide nickel ($LiNiO_2$). It is structurally more complex but costs less and its performance is similar to cobalt oxide electrodes. Manganese oxide based positive electrodes are also under research because manganese is cheaper, widely available and less toxic.

During discharge, at the negative electrode, the Li_xC discharges to produce carbon, lithium ions and electrons. The electrons will move toward the positive electrode through the external circuit, while the lithium ions move in the same direction through the electrolyte. At the positive electrode, the lithium ions recombine with the electrons and the metallic oxide to produce lithium cobalt oxide.

The features of a Li-ion battery include high specific energy, high specific power, high energy efficiency, good high temperature performance, low self-discharge and recyclable components. The Li-ion battery has features which are ideal for battery use for EV applications. However, it contains flammable electrolytes and can explode when exposed to abusive conditions.

15.2.2.5 Lithium–polymer Battery

The lithium-polymer (Li-poly) battery is similar in working to the Li-ion battery, except that it uses solid-sate electrolytes, i.e. solids capable of conducting ions but which are electron insulators.

The Li-poly battery has the highest specific energy and power, good cycle and calendar life, and all the advantages of the Li-ion battery. Also, because the electrolyte is a solid, the battery can have any desired shape. They are still not being used for EV applications, but research is ongoing. If the Li-poly battery does get commercialized for EVs, they will provide a huge space advantage. Also, because they have a solid electrolyte, they are ideally suited for moving applications. The only disadvantage is that the battery cell has to be operated in the range of 80 to 120 °C.

15.2.2.6 Zinc-Air Battery

The Zinc-air (Zn-Air) battery is similar to the conventional batteries in the sense that it generates electric power from chemical reactions (Review 2020). However, in most respects, its working is more similar to that of a fuel cell. These batteries have a gaseous positive electrode of oxygen, whereas the negative electrode is metallic zinc. The Zn-Air battery is an alkaline battery and uses potassium hydroxide (KOH) as an electrolyte. The amount of electric power generated can be controlled by controlling the flow of oxygen (air). Electrical recharging of the battery is not possible; in this sense, it is similar to a primary cell. However, once the battery is discharged completely, the zinc electrode can be mechanically replaced with fresh zinc.

Zn-Air batteries have a high specific energy. Its other inherent features such as safety and lower cost ensure that it is a promising technology for future EV applications. Although the alkaline electrolyte is still widely used, Room Temperature Ionic Liquids and Quasi-Solid Flexible Electrolytes are also being researched for use as electrolytes in Zn-Air batteries (Peng et al. 2020).

15.2.3 Battery Parameters

Battery Capacity: Battery capacity is the amount of free charge generated by the active material at the negative electrode and consumed by the positive electrode. Battery capacity is measured in ampere-hour (Ah), where 1 Ah = 3600 C and 1 C is the charge transferred by a current of 1 A in 1 s.

The theoretical capacity of a battery Q_T is given by $Q_T = xnF$ C, where x is the number of moles of the limiting reactant associated with complete discharge of the battery, n is the number of electrons produced by the negative electrode discharge reaction and $F = Le_o$. Here L is the number of molecules or atoms in a mole, also known as the Avogadro constant (6.022 × 1023) and e_o is the electron charge (1.601 × 10–19 C). Thus F is a constant whose value is equal to 96412.2 C/mol. Limiting reactant is the reactant which is consumed fully and which decides the amount of product that is formed. In terms of Ah, the theoretical capacity is given by $Q_T = 0.278\ Fm_R n/M_m$ Ah, where m_R is the mass of the limiting reactant in kg and M_m is the molar mass of the limiting reactant in grams/mol.

If the capacity of a single cell of a battery is Q_{Tcell}, and a number of cells are connected in series to form a battery, then the capacity of the whole battery is also Q_{Tcell}. Thus the capacity of a battery is the same as the capacity of an individual cell when cells are connected in series. If the capacities of individual cells are different, then the capacity of the battery will be the same as the capacity of the smallest cell in the battery.

The practical capacity of a battery is always much lower than the theoretical capacity due to practical limitations. The practical capacity can be obtained by integrating the current drawn from the battery with respect to time from the instant at which the battery is fully charged to the instant at which the battery terminal voltage becomes equal to the cut-off voltage.

Discharge rate: The discharge rate is the current at which the battery is discharged. It is expressed as Q/h rate, where Q is the rated battery capacity and h is the discharge time in hours.

Suppose the capacity of a battery is 100 Ah (1Q = 100 Ah) and the battery is discharged in 5 h, then the discharge rate is Q/5 = 100 Ah/5 h = 20 A. If the same battery is discharged in 0.5 h, the discharge rate is Q/0.5 = 200 A. Thus, the battery discharge rate will determine the amount of time for which the battery will last before it runs out of charge. The higher the discharge rate of the battery, the shorter is the time for which it can be used before recharging.

State of Charge: The state of charge (SoC) is the present capacity of the battery at any instant of time. It is the amount of capacity that remains after discharge from a top-of-charge condition. The SoC is a very important parameter for a battery. It is somewhat analogous to the amount of fuel in a conventional vehicle at any instant of time.

The SoC of a battery is decided by the amount of current that is drawn from the battery since it was fully charged. It can be determined by integrating the current drawn from a battery over a duration of time and then subtracting this integral from the theoretical capacity of the battery.

State of Discharge: The state of discharge (SoD) is just opposite to the SoC. It is a measure of the charge that has been drawn from a battery. The SoD can be obtained by integrating the current drawn from the battery over a duration of time. The state of charge of a battery is the difference between the theoretical capacity of a battery and the SoD. If the SoD of a battery is known, the SoC of the battery can be determined, and vice versa.

Depth of Discharge: The depth of discharge (DoD) is the percentage of battery capacity (rated) to which a battery is discharged. Basically, the DoD expresses the SoD as a percentage of the rated theoretical capacity of a battery.

Battery terminal voltage: Battery terminal voltage is the voltage obtained across the terminals of the battery with load applied to it. It depends on the state of charge of the battery as well as the charging/discharging current of the battery.

Cut-off voltage: The cut-off voltage of a battery is the minimum allowable voltage from a battery. If the battery terminal voltage falls below the cut-off voltage, the battery is assumed to have discharged and can be used only after recharging.

15.3 Fuel Cell

Having studied the battery in considerable detail, let us now take a look at some alternative energy sources. The first alternative energy source that we will consider is the fuel cell. The fuel cell is under research and is a promising technology for the future.

A fuel cell is an electrochemical device that produces electricity by means of a chemical reaction. Similar to a battery, it also converts chemical energy into electrical energy and uses redox reactions to achieve this end. However, batteries will produce electricity as long as there is stored chemical energy, after which they have to be recharged; whereas, in the case of a fuel cell, there is no stored chemical energy. As long as fuel is supplied to the fuel cell, it will continue to produce electricity. Once the fuel supply is stopped, the fuel cell will stop providing electrical energy.

The fuel cell has a negative terminal called the anode and a positive terminal called the cathode. It has two inlets. One inlet is for the hydrogen. Some fuel cells require pure hydrogen while some can work even if there is some impurity in the hydrogen. The other inlet is for the oxygen. There is no need to provide pure oxygen; supply of air is sufficient.

At the negative electrode, the electrode material will react with the hydrogen and split the hydrogen atom into hydrogen ions and electrons. The hydrogen (H^+) ions, which are essentially two single protons, will move through the electrolyte placed between the electrodes to the positive electrode while the electrons will move through the external circuit to constitute a current flow. At the positive electrode, the hydrogen ions will combine with the electrons and the oxygen to produce water. Thus the output of the fuel cell obtained from the outlet is water. Unreacted hydrogen is also obtained from the outlet, as is the unused air. This is the basic working of a fuel cell. As long as fuel in the form of hydrogen, as well as air, is being supplied to the fuel cell, the fuel cell will continue to produce electricity. As soon as the flow of hydrogen is stopped, the fuel cell will stop the production of electrical energy. The amount of electrical energy produced will depend on the material of the electrodes as well as the rate of flow of hydrogen. Thus, the amount of electricity produced from a fuel cell at any instant of time can be controlled by controlling the rate of flow of hydrogen. The energy conversion process is instantaneous as long as the negative electrode is maintained within an operating temperature range.

The advantages of a fuel cell over a battery for EV applications are as follows:

- There is no need to charge a fuel cell. No charging stations are required for fuel cells.
- The fuel system is similar to a petrol vehicle. Instead of petrol, hydrogen is being used as fuel. Just like a petrol vehicle, knowing the amount of hydrogen in reserve, the distance that can be traversed can be estimated. The hydrogen can be refueled as desired.
- The hydrogen capacity can be varied depending on the EV capacity and application. For smaller/short-run vehicles, the hydrogen storage capacity can be smaller, whereas for larger vehicles or vehicles designed for long runs, the hydrogen storage capacity can be appropriately increased.

In spite of these advantages, the fuel cell is still not being used for commercial EV applications. This is because there are a lot of disadvantages of using fuel cells, and the disadvantages are currently outweighing the advantages. Until the disadvantages are removed, the fuel cell cannot be used in practice. These disadvantages are as follows:

- Hydrogen is highly flammable. As such, there are safety issues in the usage of the fuel cell for commercial applications.
- Storage of hydrogen is a very big issue. It is not as simple as storage of petrol or diesel. Hydrogen storage is a refined and complex technology. Storage and

management (containment) of hydrogen is a huge hindrance that is hampering the growth of fuel cells as an energy source for EVs.

However, extensive research is being carried out for making fuel cells viable for EV applications in the future.

15.3.1 Types of Fuel Cells

15.3.1.1 Alkaline Fuel Cell (AFC)

The alkaline fuel cell uses an aqueous solution of potassium hydroxide (KOH) as an electrolyte. The original fuel cells used acidic electrolytes. The performance of an alkaline electrolyte is as good as an acidic electrolyte while being significantly less corrosive toward the electrodes. The AFC shows very high electrical efficiency as compared to other fuel cells, as high as 60%. However, they require pure hydrogen as fuel, as there can be no impurity in the fuel. The AFC operates at low temperatures (80 °C); they are therefore suitable for EV applications.

15.3.1.2 Proton Exchange Membrane (PEM) Fuel Cell

The PEM fuel cell uses a solid electrolyte instead of an acidic or alkaline solution. Nafion is an example of a solid polymer electrolyte. This fuel cell also operates at low temperatures (80 °C), and hence is suitable for EV applications. Its electrical efficiency is lower than that of the AFC, about 40%. However, the construction is rugged and simple, which is a huge advantage. They can also tolerate impurity in the fuel as compared to the AFC which require pure hydrogen.

15.3.1.3 Direct Methanol Fuel Cell (DMFC)

In case of DMFC, instead of storing hydrogen, methanol is stored. Storage and containment of methanol is much easier than storage of hydrogen. The methanol is reformed on board to supply hydrogen to the fuel cell. The DMFC works on temperatures in the range of 90–120 °C to facilitate the reformation of methanol into hydrogen. The electrical efficiency of the DMFC is low, only about 30%. The technology is still in the design and research stage because the search for a good electrocatalyst to reform the methanol efficiently and to reduce oxygen in the presence of methanol is ongoing.

15.3.1.4 Other Fuel Cell Types

There are some other fuel cell types also. One of them is the Phosphoric Acid Fuel Cell (PAFC), which is the oldest type of fuel cell. The electrolyte used is phosphoric acid and the cell operating temperature is around 200 °C. The efficiency is reasonable at around 40%. However, because of the high operating temperature, it is not considered for EV applications. It is also quite bulky.

Another type of fuel cell is the Molten Carbonate Fuel Cell (MCFC). It operates directly from coal at temperatures of around 600 °C. It requires carbon monoxide or carbon dioxide on the cathode side and hydrogen on the anode. The MCFC uses carbonate as the electrolyte. The efficiency is considerably higher at around 50%. Because the MCFC operates at very high temperatures, the excess heat can be used for cogeneration to have an improved efficiency. Cogeneration means that the excess heat can be used for other purposes also such as home heating. The very high operating temperature is, however, a deterrent in the use of MCFC for EV applications.

The solid oxide fuel cell (SOFC) uses a solid ionic conductor as an electrolyte which reduces corrosion problems. However, to achieve adequate ionic conductivity, the system must operate at very high temperatures, around 1000 °C. Hence the SOFC is also not considered for EV applications. Intermediate Temperature SOFC (ITSOFC) is also undergoing research. The intermediate temperatures are still quite high for this fuel cell to be considered for EV applications. These fuel cells are more suitable for stationery applications.

15.3.2 Reformers

Some fuel cell types, such as the DMFC, use a hydrocarbon such as methanol, ethanol, diesel, biogas or natural gas instead of pure hydrogen as a fuel. While this solves the problem of storage and transportation of hydrogen, the fuel cell has to extract hydrogen from the hydrocarbons. For this purpose, the fuel cell system must then include a reformer system (Sundén 2019).

The reformer system consists of a reformer which converts the original fuel to a hydrogen containing gas, and a clean-up system, which removes carbon monoxide and sulphur from the gas to make it appropriate for the fuel cell. The clean-up system will thus remove the unwanted gases and will provide hydrogen to the fuel cell. The hydrogen thus provided to the fuel cell will not be exactly pure as can be produced in a laboratory or industry. Hence, the efficiency of such fuel cells will be slightly less as compared to those fuel cells which utilize pure hydrogen.

15.4 Ultracapacitors

Having studied the battery and fuel cell in some detail, let us now turn our attention and briefly discuss two energy sources which are by themselves not sufficient to power an electric vehicle. However, they can be used in conjunction with a battery or fuel cell to enhance the vehicle performance. The first such energy source is the ultracapacitor.

If the battery alone is used as an energy source for an EV, the discharge profile of the battery is highly variable. The average power required from the battery is relatively low. When moving on a level surface with more or less constant speed, the power required by the vehicle is relatively low. Whereas, if the vehicle is moving up a gradient, or at the time of acceleration, the power required is much higher. The ratio of this peak power to average power can be as high as 16:1 for high performance EVs. In fact, the amount of energy required for acceleration and during transient conditions is around $2/3^{rd}$ of the total amount of energy over the entire vehicle ride. Based on present battery technology, the design of batteries has to carry out a trade-off between the specific energy, specific power and cycle life. It is difficult to simultaneously obtain high values of these three performance parameters.

A battery with high specific power will provide very fast acceleration and deceleration, but the battery will discharge very fast. On the other hand, a battery with high specific energy will last longer between discharge cycles but will be unable to provide fast acceleration. Both these requirements are contradictory. A battery designed for high specific power will have a low specific energy, while a battery designed for high specific energy will be unable to provide high power for short durations of time. Thus, both these requirements cannot be satisfied simultaneously. This difficulty has led to suggestions that EVs may be powered with a pair of energy sources, and is an option that is being explored. Out of these two sources, the main energy source, which is usually the battery, is optimized for the range. That means that the battery must last for a longer duration between charging and discharging. The auxiliary source is optimized for acceleration and going up a gradient. The auxiliary source should be such that it can be recharged from the main source during less demanding driving conditions or during regenerative braking. One of the auxiliary sources which has received wide attention is the ultracapacitor.

Ultracapacitors are derivatives of conventional capacitors in which the energy density has been increased at the expense of power density to make the device function more like a battery. Normal capacitors can charge and discharge at a fast rate; they have a high power density and a very poor energy density. In case of ultracapacitors, the power density is sacrificed so as to increase the energy density. Also, the capacitance is much higher in case of ultracapacitors or super capacitors as compared to normal capacitors. It is in farads whereas for normal capacitors it is in microfarads or millifarads. The power and energy densities of ultracapacitors are of the order of 10^6 W/m^3 and 10^4 Wh/m^3, respectively. The energy density is still much lower as compared to batteries, but the discharge times are much faster and the life cycle is much more. Because the ultracapacitor can discharge much faster, it is able to provide high power for a short duration of time.

Because the specific energy of ultracapacitors is low, they cannot be used as the sole energy source for an electric vehicle. It is not possible to have an EV powered solely by an ultracapacitor because the ultracapacitor will discharge very fast. Still there are a number of advantages that can be obtained from using the ultracapacitor as an auxiliary energy source. If the battery and ultracapacitor are used together to form a hybrid energy system for an EV, the specific energy and specific power requirements of the EV energy source can be decoupled. The specific energy requirement is now handled by the battery, whereas the specific power requirement is handled by the ultracapacitor. The battery can now be optimized for specific energy. This is nothing but a sort of load equalization for the battery, as it is obtained using a flywheel in case of variable speed electric drives. At the time of high power demand, the ultracapacitor will provide the power to drive the EV along with the battery. While the EV is running on constant speed, the ultracapacitor can be charged from the battery. It can also be charged during regenerative braking conditions. Since the battery is no longer required to provide high power, it does not have to discharge at a high current. Hence, the available energy, endurance and life of the battery is significantly increased. Also, because the time constant of the ultracapacitor is much less as compared to the battery, it can provide much faster acceleration as well as recovery of energy during regenerative braking. The battery does not have the capability to absorb power at a high speed for the short duration of regenerative braking, but the ultracapacitor does. It can accept charge at a much faster rate and can be charged to a greater extent than the battery during the short interval for which regenerative braking occurs. Because of this combined effect of load equalization and efficient energy recovery, the vehicle range can be greatly extended.

The targets set by the US Department of Energy state that the near-term specific energy and specific power for ultracapacitors should be better than 5 Wh/kg and 500 W/kg, respectively, while the advanced performance values should be over 15 Wh/kg and 1600 W/kg (Hussain 2003). So far, none of the available ultracapacitors can fully satisfy these goals. If current state-of-the-art ultracapacitors are used as an auxiliary source for EVs, the vehicle weight will increase by about 300 kg and the volume of the ultracapacitor will also be prohibitive. That is why, till now, ultracapacitors are not being used in practice. Things may change in the future with further research and development in the field of ultracapacitors.

15.5 Flywheel

Flywheels have been in use for a long time for load equalization in variable speed electrical drives. For load equalization, the flywheels are expected to have a large moment of inertia; the reason being that flywheels work on the principle of storing and releasing energy. When the flywheel is running at a speed greater than the normal speed due to the absence of load or the load being light, or when it is accelerating, it stores kinetic energy. When the full load is applied and the flywheel decelerates, the kinetic energy is released to the load. This comes in use in case of load equalization in variable speed drives, especially where the load is applied intermittently.

Would it be possible to have an EV powered solely by a flywheel? Surprisingly, there was a bus that did exactly that; that too, not today but 80 years ago. This flywheel powered bus was developed by the Oerlikon Engineering Company of Switzerland. The bus was called the 'Gyrobus'. The bus is not in use anymore, but was in actual use in the 1940s. It used a flywheel that weighed around 1500 kg and was spun at speeds up to 3000 rpm.

The flywheel was initially charged by using an electric motor, specifically the squirrel cage induction motor. Charging the flywheel implies bringing it up to its maximum speed so that it stores kinetic energy. On account of the large moment of inertia of the flywheel, this stored kinetic energy was considerable. This charging was done at a charging point, where the induction motor responsible for charging the flywheel was connected to an electrical supply. Once the flywheel was charged to its maximum speed, the electrical motor was disconnected from the supply. The flywheel now acted as a prime mover for the induction machine, so that it started working as a self-excited induction generator. This generated power was used to power the electric motor that drove the bus. As the bus moved, the flywheel basically kept releasing its kinetic energy and its speed went on decreasing. After some appropriate distance, a charging point was located to charge the flywheel again; it again discharged while driving the bus, and so on. A fully charged Gyrobus could typically travel up to 6 km on a level route with speeds ranging from 50 to 60 km/hr. Charging a flywheel took between 30 s and 3 min. The Gyrobus had its advantages: quiet operation, pollution free and did not require any rails. The disadvantage was that it required a very large flywheel which had to be repeatedly charged after short distances. As a result, the Gyrobus service was discontinued as soon as buses with IC engines became available.

In order for the flywheel to store a large amount of kinetic energy, it is not necessary that the flywheel have a large moment of inertia. Instead, the flywheel can be designed to have a small moment of inertia but a high rotating speed. Such a flywheel uses a lightweight composite rotor with tens of kg and rotates at the order of ten thousands of rpm and is called an ultrahigh-speed flywheel. It stores energy in mechanical form during periods of cruising or regenerative braking while it generates electrical energy to meet the peak power demands during periods of starting, acceleration and hill climbing.

If the flywheel is to be used as an energy source for an electric vehicle, the weight of the vehicle increases to account for the weight of the flywheel. So instead of using a flywheel with a large weight, it is better to use a flywheel that is lightweight but rotates at a faster speed. Hence ultrahigh-speed flywheels are preferable for EV applications.

Ultrahigh-speed flywheels have high specific energy, high specific power and high efficiency for conversion between electric and mechanical energies. They can recharge rapidly with high efficiency during regenerative braking leading to a remarkable improvement in the vehicle range. Note that this is the case only with ultrahigh-speed flywheels and not with conventional flywheels. The ultrahigh-speed flywheel can enhance the usable energy, endurance and cycle life of a battery when used in conjunction with it. However, even though the concept is very good, there are significant problems with the use of flywheels for EVs due to which they are still not being used for practical applications. The first such problem is gyroscopic forces. Gyroscopic forces come into play whenever an EV with a flywheel tries to deviate from its path. For example, when the vehicle is moving along a straight path initially, the flywheel tends to move along that straight path. Now if the vehicle tends to turn, the flywheel opposes this motion. The vehicle control will therefore be slightly difficult. The second problem hampering the use of flywheels for EVs is failure containment. If due to some fault, the ultrahigh-speed flywheel, rotating at a speed of 10,000 rpm, comes out of its enclosure, it will have devastating consequences. Progress needs to be done on these two issues before the flywheel can be used in practice for EV applications.

There has, however, been interest in using the flywheel as a stationery energy storage system. Instead of using the flywheel on the vehicle, it can be kept stationery and can be used as an energy storage system in a charging station. The power levels required for EV charging stations are in the MW range. This power demand is highly fluctuating and depends on the number of vehicles being charged at the station at any time as well as their SoC. This large fluctuation in the power demand at a charging station can be equalized to a certain extent by using the flywheel as an energy storage system at the charging station.

15.6 Conclusion

This chapter has considered four energy sources for EVs in brief. These sources were the battery, the fuel cell, the ultracapacitor and the flywheel. All these sources had their particular advantages and drawbacks. Notable advantages are the maturity and low cost of batteries, the outstanding specific energy and high fuel efficiency of fuel cells, the enormous specific power and instantaneous charge/discharge capability of ultracapacitors and the outstanding specific power and practically unlimited cycle life of ultrahigh-speed flywheels. Still, none of the energy sources can fulfil all the demands of EVs; no source is entirely self-sufficient. In essence, no energy source can provide high specific energy and high specific power simultaneously. However, this limitation can be overcome by the hybridization of two energy sources; one with high specific energy (battery or fuel cell) and one with high specific power (ultracapacitor or flywheel). The hybridization of three or more sources together would be quite complex and hence is not considered. The advantage of hybridization is that the responsibilities can now be divested. EV sources such as batteries and fuel cells can be optimally designed for high specific energy, while sources such as ultracapacitors and flywheels can be optimally designed for high specific power. The cycle life and production cost of these sources can be lengthened and minimized, respectively.

References

Britannica T (2021, March 1) Editors of Encyclopedia. Alessandro Volta. Encyclopedia Britannical. https://www.britannica.com/biography/Alessandro-Volta

Chan CC, Chau KT (2001) Modern electric vehicle technology. Oxford University Press

How a battery works (nd) science.org.au. https://www.science.org.au/curious/technology-future/batteries. Accessed 9 Aug 2021

Hussain I (2003) Electric and hybrid vehicles: design fundamentals. CRC Press, Boca Raton, FA

Iclodean C et al (2017) Comparison of different battery types for electric vehicle. IOP Conf Ser Mater Sci Eng 252:012058

Li Q, Chen J, Fan L, Kong X, Lu Y (2016) Progress in electrolytes for rechargeable Li-based batteries and beyond. Green Energy Environ 1(1):18–42

Mi C, Masrur MA, Gao DW (2011) Hybrid electric vehicles: principles and applications with practical perspectives. Wiley

Peng C, Keyi Z, Dejian T, Weilin L, Fancheng M, Qiuwei H, Jiehua L (2020) Recent progress in electrolytes for Zn-Air batteries. Front Chem 8

Review T (2020, April 2) Zinc-air batteries. MIT Technol Rev. https://www.technologyreview.com/2001/06/01/235591/zinc-air-batteries/

Sundén B (2019) Chapter 8—Fuel cell types—overview. In: Hydrogen, batteries and fuel cells. Academic Press, pp 123–144. ISBN 9780128169506

Vincet CA, Scrosati B (1998) Modern batteries: an introduction to electrochemical power sources, 2nd edn. Butterworth-Heinemann

Chapter 16
Wireless Charger for E-Vehicle Using Green Technology

Gitimayee Sahu, Shubham V. Kadam, and Saee P. Mane

Abstract This research work presents an overview of the technique of wireless power transmission (WPT) for electric vehicles (e-vehicles) and solar vehicles. E-vehicles are the new era in the world of automobiles aiming to modernize the fuel-free transport mechanism. WPT is an efficient mechanism for charging the e-vehicles using electromagnetic waves and renewable energy. The working model comprises two parts: the transmitting unit consists of an inductive coil placed underground and the receiving unit consists of inductive coil and rectifying circuit placed inside the vehicle. The roof of the parking lot is covered with solar cells (green energy). The sun rays gathered by the solar panel are converted to electrical energy and stored in the battery placed underground. This battery is connected to the primary winding of the transformer while the secondary winding is placed inside the vehicle. During the parking hours as the secondary winding comes in line of sight with the primary winding, energy transfer starts automatically and charges the battery. The experimental study shows that the battery of the vehicle takes 30% less time to charge compared to conventional means. The significance of this research is that the e-vehicles are charged using green technology (solar energy) instead of the conventional power grid energy.

Keywords E-vehicle · Solar cell · Inductive coil · Wireless power transmission · Green technology

16.1 Introduction

With exponentially rising concern of emission and pollution in the environment, the electric vehicles (e-vehicles) are found to be substitute for conventional automobile. The major limitation in use of e-vehicles is that it takes more time to charge and currently have finite number of charging stations. Thus, it can be used for short-range commute only (Sreedhar 2006). The inclusion of internal combustion engine

G. Sahu (✉) · S. V. Kadam · S. P. Mane
Lokmanya Tilak College of Engineering, Navi Mumbai, India
e-mail: giti.sahoo@gmail.com

A. K. Bohre et al. (eds.), *Planning of Hybrid Renewable Energy Systems, Electric Vehicles and Microgrid*, Energy Systems in Electrical Engineering,
https://doi.org/10.1007/978-981-19-0979-5_16

(IC engine) enhances the efficiency and reduces the harmful emission. The motivation for using e-vehicles is because they are environment friendly and have zero exhaust emission which reduces the pollution level in the atmosphere. By around 2022, the usage of e-vehicles will be 5 times more compared to automobiles with IC engine (KingMeter 2021; Fastin Charge 2021). The electric motor has high power efficiency of 90% while IC engine has limited thermal efficiency of 40%.

However, the drawback of longer time period of battery charging can be overcome by gasoline engine. As the liquid hydrocarbon fuel has high energy density and time required to fill in the gasoline station is in order of few minutes which is very less compared to hours of battery charging time.

E-vehicle batteries are mainly used in industrial vehicles for powering the propulsion system. The secondary battery is made up of lithium ion. Traction batteries have high ampere–hour capacity used in electric motor cycles, electric golf carts, riding floor scrubbers, trucks, vans and electric cars. E-vehicle batteries are different from starting, lightening and ignition (SLI) battery. It is constructed to provide power over a sustained period of time. Deep cycle battery is replaced with SLI battery for these applications. Batteries of e-vehicle is described by its relative high power to weight ratio, specific energy density, smaller size and low mass which improves the performance and reduces the weight of the vehicle.

Compared to liquid fuel, modern battery technology has less energy and provides maximum range of the e-vehicle. Furthermore, metal–air battery has high specific energy since the cathode is surrounded by oxygen in the air. Rechargeable battery used in e-vehicle includes lead–acid (flooded, deep-cycle, and VRLA), NiCd, nickel–metal hydride, lithium-ion (Li-ion polymer), zinc–air and sodium nickel chloride ('zebra') batteries (Kelber et al. 2003; Muralikrishnan and Kalaivani 2020). The commonly used batteries in contemporary electric cars are lithium polymer, i.e. lithium-ion due to their high energy density compared to weight. The amount of electricity (i.e. charge) stored in battery is measured either in ampere hours or coulombs, with the total energy quantified in watt hours. For battery operated e-vehicle, the battery package is significantly high in cost which is proportionate to the range. For longer range large size battery is required. From 2018, many e-cars with a range of 500 km, for example, the tesla model and TATA electric car had commercialized and are presently used in different vehicle segments.

In this research work, we are proposing a prototype of wireless battery charging system for e-vehicle by means of renewable energy using solar panel and electromagnetic inductive coil. The proposed model is environment friendly and efficient as it uses solar energy, i.e. green energy for charging the battery using wireless power charging mechanism. The solar panel is installed on the rooftop of parking area. During day time the solar energy from sun rays are collected through solar panel and converted to electrical energy and stored in the battery. The additional charging arrangement in night hours or in cloudy days is by using inductive resonance coupling at parking lots. This charging mechanism saves time, as battery of the vehicle will charge automatically without human intervention during the parking hours at night time. This kind of charging is applicable for various e-vehicles such as bus, car, van and light train. WPT uses wireless charging mechanism instead of conventional grid power system using the plug-in cable and is completely autonomous. This approach is advanced and has been used in industrial applications, but their usage in the transport sector are still emerging.

The primary issue concerned with present plug-in charging for e-vehicle is the electric shock, due to any wreckage in cable system. With WPT, it can be prevented since no physical wire or cable is needed and power transfer is done using electromagnetic waves. The secondary issue with the plug-in system is to find the battery charging points. This can be overcome by installing WPT in the parking lots.

The objective of this research is to investigate and design the prototype of WPT for e-vehicle. This technique is used in inductive resonance coupling and to measure the amount of power (voltage and current) that can transfer from transmitting coil (source) to receiving coil (load).

The paper is organized in the following manner. Section 16.1 presents introduction; Sect. 16.2, literature review; Sect. 16.3, system model; Sect. 16.4, results and discussion; and Sect. 16.5, conclusion and future scope.

16.2 Literature Review

Triviño et al. (2021) has used induction technique and capacitive methods for charging the e-vehicle using EM waves. Sovacool et al. (2020) present an extensive document regarding business model of e-vehicle and integrating it with power grid for optimizing the power consumption. Statharas et al. (2019) explain reduction of pollution by the inclusion of e-vehicle within the period of 2030 and 2050. PRIMES-TREMOVE designed by E3M Lab consists of a precise energy-economic model for the automobile sector. Bui et al. (2020) developed an energy monitoring system (EMS) for e-bikes build upon 'Energy bus'. Wei et al. (2019) describe assessment of power capacity which is essential for battery safety and for enhancing energy efficiency. It uses adaptive forgetting recursive total least-squares (AF-RTLS) algorithm to minimize the noise effect. Gonzalez-Gonzalez et al. (2020) present case studies using green energies such as wind, solar and photo-voltaic power plants combined with hybrid energy storage method. Trivinño-Cabrera et al. (2020) explain power transfer wirelessly using heterogeneous technology. Nissan corporation (2021) explains charging of the e-vehicle without cable and emphasized on auto-parking of the car in a particular location using dynamic hassle free steering. Daymak Inc. (2021) has developed sleek e-bike made up of modern technology like carbon and fiber. They installed the wireless charger on EC1 at the head quarter of Daymak, 15 Curity Ave., Toronto. The concept of transfer of electricity using air interface through wireless channel is known as WiTriCity (2021). Luo et al. (2020) use capacitive power transfer method to overcome the disadvantage of conventional Pantograph-catenary contact power supply for railway usage. Paglialunga et al. (2019) explain wireless charging of the boats using 55 kHz resonant system, which is build on the concept of relative distance and lateral shift of the coils. King meter from china (2021) develops different parts of e-bike, sensors and e-bus in 2004.

FastIn Charge (2021) demonstrates e-vehicle for urban environment by introducing a faster and compatible charging solution which enables wide usage of it by integrating with urban grid. They have built a consortium having 9 partners from France, Spain, Greece, Bulgaria, Italia and Slovakia. The consortium has done extensive research on contactless power transmission, automobile engg. and increasing energy efficiency using various innovative techniques. Eltis (2021) explains the comfortable aspects of e-buses compared to conventional buses and introduces auto charging using inductive coil on the roads. In European countries e-vehicle revolutionize the vehicle sector using low energy, cost and zero tail pipe emissions (UNPLUGGED 2021).

Nguyen (2020) explains the negotiation between e-vehicle and wireless charging–discharging lanes (WCDLs) for energy trade-off in de-centralized manner. Vincent et al. (2019) compare the inductive and capacitive wireless charging system. The inductive link establishment is costlier compared to capacitive method, which is easy to implement, low cost, low power and mainly used for short distances. Kelber et al. (2003) explains regarding automation of vehicles using drive by wire mechanism. They have selected mini Baja automobile as the platform for implementing the technique. There are several benefits of e-vehicle such as less vibration, minimum level of noise and pollution, and moreover it can be tested in the laboratory itself.

UShanmuganathan et al. (2006) explains modeling of IC engine using permanent magnet generator and simulation of the same using MATLAB14b and Simulink. The adaptive response of the IC engine in hybrid vehicle is endorsing the permanent magnet generator. Xiong et al. (2007) present unmanned heavy tracked vehicle with remote control system. They have used the method of teaching playback which depends on preview heading. It changes the position in order to minimize the path deviation by operating remotely.

Buchta et al. (2013) demonstrated and analyzed mechatronic mobile research vehicle designed at the University of Applied Sciences, Ostfalia. It has four wheels and de-centralized direct drive, active suspension and steering. It uses longitudinal and lateral dynamics control and analytically calculates the distribution on the drive, brake and the steering system. Frieske et al. (2013) developed the concept of hybrid and battery e-vehicles. They emphasize the different stages, segments of vehicle and architecture of power train. Two different key technologies are battery and electric machines. The results illustrate the usage of front and rear axle motors and the high power electric machines with battery capacity of 19Wh per Kg.. Technical parameters such as, energy capacity and power density are highlighted.

Wang et al. (2016) introduced three-layer architecture of body and brain concept of new electrical/electronic vehicle. The sense and execution layer comprises intelligent network node. The network and transmission layer exchanges information between sense, execution and decision layer. The central co-ordination mechanism analyzes the correlative action among different devices of the vehicle. It co-ordinates and supervises control mechanism of the entire automobile. Data from vehicle to everything (V2X) has real-time online diagnosis and control command for implementation.

Bomber et al. (2006) explains the principle of solar car. The energy obtained from solar panel will be stored in the battery. The battery drives the motor of the engine and moves the vehicle in forward and backward direction. The electrical tapping rheostat is connected to regulate the motor speed. This prohibits overload of current when the vehicle has to stop abruptly. It consumes less fuel compared to the normal cars.

The various kinds of wireless charging systems are RF WPT, inductive WPT, inductive resonant WPT, capacitive WPT, electromechanical (ultrasound) WPT, electric vehicle (EV) WPT and laser WPT. The different forms of radio waves are WiFi, Bluetooth, GPS and cellular phones. The received power can be calculated using Friis transmission equation

$$P_r = P_t G_t G_r \left(\frac{\lambda}{4\pi r}\right)^2 \tag{16.1}$$

where P_t is the transmitted power of the antenna, G_t is the gain of the transmitting antenna, G_r is the gain of the receiving antenna and 'r' is the distance between transmitting and receiving antenna. Inductive resonance method maximizes the coupling between two coils and enhances magnetic field interaction as shown in Fig. 16.1. It has higher efficiency, spatially independent and better flexibility in design. The near field communication (NFC) uses 13.56 MHz of frequency with 250 milli-watt to 1 W power for data and power transmission. It is used for in-band communication (Fisher et al. 2014). Though this method is complex and costly, it has many benefits such as flexible positioning, higher power levels, out of band communication (OOB) with multiple receivers per transmitter and range expansion.

The capacitive method of WPT is as shown in Fig. 16.2, where the source energy is

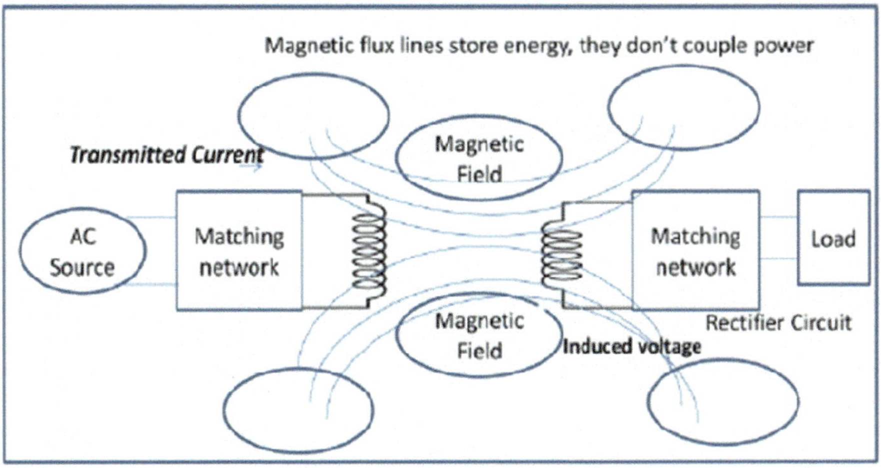

Fig. 16.1 WPT using inductive resonance method

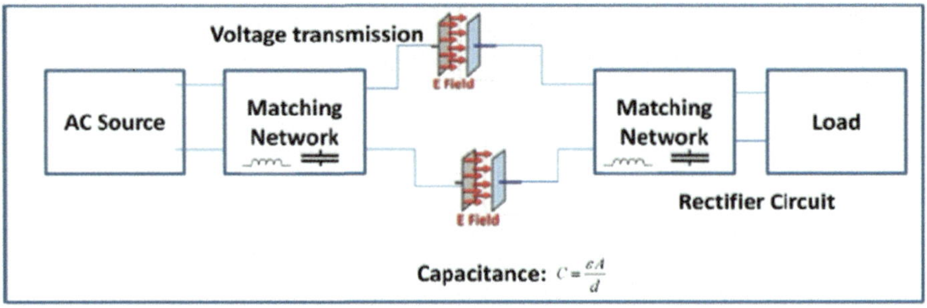

Fig. 16.2 WPT using capacitive method

Table 16.1 Power consumption of different vehicles

Sl. No.	Name of the vehicle	Power consumption
1	Audi e-tron	95 kWh
2	BMW i3	22–42 kW
3	Chevrolet Bolt/Opel Ampera-e	60 kW
4	Fiat 500e	24 kWh
5	Honda Clarity	25.5 kWh
6	Hyundai Iconiq	28 kWh
7	Nissan Leaf I	24–30 kWh
8	Mercedes-Benz EQ C	80 kWh
9	Mitsubishi V	16 kWh
10	Renault Fluence	22 Wh

transmitted through the matching network which generates electric field E(Volt/m) and is transmitted to the load through the matching and rectifier circuit. It is an efficient way of wireless power transmission.

Table 16.1 lists the power consumption of different vehicles.

16.3 System Model

The system model consists of solar panel which is used in order to generate the electrical energy from sun rays as shown in Figs. 16.3 and 16.4. The energy gathered from the panel is stored in the battery. The solar panel controller enhances the output power from the photovoltaic (PV) cell. The transmitter circuit novices the DC supply gathered from the source battery into a high-frequency AC output. It is then carried through inductive resonance coil to the load in the form of electromagnetic (EM) waves. The receiver circuit consists of inductive coil kept inside the e-vehicle. It measures the received EM wave and delivers the analogous DC output. The output is

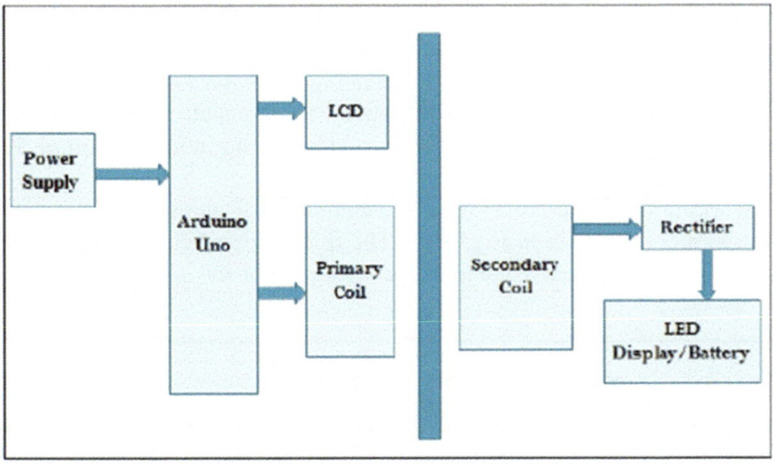

Fig. 16.3 Block diagram of the system model

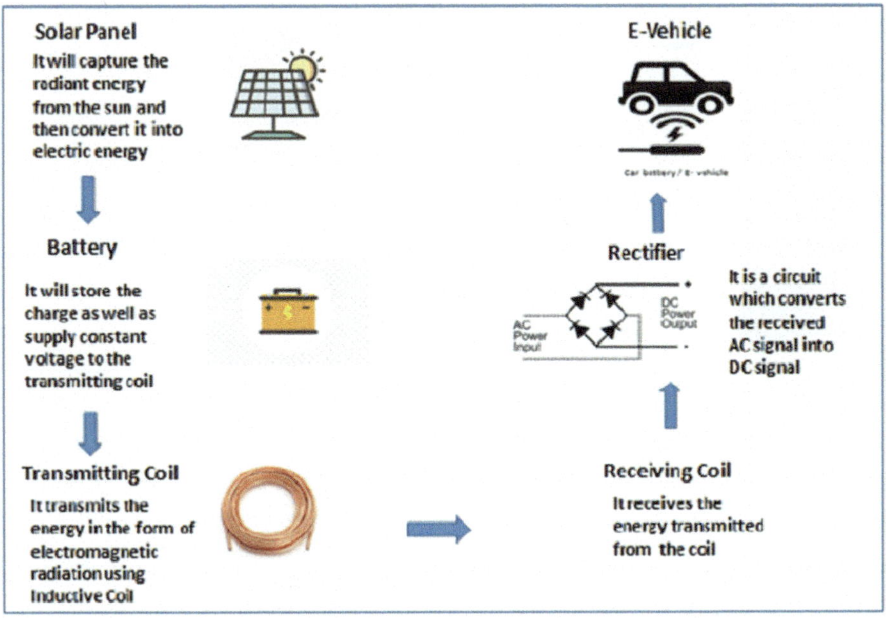

Fig. 16.4 Two different key technologies for charging the e-vehicle i. Solar panel ii. EM induction for wireless charging

used for charging the battery in the e-vehicles. The entire setup is compact, thereby minimizing the space occupancy and results in ease installation and maintenance.

The circuit diagram of the model shown in Fig. 16.5 consists of BD139 transistor, capacitor 224 J, diode 1N4007, 12 V battery, solar panel, transmitting inductive coil of 50 turns with centre tap at 25 turns and receiving inductive coil of 50 turns, Arduino-uno and LCD.

The working principle is based on the technique of mutual inductance via two port transformer as shown in Fig. 16.6. The difference of current flow in primary winding results in induced voltage at the terminals of the secondary winding due

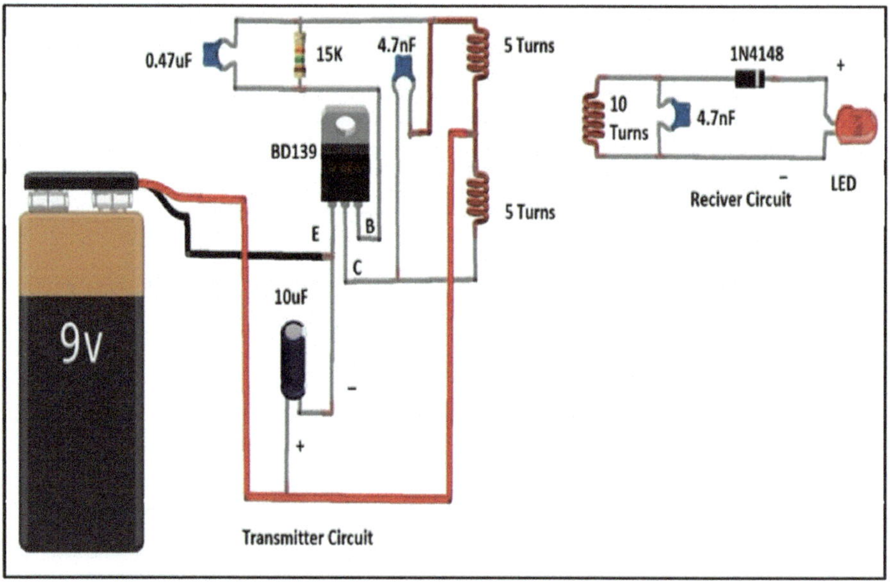

Fig. 16.5 Circuit diagram of EM induction wireless charger

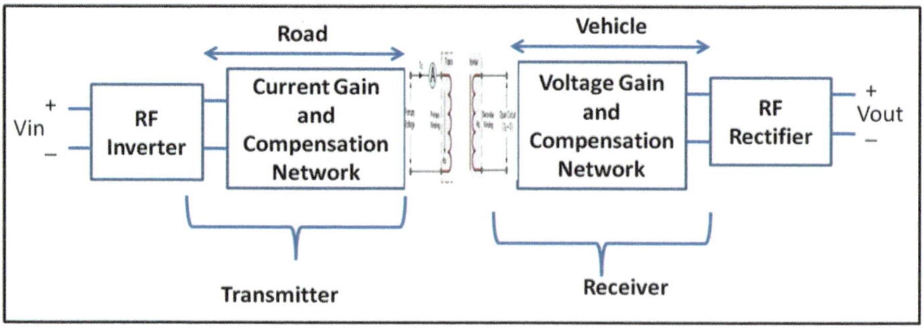

Fig. 16.6 Working principle of wireless charger

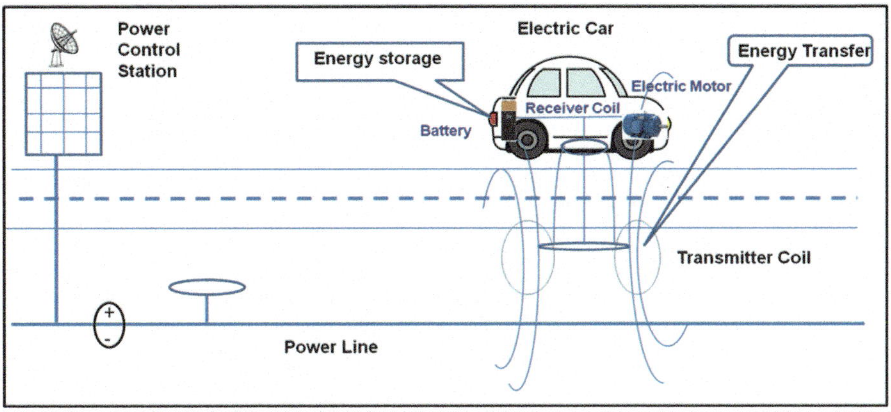

Fig. 16.7 Charging of the E-vehicle while moving on the road

to electromagnetic induction. The magnetic coupling uses the principle of electromagnetic (EM) induction to wirelessly charge the e-vehicles. The setup consists of two coils. A varying magnetic field is generated due to passing of the electric current through the transmitting coil, which induces electromagnetic waves in the receiving coil and is used to charge e-vehicles. To enhance the efficiency of inductive coupling, the primary and the secondary coil must be nearer to each other, i.e. line of sight (LOS) and must be properly aligned.

Figure 16.7 shows the pictorial representation of wireless charging of the e-vehicle using solar panel while moving on the road.

Algorithm 1 explains step by step procedure of execution of the system model. First, the solar panel converts the gathered solar power to electrical energy which is stored in battery. The battery will be connected to the inductive transmitting coil. The energy is then transferred to the secondary coil by the process of induction. The AC signal from the secondary coil is converted to the DC signal using rectifier circuit. The DC energy is stored in battery for charging of e-vehicles.

Algorithm 1: working procedure of the system model.

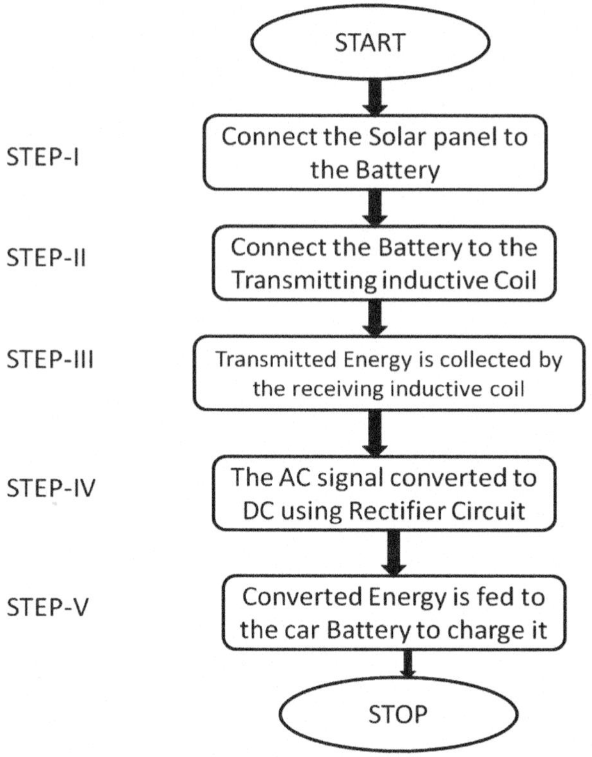

Solar energy is the sustainable green energy we get from sun rays. The demand for renewable sources of energy is increasing globally (United States Environmental Protection Agency 2020; EUROSTAT 2020). Solar panels are made from silicon and convert the solar energy to electrical energy. Solar panels need to be deployed at the top of the parking area where maximum amount of sun light falls during the day time. Fixed and tracking solar panels can be used. The tracking type dynamically changes its direction as per the movement of the sun. The charge controller regulates the charging of the battery and prevents from overcharging. The DC power generated by the battery will be applied to the primary winding of the transformer. Solar fuel, in the contour of biomass, accounts nearly 11% of fuel usage around the world. It is predicted that by the year 2030 the requirement for electricity will increase by twice and the demands for fuel and heat will enhance by 60% globally (https://physicstoday.scitation.org/doi/10.1063/1.2718755).

Assume ten solar panels which are installed and rated of 100watts each and having conversion efficiency of 18%. The sum output power of the solar system can be calculated as

$$\text{Total Power Output} = \text{Total Area} \times \text{Solar Irradiance} \times \text{Conversion Efficiency} \tag{16.2}$$

Total power output is 1000 Watts (10 panels × 100 Watts each). The solar infrared radiance of a surface perpendicular to sun rays at sea level on sunny days is 1000 Watts/sq.m. Consider the conversion efficiency of 20%.

$$\begin{aligned} &\text{Hence, } 1000\text{ Watts} = \text{Total Area} \times 1000\text{ Watts/m}^2 \times 0.2 \\ &\text{Hence, Total area of each solar panel} = 5\text{ m}^2 \end{aligned} \tag{16.3}$$

The solar panels need to be installed in a sequence. The minimum space requirement on the rooftop is estimated as 1 m * 5 m = 5 m^2. Consider the conversion efficiency of the solar panel as 100%. It converts all the solar energy into electrical energy. Hence with 100% efficiency, 1 m^2 solar panel produces 1000 watts of electrical energy (http://www.raymaps.com/index.php/how-to-calculate-the-area-required-by-solar-panels/). As a thumb rule, a 10 m^2 area is required for a 1 KW capacity of solar system. If the actual power available is 8 h per day, the energy needs to be stored and to be utilized as per requirement.

16.4 Results and Discussion

The circuit for e-vehicle uses inductive power transmission (Li 2015) which was simulated using powerSIM simulation software as shown in Fig. 16.8. The transmitter side consists of MOSFETs M1 and M2 which were connected using half-bridge topology. The receiver circuit consists of four diode rectifier. The load was a current generator circuit and produces current for battery charging. The series capacitor generate resonant frequencies in the primary and secondary coil of the transformer. The two frequencies ($f_1 \approx f_2$) must be matched for maximum power transfer.

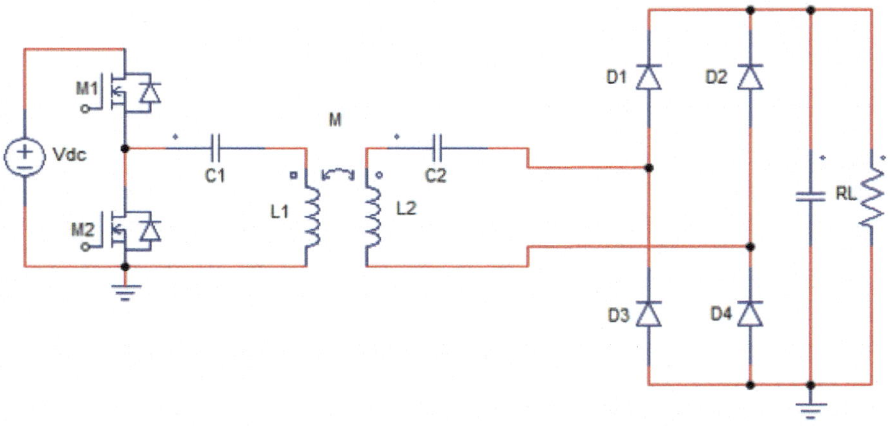

Fig. 16.8 Simulation of the inductive power transmission (IPT) model for e-vehicle using power SIM software

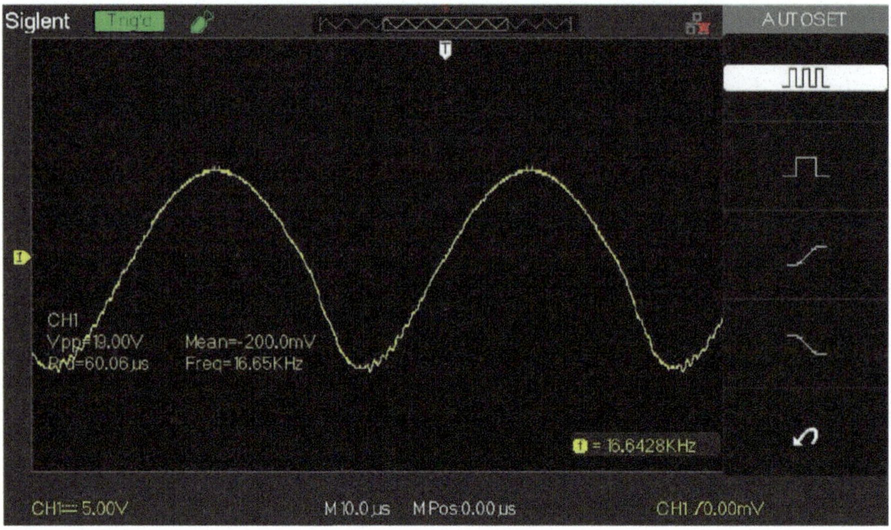

Fig. 16.9 Measurement of output power

$$f_1 = \frac{1}{2\pi\sqrt{L_1 C_1}} \tag{16.4}$$

$$f_2 = \frac{1}{2\pi\sqrt{L_2 C_2}} \tag{16.5}$$

Actual car battery requires 60 W of power for charging. The prototype produces 200 mVolt of power at 16.65 kHz of frequency as shown in Fig. 16.9. The amount of power can be enhanced by increasing number of turns in the primary and secondary winding. Instead of inductive coil, Tesla coil can also be used for increasing the power in a regulated manner. The electromagnetic interference and compatibility (EMI/EMC) effect should be taken care for minimizing inter and intra component interference.

The challenge of WPT system is low efficiency and losses during energy transmission. The deployment cost is high compared to plug-in procedure due to additional infrastructure, EMI/EMC interference and compatibility and also safety/shielding measures. Primary and secondary coils must be aligned properly and minimum air gap distance should be maintained. This method is more effective for shorter distance. Wireless charging systems (WCS) is used for high-power applications. The e-vehicle need fewer dimension for storage of costly battery and increased range of transportation.

Due to the magnetic permeability of the atmospheric materials, such as air, water and dust, the wireless charging need to be weatherproof. So that it can transfer energy in adverse climatic condition such as rain, snow or dirt. The volume of the battery will be small to reduce weight and cost. WPT can be operated by several methods such as microwave, capacitive, inductive, acoustic and luminous. The procedure is user friendly and saves battery charging time.

16.5 Conclusion and Future Scope

The use of WPT proves to be beneficial and highly efficient compared to the plug-in transmission. In coming days, large-scale setup of wireless power station is expected as per the prototype model. The use of solar power decreases the dependence of the station on the government or private players for the power input. This maximizes its adaptability and makes it suitable for positioning over all topography regions that collects substantial amount of solar radiation.

In addition, the usage of wireless data transfer is very much efficient as opposed to wired transmission. A large-scale construction of wireless power station has to be done in the near future. It improves flexibility for installing on every landscape that absorbs enough solar radiation.

The future research directions includes increasing battery efficiency, dynamic charge parking, e-vehicle solar roofs, rapid wireless charger standardization for different vehicle specifications, DC quick charging skills and autonomous charging. The next generation WPT has the capability to accomplish numerous things. With enhanced research and optimization, this technology will be able to reach a great height in the future with increased capability and efficiency.

Conflict of Interest I declare there is no conflict of interest with anyone for this research article.

References

Arulraj B, Elumalai M, Rooba M (2019) Wireless charging for electric vehicle using solar PV-wind system. Int J Adv Res Electr Electron Instrum Eng 8(Issue 3). https://doi.org/10.15662/IJAREEIE.2019.0803067

Bomber (2006) Wireless power transmission: an obscure history, possibly a bright future. Physics 464: Applied Optics

Buchta R, Liu-Henke X (2013) Energetic aspects for an integrated vehicle dynamics control for E-vehicles with de-central drives and all-wheel steering. In: 2013 international symposium on electro-dynamic and mechatronic systems (SELM). Opole-Zawiercie, Poland

Bui VT, Dow CR, Huang YC, Liu P, Thai VD (2020) A Canopen-based gateway and energy monitoring system for electric bicycles. Energies 13:3766

Daymak (2021) Daymak launches new special edition EC1 featuring built in wireless charging technology; plans to roll out this technology on certain E-bike models. https://www.prnewswire.com/news-releases/daymak-launchesnew-special-edition-ec1-featuring-built-in-wireless-charging-technology-plans-to-roll-out-this-technology-on-certain-e-bikemodels-300464874.html. Accessed 23 Jan 2021

Doug K (2015) Electric vehicle charging technology analysis and standards. FSEC Report Number: FSEC-CR-1996-15, U.S. Department of Transportation's University Transportation Centers Program Contract Number: DTRT13-G-UTC51. http://evtc.fsec.ucf.edu

Eltis (2021) Wireless charging for quiet and clean public transport in Torino (Italy). https://www.eltis.org/discover/case-studies/wireless-charging-quiet-and-clean-public-transport-torino-italy. Accessed 22 Jan 2021

EUROSTAT (2020) Trends in greenhouse gas emissions; Technical Report. EUROSTAT, Luxembourg

Fan J, Zhang C, Wang Z, Dong Y, Nino CE, Tariq AR, Strangas EG (2010) Thermal analysis of permanent magnet motor for the electric vehicle application considering driving duty cycle. IEEE Trans Magn 46(Issue 6)

Fastin charge: innovative fast inductive charging solution for electric vehicles. https://cordis.europa.eu/project/id/314284/reporting/es. Accessed 1 Feb 2021

Fisher TM, Farley KB, Gao Y, Bai H, Tse ZTH (2014) Electric vehicle wireless charging technology: a state-of-the-art review of magnetic coupling systems. Wirel Power Transf J 1(Issue 2):87–96

Frieske B, Kloetzke M, Mauser F (2013) Trends in vehicle concept and key technology development for hybrid and battery electric vehicles. Barcelona, Spain

Gonzalez-Gonzalez JM, Martin S, Lopez P, Aguado JA (2020) Hybrid battery-ultracapacitor storage system sizing for renewable energy network integration. IET Renew Power Gener 14:2367–2375

Kelber CR, Dreger RS, Comes GK, Webber D (2003) Cell-phone guided vehicle, an application based on a drive-by-wire automated system. In: IEEE IV2003 intelligent vehicles symposium. proceedings (Cat. No. 03TH8683) at Columbus, OH, USA

KingMeter (2021) Wireless charging. http://www.king-meter.com/#/product/17. Accessed 26 Jan 2021

Li S, Mi CC (2015) Wireless power transfer for electric vehicle applications. IEEE J Emerg Sel Topics Power Electron 3(No. 1)

Luo B, Long T, Guo L, Dai R, Mai R, He Z (2020) Analysis and design of inductive and capacitive hybrid wireless power transfer system for railway application. IEEE Trans Ind Appl 56:3034–3042

Muralikrishnan P, Kalaivani M (2020) IOT based electric vehicle charging station using Arduino Uno. Int J Adv Sci Technol (IJAST) 29(No. 6):4101–4106. ISSN: 2005-4238 Copyright ©2020 SERSC4101

Nguyen DH (2020) Electric vehicle—wireless charging-discharging lane decentralized peer-to-peer energy trading. IEEE Access 8:179616–179625

Nissan (2021) Wireless charging system | NISSAN | TECHNOLOGICAL DEVELOPMENT ACTIVITIES. https://www.nissan-global.com/EN/TECHNOLOGY/OVERVIEW/wcs.html. Accessed 1 Feb 2021

Paglialunga E, Marconcini P, MacUcci M (2019) Wireless charging of batteries for electric boats. In: Proceedings of the 5th international forum on research and technologies for society and industry: innovation to shape the future, RTSI, Florence, Italy, 9–12 September 2019. IEEE Inc., New York City, NY, USA, pp 325–330

Panchal C, Stegen S, Lu J (2018) Review of static and dynamic wireless electric vehicle charging system. Eng Sci Technol Elsevier Int J 21:922–937

Shanmuganathan U, Govarthanan R, Muthumailvaganan A, Imayakumar A (2006) Modeling and dynamic simulation of IC engine driven permanent magnet generator using matlab/simulink for hybrid tracked vehicle. In: 2006 IEEE conference on electric and hybrid vehicles at Pune, India

Sovacool BK, Kester J, Noel L, Zarazua de Rubens G (2020) Actors, business models, and innovation activity systems for vehicle-to-grid (V2G) technology: a comprehensive review. Renew Sustain Energy Rev 131:109963

Sreedhar V (2006) Plug-In hybrid electric vehicles with full performance. In: 2006 IEEE conference on electric and hybrid vehicles at Pune, india

Statharas S, Moysoglou Y, Siskos P, Zazias G, Capros P (2019) Factors influencing electric vehicle penetration in the EU by 2030: a model-based policy assessment. Energies 12:2739

Trivinño-Cabrera A, Gonzaález-Gonzaález JM, Aguado JAJA (2020) Wireless power transfer for electric vehicles: foundations and design approach. Springer, Cham, Switzerland, p 175

Triviño A, González-González JM, Aguado JA (2021) Wireless power transfer technologies applied to electric vehicles: a review. J Energies 14(Issue 6). https://doi.org/10.3390/en14061547

Final Report Summary—UNPLUGGED (Wireless charging for electric vehicles) CORDIS, European Commission. https://cordis.europa.eu/project/id/314126/reporting/es. Accessed 25 Jan 2021

United States Environmental Protection Agency (2020) Fast facts on transportation greenhouse gas emissions. Green Vehicle Guide, USEPA. https://www.epa.gov/greenvehicles/fast-facts-transportation-greenhouse-gas-emissions. Accessed 13 January 2021

Vaessen P (2009) "Wireless power transmission", Leonardo energy, briefing paper. Electricity

Vincent D, Huynh PS, Azeez NA, Patnaik L, Williamson SS (2019) Evolution of hybrid inductive and capacitive AC links for wireless EV charging—a comparative overview. IEEE Trans Transp Electrif 5:1060–1077

Wang J, Yang DG, Lian XM (2017) Research on electric/electronic architecture for connected vehicles. In: IET international conference on intelligent and connected vehicles (ICV 2016), on April 2017 at Chongqing, China

Wei Z, Zhao J, Xiong R, Dong G, Pou J, Tseng KJ (2019) Online estimation of power capacity with noise effect attenuation for lithium-ion battery. IEEE Trans Ind Electron 66:5724–5735

WiTricity: powering life, wirelessly. https://witricity.com/. Accessed 29 Jan 2021

Xiong G, Chen H, Gong J, Wu S (2007) Development and implementation of remote control system for an unmanned heavy tracked vehicle. In: 2007 IEEE intelligent vehicles symposium. Istanbul, Turkey

https://physicstoday.scitation.org/doi/10.1063/1.2718755

http://www.raymaps.com/index.php/how-to-calculate-the-area-required-by-solar-panels/

Chapter 17
Power Quality Issues in Smart Grid/Microgrid

S. Vijayalakshmi, R. Shenbagalakshmi, C. Pearline Kamalini, M. Marimuthu, and R. Venugopal

Abstract Microgrids and smart grids are emerging as the latest trending aspect in power industries. The smart grid integrates the technology dealing with Information and Communication in almost all aspects of power systems starting from electricity generation till consumption in order to improve the reliability of energy consumption and service, minimize the environmental impact, enable active participation of the consumers, new products and markets, improves the efficiency leading to more safe and reliable energy production and distribution. The other benefits include reducing carbon emission, supports the increased use of electric vehicles and creates wider opportunities for employment. Smart grids can be seen as a combination of microgrids and mini grids among which microgrid plays a major role in accomplishing authentic and more secure energy supply for retail load as well as distributed generation. On the other hand, microgrid can be seen as a decentralized energy system comprising distributed energy sources with demand management, storage and generations with loads which are capable of operating either in parallel or independently. Despite the benefits, smart grid as well as microgrids face several power quality-related issues and challenges which are to be met out in order to avail the entire benefits of this emerging technology. The challenges faced by the smart grid and microgrid can be categorized as two, viz., wide variations in power

S. Vijayalakshmi (✉) · C. P. Kamalini · M. Marimuthu · R. Venugopal
Saranathan College of Engineering, Trichy, Tamilnadu, India
e-mail: vijayalakshmi-eee@saranathan.ac.in

C. P. Kamalini
e-mail: pearline-eee@saranathan.ac.in

M. Marimuthu
e-mail: marimuthu-eee@saranathan.ac.in

R. Venugopal
e-mail: venugopal-eee@saranathan.ac.in

Present Address:
R. Shenbagalakshmi
SKN Sinhgad Institute of Technology and Science, Lonavala, Maharashtra, India

A. K. Bohre et al. (eds.), *Planning of Hybrid Renewable Energy Systems, Electric Vehicles and Microgrid*, Energy Systems in Electrical Engineering,
https://doi.org/10.1007/978-981-19-0979-5_17

quality which are unpredictable including, slow voltage changes, frequency deviations, harmonics, flicker and unbalance. Events including rapid voltage changes, dips, swells and interruptions. The disturbances and variations in power quality are mainly caused due to harmonic emission by power electronic devices, interference between power line carrier communication and devices, immunity of the devices and weakening of the transmission grid. This chapter aims to identify the root cause for the above-said power quality issues and challenges and investigation of various mitigation techniques.

Keywords Microgrid · Smart grid · Power quality

17.1 Introduction

The Smart Grid (SG) and microgrid (MG) power quality (PQ) problems are discussed in this chapter. Section 17.1.1 describes about the SGs, Sect. 17.1.2 explains the PQ challenges in SGs, Sect. 17.1.3 illustrates the PQ challenges in both AC and DC MGs. The flow of this chapter is as shown in the Fig. 17.1a

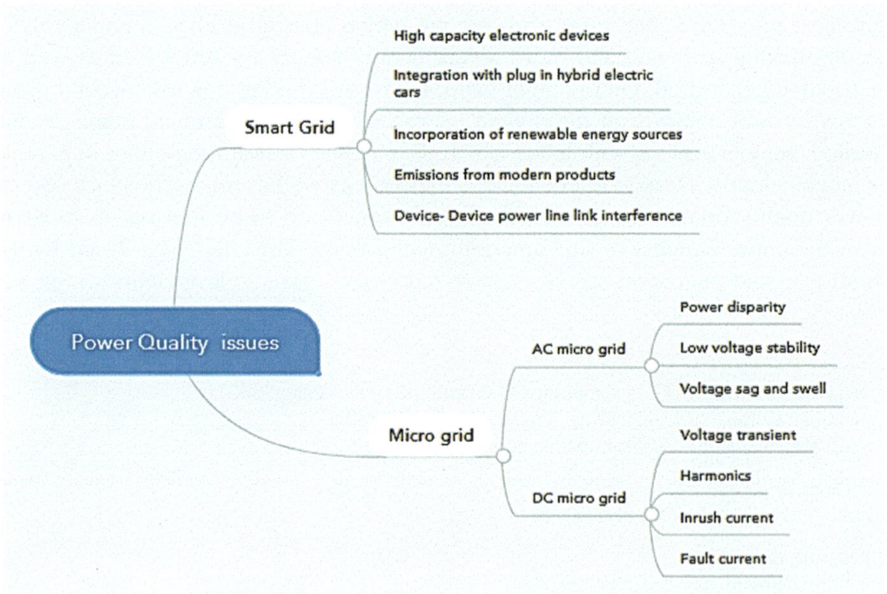

Fig. 17.1 A graphical representation of the chapter (**a**). **b** Block diagram of smart grid

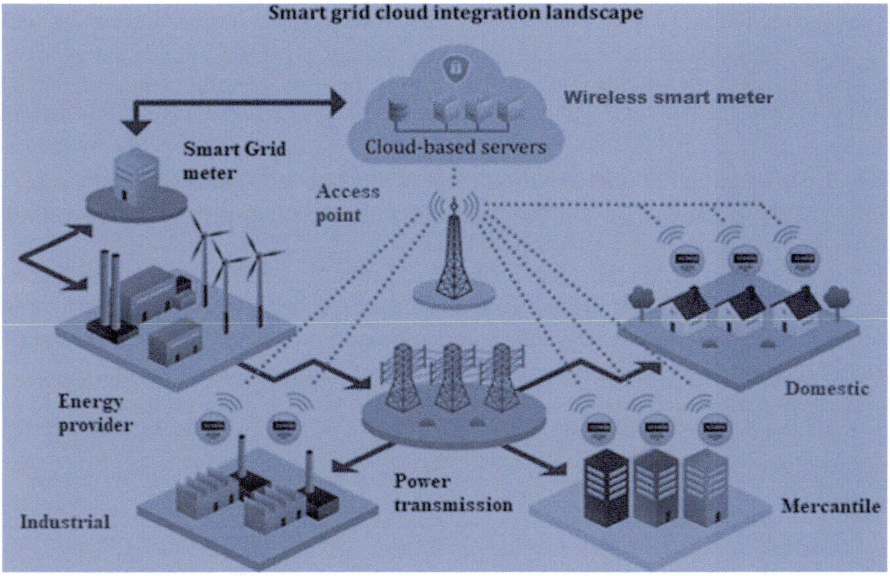

Fig. 17.1 (continued)

17.1.1 Smart Grid

- A SG is made up of claims focused on information that will allow for the seamless incorporation and advanced penetration of renewable energy. Accelerating the production and mainstream use of plug-in hybrid electric vehicles (PHEVs), in addition to their future use as grid storage, would be critical. It can be described in a variety of ways (Naderi et al. 2019):
- SG is a bright intelligent device that adds intelligence and networking capabilities to current electrical systems, such as automated metering technology, to increase the national grid's electricity efficiency.
- SG keeps track of the whole electricity chain, including transmission and delivery facilities from power plants to customers. For both consumers and utilities, it ensures advanced energy efficiency (EF). It reroutes traffic from peak hours to off-peak hours depending on goals (Gandoman et al. 2018).

However, a SG is a fully integrated grid. It will automatically react to changes in electrical parameters that are essential for the grid's smooth operation. Sensors, microcontrollers and other devices can assist with this.

Energy thefts can be stopped, correct power bills can be provided, clean energy can be encouraged, greenhouse emissions can be reduced and power losses can be reduced with the SG. The Internet age has also ushered in new developments in the power sector, with SGs now being created. It is the most recent generation of grids, using smart technologies.

SGs, in contrast to MGs, have automated knowledge and power, complex optimization of grid services, dispersed resources (similar to MGs or exactly smart MGs), request replies, request side capitals, EE possessions, smart monitoring scheme and distributed resources (like MGs or specifically smart MGs). Advanced electricity storage and smart incorporation (actual response and appropriate usage evidence). SGs also consume additional aids, faster safety, regulated auto healing, greater robustness, greater green energy, greater efficiency, higher PQ, greater re-configurability and a higher capacity factor (Jolhe et al. 2016).

Depending upon the concepts, a smart MG may be termed as a MG with some unique features that will boost the all-inclusive performance of the structure in order to sort it more environmentally efficient, gain maximum functionality by rising the energy intensity and increase the entire usage and value of current productions and transmission power. It may be utilized to rise the sources of renewable energy, boost EF to meet emerging modern demands and make the system more efficient, resilient, versatile, long-lasting and so on.

17.1.2 Power Quality Challenges in Smart Grid

SGs, like all other technologies, pose certain threats to conventional grids. Meantime, these often carry some novel techniques to boost the operations. The following paragraphs outline the major problems and resources that SGs can offer (Senthilkumar et al. 2015):

1. **High-capacity electronic devices**.
2. **Integration with plug-in hybrid electric cars**.
3. **Incorporation of renewable energy sources**.
4. **Emissions from modern products**.
   We expect development in distributed production (making at under voltage extents) and different modes of ingestion as a result of SG deployment (for instance, E-Vehicles accusing positions, extended large speed railways, etc.). Harmonic pollution is one of the PQ disruptions that some of these modern consumption devices produce. Below the frequency of 2 kHz, harmonics and inter-harmonics would be high. The release is low over the entire range above 2 kHz, where it is created through additional equipment such as energy effective drives, micro-mini generators and PV connections. It would be more difficult to quantify these small amounts of harmonics at advanced frequencies than it is now for higher levels and lower frequencies.
5. **Device-to-device and power line link interference**
   Phones, consumers, distributed generators and the grid operator would be able to communicate more easily with SGs. Because of its ease of availability, power line communication may seem to be an obvious alternative, but it could create new disruptions in the power grid, subsequently a further decrease in power efficiency. Diverged disruptions can occur relying on the incidence used for

power line communication, impeding with radio broadcasting and communication. Current strategies may hamper with power line announcement by causing an extraordinary degree of interference with the occurrence preferred for power line communication or through making a small impedance direction, essentially choosing out the signal.

6. **Increase in distorted Harmonic voltage**
   In SGs, demand volume would rise by equal development in output. Because of the continued increase in both construction and ingestion, the distorted harmonic voltage may suit unsatisfactorily high. Furthermore, the number of switching sequence may continue to rise, potentially reaching unsatisfactorily great levels. At the power system frequency, output and usage are in equilibrium, but not at harmonic frequencies.

With the above PQ concerns, it's tempting to believe it the proliferation of smooth grid technologies would exacerbate PQissues. However, here the other edge of the storey! Smooth grid expertise would aid at the reduction of PQ issues. They could only handle PQ if they could calculate it, the SG allows us to do so. Several elements of the SG work together to help utilities provide advanced excellence power to our house. Smooth metres on the supply grid aid in the management of voltage and power factor. Advanced energy metres, which are a core component of SG, provide both you and your utilities with more knowledge about the electricity supplied to your house. They have a capacitor, much like most modern gadgets, to step down the voltage for the digital electronics. To avoid interference with other electronic or networking devices, they are designed to follow stringent FCC (Federal Communication Commission) standards.

SG also guarantees that the systems connecting to it are switched and protected, resulting in network stability that is unrivalled. However, low PQ may have a significant effect on this, resulting in system malfunction and/or failure. SGs provide progressive Distributed Management System (DMS) applications such as PQ Analysis to address this. "Utilities may use Smart Grid technologies to improve service efficiency for each customer they support, resulting in a more reliable and accurate electricity supply."

SG also allows for the improvement (or prevention) of current power system output without the need for significant improvements in lines, wires, transformers and other infrastructure. Improvements may be made in terms of efficiency, voltage output or price from the customer's perspective. Figure 17.1b Shows the block diagram of SG, it could be implemented on commercial, industrial and residential, etc., is represented in the diagram.

A decrease in long-term voltage–magnitude variation is planned by rendered SG to the immediate imminent, which will increase voltage efficiency. The real-time data, utilities may practice SG technologies to improve supply efficiency for each customer they service, resulting in a more reliable and accurate power supply.

17.1.3 Microgrid

MGs are designed to provide electricity to a limited region. It's similar to supplying electricity to a rural location, a residential, government or industrial building in a community where the national grid is unavailable. The MG infrastructure is an insignificant supply system made up of output utilities and distributed energy supplies (DES) including sources of renewable energy (RE), co-generation, combined heat and power (CHP) generation, fuel cells and energy storage systems. A MG is a self-reliant electricity structure that covers a specific geographic area, such as a college campus, a hospital complex, a commercial district or a neighbourhood. MGs are driven by one or more types of distributed electricity (PV panels, wind system, shared heat and water, generators). Besides, several fresher MGs provide energy storage, mostly in the arrangement of batteries. Some now have charging points for hybrid vehicles. The MG, which is connected to nearby buildings, delivers electricity and potentially heat and cooling to its customers through sophisticated software and control systems. Figure 17.2 illustrates the applications of MG.

MGs come in a variety of shapes and sizes.

1. Institutional MGs and the campus climate.
2. MGs in the community.
3. MGs that are off-grid and off-the-grid.
4. MGs for military bases.
5. MGs for commercial and industrial (C&I) use.

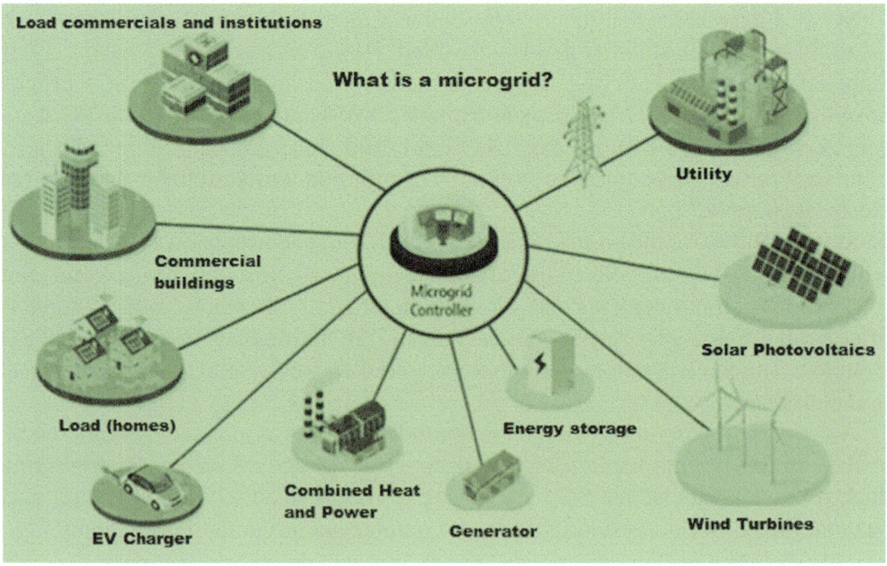

Fig. 17.2 Structure of microgrid

The combination of the erratic sort of Renewable Energy (RE) sources with progressive Power Electronics (PE) converter equipment causes a PQ problem. In addition, the existence of nonlinear and destabilizing loads in MG looks to have an effect on the PQ of the energy supply in the power delivery network.

In a (MG, three separate PQ problems are assessed: voltage decline, harmonic falsification and phase unbalance. An assorted-integer direct accumulation with harmonic load movements is proposed as a design for an energy storage procedure for MG. It has in charge of both optimizing the preparation of entire production, depository and properties of load, as well as resolving PQ problems in the tertiary extent of mastery by changing the extents of specific categories of loads inside the grid. On a theoretical experiment-case MG with housing, manufacturing and viable loads, this procedure is replicated for various situations. The findings show that the algorithm's demand-side control system will adjust the consumption behaviour of such loads effectively, ensuring that the voltage decrease, distortion in harmonic voltage and potential instability factor follow the necessary requirements in each MG node during the daytime. It is too worth noting that if energy price on the spot market rises, the MG will steadily decrease the amount of energy it buys after the efficacy grid to where it has connected.

17.1.4 Power Quality Subjects in Microgrid

A MG could work in both independent and grid-tied modes. It faces a PQ problem in both modes, where it is dissimilar together. AC MG and DC MG have also been proposed to alternatives in the literature. DC MGs have been shown to have properties that mitigate certain PQ issues. DC MG, on the other hand, has its own set of problems with power efficiency.

A. AC Microgrid Problems

An AC MG is a tiny grid structure that is associated to various kinds of distributed generators and loads. It's difficult to maintain grid reliability and power balancing with too many different types of generation and loads. There are a number of problems with AC MG. These are the problems:

i. voltage safety, ii. system reliability, iii. power output, iv. system security and control.

The following are few of the PQ problems in an AC MG:

1. **Power Disparity**

As the MG transitions from grid connected to disconnect method of action, there is a power imbalance. In grid-isolated mode, a separate micropower station connected to the MG provides electricity. Energy depository system is used to sustain power imbalances that exist throughout the MG's transformation time when these power stations have a sluggish dynamic reaction. As regular mode is restored, the phase series and voltage magnitude must be maintained in order to synchronize with the grid.

2. **Low-Voltage Stability**

Since power delivery support in distributed generation and MGs is smaller than in large grids, low-voltage stability issues arise. Since DGs and MGs are regulated by power converters, the power transmission in the converters is restricted in contrast with normal power plants. This type of converters has small energy than conventional systems at times of heavy grid demand. Traditional power plants have a lot of kinetic energy contained in their spinning turbines, so they can handle a lot of power during intermittent situations. When the huge grids are experiencing a power outage and the MG is switched to islanded approach, power-splitting care from the MG to the huge grid is nil. During the power blackout, the remote action will activate a disparity from the supply demand ratio, which will result at the grid's voltage outline being rejected.

3. When a MG experiences voltage sag or swell, the effectiveness of converter in power electronics, the distributed generation can suffer. Throughout sag periods, the grid-tied loads demand the reactive power and distributed generators based on power electronics converters attempt to inject reactive power, they encounter overcurrent in single or more phases. Another issue is an exceeding the extreme voltage cap, which could reason for the generators to trip and result in an outage.

B. Problems in DC MG

Many publications focus on PQ problems in AC MG systems, nonetheless slight consideration has given in PQ problems at DC MGs. In addition to consuming and supplying electricity to and from the grid, the DC MG can also run at grid-tied mode. It also operates in an isolated method of service. Imbalance in voltage of two-faced DC bus, voltage variations presents at the AC system, overflow current and harmonics in DC MG are some of the common PQ problems in DC MG (Ghetti et al. 2012).

1. **Voltage Transient**

This issues are common in AC MGs, nevertheless they also impact DC MGs. The key causes of voltage transients are capacitor bank swapping, distributed generation initialization and shutdown in a DC MG and load shift. The capacitor banks are used to switch so that the changes in voltage transfer from the low-voltage AC MG to DC MG through a rectifier reach up to 194 percent of the operating voltage. Thereafter, this changes in voltage stabilizes to a higher voltage level on the amount of 111 percent of the working voltage (Long et al. 2013).

2. **Harmonics at DC MG**

In a DC MG, low-level harmonics are inactive since the absence of AC–DC converter, but expanded usage of the DC–DC converter will reason for the presence of electromagnetic interference. However, several times a DC device lacks harmonics, the term harmonics refers to different frequencies of the fundamental; it is known as operating frequency of a DC–DC converter. Harmonics in a DC circuit refer to fluctuating current and voltage began by the device's working frequency. Various converters operated by PWM pulse can be used in DC MG systems in various loads and generating stations, with capacitors attached on both sides of the converter. Different resonance frequencies are caused by resistance of the DC bus and the reactive impedances of different capacitors. Whether the particular frequencies are matching with the resonant frequencies, the influence of these harmonics on the DC MG may be disastrous (Railing et al. 2004).

3. **Inrush Current**

In an AC configuration, inrush current is caused by transformers, induction motors and other strong inductive loads; however, the explanation for inrush current in a DC MG is different. Harmonics are generated when multiple loads, dispersed generating stations and different storage devices are linked via power electronics converters. As a result, sufficient filter is mounted on the load side. Additionally, Electromagnetic Interference (EMI) filters are installed on the AC edge. In a DC MG, the capacitor in this EMI filter draws a lot of inrush current. The extent of inrush current is determined by the DC system's voltage frequency, capacitor impedance and converter capacitance. Another explanation for arrival current in DC MGs are that when it is not energized loads are switched it in a loop, their big filter capacitors draw a lot of arrival current. The arrival current is so large that it could result in actual welding at the point of contact (Akagi 1994).

4. **Fault Current**

In a DC MG, fault current is generated by capacitors added to the grid by converters or various distributed generating stations, or by the EMI capacitor filter. The total liability of current is limited by the converters' ratings meanwhile entire converters in power electronic are operated in closed loop. When only a small amount of faulty current runs from a DC MG due to a transformer used in Power Electronics, safety systems have a hard time distinguishing between overload and fault conditions. As a result, designing a proper DC MG security scheme is a significant challenge. Unlike an AC system, a DC circuit wouldn't be a normal voltage and current which are crossing the zero. This means that if an arc failure happens in a DC system, it wouldn't reduce as easily as it has in an AC system. Since it is a self-supporting liability and this is does not have surge current in the DC grid, it can persist for a lengthier duration and create a significant issue in the scheme. Lower fault current, otherwise, will exacerbate voltage unbalance in a DC MG when a fault occurs (Khadkikar and Chandra 2008).

This chapter is organized into the following sections: Section 17.1 illustrates the PQ issues and challenges in SG. Section 17.2 explains the various techniques and devices used for improving the PQ in SG. Section 17.3 discusses the PQ issues and challenges and their mitigation techniques related to DC and AC MGs.

17.2 Causes and Mitigation of Power Quality Issues in Smart Grids

In this chapter, the major causes of the PQ issues and their mitigation in SGs are explained in detail.

17.2.1 *Harmonic Emission by Power Electronic Equipment*

In the field of power electronics, there have been several recent advancements which have resulted in a greater integration of power electronic devices into existing electrical grids. On the one side, these devices improve grid stability and performance while also posing new problems, such as introducing harmonic content into the grid (Naderi et al. 2019). The majority of nonlinear loads such as power electronics gadgets, switcher, information handling apparatus and high-performance lighting loads are the major reason which cases harmonic distortion. Harmonic distortion is the one in which the waveforms of voltage or current take on a non-sinusoidal structure. The waveform is defined by a set of sine waves of various magnitudes and phases with frequencies which are multiples of the grid frequency. This in turn creates a major consequence in the grid such as upsurge in the occurrence of resonance, imbalance at neutral point in three-phase systems, overheating of wires as well as equipment, deterioration of electrical machine's output, interfering with electromagnetic waves in case of communication systems, measurement inaccuracy when using DC metres and spurious thermal safety tripping (Gandoman et al. 2018).

To reduce the effect of harmonics in the SG, filters can be used. Filters are broadly categorized into active and passive filters. Passive filters are the one which uses the combination of any one of the passive components like resistors, inductors and capacitors. Active filters combine both passive and active components and uses external source for its operation (Jolhe et al. 2016). The usage of a shunt active filter mitigates current-related PQ problems alike harmonic distortion, poor power factor and consumption of reactive power. Shunt active filters are connected between the power source and the load at the Point of Common Coupling (PCC). The active filter is typically made up of power inverters and functions as a balanced source of current (Senthilkumar et al. 2015). Figure 17.3 shows the SG with shunt active filter.

17.2.2 *Interference Between Grid Connected Devices and Power Line Communication*

Power Line Communication (PLC) is the first networking technology that came out with SG, and it now offers high-speed connectivity to a variety of SG applications. Narrowband transmission, spread-spectrum transmission and DSP-processed

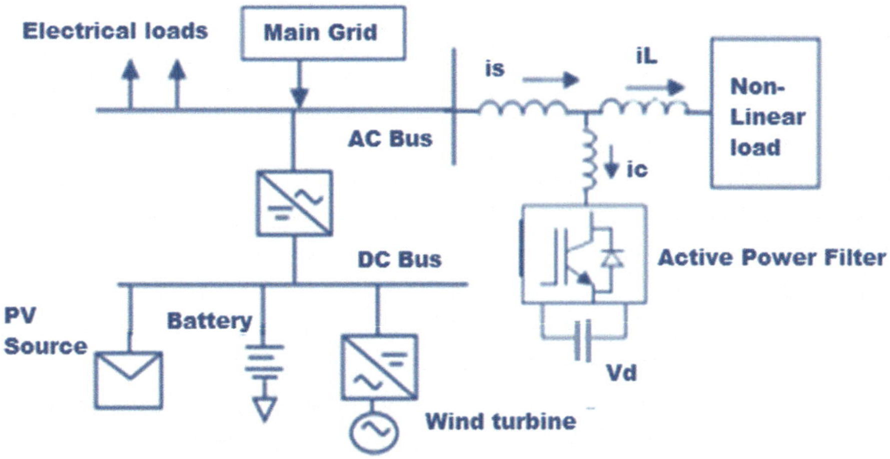

Fig. 17.3 Smart grid with shunt active filter

narrowband transmission are the three networking systems used in PLC. It seems that the spread-spectrum technology is quite disadvantageous since it diminishes the performance of PLC, when its channel is defective. On the other hand, due to the presence of the impulse noise, narrowband transmission system's performance also decreases. By integrating narrowband transmission and DSP, the results of impulse noise on narrowband transmission has been minimized and error-free communication is accomplished. DSP can also be used to resolve the narrowband transmission channel distortion issue, making PLC quite powerful. Commonly, power line communication (PLC) systems use the 1–30 MHz range to communicate a modified carrier signal over a prevailing grid. PLC employs symmetric connections which can transmit data at up to 200 Mbps. It has a range of maximum communication distances. It has a medium voltage of 3 km and a low voltage of 200 m. Until now, the greatest disadvantage of PLC has been the lack of regularity (Yigit et al. 2014).

PLC is used in many of the applications involved in SG. Some of the main applications are discussed below.

17.2.2.1 Infrastructure for Advanced Metering and Modern Monitoring Devices

PLC's two-way communication allows consumer devices and systems to communicate with one another in advanced metering infrastructure. Real-time pricing can be offered in this way by allowing service providers to communicate with metering devices. Customers will also read about their new billings. Because of its low maintenance costs, smart metering is cost-effective for utilities. Remote metre reading is accomplished using Ultra Narrowband PLC (UNB-PLC) technology. Despite the fact that UNB-PLC devices have a poor transmission rate, their connectivity range is 150 kms. PLC measures energy demand based on various time regions, for instance weekends and work hours. Further, it allows the identification of power theft as well as information gathering from another meter reading arrangements like water/gas metres. PLC's two-way data transmission often enables gas/water supply to be switched off on request once the customer is not in need of them for an extended period of time (Yigit et al. 2014).

A key parameter for understanding and optimizing system operation is monitoring in actual time and display of devices and output over a large region. Sophisticated system knowledge prevents brownouts and produces system logs in order to anticipate and prevent possible failures, produce for further decision-making, eliminate broad range disruptions and increase transmitting capacity and quality of the grid. This particular attribute marks the start of the trail that leads to PQ issues. SGs would be simply grids without these advanced metering and monitoring interfaces but activating this function will increase the performance, speed and precision of PQ issues in SGs (Naderi et al. 2019).

17.2.2.2 PLC-Based Electric Vehicle-to-Grid Connectivity

Integration of battery-operated PHEVs are made possible with the recent advancement in SG technologies. The battery of PHEV is energized by using electric vehicle supply equipment. The PHEV will communicate securely with the electric vehicle supply equipment using a PLC. The Society of Automotive Engineers has tested various PLC technologies like broadband-PLC and narrowband-PLC). In terms of applicability, narrowband-PLC has performed better (Yigit et al. 2014).

Figure 17.4 illustrates the infrastructure of the electric vehicles being charged from

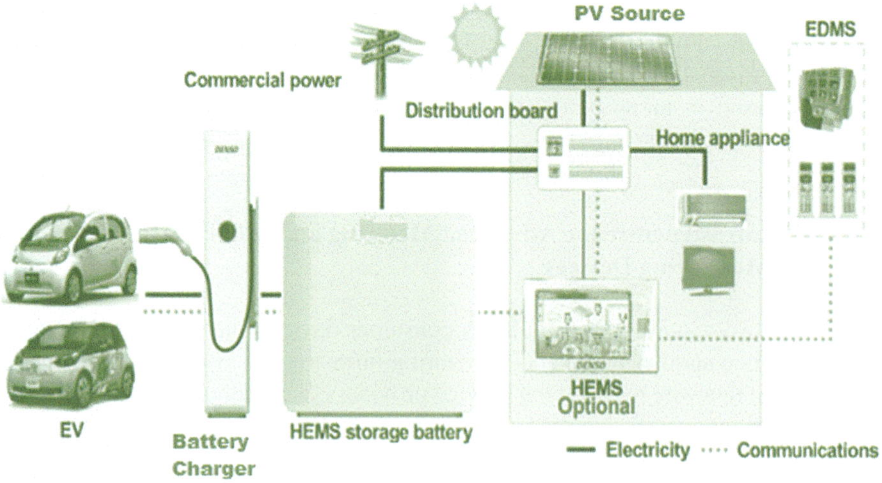

Fig. 17.4 Electric vehicle charging infrastructure (from grid to the vehicle)

the grid. As a result of the growing movement towards electric vehicles, the upcoming of the electrical grid would face a PQ issues. The incorporation of a significant number of storing elements that utilize rectifiers to energize batteries at various charging times would yield a significant consequence on the electricity grid's PQ. In addition, when adding various levels of harmonic into the electrical grid, peak demand would increase dramatically. If, on the other hand, these storage elements could be present as an appropriate demand-side management (DSM) resource, this task will be an initiative to strengthen the electrical system's reliability. This necessitates the passage of regulations governing the possession and use of electric vehicle storage devices.

17.2.2.3 Demand Response System

Demand Response (DR) responds to a variety of energy requests. DR uses a real-time pricing scheme to manage power conditions and improve device performance. Furthermore, DR can minimize peak demand and consumers have control over their energy consumption by means of a PLC between the electrical services and their home usages. A DR framework is implemented using a broadband-PLC and oblique control via a gateway. The Domestic/Building Energy Management System (HEMS/BEMS) is one of these indirect load applications. For the DR method, a less expensive solution, such as the narrowband-PLC, may be used instead of the BB-PLC (Yigit et al. 2014).

In brief, DR strategies that allow for efficient supply of electricity as well as demand management are believed to be a key component of the SG. Modelling consumer behaviour is one of the most difficult aspects of developing demand-side management models. Modelling customer relationship, constructing decision-theoretic methods, refining pricing, integrating complexities varying over time (e.g. demand fluctuations) and balancing for grid electricity restraints are altogether obstacles that must be addressed before implementing DR models.

Decision-theoretic strategies like game theory, optimization and stochastic control are needed for accurately modelling and analysing various arising DR conditions. DR is supposed to be a critical turning factor for SG implementations that are more realistic (Bari et al. 2014).

Apart from the above applications, PLCs are also used for in-home environmental applications, remote fault identification, mobile networks communication, in order to achieve peak demand elimination, reliability enhancement, diagnosing isolated fault in cables, identifying damaged insulators and to deliver greater bandwidth for voice and data communication. Though PLC's offer several benefits like, extensive coverage, cost-effective, higher flexibility, easy installation, highly stable and mobile, their disadvantages outnumber the benefits.

The major issues include the following: (i) Sources of high noise on power lines, (ii) The issue of open circuits, (iii) The signal is ameliorated and fragmented, (iv) Regulations for broadband-PLC are insufficient and (v) Compatibility is a problem (Yigit et al. 2014). Figure 17.5 illustrates the communication network for the SG.

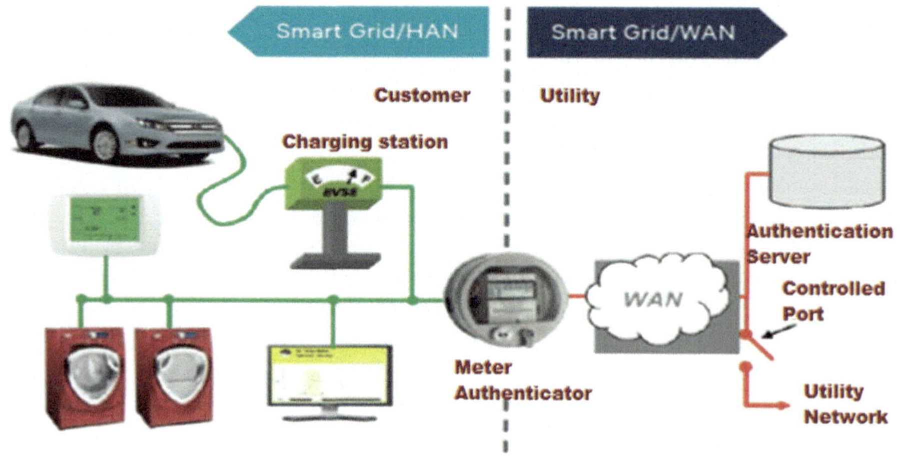

Fig. 17.5 Arrangement of communication network in a smart grid

17.2.2.4 Allotment of Restrictions of Emission on Grid by Integrating New Devices

When a new customer is tied to the controlled modern electrical grid, an estimate of the amount of pollution that is permissible from this customer without causing unreasonable amounts of voltage disruption for other customers is usually made. A so-called emission threshold is assigned to each new customer. Both current and potential customers share the cumulative amount of permissible voltage distortion. However, this means that the number of customers who will be linked in the future is understood. With modernization of the electrical grids, the quantity of demand would consume no limit as long as output grows at a similar rate. Because of the continued increase in both output and consumption, the voltage distortion due to harmonics may develop in to unusually larger threat. In addition, the number of times the switches need to be turned on and off will continue to rise, potentially reaching excessively high levels.

At the power system frequency, output and consumption are in equilibrium, but it's not the case with harmonic frequencies. Another way to look at it is that the system's intensity is now measured by the quantity of harmonic distortion being emitted by downstream apparatus, rather than the higher quantity of utilization and/or output related downstream. This will necessitate a new approach to distribution network planning. However, preliminary research has found that harmonic emission from distributed generation is very small. The majority of current end-user devices, such as computers, televisions and lamps, emit only at the lower odd integer harmonics (Bollen 2010; Pandya and Bhavsar 2018).

Immunity of Devices

An instantaneous tripping of several distributed generators can occur owing to a voltage quality disruption, or it can be said that it's because of a dip in voltage level. And this is a significant problem. Mass consumption tripping can have similar negative effects to how a SG seeks to keep a balance among generation and consumption (Bollen 2010).

Transmission Grid Deterioration

The volume of traditional generation interconnected with the transmission structure would decrease as distributed generation and huge wind parks become more common. As a result, the level of failure will be decreased, and power-quality disruptions will spread out further more. Voltage sags, rapid voltage variations (flickering in voltage) and harmonic variations would all be exacerbated. The magnitude of voltage dips has been investigated. The research concluded that even though 20% wind energy is harvested, substantial rise in the amount of voltage sags due to transmission system faults is not evident (Bollen 2010).

17.2.3 Integration of Renewable Energy Sources

The form of power generation has shifted from bulk generation units to distributed generation units as a result of sustainable energy sources like solar, wind, etc. As a result, the system's reliability, output voltage, electrical transmission expenses, damages and reliance on the primary grid have indeed strengthened. In spite of the above-said advantages, due to the stochastic aspect of sustainable energy sources such as solar or wind power, they are not completely efficient. Another disadvantage of incorporation of such sources in the grid is that they use high-power converters to transform power; as previously stated, excessive use of power electronic converters in the electricity grid would result in a lot of harmonic emissions. Recently, researchers have been working on ideas to generate these renewable energy sources infinitely customizable so that the grid's PQ would become better by the integrated power electronic converters. (Hosseini et al. 2017; Naderi et al. 2018).

17.2.4 Devices for Improving Power Quality in Smart Grid

This is important to enhance the efficiency of power supply on the customer side by using PQ control technologies and equipment. Controlling and converting electric energy to meet out the requirements of quality compliance and optimum performance is the significant aspect of PQ control. Various types of high power converters and

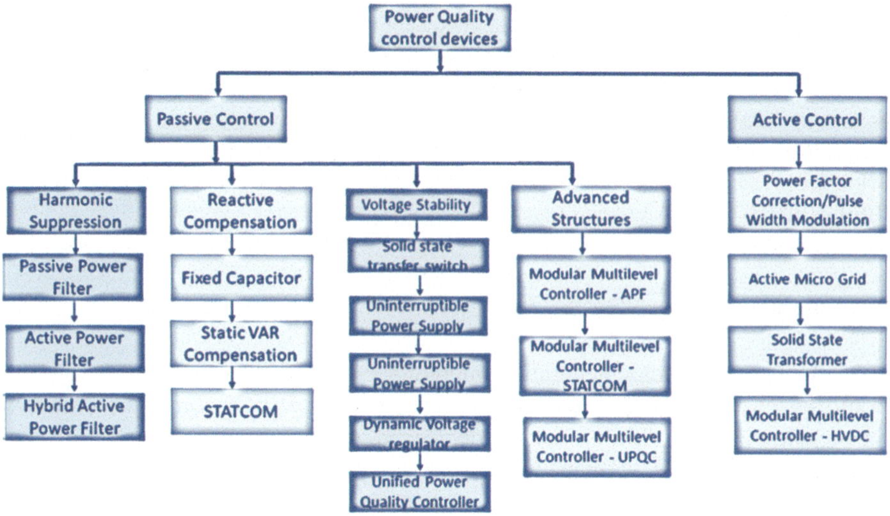

Fig. 17.6 Classification of power quality control

corresponding control topologies are key elements in achieving the above-mentioned control and conversion. Figure 17.6 depicts the category of compensators based on various types of PQ issues. There are two broad techniques available for the mitigation of PQ issues, viz. active control and passive control techniques.

17.2.4.1 Passive Components

Passive technique is described by incorporating various components to reduce or mitigate the effect of prevailing PQ issues. Currently, the passive power filter (PPF), active power filter (APF) and hybrid active power filter (HAPF) are the most common harmonic suppression strategies. Apart from their disadvantages, passive filter is implemented in certain particular applications these days due to their flexibility and a reasonable price. Its value noticing that the majority of those are applications of hybrid that use passive filters for cost cutting and improve overall device efficiency. APFs are designed to balance out and enhance power factor, mitigate harmonic distortion, balance malfunctioning and flickering of voltages and control voltage towards overcome the disadvantages of PPFs. APFs are of two types viz shunt and series APFs.

Shunt type of filters are connected in parallel for providing compensation for current harmonics through applying a harmonic current of the equal phase difference of 180 degrees and amplitude, resulting in nearly sinusoidal grid current. If an appropriate control system is used, it would act as a reactive power compensator. Harmonic voltages are compensated by incorporating harmonic voltages of the equal magnitude but opposite phase when harmonic loads are connected with series filters.

The key disadvantage of these devices is that they will have the same power rating as the load, making them a costly and unaffordable solution for high-power applications. Hybrid power filters, which combine the benefits of active and passive filters, are an appropriate option for lowering the price of utilizing high-rated active filters. In a high-voltage distribution network, it has proven to be an efficient method for mitigation of harmonic currents as well compensating the reactive power. Figure 17.7 illustrates the PPF, APF and HAPF.

The voltage disruptions and fluctuations are suppressed by the reactive power compensator, static VAR compensator (SVC), static synchronous compensator (STATCOM) and Fixed capacitors (FC) are examples of VAR compensators used in distribution networks (An et al. 2016). The STATCOM is one of these devices that is commonly used because of its various features, including grid voltage oscillation reduction and nonlinear load compensation. Due to its higher durability, consistency and specific power, extensible harmonic repression and VAR compensation systems are becoming successful. Figure 17.8 shows the devices for reactive power compensation.

Overvoltage and short disruption are the most frequently occurring types of intermittent PQ issues. The solid-state transfer switch (SSTS) will significantly minimize voltage sag complexity and length. The most powerful method for limiting voltage instability for low-power appliances in the distribution network is an uninterruptible power supply (UPS). The dynamic voltage regulator (DVR) will compensate

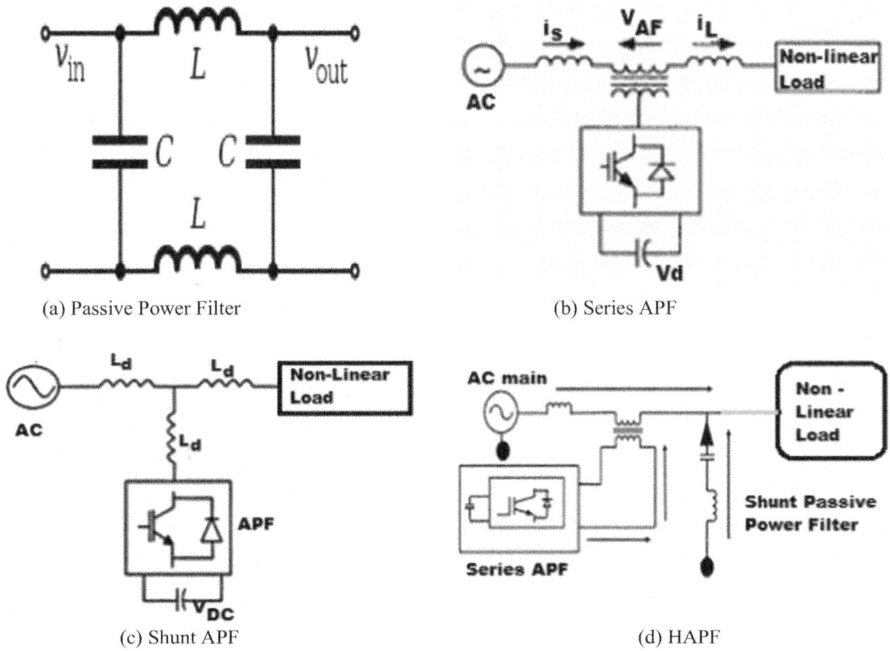

Fig. 17.7 Circuit diagram for PPF, APF and HPF

Fig. 17.8 Devices for reactive power compensation

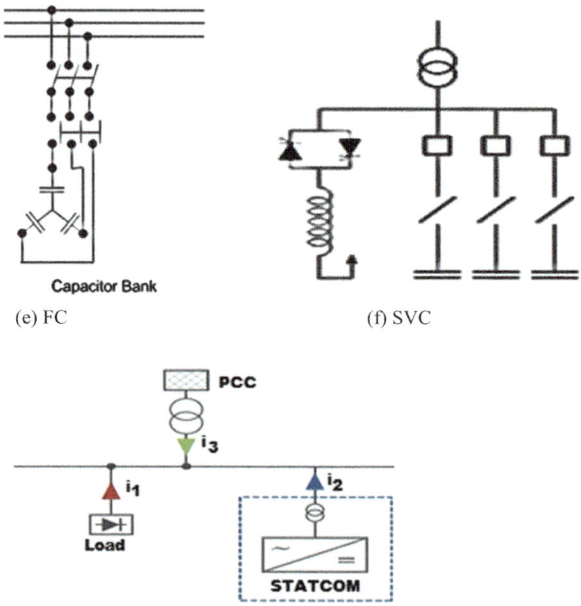

(e) FC (f) SVC

(g) STATCOM

for sudden voltage dip and swell directly and rapidly. The unified power quality controller (UPQC), which is made up of series and shunt APFs, will make comprehensive voltage and current compensation for the modern electrical grid. Figure 17.9 shows the elementary structure of DVR and UPQC.

Cascaded power converters based on modular multilevel converters (MMC) are extensively analysed in technical research and development applications in less than two decades. Modular multilevel converters significantly reduces the complexity and expense of producing high- and medium-voltage converters due to the similar module layout. MMC-based compensators have distinct advantages over conventional PQ

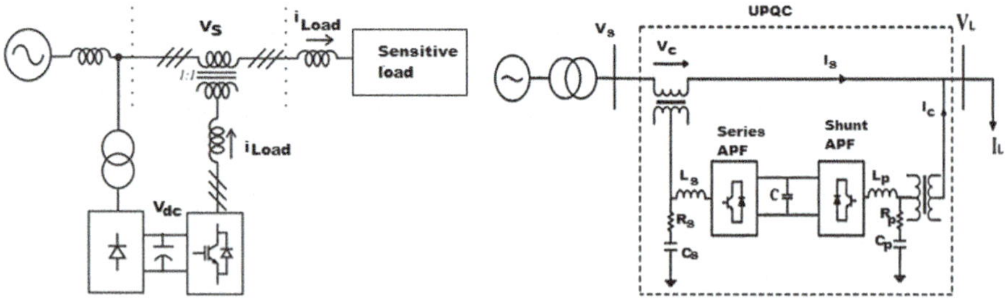

(i) DVR (ii) UPQC

Fig. 17.9 Elementary structure of DVR and UPQC

compensators in terms of normalization, adaptability, flexibility, malfunction ride-through, load demand and filtering characteristics. MMC significantly reduces the complexity and expense of producing high- and medium-voltage converters due to the similar module layout. MMC-based compensators have distinct advantages over conventional PQ compensators in terms of normalization, adaptability, flexibility, malfunction ride-through, load demand and filtering characteristics.

The versatile PQ control scheme will reduce the rated voltage of the semiconductor switches used in power converters and the unit for storing energy in the sub-system, allowing reduced losses and low-cost switching devices to be used. The cascaded framework, on the other hand, extends the use of sustainable PQ control scheme in medium- and high-voltage power transmission. Since huge amounts of data must be interpreted in a limited period of time, the multilevel structure necessitates a comprehensive control system. As a result, MMC's engineering application and advancement in the control of PQ issues is minimal. Thankfully, as digital signal processing technology advances, the implementation of complicated topologies and closed loop control are increasingly simple (Lesnicar and Marquardt 2003; Kouro et al. 2010; Ghetti et al. 2012; Long et al. 2013). Figure 17.10 describes the schematic diagram for modular multilevel converters based on APF, STATCOM and UPQC respectively.

17.2.4.2 Active Components

Most PQ issues are resolved by using active control technology to boost the intrinsic properties of electrical equipment. By means of transmission and distribution network being engineered, automated and smart, PQ issues posed through emerging electrical systems, particularly the high-power converters, would be greatly alleviated. The PQ of rectifier devices has been improved by power factor correction (PFC) techniques and pulse width modulation (PWM) methods. With the use of active control in distributed generation and MG inverters enhances the accuracy of voltage at the output and current in distributed systems, and at the same time offers additional compensation capability for the adjoining electrical grid. Consequently, the proposed solid-state transformer (SST) would prevent PQ problems from being transmitted and emitted in between end-user and the power delivery system. The MMC-type high-voltage direct current (HVDC) transmission and multi-terminal high-voltage DC technology would enhance the PQ of the entire power grid (Leung et al. 2010; Railing et al. 2004; Long et al. 2013; Akagi 1994; Khadkikar and Chandra 2008; Luo et al. 2009). Figures 17.11 and 17.12 show SST and MMC based on HVDC technology, respectively.

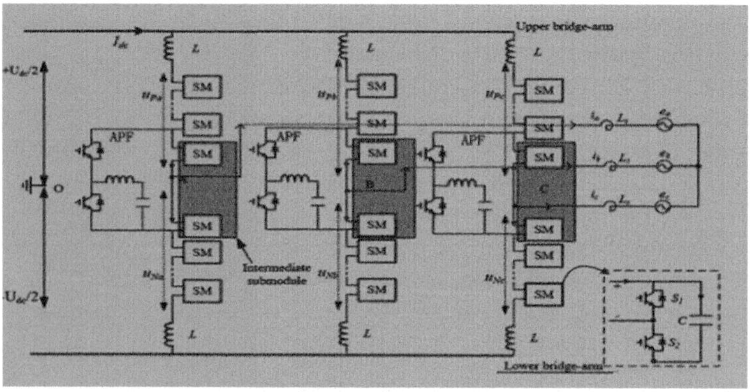

(i) MMC based on APF

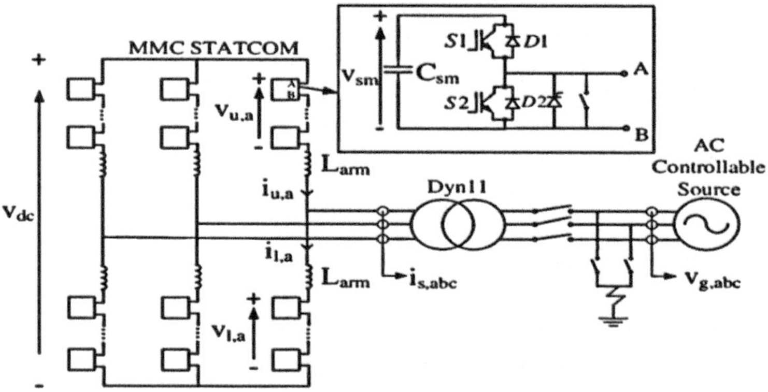

(ii) MMC based on STATCOM

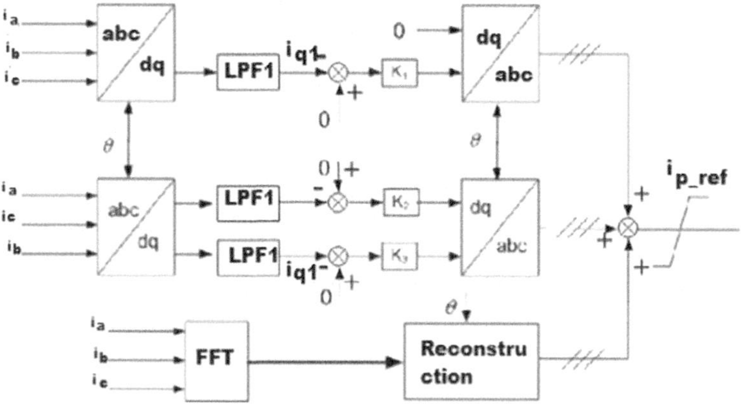

(iii) MMF based on UPQC

Fig. 17.10 APF, STATCOM and UPQC

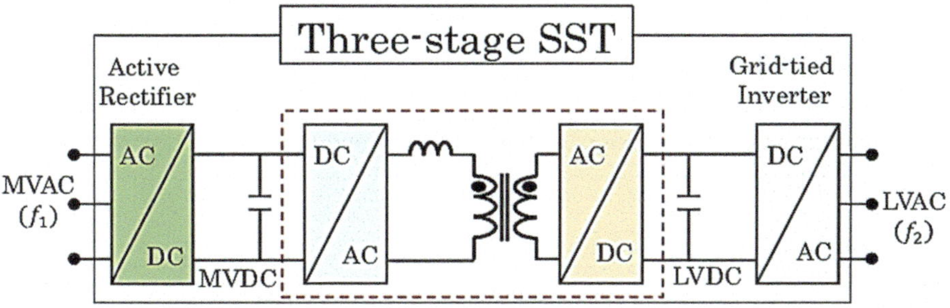

Fig. 17.11 Solid-state transformer

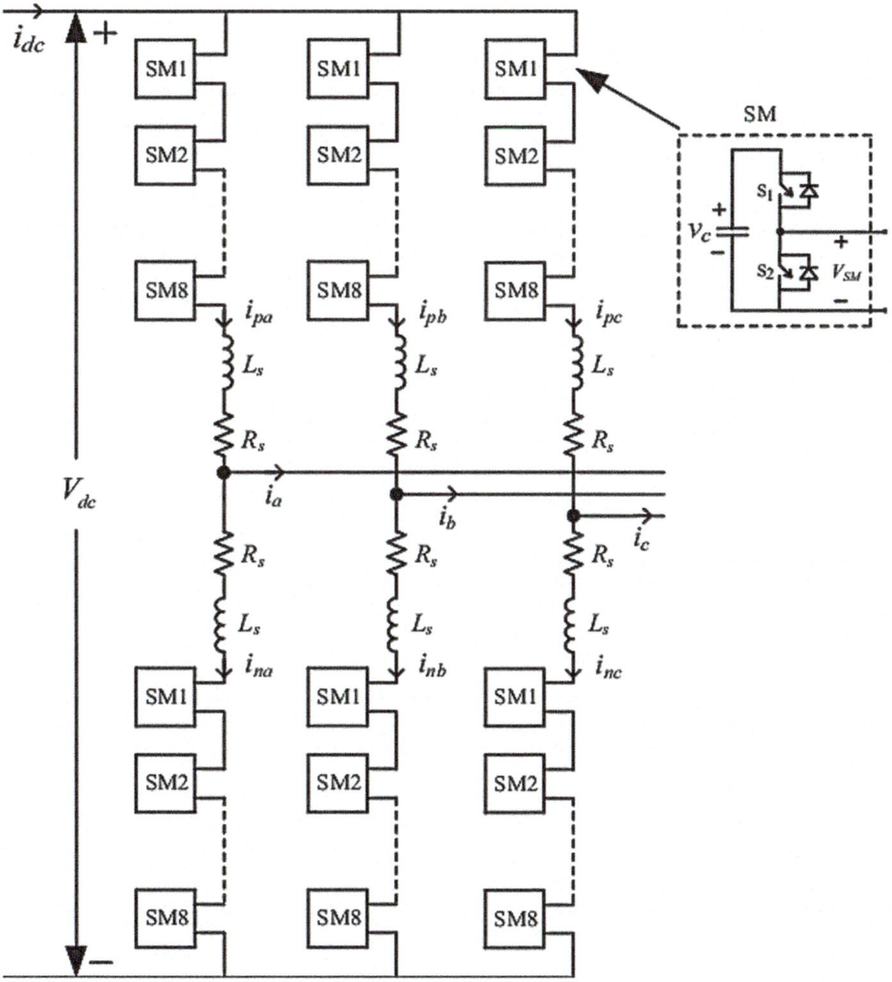

Fig. 17.12 MMC-based HVDC system Block diagram

17.3 Causes and Mitigation of Power Quality Issues in DC Microgrid

In this chapter, the major causes of the PQ issues and their mitigation in DC MGs are explained in detail.

17.3.1 Power Quality Improvement Methods for DC Microgrid

The control voltage of the DC bus is of prime importance intended for the typical function of the DC MG. To transmit effective power as well as current into the DC MG, the DC bus voltage must be controlled in a specific dimension by certain effective current-sharing techniques. Out of several techniques reported on the literature, the following are the most important and the effective one which are discussed as under:

(i) Voltage droop control method,
(ii) Hierarchical control method,
(iii) Multi-agent-based control method, and
(iv) Artificial Intelligence-based control method,
(v) Energy management strategy.

17.3.2 Voltage Droop Control Method

Droop control is a dynamic tool for improving PQ attributes like real and reactive power regulation in electrical grid as well as decentralized operations. The majority of today's current sharing approaches depend on a high-bandwidth infrastructure. As this DC MG is made up of multiple resources, implementing a high-bandwidth transmission network is not cost-effective. Figure 17.13 shows the droop control scheme. It is in turn a decentralized control scheme and it is more popularly used to enhance the MG performances. The traditional approach in this technique is to reduce the DC output voltage linearly as the output current rises. Current distribution among multiple converter topologies could be accomplished through the use of a configurable voltage variations that is confined within an appropriate range. By using high droop factor, the accuracy in current sharing can usually be improved. As a result, a larger droop factor is usually selected while limiting DC voltage variation at even the most drastic loading. This method of control is generally implemented for high-power converters, viz. DC to DC, DC to AC and AC to DC.

If V_r is the reference voltage, Io is the output current and r_d is the droop factor for the converter, the following two equations define the principle of voltage droop control method:

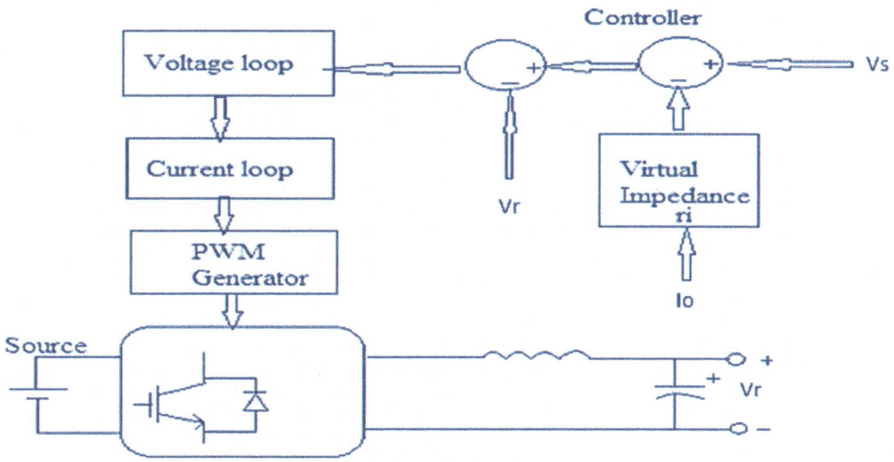

Fig. 17.13 Schematic diagram for a voltage droop control method

$$V_r = V_{r1} - r_d I_o \tag{17.1}$$

$$\Delta V_s = |V_{r1} - V_s| \le \Delta V_{smax} \tag{17.2}$$

Here, V_{r1} is the reference voltage, ΔV_s is the deviance in the DC MG voltage and ΔV_{smax} is the largest possible grid voltage deviance. In a droop-controlled DC MG, the droop factor rd is also known as virtual resistance. The major advantages of this method is that it is solely based on locally observed data and therefore does not rely on communications signal, thereby removing the challenges posed by specific location. Other benefits of the droop method include its simplicity, high efficiency, basic structure, ease of execution, unrestricted placing and multiple power ratings. Figure 17.13 illustrates the droop control implemented for a DC–DC converter (Rawat and Sathans 2018).

17.3.3 Hierarchical Control Method

Nowadays, control hierarchy is gaining popularity, because of their simplistic structure and high performance. In a nutshell, a DC MG faces complex converter-level control at the lower level but simple system-level organization at the top. Hierarchical control method plays a vital role to solve these features. In general, it comprises three control levels like, primary, secondary and tertiary levels of control strategies. Figure 17.14 shows the hierarchy control for the DC MG.

The primary control deals with the regulation of current, potential and voltage. It conducts control activities over interface control converters, following the collection

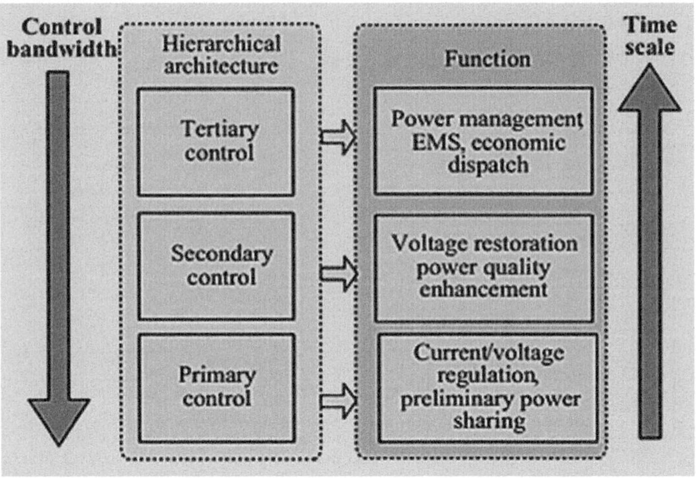

Fig. 17.14 Structure of hierarchical control

of focuses provided by upper level controllers. This primary control also focuses on the load sharing among the distributed generation units. The second level of control deals with system-level issues including control quality path, MG synchronization with the outside network for seamless reconnection and so on. The system's electrical levels are kept within the necessary limits by this central controller.

The tertiary level is the distribution management solution that also regulates the flow of power in the network system. The power flow among the DC MG along with the upper level electrical grid is regulated by this tertiary level of control. It interacts with the operator of the distribution system. The power exchange scheduling with the MG can be decided either by the distribution or transmission system's operator. The main goal of power management in DC MG is to keep the power balance between sources of energy, storage systems and loads as represented by the DC bus voltage. In order to maximize efficiency this tertiary management control determines the mode of function of the DC MG. Power sharing with the upstream side of the grid is included in interconnected mode control. The important feature of this control method is that if the load power is insufficient, the DC MG would draw power from the upper grid whereas if the power produced by the MG is in excess of what is needed, it will be transferred to the upper grid.

Depending on how secondary control is applied, these management techniques can be categorized into three main categories: centralized, decentralized and hybrid. In centralized control scheme, to overcome the voltage deviation induced by the principal controllers and also to establish balance of power among the distributed generations and loads in DC MG, centralized controller is used. This method can provide optimum control, restores the voltage level and at the same time eliminates voltage deviation which necessitates the real-time communication. In decentralized scheme, the distributed generations are independently regulated through local real-time feedback mechanisms either with communication or without communication. In the first case, i.e. decentralized control with communication, although correspondence amongst the distributed generations is essential, the operating choices are taken in a decentralized manner, typically at the level of the distributed generation. That's where the centralized secondary regulation differs significantly. In the second case of decentralized control without communication, all the required operations in this group can be completed without the use of communications. The DC bus signalling system is the most commonly used system. This approach departs from the principle of precise power sharing on the bus with limited voltage deviations.

Since the device is supposed to be power electronics-dependent, high-voltage deviations from marginal value are allowed because the line and load architectures can be configured to ultimately improve within a wider prescribed DC voltage. The key benefit is that switching between different modes of operations and altering with various appropriate control methods for power converters can be accomplished without the use of extra network communication. Cost savings and greater performance are the two added advantages. The hybrid secondary control combines the benefits of both centralized and decentralized controls in order to obtain greater performance and higher reliability. Many literatures are being reported on this hybrid control where in the basic idea is that, the first stage is activated when the communication accessibility services fail and is based on the DC bus system method. On the other hand, under normal service, the second stage is enabled in which the communication channel provides complete observability over the DC MG comprising actual bus voltage, active power flow and converter operating mode (Rawat and Sathans 2018; Natesan et al. 2014; Papadimitriou et al. 2015).

17.3.4 Multi-agent Control Technique

A multi-agent scheme combines various agents to accomplish a common goal. It consists of both software and hardware units with merely local data and limited capabilities, but it has the greatest ability to work together to achieve a global goal. The first and foremost attribute of the agents is that it can be any one of the specific physical modules. The other characteristics include the following: (i) they can interact with their surroundings and communicate with one another, (ii) without a controller it's possible for the agents to take its own decision, (iii) able to achieve specific targets by using its tools (either software or hardware), expertise and provisions. With the support of fast communication systems, multi-agent systems can be used in high-power applications more effectively and feasibly. Multi-agent control technique is shown in Fig. 17.15.

Distributed problem-solving is a subcategory of multi-agent control system that can be easily implemented on MGs and power systems, in which all the entities can cooperate to reduce operating costs and improve customer satisfaction. The basic principle is that an agent can analyse and evaluate the surrounding physical

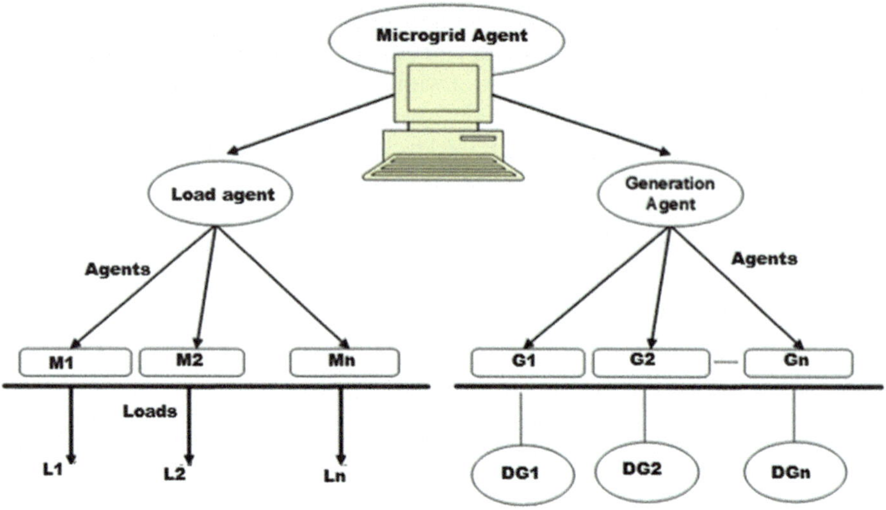

Fig. 17.15 Multi-agent control scheme

attributes, updates its local databases and responds to local events on its own. An agent can also work with other agents to coordinate and interact, negotiate offers and achieve both local and global goals. The method of designing a multi-agent system for MG power management is a step by step procedure which involves a thorough investigation of the entire power system.

First step is that complete technical details should be specifically outlined, including the comprehensive geometry of the MG, the number, form and priority of load, the total local supply generated by onsite renewable energy production and energy available in storage units and the highest possible demand that could be met in islanded mode. The specifications and goals are then defined, which are then converted into mathematical model and objective function for cost optimization. The total agents required, optimization techniques and performance metrics are then defined as per the above analyses. Finally, in both normal mode and islanded mode of operations, communications and information sharing among agents should be described. The end result is a complete model with specific behaviour of the agent which can be incorporated using a suitable agent framework like Java Agent Development Framework (Rawat and Sathans 2018).

17.3.5 Artificial Intelligence-Based Control Method

The traditional control strategy is insufficient for highly complicated or ambiguous systems. In such situations, artificial intelligence-based control has proven to be the best method for obtaining an effective control for the DC MG as well. Among many of the available methods, here we are going to discuss on the two efficient controls, viz., fuzzy logic-based control and PSO algorithm-based optimization control.

17.3.5.1 Fuzzy Logic-Based Control

The fuzzy controller was created with the help of human intelligence. It improves the capability to cope up with unpredictable or unclear data to some extent. Based on the literature, it is found that fuzzy logic controller is being used in DC and AC MGs, either stand-alone or grid-connected, for a variety of applications, including maximum power point monitoring of solar photovoltaic applications and wind energy systems, regulating battery output charge current and so on. The state of charge (SOC) of a Li-ion battery associated to a DC MG with fuel cells, photovoltaic system and wind energy systems was controlled with fuzzy logic controller. A decentralized control for gain scheduling using fuzzy logic controller was also reported in literatures to obtain balancing among the stored energy systems provided for various batteries by changing the droop coefficient of primary controllers in a DC MG. In order to supply a DC load, fuzzy logic control can be used to provide a power split between photovoltaic system and energy depository system of a battery based on the operator's expertise over a predefined set of fuzzy rules. In such cases, the controller takes three inputs such as from PV power, battery state of charge and load power. The controller's output controls the operation of the various switches, allowing for one of the three possible connections like PV power/battery, battery/load or PV power/load.

In general, when this fuzzy controller is used in MG, the control system is divided into two tiers, with control bandwidth separating them. The voltage and current controls for the DC/DC converters are located on the primary stage. The latter is configured to monitor the commanded reference values with quick disturbance rejection and a high bandwidth control loop (i.e. >500 rad/s). The fuzzy logic controller which is configured to have a lower bandwidth, represents the secondary stage. This is a general rule for nested control loops to maintain system stability. The controller regulates the flow of power in the MG as per the pre-determined regulations. Its output manipulates the primary control level's reference values (voltage or current) (Chilukuri 2021).

17.3.5.2 PSO Algorithm-Based Optimization Control

PSO (Particle Swam Optimization) algorithm can also be applied for enhancing the PQ in an individual DC MG and is being widely discussed in many literatures. This algorithm can be used to address voltage control, frequency control, sharing of power in the MG, stability, dynamic characteristic response or harmonic analysis among other PQ specifications. A real-time PSO algorithm-based self-tuning control technique can be implemented in a DC MG to enhance power efficiency. There are two feedback loops in this control technique. The first is a current mode control loop which is mainly based on a synchronous reference frames that acts as an inner control loop and the second is a power mode control loop associated with traditional Proportional–Integral (PI) controllers that serves as an outer control loop (Natesan et al. 2014).

17.3.6 Energy Management Strategy

The energy management strategy or more commonly energy management system (EMS) plays a critical role in maximizing the use of renewable energy sources to meet load requirements. Depending on the selected circumstances, an energy management system is needed to organize the energy sharing among these multiple sources connected with the MG an energy management system is a set of computer-assisted software used by electrical power system managers for monitoring, controlling and maximizing or optimizing the output of generation sources. Due to the introduction of renewable energy sources, this control strategy now plays a critical role in MGs. Higher number of power converters are employed in MG for power flow control operation. These power electronic-based high-power converters respond to grid variations in accordance with the proposed grid's energy management algorithm (Khatibzadeh et al. 2017).

The combination of energy management control system and power electronic converters simultaneously controls and manages the MG system. Photovoltaic arrays and hybrid energy storage system can be programmed to get the most power out of renewable power resources. Energy management control is commonly needed in hybrid energy storage system for deciding the number of storage units which is to be used in each cases, as well as the discharge cycles for each unit (Tazi 2019; Badwawi et al. 2016). The total power produced plus or minus the power supplied by the energy storage device should be equal to or greater than the total power consumed by the load. The control algorithm is defined by the following equation:

$$P_{gen}(t) \pm P_{enr}(t) \geq P_L \tag{17.3}$$

Here, $P_{gen}(t)$ is the instantaneous value of the generated power, $P_{enr}(t)$ is the instantaneous value of the power delivered by the energy storage system and P_L is the instantaneous value of load power.

The electrical system's energy control is critical for resolving issues in a hybrid energy storage system. The lifespan of each energy storage unit can be extended by monitoring the charge-release period. The energy management system's function here is to use the super capacitor to reduce the amount of discharge cycles of the battery during short-term load changes (Sayed and Kassem 2019).

17.4 Causes and Mitigation of Power Quality Issues in AC Microgrid

In this chapter, the major causes of the PQ issues and their mitigation in AC MGs are explained in detail.

A MG associates with the distribution network through control of AC bus. The AC bus controls the MG's assembly to and disconnect from the distribution network via the relay called circuit breaker at the Point of Common Coupling (PCC) is known as AC MG. In AC network, frequency and voltage are the dual quantities which are needed to be controlled. The frequency can be changed by changing the input mechanical power and the voltage is organized by injecting or fascinating the reactive power in AC generators (Guerrero et al. 2013).

The distribution scheme is associated to the AC bus via a circuit breaker in an AC MG, and the AC bus controls the MG's operating system via the circuit breaker at the PCC. In an AC MG, the PCC is basically a 3ϕ AC bus.

The MG control system must be capable to function reliably in two modes called grid-tied and islanded modes, with a smooth transition among them. In MGs, various control techniques like management of power, grid constraint management and PQ enhancement are carried out using multilevel controls called hierarchical control schemes (Karim Hassan Youssef 2019).

There are four different forms of MG control:

I. Output control of Converter,
II. Control for Power sharing,
III. MG supervisory (secondary) control, and
IV. The grid supervisory (tertiary) control.

Figure 17.16 depicts the various levels of power. The higher level controller provides each control level set points reached from data measurements, and contact with additional control levels (Pei et al. 2004), which are reviewed in this chapter.

17.4.1 Output Control of Converter

Voltage source inverters (VSIs) are made use in renewable energy power conditioning systems. Localized controllers monitor VSIs and their work is reached from limited measurements, which provides speedy reply than the mastery controllers in MG, and it could be capable to control in both grid-tied and islanded (standalone) modes. As a result, the system of control can be able to accommodate both modes. Figure 17.17 illustrates the formation of the converter output power.

There are two control loops in the present system. The inside loop current forms the current, while outer control loop controls the stream of active power and reactive power or controls the frequency and the output voltage of the converter depending on the modes of operation. The various types of current controllers are:

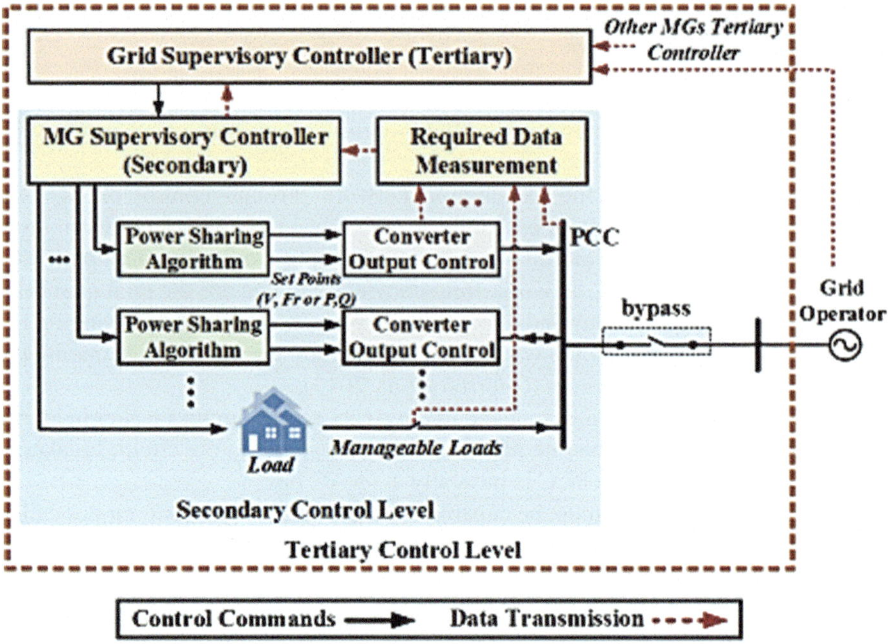

Fig. 17.16 Levels of control

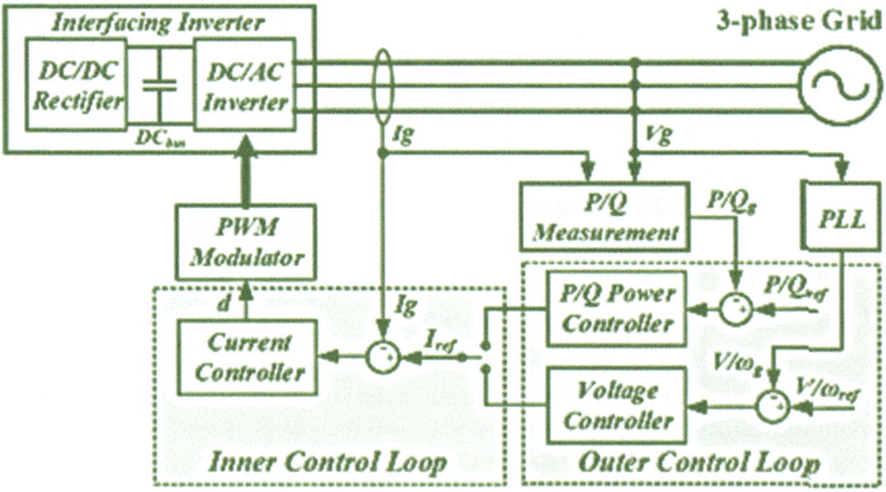

Fig. 17.17 Structure of Converter output control

(a) Conventional PI, PR and Hysteresis Band;
(b) Digital repetitive;
(c) Composite Nonlinear controller; and
(d) Composite nonlinear feedback control.

The functions of current controllers are expressed in Fig. 17.18, which shows a block diagram of various current controllers.

Integral controllers and Hysteresis band controllers are the most widely used controllers. The main advantage is that the steady state error and periodic distortion induced by periodic fluctuations are brought down. When the Overall Harmonic Disruption is low, the compensator based on Hysteresis band enhances the tracking

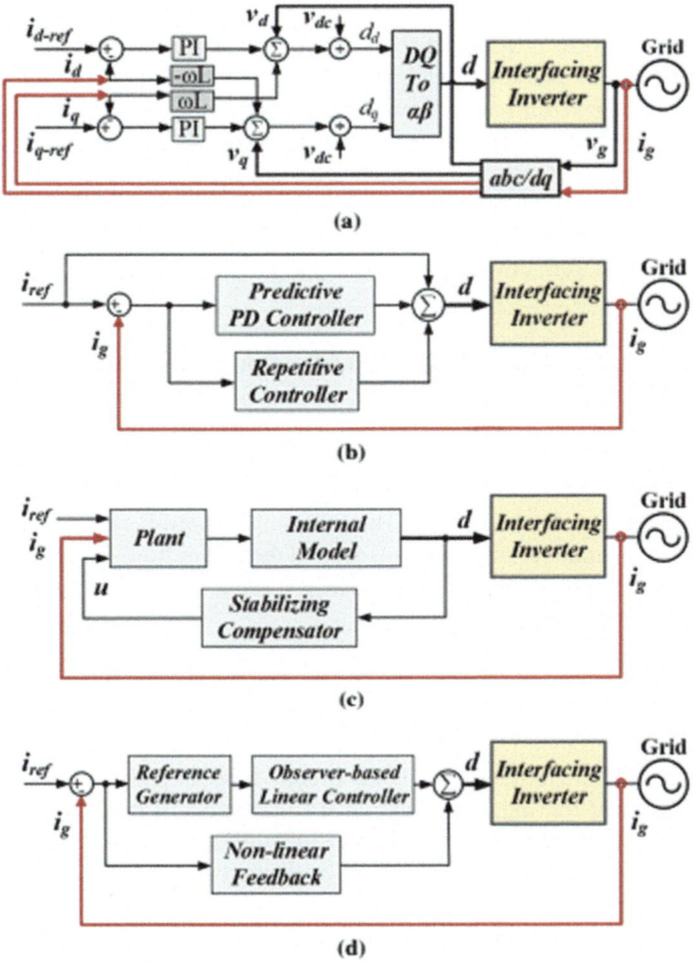

Fig. 17.18 Block diagram of various current control **a** PI controller **b** PR controller, **c** hysteresis band and **d** composite nonlinear feedback control

performance (Guerrero et al. Apr. 2013; Pei et al. 2004). The current loop's steady state and transient output are improved by the composite nonlinear feedback controller (Caldognetto and Paolo 2014).

In the Fig. 17.18, the inner current control loop can be initiated and the fix point for the inner loop is brought by the outer loop control. Hence the VSI output current VSI is controlled by using inner control loop (Guerrero et al. Jan. 2011). An external or outer current control is classified into two types which is depending upon the operating modes of power converter: They are: (1) Grid tied and (2) Stand-alone (Islanding) modes.

17.4.1.1 Mode of Operation Based on Grid Connection

Here, the primary focus is to regulate both active and reactive flow of power among the power converter and the motor/generator set. In order to regulate the active/reactive power flow, the converter has to work in the active mode, especially under current control mode to standardize the injection or absorption of the power, and a required active or reactive power is set as a reference point for the closed-loop system that must have been given by a system employed for monitoring and control (Guerrero et al. 2011). The converter will use these control schemes to execute the new synchronized control activities like effective control of voltage, power flow and unexpected failures.

17.4.1.2 Mode of Operation Based on Islanding Connection

Here, the MG is segregated from the main electrical grid and it controls both grid frequency and grid voltage. The new controller has a particular proportional-resonant control strategy along with the suppression of harmonics, which also includes a controller based on alpha–beta frame. Thus, the output control of converter is accountable to govern the transfer of output current and the power flow. With the help of this controller, the PQ issues of power of disparity is reduced.

17.4.2 Control for Power Sharing

In order to control the frequency and magnitude of the set point voltage, the control layer of the power-sharing should be added. Variation in frequency and voltage could be found from the upper control layer, and the control layer can be mentioned depending on the communication/without communication as shown in Fig. 17.19 (Pei et al. 2004).

In the Fig. 17.19, centralized current control, the controller gathers every information and at that time subject the instructions to the network. Though, this kind of method whose inverters function as source of current, and through the intermediate controller the voltage is measured. The master–slave controller is different to previous

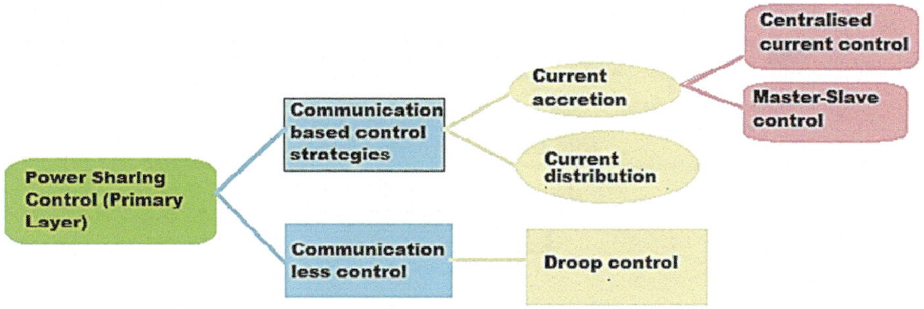

Fig. 17.19 Power control scheme organization

controller, in which 'M' no. of inverters have paralleled, among 'M' inverters, one inverter plays as a master whereas the remaining inverters plays as a role of slave. This type of controller transfers information among the master and slave. This type of controller has more economy, reliability and all information passed through common networks (Bhende and Kalam 2018).

In general, the balanced system can be controlled by distributed control system. The average current demand and the output voltage can be managed voltage controller. Droop control is the control; it has got extensive receipt since the nonappearance of the communication necessities among the inverters. The key knowledge is to control the frequency by sensing the reactive power and voltage by sensing the active power which could be sensed by the parameters.

17.4.2.1 Methods of Power Sharing in Island Mode

These are broadly classified depending on the connectivity channels as communication and as well as non-communication-based systems. These are further sub-divided in to different methods as follows:

(1) Communication-based methods
(a) System for sharing power centrally,
(b) Master–slave power sharing system,
(c) System for distributed power sharing.

Figure 17.20 depicts a block diagram for three types of contact connections. The central controller uses the high-band communications. Low-band contact is used in master–slave operation, where the master manages the voltage and the slave acts as a current controller (Dou and Liu 2013). It describes a smart master–slave communication system that enhances the transient response of the master–slave communication (Pearline and kamalini; Abinaya et al. 2019). A current sharing bus with low bandwidth communication and additional local controllers are included in the distributed sharing system to exchange information between the inverters (Alkahtani 2019).

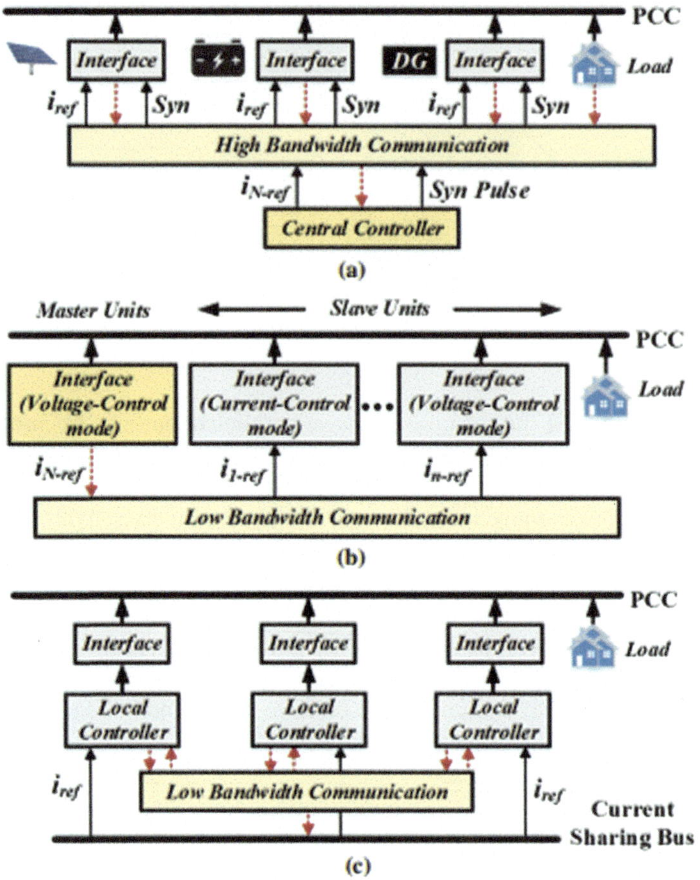

Fig. 17.20 Power sharing methods in Islanding mode. **a** central sharing, **b** master–slave and **c** distributed sharing

17.4.2.2 Non-communication-Based Method

This method uses no communication to enforce power sharing between parallel VSIs. The key benefit of this approach is its higher efficiency, lower costs and better scalability, which makes it a common choice in MGs. Droop-based control methods are also employed, i.e. P–ω droop/Q–V droop control is employed. The following are the key drawbacks of these controllers:

(1) The output impedance to resistance ratio has a high dependence on it.
(2) In data transmission, the PQ is poor.
(3) Voltage/frequency deviations.

Thus the control of power sharing is to control among the parallel inverters in which the control is responsible for the power flow among MG and utility grid. In the power sharing control, centralized current control can be obtained from the perfect current sharing even in the transient situation, no variation in the control organization and voltage regulation and various power rating of inverters could be connected with the structure. The master–slave operation control confirms the effective and consistent function of the MG in all operating condition. In the case of distributed control, which shares good current sharing even in the transient condition, minimize the circulating current and minimize the requirement of communication link. The method of droop consists of several required structures like stretch ability, compatibility, severance and flexibility. It has the advantages of very good reliability, efficient and various power-sharing and require low bandwidth (Failed 2019).

17.4.3 Secondary-Level Microgrid Supervisory

It handles the MG internal parameter guidelines to the MG and offers set points of low levels. The common control of the main grid could be attained through secondary-level control. On behalf of the inexpensive process, consistent and safe of main grid in two modes (Islanding and Grid-tied) secondary controller level or main grid Power Supervision Arrangement (PSA) is typically used. In islanding mode, the secondary control is the maximum hierarchical level in main grid with minimum bandwidth communication link and delayed control loops associated to primary control.

There are two types of controls for this. (1) Decentralized controllers using localized variables and (2) Centralized regulators relying on contact linkages to transmit data from the supervisory control to the controllers provided at the lower level.

17.4.3.1 Microgrid Supervisory Controller for Decentralized Structure

In this structure, the drawbacks in conventional droop control are overcome by augmented sags and by employing controller for compensating the distributed system.

17.4.3.2 Centralized Microgrid Supervisory Control

For collecting information and transmitting signals for controlling the power converter, these controllers depend on a high bandwidth communication networks and controller provided centrally. In the view of hierarchical management, such centralized control is implemented on the MG to ensure its reliability (Senthilkumar et al. 2015). The proposed hierarchical control structure has three levels of control: the primary level, which includes the controller for converter performance, optimizing algorithms for sharing power flow, the regulation on secondary level for maintaining the MG's potential as well as frequency, and finally tertiary level, which controls the flow of real and reactive power flow among the main electrical grid and MG.

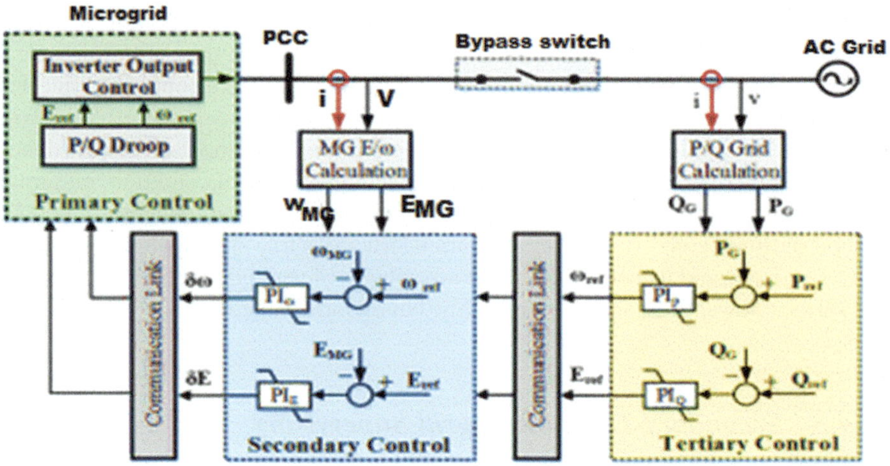

Fig. 17.21 AC hierarchical control system block diagram

The schematic diagram of the above-described hierarchical system is presented in Fig. 17.21.

In order to attain precise sharing of reactive power is an enhanced droop control scheme has been anticipated, which comprises multi-processes to progress the sharing of reactive power sharing through regularly changing the voltage bias of the simulated droop features curve and to bring back the graded rate of the output voltage. Lesser bandwidth communication links are enough for this kind of controllers.

17.4.4 Tertiary Level of Grid Supervisory Control

Tertiary level is the optimum level of control. This level acts as a current mode controller that uses communication networks for appropriately communicating with the operator available on the main grid and also with additional tertiary level controllers provided on the MGs so as to accomplish optimum reference for controllers at secondary level. This in turn comprises V/f control and real/reactive power flow control. Multi-gent scheme (MAS) is MGs' advanced control structure is developed to implement a conceptual framework depending on MAS which could manage stable potential in the meantime maximizing MG's activity at low cost. Figure 17.18 illustrates the diagram for the described control structure. It accomplishes the power flow among MG and the main grid.

From Fig. 17.22, it is seen that this control system comprises three different agents or levels of control: Unit agents or lower level agents which contains the controller of the power converter as well as sharing of power, the second agent called as the middle level coordinated agent comprises the central regulator of the MG and the third one known as the upper level agent accomplishes the collaborations among

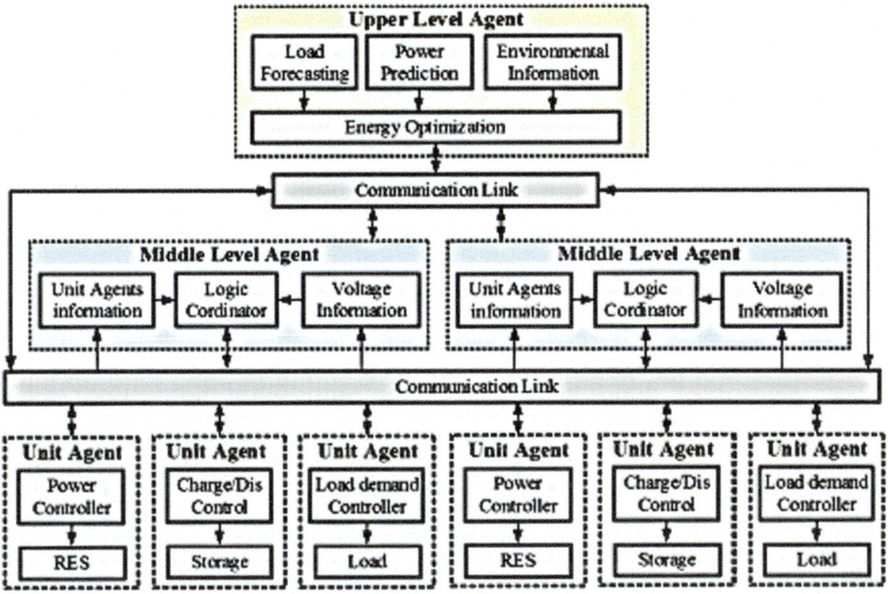

Fig. 17.22 MAS-based hierarchical control system

the main electrical grid and the MG (Dou and Liu 2013) proposed this MAS agent control level and this needs communication link to share the data among them.

The grid-tied tertiary control is used to control the magnitude and frequency of the voltage output. A tertiary level is slower and higher order of the control hierarchical, where the set points of the level of secondary are provided by this level. In grid supervisory control level, centralized control strategy is typically used. Therefore, the communication link is essential to retain the tertiary-level controllers in connect.

17.5 Conclusion

A detailed study of PQ issues and challenges in the SG and MG (both AC and DC) has been proposed. Initially, the structure of both SG and DC as well as AC MG has been explained. Various terms related to PQ issues are identified and their definitions are given in detail. In addition to the above a detailed explanation of each and every available PQ control, improvement tools and techniques are also discussed. More focus is emphasized on the various control techniques for enhancing the PQ has been given. It is to be noted that in SGs, two different technologies based on active and passive control have been discussed. Different types of filters like active, passive and hybrid power filters and other devices like Static VAR compensators, STATCOM, UPCQ, multilevel converter control, etc., are explicitly studied and their role in mitigating the voltage deviations, voltage sags, voltage swell, flicker, harmonic distortion, immunity of new devices in SG, transmission fluctuations, etc., are very well illustrated. Similarly, the DC MG control and AC MG control methods are deeply analysed. Extensive elucidation of various control techniques for improving power flow, power sharing, voltage balancing and regulation, voltage/ frequency deviation control, improved MG monitoring and control techniques, etc., are presented.

17.6 Future Scope

This work is proposed to be carried out by optimizing any one of the active devices to be implemented for mitigating the PQ issues in SG.

References

Akagi H (1994) Trend in active power line conditioners. IEEE Trans Power Electron 9(3):263–268

Alkahtani AA (2019) Power quality in microgrids including supra harmonics: issues, standards, and mitigations. IEEE Access, pp 1–19

Al Badwawi R, Issa W, Mallick T, Abusara M (2016) DC microgrid power coordination based on fuzzy logic control. In: Proceedings of 18th European conference on power electronics and applications, pp 1–11

Bari A, Jiang J, Saad W, Jaekel A (2014) Challenges in the smart grid applications: an overview. Hindawi Publishing Corporation Int J Distrib Sens Netw 1:1–11

Bhende CN, Kalam A (2018) Power quality conditioner for micro grid. IEEE Explorer, pp 505–511

Bollen MHJ, Zhong J, Zavoda F et al (2010) Power quality aspects of smart grids. In: International conference on renewable energies and power quality (ICREPQ'10) Granada (Spain), pp 1–6

Caldognetto T, Paolo T (2014) Microgrids operation based on masterslave cooperative control. IEEE J Emerg Sel Topics Power Electron 2(4):1081–1088

Chilukuri M (2021) Power quality in smart grid/micro grid. IEEE Smart Grid Resource Center

Dou C-X, Liu B (2013) Multi-agent based hierarchical hybrid control for smart Microgrid. IEEE Trans Smart Grid 4(2):771–778

Gandoman FH, Ahmadi A, Sharaf AM et al (2018) Review of FACTS technologies and applications for power quality in smart grids with renewable energy systems, vol 82, Part 1. Renewable & sustainable energy reviews, pp 502–514

Ghetti FT, Ferreira AA, Brage HAC, et al (2012) A study of shunt active power filter based on modular multilevel converter (MMC). In: Proceedings of the 10th IEEE/IAS international conference on industry applications (INDUSCON'12), Fortaleza

Guerrero JM, Chandorkar M, Lee T-L, Loh PC (2013) Advanced control architectures for intelligent microgrids—Part I: decentralized and hierarchical control. IEEE Trans Indus Electron 60(4):1254–1262

Guerrero JM, Vasquez JC, Matas J, de Vicuna LG, Castilla M (2011) Hierarchical control of droop-controlled AC and DC microgrids—a general approach toward Standardization. IEEE Trans Indus Electron 58(1):158–172

Hosseini YNSH, Zadeh SG, Mohammadi-Ivatlo B, Vasquez JC, Guerrero JM (2017) Distributed power quality improvement in residential microgrids. In: 2017 10th international conference on electrical and electronics engineering (ELECO), pp 90–94

Jolhe SP, Karalkar SMD, Dhomane GA (2016) Smart grid and power quality (PQ) issues. In: 2016 online international conference on green engineering and technologies (IC-GET), pp 1–3

Khadkikar V, Chandra A (2008) A new control philosophy for a unified power quality conditioner (UPQC) to coordinate load reactive power demand between shunt and series inverters. IEEE Trans Power Deliv 23(4):2522–2534

Khatibzadeh A, Besmi M, Mahabadi A, Haghifam MR (2017) Multi-agent-based controller for voltage enhancement in AC/DC hybrid microgrid using energy storages. Energies 10(169):1–17

Kouro S, Malinowski M, Gopakumar K et al (2010) Recent advances and industrial applications of multilevel converters. IEEE Trans Ind Electron 57(8):2553–2580

Lesnicar A, Marquardt R (2003) An innovative modular multilevel converter topology suitable for a wide power range. In: Proceedings of the 2003 IEEE Bologna power technology conference, vol 3, Bologna, pp 23–26

Leung CK, Dutta S, Baek S et al (2010) Design considerations of high voltage and high frequency three phase transformer for solid state transformer application. In: Proceedings of the 2010 IEEE energy conversion congress and exposition (ECCE'10), Atlanta, pp 1551–1558

Long YB, Xiao XN, Xu YH et al (2013) MMC-UPQC: application of modular multilevel converter on unified power quality conditioner. In: Proceedings of the 2013 IEEE power and energy society general meeting (PES'13), Vancouver, pp 21–23

Luo A, Zhao W, Deng X et al (2009) Dividing frequency control of hybrid active power filter with multi-injection branches using improved ip-iq algorithm. IEEE Trans Power Electron 24(10):2396–2405

Luo A, Xu Q, Ma F, Chen Y (2016) Overview of power quality analysis and control technology for the smart grid. J Mod Power Syst Clean Energy 4(1):1–9

Yigit M, Cagri Gungor V, Tuna G et al (2014) Power line communication technologies for smart grid applications: a review of advances and challenges. Comput Netw Int J Comput Telecommun Netw 1–18

Naderi Y, Hosseini SH, Zadeh SG et al (2018) An overview of power quality enhancement techniques applied to distributed generation in electrical distribution networks. Renew Sustain Energy Rev 93:201–204

Naderi Y, Hosseini SH, Zadeh SG et al (2019) Power quality issues of smart grids: applied methods and techniques. In: Decision making applications in modern power systems publisher. Elsevier—Academic Press, pp 89–119

Natesan C, Ajithan SK, Palani P, Kandhasamy P (2014) Survey on microgrid: power quality improvement techniques. Hindawi Publishing Corporation ISRN Renewable Energy, vol 2014, pp 1–7

Pandya RR, Bhavsar F (2018) An overview on power quality issues in smart grid. IOSR J Electr Electron Eng (IOSR-JEEE) 13(Issue1):01–04

Papadimitriou CN, Zountouridou EI, Hatziargyriou ND (2015) Review of hierarchical control in DC microgrids. Electr Power Syst Res 122:159–167

Pearline Kamalini C, Abinaya T, Eazhilarasi J, Hemadevi T (2019) Power quality improvement using series active power filter. JASC J Appl Sci Comput 6(Issue 4):43. ISSN 1076-5131

Pei Y, Jiang G, Yang X (2004) Auto-master-slave control technique of parallel inverters in distributed AC power systems and UPS. In: Proceedings of the 35th annual IEEE power electronics specialists conference, Aachen, Germany, pp 2050–2053

Railing RD, Moreau G, Ronström L et al (2004) Cross sound cable project second generation VSC technology for HVDC. In: Proceedings of the 2004 CIGRE conference, Paris, pp B4–B102

Rawat GS, Sathans (2018) Survey on DC microgrid architecture, power quality issues and control strategies. In: Proceedings of the second international conference on inventive systems and control (ICISC 2018), pp 500–505

Sayed K, Kassem AM, et al (2019) Energy management and control strategy of DC microgrid including multiple energy storage systems. In: 21st international middle east power systems conference (MEPCON). Tanta University, Egypt, pp 736–741

Senthilkumar A, Poongothai K, Selvakumar S et al (2015) Mitigation of harmonic distortion in microgrid system using adaptive neural learning algorithm based shunt active power filter. Sci Direct Proc Technol 21(2015):147–154

Senthilkumar A, Poongothai K, Selvakumar S, Silambarasan M, Ajay-D-VimalRaj P (2015) Mitigation of harmonic distortion in microgrid system using adaptive neural learning algorithm based shunt active power filter. Elsevier Ltd., pp 2212–0173

Suryavanshi RG, Korachagaon I (2019) A review on power quality issues due to high penetration level of solar generated power on the grid. In: IEEE international conference on power and embedded drive control (ICPEDC)

Tazi K, Abbou FM, Abdi F (2019) Multi-agent system for microgrids: design, optimization and performance. Artif Intell Rev 1–60

Youssef KH (2019) Power quality constrained optimal management of unbalanced smart microgrids during scheduled multiple transitions between grid-connected and islanded modes. IEEE Trans Smart Grid 8(1):457–464

Chapter 18
Comprehensive Design of Small Electric Vehicle for Powertrain Optimization for Optimum Range with Weight and Size Reduction

S. Vignesh, Yogesh Krishan Bhateshvar, Mohammad Rafiq B. Agrewale, and K. C. Vora

Abstract Installing Small Electric Vehicle (SEV) in India can potentially act as substitute for taxi/cabs in urban areas where new vehicles can't be deployed considering traffic and stringent emission norms proposed by the government. To expedite this, a quadricycle is benchmarked for retro-fitment and new design is proposed for purpose-built SEV including floor mounted battery pack. Modular electric platform is also investigated with various iterations in powertrain including the front and rear mounting possibilities of motor and battery pack, respectively, for retro-fitted one. Structural stress analysis is performed to find out the maximum possible weight of the battery pack to fit on the floor for purpose-built one. The concept design has a tubular structure for chassis and materials for the chassis are varied to mount the battery pack of weight 200 kg on the floor of SEV. The design of experiments is done on chassis materials for estimating the lowest possible curb weight of SEV. Based on modular battery pack design, a maximum of 23% weight reduction is possible following the curb weight of the Internal Combustion (IC) engine variant of the benchmarked small passenger car. In electric motor, the parameters of interest are motor speed, motor torque and motor efficiency where altering the number of poles leads to maximization in SEV performance. Vehicle parameters such as maximum speed, vehicle acceleration, final drive gear reduction ratio, and battery pack current are compared with the energy economy of the small electric vehicle. The battery pack is designed to fit under the front hood of the vehicle, whereas the motor is fitted at

S. Vignesh (✉) · M. R. B. Agrewale · K. C. Vora
Automotive Engineering, ARAI Academy, Pune, India
e-mail: vickykpm09@gmail.com

M. R. B. Agrewale
e-mail: rafiq.pga@araiindia.com

K. C. Vora
e-mail: vora.pga@araiindia.com

Y. K. Bhateshvar
ARAI Academy, Pune, India
e-mail: bhateshvar.pga@araiindia.com

A. K. Bohre et al. (eds.), *Planning of Hybrid Renewable Energy Systems, Electric Vehicles and Microgrid*, Energy Systems in Electrical Engineering,
https://doi.org/10.1007/978-981-19-0979-5_18

the rear. The driving range is estimated using Simulink and it is validated with mathematical calculation using Peukert method performed in MATLAB. It is concluded that the designed vehicle with Switched Reluctance Motor (SRM) 6/4 configuration of 15 kW, 110 Nm is showing enough capability on the replication of urban car in 2020 targets. For the betterment of range, NCA chemistry is preferred over other lithium-ion chemistries. This chapter provides a complete look of electric powertrain for SEV and its design characteristics through retro-fitted and purpose-built one and its application where electric mobility can be installed seamlessly.

Keywords Small electric vehicle · Retro-fitment · Purpose-built · Concept chassis · Switched reluctance motor · Battery swapping · Driving cycle · Range

Nomenclature

SEV	Small Electric Vehicle
SRM	Switched Reluctance Motor
kg	Kilo gram
LFP	Lithium Ferrous Phosphate
CFRP	Carbon Fiber-Reinforced Plastic
C_D	Drag coefficient
CO_2	Carbon dioxide
HWFET	High Way Fuel Economy Test
SOH	State of Health
NASA	National Aeronautics and Space Administration
LMO	Lithium Manganese Oxide
LCO	Lithium Cobalt Oxide
NEDC	New European Driving Cycle
NYCC	New York City Cycle
WLTP	Worldwide Harmonized Light vehicle Test Procedure
PMSM	Permanent Magnet Synchronous Motor
DC	Direct Current
CVT	Continuous Variable Transmission
GA	Genetic Algorithm
Ah	Ampere hour
MEET	Mahle Efficient Electric Transport
PHEV	Plugin Hybrid Electric Vehicle
ANSA	Automatic Net generation for Structural Analysis
IC	Internal Combustion
NCA	Nickel Cobalt Aluminum
BIW	Body in White
EV	Electric Vehicle
EPA	Environmental Protection Agency
SAE	Society of Automotive Engineers

UDDS	Urban Dynamometer Driving Schedule
SOC	State of Charge
BMS	Battery Management Systems
NMC	Nickel Manganese Cobalt
LTO	Lithium Titanate Oxide
TCO	Total Cost of Ownership
ISO	International Standards Organization
LA	Los Angeles
BLDC	Brush Less Direct Current
ECE	Economic Commission for Europe
AMT	Automated Manual Transmission
DCT	Dual Clutch Transmission
CNG	Compressed Natural Gas
SUV	Sports Utility Vehicle
EU	European Union
UNECE	United Nation Economic Commission for Europe
AISI	American Iron and Steel Institute

18.1 Introduction

Floor battery packs are quite common in electric vehicles. The battery pack is usually mounted in between front and rear axle for making provisions of skate board-type electric vehicle chassis architecture. There is enough ground clearance given in mounting the battery pack on the floor. Due to this, the wheelbase tends to be longer and hence the cars have lower ground clearance (Luccarelli et al. 2014). In general, the weight of the Body in White (BIW) is in the ratio of 1:4 times the curb weight of the vehicle. For example, the Renault concept has a curb weight of 800 kg and its BIW weighs 200 kg (Lesemann et al. 2013). Tesla introduced the skateboard type of arrangement where the motor is mounted on either the front or rear axle depending on the drive train configuration. The battery pack is mounted in between the front and rear axles. SEVs have a smaller wheelbase than full-size sedans. But due to lack of literatures on scalable battery pack the research is very few. Hence, this chapter brings out the need of scalable battery pack in SEV.

The chapter is organized in the following sequence as given in Fig. 18.1. SEV preparation from the scratch requires the material selection for the chassis needs to be done followed by the projected weight of the chassis and its stress analysis and corresponding mathematical validations. Road load coefficients and its significance on distance traveled by the SEV is studied. In SEV, installing a battery pack which caters a minimum of 200 km range is tedious task considering the volume availability. Hence, this chapter also deals with the proposal of modular design of battery packs which can be scaled according to powertrain requirements. In addition to this, battery chemistry for lithium ion (Lithium Ferrous Phosphate (LFP),

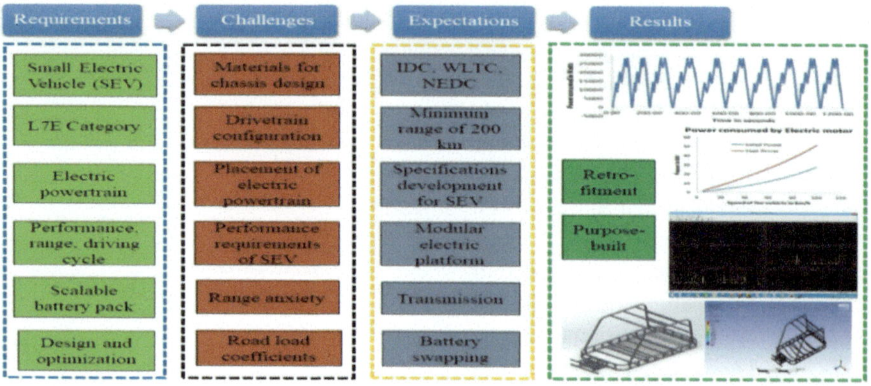

Fig. 18.1 Graphical abstract

Nickel Cobalt Aluminum (NCA)), metal air, lithium polymer, lead acid, and nickel metal hydride were considered for range estimation of the proposed SEV. This is followed by the implementation of reduction ratio to the drive axle by including the power transmission concepts applicable to electric vehicle. Motor operating voltage will have significant contribution in altering the performance of an electric vehicle. Hence, the reduction ratio applicable for various operating voltages is studied for the proposed SEV. Battery pack optimization is carried out based on the volume availability, specific energy catered by the battery pack, and expected range to be delivered by the vehicle. Once the powertrain specification is finalised for the proposed SEV, vehicle level simulation is carried out in EV reference application from MATLAB/Simulink for range estimation and performance monitoring at different driving cycles. The whole outcome of this chapter is to study the significance of modular electric powertrain design possibilities for the SEV under various iterations proposed on the battery pack. Based on this analysis, it is observed that a quadricycle can fulfil the urban mobility requirements when considered for electrification and offers a minimum range of 200 km when SRM 6/4 configuration is preferred with lithium-ion battery pack with NCA composition. The contribution and novelty of the proposed work are as follows:

- Mathematical model for the benchmarked vehicle for retro-fitment and purpose-built one.
- Concept design for purpose-built SEV chassis and its structural stress analysis.
- Empirical analysis of road load coefficients corresponding to small electric vehicle.
- Modular electric platform proposal with 48, 72, and 192 V system voltage and its corresponding gear ratio requirements.
- Scalable battery pack configuration in purpose-built SEV.

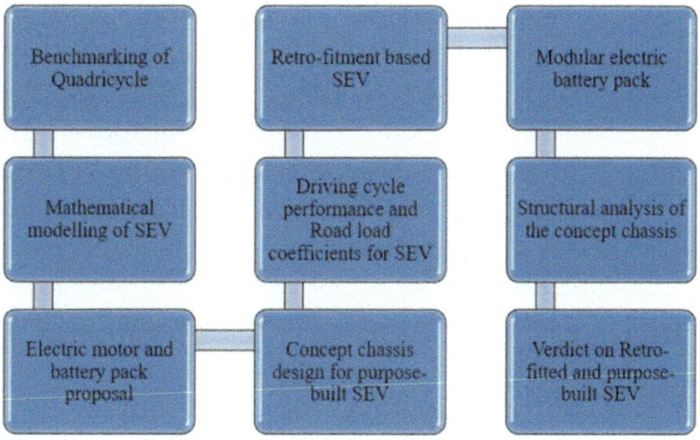

Fig. 18.2 Flow of work in this chapter

- Powertrain behavior of the proposed SEV through driving cycle Indian Driving Cycle (IDC), NEDC (New European Driving Cycle), Worldwide Harmonized Light vehicle Test Procedure (WLTP).

The flow of the proposed work is given in Fig. 18.2.

18.2 Materials for SEV Chassis

Mass of the chassis is an integral part of the curb weight of electric vehicles. In the past, chassis members are flexible. Now there is an increase in structural rigidity that demands good structural stiffness and excellent material properties. Lowering the weight of the chassis reduces inertia and simultaneously the performance of the vehicle can be improved. Ladder chassis is used in heavy vehicles and space frame (the tubular structure is preferred) is used in recent days. Chassis generally designed to keep a 200 kg battery pack on the floor. The wheelbase of 1800 mm is pre-determined to fix the length of the vehicle as 2700 mm for quadricycle having two seats. The width of the chassis is 1515 mm. It is made up of American Iron and Steel Institute (AISI) 1020 and the weight of the chassis is 84 kg (CarlosGertz et al. 2014). Stress analysis of the ladder frame is carried out using the finite element method. Alloy steel is preferred again which has a yield strength of 620 Mpa. This chassis is designed to carry 860 kg of load including motor, battery, passengers (Kristyadi et al. 2017). AISI 1018 is preferred for electric car chassis which has an ultimate strength of 634 Mpa. Chassis is analyzed for the payload of 1.5 kN. Moreover, it withstands 70 kg of the load before the actual deformation occurs (Taufik et al. 2014). The structural modifications in the chassis can be carried out by modifying the double ladder in to single ladder. But it requires the suspension should be connected strongly to the chassis made up of mild steel. One tonne of weight load applied on the carrier of the electric vehicle which is intended to carry goods, 1200 N is applied to the driver's cabin, and 300 N is applied to the battery pack region (Arun et al. 2019). Materials allocated for electric vehicle chassis can be steel, alloy steel, aluminum.

Lightweight chassis requires the material should have high yield strength at the same time low density. Aluminum is a recent entrant in the automobile industry where it is used to make body panels, bumpers. Al 6061 T4, Al 6063 T4 can be used to design the chassis for the electric vehicle. The factor of safety is pivotal in the sense of an increase in the factor of safety increases the cost of productivity. Hence, generally the factor of safety is kept between 1 and 2. If the factor of safety is low then deformation of the chassis is likely to be high (Koumartzakis et al. 2017).

Quadricycles belong to L7E offers a minimum of 150 km range with 16 kWh. When it comes to drivetrain, the requirements are as follows:

- Vehicle weight of less than 600 kg.
- Vehicle acceleration of 0–100 km/h in 10 s.
- Electric range more than 150 km.
- The energy efficiency of less than 80 Wh/km.

Based on this target, vehicle weight holds the key factor if there is a compromise between performance and range. Carbon Fibre-Reinforced Plastic (CFRP) Al space frame structure is used in the Epsilon quadricycle concept (Stein et al. 2016). While selecting the alloy steels, the ultimate strength should be given more importance. A vehicle that weighs 420 kg is designed in CATIA V5 R 18 and stress analysis is carried out using structural alloy, magnesium alloy, and aluminum alloy. Out of these three, magnesium weighs the lowest courtesy low density. It is observed that the weight of the chassis downs by a factor of 4 when it is made of magnesium alloy than steel. One hundred and seventy-five kilograms are allocated for the battery pack. The factor of safety considered is 1.25 stress values are lesser in magnesium alloy (Singh and Chauhan 2017).

Battery pack voltage is a major contributor in deciding the width of the battery pack. Packaging issues are severe when the battery pack voltage is higher. Subsequently in SEV, the floor battery pack is tedious when the chassis structure is monocoque. This is one of the major obstacles for retro-fitment of an electric vehicle with a floor battery pack. The modular design of the battery pack allows the battery pack voltage can be scalable according to the vehicle requirements. The space frame structure is more flexible than monocoque hence it is always preferred for scalable battery packs (Patel and Kumar 2017). Crashworthiness demands more strength from electric vehicle chassis in the event of a crash. High-performance composite materials can be used as chassis materials (Wismans et al. 2011).

Some of the electric vehicles are dedicatedly designed for racing applications. Shell Eco-Marathon is one such race where the electric vehicle is designed according to the guidelines given in the Shell Eco-Marathon competition. At the rear end AISI 9000 steel is used and Aluminum 6082-T6 series is preferred. Automatic Net generation and Structural Analysis (ANSA) pre-processor is used in solving the finite element model (Tsirogiannis et al. 2019).

Due to the increase of traffic in cities, shared mobility is touted as the successor among personal vehicle usage. Recent advancements in shared mobility concepts merged OLA with Mahindra for producing electric vehicles that can be used for urban mobility. The modular vehicle weighs 500 kg. A 30 kW electric motor is used which has a peak power of 90 kW. Battery capacity is 4.62 kWh and has a voltage of 77 V (Maryniuk 2017). This chassis is benchmarked for the concept design in this paper.

Hu-Go is a quadricycle which weighs 257 kg including batteries. Its energy consumption is 51 Wh/km whereas Renault Twizy consumes 70 Wh/km though placed in the same L7E category. With reduced weight, the battery capacity required is just 5.1 kWh for a 100 km range. Space constraints can be addressed by mounting the batteries on the floor (Tanik and Parlaktaş 2015).

18.3 Road Load Coefficients

Chassis dynamometer is an equipment used to measure the road load forces associated with the vehicle prone to testing where simulation of actual road profile is also possible. The results of this test give three coefficients say A, B, and C or in some works of literature it is mentioned as F0, F1, and F2. Also, there will be deviations in the vehicle specifications when the tests are carried out for getting type approval and the tests carried out in an actual vehicle. Coast-down tests are performed in order to measure car's total load. The outcomes of the coast-down test is used to predict the car's fuel economy and later it can also be used for certification purposes. The term road load can be termed as the force needed to propel at constant speeds from neutral gear on a given flat road.

Driving cycles are used to quantify the car's emission level and fuel consumption on a roller test bench. This bench consists of a dynamometer that caters driving experience applicable on a real road. The key aspect on roller test bench is that force acting on the dyno (F_{dyno}) is not equal to the force acting on the vehicle ($F_{vehicle}$). The real-time loads acting on the vehicle can be simulated by the rollers in a test bench. Usually, the resistance of the car is taken as twice in order to include vehicle losses (Norrby 2012).

18.3.1 Significance of Road Load Coefficients

The driving range of electric vehicles varies according to traffic conditions, driving cycle, rolling resistance, aerodynamic resistance, slope angle, and wind speed. If the car satisfies United Nations Economic Commission of Europe (UN ECE) 101 standards, then the vehicle is termed as a pure electric vehicle. Courant is the passenger vehicle, based on the standard version of the engine ignition. Battery charging time is 12–14 h. The economical speed to get 150 km range is 40 km/h, whereas the maximum speed is 85 km/h (Gis et al. 2012). When the vehicle is tested for fuel economy it is recommended to estimate the aerodynamic, gradient, rolling loads associated with the vehicle. Failing to do so leads to a violation of the Environmental Protection Agency (EPA) norms. Dynamometer settings are of two types. The first one deals with the estimation of road load forces under controlled test conditions. The second one involves the determination of deviations between on-road tests and chassis dynamometer tests. In other words, the second one encompasses the correction factor to compensate for the deviations between actual and simulated ones (Determination and Use of Road Load Force and Dynamometer Settings 2015). The fuel economy of an electric vehicle can be enhanced by decreasing the weight of the vehicle. Rolling resistance can be reduced by installing tires with lower rolling resistance coefficient. The value 0.009 seems to be ideal but in practical 0.011–0.015. Aerodynamic resistance can be reduced by reducing the coefficient of drag (C_D). C_D value of 0.25–0.3 seems to be perfectly aerodynamic one even though the cars whose C_D is less than 0.25 is also available in the market (Kühlwein 2016).

Coasting the vehicle is possible using the vehicle's kinetic energy. The equation of motion for deceleration is given by

$$m \cdot b \cdot \delta = -F_b \tag{18.1}$$

where m is the mass of the vehicle (kg).

b is the deceleration (m/s^2).

δ is the rotational inertia (1.03–1.04 times the mass of the vehicle).

F_b is the braking force (N).

The braking force is given by

$$F_b = G_v \cdot f \cdot \cos\alpha \pm G_v \cdot \sin\alpha + \frac{1}{2}\rho C_D A \cdot v^2 \tag{18.2}$$

where G_v is the vehicle gravitational force.

f is the rolling resistance coefficient.

α is the slope angle.

ρ is air density (kg/m^3).

C_D is the aerodynamic drag coefficient.

A is the vehicle frontal area or projected area (m^2).

v is the velocity of the vehicle (m/s).

When the vehicle is moving up the gradient, the gradient resistance is positive and vice versa for moving down the gradient (Barta et al. 2018).

While predicting the range of electric vehicles, the accuracy is of prime importance. Simulation software is employed to estimate the vehicle performance, fuel economy, and emissions tests. Such software is often used to reduce the cost instead of building prototypes to carry out the tests. Those tests were performed on Mitsubishi i-MEV, BMW i3, Nissan leaf EV, and Ford Focus EV. The road load coefficients of the same are estimated through chassis dynamometer tests and the results were compared with the simulation of such tests performed in ADVISOR (Humphries and Morozov 2016). Theoretical estimation of the coast down coefficients are also possible with certain vehicle dynamics equations. Data acquisition systems are used in coast-down tests. Acquisition errors along with errors in wind speed, road slope are the reasons for poor accuracy in coast-down results (Preda et al. 2010). Fuel consumption is usually notified in terms of litres per 100 km and CO_2 emissions are given in grams per litre. While simulating the vehicle on a roller test bench, vehicle speed is the key factor. This is performed based The road load curve gives the measure of vehicle performance. Hence, the estimation of road load curve is crucial in compliance with legislative norms (Kadijk 2012). Hence, there exists type approval and realistic specifications of the test vehicle.

18.4 Modular Battery Pack Design

Battery capacity is not the only factor that influences the range of the electric vehicle. In addition to that vehicle weight, the size of the electric motor also plays an instrumental role in determining the range of electric vehicles (Mruzek et al. 2016). Society of Automotive Engineers (SAE) J1263 norms were followed while conducting road load estimation tests. The atmospheric temperature should fall in the range of 5°–38° C. Minimum five runs should be conducted and the weighted average is taken for road load coefficients (Wishart and Diez 2015). The default driving cycle will be Urban Dynamometer Driving Schedule (UDDS) and High Way Fuel Economy Test (HWFET) according to SAE J1634. Dynamometer tests are conducted according to SAE J1263 (Implementation of SAE J1634 1997). In this paper while estimating the braking distances, thinking distance, and stopping distance is calculated at speeds in 10 km/h increments (https://www.highwaycodeuk.co.uk/answers/what-is-the-stopping-and-braking-distance-of-a-car).

Most of the vehicles built today are of monocoque chassis structure, where the body and chassis are unitized together. When mounting batteries on the floor, this type of chassis is not recommended for SEV. Therefore, there is a need to propose the front and rear mountable battery pack for SEV's.

The modularization of battery packs is the need of the hour in the electric vehicle segment. The motors selected for an electric vehicle has unique voltage requirements. Hence, to satisfy motor voltage, the battery pack should be scalable enough. Apart from basic design requirements, the cost and the life of the battery pack are the other parameters that decide the suitable battery pack. The battery pack should possess better heat transfer characteristics. The influential parameters apart from heat transfer are the design of the battery thermal management system and packaging architecture (Arora et al. 2018). Hence, there is a need for suitable battery architecture which supports the space available in SEV.

Lead–acid batteries are cheap and mature technology. Lithium-ion has several advantages than lead–acid in terms of specific energy and energy density (May et al. 2018). Lithium-ion drives the battery market today extensively for electric vehicles other than batteries for laptops, mobile phones. When it comes to the electric vehicle, the range is the primary factor. Low cost and long life are the secondary factors that drive in synchronous with electric vehicle promotion in urban mobility. The willingness to pay ratio for the additional range is very low in India compared with emerging countries. Because of space constraint and lower specific energy of lead–acid and nickel metal hydride, they are not preferred in SEV's. Hence, the potential customers to support range, low cost, and higher specific energy are lithium–sulfur, zinc-air. Cost-wise lithium–sulfur is cheaper than zinc-air. Metal-air batteries possess the capability of catering higher range than current lithium-ion battery technology. It is projected that lithium–sulfur seems to be an ideal replacement for lithium-ion, whereas metal-air batteries are still in the developing stage (Cano et al. 2018). When the electric vehicle is in use, the battery pack life deteriorates over a period. There the discussion about the second use of batteries is valid. However, due to the lack of literature about used batteries and its applications such as grid connectivity, stand-alone houses the life cycle analysis is not enough to predict the best battery pack (Bobbaa et al. 2018).

Battery performance is measured by the State of Charge (SOC), State of Health (SOH), and the number of life cycles possible in the battery before it starts to degrade. Cell-to-cell variations has to be analyzed since small changes exist in the cell that can be identified easily by the Battery Management System (BMS) (Dubarry et al. 2018). Gaussian model is available to predict the battery SOC, SOH with a root mean square value of 4.3%. National Aeronautics and Space Administration (NASA) Randomized battery sheet is used for primary data about the battery datasets (Richardson et al. 2018). Zinc-air battery can be used as a secondary power source for lithium ion. It has a longer life span than lithium ion (Sherman et al. 2018). Increasing the battery capacity of a battery pack certainly reflects an increased weight of the electric vehicle. The energy economy can be expressed by the number of Watt-hours required to travel unit km. The economic vehicles were Renault Twizzy (67.8 Wh/km), Tazzari Zero (87.9 Wh/km), and Renault Zoe ZE22 (93.6 Wh/km). Currently, Tesla Model S has a longer range courtesy to 100 kWh battery pack which weighs 600 kg. The battery pack weight is almost equal to the curb weight of SEV. However, the energy consumption of the Tesla Model S is 199.5 Wh/km. Hence, the battery pack weight influences the range of electric vehicles (Berjoza and Jurgena 2017). The battery chemistries of lithium ion involved in automotive applications are Lithium Ferrous Phosphate (LFP), Lithium Nickel Cobalt Aluminum (NCA), Lithium Nickel Manganese Cobalt (NMC), Lithium Manganese Spinel (LMO), and Lithium Titanite (LTO). The prominent chemistry in consumer applications is Lithium Cobalt Oxide (LCO). Out of the various lithium chemistries, LFP and NMC have a good balance in terms of cost, safety, specific power, specific energy, life, and performance. A 15 kWh battery pack is common in India. The cost of a 15-kWh battery pack in 2018 is reduced by 65% when compared with the cost in 2009. Further, the infrastructure required for charging the electric vehicle is in the development stage and hence while charging the electric vehicle, a part of infrastructure costs are also added to the customer. Research is still in progress, on reducing the Total Cost of Ownership (TCO) of electric vehicles (Dinger et al. 2010). Even though lithium-air has a higher theoretical specific energy of 3458 Wh/kg, achieving this in practice is difficult. It is projected that urban cars in 2020 will be having 675 kg curb weight, having a battery weight of 175 kg which fuels up to 225 km. This equates to 500 kg weight is distributed between BIW, Motor, Transmission system, and miscellaneous components in the electric vehicle. In developing countries, SEV's play a pivotal role in shifting to electric mobility. Currently, lithium-air batteries have a lower depth of discharge of 1000 mAh/g and lithium–sulfur is closer to industrialization since the C-Rate matches with lithium-ion (Lampic et al. 2016). The driving pattern of electric vehicles can impact the range as visualized by a 30% decrement in energy consumption for the NEDC driving cycle when the operating point is tilted towards the economy. Soon lithium–sulfur is one of the battery packs which propel electric vehicles (Othaganont et al. 2016).

To improve the range, battery pack design is the main factor. Employing a 24-kWh battery pack in the electric vehicle for a 50 km daily commute is like investing the surplus amount of energy into it. Hence, the appropriate size of the battery pack proposed for the right purpose makes electric vehicles successful. In modularization, the customer decides the capacity of the battery pack according to their commute (Mruzek et al. 2014). In India, a study was conducted on electric vehicle road maps and scenarios. The key findings include:

- Battery technology is gearing up for drastic improvement which involves cost reduction and improved specific energy.
- Electric 2 wheelers and cars are comparable to conventional vehicles, but lack of charging infrastructure is still a concern.
- Transport demand is expected to improve soon.
- Higher penetration of electric vehicles increases electricity demand (Shukla et al. 2014).

18.5 Energy Economy in Electric Vehicles

Electric vehicle range can also be referred as energy economy of the vehicle. When the vehicle is being tested, the parameters such as pressure, temperature, and velocity are influential in determining the fuel economy of the vehicle according to the standards published by SAE, ISO. The real-time ambience is completely different when compared to the testing standards and testing conditions. In such cases, there is a variation in the driving range is observed between the real-time conditions with respect to the tested ones under controlled conditions. In order to predict the fuel economy of the vehicle under such conditions requires an engineering model to be proposed, especially for the controlled testing conditions. O. Karabasagolu et al. used UDDS, US-06, HWFET, NYCC, and LA-92 driving cycles in order to predict the fuel economy. (Karabasoglu and Michalek 2013). Specific energy consumption of a four-seated passenger car lies around 84 Wh/km. Though the driving cycle is varying with respected to the testing conditions, it will vary the energy consumption of the vehicle. For a low-powered passenger car (power <10 kW), the range of 70–95 km is obtained at various driving cycles with the maximum speed being 117 km/h and usable battery pack capacity of 6.54 kWh (Saxena et al. 2020). In the case of urban mobility, the range of 100 km is still hovering due to electrical accessories usage consumption. Design criteria do matter a lot when it comes to low-speed or high-speed vehicle applications. Thus, the performance parameters such as maximum speed, torque, gradient, and vehicle acceleration varies according to the driving cycles (Barlow et al. 2009). Chassis dynamometer offers testing at vehicle level and also it caters a good validation by simulating the on-road conditions. Testing agencies do follow different protocols for vehicle testing along with its validation (Type I test On S.I. engines, CNG, LPG and diesel engine vehicles (Verifying the average tailpipe emission) of gaseous and particulate pollutants 2018). Increment in battery capacity will lead to increment in battery pack weight. The specific weight coefficient of the battery pack is given by

$$K = \frac{m_{\text{batt}}}{m_{EV}} \tag{18.3}$$

where, K is the specific weight coefficient,

m_{batt} is the mass of the battery pack (kg),

m_{EV} is the mass of the electric vehicle (kg).

Renault Twizzy has specific weight coefficient of 67 Wh/km, whereas Tesla Model S has specific weight coefficient of 199 Wh/km (Berjoza and Jurgena 2017). Driving cycle testing time and distance will have a significant call over CO_2 emissions (Ciuffo et al. 2017). World Harmonized Driving Cycle can be supported in having a common driving cycle which shall be considered in all terrain and driving conditions (Tutuianu et al. 2014).

A small electric vehicle having a weight of 800–900 kg is supported by a 30 kW SRM having a peak torque of 110 Nm was explored by S. Kachapornkul et al. (Kachapornkul et al. 2007). This paper closely follows the powertrain requirements fixed by S. Kachapornkul et al. Since the speed requirement is different for various driving cycles, SRM only fulfills the maximum speed criteria among BLDC, PMSM. This led to the finalization of electric model having SRM with 15 kW nominal and 30 kW peak power requirements. Battery placement is often being front hood or rear hood in SEV. Since the motor is mounted on the rear axle, the rear wheel drive configuration is preferred in mass production. LFP cells of three layers (18,650) shown in Figs. 18.1 and 18.2; 86.4 Ah, is placed in three layers under the front hood. Considering an SEV, the maximum car width is having around 1.5 m of width for battery pack placement. Driving cycles are altered using simulink platform where the battery capacity of 13.264 kWh is usable from a pack capacity of 16.58 kWh.

18.6 Electric Vehicle Transmission

Electric vehicle is gaining much attention in the automotive industry. Majority of the vehicles have single stage reduction ratio. The motors used in the vehicle have the sufficient thrust to propel basic needs of an automobile. It is the transmission system which makes the motor to operate in optimum speed and the vehicle speed can be matched to meet the driving requirements (Hofman et al. 2016). Multi-stage transmission will improve the energy economy but the vehicle purpose has to be clearly defined. It widens the range of electric vehicle operating zone. Transmission system acts as torque amplifier for electric vehicles. Motor controller is of prime importance in such a way that drivability can be improved (Zhou et al. 2012). IC engines requires gear box to amplify the torque requirements since the torque is comparatively low when compared with electric motors. Reduction ratio depends on the vehicle application and road profiles. By 2030, it is forecasted as 41% of battery electric vehicles will be on the road across the globe (https://www.ngevehicles.com/ev-technical/driving-into-2025-the-future-of-electric-vehicles.html).

18.6.1 Background in Transmission System

Two-speed transmission systems are becoming popular recent days due to their improved energy economy over single-speed one. Vehicle parameters such as mass, drag coefficient, gradient, and frontal area are instrumental in determining the power required by the motor. Battery pack defines the finite distance traveled by the vehicle in between two successive charging. Two-speed transmissions have better energy economy than single stage one. For small electric vehicles, single-speed transmission is enough to meet the driving requirements (Lu et al. 2017). Driving cycle contributes a significant part to the vehicles energy economy. Economic Commission of Europe (ECE) R 15 is such a cycle defined for urban mobility has a maximum speed of 50 km/h. Vehicle parameters can be fixed through driving cycle. The power consumed by the motor for propelling the vehicle on a gradient requires the power to be catered over a period of time. Hence, the motor should be capable of providing the required thrust for a period of time. For small electric vehicles, it is better to choose the motor with lower nominal torque and short time overloading (Prochazka et al. 2015). Downsizing of the motor and battery pack can be possible only if the vehicle have good transmission system. It indirectly correlates to the need of multiple speed transmission system. Vehicle kerb weight is also one of the deciding factors that assume the need of multiple speeds (Chander et al. 2018). Motors can generate maximum torque at low speeds and it can propel the vehicle at rest with ease than IC engine. Most of the small electric vehicles are coupled with single-speed transmission. Two-speed transmission systems can improve the functionality of DC motor-based traction system. Thus, the motor selection for small electric vehicle is critical here (Mahala and Deb 2011). Small electric vehicles sales are quite high in foreign countries like Japan, South Korea, etc. In small passenger cars segment, Nissan leaf and Renault ZOE are popular. BMW is selling smaller electric vehicles first before stepping in to SUV's and sedans but Audi and Mercedes have their own platform (https://www.ngevehicles.com/ev-technical/driving-into-2025-the-future-of-electric-vehicles.html). The automatic transmissions currently used in electric vehicles include Automated Manual Transmission (AMT), Dual Clutch Transmission (DCT), and Continuously Variable Transmission (CVT). Optimization of gear ratios can be done through Genetic Algorithm (GA) approach (Yin et al. 2017). For small electric vehicles, single-speed transmission is typical assumption. Multi-speed transmission is not recommended due to increased weight and cost (Faid 2015). Small electric vehicles have lower reduction ratio and torque from the motor is maximum. This increases the size of the motor to a certain extent. In the same page, increasing the reduction ratio to a larger value increases motor speed and power loss also increases by a margin (Isobe et al. 2011). Mileage of electric vehicles ranges from 110 to 480 km based on the battery capacity installed. In small electric vehicles the space availability of keeping the batteries is low. Motor characteristics are important in defining the reduction ratio of small electric vehicle. To ensure top speed the reduction ratio of the vehicle is kept small and making the motor to run at its maximum efficiency. The final drive ratio will be in the range of 1–4 for such small reduction ratios (Xin and Chengning 2017). Higher the speed of the vehicle, it is recommended to keep multiple gear ratios for the IC engine powered and electric vehicles. If the purpose of vehicle is meant for urban mobility, then higher speeds are not recommended. Hence, single-speed transmission is justifiable at this moment (Parkinson 2016).

18.7 Motor Voltage and Transmission-Related Optimization for SEV

In this paper, benchmarking is carried out on small passenger car. It is converted to motor powered vehicle by defining the power required by the motor. Battery calculations are performed by keeping the constraints of battery weight less than 150 kg and space availability in the front and rear boot. Retro-fitment is given higher importance hence the battery pack has to be placed either at the front or at the back. The motor voltage is varied for 48, 72, and 192 V operating system. Gear ratio is changed accordingly by keeping the maximum torque of the motor to match with Indian Driving Cycle requirements. Analysis is carried out for the small electric vehicle moving on flat road as well as at a constant gradient of 5°. This value is much higher than what is declared in the driving cycle. Even in driving cycles also the maximum gradient is applicable for finite amount of time.

Motor selection is based on rated power and peak power required by the small electric vehicle. Voltage is adjusted by matching the gear ratio with the driving cycle requirements. It is followed by the analysis of instantaneous motor voltage, torque, speed, and current for the driving cycle imposed. Battery pack is proposed for lithium ion cylindrical cell having LFP chemistry. Weight of the battery pack is compared for the motor voltages defined in this paper. Weight distribution is given higher importance since retro-fitment is proposed.

The specifications of small gasoline passenger car are given below:

- Power = 9 kW
- Engine = 220 c.c. single cylinder

- Fuel economy = 35 km/l for petrol and 43 km/kg for Compressed Natural Gas (CNG)
- Kerb weight = 450 kg
- Max speed = 70 km/h
- Dimensions = 2752 × 1312 × 1652 mm
- Projected area = 2.16 m^2
- Wheel base = 1925 mm
- Wheel track = 1143 mm
- Turning circle Radius = 3.5 m
- Front storage capacity = 60 L
- Center storage capacity = 95 L
- Rear storage capacity = 44 L
- CO_2 emitted = 66 g/Km
- Tyre size = 135/70 R12 Tubeless.

For converting this benchmarked vehicle to electric one, tractive effort calculations are carried out.

18.7.1 Rolling Resistance

$$F_r = f_r mg \cos\alpha \tag{18.4}$$

f_r is rolling resistance coefficient of tyre;
g is acceleration due to gravity (m/s^2);
α is road slope (rad);
Rolling resistance coefficient = 0.0136 × (0.40 × V^2 × 10^{-7});
Coefficient of friction = 0.4 and 0.7 for wet and dry roads;
f_r = 0.0137 for wet roads and 0.0139 for dry roads;
m = 719 kg;
g = acceleration due to gravity = 9.81 m/s^2;
Rolling resistance = 0.0139 × 719 × 9.81 = 96.631 N.

18.7.2 Aerodynamic Resistance

$$F_{\text{air}} = \frac{1}{2}\rho C_d A_f (v)^2 \tag{18.5}$$

ρ is the air density (Kg/m^3);
C_d is the aerodynamic drag coefficient;
A_f is the frontal area;
v is the vehicle speed (m/s);
C_d = 0.455;

$A_f = 2.233$ m^2;
$V = 13.88$ m/s;
$\rho =$ density of air $= 1.21$ Kg/m^3;
Aerodynamic resistance $= 0.5 \times 1.21 \times 0.455 \times 2.233 \times (13.88)^2 = 118.422$ N.

18.7.3 Gradient Resistance

$$F_g = mg \sin \alpha \tag{18.6}$$

$\alpha = 13\%$;
Gradient resistance $= 719 \times 9.81 \times \sin(13\%) = 859.158$ N.

18.7.4 Road Load

$$ma_x = F_x - F_r - F_{\text{air}} - F_g \tag{18.7}$$

a_x is vehicle longitudinal acceleration in m/s$^{2;}$
F_x is the propelling force at wheels in N;
F_r is the rolling resistance;
F_{air} is the aerodynamic resistance;
F_g is the gradient resistance.
Propelling force at wheels $= F_x$

$$F_x = f_r mg \cos \alpha + \frac{1}{2}\rho C_d A_f (v)^2 + mg \sin \alpha + ma_x \tag{18.8}$$

18.7.5 Resistance Summary

Rolling resistance = 96.631 N.

Aerodynamic resistance = 118.422 N.

Gradient resistance = 859.158 N.

Acceleration resistance = 445.061 N.

For estimating the rated power of motor, rolling resistance, aerodynamic resistance, and acceleration resistance is accounted. While computing peak power, gradient resistance is also included.

$$P_{\text{rated}} = (96.631 + 118.422 + 445.061) \times 13.88 = 9.162\ \text{kW}$$

$$P_{\text{peak}} = (96.631 + 118.422 + 445.061 + 859.158) \times 13.88 = 21.085 \text{ kW}$$

Based on the tractive effort calculations, it is observed that the rated power is 9.162 kW and peak power is 21.085 kW. This power will vary according to the driving cycle considered. The motor selected should support the rated power and peak power required by the vehicle.

18.7.6 Motor Selection

Switched Reluctance Motor (SRM) has wide speed range and high torque density. It also has reasonable power density for small electric vehicle. SRM has different configurations of stator and rotor poles. For 15 kW rated power, 6/4 and 8/6 configuration was considered in this paper. The peak torque of 8/6 is low when compared with 6/4. Hence, when 8/6 SRM is employed in real driving conditions the peak torque may not be sufficient to propel the vehicle under stiff gradients. Therefore, SRM 6/4 configuration is considered in this paper. The specifications of SRM is given below(Maryniuk 2017):

- No. of poles on Stator = 6
- No. of poled on rotor = 4
- Rated power = 15 kW
- Peak power = 30 kW
- Torque = 110 Nm
- Voltage = 192 V
- Speed = 10,000 RPM.

The driving cycle considered in this paper is Indian Driving Cycle. It is depicted in the Fig. 18.3.

18.7.7 Gear Ratio Design

$$\omega_w = \frac{v_w}{r} \tag{18.9}$$

where ω_w is angular velocity of the wheel in rad/s;
v_w is vehicle velocity in m/s;
r is wheel radius in m.

The angular velocity of the wheel is calculated instantaneously for Indian driving cycle. Wheel torque is the product of tractive force and wheel radius. Hence, it is given by

$$T_w = FTR \times r \tag{18.10}$$

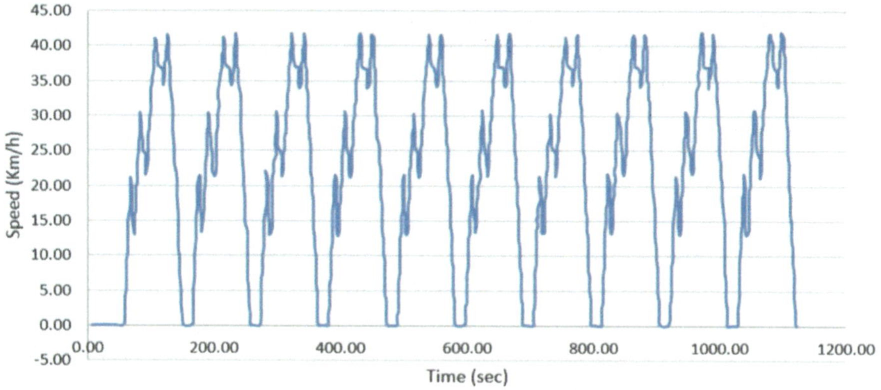

Fig. 18.3 Indian driving cycle for 2, 3 wheelers

where T_w is the wheel torque in Nm;
FTR is the tractive force in N;
r is the wheel radius in m.
Angular velocity of motor is given by

$$\omega_m = G \times \omega_w \tag{18.11}$$

where ω_m is the angular velocity of the motor in rad/s;
G is the gear ratio;
ω_w is the angular velocity of the wheel in rad/s.
Motor speed is calculated from the following equation:

$$N_m = \frac{\omega_m \times 60}{2\pi} \tag{18.12}$$

where N_m is the speed of the motor in rpm.
Motor torque is given by

$$T_m = \frac{\omega_w \times T_w}{0.98 \times \omega_m} \tag{18.13}$$

Here, the gear efficiency is assumed as 98% and
T_m is the motor torque in Nm.
Motor Power can be calculated by

$$P_m = T_m \times \omega_m \tag{18.14}$$

Here, P_m is the motor power in Watts.

For calculating the motor voltage, stroke angle and Psi peak are required. According the motor specifications, stroke angle is 0.5233 and Psi peak is 0.44.

Table 18.1 Comparison of motor system voltage

Voltage (V)	48	72	192
Gear ratio	1.25	1.9	5
Maximum constant gradient (degree)	5	5	5
Motor peak power on flat road (kW)	2.5	2.5	2.5
Motor peak power on gradient (kW)	9.6	9.6	9.6
Motor torque on flat road (Nm)	42	28	11
Motor torque on gradient (Nm)	110	110	110
Motor current on flat road (A)	50	33.54	12.35
Motor current on gradient (A)	127.48	127.48	127.48

Hence, the motor voltage is given by

$$V_m = \frac{\text{Psi peak} \times \omega_m}{\text{Stroke angle}} \tag{18.15}$$

Clearly, from Table 18.1, variation in motor system voltage doesn't affect the rated power and peak power consumed by the motor. If the system voltage is increased then in order to support the driving cycle, gear ratio have to be increased. Maximum constant gradient also doesn't depend on system voltage but it has direct relationship with motor torque and current. At level roads the scenario is different. The peak values of motor torque and current are same when the system voltage is varied. At lower voltages, increase in motor current increases the heat generation in the system. Hence, motor insulation should be given properly. Motor current and torque has a direct relationship between them. Even though motor is capable of providing high torque, increase in motor current increases the heat generation.

18.7.8 Effects on Battery Pack

As the motor voltage and battery voltage are same, placing cells in series for 48, 72, and 192 V system is applicable in small electric vehicle on both front and rear boot space available; 192 V is the maximum voltage possible for a small electric vehicle having width less than 1.5 m. Hence, the battery voltage also limited to 192 V. While connecting the cells in parallel, front boot supports up to 94.5 Ah, i.e., keeping three layers in the battery pack. In this system, the first layer consists of 40.4 Ah (15 cells in parallel), second layer contributes 32.4 Ah (12 cells in parallel), and third layer contributes 21.6 Ah (8 cells in parallel). The number of cells placed in parallel is getting reduced as no. of layers in the battery pack got increased. This is due to the phenomena of vehicle aerodynamics at the front.

At rear 216 Ah is possible with four layers of battery pack by keeping 20 cells per layer. Tata Tigor has 216 Ah battery pack. But keeping 216 Ah battery pack at the rear boot increases the battery weight as well as it affects the vehicle dynamics also. For retro-fitment purpose, keeping battery pack at the rear will change the drive configuration. Hence, rear battery pack is not recommended.

18.7.9 Weight Analysis for Battery Pack

In this analysis, the battery considered is LFP 3.6 V 2.7 Ah cylindrical cell. Motor voltage is varied and the projected battery weight is estimated is depicted in Table 18.2

For keeping battery pack at the front higher system, voltage is recommended to get a decent range. If the battery pack is placed at the rear boot, then lower system voltage is good otherwise the battery pack weight will increase and also weight distribution is affected. Indian electric cars have low system voltage but they offer a decent range through floor mounted battery pack. In retro-fitment, floor battery is not recommended as it disturbs the existing monocoque chassis structure.

18.8 Battery Pack Selection and Optimization for SEV

Fitting a battery pack in small electric vehicle which will cater a range of 200 km per charge is a challenging task considering the dimensions available for mounting the battery pack. In retro-fitment scenario, the battery pack can be mounted at the front end by taking care of vehicle aerodynamics and also the volume constraints possessed by the front hood. The battery pack can also be mounted in the rear where the free volume for fitting the battery pack is higher than the front hood. Focusing on the vehicle drivetrain, electric motors can perform efficiently and also it won't affect the drivetrain dynamics of the electric vehicle. By assuming all the challenges mentioned above, the battery pack can be optimized in such a way that it can fit various motor operating voltages, battery form factors, and high energy density cells. The motors considered for fitting in to SEV are BLDC, PMSM, SRM (8/6), and SRM

Table 18.2 Projected weight of battery pack for motor system voltage

Voltage (V)	48	72	192
Front capacity (Ah)	94.5	94.5	94.5
Front battery capacity (kWh)	4.536	6.804	18.144
Front weight (kg)	23.1	33	89.1
Rear capacity (Ah)	216	216	216
Rear battery capacity (kWh)	10.368	15.552	41.472
Rear weight (kg)	56	80	216

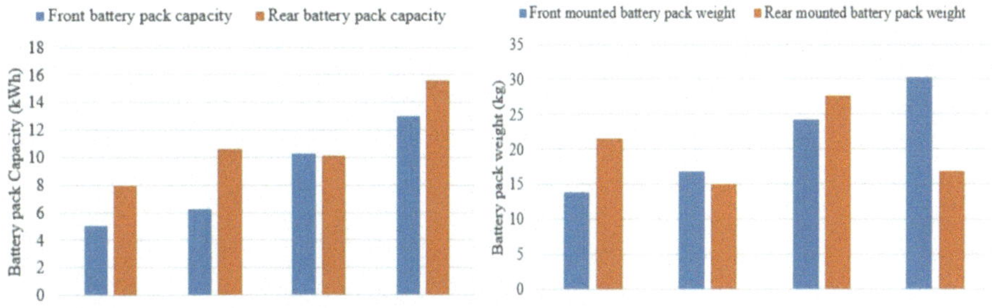

Fig. 18.4 **a** Battery pack capacity. **b** Battery pack weight

(6/4). The batteries considered in proposing the battery pack are lead–acid, Nickel–Metal Hydride, Lithium Polymer, Lithium–Sulfur, and Zinc-air. Based on the motor and battery pack selected, the optimized configuration for front- and rear-mounted battery packs which ensures a range of 200 km are given in the Fig. 18.4.

It is recommended to maintain the weight distribution of the SEV such that the center of gravity affects are negligible. The above figure shows the possible battery pack capacities which can be fitted in the SEV at front and rear ends. This also includes the projected weight of the SEV.

18.9 Retro-fitment Versus Purpose Built SEV

The benchmarked small passenger car has the following dimensions:

Length = 2800 mm,

Width = 1500 mm,

Height = 1600 mm.

Concept chassis is designed in solid works and it is exported to ANSYS for stress analysis. The concept chassis is depicted in Fig. 18.5

Volume available for battery mounting (L × B × H) = 2300 × 1300 × 200 mm.

For making the battery pack easily scalable, 12 V 35 Ah battery is considered.

The dimensions of the battery module are 175 × 104 × 165 mm and weight is 4.53 kg per module.

The maximum number of battery modules that can be mounted on the floor is 11 modules in series and 10 modules in parallel. Thus, the battery weight propels up to 500 kg on the floor. Here, the motor voltage is the key to designing scalable battery packs. For example, if the motor voltage is 72 V, then six batteries must be placed in series. Thus, by keeping ten modules in parallel, the battery rating reads 72 V 350 Ah and therefore the battery capacity reaches 25.2 kWh. The possibilities for a scalable battery pack are discussed below. Each battery module of different battery ratings is selected in such a way that it can be inserted in the floor space available. The battery modules considered in this paper are given in Table 18.3. Data for each

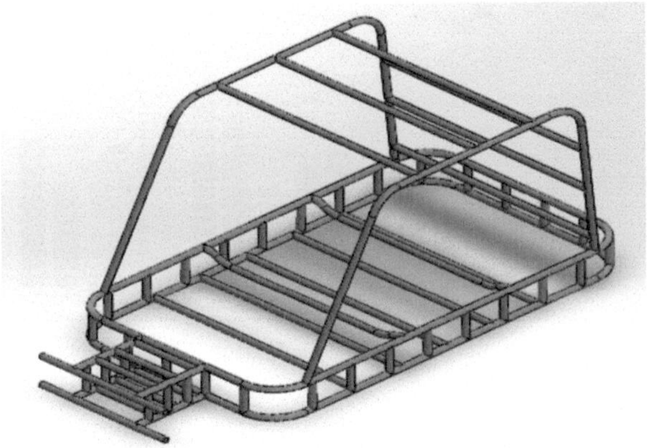

Fig. 18.5 Concept chassis for purpose built SEV

Table 18.3 Battery modules considered in this paper

Voltage (V)	12	24	48	72	96	192
Length (mm)	175	342	290	367.9	280	94.5
Width (mm)	165	205	210	183.9	200	18.144
Height (mm)	105	75	100	65.2	180	89.1
Capacity (Ah)	35	35	35	35	35	216
Weight (kg)	4.53	9	7.3	9.8	18	41.472
Battery shell	PVC	PVC	PVC	PVC	PVC	216

module is obtained from the website which is given after the reference section.

As discussed in Table 18.4, 35 Ah rating is kept constant in this paper because the battery pack benchmarked was Mahindra e2o (210 Ah). In the specifications page (Kept in Website, Reference section), 16 modules are used. Hence, the maximum length of the battery pack is obtained when 8 modules are kept parallel. Therefore, by keeping 8 modules of 35 Ah each, 210 Ah is obtained. Also, the various combinations of scaling the battery pack within the dimensional limits must be addressed. In Table 18.2 voltage scaling is done by keeping the 12 V battery module. This calculation is repeated for 24 V (shown in Table 18.5), 48 V (shown in Table 18.6), 72 V (shown in Table 18.7), and 96 V (shown in Table 18.8).

Table 18.4 Battery module having 12 V 35 Ah rating

Voltage (V)	48	72	96	120
Number of cells in series	4	6	8	10
Number of cells in parallel	6	6	6	6
Max. projected weight of battery pack (kg)	108.72	163.08	217.44	271.8
Max. projected battery capacity (kWh)	10.08	15.12	20.16	25.2

Table 18.5 Battery module having 24 V 35 Ah rating

Voltage (V)	48	72	96	120	144
Number of cells in series	2	3	4	5	6
Number of cells in parallel	6	6	6	6	6
Battery capacity for e2o (210 Ah)	10.8	15.12	20.16	25.2	30.24
Projected battery weight (kg)	108	162	216	270	324

Table 18.6 Battery module having 48 V 35 Ah rating

Voltage (V)	48	96	144	192	240	288	336
Number of cells in series	1	2	3	4	5	6	7
Number of cells in parallel	6	6	6	6	6	6	6
Battery capacity for e2o (210 Ah)	10.08	20.16	30.24	40.32	50.4	60.48	70.56
Projected battery weight for e2o (kg)	43.8	87.6	131.4	175.2	219	262.8	306.6

Table 18.7 Battery module having 72 V 35 Ah rating

Voltage (V)	72	144	216	288	360	432
Number of cells in series	1	2	3	4	5	6
Number of cells in parallel	6	6	6	6	6	6
Battery capacity for e2o (210 Ah)	15.12	30.24	45.36	60.48	75.6	90.72
Projected battery weight for e2o (kg)	58.8	117.6	176.4	235.2	294	352.8

Table 18.8 Battery module having 96 V 35 Ah rating

Voltage (V)	96	192	288	384	480	576	672	768
Number of cells in series	1	2	3	4	5	6	7	8
Number of cells in parallel	6	6	6	6	6	6	6	6
Battery capacity for e2o (210 Ah)	20.16	40.32	60.48	80.64	100.8	120.96	141.12	161.28
Projected battery weight for e2o (kg)	108	216	324	432	540	648	756	864

Weight-based optimization is carried out from the possible combinations of the battery pack with scaled voltage. The threshold weight of the battery pack is kept at 200 kg and tolerance is kept as ±50 kg. The various battery combinations with scaled voltage are shown in Tables 18.9 and 18.10.

While selecting the chassis materials, the factor of safety is kept at 5. Hence, the designed chassis can withstand 1000 kg of battery pack before failure. Initially, the maximum possible battery voltage is fixed from the calculations shown above. It is followed by the chassis materials considered in ANSYS is given in Table 18.11.

Table 18.9 Battery optimization for 200 + 50 kg

Voltage (V)	12	24	48	72	96
Individual module capacity (Ah)	35	35	35	35	35
Max battery pack voltage (V)	96	96	240	288	192
Battery capacity (Ah)	210	210	210	210	210
Projected battery capacity (kWh)	20.16	20.16	50.4	60.48	40.32
Projected battery weight (kg)	217.44	216	219	235.2	216

Table 18.10 Battery optimization for 200 − 50 kg

Voltage (V)	12	24	48	72
Individual module capacity (Ah)	35	35	35	35
Max battery pack voltage (V)	72	72	192	216
Battery capacity (Ah)	210	210	210	210
Projected battery capacity (kWh)	15.12	15.12	40.32	45.36
Projected battery weight (kg)	163.08	162	175.2	176.4

Here, the 96 V battery is not considered because the weight difference is too large based on the given boundary conditions. From the upper boundary, it is clear that the battery voltage scaled from 12 to 288 V and for lower boundary, the battery voltage scaled from 12 to 216 V. This voltage projection is applicable for the dimensions considered in designing the SEV. The proposed battery pack can be scaled for sedans, SUV's, etc., where the length and width of the volume available for mounting can be improved based on their dimensions. The materials considered for structural analysis is given in Table 18.11 where the upper and lower boundaries of battery pack is given in Tables 18.12 and 18.13.

18.9.1 Boundary Conditions

Fixed support = Bottom boundary of the chassis.

Force = 10,000 N (on the support bars for mounting the battery pack.

Output = Total deformation (mm) and Equivalent stress (Mpa).

The weight of the chassis obtained from ANSYS 2019 R1 is shown in Fig. 18.6. The results obtained from ANSYS while substituting the boundary conditions are shown in Fig. 18.7 and Fig. 18.8. Deformation and structural analysis of Al 6082 T6 and Mg alloy were shown in Figs. 18.9, 18.10, 18.11 and 18.12, respectively.

Based on the structural analysis in ANSYS, the weight of the chassis is taken. Before doing the weight distribution, the assumptions considered in this analysis are given below.

- BIW weight (W_b) = 100 kg

Table 18.11 Materials considered for chassis in ANSYS

Material	Steel	Alloy steel	Al 6082 T6	AISI 9000	Mg Alloy	Al Alloy	Al 6063 T4	Al 6061 T4	AISI 1018
Density (kg/mm^3)	7861	7700	2449.34	7076.40	1800	2770	2690	2700	8000
Youngs modulus (MPa)	205	210	72	207	45	71	69	69	210
Poisson ratio	0.29	0.28	0.33	0.29	0.35	0.33	0.33	0.33	0.3
Tensile strength (MPa)	772.56	723	270	850	540	648	80	130	634

Table 18.12 Upper boundaries of the optimized battery pack

Chassis	12 V	24 V	48 V	72 V	96 V
Steel	488.05	486.61	489.61	505.81	486.61
Alloy steel	485.94	484.5	487.5	503.7	484.5
Al 6082 T6	411.954	410.514	413.514	429.714	410.514
AISI 9000	477.152	475.712	478.712	494.912	475.712
Mg alloy	403.07	401.63	404.63	420.83	401.63
Al 6063 T4	415.344	413.904	416.904	433.104	413.904
Al 6061 T4	415.845	414.405	417.405	433.605	414.405
AISI 1018	490.17	488.73	491.73	507.93	488.73

Table 18.13 Lower boundaries of the optimized battery pack

Chassis	12 V	24 V	48 V	72 V
Steel	433.69	432.61	445.81	447.01
Alloy steel	431.58	430.5	443.7	444.9
Al 6082 T6	357.94	356.514	369.714	370.914
AISI 9000	422.792	421.712	434.912	436.112
Mg alloy	348.71	347.63	360.83	362.03
Al 6063 T4	360.984	359.904	373.104	373.304
Al 6061 T4	361.485	360.405	373.605	374.805
AISI 1018	435.81	434.73	447.93	449.13

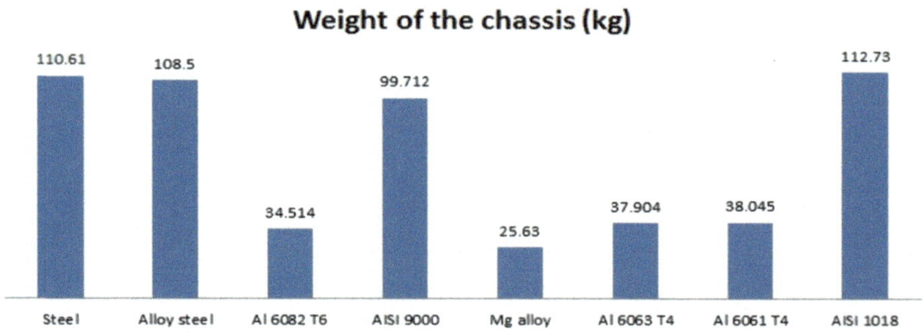

Fig. 18.6 Chassis weight comparison for purpose-built SEV

- Motor weight (W_m) = 40 kg
- Miscellaneous (W_M) = 10 kg
- Transmission (W_t) = 10 kg.

$$W_{\text{Body}} = W_b + W_m + W_M + W_t \tag{18.16}$$

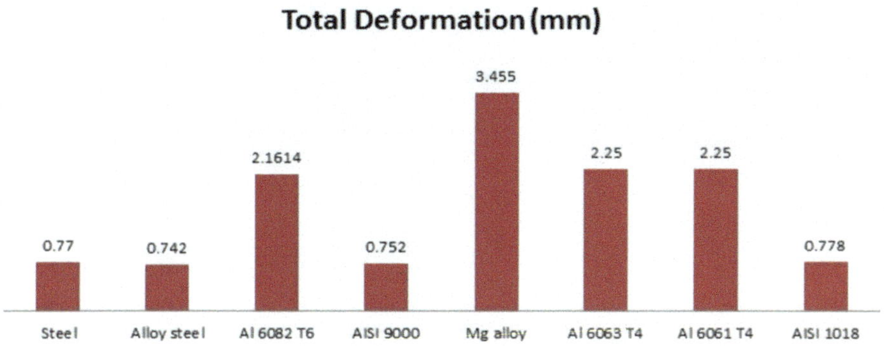

Fig. 18.7 Total deformation applicable to various chassis materials for SEV

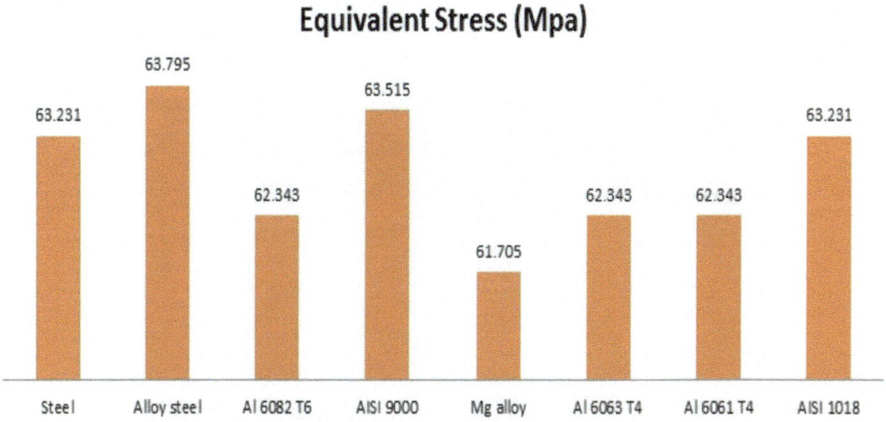

Fig. 18.8 Equivalent Stress applicable to various chassis materials for SEV

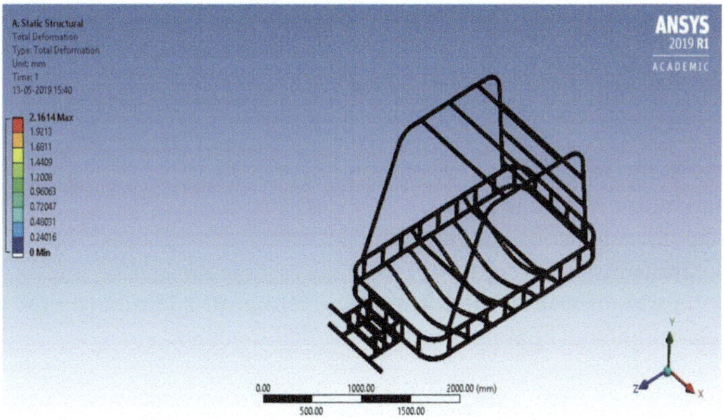

Fig. 18.9 Total deformation of the concept chassis applicable to SEV

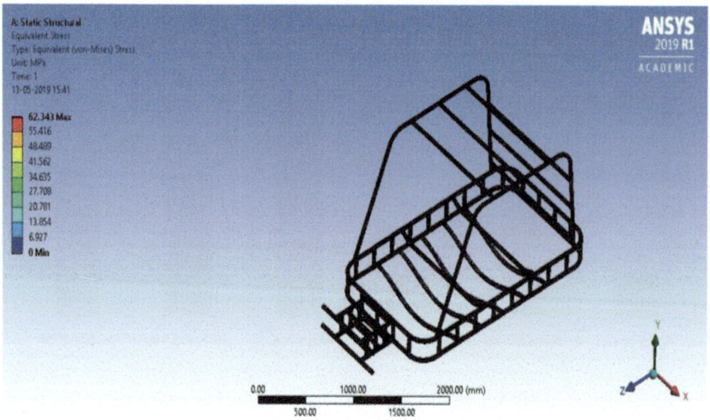

Fig. 18.10 Equivalent stress of the concept chassis applicable to SEV

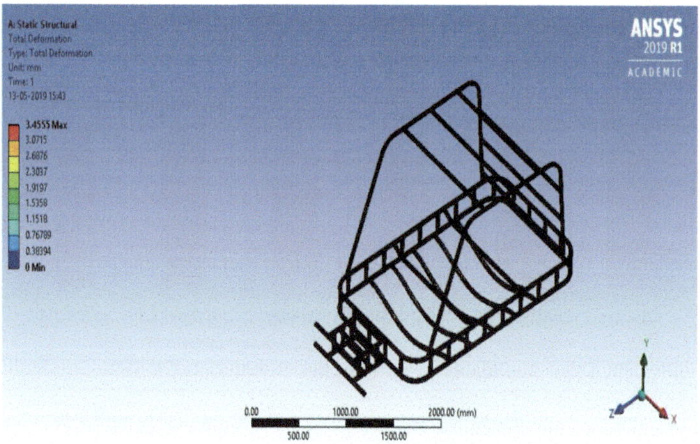

Fig. 18.11 Total deformation of the concept chassis applicable to SEV

where W_b is the BIW weight of the electric vehicle (kg).

W_m is the motor weight (kg).

W_M is the miscellaneous weight (kg).

W_t is the transmission weight (kg).

Hence, the weight to be kept constant is 160 kg. In addition to this, optimized battery pack weight and chassis weight are added together.

$$W_k = W_{\text{Body}} + W_{\text{Chassis}} + W_{\text{battery pack}} \tag{18.17}$$

where W_K is the curb weight of the electric vehicle.

W_{Body} is the body weight (kg).

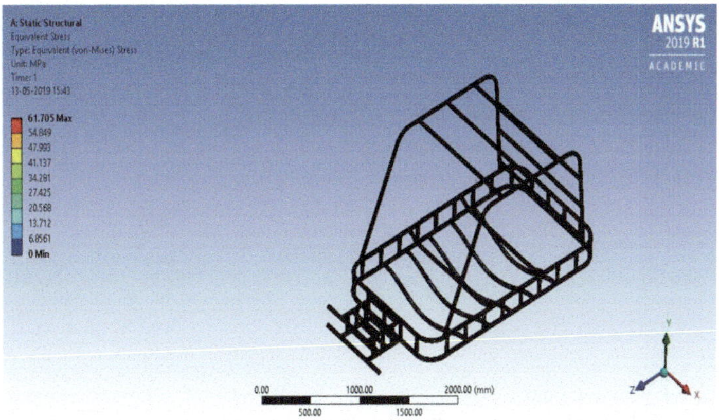

Fig. 18.12 Equivalent stress of the concept chassis applicable to SEV

W_{chassis} is the chassis weight (kg).

$W_{\text{battery pack}}$ is the battery pack weight (kg).

Based on the above design of experiments for the upper and lower boundary, the chassis should be made up of Al alloys and magnesium alloy. The lowest possible curb weight of the chassis for the upper boundary of the battery pack is 401.63 kg for magnesium alloy and for aluminum alloys the curb weight varies from 410 to 435 kg. When the lower boundary of the battery pack is considered, the lowest possible curb weight of chassis is 347.63 kg and for aluminum alloys, it varies from 356 to 375 kg. For SEV upper boundary of the battery pack capacity is quite high. Even though space is available at the floor for mounting the battery pack, a battery capacity of more than 20 kWh is not recommended. Hence, even in lower boundary also inserting 48 and 72 V modules are not recommended. The possible battery capacity which meets the requirements of SEV is 15.12 kWh from the lower boundary of the battery pack. This battery capacity is at par with Mahindra e2o, Tata Tigor, MAHLE Mahle Efficient Energy Transport (MEET) vehicles. Battery pack weight varies from 162 to 163 kg which is well inside the limit of 200 kg. Therefore, the lowest possible curb weight of the proposed SEV is 347.63 kg. Adding the passenger weight of 300 kg (4 passengers, each 75 kg) and payload of 20 kg, the gross weight of the proposed SEV is 667.63 kg. The curb weight of the IC engine variant is 451 kg. Hence, using Mg alloy chassis 22.93% weight reduction (curb weight) is possible. For aluminum alloys, it works out to be 16 to 21%.

18.10 Discussion on Road Load Coefficients

The SEV proposed in this paper will have the following specifications shown in Table 18.14.

Table 18.14 Specifications of SEV

Kerb weight (kg)	450
Pay load (kg)	20
Gross Vehicle Weight (kg)	750
Motor power (kW)	15 (SRM, 6/4)
Torque (Nm)	110
Gear ratio	6.6:1
Maximum speed (km/h)	80
Gradient	15° at 40 km/h
Battery	192 V, 86.4 Ah (LFP)

The road load coefficients obtained from previous literature for small cars are given below. The force experienced by the vehicle during coasting down test is given by

$$F = A + Bv + Cv^2 \tag{18.18}$$

where F is the Road load force in N.

A, B & C are the road load coefficients.

v is the vehicle velocity in m/s.

Then the power consumed by the electric vehicle in coasting is given by

$$P = F \times v \tag{18.19}$$

where P is the power consumed in watts.

F is the Road load force in N.

v is the vehicle velocity in m/s.

The energy stored in the battery pack is exhausted after a certain amount of time say the operation of the electric vehicle. The time can be calculated by

$$t = \frac{\text{Battery Capacity}(kWh)}{P(kW)} \tag{18.20}$$

where t is the time taken by the vehicle to exhaust full of its energy in hours.

Distance traveled by the electric vehicle is estimated by

$$D = v \times t \tag{18.21}$$

where D is the distance traveled by the electric vehicle in km.

v is the vehicle velocity in m/s.

t is the time taken by the electric vehicle to exhaust its full energy in seconds.

The road load coefficients vary according to the vehicles tested in the chassis dynamometer. The values given by the EU, SAE, and Korres et al. are given in Table

Table 18.15 Road load coefficients

Coefficients	EU	SAE	Korres et al.
A	122.2	120.4	169.6
B	−0.443	−0.207	5.12
C	0.442	0.436	0.2869

18.15. Tests were conducted on a small electric car whose gross vehicle weight ranging from 750 to 950 kg. The proposed SEV falls in the bracket and hence the road load coefficients are compared in range estimation. The vehicle considered by Korres et al. has a curb weight of 831 kg, a maximum speed of 85 km/h and a range of 150 km. The range obtained when substituting the Korres et al., EU, and SAE values of road load coefficients in Eqs. 18.1, 18.2, and 18.5–18.8 is shown in Figs. 18.13, 18.14 and 18.15, respectively.

The distance covered by the electric vehicle designed by Korres et al. is depicted in the figure. It is observed that a 200 km range is possible only when the vehicle speed

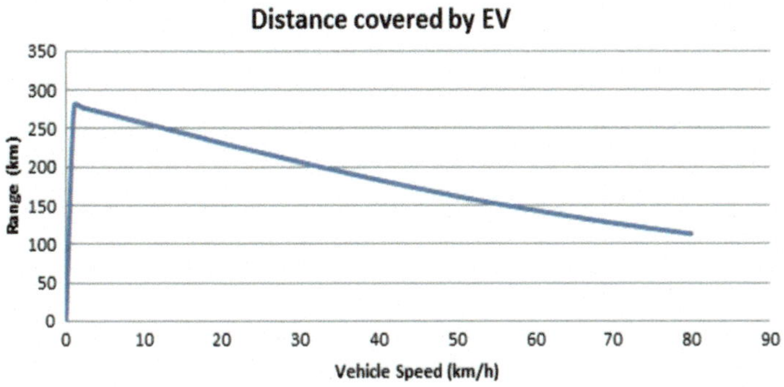

Fig. 18.13 Distance covered by the SEV under various speeds

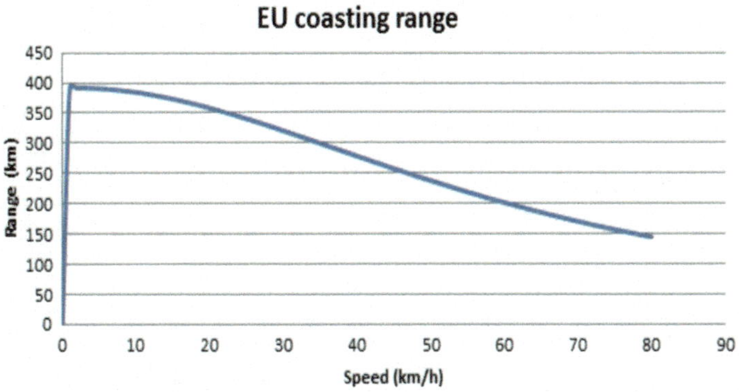

Fig. 18.14 European Union coasting range applicable to SEV

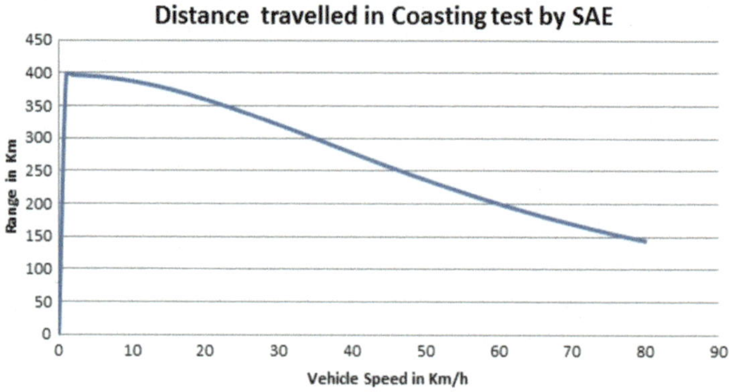

Fig. 18.15 Distance traveled in coasting test by SAE applicable to SEV

is between 30 and 35 km/h. Urban driving requires such speed ranges, especially in traffic conditions. While cruising the vehicle above 60 km/h the range drops to less than 150 km which seems to be realistic considering the battery pack provided for Mahindra e2o. The range provided by the manufacturer is 140 km for a 15-kWh battery pack. Generally, the usable energy capacity is always less than battery capacity. Assuming the usable battery capacity is 80% of total battery capacity, the usable capacity of the SEV proposed is 13.264 kWh, whereas, for Mahindra e2o, it is 12 kWh. The specific energy consumption for SEV when moving at speed of 30–35 km/h is 15.078 km/kWh and for e2o it is 11.667 km/kWh.

Euro 5 gasoline car is tested in chassis dynamometer and the coast down data for the same is given in Table 18.16.

Sample 5 is a small passenger hatchback gasoline engine car that meets Euro 5 Norms. Though the curb weight is higher than 1000 kg the road load coefficient values particularly the realistic one resembles the values declared by EU and SAE. The only change in value is the B coefficient which directly influences rolling resistance and aerodynamic resistance of the vehicle. A and C values are almost in the range when compared with the EU and SAE. From Fig. 18.16, a range of 200 km can be attained if the speed of the vehicle is in the range of 50–60 km/h (or) at an average speed of 55 km/h. This scenario is valid for a Realistic case where the actual road load forces will act on the vehicle. If the least squares fit method is implemented to estimate the coefficients after the completion of the coast down test, the average speed should be 60 km/h to obtain a 200 km range. Generally, the user ends up with 1 equation and three unknowns, therefore, the least squares fit is implemented to get the solution.

Table 18.16 Hatchback A segment coast down data

Sample 5	Type approval	Realistic	Least squares fit
A	86	123	114
B	0.612	1.008	0.3861
C	0.41472	0.41472	0.0281

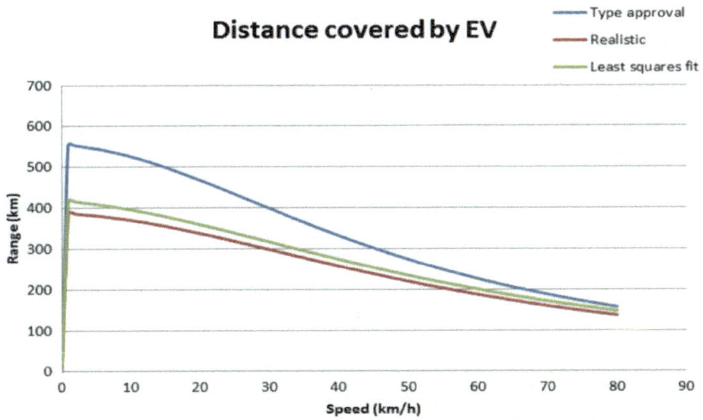

Fig. 18.16 Comparative distance covered by the proposed SEV under various coast down data

For getting type approval, the vehicle should be tested in ideal conditions, i.e., the vehicle will not experience the same in real driving conditions. In this test, the vehicle speed can reach up to 70 km/h for a range of 200 km. Based on the three cases for a sample, it is observed that maximum speed should be 55–70 km/h. This makes sense that the proposed SEV has maximum speed of 80 km/h will get a range of 200 km provided the weight of the SEV should be less than weight of Sample 5 and weight in the case of Korres et al. The curb weight of various models considered in this paper is shown in Fig. 18.16.

18.10.1 Model Validation

Here, the concept of the coast down mathematics empirical relationships was considered. The empirical relationships have a close correlation with vehicle dynamics. The expression of the vehicle deceleration during coast-down can be presented as a final function of speed is given by

$$\frac{\left[mg(f_0 + f_1 v)\cos\alpha + mg\sin\alpha + \frac{1}{2}\rho C_d A V^2 + R f_0 + R f_1 v\right]}{m_{ap}} = A + Bv + Cv^2 \tag{18.22}$$

where $A = \frac{\left[mg(f_0\cos\alpha + \sin\alpha) + \frac{1}{2}\rho C_d A V^2 + R f_0\right]}{m_{ap}}$.

$B = \frac{\left[mg(f_0\cos\alpha) + \frac{1}{2}\rho C_d A V + R f_1\right]}{m_{ap}}$.

$C = \frac{\left[\frac{1}{2}\rho C_d A\right]}{m_{ap}}$

Table 18.17 Braking results

Speed (km/h)	Distance (m)	Total stopping distance (m)	Kinetic energy (J)
80	35.99	53	177530.864
70	27.55	42	135922.067
60	20.24	32	99861.111
50	14.05	24	69347.993
40	8.99	17	44382.716
30	5.06	11	24965.277
20	2.24	6	11095.679
10	0.56	2	2773.919
0	0	0	0

$$m_{ap} = m \times \delta \tag{18.23}$$

$\delta = 1.03$ to 1.04
R_{f0} is the drive train resistance at low speeds.
R_{f1} is the increasing rate with the speed of resistance v.
$f_0 = 0.0076$.
$f_1 = 0.2$
$\alpha = 0°$ for flat roads.

While braking the vehicle, the kinetic energy at the wheels can be stored to the battery pack via regenerative braking concept. The braking distance of the vehicle moving at different speeds is given in Table 18.17.

After that, the power absorbed in the braking is estimated and the range calculated through the braking concept is depicted below. Referring to Fig. 18.17, the maximum speed range should be between 45 and 50 km/h during braking.

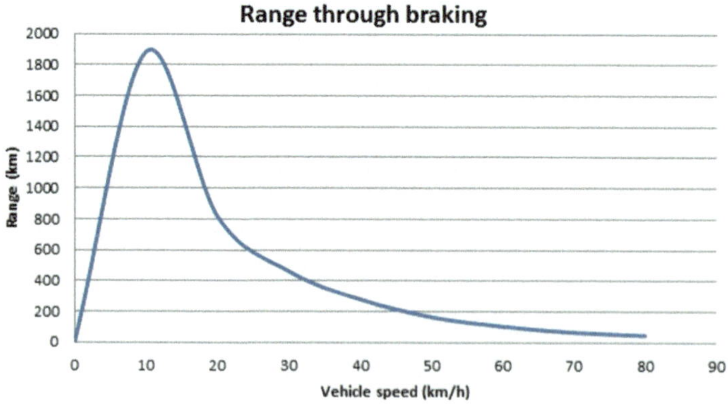

Fig. 18.17 Braking energy of the proposed SEV

18.11 Performance Analysis for SEV

Based on SEV parameters, the peak power and nominal power required by the motor is calculated from vehicle forces which is shown in Fig. 18.18.

The rated and peak power are computed at a speed of 50 km/h in the calculation. The maximum speed of the quadricycle is 70 km/h (Wismans et al. 2011). Referring to Fig. 18.18, the rated power is around 12 kW and the peak power is around 27 kW. Considering the maximum losses as 2 kW (Patel and Kumar 2017), the rated power should be 15 kW and the peak power should be 30 kW. Hence, through suitable gear reduction methods, 15 kW rated SRM is selected. The design of switched reluctance motor of 6/4 configuration is carried out according to the procedure (Patel and Kumar 2017; Tsirogiannis et al. 2019).

18.11.1 Specifications of Switched Reluctance Motor 15 kW (Patel and Kumar 2017)

Outer diameter of stator = 220 mm
Outer diameter of rotor = 116 mm
Stator pole width = 33 mm
Stator pole height = 17 mm
Rotor pole width = 33 mm
Rotor pole height = 16 mm
Shaft = 20 mm
Stator = 24 kg
Rotor = 9 kg
Accessories = 5 kg
Total weight of motor = 38 kg
Power output = 15 kW including losses

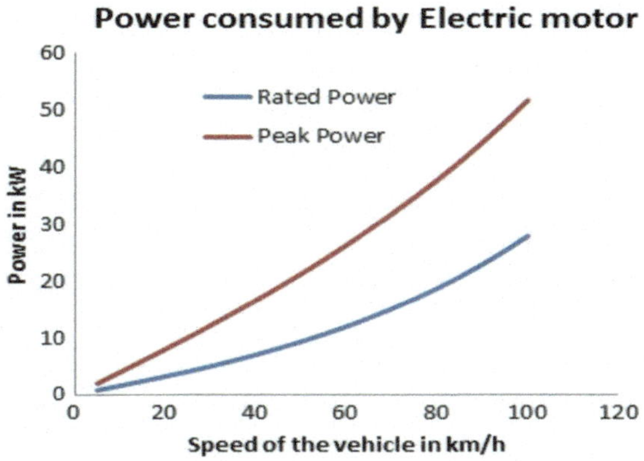

Fig. 18.18 Power consumed by SRM 6/4 for the proposed SEV

Speed = 1500 rpm
Torque = 111 Nm (calculated)
Torque = 100 Nm (expected after ripple losses).

18.11.2 Motor Losses

The losses are Copper loss, Iron loss, and windage loss.

$$q = \frac{4\sqrt{3T}}{BK\pi D^2 L} \tag{18.24}$$

where T is the output torque from the motor in Nm B is the magnetic flux density in Wb/m^2.

K is the constant which defines the saturation factor of the motor. For fully saturated condition $K = 0.75$.

D is the outer diameter of the stator in m.

L is the length of the stator (or) stack length in m.

q is the electric loading.

$$I_{\text{peak}} = \frac{q\pi D_i}{6N_{ph}} \tag{18.25}$$

where D_i is the inner diameter of the rotor in m.

N_{ph} is the number of turns per phase.

I_{peak} is the stator peak current in A.

$$I_{rms} = \frac{I_{peak}}{\sqrt{m}} \tag{18.26}$$

where m is the number of phases (usually for SRM m = Stator poles/2).

I_{rms} is the rms current in A.

For estimating the Copper loss (P_{cu}).

$$P_{\text{cu}} = 3I_{\text{rms}}^2 R \tag{18.27}$$

where P_{cu} is the copper loss in W.

R is the internal resistance in the motor in ohms.

The iron loss and winding loss is kept constant as 100 W and 5 W, respectively.

18.12 Performance Parametric Evaluation for SEV

The SEV's are intended to match the performance of urban mobility requirements. However, in particular, the Indian Driving Cycle for three-wheelers is considered with constant gradient of 10.2° and 15°. Since, the benchmarked vehicle belongs to quadricycle, the three-wheeler driving cycle justifies the power requirements regarding lower limits. Also the simulation is being performed with NEDC for benchmarking the upper limit. Recent trends show that Switched Reluctance Motor with 6/4 configuration provides 110 Nm torque at a peak power of 30 kW justifies Indian Driving Cycle with 15° gradient and also NEDC with maximum speed of 100 km/h (till phase 2 and some short range in phase 3).

Referring to the Fig. 18.19, the proposed SEV will match with IDC with top speed being 42 km/h. Urban mobility is altogether a different scenario where start and stop become the mandatory. 580 kg is the kerb weight of the vehicle benchmarked from 22 (Tanik and Parlaktaş 2015). To overcome range anxiety, the vehicle weight along with the battery capacity needs to be refined in such a way that the performance of the vehicle remains unaltered. Modern SOC estimation algorithms do prefer the datasets associated with vehicle testing parameters. Real-world test data is extremely important to benchmark a vehicle for its subsequent steps on electrification and testing protocols (Sankaran et al. 2020; Scheubner et al. 2019; Mashhoodi and Blij 2021; Petersen et al. 2019). The specifications of the proposed SEV for retro-fitment purpose is shown in Table 18.18.

The simulations performed in MATLAB/Simulink and mathematical formula from CarlosGertz et al. (2014) is used for validation. The distance traveled by the PHEV in full electric mode is given by

$$S_{\mathrm{AER}} = Zkn_{\mathrm{CD}} \tag{18.28}$$

where S_{AER} is the distance traveled by the Electric vehicle in miles.

Z is the battery swing window (Actually it resembles initial SOC of the battery, $Z = 0.8$).

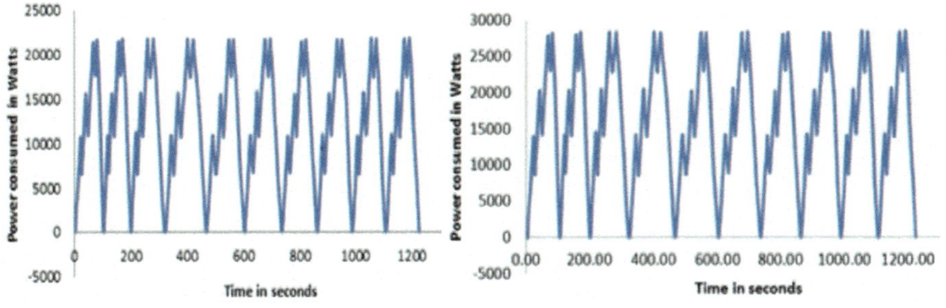

Fig. 18.19 Power consumption of the proposed SEV for a gradient of 10.2° (left) and 15° (right)

Table 18.18 Specifications of proposed SEV

Electric motor	Switched reluctance motor
Motor rated output power (kW)	15
Motor peak output power (kW)	30
Motor peak torque (Nm)	110
Battery pack output voltage	192
Battery current capacity	86.4
Battery chemistry	LFP
Theoretical battery pack capacity (kWh)	16.58
Usable battery pack capacity (kWh)	13.264
Top speed (km/h)	50
Final drive reduction ratio	8.8:1
Wheel diameter (mm)	493
Expected range (km)	200
Motor weight (kg)	50
Battery weight (kg)	80
Miscellaneous (kg)	40
Payload (kg)	20
Expected weight of BIW, seats, Instrumentation panel, wheels (kg)	492

k is the total battery capacity in kWh.

n_{CD} is the fuel economy from simulation (miles/kWh) (Fig. 18.20).

Powertrain performance parameters were estimated in Simulink simulations. Driving cycles are altered to analyze the motor and battery pack performance. Figure 18.21 shows the sample output of EV reference application in Simulink.

Usually, city-type driving cycle caters a range over 200 km. Artemis motorway 150 and HWFET were also simulated by keeping the boundary conditions constant. Driving conditions are really important in electric vehicle range prediction. The

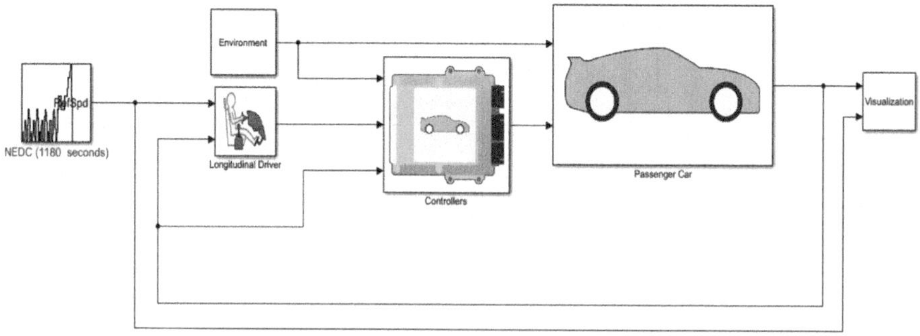

Fig. 18.20 Electric vehicle reference application from MATLAB/Simulink

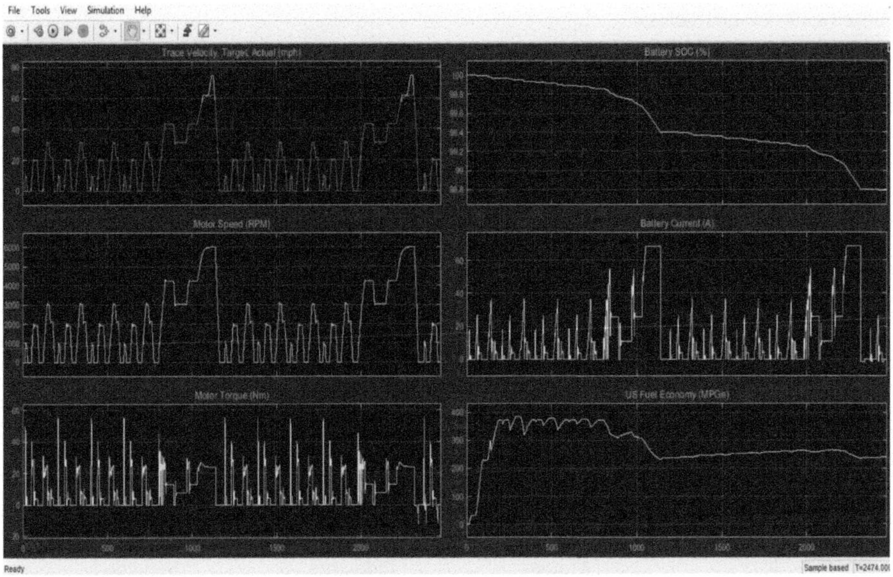

Fig. 18.21 Summary of results for the performance simulation of SEV

retro-fitted SEV is compared with MIA electric car and its specifications are shown in Table 18.19.

18.13 Conclusion

SEV requires a minimum range of 200 km in such a way that it can potentially act as an alternative to taxi's/cabs in urban mobility. The advantage of SEV can be of lower vehicle size as compared to a car. Since the SEV considered in this research is benchmarked from a quadricycle, the power to weight ratio can be quantified as 23.96 kg/kW. An SEV to be classified under WLTC cycle 1 requires minimum of 22 kg/kW and it should be able to sustain the speeds of 65 km/hr. Preferring SRM over BLDC, PMSM propels the proposed SEV to a maximum speed close to 100 km/hr. In retro-fitment scenario, fitting the battery pack in the front and motor at the rear, augurs well for maintaining stability in drivetrain dynamics and also the effective vehicle center of gravity. Lithium-ion cylindrical cells are preferred for retro-fitment case. If the SEV is designed from scratch, it can be able to accommodate the battery pack on floor. Pouch type of lithium cells can be fitted effectively on the floor which is capable of holding the battery pack weight of 200 kg in purpose built design of SEV. Also the reports on stress analysis states that aluminum alloys can be used as chassis materials for roll cage-type structure. Based on comparative study, the retro-fitted SEV performs close to the top SEV's benchmarked across the globe. In terms

Table 18.19 Comparison of proposed vehicle with MIA electric

Properties	MIA electric (Norrby 2012)	Proposed vehicle
Dimensions (mm)	3190 × 1640 × 1550	2752 × 1312 × 1652
Wheel base (mm)	1960	1925
Turning radius (m)	4.3	3.47
Number of seats	4	4
Battery type	LFP	NCA
Nominal voltage (V)	76.8	192
Battery energy (kWh)	12.3	17.23
Battery mass (kg)	145	75
Kerb weight (kg)	850	560
Gross weight (kg)	1200	900
Motor type	Asynchronous	Asynchronous
Motor	Induction	Switched reluctance
Motor power (kW)	10	9.38
Motor peak power (kW)	18	21.3
Motor torque (Nm)	65	110
Gear box	Single speed 8:1	Single speed 6.6:1
Body structure	Cold rolled steel closed sections And sheet steel joined by welding	Advanced high strength steels
NEDC vehicle range (km)	125	198.42
Specific energy consumption (Wh/km)	96	86.88
Maximum speed (km/h)	100	96.54

of performance, NEDC driving cycle up till phase 2 is applicable which guarantees a minimum speed of 80 km/hr to sustain in NEDC. In Transmission part, SRM motor having 6/4 configuration has supply voltage of 192 V and hence considering the retro-fitment and purpose built scenario, lithium-ion cylindrical cells with NCA chemistry is preferred for 200 km range.

Retro-fitted vehicles can be preferred when the vehicle population is high before vehicle electrification. However, the existing vehicles do have space constraints to be addressed by the retro-fitment kit supplier or an organization to come up with potential solutions which makes sense to vehicle electrification. The purpose-built vehicles make sense where the scalable battery packs are given the preference considering range anxiety and better control over weight distribution along with modifications in the chassis. Hence, the purpose-built vehicles can be preferred when there is a new product development which has to be started from scratch. Based on the analysis carried out in this chapter, the SEV has plenty of benefits to the urban mobility users when it is nurtured properly to cater the demands from shared mobility, vehicle electrification plans from the government and achieving the sustainability development goals. From the scalable battery pack configuration, the capability of V2G can be addressed for supporting the electricity grid when the vehicle is idle. Overall, the growth of SEV can redefine the electric mobility in all facets towards improvement.

References

Arora S, Kapoor A, Shen W (2018) Application of robust design methodology to battery packs for electric vehicles: identification of critical technical requirements for modular architecture. Batteries 30:4. https://doi.org/10.3390/batteries4030030

Arun D, Paleshwar DV, Sainath K (2019) Electric vehicle chassis design and structural analysis by using CAD and CAE techniques. Int J Res Eng Sci Manage 2(4):323–327

Barlow TJ, Latham S, Mccrae IS, Boulter PG (2009) A reference book of driving cycles for use in the measurement of road vehicle emissions. Published project report 354, TRL Limited

Barta D, Mruzek M, Labuda R, Skrucany T, Gardynski L (2018) Possibility of increasing vehicle energy balance using coasting. Adv Sci Technol Res J 12(1):228–235. https://doi.org/10.12913/22998624/86215

Berjoza D, Jurgena I (2017) Effects of change in the weight of electric vehicles on their performance characteristics. Agronomy Res 15(Special Issue 1):952–963

Bobbaa S, Mathieuxa F, Ardentea F, Blenginia GA, Cusenzac MA, Podiasd A, Pfrangd A (2018) Life cycle assessment of repurposed electric vehicle batteries: an adapted method based on modelling energy flows. J Energy Storage 19:213–225

Cano ZP, Banham D, Ye S, Hintennach A, Lu J, Fowler M, Chen Z (2018) Batteries and fuel cells for emerging electric vehicle markets. Nat Energy, Review Article. https://doi.org/10.1038/s41560-018-0108-1

CarlosGertz L, Cervieri A, AiresRodrigues AF, De Albuquerque L (2014) Chassis design for electric car prototype. SAE technical paper 2014-36-0215

Chander A, Kumar M, Nambiar S, Zaveri H (2018) Design and study of transmission system for electric vehicles. Int J Eng Sci Comput 8(3):16512–16514

Ciuffo B, Fontaras G, Id SS, Garcia MO (2017) Impact of different driving cycles and operating conditions on CO_2 emissions and energy management strategies of a Euro-6 hybrid electric vehicle. Energies 10:1590. https://doi.org/10.3390/en10101590

Determination and Use of Road Load Force and Dynamometer Settings (2015) United States Environmental Protection Agency National Vehicle and Fuel Emissions Laboratory 2565 Plymouth Road Ann Arbor, Michigan, Feb. CD-15-04 (LDV/LDT/ICI/LIMO)

Dinger A, Martin R, Mosquet X, Rabl M, Rizoulis D, Russo M, Sticher G (2010) Focus batteries for electric cars, challenges, opportunities and the outlook to 2020. Boston Consulting Group.

Driving into 2025: The future of electric vehicles. https://www.ngevehicles.com/ev-technical/driving-into-2025-the-future-of-electric-vehicles.html. Last accessed 11 June 2019

Dubarry M, Baure G, Pastor-fernández C, Fai T, Widanage WD, Marco J (2018) Battery energy storage system modelling: a combined comprehensive approach. J Energy Storage 21:172–185

Faid S (2015) A highly efficient two speed transmission for electric vehicles. EVS 28 KINTEX, Korea, May 3–6

Gis W, Żół towski A, Bocheńska A (2012) Testing of the electric vehicle in driving cycles. J KONES Powertrain Transport 19(4)

Highway code stopping distances. https://www.highwaycodeuk.co.uk/answers/what-is-the-stopping-and-braking-distance-of-a-car. Last accessed 8 Aug 19

Hofman I, Sergeant P, Van den Bossche A, Koroglu S, Kesler S (2016) Optimal design and implementation of a drivetrain for an ultra-light electric vehicle. Int J Vehicle Design 72(3):262–283

Humphries K, Morozov A (2016) A comparison of vehicle simulation software and dynamometer results for battery electric vehicles. In: Proceedings of The Canadian Society for Mechanical Engineering International Congress 2016, CSME International Congress, June 26–29, Kelowna, BC, Canada

Implementation of SAE J1634 (1997) Electric vehicle energy consumption and range test procedure. ETA–TP 003, 1–24

Isobe F, Itakura Y, Osumi Y, Li G (2011) One motor type electric vehicle drive system. Ntn Tech Rev 79:33–41

Kachapornkul S, Jitkreeyarn P, Somsiri P, Tungpimolrut K (2007) A design of 15 kW switched reluctance motor for electric vehicle applications. In: Proceeding of international conference on electrical machines and systems 2007, Oct. 8~11, Seoul, Korea, pp 1690–1693

Kadijk G, Ligterink N (2012) Road load determination of passenger cars. TNO report, TNO 2012 R10237, October 2012

Karabasoglu O, Michalek J (2013) Influence of driving patterns on life cycle cost and emissions of hybrid and plug-inn electric vehicle powertrains. Energy Policy. https://doi.org/10.1016/j.enpol.2013.03.047

Koumartzakis G, Spanoudakis P, Tsourveloudis NC (2017) Design and development of a prototype electric vehicle's chassis. In: 7th BETA CAE international conference

Kristyadi T, Putra A, Santika T, Hartawan L, Trinaldo (2017) Stress analysis of a cross over electric car chassis. IOSR J Mech Civil Eng (IOSR-JMCE) 14(5 Ver. I):13–28. e-ISSN: 2278–1684, p-ISSN: 2320–334X

Kühlwein J (2016) Driving resistances of light-duty vehicles in Europe: present situation, trends, and scenarios for 2025. White paper, ICCT, December

Lampic G, Gotovac G, Geaney H, O'Dwyer C (2016) Comparing the suitability of Lithium ion, Lithium Sulfur and Lithium air batteries for current and future vehicular applications. arXiv: 1606.06347

Lesemann M, Stein J, Malmek E-M, Wismans J, Dávila A, Monfrino G, Storer D, Coma G, Schönwald C (2013) Advanced electric vehicle architectures. ELVA SCP0-GA-2010-265898

Lu H, Zhang N, Zhou X, Wang Y, Lu E (2017) Comprehensive design and optimization of an electric vehicle powertrain equipped with a two-speed dual-clutch transmission. Adv Mech Eng 9(1):1–13. https://doi.org/10.1177/1687814016683144

Luccarelli M, Matt DT, Spena PR, Lienkamp M (2014) Purpose design for electric cars parameters defining exterior vehicle proportions. In: Bozen-Bolzano Institutional Archieve (BIA), conference publications in science and technology, pp 1–7. http://hdl.handle.net/10863/232

Mahala MK, Deb A (2011) Development of transmission specifications for an electric vehicle. In: Chakrabarti A (eds) Research into design—supporting sustainable product development. Indian Institute of Science, Bangalore, India, Published by Research Publishing ISBN: 978-981-08-7721-7, pp 978–981

Maryniuk R (2017) Development of a modular urban electric vehicle. Master thesis, Master of Applied Science, Department of Mechanical Engineering, University of Waterloo

Mashhoodi B, van der Blij N (2021) Drivers' range anxiety and cost of new EV chargers in Amsterdam: a scenario-based optimization approach. Ann GIS 27(1):87–98. https://doi.org/10.1080/19475683.2020.1848921
May GJ, Davidson A, Monahov B (2018) Lead batteries for utility energy storage: a review. J Energy Storage 15:145–157
Mruzek M, Gajdáč I, Kučera Ľ, Gajdošík T (2014) The possibilities of increasing the electric vehicle range. Procedia Eng 192:621–625
Mruzek M, Gajdáč I, Kučera Ľ, Barta D (2016) Analysis of parameters influencing electric vehicle range. Procedia Eng 134:165–174. https://doi.org/10.1016/j.proeng.2016.01.056
Norrby P (2012) Prediction of coast-down test results. Master thesis, Chalmers University of Technology
Othaganont P, Assadian F, Auger DJ (2016) Multi-objective optimisation for battery electric vehicle powertrain topologies. J Automobile Eng 231:8. https://doi.org/10.1177/0954407016671275
Parkinson S (2016) Optimum gear ratios for an electric vehicle. Undergraduate J Math Model One + Two 7(1), Article 4. https://doi.org/10.5038/2326-3652.7.1.4875
Patel R, Kumar S (2017) Analysis of packaging issues in current electric vehicles. Int J Eng Res Appl 7(8, Part-I):50–55. ISSN: 2248–9622
Petersen P et al (2019) Validation of range estimation for electric vehicles based on recorded real-world driving data. In: Bargende M, Reuss HC, Wagner A, Wiedemann J (eds) 19 Internationales Stuttgarter symposium. Proceedings. Springer Vieweg, Wiesbaden. https://doi.org/10.1007/978-3-658-25939-6_29
Preda I, Covaciu D, Ciolan G (2010) Coast down test—theoretical and experimental approach. CONAT20104030, pp 155–162
Prochazka P, Pazdera I, Cipin R, Hadas Z (2015) Optimal design techniques for small electric car operating in common urban traffic. Brno University of Technology, PRZEGLĄD ELEKTROTECHNICZNY, ISSN 0033-2097, R. 91 NR 5/2015. https://doi.org/10.15199/48.2015.05.38
Richardson RR, Osborne MA, Howey DA (2018) Battery health prediction under generalized conditions using a Gaussian process transition model. J Energy Storage 23:320–328
Sankaran G, Venkatesan S, Prabhahar M (2020) Range anxiety on electric vehicles in India—impact on customers and factors influencing range anxiety. Mater Today—Proc 33(Part 1):895–901. https://doi.org/10.1016/j.matpr.2020.06.423
Saxena S, Gopal A, Phadke A (2020) Electrical consumption of two-, three- and four-wheel light-duty electric vehicles in India. Appl Energy 115(2014):582–590
Scheubner S, Thorgeirsson AT, Vaillant M, Gauterin F (2019) A stochastic range estimation algorithm for electric vehicles using traffic phase classification. IEEE Trans Veh Technol 68(7):6414–6428. https://doi.org/10.1109/TVT.2019.2918544
Sherman SB, Cano ZP, Fowler M, Chen Z (2018) Range-extending Zinc-air battery for electric vehicle. AIMS Energy 6(1):121–145
Shukla PR, Dhar S, Pathak M, Bhaskar K (2014) Electric vehicles scenarios and a roadmap for India. UNEP DTU Partnership
Singh NP, Chauhan H (2017) Dynamic analysis & shape optimization of electric car chassis. Int J Aerospace Mech Eng 4(4):5–8
Stein J, Urban P, Seidel K, Kerschbaumer A (2016) Epsilon—small electric passenger vehicle with maximized safety and integrating a lightweight oriented novel body architecture. FORMforum 1–10
Tanik E, Parlaktaş V (2015) Design of a very light L7E electric vehicle prototype. Int J Automot Technol 16(6):997–1005. https://doi.org/10.1007/s12239-015-0102-6
Taufik AZ, Rashid N, Faruq M, Mazlan, Zahir M (2014) Electric car chassis design and analysis by using CATIA V5 R19. IOSR J Mech Civil Eng (IOSR-JMCE) 11(4 Ver. III):56–59. e-ISSN: 2278-1684, p-ISSN: 2320-334X

Tsirogiannis EC, Stavroulakis GE, Makridis SS (2019) Electric car chassis for shell eco marathon competition: design, modelling and finite element analysis. World Electric Veh J 10:8. https://doi.org/10.3390/wevj10010008

Tutuianu M, Marotta A, Steven H, Ericsson E, Haniu T, Ichikawa N, Ishii H (2014) Development of a world-wide worldwide harmonized light duty driving test cycle (WLTC). Technical report, DHC subgroup", vol 03, no January, 7–10

Type I test On S.I. engines, CNG, LPG and diesel engine vehicles (Verifying the average tailpipe emission) of gaseous and particulate pollutants, Chapter 3, MoRTH/CMVR/TAP-115/116 (Issue 4), pp 895–918 (2018)

Wishart J, Diez J (2015) Test specification—coastdown testing. Intertek company, Center for Evaluation of Clean Energy Technology (CECET)

Wismans J, Malmek E-M, Larsson R, Welinder J, Håland Y, Oldenbo M (2011) Technology needs for Safe Electric Vehicle Solutions in 2030. https://www-esv.nhtsa.dot.gov/Proceedings/22/files/22ESV-000128.pdf

Xin X, Chengning Z (2017) Optimal design of electric vehicle power system with the principle of minimum curb mass. Energy Procedia 105:2629–2634

Yin Y, Yin X, Luo W, Sun H (2017) Comparison research of electric vehicles equipped with fixed ratio gearbox and two-speed gearbox. Int J Recent Adv Mech Eng (IJMECH) 6(2)

Zhou XX, Walker PD, Zhang N, Zhu B, Ding F (2012) The influence of transmission ratios selection on electric vehicle motor performance. In: Proceedings of the ASME 2012 international mechanical engineering congress & exposition IMECE2012, November 9-15, 2012, Houston, Texas, USA IMECE2012–85906

Chapter 19
Performance Analysis of 400 kWp Grid-Connected Rooftop Solar PV System for Technical Institute

Ravindra M. Moharil, Prakash S. Kulkarni, Pranay S. Shete, and Prasad B. Joshi

Abstract In the National Development Council (NDC), India decided to enhance the renewable power share by 40% to its installed capacity by 2030. Indian government has 40 GW rooftop solar installation planned by 2022. Solar energy received by India varies from 4 kWh/m^2 during rainy season months to 7 kWh/m^2 during summer season with total sunny days of 300 per year. In 2015, Maharashtra Energy Regulatory Commission (MERC), Maharashtra state of India, permitted Rooftop Solar PV (RTSPV) system power to be connected to the distribution grid through net-metering. Industrial and commercial sectors preferred to install the SPV panels either on rooftop of building or on barren land in the premises. Various economic models of RTSPV are in place. Availability of solar energy and electricity requirements for academic institutions are coinciding with one another, which eliminates the storage requirement. Technical institutes require a large amount of power as compared to Science and Art institutes. The Maharashtra State Electricity Distribution Company Limited (MSEDCL), a distribution licensee in Maharashtra is charging an educational institute as per the tariff of HT consumer-Public Services category. As per policy of cross-subsidy, tariffs are on a higher side for the institutes. To reduce the financial burden of electricity bills, solar energy is the best option for these institutes. Variation in power generation by RTSPV occurs due to daily and seasonal variation of solar radiation. RTSPV generates active power only which results in different demand characteristics than the customer using grid supply only to meet the existing load. In this chapter, the performance analysis of a 400 kWp grid-tied RTSPV plant for the last 4 years is presented, and various challenges and issues related with interconnection of RTSPV to distribution grids are discussed. Study of system reveals that all solar power generated is consumed in the institute on all working days only on weekends and holidays export of power took place. The active power consumption

R. M. Moharil (✉) · P. S. Shete · P. B. Joshi
Yeshwantrao Chavan College of Engineering, Nagpur 441110, India
e-mail: rmm_ycce_ep@yahoo.com

P. S. Kulkarni
Visvesvaraya National Institute of Technology, Nagpur 440010, India
e-mail: pskulkarni@eee.vnit.ac.in

A. K. Bohre et al. (eds.), *Planning of Hybrid Renewable Energy Systems, Electric Vehicles and Microgrid*, Energy Systems in Electrical Engineering,
https://doi.org/10.1007/978-981-19-0979-5_19

is reduced to more than 50% but the reactive power required has increased. It was found that the payback period is low for net metering as compared to gross metering.

Keywords Rooftop solar photovoltaic System (RTSPV) · Distributed Generation Resources (DGR) · Solar Radiation (SR) · Percentage grid penetration level · Life cycle cost · Net metering · Gross metering · Payback period

Nomenclature

AMC	Annual Maintenance Cost
CEA	Central Electricity Authority
DISCOM	Distribution Company
LCC	Life Cycle Cost
MERC	Maharashtra Energy Regulatory Commission
NDC	National Development Council
RES	Renewable Energy System
SNA	State Nodal Agency
THD	Total Harmonic Distortion
APFC	Automatic Power Factor Controller
DGR	Distributed Generation Resource
IMD	Indian Meteorological Department
LCOE	Levelized Cost of Energy
MSEDCL	Maharashtra State Electricity Distribution Company Limited
O&M	Operation and Maintenance
RTSPV	Rooftop Solar Photovoltaic
SPV	Solar Photovoltaic

19.1 Introduction

Remarkable reduction in solar photovoltaic (SPV) panel cost during the last few years, the world is joining the period of cost competition of solar electricity with conventional electricity tariffs in many countries. Indian government has declared ambitious targets of installing 100 GW of SPV power with 40 GW through rooftop systems. State nodal agencies (SNA) and Distribution Companies (DISCOM) are trying to implement rooftop solar photovoltaic (RTSPV) with the help of solar rooftop project developers (National Solar Mission 2016). Manoj Kr. Behera et al. presented different forecasting models by several researchers for different RES. For developing these models' researchers, we use different meteorological parameters either in time series or statistical form. The availability of resources is predicted by developing suitable algorithms with removal of meteorological inconsistency for planners to design the suitable system (Behera et al. 2018). Moharil and Kulkarni in their paper mentioned that the output from RES is site-specific and depends on various weather parameters. It is difficult to guess the correct amount of power from these RES. Proper forecasting methodology helps in designing a suitable system with minimum imbalance between supply and demand and gap (Moharil and Kulkarni 2010). Mehmet Yesilbudak et al. explained how system operators and power system planners use these SPV power forecasting models as tools to manage the solar PV plants effectively and efficiently in their paper (Yesilbudak et al. 2018). Mandi in his paper described the solar photovoltaic plant of 500 kW installed at university campus of REVA and calculated pay back period and lifetime cost of the plant. He has also calculated specific energy generation of the plant and reduction in carbon emission (Mandi 2017). Nallapaneni Manoj Kumar et al. in their paper mention the design considerations and component selection and provided online monitored data for 95 kWp on grid PV systems and provided simulated energy performance results (Manoj Kumar et al. 2018). Wiles in his paper mentions that for better performance of SPV installations, it is necessary to select the location by using locating tool software and follow instructions given in users manual while installing components like PV modules and inverter. Good workmanship and adoption of National Electric Code helps in further improvement in performance (Wiles 2006). Yi. Guo et al. in their paper addresses the RTSPV economics for electric vehicle parking area site generation for renewable energy. Stochastic approach is used for considering uncertainty of solar energy and model predictive control (MPC) methodology for charging of electric vehicles is used for day ahead time scale pricing (Guo et al. 2016). Ahmed Ouammi in his paper presented a vehicle to building (V2B) model with the Team of Cooperating Microgrids (TCM) to lower the maximum loads with operational flexibilities of EV. MPC is used for optimal operational control of each microgrid. Presented the case study, which achieves reduction in peak load from 67 to 94% for a team of four microgrids by TCM approach (Ouammi 2021). Suresh Singh et al. in their paper done the comparative assessment of two photovoltaic plants technologies at the same location connected to the grid. Various performance parameters are compared. They have discussed various operation and maintenance issues also. They found that for comparison normalized performance indicators like specific yield and performance ratio are useful (Singh et al. 2014). Mahesh Kumar in his paper considered the interconnection issues of the SPV system with the main grid considering two d-q synchronous frames of reference (α–β and dq0) for various faults on the main grid and variation of THD for it (Kumar 2020). Ayompe et al. in their paper presented results of monitoring the RTSPV system for high-rise building in Dublin, Ireland, calculated yield parameters and compared average daily final yield with the results of the other countries (Ayompe et al. 2011). Sharma and Goel in their paper mention findings on the grid-connected solar photovoltaic system of 11.2 kWp (Sharma and Goel 2017). K. C. Rout and P. S. Kulkarni in their paper uses PVsyst V6.84 software for the design of proposed 2 kW RTSPV system in Odisha and demonstrated reduction in carbon emission, available energy, used energy, and excess energy for residential customers (Rout and Kulkarni 2020). Arun Kumar Behura et al. in their paper presented the case study of RTSPV at VIT, Vellore, using

PVSyst V6.70 software. Cost analysis with payback period is presented for two different designs with yield factors. They have also carried out the shading analysis and shading losses (Behura et al. 2021). Abhiram S. et al. in their paper focuses on impacts of 30 kW grid-tied RTSPV on secondary distribution systems and studied power quality issues with and without hybrid filters and suggested for installation of harmonic filters to maintain THD within limit (Abhiram et al. 2018). Chandrakant Dondariya et al. in their paper did the feasibility assessment of grid-connected RTSPV using different simulation software like PV*SOL, PVGIS, Solar GIS, and SISIFO. Analysis indicates that PS*SOL found suitable which is fast in operation, easy to use, and reliable as compared to other software (Dondariya et al. 2018). Subhasis Panda et al. in their paper analyzed PV renewable energy from demand side management perspective and presents different measurement index. They demonstrated that adaption of large-scale PV into the system changes load curve from spikes to smooth demand during peak period (Panda et al. 2020). Omkar K. et al. presented performance evaluation of 50 kWp RTSPV at educational institute based on IEC standard 61724 performance parameters (Omkar et al. 2015). M. Chakravarthy et al. in their paper presented the journey of a 200 kWp SPV plant from design to installation and operation at Vasavi College of Engineering, Hyderabad, India. Papers mainly present the sizing of different equipment like SPV modules, inverters, MCCB, cables, etc. They also discussed causes for poor performance and suggested solutions for them (Chakravarthy et al. 2015). Ramkrishna Kappagantu et al. presented RTSPV in smart grid project of Puducherry, India focusing on awareness of people for new technology, doing survey of availability of roof, consumption pattern, impact of government policies, and users experience (Kappagantu et al. 2015). Yash Sharma et al. in their paper presented the technical and economical analysis for the 300 kW RTSPV system installed at Rajasthan Technical University, Kota, India. Simulation is done before the installation of the system. It is concluded that high-power rating modules reduces Levelized cost of electricity (LCOE) compared to the lower power rating modules for the same system load (Sharma et al. 2019). Sandhya Prajapati and Eugene Fernandez proposed a RTSPV installations at commercial buildings for the Electric Vehicle (EV) battery charging and load in commercial building. If the charging time of EV is matched with solar power availability, then it will be more economical and efficient as compare to residential EV charging system, where solar power availability and EV battery charging times are different (Prajapati and Fernandez 2019). Shipla N. and Sridevi HR in their paper carried out the design and study of economical aspects of grid-connected PV system for an educational campus. A comparative study is done for an on-grid and off-grid PV system. Off-grid system uses PV-DG (Diesel generator) system. Proper sizing of battery bank with on-grid PV system is more efficient and economical as compared to the off-grid PV-DG for the same load (Shilpa and Sridevi 2019). Bibhu Padarbinda Mohanty and Moningi Srivalli designed the RTSPV by using System Advisor Model (SAM) and Solar Edge Software. Different operating and weather parameters are used for this designing and optimization of the output of RTSPV system (Mohanty and Srivalli 2020). Rittick Maity and Mobi Mathew in their paper studied the effect of tracking on the power generation of a rooftop PV system with the help of PVsyst simulation software. Authors modeled the SPV for monocrystalline, polycrystalline, and amorphous silicon panels. It was observed that double-axis tracking improves the energy yield of the RTSPV (Maity et al. 2018).

This chapter consider the system with a connected load of 927 kW, Contract demand of 700 kW, and RTSPV capacity of 400 kWp. The RTSPV system was installed in April 2017. For performance analysis various instruments like thermographic imager, power quality analyzer, etc., are used in addition to routine measuring devices. The details of various causes for reduction in the efficiency of the plant and methods for its improvement through proper maintenance are described. Further, economic analysis is carried out to derive the benefits of net metering. Calculations of payback period, net present value, life cycle cost, etc., are determined. Various system parameters like solar fraction, final yield, PV penetration percentage, capacity utilization factor, performance ratio, active/reactive power demand ratio, etc., are used for performance analysis. This case study helps in understanding daytime peak loads management by DISCOM and also discusses how RTSPV plants help in demand response issues with indirect benefits due to installation of RTSPV.

19.2 SPV Performance Monitoring Guidelines

Bureau of Indian Standards has published the photovoltaic system performance monitoring guidelines as IS/IEC 61724:1998, which is discussed in following paragraph and represented in Fig. 19.1.

To monitor the performance of SPV panel, IS/IEC-61724:1998 is used which has listed various meteorological and electrical parameters to be measured. Two types of irradiance data are recorded, one inclined inline with position of array for performance of the system and on horizontal plane to compare with different location

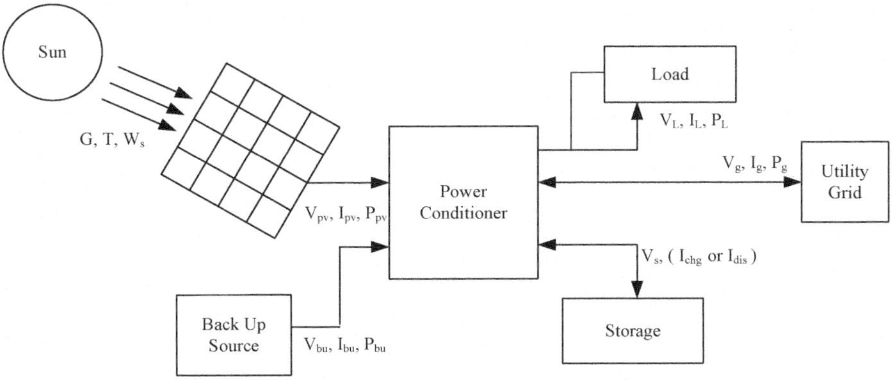

Fig. 19.1 Measurement parameters to be selected for monitoring suggested by Bureau of Indian Standards IS/IEC-61724:1998) (Bureau of Indian Standard 2010)

data. Ambient air temperature is measured with temperature sensors located in solar radiation shields. Wind speed is measured at the location of panels to consider the effects on structures of solar panels. Temperature sensors are connected on the solar panel module back surface to know the array conditions. Voltage and current on PV panel side and converter side are measured with sampling interval between 1 and 10 min. DC power is measured using DC wattmeter and AC power is measured through power sensors which take care of power factor and harmonics. Recording interval of this data is on hourly basis (Bureau of Indian Standard 2010).

19.3 CEA Grid Code

Central Electricity Authority on 7th October 2013 declared the Technical Standards for Connectivity of the Distributed Generation Resources (DGR) to the grid. The salient features of it are mentioned below:

- Planning, design, construction, reliability, protection, and safe operation of the equipment in the plant is the responsibility of the plant developer.
- Providing the facility for data storage and data communication is with the developer.
- Interconnection facilities and interconnection points with required modifications in existing systems are to be verified by appropriate licensees.
- Maximum net capacity of DGR will be based on capacity of the electric system to which it will be connected and will be finalized after considering the imbalance it will cause, affecting power quality and safety of equipment.
- Before connecting the system to the grid, it is necessary to measure DC current injection, harmonic current injection, and flickers are also to be measured. After installation also once in a year, above measurements are to be done and Flickers should be well below IEC 61000 Standard.
- Single-line diagram of the plant is to be prepared and displayed at the proper location.
- IEEE 519 limit of current injection should not be crossed by any generating station.
- At the interconnection point dc current injection should not be greater than 0.5% from DGR of the full rated output.
- Automatic synchronization device should be available at every DGR. If DGR has inverters with inherently built-in synchronizing devices, then no separate device is required.
- Electronic or electromechanical control circuit breaker shall be available with three-phase generators.
- Following protection are to be provided for DGR:
  - Overvoltage protection if voltages cross the voltage upper limit of 110% of rated voltage and under-voltage protection if the voltage drops down below the 80% of rated capacity. Fault clearing time may be up to 2 s.

  - Over frequency protection if frequency crosses 50.5 Hz frequency and under-frequency if frequency falls below 47.5 Hz with clearing time limited to 2 s. Authorities may prefer the narrower range of operation if desired.
  - Energization of the circuit should be stopped for any faulty circuit by any DGR.
  - Reconnection of DGR will be allowed only after confirming that the voltage and frequency of incoming DGR is within limits and with stable parameters.
  - Unintended islanding of DGR is to be prevented.

- Following conditions must be satisfied by generating station equipments:

  - Any circuit interrupting device should be capable of interrupting maximum available fault current at the site.
  - DGR devices safety and reliability should be such that failure of a single device should not affect the performance of a complete system. Sufficient redundancy is to be provided in the plant.
  - 220% of the rated voltage at the interconnection point should be withstood by a paralleling device of DGR.
  - Voltage fluctuation more than ±5% at the point of interconnection of DGR shall not occur after synchronization of the system with the grid.
  - Provision of the switch is to be made to manually isolate the DGR and the electricity system.

- Distribution licensee may ask the developer of DGR to provide the facility for visible verification of the plant by providing proper indicators to represent ON/OFF of the plant.
- Easy and convenient access to distribution licensee's personnel should be available (Central Electricity Authority (Technical Standards for Connectivity to the Grid) 2013).

19.4 Solar Radiation Data Modeling

The hourly output of a PV generating unit varies with time. India is a tropical country with latitude lying between 7° and 37° N. It receives an annual average intensity of solar radiation between 4.63888 and 8.12777 kW/m^2/day. There are around 300 clear day skies, indicating reception of full solar radiation.

On clear days, there is a cycle symmetrical about solar noon. Solar radiation received on the earth's surface is dependent on (a) duration of day length which depends on different geographical factors, (b) turbidity in the air, (c) solar elevation at noon, and (d) type of cloud, percentage of cloud coverage, and content of water vapour in the air. Moisture contents in the air matters much during absorption of solar radiation and varies the amount of direct or beam and diffuse or scattered solar radiation received on the earth's surface. Summation of beam and diffuse radiation gives the value of global radiation.

Every solar radiation monitoring station measures all the three types of radiations mentioned above by using various instruments. Indian Meteorological Department

(IMD) is an agency of the Ministry of Earth Sciences of the Government of India, and is accountable for meteorological observations, weather forecasting, and seismology in India. IMD measures different weather parameters like ambient temperature, wind speed, and direction and relative humidity at its station and publishes it as the statistical information. These parameters also matter for solar radiation availability on the solar panels. The graph is plotted between time in hours on X-axis and output radiation in kWh/m^2. Time is plotted from 5th hour of a day to 20th hour (from morning 5 a.m. to evening 8 p.m.). Graph is plotted as shown in Fig. 19.2 for 15 December 2018 and 22 July 2018. December being the month of winter season, the smooth variation is observed from morning 7 a.m. to 6 p.m. Midday values are higher and maximum output reaches 2.36 kWh/m^2 as shown in Fig. 19.2(a) On the other hand, the second graph represented in Fig. 19.2(b), which was plotted for 22 July has a very erratic nature as it was the month of the rainy season and the cloudy weather remarks were observed throughout the day. Also, there is a very drastic fall in output from 2 to 3 p.m. and after that output continues to be less which was due to the rainy weather remarks from 3 to 5 p.m., which directly affects the output.

The solar radiation seasonal variation analysis is shown in Fig. 19.3, which is based on 10 years data collected from IMD, Pune from January 2009 to December 2018. Here, average output variation for the rainy, winter, and summer seasons is plotted against time. The graph clearly shows the seasonal variation with less average values in the rainy season than winter and summer seasons as the number of cloudy days are more in rainy season. The day length is more in the summer season as compared to winter season, which is represented by elongated graphs over time duration for

Fig. 19.2 Hourly global solar radiation on clear day and cloudy day

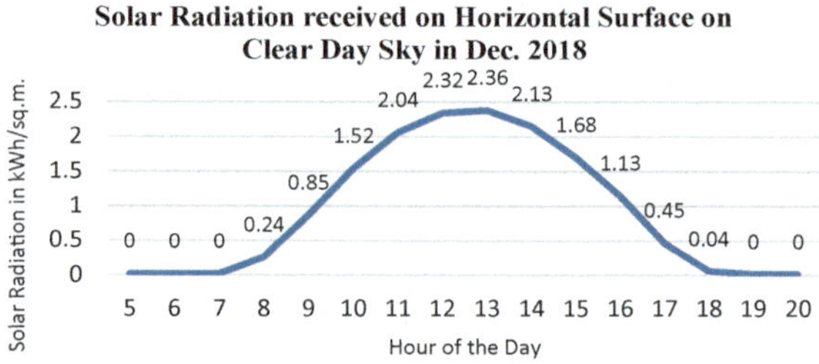

(a) Clear day sky global radiation

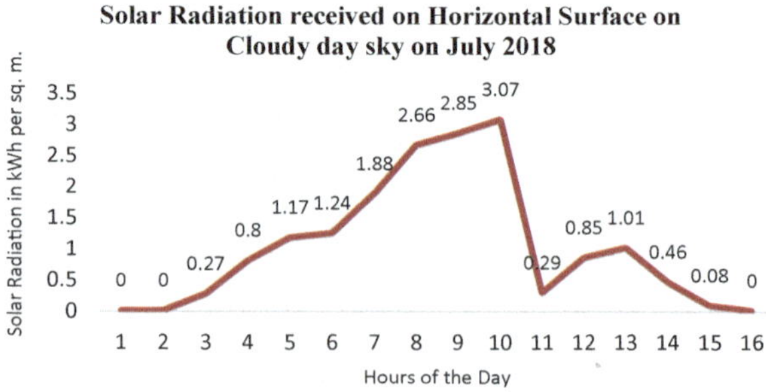

(b) Cloudy day sky global radiation

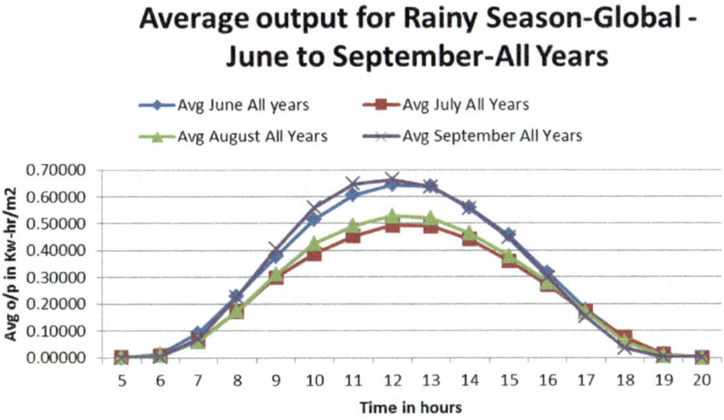

(a) – Rainy season variation in Solar Radiation

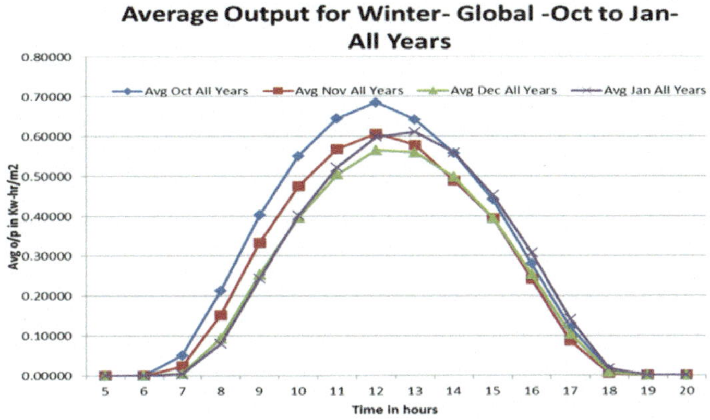

(b) – Winter Season variation in Solar Radiation

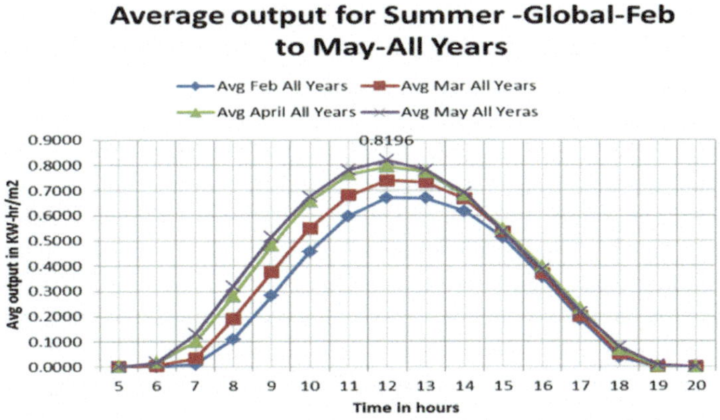

(c) – Summer season variation in Solar Radiation

Fig. 19.3 Seasonal variation of Solar radiation at the site

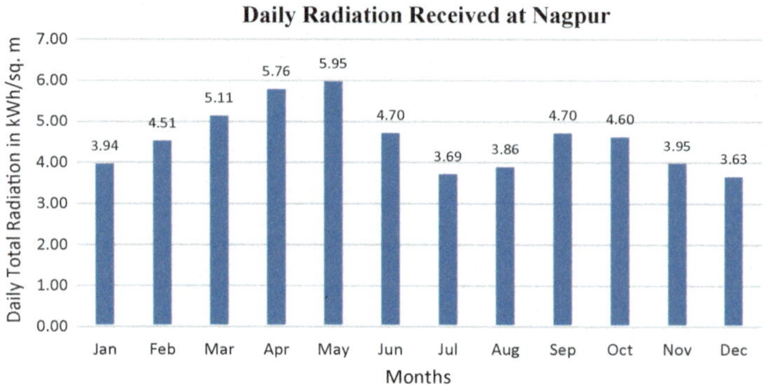

Fig. 19.4 Daily solar radiation received in different months at Nagpur

summer and rainy seasons. Daily average solar radiation in summer months is varying from 5.114 to 5.95 kW/m^2/day. Daily average solar radiation in the winter season is varying from 3.63 to 4.35 kW/m^2/day. Day temperature of Nagpur in the winter season is in the range of 25–30 °C and 40–45 °C in the summer months. The effect of temperature was observed on power output of solar PV panels during these months.

The monthly variation of solar radiation received on the earth's surface at Nagpur, which is located at latitude 21.0929° N and longitude of 78.98° E is represented in Fig. 19.4.

19.5 Site Details and Infrastructure Developed

The institute, Yeshwantrao Chavan College of Engineering (YCCE) is located in Nagpur, Maharashtra, India. The solar photovoltaic panels are installed on roofs of various buildings of the institute. Total roof coverage by PV panels is 2590.98 m^2. Figure 19.5 represents the google map of the institute representing installed solar panels. These panels were functional from April 2017. Figure 19.6 represents the predicted and power generated from date of installation to till date. It can be observed that the rainy season months are with low power generation, whereas the highest power generation took place in the summer month May, this is due to higher day length and more intensity of the beam radiation. For months of October and May the generation is very close to expected demand. In the month of April 2017, the variation in expected and generated power is large due to the teething problems of the system. Smooth power generation started from May 2017 onwards. Failure of the grid is not an issue for the institute as it has 11 kV express feeder benefit. The variation of the generated power with respect to expected power is in the range of ±20% except for a few months.

The plant rated capacity is 400 KWp. It is distributed on four different buildings with 100 kWp capacity on each. Each building has 334 panels of multi-crystalline

Fig. 19.5 Google map representing solar panel installation on various buildings at YCCE

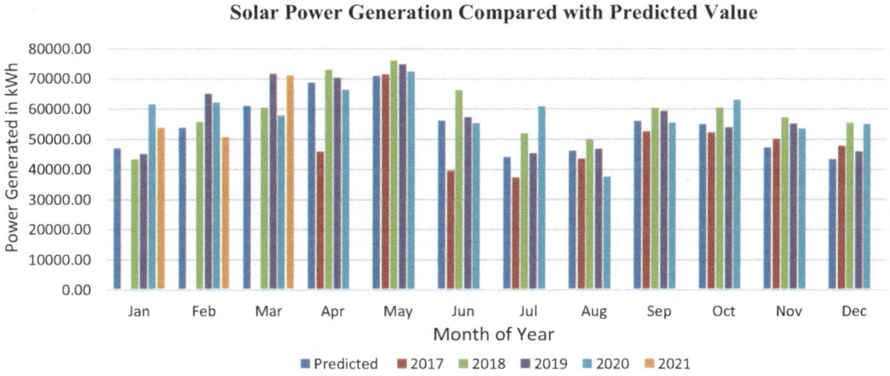

Fig. 19.6 Solar generation in various months over the years from date of installations

silicon PV modules of 300 Wp each. These panels are connected to two inverters of 50 kW each. The DG set backup is provided to take care of lighting load. Details of Solar PV panel and inverter are given in Annexure. Figure 19.7 represents installed PV panels on the science department building of the institute.

19.6 Analysis of Solar PV System and Load Profile of Institute

Table 19.1 represents the connected load of different buildings of the institute. The institute starts its functioning from 9 a.m. to 4 p.m. with a lunch break from 12 noon to 1 p.m. Administrative department works from 10 a.m. to 5 p.m. Figure 19.8 represents the typical load curve of the institute. This load curve is plotted for the actual load on

Fig. 19.7 Solar photovoltaic panels installed on Science Department Building of institute

Table 19.1 Building-wise load distribution of institute

Building	Total load (W)	Building	Total load (W)
Science Department Building	24,088	IT Department Building	35,362
Mechanical Department Building	32,497	Civil Department Building	48,948
Administrative Block	73,028	Central Computing Center	88,737
Electronic Department Building	79,262	EP & CT Department Building	110,914
Mechanical workshop	21,320	Auditorium Building	16,680
Canteen	7424	Pump load and Machine	210,740
Total campus load/connected load = 749000 W (749 kW)			

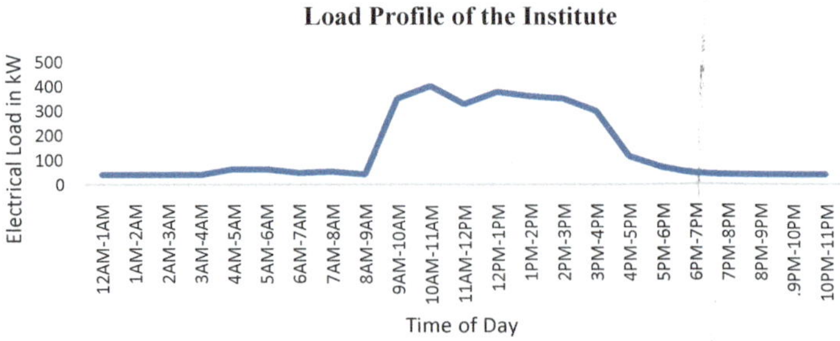

Fig. 19.8 Typical load profile of the institute on Thursday 18 January 2018

Thursday, 18 January 2018. Pump load is normally operated in the morning hours from 4 to 7 a.m. Teaching learning process begins from 9 a.m. onwards. From 4 to 5 p.m., only administrative office load is present; 5 pm onwards only essential street lighting, building lighting, and hostel load are present. When we compare the solar radiation curves for all the seasons shown in Fig. 19.3 and the load curve in Fig. 19.8, it is found that the power output of solar PV can supply the power requirement of the institute load.

Figure 19.9 represents the active power import from the distribution company before and after installation of RTSPV. It can be observed that the kWh has reduced to 50% for almost all months after installation of panels.

Table 19.2 represents the reactive power demand in kVARh lag, before and after

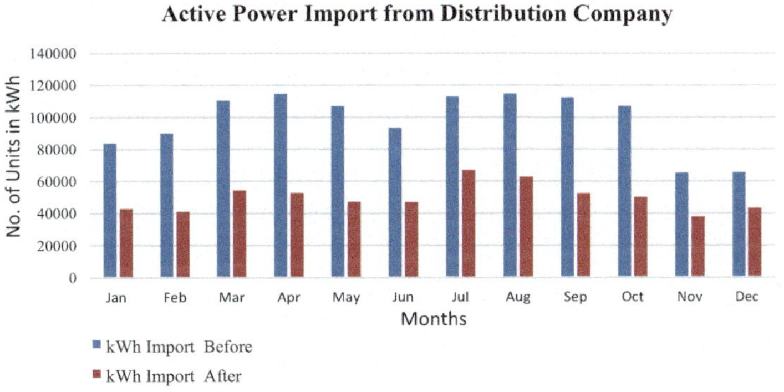

Fig. 19.9 Change in the active power import before and after installations of SPV panels

Table 19.2 Reactive power kVARh lagging drawn from distribution company

Month of the year	KVARh lag consumption		
	Before	After	After APFC installation
Jan	2240	6768	2100
Feb	2320	9415	2230
Mar	3323	11,699	2230
Apr	6423	9643	2300
May	6375	11,956	3750
Jun	3461	8655	1000
Jul	7949	15,831	1095
Aug	7430	13,518	2220
Sep	5177	13,531	1325
Oct	10,246	21,294	1260
Nov	1611	4111	820
Dec	1977	6046	365

installation of RTSPV panel. The reactive power demand has increased to 1.5 times for April to 4.1 times in the month of February. Rise in reactive power demand is highest for winter season and lowest in summer season. It can be observed that the reactive power demand has increased after the installation due to reactive power drawn from the supply by inverter and increase in harmonics due to solar power inverters. Even the power factor has reduced as active power demand has reduced but reactive power demand remains the same for load connected. To compensate for this reactive power demand, an automatic power factor controller (APFC) was installed in the control room with a capacity of 60 kVAR in addition to earlier fixed capacitors of 100 kVAR. APFC installation has reduced the reactive power demand in large amounts which is now well below before installation of RTSPV.

Figure 19.10 represents the change in kVA demand before and after the installation of RTSPV. It can be observed that kVA demand has reduced with the variation from 38% in February to 78% in June month. Figure 19.11 represents the bill demanded by the distribution company for 2016, before the installation of RTSPV and for 2019, after installation of RTSPV. The per unit charges for consumption is Rs. 9.10 for year 2016 and Rs. 9.70 for year 2019. The drop in bill demand varies from 9% for September to almost 60% in the month of February.

Table 19.3 represents the variation in solar panel voltage (V_{pv}), solar panel current

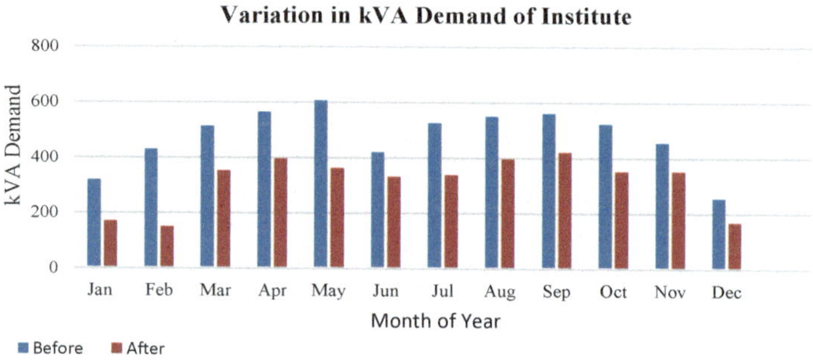

Fig. 19.10 Change in the kVA demand of institute before and after the installation of SPV panels

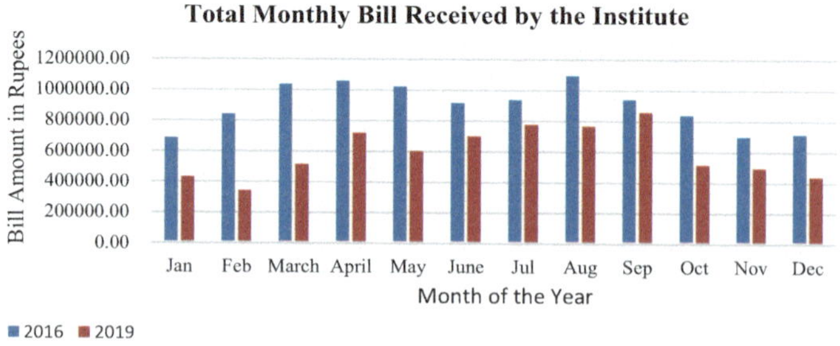

Fig. 19.11 Total electricity bill before and after installation of RTSPV at Institute

Table 19.3 Variation in parameters of inverter

S.N	Active power P_{PV} (Watts)	Array voltage	Array current	Inverter line voltage in Volts			Inverter line current in Amps			Reactive power drawn from supply in VAR		
		V_{pv} (V)	I_{pv} (A)	V_{L1}	V_{L2}	V_{L3}	I_{L1}	I_{L2}	I_{L3}	Q_1	Q_2	Q_3
01	3061	716.9	4.59	425	427	429	4.33	4.35	4.36	3	4	3
02	3124	715	4.70	427	426	433	4.43	4.37	4.34	4	4	3
03	3739	692.2	5.81	429	427	425	5.37	5.41	5.38	5	6	5
04	3749	729.9	5.52	426	424	424	4.88	4.84	4.89	6	5	6
05	3921	710.8	5.93	426	424	422	5.12	5.11	5.11	8	7	8
06	4491	700.6	6.89	434	433	431	6.49	6.43	6.48	10	8	7
07	4725	690	7.30	429	428	426	6.8	6.83	6.69	9	12	10
08	5661	710	8.57	437	433	433	8.1	8.1	8.4	12	15	17
09	7259	692	11.27	431	429	426	11	11.21	11.2	18	15	18
10	7821	670	12.55	429	428	426	8.9	9.09	9.19	22	17	19
11	23,420	625	40.28	427	426	424	8.8	8.9	8.9	54	56	54

(I_{pv}), inverter side voltages (V_{L1}–V_{L3}), and currents (I_{L1}–I_{L3}) and reactive power drawn from the grid when active power from solar panel is varying.

After observing the different parameters, line voltage, and currents from the output side of the inverter are balanced. The reactive power requirement for the inverter rises with the increase in the power output from the inverter.

$$Solar\ Fraction = \frac{Energy\ Provided\ by\ Solar\ generation\ system}{Total\ Energy\ required\ by\ the\ institute} \tag{19.1}$$

$$Final\ Yield = \frac{System\ Energy\ Output}{System\ rated\ power\ dc} \tag{19.2}$$

$$Reference\ Yield = \frac{Total\ inplant\ solar\ insolation\ (H_T\ in\ \frac{\mathrm{kWh}}{\mathrm{m}^2})}{Reference\ irradiation\ \left(1\frac{\mathrm{kWh}}{\mathrm{m}^2}\right)} \tag{19.3}$$

$$Performance\ Ratio = \frac{\text{Actual reading of plant output in kWh}}{\text{Calculated, nominal plant output in kWh}} \tag{19.4}$$

$$Capacity\ Utilization\ Factor = \frac{Actual\ Energy\ Output\ of\ PV\ system}{(No.\ of\ Hours\ in\ month)\cdot Rated\ power} \tag{19.5}$$

$$Percentage\ Grid\ Penetration\ Level = \frac{Solar\ Energy\ Exported}{Solar\ Energy\ Generated}\cdot 100\% \tag{19.6}$$

The various performance parameters are as given below and represents in Table 19.4. The solar fraction is varying from 0.51 to 0.84 indicating that more than 50%

Table 19.4 Performance parameters of solar generation system

S. N	Month	Solar fraction	Final yield	Reference yield	Performance ratio	Capacity utilization factor	% Grid penetration level (%)
01	Jan	0.74	1.21	3.94	0.87	0.190	32.77
02	Feb	0.79	1.21	4.51	0.92	0.241	45.82
03	Mar	0.75	1.23	5.11	0.90	0.252	67.52
04	Apr	0.63	0.93	5.76	0.78	0.222	22.36
05	May	0.67	0.93	5.95	0.77	0.222	30.18
06	June	0.72	1.13	4.70	0.70	0.219	25.27
07	July	0.51	1.22	3.69	0.88	0.179	18.47
08	Aug	0.60	1.39	3.86	0.92	0.215	18.76
09	Sep	0.74	1.36	4.70	0.71	0.264	10.79
10	Oct	0.73	1.32	4.60	0.95	0.243	34.66
11	Nov	0.84	1.05	3.95	0.83	0.171	54.19
12	Dec	0.70	0.98	3.63	0.88	0.142	44.19

energy need of institute is satisfied by RTSPV. Final yield is varying from 0.93 to 1.39, which indicates satisfactory performance. The values of performance ratio is varying from 0.70 to 0.95, which indicates the power loss is taking place due to temperature rise in summer seasons and better PR values for winter season. Capacity utilization factor indicates variation from 0.171 to 0.264, these calculations are based on (24 h × days of month). Percentage grid penetration level indicates that lowest grid penetration took place in September and highest penetration in the Month of March.

19.7 kWh and kVA Billing

In kWh billing, only active power consumption is charged and no charges for reactive power consumption. To reduce reactive power demand from consumer, power factor incentives and penalty are levied if power factor falls below 0.9 lagging. The highest incentives are available for maintain unity power factor. If customers draw leading power factor power, it is ignored by distribution company, which tempts customers to overcompensate during lean period by using fixed capacitors which is harmful for both distribution company and customer. Some customers used hybrid compensation of fixed and switch capacitors using automatic power factor controller.

$$kVAh = \sqrt{\sqrt{(kWh)^2 + (kVARh_{lag} + kVARh_{lead})^2}} \tag{19.7}$$

$$True\,Power\,factor = (displacement\,power\,factor) \cdot (distortion\,power\,factor)$$

$$True\,Power\,factor = cos\varnothing_i \frac{1}{\sqrt{1 + (THD)^2}} \tag{19.8}$$

In kVAh billing, the customer is charged on the customer's kVAh consumption. Customer has to maintain the power factor to unity for optimum kVAh value. Equation (19.7) is represented the calculation of kVAh. Equation (19.8) represents the effect of THD on true power factor. In this billing, no separate pf incentive is awarded. Computation of kVAh is affected due to harmonics as calculation is the rms values of parameters. The distortion power factor increases with the increase in harmonics, reduces true power factor, and increases kVAh consumption.

19.8 Maintenance and Audit of Solar PV Panels

Smooth functioning of solar PV panel and associated circuitry is possible by proper maintenance of it. It also enhances output from solar PV panels over the years.

Following are few tips to be followed for better performance of RTSPV system (Bureau of Indian Standard 2010):

- Regular visual inspection of the RTSPV for physical wellness of the panels.
- Solar cells and PV panels cleaning on a fortnightly basis with water.
- Thermal scanning of all thermal based components.
- Measurement of electrical parameters on ac and dc side of the solar system.
- Measurement of earthing resistance of solar system and electrical system.
- Collection of real-time data and its comparison with manufacturer's product catalogue on a consistent basis. Deviations may be analyzed.
- Timely performance of critical and non-critical repairs.
- Management of inventory and spares.
- Identifying and tracking key performance indicators for best performance.
- To check the interconnections of solar panels and electrical terminals for possible damages due to open to sky conditions.

Guidelines for doing the audit of Solar PV panels installed as per IS/IEC 61724:1998:

- To check the availability of various drawings/diagrams/manuals necessary for operation and maintenance of plant.
- Schedule of Module cleaning. To check the record keeping of solar panel cleaning.
- To check the availability of the operation and maintenance manual.
- To check the availability of design documents for panel mounting structure.
- To check the module connection integrity with tong tester.
- To check the junction/string combiner boxes condition (electrical and safety aspect both).
- To inspect mechanical integrity of mounting structures (Visual inspection).
- Checking fuse and cable connections
- String currents with the help of Clamp meter.
- To verify the V_{oc} and I_{sc} of an individual panel with a data sheet of manufacturer.
- To measure output power from inverter.
- Use of thermal imager to check:
  - To detect string faults with hot spot images.
  - Damages in the solar module
  - Heat loss and temperature rise issues in inverters.
- To reduce the downtime and early fault detection, cable fault locator may be used.
- Power Analyzer to check Solar PV system as per IEC 61,000 for
  - voltage flicker.
  - Harmonic variation, Voltage fluctuations, etc.
  - No DC current injection at interconnection point.
- Meter reading at point of coupling for various parameters.
- To check the labelling of different components.

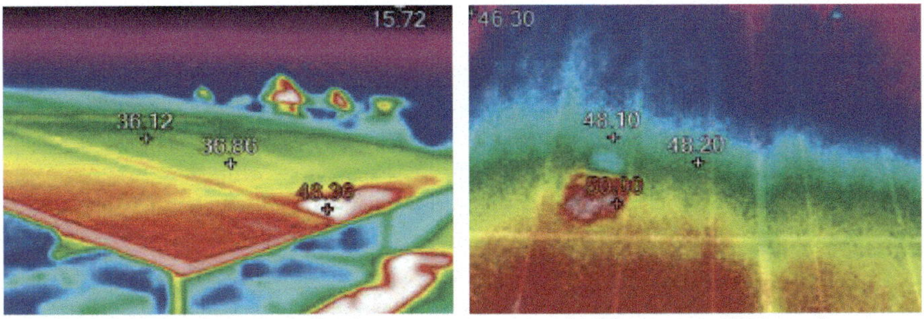

(a) Corner dirt collection (b) Dirt spot on SPV panel

Fig. 19.12 Thermal imaging of solar PV panel

- To check the various protective devices used in the system (overcurrent, overvoltage/under-voltage relay, residual current, surge and lightning protection, etc.)
- To check the earthing, insulation, and islanding condition of the system.
- To check the solar radiation recorded in the system with a solar radiation measurement device.

Figure 19.12(a) shows the thermal image of a solar PV panel. Here, the cold temperature is 15.72 °C and temperature at various locations on temperature indicates that at the corner of the panel the temperature is around 48.38 °C as compared to temperature at other spots as 36.12 °C or 36.26 °C. The reason for having higher temperature at the corner is because of deposition of dust at the corner. Figure 19.12(b) represents the spot of dirt at the middle of the panel. The spot has a temperature of 50.0 °C and cold temperature is 46.3 °C. It represents sticking dirt on the panel which did not get removed in water cleaning. After observing such a hot spot on the panels, it is the responsibility of the maintenance team to clean it with special efforts so that the corresponding part of the panel will not get permanently damaged.

19.9 Net and Gross Metering

Maharashtra state government has approved the net metering policy for individual customers through the MERC (Net Metering for Rooftop Solar Photovoltaic Systems) Regulations, 2015. The institute net meter billing is done as per this policy document. For 11 kV customers it is permitted to connect a RTSPV up to 1 MW but not more than its sanctioned load. Institute has a sanctioned load of 700 kVA and permission to install 100% of the sanctioned load, i.e. 700 kVA RTSPV. Considering the shadow free area, four buildings are identified for installation with a 100 kWp RTSPV system on each building considering availability of rooftop space. The total installed capacity is 400 kWp.

Following points from the policy are to be noted:

- High tension consumer is allowed to connect RTSPV to a low-tension busbar and the net meter is necessary to be connected to the high tension side of the consumer transformer.
- It is permissible to install RTSPV at different locations within the same premises.
- Safe operation of RTSPV with proper maintenance and necessary updates in hardware and software is the responsibility of the consumer.
- Renewable generation meters are to be maintained by distribution licensees.
- Islanding prevention for grid-connected solar inverter is to be ensured as per IEC 61000/IEEE 519 standards.
- Before injection of power into the grid, it is necessary to properly filter out harmonics and other distortions by connecting suitable filters at the inverter end so that only good quality power will be injected into the grid.
- IEEE 519 standard is to be followed for the presence of harmonic voltage and current penetrated into the grid. All meters should have advanced metering infrastructure (AMI) with RS 485 communication port.
- Before net metering if the consumer is billed as per time-of-day metering. After net metering also his consumption and generation are to be measured as per the time-of-day metering.
- Bill raised by the distribution licensee should include the following:
  - Renewable energy generation recorded by renewable energy generation meter.
  - Details of unit's consumption on time-of-day basis.
  - Active and lagging and leading var consumption during the billing period.
  - Number of units exported to distribution company, Number of units banked by distribution licensee and number of units adjusted against the solar generation units.
  - Number of units adjusted against renewable power obligations.
- If exported units are more than the imported units then after adjustment of imported units remaining units will be banked by distribution licensee and its settlement will be done at the end of financial year.
- If the imported units are more than exported units then the distribution licensee can raise the bills for units remaining after adjustment of exported units. Customer has to pay the bill before the due date.
- Payment for all the banked units in the financial year with distribution licensee will be calculated and adjusted in the first bill of the new financial year as per the tariff approved by the regulatory commission.
- For the time-of-day meter consumer, solar unit generation will be adjusted first in the same time zone. Excess unit generation will be adjusted in the off-peak period zone or banked considering generation is in the off-peak period (MERC 2015).

Institute installed its RTSPV system in April 2017. From installation till date, the number of units generated from RTSPV is 27,84,586 units. Out of these 10,44,050 units were used by the institute, 8,70,268 were exported out of that 77,651 units were adjusted, and 7,92,617 units were banked.

Presently, the distribution company has connected net meter and charges the number of units consumed at the average rate of Rs. 9.41 and purchases the solar energy generated at rate of Rs. 3.5. Considering this, saving due to solar installation is Rs. 9.41 × (10,44,050 + 77,651) + Rs. 3.5 × 7,92,617 = Rs. 1,33,29,352/- If in place of net metering if gross metering is used, then whole solar power generated is purchased by MSEDCL, a distribution licensee at Rs. 3.50, which pays (Rs. 3.5 × 27,84,586 units = Rs. 97,46,051/-. Hence, the institute will be at loss if gross metering is done to the tune of Rs. 35,83,301/-.

19.10 Life Cycle Cost of RTSPV and Tariff Decision

Regulatory bodies in India, has following guidelines for doing the calculations of life cycle cost and there on the tariff for the respective resources. Following points are to be noted for the RTSPV system calculations:

To start any renewable project requires huge funding, which normally raises through debts, equity, loan amount, etc. Regulatory commission has sanctioned a single part tariff with considerations for interest on loan and working capital, returns on equity fund and debentures, depreciation amount for developer on his own investment, O&M expenses, etc. Levelized-based tariff is decided for this. The must run priority is given to all renewable projects indicating that all the renewable energy generated from the plant will either be consumed by the customer or may be exported to the distribution licensee. When the capital cost of a renewable plant is to be considered then it includes the land cost if it is to be exclusively purchased for development of the project. In case of RTSPV, land cost investment is not required. But funds are required for infrastructure modification, plant and machinery purchased. Pre- and post-operative expenses of the project are also considered as a part of capital investment. Regulatory commission permitted for 70:30 debt: equity ratio while deciding tariff. Though the life span of project is 25 years, loan tenure is permitted up to 12 years only. Loan repayment schedule is to be considered from the first year of commercial power generation from the plant. For the first 12 years, depreciation rate of 5.83% is applicable and remaining amount will be spread over the useful life of the plant. For receiving carbon credits/renewable energy, purchase–sale obligations or any other matter related with clean energy development must be initiated by the developer. While determining tariff if the developer is getting benefits like generation-based incentive, accelerated depreciation, additional depreciation, it will be considered. Failure of renewable projects to generate guaranteed power needs to compensate for under-generation units by paying 75% of approved tariff. Excess generation than guaranteed power will be rewarded distribution licensee at the rate of 75% of tariff for excess generation.

Table 19.5, represents the costing for capital equipment. The interest rate is considered here as 8% and inflation rate 5%. The life of solar PV is considered as 25 years. The present worth of capital cost as

Table 19.5 Equipments needs capital investment

S. N	Name of equipment	Unit cost (Rs.)	No. of units	Total cost in lakhs
01	Solar panels, 300 Wp, 36 V	Rs. 16,500/-	1336	220.44
02	Inverter, 50 kW	Rs. 6,25,000/-	08	50.00
03	Module mounting structure			36.00
04	Control structure			24.00
05	Cables, lightning arrestors			04.00
06	Transportation and logistic			08.00
07	Installation charges			22.04
	Total capital cost			364.48

$$Present\ Worth = S_o = C_0\left(\frac{1+f}{1+i}\right)^n = 180.2238620\,\text{Lakhs}$$

$$Replacement\ cost = \frac{S_o}{(1+i)^n} = 26.3159\,\text{Lakhs}$$

$$Maintenance\ Cost = AMC \cdot \left(\frac{1}{i} \cdot \left(1 - \left(\frac{1}{(1+i)^n}\right)\right)\right) = 25.619\,\text{Lakhs}$$

The maintenance cost permitted by regulatory authority is 0.006 or 0.6% of the capital cost. The depreciation cost is considered for the first 15 years only as per the directives of regulatory authorities. To clean the solar panels energy is required. It is assumed that 200 units per month are required to clean the panels. Energy cost and scrap cost is assumed as zero.

$$Average\ Energy\ Generated\ per\ year = 676358.5\ units$$

$$No.\ of\ units\ used\ for\ cleaning\ panels\ per\ year = 200 \times 12 = 2400\ units$$

$$\frac{Total\ units\ to\ be\ used\ for\ billing}{25\ years = (676358.5 - 2400) \cdot 25}$$

$$= 16848962.5\ units$$

Considering the inflation and interest rate, the revenue generation from RTSPV at present rate of Rs. 9.65 per unit is

$$Total\ revenue\ generation = 16848962.5 \cdot \left((9.65) \cdot \left(\frac{1.05}{1.08}\right)^{25}\right)$$

$$= 803.96856\ \text{Lakhs}$$

Considering above factors the Life Cycle Cost (LCC) is calculated as (Abu-Rumman et al. 2017):

$$LCC = Capital\ Cost + Replacement\ Cost + Maintenance\ Cost + Energy\ Cost - Scrap\ Cost$$

$$LCC = 180.223686 + 26.3159 + 25.619 + 0 - 0 = 232.158762\,Lakhs$$

$$Levelised\ cost\ of\ energy\ (LCOE) = \frac{Life\ Cycle\ Cost\ (LCC)}{Life\ Cycle\ Energy\ Produced\ (LCEP)}$$

$$LCOE = \frac{232.158762}{803.96856} = Rs.0.2887$$

$$Pay\ back\ period = \frac{Levelised\ Capital\ Cost\ over\ 25\ years \cdot 25}{Total\ levelized\ Revenue\ generation\ in\ 25\ years} = \frac{180.2238620 \cdot 25}{803.96856} = 5.6042\,years$$

The above calculations have been carried out considering net metering. If all the generated units are purchased by the distribution company under gross metering at the rate of Rs. 3.5 per unit, then

$$Total\ revenue\ generation = 16848962.5 \cdot \left((3.50) \cdot \left(\frac{1.05}{1.08}\right)^{25}\right) = 291.5948152\,Lakhs$$

$$LCOE = \frac{232.158762}{291.5948152} = Rs.0.79616$$

$$Payback\ period = \frac{180.2238620 \cdot 25}{291.5948152} = 15.45157\,years$$

19.11 Conclusions

Institute installed RTSPV 4 years ago in April 2017. This chapter has analyzed the performance of RTSPV. Solar power generation and load curve are matching with one another. It was found that the active power drawn from the distribution company reduced to halved, similarly kVA has also reduced but the reactive power requirement increases. Installation of APFC brought down the reactive power demand also. Most of the times the exported power is adjusted with imported power and banking of power takes place during the vacation period only. Installation of RTSPV helps the institute in reducing bills. When the overall kWh consumption is reduced, other charges like wheeling charges, electricity duty, etc., also reduces. Energy bill charged by distribution licensee reduces in the range of 8–60% for various months. Grid penetration percentage varies from 10 to 60%. Though regular cleaning is done, some dirt was not removed, and special efforts are needed to remove them and save the panel from hot spot and damage. It was observed that net metering is beneficial to customers as compared to gross metering. Payback period for net metering is 5.6 years as compared to gross metering which is around 15.45 years. For further expansion in load, there is no need to take the additional sanction of load from the distribution licensee as maximum kVA demand has already dropped due to installation of RTSPV.

Annexure I

A-1 Solar panel and inverter specifications

Solar panel specification		Inverter specifications	
Watt	300 W-peak	Type of inverter	Grid tie solar inverter
Voltage	36.6 V	Model No	RPI-M 50A
Current	8.2 A	AC capacity of individual Inverter (KW)	50 kW
Type	Multi-crystalline	No. of inverters installed	8 No
Efficiency	14.3%		
Temperature	25 °C		
Dimensions (mm)	1955 × 992 × 38 mm Area of single panel = 1939.36 m^2		
Tilt angle (slope) of PV Module	18°	Total AC capacity of inverter (KW)	400 KW
Mounting	Fixed type	No. of phases	3-φ

References

Abhiram S, Ilango K, Narayanan AM (2018) The impacts of rooftop solar PV systems on secondary distribution system and advanced net metering. In: 3rd IEEE international conference on recent trends in electronics, information & communication technology (RTEICT), 18–19 May 2018, pp 189–194

Abu-Rumman AK, Muslih I, Barghash MA (2017) Life cycle costing of PV generation system. J Appl Res IndEng 4(4):252–258

Ayompe LM, Duffy A, McCormack SJ, Conlon M (2011) Measured performance of a 1.72 kW rooftop grid connected photovoltaic system in Ireland. Energy Convers Manage 52:816–825

Behera MK, Majumder I, Nayak N (2018) Solar photovoltaic power forecasting using optimized modified extreme learning machine technique. Elsevier Int J Eng Sci Technol 21(3):428–438

Behura AK, Kumar A, Rajak DK, Pruncu CI, Lamberti L (2021) Towards better performances for a novel rooftop solar PV system. Solar Energy 216:518–529

Bureau of Indian Standard (2010) Indian standard on photovoltaic system performance monitoring—for measurement data exchange and analysis

Central Electricity Authority (Technical Standards for Connectivity to the Grid) Amendment Regulations (2013)

Chakravarthy M, Ramana Murthy KV, Neelima Devi B (2015) Design, erection, testing and commissioning of 200kwp rooftop grid tied solar photovoltaic system at Vasavi College of Engineering. In: IEEE IAS joint industrial and commercial power systems/petroleum and chemical industry conference (ICPSPCIC), 19–21 Nov 2015, pp 146–152

Dondariya C, Porwal D, Awasthi A, Shukla AK, Sudhakar K, Murali Manohar SR, Bhimte A (2018) Performance simulation of grid-connected rooftop solar PV system for small households: a case study of Ujjain, India. Energy Rep 4:546–553

Guo Y, Xiong J, Xu S, Su W (2016) Two-stage economic operation of microgrid-like electric vehicle parking deck. IEEE Trans Smart Grid 7(3):1703–1712

Kappagantu R, Arul Daniel S, Venkatesh M (2015) Analysis of rooftop solar PV system implementation barrier in Puducherry smart grid pilot project. Procedia Technol 21:490–497

Kumar M (2020) Technical issues and performance analysis for grid connected PV system and present solar power scenario. In: International conference on electrical and electronics engineering (ICE3), 14–15 Feb 2020, pp 639–644

Maity R, Mathew M, Hossain J (2018) Increase in power production of rooftop solar photovoltaic system using tracking. In: 2018 international conference on power energy, environment and intelligent control (PEEIC) G. L. Bajaj Inst. of Technology and Management Greater Noida, U. P., India, 13–14 Apr 2018, pp 415–419

Mandi RP (2017) Grid interactive rooftop solar PV power plant for educational institute. In: Proceedings of 2017 international conference on smart technologies for smart nation (SmartTechCon), 17–19 Aug 2017, pp 1473–1478

Manoj Kumar N, Subathra MSP, Edwin Moses J (2018) On-grid solar photovoltaic system: components, design considerations, and case study. In: 4th international conference on electrical energy systems (ICEES), 7–9 Feb 2018, pp 616–619

Mohanty BP, Srivalli M (2020) Optimization and design of grid connected rooftop solar power plant under various operating conditions. In: IEEE international conference on computational intelligence for smart power system and sustainable energy (CISPSSE-2020), 29–31 July 2020, Odisha, India, pp 1–6

Moharil RM, Kulkarni PS (2010) Reliability analysis of solar photovoltaic system using hourly mean solar radiation data. Elsevier J Solar Energy 84:691–702

MSEDCL Commercial Circular number 258 regarding Connectivity to the Distribution network of MSEDCL for Consumers installing solar PV projects/systems below 1 MW on Rooftop or any mounting structure in their premises under the MERC (Net Metering for Roof-top Solar Photo Voltaic Systems) Regulations (2015) Procedure for application, methodology for metering & billing, etc. Dated 25/01/2016

National Solar Mission (2016) Best practices guide: implementation of state level solar rooftop photovoltaic programs in India, MNRE

Omkar K, Srikanth MV, Swaroop KP, Rama Rao PVV (2015) Performance evaluation of 50 KWp rooftop solar PV plant. In: International conference on industrial instrumentation and control (ICIC), 28–30 May 2015, pp 761–765. https://doi.org/10.1109/IIC.2015.7150844

Ouammi A (2021) Peak loads shaving in a team of cooperating smart buildings powered solar PV-based microgrids. IEEE Access 9. https://doi.org/10.1109/ACCESS.2021.3057458

Panda S, Rout PK, Sahu BK (2020) Demand side management by PV integration to micro-grid power distribution system: a review and case study analysis. Green Technology for Smart City and Society. Lecture Notes in Networks and Systems, vol 151. Springer, Singapore. https://doi.org/10.1007/978-981-15-8218-9_35.

Prajapati S, Fernandez E (2019) Rooftop solar PV systems for commercial office buildings for EV charging load. In: Proceedings of the 2019 IEEE 6th international conference on smart instrumentation, measurement and applications (ICSIMA 2019), 27–29 Aug 2019, Kuala Lumpur, Malaysia, pp 1–4

Rout KC, Kulkarni PS (2020) Design and performance evaluation of proposed 2 kW solar PV rooftop on grid system in Odisha using PVsyst. In: IEEE international students' conference on electrical, electronics and computer science (SCEECS), 22–23 Feb 2020

Sharma Y, Saxena BK, Mishra S (2019) Techno-economic analysis of rooftop solar PV system by variation in PV module capacity. In: 2nd international conference on power and embedded drive control (ICPEDC), 21–23 Aug 2019, pp 355–360

Sharma R, Goel S (2017) Performance analysis of a 11.2 kWp roof top grid-connected PV system in Eastern India. Energy Rep 3:76–84

Shilpa N, Sridevi HR (2019) Optimum design of rooftop PV system for an education campus using HOMER. In: 2019 global conference for advancement in technology (GCAT) Bangalore, India, 18–20 Oct 2019, pp 1–4

Singh S, Kumar R, Vijay V (2014) Performance analysis of 58 kW grid-connected roof-top solar PV system. In: 6th IEEE power India international conference (PIICON), 5–7 Dec 2014, pp 1–6. https://doi.org/10.1109/POWERI.2014.7117738

Wiles JC (2006) Designing and installing safe, durable, and cost-effective photovoltaic power systems. In: 4th world conference on photovoltaic energy conference, 7–12 May 2006, pp 2355–2358

Yesilbudak M, Colak M, Bayindir R (2018) What are the current status and future prospects in solar irradiance and solar power forecasting? Int J Renew Energy Res 8(1):635–648

Chapter 20
Design, Development, and Simulation Modeling of Hybrid Electric Vehicles Incorporating with BLDC Drive

Mohan P. Thakre and Nitin Kumar

Abstract Electric vehicles (EVs), including Battery Electric Vehicles (BEVs), Hybrid Electric Vehicles (HEVs), Plug-in Hybrid Electric Vehicles (PHEVs), and Fuel Cell Electric Vehicles (FCEVs), have recently become more common in the transportation sector. As the current fashion shows, the electrical transportation mode is most probably going to replace the internal combustion engine (ICE) vehicles in coming days. EVs have a notable influence on electrical power systems and other related sectors and also on environment. The current power sector can suffer enormous instability with high EV integration with the electric grid; however, with proper coordination and management, smart grid concept can be successfully implemented with EVs playing a major role. There are also opportunities for enormous economic consequences, as EVs can dramatically lower the greenhouse gas (GHG) emission levels emitted by the transportation vehicles. Though there are some significant barriers for EVs to resolve before functionalizing internal combustion engine vehicles. Industrial motor drives mainly concern with operations at motor base speed. Electric vehicle motor drives deal with operations that require sudden start and stop, low-speed operations, and base speed operations as well. The system should meet the objective of variable speed, variable torque applications like Electric Vehicles (EVs). As per studies, brushless direct-current (BLDC) motor is found to be a suitable motive element for EV application based on its suitable torque speed characteristics. In medium-sized electric vehicle, BLDC motor drive that is fed by a rectangular A.C. supply has shown much promising results than other types of motors. Permanent magnet brushless D.C motor drive is a perfect choice to be utilized in the EV propulsion system, with benefits such as removal of the brushes, capability to generate higher torque than other motors operating at the same voltage and current magnitudes, high efficiency, and power density. Since BLDC motor has been used widely in automotive, it is tested for simulation profiles and then is integrated in the HEV system model in MATLAB Simulink. The fuel economies also showed that the combination of selected motor and engine is efficient and has very high potential in market.

M. P. Thakre (✉) · N. Kumar
K. K. Wagh Institute of Engineering Education and Research, Nashik, Maharashtra, India

A. K. Bohre et al. (eds.), *Planning of Hybrid Renewable Energy Systems, Electric Vehicles and Microgrid*, Energy Systems in Electrical Engineering,
https://doi.org/10.1007/978-981-19-0979-5_20

Keywords EV · HEV · BLDC · Batteries · MATLAB simulink

20.1 Introduction

In modern years, electric vehicles (EVs) have been getting popular, and there are several reasons for all of this. The most significant is their endowment to reduction of emission of GHG levels. As per the United States Environmental Protection Agency (EPA) in 2018, the transportation sector generated the largest share of GHG emission accounting to about 28.2% of the total GHG emissions (Yong, et al. 2015; Camacho and Nørgård 2014; Chowdhary and Thakre 2020). EVs, with plenty of usage in the transport sector, have been intended to reduce a certain estimate, but that isn't the sole justification to revive this century-old, just before idea as commercially convenient and feasible product. The electric vehicles (Camacho and Mihet-Popa 2016; Marchesoni and Vacca 2007) are quiet, simple in maintaining, and therefore have no fuel costs associated as compared with traditional vehicles. It is also very useful as an urban means of transport. EVs do not utilize any stored energy or affect any emissions even during idling state, it would be responsible for frequent starting and stopping, offers the required torque from starting, but do not necessitate the excursions to the gas or petrol stations (Schaltz et al. 2008; Kramer et al. 2008; Gao and Ehsani 2010). Neither does it relate to any kind of smog that makes the environment highly polluted. EV has low infrared and silent operation and thus it is viable for military usage as well. Electrical sector is estimated via a varying phase in which non-conventional energy sources have been gaining importance. A next century power grid (Thakre et al. 2020; Miller et al. 1997; Un-Noor et al. 2017; Kim and Kum 2016), named the smart grid, is gaining momentum with EVs, considering them to be a vital contributor to the smart grid consisting of advanced grid systems and renewable generators. Because of these advantages it has led to a recommenced interest and growth throughout this transportation mode.

The concept of driving a vehicle using electric motors came to light after a series of development in direct-current (DC) motor and rechargeable batteries to operate them at high efficiency (Chan 2002). The first commercial EV was an electric taxi introduced in New York back in 1897. After few years, EVs were 28% of the total vehicles. But the ICE type (Grunditz and Thiringer 2016; Yilmaz and Krein 2013; Nakadachi et al. 2013) vehicle gained momentum later on and soon conquered the automotive sector market by becoming much more advanced with very low oil prices and EVs were lost into forgetfulness. The General Motors' EV1 concept was launched in 1996 which (Miller 2006; Kim et al. 2008) emerged as a resurrection chance for EVs. Other leading car manufacturers such as Ford, Toyota, and Honda also produced their own EVs. One of the highly successful commercial HEV Toyota Prius was launched in 1997 in Japan (Gao et al. 2005); about 18,200 units were sold in the very first year of its manufacturing. Currently, there are almost no such EVs of the twentieth century, however being an exception could be the Toyota Prius which continues to be strong in a better and more advanced manner. Now Tesla Model

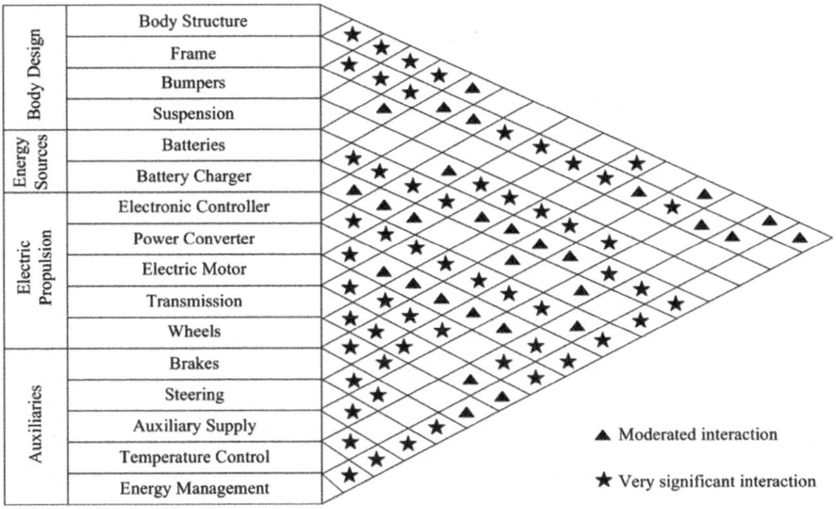

Fig. 20.1 Key component and their contribution to overall EV system

S, Chevrolet Volt, and Nissan Leaf are dominating the automotive market (Guntuk et al. 2019; Thakre et al. 2020), while BYD Auto Co., Ltd. China has taken over the Chinese market. A combination of various subsystems can be visualized as EV. These subsystems cooperate with one another to run the electric vehicle, and various mechanisms are used to run the subsystems (Chau and Chan June 2008). Figure 20.1 shows the decisive components of EV and their contribution to the overall system. Several components have to run considerably with one another as compared to others. The combined functions of all these subsystems operate an EV.a

20.2 Electric Vehicles (EV)

20.2.1 Types of Electric Vehicle (EV)

EVs run on electrical propulsion, or an internal combustion engine (ICE) can work with them. EVs working on batteries only as the power source are the basic type of EV; however, there are other EV types as well which utilizes other energy sources. These can be termed as hybrid electric vehicles (HEVs) (Gao and Ehsani 2010; Shah et al. 2010). The International Electro technical Commission's Technical Committee 69 has stated that "vehicles using two or more types of energy sources, storage and converter can be referred to as hybrid electric vehicle (HEV)". This definition enables many amalgamations of HEVs, like batteries and condensers, ICE and batteries, (Jung et al. 2007) batteries and flywheels, batteries and fuel cells, etc. The researchers have therefore begun to call for HEVs consisting of batteries and condensers as ultra-capacitator-assisted electric vehicle, and those with batteries and fuel cells as fuel cell electric vehicles (FCEVs). EVs can be distinguished as follows:

- Battery Electric Vehicle (BEV).
- Hybrid Electric Vehicle (HEV).
- Plug-in Hybrid Electric Vehicle (PHEV).
- Fuel Cell Electric Vehicle (FCEV).

20.2.1.1 Battery Electric Vehicle (BEV)

EVs in which batteries are the only power source for drive train are termed as battery electric vehicles (BEVs). BEVs rely only on battery packs with the limited stored energy, and the driving range of these vehicles therefore relies on the capacity of the batteries. On a single charge, the BEV can cover a distance of 100–250 km whereas a top-class EV model can go much further, from 300 to 500 km (Fig. 20.2).

These ranges depend on driving style, configuration of the automobile, conditions of road, weather conditions, battery type, and age. When the battery pack is discharged it requires huge time to get recharged (Naseri and Farjah May 2017). The charging time can be up to 36 h for the batteries; however, there are batteries which take less time to recharge, but none can be compared to the short time for refilling a fuel tank.

Charging time depends on the power level and design of the charger. BEVs are benefited from their easy design and functioning. They do not emit GHGs, are silent in operation, and are thus environmentally friendly. High and instant torques are furnished by electrical thrust, even at low speeds (Camacho and Mihet-Popa 2016; Marchesoni and Vacca 2007). These merits collaborated with their operating distance with single charge making them the ideal choice for use in cities; in urban traffics

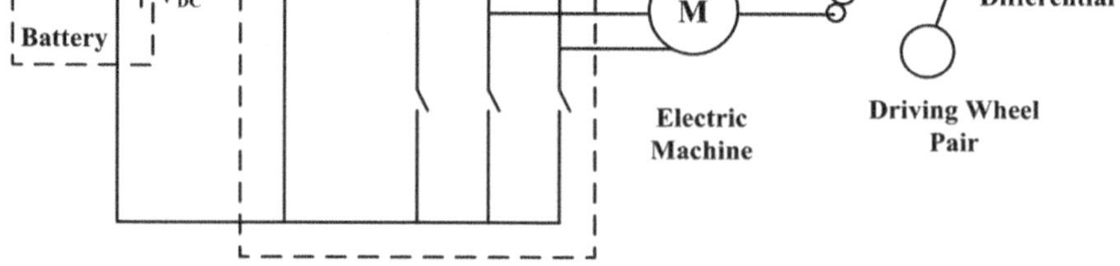

Fig. 20.2 Battery electric vehicle (BEV) configuration

vehicles need to be run at lower speeds, and these speed ranges varying from medium to slow require a lot of torque. Some of the high selling BEVs are Nissan Leaf and Tesla.

20.2.1.2 Hybrid Electric Vehicle (HEV)

HEVs are energized by both a power train and an ICE. The combinations of these components are discussed in various forms later in this chapter. The electrical propulsion system is used by an HEV when the requirement for power is less. In low-speed conditions, it is of great advantage and reduces fuel consumption and GHGs emissions, since the engine remains completely free during idling periods such as traffic jams. The HEV changeovers to ICE when speed required is high.

The two trains can also operate together to enhance performance of EV. The HEVs can regenerate the braking energy and the ICE can recharge the batteries. HEV is therefore fundamentally an ICE-driven car (Mathur et al. 2014) that utilizes an electric train to enhance the traveling distance and performance. Car manufacturers are widely adopting the HEV configurations to obtain these features. The energy flows in a fundamental hybrid electric vehicle are depicted in Fig. 20.3. When the automobile starts, the ICE can run the engine as a generator to generate power to recharge the batteries. Passing requires speed boosting, so both the engine and ICE drive the powertrain. The motor is used as a generator at the time of braking by the powertrain to recharge the batteries by regenerating braking.

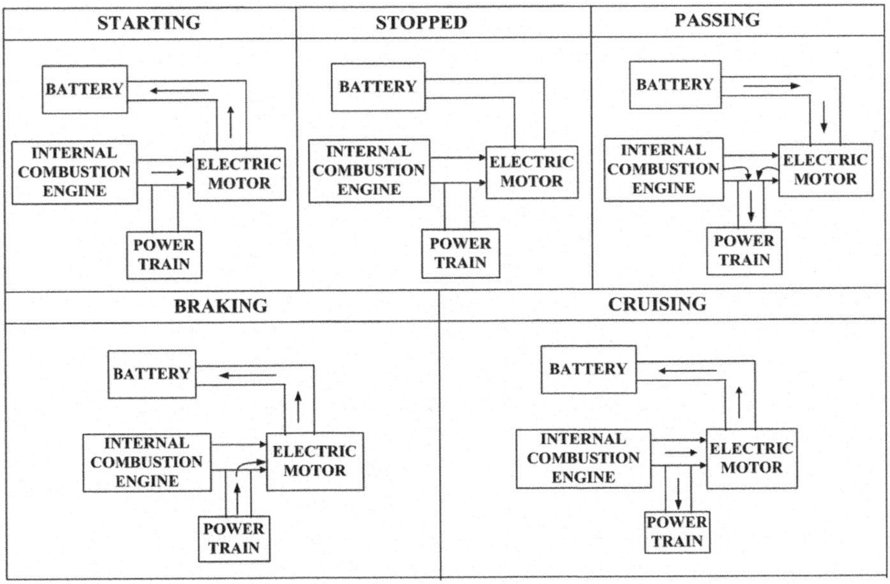

Fig. 20.3 Power flow in various stages within the basic building blocks of an HEV

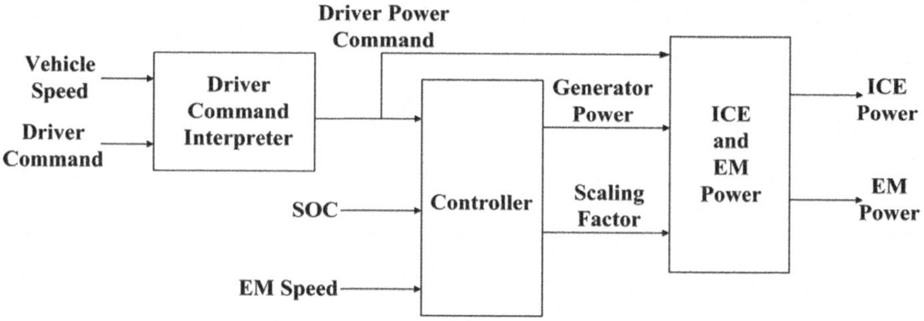

Fig. 20.4 Energy management approach in the hybrid electric vehicle

During the cruise, ICE runs the car and the engine as a generator, charging the battery. When the vehicle stops, the power flow stops. The approach of energy management systems utilized in HEVs is shown in Fig. 20.4. The one shown here splits power among the electric motor (EM) and the ICE by taking into consideration driver input, vehicle speed, engine speed, and battery charge status (SOC) to achieve full fuel efficiency (Singh et al. 2019).

20.2.1.3 Plug-In Hybrid Electric Vehicle (PHEV)

PHEV concept expanded the entire HEV concept. They use both an electric powertrain and an ICE train, such as a HEV; however, the major differentiation among them is that the PHEV's main driving force is electrical power. PHEVs run on electricity and call ICE for an increase or charge of the batteries. The ICE is utilized to expand the portfolio. The batteries in PHEVs can be charged directly from the electric grid; they can also use regenerative braking. PHEV carbon footprints are on the lower side as compared to HEVs because of their ability to operate mostly on electricity. They also consume less fuel, reducing the associated costs. These sales of Chevrolet Volt and Toyota Prius are now quite crowded on the vehicle market.

20.2.1.4 Fuel Cell Electric Vehicle (FCEV)

The core part of FCEV is the fuel cells that produce electricity by using chemical reactions. Figure 20.5 shows the basic configuration of FCEV. The fuel utilized in FCVs is hydrogen which performs these chemical reactions, so they are also known as hydrogen fuel cell vehicles. FCVs carry a high-pressure tank containing (Schaltz et al. 2008) hydrogen and an additional component in the power generation procedure is the oxygen which they absorb from the atmosphere. Electric motor drives the vehicle which is powered by the electricity produced from the fuel cells. Excess energy is stored in battery or supercapacitor storage systems.

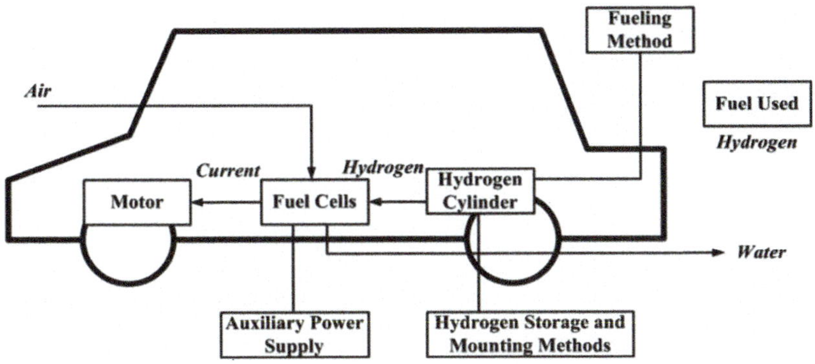

Fig. 20.5 Fuel cell electric vehicle configuration

In the fuel cell, hydrogen and oxygen react to generate electricity which drives the engine. Water is the only by-product produced.

From Fig. 20.6, it is evident that for ratio of BEV to FCEV greater than 1 fuel cell EV is advantageous over the battery EV. The only by-product produced by FCEV is water from its electricity generation process which is released from the tailpipes of the car. FCEV generates their own electricity, which does not emit carbon, and hence this type of EV has lower carbon footprint than any other type of EV. One of the major advantages is that it takes the same time to replenish fuel cell electric vehicles as taken by a conventional vehicle to fill its fuel tank, perhaps which is the most important right now. There are safety concerns when flammable hydrogen leaks from tanks. If these barriers were removed, FCVs would be the future of cars. One of

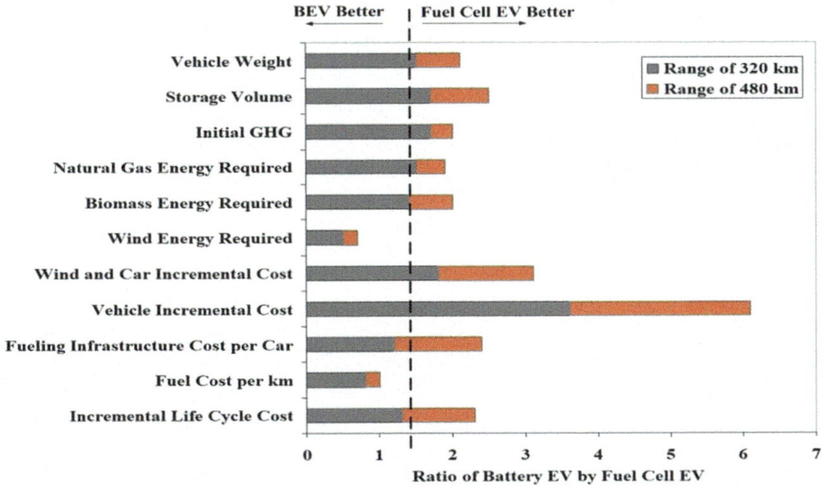

Fig. 20.6 Advanced EV attributes

Table 20.1 Description of various types of electric vehicles

EV type	Driving component	Energy source	Features	Problems
BEV	• Electric motor	• Ultra-capacitor • Battery	• Not dependent on oil • No emission • Available commercially • Range depends on type of battery	• Battery price and capacity • Range • Charging time • Availability of charging stations • High price
HEV	• ICE • Electric motor	• Battery • Ultra-capacitor • ICE	• Very little emission • Can get power from both electric and fuel • Long range • Available commercially • Complex structure containing both electric and mechanical drivetrains	• Battery and engine size optimization • Management of the energy sources
FCEV	• Electric motor	• Fuel cell	• High efficiency • Very little or no emission • High price • Not dependent on supply of electricity • Available commercially	• Feasible way to produce fuel • Cost of fuel cell • Availability of fueling capacity

the biggest drawbacks of FCEV still appears to be fuel costs, since (Lee and Ehsani 2001) there is still a lack of an inexpensive, sustainable, and environmentally friendly way to generate hydrogen, and the refueling infrastructure is behind the BEVs (Table 20.1).

20.2.2 EV Functionality

Unlike its ICE counterparts, an electric vehicle is quite flexible as it requires no complex mechanical arrangement to operate unlike a conventional vehicle. Only moving part is the motor in an EV. Figure 20.7 shows the subsystems of an EV.

Various control techniques and arrangements can be used. Such power supply is needed which can drive the engine from a range of sources. The engine and power source are placed in different positions in the vehicle and the vehicle works as long as it is connected via electric wires. An EV can then be powered only by electricity, but an electric motor and ICE may also work together to drive the vehicle. Due to

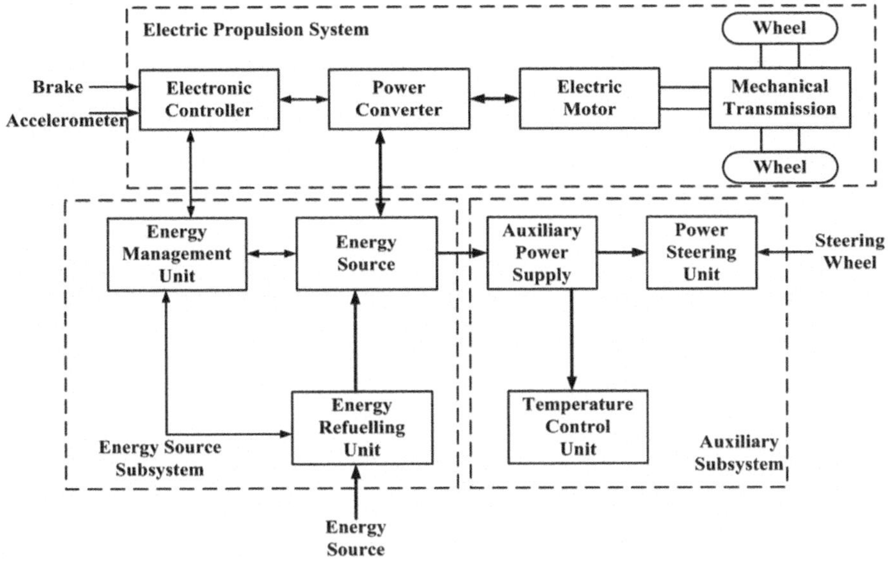

Fig. 20.7 EV subsystems

such advantages, there have been various configurations considered by this vehicle type.

An electric vehicle could be visualized as a system that incorporates three various subsystems: propulsion system, source of energy, and auxiliary system. The subsystem energy source consists of fuel system, the source, and its energy management system. The electric motor, controller, transmission, driving wheels, and power converter are part of the propulsion system. The auxiliary system contains an auxiliary supply of power, a temperature control, and a steering unit. Most EV batteries and flywheels are suitable with these methods of energy regeneration (Kumaresan and Govindaraju 2020). The in-wheel engine arrangements are convenient as by removing the connected transmission, central motor, drive shaft, universal joints, and differential they reduce the drive train's weight. They also furnish large battery space, better control, and turning operation. Both the electric propulsion system and the ICE are used by HEVs. Different ways in which electric propulsion system and ICE can be configured to rotate the wheels create various configurations which can be summarized as follows:

- Hybrid series.
- Hybrid parallel.
- Hybrid series-parallel.
- Hybrid complex.

20.2.2.1 Hybrid Series

Hybrid series is the easiest setup for a HEV to be made. Wheels are only connected to the electric motor, and an electric generator is run with the help of motor. It is positioned as an EV which is supported by an ICE generator. Figure 20.8 shows the series hybrid system drive train.

The advantages of the hybrid configuration series and their limitations are as follows:

Advantages:

- Efficient and optimized power station.
- Optimized drive line.
- Modular options for power plants.
- Speedy possibility of service exchange black box.
- Long-term life.
- Rapid reaction to mature technology.
- Able to achieve zero emissions.

Limitations:

- Large system of traction drive.
- Proper algorithms are required.
- Multiple steps in energy conversion (Table 20.2).

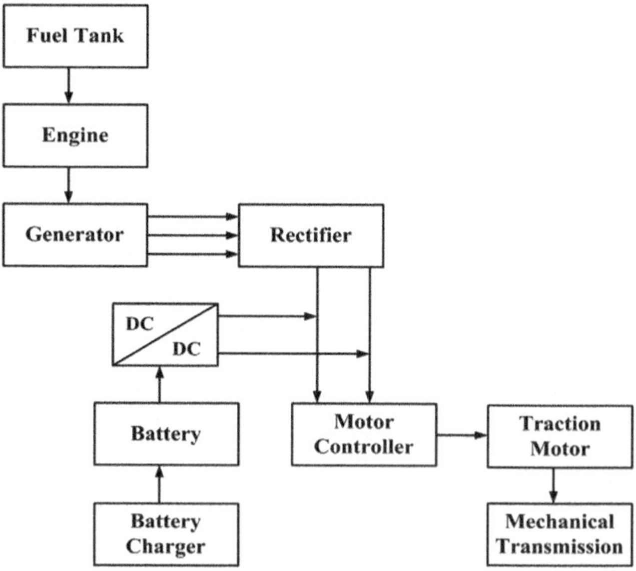

Fig. 20.8 Series hybrid system drive train

Table 20.2 Comparison of hybrid series and parallel configurations

Parameters	Parallel HEV	Series HEV
Voltage	14 V, 42 V, 144 V, 300 V	216 V, 274 V, 300 V, 550 V.900 V
Power required	3–40 kW	>50 kW
Relative gain in fuel economy (%)	5–40	>75

20.2.2.2 Hybrid Parallel

This configuration parallels the electric motor and the ICE with the wheels. Both or either one of them participates in power delivery. It is regarded as an electrically supported ICE vehicle (Gao et al. 2005). If more than the power required for driving the wheels is produced then the batteries kept in such cars are recharged by regenerative braking by the ICE or the electric motor. Figure 20.9 shows the hybrid parallel drive train.

The advantages of the hybrid configuration parallel and their limitations are as follows:

Advantages:

- Able to achieve zero emissions.
- Economic profit.
- Greater flexibility.

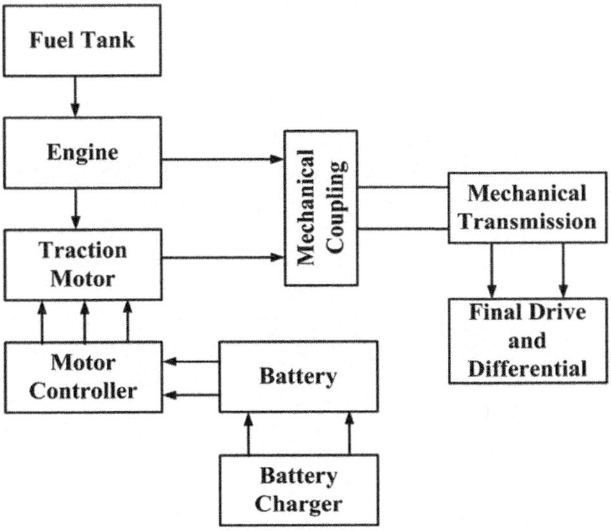

Fig. 20.9 Hybrid parallel drive train

Limitations:

- Costly.
- Complex control.
- High-voltage requirement to ensure efficiency.
- Proper algorithms are required.

20.2.2.3 Hybrid Series-Parallel

Hybrid series system requires additional generator when distinguished with the parallel type or an additional mechanical connection in comparison with the series type to combine the series with the parallel arrangement. Figure 20.10 shows the hybrid series-parallel system drive train using planetary gear unit. It offers the advantages of both hybrid series and parallel systems; however, it is very complicated and expensive. Complexities in the drive train are due to the existence of a planetary gear unit to a certain extent (Un-Noor et al. 2017). In this system, the drive wheels are connected to the rotor via the gears and the stator is attached to the engine. The engine speed is the relative speed of the rotor and stator, which alters the engine speed for each specific speed of the vehicle. Figure 20.11 shows the overall four-wheel drive HEV structure.

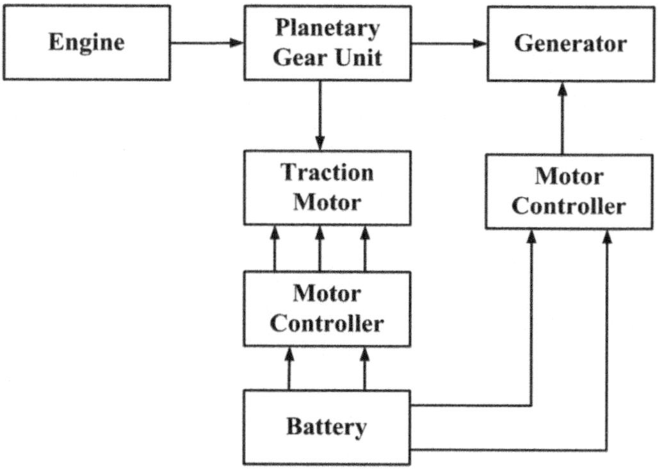

Fig. 20.10 Series-parallel hybrid system drive train

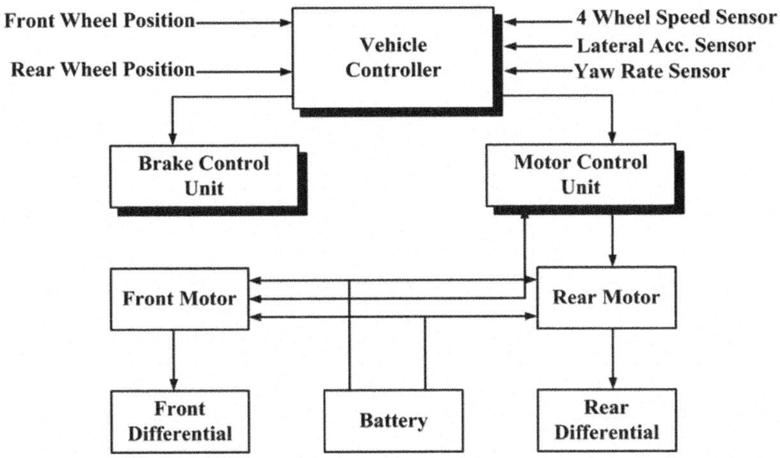

Fig. 20.11 Four-wheel drive HEV structure

20.2.3 Hybrid System Importance

In the recent years, hybrid electric vehicles have been a field of interest because of their tendency of reducing air pollution and potential for high fuel efficiency (Miller et al. 1997). The different hybrids typically change, such as how electric or mechanical energy is transmitted, or which type of energy sources/fuel can be utilized to energize the HEV. The proposed hybrid system analyzed in this chapter includes the use of high-efficient brushless DC (BLDC) drives in a predominantly electric system for propelling vehicle and to make the most of the power sources available while coordinating energy consumption. Only the supposed ICE is coupled mechanically with a permanent magnet generator is the fossil fuel component. The regeneration allows the electrical system to recover some of the mechanical energy of the vehicle. Several technological advances have led to superior power use in an electric system.

New and more efficient battery with shorter charging times, higher energy densities, and lighter physical weights are being developed. Various types of batteries have various properties which can attract a particular application. A nickel-metal hydride battery (NiMH), for example, is a better option for a hybrid system employment, given its low charging time and high overall efficiency. High-performance BLDC motors are upgraded to produce high torque from every power system per amplifier (Yilmaz and Krein 2013).

20.2.4 Hybrid System Components

20.2.4.1 Brushless DC Motor (BLDC)

Because of its high torque per amplifier capacity, high efficiency, and low maintenance requirements, brushless DC engines/generators are selected. Because of its control complexity, the uses of dq0 reference frame variables to control BLDC drives are greatly simplified. With feedback on rotor position and speed, the performance of the BLDC drive can be controlled by torque or speed control in dq0 reference frame (Grunditz and Thiringer 2016). With many diverse control schemes, power switches can be used to regulate the input voltage so as to achieve the voltage BLDC engine requires. When necessary, the BLDC can either be operated as a generator or can be operated in motor mode when braking depends on the power electronics control.

The physical performance of a BLDC engine depends on the motor's physical characteristics, like resistance of stator and rotor, motor control type, number of poles, and the DC voltage applied to the BLDC drive. These parameters are usually decided by the specifications required by the manufacturer.

20.2.4.2 Ultra-Capacitors

An ultra-capacitor utilizes the characteristics of a standard condenser to produce greater capacity than before. The condenser electrode porous surface, combined with the electrolytic solution between electrodes, contributes to a very high capacity of the ultra-capacitor. For two main reasons, the ultra-capacitor with large capacitances is beneficial in the system: the high capacity to ease voltage dips on the DC bus and high power density when used as a power source. The ultra-capacitors in hundreds of Farads can be purchased and load/discharge currents can be handled by hundreds of amperes of current. Ultra-capacitors can also have a duty cycle of up to one million before they break down. Its superior loading and unloading features and high capacitance value make the ultra-capacitor essential for quick power supplies.

20.2.4.3 Batteries

A battery is very handy in the design of an HEV as a primary or secondary power source. The power density, energy density, type of electrochemical reaction, life cycle, and cost are the basis for classification of batteries. There are many other considerations but the main considerations are listed for a hybrid system. The most efficient battery is usually also the most expensive. Tens of thousands of dollars could cost a lightweight high-energy battery.

Batteries can supply large quantities of energy over long periods (hours) but the main disadvantage is their slow charging rate. Batteries have an amp-hour rating which indicates how quickly and safely the battery can be loaded and unloaded

(Miller 2006). The rule is usually to divide the ampere hour rate by ten in order to find out the harmless charging current value. Naturally, all batteries have unlike charging times; few are faster, but these usually are very costly alternatives, mainly for larger batteries.

The battery chosen here is lead–acid battery because of their availability and low cost. The lead–acid battery sometimes has charging times in decades and normally cannot discharge less than about 80% of its rated tension. Many models of lead–acid batteries have been advanced and tested, which is another important intention for choosing platinum–acid batteries for costly solutions. Unfortunately, the battery model could not be incorporated into the final design of hybrid system.

20.2.4.4 Buck/Boost Converters

For implementation, various kinds of converters have been considered. Research has found that the most efficient design from the point of view of easy coordination and control is of DC/DC converter. This multiple DC–DC input converter is in between the inverter and source. When the BLDC is in generator mode (i.e., while braking) the converter recharges the power sources and when the BLDC is in motoring mode converter provides power from power sources (Gao and Ehsani 2010). The converter is bi-directional, i.e., it operates as a buck converter to power sources in breaking mode or it can operate as boost converter to power the DC bus. The main losses are caused by switching and driving. Due to high switching frequencies and low losses, the MOSFETs were chosen for the converters.

20.2.5 System Model Layout

The system layout is depicted in Fig. 20.12. The system's components are represented by the blocks in Fig. 20.12. The direction of flow of power is represented by direction of arrows. The overall system consists of several different subsystems. An ICE is mechanically coupled to the permanent magnet synchronous machine (PMSM). The only input to the PMSM generator is supposed to be the mechanically coupled rotor speed. To generate a DC voltage, the generator output is attached by a three-phase rectifier. The input to boost converter is DC voltage and the boost converter delivers the output voltage to the DC bus whenever necessary. The DC bus is connected with an ultra-capacitor via a bi-directional converter which powers the DC bus when operated in boost mode and powers the ultra-capacitor when operated in buck mode.

BLDC engine is powered through the DC bus. By taking note of the arrows direction in the system, the BLDC motor works as an engine, during propulsion of the vehicle and as generator during regenerative braking (Thakre et al. 2020). The power flow to the DC bus from ICE is only in one direction, as the internal combustion engine cannot recover any power which is produced during the regenerative braking. By using ultra-capacitor the energy during braking can be captured.

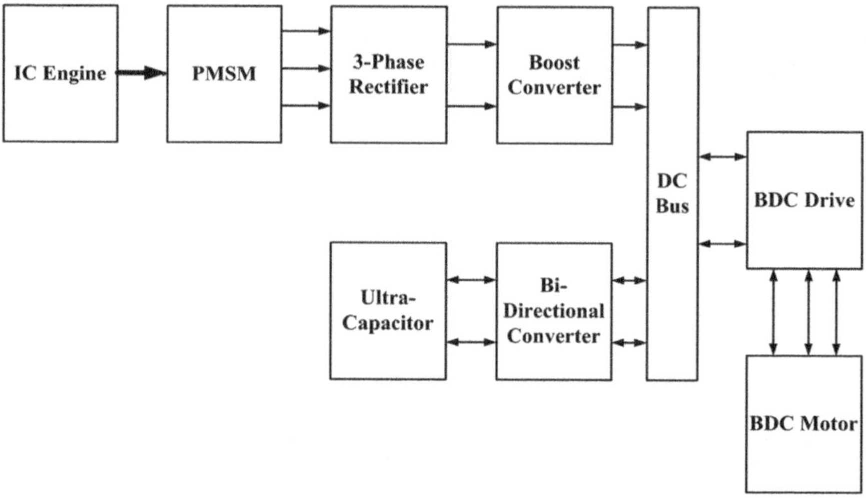

Fig. 20.12 Layout of system

20.3 Selection of Motor for Electric Vehicle

In an electric drive, the electric motor is the heart of the system. The electric motor converts the battery's electric energy into mechanical energy that moves the EV. It also behaves as a regenerative generator that returns energy to the source of energy. Different number of motors can be available on the basis of their prerequisites: the Acura NSX has three, the Toyota Prius has one, and the choice relies on the type of motor vehicle and the features it should offer. The key needs for an EV motor is high torque and power, broad speed range, reliability, high efficiency, cost-effectiveness, robustness, small size, and low noise (Schaltz et al. 2008). Direct current (DC) drives show some of the features necessary for EV applications, but their efficiency and large-scale structure, lack of reliability due to their switch or brushes, and associated maintenance requirements have made them less attractive. The advancement in control systems and power electronics has resulted in various motor types meeting the automotive sector's needs, as well as the favorite types of induction as well as permanent magnet (PM).

20.3.1 Brushed DC Motor

Such motors' rotor has brushes to supply the stator and the stator is permanent magnet (PM). The advantage of these motors is their capability to achieve maximum torque at low speed. On the other hand, the drawbacks are its low efficiency, its large structure, the heat produced by the brushes, and the related efficiency drop. As it is produced in the rotor center, the heat is difficult to eliminate (Kramer et al. 2008). For such causes, in EVs, the brushed DC motors are not used.

20.3.2 Permanent Magnet Brushless DC Motor (PMBLDC)

In PMBLDC, the rotor is made of permanent magnet (PM) (most frequently neodymium magnet (NdFeB)), while the stator is fed with an alternative current (AC) which is generated from an inverter. Because no windings are present on the rotor, no copper loss is produced thus making it highly efficient than the induction motors. BLDC motors are smaller, lighter, reliable, and high torque density. However, the constant variety is quite short because of its limited ability to weaken the field. The torque decreases with the increasing speed due to the EMF in the stator windings. The cost of BLDC motor is increased due to the use of permanent magnet. However, additional field windings can improve the speed range and improve overall efficiency. This kind of arrangements is also termed as PM hybrid motors because both field windings and permanent magnet are present. These arrangements are also restricted by physical intricacy; the rapidity ratio is not sufficient to meet the requirements of electric vehicle usage, specifically in off-road vehicles. By controlling the firing angle of the power converter, the PMBLDC motor's efficiency can be improved and the speed range to the highest fourfold base speed, although efficiency can decrease at very high speed due to the removal of PM. In addition to hybrid-PM arrangement, PMBLDCs can be mounted on the surface magnet that requires fewer magnets or can be mounted on a buried magnet that can give more air gap flux density. BLDCs with a maximum power of 60 kW are useful in small cars. Figure 20.13 represents the torque–speed characteristics of PM BLDC motor. The torque is constant at the beginning but starts decreasing exponentially after the base speed.

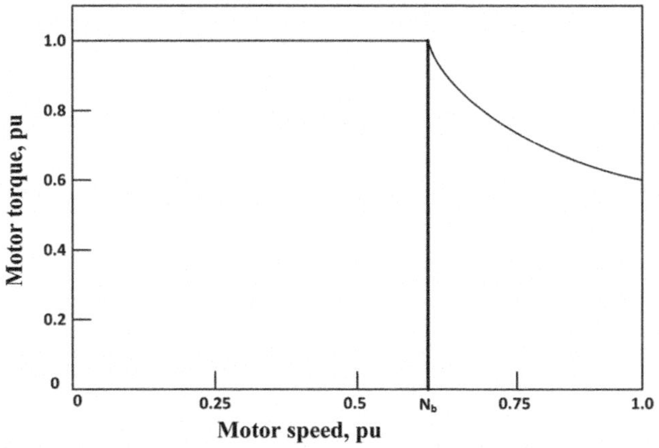

Fig. 20.13 Torque–speed characteristics of a permanent magnet brushless DC motor

20.3.3 Permanent Magnet Synchronous Motor (PM SM)

PMSM can be operated at various speed ranges without the aid of a gear system and are one of the most advanced. These motors are more efficient and portable with the feature. PMSM even at extremely low speeds offers high torque, and thus this arrangement is very suitable for in-wheel purposes. PMSMs with an external rotor can also be built without the need for rotor covers (EGG Technical Services 2002). However in in-wheel operations where the system at high speeds faces a large iron loss, the only notable disadvantage is also present. Neodymium magnet (NdFeB) permanent magnets for high energy density PMSMs have been used. The airflow connections in the airflow are sinusoidal and can be controlled by the vector control and the sinusoidal voltage. PMSM is probably the most useful motor in the battery electric vehicle and this motor technology is being utilized by more than 26 models.

20.3.4 Induction Motor (IM)

Induction motors such as the General Motor's EV1 and current models of the Tesla have been used in early EVs. Among various motor drive systems without commutators, this system is one of the most sophisticated systems. Vector control has been beneficial in enabling IM drives to fulfil EV requirements of the system. Controls for field orientation could even make an IM to act as a separately excited engine while also decoupling its control of field and torque (Kim et al. 2008). The speed range over the base speed can be increased by weakening the flux while maintaining the force constant and a range of three to five times the rated speed can be achieved by field orientation control with a correctly designed IM. The use of four polar AC motors with copper rotors in current EVs is seen in the three-phase phase. In Fig. 20.14, IM characteristics show that maximum torque has been maintained until the base speed, and then reduces significantly.

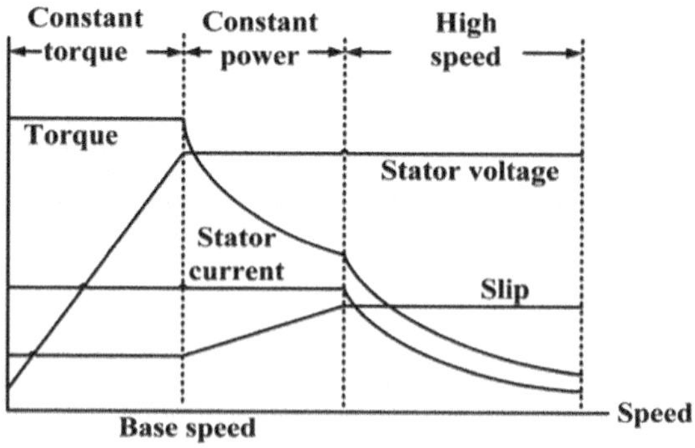

Fig. 20.14 Induction motor drive characteristics

Table 20.3 Power comparisons of motors of same size

Motor type	Power (kW)		Ease speed	Maximum speed
	HEV	BEV		
BLDC	75	110	4000	9000
SRM	42	77	2000	12,000
IM	57	93	3000	12,000

Table 20.4 Typical torque density values of some motors

Motor type	Torque/volume (N/m^3)	Torque/Cu mass (Nrn/kg Cu)
PM motor	28,860	28.7–48
IM	4170	6.6
SRM	6780	6.1

20.3.5 Switched Reluctance Motor (SRM)

Due to its sticking pole position in the stator and rotor, SRMs, also known as doubly sticking motors, are synchronous motors with unipolar inverter generated current. SRM has high speed, low cost, lower risk, high power density, and inherently long constant power supply for motor drives (Table 20.3).

Although these machines have simpler construction, their control and design is not simple because of the slots, poles, and high pole saturation. These machines have not advanced as much as induction machines or PM because of these disadvantages. However, the interest in SRMs is increasing due to the high cost of rare earth metals required in PM machines. Torque and noise reduction are the major concerns in SRM research. Total losses can be reduced by finite element analysis; sliding mode control can also be used to decrease the chatter controls and non-linear engine management (Table 20.4).

20.3.6 BLDC Motor as Most Preferable Option

PMBLDC motor has been the classic way for EV propulsion systems to advantages like no brushes, the ability to manufacture a higher value torque than the other motors at the same voltage and current magnitudes, high efficiency, and high power density. BLDC motors are DC motors where commutation is done electronically, not by brushes. Therefore, a BLDC motor needs less maintenance, has lower noise susceptibility and lesser power dissipation in the air gap compared to a brushed DC motor due to the absence of the brushes. The magnet density is higher with alloy magnets like ferrite and boron (NdFeB), neodymium (Nd), and samarium cobalt (SmCo). They produce more torque for the same volume than ferrite magnets, which

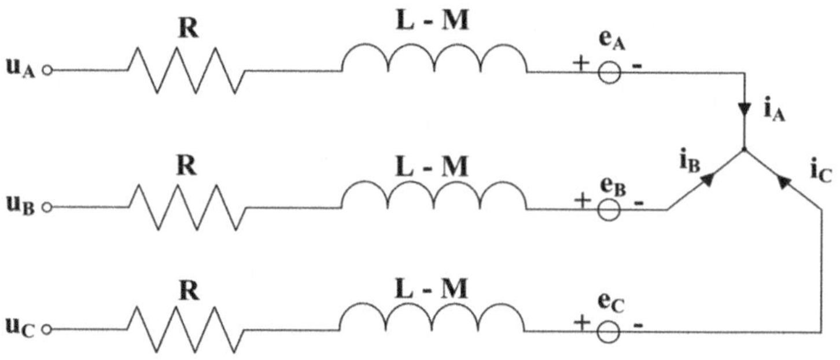

Fig. 20.15 BLDC simplified model

increases the BLDC engine power-to-size ratio. Consider a three-phase, four pole, star connected to the trapezoidal EMF permanent magnet synchronous engine. The trapezoidal back EMF means that the mutual inductance between the stator and the rotor is non-sinusoidal and that a variable "abc" model is more applicable than a d-Q axis model. Figure 20.15 shows the BLDC simplified model. The mathematical equations and the overall BLDC motor model are therefore simplified according to the following presumptions:

- Saturation of the magnetic circuit has been neglected.
- Stator resistance, self-induction as well as mutual inductances have been equal and constant in all three phases.
- Losses of hysteresis as well as eddy current have been removed.
- Semiconductor inverter switches have been ideal.

The simple analytical as well as electrical formulae of the BLDC motor could be written in the form:

$$V_a = Ri_a + (L - M)\frac{di_a}{dt} + E_a \tag{20.1}$$

$$Vb = Ri_b + (L - M)\frac{di_b}{dt} + E_b \tag{20.2}$$

$$V_c = Ri_c + (L - M)\frac{di_c}{dt} + E_c \tag{20.3}$$

$$T_e = T_{ea} + T_{eb} + T_{ec} \tag{20.4}$$

$$T_e - T_l = J\frac{d^2\theta_m}{dt^2} + \frac{d\theta_m}{dt} \tag{20.5}$$

$$\theta_e = \frac{P}{2}\theta_m \tag{20.6}$$

$$\omega_m = \frac{d\theta_m}{dt} \tag{20.7}$$

The "abc" phasor equations for emf and torque are given below:

$$E_a = K_e \omega_m F(\theta_m) \tag{20.8}$$

$$E_b = K_e \omega_m F(\theta_m - \frac{2\pi}{3}) \tag{20.9}$$

$$E_c = K_e \omega_m F(\theta_m + \frac{2\pi}{3}) \tag{20.10}$$

$$T_{ea} = K_{tia} F(\theta_m) \tag{20.11}$$

$$T_{eb} = K_{tib} F(\theta_m - \frac{2\pi}{3}) \tag{20.12}$$

$$T_{ec} = K_{tic} F(\theta_m + \frac{2\pi}{3}) \tag{20.13}$$

Where L is the phase self-inductances, R is the stator resistance, and M is the phase mutual inductance, i is the phase current, E is the back emf, Te is the electrical torque developed, Tl is the load torque, J is the moment of inertia, ωm is the speed of the motor, Ke and Kt are back emf constants, θe and θm are the electrical and mechanical angle of the motor, and F(θ) is the rotor position function. Back-EMF signaling of the BLDC motor is generated by electrical rotor degree at each phase and applied to the input voltages as negative feedback. This approach simplifies and makes the BLDC motor model more convenient for various control techniques.

The brushless DC motor looks like a DC motor from modeling perspective; however, it is actually a synchronous motor with a linear relation between torque and current, rpm, and voltage. It is controlled electronic switching system rather than a mechanical switching system that is usually of brushed motors. This addresses the problem of transferring current to a moving frame. To do that, a smart electronic controller is substituted for the brush system/switch assembly that carries out the same power distribution as a brushed DC engine. BLDC engines have several benefits surrounded by brushed DC motors as well as induction motors like good torque and speed, high efficiency, high reliability and dynamic response, long service life (no brush erosion), rustle operation, large speed ranges, and electromagnetic interference reduction (EMI).

The torque-to-motor ratio is higher and is useful for applications where weight and space are essential particularly in aircraft industry and electrical applications. Thanks to its high efficiency, power density, no maintenance characteristic, and quiet operation. PM motors are widely utilized in industrial automation systems, computers, aircraft industries, military combat vehicles, and automotive industries.

20.4 Study System Operation and Performance

The comprehensive system is made up of an ultra-capacitor as well as a generator in such a hybrid series arrangement to supply power to a BLDC propulsion machine via converters to the DC bus, which is depicted in Fig. 20.16.

The BLDC motor progress is measured for data collection to investigate rotor speed and error, supply DC voltage, as well as current injected from two power sources. The objectives are to illustrate the following:

- Due to limitations on implanted currents, the ultra-capacitor has only complexity providing all of load power.
- The PMSM generator could even supply this same structure with sufficient power, but as the generator speed has been ratcheted up, the voltage sags originally.
- The ultra-capacitor would then endorse the DC bus voltage at low velocity as well as torque.
- The energy from of the PMSM generator has been required at greater velocity for better application performance.
- The DC bus voltage would be near to the rated voltage of the system whenever the BLDC motor is in generation mode and hence regeneration occurs.

The first test series has been intended to verify the process of each network element under a defined set of control speeds and mechanical loads implemented to the BLDC motor.

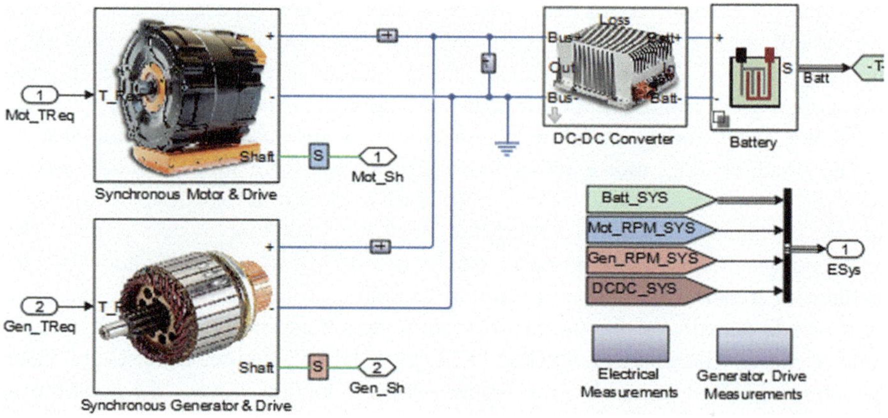

Fig. 20.16 Integration of motor and generator framework based in MATLAB

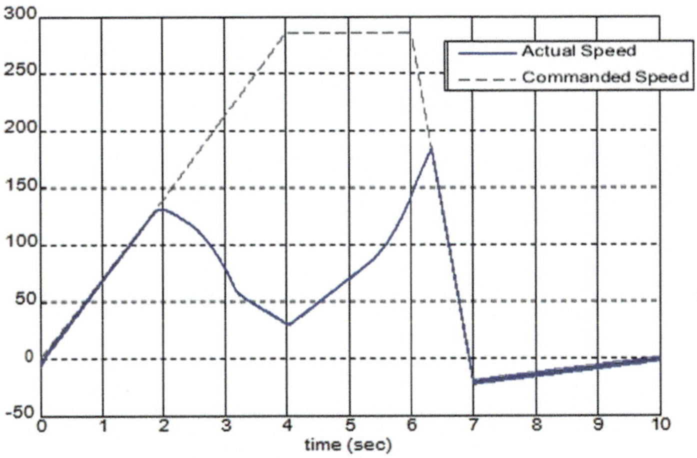

Fig. 20.17 Ultra-capacitor only rotor speed

20.4.1 Ultra-Capacitor Operation

Since for power supply only ultra-capacitor is used, simulation results demonstrate that the ultra-capacitor (UC) behaves adequately to supply power from the system which is depicted in Fig. 20.17 at lower speed and torque requirements from the BLDC drive. Whenever the mechanical torque is applied to a motor at higher levels and the BLDC motor requires more speed, the error begins to peak at maximum load torque and maximum control speed. The whole error as well as correlates to a lower unlike required voltage value of BLDC current ultra-capacitor details mostly on DC supply bus. To evaluate average currents both to and from sources, low-pass filtering of the current contributions of the power sources has been essential. Approximately, 875 W is just the peak power supplied by the ultra-capacitor.

20.4.2 Generator Operation

The presumption would be that the generator has been functioned by a speed-controlled, mechanically coupled IC engine. The PMSM generator's speed would be proportional to the input power to the system. Because of the speed control, it is believed that the speed of a DC generator must be restricted to make this process more accurate (i.e., speed of the generator rotor cannot accelerate to the rated speed instantaneously). The full angular velocity of rotor is reduced to 314 rad/s rated mechanical speed, and the speed change of the rotor is supposed to be 2.5 s from 0 rad/s to rated speed (3000 RPM) and 1 s from 0 rad/s to rated speed. Depending on all this, the PI-controlled mechanical input speed controller has been done to receive the DC supply bus voltage a negligible overshoot.

Due to the low DC bus voltage, the current speed sags within the first second. The generator would be able to supply sufficient power to maintain the voltage level required by the BLDC machine after the generator starts to boost to the rated speed. The generator's peak power supply is (14 Nm) * (290 rad/s) (i.e., 4.06 kW) the velocity error begins at a time when the mechanical load torques as well as the controlled velocities are considerably lower. At such a time when the DC supply bus voltage has been lower than that required by the BLDC driver, this same rotor speed error peaks. The error is reduced after generator takes adequate time to boost up to speed. Figure 20.18 demonstrates the mechanical load torque.

In Fig. 20.19, generator currents have been highlighted. The generator supply current indicates the current that is provided to the DC bus from the generator. The voltage haven't ever exceeds 100 A. Figure 20.20 treats the low-pass filtered supply

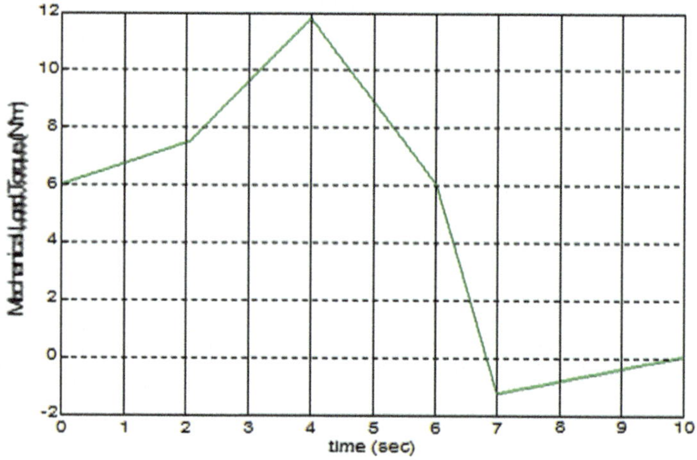

Fig. 20.18 Generator only—mechanical load torque

Fig. 20.19 Generator only—current contribution

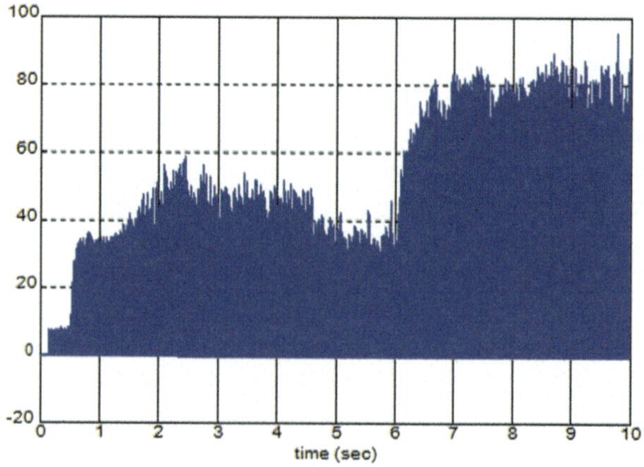

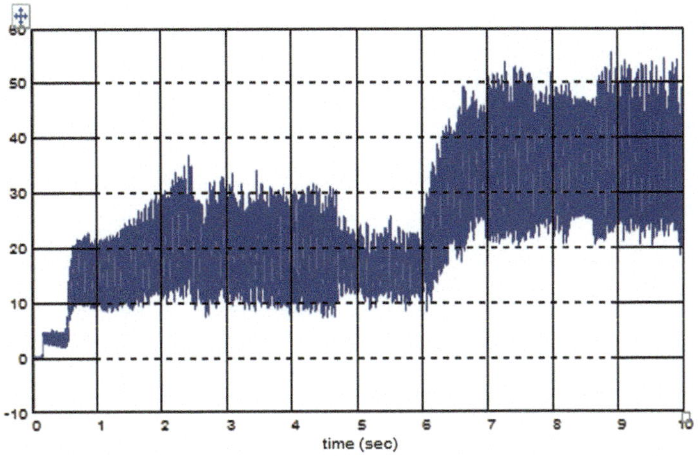

Fig. 20.20 Generator only—filtered current contribution

current. The filtered parts requiring demonstrate that throughout any voltage sag, the generator increases current infusion into the DC bus. At first, the voltage of the DC bus drops because the generator hasn't really reached speed. The initial voltage was indeed due to the DC bus capacitor's presumed initial charge to 160 V. The DC bus voltage remains constant at magnitude which is under 400 V after generator speeds up. The drop in voltage starting at t = 6 s. is because of attaining the intended DC bus voltage. The generator then slows down in velocity, after which generator should speed up and increase the DC voltage of a bus so because BLDC motor even now requires power. The above contributes for the larger current toward the end of simulation from the generator.

20.4.3 Software and Control Logic

Figure 20.21 describes the commutation scheme included. By analyzing the commutation scheme, Eq. (20.14) of the system has been extrapolated and with the help of the control logic combinational system has been modeled. Hall position sensor signals and a master PWM are the inputs to this combinational system and the outputs are the PWM signals.

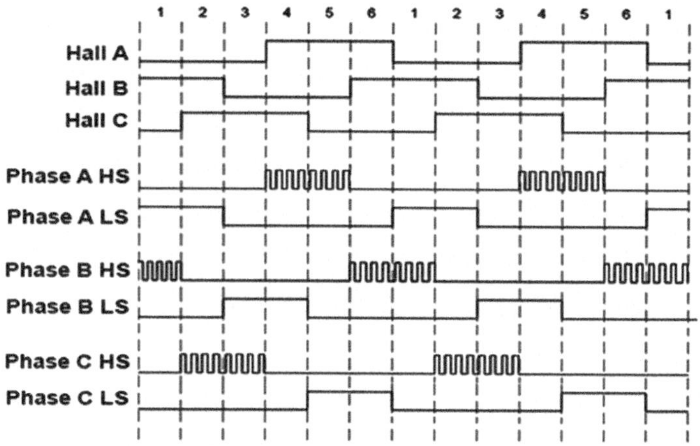

Fig. 20.21 PWM commutation scheme

$$
\begin{aligned}
A_{HS} &= PWM.(HallA.\overline{HallB}) \\
B_{HS} &= PWM.(HallB.\overline{HallC}) \\
C_{HS} &= PWM.(HallC.\overline{HallA}) \\
A_{LS} &= HallB.\overline{HallC} \\
B_{LS} &= HallC.\overline{HallB} \\
C_{LS} &= HallA.\overline{HallC}
\end{aligned}
\tag{20.14}
$$

Following causes are taken into consideration for the choice of the commutation strategy:

- Integration simplification.
- Presented motor speed performance relating to a mean speed of 20–25 km/h as well as an average speed of 20–25 km/h.
- This same amount of CPU load calculated in conjunction with both the secondary functions by a motor control feature.

The software flow model is illustrated in Fig. 20.22.

Figure 20.22 has been organized into two threads, one for each primary purpose accomplish. Two separate timers govern the threads; with the motor control thread holds the greatest priority.

Using the MATLAB/Simulink environment simulation model, the design is simulated. The simulation took into consideration the commutation table shown in Fig. 20.21 and that is characterized by Eq. (20.14). In Fig. 20.23, the Simulink approach is formulated. The simulated results are shown in Fig. 20.24 in the following sequence (Fig. 20.25).

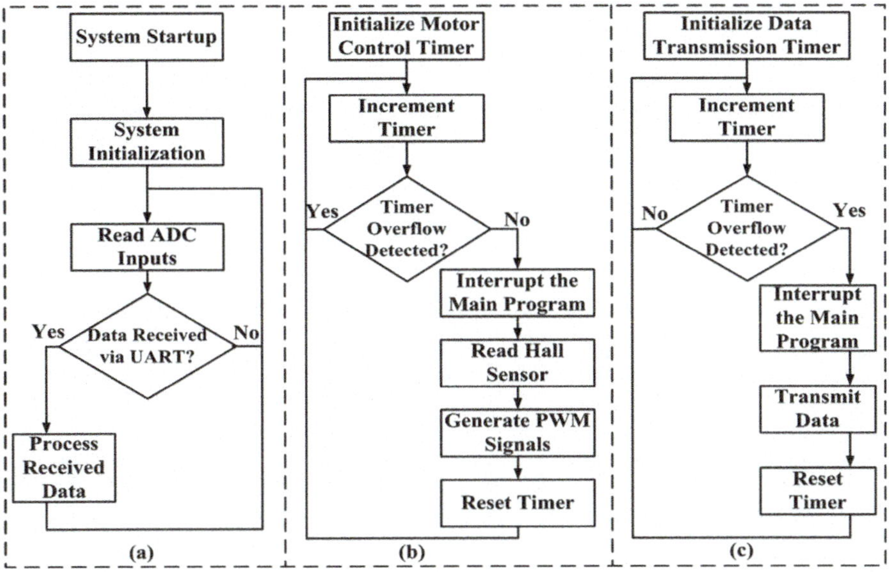

Fig. 20.22 Software flow diagram for the electronic control unit: **a** main program; **b** motor control thread; **c** data transmission thread

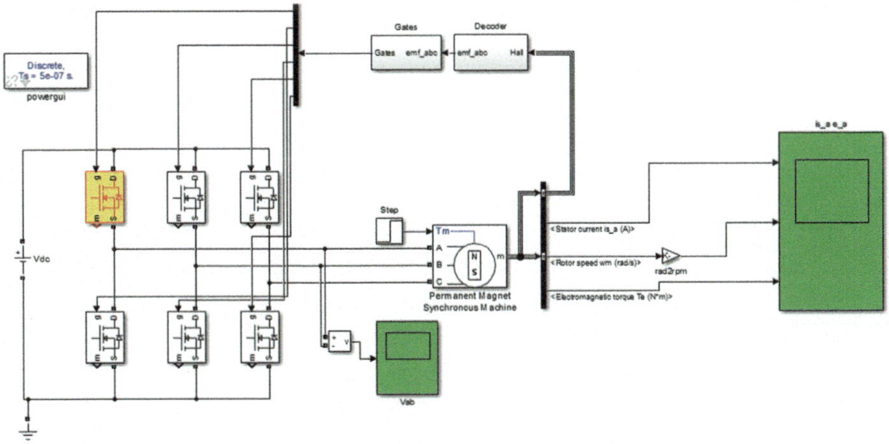

Fig. 20.23 Simulink model of proposed system

- the waveform of a stator current,
- electromagnetic torque ripple, and
- rotor speed.

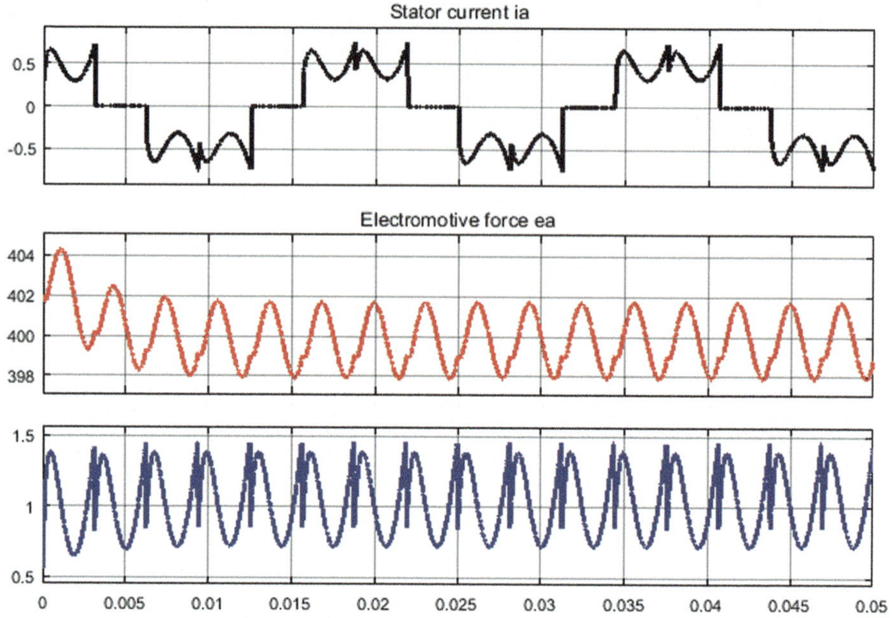

Fig. 20.24 Simulation results of proposed system

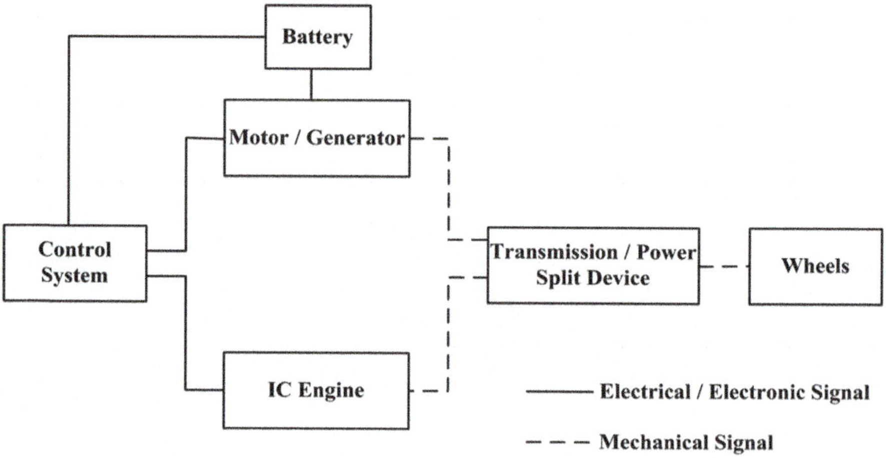

Fig. 20.25 Proposed system for HEV

Table 20.5 Engine model parameter

Parameter	Value
Power	5 7 kW (3)5 000 RPM
Mill speed	1000 rpm
Max speed	4500 rpm
Torque	115 Nm (H > 4200 RPM

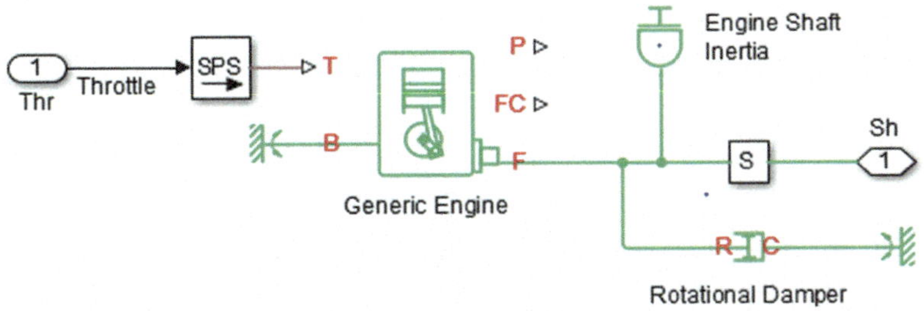

Fig. 20.26 Generic Simscape engine

20.4.4 Simulink Subsystem Used in HEV

20.4.4.1 Engine Model

The following requirements apply to the functionality of this module (Table 20.5 and Fig. 20.26).

The engine model is now a mechanical system used during Simulink that offers torque as output as well as uses throttle as input.

20.4.4.2 Planetary System

It includes a small planetary system. The carrier has been linked to an engine, the system is linked to the ring, and the sun has been attached to the generator (Fig. 20.27).

20.4.4.3 Battery Model

This battery does have an installed voltage of 8.1 Ah, a voltage of 200 V, and a total power of 1.62 kWh. The chosen GM HEV system does have a kerb weight of 1325 kg, a total range of 870 km, and a power range of 18 km (Figs. 20.28, 20.29 and 20.30).

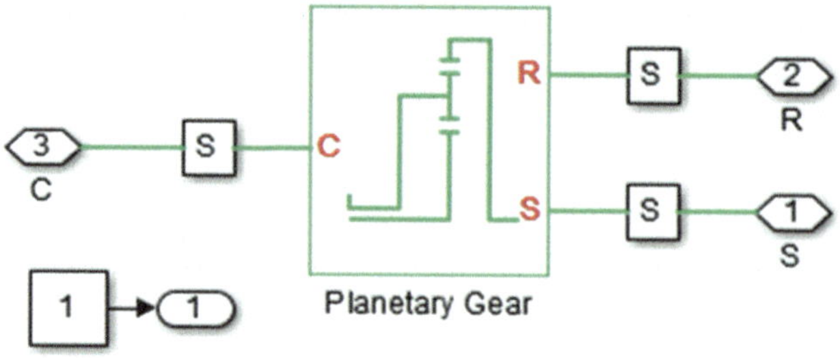

Fig. 20.27 Planetary gear system

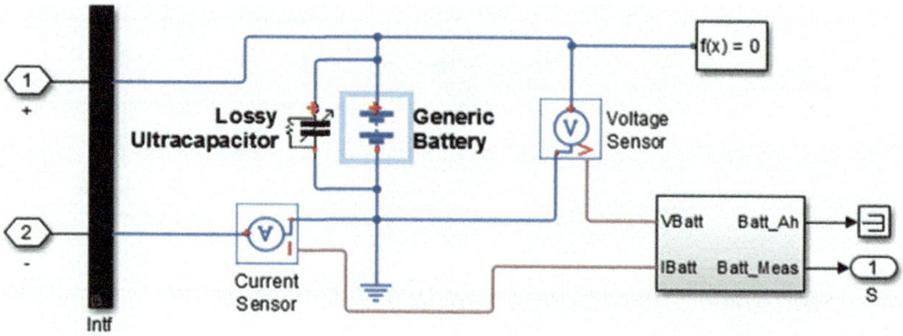

Fig. 20.28 Battery system

Engine System Requirements

The following requirements apply to the functionality of this module (Fig. 20.31 and Table 20.6).

Fuel Consumption

The following requirements apply to the fuel consumption (Table 20.7).

Speed Controller Module Requirements

For the speed controller module, the preceding rule applies:

i. A minimum of integral and proportional control will be enacted by the controller module.
ii. The system must be within 5% of the final value in 0.1 s after a change in angle (Settling Time).
iii. The system must achieve 10 percent of the final value within 0.7 s after a 10% change (Fig. 20.32).

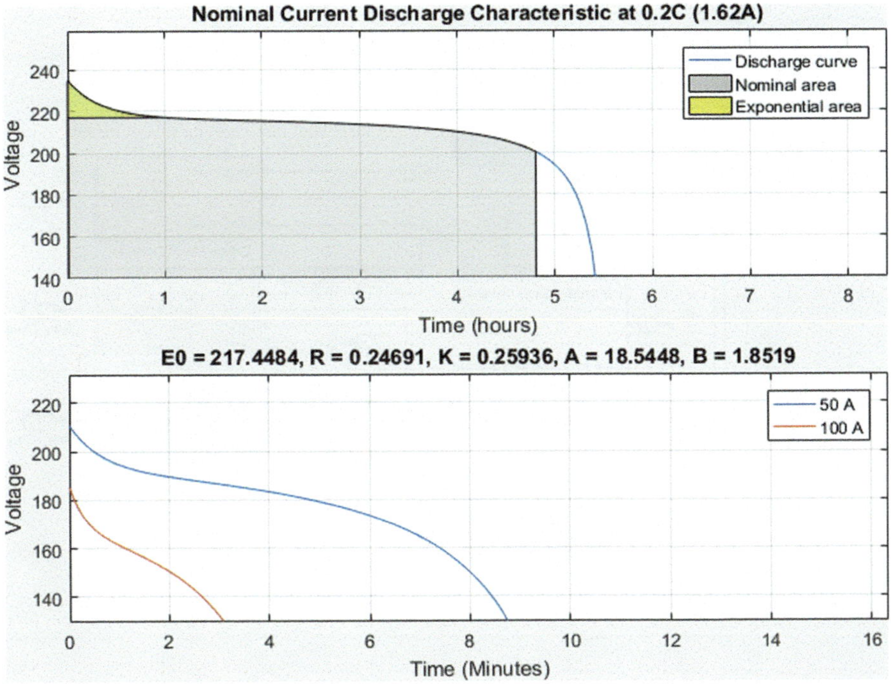

Fig. 20.29 Battery characteristics

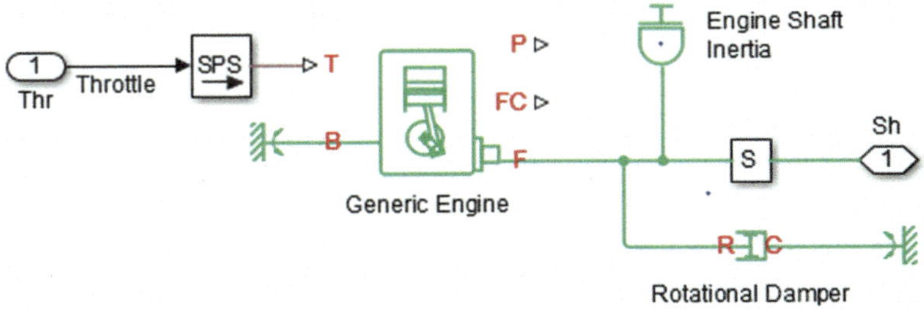

Fig. 20.30 HEV system level layout

20.5 Simulation Results

20.5.1 Drive Cycle 1

See Figs. 20.33, 20.34, 20.35 and 20.36.

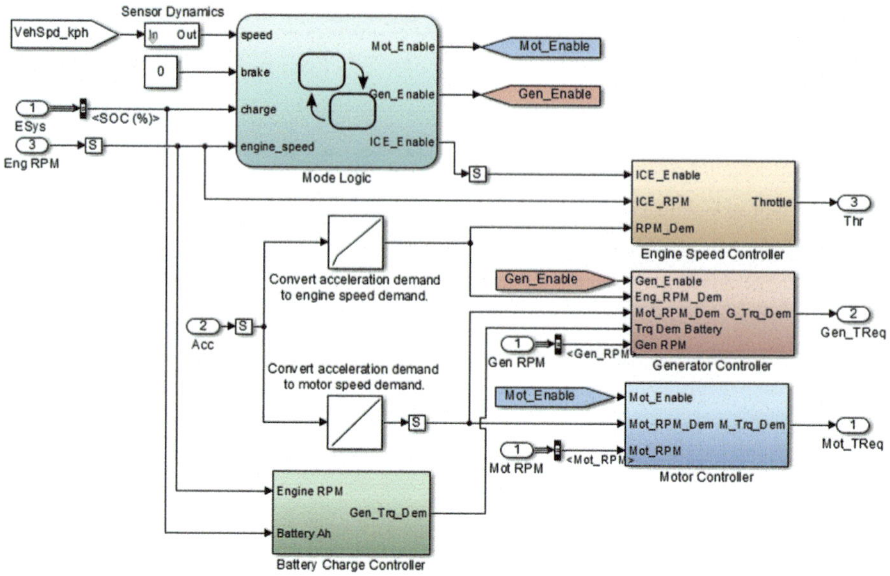

Fig. 20.31 Generic Simscape engine

Table 20.6 Engine system parameters

Parameter	Value
Power	57 kW (3)5000 RPM
Mill speed	1000 rpm
Max speed	4500 rpm
Torque	115 Nm (2) 4200 RPM

Table 20.7 Requirements for fuel consumption

Regular gas	
City	51 MPG
Highway	49 MPG
Combined	50 MPG
Electric + gas	
Combined	95 PG-e

20.5.2 Drive Cycle 2

See Figs. 20.37, 20.38, 20.39 and 20.40 and Table 20.8.

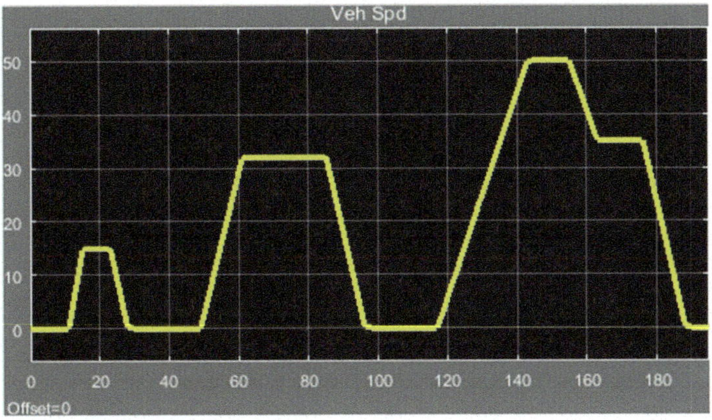

Fig. 20.32 HEV control module layout

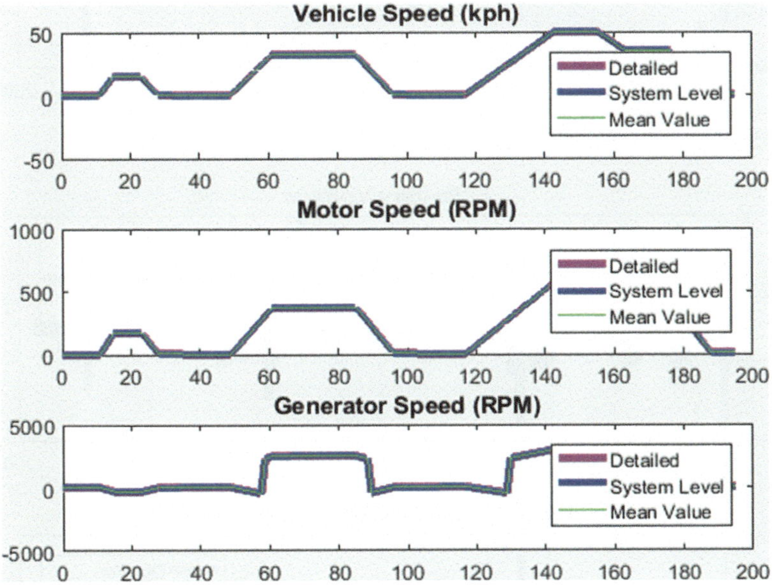

Fig. 20.33 Vehicle speed

20.6 Conclusion

In the E.V.'s, a Direct Current (DC) drive is being used extensively because it provides simple speed control and optimal specifications for torque speed. The perfect torque matches the requisites of terrain and traction in the E.V. Ones brushes and commutators end up making others to less reliable. So, for such a maintenance-free feature, this is not appropriate. A.C. motor drives with IM or PMSM are far more widely known

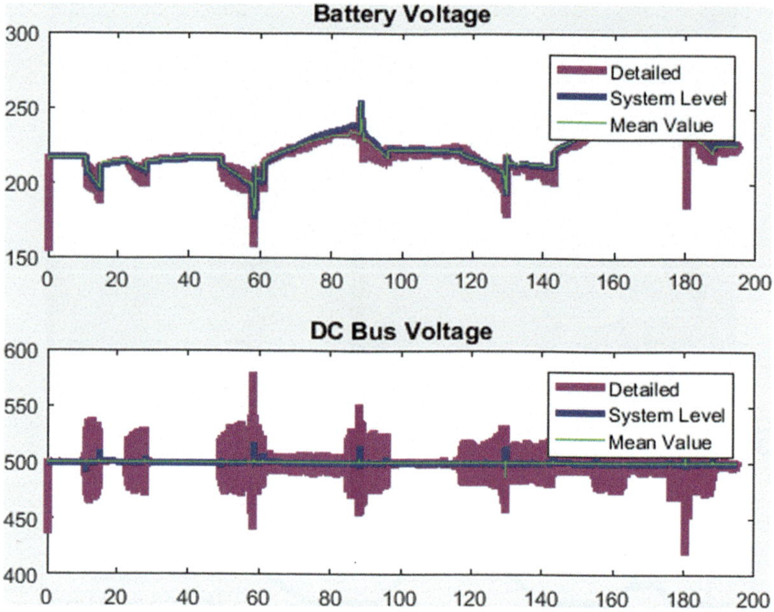

Fig. 20.34 Speeds from Urban Cycle 1

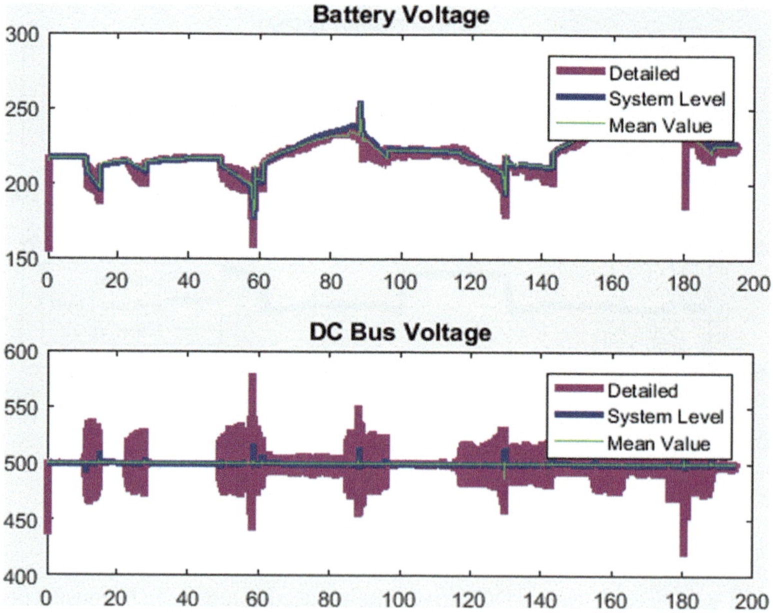

Fig. 20.35 Voltage from Urban Cycle 1

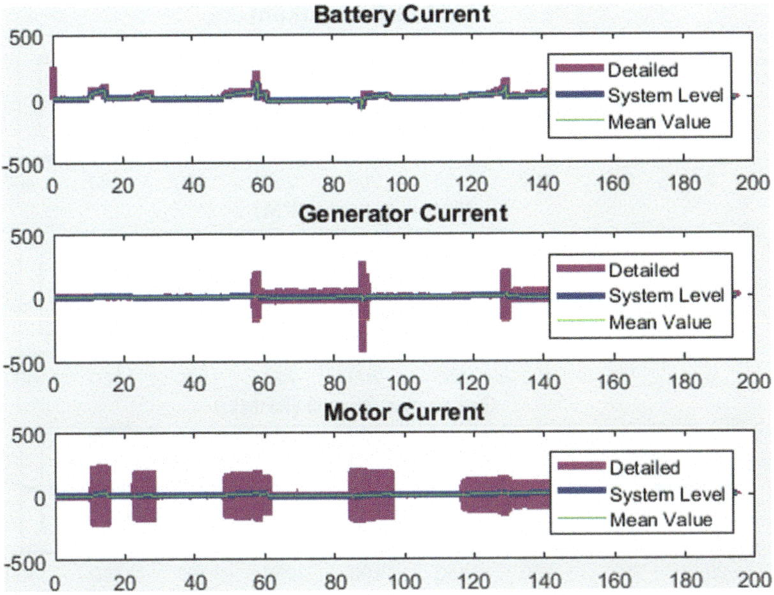

Fig. 20.36 Currents from Urban Cycle 1

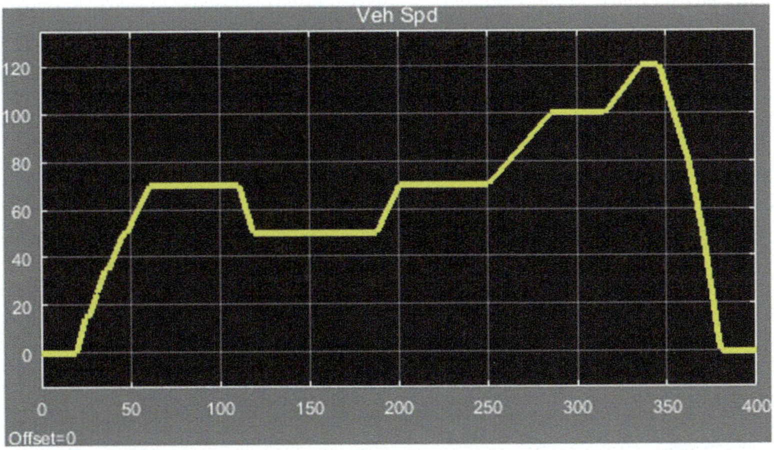

Fig. 20.37 Vehicle speed

to advance in power electronics than what a D.C. drive with robustness, less maintenance, higher efficiency, and high power density benefits. For a specified power value, the PMSM provides a general reduction in weight and volume. The efficiency is also much greater because with no loss of rotor copper. The accuracy seems to be fairly high.

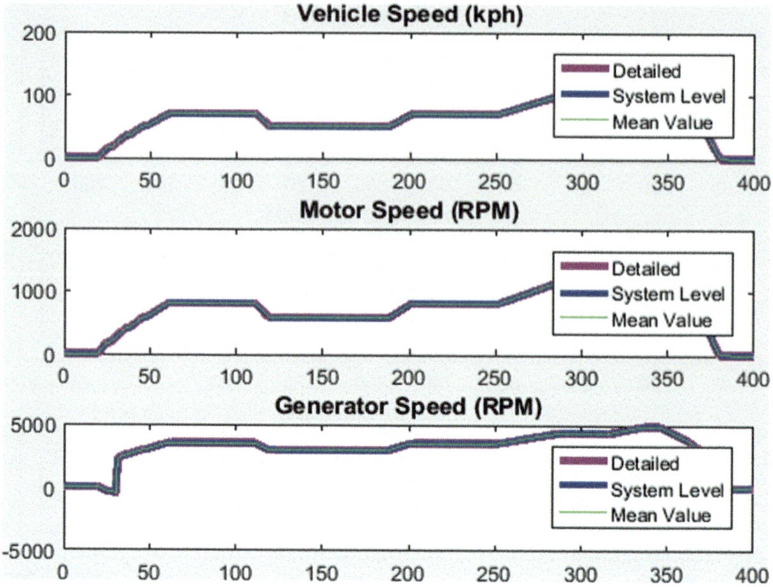

Fig. 20.38 Speeds from Urban Cycle 2

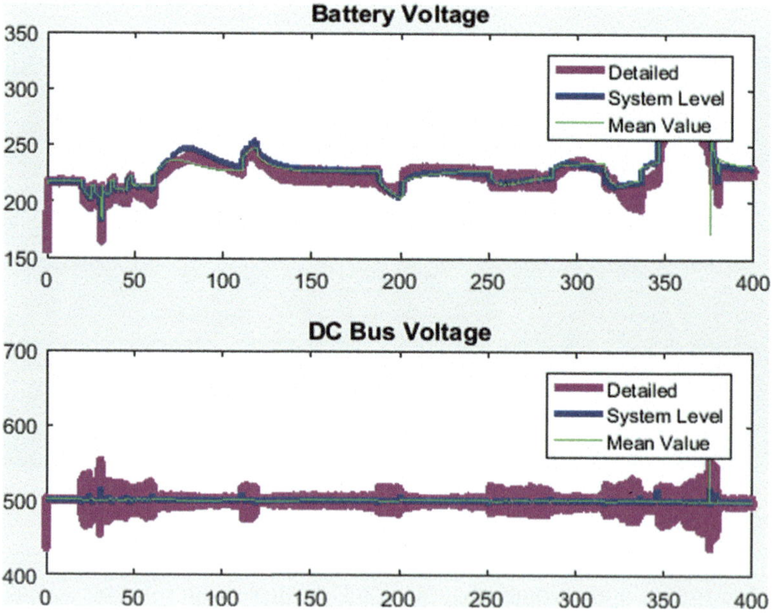

Fig. 20.39 Voltages from Urban Cycle 2

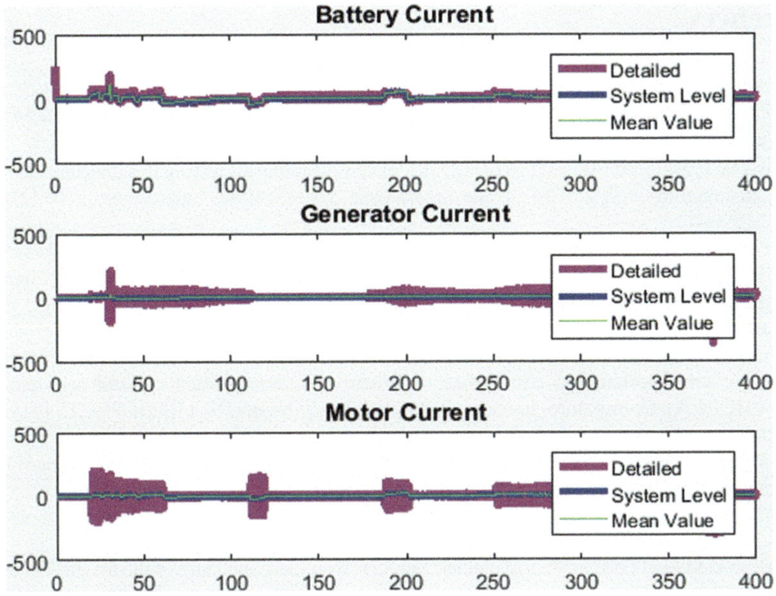

Fig. 20.40 Currents from Urban Cycle

Table 20.8 Fuel economy comparison

Sr. no.	Level used	Fuel economy (km/L)
1	System level	49.52
2	Mean level	48.66
3	Detailed level	47.53

But the winner in this medium-sized electric vehicle recruitment process stands out as BLDC motor drive that is fed by a rectangular A.C. supply. P.M. brushless D.C. motor drive is a perfect option to be utilized in the EV propulsion system, with benefits such as ability to generate a high torque than the others at the same voltage and current magnitudes, removal of the brushes, high efficiency, and power density.

Since BLDC motor has been used widely in automotive, it was tested for simulation profiles and then is integrated in the HEV system model in Simulink. The fuel economies also showed that the combination of selected motor and engine is efficient and has very high potential in market.

References

Camacho OMF, Mihet-Popa L (2016) Fast charging and smart charging tests for electric vehicles batteries using renewable energy. Oil Gas Sci Technol 71:13–25. https://doi.org/10.2516/ogst/2014001

Camacho OMF, Nørgård PB et al (2014) Electrical vehicle batteries testing in a distribution network using sustainable energy. IEEE Trans Smart Grid 5:1033–1042. https://doi.org/10.1109/TSG.2014.2299064

Chan CC (2002) The state of the art of electric and hybrid vehicles. Proc IEEE 90:247–275

Chau KT, Chan CC et al (2008) Overview of permanent-magnet brushless drives for electric and hybrid electric vehicles. IEEE Trans Ind Electron 55(6):2246–2257. https://doi.org/10.1109/TIE.2008.918403

Chowdhary PK, Thakre MP (2020) MMC based SRM drives for hybrid EV with decentralized BESS. In: 4th International conference on electronics, communication and aerospace technology (ICECA), Coimbatore, India, pp 319–325. https://doi.org/10.1109/ICPECTS49113.2020.9337029

EG&G Technical Services, Inc (2002) The fuel cell handbook, 6th edn. U.S. Department of Energy: Morgantown, WV, USA

Gao Y, Ehsani M (2010) Design and control methodology of plug-in hybrid electric vehicles. IEEE Trans Ind Electron 57:633–640. https://doi.org/10.1109/TIE.2009.2027918

Gao Y, Ehsani M et al (2005) Hybrid electric vehicle: overview and state of the art. In: Proceedings of the IEEE international symposium on industrial electronics, Dubrovnik, Croatia, pp 307–316, 20–23 June 2005. https://doi.org/10.1109/ISIE.2005.1528929

Grunditz EA, Thiringer T (2016) Performance analysis of current BEVs based on a comprehensive review of specifications. IEEE Trans Transp Electr 2:270–289. https://doi.org/10.1109/TTE.2016.2571783

Guntuk VS, Bakle UV et al (2019) A novel 4-level converter for switched reluctance motor drive in plug-in HEVs. In: 2019 International conference on intelligent computing and control systems (ICCS), Madurai, India, pp 626–631. https://doi.org/10.1109/ICCS45141.2019.9065527

Jung TU, Lee SH, Park SJ et al (2007) BLDC motor drive system of air-condition of hybrid electric vehicle. Int Conf Electr Mach Syst (ICEMS) 2007:636–641. https://doi.org/10.1109/ICEMS12746.2007.4412040

Kim H, Kum D (2016) Comprehensive design methodology of input- and output-split hybrid electric vehicles. Search of optimal configuration. IEEE/ASME Trans Mechatron 21:2912–2923. https://doi.org/10.1109/TMECH.2016.2579646

Kim D, Hwang S, Kim H (2008) Vehicle stability enhancement of four-wheel-drive hybrid electric vehicle using rear motor control. IEEE Trans Veh Technol 57:727–773. https://doi.org/10.1109/TVT.2007.907016

Kramer B, Chakraborty S, Kroposki B (2008) A review of plug-in vehicles and vehicle-to-grid capability. In: Proceedings of the 34th IEEE Industrial electronics annual conference, Orlando, FL, USA, pp 2278–2283, 10–13 November 2008. https://doi.org/10.1109/IECON.2008.4758312

Kumaresan J, Govindaraju C (2020) Development of a power management algorithm for PV/battery powered plug-in dual drive hybrid electric vehicle (DDHEV). Electric Power Comp Syst 48:1–2, 70–85. https://doi.org/10.1080/15325008.2020.1736212

Lee BK, Ehsani M (2001) Advanced BLDC motor drive for low cost and high performance propulsion system in electric and hybrid vehicles. In: IEMDC 2001. IEEE International electric machines and drives conference (Cat. No. 01EX485), pp 246–251. https://doi.org/10.1109/IEMDC.2001.939307

Marchesoni M, Vacca C (2007) New DC–DC converter for energy storage system interfacing in fuel cell hybrid electric vehicles. IEEE Trans Power Electron 22:301–308. https://doi.org/10.1109/TPEL.2006.886650

Mathur A, Parashar T et al (2014) A hybrid electric vehicle with incorporation of VaReB technology. In: 2014 IEEE 6th India International conference on power electronics (IICPE), 2014, pp 1–6. https://doi.org/10.1109/IICPE.2014.7115847

Miller JM (2006) Hybrid electric vehicle propulsion system architectures of the e-CVT type. IEEE Trans Power Electron 21:756–767. https://doi.org/10.1109/TPEL.2006.872372

Miller JF, Webster et al (1997) Testing and evaluation of batteries for a fuel cell powered hybrid bus. In: Proceedings of the energy conversion engineering conference, Honolulu, HI, USA, vol 2, pp 894–898, 27 July–1 August 1997. https://doi.org/10.1109/IECEC.1997.661887

Nakadachi S, Mochizuki S et al (2013) Bidirectional contactless power transfer system expandable from unidirectional system. In: Proceedings of the 2013 IEEE Energy conversion congress and exposition, Denver, CO, USA, pp 3651–3657, 15–19 September 2013. https://doi.org/10.1109/ECCE.2013.6647182

Naseri F, Farjah E et al (2017) An efficient regenerative braking system based on battery/supercapacitor for electric, hybrid, and plug-in hybrid electric vehicles with BLDC motor. IEEE Trans Veh Technol 66(5):3724–3738. https://doi.org/10.1109/TVT.2016.2611655

Schaltz E, Khaligh A, Rasmussen PO (2008) Investigation of battery/ultracapacitor energy storage rating for a fuel cell hybrid electric vehicle. IEEE Vehicle Power Propul Conf 2008:1–6. https://doi.org/10.1109/VPPC.2008.4677596

Shah V, Chaudhari R, Kundu P (2010) Performance analysis of hybrid energy storage system using hybrid control algorithm with BLDC motor driving a vehicle. In: 2010 Joint international conference on power electronics, drives and energy systems & 2010 Power India, 2010, pp 1–5. https://doi.org/10.1109/PEDES.2010.5712540

Singh NK, Koley C et al (2019) An implementation perspective of hybrid electric vehicle. J Inf Opt Sci 40(8):1693–1708. https://doi.org/10.1080/02522667.2019.1704520

Thakre MP, Mahadik YV et al (2020) Fast charging systems for the rapid growth of advanced electric vehicles (EVs). In: International conference on power, energy, control and transmission systems (ICPECTS), Chennai, India, pp 1–6. https://doi.org/10.1109/ICPECTS49113.2020.9336979

Thakre M, Mane J, Hadke V (2020) Performance analysis of SRM based on asymmetrical bridge converter for plug-in hybrid electric vehicle. In: International conference on power, energy, control and transmission systems (ICPECTS), Chennai, India, pp 1–6. https://doi.org/10.1109/ICPECTS49113.2020.9337059

Un-Noor F et al (2017) A comprehensive study of key electric vehicle (EV) components, technologies, challenges, impacts, and future direction of development. Energies 10:1217. https://doi.org/10.3390/en10081217

Yilmaz M, Krein PT (2013) Review of battery charger topologies, charging power levels, and infrastructure for plug-in electric and hybrid vehicles. IEEE Trans Power Electr 28:2151–2169. https://doi.org/10.1109/TPEL.2012.2212917

Yong JY, Ramachandaramurthy et. Al (2015) A review on the state-of-the-art technologies of electric vehicle, its impacts and prospects. Renew Sustain Energy Rev 49:365–385. https://doi.org/10.1016/j.rser.2015.04.130

Chapter 21
Charging Techniques of Lead–Acid Battery: State of the Art

Aarti S. Pawar and Mahesh T. Kolte

Abstract Battery charging is a very critical activity for using its electric storage capability and incorrect procedure affects its efficiency and health. The charging process plays a key role in evaluating the life of the battery. Overcharging results in high temperature in the battery, which degrades the chemical composition of the electrolyte. The conventional charging techniques such as constant current, constant voltage, and constant current-constant voltage (CC-CV) charging techniques are used for charging a battery but the problem like gas formation, grid corrosion, and sulfation is faced in extending the life of the battery. The various parameters such as ensuring battery full-service life, temperature rise, and gas evolution during charge, state of charge (SOC), charging efficiency in AH and WH, and charging time are to be considered when designing a battery charger. In this paper, the charging techniques have been analyzed in terms of charging time, charging efficiency, circuit complexity, and propose an effective charging technique. This paper also includes development in lead–acid battery technology and highlights some drawbacks of conventional charging techniques.

Keywords Constant current-constant voltage charging techniques (CC-CV) · Charging efficiency · Lead–acid battery · Grid corrosion · Sulfation

A. S. Pawar (✉)
Research Centre, Pimpri Chinchwad College of Engineering, Savitribai Phule Pune University, Pune 411044, India
e-mail: aarti.pawar@pccoepune.org

M. T. Kolte
Mahesh T.Kolte, Research Centre, Pimpri Chinchwad College of Engineering, Savitribai Phule Pune University, Pune 411044, India
e-mail: mahesh.kolte@pccoepune.org

A. K. Bohre et al. (eds.), *Planning of Hybrid Renewable Energy Systems, Electric Vehicles and Microgrid*, Energy Systems in Electrical Engineering,
https://doi.org/10.1007/978-981-19-0979-5_21

Nomenclature

VRLA	Valve-regulated lead–acid
CC	Constant current
CC-CV	Constant current-constant voltage
PPC	Pure lead punching carbon technology
LEGF	Lead electroplated graphite foil
BE	Bipolar electrodes
PAM	Positive active material
AGM	Absorptive glass mat
CV	Constant voltage
ICC	Interrupted charge control
CV	Cyclic voltammetry
HRPSOC	High-rate partial state of charge
NAM	Negative active material
LAB	Lead–acid battery

21.1 Introduction

In the case of power supply cuts, energy storage systems now play a significant role in providing an alternative source to provide electricity to various industrial and domestic applications. The energy storage device plays a significant part in applications for renewable energy in off-grid as well as load leveling and frequency regulation for on-grid systems. Batteries of lead–acid are extensively used in diverse applications like automotive industries, telecommunications systems, hospitals, emergency lighting, power tools, alarm systems, material handling, railway air-conditioning and coach lighting, and so on. Because of its enormous use, ease, and efficiency of commercial recycling, lead–acid batteries are available in various designs, capacities, and voltage configurations. Because of its wide versatility, good efficiency, and life characteristics, the production of lead–acid batteries has reached a very high level of mass production and thus the cost of the lead–acid battery is quite low. The charging time, overcharging and undercharging, operating temperature, charging process, charging state, electrolyte condition, system capacity, battery design, and application area are factors that affect the battery life in the applications listed above (Bhatt et al. 2005). There are different methods available for charging a battery such as by the use of a photovoltaic system or by converting grid AC to controlled DC for charging. Its efficiency and health will depend on the proper charging procedure.

Sulfation of plates and stratification of electrolyte occurs due to long-term undercharging which degrades the battery's performance and reduces its service life. Battery gassing and grid corrosion of plates is a result of frequent or extended overcharging which also reduces the battery life. Three-stage charging, intermittent charging, and interrupted charge control were introduced in 2008, and their output is monitored at a high battery charging level. It increased the battery life cycle (Armstrong et al. 2008).

The improper charging method is responsible for reducing battery life if the process is unable to manage to charge time, charging temperature, and charging current as per the progressive state of charge during charging. Such a process cannot protect against battery overcharge and undercharge. Common charging methods are boost charging with stepped constant current (CC) at starting and finishing rates, constant voltage (CV), and constant current-constant voltage (CC-CV). They may have some drawbacks, like longer charging time in CC-CV mode or somewhat higher temperatures in boost charging, and sometimes result in overcharging or undercharge if the current, voltage, and cut-off settings are not proper.

21.1.1 Contribution of the Work

- The purpose of this chapter is to provide enough information to researchers and academicians to comprehend how critical it is to describe the function of charging techniques in enhancing the performance of lead–acid batteries.
- The advantages and limitations of all charging techniques are highlighted.
- Nowadays, energy storage devices play a critical role in obtaining power for a variety of purposes. This review also looked at recent advancements in lead–acid batteries, such as electrode material, electrolyte placement, the use of current collectors, and lead–acid battery recycling. Researchers from a variety of disciplines will be required to solve different difficulties in energy storage applications, and their success will help to the creation of next-generation environmentally pollution-free and sustainable energy systems.

21.1.2 The Flow of the Work

This article is arranged in six parts, systematically. The introduction contains the first part. The second segment deals with the operating principle of the lead–acid battery during charging and discharging. The third section provides a thorough description of each charge technique. The fourth deals with developments in the lead–acid battery. The fifth and sixth parts deal with arguments and conclusions. The flow of the work is mentioned in Fig. 21.1.

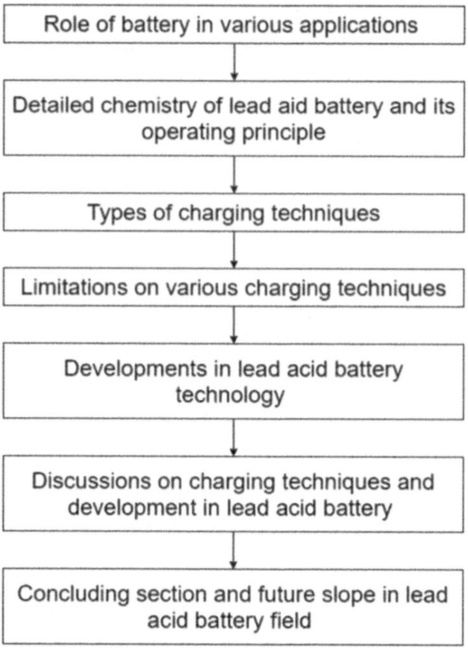

Fig. 21.1 Flow of the work

21.2 Lead–Acid Battery Chemistry and Its Working

A battery is an energy storage device. Here the lead–acid battery's working theory is discussed. It's rare in the world of rechargeable or secondary batteries. The positive plate contains lead dioxide (PbO_2), the negative plate contains sponge lead (Pb), and the electrolyte is dilute sulfuric acid (H_2SO_4). The diluted sulfuric acid is the combination of water and acid in the proportion of 3:1 ratio. It takes part in the electrode reactions. The chemical reactions which generate electricity take place at the two electrodes. Charging and discharging are the states of chemical reactions in the battery. Figures 21.2 and 21.3 depict the charge/discharge of a lead–acid battery, respectively.

The chemical reaction takes place at the electrodes during charging. On charge, the reactions are reversible. When cells reach the necessary charge and the electrodes are reconverted back to PbO_2 and Pb, the electrolyte's specific gravity rises as the sulfur concentration is enhanced. Water loss can occur as water is electrolyzed to produce oxygen and hydrogen; however, the overpotential where this occurs is significant enough that gas leakage can be regulated by changing the charging voltage. At this time, the molecules of sulfuric acid are broken into positive $2H^+$ ions and negative SO_{4-} ions. The hydrogen transfers electrons with the cathode and becomes hydrogen, and this hydrogen interacts with the $PbSO_4$ in the cathode, forming sulfuric acid (H_2SO_4) and lead (Pb). The SO_{4-} ions move to the anode and transfer their electrons. It becomes radical SO_4. This radical SO_4 cannot exist by itself and hence it reacts

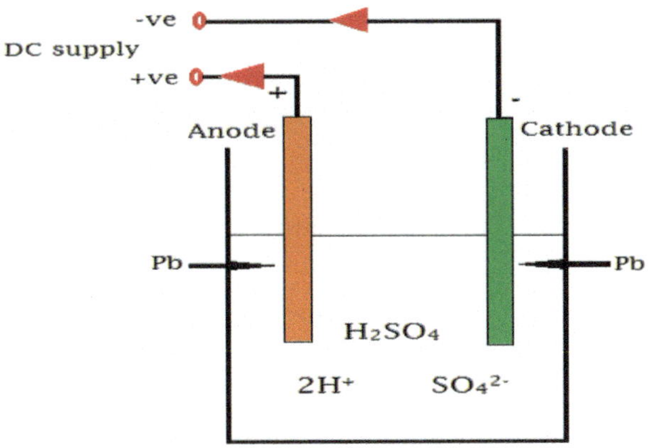

Fig. 21.2 Charging of lead–acid cell

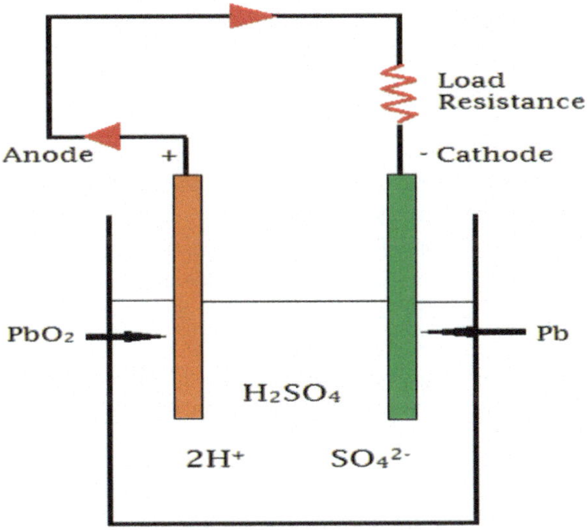

Fig. 21.3 Discharging of a lead–acid cell

with anode $PbSO_4$ and induces PbO_2 and sulfuric acid (H_2SO_4). During battery charging, the following is the chemical reaction:

$$PbSO_4 + 2H_2 + SO_4 \rightarrow PbO_2 + 2H_2SO_4 \tag{21.1}$$

The chemical reactions are again involved during the discharge of a lead–acid battery. When the loads are bound across the electrodes, the sulfuric acid splits again into two parts, such as positive $2H^+$ ions and negative SO_4 ions. With the PbO_2 anode, the hydrogen ions react and form PbO and H_2O water. The PbO begins to react with H_2SO_4 and generates $PbSO_4$ and H_2O. The SO_{4-} ions give electrons to the pure Pb plate and create radical SO_4 which interacts further with the Pb that creates $PbSO_4$. The reverse chemical reaction occurs during the charging process.

Due to the electron exchange along anode and cathode, the electrons pass through the charge and the battery gets discharged, the electron balance between the plates is affected. The gravity of diluted sulfuric acid decreases during this discharge. Even the cell's potential gap decreases at the same time. The following is the chemical reaction when it is discharged:

$$PbO_2 + 2H \rightarrow PbO + H_2O \tag{21.2}$$

$$PbO + H_2SO_4 \rightarrow PbSO_4 + H_2O \tag{21.3}$$

Combining Eqs. (21.2) and (21.3), we get

$$PbO_2 + 2H + H_2SO_4 \rightarrow PbSO_4 + 2H_2O \tag{21.4}$$

The right grid alloy and charging parameters for flooded batteries keep water loss to a minimum, requiring only occasional water addition for battery servicing. The use of an absorptive glass mat (AGM) separator and gel in the electrolyte is permitting to the passage of oxygen gas to the negative electrode, where it interacts with Pb to form $PbSO_4$.

In a gas electrolyte, gaseous exchange from the positive to the negative is much quicker than in a liquid electrolyte. As the plate depolarizes, oxygen is naturally integrated to form PbSO4, the typical discharged form, which is then recharged back to Pb in the same way as the charge method. Other criteria for shielded recombinant cells include purified materials for both active materials and grids, in general, as well as grid materials of high hydrogen over there to reduce hydrogen evolution just at negative. Valve-regulated lead–acid is designed to allow gas to exit the cell at a predetermined pressure while preventing outside air from entering.

21.3 Charging Techniques

This research article does deal with 11 charging methods, however. Some essential principles concerning the charging strategies are as follows:

A. Overcharging: During overcharging the temperature of the battery increases resulting in thermal stress on plates and the excess charging current also results in electrolysis or gassing of electrolyte resulting in loss of water from the excess charge current attacks and oxidizes the grids of positive plates which causes early failure of the battery.

B. Undercharging: When the battery terminal voltage is less than 20 percent of the maximum charge, the battery sulfates the electrodes. This phenomenon clogs the active surface area and reduces the conductivity. Therefore, the battery cell loses charge acceptance and thus capacity is reduced. This problem can be solved using superimposed pulse frequency and intermittent charge control techniques (Praisuwanna and Khomfoi 2013).

C. Operating Temperature during charging: In general, the battery's operating temperature is also a major factor responsible for prolonging its life. If the charging current is above prescribed limits or charging is continued after the battery is fully charged or if the charging current is not controlled as the SOC level builds up, these will increase the operating temperature in the battery. As per IEEE publications, the increase in operating temperature at every 8 to 10 degree Celsius above the standard temperature of 27 degree Celsius decreases the battery life by half. The large temperature change during the charging process is important to avoid (Bhatt et al. 2005).

D. Storage: If the battery is not used for long periods, because of the self-discharge phenomenon, the battery gets discharged and if it remains in this condition for long, the sulfation process takes place on the battery plates (mostly negative).

E. State of Charge: Battery efficiency is also measured by calculating the state of battery charging. Continued to improve battery life is helpful (Ng et al. 2008).

F. Charging Time: The charging time determines the battery life and should be minimal. Thereby new charging systems should be invented. (Extended charging at low currents is good for the battery but high or normal current extended charging is bad.)

Normal charging techniques, two-step current charging, and pulse charging technique are stated in this paper. Besides these methods, it discusses in detail the negative pulse discharge, intermittent charging, reflex charging method, and superimposed method.

21.3.1 Constant Current Technique

This is the conventional charging technique for charging the lead corrosive battery. The battery is charged by making the current consistent. It is a basic technique for charging batteries. The charging current is set roughly 10% of the greatest battery rating. It is constant current @ 10% or 12% of rated capacity in Amps which is reduced to 50% of that as soon as gassing voltage is reached. The battery requires enormous time for charging which makes the battery warmed. Due to this, the temperature of the battery may increment and a warm runway issue has happened. If it is over-charged, battery substitution occurs. Figure 21.4 shows the current and voltage curve of the battery.

21.3.2 Constant Voltage Technique

It is the most widely recognized customary technique for charging the battery. In this strategy, the battery is charged by keeping its voltage consistent. It requires less

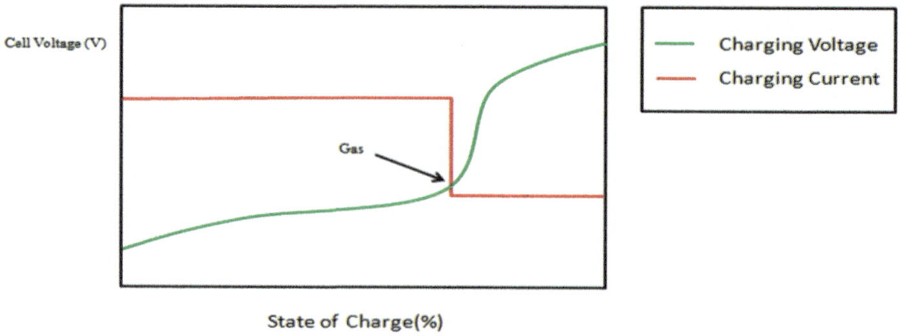

Fig. 21.4 Battery's voltage and current graph at CC method

an ideal opportunity for charging the battery. Hence, the battery limit is expanded up to 20%. Simultaneously, productivity is decreased around by 10%. The battery is initially charged at a constant current rate. When the battery terminal voltage reaches the gassing point, the charging voltage is kept constant and since the potential diff reduces, the current decreases gradually.

By using this technique, the batteries with different degrees of discharge and different capacities are to be charged. At the beginning of the charge is required the higher charging current which is of relatively short duration. The charging current drops to almost zero at the end of the cycle when the voltage of the battery is almost equal to the voltage of the supply circuit, and a potential difference is NIL (Hua and Lin 2000; Serhan and Ahmed 2018). Figure 21.5 displays the battery current and voltage curve.

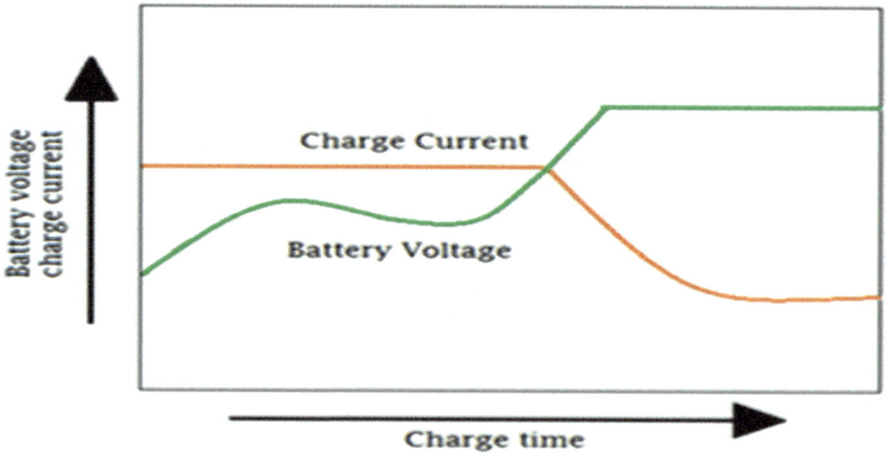

Fig. 21.5 Battery's voltage and current graph at CV method

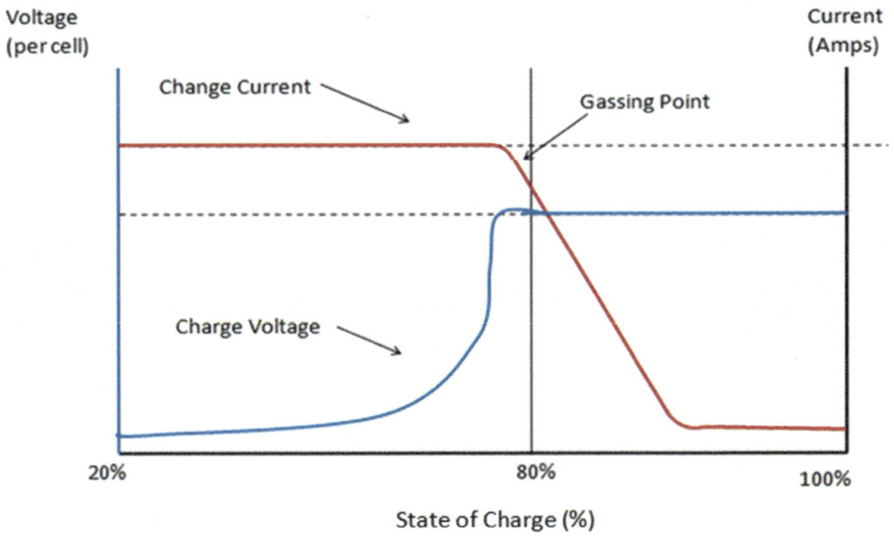

Fig. 21.6 Battery's voltage and current graph at CC-CV method

21.3.3 Constant Current-Constant Voltage Technique

This approach involves a combination of constant current and constant voltage. This process is broken down into two parts. The charger limits the amount of current in the first step to a pre-set level before the battery reaches a pre-set voltage level. Battery charging in the second stage is accomplished by a constant voltage charging system. The current decreases in the second level until the battery is fully charged (Hua and Lin 2000; Serhan and Ahmed 2018). Figure 21.6 displays the battery current and voltage curve.

21.3.4 Two-Step Current Charging Technique

Constant current and pulse charging techniques are used in this procedure. The battery is charged in the first stage by providing a consistent battery current until the voltage of the battery exceeds its pre-set value (float voltage). The battery is usually discharged by itself. At the second stage, to account for the actual self-discharge, current charging pulses move through the battery. The temperature of the battery elevates throughout the charging process. The pulse current is used to prevent overwork and self-discharging which decreases the battery life. Figure 21.7 displays the charging curve in two phases for current charging.

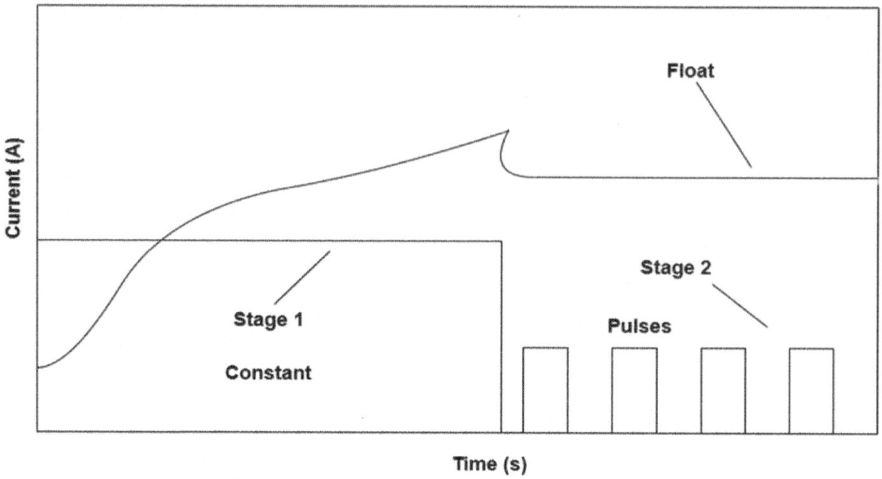

Fig. 21.7 Battery's voltage and current graph

21.3.5 *Pulse Charge Technique*

In this process, the current is added regularly to the battery in pulses (Hua and Lin 2000; Lam et al. 1995). In this process, a short rest time between current pulses is used during battery charging to allow the chemical actions in the battery. It is maintained before restarting the charge by balancing the reaction in the bulk of the electrode. The charge rate depends on the width of current pulses (based on the average current). The electrochemical reaction and battery-internal electrolyte neutralization are helped to improve the battery life cycle. Using a broad pulse current the battery would take less time to charge. This method will reduce the formation of gas, the growth of crystals, and passivation. The downside of this approach is its charge efficiency. Figure 21.8 displays the existing pulse charging approach waveform.

21.3.6 *Reflex Charge Technique*

Using this method (Horkos et al. 2015; Hua and Lin 2000), a modification is made to the pulse charging system. A charging time consists of a positive pulse, a negative pulse, and an interval of rest. To provide service, this system employs three different modes. The initial phase of active charging is when the current (high positive pulse) goes into the battery. The active discharge phase occurs when the electrolyte electrode depolarizes while current flows from the battery to the output (negative high voltage). Finally, a stability mode is the final method. This mode is used to locate the ions from the plate surfaces at an appropriate distance. The size of the positive current is equal to that of the negative. The bipolar triangle pattern is repeated frequently until

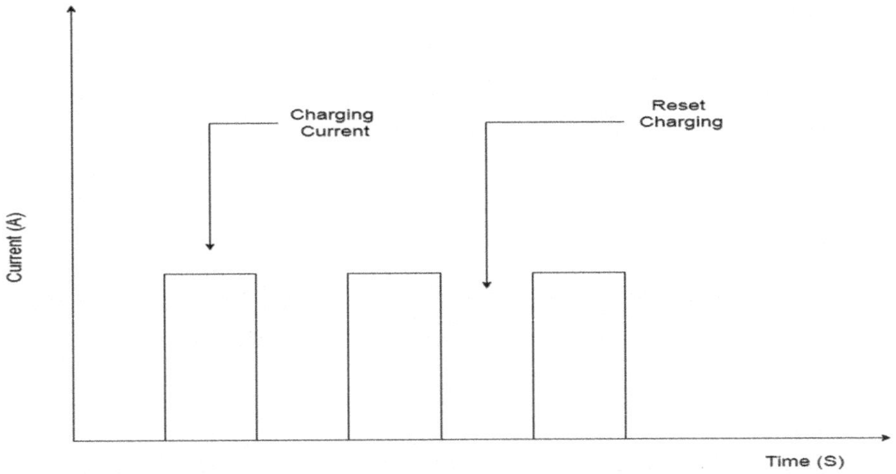

Fig. 21.8 Pulse current waveform

a sufficient amount of power is transferred to the mounted batteries. The charging efficiency of the reflex charging system is increased by the use of the quick charging circuit where the power conversion in bi-direction facilitates the power transfer in either direction between the power source and the battery. The reflex charging circuit and current charging system curve are shown in Fig. 21.9a and b.

The circuit utilizes MOSFETs, diode, capacitor, and diode. It acts as a buck converter when it is in the forward power flow model. It acts as a boost converter in the reverse power flow model.

21.3.7 Trickle Charging Technique

In this method fully charged battery is charged at a rate equal to its self-discharge rate, allowing the battery to stay at its fully charged level. This condition happens almost exclusively when the battery is not charged, as trickle charging is said to be float charging because a battery float voltage charge is not held. Trickle compensated in case a load draws current. A battery under continuous charge is designed to offset the battery's self-discharge (Jose 2019).

21.3.8 Negative Pulse Discharge Technique

Battery charging is a crucial activity. When designing a charger, attention must be given to charging speed and charging time to extend the battery life. The charging

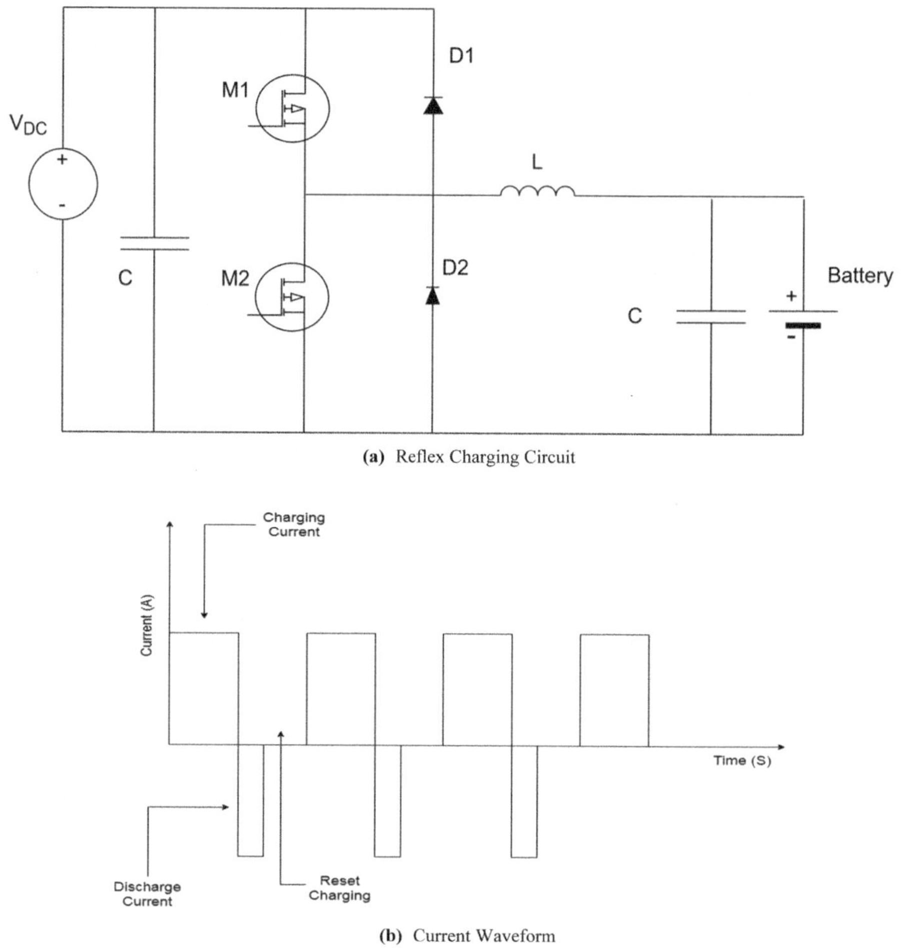

Fig. 21.9 **a** Reflex charging circuit. **b** Current waveform

velocity indicates the charge time. The charging time determines battery health. From this point of view, the negative method of discharge of pulses is stated (Yifeng and Chengning 2011). The current of battery charging directly affects the impact of charging.

The charging is to be increased by increasing the charge current rate. The negative form of discharge was studied in 2011 (Yifeng and Chengning 2011). In this process, when introducing a negative pulse for discharge, the majority of the charge current is used for charging reaction. That will increase the speed of charging. The discharge plays an important role in stirring the electrolyte, which helps to ease polarization concentration and the depolarization of the battery can be strengthened so that the battery is charged with a greater charge present.

The current and rate of charges are commensurate, as has previously been established. There are also some limitations on each battery's permissible charging current. The appropriate discharge time and battery capacity will increase the battery's charge acceptance current. Those two parameters have to be defined for this purpose. The current of the appropriate charge is a decrease in exponentiality. The capacity for release is (Yifeng and Chengning 2011)

$$C = \left[\frac{i}{D.\log(d.I_{dis})} \right] \tag{21.5}$$

where

- D 15, constant;
- d 10, constant;
- I_{dis} current during discharge; and
- i battery mark defined current

The appropriate discharge time is as follows:

$$T = \frac{C}{I_{dis}} \tag{21.6}$$

Some tests have been performed to demonstrate the efficacy of the negative pulse discharge (Yifeng and Chengning 2011). For quick charging, the parameters of negative pulse discharge are determined in this process. The amplitude of the negative pulse should therefore be taken as 85–115% of the battery power. Negative pulse time length should be taken between 100 and 600 ms. This approach has some benefits as well as some disadvantages. For types of batteries with their various capacities, the parameters of this approach vary.

21.3.9 Intermittent Charging Technique

Owing to positive grid corrosion and thermal runway from overcharging the battery life is shortening. When the battery charger is being built, battery life is a very important factor in determining its performance. An approach to increase the life span of standby lead–acid batteries is to reduce the deterioration of the positive grid. The use of a low potential on a positive electrode allows it to be within the minimum zone of corrosion (Armstrong et al. 2008; Bhatt et al. Oct. 2005).

To solve this difficulty, Reid (Reid and Glasa 1984) has developed a new intermittent charging method in which the battery is charged intermittently to its maximum charging capacity and then held in an open-circuit condition. This prevents potentially harmful electrochemical reactions. The collection of intermittent charging characteristics such as voltage levels and charging strategy, with the possibility of undercharging the negative electrode, is a challenge in implementing the intermittent charging technique in a realistic application (Muneret et al. 2000).

21.3.10 Interrupted Charge Control

An intermittent charging control method is an effective charging system whereby the risk of undercharging is decreased by ensuring maximum battery charging return. The undercharging effect clearly shows the need for battery charging to the full. The method of interrupting charge control restrains the VRAL battery from its float charging mode.

The ICC procedure is divided into two stages. The battery is first charged with a steady current to an upper voltage threshold, then with the pulsed current until the charge is complete, guaranteeing maximum charge return. As charging substantially extends the battery life, accurate estimation of the battery's voltage, current, and temperature is crucial. In this technique, choosing the correct threshold settings is critical (Bhatt et al. 2005).

Figure 21.10 shows the circuit diagram for the interrupted charge control system. The topology for flyback is used below 150 W. The flyback converter is managed via current mode control. This regulation has overcurrent safety and delivers current

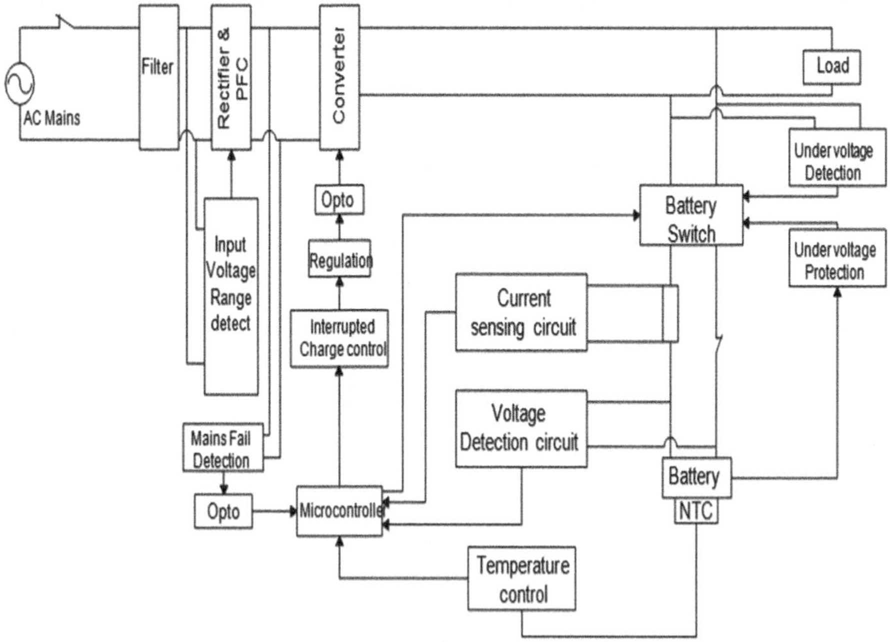

Fig. 21.10 Interrupted charge control (Bhatt et al. 2005)

pulse-to-pulse restriction. Using the voltage detection circuit, the battery voltage is continuously tracked and is entered into the microcontroller. Despite this, according to the referenced values, the microcontroller controls the voltage and current (Bhatt et al. 2005).

The experimental test was conducted to validate that this method is shielding the battery from undercharging. It showed the ampere-hour is very high at the end of the charge period. Recharging the battery is high enough which removed the risk of undercharging. Lastly, this system has advantages from the intermittent charging method and pulsed current charging method. The technique's self-discharge rate is low. Using rest periods and high pulsed current is reducing the risk of thermal runaway and grid corrosion.

21.3.11 Superimposed Pulse Frequency Technique

It is a new technique for battery charging. The main emphasis is on prolonging battery life. Sulfation is the major motivator that will destroy the battery entirely. The technique was developed from this perspective (Praisuwanna and Khomfoi 2013). The multilevel converter is used in this technique to produce superimposed pulse frequencies that range from 100 Hz to 1 kHz. This pulse generates mechanical resonance at the electrode plates. Figure 21.11 shows the circuit diagram of a multilevel converter charger. This communicates the vibration at the surfaces. As a consequence, the pore that is attached to the plates is removed due to sulfation. This increases the battery life and improves the charging efficiency (Praisuwanna and Khomfoi 2013). The battery is extended for almost a year after sulfation and thus increases the frequency of charging (Praisuwanna and Khomfoi 2013). It displays the findings in Fig. 21.12 (Praisuwanna and Khomfoi 2013).

21.4 Developments in Lead–Acid Battery

More than 160 years ago, a scientist, Gaston Plante, invented the lead–acid battery. He was probably unaware of recent developments in the battery industry. Lead–acid batteries have a smaller storage density than most batteries. The materials needed for a lead–acid battery are less costly. Lead–acid batteries' long-term sustainability is often questioned. Many have claimed that only the lead–acid battery has no future, but this is nothing new, and amid decades of predictions to the contrary, the lead–acid battery continues to dominate the global battery energy storage market.

Lead–acid batteries have issues with accelerated corrosion of the battery plates, faster self-discharge, rapid water loss, gas formation, and significant internal resistance variance. In reaction to the above requirements and concerns, improvements in the lead–acid battery era have been developed, such as the incorporation of carbon to the negative active mass and positive active mass of batteries, as well as the use

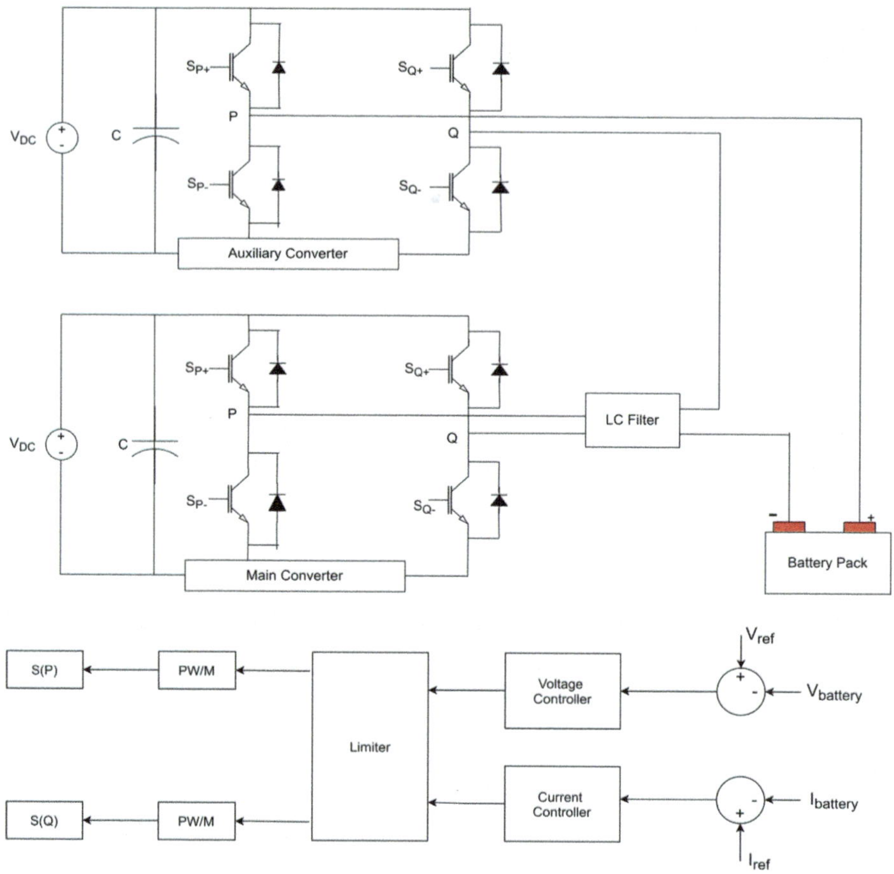

Fig. 21.11 Multilevel converter charger

of carbon in current collectors and negative electrodes. To address the above issues, pure lead punching carbon technology (PPC) is also used (Shimoura and Wei 2018). Here are some recent advancements in lead–acid battery technology.

21.4.1 Pure Lead Punching Carbon Technology

Power, high discharge rate, battery life, and environmental suitability are the four most critical parameters of a lead–acid battery. Improving these variables is a difficult task. These parameters have been improved by using a new construction process, new alloy content, and carbon as the negative active material.

This technology makes use of a Pb-Sn alloy to reduce grid corrosion and extend battery life. Grid corrosion is less in the Pb-Sn alloy than in the Pb-Ca-Sn alloy

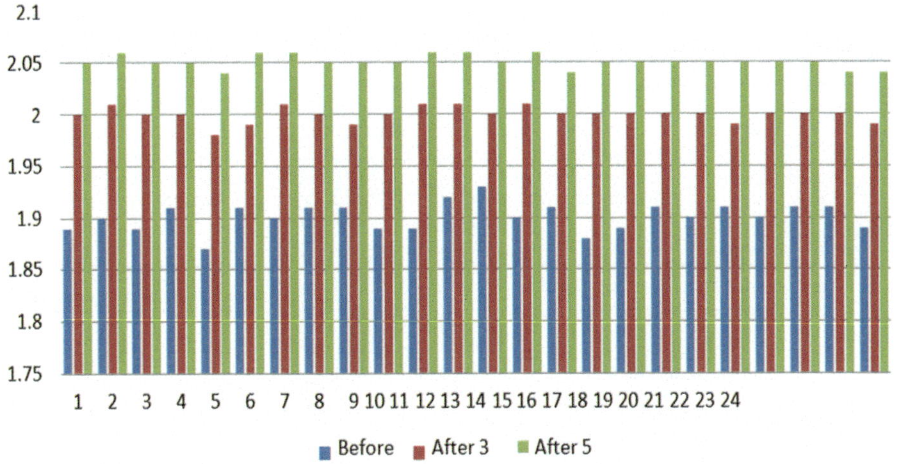

Fig. 21.12 Comparative analysis of the battery's terminal voltage before and after SPF charge (Praisuwanna and Khomfoi 2013)

grid. The PPC technology alters the battery structure to improve lead–acid battery high-rate discharge performance. The COS (cast on the trap) construction method shortens the conductive road. The internal resistance is then reduced by increasing the contact area of the penetrating welding segment between cells. The installation of an automated welding facility mitigates the impact of manual welding's faulty welding lead flow and fusing. The amount of heat produced at discharge in the strap is reduced by 60% and that in the terminal section by 58% of the structural change. As a result of this reduction, the output during high-rate discharge is improved. Using continuous lead sheet manufacturing and punching techniques to reduce weight and thickness, a thinner grid with a smaller mesh opening can be developed. As a result, conductivity is improved. The number of electrode plates that can be installed in the same size battery has increased as a result of this arrangement. As a result, the battery's capacity to store energy changes. It's depicted in Fig. 21.13.

Other than that, the addition of electrically conductive carbon to the negative energetic material improves the electrode plate's charge and discharge reputation. The battery capacity is improved in this technology because the amount of active material per electrode plate is increased. As a result, when compared to current battery technology, the discharge capacity of the PPC battery has increased by 25% in a 15 min discharge rate and by 43% in a 5 min discharge rate. This technology's production process is low impact. The CO_2 cycle is slowed down. Figure 21.12 depicts the entire PPC operation.

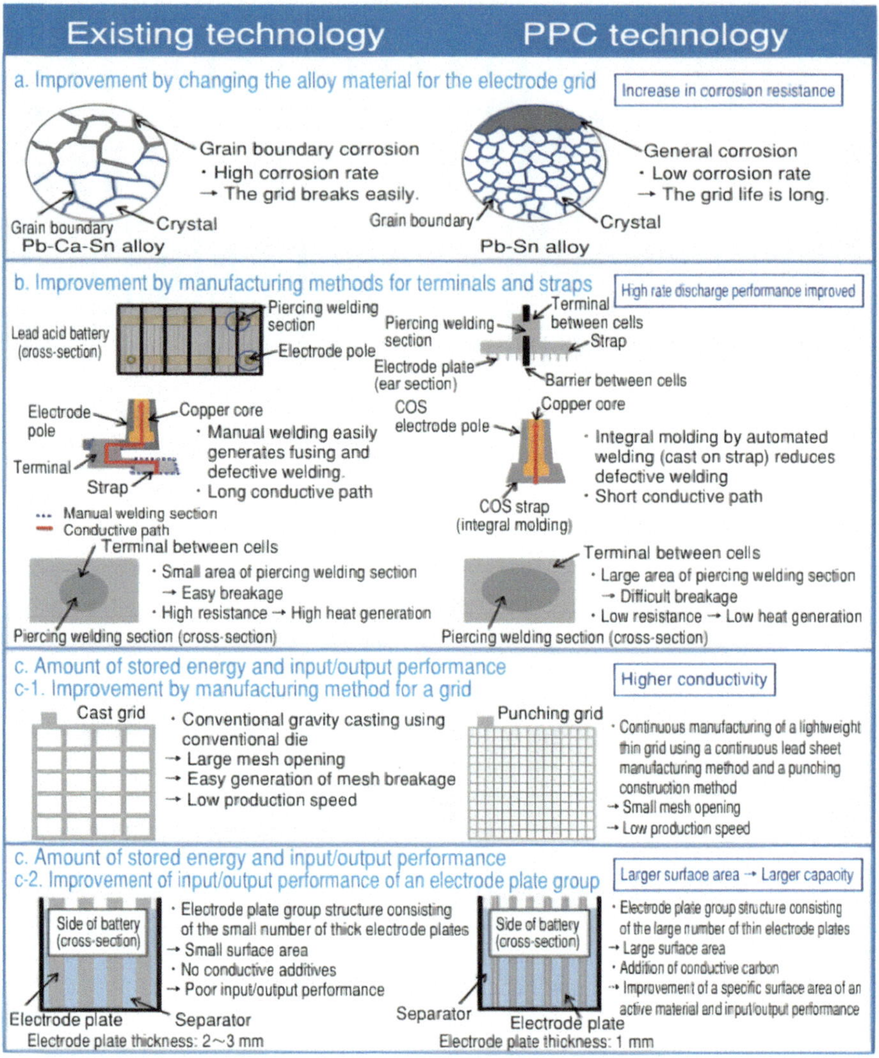

Fig. 21.13 Complete process of PPC operation (Shimoura and Wei 2018)

21.4.2 Carbon Negative Electrode

Carbon materials are typically used as a negative active material (NAM) additive because they improve battery cycle existence and charge reputation, specifically in an excessive-charge partial state of rate situations, which can be not unusual in hybrid and electric powered automobile (Lach et al. 2019). It is observed in a variety of sorts, together with graphite and carbon black.

The majority of carbon additive research is focused on retrospective research that indicates improved lifetime. Cyclic voltammetry was used to examine the presence of carbon in the negative plate of a lead-acid battery during high-rate processing using Pb and carbon electrodes. The capacitive contribution to Faradaic processes can be isolated using CV, which has been demonstrated experimentally. The discharge and charge capacities of capacitive efficiency are further divided. This is important for lead-acid batteries in general, but particularly for Carbon Lead Acid batteries, where charge acceptance is crucial (Jaiswal and Chalasani 2015). It's miles feasible to absolutely replace the NAM with carbon, wherein case the electrode acts as a capacitor as opposed to a faradaic strength storage device, and with carbon materials of appropriate bodily shape and high precise floor area, the electrode features as a supercapacitor. Combining this with standard PbO_2 positive plate results in an asymmetric supercapacitor. This type of device has smaller energy efficiency and a much steeper discharge curve than a lead-acid battery, however, it has a longer cycle lifestyle. It can also be charged up easily. Axion Power in the United States developed this idea as an energy storage device (Volfkovich and Serdyuk 2002; Buiel et al. 2010; http://www.axionpower.com).

21.4.3 Enhancement in Carbon

Micro-hybrid technology, which has been embraced by the automotive industry and includes stop-start features, improves fuel economy while also lowering emissions. This is essential to improve lead-acid battery capacity in situations during which a significant number of engines start at the beginning of a journey and are subjected to shallow charge-discharge cycles. The high-rate partial-state-of-charge (HRPSoC) process is the name for this type of operation. vulnerable dynamic charge reputation on the negative plate is a sign of this interest.

The factors that contribute to this shear failure are addressed to become aware that supports the charge reaction as well as opposing secondary reactions such as redox reactions (for VRLA designs), electrolytic, and capacitive charge aggregation that can occur at the plate during charge.

Researchers must investigate facets of the charge response that have previously raised concerns, such as local reactions within the pores including its plate active material, the effect of acid awareness on localized solvation of $PbSO_4$, and possible plate inhomogeneity, to fully comprehend the processes that underpin HRPSoC behavior. It is also important to determine the possible benefits of biomass, grid configuration changes (twin tab), and the UltraBatteryTM design. The negative plate's charge acceptance loss is the source of the battery's poor efficiency.

It's difficult to completely charge lead sulfate; it builds up during cycling and prevents plate activity. Since the battery cannot accept charge and restore energy when completely charged, the issue is exacerbated. In the industry, extensive research has been conducted to develop new flooded and VRLA battery architectures that can match the needs. The discovery of new carbons that alter the nature of negative plates has been the secret to this. Many of the technologies developed for automobile batteries can be extended to industrial batteries (Jaiswal and Chalasani 2015), (Moseley et al. 2015b). Similar challenges confront energy systems, especially solar photovoltaic and utility support systems. The method of adding a carbon portion to the lead-acid battery's negative plate to increase its performance has been the subject of numerous studies. There have been eight different carbon roles proposed (Shaffer 2016), but only three of them have a significant effect on their own.

Carbon has several effects on a lead-acid battery's negative plate production. The three types of effects are the capacitive, surface area of electrochemical processes on which electrical and chemical charge/discharge processes occur, as well as side effects.

(a) Carbons with large specific surface areas and close contact with the grid as a current collector, as well as the active mass's spongy lead matrix, prefer capacitive effects. On the other hand, sponge lead does not require thorough carbon mixing.
(b) The carbon must be active and in alignment mostly with a current collector to have an effect on the surface area, and for the reason that carbon serves bulk instead of surface methods, the surface area can indeed be underneath that of the capacitive method.
(c) The carbon in the physical process does not need to be active; however, it must be closely combined with that of the sponge lead and have a particle size significant enough to keep the mechanism from deteriorating over time.

Since there are so many different carbon materials, these expectations conflict to some degree. Although much research has gone into finding the best combinations of carbons, there is still a lot of basic research to be conducted to explain the underlying mechanism (Moseley et al. 2015b). As a result, lead batteries' capability for a variety of energy storage applications will continue to grow.

21.4.4 Carbon Negative Current Collectors

Sulfation prevents small amounts of those carbons from being added to the original NAM, but an evolution of hydrogen is accelerated. Carbon may be used to consume one or more metallic elements of the negative grid. Tough carbon polymers, LEGF, as well as light carbon felt, have also been used to test various concepts. The flexible carbon polymers were durable, had a large surface area for electrochemical reactions, and had a high active mass consumption (http://www.fireflyenergy.com), but their brittle nature made manufacturing difficult. The negative and positive grids have been replaced by LEGF and extended titanium mesh coated with SnO2. Reversible hydrogen storage, with activated carbon as the basic element of the NAM, is the perfect energy storage device. Titanium as an alternate current collector in a partial state of charge (PSoC) cycling applications has several benefits, including higher power output and a battery lifespan of more than 3000 equivalent cycles. In PSoC cycling, LEGF had a lower likelihood of being used but a longer lifetime, implying that lead sulfate formation was avoided (Lannelongue et al. 2017). ArcActive in New Zealand has developed a more promising model that substitutes the Pb grid with carbon felt that is powered by an electric arc under realistic circumstances (Christie et al. 2017). The active material is infused into the felt, which is then attached with existing lead alloy collectors. This structure performs extremely well in PSoC action. Although it was designed for vehicle industries, it has a bright future for electrochemical devices in wider applications, particularly because the high-rate capacity required for automobile operation isn't required in most energy production duty cycles.

21.4.5 *Additives for Positive Electrodes*

In a lead–acid battery with an HRPSOC, sulfation on the negative electrode will occur. The sulfation problem of a lead–acid battery's negative electrode can be easily solved by adding carbon material to the negative electrode. As a result, the "Lead–Carbon" battery is developed (Moseley et al. 2015b). Since the negative electrode problem was solved, the positive electrode's strength has decreased. A lot of studies are focused on increasing the performance of the positive electrode. This can be accomplished by adding additives to positive active material (PAM). Positive additives were categorized into different classes based on their chemical properties and effect on lead–acid battery production. These groups included conductive additives, porous additives, binder additives, and nucleating additives. Figure 21.14 depicts the causes of failure and methods for reinforcing the positive electrode battery. These additives are approved for modifying the conductivity, porosity, mechanical strength,

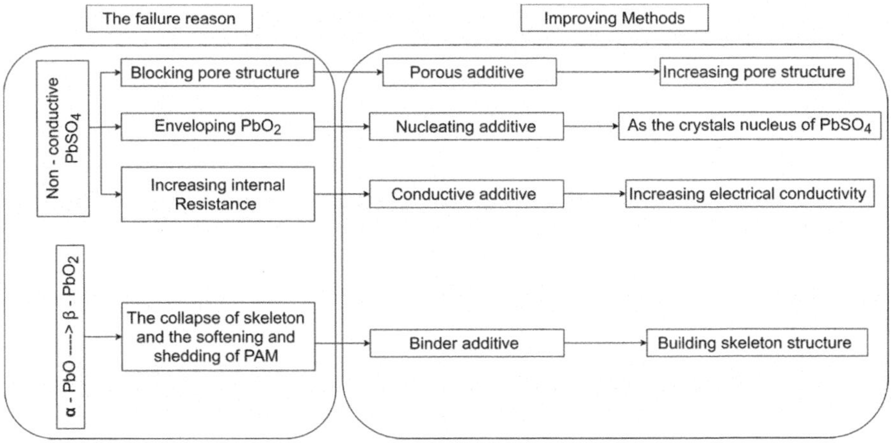

Fig. 21.14 Causes of failure and methods for reinforcing the positive electrode battery

crystal morphology, and geometry of the active material to improve its performance (Dietz et al. 1985).

(a) Conductive Additive: Its role is to make sure the plate charges and discharges correctly. When a plate is formed, conductive materials are often introduced as a conduit to link isolated PbO_2 to the Pb grid, allowing a tiny current to build between the active materials and the grid (Ponraj et al. 2009). Simultaneously, the frequency of hydrogen ion migration inactive materials can be significantly improved, increasing the electrode's ability to charge and discharge.
(b) Porous additive: The porous additive's pore structure has the potential to adsorb sulfuric acid and speed up electrolyte ion migration. It has excellent ionic conductivity. According to the researchers, a porous additive enhanced HSO_{4-} diffusion and produced regional concentrations of the ion within the pores (Metzendorf 1982).
(c) Nucleating additive: $PbSO_4$ expands due to the use of sulfate ions as nuclei in the discharge, specifically $CaSO_4$ and $NaSO_4$. This sort of additive might inhibit the development of bigger particles by generating an evenly distributed tiny growth nucleus in the PAM. The use of this type of additive improves the discharge phase and increases discharge capability.
(d) Binder additive: Binders (Pavlov and Ruevski 2001; Ferg et al. 2017): Positive plates last longer with these additives. When a portion of PAM, including its PbO_2 particles, is limited during discharge, the relationships with the residual PbO_2 particles near the restriction site change. Certain particle recovery processes in granules are governed by surface forces. These systems are influenced by the ions Sb, Bi, and As, which promote the regeneration of PbO_2 particles in granules. Binders have been shown to affect the processing of tubular electrodes equipped by PbO_2 powder and salts of the three elements listed above. In the existence of Sb_2O_5 or Bi_3 ions, the potential of the electrodes under evaluation increases at a faster rate as the number of cycles increases. Batteries containing 0.05 % weight of Bi_3 in PAM include an 18% longer battery life than batteries that do not contain bismuth (Lang et al. 2017). Moseley conducted a study of all additives tested for enhancing PAM conductivity.

Sulfations of the negative electrode, as well as weakening and flaking of the PAM, have been the main issues restricting the usage of lead–acid batteries (Guo et al. 2009). According to recent research, the failure mode of lead–acid batteries is PAM weakening and shedding, and the battery lifespan is primarily confined to the positive electrode. As a consequence, the lead–acid battery has hit a stumbling block that must be addressed to improve the PAM of the lead–acid battery's efficiency. As a result, the battery's overall performance can be greatly improved when suitable PAM additives are used. It is predicted that the lead–acid battery energy storage system modified by positive electrode active material additives would achieve better service efficiency as a result of the researchers' thorough investigation. Moreover, certain electrode additions might be investigated in combination with positive additives to increase the battery's efficiency. In the fields of start–stop systems, power batteries, solar energy storage, and other sectors, lead–acid batteries are expected to see increased usage and development. As a consequence, the positive plate's shape and several additives should be selected with the discharge rate in mind. Positive plates with additives that increase PAM's electrical properties should be used in batteries with low discharge currents, while additives that enable ionic flow diffusion in PAM's pores should be used in batteries with high-rate discharge duties.

21.4.6 Supercapacitor/Battery Hybrids

The system is referred to as a "Composite" negative electrode, and it is known as an ultrabattery device when the carbonaceous material (specifically activated carbon) content of the negative electrode is increased further, or if a typical negative plate is paired in conjunction with a porous carbon supercapacitive electrode and placed versus a conventional positive electrode. It outperforms unbalanced lead-based electrochemical capacitors and conventional lead–acid batteries in a variety of ways (Lam et al. 2012; Cooper et al. 2009). Enhanced high-rate charge adoption, enhanced cell self-balancing in series strings, a discharge energy density and voltage profile comparable to a lead–acid battery, prevention of permanent sulfation on the negative electrode in PSoC cycling, and the removal of the need for long-term charging cycles are just a few of the benefits. During a Neighborhood Electric Vehicles (NEV) driving conditions, Charles Kagiri investigated a suitably operated supercapacitor-centered battery hybrid approach that is used to minimize the effects of high current demands on the battery. The integrated storage device demonstrates that it can lower the peak current drawn from the battery, which is a key factor in deciding the battery's effective performance per single charge and battery capacity. This rise would result in lower NEV operating costs leading to decreased energy losses and better battery life. High-capacity battery electrodes and high-rate capacitor electrodes are commonly used in battery supercapacitor hybrid systems. Many studies are being conducted to increase the energy density for lead–acid batteries (Kagiri and Xia 2017).

21.4.7 Bipolar Lead–Acid Batteries

Increased energy and power density are possible with bipolar lead–acid batteries (bipolar LAB). Purushothama Rao and Jing-Yih Cherng patented a bipolar LAB of increased capacity as a result of the use of a central bi-negative or bi-positive plate, which allows for increased capacity without increasing plate size. For many years, bipolar structures have been studied, and a few concepts have recently shown greater significance in terms of technological and economical success. The bipolar method increases the basic power of the battery. Aside from the end plates, a bipolar battery's plates provide one positive side as well as the other negative sides, separated by an impervious, discover the connection, and corrosive environment membrane.

Choosing the right substrate of lead–acid batteries is critical, as is forming solid edge seals around the substrate for both electrodes on both faces. Bipolar lead–acid batteries have a lower mass/volume ratio than conventional lead–acid batteries, resulting in higher energy densities in both dimension and mass. However, due to a lack of suitable materials to seal dissolved H_2SO_4 in a single cell, long-life bipolar LAB has proven difficult to achieve. In the 1990s, reports of technically viable bipolar lead–acid batteries became more common. LaFollette et al. (Lafollette 1990, 1995) presented the fundamentals of bipolar lead–acid battery design, assembly, and development. Their contributions are primarily concerned with the creation of basic models and packaging techniques. As a result, the substrate, which acts as both the cathode and anode in bipolar LAB, is responsible for the majority of the problems. Power efficiency, cycling longevity, battery size, and production costs are all influenced by the chemical and physical properties of bipolar substrates (Coux et al. 1999). Corrosive environment stability and overpotential (H_2–O_2) are particularly important throughout the electrolytes of aqueous sulfuric. For this reason, bipolar LAB substrate materials have been widely explored (Karami et al. 2007; Bagshaw 1995; Lun-Shu et al. 1987; Bullock 1995; Kao 1998). The bipolar LAB substrate material used by Karami et al. was a lead–tin mixture (Karami et al. 2007). Tin in alloys will increase the water splitting overpotential under strongly oxidizing conditions, resulting in a somewhat more robust sulfuric acid interface. Y. R. Lun-Shu et al. used a thermal treatment to improve the lead–titanium bonding after electroplating a lead coating on a titanium substrate (Lun-Shu et al. 1987). Lang et al. investigated the bipolar lead–acid battery's substrate in a set of research (Lang et al. 2014a, b, 2018, 2017b). Pyrolytic carbon and TiO2x were added to the titanium foil to increase its corrosion resistance in an H_2SO_4 electrolyte. The effectiveness of these different techniques and methods in extending lifespan have been discovered, showing the need for an inert layer between the substrate as well as the electrolyte in each case. Anti-corrosive powders against the acid electrolyte were produced in bipolar substrates in addition to the requisite surface treatments. Paleska et al. identified BaPbO3 as a low-weight substrate material. It is a high-conductivity perovskite ceramic material (Paleska et al. 2004). Using cyclic voltammetry, a lead layer was electrodeposited onto the surface. Many transition metal oxides, such as TiO2x, Mo2O3x, WO3x, Ni2O5x, and V2O5x, were scattered into polypropylene membranes by Reichman et al. (Reichman 1991), effectively raising the ductility of the polypropylene film. Bipolar lead–acid battery substrates are suitable for this form of film. Ellis et al. (Ellis et al. 2004) constructed a bipolar substrate using a resin-bonded composite. The mixture included TiO2x powder, fibers, and adhesives. Thermal pressure was used to force the resulting slurry into the substrate, which can be easily scaled up in the industrial field. The industry group is very interested in bipolar lead–acid batteries during the commercialization period. Ellis et al. proposed a substrate which was later produced by Atraverda Company in the United Kingdom and registered as an Ebonex trademark. Ebonex technology would be another title for the technology used to prepare this layer. Using an Ebonex substrate, Loyns et al. designed a bipolar LAB. Bipolar LAB was more successful and lasted longer (Wertz 2001; Ghaemi et al. 2006). OPTIMA of Sweden and VOLVO Automobile Company collaborated to create Effpower, a bipolar LAB system for the era of the integrated vehicle. In the United States, BPC organization has been interested in bipolar lithium–ion battery technology, resulting in a bipolar LAB device with a 180 V, 60 Ah specification, as well as improved specific features and durability. This empirical illustration proves that bipolar lead–acid batteries are cost-effective and high-efficiency power source for electric vehicles. An effective bipolar lead–acid design would be a viable energy storage device.

21.4.8 Bipolar Electrodes for Lead–Acid Battery

The invention of powerful backup batteries offers a fantastic opportunity among applied research scholars that collaborate on challenging scientific and technological problems while still meeting a critical energy storage need. Aside from novel battery chemistries, which are commonly studied by scientists, battery-based structures resulting from technological advancements that improve battery efficiency should be considered. Bipolar electrodes (BEs) can enhance basic strength, strengthen cell components, and lower rechargeable battery manufacturing costs throughout this case. Kaptiza et al. designed the first bipolar electrodes, which were used in lead–acid batteries in the early 1920s (Kapitza and Heath 1923). However, some issues exist, such as cell sealing and severe substrate corrosion. It is the cause of battery failure. Biddick and Nelson's work reinforced the need for seals and vents in bipolar lead–acid batteries until the 1960s (Kao 1998). Since the 1960s, several scientists have been researching how to manipulate bipolar substrates to overcome cell trapping as well as substrate corrosion. The cell design is created from scratch. Nickel–metal hybrid batteries, lithium–ion batteries, as well as post-lithium–ion batteries are now all made with BEs. Bipolar electrodes first appeared in post-lithium–ion batteries in 2019, such as lithium–sulfur batteries (Kim et al. 2019) and sodium–ion batteries. In the next generation of rechargeable batteries, bipolar electrodes will be employed, as shown in Fig. 21.15.

Rechargeable batteries with bipolar electrodes have also seen a large rise in volumetric/gravimetric energy density. Furthermore, depending on the application-oriented nature, the battery shape can be easily changed, resulting in better battery storage available space within targeted devices. As a result, BE's advantages are particularly appealing in the design of rechargeable batteries for mobile devices as well as electric vehicles. BEs remove the need for several tabs and a cooling system in rechargeable batteries. Aside from the advantages of BEs, certain conditions must be met, such as the substrate's position, the electrolyte's segregation, the fact that neither chemical processes of the substrate systems to deliver in the cathode or anode, the same amount of electron transfer during the charging–discharging process, as well as the cost of the substrate material. BEs aren't without their drawbacks. The BEs permit a standard chain arrangement without the use of external wires. When it comes to determining cell voltage as well as preserving the health of bipolar batteries, battery control becomes increasingly important.

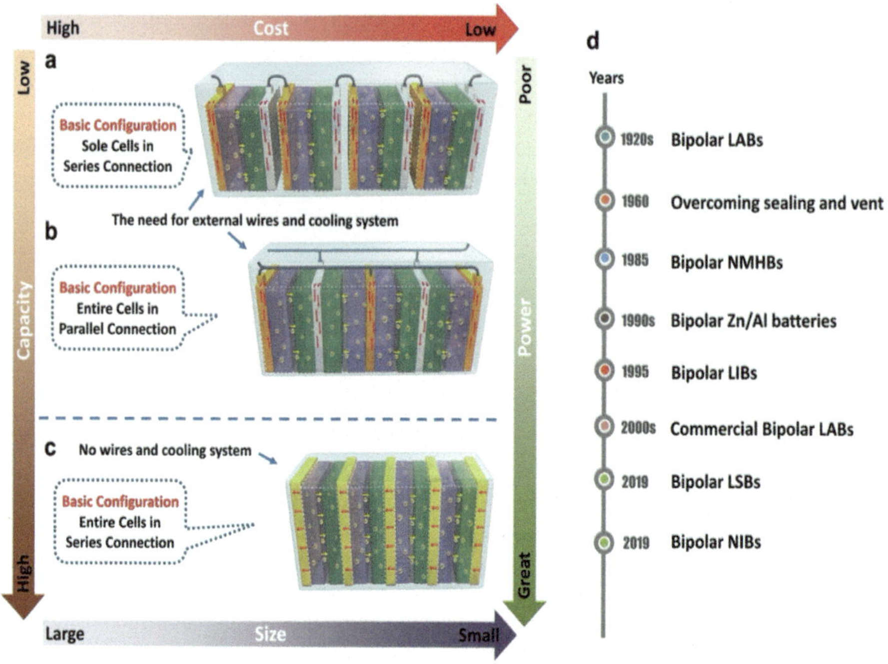

Fig. 21.15 Bipolar electrodes for next-generation rechargeable batteries (Liu et al. 2020)

A significant technological gap provides an opportunity for researchers and research teams to collaborate to accelerate the production of lead–acid batteries for electricity grid purposes, where no single energy storage device can currently satisfy all of the technical and economic demands. Even though the battery's operating principle has not changed, manufacturers have advanced technology while increasing the efficiency of the electrode, which includes active components, especially for the automotive industry.

The effectiveness in which the active material is used for the electrode determines the battery's output. The choice of active material has a significant impact on the battery cycle life and calendar life. The degradation of electrode grids made of pure Pb, Pb-Ca including Pb-Sn materials is facilitated by the continuous evolution of electrode morphology and microstructure, reducing battery life span as well as material usage quality.

As maintaining total electrode surface area provides effective charge–discharge processes, the design of battery component materials, surface electrochemistry, and area development of energy materials at electrochemistry terminals all have an impact on battery performance. Reactions that breakwater is undesirable. Lead as well as lead dioxide is inadequate facilitators for these reactions, with high overpotentials that kinetically limit them while charging rapidly at high voltages.

Many of the risks associated with this technology, like those associated with any other, can be mitigated by proper material handling, efficient manufacturing processes, and committed sewage treatment. The 99 percent recyclable content of lead–acid batteries, as well as strict environmental limitations on Pb emissions, substantially decreases Pb exposure. (SmithBucklin Statistics Group 2019). However, in the short term, the absence of economically feasible recycling options for lithium–ion battery technology, along with the projected rise in the number of battery cells reaching their dying process, raises the danger of pollution from disposable lithium–ion battery devices. Lithium–ion batteries are dangerous in lead recycling, demanding improved safety and recycling processes.

The variety of experimental and theoretical approaches and methodologies established over the last 30 years can be used to further extend and demonstrate the research regarding lead–acid batteries. More involvement from US National Lead Acid Battery Laboratories and academics on these topics would be extremely beneficial (Davidson 2019; https://www.batteryinnovation.org).

21.5 Discussions on Charging Techniques and Developments of Lead–Acid Battery

Various charging techniques are used to charge a lead–acid battery. Each technique is having some pros and cons. But as such, there is no perfect technique to charge it. In the constant current charging technique, it is observed that the current at the starting is less than the current of a reasonable charge. Hence, the longer charging time this technique offers. This causes the temperature to rise during charging. This is the drawback of the constant current charging technique. Higher gas evolution problem has occurred in the constant voltage charging technique method which affects its functioning (Horkos et al. 2015; Valeriote et al. 1994). The problem of controlling the initial current in the constant current constant is solved in the CC-CV method. But it tackles the issue of increasing temperatures and battery overcharging (Serhan and Ahmed 2018). Similarly, overcharging and discharging are avoided in the two-step current charging method which is similar to the constant current method. The charge efficiency of the pulse charge technique is extremely low (Cheng and Chen 2002). To improve the charge efficiency, the fast-charging technique is developed.

Sulfation is the main issue that occurred in the deep battery discharge. A recently developed charging technique known as the superimposed pulse frequency technique is used to remove it from it. Because of this technique battery life is extended and we can use the battery for applications. This technique is highly effective in charging. This technique is undergoing research (Horkos et al. 2015; Praisuwanna and Khomfoi 2013). Table 21.1 includes the comparative study of the charging techniques. From the comparative analysis, it is observed that the charging techniques have scope to work. Research is underway to boost the battery life cycle. Some converters are also designed to keep the battery's charging profile. This will extend battery life (Hakim

Table 21.1 Comparison of charging techniques

Sr.no.	Charging techniques	Charge efficiency	Charging time	Limitations
1	Conventional charging techniques (CC, CV, CC-CV)	Medium	Medium	Thermal runaway, overcharging, grid corrosion
2	Pulse charging	Poor	High	Overcharging
3	Reflex charging	Better	Less	This method has a high temperature than the CC-CV method
4	Negative pulse discharge	Better	Medium	Controlling the negative pulse is a crucial task. Thermal runaway problem has occurred
5	Intermittent charging	Medium	Medium	Undercharging
6	Interrupted charge control	Better	Less	Circuit complexity
7	Superimposed pulse frequency	High	Less	Circuit complexity

et al. 2016; Shafiei et al. 2016). There is some development of the charging techniques (Valeriote et al. 1994; Alaoui and Salameh 2003; Bayya and Rao 2018).

In today's world, combined battery chemistry is used to meet energy storage needs in automobiles and other fields. Lead material is used to fabricate anodes for lithium–ion batteries at Argonne National Lead Acid Battery laboratory. This lead-based anode can store twice as much energy as current graphite anodes while maintaining stable performance during cycling. The advanced lead battery development group at the Battery Innovation Center is attempting to meet potential technological demands for clean mobility and clean energy storage (https://www.batteryinnovation.org).

21.6 Conclusion

It is commonly used in various fields due to numerous advantages of lead–acid batteries such as low cost, lead recycling, etc. This paper explores the thorough functioning of charging methods over time along with voltage and current graphs. In this paper, we have presented various charging techniques like the conventional charging techniques, two-current step, pulse, reflex charging, negative pulse discharge, intermittent charge, and interrupted charge control techniques. The superimposed pulse technique is also introduced here in brief.

The battery life and charging time are the key parameters focused on by most researchers. Through these techniques, it comes to know that there is scope for developing a proper charging technique to boost battery life in the current situation and prevent overcharging, undercharging, and deep charging problems.

Based on existing technological expertise and a proven manufacturing sector with outstanding safety and recycling records, lead–acid batteries will seek to function as part of a potential arsenal for energy storage systems with strong economic, technical, and environmental support. The influence of battery chemistry, correct use of active material for electrodes, placement of current collectors, and electrolyte on the efficiency of lead–acid batteries will be addressed in future research. Researchers, academicians, and industry professionals will have the opportunity to work on aging phenomena, weight, maintenance, temperature issues, and dead battery disposal.

References

Alaoui C, Salameh ZM (2003) Experiments in fast charging lead-acid electric vehicle batteries. In: Vehicular technology conference, vol 5, pp 3326–3331

Armstrong S, Glavin ME, Hurley WG (2008) Comparison of battery charging algorithms for stand-alone photovoltaic systems. In: PESC Record—IEEE Annual power electronics specialists conference, pp 1469–1475. https://doi.org/10.1109/PESC.2008.4592143

Bagshaw NE (1995) Lead alloys: past, present, and future. J Power Sources 53(1):25–30

Bayya M, Rao UM (2018) Comparison of voltage charging techniques to increase the life of lead-acid batteries. In: IEEE Symposium on a smart electronic system

Bhatt M, Hurley WG, Wolfle WH (2005) A new approach to intermittent charging of valve-regulated lead-acid batteries in standby applications. IEEE Trans Ind Electron 52(5):1337–1342

Buiel E, Dickinson E, Stoermer A, Schaeck S (2010) Dynamic charge acceptance of lead-Acid batteries in micro-hybrid board net. In: 12th European lead battery conference, Istanbul

Bullock KR (1995) Progress and challenges in bipolar lead-acid battery development. J Electrochem Soc 142(5):1726–1731

Cheng PH, Chen CL (2002) A high-efficiency fast charger for lead. A high-efficiency fast charger for lead-acid batteries. In: 28th Annual conference of the industrial electronics society, vol 2, pp 1410–1415

Christie S, Wong YS, Titelman G, Abrahamson J (2017) US Patent 9 543 589, 10 January 2017

Cooper A, Furukawa J, Lam LT, Kellaway M (2009) J Power Sources 188:58

Coux M, Muneret X, Lenain P, Wojkiewicz JL, Renard J (1999) Bipolar lead/acid batteries: effect of membrane conductivity on performance. J Power Sources 78(1–2):115–122

Davidson A (2019) Consort Batter Innov

Dietz H, Garche J, Wiesener K (1985) The effect of additives on the positive lead-acid battery electrode. J Power Sources 14(4):305–319

Ellis K, Hill A, Hill J, Loyns A, Partington T (2004) The performance of Ebonex® electrodes in bipolar lead-acid batteries. J Power Sources 136(2):366–371

Ferg EE, Billing DG, Venter AM (2017) Thermal characterization of tetrabasic lead sulfate used in the lead-acid battery technology. Solid State Sci 64

Ghaemi M, Ghafouri E, Neshati J (2006) Influence of the nonionic surfactant Triton X-100 on electrocrystallization and electrochemical performance of lead dioxide electrode. J Power Sources 157(1):550–562

Guo Y, Tang S, Meng G, Yang S (2009) Failure modes of valve-regulated lead-acid batteries for electric bicycle applications in deep discharge. J Power Sources 191(1):127–133

Hakim MS, Latif F, Khan MI, Basir A (2016) Design and implementation of three-stage battery charger for lead-acid battery. In: 2016 3rd International conference on electrical engineering and information communication technology (ICEEICT), Dhaka, 2016, pp 1–5

Horkos PG, Yammine E, Karami N (2015) Review on different charging techniques of lead-acid batteries. In: Proceedings of Third international conference on technological advances in electrical, electronics and computer engineering (TAEECE), Beirut, pp 27–3
http://www.axionpower.com
http://www.fireflyenergy.com.
http://www.batteryinnovation.org/combining-strengths-lead-anodes-for-lithium-ion-batteries/.
https://www.batteryinnovation.org/lead-battery-innovation-and-the-year-ahead/
Hua CC, Lin MY (2000) A study of charging control of lead-acid battery for electric vehicles. In: IEEE International symposium on proceedings of the 2000 industrial electronics, Cholula, Puebla, vol 1, pp 135–140
Jaiswal A, Chalasani SC (2015) The role of carbon in the negative plate of the lead-acid battery. J Energy Storage
Jose PJ (2019) 12V DC trickle charger. https://doi.org/10.13140/RG.2.2.21832.37127
Kagiri C, Xia X (2017) Optimal control of a hybrid battery/supercapacitor storage for neighborhood electric vehicles. Sci Direct
Kao W-H (1998) Substrate materials for bipolar lead/acid batteries. J Power Sources 70(1):8–15
Kapitza P, Heath HF (1923) GB226857
Karami H, Shamsipur M, Ghasemi S, Mousavi MF (2007) Lead-acid bipolar battery assembled with primary chemically formed positive pasted electrode. J Power Sources 164(2):896–904
Kim SH, Kim JH, Cho SJ, Lee SY (2019) All-solid-state printed bipolar Li–S batteries. Adv Energy Mater 9:1901841
Lach J, Wróbel K, Wróbel J, Podsadni P, Czerwiński A (2019) Applications of carbon in lead-acid batteries: a review. J Solid State Electrochem
Lafollette RM (1990) Design fundamentals of high power density, pulsed discharge, lead-acid batteries: II. Modeling. J Electrochem Soc 137(12):3701
Lafollette RM (1995) Proceedings of the tenth annual battery conference on applications and advances, IEEE, Piscataway, NJ, p 43
Lam LT, Ozgun H, Lim OV (1995) Pulsed-current charging of lead/acid batteries – a possible means for overcoming premature capacity loss? J Power Sources 53:215–228
Lam LT, Haigh NP, Phyland CG, Rand DAJ (2012) US Patent 8 232 006, 31 July 2012
Lang X, Xiao Y, Cai K, Li L, Zhang Q, Yang R (2017b) High-performance porous lead/graphite composite electrode for bipolar lead-acid batteries. Int J Energy Res 41:1504
Lang X, Li Y, Cai K, Li L, Zhang Q, Wu H (2018) Preparation of bipolar lead-carbon electrode and study on its electrochemical performance. Ionics 24:935
Lang X, Wang D, Zhu J (2014a) Modified titanium foil's surface by high-temperature carbon sintering method as the substrate for bipolar lead-acid battery. J Power Sources 272:176–182
Lang X, Wang D, Tang S, Zhu J, Guo C (2014b) Study on titanium foil coated with partial reduction titanium dioxide as bipolar lead-acid battery's substrate. J Power Sources 271:354–359
Lang X, Zhao Y, Cai K, Li L, Zhang Q, Wu H (2017) Ionics
Lannelongue J, Cugnet M, Guillet N, Kirchev A (2017) Electrochemistry of thin-plate lead-carbon batteries employing alternative current collectors. J Power Sources
Liu T, Yuan Y, Tao X, Lin Z, Lu J (2020) Bipolar electrodes for next-generation rechargeable batteries. Adv Sci
Lun-Shu YR, Grenness M, Fuller RJ (1987) US4637970
Metzendorf H (1982) The capacity limiting the role of the electronic conductivity of the active material in lead-acid batteries during discharge. J Power Sources 7(3):281–291
Moseley PT, Rand DAJ, Peters K (2015b) Enhancing the performance of lead-acid batteries with carbon—in pursuit of an understanding. Power Sources
Muneret X, Coux M, Lenain P (2000) Analysis of the partial charge reactions within a standby VRLA battery leading to an understanding of intermittent charging techniques. In: Telecommunications energy conference, Phoenix, AZ, pp 293–298
Ng KS, Moo CS, Chen YP, Hsieh YC (2008) State-of-charge estimation for lead-acid batteries based on dynamic open-circuit voltage. In: Power and Energy Conference, Johor Bahru, pp 972–976

Paleska I, Pruszkowska-Drachal R, Kotowski J, Rogulski Z, Milewski JD, Czerwinski A (2004) Electrochemical behaviour of barium metaplumbate as a lead carrier. J Power Sources 129(2):326–329
Pavlov D, Ruevski S (2001) Semi-suspension technology for preparation of tetrabasic lead sulfate pastes for lead-acid batteries. J Power Sources 95:191–202
Ponraj R, Mcallister SD, Cheng IF, Edwards DB (2009) Investigation on electronically conductive additives to improve positive active material utilization in lead-acid batteries. J Power Sources 189(2):1199–1203
Praisuwanna N, Khomfoi S (2013) A seal lead-acid battery charger for prolonging battery lifetime using superimposed pulse frequency technique. In: 2013 IEEE Energy conversion congress and exposition, Denver, CO, pp 1603–1609
Reichman B (1991) US5017446
Reid DP, Glasa I (1984) A new concept: intermittent charging of lead-acid batteries in telecommunication systems. In: Proceedings of IEEE INTELEC'84, pp 67–71
Serhan HA, Ahmed EM (2018) Effect of the different charging techniques on battery life-time: review. In: 2018 International conference on innovative trends in computer engineering (ITCE), Aswan, pp 421–426
Shaffer EO II (2016) Advancing the energy power envelope of lead-acid batteries with bipolar design. In: 15th European lead battery conference, Valletta Malta
Shafiei N, Arefifar SA, Saket MA, Ordonez M (2016) High-efficiency LLC converter design for universal battery chargers. In: 2016 IEEE Applied power electronics conference and exposition (APEC), Long Beach, CA, pp 2561–2566
Shimoura I, Wei C-T (2018) New technology for industrial lead acid battery. Hitachi chemical technical report no. 60. Hitachi Chemical Energy Technology Co. Ltd
SmithBucklin Statistics Group (2019) National recycling rate study
Valeriote EM, TG Jochim DM (1994) Fast charging of lead-acid batteries. In: Battery conference on applications and advances, Proceedings of the ninth annual, Long Beach, CA, pp 33–38
Volfkovich YM, Serdyuk TM (2002) Russ J Electrochem 38:935
Wertz JA (2001) Proceedings of sixteenth annual battery conference on applications and advances, IEEE, Piscataway, NJ, USA, p 105
Yifeng G, Chengning Z (2011) Study on the fast charging method of lead-acid battery with negative pulse discharge. In: Power electronics systems and applications, 4th International conference, Honk Kong, pp 1–4

Chapter 22
A Recursive PID Tuning Approach for the Inherently Unstable System

Pankaj Swarnkar and Harsh Goud

Abstract PID (proportional–integral–derivative) control is the quicker control approaches. It has an uncomplicated control arrangement which was grasped by plant engineers and which they got comparatively manageable to tune. As various control processes employing PID have confirmed adequate, it still has a broad range of applications in automated control. Meanwhile, various researchers in the area of process control system observed that the design of PID controller is very tedious if the plant is highly nonlinear. Controller synthesis is a very difficult task for an unstable system because there are closed-loop performance and specific configuration constraints that range and narrow available solutions. These restrictions reveal peaks in sensitivity functions, overshoots, and overall system bandwidth. Selection of PID control design approach and arrangement is based on considerably a previous theory of the plant demands and process dynamics. The two most common conventional PID tuning procedures were the frequency response (cycling) and time response (reaction curve) experiment followed by proportional control. This chapter mainly focuses on the design and analysis of PID tuning for an unstable system. A related investigation of conventional (Pessen integral rule) and real-time online tuning techniques is also represented based on individual simulation examinations. The study confirms the effectiveness of suitable tuning techniques to regulate the unstable system for getting the desired performance.

Keywords PID controller · Controller tuning · Unstable system · Online PID tuner · SISO tool

P. Swarnkar (✉)
Pankaj Swarnkar-1 Maulana Azad National Institute of Technology, Bhopal, Madhya Pradesh, India

H. Goud
Indian Institute of Information Technology, Nagpur, Maharashtra, India
e-mail: hgoud@iiitn.ac.in

A. K. Bohre et al. (eds.), *Planning of Hybrid Renewable Energy Systems, Electric Vehicles and Microgrid*, Energy Systems in Electrical Engineering,
https://doi.org/10.1007/978-981-19-0979-5_22

Nomenclature

PIR	Pessen Integral Rule
PID	Proportional–Integral–Derivative
ZN	Ziegler–Nichols
LQG	Linear–Quadratic–Gaussian

22.1 Introduction

Real-world systems are naturally unstable when analyzed over the broad functioning range while various of the systems are supposed to "act" as linear in the environment of the specific operating condition at moderate speeds under particular consideration (Albertos et al. 1997; Zarei 2020; Precup et al. 2020; Priyanka et al. 2020; Boonpramuk et al. 2019; Fiser and Zítek 2019). Numerous physical process systems are denoted by nonlinear models. Examples incorporate drag on a vehicle in movement, chemical reactor, coulomb friction, robotic manipulator, electrostatic and gravitational attraction, electrical or electronics characteristics, etc. Recently, many researchers from such abovementioned broad area have revealed an intense concern in the modeling and examination of nonlinear control strategies. However, the main challenges force constantly striking into nonlinearities (Iqbal and Ullah 2017). Main causes behind the increasing concern in nonlinear control cover are investigations of strong nonlinearities, advancement of linear control systems, and requirement of dealing with model changes and design simplicity. PID controller is broadly employed in several industrial applications due to its simple design and structure. However, it is a challenge to obtain the desired control performance in the appearance of time delays, anonymous nonlinearities, disturbances as well as variation in system parameters. In this era, several control techniques have been invented for the industrial application. Though PID controller appeared as the simplest and usually accepted classical controller (Goud and Swarnkar 2019a, 2018; Rajesh and Dash 2019; Khan et al. 2019; Gaidhane et al. 2019; Salman and Jafar 2019; Khanduja et al. 2019). PID control tuning is the mathematical and practical procedure for obtaining the optimized value of controller parameters. Conventional controller tuning provides reliable and adequate action for linear time-invariant systems (Dubey et al. 2021; Goud et al. 2021; Swarnkar and Goud 2021; Ekinci and Hekimoglu 2019; Goud and Swarnkar 2019b).

Though, as the process is conducted in the limit enough nearly to a given set point, it cannot guarantee to perform well with fluctuations in set points or environmental conditions. Controller using online tuning for nonlinear systems should adjust variations in the variable and transfer them to control technique such that the performance of the control system is not degraded in spite of variations in the environment changes (Erol 2021; Feng et al. 2021; Jayaswal et al. 2021; Ghosh et al.

2021; Goud and Swarnkar 2019c; Swarnkar and Goud 2020; Noordin et al. 2021; Guo et al. 2021; Pongfai et al. 2021; Gopi and Reddy 2015).

In this investigation, an attempt is made to carry out a comparative analysis of offline and online PID tuning techniques for the unstable system. Pure unstable modes normally occur in the system such as an inverted pendulum system and flight system. Both classical and advanced online PID tuning methods are employed on unstable system to obtain satisfactory performance. The rest of this book chapter is ordered as follows: Sect. 22.2 illustrates the unstable system model. Section 22.3 explains the classical control action. Section 22.4 summarizes classical modes of PID tuning. Section 22.4 describes the online controller tuning for unstable system. Eventually, a short conclusion is discussed in Sect. 22.5.

22.2 Unstable System Model

Controller design of unstable systems could be tried applying many of the techniques suitable to the usual stable system but the outcomes may not be adequate. Controller design for stable system is based on various linear opinions when applied to unstable system practically may fail. This situation, accordingly, requires the sensible and accurate design of the controller. The system, which is now used for the investigation of controllers, is the third-order unstable system with feedforward transfer function shown in Eq. (22.1) (Swarnkar and Goud 2021, 2020).

$$G(s) = \frac{1000}{(1 + 0.01s)(0.01s^2 + s)} \tag{22.1}$$

and feedback transfer function H(s) = 1.

The step response of the plant without any controller is exhibited in Fig. 22.1.

It can be remarked that oscillations of the system are rising with time and becoming the system unstable. In an unstable system, the amount of overshoot is closely related to rise time. Unexpectedly, the higher the rise times more the overshoots, which is opposed to the cause of stable plants. Bode plot is given for the loop transfer function in Fig. 22.2. The gain crossover frequency (ωg) is calculated as 200 rad/sec whereas the phase crossover frequency (ωp) is calculated as 100 rad/sec. The phase and gain margins measured by the plot are −37° and −13.9 dB, respectively. A negative sign of margins reveals the instability of the system. To check the stability range with respect to feedforward gain of the system, the root locus plot is also shown in Fig. 22.3. This plot shows very small stability range for the plant under consideration. System becomes marginally stable for feedforward gain of 0.2 only, and beyond this system becomes unstable.

The traditional and advanced online control techniques are now used to examine their applicability and usefulness for such uncontrolled plant (Swarnkar and Goud 2021, 2020).

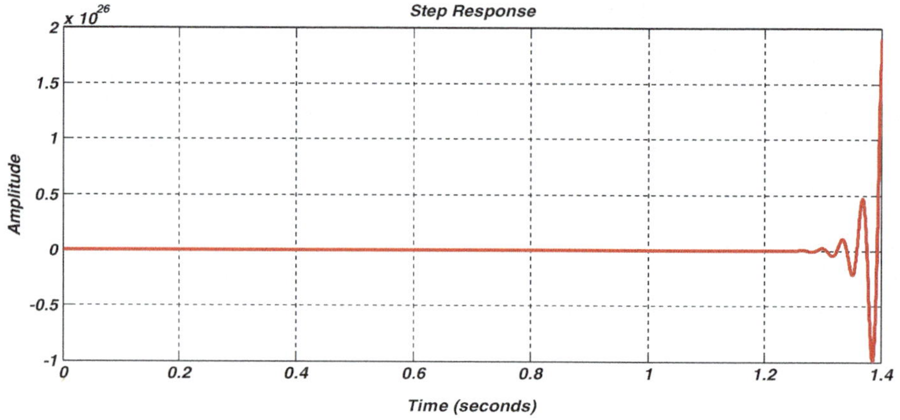

Fig. 22.1 Time response characteristic of third-order unstable system (Swarnkar and Goud 2021, 2020)

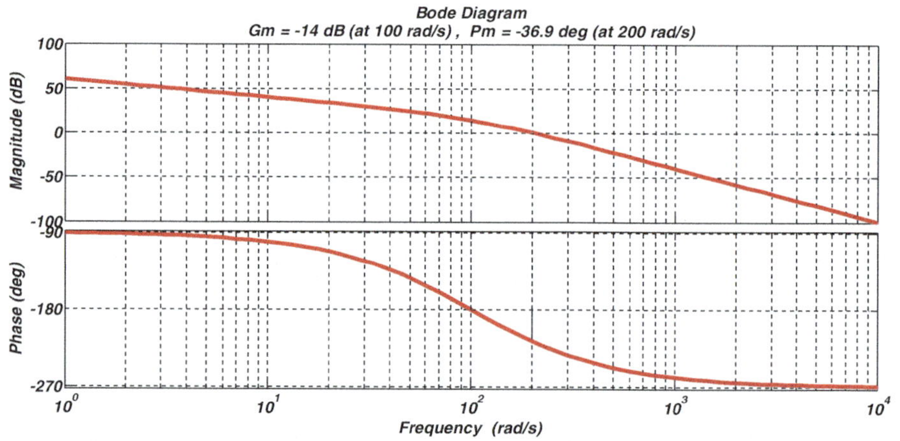

Fig. 22.2 Bode plot for third-order unstable system (Swarnkar and Goud 2021, 2020)

22.3 Classical Control Action

All standard design control methods are beneficial for PID controllers. Some novel approaches are also revealed to evaluate parameters for PID controllers and these approaches are often called tuning techniques. The most popular offline tuning methods are invented by Ziegler–Nichols (ZN). By using proper tuning techniques, errors can be overcome and output can move toward given desired value. In some circumstances, where the system has a notable amount of instability, the traditional tuning techniques are unsuccessful to give an excellent solution. A schematic diagram of PID control operation is revealed in Fig. 22.4.

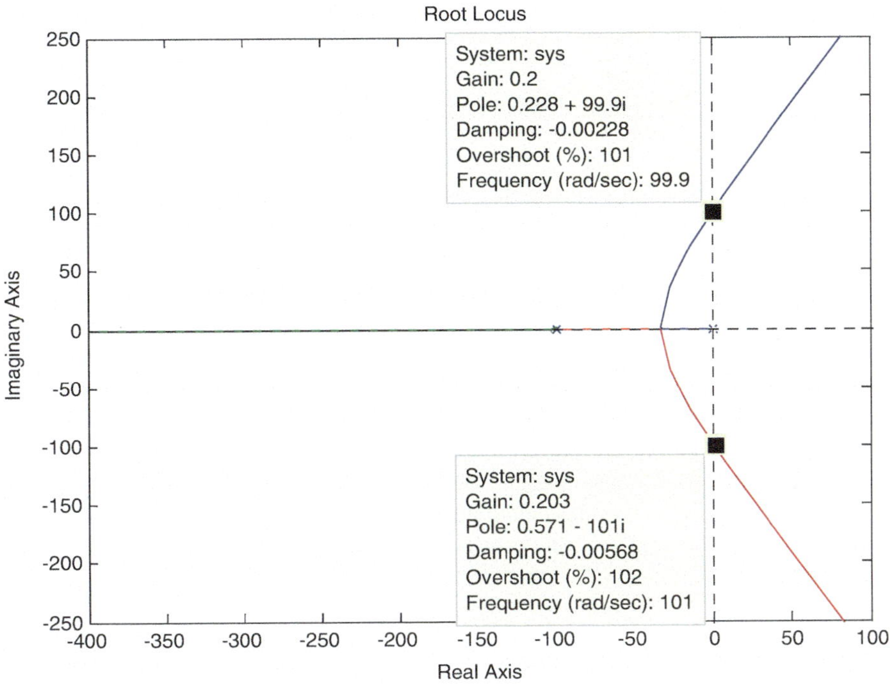

Fig. 22.3 Root locus plot for third-order unstable system (Swarnkar and Goud 2021, 2020)

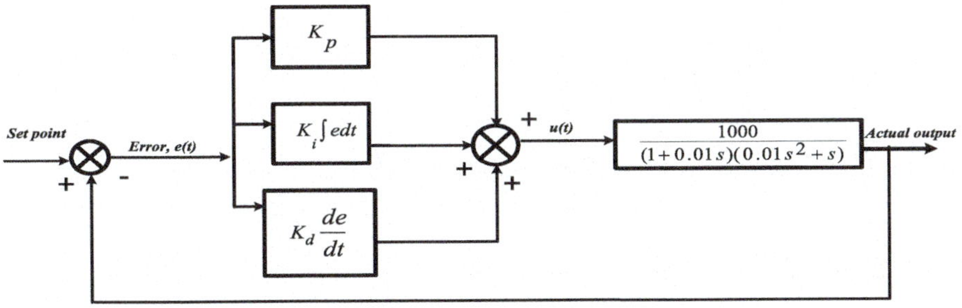

Fig. 22.4 A schematic diagram of PID control action (Swarnkar and Goud 2020, 2021)

In conventional control, once these parameters are decided then they remain fixed throughout the process irrespective of the variations in the system. In PID controller, the control signal generated by the controller is defined by Eq. (22.2).

$$u(t) = K_p e(t) + K_i \int e(t)dt + K_d \frac{de(t)}{dt} \tag{22.2}$$

The transfer function of PID controller is defined in Eq. (22.3).

Table 22.1 PID tuning using Pessen integral rule (PIR) (Pongfai et al. 2021)

Rule	Controller	k_p	τ_i	τ_d
Pessen integral rule (PIR)	PID	$0.7 \times k_u$	$\frac{T_u}{2.5}$	$\frac{3 \times T_u}{20}$

$$G_{PID}(s) = \frac{U(s)}{E(s)} = K_p + \frac{K_i}{s} + sK_d \tag{22.3}$$

where K_p is the proportional controller parameter, K_i is the integral controller parameter, and K_d is the derivative controller parameter, e(t) is the error signal, and u(t) is the control signal. These gains fix the role of each individual controller in the process. Correct choice of these gains is actually the base for the perfect controlling (Albertos et al. 1997; Fiser and Zítek 2019; Goud and Swarnkar 2019b). For the nonlinear system, the measurement of these parameters is a challenging assignment. This book chapter highlights the importance of classical and advanced (online) tuning techniques. As these techniques calculate the values of Kp, Ki, and Kd based on their previous values, the next section describes the mathematical model of its execution.

22.3.1 *Classical PID Tuning*

A traditional tuning method for PID controller was invented by ZN known as Pessen Integral Rule (PIR). The procedure for tuning PID gains using PIR is similar to the procedure of ZN PID tuning. Traditional techniques are offline methods, once the controller gains are set cannot be adjusted during the process which can be overcome by using smart or online techniques.

The steps for getting controller gains using PIR are as follows:

- Put $K_d = 0$ and $K_i = 0$, system is proportional.
- Take the characteristic equation and by using Routh criteria find the value of $K_p = K_u$.
- Using characteristic equation and Routh criteria find the value of Tu.
- Find out the values of controller gains using Table 22.1 (Pongfai et al. 2021).

22.3.2 *System Response with Classical PID Tuning*

The time response characteristic of the unstable system controlled with PIR technique is shown in Fig. 22.5. Remarkable improvement in system characteristic can be examined with the use of PID controller. The performance of the unstable system is examined in terms of rise time, peak time, settling time, and overshoot. By using the PIR-PID tuning approach highest overshoot is reduced to 56.9987% and settling time to 0.1357 s. which is comparatively better than unstable system. Table 22.3 shows

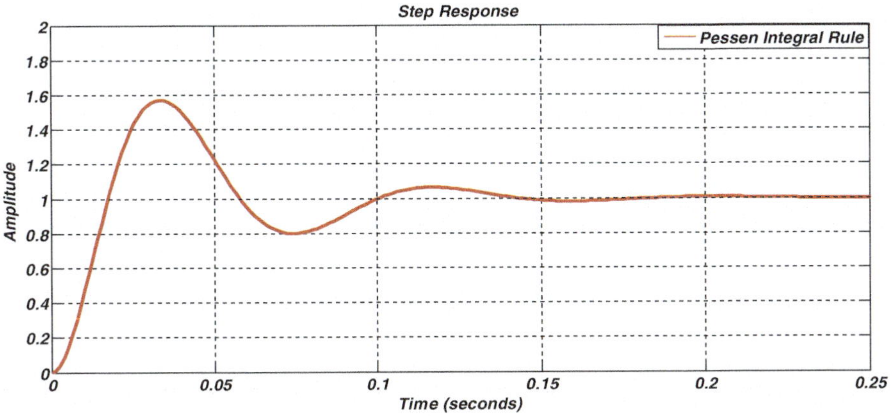

Fig. 22.5 Time response characteristic of third-order unstable system using PIR tuning technique

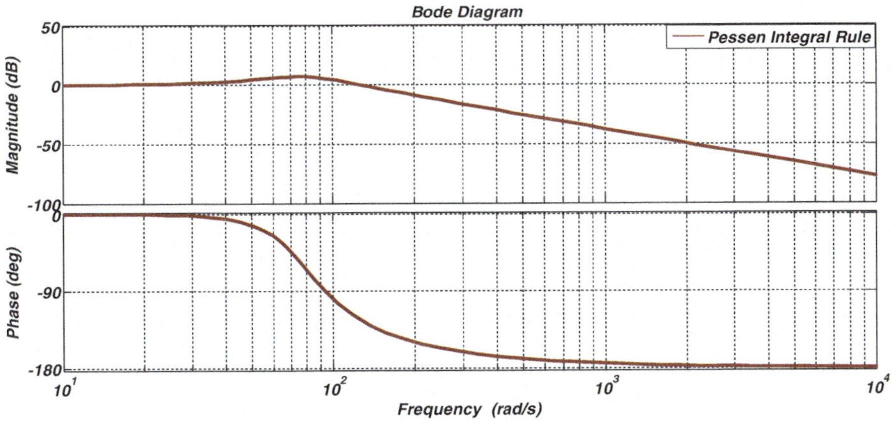

Fig. 22.6 Bode plot for third-order unstable system using PIR tuning technique

the values of the PID controller parameters obtained by various tuning techniques. Figure 22.6 shows the Bode plot of system with PIR technique in which calculated values of phase margin and gain margin are positive that shows system is stable and that parameters are summarized in Table 22.2.

22.4 Online Controller Tuning for Unstable System

The offline conventional method has improved the performance of the system under consideration which can be further improved by advanced online tuning techniques. This section presents the controlling of unstable system using advanced online tuning

Table 22.2 Bode plot parameters of different PID tuning

Parameter	Unstable system	PIR	LQG synthesis
Phase margin	−37°	55.8°	16.3°
Delay margin	0.0282 s at frequency 200 rad/s	0.0077 s at frequency 127 rad/s	0.000765 s at frequency 371 rad/s
Peak gain	460 dB at frequency 1e-20 rad/s	6.95 dB at frequency 73.2 rad/s	4.58 dB at frequency 201 rad/
Closed-loop stability	No	Yes	Yes

Table 22.3 PID parameters obtained by advance (online) and classical PID tuning for unstable system

Parameters	Tuning method		
	Advanced tuning (online)		Classical PID tuning
	With PID tuner	With signal constraint box	Pessen integral rule
K_p	0.099254	0.0325	0.1400
K_i	0.090928	−0.0069	5.5704
K_d	0.0019233	2.2402e-054	0.0013

methods and "SISOTOOL" in the Control System Toolbox, PID tuner block, and signal constraint box. Finally, results will be compared for all tuning methods.

22.4.1 Controller Design with SISOTOOL:

"SISOTOOL" is one of the advanced features in the Control System Toolbox in MATLAB that facilitates software-based controller configuration for single-input single-output (SISO) systems. A typical design workflow with the SISOTOOL involves the following steps:

- Run the transfer function of an inherently unstable system in the editor or command window of MATLAB.
- Type "SISOTOOL" at the command prompt.
- A new window SISO design task is open.
- Import the system model (transfer function) of an inherently unstable system.
- Design controllers using Linear–Quadratic–Gaussian (LQG) synthesis design method in automated tuning and update compensator.
- Now Eq. (22.4) of compensator is found.

Now click on show analysis plot and investigate control system designs using frequency-domain responses and time domain such as step response (Fig. 22.7) and bode plot (Fig. 22.8). Time response characteristic reveals the remarkable change in performance with the higher overshoot as decrease as 46.2%. In Bode plot the gain

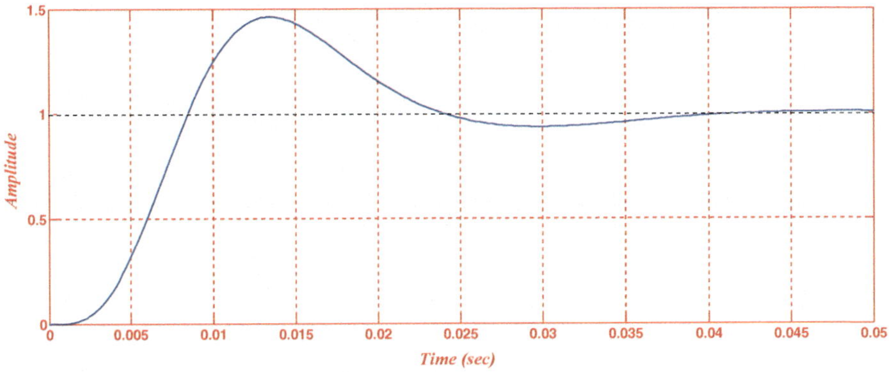

Fig. 22.7 Time response graph with tuned LQG synthesis tuning technique

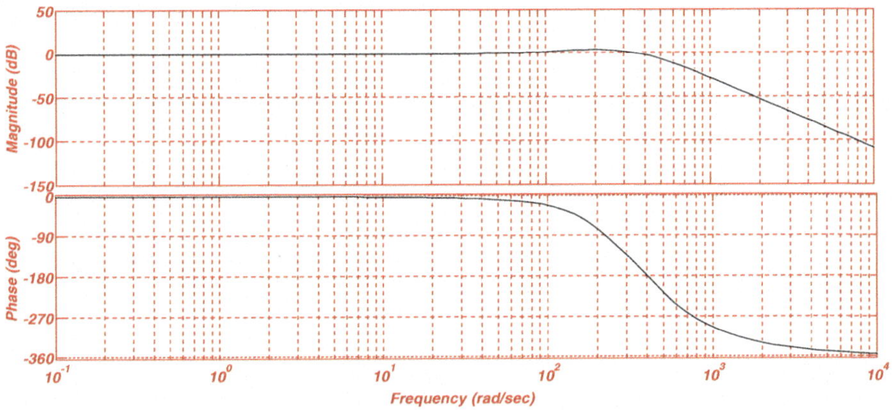

Fig. 22.8 Bode plot with tuned LQG synthesis tuning technique

and phase margins calculated by the graph are positive which indicate the stability of the system. Observations by Bode plot are summarized in Table 22.2, where the results are compared with conventional tuning method also.

$$Compensator = 0.40446 \times \frac{(1+s)(1+0.012s+(0.0064s)^2)}{s(1+0.0015s)(1+0.0014s+(0.0017s)^2)} \quad (22.4)$$

22.4.2 PID Tuning with Signal Constraint Box

In this method of online tuning, the desired output characteristic is defined in advance and controller is tuned for obtaining this characteristic. Following steps are adopted for performing tuning using signal constraint box in MATLAB.

- Open PID controller box and assign the controller gains in the form of variables only, i.e., Kp, Ki, Kd.
- Open signal constraint box and set the limits of output waveform as shown in Fig. 22.9.
- Go to optimization block then Tuned parameters and add the variables one by one.
- Assign the initial guess to each variable.
- Run the system and start the optimization.

The system performance is checked for different set of controller gains. After iteration the best possible set of controller gain is chosen for getting the desired performance of the system. Figure 22.9 shows the number of iterations with the time response characteristics for different sets of controller gains. Finally, the feasible or optimal solution provided by this tuning within the specified tolerances defines the controller gains as shown in Table 22.3.The time response characteristic with this optimal set of controller gains is shown in Fig. 22.10. It is obvious from the characteristic that system becomes stable with reasonable specifications. The maximum overshoot is limited to 11.35% with settling time 0.225 s which shows the great improvement in performance as compared to conventional offline tuning method.

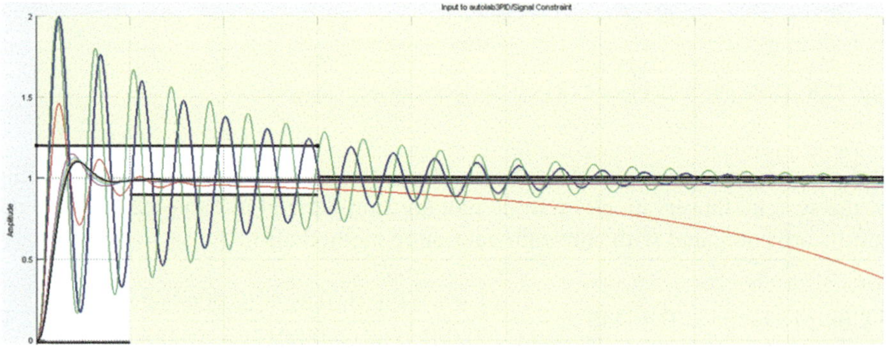

Fig. 22.9 Time response characteristics for different sets of controller gains

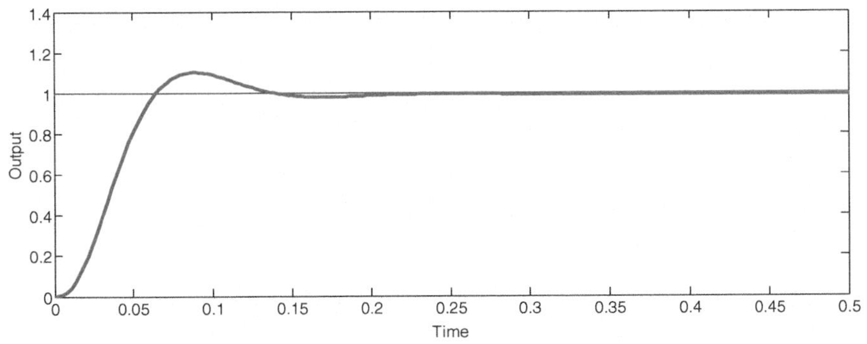

Fig. 22.10 Time response characteristic for optimal values of controller gains

22.4.3 *Controller Design with Online PID Tuner*

PID tuner in MATLAB gives a quick and extensively used single-loop PID tuning approach for the Simulink. This technique is employed to tune controller parameters to obtain a strong PID controller design with the required reaction time. A common design step with the online tuner includes the subsequent steps:

- Start the PID tuner.
- When import, the MATLAB software calculates a model from the designs of Simulink model and primary controller.
- Tune the parameters in the Graphical User Interfaces (GUIs), by manually fixing design rules in two design approaches.
- The tuner calculates controller parameters that strongly sustain the system.
- Export the parameters of the designed controller following the PID controller and validate controller performance in Simulink (Gopi and Reddy 2015).

After performing the tuning using PID tuner, controller gains are obtained as shown in Table 22.3.The time response characteristic of the given unstable system with this tuned PID controller is shown in Fig. 22.11. Characteristic shows the remarkable improvement in performance with maximum overshoot as low as 7.13%. This not only makes the system stable but making it convenient for all practical applications. Bode plot of such system is revealed in Fig. 22.12. Graph shows that the system is having both the margins positive with sufficient magnitudes.

Unstable system is not only showing the bonded output but performance is also quite satisfactory with the adaptive schemes. The performance of the system based on time response specifications is summarized in Table 22.4 which shows the importance of suitable tuning of controller for obtaining the optimal performance of uncontrolled system. Performance of such conventional controllers can be used to understand complex behaviors of the system. PIR response yields higher overshoot which can be slightly overcome by using tuning using Control System Toolbox. The analysis of Control System Toolbox for the design of compensator used in LQG synthesis

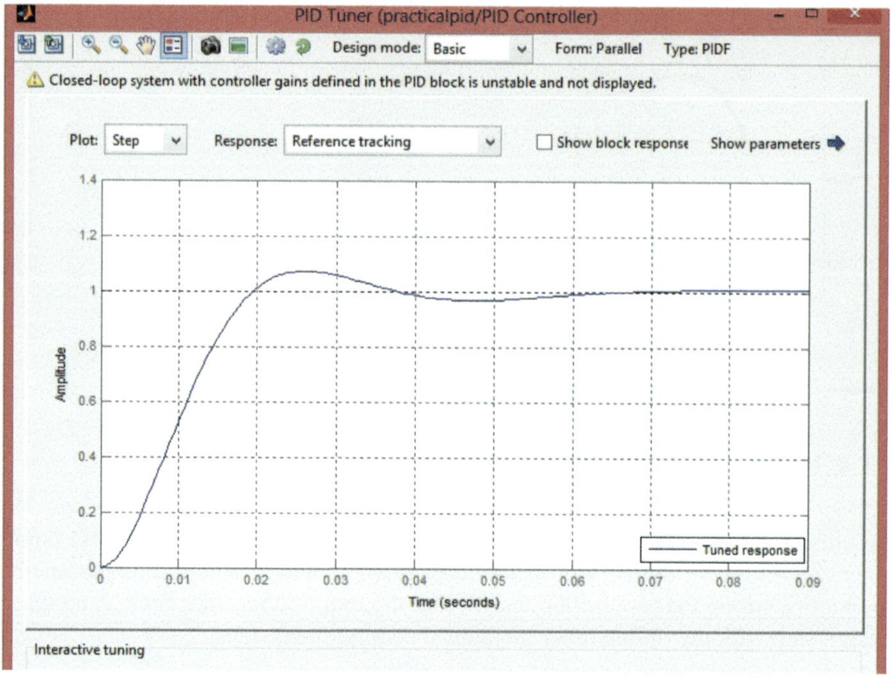

Fig. 22.11 Time response of unstable system with auto-tuned PID controller

reveals that tuning technique used is robust toward uncertainty in system parameters and it can be strongly executed to any process industries.

22.5 Conclusion

Conventional PID controllers are suitable for performance improvement of stable system while controlling the unstable system which is the challenging job for controllers. If the calculated values of phase and gain margins are negative with large magnitudes for unstable systems, then the performance improvement of such system is not feasible for traditional rigid gain controllers. Tuning of PID controller is a tedious task for such an unbounded plant. With detailed simulation studies this work presents the design and implementation of traditional and advanced online tuned PID controller for controlling the highly unstable system. It is found that an unstable system is not only giving the bonded response but performance is also quite satisfactory with the suitable adaptive control. The performance of such traditional controllers can be employed to examine the complicated behavior of the system. PIR response produces a larger maximum overshoot which can be insignificantly reduced by applying LQG synthesis. The settling time (0.0379 s) and overshoot (46.2%) are

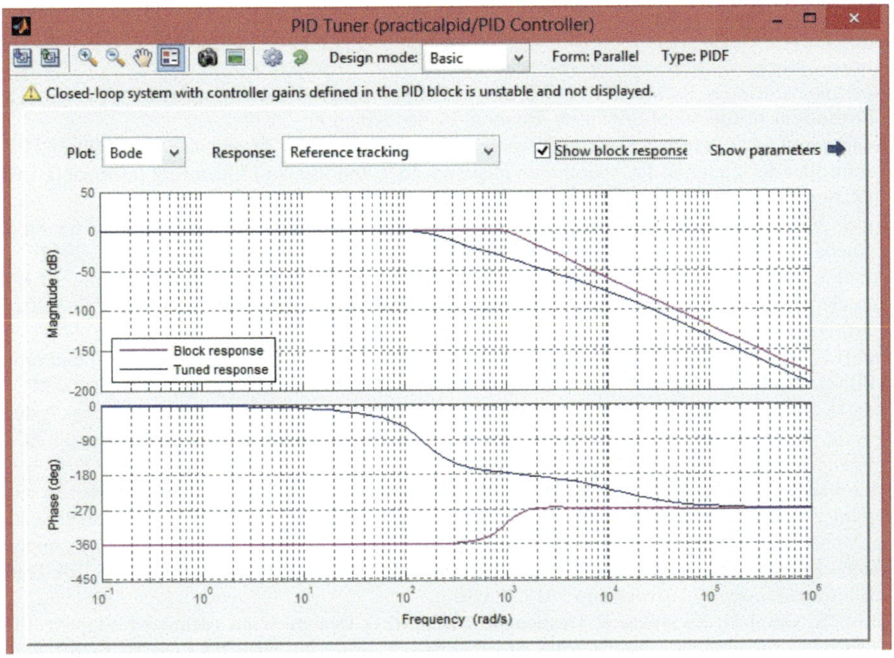

Fig. 22.12 Bode plot for the system tuned with online PID tuner

Table 22.4 Step response specification of different PID tuning

Specifications	Tuning method			
	Pessen integral rule	Advance (online)		SISOTOOL
	Pessen integral rule	With online PID tuner	With signal constraint box	LQG synthesis
Rise time (s)	0.0118	0.0129	0.0618	0.00458
Peak time (s)	0.0332	1.07	0.085	0.0135
Maximum overshoot (%)	56.9987	7.13	11.35	46.2
Settling time (s)	0.1357	0.0561	0.225	0.0379

under the prescribed limit with LQG synthesis tuning method. The performance of the unstable system is further enhanced with the employment of online advanced tuning techniques. The online PID tuning techniques exhibit that appropriate tuning procedure can not only drag one unstable system toward stability but can also improve its performance to large extent.

References

Albertos P, Strietzel R, Mort N (1997) Control engineering solution-a practical approach. The Institutions of Electrical Engineers, London, United Kingdom

Boonpramuk M, Tunyasirut S, Puangdownreong D (2019) Artificial intelligence-based optimal PID controller design for BLDC motor with phase advance. Indonesian J Electr Eng Informat (IJEEI) 28. https://doi.org/10.52549/ijeei.v7i4.1372

Dubey V, Goud H, Sharma P (2021) Comparative analysis of PID tuning techniques for blood glucose level of diabetic patient. Turk J Comput Math Educ (TURCOMAT) 12:2948–2953

Ekinci S, Hekimoglu B (2019) Improved kidney-inspired algorithm approach for tuning of PID controller in AVR system. IEEE Access 7:39935–39947. https://doi.org/10.1109/ACCESS.2019.2906980

Erol H (2021) Stability analysis of pitch angle control of large wind turbines with fractional order PID controller. Sustain Energy Grids Netw 26. https://doi.org/10.1016/j.segan.2021.100430

Feng H, Ma W, Yin C, Cao D (2021) Trajectory control of electro-hydraulic position servo system using improved PSO-PID controller. Autom Constr 127. https://doi.org/10.1016/j.autcon.2021.103722

Fiser J, Zítek P (2019) PID controller tuning via dominant pole placement in comparison with ziegler-nichols tuning. IFAC-Papers OnLine 1 2019. https://doi.org/10.1016/j.ifacol.2019.12.204

Gaidhane PJ, Nigam MJ, Kumar A, Pradhan PM (2019) Design of interval type-2 fuzzy precompensated PID controller applied to two-DOF robotic manipulator with variable payload. ISA Trans 1. https://doi.org/10.1016/j.isatra.2018.12.030

Ghosh S, Goud H, Swarnkar P, Deshpande DM (2021) Design of an optimized adaptive PID controller for induction motor drive. Mechatr Syst Control 49. https://doi.org/10.2316/J.2021.201-0197

Gopi P, Reddy PotlaLinga (2015) A robust decentralized controller design for interconnected power system with random load perturbations using SDO software. Proc Technol 21:406–414. https://doi.org/10.1016/j.protcy.2015.10.060

Goud H, Swarnkar P (2018) Signal synthesis model reference adaptive controller with artificial intelligent technique for a control of continuous stirred tank reactor. Int J Chem Reactor Eng 17. https://doi.org/10.1515/ijcre-2018-0145

Goud H, Swarnkar P (2019b) Signal synthesis model reference adaptive controller with genetic algorithm for a control of chemical tank reactor. Int J Chem Reactor Eng 17. https://doi.org/10.1515/ijcre-2018-0199

Goud H, Swarnkar P (2019c) Analysis and simulation of the continuous stirred tank reactor system using genetic algorithm. In: Harmony search and nature inspired optimization algorithms. Springer, Singapore. https://doi.org/10.1007/978-981-13-0761-4_106

Goud H, Sharma PC, Swarnkar P, Shankar VG, Sharma VP, Sahu AK (2021) A comparative analysis of conventional PID tuning techniques for single link robotic arm. Solid State Technol 64

Goud H, Swarnkar P (2019a) Investigations on metaheuristic algorithm for designing adaptive pid controller for continuous stirred tank reactor. Mapan 34:113–119. https://doi.org/10.1007/s12647-018-00300-w

Guo B, Zhuang Z, Pan JS, Chu SC (2021) Optimal design and simulation for PID controller using fractional-order fish migration optimization algorithm. IEEE Access 9:8808–8819. https://doi.org/10.1109/ACCESS.2021.3049421

Iqbal J, Ullah M, Khan SG, Cukovic BKS (2017) Nonlinear control systems—a brief overview of historical and recent advances. Nonlinear Eng 6:301–312. https://doi.org/10.1515/nleng-2016-0077

Jayaswal K, Palwalia DK, Kumar S (2021) Performance investigation of PID controller in trajectory control of two-link robotic manipulator in medical robots. J Interdisc Math 24:467–478. https://doi.org/10.1080/09720502.2021.1893444

Khan IA, Alghamdi AS, Jumani TA, Alamgir A, Awan AB, Khidrani A (2019) Salp swarm optimization algorithm-based fractional order PID controller for dynamic response and stability

enhancement of an automatic voltage regulator system. Electronics 8:1472. https://doi.org/10.3390/electronics8121472
Khanduja N, Bhushan B (2019) CSTR control using IMC-PID, PSO-PID, and hybrid BBO-FF-PID controller. In: Applications of artificial Intelligence techniques in Engineering 2019. Springer, Singapore. https://doi.org/10.1007/978-981-13-1822-1_48
MATLAB optimization toolbox, the MathWorks, Inc., Natick, MA, USA
Noordin A, Basri MA, Mohamed Z, Lazim IM (2021) Adaptive PID controller using sliding mode control approaches for quadrotor UAV attitude and position stabilization. Arab J Sci Eng 46:963–981. https://doi.org/10.1007/s13369-020-04742-w
Pongfai J, Angeli C, Shi P, Su X, Assawinchaichote W (2021) Optimal PID controller autotuning design for MIMO nonlinear systems based on the adaptive SLP algorithm. Int J Control Autom Syst 1:392–403. https://doi.org/10.1007/s12555-019-0680-6
Precup RE, Teban TA, Albu A, Borlea AB, Zamfirache IA, Petriu EM (2020) Evolving fuzzy models for prosthetic hand myoelectric-based control. IEEE Trans Instr Measur 69. https://doi.org/10.1109/TIM.2020.2983531
Priyanka EB, Maheswari C, Thangavel S, Bala MP (2020) Integrating IoT with LQR-PID controller for online surveillance and control of flow and pressure in fluid transportation system. J Ind Inf Integr 1. https://doi.org/10.1016/j.jii.2020.100127
Rajesh KS, Dash SS (2019) Load frequency control of autonomous power system using adaptive fuzzy based PID controller optimized on improved sine cosine algorithm. J Ambient Intell Humaniz Comput 10:2361–2373. https://doi.org/10.1007/s12652-018-0834-z
Salman GA, Jafar AS (2019) Application of artificial intelligence techniques for LFC and AVR systems using PID controller. Int J Power Electron Drive Syst 10:1694. https://doi.org/10.11591/ijpeds.v10.i3.1694-1704
Swarnkar P, Goud H (2020) Design of fuzzy adaptive pi controller for inherently unstable system. SSRN 3623747
Swarnkar P, Goud H (2021) Design and tuning of PID controller for an inherently unstable system. In: Smart computing. https://doi.org/10.1201/9781003167488
Zarei M (2020) A physically based PID controller for the power maneuvering of nuclear reactors. Prog Nucl Energy 1. https://doi.org/10.1016/j.pnucene.2020.103431

Chapter 23
A Review on Motor and Drive System for Electric Vehicle

Suchita Roy and Rahul Pandey

Abstract Due to the increasing adverse effect of global warming and the serious concern related to the depletion of fossil fuels our society is moving rapidly towards the use of electric vehicle. Nowadays, electric vehicles are preferred since they are environment friendly as they emit less greenhouse gases and less pollutant than conventional drives. The improvements in power electronic technologies have made it possible to achieve optimum performance of electrical vehicles for different load conditions. Freshly adopted techniques for driving the electric vehicles driven by induction motor which is utilized by the industries and the civilized society as well as associations that moves into viable expertise and recommended choice for the use of electric vehicles. In this paper investigation has been done over drive systems and different loss minimization techniques for EVs so that the controller can be designed for enhancing the efficiency of induction motor in every load condition by inclusion of core losses. This paper gives an evaluation and assessment over various loss minimization techniques used while driving the induction motor. It would help to operate the motor smoothly at optimum torque by elevating the efficiency/performance of the system during high load as well as in light load conditions. The various loss minimization methods generalize the system operation by reducing the copper loss and iron loss, reduced torque and speed control of the system. As the benefits of electric vehicle, countries can meet the environmental targets as well as provide the best alternative for the conventional drives by achieving the desired performance as required by the society and industries.

Keywords Electric vehicle · Motor drive system · Induction motor · Loss minimization · Torque · Efficiency

S. Roy (✉) · R. Pandey
Shri Shankaracharya Group of Institution, Bhilai C.G., India

A. K. Bohre et al. (eds.), *Planning of Hybrid Renewable Energy Systems, Electric Vehicles and Microgrid*, Energy Systems in Electrical Engineering,
https://doi.org/10.1007/978-981-19-0979-5_23

Nomenclature

EV	Electric Vehicle
ICE	Internal Combustion Engine
PHEV	Plug-in Hybrid Electric Vehicle
LMC	Loss Minimization Controller
LMT	Loss Minimization Technique
HEV	Hybrid Electric Vehicle
FEV	Full Electric Vehicle
IM	Induction Motor

23.1 Introduction

Electrical machines are playing a leading role for over a decade in industrial application. Their positions in industries are due to the versatility of their application implementation and the importance of the technologies that can be built into them. Beyond 60 percent of the electrical energies are used by the industries which are used by electric motors, 90 percent of which are asynchronous devices, suggesting that electromagnetic drives are the foremost motive force of industrial use.

The electrical motors are also the principal motive force of the electric domestic stuffs. Therefore, they suggest the key priorities to achieve energy savings. A successful effort to protect the environment and energy saving is the electric vehicle (EV). The application and investigation of electric propulsion system have quickly gone into full swing over the previous few years (Un-Noor et al. 2017). The basic scheme for driving an electric vehicle is shown in Fig. 23.1. Recently, many efforts are being taken to drive such type of vehicle even on light load conditions by minimizing the losses.

23.2 Electric Propulsion System

Definition Electric propulsion is a device that uses electrical power to accelerate a propellant. As long as the propeller is well-balanced, the operation of electric motor is without vibration. An electric propulsion system must have the following features:

1. High efficiency over wide torque ranges.
2. Fast torque response.
3. Wide speed range.
4. High torque at low speed.
5. Reasonable cost.

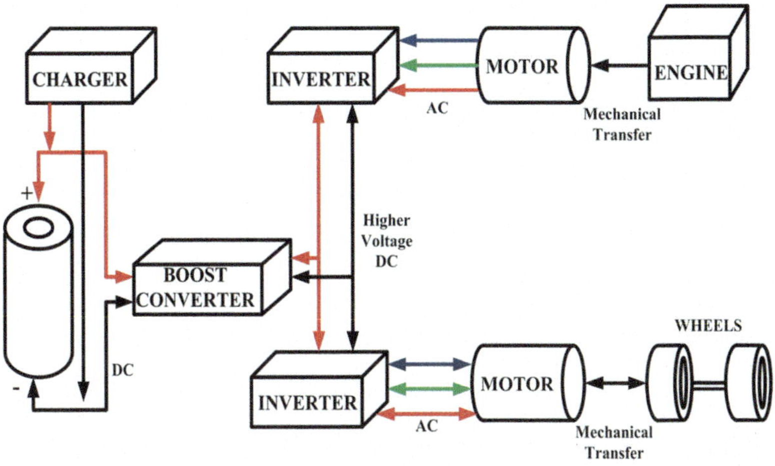

Fig. 23.1 Basic block diagram of electric vehicle

23.2.1 Comparison of Electrical Vehicle (EV) with Traditional Vehicle

The following Table 23.1 shows the comparison between traditional vehicle and electric vehicle (Bhatt et al. 2020; Li et al. 2017):

23.2.2 Types of Electric Propulsion System

Basically, electric vehicle (EV) is the type of vehicle that uses electric drive system as a propulsion mechanism and they do not depend on ICE for propulsion. Mainly, electric propelled vehicle are of two types namely, Plug-in EV and Hybrid EV. The Plug-in type EV is of two types-

a. Plug-in Hybrid EV
b. Full Electric Vehicle

Figure 23.2 classifies the different types of electrically propelled vehicles.

i. ***Plug-in Hybrid Electric Vehicle (PHEV)***: A plug-in hybrid electric vehicle is a hybrid electric vehicle with a battery that can feasibly be energized both internally and externally by plugging a charging cable into an external electric power source, as well as by its on-board internal combustion engine-powered generator. The main distinction is the charging method, which allows the batteries to be charged from the power grid through the vehicle's plug-in addition to regenerative braking and energy from the combustion engine. The combustion engine replaces them after batteries are discharged. The vehicle's batteries

Table 23.1 Comparison of traditional vehicle and EV

S. no.	Parameters	Traditional vehicle	Electric vehicle
1	Sources	Nonrenewable	Renewable
2	Power transmission	Mechanical only	Mechanical as well as electrical
3	Braking system	Friction braking	Regenerative braking
4	Prime mover	I.C. engine	Electric motor
5	Gear system	Complex gear system required	Requires only one gear
6	Power	Petrol, diesel	Charged battery
7	Specific energy	High specific energy of propellant	Low specific energy of battery
8	Power density	High	Low
9	Initial cost	Average	High
10	Maintenance cost	High	Low
11	Running cost	High	Low
12	Refuelling cost	High	Low
13	Noise	Produces noise	Noiseless operation
14	Emission	Emits greenhouse gases	No greenhouse gases emission
15	Maximum torque	Need to raise the speed to yield maximum torque	As soon as the motor starts maximum torque is produced

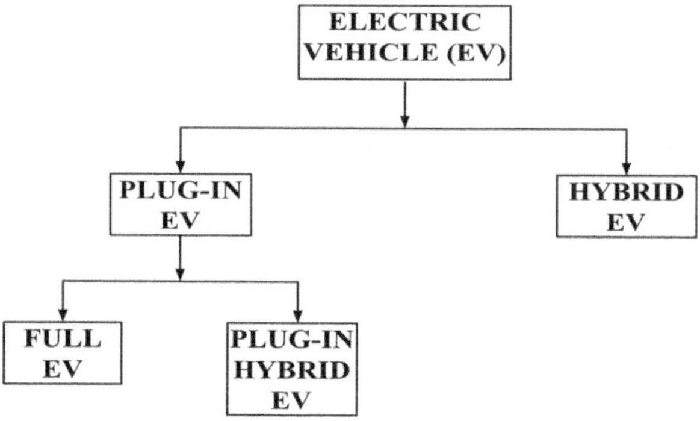

Fig. 23.2 Flowchart of classification of electric vehicle

are significantly larger and can provide several kilometres of purely electric driving. In comparison to HEVs, this vehicle has a larger battery that can be recharged using household outlets or charging stations. The cost, weight and size of a large battery are the disadvantages of PHEVs (Crisostomi et al. 2018; Rajakaruna et al. 2015; Lausenhammer et al. 2016).

ii. ***Full Electric Vehicle*** (***FEV***): A full electric vehicle uses electric drive mechanism to propel. The use of this type of vehicle is beneficial and considered as filter for turning dirty into clean energy (Jose et al. 2017).

iii. ***Hybrid Electric Vehicle* (*HEV*)**: A hybrid electric vehicle incorporates a traditional internal combustion engine with an electric propulsion system (drive train). An inclusion of electric power train is deliberated to attain either improved propellant economy or improved efficiency than traditional vehicle (Rahman et al. May 2016). The electric motor reduces idling and enhances the vehicle's stopping and going performance, which is especially useful in metropolis traffic. Basically, there are three types of HEVs

   a. ***Series HEV***: The vehicle is powered by both internal combustion and drive train. However, in this type, the combustion engine alone would not be able to propel the vehicle because its primary purpose is to produce energy when the vehicle's battery runs out. As a result, the combustion engine performs the role of a range extender.
   b. ***Parallel HEV***: This type of vehicle uses internal combustion and electric combustion engine at the same time to fuel the vehicle. This implies that both forms can completely activate the vehicle's wheels.
   c. ***Twin drain***: In this type each drive axle is driven differently, one is powered by a combustion engine and the other by an electric motor. The gearbox is used in the drive axle driven by an internal combustion engine, while the drive axle powered by an electric engine does not have gearbox. The electric and internal combustion engine drives are kept completely independent from each other and the vehicle can move between them, or both can be used simultaneously.

23.2.3 Factors Affecting Use of Electric Vehicle

The main factors affecting the frequent use of electric vehicle from user's perspectives are (Habib et al. 2015; Shareef and Islam 2016; Camacho and Mihet-Popa 2016):

a. **Cost**: The most important factor that is hindering the popularity of electric vehicle among the people is the cost. The costs of EV are generally higher than the cost of traditional vehicle because of the high cost of battery. Due to this reason, it is impossible for manufacturers to meet their profit margins while also striving to have a low-cost vehicle. As a result, electric vehicles are more costly than their traditional vehicles. But it is expected that within few years the battery cost can be reduced to affordable price as claimed by some companies.
b. **Range**: The range of an electric vehicle on one time charge is far less than that of a completely tanked traditional vehicle, which is a major deterrent to purchasing one. The maximum running range of electric vehicle is mainly less than 500 km in favourable condition and this range is too affected if blower or air conditioner is been used. This range reduces when travelling on the highway at speeds greater than 130 km/h, when the vehicle requires more fuel. Hence, the range of EV becomes very unpredictable.

c. **Vehicle size**: To minimize energy consumption, electric vehicles are often designed in small size and have restricted baggage space. This can be a challenge for potential vehicle buyers who want to carry large loads from time to time.
d. **Charging time**: On an average the charging time of plug-in EV is nearly 30 min which is not acceptable for many users. Since in large cities the availability of charging point is a real concern, especially in large apartment block. Most of the electric vehicle owners (about 90%) charge their vehicle's batteries at night while they are not using them. Regardless, not everyone in the population has the privilege of parking in a garage where they can simply plug their car in for charging. Thus, due to long charging time and lack of charging points for EV, it hinders the customers to buy an electric vehicle.

23.2.4 Trait Encouraging Electric Propulsion System

The governments of various countries around the world are attempting to stimulate consumer demand for any kind of electric vehicle in order to popularize EV purchase and reduce the typically higher retail price of these vehicles. There are mainly three traits that focus to increase the sale and purchase of electric vehicles which are described below (Yao et al. 2017; Singh and Karandikar 2017):

i. ***Tax incentives***: For electric vehicles, this includes a lower purchase and annual tax. This can be grouped into three types:

    a. ***Value Added Tax***—This type of tax is applied while purchasing any type of electric vehicle. In general, the base price of any newly purchased vehicle will vary anywhere from 5 to 25% globally.
    b. ***Registration tax***—In certain markets, this is in addition to the VAT on the vehicle's selling price, whereas some countries government prohibit the EV's owner to pay this type of tax which is a motivation to purchase EVs.
    c. Local governments in certain countries often owe car owners an annual circulation ownership fee on a quarterly basis. Some markets are attempting to incentivize the industry by exempting this form of tax in the case of an electric vehicle.

ii. ***Reduction of usage of fuels***: These occur as a result of lower electricity prices than fossil fuel prices, as a result of lower taxes and lower energy costs, and as a result of EVs' higher performance. This could be more beneficial and more appealing to customers who travel a lot each year and want to take full advantage of the EV's performance.
iii. ***Direct subsidies***: The direct subsidies, which can be thought of as a one-time incentive for purchasing a new electric vehicle. This form of subsidy is still uncommon nowadays, with only a few countries offering such a direct subsidy to prospective buyers.

23.2.5 Advantages of Electric Propulsion System

- EVs produce no noise as compared to petrol or diesel vehicles, thus it produces less noise pollution.
- There is negligible emission from the electric vehicle which is good for the environment and thus it causes less or no air pollution.
- Maintenance and running cost of such vehicle are low.
- Refuelling of EV is low as compared to conventional vehicle as electricity is cheaper than the fuels.
- Tax credits are provided to the owner of electric vehicle.
- Electric propulsion systems are eco-friendly vehicles.

23.3 Drive System for Electric Vehicle

One of the crucial parts of an electric vehicle is the motor drive mechanism. To meet the performance specifications of electric vehicles, such as starting, accelerating and other driving states, the drive system must have high torque at low speeds, high power at high speeds and a large speed control range. Front-wheel drive, rear-wheel drive and wheel motor drive are the different topologies options for electric vehicles drive systems. Drive system for electric vehicle includes motor, power driven devices and control algorithms. The transmission gear ration and transmission performance can estimate the power requirement of the drive system (Bazzi 2010; Krishnan 2015; Lulhe and Date 2015).

Basic definitions of drive system:

- A mechanism that is used to direct the movement of an electrical machine is known as an electric drive. A prime mover is used to supply mechanical energy to the drive, which is used to control the motion.
- An electrical drive is a type of industrial device that transforms electrical energy into mechanical energy or vice versa for the purpose of controlling and running various processes.
- An electrical drive is an electromechanical system that converts electrical energy into mechanical energy in order to provide mobility to various machineries and mechanisms for process control.

23.3.1 Components of Drive System

The basic modules of electrical drive system consist of Sato et al. (2016):

- Power source.
- Power modulator.
- Motor.

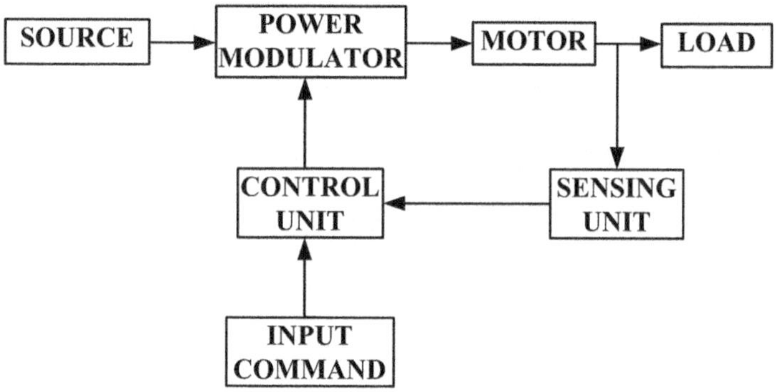

Fig. 23.3 Basic components of electrical drive system

- Load.
- Control unit.
- Sensing unit.

The basic components of electrical drive system are shown through a block diagram in Fig. 23.3.

Power Source:

The power source module is used to provide necessary excitation to the system. The power source communicates with both the converter and the motor to provide variable voltage, frequency and current to the motor.

Power Modulator:

This is used to regulate the supply's output power. The motor's power can be governed in such a way that the electrical motor sends out the speed-torque feature that is required by the load. Extreme current would be drawn from the power source during the short-term operations. The current withdrawn from the power source may be excessive, resulting in a voltage drop. As a result, the power modulator restricts both the motor current and the source current.

Motor:

The electric motor that is best suited for a given application can be chosen based on various factors such as cost, achieving the load's required level of power and efficiency in both the stable and active states.

Load:

The mechanical load can be determined by the environment industrial process and the power source can be determined by the location's available power source. Other electric parts, such as the electric motor, controller and converter, can be chosen.

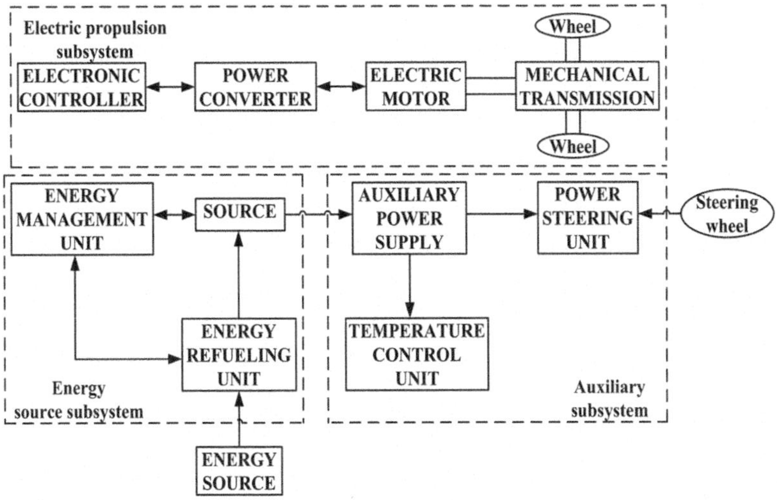

Fig. 23.4 Basic components of Electric Vehicle subsystems

Control Unit:

The power modulator is controlled by the control unit, and it can work at high power levels as well as low voltage. This unit generates the command for the motor's safety as well as for the power modulator. The input control signal governs the working point of the drive from input to the control unit.

Sensing Unit:

The sensing device detects specific drive factors such as speed, torque, motor current, voltage, position and temperature. This device is primarily used for closed-loop control and safety in the event of a power failure.

The block diagram of electric vehicle (EV) subsystem's components is shown in Fig. 23.4.

23.3.2 Classification of Drive System

Basically, there are two types of electrical drive system, namely, AC drive and DC drive (Tolochko et al. 2017a).

AC Drive:

Speed, torque, acceleration, deceleration and rotation direction are regulated by AC drives. The power delivered by AC drives is tailored to meet the actual load requirement, thus saves energy. In certain cases, AC drives are an alternative for mechanical gearboxes, belts, hydraulic couplings, servo-drives or eddy current drives. In a highly reliable, low-maintenance, compact size bundle, AC drives provide an accurate and large range of smooth and step-less regulation of motor speed and torque. AC drives are also free of mechanical inertia and resonance and have a variety of built-in features, complex control and networking capabilities.

DC Drive:

To operate a DC motor, DC drive converts the alternating current into direct current. The output current fluctuates when the voltage to the gate is applied at a distinct angle of the input phase, allowing the drive to control the motor speed. DC drives are further categorized into two types: non-regenerative drives and regenerative drives.

- ***Non-regenerative DC drive***—They control the motor speed and torque in only one direction. This is the conventional type DC drive.
- ***Regenerative DC drives***—This type is proficient in directing the speed and regulation of motor rotation as well as direction of motor torque.

23.3.3 Microprocessor-Based Drive System for Electric Propulsion System

The foremost aim of an electrical drive is to regulate the motion of mechanisms. The electrical drive is a multi-feedback automatic control system that employs various automatic control concepts such as error-driven feedback control, model-based control, logical binary control and fuzzy logic control methods. Distinct sensors for calculating currents, voltages, velocity, acceleration, torque and other parameters in the electrical drive are utilized depending on the technological solution and control theory chosen (Lei et al. 2014; Farhani et al. 2017; Kumar et al. 2015; Kim and Kum 2016; Nasri et al. 2016) (Fig. 23.5).

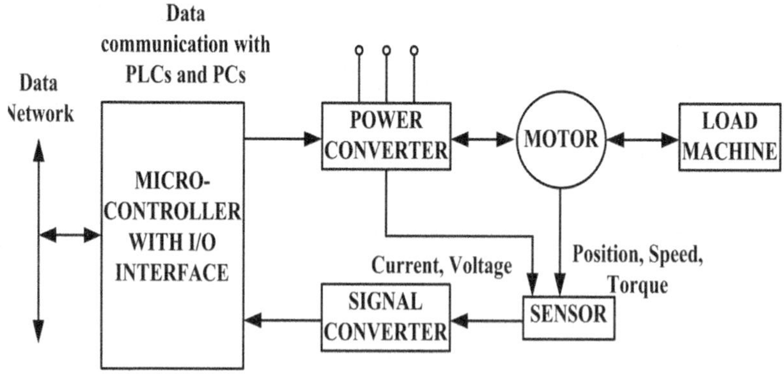

Fig. 23.5 Microprocessor based Electric Propulsion System

23.3.4 Merits of Drive System

- In comparison to another drive device, the price is too low.
- These drives have a variety of speed, power and torque range.
- The operation is simple and smooth.
- Owing to the use of electric braking, they have a versatile characteristic.
- It can be started immediately without any time delay.
- Its maintenance cost is relatively inexpensive.
- Remote control capability is available.
- They have the ability to work in all four quadrants of the speed-torque plane.

23.3.5 Demerits of Drive System

- The resulting drive output power is low.
- The whole system comes to a halt due to a power outage.
- This drive will not work if the power supply is not available.
- Drive system has a low dynamic response.

23.3.6 Applications of Drive System

- ***Electric traction***: This is the primary application of drive system. DC series motors are appropriate for almost all types of services, but especially this is favourable for sub-urban services where a high rate of acceleration is needed. For main line work, single phase AC compensated series motors with speed-current and speed-torque characteristics identical to DC series motors are widely used.
- ***Lift***: High accelerating torque, high overload power, pull out torque, high degree of silence and moderate speed are the crucial need for this system.
- ***Hoist motor and cranes***: A crane or hoist motor must have a high starting torque and be able to handle a large number of switching operations as a minimum requirement. Hoist motors are equipped with special electromechanical brakes, by the use of springs that are designed to hold the load in the event of a power outage.
- ***Frequency converters***: Converters are powered by squirrel cage induction and synchronous motors. Static frequency changers are becoming more common as a result of recent developments in the field of silicon controlled rectifiers.
- ***Machine tool drives***: An easy, more or less constant speed drive is needed by many machine tools. Gear boxes and stepped pulleys are used to transmit power to cutting tools or work parts at a finite number of speeds.
- ***Belt conveyors***: Sand and gravel are transported using belt conveyors. Acceleration is needed for heavy loads. Double cage induction motors with direct-on-line

starters or wound rotor induction motors are used with normal starting current and high starting torque.

- ***Ship propulsion***: The prime mover in electric ship powers the generator, which supplies power to an AC or DC motor mounted on the propeller shafts. Steam turbines with AC equipment are used on large ships needing more than 2,500 kW of power, while diesel engines with DC equipment are used on smaller ships.
- ***Electric vehicle***: These types of vehicles are encouraged in order to save environment from harmful emissions. Hyundai Kona, Tata Nexon and MG ZS are the popular type of EVs in the market.
- ***Petrochemical industry***: Fluid handling equipment is widely used in the petroleum and chemical industries. Pumps have been powered by induction motors, with flow control provided by throttling valves when required. Gas liquefaction, compression, refrigeration and heat recovery are all critical activities in the petrochemical industry's processes.
- Electrical drives are also widely used in domestic and industrial applications including engines, transportation systems, factories, pumps, fans, etc.

23.4 Loss Minimization Technique for Electric Vehicle

Basically, in rated condition each and every type of motor work very efficiently and have high efficiency without any problem but for low load condition, losses are unbalanced and whenever the load fluctuates, it is difficult to maintain a constant speed in the motor. Losses in electric motors include grid loss, mechanical loss, converter loss (stray and commutation losses), winding loss (copper losses), motor loss and transmission loss. These losses can be minimized by using different loss minimization techniques. A loss minimization technique (LMT) usually requires that the voltage, current, flux, torque and speed can be controlled.

Modifications are done in materials, design and construction technique in an attempt to improve performance of the drives. Converter loss and motor loss, however, are even now highly influenced by the control techniques, chiefly when the motor works at low load condition. The flux must be decreased in order to increase motor performance, achieving balance between copper and core losses.

23.4.1 Selection of Motor

Mainly, three types of motor are been considered for driving the electrical vehicle, namely, DC motor, induction motor (IM) and permanent magnet synchronous motor (PMSM). The motors are selected in such a way that there is a proper balance between motor performance and the system price. The first step in designing an electric vehicle is to choose the motor type based on vehicle demand and other boundary conditions. The basic trait of these three types of motor is discussed below:

DC Motor: This is the simplest type of motor from motor class that converts the DC electrical energy into mechanical energy. The speed of such motor can be altered over a broad range by fluctuating the supply voltage or modifying the current intensity in the field windings. Mechanical brush commutation is used in DC motors, which is susceptible to depreciation, chances of inadequacy are high and are difficult to maintain. DC motors have a wider drive mechanism, a limited speed control range and a low energy density, all of which restrict their use in electric vehicles.

Induction Motor: An induction motor is an electric motor that produces torque by electromagnetic induction from the stator winding's magnetic field. Because of high performance, excellent speed control and absence of commutators, three-phase induction motors are frequently used in electric vehicles. Electronic brush commutation is used in induction motors to solve the drawbacks of mechanical brush commutation and sine wave control mode is used to introduce high accuracy torque control.

Permanent Magnet Synchronous Motor: PMSM uses permanent magnet which is embedded in the steel rotor to produce a constant magnetic field. These types of motors have the capability of delivering high torque-to-current ratios, high power-to-weight ratios, high accuracy and stability which make it suitable for electric vehicle application.

23.4.2 Control Strategy for Induction Motor Drives

Centred on scalar control or vector control of IM drives, there are many loss minimal control systems. The rotor-flux-oriented control, stator-flux-oriented control and magnetizing-flux-oriented control techniques are studied in various vector control techniques (Bazzi and Krein 2010; Stumper et al. 2013). It is possible to split the control strategy to boost motor effectiveness into two categories:

(a) **Search controller**: By changing the amount of flux in the motor, the optimization algorithm iteratively decreases the input power until the difference between the input and output power is fulfilled. Slow convergence and torque ripples are significant downsides of the search controller.
(b) **Loss model-based controller**: While utilizing the system model, the model-based controller measures losses and determines a flux level that reduces the losses. Loss model-based controller is quick and does not generate a ripple of torque.

The basic block diagram of induction motor drive operated under loss minimization technique is shown in Fig. 23.6.

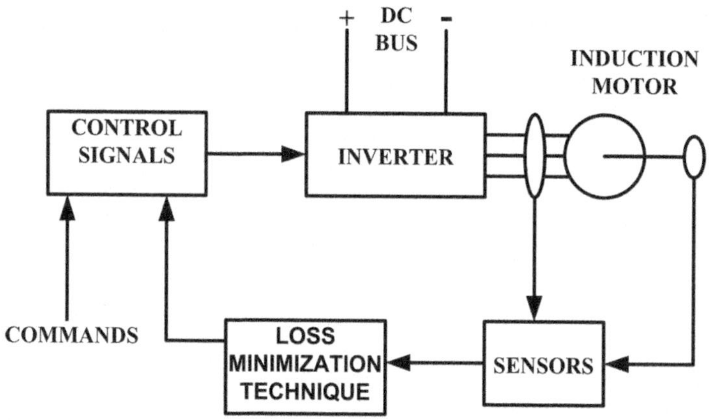

Fig. 23.6 Induction machine as governed by LMT

23.4.3 Classification of Loss Minimization Techniques (LMT)

The distinct types of methods are followed by existing loss minimization techniques and chosen according to the application or type of drive used. The application is connected to the required speed of convergence, responsiveness of the parameter and error of convergence, while the drive sets the optimization variables that are available. Basically there are two types of LMTs (Verrelli et al. 2014):

(1) Offline techniques.
(2) Online techniques.

The classification of loss minimization technique is shown in Fig. 23.7.

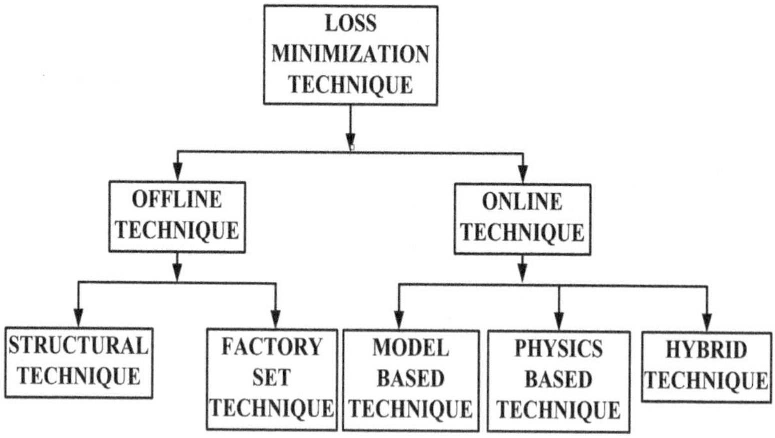

Fig. 23.7 Flowchart of classification of LMT

23.4.3.1 Offline Techniques

By setting the optimum operating point that depends on preset values, offline method aims to minimize power losses. Offline strategies cannot further change losses when the motor is working. Offline techniques are categorized into two groups:

i. Structural technique: utilizing electromagnetic analysis.
ii. Factory-set technique: Based on planned operating conditions, sets an optimal motor drive command.

23.4.3.2 Online Techniques

While the motor is running, online or real-time methods aim to attain minimal losses. Online techniques are categorized into three groups:

i. ***Model-based technique:***

   It depends on power-loss models and motor parameters.

   Advantages: The pros of this technique are as mentioned below:

   a. It has the ability to operate near the optimum point.
   b. Its concurrence to the optimum operating point is at high speed.
   c. It is highly responsive to the variance of motor specifications, particularly at light load condition.

   Disadvantage: It depends on the motor parameters is the main pitfall of this technique.

ii. ***Physics-based technique***:

   To drive the control input, it utilizes electromechanical or mathematical principles.

   Advantages: Its ease of implementation is the benefit of this technique.

   Disadvantage: This technique is slower than model-based approach as in order to change their control variable, they hold back to estimate a steady-state power loss which is the major drawback of this technique.

iii. ***Hybrid technique***:

   It requires a system model to examine for the minimal power loss.

   Advantages: This technique is greatly preferred due to its fast convergence speeds than physics-based technique and reduced specification dependency than model-based technique.

   Disadvantage: The implementation of this technique is comparatively complex.

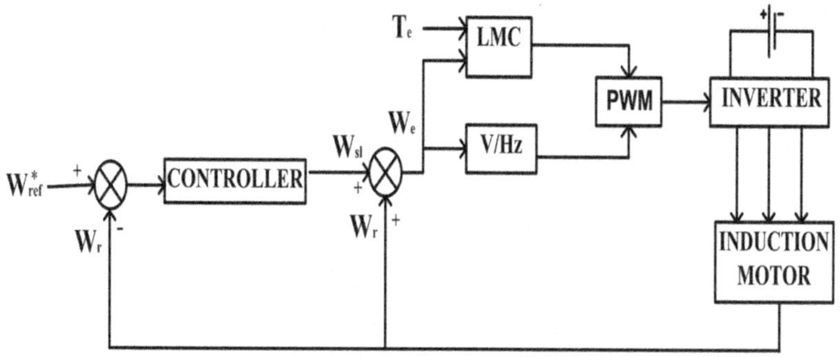

Fig. 23.8 Block diagram of scalar-controlled IM drive governed by LMT

23.4.4 Scalar Controller for IM Drives

Scalar control is due to the magnitude variance of the control variables and neglects the machine's coupling effect. Scalar controlled drives produce lower quality performance but are simple to implement. They have been widely used in industry. However, because of the superior performance of vector-driven drives, which are used in many applications, the significance of scalar-controlled drives has recently decreased.

Loss Model:

This control technique maintains V/f constant so as to keep the flux constant in the system. A scalar drive loss minimization controller is depending on calculating the optimum flux for loss minimization. The self-governing parameters: speed, end voltage, end frequency and the specifications of the machine and its power source define the behaviour of the system. An optimum flux that satisfies the operating point requirement and reduces the total losses which can be found at any operating point is defined by speed and torque. The speed and torque are taken as inputs by the controller and the optimum flux value is measured and the LMC generates the optimum modulation index value (Tolochko et al. 2017b). To avoid excess currents, the optimal value of flux has a minimum value. The block diagram of scalar-controlled induction motor drive operated under loss minimization technique is shown in Fig. 23.8.

23.4.5 Vector Controller for IM Drives

Vector controller is also known as Field-Oriented Controller (FOC), initially implemented by Blaschke, the principle of field-oriented control made induction motors ideal for speedy functioning of servo applications. The probability of self-governing modulation of the quadrature stator current, which directly determines the torque of the motor and the rotor flux control, is the main focus of the FOC induction motor control. Independent quadrature current and flux power converts the system with independent excitation to a DC motor. Direct Field-Oriented Control and Indirect Field-Oriented Control are two main strategies used in FOC. These methods depend on the premise that the regulated flux vector is associated with the rotating reference frame's d-axis. In consideration of this inference, the stator current can be dissociated into two units, namely torque generating components and flux generating components (Tolochko and Kaluhin 2019).

Loss Model:

Vector control includes the potential to control the IM's electromagnetic torque easily and accurately. However, even at low torque values, the traditional vector-driven drive mechanism operates at nominal flux. The downside of this approach is that when the torque varies and the flux is held at the nominal point, performance decreases. This reduces the induction motors efficiency when it works under conditions of light load.

There are various ways of managing the mentioned drawback. There are systems to reduce different kinds of losses in the motor and converter, in addition to maximization of efficiency and power factor, among the known solutions. The simplest among them are approaches that focus on loss models. In the possibility of minimal losses, supposition validated in optimization and the structural execution of the algorithm acquired, they vary from each other (Sreejeth et al. 2012; Iffouzar and Amrouche 2017; Zhao et al. 2013).

By measuring the optimum air gap flux or equivalent magnetizing current, the power factor and performance of the field-oriented operated induction machine can be improved. Motor performance can also be increased by regulating the d-axis current in the d-q reference frame in vector-controlled drives by running the machine at optimum flux. By measuring the ideal air gap flux or equivalent magnetizing current, the loss model-based controller will enhance the performance and power factor of the field-oriented regulated induction machine.

The equivalent circuit diagram of induction motor in d-axis is shown in Fig. 23.9.
The equivalent circuit diagram of induction motor in q-axis is shown in Fig. 23.10.

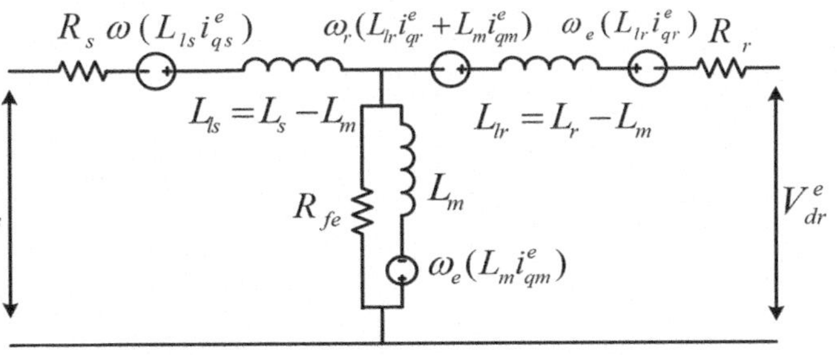

Fig. 23.9 d-axis equivalent circuit

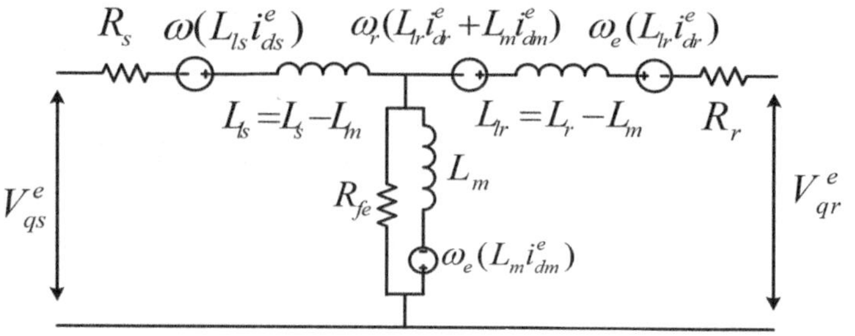

Fig. 23.10 q-axis equivalent circuit

23.4.6 Direct Torque Controller for IM Drives

The best magnificent control technique accessible to regulate the torque of the induction motor is direct torque control. It uses the stator flux as a control variable and the flux level is chosen in accordance with the torque demand of the propulsion drive to attain the desired performance. It depends on a direct flux and torque control system. However, the DTC's low switching frequency gives rise to high flux and torque ripples, leading to an acoustic noise that degrades the efficiency of the control, particularly at low speeds. The goal is to do an analytical study of these procedures in respect of depletion of ripples, monitoring of speed, lack of switching, complication of algorithms and responsiveness of parameters. The minimization of loss in the induction machine is directly related to the flux level option. DTC has the benefit of not needing encoders for speed or location and uses only measurements of voltage and current. Since the proportional-integral (PI) current controller is absent, it also has a quicker dynamic response. The block diagram represented in Fig. 23.11 shows

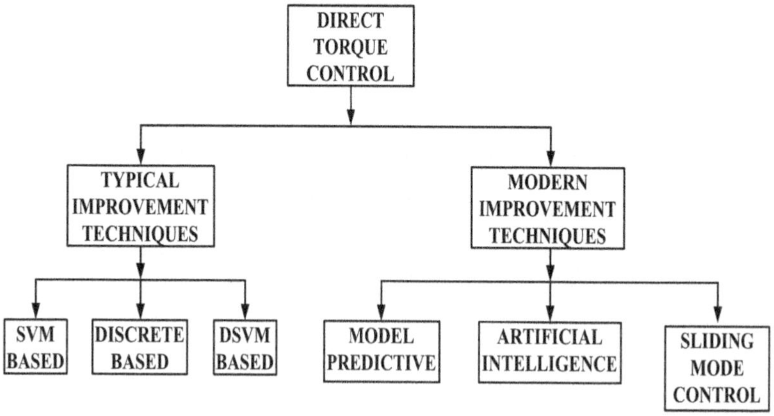

Fig. 23.11 Flowchart of classification of DTC

the classification of the direct torque controller (DTC) method for induction machine drives (Ouanjli et al. 2019; Sutikno et al. 2014).

Loss Model:

Its theory is derived from the immediate choice of the control pulses of the voltage inverter switches. DTC control makes certain operation with high performance and gives precise and rapid torque dynamics. DTC theory is rooted on the direct implementation of the control sequence to the voltage inverter switches mounted in front of the machine. Using a switching table and two hysteresis regulators, whose function is to monitor and direct the electromagnetic torque and the system flux in a decoupled manner, the choice of this sequence is made.

As shown in Fig. 23.12, the electromagnetic torque is directed through a comparator of three levels of hysteresis. By utilizing a two-stage hysteresis comparator, the stator flux is regulated. For determining the switching table, the outcomes of these comparators, in addition to the flux vector information, are used. The implementation of hysteresis controllers induces high flux and electromagnetic torque ripples that produce mechanical disturbances and unwanted acoustic noise, thus degrading the efficiency of the system, resulting in variable switching frequencies and current distortions that can deteriorate the output power trait.

Immense work has been done in traditional DTC to overcome the troubles that were faced in conventional methods. These techniques are build on the identical concept of regulation of instantaneous torque and stator flux and the immediate value from a switching table of the inverter control signals. These control techniques are usually split into two groups: modern and typical methods. The purpose of this

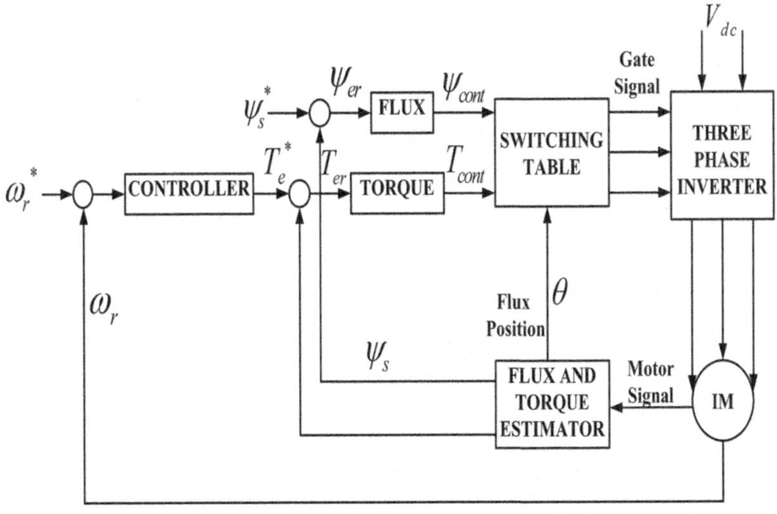

Fig. 23.12 Basic direct torque controller. Adapted with permission from Aktas et al. (2020). Copyright 2020, Elsevier

enhancement is, on the one hand, to decrease the distortion of the couple and the flux of the IM and, on the other hand, to reduce the switching frequency of the inverter (Ouhrouche et al. 2016; Ghezouani et al. 2018; Ammar et al. 2017; Naik et al. 2016; Lascu et al. 2017; Bermudez et al. 2017; Aktas et al. 2020; Alagna et al. 2016; Garcia et al. 2016; Xie et al. 2018; Tazerart et al. 2015).

23.4.7 Indirect Field-Oriented Controller for IM Drives

By dissociating the rotor flux and torque generating current constituent of stator current, indirect field-oriented control is known to achieve high efficiency in induction motor drives. The proposed observer correctly obtains the rotor flux and the magnetizing current, and the PI controller and SVM technique are used to achieve good results. To conduct IFOC of IM drives using SVM, the initial angular frequency, stator voltages, stator currents, magnetizing currents, rotor fluxes and rotor speed must all be measured.

Indirect field-oriented controller uses the slip relation to calculate the angle of the rotor flux respective to the stator after computing the rotor orientation. This is a high potential technique that controls the stator currents to regulate the induction motor. It functions on the principle of converting three-phase stator currents into two-phase constituents, i.e. d-axis and q-axis. The d-component of the current i_{sd} controls the flux, while the q-component i_{sq} directs the torque of motor. In this technique, the reference values i_{sd} and i_{sq} are compared, and then the two PI controllers process the error signals to determine the reference values of the voltage components V_{sd} and V_{sq}. The reference voltages are transformed into a stationary reference frame using the Clark transformation. To provide the appropriate voltage to the induction motor, SVPWM and it is also used to measure the necessary switching signals for the VSI. DC voltage from the battery powers the three-phase inverter. Since function of such drives is simple in closed-loop mode throughout the speed range from zero to high in the field weakening, indirect vector control is more widely applied.

Loss Model:

As shown in Fig. 23.13, the optimal regulation of d-axis stator current (i_{ds}) that supervises the magnitude of rotor flux reduces the governable electrical loss in the induction motor drive system. By deteriorating the rotor flux, the motor's performance is improved, and core losses are reduced. The difference between the commanded speed and the real motor speed is used by the speed control loop to produce the torque generating portion of current. The slip frequency is calculated using the feed forward method from i_{qs}^* and i_{ds}^*. Using the Park's Inverse Transformation, the reference stator phase currents,i_a^*, i_b^* and i_c^* are calculated via reference d and q-axes currents and rotor pole orientation. In order to obtain the unit vector rotating at ω_e, the slip frequency is summed to the rotor speed. To acquire rotor pole position, ω_e is integrated. The hysteresis PWM tries to match the real motor currents (i_a,i_b,i_c) to the reference currents. To control the PWM inverter, the difference between the

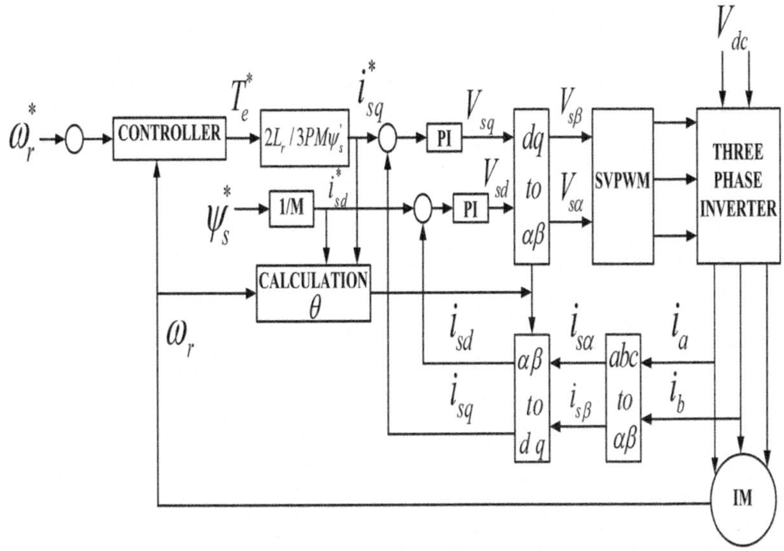

Fig. 23.13 Basic indirect torque controller. Adapted with permission from Aktas et al. (2020). Copyright 2020, Elsevier

reference and true currents is employed. The stator of an induction motor receives the inverter's output (Alagna et al. 2016).

23.5 Environmental Impact of Electric Vehicle

Electric vehicles (EVs) are being promoted by governments and manufacturers all over the world as a crucial technology for reducing oil consumption and combating climate change. Electric vehicles run merely on electricity; depending on how the electricity is generated, EV can run entirely on sustainable, renewable energy (Lu 2016; Calzada-Lara 2016). When assessing the environmental effect of electric vehicles, four points must be considered:

i. Tailpipe emissions.
ii. Well-to-wheel emissions.
iii. Energy source that charges the battery.
iv. Vehicle's power.

Since electric vehicles are powered by electricity there are no tailpipe emissions. This tailpipe emission is also popularly known as direct emission. Smog-forming contaminants (e.g. nitrogen oxides), other toxic emissions detrimental to human health, and greenhouse gases, mainly carbon dioxide, are all instances of direct pollution. All electric vehicles release no direct pollution, which boost the air quality in cities. EVs are much more environmentally friendly than today's gasoline-powered vehicles when measured solely on this factor. Electric vehicles are becoming progressively enticing in the area where the population is bothered about diminishing carbon emissions and pollution. Electric vehicles have been shown to be more environmentally friendly in studies. They release less greenhouse gases and toxins than a gasoline or diesel vehicle. Over the course of a year, only a single electric vehicle on the road can approximately save 1.5 million grams of carbon dioxide.

Well-to-wheel emission is a broad concept that encompasses greenhouse gas emissions and air pollution related with the production and distribution of energy used to power automobiles. EVs emit about 4,400 pounds of carbon dioxide equivalent per year when well-to-wheel emissions are considered. Depending on the resource, electricity output produces varying amounts of pollution. If the combination of energy is used to generate the power to drive an electric vehicle, it will produce comparatively less well-to-wheel emissions.

Electric vehicles can also help to mitigate noise pollution. Electric vehicles are much quieter than traditional vehicles, resulting in a more peaceful surrounding for all of us. The parts of this type of vehicle can be dismantled and recycled at the end of a vehicle's life thus; waste is minimized hence promoting eco-friendly manufacturing. The batteries that are used to power electric vehicle can be recycled as well. Electric vehicle drivers can diminish the life cycle emissions even further by using electricity produced from renewable sources such as solar, wind and so on.

Modern electric vehicles have a comprehensive information system that is constantly updated and monitors internal and external parameters to achieve optimum energy savings. EVs are more fuel efficient and require no or less fuel to operate thus reducing the overall fuel cost (Bortoni et al. 2013).

23.5.1 Comparison of Electric Vehicle with Other Technologies

The procedure of transforming potential energy into kinetic energy is the basic difference between traditional vehicle, thermal vehicle and electric vehicle. This energy is preserved in a chemical form in thermal vehicles and discharged through a chemical reaction within the engine. Electric vehicles, contrarily, despite having chemically stored electricity, release it electrochemically rather than by combustion. This assures that no fuel is utilized, and therefore no carbon dioxide is released into the atmosphere when driving. This means that EVs emit less greenhouse gases and other pollutants that lead to climate change and smog than traditional vehicles. They are even more fuel efficient than traditional automobiles.

About 77 percent of the electrical energy from the grid is converted to electricity at the wheels of electric vehicles. Traditional fuel vehicles transform only about 12–30% of the energy contained in gasoline to electricity at the wheels. A substantial reduction in consumption costs is another significant benefit of any fully electric vehicles.

Petrol stations are a health and safety threat since they require bulk storage of volatile hazardous fuels. Refuelling stations for EVs can be set up almost anywhere for a low cost and are simple to use without any supervision requirement (Changjian et al. 2015). Although electric vehicles have some environmental effects, they are still a great way to reduce the carbon footprint as compared to traditional gasoline or diesel engines.

There is no way to convert the vehicle's latent momentum into a force that can be used to accelerate further without significantly increasing the vehicle's weight, which would negate the benefits of regeneration in conventional vehicles whereas in EV during braking, the vehicle's latent momentum can be used to regenerate power.

EVs are frequently more digitally connected than traditional vehicles, with many EV charging stations allowing charging to be managed through android mobile application. Modern navigation technologies, batteries and protection devices can be integrated into our own electric vehicles, at high-traffic areas. This has the ability to track and control charge to grid operations, as well as monitor and control protection and flow. Electric vehicle can be charged and it can be ready to drive very next moment. Since the electric grid is nearly present in every spot, there are a number of charging options such as at home, at work, or on the road. Thus, charging it frequently one does not need to go to fuel station again and again for refuelling.

Charge from electric car batteries can also be fed back into the grid through charging stations. This ability can be used as a backup in the event of a major plant outage, as well as load balancing on the grid to help feed stored renewable produced power back into the grid at night (Abhishek et al. 2017; Liao and Lu 2015).

Electric vehicle and traditional vehicle comparisons in every aspect are difficult since they cannot be compared on the basis of single factor. Comparison is being affected by the vehicle's size, the precision of the fuel efficiency calculations used, how energy emissions are measured, presumed driving habits and even the weather in the regions where the vehicles are used. This means that there is no one-size-fits-all estimate that can be used anywhere.

23.5.2 Opportunities of Business Expansion

Electric vehicles are expected to have massive market penetration in the future all over the world. With the current expansion in the automotive industry, the market for electric vehicles has seen rapid growth. To plan for the anticipated EV boom, automakers are forming separate EV business units. Nevertheless, the increase in EV demand would result in a major increase in the demand for charging framework, as well as the implementation of safety regulations and standards (Yilmaz and Krein 2013). Economies of scale would support electric vehicle production by allowing for improved infrastructure, more efficient manufacturing processes, recycling options and a reduction in the need for new material mining.

In terms of electric vehicles, India is still a developing market; however, OEMs have begun demonstrating and prepare for the obligatory—an electric future. Electric vehicle market was first entered by Tata Motors and Mahindra & Mahindra. Maruti Suzuki intends to enter the market with a small mini EV as its first product in the near future. Following China, India's electric two-wheeler market has a lot of promise, owing to the fact that it is the world's largest two-wheeler market in terms of sales and demand.

a. ***Using a network business model, include public EV charging***:

   The accessibility of EV chargers is one of the aspects that will give rise to the electric vehicles. The more electric vehicle chargers there are, the more electric mobility connectivity there is, and therefore more EVs will be offered. Increased EV charger utilization in the future could result in higher desire for the facility, creating a good business opportunity. By using the Platform Business Model to include EV chargers, customers are saved from paying for electricity. This is an excellent way for fuel marketers to mitigate the risk of lower gasoline demand as electric vehicles become more common.

b. ***Join forces with charging stations to create a one-stop-shop for the 'Charge and Chill' experience***:

   Charging an electric vehicle requires time. Tesla has developed plans to construct Tesla supercharging stations. Their supercharging may be called cutting-edge technology in today's world. In spite of this also it took 30–40 minutes to fully charge the vehicle. Since it takes minutes to charge it allows users to 'Charge and Chill,' which are essentially another profitable business opportunity.

c. ***Dispose of batteries***:

   The new EV battery technology is the same as the Lithium-ion batteries used in our smartphones. Due to environmental concerns, this is a regulated substance that requires careful disposal. The EV battery must be disposed of after it has reached the end of its useful life and requires substitution. Due to the rise of electric vehicles, specialized battery disposal services are another business opportunity. According to surveys, there will be a high demand for battery disposal facilities in the future. The good news is that battery recycling facilities will be available for both electric vehicles and other electronic devices as well.

23.5.3 Challenges to Overcome in Electric Vehicle

The manufacturing of electric vehicles has its own set of consequences. Since the batteries are so large and heavy, designers strive to make the rest of the vehicle as light as possible. As a result, many lightweight materials used in electric vehicle components need a lot of energy to manufacture and process. The rare earth metals are used in electric vehicle's motor in order to enhance the output power of these motors. A lot of energy is required for processing and mining of the metals like copper, nickel, and lithium which emits poisonous gas in the environment. Mineral mining can increase risks even more in developing countries with poor regulation and compliance. As a result, the local public may be endangered by poisonous pollutants which contaminate the air and underground water. To overcome this problem more advanced technology is needed to manufacture environmental friendly batteries.

Usually charging the battery pack takes 3–12 h for full charge. Paradoxically, a fast charge to 80% capacity can take up to 30 min. EV batteries are made to last a long time. According to a study conducted by the Department of Energy's National Renewable Energy Laboratory, urged that these batteries could last 11–15 years in mild climates and 7–12 years in harsh climates. These batteries, on the other hand, are expensive and replacing them if they malfunction is quite pricey.

In a broader vision from customer point of view, there is a lack of knowledge about electric vehicles and their mechanical systems, so if breakdown occurs in this type of vehicles one cannot go to the normal mechanic for fixing the problem rather they need to contact authorized service personnel. Moreover, for these types of vehicles, the precise maintenance costs have yet to be determined. Many service centres have a reputation for deceiving consumers by claiming that their vehicles have fake problems. Electric vehicles would only exacerbate the issue. All these issues sometimes cause serious inconvenience to the customers.

Another vital challenge for electric vehicles is the country's infrastructure funding for them to be trouble-free and become popular among users. This primarily leads to the charging stations, which will be critical for EVs. At the moment, electric vehicles cannot be used on highways because there are no charging stations along the highways or in the countries.

In some countries like Europe, government has welcomed EV technology full-heartedly by reducing tax credits for the EV's owner and also providing subsidies in many different ways, and to attract consumers, they have simple rules and policies designed specifically for electric vehicles. In contrast to this, in many countries, including India government does not have clear policies and special privileges to the owner of EVs. This has caused manufacturers to be perplexed about their electric vehicle plans. Besides all this, the boon is that the industrialist is developing vehicles like the Hyundai Kona, Tata Nexon and MG ZS that are dedicated to being future-ready and prioritizing them. Hopefully, the government recognizes this and is able to address all of the EV-related issues.

23.6 Conclusion

A number of modern improvement strategies for minimizing loss in induction motor drives are reassessed. The main target of these improvements is to minimize the losses and to increase the efficiency in light load condition by the inclusion of core losses resistance in the equivalent d-axis and q-axis diagram of induction motor. An overview of offline and online LMTs, and practice of examining power losses were also proposed. The outlines of each of the three online LMT techniques were also presented which are suitable loss minimization techniques for propulsion drives at low load. The choice of strategy is determined by the type of application, expense, accuracy, precision and usability of hardware for which they are being used. Such systems are ideal for the use of propulsion drives with a small torque load in a very long operation, for example- escalator; conveyors; mine lifting installations and turbo-mechanics. This analysis is intended to provide all investigators, research workers and industries working on electric machine controller drives with useful information.

References

Abhishek A, Karthikeyan V, Sanjeevikumar P, Rajasekar S, Blaabjerg F, Asheesh KS (2017) Optimal planning of electric vehicle charging station at the distribution system using hybrid optimization algorithm. Energy 133:70–78

Aktas M, Awaili K, Ehsani M, Arisoy A (2020) Direct torque control versus indirect field-oriented control of induction motors for electric vehicle applications. Eng Sci Technol Int J

Alagna S, Cipriani G, Corpora M, Di Dio V, Miceli R (2016) Sliding mode torque control of an induction motor for automotive application with sliding mode flux observer. In: IEEE International conference on renewable energy research and applications (ICRERA), pp 1207–1212

Ammar A, Bourek A, Benakcha A (2017) Nonlinear SVM-DTC for induction motor drive using input-output feedback linearization and high order sliding mode control. ISA Trans 67:428–442

Bazzi AM (2010) Designing better induction motor drive systems from efficiency, reliability, and power electronics perspectives. PhD dissertation, Doctoral dissertation, University of Illinois at Urbana Champaign

Bazzi AM, Krein PT (2010) Review of methods for real-time loss minimization in induction machines. IEEE Trans Ind Appl 46(6)

Bermudez M, Prieto IG, Barrero F et al (2017) Open-phase fault-tolerant direct torque control technique for five-phase induction motor drives. IEEE Trans Ind Electron 64(2):902–991

Bhatt P, Long C, Saiyad M (2020) Review of the impact of vehicle-to-grid schemes on electrical power systems. Springer Nature Singapore Pte Ltd. Advances in electric power and energy infrastructure, Lecture notes in electrical engineering. https://doi.org/10.1007/978-981-15-0206-4

Bortoni EC, Nogueira LAH, Cardoso RB, Haddad J, Souza EP, Dias MVX, Yamachita RA (2013) Assessment of the achieved savings from induction motors energy efficiency labeling in Brazil. Energy Convers Manage 734–740, 00005. https://doi.org/10.1016/j.enconman.2013.08.034

Calzada-Lara (2016) Energetic consumption improvement in induction motors with possible mechatronics applications. Mech Based Des Struct Mach, Int J 44(1–2):137–145

Camacho OMF, Mihet-Popa L (2016) Fast charging and smart charging tests for electric vehicles batteries using renewable energy. Oil Gas Sci Technol 71:13–25

Changjian H, YiminG, Alex QH (2015) Power management strategy of hybrid electric vehicles based on particle swarm optimization. In: IEEE Transportation electrification conference and expo (ITEC), 1–6

Crisostomi E, Shorten R, Studli S, Wirth F (2018) Electrical and plug-in hybrid vehicle network. CRC Press

Farhani F, Zaafouri A, Chaari A (2017) Real time induction motor efficiency optimization. J Franklin Inst

Garcia C, Rodriguez J, Silva C, Rojas C, Zanchetta P, Abu-Rub H (2016) Full predictive cascaded speed and current control of an induction machine. IEEE Trans Energy Convers 31(3):1059–1067

Ghezouani A, Gasbaoui B, Ghouili J (2018) Modeling and sliding mode DTC of an EV with four in-wheel induction motors drive. In: International conference on electrical sciences and technologies in Maghreb (CISTEM), pp 1–9

Habib S, Kamran M, Rashid U (2015) Impact analysis of vehicle-to-grid technology and charging strategies of electric vehicles on distribution networks—a review. J Power Sources 277:205–214

Iffouzar K, Amrouche B, Cherif TO, Benkhoris M-F, Aouzellag D, Ghedamsi K (2017) Review article: improved direct field oriented control of multiphase induction motor used in hybrid electric vehicle application. Int J Hydrog Energy Publ

Jose CP, Meikandasivam S (2017) A review on the trends and developments in hybrid electric vehicles in innovative design and development practices in aerospace and automotive engineering. Springer, Singapore, pp 211–229

Kim H, Kum D (2016) Comprehensive design methodology of input- and output-split hybrid electric vehicles: in search of optimal configuration. IEEE/ASME Trans Mechatron 2912–2923

Krishnan R (2015) Electric motor drives modelling, analysis and control. Pearson Education, Inc., India

Kumar N, Raj Chelliah T, Srivastava SP (2015) Adaptive control schemes for improving dynamic performance of efficiency-optimized induction motor drives. ISA Trans 57:301–310. https://doi.org/10.1016/j.isatra.02.011

Lascu C, Jafarzadeh S, Fadali MS et al (2017) Direct torque control with feedback linearization for induction motor drives. IEEE Trans Power Electron 32(3):2072–2080

Lausenhammer W, Engel D, Green R (2016) "Utilizing capabilities of plug in electric vehicles with a new demand response optimization software framework. Okeanos Int J Electr Power Energy Syst 75:1–7

Lei G, Wang T, Guo Y, Zhu J, Wang S (2014) System-level design optimization methods for electrical drive systems: deterministic approach. IEEE Trans Ind Electron 61(12):6591–6602

Li Y, Yang J, Song J (2017) Nano energy system model and nanoscale effect of graphene battery in renewable energy electric vehicle. Renew Sustain Energy Rev 69:652–663

Liao YT, Lu CN (2015) Dispatch of EV charging station energy resources for sustainable mobility. IEEE Trans Electr 1(1):86–93

Lu S-M (2016) A review of high-efficiency motors: Specification, policy, and technology. Renew Sustain Energy Rev 59:1–12. https://doi.org/10.1016/j.rser.2015.12.360

Lulhe AM, Date TN (2015) A technology review paper for drives used in electrical vehicle (EV) & hybrid electrical vehicles (HEV). In: International conference on control, instrumentation, communication and computational technologies (ICCICCT), Kumaracoil, India, 18–19 December 2015

Naik V, Panda A, Singh SP (2016) A three-level fuzzy-2 DTC of induction motor drive using SVPWM. IEEE Trans Ind Electron 63(3):1467–1479

Nasri A, Gasbaoui B, Fayssal BM (2016) Sliding mode control for four wheels electric vehicle drive. Proc Technol 22:518–526

El Ouanjli N, Derouich A, El Ghzizal A, Motahhir S, Chebabhi A, El Mourabit Y, Taoussi M (2019) Modern improvement techniques of direct torque control for induction motor drives—a review. Protection and control of modern power systems

Ouhrouche M, Errouissi R, Trzynadlowski AM et al (2016) A novel predictive direct torque controller for induction motor drives. IEEE Trans Ind Electron 63(8):5221–5230

Rahman I, Vasant PM, Singh M, Abdullah-Al-Wadud M, Adnan N (2016) Review of recent trends in optimization techniques for plug-in hybrid, and electric vehicle charging infrastructures. Renew Sustain Energy Rev 58:1039–1047

Rajakaruna S, Shahnia F, Ghosh A (2015) Plug in electric vehicles in smart grids, 1st edn. Springer Science and Business Media Singapore Pte Ltd., Singapore

Sato M, Yamamoto G, Gunji D, Imura T, Fujimoto H (2016)Development of wireless in-wheel motor using magnetic resonance coupling. IEEE Trans Power Electron 5270–5278

Shareef H, Islam MM (2016) A review of the stage-of-the-art charging technologies, placement methodologies, and impacts of electric vehicles. Renew Sustain Energy Rev 64:403–420

Singh A, Karandikar PB (2017) A broad review on desulfation of lead-acid battery for electric hybrid vehicle. Microsyst Technol 23:1–11

Sreejeth M, Singh M, Kumar P (2012) Efficiency optimization of vector controlled induction motor drive. In: 38th Annual conference on IEEE industrial electronics society (IECON), pp 1746–1753
Stumper JF, Dotlinger A, Kennel R (2013) Loss minimization of induction machines in dynamic operation. IEEE Trans Energy Convers 28(3):726–735
Sutikno T, Idris NRN, Jidin A (2014) A review of direct torque control of induction motors for sustainable reliability and energy efficient drives. Renew Sustain Energy Rev 32:548–558
Tazerart F, Mokrani Z, Rekioua D, Rekioua T (2015) Direct torque control implementation with losses minimization of induction motor for electric vehicle applications with high operating life of the battery. Int J Hydrogen Energy 40(39):13827e38
Tolochko O, Bovkunovith V, Kalugin D (2017b) Structural realization of the maximum torque per ampere strategy for the vector speed control system of induction motors. Sci Works Donetsk Natl Tech Univ Electr Eng 1(18):23–30
Tolochko DK, Kaluhin D (2019) Speed vector control of induction motor with copper and iron losses minimization. In: IEEE 2nd Ukraine conference on electrical and computer engineering
Tolochko O, Sopiha M, Melnyk A (2017a) Heat loss minimization field control of motionless induction motors in pause of intermittent duty. In: IEEE First Ukraine conference on electrical and computer engineering (UkrCon), pp 442–447
Un-Noor F, Padmanaban S, Mihet-Popa L, Mollah MN, Hossain E (2017) A comprehensive study of key electric vehicle (EV) components, technologies, challenges, impacts, and future direction of development
Verrelli CM, Savoia A, Mengoni M, Marino R, Tomei P, Zarri L (2014) On-line identification of winding resistances and load torque in induction machines. IEEE Trans Control Syst Technol 22(4):1629–1637
Xie Y, Tang X, Zheng S, Qiao W, Song B (2018) Adaptive fractional order PI controller design for a flexible swing arm system via enhanced virtual reference feedback tuning. Asian J Control 20(3):1221–1240
Yao L, Lim WH, Tsai TS (2017) A real-time charging scheme for demand response in electric vehicle parking station. IEEE Trans Smart Grid 8:52–62
Yilmaz M, Krein PT (2013) Review of the impact of vehicle-to-grid technologies on distribution systems and utility interfaces. IEEE Trans Power Electron 28(12):5673–5689
Zhao X, Liu H, Zhang J, Zhang H (2013) Simulation of field oriented control in induction motor drive system. Telkomnika Indonesian J Electr Eng 11(12):7555–7563

Chapter 24
Control Architectures for Low Voltage DC (LVDC) Microgrid

S. P. Gawande, Pranay S. Shete, and Pradyumn Chaturvedi

Abstract LVDC microgrid is considered as the desired solution against the continuous increase of load demand which is powered by renewable energy sources (RESs) which upholds stability between energy needs and supply. The LVDC may escalate the trustworthy and energy-efficient electrical network compared with the existing AC network in many aspects. This chapter discusses the different possible and most efficient control architectures available for the stable operation of DC microgrids. The controls are categorized as decentralized, centralized, and distributed control, which is used for overall control, and communication purpose. In decentralized control, the adaptive control equipped with the droop coefficient, power line signal, and data bus signal is mainly used. In centralized control, the digital communication network will be set up which will allow connecting the source and load controller and will be controlled by central control. Whereas in distributed control, the digital communication link will be set up through a set of local controls which will provide the DC bus voltage at a constant magnitude and may enhance the output current sharing with a proportional-integral (PI) controller. The Local controls are based on voltage control, current control, and droop control. This chapter emphasizes the detailed design architecture of control techniques, their key aspects, communication dynamics, functionalities, and suitable applicability. The chapter also illustrates various other control techniques used in DC microgrids such as droop control, inverse droop control, modified droop control, adaptive droop control, master-slave control, virtual impedance method, etc. which will be engaged in the direction of the power, load sharing, electrical power flow control among the several other inline DC microgrid, and the trouble-free switching and significant drop in power loss.

Keywords Coordinated control · Control architecture · Communication · Low voltage DC (LVDC) microgrid · Stability

S. P. Gawande (✉) · P. S. Shete
Yeshwantrao Chavan College of Engineering, Nagpur, India

P. Chaturvedi
Visvesvaraya National Institute of Technology, Nagpur, India

A. K. Bohre et al. (eds.), *Planning of Hybrid Renewable Energy Systems, Electric Vehicles and Microgrid*, Energy Systems in Electrical Engineering,
https://doi.org/10.1007/978-981-19-0979-5_24

Nomenclature

LVDC	Low Voltage Direct Current
RESs	Renewable Energy Sources
PI	Proportional Integral
DCMG	Direct Current Microgrid
GHGs	Green House Gases
ESSs	Energy Storage Systems
T&D	Transmission and Distribution
PE	Power Electronics
MVDC	Medium Voltage Direct Current
HVDC	High Voltage Direct Current
MS	Micro Source
DERs	Distributed Energy Resources
VSC	Voltage Source Converter
MPPT	Maximum Power Point Tracking
PCC	Power Control Center
CV	Constant Voltage
CCS	Current Controlled Source
MAS	Multi-agent system
CRM	Closed-loop reference model
DG	Distributed Generation
SoC	State of Charge
WESs	Wind Energy Systems

24.1 Introduction

DC Microgrid (DC MG) is now an emerging area in Research as its applications are not only limited to industrial, commercial, and domestic sectors but also can be utilized for agricultural and municipal loads. Microgrids can be distinguished as AC, DC, and hybrid microgrids depending upon their application. Microgrids can be defined as a reduced in size version of the centralized power system. Historical overview of a microgrid is in 1882, the Manhattan Pearl Street Station, built by Thomas Edison, was the USA's first commercial electrical power plant (Bhattacharya 2018). The thermal power station initially served few customers in a few blocks, powering a few hundred resistive loads (e.g., lamps) (Bhattacharya 2018). Two years later it was up to 510 customers approximate with 10,000 lamps approximate, in 2004 Japan's Sendai system, in 2005 the Shimizu microgrid, the gas microgrid, the Labein microgrid, the National Laboratories and last but not the least in 2006, Germany's Manheim microgrid. Microgrids can be able to generate, distribute and control electrical power for dedicated load or small locality (Bharath et al. 2019). Aspects of DC MG are pliable and reliable as to uphold the steadiness between the supply and load by the microgrids which are usually designed to fulfill the load demands. In the case of DC MG, one should always look after the change in supply voltage in the DC MG. It is more sheltered because it is small in size, its generation and controlling can be locally possible hence it can be safe from any cyber-attack. It is healthy as the operation and control of the DC MG are not dependent on the conventional grid. Many a time, DC MG is used to fulfill the essential load demands when the AC supply gets disconnected from the conventional grid. It is cost-effective as it can help to optimize the power usage which intern saves money. It can store and incorporate renewable energy hence can be able to reduce greenhouse gases (GHGs) gas emissions as per government guidelines and regulations. A modern microgrid may comprise a lot many things from a small group of households to a whole community. A microgrid can work as an off-grid mode that is usually independent and empowers areas located in a remote place to get linked using marginal investment and can connect rapidly to a larger grid. A microgrid maybe works as a cybernetic model which can be installed physically as an energy distribution grid that can't separate but which may be able to operate within the energy market as an independent or as a grid-connected mode where it can be "islanded" to run on its own.

In this mode, it is set in the widespread grid to authorize other power sources to balance the loads by adapting the revolutionary communication and, IT skills for electric power systems. Nowadays, DC MG has such a capability that it is agile, free from the occurrence of frequent interruptions, secure, and more economical if used in a smart grid application also it can be possible that the smart interaction amongst all links by emerging and make them known to most recent innovative control technologies, which may further lead to optimizing the production, transmission, and electrical energy distribution. In the smart grid, the power distribution has to alter from pliant to agile mode, to optimize the operation of the system, it ropes the DG for real-time involvement in the production and distribution of electrical energy. The microgrid comprises a dynamic distribution network, which enables the combination of DG and conversion from the conventional grid to a versatile smart grid on a large scale. The usage of highly recommended Energy Storages Systems (ESSs) and DGs in the microgrid serves the justifiable energy saving, reduction in the emission of toxic gases like SO_x, NO_x, etc., and may motivate to employ the suitable strategy. Feeder losses reduce to some extent by using new and RESS DGs, which may direct toward the chances of saving eye-catching investment on transmission and distribution (T&D) networks effectively. To do so, the shared support with the microgrid, the utilization of resources to satisfy the desires, and the equipment is needed. The said approach is considered as reliable as well as maintaining the quality in that way energy efficiency and grid security may increase. From Fig. 24.1 there are various kinds of DGs such as solar PV, windmill, super-capacitor, and battery pack used as a main energy source in the DC microgrid. Here all the sources are not sustainable, as one or the other factor will hamper the continuity of the power supply as energy generation from the solar PV may vary depending on the solar irradiation, windmill performance may be hampered due to the wind speed, the super-capacitor, and the battery pack may not last long for longer duration. The generation from all the sources is of DC hence, to maintain the overall voltage level equal to the DC

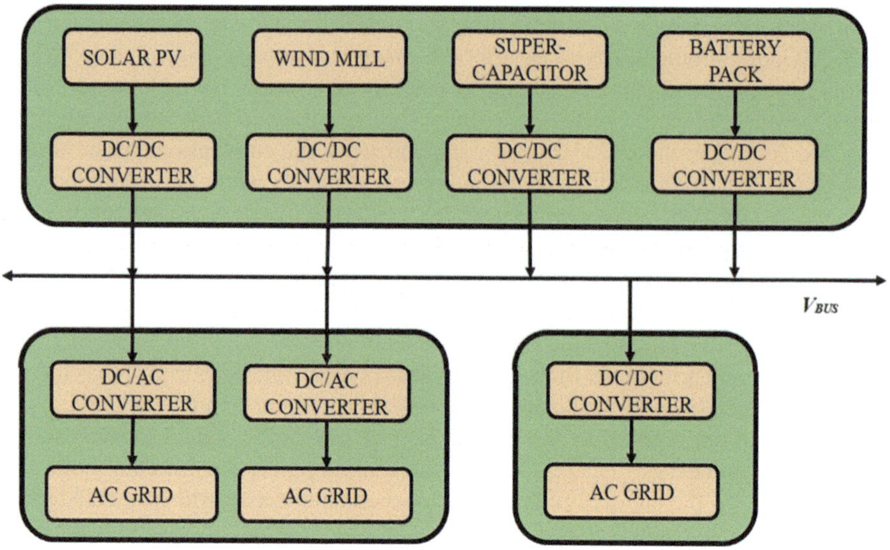

Fig. 24.1 Architecture of DC microgrid

grid voltage level it is necessary to employ DC/DC converter in the system. Most loads are of AC so an inverter is needed to invert the fixed DC voltage into variable AC voltage and there is DC/DC converter at the distribution end to accomplish the power requirement of DC loads. The recent technological advancement in renewable energy empowers the user to utilize the common facility such as the rooftop solar photovoltaic panel for home, terrace of commercial complex and industry for generation of electricity which then injected into the utility grid simultaneously the user can use the power received from the grid. The system must be well equipped with advanced power electronics components which allow the user to operate, coordinate, and optimize the size of the microgrid. When the utility grid is interconnected with the system and can operate as a primary energy source (Guerrero et al. 2011). In a microgrid, highly effective storage batteries are used, which can store the electrical energy in the form of electrical charges and can be used as a backup energy source to meet any future contingencies. The DC MG provides the desired voltage magnitude and the variable voltage and frequency magnitude is provided in the AC system with tolerance. To deal with the microgrid, one should have the practice to enable the use of a microgrid management system that coordinates with the utility grid during outages. This can be done by segregating the load into essential and non-essential loads. The uninterrupted power supply must always be available for the essential loads whereas, non-essential loads can be thrown off or get connected as per requirement. Many times, load scheduling will reduce the occurrence of frequent outages of both AC and DC microgrids. In the distribution network, microgrids can be acts as a power source, or sometimes they may act as a virtual load. Hence, microgrids may be able to maintain the power balance in a network. One can achieve the optimal energy distribution through a microgrid in a network (Bharath et al. 2019, 2011).

To achieve and overlook the above-mentioned facts, this chapter

- Discusses the different possible and most efficient and latest control architecture available for stable operation of DC Microgrid
- Emphasizes the detailed design architecture of control techniques, its key aspects, communication dynamics, functionality, and suitable applications

The chapter is structured and divided into different sections. Section I introduces the fundamental aspects of DC microgrid. Section 24.2 elaborates exclusively on the low voltage DC microgrid. Section 24.3 emphasizes the DC microgrid control techniques including primary, secondary, centralize, decentralize and droop control. Section 24.4 discusses the key aspects of low voltage DC microgrid such as utilization, stability issues and challenges to be faced. Further, the chapter is followed by a conclusion and references.

24.2 Low Voltage DC (LVDC)

An LVDC uplifts the capacity of the existing electricity distribution network due to its capabilities. In a new era, due to technological and economic growth in advanced power electronic elements allows the user to use them in the LVDC network to fulfill the need of the consumer demands. In this regard, the LVDC distribution network concentrates on the DC/AC interface which is directed toward the implementation of the utility-end inverter. An LVDC energy distribution network consists of fast-acting power electronic (PE) converters and a DC link flanked by the modified converters. In the distribution network, the converter usually positioned adjacent to a medium voltage (MV) electrical distribution (Hatziargyriou 2013; Gao et al. 2019; Khorsandi et al. 2014; Javed and Chen 2018; Hajian et al. 2013).

The DC/AC conversion is situated at unlike places, and which may reliant upon the place, the LVDC network can be either high voltage direct current (HVDC) distribution or an LVDC where the DC/AC conversion can be made in power distribution network at the utility end. With technological advancements, the existing AC distribution network can easily be replaced with a modernized LVDC network. In Fig. 24.2, the LVDC network architecture is as shown. Here, the solar PV and fuel cells are acting as an energy source. To convert the DC power with a suitable voltage limit, DC/DC converter is used. LVDC and MVDC may be employed for railway traction, telecom centers, and vehicle power systems (AlLee and Tschudi 2012). The increase in DC loads depends upon the installed capacity of RESs. The applications of power distribution system in residential and commercial are capable enough to fulfill the load demand. The various DC Microgrid architectures may propose the necessity of specific design, its applications, and investigation of the literature (Kwasinski 2011; Lindman and Thorsell 1996; Dragicevic et al. 2014; Park et al. 2013). From a technical point of view, the LVDC network is always on the upper

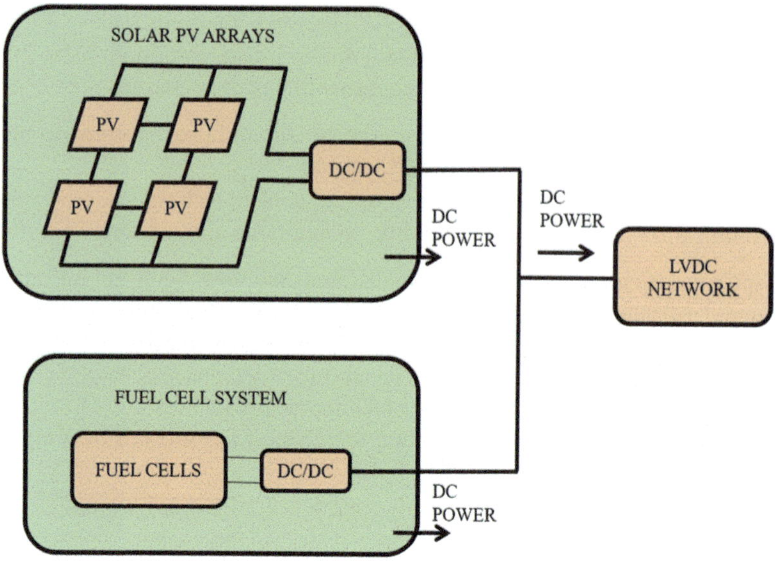

Fig. 24.2 LVDC network architecture

hand and has noteworthy advantages over the conventional AC network. The conversion steps for the electronic equipment present within the premises may reduce the cost and may improve the overall efficiency. It can be said that the removal of every conversion step will increase 2.5% efficiency. In addition to that, the filtering process of harmonics within buildings may be facilitated by the LVDC distribution network in combination with the centralized AC/DC converter. The LVDC network surely is more effective when it gets connected with RESs with low carbon emissions to the distribution grid. The decentralized RESs such as solar PV, rechargeable batteries used in electric vehicles, fuel cells, super-capacitors, and many more RESs generate DC energy and somehow require a midway conversion stage to connect the conventional grid. Hence, RESs can be linked to DC MG in the LVDC network. Ultimately, the probabilities of power losses during the conversation stages will reduce and the capital cost of the overall system decreases. The LVDC network has a higher power transmission capacity than the conventional AC network with 400 VAC.

In the case of LVDC, the power transmission capacity can be more than 16 times compared to the conventional 400 VAC network at the voltage drop limit and the thermal limit is more than 4 times compared to the conventional 400 VAC network. The coefficient of transmission capacity is contingent on the DC voltage level in the LVDC network. Here, the voltage quality may improve which results in the implementation of the inverter's active voltage control which may help to reduce the voltage fluctuations in the DC network and may be able to restore to their original operating voltage.

The rebuilding of voltage dip depends upon the capacitance allied to the DC network. The high transmission capacity of the LVDC network permits the substitute as a medium voltage direct current (MVDC) network. The suitable protection zone for the DC network is created by the LVDC network. The faults may take place in the DC line which may originate from the outages planned for the consumers in DC area only. The reliability of DC network elevates while MV line reliability suppresses (Wang et al. 2015; Paul 2002; Cuzner and Venkataramanan 2008; Salomonsson et al. 2009). Hence, it can be said that the higher the transmission capacity the smaller is the cable cross-section area. The higher power transmission dimensions may cause smaller currents to flow through the system which may help to reduce the system power losses for the same cable sizes in both networks. The actual difference in the power loss can however be smaller than transmission capacity presumes that if the cable sizes are not equal then the LVDC network may replace the number of AC distribution transformers easily as it includes only one LV transformer at the commencement of the LVDC network. In addition to that, the new LV distribution network has to acquaint with challenges as well (Emhemed et al. 2017; Tokuyama et al. 1985; Pauli et al. 1988). The DC network is more complicated than the conventional AC network due to which operational challenges may occur. The LVDC network has challenges to electrical safety. The voltage levels of the LVDC network can cause high voltages in difficult grounding conditions which may require an ungrounded IT grid. The use of PE devices for the distribution network also the usage of PE converter can cause switch faults and operation of complicated switchgear used for protection. The biggest challenge in the converter is that it has to operate with a sufficient fault current capability. The operating limit of suitable power electronic devices may be set as one-third of the widely used power electronic devices. The short operating lifespan of converters may accentuate maintenance issues and increase costs.

24.3 DC MG Control Techniques

The DC MG Control techniques promise that the control will be improved, steady, and efficient. The PE converters act as an interface between the grid and the load which may provide proper control to the microgrid with modified voltage regulation, and better distribution of current (Zhang et al. 2016). This interface may simplify the connections of several units available in the DC MG (Hatziargyriou 2013). A wise control technique is needed to develop and lower down the distortion effect generated by the power electronic converters which is due to their fixed power actions. The sudden increase in distorted loads and DGs results in complexity of the control. Necessities of the various control employed for smooth shifting from standalone to the grid integrated operation which usually consists of regulation of voltage and controlled current sharing, and steady operation with non-linear as well as a constant power load. To control the power flow of the MG, the Energy Management Scheme (EMS), efficient load power-sharing, and the proper communication medium between DERs is used (Zhang et al. 2018). To regularize the proper control mechanisms, it is required to avoid the grid failure, avoid black start, possible reduction in the transmission losses and, try to enhance the capability of DERs. An uninterrupted

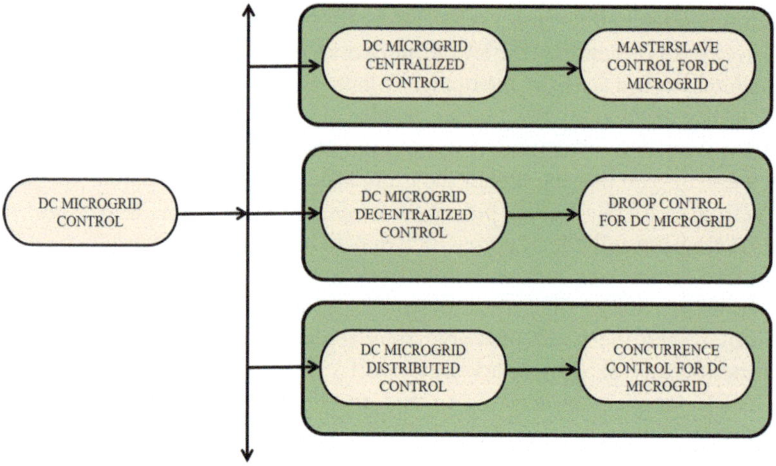

Fig. 24.3 Control strategies of DC Microgrid

power supply will be made available to the critical loads (e.g., hospitals, industries, and other key utilities) the proposed microgrids need to provide the desired voltage magnitude in the DC system. The DC microgrid control is discussed in Fig. 24.3. Here, the centralized control will be extended to the master-slave, the droop, and the concurrence control. There may be an uncontrolled situation that has occurred when the source of energy is renewable energy, and which is hampered due to the change in weather. The use of modern controls may help to optimize the energy use depending upon the priorities and the critical requirement within the premises.

To deal with the DC MG, one should have the practice to get acquainted with the use of a microgrid management system that may coordinate with the utility grid during outages. This can be done by segregating the load into essential and non-essential loads. The uninterrupted power supply must always be available for the essential loads whereas, non-essential loads can be thrown off or get connected as per requirement. Many times, load scheduling will reduce the occurrence of frequent outages of both AC and DC microgrids. In the distribution network, microgrids can be acts as a power source, or sometimes they may act as a virtual load. Hence, microgrids usually maintain the power balance in a network. One can achieve the optimal energy distribution through a microgrid in a network (Bharath et al. 2019). From Fig. 24.4, we may refer to the control architecture of the DC MG. Here, various controls are discussed such as tertiary, secondary, and primary control (Hatziargyriou 2013). Necessities of the various control are, they have smooth switching operating characteristics from standalone to the grid integrated operation, and there must be voltage regulation and current distribution. The effective control should be efficient during the load power-sharing operation and there must be a proper communication medium between distributed energy resources (DERs). The use of effective control may help to optimize the cost and economic dispatching of loads can be possible. To maximize the abilities of DERs and to reduce the transmission losses a proper

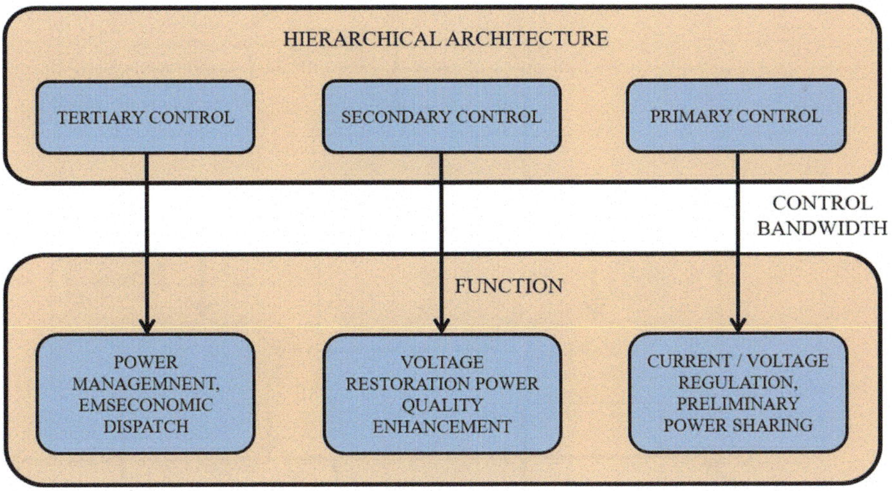

Fig. 24.4 Hierarchical architecture of DC Microgrid

control plays a vital role and may have the ability to make available an uninterrupted power supply to loads considered as critical.

24.3.1 Primary and Secondary Control of DC MG

Due to numerous advantages of LVDC systems stated as the reduction in the losses and it is easy to integrate with ESRs, that is why the DC MGs may gain the attention of researchers. The challenge in this DC MG is to offer the voltage support and power-sharing performance in a better way. The control strategy shows a significant role in confirming MG's power quality and efficiency, it is necessary to conduct a complete review of the state-of-the-art control and their approaches in DC MGs (Gao et al. 2019).

24.3.2 Primary and Secondary Control of DC MG

The ESs are placed to accomplish intellect which is organized by a microcontroller or with a server. To have a better communication medium is the essential component of such a scheme, which helps in the easier functioning of the centralized system. Some of the advantages of the centralized controller are listed as it should have strong controllability, there must be a single controller, it can outline comprehensive approaches for regulatory operation of the system, and it should have good observability. Form Fig. 24.5, the master-slave control is shown and, which is a prominent

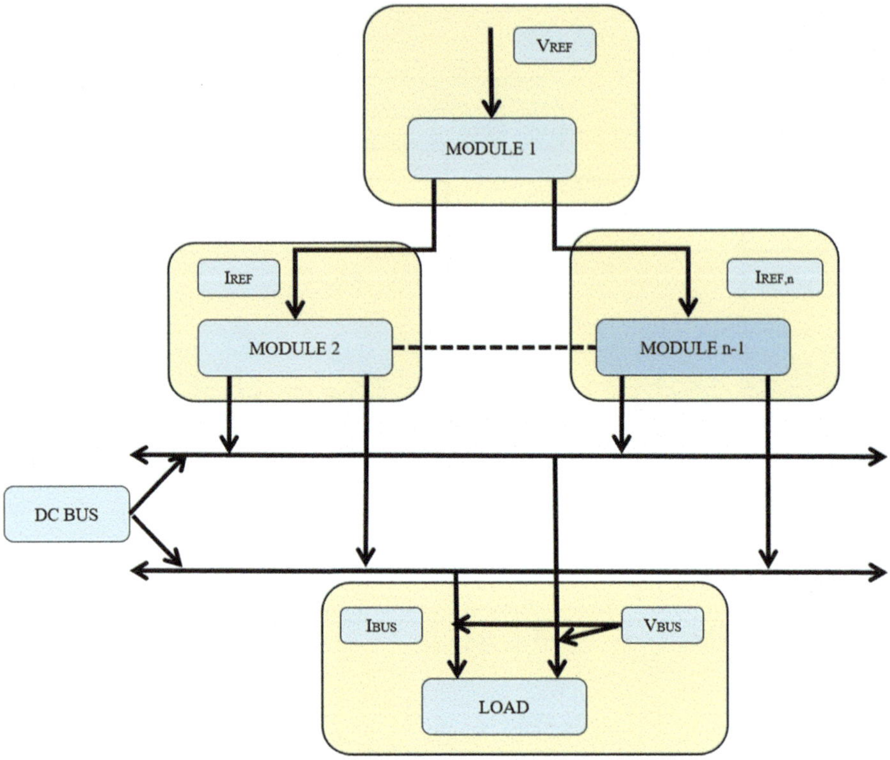

Fig. 24.5 Master–slave controller for DC MG

technique used to attain the corresponding operation of several sources. Here in this scheme, one converter turned as a voltage source converter (VSC), which is used as a master and instruct slave elements for DC bus voltage regulation. The master-slave converter function so that the grid voltage is retained in the tolerable limit, and the persistent converters that act as slaves may be backing the master due to which the set point may be achieved.

The control mechanism relied on a quick response communication system. The overall performance of the system may affect when the system is subjected to any failure in communication, which will result in a total black-out of the entire system. The drawback of this said control approach is the reduction in the battery life, low scalability, the necessity of intelligent supervisory control, luxury, which has poor fault tolerance capability (Tamura et al. 1980)–(Ferreira et al. 2012). A phase-shift controller is used in the master-slave control algorithm which is used in series and parallel full-bridge converter as an input and output. It has an inefficient range of voltage stability which is compensated using the master-slave supervisor. In some cases, it is more flexible, and easier to implement, which facilitates it for high power applications. The master-slave control used for the electric bus application is assumed to have high efficiency of up to 3%.

24.3.3 Decentralized Control for LVDC

For stable operation of the DC MG, one should have to keep the power balance in between the DGs and loads (Bharath et al. 2019; Hatziargyriou 2013; Khorsandi et al. 2014). The operating modes of DC MG are as follows:

(1) **Mode I**: Here, the DC MG is linked to the system and can work as an AC grid with AC/DC converter. The converter regulates the DC bus voltage by extracting the excess power from the grid and may act as a voltage source. To extract the peak power from the solar PV panel, the DC/AC converter works with a MPPT scheme. MPPT working as a current source here. The battery converts the energy into a safe charge current. The battery converters are detached from the system when the batteries are fully charged. Hence, the output currents of the DG units are used to estimate the cable resistances. The cable resistances can be assessed with the Eq. (24.1) given below

$$r_{es,i} = \left| \frac{V_{nor} - V_{out,i}}{I_{out,i}} \right| \tag{24.1}$$

where V_{nor} is nominal voltage. Whereas $R_{\text{es, i}}$ is the cable resistance of the *i*th cable, $V_{\text{out, i}}$, and $I_{\text{out, i}}$ are the output voltage and current of the ith DG unit, respectively. To estimate the cable resistances, the charging and discharging droop gains must be taken into consideration.

(2) **Mode II**: When a fault has occurred in the AC grid, the DC converter is detached from the system, and in this case, the DC MG functions in the standalone operation (Saleh and AL-Ismail, 2021; Kumar et al. 2020). The solar PV converter works with MPPT, and the load power is greater than the solar PV power. When working with the voltage sources, the battery converters may regulate the battery discharging current. To access this, the discharge droop gains are used such that at the power control center (PCC), the system voltage extends up to their minimum suitable limit (Tamura et al. 1980; Meyer et al. 2004). Here, the discharge current of the battery pack is also maximum.

(3) **Mode III**: Here, the DC MG functions in an islanded operation, and the solar PV converter works with the MPPT algorithm. Since the solar PV power generated is larger than the load demand, and hence the PCC voltage rises. The batteries are not fully charged in this case, as the excess amount of power is required to accomplish the operation. Therefore, it is necessary to regulate the dc bus voltage through the battery charging operation.

(4) **Mode IV**: In such a mode, the DC MG works in an standalone operation, and the batteries need to remain fully charged. As the PCC voltage increases which in turn increases the power generation through the solar PV which is greater than the load power.

In a decentralized controller, to control the scattered elements are required to use the local controllers through non-dependent local variables, and to do so no communication medium is required. This can be considered as the most reliable control strategy. A switching of current is based on a decentralized controller, which is proposed where the VCS is replaced by a CCS. Here, the proposed controller gives a improved transient response, plug-in and plug-out facility, enhanced voltage regulation, and satisfactory current sharing. Some of its advantages may comprise of healthier fault tolerance ability, more flexibility, suitable scalability, and ease of implementation.

24.3.4 Droop-Controller for DC Microgrid

The droop-controller shown in Fig. 24.6 may effectively work when a current is supplied to the grid it generates a DC grid voltage. The decentralized control scheme was accepted to remove the circulating current present in the converters (Bharath et al. 2019; Ferreira et al. 2012). The Eq. (24.2) indicating the operation is given as

$$V_{ref} = V_o + \left(I_{ref}.R_{droop}\right) \tag{24.2}$$

where V_{ref} is the DC grid reference voltage, V_{o} is the PCC voltage, I_{o} is load current and R_{droop} is the virtual resistance.

In the conventional droop-control strategy, the way that the incorrect selection of droop resistance gives unsatisfactory load distribution, poor voltage regulation, and unefficient performance of RESs.

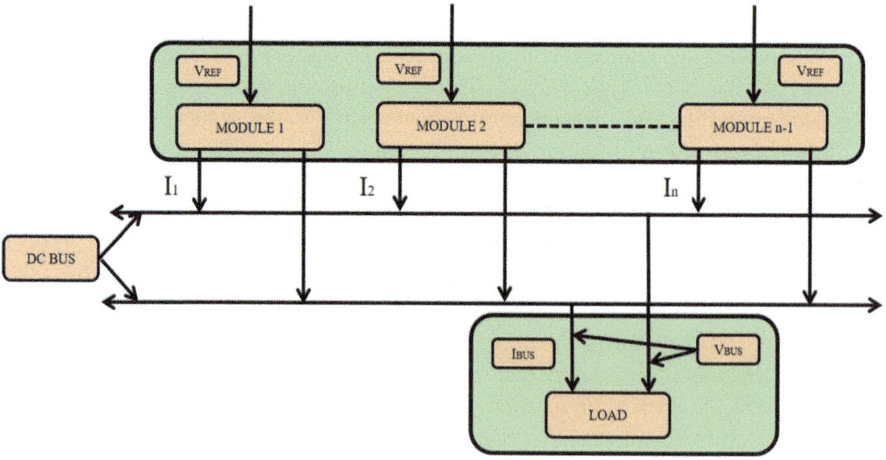

Fig. 24.6 Droop-controller for DC MG

24.3.5 Cybernetic Resistance-Based Droop Control

The proposed virtual droop resistance may solve the disadvantage of the droop. The said method introduces a cybernetic Resistance (R_{droop}) which is a function of the terminal voltage as given in (24.3) and (24.4).

$$I_{ref} = \frac{\left(V_{ref} - V_o\right)}{R_{droop}} \tag{24.3}$$

$$R_{droop} = func(V_o) \tag{24.4}$$

To non-linearity in the droop, the controller can be achieved and the voltage regulation of the system may improve with an adaptive droop-control strategy applicable for DC MG. For effective power-sharing, the charging and discharging of the BESS units are controlled with the help of bidirectional adaptive droop-control scheme. An adaptive droop-control strategy is used in a closed-loop reference model (CRM) to comprehend the DC MGs control. To achieve voltage stability and current sharing, a time-varying model is required (Ahmadi et al. 2597).

24.3.6 Distributed Control for LVDC

Nowadays, controllers are equipped with an energy source to form a distributed controller, which assist in appropriate distribution of load to regulate grid voltage. The communication through a communication medium controller for each PE device is similar to that of decentralized control (Bhargav et al. 2020). This may be carrying out an important operation (like maintaining the balance of state of charge (SOC), voltage maintenance, etc.). Any variations in the distributed generation (DG) make it complex for the implementation of centralized control. During such a difficult situation, the distributed controller is having the advantage as a controller restricts the chances of occurrence of failure of communication link hence, the system become invulnerable. The improvement of voltage regulation and finest power using a non-linear optimal controller can be done using modified droop-control architecture.

An extensive comparison between the various control schemes is provided in Table 24.1.

Table 24.1 Summary and evaluation of DC MG algorithm

Sr. no.	Control scheme	Method	Load current sharing	Complexity	Voltage regulation	Additional features	Claims
1	LVDC centralized control	Phase control	Impartial	Modest	Decent	(a) More supple (b) Fault-tolerant	HV and power DC Grid
		Relative integral control	Impartial	Modest	Decent	(a) Healthy (b) Better efficiency (c) Rejects extra wiring	LVDC Electric bus
		Multi-level control	Decent	Complex	Improved	(a) Transient response is faster (b) Load shedding minimizes (c) Operational cost reduced	Islanded DC MG with Plug-in EVs
2	LVDC decentralized control	Predictable PI control	Unfortunate	Easy	Decent	(a) More supple (b) Ascendable	DC System
		Adaptive-droop-control strategy	Decent	Easy/modest	Improved	(a) Healthy (b) Fit in RESSs	only DC system
		Closed-loop reference model (CRM)	Improved	Easy/moderate	Improved	(a) Healthy under noise instabilities and misgivings (b) Improved power-sharing (c) More operative than linear control	HVDC MG
		Adaptive droop-controller	Improved	Modest	Improved	(a) More commercial (b) More consistent	LVDC MG

(continued)

Table 24.1 (continued)

Sr. no.	Control scheme	Method	Load current sharing	Complexity	Voltage regulation	Additional features	Claims
3	LVDC distributed control	Non-linear optimal Control	Decent	Modest	Improved	(a) More stable (b) Strong	DC MG
		HSS using a PI controller	Decent	Modest/Complex	Improved	(a) Supple and pliable (b) Substantial reduction in the use of energy storage elements	DC MG
		Adaptive-droop control using a PI controller	Decent	Modest	Improved	(a) Proportionate load sharing (b) Plug-and-play compatibility	LVDC MG

24.4 DC MG Utilization, Stability Issues, and Challenges

24.4.1 RESs and ESS Utilization

The use of RESs is a need of the current scenario. The operation of numerous components, for example, the BESS, the RESSs, and the loads using a centralized controller in a synchronized operation is considered here. The tasks performed by the controller are equalizing of SOC, stoppage of overcharging of battery by an real power break, and load shedding due to which battery discharging may be avoided. The proposed system results in improving the life duration of energy storage devices (Ruiz-Martinez et al. 2182; Wang et al. 2021).

24.4.2 Microgrid Protection and Control

The presence of a zero-crossing current is considered a risky feature in the DC system compared to AC systems. Many research gaps exist in the current scenario and researchers are still involved in designing a better protection scheme for DC MG that leads to healthier operation. In addition to that the safety of the operator is a major issue and which can't be tolerated (Park et al. 2013). However, there are some modified and tested switchgears devices with faster response time, but found to be costly due to technological advancements. A quick-acting protection scheme is required for the bidirectional communication schemes, for EVs and MG (Salomonsson et al. 2009). There are few other areas the changes getting introduce the advanced research to enhance the capability of the system. The system reliability may ensure through a stability analysis. Dring the occurrences of sudden disturbances in the grid, it is required that the voltage must be stable for better operation and reliability (Meyer et al. 2004).

24.4.3 Microgrid Protection and Control

To meet the numerous practical challenges existing in the DC grid, the centralized controller will not be able to serve the purpose without a proper communication medium. A suitable communication link must be employed to attaining improved dependability and functioning of the system. The communication link should be best, inexpensive, more practicable, and must be less difficult. Additional required important feature includes improved load current distribution, and good regulation of voltage, which is considered a main problem for conventional controllers.

24.4.4 Stability Issues of DC Microgrid

The load with a constant magnitude generally faces the challenge of instability of DC microgrid. Therefore, such loads can be attached with a system that has negative impedance features. The resonant conditions may arise due to the replicated negative inductance which may likely to induced during the constant power load operation and has an unfavorable effect on a DC microgrid from a stability point of view. It is common practice that there is a rise in the implementation of damping mechanisms that may adopt to overcome the unwanted situation. But this may result in a voltage sag up to some extent. A right trade-off is mandatory in between allowable voltage swing and voltage sag duration which may help to meet the applicable standards.

24.4.5 Challenges of DC Microgrid

The fundamental conventional control schemes are analyzed by adopting the advanced microprocessors technique which is already proven as faster as the response time is very less in a protection device. In a DC MG, the scheme of circuit breakers in general practice is a big encounter as the scheme must be economic and effective as far as its operation is concerned. Here, the DC voltage level should not fall to zero magnitudes hence the magnitude of current also never be equal to zero this can be considered as one of the major interferences in DC microgrid protection. The security of an operator is one of the prime requirements. It is a real challenge to lower down the DC stray current within the limit, to achieve the same the proper grounding may help a lot. The unified controllers used as a local control along with the other related control tasks should certify that the system is firm during islanding and grid integration. The selection of an improved control scheme for the battery operation must be accomplished with extreme attention and appropriate methods which may safeguard against overcharging and undercharging. The battery packs should have to maintain the SOC to avoid damages that may occur in the future.

24.5 Conclusion

This chapter mainly focuses on the low voltage DC Microgrid structure, control architecture, and the other associated aspects. The chapter essentially discusses the different post efficient control architectures available for the stable and smooth operation of the low voltage DC Microgrid. An extensive comparison between the various control schemes is provided in the tabular form. Further, the paper satisfactorily describes the detail design architecture of control schemes, its key aspects, the communication dynamics, functionality, and suitable applications of these controls in DC Microgrid.

References

AhmadiS, Sadeghkhani I, Shahgholian G, Fani B, Guerrero JM (2021) Protection of LVDC micro-grids in grid-connected and islanded modes using bifurcation theory. IEEE J Emerg Sel Topics Power Electron 9(3):2597–2604. https://doi.org/10.1109/JESTPE.2019.2961903

AlLee G, Tschudi W (2012) Edison Redux: 380 Vdc brings reliability and efficiency to sustainable data centers. IEEE Power Energy Mag 10(6):50–59. https://doi.org/10.1109/MPE.2012.2212607

Bharath KR, Krishnan Mithun M, Kanakasabapathy P (2019) A review on DC microgrid control techniques, applications and trends. Int J Renew Energy Res 9:1–11. https://www.ijrer.org/ijrer/index.php/ijrer/article/view/9671

Bhargav R, Bhalja BR, Gupta CP (2020) Novel fault detection and localization algorithm for low-voltage DC microgrid. IEEE Trans Ind Inf 16(7):4498–4511. https://doi.org/10.1109/TII.2019.2942426

Bhattacharya A (2018) Control of DC microgrid system—overview of microgrids, NPTEL

Cuzner RM, Venkataramanan G (2008) The status of DC micro-grid protection. In: Proceedings of 2008 IEEE Industry applications society annual meeting, pp 1–8. https://doi.org/10.1007/978-3-642-38786-9_38

Dragicevic T, Vasquez JC, Guerrero JM, Skrlec D (2014) Advanced LVDC electrical power architectures and microgrids: a step toward a new generation of power distribution networks. IEEE Electronic Mag 2(1):54–65. https://doi.org/10.1109/MELE.2013.2297033

Emhemed AAS, Fong K, Fletcher S, Burt M (2017) Validation of fast and selective protection scheme for an LVDC distribution network 32(3):1–8. https://doi.org/10.3390/en13123167

Ferreira RAF, Braga HAC, Ferreira AA, Barbosa PG (2012) Analysis of voltage droop-control method for dc microgrids with Simulink: modelling and simulation. In: 2012 10th IEEE/IAS International conference, Fortaleza, Brazil, pp 1–6. https://doi.org/10.1109/INDUSCON.2012.6452563

Gao F, Kang R, Cao J, Yang T (2019) Primary and secondary control in DC microgrids: a review. J Mod Power Syst Clean Energy 7(2):227–242. https://doi.org/10.1007/s40565-018-0466-5

Guerrero JM et al (2011) Hierarchical control of droop-controlled AC and DC microgrids—A general approach toward standardization. IEEE Trans Ind Electron 58:158–172. https://doi.org/10.1109/TIE.2010.2066534

Hajian M, Jovcic D, Wu B (2013) Evaluation of semiconductor based methods for fault isolation on high voltage DC grids. IEEE Trans Smart Grid 4(2):1171–1179. https://doi.org/10.1109/TSG.2013.2238260

Hatziargyriou N (2013) Microgrid: architecture and control. Wiley & Sons Publications, IEEE Press

Javed W, Chen D (2018) Low voltage DC microgrid protection system—a review. In: 53rd International universities power engineering conference (UPEC), pp 1–6. https://doi.org/10.1109/UPEC.2018.8541944

Khorsandi A, Ashourloo M, Mokhtari H (2014) A decentralized control method for a low-voltage DC microgrid. IEEE Trans Energy Convers 29(4):793–801. https://doi.org/10.1109/TEC.2014.2329236

Kumar J, Agarwal A, Singh N (2020) Design, operation and control of a vast DC microgrid for integration of renewable energy sources. Renew Energy Focus 3(4):1–36. https://doi.org/10.1016/j.ref.2020.05.001

Kwasinski A (2011) Quantitative evaluation of DC microgrids availability: effects of system architecture and converter topology design choices. IEEE Trans Power Electron 26(3):835–851. https://doi.org/10.1109/TPEL.2010.2102774

Lindman P, Thorsell L (1996) Applying distributed power modules in telecom systems. IEEE Trans Power Electron 11(2):365–373. https://doi.org/10.1109/63.486187

Meyer C, Schroder S, DeDoncker RW (2004) Solid-state circuit breakers and current limiters for medium-voltage systems having distributed power systems. IEEE Trans Power Electron 19(5):1333–1340. https://doi.org/10.1109/CIEP.2002.1216642

Park J, Candelaria J, Ma L, Dunn K (2013) DC ring-bus microgrid fault protection and identification of fault location. IEEE Trans Power Deliv 28(4):2574–2584. https://doi.org/10.1109/TPWRD.2013.2267750

Paul D (2002) DC traction power system grounding. IEEE Trans Ind Appl 38(3):818–824. https://doi.org/10.1109/TIA.2002.1003435

Pauli B, Mauthe G, Ruoss E, Ecklin G, Porter J, Vithayathil J (1988) Development of a high current HVDC circuit breaker with fast fault clearing capability. IEEE Trans Power Deliv 3(4):2072–2080. https://doi.org/10.1109/61.194019

Ruiz-Martinez OF, Mayo-Maldonado JC, Escobar G, Frias-Araya BA, Valdez JE (2020) Data-driven control of LVDC network converters: active load stabilization. IEEE Transa Smart Grid 11(3):2182–2194. https://doi.org/10.1109/TSG.2019.2949259

Saleh F, AL-Ismail (2021) DC microgrid planning, operation, and control: a comprehensive review. IEEE Access 10(9):36154–36172

Salomonsson D, Soder L, Sannino A (2009) Protection of low-voltage DC microgrids. IEEE Trans Power Deliv 24(3):1045–1053. https://doi.org/10.1109/TPWRD.2009.2016622

Tamura S et al (1980) Parallel interruption of heavy direct current by vacuum circuit breakers. IEEE Trans Power Apparatus Syst 99(3):1119–1129. https://doi.org/10.1109/TPAS.1980.319742

Tokuyama S, Arimatsu K, Yoshioka Y, Kato Y, Hirata K (1985) Development and interrupting tests on 250KV 8KA HVDC circuit breaker. IEEE Power Eng Rev 5(9):42–43. https://doi.org/10.1109/TPAS.1985.318990

Wang Y, Yu Z, He J, Chen S, Zeng R, Zhang B (2015) Performance of shipboard medium-voltage DC system of various grounding modes under monopole ground fault. IEEE Trans Ind Appl 51(6):5002–5009. https://doi.org/10.1109/ISGT-Asia.2017.8378426

Wang L, Wong M-C, Zhou X, He Z, Xu, Qianming, Zhou L (2021) A selective power droop control for hybrid interlinking converter in AC/LVDC microgrid. IEEE Trans Ind Electron 68(10):9046–9057. https://doi.org/10.1109/TIE.2020.3020017

Zhang Y, Ravishankar J, Fletcher J, Li R, Han M (2016) Review of modular multilevel converter based multi-terminal HVDC systems for offshore wind power transmission. Renew Sustain Energy Rev 61:572–586. https://doi.org/10.3390/app10217719

Zhang L, Zhang W, Zeng F, Yang X (2018) A review of control strategies in DC microgrid. In: First international conference on advanced algorithms and control engineering, pp 1–9. https://doi.org/10.1088/1742-6596/1087/4/042035

Chapter 25
Multilevel Planning for Smart Charging Scheduling for On-Road Electric Vehicles Considering Seasonal Uncertainties

Sourav Das, Arnab Pal, Parimal Acharjee, Ajoy Kumar Chakraborty, and Aniruddha Bhattacharya

Abstract In this chapter, a complete charging solution is proposed, which consists of optimal allocation of Electric Vehicles Charging Station (EVCS), optimal assignment of charging station to each EV, which is followed by an optimal charging strategy to reduce the daily cost of charging considering Grid to Vehicle (G2V) and Vehicle to Grid (V2G) dual-mode of operation. In this respect, a multilevel algorithm is proposed, where, in the first level, the optimal allocation of EVCS has been done at IEEE 33-bus system with the objective to minimize the power loss in the system by considering various constraints. In the second level, apt charging station for each and every EV has been identified where they can reach at minimal battery energy consumption. This is followed by a smart charging strategy considering the coordination of both G2V and V2G dual-mode of maneuver with the objective to maximize the profit by reducing the daily cost of charging for both CSO and EV owners, considering various practical constraints, in the third level of algorithm. But, while performing all these tasks, several driving cycle features like daily mileage, arrival time, departure time, initial SOC (State-of-Charge), Departure SOC, ensuing trip length after leaving the charging station are required, which is uncertain in nature for each and every vehicle. Day to day basis, it could vary. Hence, seasonal variation is one of the major factors which needs to be incorporated while doing such work. To

S. Das (✉) · P. Acharjee · A. Bhattacharya
Department of Electrical Engineering, National Institute of Technology Durgapur, Durgapur 713209, India
e-mail: sd.18ee1103@phd.nitdgp.ac.in

P. Acharjee
e-mail: parimal.acharjee@ee.nitdgp.ac.in

A. Bhattacharya
e-mail: aniruddha.bhattacharya@ee.nitdgp.ac.in

A. Pal · A. K. Chakraborty
Department of Electrical Engineering, National Institute of Technology Agartala, Agartala 799046, India
e-mail: arnabee.sch@nita.ac.in

A. K. Chakraborty
e-mail: akcall58.ee@nita.ac.in

A. K. Bohre et al. (eds.), *Planning of Hybrid Renewable Energy Systems, Electric Vehicles and Microgrid*, Energy Systems in Electrical Engineering,
https://doi.org/10.1007/978-981-19-0979-5_25

deal with uncertainties related with driving cycles, a 2m Point Estimation Method (2m PEM) is chosen. Later, using statistical analysis (i.e., Wilcoxon Signed Rank Test and Quade Test) the robustness and consistency of the proposed algorithm is established.

Keywords Electric vehicles · Charging scheduling · Charging station allocation · Grid to vehicles · Vehicle to grid · Point estimation · Optimization · Seasonal variation · Uncertainty

Nomenclature

AER	All Electric Range
BEV	Battery Electric Vehicle
DE	Differential Evolution
EV	Electric Vehicle
ETL	Ensuing Trip Length
EVCS	Electric Vehicles Charging Station
G2V	Grid to Vehicle
HGSO	Henry's Gas Solubility Optimization
ICE	Internal Combustion Engine
ILP	Integer Linear Programing
PEM	Point Estimation Method
PHEV	Plug-in Hybrid Electric Vehicle
PLO	Parking Lot Operator
QT	Quade Test
RDN	Radial Distribution Network
RTP	Real-Time Pricing
SOC	State of Charge
V2G	Vehicle To Grid
WSRT	Wilcoxon Signed Rank Test
WPDF	Weibull Probability Distribution Function

25.1 Introduction

In this Anthropocene period, global warming is a huge issue to deal with. From literature, it can be seen that the main reason behind this global warming is air pollution created by Internal Combustion Engine (ICE) vehicles. Thus, slowly, electrification of transportation is happening and consequently, the new asset of transportation, i.e., electric vehicles (EV) are coming into pictures, which can drastically minimize the air pollution because of its zero-tail pipe emission. Moreover, EVs are noiseless. The predominant use of EVs is an operative means of plummeting CO_2 emissions, which will be beneficial from environmental aspects. Moreover, scarcity of fossil fuels is happening. Introducing EVs may cause less consumption in petrol and diesels, which leads to lesser air pollution (Zeng et al. 2020; Almehizia and Snodgrass 2018).

In recent market, besides the environmental impact, EV is getting popular for some of its other features also. Recently, lots of countries are experiencing growth of electricity generations due to the rapid increase of loads. To handle this ramping in loads, electric vehicles are grabbing attention, since it can deliver power back to the grid, when the grid is in peak load condition. This feature can be named as Vehicle to Grid mode, which is making EV more popular. Therefore, from the grid's perspective, EVs can act as spinning reserves and as well as load. Though proper systematic approach is needed by scheduling the charging/discharging mode of EVs to deal with grid's loading condition (Amini et al. 2014).

But each and every emerging technology has its own challenges. In the above discussion, it is pretty much clear that, why in today's environment, involvement of EVs is required. But few major cons are also there, which is resisting the EVs for coming into the automobile industry. This problem can be defined from three perspective; one from grid's perspective and another is from user's perspective and third one from the investor's perspective.

25.1.1 Grid's Perspective

EV's are gradually grabbing attention because of its G2V and V2G mode of operation. In G2V mode of operation, the EV will draw power from the grid for charging its battery. In V2G mode of operation, the exact opposite scene occurred. In V2G mode, the battery will discharge and deliver its power back to the grid during peak load condition for peak shaving. When the EV market will flourish, the load demand will consequently increase. Huge amount of current will be drawn by EV, which can damage the transmission network and as well as distribution network. Thus, before penetration of huge amount of EVs in the market, it is essential to increase the ratings of transformers, cables and other auxiliary equipment. Moreover, by forecasting the number of EVs in a state/country, the grid authority should check the requirement of transmission and distribution network expansion. Otherwise, the system can be collapsed due to huge power drawn by EVs. Moreover, due to the large battery capacity, charging of EVs will automatically increase the power losses in the distribution network. In most of the cases, EV's used to get charge from charging stations and parking lots. Thus, selection of location for installing charging stations/parking lots in the distribution network is another crucial matter for power system engineers to deal with. In this way, finding out the optimal location for placing the charging stations is required, so that least amount of power losses (both active and reactive power loss) will occur. Moreover, due to the presence of various power electronics components in the EV, it can introduce harmonics in the network. Thus, grid authority should consider this factor before deciding the ratings of various grid equipment.

25.1.2 User's Perspective

The main issue from user's perspective is the range anxiety, charging time and "When" and "where" to charge the EVs while cruising. These three factors are slowing down the EV market growth. Most of the time, drivers are unaware of the distance an EV could cover, provide from current battery status. Consequently, in the middle of the road, the breakdown of car may occur. Even if they are aware about the status of the battery and having the prior knowledge regarding the amount of distance it could cover with it, then also, due to lack of knowledge regarding the location of nearest charging station, the EV may breakdown at the middle of the road. Again, if it can be assumed that both the charging status of the battery and the location of the charging station is known to the EV user, then also, they may unsure about the stipulated charging time required for that EV, on that very charging station. They are unsure about the unavailability of the charging slots. Now these three issues are strongly correlated with each other. Thus, power system engineers should raise this challenge and provide some solution for it.

25.1.3 Investor's Perspective

Investor's do investment only if they find the notion of profit in it. On the other hand, for installing an EVCS, the involvement of investors is important. But due the huge anxiety among people, most of the time there is least interest in EV. Lesser amount of EVCS is one of the major factors behind this (Almehizia and Snodgrass 2018). Thus, EV selling become sluggish. At the same time, due to the less selling of EV, investors are not so interested to invest in EVCS. It is like a chicken and egg problem while dealing with it. From critical observation, it can be said that, before EVs huge market penetration, it is essential to invest on charging stations, and in order to do this, government should take the initiative so that in subsidized rate, these charging station can be installed. Moreover, not only the private investors, government also needs to take some vital steps by themselves to promote EV in recent days for creating a better charging infrastructure.

Though problems are technical and can be handled by the power system engineers but the social constraints must be tackled with the help of government. Otherwise, EV will face many hurdles to come into the auto mobile industries. Government should provide subsidy to the common people for buying EV. This will provoke people to buy an EV. At the same time, by providing subsidy to the investors, needs to be promoted for the installing EVCS (Almehizia and Snodgrass 2018).

From the above comprehensive discussion, it can be briefed that power system engineers need to take challenges for finding out the appropriate location in distribution network to install EVCS. Thereafter, there should be some intelligent strategy, which will take all the necessary inputs from the EV and will show the appropriate charging station, where the charging operation can be done at minimum time and energy consumption. After these tasks, it is also very crucial to decide at what time the EV will participate in G2V mode of operation and at what time the EV should participate in V2G mode of operation and when the EV will stay idle. These can be possible only when the grid status is known. Accordingly, some strategy can be set, so that both the charging station owner and EV user will get profit. Therefore, it can be said that the role of government and power system engineers both are equally important and strongly associated to promote EV in recent days.

By following the above discussion, in this chapter, three major aspects have been considered.

- Optimal Allocation of Charging station.
- Optimal Selection of apt charging station for individual EV.
- Optimal charging strategy of EVs considering G2V, V2G and idle mode of operation in such a way that both Charging station operator and EV owner will be in a win-win situation.

25.1.4 Related Literatures and Major Contributions

In Zeng et al. (2020) and Almehizia and Snodgrass (2018), the parking lots (PLO) have been allocated in a test distribution network using genetic algorithm (GA). Here, the authors have assigned charging station for PHEVs in a 30-bus system to minimize the power loss using GA (Almehizia and Snodgrass 2018). Enhancement of reliability has been performed in Amini et al. (2014), Liu et al. (2020), Mehta et al. (2019), for allocating charging station in a distribution network. In Liu et al. (2020), optimal allocation of EVCS (Electric Vehicles charging station) and DG (Distributed Generation) has been done in RDN (Radial Distribution Network) in standard IEEE 33 and 69 bus test system. In Mehta et al. (2019), the author performed 2-layer optimization for integrating charging station in IEEE 33 bus test system. Again, few authors in Shojaabadi et al. (2016), Liu et al. (2013), Zhang et al. (2019), Pal et al. (2021) planned to allocate the charging station in IEEE 123 node test system with the objective to minimize different cost associated with it and as well as the power loss of the system. Some of the paper considered EVCS installation cost, grid operational cost, and EV users expenses as objective function while installing EVCS in optimal way (Zhang et al. 2019). In Kong et al. (2019), Liu et al. (2018), Chen et al. (2020), Awasthi et al. (2017), slow and fast-charging stations have been allocated in a 33-bus distribution network and energy loss has been minimized using DE and HHO.

But these papers don't consider traffic congestions and few other practical constraints, while performing optimal allocation of EVCS. Again, in Zhang et al. (2019a), by using hybrid optimization, allocation of EVCS has been performed in

RDN of Allahabad city with the objective to minimize the complete set up cost and power quality improvement by considering traffic congestions. Again in Kong et al. (2019), Kong et al. (2019), Liu et al. (2018), Zhang et al. (2019b), a traffic network corresponding to the RDN has been considered for optimal placement of EVCS using Particle Swarm Optimization and Cross Entropy solver. But in these articles, comprehensive analysis of traffic congestion is missing and as well as few other essential features like seasonal variations, uncertainties related with driving cycles of EVs in CS have not been considered. Moreover, a complete solution in terms of promoting EV in recent market scenario is not found.

Yet again, in Deb et al. (2021) some charging management based on window optimization scheme has been proposed where real-time dynamic data can be fetched to determine the optimal scheduling at lower cost. A context for G2V/V2G scheduling of electric vehicles considering variation on spot tariffs of electricity and driving pattern of EV is shown in Sun et al. (2015), Zhang et al. (2019c), González-Garrido et al. (2019). This work promotes night day pricing, whereas day time charging can be expensive due to high tariff at day time. A supervised energy management scheme is shown in Shenghua Zhou et al. (2021). Here, the main objective is to uplift the fuel consumption by considering historical traffic data and compared it with other energy management scheme having no traffic information and static traffic information. In few literatures (Hussain et al. 2020; Khan and Ahmad 2019; Karmaker et al. 2019; Girade et al. 2021) online energy optimization has been performed to reduce the charging time at optimum cost. By increasing the charging current and voltage, the increased power has been fed to the charging station to minimize the charging time. Furthermore, minimization of total cost of EVs is performed for the scheduling of local group of EVs. EVs are optimally scheduled by a decentralized algorithm in Deb et al. (2021), where the goal is to fill the valleys and reduce the peaks of the load profile. An energy management system is developed in Sun et al. (2015) to keep cost of energy low and battery banks are used to manage renewable generations. High power transfer to EV with higher voltage and current rating is another solution (Shafiq et al. 2020; Chhawchharia et al. 2018; Sharma and Sharma 2019). Conversely, constant charging strategy is applied in an algorithm for fast charging of the EVs (Bendary and Ismail 2019; Hadi Amini et al. 2016). Two methods are taken in that algorithm, i.e., constant current and constant voltage. In Hadi Amini et al. (2016), ultra-capacitors are used as the energy storage device in the fast-charging station to reduce charging time and long persistence stress on the power system.

From literature (Amini et al. 2018; Saleem et al. 2019), it can be observed that most of the authors focused in energy management of EV charging to reduce the charging time by varying the current or voltage or sometimes both. But none of the author considered the appropriate selection of charging station, which can be a great way to reduce the charging time. Moreover, if the EV driver knows about the free charging slot in some charging station, in that way also, the charging time can be reduced. Thus, this very basic research gap has been found in the critical literature survey, which needs to be incorporated in the proposed book chapter.

After successful allotment of EV, the next goal is to perform an intelligent charging scheduling. In previous literatures (Khaligh and D'Antonio 2019), it can be seen that many authors only focused on G2V mode of operation. As per the real-time pricing, most of the time at night they suggested for charging of EVs. On the other hand, few of them considered both G2V and V2G mode of operation for charging scheduling purpose (Teng et al. 2020; Susowake et al. 2018). In those literature, it can be observed that the EV is working on G2V mode of operation continuously for several durations and again V2G mode of operation for several hours, to minimize the charging cost. Though comprehensive analysis regarding the charging cost has not been done. Very few papers have executed charging scheduling with the historical data (Lam et al. 2018). Thus, for very obvious reason uncertainties related to driving cycles have not been considered which have a strong correlation with charging scheduling. A limited paper has considered seasonal variation, but the uncertainties related to the seasonal variation has not been considered in many works as per authors knowledge (Khorramdel et al. July 2015).

Hence, from the above literature survey, it can be observed that there are few basic and vital factors and practical constraints are there to perform a multilevel scheduling.

Thus, in this chapter, the major contributions are as follows:

- Comprehensive analysis of EV technology from techno-economic-social aspects has been done.
- The optimal allocation of EVCS by considered the seasonal variation and its worst-case scenario with the objective to minimize the overall power loss of the system. Here detailed analysis due to seasonal variation has been considered during optimal allocation of EVCS.
- After successful allotment of EVCS, the apt EVCS for each and every EVs has been as been selected considering road's traffic congestions. Impact of traffic conditions are mostly avoided in most of the works.
- Variable charging rates and seasonal uncertainties associates with driving cycles of EV have been considered for optimal charging scheduling of EVs considering V2G (−1), G2V (+1) and idle mode (0) of operation, where zero power transaction has been considered.

25.2 State of the Art of the Problem

"Anxiety" is a major obstacle between the transition of ICE vehicles to EV in automobile industry. There are two major aspects of anxiety. As previously mentioned, one is from investors perspective and another is from common people's perspective. Common people always hesitate to buy an EV because the charging infrastructure for electric vehicles are not developed fully. There are very few charging stations are available in most of the places. Besides this, those who already bought EV, they are also facing apprehension regarding "when" and "where" to charge the EV. Again, from investor's point of view, it is also true that if they found that EV selling is not satisfactory, in that case, investing on EVCS may not be possible. This scenario is analogous to the famous chicken and egg problem.

To deal with such situation and to promote EV in the recent market, as a researcher, it is our duty to find some fruitful way-out through which both the EV owner and CS operator will be in some win-win situation. In order to do so, the first initiative should be initiated by the government by subsidizing the building cost of EVCS. If government can assure the economic platform, then researcher can build the charging infrastructure in the existing power system. It is true that if large number of EV suddenly started to penetrate the transportation network, then the overall grid system may collapse and thus, the placement of charging station should be optimal where the overall power loss of the system will be minimized (Pal et al. 2019). In order to do so, it is much essential to observe the driving cycles of the EVs and also the variation in driving cycles in a whole year. Thus, it is much essential to consider the seasonal variation and its impact in driving cycles. By selecting the worst scenario, optimal placement of charging station should be done, which will be more robust and more practical from techno-economic aspects.

Again, from common people's perspective also, there are some major challenges which researcher need to address. Among them the first is "when" and "where" to charge the EV at least time. It is never compulsory that adjacent charging station (CS) is always relevant for charging. It might be possible that there are huge traffic congestion while reaching to the nearest EVCS, which result the breakdown of EV at the middle of the road or maybe there are no slots available for which the EV needs to stay in que for long to perform charging. In the contrary, EV could reach another EVCS with the stipulated charge, which is situated little far but the overall charging process can be done smoothly because of less traffic congestion and availability of slots. Thus, there should be some intelligent algorithm through which the from earlier an EV could reach to its apt charging station and at minimum time and at minimum energy consumption, to complete the charging process. In order to do so, ILP (Integer linear Programming) can be applied with the objective to reach the appropriate charging station at minimum energy consumption (Das et al. 2020a).

After reaching to the apt charging station, next challenge is to decide how much energy it required to reach the subsequent destination after leaving the charging station. In order to decide it, from some input attributes related to driving cycles are required for each EV. In this chapter, few assumptions have been made to establish the proposed strategy. Among them, the first and foremost assumption is, the charging station should be a slow charging station. Unless and until it's a slow charging station, it is not possible to incorporate V2G mode of operation with it. It should be noted that V2G mode of operation cannot possible in fast-charging station, since during V2G mode of operation, slow charging rates are required (Das et al. 2020a).

Thus, by incorporating G2V, V2G and idle mode of operation and as per the real-time tariff of energy, the charging scheduling algorithm is proposed in this chapter where seasonal variation and its uncertainty correlated with driving cycles of EV have been tackled by 2m Point Estimation Method (2m PEM), to make the scheduling more practical and robust (Pal et al. 2019; Das et al. 2020b). The main objective is to reduce the charging cost in such a way that both EV owner and Charging station operator will be in some win-win situation. To optimize the cost, a recent optimization technique, named Henry Gas Solubility optimization has been applied (Hashim et al. 2019).

In this way, a complete charging solution in multiple level is proposed to promote the electrification of transportation as a decarbonization tool to keep ecological balance in the environment. The schematic diagram of the proposed method has been shown in Fig. 25.1a and the working flow is elaborated through flowchart in Fig. 25.1b. In the next section, the problem formulation for the proposed method has been discussed.

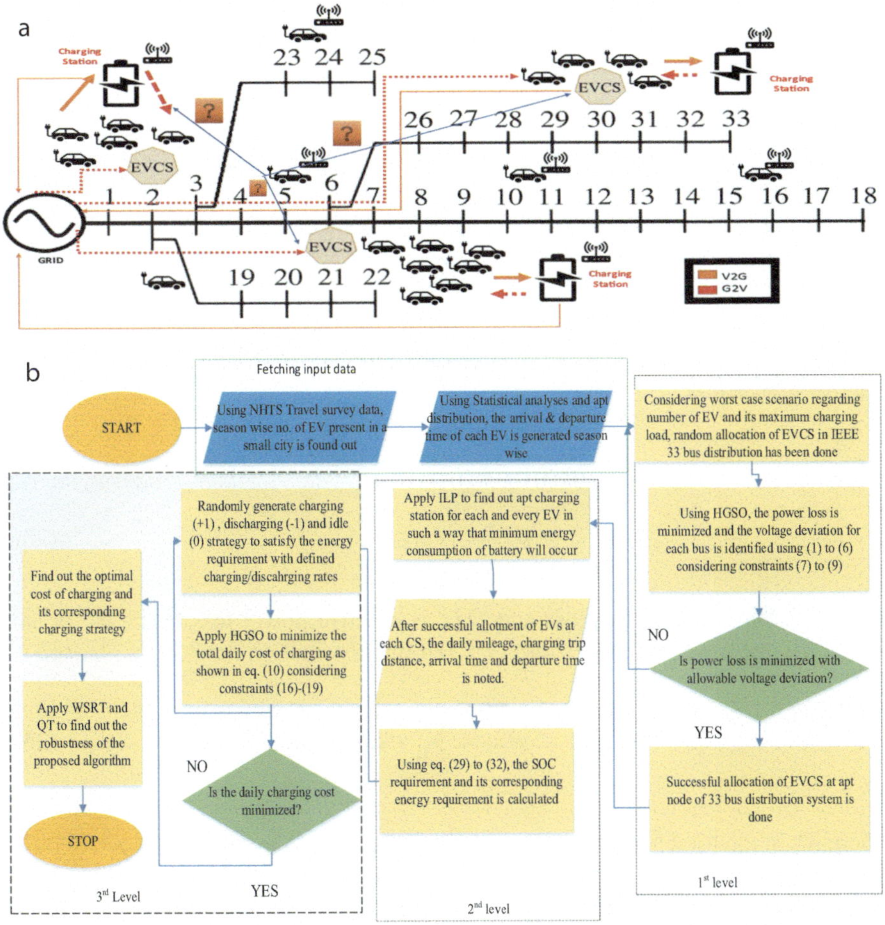

Fig. 25.1 **a** Schematic diagram of the proposed multilevel charging solution. **b** The overall working flow and methodology of the proposed multilevel algorithm

25.3 Problem Formulation

25.3.1 First Level (Optimal Placement of EVCS)

25.3.1.1 Objective Function

In the first level optimization, a single objective problem has been taken where the 33-bus radial distribution system's the real power loss is minimized (Pal et al. 2019). For a generalized problem formulation, a distribution network with b number of branches can be taken. Where the parameters of bth branch are as, I^b is current, Z^b is the impedance, R^b is resistance, and X^b is reactance. Therefore, Z^b can be calculated as

$$Z^b = R^b + jX^b \tag{25.1}$$

The total apparent power loss (S^b) at bth branch can be expressed as

$$S^b = \left(I^b\right)^2 \times Z^b \tag{25.2}$$

Total apparent power loss (S^b) can be separated into active power loss (PL^b) and reactive power loss (QL^b) by

$$PL^b = Re\left(S^b\right) \tag{25.3}$$

$$QL^b = Img\left(S^b\right) \tag{25.4}$$

where PL^b is active power loss and QL^b is reactive power loss in the bth branch. Therefore, the total real power loss (PL) of the distribution network is planned as

$$PL = \sum_{b=1}^{nb} PL^b \tag{25.5}$$

where nb in the total branches. The objective function to minimize the active power loss, can be written as

$$Min(f) = Min(PL) \tag{25.6}$$

- **Constraints**

The power supply should be balanced with load and losses as

$$P^{SS} = \sum_{bs=1}^{Nbs} P^{bs} + PL \tag{25.7}$$

where P^{SS} is total delivered power from the grid, P^{b} is loading of bs^{th} bus, Nbs is bus's total number. All the bus voltages of the distribution network are to be maintained within permissible limits as

$$V^{min} \leq V^{bs} \leq V^{max} \tag{25.8}$$

where V^{min} and V^{max} is minimum and maximum voltage limit. V^{bs} is the genuine voltage of the bs^{th} bus.

The branches have maximum current carrying capacity. The current flow capacity in kept within the limits by using Eq. (25.9)

$$\left|I^{b}\right| \leq I^{max} \tag{25.9}$$

where I^{max} is current carrying's maximum capacity of all the branches. I^{b} is the current flowing through branch b.

25.3.2 Second Level (Optimal Charging Scheduling)

Here, the objective function is to minimize the daily cost of charging (incudes G2V, V2G, and idle mode of operation and battery degradation cost) shown by (25.10). The cost associates with it is shown in (25.11), where $Cost^{ch}$ is the overall costing considering grid to vehicle, vehicle to grid and idle mode of operation and $Cost_{batt.deg}$ is the battery degradation cost (Das et al. 2021). This is a vital factor which needs to be considered while performing the scheduling, since, dynamically, the battery lifecycle is reduced due to charging discharging cycle and thus, to the aggregator should consider the battery degradation cost in terms of incentives for the EV users to promote V2G mode of operation.

$$min\{Cost^{all}\} \tag{25.10}$$

$$Cost^{TOT} = Cost^{oc} \cup Cost_{batt.deg} \tag{25.11}$$

Thus, overall cost of charging ($Cost^{ch}$) consists of the cost for G2V mode of operation ($Cost^{G2V}$) and earning s coming from V2G mode of operation ($Earn^{V2G}$), as shown in (25.12).

$$Cost^{oc} = Cost^{G2V} \cup Earn^{V2G} \tag{25.12}$$

Again, to evaluate these$Cost^{G2V}$,$Earn^{V2G}$, and $Cost_{batt.deg}$, (25.13), (25.14), (25.15), can be used, where, (tot_slot) is the total slot number vector and $(slot \in tot\_slot)$.

$$Cost^{G2V} = \sum_{slot=1}^{tot\_slot} \left(\sum_{PEV\_no}^{tot\_PEV} (C^{slot}_{PEV\_no} - DC^{slot}_{PEV\_no}) * ch^{rate,PEV\_no} \right) * RTT(slot) \tag{25.13}$$

$$Earn^{V2G} = \sum_{slot=1}^{tot\_slot} \left(\sum_{no=1}^{tot\_PEV} (C^{slot}_{PEV\_no} - DC^{slot}_{PEV\_no}) * ch^{rate,PEV\_no} \right) * RTT(slot) \tag{25.14}$$

$$Cost_{batt.deg} = \sum_{PEV\_no=1}^{tot\_PEV} \left(\frac{(batt_{cost,PEV\_no} * Batt^{cap,PEV\_no} + Cost_{lab}) * |Eng^{dis}|}{(L^{batt} * Batt^{cap,PEV.no} * DOD)} \right) \tag{25.15}$$

where battery manufacturing cost is $batt_{cost,PEV\_no}$. The labor cost to install the battery is $Cost_{lab}$. L^{batt} is the battery lifecycle and the depth of discharge of the battery is presented by DOD (Bendary and Ismail 2019) and the total discharged energy due to V2G mode of operation is implied by Eng^{dis}.

25.3.2.1 Constraints in Charging Scheduling

During execute this multilevel algorithm, in this level, few constraints are there which are related to the driving cycles and uncertain in nature. These are, SOC of the battery, requirements of energy, number of charging stations and number of cars. These attributes are need to be well-thought-out for performing coordinated charging scheduling.

Charging (G2V) and Discharging (V2G) Rate

The G2V/V2G mode's charging/ discharging should not exceed the rated power limit of charging slots, which is shown in Eq. (25.16). The controller for charging control can regulate the rate of charging in a predefined charging limits as shown in (25.17) (Hadi Amini et al. 2016).

$$ch^{rate,PEV\_no} < PW_{rated} \tag{25.16}$$

$$ch^{rate,PEV\_no}{}_{min} \leq ch^{rate,PEV\_no} \leq ch^{rate,PEV\_no}{}_{max} \tag{25.17}$$

Requirements of Energy

The required energy must be fulfilled within predefined time interval with apt rate of charging.

$$\sum_{slot=slot\_in}^{slot.out} ST_{PEV\_no}^{slot}.ch^{rate,PEV\_no} = Eng^{req} \tag{25.18}$$

SOC Limit of the Battery

The SOC of the battery should not surpass its maximum limit or should not go underneath the minimum battery energy capacity. The lower limit of the SOC should not fall below 20%.

$$SOC^{min} \leq SOC_{PEV\_no}^{slot} \leq SOC^{max} \tag{25.19}$$

Number of CS and Number of EVs

The number of CS should be lesser than the number of EVs.

25.3.2.2 Charging and Discharging Strategy for Electric Vehicles

Here, C_{PEV_no} is the charging strategy vector, DC_{PEV_no} is the discharging strategy vector, which is the subset of complete strategy vector, i.e., ST_{PEV_no} (where, $C_{PEV\_no}^{slot}$, $DC_{PEV\_no} \subseteq ST_{PEV\_no}$ and $C_{PEV\_no}^{slot} \in C_{PEV\_no}$,$DC_{PEV.no}^{slot} \in DC_{PEV\_no}$

$$C_{PEV\_no} = [C_{PEV\_no}^{1}, \ldots.., C_{PEV\_no_x}^{slot\_in}, \ldots., C_{PEV\_no}^{slot\_out}, \ldots., C_{PEV\_no}^{tot\_slot}] \tag{25.20}$$

where,

$$C_{PEV\_no}^{slot} = \begin{cases} 1, \quad \text{if } ST_{PEV_{no}}^{slot} = 1, \forall slot : P^{slots,PEV_{no}}, \forall PEV\_no \in \vec{Z} \\ 0, \quad orelse \\ 0, \forall slot \notin P^{slots,PEV_{no}}, \forall PEV\_no \in \vec{Z} \end{cases} \tag{25.21}$$

$$DC_{PEV\_no} = \left[DC_{PEV_{no}}^{1}, .., DC_{PEV_{no}x}^{slot_{in}}, \ldots, DC_{PEV_{no}}^{slot_{out}}, .., DC_{PEV_{no}}^{tot_{slot}}\right] \tag{25.22}$$

where

$$DC_{PEV.no}^{slot} = \begin{cases} -1, \text{ if } ST_{PEV_{no}}^{slot} = -1 \forall slot : P^{slots,PEV\_no} \forall PEV\_no \in \vec{Z} \\ 0, \; orelse \\ 0, \forall slot \notin P^{slots,PEV\_no}, \forall PEV\_no \in \vec{Z} \end{cases} \tag{25.23}$$

$$ST_{PEV\_no} = [ST_{PEV\_no}^{1}., ST_{PEV\_no}^{slot\_in}, \ldots .ST_{PEV\_no}^{slot}.., ST_{PEV\_no}^{slot.out}, .., ST_{PEV\_no}^{tot\_slot}] \tag{25.24}$$

and

$$ST_{PEV\_no}^{slot} = \begin{cases} 1, charging \\ -1, discharging \\ 0, idle \end{cases}, \quad \forall PEV\_no \in \vec{Z} \tag{25.25}$$

Again, the vector of parking time interim for PEV_no^{th} is shown by (25.26),

$$P^{slots,PEV} = \left[t_{in,PEV\_no}, \ldots.., t_{PEV\_no}, \ldots..., t_{out,PEV\_no}\right] \tag{25.26}$$

The scheduling is done for 24 h at 30 min interval, thus total 48 slots has been considered for each electric vehicle. The horizon of time is defined by $\vec{H}$,

$$\vec{H} = \{\overline{Slot}\} \tag{25.27}$$

The number of EVs reaching at CS is implied by $\vec{Z}$

$$\vec{Z} = [1, .., PEV\_no, ..tot\_PEV \tag{25.28}$$

Three possible operating mode can be occurred while satisfying the required energy, i.e., charging mode (+1), discharging mode (-1), and idle (0) mode of operation.

25.3.2.3 Energy Modeling

To satisfy the constraints, as shown in (25.18), the requirement of energy (for charging/discharging) for each EV needs to be determined by using (25.29) (Das et al. 2020b)

$$eng^{req}(\text{or} eng_{dis}{}^{req}) = \frac{SOC^{req}.b_{cap}}{\eta^{eff}} \tag{25.29}$$

When, in G2V mode, $\eta^{eff} = \eta^{ch}$ and in discharging/ V2G mode, $\eta^{eff} = 1/\eta^{dis}$. Where, η^{eff} is the charging (η^{ch}) / discharging (η^{dis}) efficiency of EVs.

Again, to evaluate the SOC^{req} for each EV, Eq. (25.30) is used,

$$SOC^{req} = \begin{cases} 1 - SOC^{arr}, when, SOC^{dep} > 1 \\ \left(SOC^{dep} - SOC^{arr}\right), when, SOC^{arr} < SOC^{dep} < 1 \\ 0, when SOC^{arr} = SOC^{dep} \\ -\left(SOC^{arr} - SOC^{dep}\right), when 0.2 < SOC^{dep} < SOC^{arr} \end{cases} \quad (25.30)$$

where (*SOC* during arriving at charging station) (SOC^{arr}Zhang et al. 2019c) and SOC^{dep} (*SOC* during departing from charging station) (Zhang et al. 2019c) can be calculated from (25.31) to (25.32). Where ETL is the ensuing trip length, d^f is the charging trip distance and AER is the All-Electric Range (AER) of each EV.

$$SOC^{arr} = 1 - \left({}^{df}/_{AER}\right) \quad (25.31)$$

$$SOC^{dep} = \left[\left({}^{ETL}/_{AER}\right) + 20\%\right] \quad (25.32)$$

Though here SOC^{arr} and SOC^{dep} is being calculated, but in real time, the SOC status after can be gauged from the telematic system of EV. Besides this, complete charging of battery may lessen the lifecycle of battery (Hadi Amini et al. 2016), therefore, it's evaded while formulating the proposed multilevel algorithm.

Now to evaluate the SOC during the arrival and departure of the vehicles, the charging trip distance, daily mileage, it's arrival time and desired departure time are required which are not available in real time. Moreover, these factors are stochastic. Thus 2m PEM is used to generate data.

25.4 State of the Art for Modeling Uncertainty

25.4.1 2m PEM (2m Point Estimation Method)

Various characteristics of driving cycle used in the proposed algorithm are of uncertain characteristics and have been disparagingly tackled using 2m PEM, which has certain advantages, such as it utilizes deterministic sequences to grip the probabilistic contexts and also has lesser computational complexity and efficiency to grip larger quantity's stochastic attributes. The efficiency or performance of optimization technique in case of 2m PEM estimation, is measured by the average value of objective function. Since the number of uncertain variables is large in this case and the number of uncertain variable set is generated by 2m PEM to solve the problem is twice the number of uncertain variables, so the objective function is not represented by the actual value of the individual set rather, it is calculated individually using each set, out of 2m sets of uncertain variables. So, for any particular deterministic solution set,

2m number of cost value can be determined (Das et al. 2020c). So as a result of it, it is not wise to deal with the best value out of it, rather, while doing the optimization, it should be dealt with the mean value of the all these 2m costs obtained for a particular deterministic set. Since, for each and every deterministic solution set (population sets solution) performance analysis, the mean value of the 2m sets solution is being used, therefore 2m PEM approach is much more realistic while working with the stochastic nature of driving cycles.

Methodically, the deterministic routine of daily average cost of charging minimization problem can be shown as (25.33), where f transferences the randomness from input random variables (*IRV*) to output variables (Hashim et al. 2019; Das et al. 2021).

$$C = f(r_i) \tag{25.33}$$

$$C = f(C_r, Z_1, Z_{2,} \ldots Z_l) \tag{25.34}$$

Here, $Z_l(l = 1, 2, 3 \ldots, l)$, are *IRV* under imprecision with the PDF f_{Z_l}. Statistical instant's output of input random variables using first four central moments, i.e., mean(μ_{zl}), standard deviation (σ_{zl}), skewness $(\lambda_{Zl,3})$ coefficients and kurtosis$(\lambda_{Zl,4})$ coefficient, have been determined. This method produces two fixed values for each IRVZ_l, as $(Z_{l,1} W_{l,1})$ and $(Z_{l,2} W_{l,2})$. The $Z_{l,pos}(position = 1, 2)$ is named as pos^{th} location of Z_l and $W_{l,pos}$ $(pos =$ 1,2) is a weighting factor which signifies the corresponding points in measuring the statistical instants of the output random variable. Here $Z_{l,pos}$ can be determined as

$$Z_{l,pos} = \mu_{z_l} + \xi_{Z_{l,pos}} * \sigma_{Z_l}, \; pos = 1, 2 \tag{25.35}$$

In Eq. (25.35), to asses $\xi_{Z_{l,pos}}$ for the pos^{th} standard location:

$$\xi_{Z_{l,pos}} = \lambda_{Zl,3}/2 + (-1)^{(3-pos)} \sqrt[2]{\lambda_{Zl,4} - (\sqrt{3}\lambda_{Zl,3}/2)^2} \tag{25.36}$$

In Eq. (25.36) $\lambda_{Zl,3}$ is the skewness (Amini et al. 2018) and $\lambda_{Zl,4}$ is the kurtosis (Amini et al. 2018) of Z_l, respectively, and been assessed by

$$\lambda_{zl,3} = {}^{m_3}/_{\lambda^3_{zl,3}} \tag{25.37}$$

where

$$m_3 = E\left(Z_l - \mu_{z_l}\right)^3 = \sum_{j=1}^{N} \left(Z_l - \mu_{z_l}\right)^3 .Prob\left(Z_{l,j}\right) \tag{25.38}$$

$$\lambda_{zl,4} = {}^{m_4}/_{\lambda^4_{zl,4}} \tag{25.39}$$

where

$$m_4 = E(Z_l - \mu_{z_l})^4 = \sum_{j=1}^{N} (Z_l - \mu_{z_l})^4.Prob(Z_{l,j}) \quad (25.40)$$

In Eq. (25.38) and (25.40), the expected value is *E, N* belongs to the observation number of Z_l and $P(Z_{l,j})$ is the probability of every input random variable $Z_{l,j}$.

One *IRV* is kept fixed at one location and the all-other input random variables are assumed to be equal to their mean, in each iteration of 2m PEM,

$$C_{(l,pos)} = fn(C_r, \mu_1, \mu_{2,} \ldots\ldots\ldots z_{l,pos}, \ldots \mu_{Zm}) \quad (25.41)$$

Here, *pos* $\epsilon\{1,2\}$ and $l = \{1, 2, 3, \ldots, m\}$.

Couple of allowance features of Z_l has been determined by (25.42):

$$W_{l,pos} = \frac{(-1)^{pos}}{m} \times \frac{\xi_{Z_{l,3-pos}}}{\xi_{Z_{l,1}} - \xi_{Z_{l,2}}} \quad (25.42)$$

The first and second moment of the *ORV* (Output Random Variable) can be calculated as

$$E(C^y) = \sum_{l=1}^{m} \sum_{pos=1}^{2} (W_{l,pos} \times ((C_{l,pos}))^y), \; y = 1, 2 \quad (25.43)$$

This 2m PEM method can be useful to all ambiguous attributes of the delinquent using (25.30–25.38). Henceforth, the average and standard deviation of the output random variable can be determined by (25.44):

$$\mu_c = E(C^1) \;\&\; \sigma_C = \sqrt{(E(C^2)) - (E(C^1))^2} \quad (25.44)$$

25.5 Solution Strategy of "The Multilevel Planning for Charging Scheduling"

25.5.1 Level 1: Solution Strategy for Selection of Appropriate Charging Station

In this stage, after successful findings of nodes for installation of EVCS, using soft-computing technique (Will be discussed in next section), Integer Linear Programming has been used to choose or select the relevant charging station for each EV has been

Table 25.1 Structure of assignment matrix

	EV_1	EV_2		EV_e	Supply
CS-1	E_con (1,1, t)	E_con (1,2, t)		E_con (1, e, t)	…
CS-2	….				…
…	….				…
CS-F	E_con (f,1, t)	E_con (f, 2, t)	…	E_con (f, e, t)	…
Demand	1	1		1	

chosen, where the EV could reach at minimum energy consumption. Thus, it is clear that the successful allotment of CS for each EV has been done by consuming optimum amount of energy. Since, in real time, it is never compulsory that the nearest charging station is always relevant (Das et al. 2020a).

Let e number of EVs and f is the number of CSs, where e, f are the positive integers ($e >> f$). Now, C is the assignment matrix which can be well-defined as $C(f, e)$ where each EV e can be assigned to CS_f. This problem is following the well-known benchmark problem, i.e., "Assignment problem" with the mixture of "Transportation Problem" as shown in Table 25.1 (Das et al. 2020a).

Consumption of energy depends on various factors such as vehicles speed ($v(f, e, t)$), distance between the location of vehicle and the charging station ($d(f, e, t)$), the AER(e) of individual vehicles, its battery capacity $\left(b_{cap}(e)\right)$, vehicles flow $\left(v_q(f, e, t)\right)$ and traffic density $\left(j_{jam}(f, e, t)\right)$. The traffic density is assumed to be same in every route and the vehicles flow is considered constant (González-Garrido et al. 2019; Pal et al. 2019).

The mandatory time to reaching the charging station's distance ($d(f, e, t)$) by an EV can be shown as

$$T = {}^{d}/_{v}, \quad where,\ T, d, v \in fn(f, e, t) \tag{25.45}$$

Again, from "Greenshields Model", it has been proven that the flow of vehicle (v_{fq}) is proportional to the vehicle's density (D_{jam}) (Hadi Amini et al. 2016)

$$v_{fq} \propto D_{jam} \tag{25.46}$$

$$or,\ v_q = k_1.D_{jam},\ where,\ k_1 = \frac{1}{v} and v_{fq}, D_{jam} \in fn(f, e, t) \tag{25.47}$$

The consumed energy E_{con} under normal conditions is directly proportional to its battery capacity, the distance it covers and inversely proportional to its AER or battery autonomy.

$$E_{con} = \frac{b_{cap}.d}{AER}, \text{ where, } C, AER \in fn(e) \text{and} d \in fn(f, e, t) \tag{25.48}$$

So, by bearing in mind the flow of EV and traffic congestions it can be redefined as

$$E_{con} = \frac{b_{cap}.T.v_q}{AER.j_{jam}} \tag{25.49}$$

Here, 3 CS has been considered and the demand of EV is to get a single charging station at a particular time, thus, the demand is 1.

In this ILP, keeping the energy consumption of battery at a minimum is the main objective, i.e.,

$$min\left\{\sum_{f=1}^{f}\sum_{e=1}^{e}(E_{con}(f,e,t).x(f,e,t))\right\} \tag{25.50}$$

Subjects to, 1. $\sum_{f=1}^{cs} x(f,e,t) = 1, \forall EV_e(\text{eachvehicle})\text{where}1 \le e \le EV$, i.e., each EV can visit only one CS at a time.

2. $\sum_{e=1}^{ev} x(f,e,t) \le CS_f, \forall CS_f\text{where}1 \le f \le CS$, i.e., the slot capacity of each CS must not surpass its maximum limit.

3. First trip length d^f should not be the lesser than the distance of EVs from charging stations dis^{e-f}.

i.e.,

$$dis^{e-f} \le d^f \tag{25.51}$$

This Integer Linear programming problem (ILLP) has been unraveled in MATLAB interface. In this ILLP problem, the decision variable can be stated as

$$x(f,e,t) = \begin{cases} 1, & |\text{if } EVLe \text{ is selected for } CS_f \text{ at time } t \\ 0, & |\text{or else} \end{cases} \tag{25.52}$$

25.5.2 Second Level: Henry's Gas Solubility Optimization:

To optimize Power loss and as well as the daily total charging cost has been optimized by Henry's Gas solubility optimization (HGSO) (Hashim et al. 2019).

25.5.2.1 Henry Gas Solubility Optimization (HGSO)

To solve the real-world problem, several metaheuristic optimization algorithms have been developed. Among the Henry Gas Solubility optimization is a very recent optimization proposed by Fatma. A. Hashim and her team in 2019 (Hashim et al. 2019). This is an NSA (Nature Science based Algorithm), which takes the inspiration from the Henry's Gas solubility law, which states that "At constant temperature the amount of a provided gas that dissolves in a particular type and volume of liquid is directly proportional to the partial pressure of that very gas in equilibrium with that liquid" (Hashim et al. 2019). Therefore, from the definition it can be observed that the gas's solubility is highly dependent on the temperature and thus, the solubility of the gas is directly proportional to the partial pressure of the gas as shown in Eq. (25.52–25.58). This HGSO algorithm emulates the clustering nature of gas to balance between exploration and exploitation. Thus, local optima can be avoided. This HGSO mimics the Henry's gas law to solve typical optimization problems. Here, in this paper, HGSO is applied to minimize the total daily charging cost of electric vehicles

$$sol^{gas} \propto p^{gas},\ p^{gas} \in \ partial\ pressure\, of\ the\ gas \tag{25.53}$$

$$sol^{gas} = k.p^{gas} \tag{25.54}$$

where, $k = H^{gas}$, here,$H^{gas} \in Henry^{'} sconstant$

Moreover, the effect of temperature reliance on Henry's law constant should be taken into consideration. The change in variation of temperature can be shown and described by the Van't Hoff equation as shown in Eq. (25.55):

$$\frac{\partial(2.303\log H)}{\partial(\frac{1}{T^k})} = \frac{-\nabla_{sol}E}{R_{molar}} \tag{25.55}$$

Here, Henry's constant H is the function of T^k. $-\nabla_{sol}E$ is the enthalpy of the disbanding solution, R_{molar} is the molar gas constant. After integrating the Eq. (25.55) and simplifying it, Eq. (25.56) can be determined.

$$H\left(T^k\right) = \exp\left(\frac{M}{T^k}\right) * N \tag{25.56}$$

Here, M and N are dependent parameter of $H\left(T^k\right)$. The value T^k is considered to be 298.15 k.

$$H\left(T^k\right) = \tilde{\mathrm{H}}.\ exp\left(\frac{-\nabla_{sol}E}{R_{molar}}\left(\frac{1}{T^k} - \frac{1}{\tilde{T}}\right)\right) \tag{25.57}$$

Now, in case of Van't Hoff's equation, the enthalpy must be constant. So, considering it, Eq. (25.57) can be simplified and written as Eq. (25.58)

$$H\left(T^k\right) = \tilde{\mathrm{H}}\ x exp\left(-\tilde{C}\left(\frac{1}{T^k} - \frac{1}{\tilde{T}}\right)\right) \tag{25.58}$$

25.5.2.2 Mathematical Model and Algorithm for HGSO

Initialization Process:

The random population can be initialized by Das et al. (2020c).

$$Y^{i}(k+1) = Y^{k}_{min} + rand(Y^{k}_{max} - Y^{k}_{min}) \tag{25.59}$$

Here, the function "rand" generates arbitrary values between 0 and 1 interval, i.e., [0, 1] interval. In the suffix of Y^{i}, $i = 1, 2, \ldots$, N_{pop}, where size of the population *is* N_{pop}. Y^{k}_{max}, Y^{k}_{min} are the lower and upper limit of the algorithm where k is the iteration time. The total number of gases are i and "Henry's constant" value can be expressed in $(H^{l}(k))$, which is of type l. By $p^{i,l}$ of gas i of cluster l, the partial pressure of gas can be expressed. The value of constant $\tilde{C}^{l}$ of cluster type l, the partial pressure and the Henry's constant can be initialized at the beginning and shown by Eq. (25.60):

$$\left.\begin{aligned} H^{1}(k) &= C^{1} * rand\ (0, 1) \\ p^{i,1} &= C^{2} * rand\ (0, 1) \\ \tilde{C}^{l} &= C^{3} * rand\ (0, 1) \end{aligned}\right\} \tag{25.60}$$

Here, the value of C^{1}, C^{2}, andC^{3} are the constants with the values 5E-02,100 and 1E-02, respectively.

Clustering and Evaluation

Clustering in done where each cluster l contains same type of gases and with the same Henry's constant value. The total number of populations are separated in few clusters as per the gas type's number. After dividing into clusters, each of them is evaluated to choose the best gas that achieves the balance of equilibrium state from other gases in each cluster l. After that, these gases are ranked to determine the optimal gas in the entire population of swarm.

Update of Henry's Coefficient and Corresponding Solubility

$$(H^{l}(k+1)), = (H^{l}(k))xexp\left(-\tilde{C}^{l}\left(\frac{1}{T(k)} - \frac{1}{\tilde{T}}\right)\right), , T(k) = \exp\left(-\frac{k}{iter}\right) \tag{25.61}$$

$$sol^{i,l}(k) = \alpha.H^{l}(k+1).p^{i,l}(k) \tag{25.62}$$

By Eq. (25.61), the Henry's coefficient can be updated which will be followed by the update of solubility of the gas i in cluster l as shown in Eq. (25.62). $p^{i,l}$ is the partial pressure of gas i of cluster l. Here, α is a constant.

Update of molecule's Position

The positions can be updated using the below mentioned Eqs. (25.63) and (25.64)

$$\begin{aligned} Y^{i,l}(k+1) =& Y^{i,l}(k) \\ & + F * rand * \sigma * \left(Y^{i,best}(k) - Y^{i,l}(k)\right) + F * rand * \mu * (sol^{i,l}(k) \\ & * Y^{best}(k) - Y^{i,l}(k)) \end{aligned} \quad (25.63)$$

$$\sigma = \beta.e^{(-F^{best}(k)+\epsilon/F^{i,l}(k)+\epsilon)}, where\epsilon = 0.05 \quad (25.64)$$

Here, the position of gas i in cluster l is denoted as $Y^{i,l}$. $Y^{i,best}$ signifies the ith gas is the best gas in cluster l. σ is the gas's ability in cluster l to intermingle with other gases of its own cluster.

Moreover, μ is the influence of other gases of different cluster to interact with the gas i in cluster l. Its value is considered to be 1. Here, another constant is β.

In contrast, $C^{i,l}$ is the fitness of the gas i in cluster l and C^{best} is the fitness value of best gas in that very cluster.

Here, F is the factor or rather flag which helps the search space for exploration in the search space for the swarm (search agent) and it provides the diversity = $\pm$.

$Y^{i,l}$, and $Y^{i,best}$ are two parameters to make balance between exploration and exploitation.

Escaping from Local Optima and Updating the Worst gas's Position

$$N^{worst} = N * \left(rand\left(r^2 - r^1\right) + r^1\right) \quad (25.65)$$

$$G^{i,l} = G^{\min(i,l)} + rand(G^{\max(i,l)} - G^{\min(i,l)})) \quad (25.66)$$

Equation (25.65) is used to escape from the local optima of the search space and make the search agent for the exploration. These gases are aggregated as per their fitness value and the worst agents are identified and again using Eq. (25.66), their positions has been updated and accordingly the fitness value is determined.

$G^{\max(i,l)}$, $G^{\min(i,l)}$ are the minimum and maximum value of the worst agent of gas i in cluster l.

25.6 Case Studies and Input Data

As mentioned in introduction, to establish or install a charging station and to perform scheduling various practical constraints and factor needs to be considered. Many works have either ignored these factors or didn't discussed in elaborative manner. Among them, one of the major factors is the seasonal variation. Due to global warming, one of the major issues is climate change. Earth is getting warmer day by day. This change in climate is directly correlated with the power system loads. Consequently, it is necessary to consider this while performing the optimal allocation of charging station. Moreover, as per the weather, driving cycles are also changed. Thus, reflects on the charging scheduling. Therefore, in charging scheduling also, seasonal variation is needs to be considered. On the other hand, since optimal allocation of EVCS and charging scheduling is indirectly correlated, thus, in the proposed algorithm, the data has been considered considering seasonal variation and according the execution process has been carried out. While executing the process, few assumptions and factors has been established.

- A small urban area in a summer-based country has been selected in Asia. The reason behind selecting a summer base country lies in the battery of EV. EV battery is very much correlated with the temperature and thus, intentionally this kind of place has been chosen to verify our algorithm at extensive condition.
- In this area, broadly 3 kinds of season can be observed. Dry season, monsoon season and the winter season.
- 33-bus radial distribution network has been chosen in that area, for installing EVCS (Pal et al. 2019, 2021).
- The homogenous traffic scenario has been considered for allotting apt charging station to each EV (Das et al. 2020a).
- Three kinds of EV, such as BEV, PHEV-30, and PHEV-40 has been selected for the experiment (Das et al. 2020b).
- The charging stations are slow charging station. Because, to implement V2G mode of operation, it is essential to have charging station with lower charging rates.

25.6.1 Input Data

25.6.1.1 No of Cars Arriving in Charging Station:

Season wise, the number of EVs in the road may vary. Especially during monsoon and winter, many people try to travel less by driving its own car or drive their car for smaller distances. According to the data found from National Household Travel Survey (NHTS) (Pal et al. 2021), from statistical analysis, it has been seen that the arrival of EV at charging station generally follows the normal distribution (Pal et al. 2021). This value varies from season to season. The mean and standard deviation of no of cars arriving at charging station in 24 h is shown in Table 25.2.

Table 25.2 Parameter for normal distribution for the number of cars arriving at charging station (Pal et al. 2021)

Season	Mean (μ)	Standard deviation (σ)
Summer	80	15
Monsoon	45	15
Winter	60	20

During performing the experiment, it has been assumed that total 300 EVs are present in that area and after considering meteorological scenario, the worst condition among these three seasons (maximum number of cars arrived at charging station) has been considered for further analysis. It has been seen that in summer, maximum 100 eV arrived at charging station. Similarly, during monsoon, it is only 55 and in winter it is 80. But before installing a charging station, it is assumed that maximum 100 eV can be accommodated in each charging station. And based on its maximum available capacity, the charging station needs to be placed in 33 bus radial distribution network based on the worst-case scenario.

25.6.1.2 Arrival and Departure Time of EV in Charging Station

The arrival and departure time of EVs vary as per meteorological condition and from literature, it has been observed that most of the time these two attributes can be modeled using two parameter Weibull distribution function (WPDF) (Pal et al. 2021). The seasonal details regarding the parameter (shape factor and scale factor) have been shown in Table 25.3.

Using these parameters, it has been observed that mean time to arrive/depart, to/from charging station is 7.00 am and 8 pm. For monsoon it is around 10 am and 5 pm and for winter, it is around 9 am and 6 pm. This variation is happening due to bad climate in monsoon and winter. During monsoon, heavy rainfall causes traffic congestions and as well as water logging, which make resistance for cruising of EV. At the same time, in winter season, high windy weather and smog causes difficulties in cruise. Thus, most of the time, the EV reach the charging station in late morning and it also leave the charging station in early evening.

Table 25.3 The shape and scale factor of WPDF

	Arrival time				Departure time			
Season	Mean shape factor (α)	Standard deviation (σ) of α	Mean scale factor (β)	Standard deviation (σ) of β	Mean shape factor (α)	Standard deviation (σ) of α	Mean scale factor (β)	Standard deviation (σ) of β
Summer	1.01	0.02	16.49	0.45	5.24	0.17	50.09	0.38
Monsoon	0.99	0.01	16.80	0.25	4.85	0.10	50.00	0.33
Winter	1.00	0.02	16.41	0.40	4.89	0.12	50.07	0.35

25.6.1.3 Daily Mileage and Charging Trip Distance of EV

As per the literature, it can be observed that the daily mileage of EV can be modeled using normal distribution (Saleem et al. 2019). This mileage can vary due to seasonal factor. Considering this factor, in Table 25.4, the mean daily mileage and standard deviation has been shown, which has been considered in 2m PEM, for execution purpose. Moreover, from literature, it also been observed that almost 90% of EV make at least 3 trips per day among which, each trip is in average of 35 km, irrespective of all season.

25.6.2 *Type of EVs*

In recent scenario, basically 3 types of EV have been assumed which usually cruise in that area. These are PHEV-30(Das et al. 2020a), PHEV-40 (Pal et al. 2019) and BEV (Battery Electric Vehicles) (Pal et al. 2019), whose specifications are given in Table 25.5.

25.6.3 *Seasonal Real-Time Pricing (Seasonal RTP)*

The statistics about seasonal RTP are highly unstable. The data is fetched from NEMS of 2019 (http://www.emcsg.com), where RTP of every 30 min interval are shown. The whole data of real-time pricing is segregated season wise and the average real-time pricing has been considered for charging scheduling purpose. To justify the RTP, the corresponding seasonal average real-time load at 30 min. interval is shown in Figs. 25.2, 25.3 and 25.4.

Table 25.4 Statistical value for daily mileage and standard deviation

Season	Mean daily mileage (μ) (km)	Standard deviation (σ)
Summer	55	10
Monsoon	29.5	10
Winter	35	10

Table 25.5 EV's specifications (Hashim et al. 2019; Das et al. 2021, 2020c)

Vehicle type	AER (km)	Battery capacity (kWh)	Battery type
PHEV-30	48	13.8	Lithium ion
PHEV-40	64	18.4	Lithium ion
BEV	188	28	Lithium ion

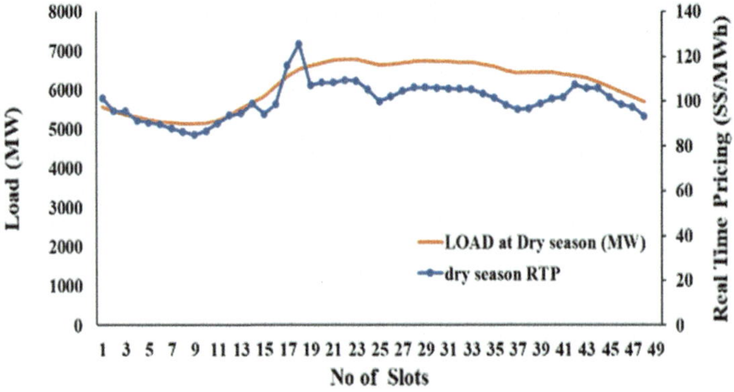

Fig. 25.2 Real-time pricing and Load of summer (http://www.emcsg.com)

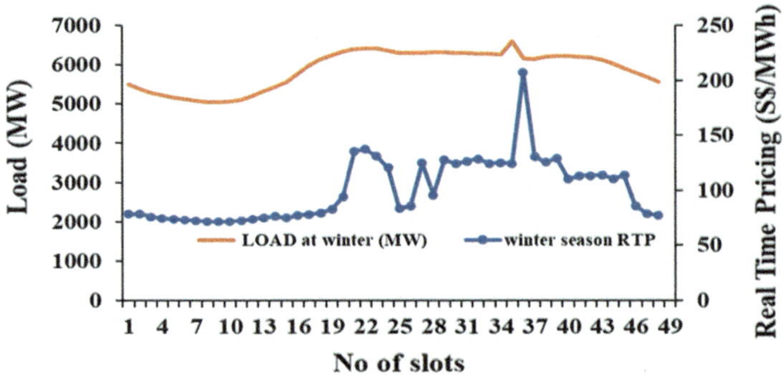

Fig. 25.3 Real-time pricing and Load of Winter (http://www.emcsg.com)

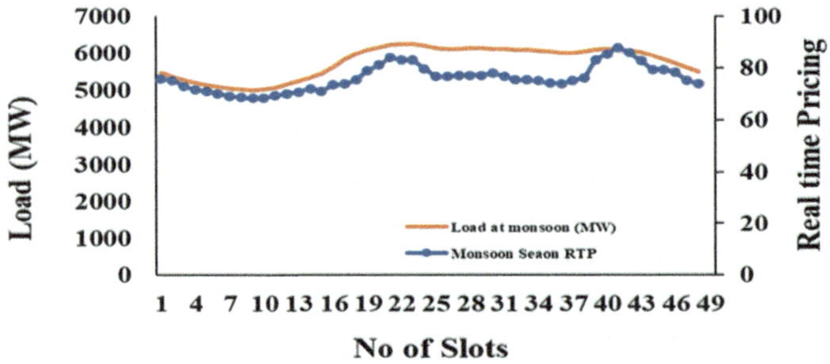

Fig. 25.4 Real-time pricing and Load of monsoon (http://www.emcsg.com)

25.7 Execution of Work, Results, and Discussion

The execution of work has been done in three operational level. In each level, the meteorological factor has been considered. As mentioned earlier, the multilevel strategy for smart charging scheduling has been done considering charging (G2V), discharging (V2G) and idle mode of operation. The detailed execution procedure has been described as follows.

25.7.1 Executing Multilevel Algorithm

25.7.1.1 First Level: Optimal Allocation of EVCS

Three charging stations have been considered for this work. Optimal allocation of EVCS has been performed on IEEE 33 bus radial distribution system. The network has been divided into three regions according to EV charging demand as shown in Fig. 25.5. Region wise optimal locations have been found out based on power loss minimization.

Considering all the seasonal variation, the worst case has been selected for placement. As there are 300 EVs are present in the study area, therefore 100 EVs in each charging station have been considered for the allocation problem. The optimization problem has been solved by HGSO technique. The optimal locations have been presented in Table 25.6.

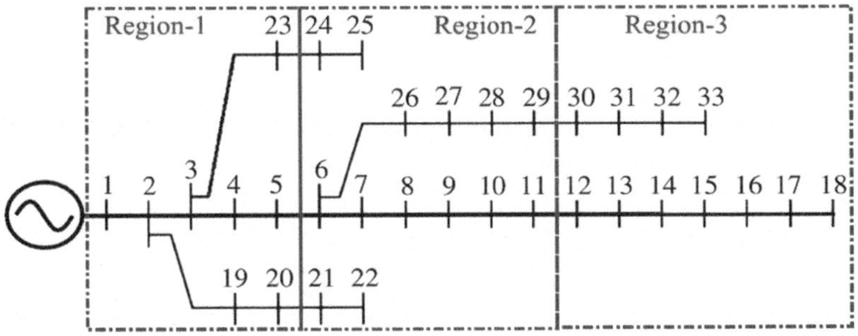

Fig. 25.5 Region wise divisions of the distribution system (Pal et al. 2019, 2021)

Table 25.6 Optimal allocation of EVCS considering worst-case scenario

Place in region-1	Place in region-2	Place in region-3	Mean power loss (kW)	SD power loss (kW)
2	21	30	240.4525	2.3568

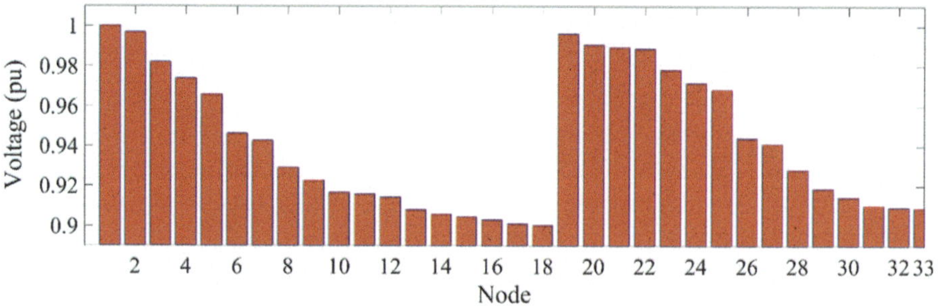

Fig. 25.6 Voltage profile of the distribution system

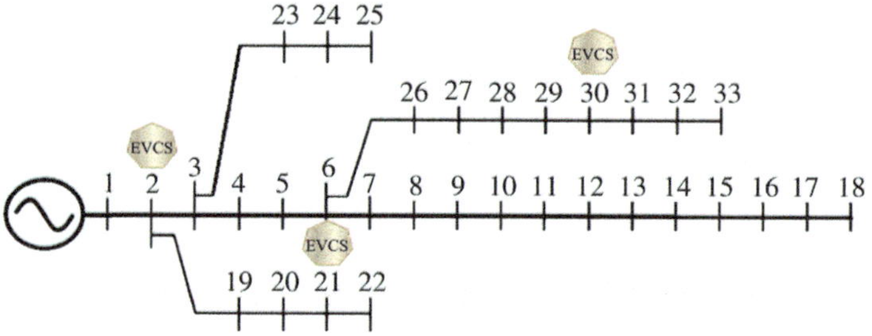

Fig. 25.7 Locations of EVCS in IEEE 33 distribution network (Pal et al. 2019)

It can be observed in table 25.6 that three optimal locations are selected from each region. Minimized mean power loss obtained by the optimization technique is 240.4525 kW by taking the uncertainties using 2m PEM. The voltage profile of the distribution system has been illustrated in Fig. 25.6 and the graphical allocation in the IEEE 33 node distribution system has been shown in Fig. 25.7. It can be easily noticed in Fig. 25.7 that the EVCSs are scattered in the area to attend all the EVs equally.

25.7.1.2 Second Level: Selection of Apt Charging Station

After fixing the optimal location in 33 bus radial distribution network, next motto is to send the EVs (which required charging/want to participate in V2G mode) to its apt charging station at minimum battery energy consumption. Here, it is considered that this is not always essential that the adjacent CS will always be the relevant. To break this myth, this kind of analysis has been executed using ILP. This is executed in three main seasons, i.e., dry season (summer), monsoon (rainy) and winter (or fall). As described earlier, it is very obvious that, in these three seasons, the distance traveled by each car in daily basis will vary and accordingly, the battery state (battery energy) will also vary and the number of EVs on road will also differ. Keeping this in mind, using normal distribution, three scenarios has been considered and one scenario is for each season. This scenario is considered to be the wort scenario in that very season. Based on this, it has been seen that most EV has been found in road in summer, then little less in winter and very less in rainy season. The number of EVs available on road in these three seasons, have been shown in tabular form in Table 25.7. Thereafter, using ILP, as the availability and requirement how many EVs went to each charging station, and in what ratio, which type of car has reached in that very charging station, that also been shown in Table 25.7. It is assumed that each charging station can accommodate maximum 100 EVs.

25.7.1.3 Third Level: Optimal Charging Scheduling

After successful allotment of CS for each EV, the ultimate aim is to perform the optimal scheduling of these EVs in each charging station at minimum cost considering G2V, V2G, and idle mode of operation. From the previous allotment it has been observed that in each charging station contains three types of EVs, i.e., PHEV-30, PHEV-40, and BEV at certain ratio. In some CS, the higher battery capacity EVs (BEV) are in greater ratio and somewhere, lower battery capacity EVs are in greater ratio. Besides this, it has been observed that, season wise, this percentage is varying. From this observation, it is pretty prominent that this will keep an impact on the charging cost. As well as the charging scheduling criteria will also differ for different seasons. Keeping this in mind, the charging scheduling of EVs has been performed and the assumed statement has been verified. The charging scheduling for each charging station of each season has been shown in Figs. 25.8, 25.9, 25.10, 25.11, 25.12, 25.13, 25.14, 25.15 and 25.16. Its corresponding cost is also depicted in tabular form in Table 25.8.

Moreover, to establish the consistency of the proposed algorithm, applying another soft-computing technique, i.e., DE (Differential Evolution), the results for level 3 have been determined and compared. From Table 25.8, it has been seen that the using both the soft-computing technique, the proposed algorithm can determine the charging cost with higher accuracy.

25.7.1.4 Wilcoxon Signed Rank Test and Quade Test

Wilcoxon Signed-Rank Test (WSRT) (Das et al. 2020c)

The Wilcoxon signed-rank test is a non-parametric test for hypothesis analysis. For comparison of two liked samples, coordinated samples, or repetitive calculations on a single sample and to assess whether their population mean ranks differ, this can be done with this test. In this chapter, Wilcoxon Signed rank test has been done on 30 repeated measurements on a single sample and two hypotheses we have made, i.e.,

H_0 (Null Hypothesis): No significant difference among those 30 values.

H_1 (Alternative hypothesis): significant difference among those 30 values.

Table 25.7 Season wise number of EVs available on road and arriving in charging station

Season	EVs on Road	EVs need charge	CS1	PHEV-30 (%)	PHEV-40 (%)	BEV (%)	CS2	PHEV-30 (%)	PHEV-40 (%)	BEV (%)	CS3	PHEV-30 (%)	PHEV-40 (%)	BEV (%)
Summer	100	**55**	**27**	20	30	50	**15**	35	25	40	**20**	55	20	25
Monsoon	65	**45**	**27**	50	35	15	**12**	20	30	50	**6**	20	50	30
Winter	80	**50**	**9**	30	25	45	**22**	10	30	60	**19**	25	10	6

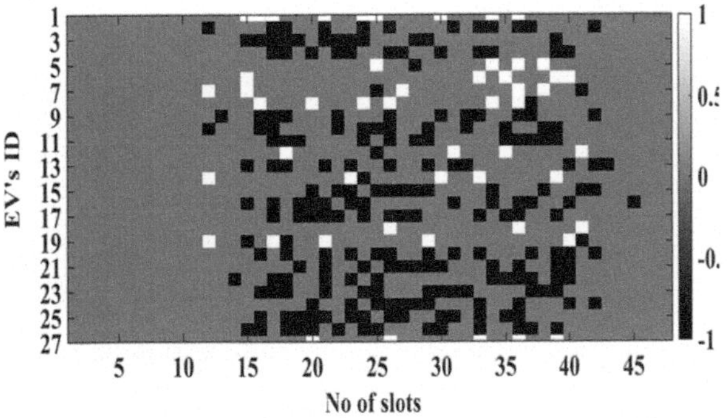

Fig. 25.8 Charging strategy of summer in CS1

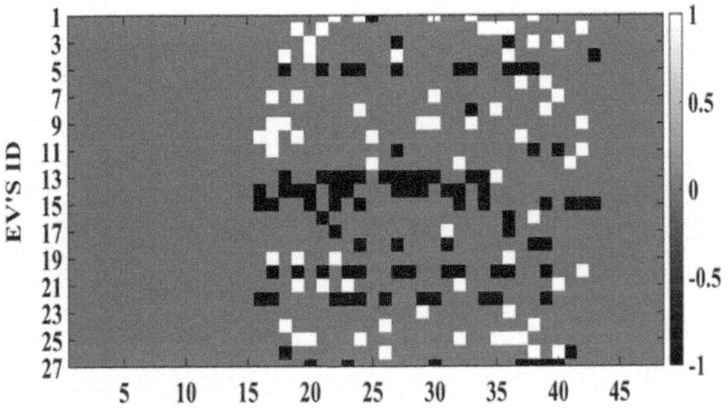

Fig. 25.9 Charging strategy of monsoon in CS1

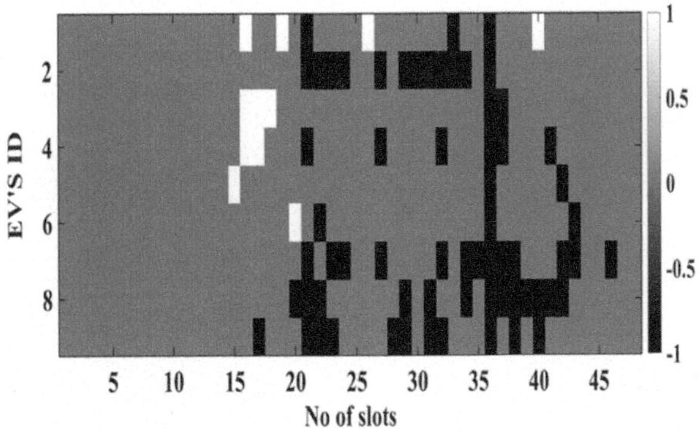

Fig. 25.10 Charging strategy of winter in CS1

From Table 25.9, it is observed that, at 95% significance level for 30 of samples, the value of test statistic is 216 for HGSO and 198 for DE. But as per α-distribution table, the value corresponds to 30 sample is 137.Since, the absolute value, i.e., 137 is lesser than both test statistic values; therefore, it can be said that there is no significance difference between these 30 repeated measurements and thus the null hypothesis, i.e., H_0 can be accepted. This implies that the proposed multilevel algorithm for charging scheduling is robust enough to give consistent output.

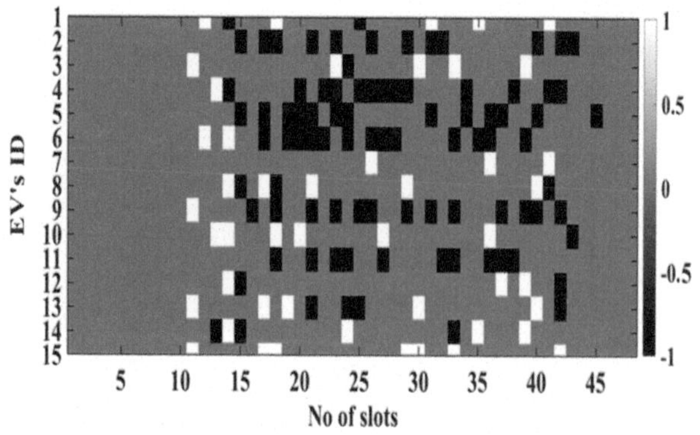

Fig. 25.11 Charging strategy of summer in CS2

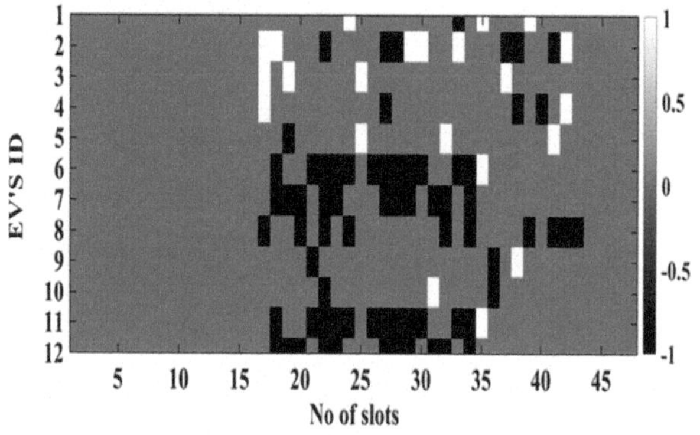

Fig. 25.12 Charging strategy of monsoon in CS2

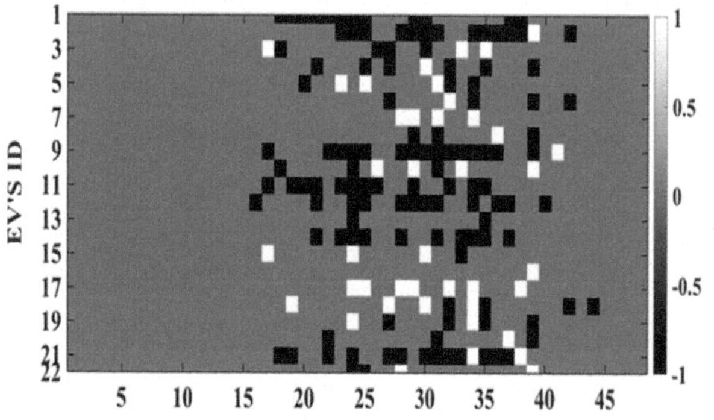

Fig. 25.13 Charging strategy of winter in CS2

Quade Test (Das et al. 2020c)

This is another the non-parametric test by which the significant difference between two or more than two groups can be determined. The Quade test is used for analyzing randomized complete block designs. Here, the level of significance is $\alpha = 5\%$ (95% confidence interval). From Table 25.10, it can be said that for both HGSO and DE,

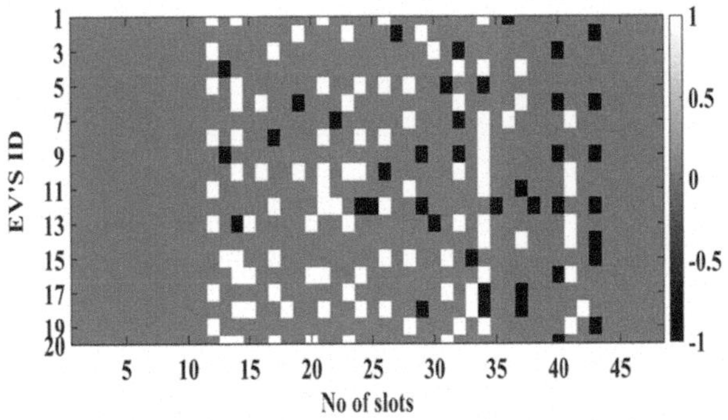

Fig. 25.14 Charging strategy of summer in CS3

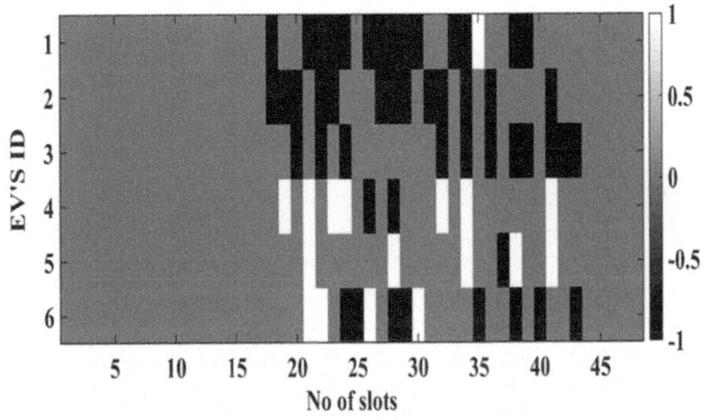

Fig. 25.15 Charging strategy of monsoon in CS3

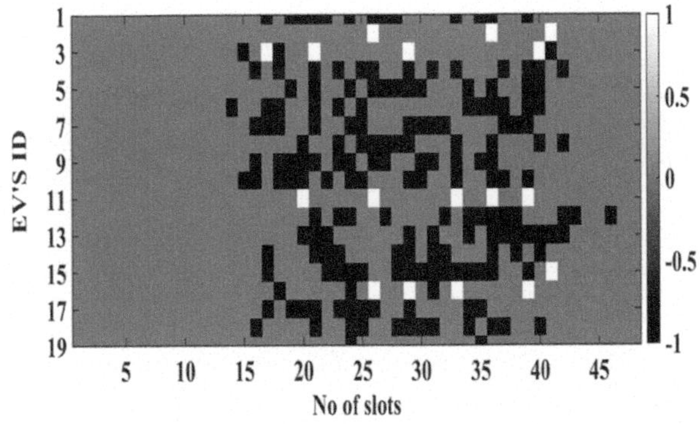

Fig. 25.16 Charging strategy of winter in CS3

the absolute value (which can be obtained from *f*-distribution table) is less than the test statistic value. Thus, null hypothesis can be denied. Alternatively, it implies that there are significant differences between the results coming after applying HGSO and DE.

Table 25.8 Season wise charging scheduling cost for EVs in 24 h

Season	EVs on Road	EVs need charge	EV@ CS1	Best daily charging cost (S$), SD ($\sigma$)		EV@ CS2	Best daily charging cost (S$), SD ($\sigma$)		EV@ CS3	Best daily charging cost (S$), SD ($\sigma$)	
				HGSO	DE		HGSO	DE		HGSO	DE
Summer	100	**55**	**27**	02.2698, 0.235	3.2348, 0.342	**15**	5.5896, 0.328	6.5664, 0.447	20	14.5774, 0.225	16.0014, 0.305
Monsoon	65	**45**	**27**	15.8978, 0.221	16.6998, 0.302	**12**	1.0599, 0.215	2.2558, 0.335	6	1.1558, 0.166	2.5448, 0.2115
Winter	80	**50**	**9**	00.6899, 0.114	1.0223, 0.192	**22**	2.0018, 0.056	3.0244, 0.119	19	1.9878, 0.3645	3.0012, 0.449

Table 25.9 WSRT For multilevel algorithm

Optimization	No of samples	Value of α (level of significance)	Test statistic value	Absolute value (from α-distribution table)	Hypothesis
HGSO	30	0.05 (95%)	216	137	Accepting H_0
DE	30	0.05 (95%)	198	137	Accepting H_0

Table 25.10 QT result for multilevel charging scheduling algorithm

No. of samples	No of groups	Value of α (level of significance)	Test statistic value	Abs value	Hypothesis
30	2	0.05 (95%)	3.89	2.04	Reject H_0

25.7.2 Discussions

From first level, it has been realized that, irrespective of seasonal variation and other constraints, the optimum location to install the charging station is at node 2, 21 and 30. Placing in this location will minimize the power loss by 240.4525 kW. Since the worst condition among these three seasons have been considered, thus it can be said that, under any circumstances, these locations are optimum and the maximum power minimization can be achieved throughout the year.

Thereafter, the apt charging station has been assigned for the required EVs. While assigning this charging station, it has been observed that, in each season, the number of cars available on road is varying. Besides this, in some charging station, the number of lower capacities EVs are more. Again, in some places, the higher battery capacity EVs are more. While performing the successful allotment, the dynamic vehicle congestion has been considered and it can be seen that, in some cases, the nearest charging station is not always the apt one.

After successful allotment, in the final level, the combined scheduling of EVs considering G2V, V2G, and idle mode of operation has been performed. Here idle mode played an important role.

- When Real-time tariff for purchasing the electrical energy from grid is high (so it cannot go for G2V mode of operation) and both CSO and EV owners wants to be benefited and earn some revenue, in that scenario, in those hours, they will try to keep the EV in idle condition.
- When there is no sufficient amount charge in the battery of the vehicles and so, it cannot go for V2G mode of operation, in that case also, the EV may remain in idle mode.
- Otherwise, when the cost of selling the energy is less, but the cost of purchasing power as well as is high, but sufficient amount of power SOC is there in the EV, then also, EV owner may keep their EV in the idle condition, so that it may make profit in the later hours, when the selling price of energy will be higher.

Based on this rules and concept, the coordinate charging scheduling of EVs at each charging station has been performed for every individual season. From Figs. 25.8, 25.11 and 25.14, it can be seen that when higher battery capacity EVs are in majority and lower battery capacity EVs are less, in that case, when the real-time pricing is higher, the EVs are contributing in V2G mode of operation. Again, when real-time pricing is decreasing, the EV are contributing in G2V mode of operation. This results reduction in the overall daily charging scheduling cost of EVs. In contrast, when the scenario is opposite, i.e., the lower battery capacity EVs are more and Higher battery capacity EV is lesser, in that case, sometimes besides being the higher tariff rate, EVs are participating in G2V mode of operation. Such incident is occurring due to limited battery capacity of the EVs and uncertainties. This results more cost in charging process.

Again, from Figs. 25.2, 25.3 and 25.4, it can be observed that in summer, the average RTP is higher than the RTP of Winter and Monsoon. From Table 25.8, it can be seen that when majority of EVs connected with the grid is of lower battery capacity, in that case, the daily cost of charging is increasing by 80% in average irrespective of any season. Elaborately, it can be said that though the average RTP of monsoon is lower than the summer, then also, for equal number of EV (27 eV in this case), the daily cost of charging is more than the daily charging cost of summer. Such incident is occurred because, even though the number of EV is equal, but in case of summer, Higher battery capacity EV is more and in monsoon the Lower battery capacity EV is more. This is why, the cost of charging in monsoon is increased about 7 times than the charging cost of summer. With compared to the winter season also, the scenario is same. Hence, it can be said that, for proper economic coordinated charging process, greater number of higher battery capacity EVs are needs to be connected with the grid. This will decrease the charging cost, moreover will increase the V2G facilities, which will help in peak shaving of load. This is how, both the utility and EV users can be in some win-win situation.

Apart from this, it can be seen that, during rainy season and sometimes in winter also, the number of cars arriving in CS is less because of many constraints, natural calamities and as well as the tariff rate is also less in average (Figs. 25.9 and 25.10) at that time. This is resisting EVs users to participate in V2G mode of operation. Therefore, to promote V2G technology and as well as to endorse coordinated charging process for the whole year, the utility should think of providing more incentives in terms of money or in terms perks, during monsoon and winter, which will insist EV users to participate more in V2G mode of operation. Consequently, this will help to reduce unwanted peak loading condition during in these two seasons, especially in evening time.

Moreover, using non-parametric analysis, i.e., WSRT, it can be observed that, for both HGSO and DE, the proposed multilevel charging scheduling algorithm is robust and consistent enough. Furthermore, after performing the Quade test, it can be seen from Table 25.10, that a significant difference is there in the results are there after applying HGSO over DE. HGSO is performing better than DE. Though both of the soft-computing techniques are robust enough to provide consistent output.

25.8 Conclusions

In this book chapter, an innovative multilevel charging scheduling algorithm has been proposed, which provides a complete charging solution in a smart and optimal way considering seasonal uncertainties related to the driving cycles of EVs. This chapter discussed mainly Socio-techno-economic perspective and various perspectives of EVs. In the first level, the optimal placement of EVCS has been performed in such a way that it could be able to cover the complete distribution network for attending all the EVs equally by consuming minimum amount of power loss. Thereafter, for each and every EV, the relevant charging station has been assigned in such a way that it could reach corresponding EVCS by overriding minimum battery energy. Subsequently, for coordinated charging scheduling of EVs, G2V, V2G, and the idle mode of operation has been utilized in such a way that the mean daily charging cost of the EVs will be minimized. In all these three levels, the uncertainties related to driving cycles, due to seasonal changes, have been considered and handled using 2m PEM method. From the nature of charging scheduling, it can be realized that, for better coordination of G2V, V2G, and idle mode of operation, a greater number of EVs with higher battery capacity are required. This could result in V2G mode of operation, when Real-time pricing is higher and G2V mode of operation when Real-time pricing is lower. But using lesser number of EV with lower battery capacity is not apt to serve this purpose and consequently the charging cost will become higher. On the other hand, it can be said that, to promote V2G for all year along, the utility should provide more incentives to the EV users during monsoon and winter season for better coordination between the grid and EVs. Above and all, the proposed methodology along with the algorithm is a complete solution, through which all the three major sectors, i.e., Grid, EVCS, and EV users will be in a beneficial position.

References

Almehizia A, Snodgrass JM (2018) Investigation of V2G economical viability. IEEE Texas Power Energy Conf (TPEC) 2018:1–6. https://doi.org/10.1109/TPEC.2018.8312048

Amini MH, Islam A (2014) Allocation of electric vehicles' parking lots in distribution network. In: ISGT 2014, Washington, DC, pp 1–5. https://doi.org/10.1109/ISGT.2014.6816429

Amini M, Boroojeni K, Iyengar S, Pardalos P, Blaabjerg K, Madni A () Sustainable interdependent networks. Studies in systems. Decision and control, vol 145. Springer, Cham. https://doi.org/10.1007/978-3-319-74412-4_1

Awasthi A, Venkitusamy K, Padmanaban S, Selvamuthukumaran R, Blaabjerg F, Singh AK (2017) Optimal planning of electric vehicle charging station at the distribution system using hybrid optimization algorithm. Energy 133:70–78. https://doi.org/10.1016/j.energy.2017.05.094

Bendary AF, Ismail MM (2019) Battery charge management for hybrid PV/wind/fuel cell with storage battery. Energy Proc 162:107–116. https://doi.org/10.1016/j.egypro.2019.04.012

Chen R, Qian X, Miao L, Ukkusuri SV (2020) Optimal charging facility location and capacity for electric vehicles considering route choice and charging time equilibrium. Comput Oper Res 113. https://doi.org/10.1016/j.cor.2019.104776

Chhawchharia S, Kumar Sahoo S, Balamurugan M, Sukchai S, Yanine F (2018) Investigation of wireless power transfer applications with a focus on renewable energy. Renew Sustain Energy Rev 91:888–902. https://doi.org/10.1016/j.rser.2018.04.101

Das S, Bhattacharya A, Acharjee P (2020c) Charging scheduling of electric vehicles (EV) in probabilistic scenario considering grid-to-vehicle (G2V) and vehicle-to-grid (V2G). IEEE Calcutta Conf (CALCON) 2020:467–472. https://doi.org/10.1109/CALCON49167.2020.9106552

Das S, Acharjee P, Bhattacharya A (2020a) Allocation of appropriate charging station and intelligent charging scheduling for on-road electric vehicles. Special Issue on Water, Energy and Environmental Sustainability (WEES 2020a). J Indian Chem Soc (JICS) 97:1–17. ISSN: 0019-4522

Das S, Acharjee P, Bhattacharya A (2020b) Charging scheduling of electric vehicle incorporating grid-to-vehicle (G2V) and vehicle-to-grid (V2G) technology in smart-grid. In: IEEE International conference on power electronics, smart grid and renewable energy (PESGRE2020b), pp 1–6. https://doi.org/10.1109/PESGRE45664.2020b.9070489

Das S, Acharjee P, Bhattacharya A (2021) Charging scheduling of electric vehicle incorporating grid-to-vehicle and vehicle-to-grid technology considering in smart grid. IEEE Trans Ind Appl 57(2):1688–1702. https://doi.org/10.1109/TIA.2020.3041808

Deb S, Gao X-Z, Tammi K, Mahanta KKP (2021) A novel chicken swarm and teaching learning-based algorithm for electric vehicle charging station placement problem. Energy 220:119645. https://doi.org/10.1016/j.energy.2020.119645

Energy Market Company Price Information. http://www.emcsg.com/MarketData/PriceInformation

Girade P, Shah H, Kaushik K, Patheria A, Xu B (2021) Comparative analysis of state of charge based adaptive supervisory control strategies of plug-in Hybrid Electric Vehicles. Energy 230. https://doi.org/10.1016/j.energy.2021.120856

González-Garrido A, Thingvad A, Gaztañaga H, Marinelli M (2019) Full-scale electric vehicles penetration in the Danish Island of Bornholm—Optimal scheduling and battery degradation under driving constraints. J Energy Storage 23:381–391. https://doi.org/10.1016/j.est.2019.03.025

Hadi Amini M, Parsa Moghaddam M, Karabasoglu O (2016) Simultaneous allocation of electric vehicles' parking lots and distributed renewable resources in smart power distribution networks. Sustain Cities Soc 28:332–342. https://doi.org/10.1016/j.scs.2016.10.006

Hashim FA, Houssein EH, Mabrouk MS, Al-Atabany W, Mirjalili S (2019) Henry gas solubility optimization: a novel physics-based algorithm. Fut Gener Comput Syst 101:646–667. https://doi.org/10.1016/j.future.2019.07.015

Hussain A, Bui V, Kim H (2020) Optimal sizing of battery energy storage system in a fast EV charging station considering power outages. IEEE Trans Transp Electr 6(2):453–463. https://doi.org/10.1109/TTE.2020.2980744

Karmaker AK, Ahmed MR, Hossain MA, Sikder MM (2019) Feasibility assessment & design of hybrid renewable energy based electric vehicle charging station in Bangladesh. Sustain Cities Soc 39:189–202. https://doi.org/10.1016/j.scs.2018.02.035

Khaligh A, D'Antonio M (2019) Global trends in high-power on-board chargers for electric vehicles. IEEE Trans Vehicul Technol 68(4):3306–3324. https://doi.org/10.1109/TVT.2019.2897050

Khan W, Ahmad F (2019) Fast EV charging station integration with grid ensuring optimal and quality power exchange. Eng Sci Technol Int J 22(1):143–152. https://doi.org/10.1016/j.jestch.2018.08.005

Khorramdel B, Khorramdel H, Aghaei J, Heidari A, Agelidis VG (2015) Voltage security considerations in optimal operation of BEVS/PHEVS integrated microgrids. IEEE Trans Smart Grid 6(4):1575–1587. https://doi.org/10.1109/TSG.2015.2394499

Kong W, Luo Y, Feng G, Li K, Peng H (2019) Optimal location planning method of fast charging station for electric vehicles considering operators, drivers, vehicles, traffic flow and power grid. Energy 186. https://doi.org/10.1016/j.energy.2019.07.156

Lam AYS, Yu JJQ, Hou Y, Li VOK (2018) Coordinated autonomous vehicle parking for vehicle-to-grid services: formulation and distributed algorithm. IEEE Trans Smart Grid 9(5):4356–4366. https://doi.org/10.1109/TSG.2017.2655299

Liu Z, Wen F, Ledwich G (2013) Optimal planning of electric-vehicle charging stations in distribution systems. IEEE Trans Power Deliv 28(1):102–110. https://doi.org/10.1109/TPWRD.2012.2223489

Liu J, Zhang T, Zhu J, Ma T (2018) Allocation optimization of electric vehicle charging station (EVCS) considering with charging satisfaction and distributed renewables integration. Energy 164:560–574. https://doi.org/10.1016/j.energy.2018.09.028

Liu L, Zhang YDC, Huang Z, Wang M (2020) Optimal allocation of distributed generation and electric vehicle charging stations based on intelligent algorithm and bi-level programming. Int Trans Electr Energ Syst 30:12366. https://doi.org/10.1002/2050-7038.12366

Mehta R, Verma P, Srinivasan D (2019) Double-layered intelligent energy management for optimal integration of plug-in electric vehicles into distribution systems. Appl Energy 233–234:146–155. https://doi.org/10.1016/j.apenergy.2018.10.008

Pal A, Bhattacharya A, Chakraborty AK (2021) Placement of public fast-charging station and solar distributed generation with battery energy storage in distribution network considering uncertainties and traffic congestion. J Energy Storage 41. https://doi.org/10.1016/j.est.2021.102939

Pal A, Bhattacharya A, Chakraborty AK (Mar. 2021) Allocation of electric vehicle charging station considering uncertainties. Sustain Energy Grids Netw 25. https://doi.org/10.1016/j.segan.2020.100422

Pal A, Bhattacharya A, Chakraborty AK (2019) Allocation of EV fast charging station with V2G facility in distribution network. In: 2019 8th International conference on power systems (ICPS), pp 1–6. https://doi.org/10.1109/ICPS48983.2019.9067574

Saleem Y, Crespi N, Rehmani MH, Copeland R (2019) Internet of Things-aided smart grid: technologies, architectures, applications, prototypes, and future research directions. IEEE Access 7:62962–63003. https://doi.org/10.1109/ACCESS.2019.2913984

Shafiq S, Irshad UB, Al-Muhaini M, Djokic SZ, Akram U (2020) Reliability evaluation of composite power systems: evaluating the impact of full and plug-in hybrid electric vehicles. In: IEEE Access 8:114305–114314. https://doi.org/10.1109/ACCESS.2020.3003369

Sharma A, Sharma S (2019) Review of power electronics in vehicle-to-grid systems. J Energy Storage 21:337–361. https://doi.org/10.1016/j.est.2018.11.022

Shojaabadi S, Abapour S, Abapour M, Nahavandi A (2016) Optimal planning of plug-in hybrid electric vehicle charging station in distribution network considering demand response programs and uncertainties. IET Gener Transm Distr 10(13):3330–3340. https://doi.org/10.1049/ietgtd.2016.0312

Sun C, Sun F, Hu X, Hedrick JK, Moura S (2015) Integrating traffic velocity data into predictive energy management of plug-in hybrid electric vehicles. In: 2015 American control conference (ACC), 2015, pp 3267–3272. https://doi.org/10.1109/ACC.2015.7171836

Susowake Y, Yongyi H, Senjyu T, Howlader AM, Mandal P (2018) Optimum operation plan for multiple existing EV charging stations. In: 2018 IEEE PES Asia-Pacific power and energy engineering conference (APPEEC), pp 611–615. https://doi.org/10.1109/APPEEC.2018.8566620

Teng F, Ding Z, Hu Z, Sarikprueck P (2020) Technical review on advanced approaches for electric vehicle charging demand management, Part I: Applications in electric power market and renewable energy integration. IEEE Trans Ind Appl 56(5):5684–5694. https://doi.org/10.1109/TIA.2020.2993991

Zeng B, ZhuZ, Xu H, Dong H (2020) Optimal public parking lot allocation and management for efficient PEV accommodation in distribution systems. IEEE Trans Ind Appl 56(5):5984–5994. https://doi.org/10.1109/TIA.2020.2986980

Zhang Y, Zhang Q, Farnoosh A, Chen S, Li Y (2019b) GIS-based multi-objective particle swarm optimization of charging stations for electric vehicles. Energy 169:844–853. https://doi.org/10.1016/j.energy.2018.12.062

Zhang H, Tang L, Yang C, Lan S (Aug. 2019) Locating electric vehicle charging stations with service capacity using the improved whale optimization algorithm. Adv Eng Inform 41. https://doi.org/10.1016/j.aei.2019.02.006

Zhang H, Tang L, Yang C, Lan S (2019a) Locating electric vehicle charging stations with service capacity using the improved whale optimization algorithm. Adv Eng Informat 41:100901. https://doi.org/10.1016/j.aei.2019a.02.006

Zhang S, Dou W, Zhang Y, Hao W, Chen Z, Liu Y (2019c) A vehicle-environment cooperative control based velocity profile prediction method and case study in energy management of plug-in hybrid electric vehicles. IEEE Access 7:75965–75975. https://doi.org/10.1109/ACCESS.2019c.2921949

Zhou S, Ng ST, Yang Y, Xu JF (2021) Integrating computer vision and traffic modeling for near-real-time signal timing optimization of multiple intersections. Sustain Cities Soc 68:102775. https://doi.org/10.1016/j.scs.2021.102775

Chapter 26
Layout Optimization Planning of Hybrid Offshore Wind-Solar PV Power Plants

Santanu Paul, Syed Raahat Ara, and Zakir Hussain Rather

Abstract The integration of renewable energy, particularly wind and solar, is being done on a large scale in the modern power system. The installation of these technologies was earlier limited to onshore, but with advancements in technology and increasing land requirements, these renewable energy generations are gradually shifting offshore. There are multiple advantages associated with offshore renewable power generations, such as the proper utilization of the potential of renewable resources without hindrance. Improvement of the annual capacity factor of renewable power plants is another major factor in moving offshore locations. Wind and solar resources are often complementary in nature; hence, with many wind power plants already in place, it might be a good option to install solar PV with the existing infrastructure, which will reduce its seasonal intermittency and also increase the capacity factor. For the maximum utilization of these sources, optimal placement of wind turbines (WTs) and solar PV panels in an offshore location is an inevitable part of planning for setting up hybrid wind and solar PV offshore power plants. This chapter mainly focuses on the layout optimization of offshore hybrid wind and solar PV plants to improve system-level planning to maximize the energy output. The generation from the offshore hybrid plants needs to be optimized considering wake effect and tower shadow effect loss on wind turbines and solar panels, respectively, to improve the overall efficacy of the hybrid offshore plants. This chapter also deals with different aspects of mathematical modeling of the wind and solar PV systems

S. Paul and S. R. Ara both have contributed equally to this chapter.

S. Paul (✉)
Department of Electrical Engineering, Indian Institute of Science Bangalore, Bangalore 560012, India

S. R. Ara (✉)
Department of Applied Sciences and Humanities, Faculty of Engineering, Jamia Millia Islamia, New Delhi 110025, India

Z. H. Rather
Department of Energy Sc. and Engineering, Indian Institute of Technology Bombay, Mumbai 400076, India
e-mail: zakir.rather@iitb.ac.in

A. K. Bohre et al. (eds.), *Planning of Hybrid Renewable Energy Systems, Electric Vehicles and Microgrid*, Energy Systems in Electrical Engineering,
https://doi.org/10.1007/978-981-19-0979-5_26

to calculate the wake and tower shadow losses while determining the optimal layout of large-scale hybrid offshore wind-solar PV plants.

Keywords Wind energy · Hybrid solar PV wind · Layout optimization · Particle Swarm optimization · Wake modeling · Shadow analysis

Nomenclature

C_t	Coefficient of thrust
G^q_{best}	Global best position up to the qth iteration
$P^q_{i,best}$	Local best position up to the qth iteration for the ith particle
C_A	Annual loan repayment
η_1	Panel efficiency
η_2	Performance ratio
CSP	Concentrated Solar Power
GHI	Global Horizontal Irradiance
MPP	Maximum Power Point
PDF	Probability Distribution Function
WF	Wind Farm
ψ_1, ψ_2	Cognitive and Social learning constants
$rand_1$, $rand_2$	Random values between 0 and 1
u_{ci} and u_{co}	Cut-in wind speed and Cut-out wind speed
L_x, L_y	Projection of WT on x and y axis
S_x, S_y	Coordinates of elliptical shadow along the x and y-axis
C_x, C_y	Coordinates of circle center or coordinates of WT
CF	Capacity Factor
LCoE	Levelized Cost of Energy
PV	Photo Voltaic
PSO	Particle Swarm Optimization
WT	Wind Turbine

26.1 Introduction

The apparent rise in the electricity demand, the rapid exhaustion of fossil fuels, and the increasing environmental concerns along with the stringent environmental laws throughout the world have led to the deployment of renewable energy sources on a large scale, especially wind and solar. Even though, since 2008, the world economic crisis has been affecting the world economy, significant investments in renewable energy have continued nevertheless. As of the latest data, wind and solar generation account for about 24% of India's installed capacity, generating 10.70%

of total electricity throughout the country (India 2020). In the Paris Agreement, the Government of India has committed to meeting a target of 40% of its total production of electricity from renewable sources by 2030. The goal was to install 175 GW of renewable energy power by 2022, including 100 GW of solar energy and 60 GW of wind and rest from other renewable sources (The Indian Express 2015).

It is seen that new investments in wind and solar systems have been at the forefront. The installation of these technologies was earlier limited to land, but land being a premium commodity and their huge requirement for it is making these technologies go offshore. This move has presented itself with its own merits and demerits. For solar, the merits include fewer obstacles to block sunlight, less dust effect, the negative coefficient efficiency of the PV junction, higher energy efficiency due to lower temperature underneath the panels. On the other hand, for wind, it is the higher and more consistent wind speeds, thus higher power generation. The demerits continue to be transportation, installation, evacuation of power, threats like sea waves, cyclones, high tides, storms, and tsunami, increased corrosion of the metallic structure, increased maintenance, effect on fishing, and other transportation activities depending on the selected site (Sahu et al. 2016).

Nonetheless, with the advances in technologies, there has been significant growth in offshore wind. And now, offshore solar is also making its way (Snieckus 2019). Onshore hybrid solar PV-wind is not a new concept. Due to the unpredictable nature of these sources, they are not entirely reliable when explored independently. Their complementary nature, however, makes their use as a hybrid system more cost-effective and dependable.

26.2 Brief History Associated with Wind and Solar PV Power Plants

The renewable energy industry is expanding throughout the world, developing reliable and cost-competitive systems using a variety of technologies. Various hybrid systems have been developed, but the wind and solar hybrid systems dominate the charts. These two technologies, due to their complementary nature, have been used together for some time. When used together, it increases the reliability and provides a more constant output.

The first known reference or some technology resembling a windmill dates back to 1st BC or 1st AD and is from "Hero of Alexandria" in his work "Pneumatics." Windmills made their first recorded appearance in twelfth century Northern Europe (England) but are believed to have arrived in the 10th or 11th-century (Manwell et al. 2010). On the other hand, solar PV technology was born in 1954, when Calvin Fuller, Daryl Chapin, and Gerald Pearson created the first silicon PV cell at Bell Labs in the United States. It is reported that Willoughby Smith, in 1873, discovered the photo-conductivity of selenium which led to the various discoveries that later lead to the invention of the first solar PV cell (Perlin 1999).

26.3 Recent Development and Potential of Wind and Solar Power at Off Shore Locations

The cost of building solar PV and wind power plants is continuously falling. Hence, a significant scale-up of renewable generation has become feasible for the developing world. Thousands of villages in many parts of the globe are still being exiled from electricity, and energizing these villages with conventional generation alone by extending grid infrastructure will be uneconomical. In 2019, renewable energy accelerated its growth rate, growing globally by 12.2% relative to 2018. Renewable energy demand has risen at an average annual rate of 13.7% over the past decade. Renewables have been the only energy group that has risen at double-digit rates globally over the past decade. The planet used 8.2 exajoules of green energy in 2009 for comparison. That has risen dramatically to 29.0 exajoules in 2019 (Rapier 2020).

India has a high potential for renewable generation. In the last few years, the share of installed renewable generation capacity has grown many folds. India has set an ambitious target of generating 175 gigawatts (GW) of electricity through renewable resources, especially wind (60 GW), and solar (100 GW), by 2022. Of this, around 38.80 GW of wind and 40 GW of solar have already been installed till 31st March 2021. The rest of the target of 175 GW would be mainly through hydro and biomass. This implies an enormous potential in energy generation with current renewable energy technologies. Various countries and entities around the world have already started to shift their focus from offshore wind to offshore hybrid systems. The firms, including Denmark's Floating Power Station, UK Marine Power Systems, and Norway's Pelagic Power, are all promoting the concept of using floating wind turbine platforms for a range of generation properties, from wind and wave to solar and ocean thermal power. Experts claim that using a single multi-technology floating platform will help increase the energy yield per unit of area, thus reducing the total cost of electricity.

26.4 Studies Related to Hybrid Offshore Wind-Solar PV Plants

Several studies have been reported in the literature related to optimal planning and operation of hybrid renewable resource power plants in recent times. Techno-economic evaluation is an essential parameter to assess the feasibility of such hybrid renewable systems (Ma et al. 2018). Optimization of different performance indicators such as net present value, fuel cost, operation cost, and cost of energy for a standalone hybrid PV- wind- battery- diesel power system has been reported in Masrur et al. (2880), Sethi 2020). The authors in Alharthi et al. (2019), Chong et al. 2011) have studied performance evaluation of hybrid wind- PV systems connected to the utility grid considering several economic and environmental factors. A new planning approach for optimal sizing of standalone PV- wind hybrid system has been proposed in Belmili et al. (2014) from an economic and reliability point of view. Sizing of hybrid renewable generation systems such as wind and solar PV always needs to be done considering maximum utilization of their capacity (Celik 2003). In Islam (2018), the authors are focused on developing an efficient model of the renewable energy system to minimize the consumption of electricity from the grid by producing maximum electric power for a large office building. On the other hand, accurate mathematical modeling and optimal sizing of standalone solar PV-wind-biogas energy systems are determined considering economic and reliability issues for a large building in Mudgal et al. (2019). The study reported in Aguilar-Jiménez et al. (2018) shows that the hybrid combination of solar PV and concentrated solar power (CSP) plants perform economically better compared to the solar PV-battery combination for large-scale isolated microgrids. A new planning framework has been provided in Blechinger et al. (2016) to assess the techno-economic benefit of solar PV, wind, and battery systems over conventional generating units on small islands. A methodology to determine optimal sizing of the hybrid system consisting of solar PV, wind, biogas, and fuel cell for both on-grid and off-grid operation for small villages has been discussed in Rad et al. (2020). A comprehensive analysis of the utilization of solar PV- diesel and battery systems from a socio-techno and environmental point of view have been provided in Shaahid and Elhadidy (2008). A multi-objective planning framework to optimize the hybrid renewable energy systems considering the socio-techno-economic issue has been developed in Sawle et al. (2018).

Many studies have been carried out related to the planning and techno-economic aspect of hybrid power plants. Large-scale hybrid offshore power plants are an emerging area of research, but comprehensive study on layout planning issues is missing in this domain. Against this backdrop, in this chapter, a comprehensive planning methodology has been proposed to optimize the layout of offshore wind-solar PV power plants to maximize the generation from renewable resources.

26.5 Problem Description and Overall Proposed Approach

Proper planning of any project is a prerequisite for its efficient operation. Planning of off- shore hybrid wind-solar PV power plants can be divided into various categories like layout optimization, sizing of electrical components, techno-economic performance evaluation, etc. In this chapter, the optimal layout design of a hybrid offshore wind-solar PV plant has been carried out.

Layout optimization is an important aspect to maximize the power generation of the hybrid offshore wind-solar PV plant. Power generation from WTs is affected by the wake effect losses; if appropriate spacings among the WTs are not maintained during the construction stage, these losses may increase. On the other hand, generation from solar PV panels is reduced due to the shadow effect. Hence, an optimization methodology has been proposed to determine the optimal layout of the power plant considering the wake and tower shadow effect on WTs and solar panels, respectively.

26.5.1 *Requirements for the Optimum Configuration of Offshore Hybrid Wind-Solar PV Power Plants*

Knowledge of the local weather pattern in addition to economic and reliability analysis is very critical for optimal sizing of the hybrid offshore wind-solar PV network. To study the performance analysis, a proper mathematical model of different components is essential. A few key factors for planning and sizing offshore hybrid wind-solar PV power plants have been discussed below.

26.5.1.1 Meteorological Data

The power generation from offshore hybrid wind-solar PV plants is dependent on the climatic conditions of a place. Therefore, weather data of the area is very important for a feasibility study or optimal sizing of the hybrid systems. Analysis of the climatic condition is important before setting up a plant. This data is available mainly at the local weather stations. The significant parameters necessary for such a study are solar radiation, ambient temperature, and wind speed. Wind data is very site-specific, but solar data does not vary much within a few hundred kilometers of a site. The data used can be time-series data or statistically generated probability distribution. The data might not be available in some places for the entire necessary duration of the study; in that case, the statistical methodologies are used to produce the data. If the data is available for a nearby venue, it can be extrapolated by making appropriate changes to the site (Wahab and Essa 1998).

26.5.1.2 Modeling of Photovoltaic System

There are different mathematical models available for modeling solar PV systems. To capture maximum possible solar irradiation, it is essential to consider the effect of the tilt angle of solar panel on the calculation (Lin et al. 2013). Some studies have modeled power output from solar PV panels as a function of short circuit current and open-circuit voltage (Alsayed et al. 2013). Modeling of solar PV output as the function of area and panel efficiency (Solanki et al. 2017) has also been adapted in various studies. To estimate the power output of PV panels, it is important to consider the effect of diffused, reflected, scattered, and incidental radiation on an inclined surface (Mahesh and Sandhu 2015).

26.5.1.3 Modeling of Wind Energy System

Mathematical modeling of the wind energy system includes power curve modeling of WT. The power curve of WT represents maximum power point (MPP) operation at a particular wind speed. According to the available studies, the power curve of

WT has been modeled as linear, quadratic, cubic, or a piecewise linear function with few nodes. On the other hand, studies have considered WT power output in terms of rotor swept area (Mahesh and Sandhu 2015). Manufacturers of WT provide a power output of individual WT at different wind speed; using those data appropriate power curve can be reconstructed using cubic spline regression technique (Paul and Rather 2016). WT output in a WF is greatly influenced by the downstream wake generated by the upstream WTs. Previously, many mathematical models were used to estimate the wake-effect on the WF performance, including Jensen's, Frandsen- Gaussian wake model (Tao et al. 2019) etc. Jensen's wake model being popular among all and used for wake effect estimation (González-Longatt et al. 2012) (Paul and Rather 2016). In a large WF interaction of multiple wakes among different WTs take place, the multiple wake effect can be calculated by geometric analysis.

26.5.2 *The Complete Process*

Different inputs to this optimization problem are given in Fig. 26.1. For better understanding the process of optimization of the layout of an offshore hybrid wind-solar PV power plant, the entire methodology is represented by a flowchart in Fig. 26.2.

26.6 Accessing Meteorological Data

The weather data for the study was taken from the Photovoltaic geographical information system (PVGIS), which provides the solar data, wind data, and temperature of different locations on land. However, the solar data does not vary much for a few hundred kilometers, as can be found in various solar potential maps (Solanki et al. 2017). So, the solar data of a nearby coastal area of the Indian state of Gujarat was used for the analysis. As the wind data varies a lot from onshore to offshore, the data of the nearby onshore location could not be used. So, the wind data of that specific

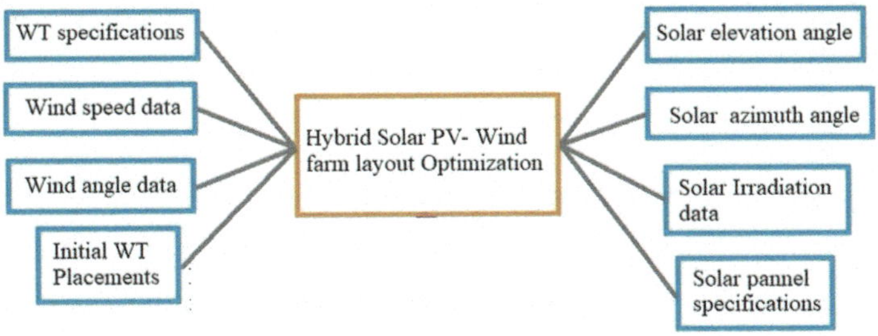

Fig. 26.1 Inputs for the optimization problem

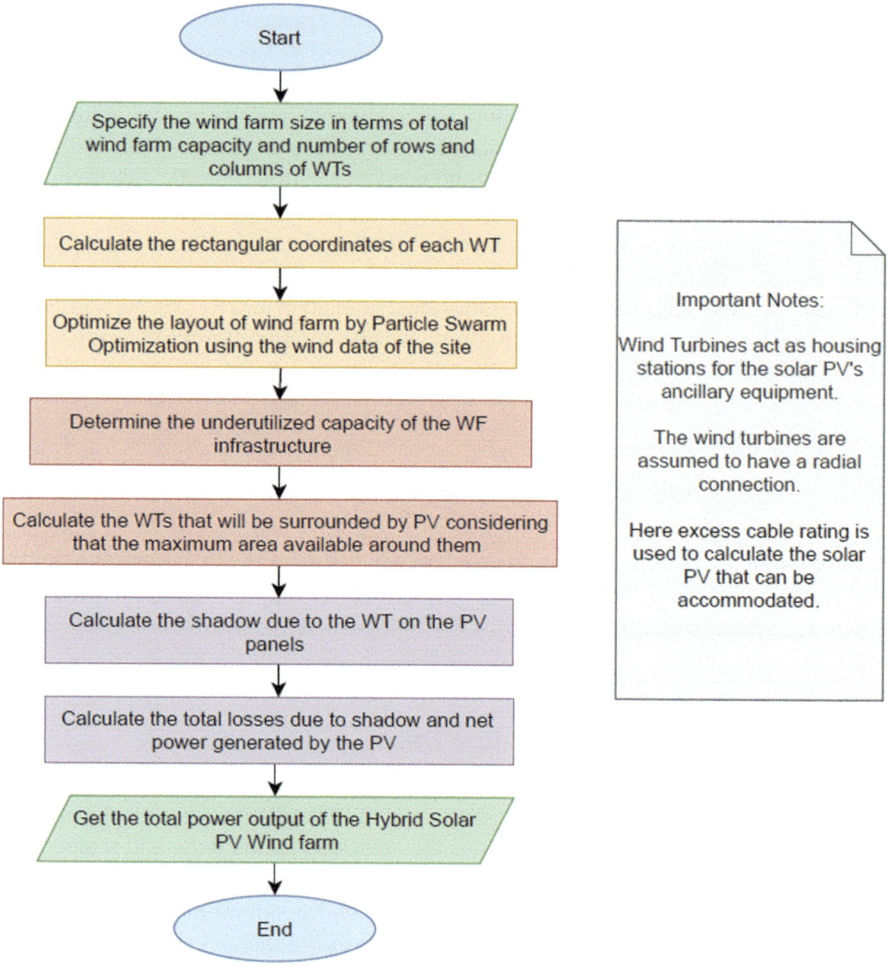

Fig. 26.2 Flowchart of the Entire process

offshore location near the coast of Gujarat was accessed from a study referenced as (COWI 2017).

26.7 Modeling of Wind Energy System

WTs extract energy from the wind to generate electricity; therefore, the wind leaving the turbine has a lower energy content than the wind upstream of the turbine. This effect is known as the wake effect. As the wind flows further downstream, this wake will begin to spread and slowly return to the conditions of free streams. Down-stream

WT is said to be shadowed by the turbine producing the wake if the wake intersects with the swept area of that downstream WT. The two main drawbacks of the wake effect are (i) a reduction in the wind farm's energy production and (ii) an increase in the mechanical loading on downwind turbines due to the turbulence in the wind. In this section, a detailed model of the wake effect has been discussed.

26.7.1 Mathematical Modeling of Wake Effect

In designing a wind farm, it is crucial to consider the wake effect losses in order to maximize the WT's energy output and lifetime. Model development to describe the wake effect began in the 1980s. Since then, a significant number of numerical models have been developed. These are of varying depths and complexities and can generally be classified as either explicit or implicit (González-Longatt et al. 2012).

The explicit or kinematic wake models are the earliest. Some of these models include "Ainslie's model," "Frandsen's model," and "Jensen's model." They are simple and need fewer measurements. However, these methods tend to deliver fairly accurate results despite their relative simplicity. Implicit models of wake were developed as elaborate alternatives to explicit models of the wake effect. They are based either on approximations of the "Navier–Stokes equations" or "vorticity transport equations." The implicit models are generally believed to give the most comprehensive and accurate simulation results. However, these models are also the most complex models, which are not suitable for optimizing layouts (Hu 2016). The wake models commonly used are based on simplified "aerodynamic equations" and "empirical relationships." They simulate a less precise but much faster wake-effect. Moreover, numerous studies show that there is not much difference between the complex and simplistic versions in terms of precision. In this chapter, the wake is calculated using Jensen's model.

26.7.1.1 Jensen's Single Wake Model

A simple single-wake model is the N.O. Jensen wake model. In Ref. González-Longatt et al. (2012) and Molina (2014) the model is well documented. This model is based on the assumption that the diameter of a wake expands linearly. Assuming a linear expansion of the wake, a cone represents the direction taken by the wind that has passed through the turbine blades. This idea is represented in Fig. 26.3.

The wake velocity can be obtained from (26.1) at a distance x from the wind turbine.

$$V_1 = V_0 + V_0\left(\sqrt{\{1 - C_T\}} - 1\right)\left(\frac{r_0}{r}\right)^2 \tag{26.1}$$

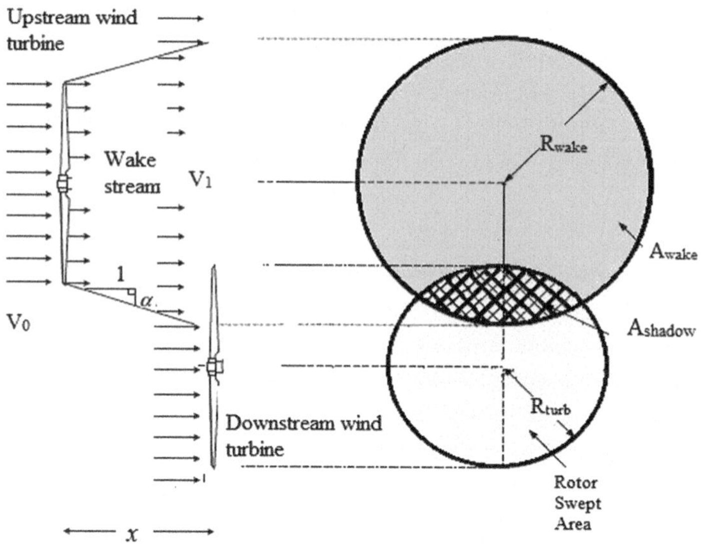

Fig. 26.3 The Jensen's single wake model

where V_1 is the downstream wind velocity, V_0 is the incoming wind velocity, C_T is the coefficient of thrust, r_0 is the radius of rotor blades, and r is the radius of the wake at distance x behind the turbine. As a function of incoming wind speed, Eq. 26.1 provides the wake speed at the downwind spot where r can be set to (26.2).

$$r = r_0 + \alpha X. \tag{26.2}$$

The dimensionless scalar α, known as the coefficient of wake expansion, determines how rapidly the wake expands over space. Its value is deemed to be 0.08.

26.7.1.2 Multiple Wake Model

In a WF, where a large number of WTs are placed, wake created by multiple WTs interacts, as shown in Fig. 26.4. The results of all the different wakes are combined into a single effect to measure the multiple wake interaction within a WF. The wake shadow created by upstream WTs measures the degree of overlap between the area spanned by the wake shadow cone (A_{shadow}) and the shadowed downstream turbine's swept area (A_0), an example of wake shadowing is shown in Fig. 26.3. There are four possible distinct wake shadows, namely: "complete shadowing," "quasi-complete shadowing," "partial shadowing," and "no shadowing." Provided that all wind turbines have the same diameter ($2r_0$), the shadowed turbine area can be determined using (26.3), (Paul and Rather 2016).

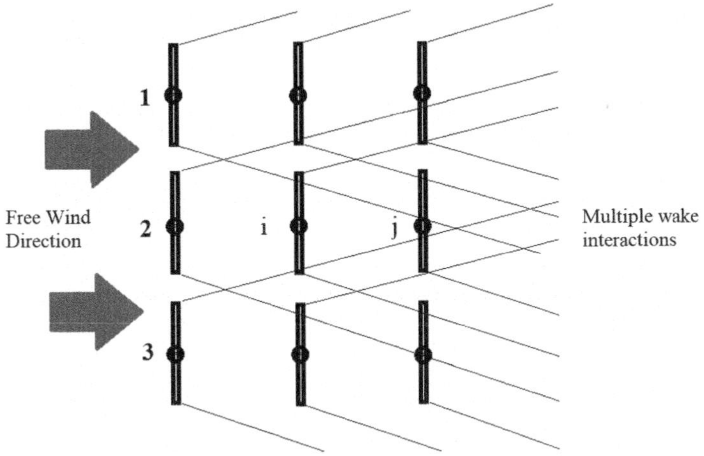

Fig. 26.4 The multiple wake model

$$V_m = \sqrt{\left\{V_0^2 + \sum_{j=1}^{N} \left\{V_{jm}^2 - V_0^2\left(\frac{A_{shadow}}{A_0}\right)\right\}\right\}} \quad (26.3)$$

where V_m is the wind velocity faced by the m-th turbine, j is the turbine number, A_0 is the circular area covered by rotation of blades, A_{shadow} is the blade area shaded by the wake, V_{jm} is the velocity of m-th turbine when j-th turbine is producing the wake, and N is the total number of turbines.

26.7.2 Inclusion of Wind Direction

The wake impact depends on the distance between the turbines in a WF, and as the wind direction changes, so does the interspacing distance between the WTs as seen by the observer sitting in the direction of wind (Paul and Rather 2018). Hence, wind direction influences the WF's power production. To take wind direction into account coordinate transformation-based method is used (Paul and Rather 2016). Coordinates of each WT are first determined in Cartesian coordinate for 0° wind direction. If initially the coordinates are considered to be (x_i, y_i), then the transformed coordinates represented by (x_f, y_f) can be calculated using (26.4).

$$\begin{bmatrix} x_f \\ y_f \end{bmatrix} = T \begin{bmatrix} x_i \\ y_i \end{bmatrix} \quad (26.4)$$

where T is the transformation matrix given by (26.5).

$$T = \begin{bmatrix} cos\theta & sin\theta \\ -sin\theta & cos\theta \end{bmatrix} \quad (26.5)$$

where θ is the "direction of the wind with respect to the X-axis taken anticlockwise," as shown in Fig. 26.5. The figure shows the layout of transformed coordinates (solid line) of an example WF with 9 WTs as seen by the observer placed in the wind direction when the wind flows at an angle of 0°, 45°, 135°, 225°, and 315°.

Thus, for every wind direction, the wake pattern hence the wake loss, changes. The wind directions have a significant impact on WF performance, as changes in wind direction change the wake interaction (overlapping area) and consequently their effects on the power output of each individual turbine.

26.8 Wind Farm Layout Optimization Using Particle Swarm Optimization (PSO)

26.8.1 Importance of Layout Optimization

Layout optimization of the hybrid offshore wind-solar PV plant is a critical factor in maximizing power generation. Power generation from WTs is affected if appropriate spacing among the WTs is not maintained during the construction stage. On the other hand, generation from solar PV panels is reduced due to the shadow effect. This chapter proposes an optimization technique to determine the optimal layout of hybrid offshore wind-solar PV plants, considering the wake and tower shadow effect on WTs and solar panels, respectively. It has been assumed that the foundation of solar panels is surrounded, considering the WT foundation as the center point.

This chapter will discuss the approach of optimal placement of WTs and solar PV (the solar PV panels are placed around WTs) in an offshore location, which will be determined considering the wake effect of wind and tower shadow effect on solar panels. The aim is to maximize the generation from the hybrid power plant. The major goals of the study presented in this chapter are summarized below,

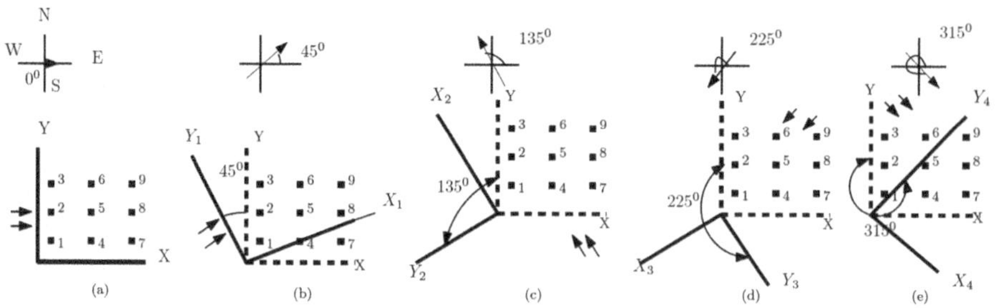

Fig. 26.5 Coordinate transformation approach for different wind directions (Paul and Rather 2016)

- Determination of the optimized layout of the hybrid wind-solar PV plant.
- Mathematical modeling of the wake effect on WT and WT shadow effect on solar PV panels.

26.8.2 Particle Swarm Optimization

A stochastic optimization technique introduced by Eberhart and Kennedy in 1995, the Particle Swarm Optimization (PSO) algorithm imitates the social behavior of animals, including insects, herds, birds, and fish. These swarms stick to a collaborative way of finding food, and each participant in the swarm adjusts the search pattern based on the learning experiences of their own and other nearby participants. The fundamental idea of the algorithm can be explained as follows. Individual participant in the swarm is represented by a point in the Cartesian coordinate system with a randomly assigned initial velocity and position. Every participant referred to as a particle is a potential solution of the optimized problem in an N-dimensional search space. All particles memorize the best position of the swarm and that of its own, along with its velocity. Particles update their locations and velocities according to the change in the environment so that it meets the proximity and consistency specifications. Particle information is merged in each generation to change the velocity of each dimension and is used to measure the new direction of the particle. The algorithm operates so that all particles achieve the same velocity at the optimal point and move in the same direction. Particles in the multi-dimensional search space continuously change their states before they attain stability or optimal state. Through the objective functions, specific connections between various dimensions of the problem space are introduced.

So PSO is a computational approach that optimizes a problem by iteratively attempting to improve a candidate solution on a given quality measure. It finds the optimal solution by having a population of candidate solutions (particles) and moving these particles over the position and velocity of the particle in the search space according to simple mathematical formulae, given in (26.6) and (26.7).

$$v_i^{q+1} = wv_i^q + \psi_1 rand_1\left(P_{\{best,i\}}^q - Y_i^q\right) + \psi_2 rand_2\left(G_{\{best\}}^q - Y_i^q\right) \tag{26.6}$$

$$Y_i^{q+1} = Y_i^q + v_i^{q+1} \tag{26.7}$$

where w is the inertial weight, ψ_1 and ψ_2. are cognitive and social learning constants. The best position is defined by $G^q{}_{best}$ and $P^q{}_{best.i}$ is the best position up to the qth iteration for the ith particle. The $rand_1$ and $rand_2$ produce random numbers. The movement of each particle is determined by its best-known local location (local best) but is also directed toward the best-known search space positions (global best), which are modified as other particles find better positions. This is supposed to move swarm toward the optimal solutions (Wang et al. 2018).

Analyzing the velocity update formula, we can see that the first component is the effect of the previous velocity of the particle. This suggests that the particle has confidence in its current state and conducts inertial movement according to its present velocity, so the parameter is called inertia weight. The second component, called the "cognitive" factor, depends on the distance between the current location of the particle and its own ideal position. It implies the cognition of the particle itself, i.e., the action of the particle arising from its own experience. Therefore, the parameter ψ_1 is named the cognitive learning factor (also termed as cognitive acceleration factor). The third element depends on the distance between the present location of the particle and the optimum global (or local) position in the swarm and is called the 'social' component. It implies the exchange of knowledge and interaction between particles, namely the movement of particles from the experience of other particles in the swarm. This stimulates the flow of good particles through learning, so the ψ_2 parameter is called the element of social learning (also called the social acceleration factor).

Since the birth of the PSO algorithm, it has obtained great interest because of its intuitive context, convenient, and simple implementation, as well as broad adaptability to various types of functions. Both the theory and implementation of the PSO algorithm have made tremendous progress in the past two decades. Researchers had a reasonable grasp of the theory, have carried out its implementation in various domains. A lot of scientific research has proven that this algorithm is an efficient method for optimization. It is possible to outline the benefits as follows: It does not require differential, derivative, and continuous optimized functions; its convergence rate is rapid and very easy to implement (Fig. 26.6).

26.8.3 Implementation of PSO for WF Layout Optimization

In order to implement PSO for this problem, the placement of all WTs in the WF is considered as a single particle. Hence, each particle is embedded with the two-dimensional coordinate position of all WTs, along with their velocity over each axis of the coordinate. Each iteration produces many possible WF layouts (depending on the number of particles) in terms of individual WT coordinates. The net energy generated by the entire WF, which depends on each WT's position, is the objective function considered in this chapter. For a better understanding of the problem, every aspect is further discussed in detail.

26.8.4 Calculation of Objective Function for the Optimization

In this chapter, the power output of the offshore WF is taken as the objective function. The impact of the wake effect has been taken into account in the optimization model. The cumulative influence of multiple wakes created by upstream WTs to the

Fig. 26.6 Flowchart of PSO

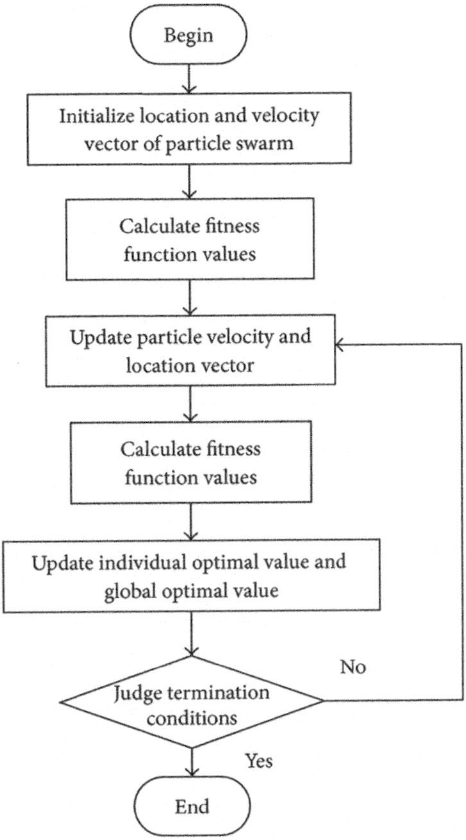

downstream WTs, and the effect of wind direction has been considered to calculate the wake effect.

26.8.4.1 Calculation of Power Output of a Wind Farm

The power generated at hub height by a wind turbine is a function of the wind speed. For any WT, this relation is generally given by the power curve, which represents MPP operation at any particular wind speed. The manufacturer provides the WT power output for different wind speeds. These data can be interpolated to reconstruct the power curve that can be used to calculate power output from WTs. In this chapter, piece-wise cubic Hermite interpolation polynomial (PCHIP) and the result is shown in Fig. 26.7.

Along with the power curve, the probability density function (PDF) of the yearly wind speed of the site at the hub height (denoted as frequency (U)) is also required

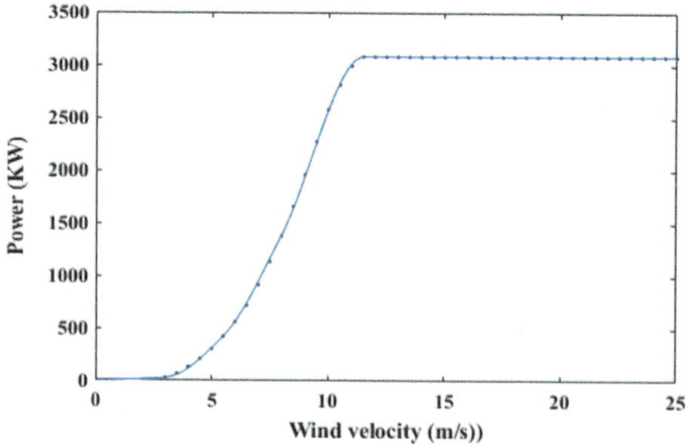

Fig. 26.7 Power curve of Vestas V90 WT after reconstruction

to be calculated. The PDF is represented by the Weibull distribution. The histogram of the wind speed distribution is shown in Fig. 26.8.

Finally, a wind turbine's average power output can be determined by (26.8),

$$P_{turbine} = \int_{u_{ci}}^{u_{co}} P(u) freq(u) du \tag{26.8}$$

where $P_{turbine}$ is the "power output of the turbine", *freq(u)* is the "frequency of w" $P(u)$is the" power output for the velocity u", u_{ci} and u_{co} are the "cut-in wind speed" and "cut-out wind speed", respectively.

26.8.4.2 Average Power Output and Annual Energy Yield of a Wind Farm Considering Wind Direction

The annual energy yield of the entire WF is the summation of the annual power output of individual WT (Fig. 26.8). Different wind speeds and wind directions will cause different wake scenarios in the WF for a WF of a given layout. The wind speed distribution and direction for the site are shown using the wind rose in Fig. 26.9.

Figure 26.9 shows two-dimensional wind rose to depict the probability distribution of the wind speed for different wind directions for an offshore location near the coast of the Indian state of Gujarat. The wind speed shown in the wind rose is the undisturbed wind speed at the height of 10 m above sea level. The relation between height and the wind speed can be represented approximately by (26.9),

$$u_{h2}/u_{h1} = (h2/h1)^a \tag{26.9}$$

where $h2$, $h1$ are the heights above sea level, $u_{h2,\ h1}$ are the wind velocities at those heights respectively, and a is a constant with the value of 0.1 (for calm sea surface).

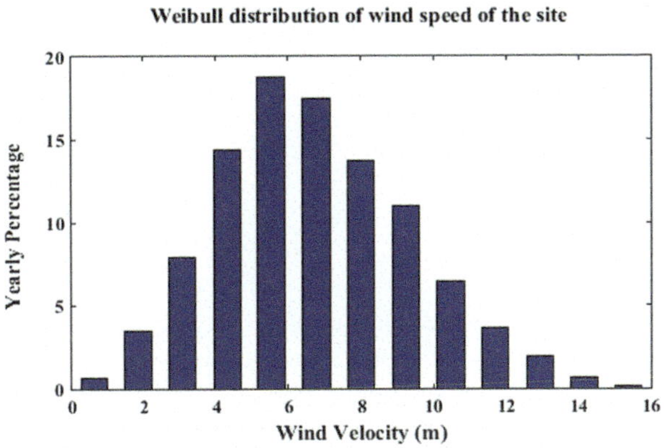

Fig. 26.8 Histogram of the wind speed data

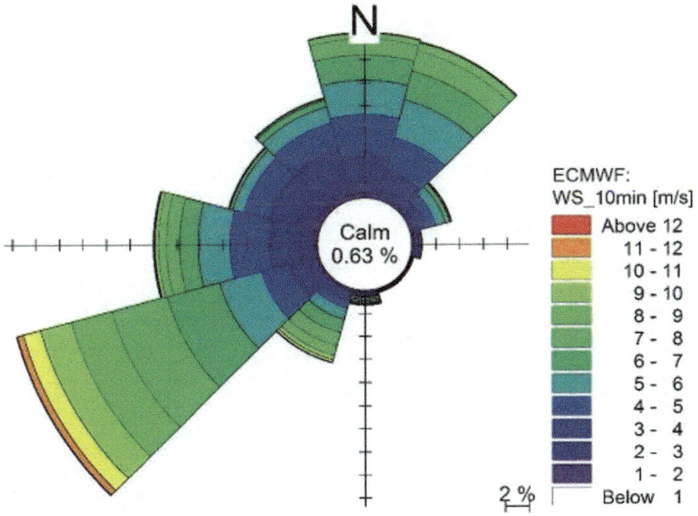

Fig. 26.9 Rose plot of wind speed (at the height of 10 m above sea level) (COWI 2017)

The average power output and annual energy production is given by (26.10),

$$E_{total} = \sum_{\theta=0^0}^{360^0} \left[\sum_{i=1}^{N} \left(\int_0^{u_{co}} P(u,\theta) freq(u,\theta) du \right) \cdot T \right] \tag{26.10}$$

where E_{total} is the energy output for a year, $freq\,(u, \theta)$ is the frequency of wind velocity (u) along wind direction θ, N is the total number of wind turbines, $P\ u, \theta$ is the power output for the velocity u and direction θ, T is the number of hours per year and u_{co} is the cut-out wind speed. Capacity factor (CF) of the farm is given by (26.11),

$$CF = \frac{E_{total}}{P_r.T} \tag{26.11}$$

where P_r is the rating of the WF.

26.8.4.3 Constraints

The constraint for this study is the minimum distance requirement between any two successive WTs. The distance between the two turbines should not be less than four rotor diameters (RD) as the rule of thumb to avoid erosion of turbine blades caused by wake turbulence. The net area in which the turbines are allowed to move is shown by the boxes as given in Fig. 26.10 (Paul and Rather 2019).

26.9 Mathematical Modeling of Solar PV System

The photovoltaic system has been modeled in PVGIS. A study related to the contrast between photovoltaic systems on land and at sea shows that the relative annual average energy production is around 12.96% higher at sea compared to land, and in some months, it increases up to 18% (Snieckus 2019). Several studies show that the effect of ocean waves changing the optimum tilt angle of solar PV panels doesn't affect the output much; in fact, the increased solar irradiation and less temperature increase the output. The Energy output of PV (per day) is given by (26.12),

$$E_{day} = GHI\,Area\;n_1 n_2 \tag{26.12}$$

$$Area = \frac{P_{rated}}{n_1}$$

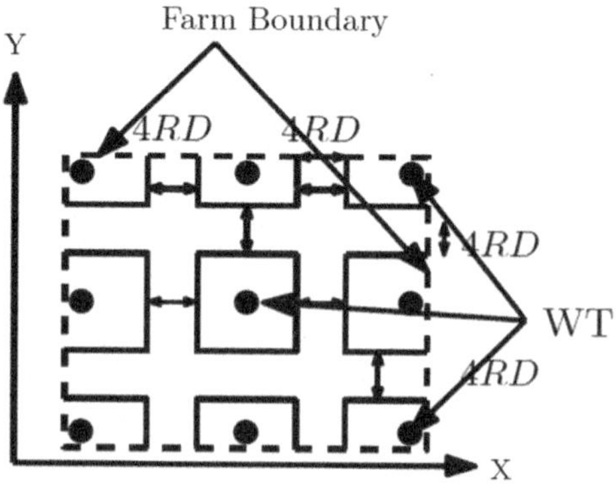

Fig. 26.10 An example of area constraint for the optimization problem (Paul and Rather 2019)

where E_{day} is the energy output of PV (per day), $GH\ I$ is the Global Horizontal Irradiance,

η_1 is the panel efficiency, and η_2 is the Performance ratio and P_{rated} is the nominal power or peak power, which is the power claimed by the manufacturer of the module. It is the power output of the module(s) measured at Standard Test Conditions (STC). The capacity factor of the solar PV is given by (26.13),

$$CF = \frac{E_{total}}{P_{rated} 24.365} \tag{26.13}$$

26.9.1 Mathematical Modeling of Tower Shadow Effect on Solar PV Panels

WTs exert shadow on solar panels placed around them and cause a decrease in power output. In order to calculate the losses due to the shadow of the WT, it is first required to determine the area of shadow around a single turbine. The elevation and azimuth angle of the sun is taken from University of Oregon (2020) and the method for estimating the shading on PV by using sun path charts is given in Vignola (2004). The WTs are placed far apart, so the shadow of one turbine does not fall on PV surrounding the other WTs. Figure 26.11 shows the solar elevation path throughout a year for a particular site. The shadow is calculated using the solar elevation angle and azimuth angle of an hour, and this shaded area is considered as the average shaded area for that hour. To calculate the shadow, basic trigonometry is used, and the intersecting area is calculated. A few assumptions to calculate the shadow area are provided below,

- Solar panels are fixed around the wind turbine in the form of a circle.
- Tower has a uniform width from top to bottom, and shadow has the same width as the tower.
- The swept area of the blade as a solid disc.
- The sun is always behind the rotating blades (never sideways).
- Output of the shaded panels is decreased by 80%.
- Panels have no inclination.

The shadow of the tower is divided into two parts:

(a) Rectangular shadow (shadow area of the tower)
(b) Elliptical shadow (shadow area of the blades)

Here, the tower is referred to as the actual tower length minus the blade length. The shadow length throughout the day presents us with three main scenarios as shown in Fig. 26.12. First, the total length, i.e., the maximum length of the tower shadow with the blades, is less than the radius of the circle in which PV panels are placed as shown in Fig. 26.12a. In this case, the area of the shadow is calculated

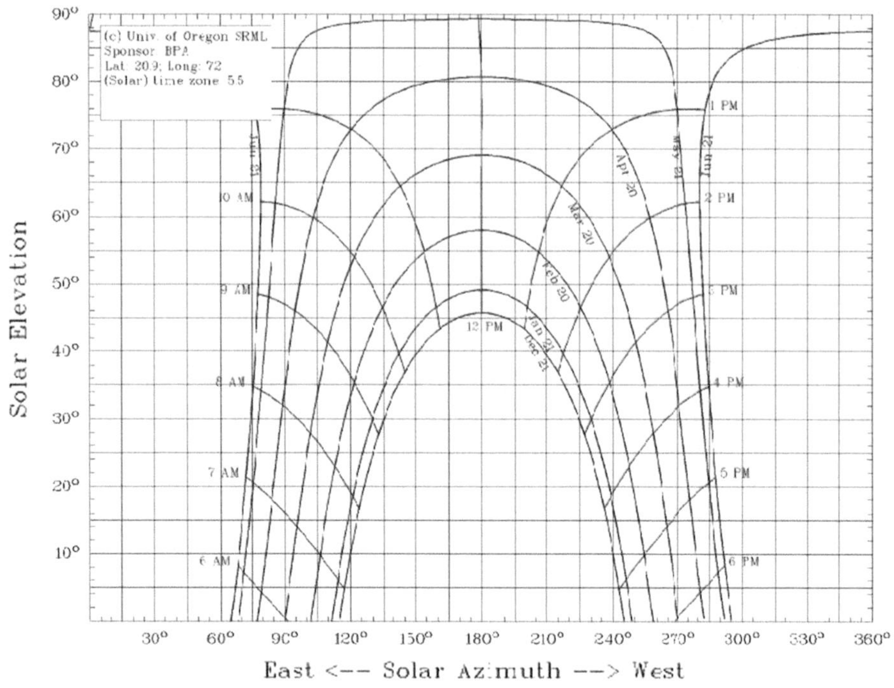

Fig. 26.11 Sun Path chart for the coastal area of Gujarat (University of Oregon 2020)

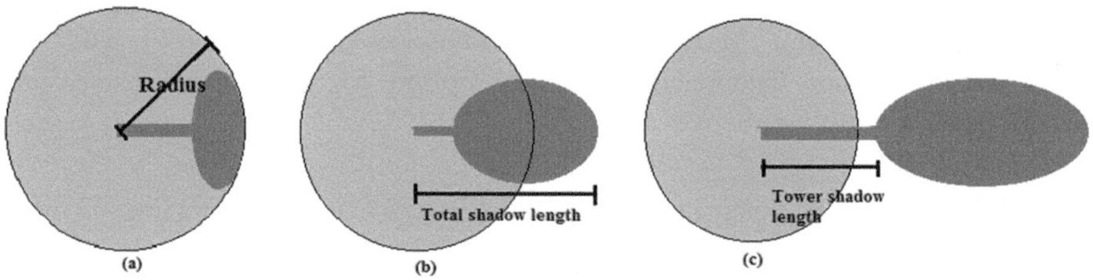

Fig. 26.12 Shadow of the WT at different times of the day

considering the area of the elliptical shadow (for WT blades) plus the rectangular shadow (WT tower). Second, when the total length of the shadow is greater than the radius of the PV panel circle, but the length of the tower shadow is less than the radius of the circle, as shown in Fig. 26.12b. In this case area of shadow consists of the complete rectangle and partial ellipse to be calculated. Third, when the length of the tower shadow is greater than the radius of the circle in which the PV panels are placed, as shown in Fig. 26.12c. In this scenario, the shaded area will be the partial rectangular area of the tower only. To calculate the partial elliptical shadow, the points of intersection of ellipse and circle are calculated first (Fig. 26.14), then the intersected area is calculated.

26.9.2 Calculation of Shadow

In Fig. 26.13, a pole of length u in the y–z plane inclined at an angle α and angled at an angle θ is shown. The shadow length P is "co," Px and Py are given in (26.14) (Groumpos and Khouzam 1988).

$$P_x = u.\cos\alpha.sin\theta + u.sinZ.cot\beta.sin\alpha \tag{26.14}$$

$$P_y = u.cos\alpha.cot\theta + u.cosZ.cot\beta.sin\alpha$$

where Z is the azimuth angle of the sun and β is the elevation angle of the sun. For calculating the area of a complete ellipse, we consider one axis of the ellipse (axis parallel to the surface of the sea) to be of the fixed length and equal to the rotor diameter. The perpendicular axis changes the length with the elevation of the sun. The total area of the ellipse and the length of its axis is calculated using (26.15),

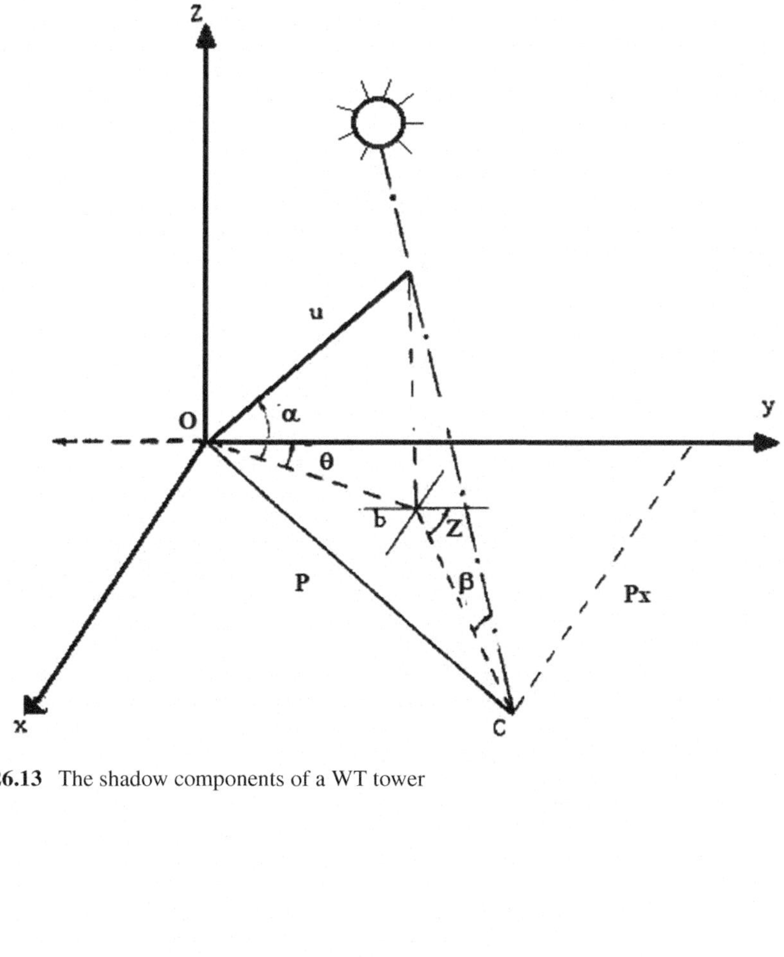

Fig. 26.13 The shadow components of a WT tower

$$S_x = RD/tan\beta$$

$$S_y = RD \tag{26.15}$$

$$Area = \frac{\pi}{4} S_x S_y$$

RD is the rotor diameter, S_x, S_y is the axis of elliptical shadow along the x and y-axis. For calculating the shadow area when part of the blade shadow shades the PV panels, we first calculate the points of intersection between the circle and ellipse by solving their equations. Equation of circle is given by:

$$C_x^2 + C_y^2 = radius^2 \tag{26.16}$$

Equation of ellipse is given by:

$$x_0 = h_{hub}/tan\beta$$

$$\frac{(x - x_0)^2}{S_x^2} + \frac{y^2}{S_y^2} = 1. \tag{26.17}$$

solving the above two equations give the points of intersection. Using the point in the first quadrant of Fig. 26.14 (x_1, y_1), we find the area under ellipse, integrate the ellipse from $x = x_{min}$ to the point of intersection (x_1) and integrate the circle from x_1 to *radius*.

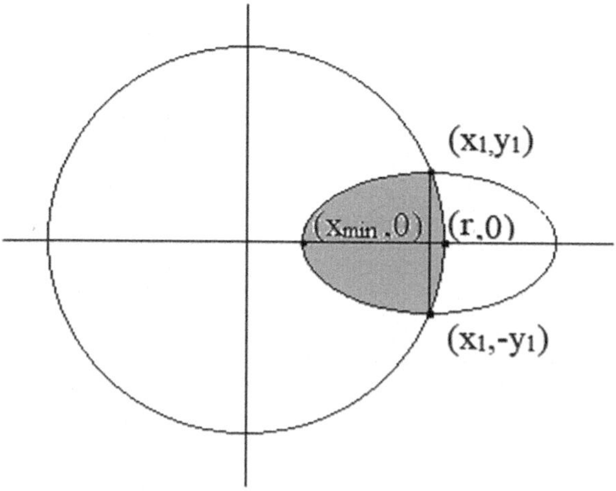

Fig. 26.14 Diagram for calculation of shadow area

26.10 Application of the Methodology for a Practical Site and Observations

When an offshore WF is set up, sometimes a margin of the capacity of its electrical infrastructure remains un-utilized (keeping in mind partially possible future expansion), such as the capacity of cable, transformer, and other associated electrical components. Some portion of these unused capacities can be utilized without losing operational stability to generate electricity by adding up solar PV panels.

In this chapter, 80% of the excess capacity of the cable has been utilized to determine the penetration of the solar PV that could be set up. An offshore hybrid plant of 230 MW wind capacity and about 13 MW additional solar (which is the maximum capacity of solar that can be set up using the un-utilized electrical infrastructure) is considered, and the meteorological data of a site near the coast of Gujarat (Gulf of Khambhat) in India is used for the calculations. From the Fig. 26.15, it is evident that the wind and solar resources are complementary, and hence the seasonal variation in outputs can be reduced by their combined utilization. The net capacity factor of the offshore hybrid plant after optimization comes out to be 32%.

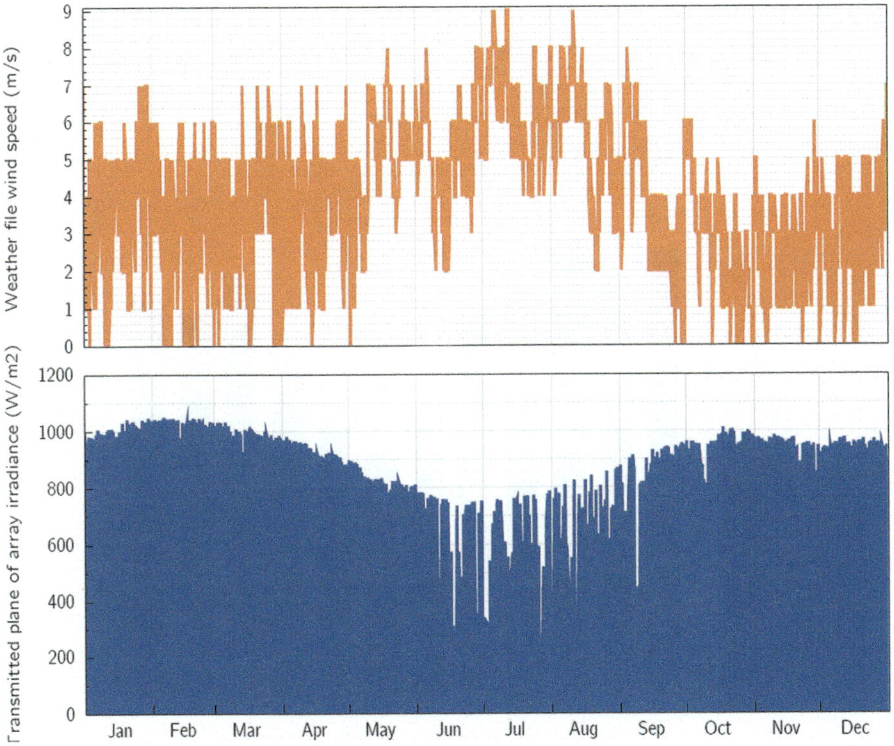

Fig. 26.15 Seasonal variation of wind speed and solar irradiance data for the Gujarat site

26.10.1 Wind Farm Optimization Using PSO

WF having eight rows and five columns has been considered for this study, and the results after optimization are given in Fig. 26.17. The green dots in Fig. 26.17 represent the position of WT for different iteration counts of PSO. The black dots represent the final position of WTs (Fig. 26.16).

after optimization. It can be seen that WTs settle at the corners of the individual rectangular area in which they are allowed to move. The convergence of the PSO algorithm is shown in (Fig. 26.18) (for better clarity, the convergence of three particles is shown in Fig. 26.18).

26.10.2 Shadow Analysis

The average solar irradiance for the Gujarat site is given in Fig. 26.16. The table shows the result of the shadow analysis. Here, we are considering the area surrounded by PV to be the maximum possible area taking into account the 4RD constraint as discussed in Sect. 10.3.3. The average loss of energy in the PV system due to shading by the WT is around 3% (Figs. 26.17, 26.18 and Table 26.1).

An average of about 5.418 MWh is lost to shading in one day if shaded panels are considered to be producing no output. If we consider panels generating 20% of the output in un-shaded conditions, the average loss in a day is 4.33 MWh. 151.8 MWh is the total energy output of the 46,000 panels throughout the day, considering

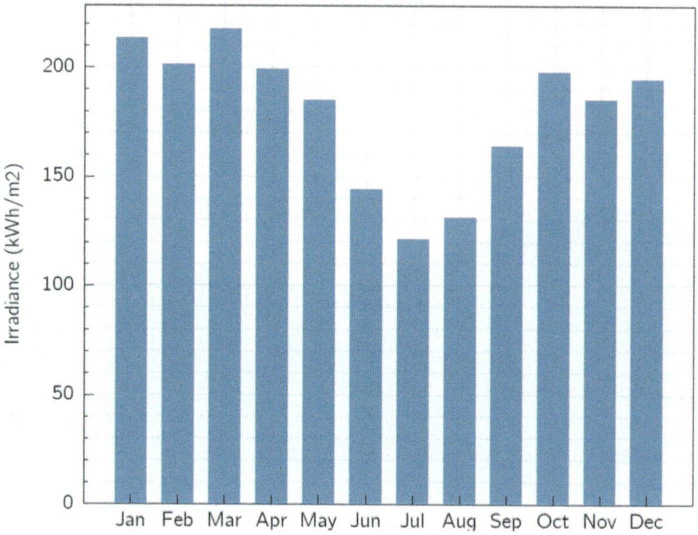

Fig. 26.16 Average monthly irradiance for the site of Gujarat

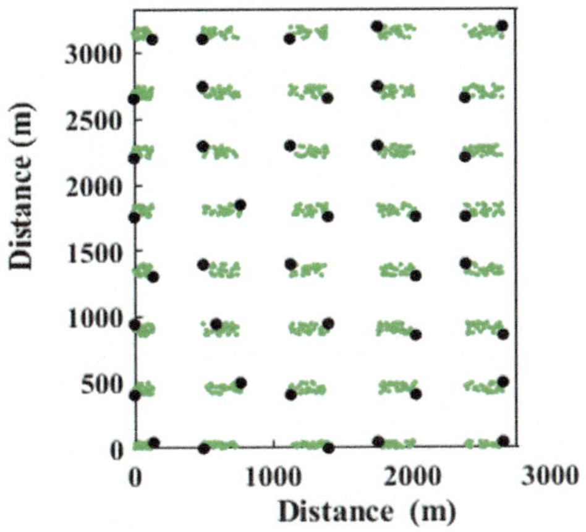

Fig. 26.17 Offshore wind farm layout after optimization

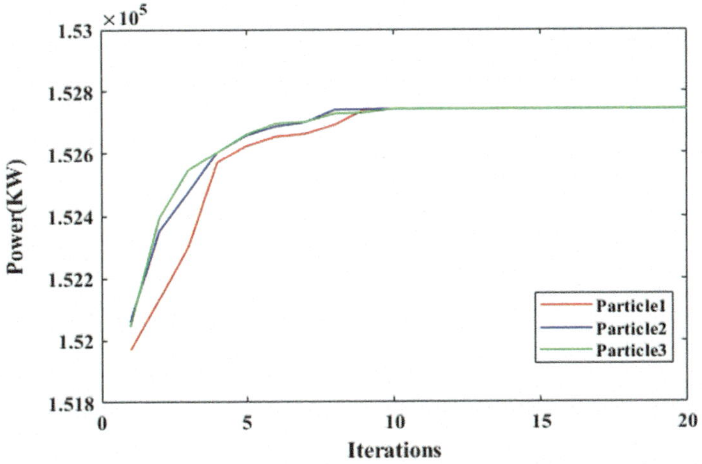

Fig. 26.18 Plot of power output of the WF with every iteration

STC for the 11 h of sunlight. The hybrid offshore wind-solar PV plants are especially beneficial for countries having relatively lower wind velocity and higher solar irradiance. As solar is far cheaper than wind, often the Levelized cost of energy of offshore wind is higher than a hybrid offshore wind-solar PV plant.

Table 26.1 Number of shaded panels at different times of the day

Time (Hour)	7	8	9	10	11	12	13	14	15	16	17
No. of shaded panels	380	2481	3309	2165	1390	0	1390	2165	3309	2481	380

26.11 Conclusions

This chapter discusses the layout planning of offshore hybrid wind-solar PV power plants. In a region with lesser wind speed and higher solar irradiance, wind and solar together improve the efficacy of the plant. The layout optimization further enables this better utilization of the renewable resource as the power output increases. The presented planning framework is based on parameterized models of various equipment and renewable resource profiles. In this chapter, different phenomenon such as the wake effect and tower shadow loss which impact the overall generation from the offshore hybrid plant, have been model mathematically. Since the cost of the generated electricity is the deciding factor, the planning framework discussed can be customized to compare the cost of generation from wind and wind-solar hybrid plants for any site. In the future, the presented framework could be extended to determine the optimal share of wind and solar PV to obtain a minimum Levelized cost of energy.

References

Aguilar-Jiménez JA, Velázquez N, Acuña A, Cota R, González E, González L, López R, Islas S (2018) Techno-economic analysis of a hybrid pv-csp system with thermal energy storage applied to isolated microgrids. Sol Energy 174:55–65

Alharthi YZ, Siddiki MK, Chaudhry GM (2019) Techno-economic analysis of hybrid pv/wind system connected to utility grid. In: 2019 IEEE texas power and energy conference (TPEC). IEEE, pp 1–6

Alsayed M, Cacciato M, Scarcella G, Scelba G (2013) Multicriteria optimal sizing of photovoltaic-wind turbine grid connected systems. IEEE Trans Energy Convers 28(2):370–379

Belmili H, Haddadi M, Bacha S, Almi MF, Bendib B (2014) Sizing stand-alone photovoltaic–wind hybrid system: Techno-economic analysis and optimization. Renew Sustain Energy Rev 30:821–832

Blechinger P, Cader C, Bertheau P, Huyskens H, Seguin R, Breyer C (2016) Global analysis of the techno-economic potential of renewable energy hybrid systems on small Islands. Energy Policy 98:674–687

Celik AN (2003) Techno-economic analysis of autonomous pv-wind hybrid energy systems using different sizing methods. Energy Convers Manag 44(12):1951–1968

Chong WT, Naghavi MS, Poh SC, Mahlia TMI, Pan KC (2011) Techno-economic analysis of a wind–solar hybrid renewable energy system with rainwater collection feature for urban high-rise application. Appl Energy 88(11):4067–4077

COWI (2017) Metocean data reqirements - Offshore Wind Foundations. First offshore wind project of India –metocean study. https://www.cecpeu.in/uploads/documents/fowpi/1_fowpi_metocean_study_87386800442

González-Longatt F, Wall P, Terzija V (2012) Wake effect in wind farm performance: steady-state and dynamic behavior. Renew Energy 39(1):329–338
Groumpos PP, Khouzam K (1988) A generic approach to the shadow effect of large solar power systems. Solar Cells 22(1):29–46
Hu B (2016) Design of a simple wake model for the wind farm layout optimization considering the wake meandering effect. Master Thesis, TU Delft
India CEA (2020) All India installed capacity of utility power stations. https://cea.nic.in/wp−content/uploads/installed/2020/11/installed_capacity
Islam MS (2018) A techno-economic feasibility analysis of hybrid renewable energy supply options for a grid-connected large office building in southeastern part of France. Sustain Cities Soc 38:492–508
Ma W, Xue X, Liu G (2018) Techno-economic evaluation for hybrid renewable energy system: application and merits. Energy 159:385–409
Manwell JF, McGowan JG, Rogers AL (2010) Wind energy explained: theory, design and application. Wiley
Masrur H, Howlader HO, Elsayed Lotfy M, Khan KR, Guerrero JM, Senjyu T (2020) Analysis of techno-economic-environmental suitability of an isolated microgrid system located in a remote island of Bangladesh. Sustainability 12(7):2880
Mahesh A, Sandhu KS (205) Hybrid wind/photovoltaic energy system developments: critical review and findings. Renew Sustain Energy Rev 52:1135–1147, 2015
Molina DL (2014) Modeling of wakes behind wind turbines. PhD thesis, Valencia Polytechnic University
Mudgal V, Reddy KS, Mallick TK (2019) Techno-economic analysis of standalone solar photovoltaic-wind-biogas hybrid renewable energy system for community energy requirement. Futur Cities Environ 5(1)
Paul S, Rather ZH (2016) A new approach for selection of a suitable wind turbine for a wind farm. In 2016 IEEE international conference on power electronics, drives and energy systems (PEDES), IEEE, pp 1–6
Paul S, Rather ZH (2018) A pragmatic approach for selecting a suitable wind turbine for a wind farm considering different metrics. IEEE Trans Sustain Energy 9(4):1648–1658
Paul S, Rather ZH (2019) A new bi-level planning approach to find economic and reliable layout for large-scale wind farm. IEEE Syst J 13(3):3080–3090
Perlin J (1999) From space to earth: the story of solar electricity. Earthscan
Rad MA, Ghasempour R, Rahdan P, Mousavi S, Arastounia M (2020) Techno-economic analysis of a hybrid power system based on the cost-effective hydrogen production method for rural electrification, a case study in Iran. Energy 190:116421
Rapier R (2020) Renewable energy growth continues at a blistering pace. Forbes Mag. https://indianexpress.com/article/india/india-news-india/here-are-indias-indc-objectives-and-how-much-it-will-cost/
Sahu A, Yadav N, Sudhakar K (2016) Floating photovoltaic power plant: a review. Renew Sustain Energy Rev 66:815–824
Sawle Y, Gupta SC, Bohre AK (2018) Socio-techno-economic design of hybrid renewable energy system using optimization techniques. Renew Energy 119:459–472
Sethi R (2020) Techno-economic analysis of a Microgrid with solar PV, Battery Energy Storage and Power to Hydrogen (P2H) System. PhD thesis, UC San Diego
Shaahid SM, Elhadidy MA (2008) Economic analysis of hybrid photovoltaic–diesel–battery power systems for residential loads in hot regions—a step to clean future. Renew Sustain Energy Rev 12(2):488–503
Snieckus D (2019) World's first offshore solar array rides out storm ciara off netherlands. Recharge News. https://www.rechargenews.com/transition/worlds-first-offshore-solar-array-rides-out-storm-ciara-off-netherlands/2-1-757022
Solanki C, Nagababu G, Kachhwaha SS (2017) Assessment of offshore solar energy along the coast of India. Energy Procedia 138:530–535

Tao S, Kuenzel S, Xu Q, Chen Z (2019) Optimal micro-siting of wind turbines in an offshore wind farm using Frandsen–Gaussian wake model. IEEE Trans Power Syst 34(6):4944–4954

The Indian Express (2015) Here are india's indc objectives and how much it will cost. https://indianexpress.com/article/india/india-news-india/here-are-indias-indc-objectives-and-how-much-it-will-cost/

University of Oregon. Solar radiation monitoring laboratory (2020). *h*tt ps://solar dat.uor egon.edu

Vignola F (2004) Using sun path charts to estimate the effect of shading on PV arrays. In: Proceedings of the solar conference. American Solar Energy Society; American Institute of Architects, pp 27–32

Wahab MA, Essa KSM (1998) Extrapolation of solar irradiation measurements: case study over Egypt. Renew Energy 14(1–4):229–239

Wang D, Tan D, Liu L (2018) Particle swarm optimization algorithm: an overview. Soft Comput 22(2):387–408

Xu L, Ruan X, Mao C, Zhang B, Luo Y (2013) An improved optimal sizing method for wind-solar-battery hybrid power system. IEEE Trans Sustain Energy 4(3):774–785

Chapter 27
Analysis of Acoustic Noise and Vibration of PMSM Coupled with DC Generator for Electric Vehicle Applications

Rajesh M. Pindoriya, Rishi K. Thakur, Bharat S. Rajpurohit, and Rajeev Kumar

Abstract In recent years, the market of the brushless Permanent Magnet (PM) motors, such as Permanent Magnet Synchronous Motor (PMSM) and Brushless Direct Current Motor (BLDCM) drives, has become huge due to demand of the Electric Vehicles (EVs) and Hybrid Electric Vehicles (HEVs). However, brushless PM drives are less robust compared to other types of motor drives due to the high acoustic noise, vibrations, and de-magnetization risk of the PM (Chan. Proc IEEE 95:704–718, 2007; Report, Implementing Agreement for Co-operation on Hybrid and Electric Vehicle Technologies and Programmed. International Energy Agency, 2016). These shortcomings pose important restrictions for critical applications. Initially, to run PMSM, Sinusoidal Pulse Width Modulation (SPWM) is implemented. But this technique generates current harmonics and high torque ripples, which ultimately leads to Acoustic Noise and Vibration (ANV) in PMSM drive. Hence, for analysis purpose, a framework based on lumped model along with effective mass and mass participation factor technique for prediction of torsional vibration in case of SPWM technique is elaborated to show detailed methodology for vibration response caused by high torque ripples. This framework is generalized in a way that can be easily extended to any mechanical power transmission system having shaft-coupler or geared system especially for EV and HEV application. Also, vibration prediction modelling is integrated with optimum number of modes or degree of freedom selection technique, which help to enhance the accuracy of model along with computationally efficient, which is the novelty of present work, which usually researchers took earlier randomly based on their setup and mass distribution without any specific technical justification. The vibration analysis reveals high torsional twisting and untwisting of shaft in case

R. M. Pindoriya (✉) · B. S. Rajpurohit
School of Computing and Electrical Engineering, Indian Institute of Technology Mandi, Kamand, Himachal Pradesh, India
e-mail: rajeshpindoriya@gmail.com

B. S. Rajpurohit
e-mail: bsr@iitmandi.ac.in

R. K. Thakur · R. Kumar
School of Engineering, Indian Institute of Technology Mandi, Kamand, Himachal Pradesh, India
e-mail: rajeev@iitmandi.ac.in

A. K. Bohre et al. (eds.), *Planning of Hybrid Renewable Energy Systems, Electric Vehicles and Microgrid*, Energy Systems in Electrical Engineering,
https://doi.org/10.1007/978-981-19-0979-5_27

of SPWM, because of high source torque ripple. Henceforth, a Random Pulse Width Modulation (RPWM) technique for reduction of ANV is discussed in this chapter. The proposed RPWM method brings a significant reduction in torque ripples which directly influence ANV in the motor, thereby enhancing the performance of the complete drive system under operation. The relationships between the stator current harmonics feed by drive and non-sinusoidal magnetic field flux distribution, with torque ripples is developed and detailed analysis is discussed in this chapter. An extensive simulation and experimental work are carried out on a 1.07- kW, 4-poles, 36-slots, 3-phase PMSM drive for validation of proposed control strategy. In the end, experimental validation part is presented for all analytical modelling and simulation results presented in this chapter.

Keywords Acoustic Noise and Vibration (ANV) · Electric Vehicles (EVs) · Hybrid Electric Vehicles (HEVs) · Random Pulse Width Modulation (RPWM) · Permanent Magnet (PM) motors · Resonance · Mode shape · Modal analysis · Orthonormality · Effective modal mass · Mass Participation Factor (MPF)

Nomenclature

ANV	Acoustic Noise and Vibration
EVs	Electric Vehicles
FPGA	Field Programmable Gate Array
FEM	Finite Element Method
HEVs	Hybrid Electric Vehicles
MMF	Magneto-Motive Force
MPF	Mass Participation Factor
PMSM	Permanent Magnet Synchronous Motor
PM	Permanent Magnet
RPWM	Random Pulse Width Modulation
SPWM	Sinusoidal Pulse Width Modulation
VSI	Voltage Source Inverter

27.1 Introduction

Electric Vehicles (EVs) are the solution of energy crises and toxic emissions but its cost, range anxiety, long charging times, limited cargo and slow acceleration limits its usability (Chan 2007). Traction motors such as, Permanent Magnet Synchronous Motor (PMSM) and Brushless Direct Current Motor (BLDCM) drives, reduces these problems to some extent. In recent years, the market of Permanent Magnet (PM) motors, such as (Permanent Magnet Synchronous Motor (PMSM) and Brushless

Direct Current Motor (BLDCM)) drives, has become huge due to demand of the EVs and Hybrid Electric Vehicles (HEVs) (Report 2016). During last decade, design and control of PMSM and BLDCM have been the focus of significant research efforts due to their superior performance in key aspects as high efficiency, high power density, low moment of inertia, and high torque to power ratio compared to other motor types (Dorrell et al. 2011). Superior features have made PMSM and BLDCM inherently suitable for many applications such as automotive, robotics, servo drive system, military, and aerospace applications.

But they have to face mechanical anomalies like Acoustic Noise and Vibration (ANV), de-magnetization risk of the PM (Zou et al. 2017; Bosing 2012). These shortcomings pose important restrictions for critical applications. For safe and better performance, analysis of acoustic noise, vibration, condition monitoring and fault diagnosis are key aspects of being carefully looked for preventive measures for industrial applications. The space harmonics in the airgap flux density is the main cause of ANV which depends on the number of stator slots and type of machine winding of the PMSM. Also, electromagnetic torque ripple and cogging torque are the main cause of ANV, which is produced by the interaction between time and space harmonics inside the airgap of PMSM (EL-Refaie 2010; Besnerais et al. 2009; Knopik and Binder 2011; Chattopadhyay et al. 2014).

Extreme vibration especially at resonance leads to adjoining components under impact loading causes comfortless, reduction in power transmission efficiency, further noise and catastrophic shaft and bearing failures (Huang et al. 2019a*)*. Therefore, need of analysis of design is there which will help in revealing the cause of noise and vibration, which ultimately improve the effectiveness of vibration reduction tactics, hence enhance the reliability and performance of EVs.

Xiaohua et al. (2020) presented an analytical mathematical model of electromagnetic forces and analysis the characteristic parameters of generation of ANV by PMSM drive for EVs applications.

All components under rotational motion are under torsion to some extent due to source power fluctuation especially during starting, stopping and running condition. As Zhang et al. (2020) presented that source of ripple in torque is harmonics present in three-phase voltage source inverter which induces torsion of components and sometimes creates defects or even catastrophic failure of the transmission shaft (Zhang et al. 2020). Manguelle et al. (2018) reported relation between current and voltage harmonics with torque at the output. Similarly, Han et al. (2013) give shaft life prediction using power harmonics. Rabbi et al. presented multi-degrees of freedom lumped model for torsional vibration analysis of shaft of submersible pump during starting and stopping of motor shaft (Rabbi et al. 2020). Shen et al. present vibration analysis of two stage gear system considering shaft crack into consideration to make system reliable (Shen et al. 2020). Hence, to make system power efficient and reliable, it is very important to analyse the torsional vibration and stresses induced due to harmonics in power supply and ripple in source torque. So, for enhancing system reliability, in present chapter:

- A detailed generalized framework based on multi-Degrees of Freedom (DOF) lumped modelling for dynamic torsional vibration response is presented, which can be easily extended for any type of mechanical power transmission shafts, couplers and geared system especially in EVs and HEVs applications.
- The important step associated with modelling is number of DOF to be considered. Less number of DOF leads to inaccuracy and more number of DOF leads to increase in computational cost. The present methodology is integrated with optimum number of DOF or mode selection technique to enhance accuracy along with computational advantage.
- A small-scale experimental setup driven by PMSM which is run by vector control based SPWM technique is developed, for all experimental analysis related to ANV.

Schematic layout of EVs and HEVs is shown in Fig. 27.1. The schematic layout of EVs and HEVs are become combinations of power electronics components, batteries and control logic.

The main objective of this book chapter is to present detailed elaborative study of ANV of PM based electric drive using simulations and experimental results. Time-domain and frequency-domain analysis is performed to show the effects of changing the operating parameters of the power electronic converters on the ANV. The focus of this chapter is to develop a better understanding and to draw an operational concept of ANV produced by PM electric drive from deduced theoretical and experimental investigations. The experimental investigations are performed on a low-cost laboratory setup of PM electric drive and acoustic chamber to validate the theoretical deductions.

The organization of this chapter is as follows. Sources of acoustic noise and vibration in permanent magnet based electric drive are given in Sect. 27.2. SPWM based control strategy and operation characteristics of PM based electric drive is discussed in Sect. 27.3. Small-scale experimental setup driven by PMSM transmitting power to load DC generator which is required for validation of simulation results is explained in Sect. 27.4. Further, a detailed and generalized modelling of

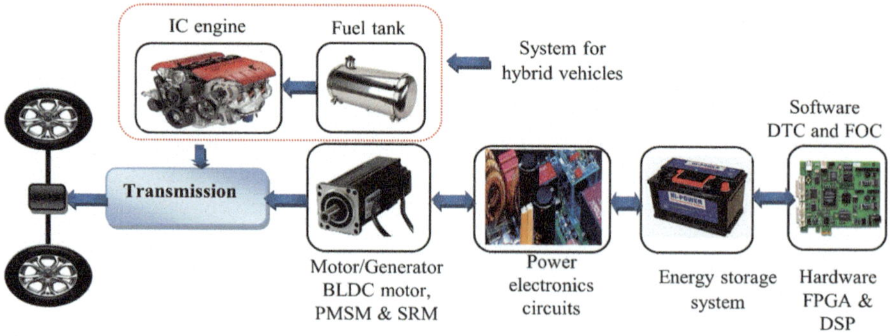

Fig. 27.1 Schematic layout of electric and hybrid electric vehicles

torsional vibration induced in the same experimental setup is presented in Sect. 27.5. Experimental validation of simulation results and key finding of results are discussed in Sect. 27.6. Techniques for reduction of acoustic noise and vibration is given in Sect. 27.7. Conclusion of this work is given in Sect. 27.8.

27.2 Sources of Acoustic Noise and Vibration of PM Based Electric Drive

Electric vehicles have acoustic noise as an audible sound which is unwanted. Vibrations may be supposed to state forward where they are transmitted to the body through. The ANV are key factors to select an electric machine for industrial and commercial applications. The power electronic converters are the heart of an electric drive which operates at very high switching frequency range, during the normal operation of drives. Due to the non-sinusoidal output and high switching frequency of the power electronic converters, it produces time harmonics as well as ANV in the electric drives. Also, vibration due to design parameters and power supply fluctuations causes collision, bending and twisting issues of shafts of EVs which leads to acoustic noise. For rotating electrical machines, ANV has been classified in the following three categories: aerodynamic noise, mechanical noise and electromagnetic noise as given in Fig. 27.2 (Gieras et al. 2006). In Basu et al. (2009), Pindoriya et al. (2018), authors had presented a classification of ANV sources in electrical machines and its communication pathway to airborne sound. The Field Programmable Gate Array (FPGA) based architecture of the hall sensor-based control system for PMSM drive is shown in Fig. 27.3. The control system features the cascaded control principle included as the inner loop current controller and outer loop speed controller with a sinusoidal PWM technique for voltage source inverter-fed PMSM drive.

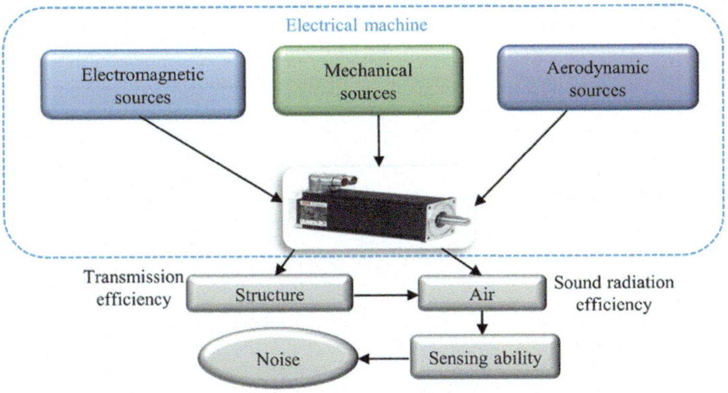

Fig. 27.2 Noise generation and propagation in electrical machines

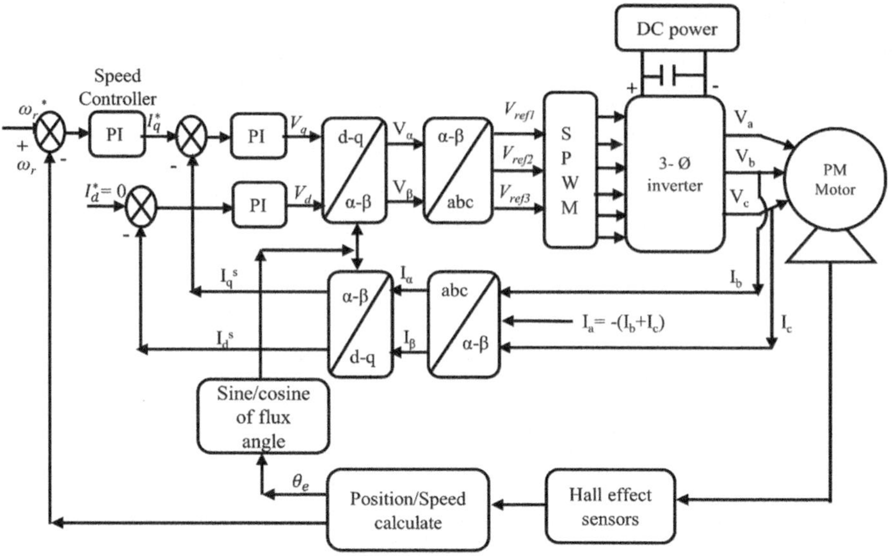

Fig. 27.3 The control block diagram of a hall sensor based PMSM drive

The mechanical structure of the stator and rotor of electric machines and design parameters of power transmission line in Electric Vehicles are also the key factors which influence the generation of ANV. The three-phase (3-ɸ) star connected stator winding of PMSM has its own natural frequency. When the frequencies of the exciting forces match with the natural frequencies of three-phase star connected stator winding, then resonance occurs, which produces a high level of acoustic noise in PMSM drive. Therefore, it is essential to calculate the natural frequency of the stator. There are many techniques reported in the literature to find out natural frequencies of the stator of PMSM (Rik et al. 2016; Verma et al. 1989). Similarly, the effect of fluctuation in operating characteristics of PM drive on power transmission line is to be considered, to find out induced dynamic vibration response, which helps in improving effectiveness of reduction of vibration and hence enhance system reliability.

27.3 Control Strategy and Operation Characteristics of PM Based Electric Drive

Initially, vector control based SPWM technique is used to run PMSM. Error in PMSM speed is calculated by comparing actual speed with reference speed and processed through the Proportional and Integral (PI) controller. The output of the PI controller generates the current reference i_q. With the help of theta (θ), i_{dq} is converted to i^*_{abc}. Reference current i^*_{abc}, actual current (i_{abc}) and carrier wave signals are fed to pulse generator and pulse generator, generated pulses are fed to the three-phase inverter.

The electric drives use solid-state power electronic converters, which produces non-sinusoidal signals as supply voltage and current to electric drives. Depending on the system's configuration, the frequency of the generated harmonics may be of any order. The order of the important harmonics having higher magnitudes are *3rd, 5th, 7th,* and *11th* only. Gieras et al. (2006) had presented a mathematical model of electromagnetic noise and vibration which is generated due to electromagnetic fields. A three-phase balanced stator has three similar windings shifted in space by *120°* electrical. For a three-phase balanced stator (I_a, I_b and I_c), the amplitude of the input stator current of each phase is $\sqrt{2}I$ and *120°* electrical phase shift, which are given by (27.1)–(27.3):

$$I_a = \sqrt{2}I cos(\omega t) \tag{27.1}$$

$$I_b = \sqrt{2}I cos\left(\omega t - \frac{2\pi}{3}\right) \tag{27.2}$$

$$I_c = \sqrt{2}I cos\left(\omega t - \frac{4\pi}{3}\right) \tag{27.3}$$

The Magneto-Motive Force (MMF) of a single-phase winding changing in space and time can be resolved into two waves rotating in opposite directions, which can be found by (27.4).

$$\begin{aligned} f(x,t) &= \sum_{\upsilon=1,3,5}^{\infty} f_{m\upsilon} cos(\omega t) cos\left(\upsilon\frac{\pi}{\tau}x\right) \\ &= \frac{1}{2}\sum_{\upsilon=1,3,5}^{\infty} f_{m\upsilon} cos\left(\omega t - \upsilon\frac{\pi}{\tau}x\right) \\ &+ \frac{1}{2}\sum_{\upsilon=1,3,5}^{\infty} f_{m\upsilon} cos\left(\omega t + \upsilon\frac{\pi}{\tau}x\right) \end{aligned} \tag{27.4}$$

where $f(x,t)$ is the variation of the MMF with the linear coordinate x-axis with respect to time (t), f_{mv} is the magnitude of the vth harmonic of the MMF per phase, v is the no. of the stator vth space harmonics or Poisson's ratio, τ is the pole pitch and ω is the angular frequency.

Based on (27.4) the space harmonics of the MMF produced by each phases winding which are written by (27.5)–(27.7):

$$f_{A\upsilon} = \frac{1}{2} f_{m\upsilon} cos\left(\omega t - \upsilon\frac{\pi}{\tau}x\right) + \frac{1}{2} f_{m\upsilon} cos\left(\omega t + \upsilon\frac{\pi}{\tau}x\right) \tag{27.5}$$

$$f_{B\upsilon} = \frac{1}{2} f_{m\upsilon} cos\left[\left(\omega t - \upsilon\frac{\pi}{\tau}x\right) + (\upsilon - 1)\frac{2\pi}{3}\right] + \frac{1}{2} f_{m\upsilon} cos\left[\left(\omega t + \upsilon\frac{\pi}{\tau}x\right) - (\upsilon + 1)\frac{2\pi}{3}\right] \tag{27.6}$$

$$f_{C\upsilon} = \frac{1}{2} f_{m\upsilon} cos\left[\left(\omega t - \upsilon\frac{\pi}{\tau}x\right) + 2(\upsilon - 1)\frac{2\pi}{3}\right] + \frac{1}{2} f_{m\upsilon} cos\left[\left(\omega t + \upsilon\frac{\pi}{\tau}x\right) - 2(\upsilon + 1)\frac{2\pi}{3}\right] \tag{27.7}$$

According to (27.5)–(27.7), harmonics waves of the MMF of each phase are shifted by the angle of $(\upsilon \mp 1)(2\pi/3)$ one from each other. From (27.5)–(27.7) it can be easily identified that which order and what magnitude of the space harmonics in MMF are present in the three-phase voltage source inverter-fed PMSM drive. The Voltage Source Inverter (VSI) supplies the current, which is not pure sinusoidal and rich in harmonics. The existing harmonics are further distinguished into, set of time and space harmonics. These harmonics are the cause of vibrations in the stator core of PMSM. The ripple produced in the torque finally leads to acoustic or electromagnetic noise. Bolton et al. (1984), Le-Huy et al. (1986) presented a general relation for the electromagnetic torque developed by a converter-fed PM synchronous motor is analyses for harmonic rich current waveform and the magnetic field flux distribution. The torque harmonics frequencies depend upon stator phases and can be calculated by the stator MMF harmonics rotational speeds with respect to the synchronous rotor speed.

The stator supply phase current is assumed to be symmetrical and with no even harmonics, then the phase current in the Nth phase is given by (27.8) (Le-Huy et al. 1986; Bolton and Ashen 1984):

$$i_N = \hat{I}\left[sin\left(\omega t + (N-1)\frac{2\pi}{N}\right) + K_{I3} sin3\left(\omega t + (N-1)\frac{2\pi}{N}\right) + \cdots\right] \quad (27.8)$$

where $\hat{I}$ is peak stator phase current, ω is supply frequency, i is instantaneous current and K_{IP} is peak value relative to fundamental of pth harmonic in current waveform.

Hence, the torque for 3-ϕ PMSM is shown as (27.9) (Le-Huy et al. 1986; Bolton and Ashen 1984):

$$\begin{aligned} T = {} & 1.5\hat{I}\hat{B}mDLPK_{w1}\times \\ & \left[\begin{array}{l} 1 + \dfrac{K_3K_{w3}K_{I3}}{K_{w1}} + \dfrac{K_5K_{w5}K_{I5}}{K_{w1}} + \ldots + \\ \left(\begin{array}{l} K_{I7} - K_{I5} + \dfrac{K_3K_{w3}K_{I9}}{K_{w1}} - \dfrac{K_3K_{w3}K_{I3}}{K_{w1}} \\ \dfrac{K_5K_{w5}}{K_{w1}} + \dfrac{K_7K_{w7}}{K_{w1}} + \dfrac{K_9K_{w9}K_{I3}}{K_{w1}} + \ldots \end{array}\right)\cos 6\omega t \\ \left(\begin{array}{l} K_{I13} - K_{I11} + \dfrac{K_3K_{w3}K_{I9}}{K_{w1}} - \dfrac{K_5K_{w5}K_{I7}}{K_{w1}} - \\ \dfrac{K_7K_{w7}K_{I5}}{K_{w1}} - \dfrac{K_9K_{w9}K_{I3}}{K_{w1}} + \ldots \end{array}\right)\cos 12\omega t \end{array}\right] \end{aligned} \quad (27.9)$$

Torque ripple frequency will be 6, 12, 18... cycles per pole-pair of rotor position. From (27.9), for the ideal case, if the field flux and the stator currents are perfect sinusoidal, then the torque is constant and torque ripple will be zero. In practice, due to the motor construction and the step field distribution, the induced EMF's are non-sinusoidal and contain high-order harmonics. The fundamental torque T_1 due to the fundamental of the stator ampere-conductor and field distribution is given by putting with $p = 1$,

$$T_1 = 0.5N\hat{I}\hat{B}mDLPK_{w1} \tag{27.10}$$

For three-phase PMSM, $N = 3$ and the values of X (p, q) and Y (p, q) are X (p, q) = 1 for |p-q| = 0, 3, 6… and Y (p, q) = 1 for $p + q = 3, 6, 9\ldots$ and zero otherwise.

Since the torque harmonics/ripples are a function of stator current time-harmonics/ripples which in turn contributes to the production of PMSM ANV. Hence, once the magnitude and order of harmonics are known then suitable techniques like spread harmonics spectrum over the time range through random PWM techniques can be easily implemented for ANV reduction for PMSM drive.

The harmonics in the current–voltage inverter output with high ripples produces corresponding torque with high ripples causes twist and untwist or angular acceleration or deceleration of shaft about axis of rotation, may lead to fracture of shaft. Torque response (τ) at the motor shaft of PMSM is shown in Fig. 27.4 and measured experimentally using in-line torque sensor (Torque constant = 0.49 N-m, range 0–5 N-m for 12,000 rpm, manufactured by FUTEK) placed over PMSM and DC generator for vibration analysis. Using waveform from Fig. 27.4, mathematically torque waveform can be expresses as (27.11).

$$\tau = 1.06 + 0.117 \sin 314t \tag{27.11}$$

The torque waveform as shown in Fig. 27.4 acts as initial input for vibrational analysis and hence leads to acoustic noise.

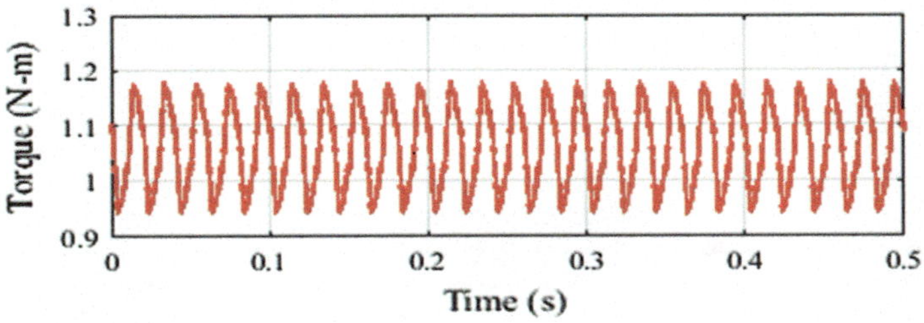

Fig. 27.4 Torque ripple measured at shaft (S_1) of PMSM drive

27.4 Small-Scale Experimental Setup of PMSM Coupled with DC Generator

A three-phase star-connected inverter fed PMSM drive is used for the analysis of ANV. For modelling, analysis and results validation of ANV, a small-scale experimental setup is considered as shown in Figs. 27.5, 27.6, 27.7, 27.8, and 27.9, in which PMSM is coupled with DC generator through end to end coaxially coupled stepped shafts. A schematic layout and pictorial view of the experimental set-up are shown in Figs. 27.5 and 27.6, respectively. For measurement of acoustic noise and vibration of electric machines, a highly effective low-cost acoustic chamber has been designed with the highest cut-off frequency of 40 Hz. The whole setup is placed inside low-cost acoustic chamber to measured exact acoustic noise of PMSM drive as shown in Fig. 27.7. Also, Table 27.1 summaries the experimental system parameters of the PMSM drive used for this study.

The real-time National Instrument (NI) Compact Reconfigurable Input–Output (c-RIO) 9081 from National Instruments Pvt. Ltd. use as a NI-DAQ. Compact-RIO controllers are embedded real-time controllers that are ideal for advanced control and monitoring applications. The controller with a compatible compact chassis. The 8-slot, 1.06 GHz Dual-Core CPU, 2 GB DRAM, 16 GB Storage, Xilinx Spartan-6 LX75 Field Programmable Gate Array (FPGA). For the realization of measurements,

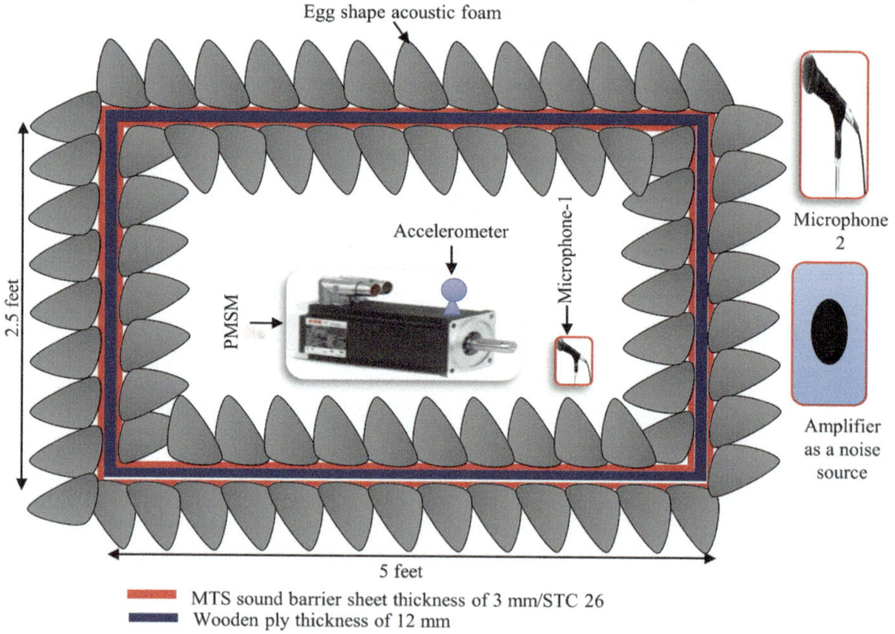

Fig. 27.5 Schematic layout of a low-cost acoustic chamber for measure of acoustic noise of electric motors

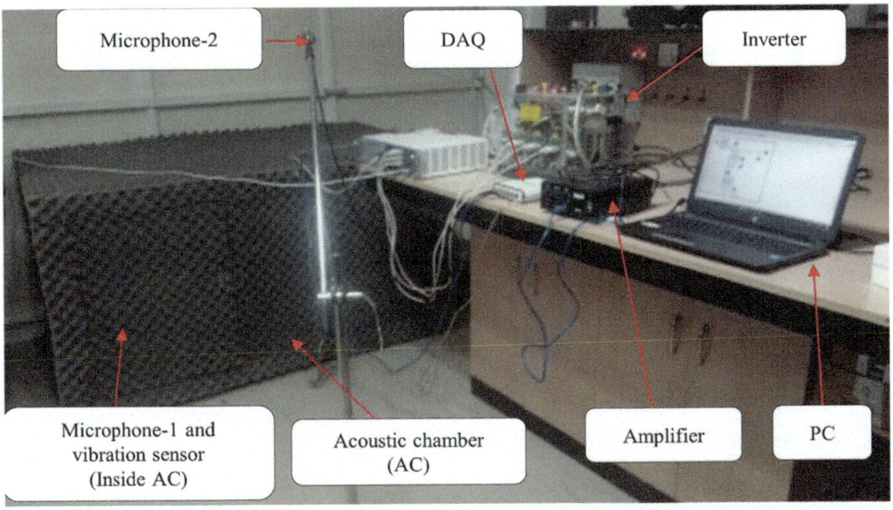

Fig. 27.6 An experimental setup for characterize of a low-cost acoustic chamber

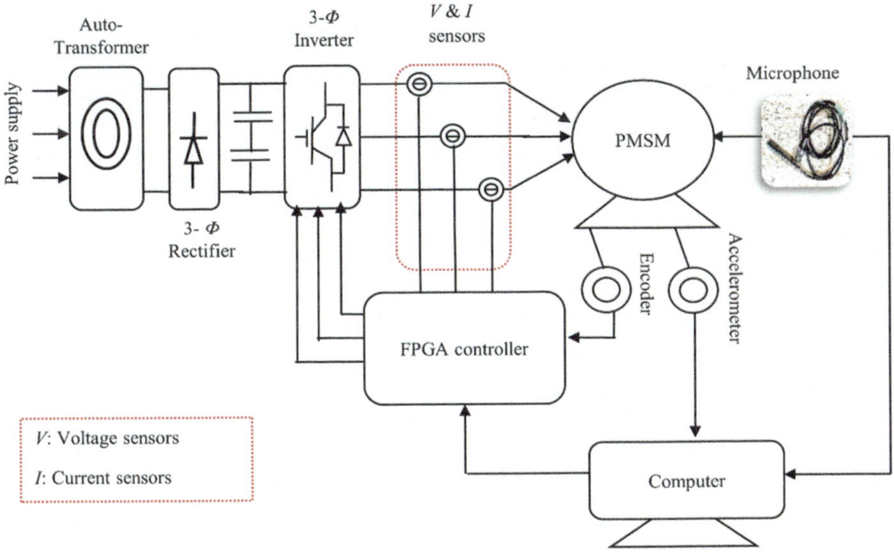

Fig. 27.7 Schematic diagram of experimental setup for analysis of ANV of PMSM drive

a directive sound source was used, placed microphones in two different locations. Microphone-1 put inside the chamber and microphone-2 put outside the chamber, as shown in Fig. 27.6.

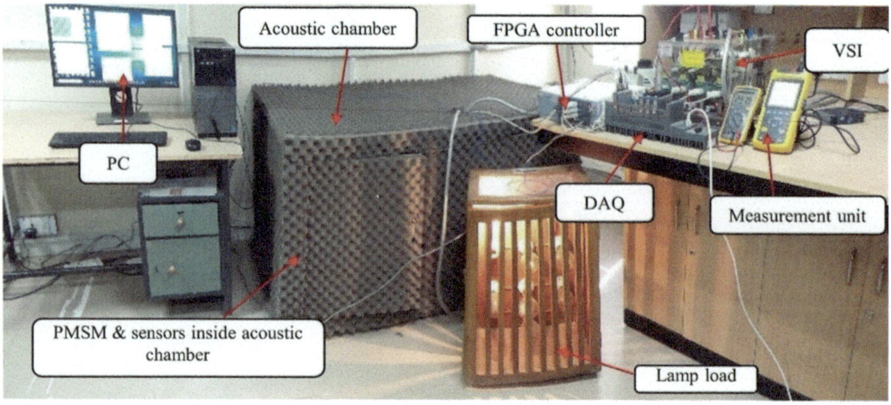

Fig. 27.8 An experimental setup for analysis of ANV of PMSM drive

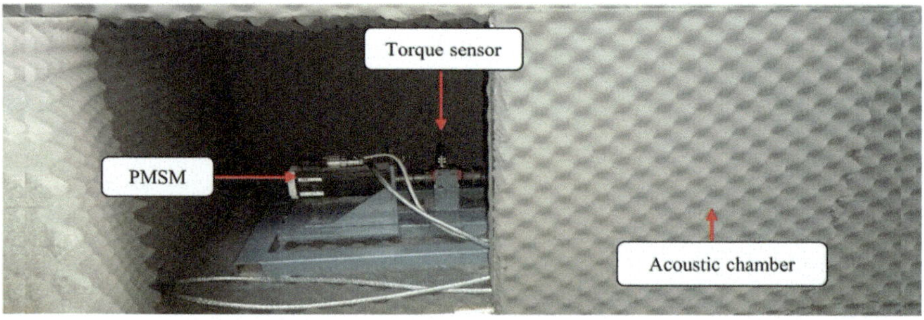

Fig. 27.9 Acoustic chamber with sensors and PMSM drive

The complete setup having PMSM coupled with DC generator through end to end coaxially coupled shaft (placed inside acoustic chamber as shown in Fig. 27.9) which is required for vibration analysis is shown in Fig. 27.10.

Schematic layout of PMSM driven experimental setup is shown in Fig. 27.11. Power from motor shaft (S_1) used to drive generator shaft (S_3) through shaft (S_2). Generally axial, transverse and torsional vibrations are present, if axial motion is arrested as in most of the cases, only transverse and torsional vibration are there in electric vehicles power transmission system. Transverse vibration is usually found where aspect ratio is high and span between end supports of shaft is high. Roller support (B_2) is attached over shaft S_2, to reduce the chances of shaft whirl because of high aspect ratio and eccentricity from support, therefore only torsional vibration is considered and lateral or transverse vibration is neglected. Claw coupling (A) with discs (C_1) and (C_2) are attached with shaft S_1 and S_2, and claw coupling (B) having disc (C_3) and (C_4) are attached with shaft S_2 and S_3, respectively. Claw couplings are flexible coupling easily transmit power between non-coaxial shaft also. Inbuilt

Table 27.1 Experimental setup specification of PMS drive

Specification item	Value	Unit
Power	1.07	kW
Rated speed	3000	rpm
No. of poles	04	
Torque cont. stall	3.6	N-m
Current cont. stall	6.29	A
Rated bus voltage	300	V
Per phase resistance	3.07	Ω
Per phase inductance	6.57	mH
Rotor inertia	1.4–1.8	kg-m^2
Spartan 3AN FPGA kit	20	MHz Clock frequency
Peak current	16	A
Torque constant	0.49	N-m/A
IGBT based inverter stack	600, 30	V, A
Microphone (1/2″ free-field)	10	mV/Pa
Accelerometer sensitivity	10	mV/g

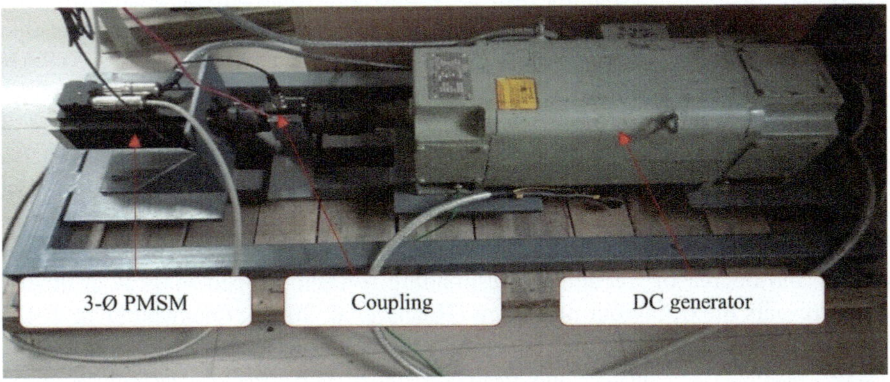

Fig. 27.10 Small-scale experimental setup with PMSM driven including coupled shaft and load DC generator

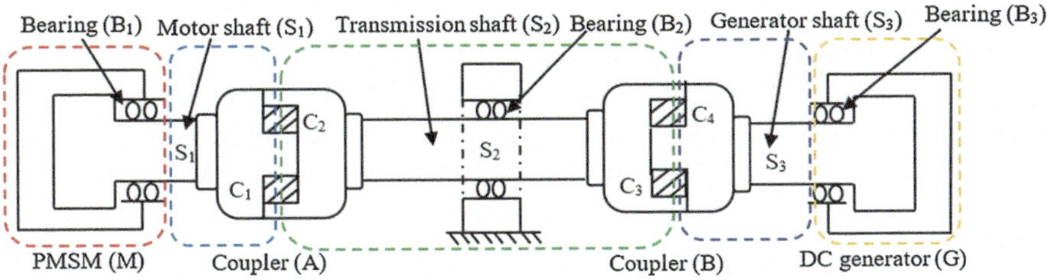

Fig. 27.11 Schematic layout of PMSM drive experimental setup

bearing (B_1) is present in between shaft S_1 and motor hub, similarly, B_3 is inbuilt bearing present in generator shaft S_3 and its hub.

27.5 Generalized Framework of Dynamic Response of Torsional Vibration in PMSM Drive Coupled with Load DC Generator

Torsional vibrations in end to end coaxially coupled shaft are produced due to high ripples in torque as shown in Fig. 27.4 which act as input for modelling of torsional vibration. The detailed generalized procedure and analysis for steady state response of torsional vibration is as follows.

Step 1: Calculation of optimum number of lumped elements, Natural frequencies and mode shape at resonance

To analyse dynamic response of torsional vibration various methods are used in literature. Analytic methods in which system is considered as continuous, results in partial differential equations and especially for such complex system, complex Partial Differential Equation (PDE) are formed, which are computationally expensive and sometime unsolvable. To solve these differential equation, approximate methods like Finite Element Method (FEM) and various software based on FEM like NASTRAN, SAP, ADINA, ABAQUS, ANSYS etc., are introduced which convert these PDE's to set of linear equation (Chengsheu and Wenchen 2000; Junhui 2019). But because of limitation such as lengthy pre-processing like geometry preparation, material properties, meshing, boundary condition as input from user make it time consuming and computationally expensive (Seung et al. 2016). So, a lumped modelling method is introduced in which whole structural domain is discretised into some finite lumped masses and degree of freedom corresponding to each lumped mass is evaluated, which reveals the behaviour of system (Huang et al. 2019b; Qianwen et al. 2017). External loading conditions may be thermal, electrical, acoustic, magnetic or a multi-physics problem (Chengsheu and Wenchen 2000). This method is computationally efficient with a level of accuracy (Rao 2016). Also, some researchers integrated FEM with lumped modelling like for vibration analysis problem. Sometime, lumped model is not able to incorporate various geometrical or boundary conditions like stepped beam, eccentric loading. So, researchers had done modification work in conventional lumped modelling to make it suitable for specific problem. In literature, an improvement in lumped model is presented, for transverse vibration analysis using piezoelectric sensors. Lumped method is still a preferred choice in latest research paper for vibration analysis and in other areas (Huang et al. 2019b ; Qianwen et al. 2017), reveals its computational effectiveness.

In lumped model, system is discretized, and finite lumped masses are created and the torsional vibrational characteristics like angular displacement, natural frequencies etc., are calculated for respective lumped element. Lumped model has less

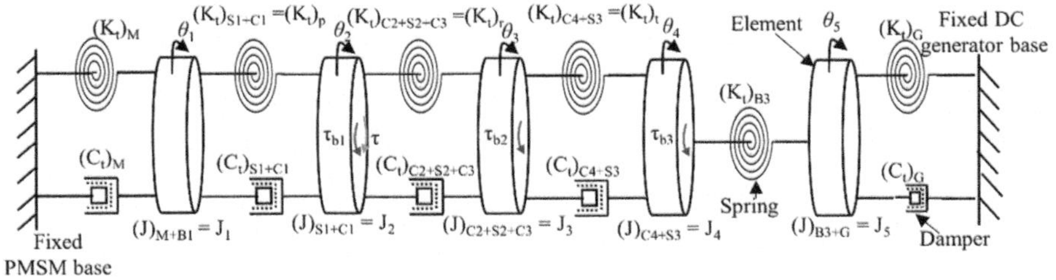

Fig. 27.12 Lumped parametric modelling of torsional vibration in PMSM coupled with DC generator

accuracy compared to FEM but various accuracy enhancement techniques are there which will make lumped model accurate along with advantage of computationally inexpensiveness (Rahman and Gupta 2020; Papadrakakis et al. 2000; Cremer and Heckl 1988). One such method is effective mass and mass participation factor method which helps to decide optimum number of lumped elements which is to be considered (Rahman and Gupta 2020; ABB Manual 2015; Papadrakakis et al. 2000). According to this method, modes with relatively high effective modal masses are more likely to be excited using external electromagnetic torque and the number of modes to be considered in such a way that the total effective model mass of lumped model is at least 95% of the actual mass (Papadrakakis et al. 2000). In the present case, we have torsional vibration so actual mass is combined polar moment of inertia of the system. So initially a guess of five number of lumped elements are considered as shown in Fig. 27.12.

Step 1.1: Modelling of Torsional vibration equation corresponding to each lumped element

In lumped parametric modelling, each component including PMSM and DC Generator is replaced by torsional spring, damper and polar moment of inertia (Rao 2016; Irvine 2015). PMSM (M) is considered as first lumped element in the present system. S_1 shaft along with disc C_1 of claw coupler A as 2nd element, C_2-S_2-C_3 as 3rd, C_4-S_3 as 4th and generator as 5th element respectively (Manjibhai et al. 2021). In this way, 5 degrees of freedom lumped system is established. θ_1 to θ_5 and J_1 to J_5 are degree of freedom in radians and inertia of each lumped mass. $(K_t)_M$ and $(C_t)_M$ are torsional stiffness and coefficient of damping of PMSM used in experimental setup. $(K_t)_{S1+C1}$ $((K_t)_p)$, $(K_t)_{C2+S2+C3}$ $((K_t)_r)$ and $(K_t)_{C4+S3}$ $((K_t)_t)$ are stiffness and $(C_t)_{S1+C1}$ $((C_t)_q)$, $(C_t)_{C2+S2+C3}$ $((C_t)_s)$ and $(C_t)_{C4+S3}$ $((C_t)_u)$ are damping of lumped mass 2nd, 3rd and 4th, respectively. $(K_t)_{b3}$ is stiffness of B_3 bearing. $(K_t)_G$ and $(C_t)_G$ are stiffness and coefficient of damping of load DC generator, respectively. τ, τ_{b1}, $\tau_{b2,}$ and τ_{b3} are source torque and resisting torque at B_1, B_2, B_3, bearings, respectively.

For jth arbitrary lumped mass, all possible torque and their directions is shown in Fig. 27.13 that further used in place of any lumped element.

where, θ_j, J_j, K_{tj}, and C_{tj}, are angular displacement, inertia, stiffness and coefficient of damping of arbitrary jth element. τ_j is the source torque having ripples due to

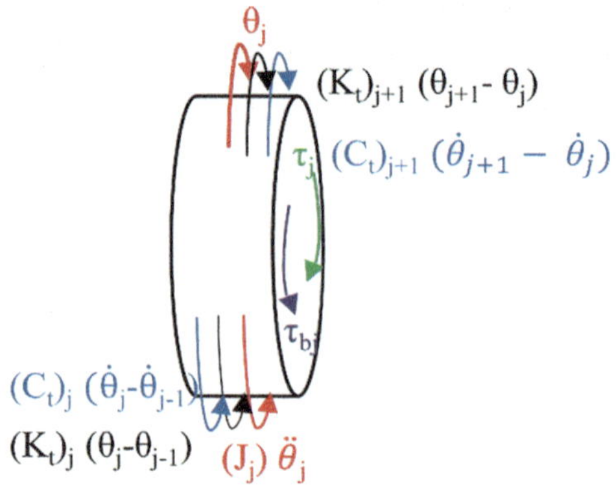

Fig. 27.13 Generalized free body diagram of jth lumped element

power harmonics, and τ_{bj} is frictional torque at each bearing. $(K_t)_{j+1}$ and $(C_t)_{j+1}$ are stiffness and coefficient of torsional damping of successive (j+1)th lumped mass. Similarly, θ_{j-1} is the DOF of preceding (j-1)th lumped mass. $(C_t)_j$ $(\dot{\theta}_j - \dot{\theta}_{j-1})$ and $(K_t)_j$ $(\theta_j - \theta_{j-1})$ are torsional torque exerted by equivalent spring and damper of jth lumped mass. Similarly $(C_t)_{j+1}$ $(\dot{\theta}_{j+1} - \dot{\theta}_j)$ and $(K_t)_{j+1}$ $(\theta_{j+1} - \theta_j)$ are torsional torque due to damping and spring exerted by damper and equivalent spring of (j+1)th lumped mass.

Using Newton's 2nd law of motion for rotational system, the differential equation representing state of motion of *j*th lumped mass as shown in Fig. 27.13, is given by (27.12).

$$J_j\ddot{\theta}_j + (K_t)_j(\theta_j - \theta_{j-1}) + (C_t)_j(\dot{\theta}_j - \dot{\theta}_{j-1}) - (K_t)_{j+1}(\theta_{j+1} - \theta_j) - (C_t)_{j+1}(\dot{\theta}_{j+1} - \dot{\theta}_j) = \tau_j - \tau_{bj} \tag{27.12}$$

Table 27.2 represents various parameters as in (27.12) corresponding to respective

Table 27.2 Values of generalized coefficient corresponding to respective lumped element

Element no.	1	2	3	4	5	jth
J_j	J_1	J_2	J_3	J_4	J_5	J_j
θ_j	θ_1	θ_2	θ_3	θ_4	θ_5	θ_j
$(K_t)_j$	$(K_t)_M$	$(K_t)_p$	$(K_t)_r$	$(K_t)_u$	$(K_t)_{b3}$	$(K_t)_j$
θ_{j-1}	0	θ_1	θ_2	θ_3	θ_4	θ_{j-1}
$(C_t)_j$	$(C_t)_M$	$(C_t)_q$	$(C_t)_s$	$(C_t)_v$	0	$(C_t)_j$
$(K_t)_{j+1}$	$(K_t)_p$	$(K_t)_r$	$(K_t)_u$	$(K_t)_{b3}$	$(K_t)_G$	$(K_t)_{j+1}$
θ_{j+1}	θ_2	θ_3	θ_4	θ_5	0	θ_{j+1}
$C_t)_{j+1}$	$(C_t)_q$	$(C_t)_s$	$(C_t)_v$	0	$(C_t)_G$	$C_t)_{j+1}$
τ_j	0	τ	0	0	0	τ_j
τ_{bj}	0	τ_{b1}	τ_{b2}	τ_{b3}	0	τ_{bj}

lumped element.

Using Tables 27.2 and (27.12), equation of motion for any number of lumped elements are given as.

For motor as 1st element after simplification,

$$J_1\ddot{\theta}_1 + \left((C_t)_M + (C_t)_q\right)\dot{\theta}_1 - (C_t)_q\dot{\theta}_2 + \left((K_t)_M + (K_t)_p\right)\theta_1 - (K_t)_p\theta_2 = 0 \tag{27.13}$$

Similarly, for 2nd, 3rd, 4th and 5th lumped masses, equations of motion shown as (27.14), (27.15), (27.16) and (27.17).

$$J_2\ddot{\theta}_2 + \left((C_t)_q + (C_t)_s\right)\dot{\theta}_2 - (C_t)_q\dot{\theta}_1 - (C_t)_s\dot{\theta}_3 + \left((K_t)_p + (K_t)_r\right)\theta_2 - (K_t)_p\theta_1 - (K_t)_r\theta_3 = \tau - \tau_{b1} \tag{27.14}$$

$$J_3\ddot{\theta}_3 + \left((C_t)_s + (C_t)_r\right)\dot{\theta}_3 - (C_t)_s\dot{\theta}_2 - (C_t)_q\dot{\theta}_4 + \left((K_t)_r + (K_t)_q\right)\theta_3 - (K_t)_r\theta_2 - (K_t)_r\theta_4 = -\tau_{b2} \tag{27.15}$$

$$J_4\ddot{\theta}_4 + \left((C_t)_r + (C_t)_{b_3}\right)\dot{\theta}_4 - (C_t)_s\dot{\theta}_3 - (C_t)_{b_3}\dot{\theta}_5 + \left((K_t)_u + (K_t)_{b_3}\right)\theta_4 - (K_t)_u\theta_3 - (K_t)_{b_3}\theta_5 = -\tau_{b3} \tag{27.16}$$

$$J_5\ddot{\theta}_5 + (C_t)_G\dot{\theta}_5 + \left((K_t)_{b_3} - (K_t)_G\right)\theta_5 - (K_t)_{b_3}\theta_4 = 0 \tag{27.17}$$

Similarly, for jth element, an equation can be written as:

$$J_j\ddot{\theta}_j + (K_t)_j(\theta_j - \theta_{j-1}) + (C_t)_j(\dot{\theta}_j - \dot{\theta}_{j-1}) - (K_t)_{j+1}(\theta_{j+1} - \theta_j) - (C_t)_{j+1}(\dot{\theta}_{j+1} - \dot{\theta}_j) = \tau_j - \tau_{bj} \tag{27.18}$$

These equations in matrix form are written as:

$$[J]_{j\times j}\{\ddot{\theta}\}_{j\times 1} + [C_t]_{j\times j}\{\dot{\theta}\}_{j\times 1} + [K_t]_{j\times j}\{\theta\}_{j\times 1} = \{\tau\}_{j\times 1} \tag{27.19}$$

This equation will give j number of natural frequencies, mode shapes and dynamic response of torsional vibration. As initially five elements ($j = 5$) are considered, so (27.19) can be written using (27.13) to (27.17) in expanded form as (27.20).

$$\begin{bmatrix} J_1 & 0 & 0 & 0 & 0 \\ 0 & J_2 & 0 & 0 & 0 \\ 0 & 0 & J_3 & 0 & 0 \\ 0 & 0 & 0 & J_4 & 0 \\ 0 & 0 & 0 & 0 & J_5 \end{bmatrix} \begin{bmatrix} \ddot{\theta}_1 \\ \ddot{\theta}_2 \\ \ddot{\theta}_3 \\ \ddot{\theta}_4 \\ \ddot{\theta}_5 \end{bmatrix} + \begin{bmatrix} (C_t)_M + (C_t)_q & -(C_t)_q & 0 & 0 & 0 \\ -(C_t)_q & (C_t)_q + (C_t)_s & -(C_t)_s & 0 & 0 \\ 0 & -(C_t)_s & (C_t)_s + (C_t)_v & -(C_t)_v & 0 \\ 0 & 0 & -(C_t)_v & (C_t)_v & 0 \\ 0 & 0 & 0 & 0 & (C_t)_G \end{bmatrix} \begin{bmatrix} \dot{\theta}_1 \\ \dot{\theta}_2 \\ \dot{\theta}_3 \\ \dot{\theta}_4 \\ \dot{\theta}_5 \end{bmatrix}$$

$$+\begin{bmatrix} (K_t)_M+(K_t)_p & -(K_t)_p & 0 & 0 & 0 \\ -(K_t)_p & ((K_t)_p+(K_t)_r) & -(K_t)_r & 0 & 0 \\ 0 & -(K_t)_r & ((K_t)_r+(K_t)_u) & -(K_t)_u & 0 \\ 0 & 0 & -(K_t)_u & ((K_t)_u+(K_t)_{b3}) & -(K_t)_{b3} \\ 0 & 0 & 0 & -(K_t)_{b3} & ((K_t)_{b3}+(K_t)_G) \end{bmatrix}\begin{bmatrix}\theta_1\\ \theta_2\\ \theta_3\\ \theta_4\\ \theta_5\end{bmatrix}=\begin{bmatrix}0\\ \tau-\tau_{b1}\\ -\tau_{b2}\\ -\tau_{b3}\\ 0\end{bmatrix} \tag{27.20}$$

Step 1.2: Calculation of various geometric parameters required in lumped model

As mentioned in (27.20), along with external loading, various geometrical and material parameters are required to find out the dynamic response of torsional vibration of system. This step is also generalized by considering a stepped generalized element as shown in Fig. 27.14.

To find out various parameters required in (27.20) like J_j, $(K_t)_j$, $(C_t)_j$ which are inertia, stiffness, coefficient of damping and τ_j is source torque and τ_{bj} is frictional torque at bearings for *j*th element respectively, the methodology used is as follows.

Equivalent torsional stiffness ($(K_t)_j$): For equivalent torsional stiffness of generalized *j*th element which is assumed as Fig. 27.14, is a series combination of n number of shaft and coupler which are in the form of a solid circular disc having varying diameter and length is given as

$$(K_t)_j = \frac{1}{(K_t)_{j1}} + \frac{1}{(K_t)_{j2}} + \ldots\ldots\frac{1}{(K_t)_{jn}} \tag{27.21}$$

where, $(K_t)_{jn}$ is nth disc in jth element.

$$(K_t)_{jn} = \frac{G_j(J_a)_{jn}}{L_{jn}}, n \text{ varies from 1 to } n. \tag{27.22}$$

where, G_j is the Shear modulus of components (for stainless steel (Grade-304), G_j is 77×10^3 MN/m^2) and its density is 7780 kg/m^3 (Cremer and Heckl 1988; Bachmann and Ammann 1995). $(J_a)_{jn}$, L_{jn} is inertia and length of *n*th disc in *j*th element, respectively. where,

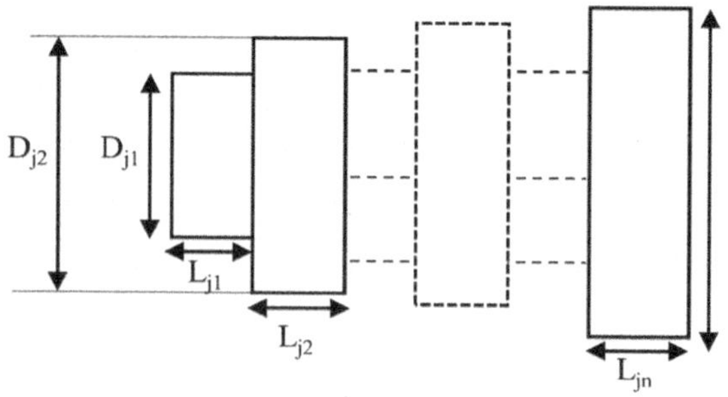

Fig. 27.14 Generalized jth element

$$(J_a)_{jn} = \frac{\pi d_{jn}^4}{32}; n \text{ varies from 1 to } n. \tag{27.23}$$

where, d_{jn} is the diameter of various coaxial shaft & coupler present in *j*th element as shown in Fig. 27.14. Using (27.23) in (27.22), and then (27.22) in (27.21), equivalent torsional stiffness $(K_t)_j$ of a generalized jth element is calculated.

Mass polar moment of inertia (J_j): Mass polar moment of inertia of assumed *j*th element is given as

$$(J)_j = \frac{m_{j1} d_{j1}^2}{8} + \frac{m_{j2} d_{j2}^2}{8} + \ldots\ldots \frac{m_{jn} d_{jn}^2}{8} \tag{27.24}$$

where, m_{jn} is mass of *n*th disc in *j*th element.

This generalized methodology is used for small scale experimental setup of initially assumed five number of lumped elements as follows.

For PMSM and DC Generator (lumped element 1 and 5), the value of torsional stiffness (K_t) and polar moment of inertia (*J*) is directly taken from the manufacturer catalog (ABB Manual 2015) and given in Table 27.3.

For lumped element 2, 3 & 4, value of torsional stiffness (K_t) and inertia (*J*) is calculated using above stated methodology and various geometrical parameters and calculations are in tabular form given as follows.

For lumped element -2

See Fig. 27.15 and Table 27.4.

For lumped element -3

Table 27.3 PMSM and load DC generator parameters

Element No	Inertia (J)	Inertia (kg-m^2)	Stiffness (K_t)	Stiffness (N-m/rad)
1 (PMSM)	J_1	1.6	$(K_t)_M$	0.0434
5(DC generator)	J_5	2.4	$(K_t)_G$	0.642

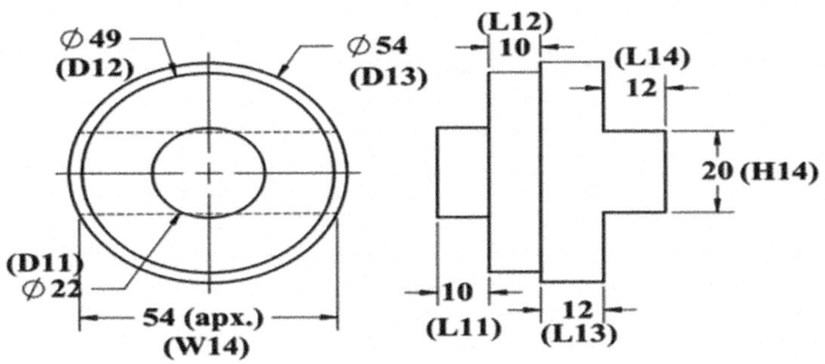

Fig. 27.15 Parameters of lumped mass-2 (Scale 1:1)

Table 27.4 Parameters of lumped mass-2

Parameters	Values (m) $\times 10^{-3}$	Mass (kg) $\times 10^{-3}$	J (kg-m^2) $\times 10^{-7}$	$(K_t)_p$ N-m/rad
D11, L11	22,10	22.56	17.88	4.36×10^5
D12, L12	49, 13	190.63	572.11	
D13, L13	54,12	213.7	778.95	
L14, H14, W14	12, 20, 54	100.8	278.59	
$J_2 = \sum J = 1647.53 \times 10^{-7}$				

See Fig. 27.16 and Table 27.5.

<u>**For lumped element -4**</u>

See Fig. 27.17 and Table 27.6.

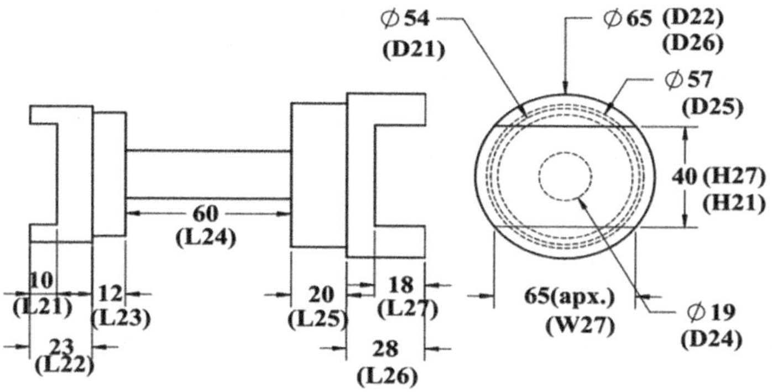

Fig. 27.16 Parameters of lumped mass-3 (Scale 1:2)

Table 27.5 Parameters of lumped mass-3

Parameters	Values (m) $\times 10^{-3}$	Mass (kg) $\times 10^{-3}$	J (kg-m^2) $\times 10^{-7}$	$(K_t)_r$ N-m/rad
L21, H21, W21 (Cavity)	10, 40, 54	−168.048	−632.42	0.146×10^5
D22, L22	54, 23	409.60	1492.99	
D23, L23	49, 12	175.963	528.10	
L24, D24	60, 19	132.28	59.69	
L25, D25	20, 57	396.85	1611.70	
L26, D26	28, 65	722.99	3815.65	
L27, H27, W27	18, 40, 65	−364.104	−1767.4	
$J_3 = \sum J = 5108.29 \times 10^{-7}$				

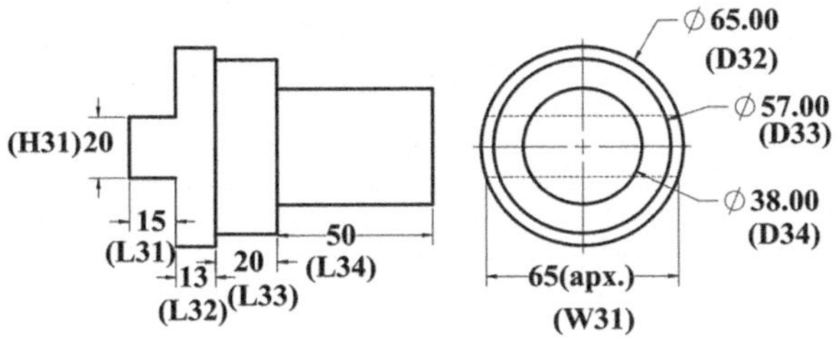

Fig. 27.17 Parameters of lumped mass-4 (Scale 1:1)

Table 27.6 Parameters of Lumped Mass-4

Parameters	Values (m) $\times 10^{-3}$	Mass (kg) $\times 10^{-3}$	J (kg-m^2) $\times 10^{-7}$	$(K_t)_u$ N-m/rad
L31, H31, W31	15, 20, 65	151.71	584.71	2.56×10^5
D32, L32	65, 13	335.44	1771.54	
D33, L33	57, 20	396.85	1611.70	
L34, D34	50, 38	440.94	795.89	
$J_4 = \sum J = 4763.84 \times 10^{-7}$				

Step 1.3: Natural frequencies and mode shape calculations

For natural frequencies or frequencies at resonance and mode shape calculations, (27.19) rewrite as (27.25).

$$[J]_{j\times j}\{\ddot{\theta}\}_{j\times 1} + [K_t]_{j\times j}\{\theta\}_{j\times 1} = \{0\} \tag{27.25}$$

By considering sinusoidal solution whose resonance frequency is ω_n and vibration magnitude is $\{\theta_0\}$, (27.25) rewrite as (27.26).

$$([D] - \lambda[I])\{\theta_0\} = \{0\} \tag{27.26}$$

where $[D]$ is given as $[J]^{-1}[K_t]$ is called as dynamic matrix, λ is ω_{n2} called as eigen values of system. Equation (27.26) gives j no. of natural frequency (ω_{nj}) and mode shape (θ_{0j}). Using the parameters from Step 1.2 in (27.26), five natural frequencies are obtained and given in Table 27.7.

Table 27.7 Resonance frequency corresponding to each lumped mass

Mode number	1	2	3	4	5
Natural frequency (Hz)	23.78	34.6	3.133×10^3	7.855×10^3	8.33×10^3

Putting these natural frequencies in (27.26), mode shape (θ_{0j}) can be obtained. For plotting purpose, all mode shapes are converted to scale of 1, called as displacement normalized (θ_{Ni}) and shown in Fig. 27.18.

For dynamic behaviour of the system, a modal analysis approach is used. Further, these, θ_{Nj} are ortho-normalized. θ_{Nj} are already orthogonal but are not inertia normalized. A_{j} is modal mass multiplied with θ_{Nj} to make them inertia-normalized.

where

$$A_j = \sqrt{\frac{1}{[\theta_{Nj}]^T [J][\theta_{Nj}]}} \tag{27.27}$$

where j varies from 1 to 5.

Hence,

$$A_1 = 0.7587,\ A_2 = 0.6194,\ A_3 = 42.683,\ A_4 = 44.184 \text{ and } A_5 = 77.8274. \tag{27.28}$$

Hence, modes become orthogonal and also inertia normalized by multiplying A_{i}, So,

$$\begin{aligned}
\{\theta_{ON1}\} &= [0.75;\ 0.74;\ 0.22;\ 0.19;\ 0.18] \\
\{\theta_{ON2}\} &= [0.22;\ 0.19;\ -0.56;\ -0.60;\ -0.61] \\
\{\theta_{ON3}\} &= [0.001;\ -1.61;\ -42.68;\ -12.002;\ 0.01] \\
\{\theta_{ON4}\} &= [-0.0003;\ 3.36;\ 11.55;\ -44.18;\ 0.006] \\
\{\theta_{ON5}\} &= [0.007;\ -77.82;\ 1.38;\ -1.66;\ 0.0002]
\end{aligned} \tag{27.29}$$

Hence, ortho-normalized modal matrix.

$$[\theta_{\mathrm{ON}}] = [\theta_{\mathrm{ON1}}, \theta_{\mathrm{ON2}}, \theta_{\mathrm{ON3}}, \theta_{\mathrm{ON4}}, \theta_{\mathrm{ON5}}] \tag{27.30}$$

Step 1.4: Optimum number of lumped elements using effective modal mass and mass participation factor method.

In this step, the initial guess of five number of lumped elements is tested using effective modal mass and mass participation technique. For this, orthonormalized mode are required as calculated in step 1.1 to step 1.3.

Step 1.4.1: Calculation of influence vector {r}

Influence vector represents the value of the degree of freedom (angular displacement) of all lumped elements when the whole body is assumed as a rigid body and fixed supports are rotated by unit degree. So as the body is rigid for calculating the influence

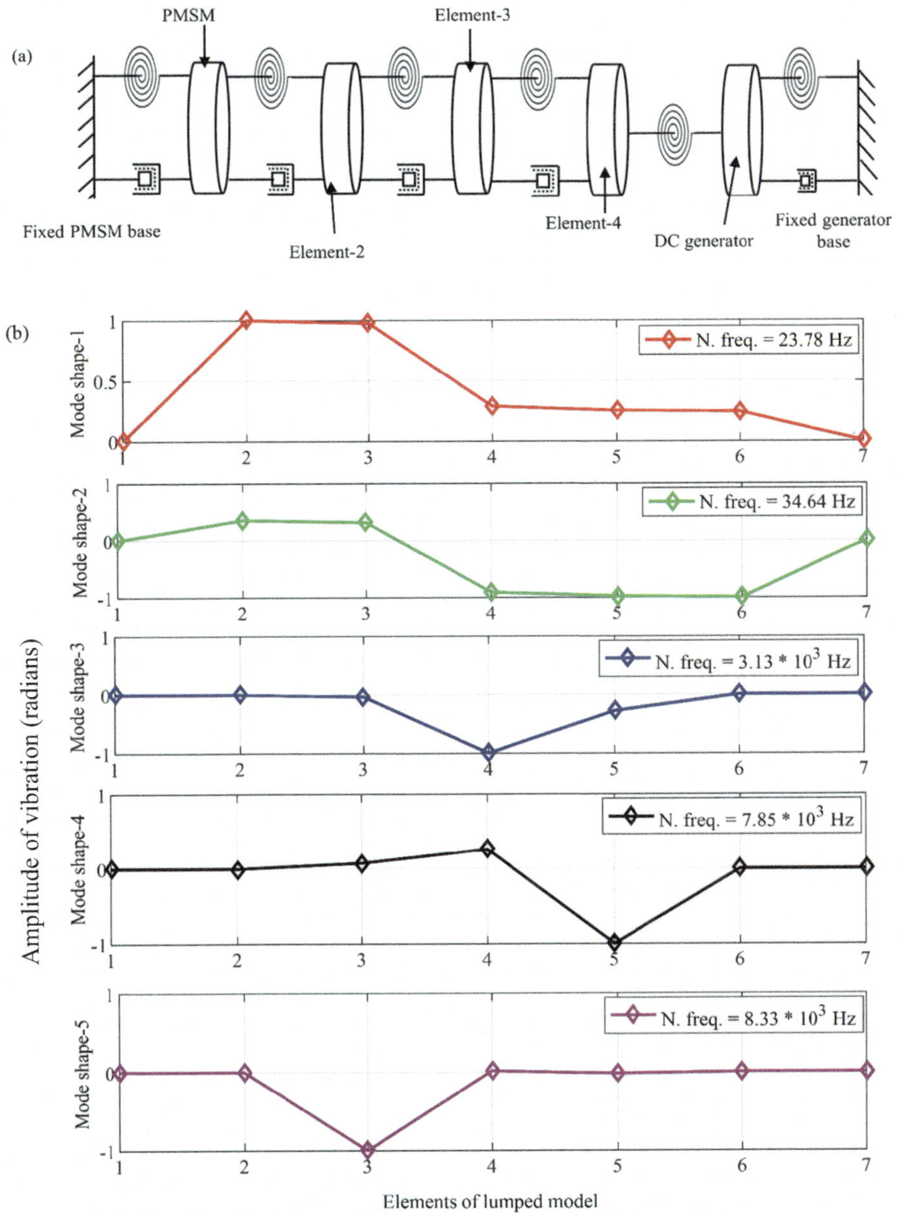

Fig. 27.18 Mode shapes corresponding to each lumped mass: (**a**) lumped model and (**b**) mode shapes for all lumped masses at resonance

vector, so all lumped elements are displaced by unity (i.e., 1 rad). Let initial guess of 5 number of mode or indirectly 5 number of degrees of freedom.

Hence,

$$\{r\} = [1;\ 1;\ 1;\ 1;\ 1] \tag{27.31}$$

Step 1.4.2: Calculation of coefficient vector {L}

$$\{\mathrm{L}\} = [\theta_{\mathrm{ON}}]^T [J]\{r\} \tag{27.32}$$

$$\{L\} = 10^{-2}\begin{bmatrix} 75.9 & 22.2 & 0.1 & -0.04 & 0.78 \\ 74.13 & 19.64 & -161.2 & 336.8 & -7782.7 \\ 22.1 & -56.28 & -4268.4 & -4418.8 & -166.32 \\ 19.02 & -60.62 & -1200.2 & -4418.48 & -166.32 \\ 18.11 & -61.94 & 1.08 & 0.63 & 0.02 \end{bmatrix}^T$$

$$10^{-4}\begin{bmatrix} 16000 & 0 & 0 & 0 & 0 \\ 0 & 1.647 & 0 & 0 & 0 \\ 0 & 0 & 5.108 & 0 & 0 \\ 0 & 0 & 0 & 4.763 & 0 \\ 0 & 0 & 0 & 0 & 24000 \end{bmatrix}\begin{Bmatrix} 1 \\ 1 \\ 1 \\ 1 \\ 1 \end{Bmatrix}$$

$$\{L\} = 10^{-7}\begin{Bmatrix} 16489983.6 \\ -11322567.7 \\ 26.67 \\ 2.45 \\ 0.81 \end{Bmatrix} \tag{27.33}$$

Step 1.4.3: Calculation of Model Participation Factor (MPF)

The MPF for jth mode is given as

$$MPF_j = \frac{L_{j1}}{(J_g)_{jj}} \tag{27.34}$$

As modal vectors are orthonormalized so the generalized mass polar moment of inertia matrix ($[\mathrm{J_g}]$) is an identity matrix. so $\left(J_g\right)_{jj} = 1$. It means.

$$MPF_j = L_{j1} \tag{27.35}$$

The effective modal mass for jth mode ($(\mathrm{J_{eff}})_j$) is given as

$$\left(J_{eff}\right)_j = \frac{\text{MPF}_j^2}{\left(J_g\right)_{jj}} \tag{27.36}$$

On calculating.
$(J_{eff})_1 = 2.71$ kg-m^2
$(J_{eff})_2 = 1.28$ kg-m^2
$(J_{eff})_3 = 7.1 \times 10^{-12}$ kg-m^2
$(J_{eff})_4 = 6 \times 10^{-14}$ kg-m^2
$(J_{eff})_5 = 6 \times 10^{-15}$ kg-m^2
Total actual polar moment of inertia of system is given as

$$J_{actual} = \sum_{j=1}^{5} J_J = J_{11} + J_{22} + J_{33} + J_{44} + J_{55}$$

$$J_{actual} = 10^{-4} \times (16000 + 1.647 + 5.108 + 4.763 + 24000)$$

$J_{actual} = 4.0016$ kg-m^2
Contribution of the first mode ineffective mass

$$\%\left(J_{eff}\right)_1 = \frac{\left(J_{eff}\right)_1}{J_{actual}} \times 100$$

$$\%\left(J_{eff}\right)_1 = \frac{2.71}{4.0016} \times 100 = 67.72\%, < 95\%$$

Contribution of first two modes ineffective mass

$$\%\left(J_{eff}\right)_2 = \frac{\left(J_{eff}\right)_1 + \left(J_{eff}\right)_2}{J_{actual}} \times 100$$

$$\%\left(J_{eff}\right)_2 = \frac{2.71 + 1.28}{4.0016} \times 100 = 99.71\%, > 95\%$$

As the minimum modes number is taken in a way that their sum of effective mass is at least 95% (Song and Su 2017). So, in the present case, the number of significant modes (m) is equal to 2. For Rayleigh damping and dynamic response calculation (Song and Su 2017), a number of modes or degree of freedom are 2.5 times of significant number of modes (m). Therefore, a number of modes to take advantage of less computational time along with ensuring accuracy is (2.5 × 2 = 5). Hence, five number of modes are optimum for the considered system and further dynamic response of torsional vibration calculation is done by considering five lumped elements.

Step 2: Dynamic response of torsional vibration of lumped masses of experimental setup

As in step 1.3, modes become orthonormalized which helps in ease in computation during torsional vibration calculations. Orthonormalized mode shape satisfy the conditions as given in (27.37).

$$\left[\theta_{\mathrm{ONj}}\right]^{\mathrm{T}}[\mathrm{J}][\theta_{\mathrm{ONi}}] = 0 \text{ and } \left[\theta_{\mathrm{ONj}}\right]^{\mathrm{T}}[\mathrm{K}][\theta_{\mathrm{ONi}}] = 0 \tag{27.37}$$

$$\text{Now, } \left[\,J\,\right]\{\ddot{\theta}\} + [C_t]\{\dot{\theta}\} + \left[\,K_t\,\right]\{\theta\} = \{\tau\} \tag{27.38}$$

By principal coordinate $q_{\mathrm{i}}(\mathrm{t})$ these all equation of motion represented by differential equation become independent or uncoupled. By theorem of expansion $q_{\mathrm{i}}(\mathrm{t})$ is expressed as (27.39).

$$\{\theta\} = [\theta_{\mathrm{ON}}]\{q\}and\{q\}[q_1; q_2; q_3; q_4; q_5] \tag{27.39}$$

where q_1, q_2, q_3, q_4, and q_5 are principal coordinates for each lumped mass. Using (27.39) in (27.38) further multiplying $[\theta_{\mathrm{ON}}]^{\mathrm{T}}$, equation become

$$[\theta_{\mathrm{ON}}]^T\left\{[J][\theta_{\mathrm{ON}}]\{\ddot{q}\} + [\theta_{\mathrm{ON}}]^T[C_t][\theta_{\mathrm{ON}}]\{\dot{q}\} + [\theta_{\mathrm{ON}}]^T[K_t][\theta_{\mathrm{ON}}]\{q\}\right\} = [\theta_{\mathrm{ON}}]^T\{\tau\} \tag{27.40}$$

Also, as modes are inertia normalized, so $[\theta_{\mathrm{ON}}]^{\mathrm{T}}$ $[\mathrm{J}][\theta_{\mathrm{ON}}]$become identity matrix called generalized inertia matrix ($[\mathrm{J_g}]$) and $[\theta_{\mathrm{ON}}]^{\mathrm{T}}$ $[\mathrm{K}][\theta_{\mathrm{ON}}]$ become diagonal matrix with ωnj^2 as diagonal element, named as generalized stiffness matrix ($[\mathrm{K_g}]$) where, ω_{nj} is natural frequency. $[\theta_{\mathrm{ON}}]^{\mathrm{T}}$ $[\mathrm{C_t}][\theta_{\mathrm{ON}}]$ is the generalized damping matrix ($[C_{\mathrm{g}}]$).

Step 2.1: Calculation of generalized damping matrix ($[C_{\mathrm{g}}]$)

In this step, the procedure for calculation of modal damping ratio (ξ) and damping matrix is presented which is generalized for any shaft coupler system. For torsional damping, the present system is in such a way that its damping matrix is a linear combination of inertia and stiffness matrix. This mentioned in (Song and Su 2017), that this assumption can be used where inertia and stiffness matrix is symmetric and in present case these matrices are symmetric. i.e.,

$$C_t = \alpha[J] + \beta[K_t] \tag{27.41}$$

where α and β are constants. This is called Rayleigh damping, classically damped or proportional damping.

This is used in generalized damping matrix ($[C_{\mathrm{g}}]$) i.e.,

$$[C_g] = [\theta_{\mathrm{ON}}]^T[C_t][\theta_{\mathrm{ON}}] \tag{27.42}$$

Using (27.42) in (27.41)

$$[C_g] = \alpha[\theta_{ON}]^T[J][\theta_{ON}] + \beta[\theta_{ON}]^T[K_t][\theta_{ON}] \tag{27.43}$$

Using property of orthonormality of modes as in paper, i.e.,

$$[\theta_{ON}]^T[J][\theta_{ON}] = \text{I (Identity matrix)} \tag{27.44}$$

And

$$[\theta_{ON}]^T[K_t]\,[\theta_{ON}] = \text{diag}\left(\omega_{nj}^2\right)\left(\text{diagonal matrix having diagonal element}\omega_{nj}^2\right) \tag{27.45}$$

Using (27.44) and (27.45) in (27.43),

$$\left[C_g\right] = \alpha[\text{I}] + \beta\text{diag}\left(\omega_{ni}^2\right) \tag{27.46}$$

And as in (27.40),

$$\alpha + \beta\text{diag}\left(\omega_{ni}^2\right) = 2 \times \xi_i \times \omega_{ni} \tag{27.47}$$

where ξ_j is modal damping ratio.

$$\xi_i = \frac{\alpha}{2 \times \omega_{ni}} + \frac{\beta \times \omega_{ni}}{2} \tag{27.48}$$

To find α and β, by using (Papadrakakis et al. 2000; Irvine 2015). It seems from (27.48) that initially for a small value of ω_{nj} vary from 0.5–8.5 rad/s as given in (Papadrakakis et al. 2000) $\frac{\alpha}{2\times\omega nj}$ factor dominates and variation of ξ_j vs ω_{nj} is not linear. But for, $\omega_{ni} > 8.5$ rad/s the variation become linear. In the present case the minimum natural frequency i.e., first natural frequency is 1.495×10^2 rad/s. Hence, for the present case and also it is said in Rao (2016), in most of the cases, linear variation is assumed between ξ_j vs ω_{nj}. From (Bachmann and Ammann 1995, ISO 2001), for carbon steel (Grade -304), the upper and lower bond of damping ratio varies from 0.001 to 0.002. The minimum value of ξ is considered at a first natural frequency and maximum value at a natural frequency of the last mode which is significant, i.e., in present case up to 2nd mode Therefore, the highest value of ξ is at a natural frequency corresponding to the 2nd mode. Hence for mode 1, modal damping ratio $\xi_1 = 0.001$, and for the last significant mode (in this case 2nd mode) $\xi_2 = 0.002$. Further, for calculation of remaining ξ_i, the following methods are followed-

Method 1- Considering linear variation between ξ_i and ω_{ni} as an angular frequency in the present case is higher.

Using linear interpolation

$$\frac{\xi_i - \xi_1}{\omega_{n,i+1} - \omega_{nm}} = \frac{\xi_m - \xi_1}{\omega_{nm} - \omega_{n1}}, \text{ for, m} < \text{i} < 2.5\text{m} \quad (27.49)$$

where m is a number of significant modes. By using (27.49), ξ_3, ξ_4, and ξ_5 are calculated i.e., modal damping ratio corresponding to 3rd, 4th, and 5th modes are calculated.

Method 2- Using data points as mentioned above $\xi_1 = 0.001$, $\xi_2 = 0.002$, $\omega_{n1} = 1.495 \times 10^2$ rad/s and $\omega_{nm} = 2.177 \times 10^2$ rad/s (in present case m = 2) in (27.50)

$$\beta = \frac{(2 \times \xi_1 \times \omega_{n1}) - (2 \times \xi_m \times \omega_{nm})}{\omega_{ni}^2 - \omega_{nm}^2} \quad (27.50)$$

where ξ_1, ξ_m are modal damping ratios for 1st and last significant mode (i.e., m = 2). And ω_{n1} and ω_{nm} are natural frequency in rad/sec for 1st and last significant mode (m = 2).

(27.50) gives β and using (27.51) i.e.,

$$\alpha + \beta\left(\omega_{n1}^2\right) = 2 \times \xi_1 \times \omega_{n1} \quad (27.51)$$

Equation (27.51) gives α. Using (27.52).
i.e.,

$$\alpha + \beta\left(\omega_{ni}^2\right) = 2 \times \xi_i \times \omega_{ni} \quad (27.52)$$

It gives ξ_3, ξ_4, and ξ_5 i.e., modal damping ratio corresponding to 3rd, 4th, and 5th modes are calculated.

Method 3- Using data points as mentioned above ξ_1 and $\xi_{2.5\,m}$ as calculated in the previous step and $\omega_{n1} = 1.495 \times 10^2$ rad/s and $\omega_{n,2.5\,m} = 5.237 \times 10^4$ rad/s (in present case m = 2) in (27.53)

$$\beta = \frac{(2 \times \xi_1 \times \omega_{n1}) - \left(2 \times \xi_{2.5m} \times \omega_{n,2.5m}\right)}{\omega_{ni}^2 - \omega_{n,2.5m}^2} \quad (27.53)$$

where ξ_1, $\xi_{2.5\,m}$ are modal damping ratios for 1st and last mode (i.e., 2.5 × m = 5). And $(\omega_n)_1$ and $(\omega_n)_{2.5\,m}$ are natural frequencies in rad/s for 1st and last mode (i.e., 5th mode).

(27.53) gives β and using (27.54) i.e.,

$$\alpha + \beta\left(\omega_{n1}^2\right) = 2 \times \xi_1 \times \omega_{n1} \quad (27.54)$$

(27.4) gives α. Using (27.55).
i.e.,

$$\alpha + \beta\left(\omega_{ni}^2\right) = 2 \times \xi_i \times \omega_{ni} \tag{27.55}$$

It gives ξ_3, ξ_4, and ξ_5 i.e., modal damping ratio corresponding to 3rd, 4th, and 5th modes are calculated.

Method 4- In this method, the average value of ξ_i obtained in method -2 and method -3 are considered. So, in actual practice which method seems to be near to linear interpolation values that method is followed. This is the procedure of finding modal damping ratio ξ_i. But as in Rao (2016), to avoid such a long procedure, authors considered a constant value of modal damping ratio within the limit of accuracy. Hence, ξ_i is equal to 0.0015, constant for all modes, and found that this value best fits the values that come from the linear interpolation method. In this way, the torsional damping coefficient used in the generalized torsional damping matrix ($[C_g]$)is calculated.

Step 2.2: Calculation of resisting bearing torque (τ_{bj}): τ_{bj} is resisting torque offered by ball bearing and its empirical relation (27.56) is taken from manufacturer catalogue.

$$\tau_b = 0.5 \times 0.0015 \times \text{radial force (N)} \times b\text{earing bore dia (m)}. \tag{27.56}$$

where the radial force (N) is calculated using the shear force diagram method and shown in Table 27.8.

Putting all these in (27.40) and the uncoupled five second order differential equation in principal coordinates are obtained as (27.57)–(27.61)

$$\ddot{q}_1 + 0.45\dot{q}_1 + 22337.8q_1 = 0.78 + 0.87sin314t \tag{27.57}$$

$$\ddot{q}_2 + 0.65\dot{q}_2 + 47380.5q_2 = 0.21 + 0.02sin314t \tag{27.58}$$

$$\ddot{q}_3 + 59.0\dot{q}_3 + 38751.2 \times 10^4 q_3 = -1.7 - 0.18sin314t \tag{27.59}$$

$$\ddot{q}_4 + 148.1\dot{q}_4 + 24.36 \times 10^8 q_4 = 3.51 + 0.394sin314t \tag{27.60}$$

$$\ddot{q}_5 + 157.1\dot{q}_5 + 27.42 \times 10^8 q_5 = -82.45 - 9.1sin314t \tag{27.61}$$

Table 27.8 Resisting torque offered by bearings

Sr. No	Name of bearing	Radial force (N)	Bore diameter (m)	Resisting torque (τ_{bj}) (N-m)
1	B_1	13.37081	22×10^{-3}	2.265×10^{-4}
2	B_2	1.25	19×10^{-3}	1.7281×10^{-5}
3	B_3	16.06351	38×10^{-3}	4.5781×10^{-5}

These equations can be generalized as

$$\ddot{q}_i + C_i\dot{q}_i + D_iq_i = E_i + B_i sin\omega_i t \tag{27.62}$$

whose solution is given as (27.63)

$$q_i(t) = \frac{E_i}{D_i} + \frac{B_i/D_i \times sin(\omega_i t - \varnothing_i)}{\sqrt{\left(1 - r_i^2\right)^2 + (2\xi_i r_i)^2}} \tag{27.63}$$

$$\text{where } \varnothing_i = \tan^{-1}\left(\frac{2\xi_i r_i}{1 - r_i^2}\right) \text{ and } r_i = \frac{\omega_i}{(\omega_n)_i} \tag{27.64}$$

By putting all values of E_i, D_i, and B_i from (27.57)–(27.61) in (27.63, 27.64), the steady state response in principal coordinate are:

$$q_1 = 3.516 \times 10^{-5} + 1.137 \times 10^{-5}\sin(314t - 3.138) \tag{27.65}$$

$$q_2 = 4.393 \times 10^{-6} + 4.486 \times 10^{-7}\sin(314t - 3.135) \tag{27.66}$$

$$q_3 = -4.405 \times 10^{-9} - 4.868 \times 10^{-10}\sin\left(314t - 4.784 \times 10^{-5}\right) \tag{27.67}$$

$$q_4 = 1.465 \times 10^{-9} + 1.617 \times 10^{-10}\sin\left(314t - 1.907 \times 10^{-5}\right) \tag{27.68}$$

$$q_5 = -3.006 \times 10^{-8} - 3.319 \times 10^{-9}\sin\left(314t - 1.975 \times 10^{-5}\right) \tag{27.69}$$

Putting Eqs. (27.65–27.69) in (27.39), steady state response of torsional vibration is $\{\theta\} = [\Theta_{ON}]\ \{q\}$ and is plotted as Fig. 27.19.

27.6 Experimental Validation of Simulation Results and Finding of Results

For validation of modelling results (Table 27.7), accelerometers are placed on PMSM and load DC generator. Initially PMSM is run at rated rotational speed, and then power off causes reduction in speed to zero. In between, the frequency response is noted as shown in Fig. 27.20. In present setup, the maximum rotational speed of PMSM is 3000 rpm. So, out of five natural frequencies, first two natural frequencies i.e., 23.78 Hz and 34.64 Hz, respectively are within range of maximum rated speed and are possible to excite during starting or stopping of motor. So, only first two modes are significant and need to be analysed, remaining modes are not able to excite as

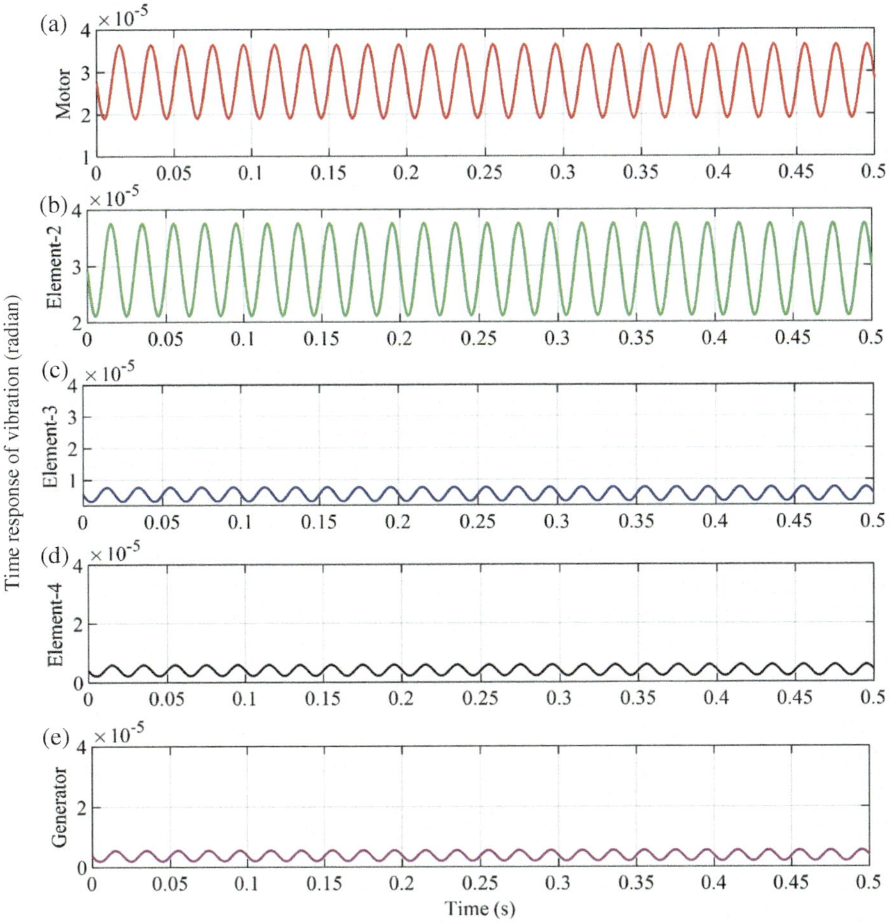

Fig. 27.19 Dynamic torsional vibration response corresponding to lumped masses: **a** PMSM Drive, **b** S_1-C_1, **c** C_2-S_2-C_3, **d** C_4-S_3, and **e** load DC generator

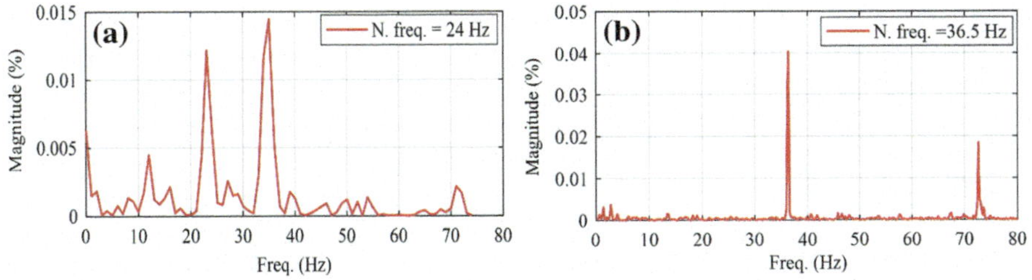

Fig. 27.20 Experimental validation of resonance response: **a** PMSM and **b** load DC generator

their natural frequencies are far away from rated speed. That is why, only first two modes can be validated experimentally and results are presented as Fig. 27.20.

It is observed from Fig. 27.20a, that the first peak of frequency response in case of accelerometer attached over PMSM is 24 Hz, close to the modelling result i.e., 23.78 Hz, which is resonance frequency of PMSM. Also, by Fig. 27.20b, for load DC Generator peak of frequency response comes at 36.5 Hz, close to modelling result i.e., 34.6 Hz, which is resonance frequency of load DC generator. Hence analytically, experimental results are validated.

Also, using MATLAB, the vibration response at resonance called mode shape (here modes are displacement normalised i.e., reduced to scale 1) is obtained corresponding to first two natural frequencies. From Fig. 27.21a, it can be easily noticed that all the lumped elements are twisted in one direction when resonance at frequency 23.78 Hz will occur, and maximum amplitude of vibration is in PMSM so this verify that 23.78 Hz is natural frequency of PMSM. To analyse relative twisting force on each element, the amplitude of angular displacement of that element is subtracted from previous one as twisting force is proportional to difference in angular displacement. So, from Fig. 27.21c, it can be easily visualised that maximum twisting occur at fixture of PMSM. From Fig. 27.21b, it can be noticed that at resonance frequency 34.64 Hz, PMSM and lumped element-2 is twisting in one direction while lumped element 3,4 and 5 twisted in other direction, means maximum torsion is to be occur between lumped element 2 and 3 i.e., coupler A. And Fig. 27.21d justify this, since lumped element 2 has highest relative twisting force. Also Fig. 27.21b shows that maximum amplitude of vibration at 34.64 Hz occurs at DC generator, so this justifies experimental and analytical result. Hence these numerical results are also consistent with experimental results.

The sinusoidal carrier-based PWM technique has been implemented to observe and analyses ANV. The simulation results of the PMSM presented in Fig. 27.22.

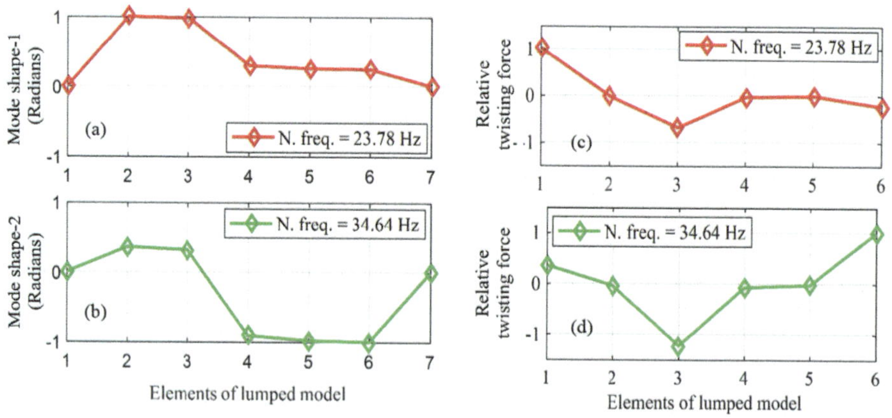

Fig. 27.21 Numerical result of vibration response of lumped elements: **a** First mode shape, **b** Second mode shape, **c** Twisting force corresponding to lumped elements at first resonance and **d** Twisting force corresponding to lumped elements at second resonance

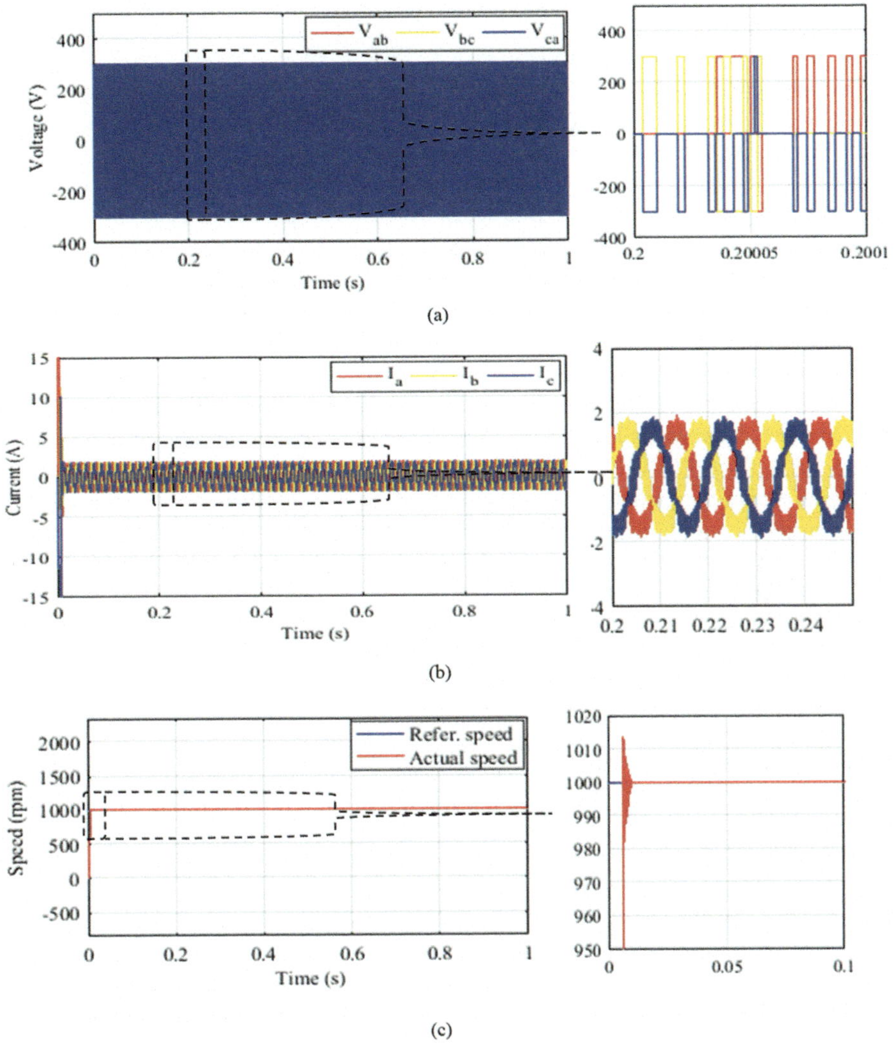

Fig. 27.22 Simulation results of PMSM drive: **a** three-phase stator voltage response, **b** three-phase stator current response and **c** steady-state speed response

Three-phase stator voltage V_{ab}, V_{bc} and V_{ca} of PMSM drive are shown in Fig. 27.22a. Phases are 120° shifted with from each other. 3-ϕ stator current response of PMSM is also presented in Fig. 27.22b. Waveforms of three-phase stator current are pure sinusoidal but with ripple value which is not acceptable in the industry. The steady-state speed response of the PMSM drive is shown in Fig. 27.22c. The PMSM drive is operating while taking a reference speed of 1000 rpm, as shown in Fig. 27.22c. The speed controller forces the motor to follow reference speed. The settling time for speed response is less than 0.001 s. Settling time for a speed response is one of the characteristics of a high-performance PMSM drive.

Electromagnetic torque produced by the motor is shown in Fig. 27.23. Torque is a component of current, so harmonics in current produces ripple in torque. Therefore, as shown in Fig. 27.23 the ripple or harmonic component in the torque is touching above the 1.36 N-m value in the graph, although peak to peak ripple is 1.2 N-m. Figure 27.23 presents a time and frequency domain vibration spectrum of the PMSM drive obtained from shaft modelling of PMSM in MATLAB/Simulink at I = 1.95 A, N = 1000 rpm with an average magnitude of vibration at a speed of 1000 rpm is observed 0.78 gravitation (values of vibration in m/s^2 is 7.644), and maximum vibration occurs at 160 Hz with the magnitude of 80% gravitation.

The experimental results of hall-sensor based PMSM drive are shown in Fig. 27.24. The three-phase stator input voltage with the amplitude of 300 V of PMSM drive is shown in Fig. 27.24a. The three-phase stator current with the amplitude of 2. 1 A of PMSM drive is shown in Fig. 27.24b. The waveforms of stator current are pure sine wave with 120° phase shift to each other. But the three-phase stator current contained too much ripple due to Maxwell forces. The steady-state speed response of the PMSM drive is shown in Fig. 27.24c. The settling time for speed response is less than 0.001 s. The experimental speed response shows very less steady-state error.

The steady-state torque response of SPWM based PMSM drive is shown in Fig. 27.24d. The observed average torque response is around 1.6 N-m. The observed torque ripple is around 0.8 N-m, which is very high and not acceptable in the industry. The generated torque ripple in PMSM drive may be causes of the internal structure of PMSM, the configuration of the stator winding, the shape of the stator slots and power electronics control algorithm.

Experimentally obtained acoustic noise and vibration of SPWM based PMSM drive are shown in Fig. 27.24e, f. The experimental result of vibration was measured using the accelerometer sensor with an output sensitivity of 10 mV/*g*. The accelerometer sensor was deployed horizontally on the end body of the PMSM drive to measure vibration. A microphone with an output sensitivity of 10 mV/Pa is located in line with the radial-axial of one of the stator poles to measure acoustic noise. To

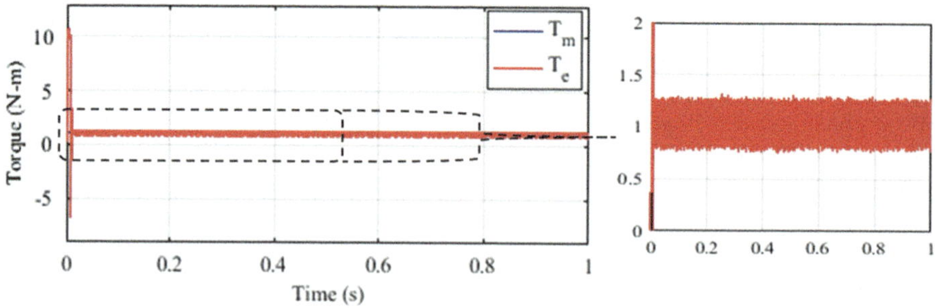

Fig. 27.23 Electromagnetic torque response of PMSM drive

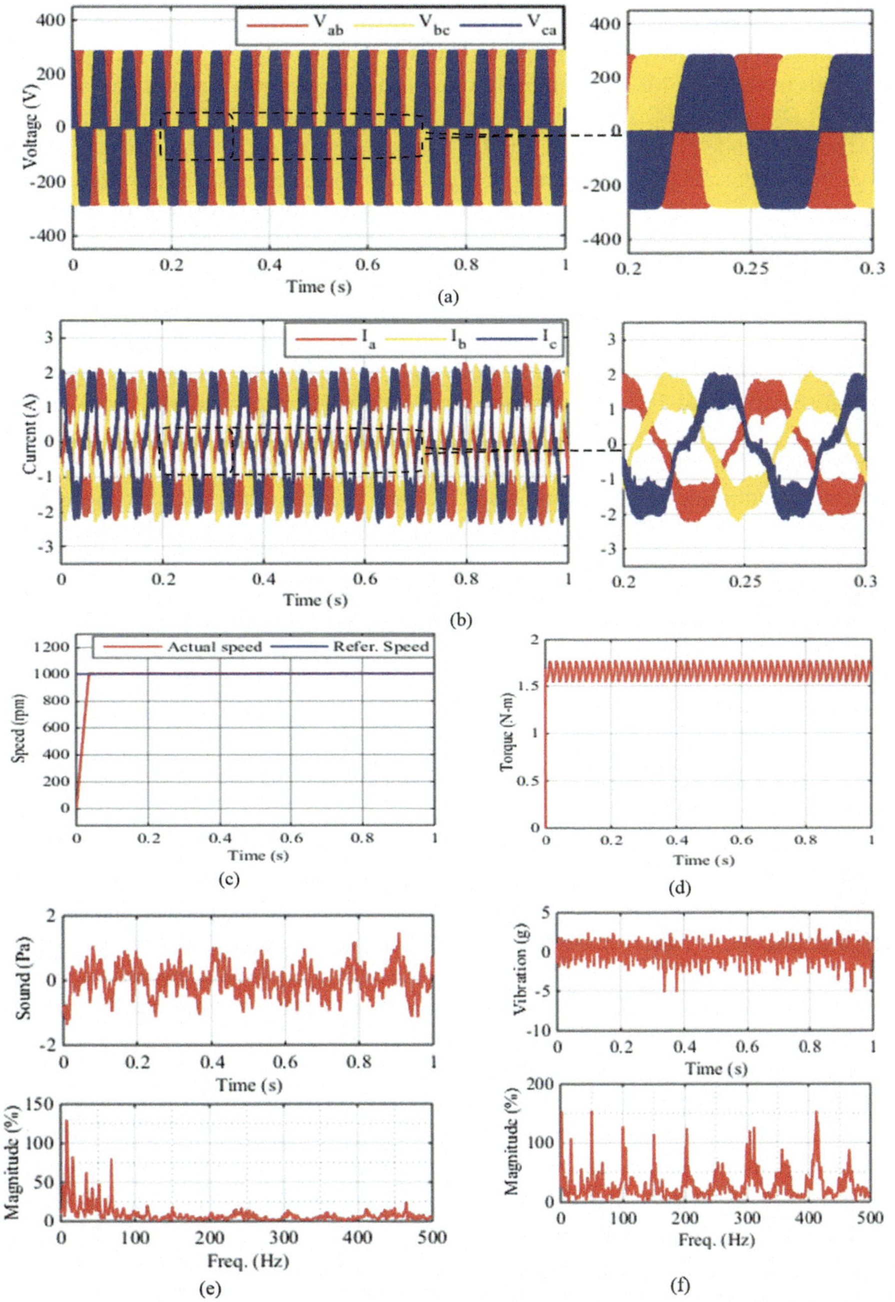

Fig. 27.24 Experimental results of PMSM drive: **a** rated input supply voltage, **b** current response, **c** steady-state speed response (refer. speed 1000 **rpm**), **d** torque response, **e** time and frequency domain response of acoustic noise and **f** time and frequency domain response of vibration

measure acoustic noise of PMSM drive, a half-inch free-field microphone provided by National Instruments (NI) is used. The microphone placed 5 cm far away from the shaft of the PMSM. The sampling rate of the sound and vibration data recorded through DAQ is 50000.

The amplitude of acoustic noise is 1.2 Pa measured at I = 2.1 A and N = 1000 rpm is shown in Fig. 27.24e. The generated acoustic noise has a direct correlation with electromagnetic torque ripple of PMSM drive. Also, it depends on the switching frequency of the power electronics converters. Here, measured acoustic noise at a frequency range of 500 Hz. But the magnitude of acoustic noise varies according to switching frequency of power electronics. The time and frequency spectrum of vibration signal for PMSM drive is shown in Fig. 27.24f. Vibration sensors were located horizontally at the end of the body of the PMSM. One accelerometer with an output sensitivity of 10 mV/g was attached with beeswax at a point directly behind any stator poles to measure vibration. Figure 27.24f presents a time, and frequency domain vibration spectrum of the PMSM drive at I = 2.00 A, N = 1000 rpm with vibration magnitude of 2.13 gravitation ($2.1 * 9.8 = 20.58\ \text{m/s}^2$) and maximum vibration occurs at 50 Hz with magnitude 150% gravitation, it could be due to mechanical resonance of mechanical parts like fan blades or the end cap. The vibrations are well in range as per the ISO 7919–2: 2001 (2001) guidelines. However, acoustic noise signals were almost double as per the prescribed range according to the IEEE Standard 85–1980 (1980).

27.7 Acoustic Noise and Vibration Reduction Techniques

The relation between the flux density and radial forces in the airgap of the PMSM drive is defined as; the radial forces are directly proportional to the square of the flux density in the air gap. The easy and fast ways to mitigate ANV would decrease the flux density by increasing the air gap. Skewing the stator slots decreases the radial forces of PMSM drive, which results in a decreased level of acoustic noise and vibration (Basu et al. 2009). Various guidelines are available for how to design stator slot and how to select a number of slots to mitigate unbalanced magnetic pull and locking between stator slots and rotor of PMSM drives, it can be helpful in reducing the ANV of PMSM drive (Preindl and Bolognani 2013; Gamoudi et al. 2018).

Many methods are available in the literature to eliminate the space harmonics, which is generated by power electronic converters (Basu et al. 2009; Pindoriya et al. 2018; Qu et al. 2021). The popular methods are Random Pulse Position (RPP), random switching operation, and a random switching frequency of power electronic semiconductor switches of three-phase voltage source inverters. The switching strategies of three-phase voltage and current source inverter will fix the voltage and current harmonic content. The supply voltage harmonics contain harmonics linked to the three-phase inverter PWM switching frequency like, (f_{sw}, $2f_{sw}$, $3f_{sw}$ and so on) and harmonics link to the fundamental switching frequency of power semiconductor switches (Qu et al. 2021). The largest harmonics and vibration can occur at once or twice the switching frequency depending on the PWM strategy and torque-speed operating point of electric machines (Pindoriya et al. 2018). Increase in the switching frequency of the three-phase inverter, reduces the ANV of electric machines drives but, increasing the power semiconductor switching losses. To eliminate the complete odd-order PWM frequency vibration, a dualbranch three-phase PMSM with carrier phase-shift implemented in Zhang et al. (2020). The comparison of different random PWM techniques is given in Fig. 27.25 (Zhang et al. 2020).

The abbreviations of spread spectrum PWM techniques are as following; Random Pulse Position Modulation (RPPM), Random Pulse Width Modulation (RPWM), Randomized Carrier Frequency Modulation with Fixed Duty-ratio (RCFMFD), Randomized Carrier Frequency Modulation with Variable Duty-ratio (RCFMVD), Randomized Duty-ratio and RPPM with Fixed Carrier Frequency (RDRPPMFCF), Randomized Carrier Frequency and RPPM with Fixed Duty-ratio (RCFRPPMFD), Randomized carrier frequency and a Randomized duty ratio, with RPPM (RRRM), Random Lead-Lag Modulation (RLLM), Random Phase Shift Modulation (RPSM), Asymmetric Carrier Modulation (ACM), Variable Delay Modulation (VDM), Fractal-Based Modulation (FBM), and Separately Randomized Pulse Position Modulation.

27.8 Conclusions

In this chapter, a detailed mathematical, numerical and further experimental validation of ANV in case of PMSM is presented. A mathematical framework suitable for torsional vibrational prediction of PMSM based drive by implementing lumped model along with effective mass method is proposed, which further used for any

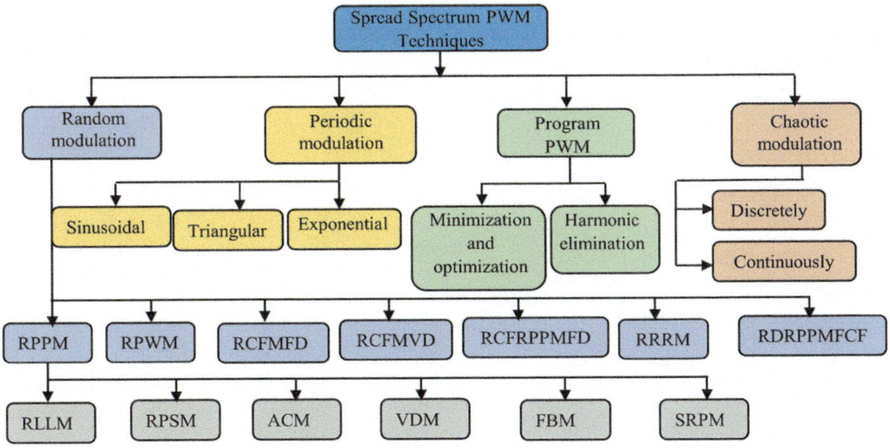

Fig. 27.25 Spread spectrum PWM techniques

power transmission system used in EV and HEV applications. Further, this methodology provides a detailed elaboration about selection of optimum number of modes or DOF which usually researchers took randomly based on their setup and mass distribution without a technical justification. This method having computational easiness along with accuracy and required minimal input in term of geometrical, material and source of vibration i.e., waveform of torque only. The analytical results verified over a small-scale laboratory setup and further extended for dynamic torsional vibration response.

From analytical and simulation results related to dynamic response of torsional vibration, quantitative observation is that the amplitude of vibration is maximum in PMSM which gradually fall down towards load DC generator end, reason is material own property to dampen the vibration. Also, as maximum twist angle lies between C_1 and C_2 disc of coupler A, so, coupler A is under highest fatigue load causes failure of component. The whole methodology is presented in generalized way so that it can be implemented to such systems used in electric vehicle applications and helps in vibration analysis and set design requirement for reliable design.

Also, an introductory idea of further reduction in ANV using RPWM also presented in this chapter which will be future scope of study. Another way of controlling the vibration is use of smart materials, for that also the present study act as pre-requisite to decide number of patches, patches dimension etc., which will precisely and efficiently reduce vibration. Vibration control tactics enhance the efficiency of mechanical power transmission, system reliability and reduces the ANV to a significant extent.

Appendix

A detailed graphical representation of methodology including SPWM driven PMSM along with setup of acoustic noise and torsional vibration measurement sensors is shown in Fig. 27.26.

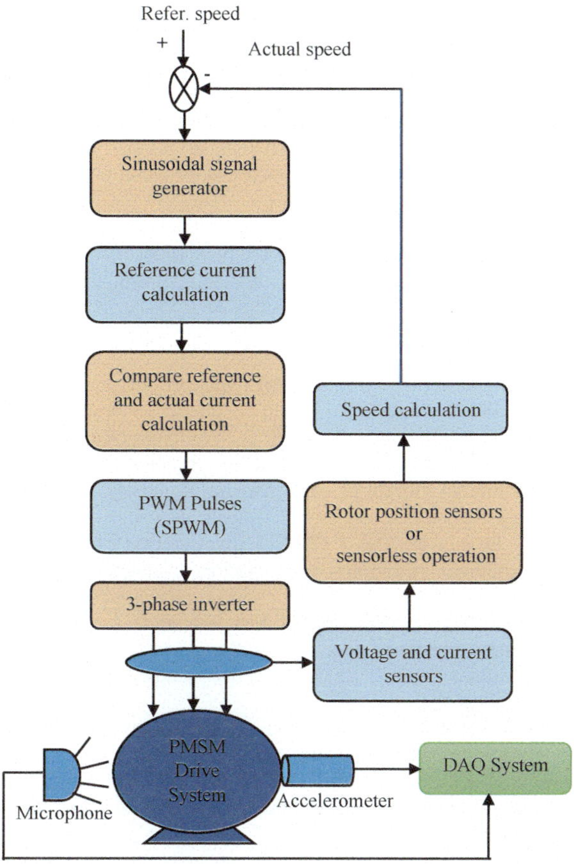

Fig. 27.26 Flow chart for analysis of acoustic noise and torsion vibration of PMSM drive system

References

ABB Manual (2015) AC Servo Motors BSM Series, pp 1–99. https://library.e.abb.com/public/

Bachmann H, Ammann WJ (1995) Vibration problems in structures. Birkhauser Verlag, Berlin

Basu K, Prasad JSS, Narayanan G (2009) Minimization of Torque Ripple in PWM AC Drives. IEEE Trans Ind Electron 56(2):553–558, Feb

Bosing M (2012) Acoustic modeling of electrical drives—Noise and vibration synthesis based on force response superposition, Ph. D. dissertation, Power Electronics and Electrical Drives, RWTH Aachen University, Aachen, Germany

Bolton HR, Ashen RA (1984) Influence of motor design and feed-current waveform on torque ripple in brushless DC drives. IEEE Proc B Electr Power Appl 131(3):82–90, May

Chan CC (2007) The state of the art of electric, hybrid, and fuel cell vehicles. Proc IEEE 95(4):704–718, Apr

Chengsheu H, Wenchen L (2000) A lumped mass model for parametric instability analysis of cantilever shaft-disk systems. J Sound Vib 234(2):331–348

Chattopadhyay S, Chattopadhyaya A, Sengupta S (2014) Measurement of harmonic distortion and skewness of stator current of induction machine at crawling in clarke plane. IET Sci Measur Technol 8(6):528–536

Cremer L, Heckl M (1988) Structure-borne sound. Springer-Verlag, New York

Dorrell DG, Hsieh MF, Popescu M, Evans L, Staton DA, Grout V (2011) A review of the design issues and techniques for radial-flux brushless surface and internal rare-earth permanent magnet motors. IEEE Trans Industr Electron 58(9):3741–3757, Sep

EL-Refaie M (2010) Fractional-slot concentrated-windings synchronous permanent magnet machines: opportunities and challenges. IEEE Trans Industr Electron 57(1):107–121, Jan

Empirical relation to calculate frictional torque offered by radial ball bearing. https://www.smbbearings.com/

Gamoudi R, ElhakChariag D, Sbita L (2018) A review of spread-spectrum-based PWM techniques—A novel fast digital implementation. IEEE Trans Power Electron 33(12):10292–10307, Dec

Girgis RS, Vermas SP (1981) Method for accurate determination of resonant frequencies and vibration behavior of stators of electrical machines. IEE Proc B - Electric Power Appl 128(1):1–11, Jan

Gieras JF, Wang C, Lai JCS (2006) Noise of polyphase electrical motors. Taylor & Francis Group

Han X, Palazzolo AB (2013) VFD machinery vibration fatigue life and multilevel inverter effect. IEEE Trans Ind Appl 49(6):2562–2575, Nov–Dec

Huang Q, Liu H, Cao J (2019a) Investigation of lumped-mass method on coupled torsional-longitudinal vibrations for a marine propulsion shaft with impact factors. J Marine Sci Eng 7(4):1–16

Huang Q, Liu H, Cao J (2019b) Investigation of lumped-mass method on coupled torsional-longitudinal vibrations for a marine propulsion shaft with impact factors. J Mar Sci Eng 7:95

IEEE standard 85–1980. IEEE Test Procedure for Airborne Sound Measurements on Rotating Electric Machinery, New York, USA, pp 1–21, Dec

ISO standard 7919–2: 2001, Reference standards for vibration monitoring and analysis, pp 1–10, 2001

Irvine T (2015) Effective modal mass and modal participation factors. Vibration data. http://www.vibrationdata.com/tutorials2/ModalMass.pdf

Junhui Y, Xu L, Wang H, Xie P, Huang S, Liu H, Yang Z, Li B (2019) Accurate and fast three-dimensional free vibration analysis of large complex structures using the finite element method. Comput Struct 221:142–156

Knopik T, Binder A (2011) Measurement proven analytical and numerical models for calculation of the teeth flux pulsations and harmonic torques of skewed squirrel cage induction machines. In 2011 IEEE energy conversion congress and exposition, Phoenix, AZ, pp 162–169

Le Besnerais J, Lanfranchi V, Hecquet M, Brochet P (2009) Optimal slot numbers for magnetic noise reduction in variable-speed induction motors. IEEE Trans Magn 45(8):3131–3136, Aug

Le-Huy H, Perret R, Feuillet R (1986) Minimization of torque ripple in brushless DC motor drives. IEEE Trans Ind Appl IA-22(4): 748–755, Jul

Manguelle JS, Ekemb G, Mon-Nzongo DL, Jin T, Doumbia ML (2018) A theoretical analysis of pulsating torque components in AC machines with variable frequency drives and dynamic mechanical loads. IEEE Trans Industr Electron 65(12):9311–9324, Dec

Manjibhai RP, Thakur RK, Bharat Singh R, Kumar R (2021) Numerical and experimental analysis of torsional vibration and acoustic noise of PMSM coupled with DC generator. IEEE Trans Ind Electron 69(4):3345–3356, April 2022. https://doi.org/10.1109/TIE.2021.3076715

Papadrakakis M, Lagaros N, Plevris V (2000) Optimum design of structures under seismic loading. In European Congress on Computational Methods in Applied Sciences and Engineering, Barcelona

Pindoriya RM, Mishra AK, Rajpurohit BS, Kumar R (2018) An analysis of vibration and acoustic noise of BLDC motor drive. In 2018 IEEE power & energy society general meeting (PESGM), Portland, pp 1–5

Preindl M, Bolognani S (2013) Model predictive direct torque control with finite control set for PMSM drive systems, part 1: Maximum torque per ampere operation. IEEE Trans Ind Inf 9(4):1912–1921, Nov

Qianwen H, Xinping Y, Wang Y, Zhang C, Wang Z (2017) Numerical modeling and experimental analysis on coupled torsional-longitudinal vibrations of a ship's propeller shaft. Ocean Eng 136: 272–282

Qu J, Zhang C, Jatskevich J, Zhanga S (2021) Deadbeat harmonic current control of permanent magnet synchronous machine drives for torque ripple reduction. IEEE J Emerg Select Topics Power Electron [Early Access]

Rahman MH, Gupta C (2020) Computation of Rayleigh damping co-efficient of a rectangular submerged floating tunnel (SFT). SN Appl Sci 2:936

Rao SS (2016) Mechanical vibration (6th ed). Pearson India Education Services, Pvt. Ltd.

Rik HC, Doncker WD, Kennel R (2016) Acoustic modeling of electrical drives—Noise and vibration synthesis based on force response superposition. Thesis

Rabbi SF, Kahnamouei JT, Liang X, Yang J (2020) Shaft failure analysis in soft-starter fed electrical submersible pump systems. IEEE Open J Ind Appl 1:1–10. https://doi.org/10.1109/OJIA.2019.2956987

Report (2016) Implementing agreement for co-operation on hybrid and electric vehicle technologies and programmed. International Energy Agency, Jun

Seung Cho D, Hee Kim B, Kim J-H, Choi TM, Vladimir N (2016) Free vibration analysis of stiffened panels with lumped mass and stiffness attachments. Ocean Eng 124:84–93

Song Z, Su C (2017) Computation of Rayleigh damping coefficients for the seismic analysis of a hydro-powerhouse. In Hindawi Shock and Vibration, vol. 2017, pp 1–11

Shen Y et al (2020) Comparative study on dynamic characteristics of two-stage gear system with gear and shaft cracks considering the shaft flexibility. IEEE Access 8:133681–133699. https://doi.org/10.1109/ACCESS.2020.3009398

Xiaohua L, Zhang L, Hongliang Y, Surong H, Zhang Q (2020) Study of suppression of vibration and noise of PMSM for electric vehicles. IET Electr Power Appl 14(7):1274–1282

Verma SP, Williams K, Singal RK (1989) Vibrations of long and short laminated stators of electrical machines, part I: theory, experimental models, procedure, and set-up. Sound Vibr 129:1–13

Zhang W, Xu Y, Huang H, Zou J (2020) Vibration reduction for dual-branch three-phase permanent magnet synchronous motor with carrier phase-shift technique. IEEE Trans Power Electron 35(1): 607–618, Jan

Zhang Q, Lu G, Xu Y, Zhang C (2020) Torsional vibration analysis of shaft in an induced draft fan due to variable frequency drive. IEEE Access 8:174723–174735

Zou J, Lan H, Xu Y, Zhao B (2017) Analysis of global and local force harmonics and their effects on vibration in permanent magnet synchronous machines. IEEE Trans Energy Convers 32(4):1523–1532, Dec

Chapter 28
Integration of Renewable Sources and Energy Storage Devices

Dipanshu Naware, Ram Babu Thogaru, and Arghya Mitra

Abstract With the ever-increasing penetration of renewable energy sources, solar and wind are emerging as eco-friendly generating resources in modern-day power systems. Due to their highly unpredictable nature, the energy storage system is frequently being used in coordination with these sources. This chapter focuses on the overview of the integration of solar, wind, and energy storage system in the present-day power system along with the challenges and control strategies. Photovoltaic systems are used to extract the maximum amount of energy from the available solar intensity. The most commonly used configurations are grid-connected solar PV and stand-alone PV with an energy storage device. Similarly, wind energy has been there for thousands of years for sailing ships, water pumping applications, and so on. Based on the geographical locations, the potential of wind can be utilized to generate energy. Among the available wind generators, doubly-fed induction generator and permanent magnet synchronous generator become popular because of their features of working at variable speed in both super-synchronous and sub-synchronous regimes while extracting maximum power from wind. With the advent of new technologies, modern-day energy storage systems are cost-effective and more efficient. An energy storage system, when integrated with a renewable energy source, plays a vital role as it absorbs energy during periods of high generation and acts as a source during periods of high demand. The storage device thus can be used to reduce the fluctuations in power generated by the renewable sources that are being exchanged with the grid. To improve the resiliency of the modern-day grid, it can also be used as an emergency backup to satisfy the critical loads in the presence of any disturbance. An array of energy storage options is available in the market such as super-capacitors, superconducting magnetic energy storage, compressed air, pumped

D. Naware (✉) · R. B. Thogaru · A. Mitra
Visvesvaraya National Institute of Technology, Nagpur (M.S), India
e-mail: dipanshunaware@students.vnit.ac.in

R. B. Thogaru
e-mail: ramthogaru@students.vnit.ac.in

A. Mitra
e-mail: mitraarghya@eee.vnit.ac.in

A. K. Bohre et al. (eds.), *Planning of Hybrid Renewable Energy Systems, Electric Vehicles and Microgrid*, Energy Systems in Electrical Engineering,
https://doi.org/10.1007/978-981-19-0979-5_28

hydro storage, flywheels, and rechargeable batteries. This chapter covers the basics of solar, wind, and energy storage device, especially superconducting magnetic energy storage and battery energy storage system, with schematic illustrations such as electrical equivalent circuits, block diagrams, and control strategies. Apart from this, a novel approach for hybrid integration of solar PV, wind energy, and energy storage devices is proposed here. Specifically, the several possible configurations incorporating solar PV, a doubly-fed induction generator, a permanent magnet synchronous generator, a SMES, and a battery energy storage system are presented and compared based on cost, reliability, stability, and environmental impact. This chapter aims to induce knowledge of new-age power generating options and their integration with the utility grid, increase the share of clean and green energy and encourage contribution towards sustainable development.

Keywords Converters · DFIG · Energy storage system · Grid integration · Microgrid · PMSG · Solar PV · Wind energy

Nomenclature

BESS	Battery energy storage system
C-rate	Charge/discharge rate
CSC	Current source converter
DOD	Depth of discharge
DFIG	Doubly fed induction generator
ESS	Energy storage system
GSC	Grid side converter
IGBT	Insulated gate bipolar transistor
PMSG	Permanent magnet synchronous generator
PV	Photovoltaic
RES	Renewable energy sources
RSC	Rotor side converter
SOC	State of charge
STC	Standard test conditions
SMES	Superconducting magnetic energy storage
VSI	Voltage source inverter
WECS	Wind energy conversion system

28.1 Introduction

The growth of renewable energy sources (RES) in the current scenario is enormous and widely accepted all over the world due to their eco-friendly nature. It is the fastest growing industry of recent times by establishing new technologies in the modern-day power systems. Due to the unfavourable impact of fossil fuels on the environment, people are moving towards green energy generation such as solar photovoltaic, wind energy, fuel cells, and bio-generation, to reduce carbon emissions. The chapter begins with the fundamentals of RES followed by recent developments and includes integration of these sources with various system configurations and challenges associated with it. Also, few energy storage devices are discussed and finally, stress on the economic operation of grid integrated RES is given focusing to minimize the gap between the existing and future smart grid.

Solar is a free source of energy available in abundance. Photovoltaic systems are used to extract the maximum amount of energy from the available solar intensity. Latitude of the location, clearness index, sunrise and sunset hour angle, etc. are responsible for the amount and quality of radiation reaching the earth's surface. A Series combination of PV cells forms a module while an array is formed when multiple modules are connected in series or parallel depending upon the requirement (Masters 2005). The most commonly used configurations, grid-connected solar PV and stand-alone PV with an energy storage device are discussed in this chapter.

Wind energy has been there for thousands of years for sailing ships, water pumping applications, and so on. Based on the geographical locations, the potential of wind can be utilized to generate energy. Wind turbines convert the available wind into mechanical energy and mechanical to electrical energy conversion takes place with the help of a generator. Synchronous and asynchronous generators are being widely used in the modern-day application. The available configurations are a doubly-fed asynchronous generator and a permanent magnet synchronous generator. They become popular because of their features of working at variable speed in both super-synchronous and sub-synchronous modes for the extraction of power. Various evolutionary nature-inspired optimization algorithms can be used to extract the maximum power from both configurations (Vasavi Uma Maheswari et al. 2021; Dursun et al. 2021; Dursun and Kulaksiz 2020).

Intermittency issues related to RES need to be addressed for reliable operation of the system. As uncertainty looms with PV and wind generation, an energy storage system (ESS) is the viable solution. In the absence of PV/wind, ESS will serve the purpose of power generating source by satisfying the load demand, thus maintaining a reliable and uninterrupted power supply (Pamu et al. 2020; Rana et al. 2021). Nowadays, more and more cost-effective storage options are available to deal with these scenarios (Korjani et al. 2020). Superconducting magnetic energy storage (SMES) is an emerging technology due to its high efficiency, faster response, and limitless charging/discharging cycles (Mukherjee and Rao 2019a). On the other hand, a battery energy storage device (BESS), also known as a rechargeable battery, is frequently used in a modern-day microgrid. A lead-acid battery is the cheapest and the oldest one while a lithium-ion battery is under development stage. Although lithium-ion is costlier than lead-acid for the same size of the battery, the cyclelife feature outclasses the latter one. In this chapter, the focus is limited to SMES and rechargeable batteries applications. The proposed approach in a graphical representation is depicted in Fig. 28.1.

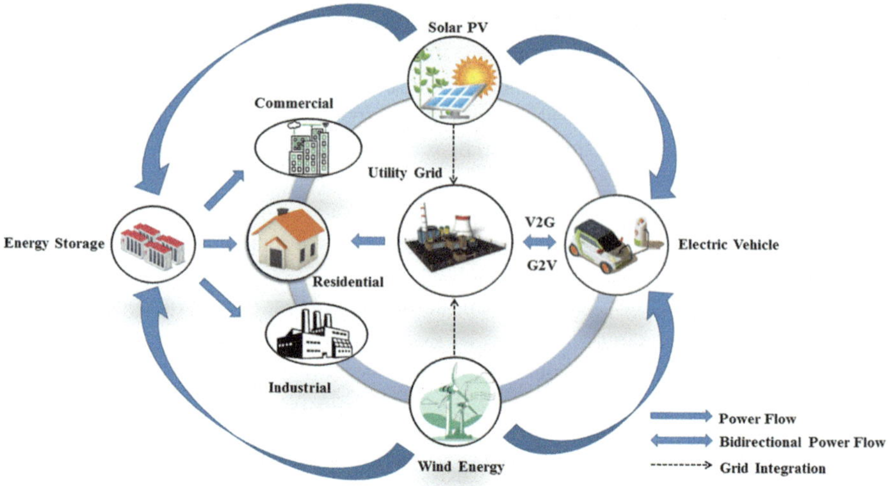

Fig. 28.1 Graphical representation of the proposed approach

The major contributions of the study are listed below:

- The hybrid integration of solar PV, wind, and ESS is presented.
- Few possible configurations incorporating BESS and SMES with PV and wind are proposed.

The chapter is structured as follows: Sect. 28.2 gives the overview of the solar photovoltaic system, its mathematical modelling, and two commonly used configurations; a gentle introduction to doubly fed induction generator and permanent magnet synchronous generator wind energy conversion systems; ESS with classifications and overview of power electronic interfaces; Sect. 28.3 covers various possible configurations for grid integrated hybrid RES with associated challenges and their comparison and Sect. 28.4 gives conclusive remarks and probable future scope.

28.2 Overview of Solar, Wind, Energy Storage System and Power Electronic Interfaces

This section covers the basics of solar PV, its mathematical modelling along with the description of on-grid and off-grid configuration; wind energy conversion system, its mathematical modelling, and associated configurations; ESS along with schematic illustrations, block diagram, mathematical modelling, applications, and various power electronic interfacing components associated with hybrid configurations. The objective of this section is to lay the foundation for RES and induce basic knowledge to the fraternity working in this field.

28.2.1 Solar Photovoltaic System

The abundance of available sunlight makes photovoltaic technology the most suitable candidate for renewable energy generation. It makes use of a solar cell to convert sunlight into electricity. The most commonly used material is silicon. Several solar cells are arranged in series and/or parallel to get the required rating, known as a photovoltaic module. The current produced is directly proportional to the solar intensity received on the panel. Similarly, a series or parallel combination of modules forms an array (Elbaset et al. 2014; Messenger and Abtahi 1999; Zhao et al. 2017). Figure 28.2 shows the basic electric circuit of a solar cell.

From the above equivalent circuit, the load current is given by (28.1),

$$I = I_{ph} - I_s\left\{e^{q\left[\frac{V+IR_s}{\alpha KT}\right]} - 1\right\} - \left[\frac{V + IR_s}{R_{sh}}\right] \tag{28.1}$$

The photo current, I_{ph} as a function of the temperature and the solar insolation can be expressed as follows

$$I_{ph} = \left(\frac{G}{G_{STC}}\right)\left[I_{ph,atSTC} + K_i(T - T_{STC})\right] \tag{28.2}$$

The diode saturation current as a function of PV temperature is given by

$$I_s = I_{s,atSTC}\left(\frac{T}{T_{STC}}\right)^3 e^{\left[\left(\frac{qE_g}{\alpha K}\right)\left(\frac{1}{T_{STC}} - \frac{1}{T}\right)\right]} \tag{28.3}$$

where,

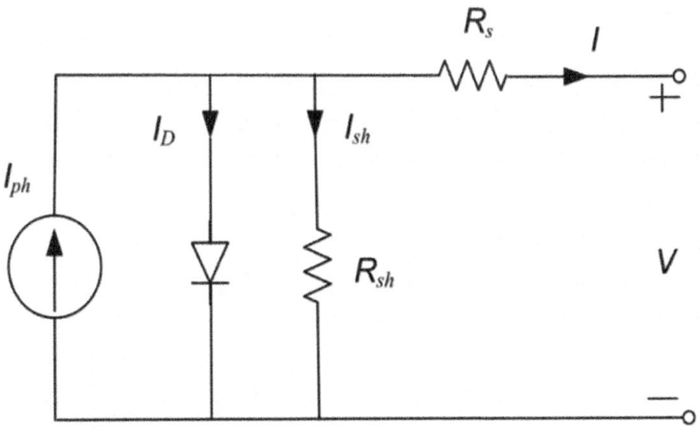

Fig. 28.2 Single solar cell circuit

I	load current (A)
V	load voltage (V)
I_{ph}	photo current (A)
I_s	cell reverse saturation current (A) at the standard test conditions (STC)
K	Boltzmann constant $\left(1.38 * 10^{-23} \text{J/K}\right)$
T	cell working temperature (K)
α	diode ideality factor
R_s	series resistance (Ω)
R_{sh}	shunt resistance (Ω)
G	is the solar irradiance (KW/m^2)
G_{STC}	is the solar irradiance at STC $\left(1\ \text{KW/m}^2\right)$.

The two most commonly used configurations are on-grid and off-grid. Grid-connected systems are frequently used worldwide. Here, the solar PV is linked to the utility grid and AC loads via power conditioning unit, such that during the presence of PV power, it feeds the AC loads via DC–AC converter, whereas excess generation can be injected into the utility grid facilitating the bidirectional flow of power. Similarly, during the absence of PV generation, the utility grid will take care of the loads as shown in Fig. 28.3. Inadequacy of PV imposes stress on the utility grid which may upset the bus voltage and deteriorates the system performance causing reliability issues. Hence there is a need to encompass a backup storage device that will share the deficit energy demand along with the utility grid and enhance the overall efficiency.

The challenge possessed by grid-connected PV systems can be addressed in standalone mode solar PV which is equipped with an energy storage device while the utility

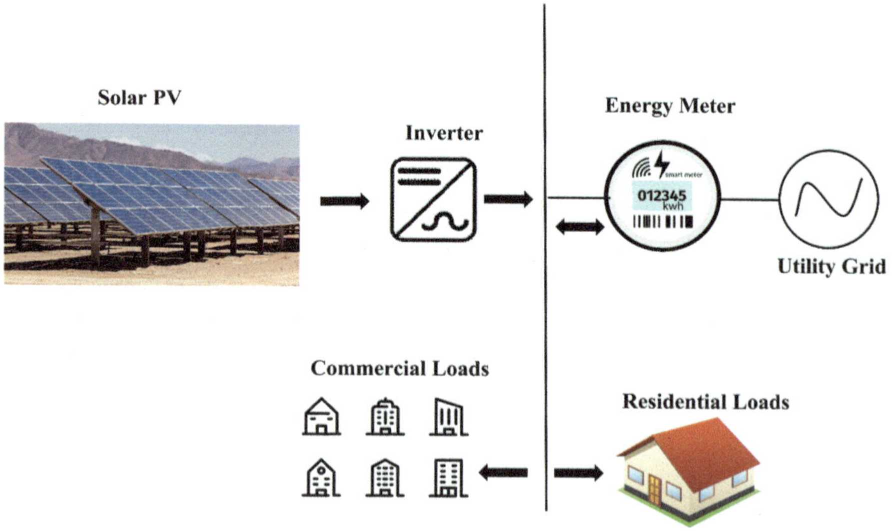

Fig. 28.3 Grid connected solar PV system

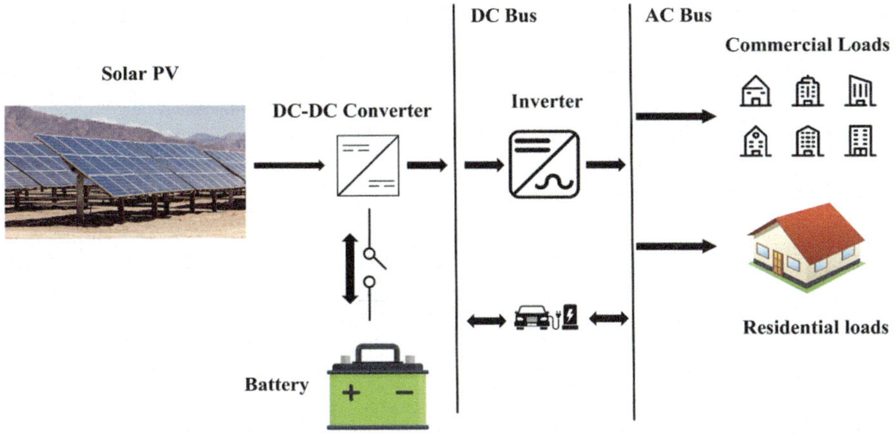

Fig. 28.4 Stand-alone solar PV system

grid is absent. During the periods of high solar intensity, solar PV generates power that can be fed to the loads and also can be stored in ESS while in the absence of solar PV power, ESS will deliver current to the load as depicted in Fig. 28.4. The total amount of deficit energy demand will be taken care of by rechargeable batteries. The selection of BESS is crucial and may be application-specific. Poor choice of storage device may lead to uneconomical operation causing faster battery degradation and lesser lifetime. Hence appropriate modelling of the battery aging phenomenon is a prerequisite.

28.2.2 Wind Energy Conversion System (WECS)

Energy extraction from wind is one of the oldest and cheapest forms of RES used to harness power when the earth's surface is non-uniformly heated by the sun. Over the past few years, it has been emerged significantly in the power industry (Steiger 1988). Wind turbines are used to alter the kinetic energy of wind into mechanical energy. Power generated by the wind turbine is proportional to the swept area and also to the cube of the wind speed. Thus larger the swept area, the larger will be the power production while double the wind speed produces eight times power as given in (28.4),

$$P_w = \frac{1}{2} C_p \rho A v^3 \tag{28.4}$$

where P_w is the power extracted from the wind (W), C_p is the power coefficient, ρ is the air density $\left(\mathrm{kg/m^3}\right)$, A is the cross-sectional area intercepted by the turbine blades $\left(\mathrm{m}^2\right)$ and v is the wind speed $(\mathrm{m/s})$.

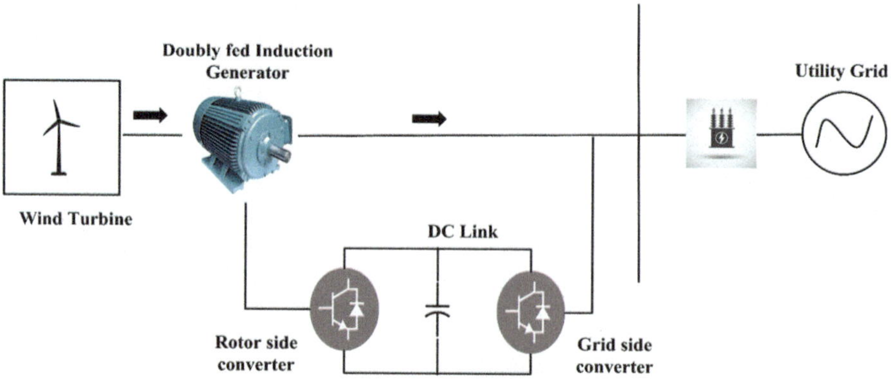

Fig. 28.5 Schematic of doubly fed induction generator based wind energy conversion system

Due to the uncertainty of wind, fixed-speed wind turbines are incapable of extracting maximum power. To attain variable speed, doubly-fed asynchronous generators and permanent magnet synchronous generator-based wind turbines are used by incorporating the merits of advanced power electronic devices. As the output of these generators is a function of variable wind speed, their integration with the utility grid is challenging. Both above-mentioned configurations are discussed below.

Doubly fed Induction Generator based WECS

As the name suggests, this type of wind generator is fed from both the ends, i.e. stator and rotor as shown in Fig. 28.5. The wound-rotor induction generator is connected to the wind turbine through a gearbox. The stator of the generator is directly connected to the utility grid while the rotor is connected via back-to-back converters, i.e. rotor side converter (RSC) and grid side converter (GSC). In this way, decoupling is achieved between rotor frequency and grid frequency. This configuration belongs to a Type 3 wind turbine generator (Mitra and Chatterjee 2016).

The shafts of the wind turbine and generator are connected through gears. These three arrangements (wind turbine, generator, and gear) are mechanically coupled and thereby arising some torsional oscillations. This 3-mass system may be represented by a 5th order model involving the dynamics related to each of these three masses along with that of the connected shafts. Nevertheless, it is reasonable to consider that gearbox inertia is negligibly small when compared with that of the turbine and/or generator. Thus, the high-speed shaft is considered to be rigid with respect to the low-speed shaft, which results in a lumped inertia of the combined gearbox and generator. So, the system behaves as a two-mass model and therefore may be represented using a 3rd order dynamic equation. When the generator and the turbine masses are assumed to be lumped to give a single rotating mass, the drive train can be expressed by a 1st order dynamic equation.

As the shaft connecting the gearbox in a wind farm is lean, it results in a low mechanical stiffness while compared with electrical stiffness. So, an equivalent

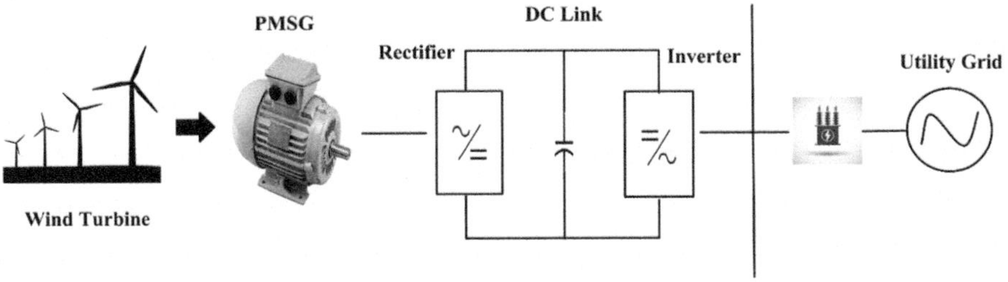

Fig. 28.6 Permanent magnet synchronous generator based wind energy conversion system

lumped mass model with a first-order differential equation can be considered as an over-approximation. On the other hand, three mass model is supposed to be best fitted when a detailed stability study of the wind farm itself is needed which forces us to consider each of the wind turbo-generator sets individually. It is to be mention here that, while the study is focusing on the power system stability, representation of the drive train with two mass model may be suitable.

Permanent Magnet Synchronous Generator based WECS

A permanent magnet synchronous generator (PMSG) can be used to generate variable voltage and variable frequency power as visible from Fig. 28.6. The PMSG integrated wind turbine is connected to the utility grid via back-to-back converters in series. The major drawback of this system is visible from the figure as both the stator and rotor side converters are in series with PMSG and thus they have to carry total power making this configuration costlier than DFIG-WECS. This configuration is known as a Type 4 wind turbine generator.

To extract maximum power, the speed of the wind turbine is controlled with the help of a rotor side converter. The rotor speed should be adjusted as per the wind variations. Similarly, the role of the grid side converter is to maintain the dc-link voltage constant to assure smooth active power flow.

Back-to-back Converters

The purpose of a back-to-back AC-DC-AC converter in both the above-mentioned configurations is to attain smooth integration with the utility grid. In DFIG, the converter connected to the rotor is known as rotor side converter (RSC) while the one connected to the utility grid is grid side converter (GSC). Depending upon the modes of operation, i.e. sub-synchronous or super-synchronous, the converters will act as rectifiers or inverters.

Based on the theory of rotating magnetic field, following active (P) and reactive power (Q) flow equations associated with WECS are developed. It represents the total power delivered to the grid as given in (28.5) and (28.6),

$$P_{dg} = V_{ds}i_{ds} + V_{qs}i_{qs} + V_{dr}i_{dr} + V_{qr}i_{qr} \tag{28.5}$$

$$Q_{dg} = V_{qs}i_{ds} - V_{ds}i_{qs} \tag{28.6}$$

where V_d and i_d are d-axis voltage and current respectively, V_q and i_q are q-axis voltage and current respectively.

28.2.3 *Energy Storage System*

With the increasing popularity of renewable energy sources in the modern power system such as solar, wind, fuel cells, bio-generation, etc. the research and development of energy storage system is achieving a peak concerning environmental protection. Due to the intermittent nature of these sources, there is a need to overcome the reliability issues that can be fulfilled by an energy storage device that is actively adopted nowadays. An energy storage system, when integrated with a renewable energy source, plays a vital role as it absorbs energy during periods of high generation and acts as a source during periods of high demand. To improve the resiliency of the modern-day grid, it can also be used as an emergency backup to satisfy the critical loads in the presence of any disturbance. An array of energy storage options is available in the market such as super-capacitors, compressed air, pumped hydro storage, flywheels, and rechargeable batteries. Each has its own merits and demerits.

Superconducting Magnetic Energy Storage

Superconducting magnetic energy storage (SMES) system combines the advantage of circulating current within the superconducting coil at critically low temperature and magnetic field for energy storage. At such a low-temperature resistive loss are negligible, hence it is one of the top contenders for future ESS in the smart grid. Mercury, vanadium, and niobium-titanium are the most commonly used superconducting materials. It consists of a superconducting coil magnet, a cryogenic refrigerator, a cryostat, and a passive filter circuit as shown in Fig. 28.7.

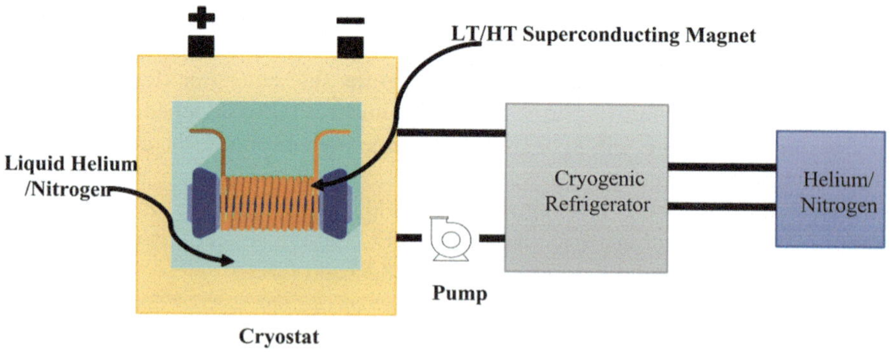

Fig. 28.7 Schematic of superconducting magnetic energy storage

A superconducting coil forms the heart of the SMES as it stores energy generated by the circulating current magnetic field. The large size of the coil ensures more amount of stored energy. To maintain the critically low temperature of SMES coil, a cryogenic refrigerator is used with helium as a coolant. Due to its rapid discharge capabilities, it can be used to address the transient stability issues, in modern-day power systems (Mukherjee and Rao 2019b; Hashem et al. 2021). Furthermore, the role of SMES is studied in Salama and Vokony (2020) for electric vehicle integration approaches whereas, at the distribution level, the control strategy of SMES in coordination with RES integration is presented in Yang et al. (2020), Said et al. 2020).

The SMES, in simulation studies, may be represented as a pure inductor of high value with zero resistance. The SMES is integrated with the utility grid in two possible ways; one through a voltage source converter (VSC) in series with a DC-DC converter, second through a current source converter (CSC) (Mitra 2015).

The VSC-based SMES is shown in Fig. 28.8. The VSC and the DC-DC converter are connected by a DC link capacitor (CDC). The control here is complicated and the reliability is also less because of the presence of two converters in series. The circuit connection of an IGBT-based CSC integrating the SMES to the utility grid is shown in Fig. 28.9. Because of the presence of a single converter, the control is

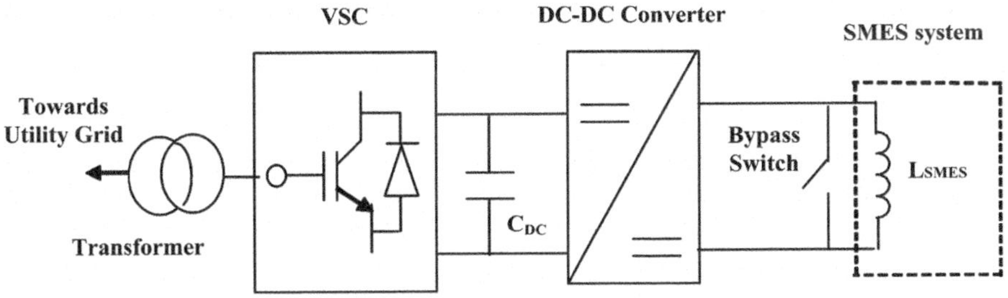

Fig. 28.8 SMES integrated with the grid-based on VSC

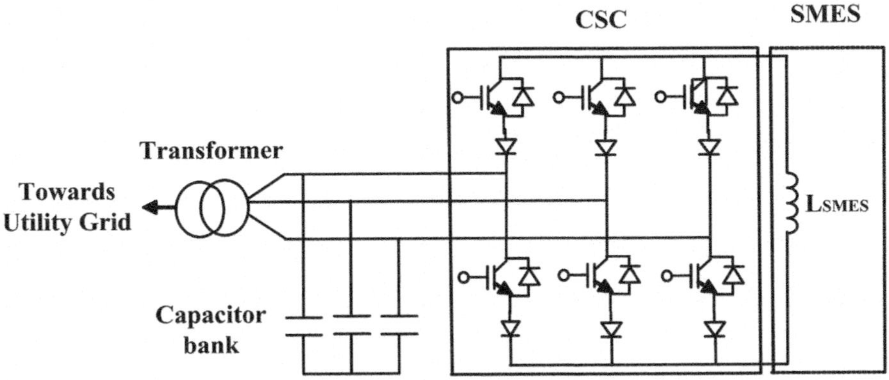

Fig. 28.9 CSC based SMES integrated with the grid

easier with more reliability when compared with VSC-based SMES.

Battery Energy Storage System

The most widely used option is rechargeable batteries due to their prolonged life and cost-effective nature. A rechargeable battery facilitates chemical energy to electrical energy conversion during the discharging process and vice versa during the charging process (Srujan et al. 2019; Fatnani et al. 2020).

Terminal voltages of BESS during charging and discharging modes are given as,

$$V_{ch} = E_0 + V_{op+} + V_{op-} + I R_0 \tag{28.7}$$

$$V_{dic} = E_0 - V_{op+} - V_{op-} - I R_0 \tag{28.8}$$

where,

V_{ch} and V_{dic}	Cell charging and discharging voltage respectively,
E_0	Open-circuit voltage,
V_{op+} and V_{op-}	Voltages at positive and negative electrodes respectively,
I	cell current,
R_0	Polarization resistance.

Shepherd model is considered to be the best voltage-current model for constant battery current. The battery voltage may be represented as in (28.9), or as in (28.10),

$$V_{bat} = E_0 - \left(\frac{K}{Q - It}\right)I - R_0 I \tag{28.9}$$

$$V_{bat} = E_0 - \left(\frac{K}{SOC}\right)I - R_0 I \tag{28.10}$$

where

V_{bat}	battery terminal voltage (V),
K	Polarization resistance coefficient (Ω),
Q	Capacity of the battery (Ah); It = ∫Idt.

Battery Basics and Terminologies

Basic terminologies related to the battery are discussed below.

- **Cells and Modules**—A cell is the smallest form of a battery in the range of fewer than 10 V whereas a series or parallel combination of cells forms a module.

Similarly, when modules are connected in series or parallel, it is known as the battery pack.

- **State of Charge (SOC) (%)**—It indicates the present available battery capacity as a percentage of nominal capacity (Ah_{rated}) as given in (28.11). It is usually bounded in the range $soc_{min} \leq soc \leq soc_{max}$ depending on the type of battery.

$$soc(t) = \frac{I_{bat} * \Delta t}{Ah_{rated}} \tag{28.11}$$

- **Depth of discharge (DOD) (%)**—It represents the percentage of the battery capacity that has been discharged.
- **Charge/Discharge Rate (C-rate)**—The rate at which a battery charges/discharges with respect to the nominal capacity is known as C-rate which is the ratio of charge/discharge current to the rated capacity of the battery given in (28.12).

$$C_{rate} = \frac{I_{bat}[A]}{Nominal\ Capacity[Ah]} \tag{28.12}$$

- **Cyclelife**—During its entire lifetime, a battery may get exposed to several charging or discharging cycles, known as cyclelife. It depends on various factors such as ambient temperature, SOC, C-rate, etc. A lower value of DOD yields more cyclelife.
- **Open Circuit Voltage (V)**—The voltage appearing between the battery terminals without any load is known as open-circuit voltage.
- **Nominal Voltage (V)**—The voltage specified by the manufacturer, also known as normal voltage.
- **Cut-off Voltage (V)**—The minimum voltage allowed by the battery.
- **Nominal Capacity (Ah)**—It represents the coulometric capacity in ampere-hours. It is calculated as a product of discharge current and discharge time.
- **Internal Resistance (Ω)**—It represents the resistance within the battery. It is inversely proportional to cell temperature.

Classification and Design Considerations

The classification of various rechargeable batteries is depicted in Fig. 28.10. Lead-Acid and Lithium-ion are the most widely used batteries in renewable energy applications while Nickle batteries are rarely used due to their high cost and less eco-friendly nature. Lead-acid is the oldest and cheapest of all batteries available whereas lithium-ion technology is rising and finding its applications in electric vehicles due to its longer life span but at a higher cost. Even though the initial cost of a lithium-ion battery is high, it offers low maintenance cost, high energy and power density, higher cyclelife, and higher efficiency at lower C-rates as compared to its counterpart batteries.

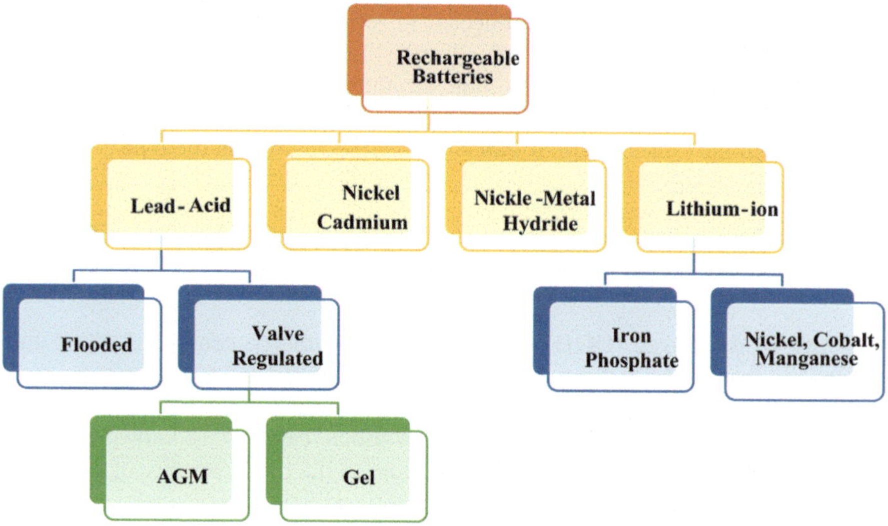

Fig. 28.10 Classification of rechargeable batteries

Total cost, lifetime, reliability, and environmental impact are the major variables considered while designing BESS. Lithium-ion has a higher cyclelife due to deep discharge application as compared to lead-acid (LA) battery as shown in Figs. 28.11 and 28.12 (Zhou et al. 2011). In Fig. 28.11, the y-axis is on a logarithmic scale. It is clear from both the figures that for lower DOD, lithium-ion outclasses lead acid with a higher number of cycles.

Similarly, operating BESS at higher temperature and C-rate cause faster degradation, and the BESS reaches its end of life when the capacity fades to 80% of rated as depicted in Figs. 28.13 and 28.14 respectively. It is clear that due to the Arrhenius

Fig. 28.11 Cyclelife versus DOD, Li-ion

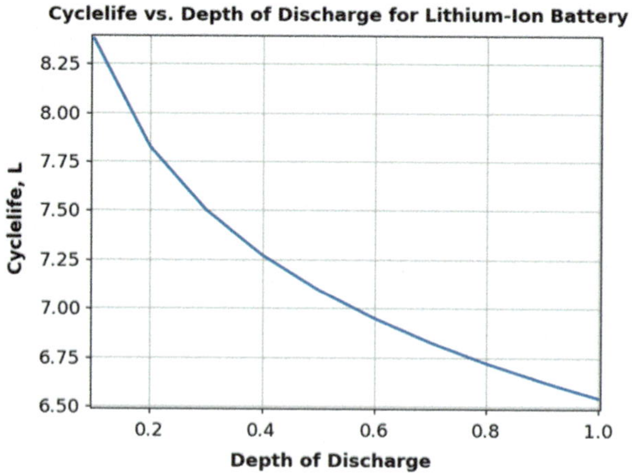

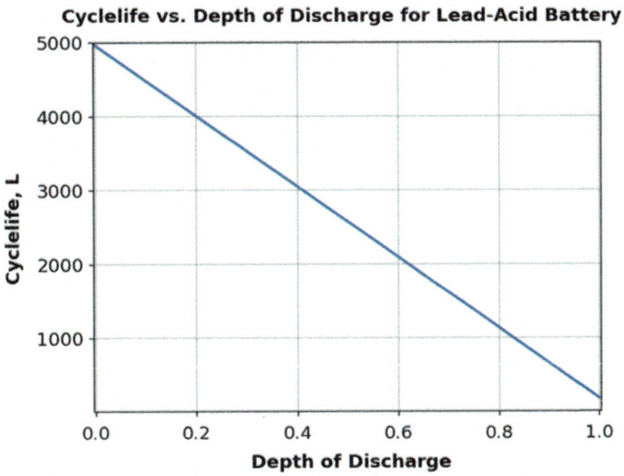

Fig. 28.12 Cyclelife versus DOD, lead acid

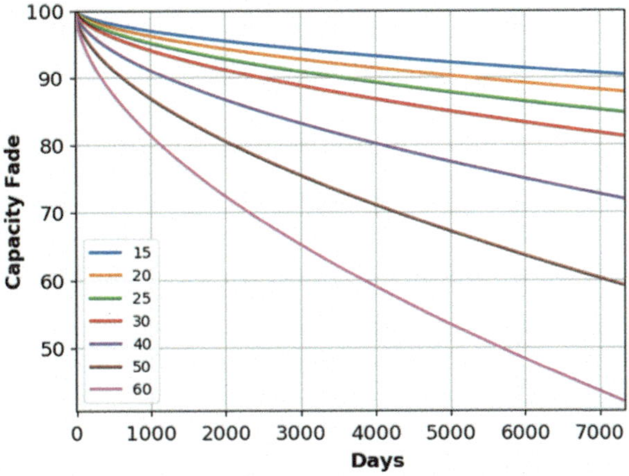

Fig. 28.13 Capacity retention w.r.t. temperature

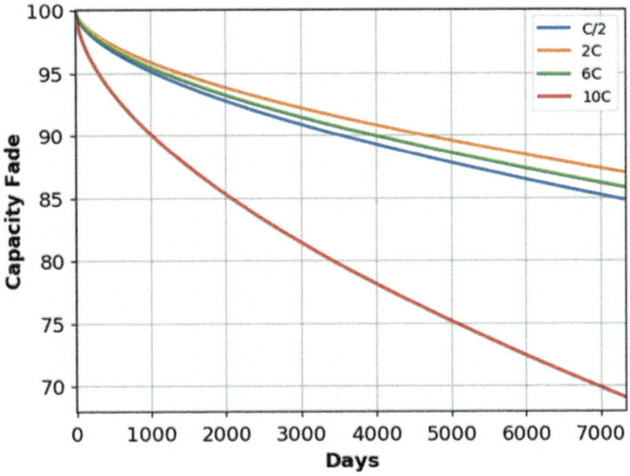

Fig. 28.14 Capacity retention w.r.t. C-rate

dependence (Eq. 28.13) of reaction rate (k) of a chemical process with temperature, the useful life of the battery is reduced at higher values of temperature and the capacity loss is faster. Moreover, with the increasing charge/discharge rate, cyclelife reduces to a great extent. Fast charging/discharging means, the shorter life span of the BESS (Suri and Onori 2016; Wang et al. 2011; Zhang et al. 2019).

$$k = (\alpha * SOC + \beta) * \exp\left(\frac{E_a}{RT}\right) \tag{28.13}$$

where, E_a is the activation energy $\left(\text{Jmol}^{-1}\right)$, R is the universal gas constant, T is the temperature in Kelvin, and α, β are the fitting coefficients.

The selection of battery plays a crucial role for a particular type of application. For economic operation, the optimal size and cost of the battery need to be evaluated by a suitable optimization technique. This can be achieved with a mathematical battery aging model subjected to a range of temperature, SOC, and C-rate. The optimal cost of the battery ensures minimum degradation cost and prolonged lifetime of the battery. Hence appropriate modelling of the battery aging phenomenon is essential.

28.2.4 Power Electronic Interfaces

AC-DC Converters (Three Phase Controlled Rectifiers)

Three-phase controlled rectifiers find their usage ranging from low voltage to high voltage DC transmission. It provides a controllable dc output voltage from the three-phase ac input by adjusting the conduction interval of each thyristor. Figure 28.15 shows the basic arrangement. Load current completes its path when at least one thyristor, each from the top and bottom group conducts. This converter works as a rectifier for firing angle $\alpha < 90°$ while if α is made larger than 90°, the direction of power flow reverses, known as inverter operation.

The r.m.s. component of output voltage and current is given by,

$$V_{rms} = \sqrt{3}V_m\sqrt{\frac{1}{2} + \frac{3\sqrt{3}}{4\pi}\cos 2\alpha} \tag{28.14}$$

$$I_{rms} = \frac{V_{rms}}{Z} \tag{28.15}$$

where, V_m is the peak phase voltage, α is the firing angle and z is the load impedance.

DC-DC Converters (Choppers)

A boost converter is the most commonly used DC-DC converter in microgrid applications as shown in Fig. 28.16. It is one of the simplest non-isolated DC-DC converters

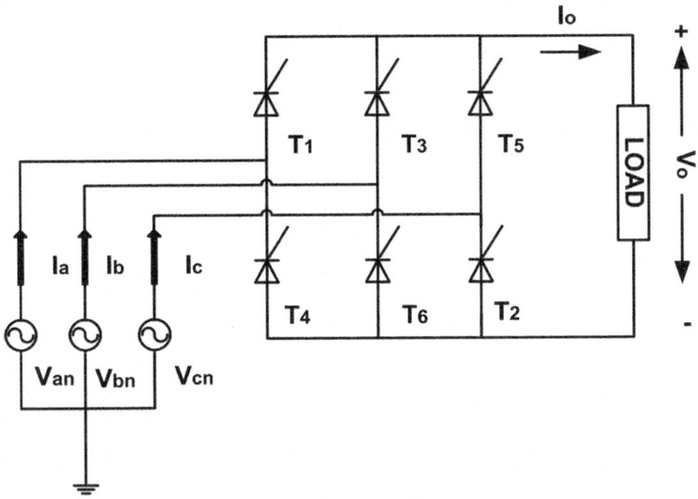

Fig. 28.15 Three phase controlled rectifier

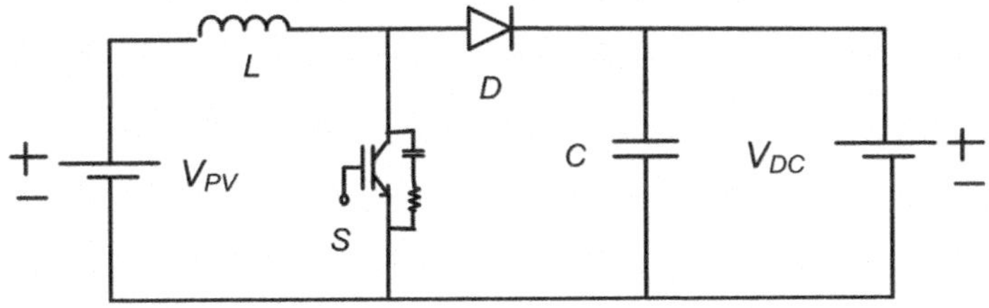

Fig. 28.16 DC-DC boost converter

used to step up DC voltage. It consists of dc input source, diode D, controlled switch S, capacitor C, and boost inductor L. When the switch is on and the diode is off, it creates a short circuit and the current flows through the inductor storing energy disconnecting the rest of the circuit. On the other hand, when the switch is off and the diode is on, the reverse polarity of the inductor causes energy to be released to the load and also steps up the output voltage. Basic equations governed by the DC-DC converters are given below (Irwin 2011).

For a given duty ratio d, the transfer function equation for boost converter is as given below,

$$\frac{V_{dc}}{V_{pv}} = \frac{1}{1-d} \tag{28.16}$$

The boost inductance L, when operated in continuous conduction mode is given as,

$$L = \frac{(1-d)^2 * d * R}{2 * f} \quad (28.17)$$

The minimum value of filter capacitance C_{min}, for ripple voltage V_r, is given by (28.18),

$$C_{min} = \frac{d * V_{dc}}{V_r * R * f} \quad (28.18)$$

where R is the load resistance and f is the switching frequency.

DC-AC Converters (Three Phase Voltage Source Inverters)

A grid-connected inverter plays a vital role in the transfer of power to the grid in an effective manner. Figure 28.17 shows the adopted topology. A decoupled control technique is a prominent feature in voltage source inverter (VSI) for controlling the flow of power between the inverter and grid independently (Lakshmi and Hemamalini 2018).

The total active and reactive power delivered to the utility grid can be written as,

$$P = \frac{3}{2}\left[V_{sd} I_d + V_{sq} I_q\right] \quad (28.19)$$

$$Q = \frac{3}{2}\left[-V_{sd} I_q + V_{sq} I_d\right] \quad (28.20)$$

Assuming that for the grid synchronization, the component V_{sq} is regulated to zero, this gives the expressions,

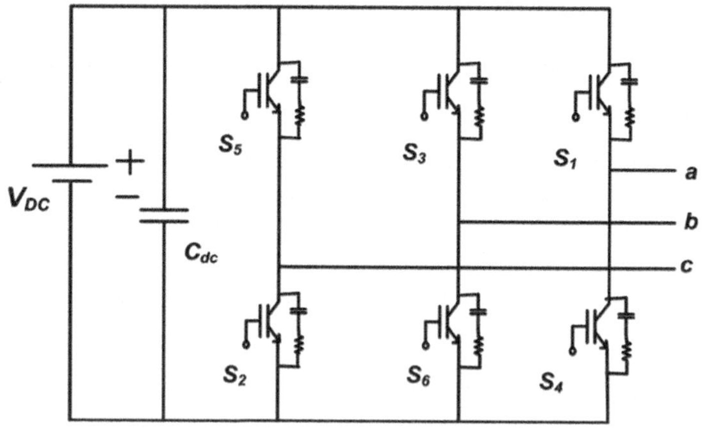

Fig. 28.17 Three phase voltage Source Inverter

$$P = \frac{3}{2}[V_{sd} I_d] \tag{28.21}$$

$$Q = \frac{3}{2}\left[-V_{sd} I_q\right] \tag{28.22}$$

From the above expression, it is clear that the direct axis component of the current (I_d) and the quadrature axis current (I_q) are used to control the active and reactive power respectively.

28.3 Hybrid Integration of Renewable Energy Sources

As fossil fuel is regularly being exhausted, RES is gaining popularity in the present-day microgrid. Although RES has several advantages of reduced transmission and distribution losses, customer participation in the retail electricity market, net metering, feed-in-tariff facility, etc., integration of distributed energy resources with the utility grid invites more challenges such as energy trading, system stability, the control strategy for converters, reliability, efficiency, and security. With the advent of new technologies in the power electronic domain, modern-day power conditioning units and high-efficiency controllers allow the smooth operation of the system.

The idea of hybrid integration comes with the fact that the potential of solar and wind can be utilized maximally. Given the fact that two adjacent wind turbines are located at a safe distance, the space between them can be used to install solar PV panels. This configuration has the advantage of generating PV power during the presence of insolation while power from wind turbines can be extracted during the absence of PV. Also, this operation can be supported by assimilating ESS to undertake the fluctuations in power generated by the renewable sources that are being exchanged with the grid (Al-Shetwi et al. 2020; Shi et al. 2020; Mbungu et al. 2020).

This section covers various possible configurations of RES, schematic diagrams, mathematical modelling, and challenges associated with them.

28.3.1 Case-I: Hybrid Integration of Solar PV, PMSG, and BESS

Figure 28.18 shows the schematic of hybrid integration of solar PV, PMSG, and BESS. Here, a solar PV along with PMSG and BESS are connected to the utility grid via power conditioning units. The DC power produced by solar PV is stepped up with the help of DC-DC converters which feed to the DC bus. Here static, dynamic, and constant power loads can be connected along with the electric vehicle. Also,

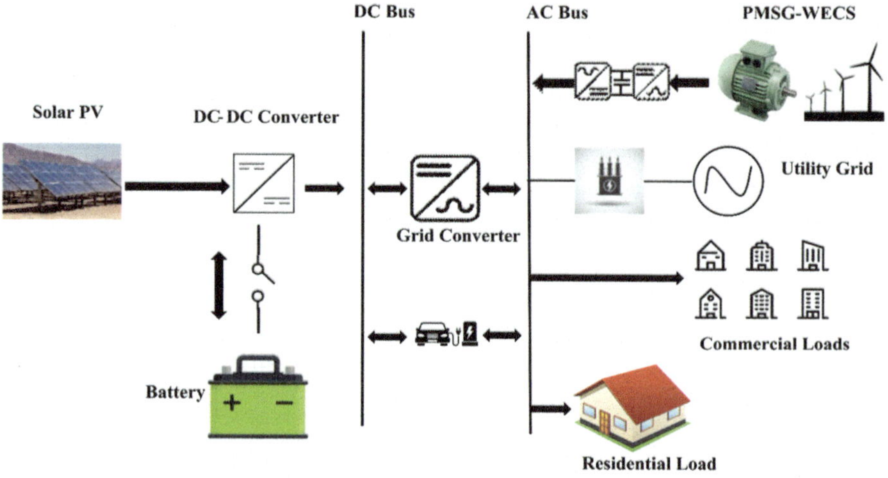

Fig. 28.18 Hybrid solar PV, PMSG, and BESS

a PMSG based wind turbine is directly connected to the AC bus via back-to-back converters in series.

Since PMSG and its associated power electronic converters have to carry total power, the cost of the converter is high. Again, as permanent magnets are not available easily, the overall cost becomes comparatively higher than that of the DFIG. Also, hybrid integration of these sources is unable to address the short-term fluctuations as BESS response time is small, hence the improvement of transient stability with this configuration remains a challenge.

28.3.2 *Case-II: Hybrid Integration of Solar PV, DFIG, and BESS*

Figure 28.19 shows the schematic of hybrid integration of solar PV, DFIG, and BESS. Here, a solar PV along with DFIG and BESS are connected to the utility grid via power conditioning units. Moreover, a DFIG based wind turbine is connected via back-to-back converters. The power generated by DFIG will vary as per the available wind speed. Along with these RES, a BESS is equipped to take care of the fluctuations in the grid.

As we are aware that during the daytime, solar PV will harness maximum power from the available solar intensity and will satisfy the load demand while surplus generation will be used to charge the BESS and further injected into the utility grid. Similarly, the potential of available wind will be utilized by the wind turbines for the energy conversion process with the help of DFIG. To overcome the issue of variable wind speed, the rotor of DFIG operates at speeds both above (super-synchronous)

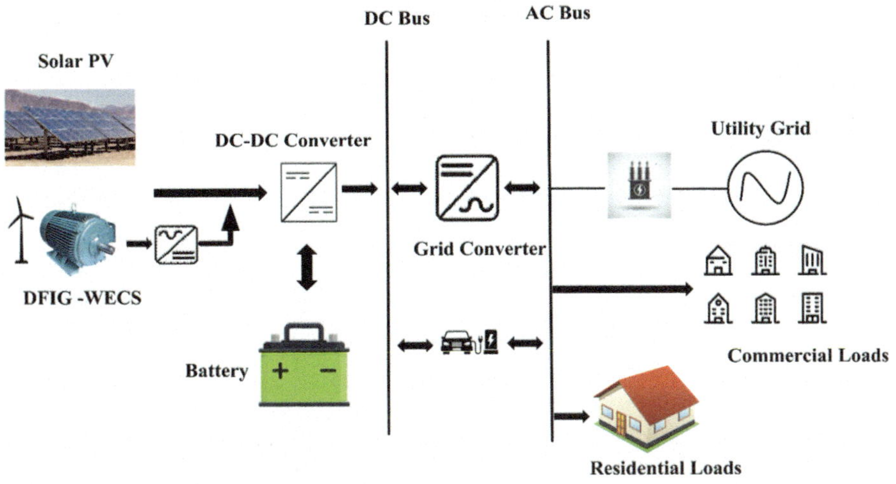

Fig. 28.19 Hybrid solar PV, DFIG, and BESS

and below (sub-synchronous) the synchronous speed. As the stator and rotor are decoupled with each other, the total power carried by converters is relatively small, hence it is preferred over PMSG. It also offers reduced inverter cost and improved system efficiency.

The role of BESS in this system is to absorb energy during the periods of high generation from PV and wind while acting as a source when either of them is unavailable. As both PV and wind are intermittent, the fluctuation caused by these will impact the stability of the system, hence BESS is a viable option to handle long-term fluctuations but enhancement of transient stability of such a hybrid system with BESS is a challenging task.

28.3.3 Case-III: Hybrid Integration of Solar PV, DFIG, BESS, and SMES

To address the short-term fluctuations arising due to the stochastic nature of solar and wind, another possible hybrid configuration is shown in Fig. 28.20. A solar PV and a DFIG based wind turbine are incorporated with the utility grid along with BESS and SMES.

As mentioned in the previous sub-section, BESS is comfortable with long-term fluctuations but suffers from several drawbacks. Its charging/discharging is governed by load current, ambient temperature, state-of-charge, depth-of-discharge, and C-rate. These factors are responsible for the aging of the battery; hence battery degradation cost is high. Due to its high efficiency, faster response time, and effectively zero resistance, SMES is a suitable candidate to overcome short-term fluctuations

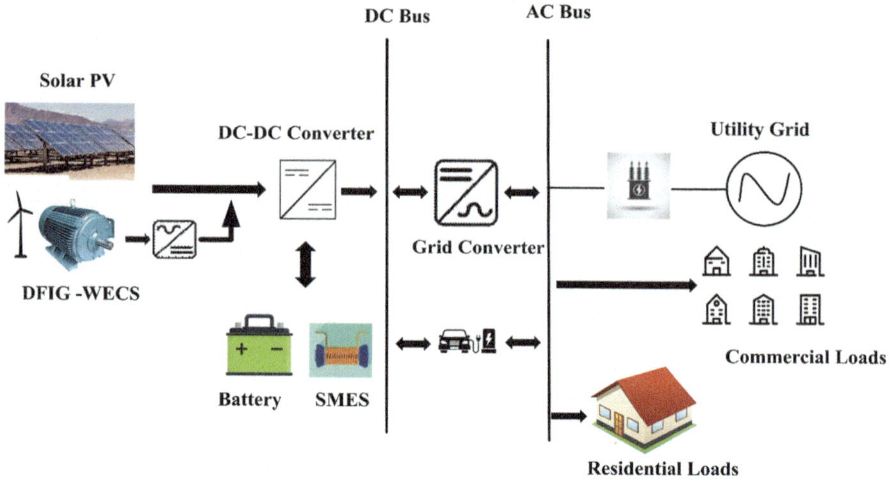

Fig. 28.20 Hybrid solar PV, DFIG, BESS and SMES

caused by solar and wind or may be due to disturbance as it offers immediate power requirement.

28.3.4 Comparison Between Different Configurations

This sub-section compares the various configurations of hybrid integration discussed above based on several factors such as cost, reliability, stability, and environmental impact. Table 28.1 depicts the overall comparison.

Large-scale integration of hybrid renewable sources for the probable improvement of the transient stability of multi-machine power systems is proposed in Verma and Mitra (2018). The study shows that a modification in the control strategy of the

Table 28.1 Overall comparison of various configuration

Particulars	Case-I	Case-II	Case-III
Cost	High	Low	High
Reliability	Yes	Yes	Yes
Stability	Can handle Long-term fluctuations, hence steady-state stable	Can handle Long-term fluctuations, hence steady-state stable	Can handle both long and short term fluctuations, hence improved transient stability
Environmental Impact	Yes, due to BESS	Yes, due to BESS	Mostly due to BESS, as SMES is hazard-free

GSC of both solar PV farm and DFI-based wind farm can be made during a fault aiming to support the reactive power and thereby improving the magnitude of the terminal bus voltage of the system, where the hybrid sources are integrated. This strategy ultimately helps to improve the overall transient stability of the system. The hybrid sources, comprising of solar PV and wind farm, not only use the advantages of both the sources, but also reduce the capacity of the energy storage used for long term support to decrease fluctuations in the injected power to the grid because the availability of power through such sources are balancing each other in the majority of the times in one solar day.

So, not only at the microgrid level, large-scale implementation of hybrid renewable sources is the future of the power system towards sustainable energy supply.

28.4 Conclusion and Future Scope

This chapter aims to provide an overview of grid integration of renewable energy sources. The basics of various RES such as solar PV and wind energy along with battery energy storage systems and superconducting magnetic energy storage are covered in detail with schematic illustrations and building-block diagrams. Each RES with its equivalent circuit, mathematical modelling, various configurations, and their applications is presented here in a comprehensive manner. The role of an energy storage system in modern-day microgrids such as BESS and SMES is highlighted and compared in this chapter. The chapter also discusses the various power electronic interfaces which are widely used for the integration of renewable sources and energy storage systems with the power system grid.

The hybrid integration of RES with the utility grid and the challenges associated with these configurations is reported here. Configuration I deals with the integration of solar PV, PMSG, and BESS with the utility grid with improved reliability as both solar and wind are acting during the day while BESS takes care of long-term fluctuations but the overall system cost is high due to series-connected back-to-back converters. Moreover, in configuration II, PMSG is replaced by DFIG to address the cost-related issue as converters do not carry total power. Although the short-term fluctuations are still untouched due to the slow response time of BESS and it also causes environmental hazards. On the other hand, in configuration III, there is a trade-off between cost and environmental impact caused by the inclusion of SMES along with BESS. The transient stability is enhanced with the help of SMES due to its faster response time while BESS continues to handle long-term fluctuations. The overall comparison shows that the hybrid integration of renewable sources and energy storage devices is practically feasible with some trade-offs between various comparison parameters.

As we are moving towards the deregulated environment, the share of RES is exponentially increasing and the results are overwhelming. Also with the advancement in the field of the energy storage system, fluctuations arising due to the stochastic nature of these sources can be suppressed thus making the modern-day grid reliable, efficient, secure, and hazard-free. This chapter makes a valuable contribution in the area of power systems and will be useful to the researchers, academicians, and fraternity working in this field.

References

Al-Shetwi AQ, Hannan MA, Jern KP et al (2020) Grid-connected renewable energy sources: review of the recent integration requirements and control methods. J Clean Prod 253:119831. https://doi.org/10.1016/j.jclepro.2019.119831

Dursun EH, Koyuncu H, Kulaksiz AA (2021) A novel unified maximum power extraction framework for PMSG based WECS using chaotic particle swarm optimization derivatives. Eng Sci Technol An Int J 24:158–170. https://doi.org/10.1016/j.jestch.2020.05.005

Dursun EH, Kulaksiz AA (2020) Second-order sliding mode voltage-regulator for improving MPPT efficiency of PMSG-based WECS. Int J Electr Power Energy Syst 121:106149. https://doi.org/10.1016/j.ijepes.2020.106149

Elbaset AA, Ali H, Abd-El Sattar M (2014) Novel seven-parameter model for photovoltaic modules. Sol Energy Mater Sol Cells 130:442–455. https://doi.org/10.1016/j.solmat.2014.07.016

Fatnani M, Naware D, Mitra A (2020) Design of solar PV based EV charging station with optimized battery energy storage system. In: 2020 IEEE first international conference on smart technologies for power, energy and control (STPEC). IEEE, Nagpur, Maharashtra, India, pp 1–5

Hashem M, Abdel-Salam M, El-Mohandes MT et al (2021) Optimal placement and sizing of wind turbine generators and superconducting magnetic energy storages in a distribution system. J Energy Storage 38:102497. https://doi.org/10.1016/j.est.2021.102497

Irwin JD (2011) Power electronics academic press series in engineering

Korjani S, Serpi A, Damiano A (2020) A genetic algorithm approach for sizing integrated PV-BESS systems for prosumers. In: 2020 2nd IEEE international conference on industrial electronics for sustainable energy systems (IESES). IEEE, pp 151–156

Lakshmi M, Hemamalini S (2018) Decoupled control of grid connected photovoltaic system using fractional order controller. Ain Shams Eng J 9:927–937. https://doi.org/10.1016/j.asej.2016.06.002

Masters GM (2005) Renewable and efficient electric power systems. Choice Rev Online 42:42-3448–42–3448. https://doi.org/10.5860/choice.42-3448

Mbungu NT, Naidoo RM, Bansal RC et al (2020) An overview of renewable energy resources and grid integration for commercial building applications. J Energy Storage 29:101385. https://doi.org/10.1016/j.est.2020.101385

Messenger RA, Abtahi A (1999) Photovoltaic systems engineering. CRC Press

Mitra A (2015) Impacts of integration of doubly fed induction generator based wind farms on the transient stability of power systems. Indian Institute of Technology Kharagpur

Mitra A, Chatterjee D (2016) Active power control of DFIG-based wind farm for improvement of transient stability of power systems. IEEE Trans Power Syst 31:82–93. https://doi.org/10.1109/TPWRS.2015.2397974

Mukherjee P, Rao VV (2019a) Design and development of high temperature superconducting magnetic energy storage for power applications—A review. Phys C Supercond Its Appl 563:67–73. https://doi.org/10.1016/j.physc.2019.05.001

Mukherjee P, Rao VV (2019b) Superconducting magnetic energy storage for stabilizing grid integrated with wind power generation systems. J Mod Power Syst Clean Energy 7:400–411. https://doi.org/10.1007/s40565-018-0460-y

Pamu S, Thogaru R, Mitra A (2020) Resiliency analysis of grid connected solar PV system with battery support. In: 2020 IEEE first international conference on smart technologies for power, energy and control (STPEC). IEEE, pp 1–6
Rana MM, Romlie MF, Abdullah MF et al (2021) A novel peak load shaving algorithm for isolated microgrid using hybrid PV-BESS system. Energy 234:121157. https://doi.org/10.1016/j.energy.2021.121157
Salama HS, Vokony I (2020) Comparison of different electric vehicle integration approaches in presence of photovoltaic and superconducting magnetic energy storage systems. J Clean Prod 260:121099. https://doi.org/10.1016/j.jclepro.2020.121099
Said SM, Aly M, Balint H (2020) An efficient reactive power dispatch method for hybrid photovoltaic and superconducting magnetic energy storage inverters in utility grids. IEEE Access 8:183708–183721. https://doi.org/10.1109/ACCESS.2020.3029326
Srujan D, Thogaru R, Mitra A, Borghate VB (2019) Energy management in grid assisted BESS integrated solar PV based smart charging station. In: IECON 2019 - 45th annual conference of the IEEE industrial electronics society. IEEE, pp 6363–6368
Shi R, Li S, Zhang P, Lee KY (2020) Integration of renewable energy sources and electric vehicles in V2G network with adjustable robust optimization. Renew Energy 153:1067–1080. https://doi.org/10.1016/j.renene.2020.02.027
Steiger J (1988) Renewbale energy resources. In: Renewable Energy Resources in ASEAN. ISEAS Publishing, pp 4–10
Suri G, Onori S (2016) A control-oriented cycle-life model for hybrid electric vehicle lithium-ion batteries. Energy 96:644–653. https://doi.org/10.1016/j.energy.2015.11.075
Vasavi Uma Maheswari M, Ramana Rao PV, Jayaram Kumar SV (2021) Tracking maximum power point of a grid-connected DFIG wind turbine systems using AI and evolutionary controllers. Springer, Singapore, pp 261–272
Verma PG, Mitra A (2018) Impact of hybrid renewable sources on the transient stability of multimachine power system. In: 2018 IEEE international conference on power electronics, drives and energy systems (PEDES). IEEE, pp 1–5
Wang J, Liu P, Hicks-Garner J et al (2011) Cycle-life model for graphite-LiFePO4 cells. J Power Sources 196:3942–3948. https://doi.org/10.1016/j.jpowsour.2010.11.134
Yang B, Wang J, Zhang X et al (2020) Control of SMES systems in distribution networks with renewable energy integration: a perturbation estimation approach. Energy 202:117753. https://doi.org/10.1016/j.energy.2020.117753
Zhang S, Hu X, Xie S et al (2019) Adaptively coordinated optimization of battery aging and energy management in plug-in hybrid electric buses. Appl Energy 256:113891. https://doi.org/10.1016/j.apenergy.2019.113891
Zhao B, Wang C, Zhang X (2017) Grid-integrated and standalone photovoltaic distributed generation systems. John Wiley & Sons Singapore Pte. Ltd, Chichester, UK
Zhou C, Qian K, Allan M, Zhou W (2011) Modeling of the cost of EV battery wear due to V2G application in power systems. IEEE Trans Energy Convers 26:1041–1050. https://doi.org/10.1109/TEC.2011.2159977

Chapter 29
Optimal Allocation of Electric Vehicles Charging Station in Distribution Network Beside DG Using TSO

Jitendra Singh Bhadoriya, Atma Ram Gupta, Mohamed Zellagui, Nitin Kumar Saxena, Aadesh Kumar Arya, and Aashish Kumar Bohre

Abstract Increasingly, there is growth in electric cars globally, and it will keep rising owing to increasing knowledge and interest on the part of people, all while considering the significant environmental and financial effects. Installation in the distribution system of rapid electric vehicle charging stations that meet the increasing charging demand for electric vehicles. Although fast-charging stations are placed in the distribution system, the implementation of these stations leads to adverse effects such as higher power loss and a more inferior voltage profile. To minimize these adverse effects, one must strategically locate charging stations and allocate dispersed generation appropriately across the distribution system. In the article, the negative impacts of charging stations on radial distribution systems and the positive impacts of distributed generation on unbalanced systems to balance single and multiple distributed generation to reduce active and reactive power loss and enhance voltage stability were evaluated for IEEE-25 unbalanced radial distribution systems. The distribution system with sufficient active and reactive power injection has the distributed generation with unity, fixed, and optimum power factor assigned to enhance it. Various optimization strategies were proposed for EVCS allocation,

J. S. Bhadoriya (✉) · A. R. Gupta
Department of Electrical Engineering, NIT Kurukshetra, Kurukshetra, Haryana, India

A. R. Gupta
e-mail: argupta@nitkkr.ac.in

M. Zellagui
Department of Electrical Engineering, University of Batna, 2 Fesdis, Batna, Algeria
e-mail: m.zellagui@univ-batna2.dz

N. K. Saxena
Department of Electrical and Electronics Engineering, KIET Group of Institutions, Ghaziabad, India

A. K. Arya
Department of Electrical Engineering, College of Engineering Roorkee, Roorkee Haridwar, India

A. K. Bohre
Department of Electrical Engineering, NIT Durgapur, Durgapur, West Bengal, India
e-mail: aashishkumar.bohre@ee.nitdgp.ac.in

A. K. Bohre et al. (eds.), *Planning of Hybrid Renewable Energy Systems, Electric Vehicles and Microgrid*, Energy Systems in Electrical Engineering,
https://doi.org/10.1007/978-981-19-0979-5_29

but a novel physics-based meta-heuristic algorithm, Transient search optimization (TSO), which had just been created, was used for multi-objective functions. Based on the modeling findings, the voltage profile improved, and power loss was reduced in every scenario. The convergence features have emerged regarding the new algorithm recently created to coordinate all scenarios with the findings of the outcomes.

Keywords Electric vehicle · Electric vehicle charging station · Transient search optimization · Power loss · Voltage profile

Nomenclature

ANN	Artificial neural network
APL	Active power loss
CO_2	Carbon dioxide
CRAE	Correlation attribute evaluation
DG	Distributed generation
EVCS	Electric vehicles charging stations
EVs	Electric vehicles
EVSE	Electric vehicle charging station expenditure
GA	Genetic algorithm
PSO	Particle swarm optimization
RAPL	Reduction in active power loss
TAPL	Total active power loss
TSO	Transient search optimization
URDS	Unbalanced radial distribution system

29.1 Introduction

29.1.1 Overview

The road transportation industry is a significant source of greenhouse emissions (Verma and Kumar 2013). Because Electric vehicles (EVs) have the potential to decrease greenhouse gas emissions, they have emerged as an ecologically friendly alternative to conventional Internal Combustion Engine powered cars. However, widespread adoption of EVs may pose a significant danger to the electric power system, owing to the rise and variability in power consumption caused by electric vehicle charging stations (EVCS). Indeed, voltage stability and reliability indices are degraded, reserve margins are decreased, and power losses rise due to the inappropriate placement of EV charging stations in the electric power distribution network

(Chen et al. 2021). The electrification of cars has been prompted by the economic and environmental issues associated with fossil fuel transportation. Many nations have set goals to achieve 100% EV penetration over the next few years, with Norway having achieved 28.8% market share, the Netherlands having 6.4%, and China getting 1.4% (Sharma et al. 2019). A worldwide total of 35 million electric vehicles are projected to be sold by 2022. When EVs are fuelled by renewable energy, the expanding EV sector indicates the possibility of zero emissions. In reality, it is essential to power these EVs with renewable energy sources to the greatest extent feasible since the high charging loads would have negative environmental consequences if they are fuelled by fossil fuel power plants (Deb et al. 2018). The number of EVs is projected to multiply over the next several years, owing to the waste of oil and the environmental damage connected with their usage. As a result, efforts to decrease urban pollution and greenhouse gas emissions tend to be coordinated. At the moment, one of the most severe issues confronting EV growth is the scarcity of charging infrastructure. Drivers may charge their EVs at home, but it takes a long time. To encourage the growth of electric vehicles, it is essential to build fast-charging stations that can fully charge an EV battery in around 15 min. The downside of rapid charging, on the other hand, is the increased power consumption and its effect on the grid. Renewable energy sources and storage technologies may be added to these stations to solve this issue. A review of research on the suitable system design for the installation and operation of an EV charging station can be found in Drossinos et al. (2016). Environmental pollution and the fossil fuel problem are the primary reasons for the electric vehicle's attractiveness. Governments are pushing EV usage in every way possible and encouraging businesses to establish charging stations. The need for EV charging presents enormous difficulties to power system experts. Optimized planning is required to install EVCS in optimum locations to service customers in crowded regions with high charging demand. Furthermore, power losses must be maintained to a minimum and voltage deviations within statutory limitations on the power system network. Additionally, the cost of land is a significant issue in constructing a charging station. Dealing with the uncertainties associated with electric vehicles is also a challenging task when planning the construction of charging stations. Due to the environmental issues connected with greenhouse emissions and fossil resources running out, EVs are quickly replacing gasoline-fueled vehicles. As EVs quickly increase, peak load demand will rise, and feeder currents will also increase. This will cause voltage profiles to deteriorate in the distribution system. Inappropriate placement of distributed generation (DG) produces a higher amount of power loss and a decrease in voltage profile magnitude. Shunt capacitors are often used in the industry to enhance voltage profiles on distribution systems. To address the risks caused to the existence of EVs and DGs, distribution system planners will need to establish best practices for EV, DG, and capacitor size and location (Rao et al. 2020). EVs are a viable alternative to conventional gasoline and diesel cars. When it comes to the benefits of electric vehicles, it is essential to note that they decrease carbon dioxide emissions, air pollution, and noise. The existing charging infrastructure may be a constraint on the commercial potential of electric cars soon. One of the issues that may hinder the adoption of electric cars in cities is the lack of infrastructure to restrict the duration of trips that customers could take in their vehicles. Apart from that, the charging time is another issue that may impact the adoption of electric vehicles. However, even if the cost of charging an electric car is cheaper than the cost of filling a tank, consumers will choose the quick charge of conventional vehicles if charging periods are prolonged. For instance, it may take between 4 and 8 h to charge a battery fully (Injeti and Thunuguntla 2020).

29.1.2 Previous Work

It was shown in Reddy and Selvajyothi (2020) that the placement issue was formulated by taking into account three objective functions: EV flow, voltage variation, and power loss. Cost, annualized traffic flow, and energy losses were included as objective factors in the model developed for this issue. As previously stated in Johari et al. (2013), the cost was considered the primary goal function of the placement issue, and the placement problem was solved using the Binary Firefly Algorithm. When it comes to placement schemes for public charging stations, the authors of Jacobs (2018) presented a system that used cost as the goal function. In Tang et al. (2018), the authors modeled the placement problem using a multi-objective framework, with the objective functions being cost, real power loss reduction index, reactive power loss reduction index, and voltage profile improvement index. The cost, real power loss reduction index, and voltage profile improvement index were used as the objective functions. In order to deal with this issue, a hybrid GA PSO method was developed. According to Deb and Tammi (2018), the authors proposed an optimum placement strategy for charging stations that took stochastic charging into consideration. A voltage sensitivity index-based optimum design method for EV charging stations was presented in Roosta and Khooban (2019) and it was based on the authors' findings. In Moghaddam et al. (2017), the authors proposed an optimum placement and charging strategy for electric vehicle charging stations in a smart distribution grid that was subjected to a variety of contingency circumstances. The authors proposed a scenario-based planning model for charging station placement that took into consideration network reconfiguration, and they used a coevolutionary method to address the issue at hand. According to Xiong et al. (2018), the authors developed a multi-objective framework for charging station location that considered voltage variation, power losses, and electric vehicle traffic.

Optimal planning of EV charging stations has been accomplished via the use of a variety of methods and goals. The use of EVs as a spinning reserve to provide peak demand and improve system performance may aid in the optimum design of charging station locations. In order to achieve better economics and key parameter improvement, such as loss reduction and voltage deviation minimization EVs are being developed. It is discussed in Satish et al. (2020) how particle swarm optimization is applied to the issue of locating charging stations. To design EV charging stations, the authors of Ahmadi et al. (2021) have considered CO_2 emissions. It is shown in Shariff et al. (2020) that maximizing interaction among parking lots using the K-means clustering method may result in profit maximization for a parking lot owner. Furthermore, by arranging parking spaces with charging stations most efficiently, it is possible to reduce power loss and voltage variation while increasing network dependability. Ghatak et al. (2019) proposes a novel method for mitigating the negative impacts of high solar penetration and charge stations by strategically placing and situating both at the most advantageous locations. In the same way, researchers have utilized sun photovoltaic production to enhance voltage profile to mitigate the detrimental effects of EV parking in Grahn and Söder (2013). In order to expand the usage of electric vehicles, a well-planned charging infrastructure is needed to make them more accessible to users (Pillai et al. 2018). Despite this, their cost is very expensive, and if appropriate planning is not undertaken, they have the potential to overwhelm the electrical power grid. Jacobs (2018) describes how to design fast-charging stations while considering the costs and traffic in a distribution network. In Hadian et al. (2020), the Nash bargaining theory is used to maximize the profit of operators by locating and sizing fast-charging stations in the most profitable locations. It is shown in Gong et al. (2019) that a complete strategy for optimizing the location and size of fast-charging stations on metropolitan roadways is developed. This plan considers electric vehicle and power grid failures, and it recognizes them as key considerations when choosing the location and size of charging stations. Nonetheless, none of the studies examined has addressed maximizing and evaluating the advantages of utilizing more than one kind of charging station, even though this is common. It is also not thoroughly investigated the potential advantage of DG allocation in terms of preserving voltage in the distribution system in the presence of charging stations.

Authors may deal with the question of sites for these EV stations and their size but simplify matters by just talking about the quantity of electricity needed and how it is supplied. In order to identify locations for fast-charging stations, a team of researchers used a mixed-integer non-linear optimization method. They first determined the monetary, energy, and power grid losses of EVs and the development and electrification expenses. In the model, just the number of EVs that reach the station was specified, but they did not include the time it takes for them to arrive or whether or not people are using the station. Using a multi-objective planning model, they proposed a new approach to power losses and voltage variations in the distribution system that reduced power losses and voltage deviations. A pre-set demand was specified, and the functioning of the charging process was ignored. The location of the EV charging station was found by using a geographical and temporal model of EV charging demand. The researchers utilized a fluid dynamic traffic model and queueing theory to simulate traffic flow. With regards to the placement and size of charging stations, the team created a model. The team also considered interactions with the power system but did not include alternative sources of energy. As some other researchers have shown, it is possible to derive models for the design of EV charging stations, where the design incorporates various simplifications. Having no connection to the grid and thus cannot determine the size of a charging station using renewable energy. Designed an EV charging station that minimizes the total cost of ownership. Renewable energy, connection to the grid, and batteries were all investigated, but their algorithm did not factor in arrival time. Using PV and wind generators, they demonstrated the electrical architecture of an EV charging station, but they did not examine the space required for the station. Converters and control algorithms were the emphases of his design. Previously authors constructed a daily demand curve, solar and wind power production, by using a logarithmic equation. Since the demand models utilized for EV charging stations are straightforward, constant demand and load profiles are assumed. In terms of EV charging, authors are motivated by various aims, including maximizing revenues, minimizing overall energy costs for consumers, decreasing network power losses, limiting generating costs, and preventing congestion in the distribution network. For the most part, prior studies have examined the operational and impact aspects of EV charging, while just a few papers concentrate on EV charging station design.

Design papers do not address the charging dynamic because they must customize the input data to programming like Homer, which is not designed to handle this issue (Domínguez-Navarro et al. 2019), also describes the optimum design of EVCS for the IEEE 30 bus system. In Chen et al. (2021), a hybridized optimization method was utilized to appropriately distribute EVCS in Allahabad's distribution system while minimizing start-up costs and improving the gird's power quality. The authors in Reddy and Selvajyothi (2020) optimized the charging station's integration into a 33-bus distribution system utilizing two-layer optimization. The authors of Battapothula et al. (2019) attempted to serve the most significant number of vehicles with the least amount of power loss and voltage variation. In Gong et al. (2019), charging stations were installed across the Beijing area, reducing costs and power loss.

Using various optimization methods, researchers designed EVCS with renewable integration in order to optimize various goals such as power loss and cost. The literature on EVCS allocation is sparse since research on electric vehicles and charging stations is still in its infancy. There is a dearth of literature on the allocation of EVCS in the context of EV uncertainty. Additionally, most study articles do not consider the distribution network in conjunction with the road network when allocating EVCS. Numerous research studies have placed charging stations in nearby places, which may not be sufficiently robust to service all customers in a region. While most research papers have attempted to reduce power loss, power loss fluctuates as load changes with EV flow. As a result, reducing power loss may not result in optimal allocation. Numerous research studies have examined just fast-charging stations for allocation; however, not all EVs are compatible with fast charging. As a result, the charging station should provide both slow and rapid charging.

It is also apparent from literature that researchers have used a wide range of meta-heuristics and conventional optimization methods to deal with charging station location to solve it. When dealing with complicated non-linear issues like the charging station location problem, heuristics or meta-heuristics may provide near-optimal solutions in less time than analytical techniques Lelièvre et al. (2016), which can save time and money. As a result, there is a continuing need for the development of efficient and quick meta-heuristics. It is also apparent that researchers have used a wide range of meta-heuristics and conventional optimization methods to deal with charging station location to solve it. In the proposed work, a new transient search optimization technique is implemented for EVCS allocation.

29.2 EVCS

There is an urgent need to provide an adaptable charging infrastructure for various car segments to increase EV adoption. The most critical enabler in the whole EV value chain is the charging infrastructure. The examination of various pricing models concerning the local Conditions must be created to expedite the adoption of electric cars in the nation. An electric vehicle charging station expenditure (EVSE) is a wall-mounted module that provides electricity for recharging electric car batteries. An essential element of EVSEs is a safety mechanism that prohibits the flow of electricity until the plug is physically connected to the vehicle. It is possible to add customization options to EVSEs such as:

- Verification (such as driver registration)
- Integration with payment gateways and,
- software for remote monitoring.

Several industry standards and recommendations have arisen due to increasing electric car charging technologies. A quick perspective of the technology, standards, and nomenclature for the charging infrastructure is given here. Charging power, which dictates how long it takes to charge a car, varies by order of magnitude between charge stations. A modest home outlet can charge at a rate of up to 1.2 kW, while the most sophisticated fast-charging stations can charge at speeds of up to 350 kW. The charging infrastructure is classified generally according to its speed: Level 1, Level 2, and direct current (DC) rapid charging (sometimes referred to as Level 3). Table 29.1 lists the three different kinds of chargers that may be used to refuel

Table 29.1 Charging station details

Charger Type	Charger connectors	Location	No. of charging point	Minimum power and voltage range	Charging time (hours)
Fast (Level 3)	CCS	Commercial/Industrial	1	50 kW, 200–1000 V	0.5–1
	CHAdeMO		1	50 kW, 200–1000 V	
	Type-2 AC		1	22 kW, 380–480 V	
Slow/Moderate (Level 1 and 2)	Bharat DC-001	Residential/Commercial	1	15 kW, 72–200 V	6–8
	Bharat AC-001		3	10 kW, 230 V	11

electric vehicles. Level 1 chargers are low-power chargers that are most often seen in residential regions of the world. Level 2 chargers are preferred over level 1 chargers when it comes to minimizing charging time. Level 2 chargers, on the other hand, need an increase in protection if they are used inside a residential building. Level 3 chargers are intended for use at the commercial level, as they can completely charge an EV battery in one hour or less. As a result of their decreased charging time, fast-charging stations are gaining in popularity.

Private EVCS: charging the batteries of privately owned automobiles through household charging stations. Billing is a significant component of home/domestic metering.

AC "Slow" EVCS: Generally, home private chargers are used with a 230 V/15A single-phase socket capable of delivering up to about 2.5KW of power. The EVSE provides alternating current to the vehicle's onboard charger, converting it to a direct current to charge the battery.

Public EVCS: Outside of the house, electric power must be invoiced, and payment collected. Occasionally, the power consumed by these chargers may need to be controlled.

DC "Fast" EVCS: Direct current is delivered straight to the electric vehicle's battery through the charging connector. FC chargers (often 50 KW or more) may provide up to 100 kms of range per hour.

Fast chargers are often used to supplement rather than charge cars entirely. These are critical for taxi firms and corporate users with an electric vehicle fleet. Two- and three-wheeler electric vehicles need slow charging, while light four-wheeler electric vehicles may be charged both slowly and quickly. Heavy four-wheeled vehicles (buses) need just rapid charging at public stations. Table 29.1 provides details of the charging station.

29.3 Objectives

The optimization method may be used to determine the optimum value or solution. The optimization issues may include determining a maximum or minimum value or using a single or many objectives. Multi-objective optimization is a term that refers to problems that have more than one goal (Gunantara 2018). Table 29.2 enlists the objective function formulated previously.

The following single-objective function is used to form multi-objective functions.

29.3.1 Total Active Power Loss (TAPL)

With the high R/X ratio and radial construction of the distribution network, active power loss is high compared to the transmission system. The TAPL is the difference between power given by generators and demanded by the load. If all the power given

Table 29.2 Objective considered previously

Reference	Bus system	Maximization	Minimization
Hadian et al. (2020)	IEEE-69 Bus	DNO profit, EVCS profit	
Pashajavid and Golkar (2013)	Practical 11 Bus		Voltage deviation, Electrical energy loss
Moghaddam et al. (2017)	Transport network		Total travel time, Recharging cost
Reddy and Selvajyothi (2020)	IEEE-19 and 25 Bus		Active power loss
Pal et al. (2021)	Practical 33 Bus		Energy loss
Proposed	IEEE-25	Voltage profile	Active power loss

by the generator meets the demand of load, then no loss of power will take place, but generally, the difference gets increases because loads do not demand all the time power in all three phases, but since power cannot provide storage, so excess power is generated for fulfilling the demand of load at any time. The power loss occurs in a branch or distribution line connected with two adjacent buses, whereas the voltages change occurred on the bus of the distribution system. The primary objective is to reduce its TAPL to augment the distribution system's capability, which will smoothen the operation and improve its performance. The APL is determined using the branch current loss formula given by the following equations for all three phases.

$$P^a_{Loss,br(xy)} = \text{Real}\{(U^a_x - U^a_y)(I^a_{xy})^*\} \tag{29.1}$$

$$P^c_{Loss,br(xy)} = \text{Real}\{(U^c_x - U^c_y)(I^c_{xy})^*\} \tag{29.2}$$

$$P^c_{Loss,br(xy)} = \text{Real}\{(U^c_x - U^c_y)(I^c_{xy})^*\} \tag{29.3}$$

where $P^a{}_{Loss,br(xy)}$, $P^b{}_{Loss,br(xy)}$, $P^c{}_{Loss,br(xy)}$ represent the active power loss in branches for phases *a, b*, and c, respectively. *x* and *y* are the distribution bus, *xy* is the branch or distribution line connecting buses *x* and *y*. U_x, U_y is the voltage of bus *x*, *y*. I_{xy} is the current flowing through the branch *xy*.

The TAPL determined by the summation of all phase losses of distributed lines is given by the following equation.

$$TAPL = P^T_{Loss} = \sum \left(P^a_{Loss,br(xy)} + P^b_{Loss,br(xy)} + P^c_{Loss,br(xy)}\right) \tag{29.4}$$

So, the first single-objective function is subjected to the minimization of TAPL of the distribution system, including all three phases.

$$F_1 = \min(TAPL) \tag{29.5}$$

29.4 Constraints

The following constraints are checked in every iteration for the feasible solution according to the DG allocation optimization problem's obligation.

29.4.1 Equality Constraints

The algebraic sum of the power in the electrical distribution network and loss should be identical to the DG's power. Since the non-optimal allocation of DG could affect the distribution system's operation, that condition is restricted by applying the below constraints.

$$P_{Loss}^{T} + \sum_{l=1}^{Nl} P_D = \sum_{i=1}^{NDG} P_{DGi} \tag{29.6}$$

$$Q_{Loss}^{T} + \sum_{l=1}^{Nl} Q_D = \sum_{i=1}^{NDG} Q_{DGi} \tag{29.7}$$

where P_D active power demand raised by the load, *Nl* total number of distribution lines, *NDG* number of DG, P_{DG} active power injected by DG, Q_D reactive power demand raised by the load, Q_{DG} reactive power injected by DG.

29.4.2 Inequality Constraints

a. *Voltage Constraint*:

   The magnitude of bus voltage after every single iteration should be intolerance limit; otherwise, it leads to the problem of unbalancing in voltage magnitudes, since if there will be significant fluctuation in bus voltage, then it will inject a higher amount of current through the branches, that will make solution unachievable.

$$U_i^{\min} \le U_i \le U_i^{\max} \tag{29.8}$$

   where i = *1,2,3 … nbus, nbus* is the total number of the bus in the distribution system.

b. *Loss Constraint*:

When EVCS is placed in the distribution network, then the power loss started to increase; while finding the best location, it is essential to limit the value of active power loss.

$$P_{loss,i}^{\min} \leq P_i \leq P_{loss,i}^{\max} \quad (29.9)$$

c. *DG Size Constraint*:

The power injected by DG should be within described limits to operate the distribution system. The power supplied by DG with UPF is controlled by real power constraints; however, the power supplied through DG with power factor (PF) should be within reasonable limits, for the efficient functioning of the power supplied by it is controlled by additional reactive power constraint.

$$P_{DG}^{\min} \leq P_{DG} \leq P_{DG}^{\max} \quad (29.10)$$

$$Q_{DG}^{\min} \leq Q_{DG} \leq Q_{DG}^{\max} \quad (29.11)$$

d. *Thermal Limits Constraint*:

The magnitude of the current flowing in the branch of the distribution system should not surpass the thermal limit or ultimate point of line outage; for imposing this situation, the following thermal limit constraint is applied during the optimization of the objective function.

$$I_{xy} \leq I_{\max} \quad (29.12)$$

where I_{max} is the maximum current flowing in the branch without line outage.

29.5 Optimization Techniques

29.5.1 Chicken Swarm Optimization (Deb et al. 2020)

CSO replicates the chicken swarm's behavior and food-seeking method. The group is split into dominant roosters, hens, and chicks according to the chickens' status. Roosters are at the top of the food chain, hens are in the middle, and chicks are at the bottom. The algorithm's random assignment of mother–child relationships inside the swarm is another noteworthy aspect. The hierarchical order and mother–child connection are modified after every step. The program effectively exploits hens' innate proclivities for following their groupmate rooster and chicks' proclivities for following their mother in food quest. Additionally, this algorithm considers that hens may attempt to take food discovered by others, resulting in a rivalry for food within the group. CSO's general and algorithm-specific parameters are specified during the initialization step. In multi-objective CSO, the population is classified as rooster, hen, or chick based on rank, rather than fitness value, as in single-objective CSO. The concept described is used to determine the rank of all the people in the population. A hierarchical order is created based on the rank of the individuals in the population.

29.5.2 Teaching Learning-Based Optimization Algorithm (Vadhera 2020)

TLBO is an evolutionary algorithm based on population genetics inspired by the interactive process of teaching and learning. The population of TLBO is made up of learners. The teacher is an educated scholar who imparts information to his students. The learners' performance is contingent upon the teacher's knowledge and teaching skills. The algorithm is split into two phases: the instructor phase, during which students learn directly from the teacher, and the Learner phase, during which students learn from one another via mutual contact. The student with the highest rating in a randomly formed population is usually given the instructor's position in multi-objective TLBO. Each student gains knowledge from the instructor. The learner acquires knowledge via reciprocal contact with other learners.

29.5.3 Harris Hawks Optimizer (Selim et al. 2020)

HHO is a population-based approach that is now being used by a hybridized process of exploration and exploitation. The Harris hawks primary goal is to hunt prey such as rabbits. Therefore, first of all, the hawks are seeking the rabbit. Exploration may be done utilizing two different approaches. This concept says that hawks should be placed near people and prey since this knowledge would benefit the family. But, the second approach considers the hawks randomly distributed throughout the forest. During the pursuit, the rabbit's fleeing energy was utilized to transition between the exploratory and exploitative phases of the HHO. The HHO algorithm can transition from exploration to exploitation and then switch between several exploitative actions depending on the prey's escape energy. A prey's energy level drops significantly during its evasion activity. The HHO's computing complexity is mostly determined by three processes: initialization, fitness assessment, and hawk updating. Notably, the computing cost of the initialization procedure with N hawks is O. (N). The updating method has a computational complexity of O(TN) + O(TND), which is comprised of searching for the optimal position and updating the location vectors of all hawks, where T denotes the maximum number of iterations and D denotes the dimension of particular issues. As a result, HHO has a computational complexity of O(N(T + TD + 1)).

29.5.4 Genetic Algorithm (Bohre et al. 2016)

The fundamental idea behind GA Optimization is synonymous with survival of the fittest.' The optimization method for GA solutions contains just one strong solution since only the robust solution can survive the process, while the weak solution cannot. The GA is capable of establishing an initial population associated with potential optimal solutions. Following that, GA recombines these people to direct their specific search to the most favorable results in the search space. The potential outcome is stored as a string or chromosome, and each chromosome is equipped with a fitness function or objective function for evaluating fitness. The fitness of a chromosome determines its capacity to survive and reproduce. Throughout the process, the GA maintains a limited population of chromosomes. The GA makes use of probabilistic concepts to develop the population in successive generations. In successive generations, new solutions are produced by GA recombination operators. These operators include those for selection or reproduction, crossover, and mutation. The crossover operator combines the 'fittest' chromosomes that give superior genes to the following generation. The mutation ensures that the whole search space is likely to be explored simultaneously, ensuring that populations are free of local minima. The population size, the assessment of the fitness function, the crossover technique, and the mutation rate are all critical factors in genetics algorithms. Generate a population of string or chromosomes as a starting point. Step 2: Determine the fitness level of each member of the population concerning the issue category (minimization or maximization). Step 3: Using reproduction, crossover, and mutation, generate offspring strings and then assess. Calculate the fitness value for each string in Step 4. Step5: Verify convergence to determine whether the necessary solution was found, or the requisite number of generations was reached.

29.5.5 Particle Swarm Optimization (Pandey and Bhadoriya 2014)

Kennedy and Eberhart proposed the PSO for the first time in 1995 (Kennedy and Eberhart 1995). The PSO is a well-known population-based optimization method. Throughout the PSO, each particle inside the search space has a unique velocity and inertia about the generations to which it belongs. The particle's velocity is modified in both direction and speed depending on its prior best experience (self) and the particle's historical best experience in its neighborhood (social). As a result, the particle will gravitate toward a promising region in the search space. Each individual flies in the whole search space with a unique velocity that is updated based on its own flight experience and that of its friends. The PSO optimization method initializes a random population of particles and continuously changes their positions depending on personal and neighboring experiences. The population's updated value depends on the velocity's updated values in each generation (or iteration). The population of particles in the next iteration is updated by adding the population from the previous iteration and the current iteration velocity.

29.5.6 Grasshopper Optimization Algorithm (Rao et al. 2020)

Grasshopper swarming activity is emulated by using an evolutionary computing algorithm and imitating grasshopper swarming behavior. It's utilized as an optimization tool for getting the best size and allocation of digital goods units and service classes for reaching various goals. The mathematical equations for GOA are created using the eating habits of grasshopper swarms as a reference. The swarming behavior of grasshoppers is affected by gravitational force, wind advection, and social interaction among individuals. Grasshoppers are a kind of bug. They are classified as a pest because they wreak havoc on agricultural production and agriculture. The grasshopper's life cycle. Although grasshoppers are often observed alone in the wild, they form one of the enormous swarms of any species. The swarm's size may be continental in scope, creating havoc for agriculture. The grasshopper swarm is unusual because it exhibits swarming behavior in both nymph and adult stages. Hundreds of thousands of nymph grasshoppers leap and move in the manner of rolling cylinders. They consume nearly all vegetation in their path. When they reach adulthood, they develop this behavior and form a swarm in the air. This is the method through which grasshoppers travel long distances. The swarm's primary feature during the larval period is the grasshoppers' sluggish mobility and tiny steps. In maturity, by contrast, the swarm's primary characteristic is long-range and sudden movement. Another essential feature of grasshopper swarming is the search for food sources. As said before, nature-inspired algorithms split the search process logically into two tendencies: exploration and exploitation. The search agents are urged to move suddenly during exploration while moving locally during exploitation. Grasshoppers naturally execute these two tasks, as well as target finding.

29.5.7 Rao Algorithm (Rao 2020)

When it comes to optimization, different metaphor-based methods are often used. All algorithms, however, are not flawless, and the computational time is costly. Therefore, there is a need for metaphor-less optimization algorithms to handle complicated problems accurately. It is a simple idea that does not need fine-tuning in artificial particles to make implementation easier. Successful implementation of the method for multio-bjective engineering optimization implemented, finishing in encouraging results. An advantage of the RAO algorithm is that it does not use any metaphor to optimize multio-bjective problems, which results in it being optimal. This method has two steps. In the first step, the best and worst candidates in the problem-solving process are optimized, and at random, these candidates interact with one another. It requires population size, iteration numbers, and control parameters in general. Nothing else than the algorithm's general parameters are being used. This method has previously been thoroughly evaluated on several benchmarks with restricted and multi-modeling properties, demonstrating its applicability in diverse optimization problems. The most important thing to do in the beginning is to populate the first population to reduce the objective function (OF). In every cycle, there are (n-m) decisions, and therefore, (n-m) potential solutions. Assign the OF's greatest value to the best agent; assign the OF's lowest value to the worst agent. By taking the difference between the best and worst answers, the RAO algorithm enhances the outcome. When a variable's new value is achieved, it is calculated by adding the difference between the best and worst values multiplied by a randomly generated number, with the iteration itself determining the iteration number. By adjusting the best and worst solutions and candidate solutions' random interactions, the algorithm enhances the outcome. The RAO algorithm stores all the function values that were approved after each iteration. The values used in the following iteration are formed from these. The randomly generated number is used in the fantastic exploration of the search space since it is applied to measure the gap between the best and worst answers. The RAO algorithm has shown superior competitive outcomes due to these factors.

29.5.8 Transient Search Optimization (TSO) (Bhadoriya and Gupta 2021)

The transient responses of switched electric circuits that involve storage components, such as inductance plus capacitance, stimulate this algorithm. Exploring and exploiting the TSO algorithm's capacity tested by optimizing many mathematical benchmark functions and solving problems with engineering design optimization, and results are associated with the new algorithms of optimization. The complete response of the circuits to resistive and energy storage variables such as condensers and inductors and a transient response combined with a steady-state response are interdependent. The system cannot be instantly switched to the next constant condition by altering the first and second-order circuits, which need time for the condenser or inductor to charge or discharge before the steady-state value can be calculated. The TSO algorithm is composed of three sequential steps: random solution initialization, exploration, and exploitation to find a balance between the various solutions obtained after each scenario in order to obtain the optimal solution, followed by updating steps for the various iteration counts chosen to terminate the algorithm. The search agents are initialized randomly inside the bottom and topmost limits of the search area. The search agents' initialization is created randomly. Exploration finds the optimal solution, while exploitation seeks the optimal or steady-state solution. TSO's discovery exhibited improved behavior due to second-order circuit oscillations near the null point. As a result, the use of TSO enables the exponential decay of the first-order discharge of storage components into electrical circuits. The TSO algorithm's exploitation and exploratory computational modeling, which are inspired by the equation's full response (transient and steady-state) of the first and respectively, are analogous to an electrical circuit's final state. The optimal position of the decision variable is determined by the computational complexity O of the proposed method, which is dependent on the product of the various number of search agents N. The flow chart is given in Fig. 29.1.

29.6 Results and Discussion

Figure 29.2 demonstrates the one-line diagram for DG and EVCS allocation of the IEEE-25 unbalanced radial distribution system (URDS), modeling data for active and reactive power demand, and branch taken from Ramana et al. (2010). The base kV and base MVA of the system are 11 kV and 1 MVA, respectively. The approach for optimizing the best distribution generator allocation in an unbalanced distribution system in the presence of EVCS optimal position with constant capacity. For optimization of the single-objective function, TAPL is chosen. The multi-objective function included TAPL, TRPL, and VSI to analyze the distribution network's characteristics. Single and multiple DG allocated for single as well as multi-objective function formulated. The proposed algorithm's convergence characteristics for all cases under every scenario formed are presented. The negative impact of EVCS on distribution network managed with optimal allocation of DG The following cases reduced TAPL, TRPL, and improved VSI.

Case 1: DG with unity power factor (UPF)

Case 2: DG with fixed power factor (FPF)

Case 3: DG with optimal power factor (OPF).

29.6.1 Single Objective

The active power loss minimization aimed with optimal allocation of EVCS, and DG The active power loss is 150.12 kW for the base case when EVCS is absent in the distribution network. The three scenarios are considered for this study for optimal placement of one, two, and three EVCS depicted in Table 29.3. The insertion of EVCS increases the power loss in the distribution network; therefore, optimal placement

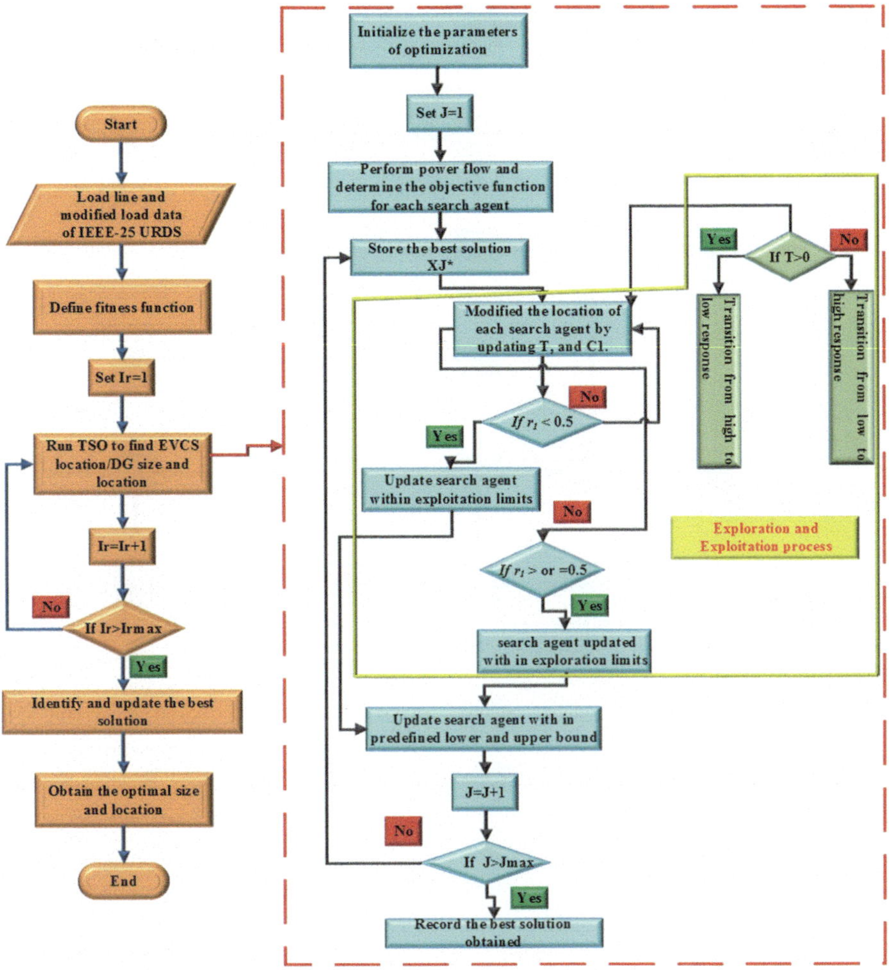

Fig. 29.1 Flowchart of TSO

is needed. The losses are increased from 150.12 kW to 223.82 kW after placement of three EVCS on bus locations 3, 18, and 23; The corresponding voltage profile is degraded, which needs to be improved.

29.6.1.1 Single DG Allocation

The cases, as mentioned earlier, are applied for optimal placement of DG in the presence of EVCS. All the scenarios of EVCS are considered. Table 29.4 summarizes the results obtained after optimal allocation of single DG for each scenario; the TAPL is reduced to 87.93 kW, under case 1 when DGs of 744.46 kW, 712.14 kW, and

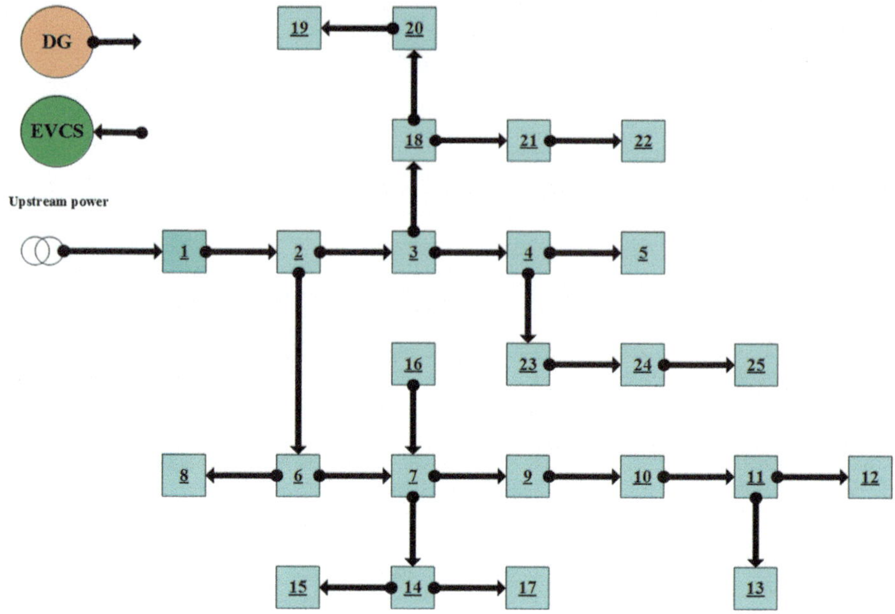

Fig. 29.2 Single line layout of IEEE-25 unbalanced radial distribution system

Table 29.3 The optimal location of EVCS

Scenario	EVCS	Location (bus)	TAPL (kW)	Relative Increment (%)
Base case	–	–	150.12	–
1	1	3	167.47	11.55
2	2	3–23	192.61	28.30
3	3	3–18-23	223.82	49.09

756.18 kW are placed on phases a, b, and c, respectively, of bus 7 for scenario 1. The RAPL is further reduced by 30.96% and 17.05% for scenarios 2 and 3. For case 2, the RAPL are 59.78%, 50.35%, and 37.37% for scenario 1, 2, and 3, respectively. The maximum reduction in TAPL is found in case 3 and the RAPL is 66.07%, 55.95%, and 42.72% for scenarios 1, 2, and 3. The reduction in power loss is obtained after optimal single DG allocation in all the phases of optimal bus location.

a. Voltage Profile

The voltage magnitude updates in every case for all test system phases. The phase-wise voltages profile for all the scenarios is shown in Fig. 29.3. It is viewed from the voltage profile that after placement of EVCS, degradation occurs in voltages on several buses causing lower voltage, ultimately introducing the significant difference in voltages of two different buses compared to the base case. This situation causes voltage collapse in the distribution network.

Table 29.4 Single DG allocation

Scenario	Location (bus)	DG size (kW)			TAPL (kW)	RAPL (%)
		Phase A	Phase B	Phase C		
Case 1 (UPF)						
1	7	744.46	712.14	756.18	87.93	41.42
2	7	771.27	764.07	835.74	103.64	30.96
3	7	844.63	829.07	816.72	124.51	17.05
Case 2 (FPF)						
1	7	785.96	845.35	774.52	60.37	59.78
2	7	873.33	853.38	875.08	74.52	50.35
3	6	890.96	898.72	899.2	94.08	37.37
Case 3 (OPF) (PF1 = 0.80)						
1	7	738.94	755.05	747.99	50.93	66.07
2	7	749.28	725.50	740.08	66.12	55.95
3	6	818.73	842.98	824.33	85.98	42.72

After integrating DG, the voltage profile becomes closer and more circular with an enhanced magnitude of the voltage produced in a better voltage profile with lower voltage magnitudes on successive buses. The voltage on each bus is within acceptable limits. The improvement in voltage profile balances the reactive power requirement in the distribution network.

b. Convergence Characteristics

The convergence criterion is an essential aspect of optimization algorithms. The success or failure of optimization algorithms depends on rapid or slow convergence characteristics for a specific objective function that will directly affect the favorable or unfavorable selection of convergence criteria in a certain number of iterations during the independent run. The convergence characteristics are shown in Fig. 29.4, revealing that the placement of EVCS converged with minimum increment in power loss reference to the base case, whereas the DG with various power factors converges with lower power loss. The convergence curve of all the cases was obtained within 300 iterations, with the global minima of objective function subjected to various operational constraints.

29.7 Conclusion

The electric vehicle charging station is optimally placed with a minimum negative impact on the distribution network. The URDS IEEE-25 analyzed with the optimal allocation of DG with UPF, FPF, and OPF in the presence of EVCS, reduced the

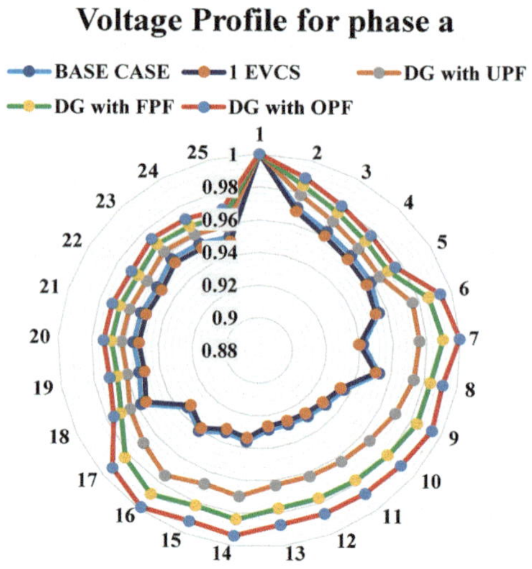

(a)

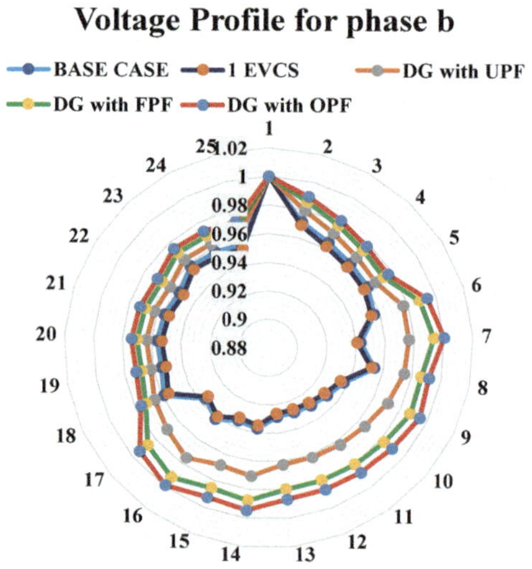

(b)

Fig. 29.3 The phase-wise voltage profile for (**a**) phase a, (**b**) phase b, and (**c**) phase c

power loss with improvement in voltage profile utilizing a new recently developed TSO algorithm. The main findings are as follows:

- The active power loss reduced to 66.07%, 55.95%, and 42.72% after allocation of DG.
- In the fitness function, the active and reactive power loss is 50.93 kW, compared to 167.47 kW, for optimally allocated EVCS.

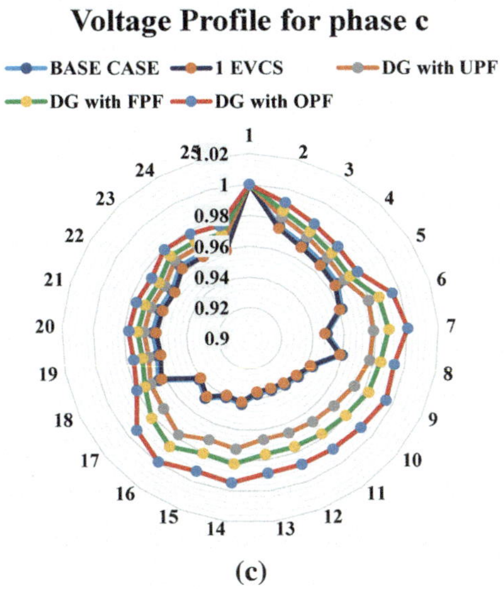

Fig. 29.3 (continued)

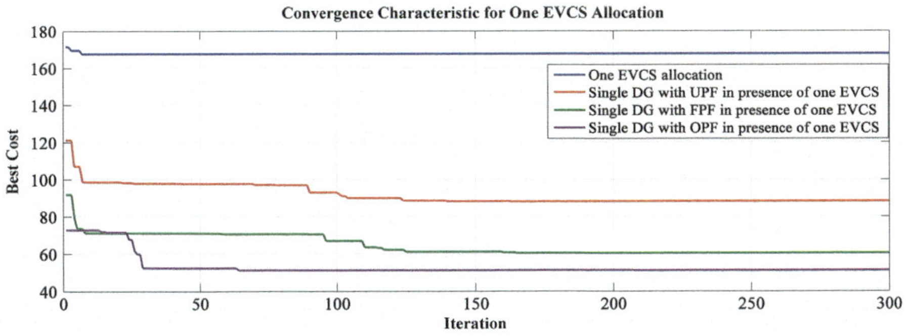

Fig. 29.4 The convergence characteristic of TSO

- The proposed algorithm's convergence characteristic obtained global minima for multi-objective functions when DG with OPF is allocated optimally.
- The voltage profile enhanced significantly after allocation of different power factor-based DG in the presence of three EVCS; however, for every case, it is improved and within limits.

A uniform circular voltage profile indicates a resilient distribution network with lower active and reactive power loss. This study will help distribution network operators plan the distribution system with a more significant number of EVCS done simultaneously with DG allocation for smooth operation. The fixed and uncertain schedule for charging electric vehicles may be considered with EVCS to analyze the distribution network's performance for residential, commercial, and industrial zones implemented with mentioned optimization algorithms.

References

Ahmadi M, Adewuyi OB, Danish MSS, Mandal P, Yona A, Senjyu T (2021) Optimum coordination of centralized and distributed renewable power generation incorporating battery storage system into the electric distribution network. Int J Electr Power Energy Syst 125(August 2020):106458. https://doi.org/10.1016/j.ijepes.2020.106458

Battapothula G, Yammani C, Maheswarapu S (2019) Multi-objective simultaneous optimal planning of electrical vehicle fast charging stations and DGs in distribution system. J Mod Power Syst Clean Energy. 7(4):923–934. https://doi.org/10.1007/s40565-018-0493-2

Bhadoriya JS, Gupta AR (2021) A novel transient search optimization for optimal allocation of multiple distributed generator in the radial electrical distribution network:1–23

Bohre AK, Agnihotri G, Dubey M (2016) Optimal sizing and sitting of DG with load models using soft computing techniques in practical distribution system. IET Gener Transm Distrib 10(11):2606–2621. https://doi.org/10.1049/iet-gtd.2015.1034

Chen L, Xu C, Song H, Jermsittiparsert K (2021) Optimal sizing and sitting of EVCS in the distribution system using metaheuristics: a case study. Energy Rep 7:208–217. https://doi.org/10.1016/j.egyr.2020.12.032

Deb S, Member S, Tammi K, Gao X, Kalita K, Mahanta P (2020) A hybrid multi-objective chicken swarm optimization and teaching learning based algorithm for charging station placement problem. IEEE Access:92573–92590. https://doi.org/10.1109/ACCESS.2020.2994298

Deb S, Tammi K, Kalita K, Mahanta P (2018) Impact of electric vehicle charging station load on distribution network. Energies 11(1):1–25. https://doi.org/10.3390/en11010178

Deb S, Tammi K (2018) Distribution network:1–25. https://doi.org/10.3390/en11010178

Domínguez-Navarro JA, Dufo-López R, Yusta-Loyo JM, Artal-Sevil JS, Bernal-Agustín JL (2019). Design of an electric vehicle fast-charging station with integration of renewable energy and storage systems. Int J Electr Power Energy Syst 105(March 2018):46-58. https://doi.org/10.1016/j.ijepes.2018.08.001

Drossinos Y, Zambelli P, Thiel C (2016) Optimal allocation of electric vehicle charging infrastructure in cities and regions irst main title line first line second main title line third main title line Alyona. https://doi.org/10.2790/353572

Ghatak SR, Sannigrahi S, Acharjee P (2019) Multi-objective approach for strategic incorporation of solar energy source, battery storage system, and DSTATCOM in a smart grid environment. IEEE Syst J 13(3):3038–3049. https://doi.org/10.1109/JSYST.2018.2875177

Gong D, Tang M, Buchmeister B, Zhang H (2019) Solving location problem for electric vehicle charging stations—A sharing charging model. IEEE Access 7:138391–138402. https://doi.org/10.1109/ACCESS.2019.2943079

Grahn P, Söder L (2013) Electric vehicle charging impact on load profile. http://urn.kb.se/resolve?urn=urn:nbn:se:kth:diva-116145

Gunantara N (2018) A review of multi-objective optimization: methods and its applications. Cogent Eng 5(1):1–16. https://doi.org/10.1080/23311916.2018.1502242

Hadian E, Akbari H, Farzinfar M, Saeed S (2020) Optimal allocation of electric vehicle charging stations with adopted smart charging/discharging schedule. IEEE Access. 8:196908–196919. https://doi.org/10.1109/access.2020.3033662

Injeti SK, Thunuguntla VK (2020) Optimal integration of DGs into radial distribution network in the presence of plug-in electric vehicles to minimize daily active power losses and to improve the voltage profile of the system using bio-inspired optimization algorithms. Prot Control Mod Power Syst 5(1). https://doi.org/10.1186/s41601-019-0149-x

Jacobs B (2018) Optimal electric vehicle fast charging station placement based on game theoretical framework. IEEE Trans Intell Transp Syst 19(8):21–38. https://doi.org/10.1007/978-981-15-3525-3_2

Johari NF, Zain AM, Mustaffa NH, Udin A (2013) Firefly algorithm for optimization problem. Appl Mech Mater 421(April 2014):512–517. https://doi.org/10.4028/www.scientific.net/AMM.421.512

Kennedy J, Eberhart R (1995) Particle swarm optimisation. Stud Comput Intell 927:5–13. https://doi.org/10.1007/978-3-030-61111-8_2

Lelièvre PG, Bijani R, Farquharson CG (2016) Optimization methods. In 78th EAGE conference exhibit 2016—Work program. 1–62. https://doi.org/10.7551/mitpress/2687.003.0008

Moghaddam Z, Ahmad I, Habibi D, Phung QV (2017) Smart charging strategy for electric vehicle charging stations. IEEE Trans Transp Electrif 4(1):76–88. https://doi.org/10.1109/TTE.2017.2753403

Pal A, Bhattacharya A, Chakraborty AK (2021) Allocation of electric vehicle charging station considering uncertainties. Sustain Energy, Grids Netw 25:100422. https://doi.org/10.1016/j.segan.2020.100422

Pandey D, Bhadoriya JS (2014) Optimal placement & sizing of distributed generation (DG) to minimize active power loss using particle swarm optimization (PSO). Int J Sci Technol Res 3(7):246–254

Pashajavid E, Golkar MA (2013) Optimal placement and sizing of plug in electric vehicles charging stations within distribution networks with high penetration of photovoltaic panels. J Renew Sustain Energy 5(5). https://doi.org/10.1063/1.4822257

Pillai RK, Suri R, Kundu S, Singh H, Sarkar Roy S, Dhuri S (2018) Electric vehicle charging stations business models for India. ISGF Report. 1–20. www.indiasmartgrid.org

Ramana T, Ganesh V, Sivanagaraju S (2010) Cogeneration & distributed distributed generator placement and sizing in unbalanced radial distribution system. August 2014:37-41. https://doi.org/10.1080/15453661009709862

Rao RV (2020) Rao algorithms: three metaphor-less simple algorithms for solving optimization problems. Int J Ind Eng Comput 11(1):107–130. https://doi.org/10.5267/j.ijiec.2019.6.002

Rao S, Jasthi K, Goli P, Das D, Bansal RC (2020) Grasshopper optimization algorithm based two stage fuzzy multiobjective approach for optimum sizing and placement of distributed generations, shunt capacitors and electric vehicle charging stations. J Energy Storage 27(Nov 2019):101117. https://doi.org/10.1016/j.est.2019.101117

Reddy MSK, Selvajyothi K (2020) Optimal placement of electric vehicle charging station for unbalanced radial distribution systems. Energy Sources, Part A Recover Util Environ Eff 00(00):1–15. https://doi.org/10.1080/15567036.2020.1731017

Roosta A, Khooban HEM (2019) Optimization of radial unbalanced distribution networks in the presence of distribution generation units by network reconfiguration using harmony search algorithm. Neural Comput Appl 31(11):7095–7109. https://doi.org/10.1007/s00521-018-3507-0

Satish M, Reddy K, Selvajyothi K (2020) Environmental Effects Optimal placement of electric vehicle charging station for unbalanced radial distribution systems. Energy Sources, Part A Recover Util Environ Eff 00(00):1–15. https://doi.org/10.1080/15567036.2020.1731017

Selim A, Kamel S, Alghamdi AS, Jurado F (2020) Optimal placement of DGs in distribution system using an improved Harris Hawks optimizer based on single- and multi-objective approaches. IEEE Access. 8:52815–52829. https://doi.org/10.1109/ACCESS.2020.2980245

Shariff SM, Alam MS, Ahmad F, Rafat Y, Asghar MSJ, Khan S (2020) System design and realization of a solar-powered electric vehicle charging station. IEEE Syst J 14(2):2748–2758. https://doi.org/10.1109/JSYST.2019.2931880

Sharma A, Kapoor A, Chakrabarti S (2019) Impact of plug-in electric vehicles on power distribution system of major cities of India: a case study

Tang H, Wu J, Wu Z, Chen L (2018) Two-stage optimization method for power loss and voltage profile control in distribution systems with DGs and EVs using stochastic second-order cone:501–517. https://doi.org/10.3906/elk-1602-65

Vadhera S (2020) TLBO-based approach to optimally place and sizing of energy storage system for reliability enhancement of radial distribution system. July 2019:1–20. https://doi.org/10.1002/2050-7038.12334

Verma YP, Kumar A (2013) Potential impacts of emission concerned policies on power system operation with renewable energy sources. Int J Electr Power Energy Syst 44(1):520–529. https://doi.org/10.1016/j.ijepes.2012.03.053

Xiong Y, Gan J, An B, Miao C, Bazzan ALC (2018) Optimal electric vehicle fast charging station placement based on game theoretical framework. IEEE Trans Intell Transp Syst 19(8):2493–2504. https://doi.org/10.1109/TITS.2017.2754382

Chapter 30
Solar Power Charging of Electric Vehicle Along with the Implementation of Vehicle-To-Grid (V2G) Technology

Shreastha Varun and Sandeep Bhongade

Abstract Reducing pollution in the environment is the most important topic of the present times. Pollution mainly arises from the burning of fossil fuel which emits CO_2 and other harmful gases in the environment. The transportation sector is introducing technologically advanced Electric Vehicles (EVs) which are eco-friendly. EV integration to the grid is the attraction of many researchers and engineers around the world due to the rapidly growing numbers of EVs in the global market. The use of EVs in the grid will play an important role in future smart grid technology. But EVs charging puts significant load demand on the grid as well as uneven and uncoordinated charging creates disturbance in the grid. In order to reduce the burden on the grid, the EVs can be charged by Renewable Energy Resources (RESs). Solar is the most widely available renewable energy source. Solar charging stations can be installed at homes, offices, parking lots, public areas and isolated areas. The proposed system in this chapter is designed in the MATLAB Simulink environment in which Electric Vehicle (EV) is charged in a standalone solar power charging station and this EV is integrated into the grid when the solar power is unavailable. The use of a DC–DC bidirectional converter with appropriate controlling technique enables the transfer of power from EV to the grid whenever there is demand. The benefits and limitations of integrating EV in the power systems have been discussed in this chapter.

Keywords Electric Vehicle (EV) · Solar-to-Vehicle (S2V) · Vehicle-to-grid (V2G) · DC–DC bidirectional converter · DC–AC converter

Nomenclature

CCS	Combined Charging System
EV	Electric Vehicle
EVs	Electric vehicles

S. Varun (✉) · S. Bhongade
SGSITS, Indore, India

A. K. Bohre et al. (eds.), *Planning of Hybrid Renewable Energy Systems, Electric Vehicles and Microgrid*, Energy Systems in Electrical Engineering,
https://doi.org/10.1007/978-981-19-0979-5_30

MPPT	Maximum Power Point Tracker
P&O	Perturb & observe
PLL	Phase-Locked Loop
RESs	Renewable Energy Resources
S2V	Solar to Vehicle
SAE	Society of Automotive Engineers
SLM	Smart Load Management
SoC	State of Charge
SRF	Synchronous Reference frame
V2G	Vehicle-to-Grid
VPP	Virtual Power Plant

30.1 Introduction

From Electric vehicles (EVs) such as Electrobat (1895), General Motors EV-1(1996–1999), Toyota Prius (1997) to the present-day Tesla Model 3 (2017), Electric vehicles (EVs) that we see today comes from a series of development in the field of electric locomotives over the past two centuries.

Due to the growing concern of the environment and the depletion of fossil fuels, Electric Vehicles (EVs) are now seen as the future of the transportation sector (Sneha Angeline and Newlin Rajkumar 2020).

The innovations in the technologies of EVs have significantly increased the market shares of EVs around the world today. There are mainly following types of Electric vehicles:

1. Hybrid electric vehicles (HEVs) run on an internal combustion engine and an electric motor, which uses energy stored in batteries.
2. Plug-in hybrid electric vehicle (PHEV) is a hybrid electric vehicle whose battery can be recharged by plugging it into an external source of electric power, as well as through regenerative braking.
3. Battery electric vehicles (BEVs) are fully-electric vehicles with rechargeable batteries. BEVs use an external source for charging the battery.
4. Fuel cell electric vehicle (FCEV) works on fuel cell which uses oxygen and compressed hydrogen to generate electricity to run the electric motor.
5. Solar electric vehicle (in future), work is going on in developing an electric vehicle that can be powered partially or fully by direct solar energy.

Mainly PHEVs and HEVs are more popular in the EV market. Lithium-ion batteries are now largely used in EVs due to their very high efficiency and energy density as compared to Lead-acid batteries or any other chemical batteries that are available in the market.

EVs charging technology plays an important role in the growth of the EVs market. There are a charging levels of EV such as shown in the Fig. 30.1, the AC level 1

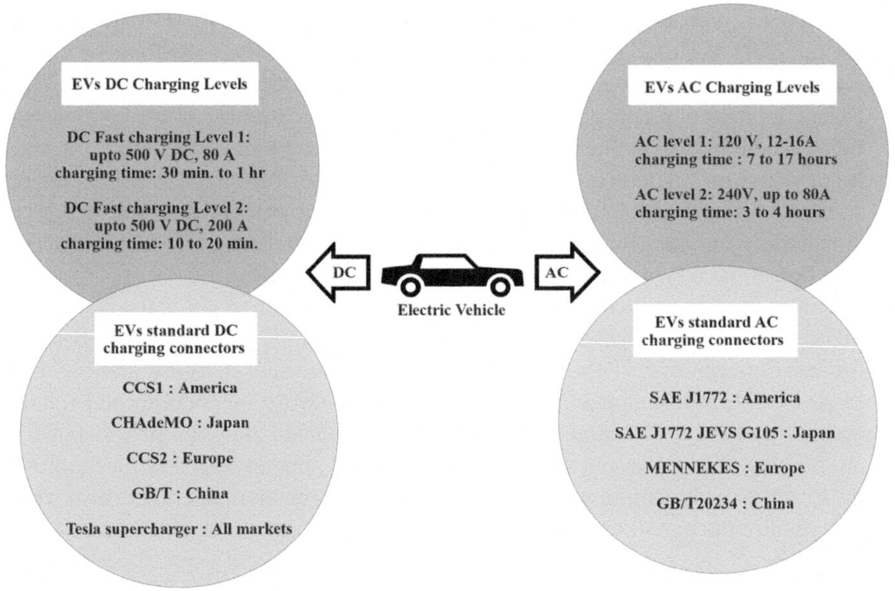

Fig. 30.1 Electric vehicle charging levels and standard electrical connectors

and AC level 2 both uses on -board EV chargers and are slow charging levels. AC level 1 charging can be done in households, whereas AC level 2 mainly works for public areas like offices, movie theatres, shopping malls, institutions, etc. DC fast charging uses off-board chargers, and therefore, requires DC charging stations. EVs that support fast charging can only be charged by DC charging stations. DC fast charger connectors can be separate DC only types like CHAdeMO or both AC and DC combined type such as combo-charging system (CCS/SAE combo) which supports both slow as well as fast charging of EV. Tesla models come with adapters to join any type of charging power outlet (Mohammed and Jung 2021).

In the present scenario, the penetration of both EVs and renewable energy sources (RESs) plays crucial roles in the power sector (Patil and Kalkhambkar 2021; Shahrukh Adnan Khan et al. 2019) which has been discussed below.

- Major pollution is caused by the emissions from fossil fuel-based transportation vehicles. Especially in the densely populated areas like metropolitan cities, there is a high risk of health to the people as pollution has reached an alarming level. The transportation sector is, therefore, rapidly introducing technologically advanced EVs which are eco-friendly. By 2030, India is planning to be an EV nation and hence coming up with innovative schemes such as Faster Adoption and Manufacturing of Hybrid and Electric vehicle (FAME) phase II for promoting the sales of EV by the end of 2024.

- Though Carbon emissions due to the transport sector have been reduced by adopting EVs, still EVs are not considered as absolutely Zero-Emission Vehicles as the batteries of EVs are charged from the electricity produced by coal or natural gas power plants which cause emissions. Large number of EVs will require more production from these coal or natural gas power plants which in turn mean more emissions. Therefore, if EVs are charged from renewable or clean energy sources such as hydroelectric, wind, solar, or nuclear power then there will be no emissions from the electricity production to transportation cycle and hence EVs will be completely well-to-wheel Zero-Emission Vehicles.
- The most important incentive for the large-scale adoption of EV is the installation of large numbers of easily accessible charging stations at appropriate distances for the convenience of the EV drivers. Grid-connected charging stations have to be installed with careful planning in order to ensure the smooth and reliable operation of the grid. Unpredictable and uncoordinated charging patterns of EVs at charging stations may cause voltage instability, power loss and harmonics in the grid. Increasing numbers of EVs will put significant demand on the grid which will require the installation of additional generation plants, thereby increasing the efficiency of the existing electrical network for a safe and smooth operation of the grid (Deb et al. 2019).
- The inclusion of renewable energy sources can help reduce the congestion on the grid, minimizing the cost and emission from thermal units but large penetration of renewable energy sources will impact the grid due to the intermittent nature of these sources. The off-grid charging of EV through renewable energy sources can help in reducing the burden on the grid as well as help in the integration of RE to the grid. EV can be considered as decentralized battery storage units for renewable energy sources which will help in controlling and limiting the amount of power injected into the grid (Preetha Yesheswini et al. 2020).
- In the smart grid technology, EVs along with RESs are seen as Virtual Power Plant (VPP) that can provide Smart Load Management (SLM) and power regulation services. EVs behave as a distributed spinning reserve for unexpected frequency fluctuations mostly caused due to renewable energy sources in the grid (Prem et al. 2020).

The main contribution of this chapter is to evaluate the standalone solar-based charging station for EV and utilize the storage capability of EV in meeting the demand requirement of the grid. Figure 30.2 shows the scheme of the proposed system. The main content of this chapter is summarized as follows:

- Standalone solar-based sustainable charging station as an alternative to the grid-connected charging station for charging of the EVs.
- Mitigating the impact of integrating renewable energy sources into the grid with the help of EV.
- Utilizing the V2G capability of EV to provide ancillary service to the grid.

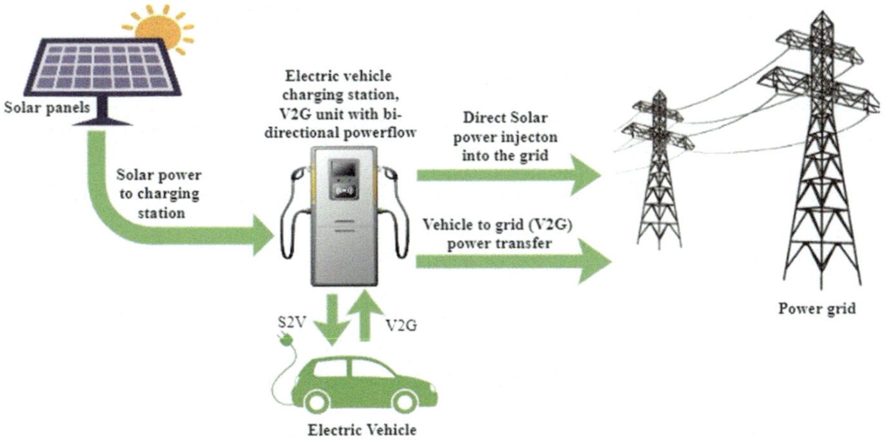

Fig. 30.2 Scheme of the proposed system showing S2V and V2G operation of EV

30.2 System Configuration

The main objective of the proposed system is to demonstrate and observe the integration of electric vehicles with the power system. The solar power is used to charge the vehicles in the Solar to Vehicle (S2V) concept (Shariff et al. 2020), and the energy stored in the EV batteries can be delivered to the grid when required in the Vehicle-to-Grid (V2G) concept. The system is designed to implement S2V and V2G concepts as shown in the block diagram in Fig. 30.3. The MATLAB/SIMULINK environment is used to design the system.

30.2.1 System Components

30.2.1.1 MPPT Solar Charge Controller

Output power of a solar panel varies due to irradiance level on PV panel, temperature level and current drawn from the cells. Maximum Power Point Tracker (MPPT) is mainly required to extract the maximum power from the solar panel. The maximum power point tracker sweeps through the panel voltage to find the best combination of voltage and current to produce maximum power. MPPT continuously tracks and adjusts the voltage to get maximum power. This makes MPPT very important especially in bad weather conditions.

There are different control techniques to get maximum power from the solar panel such as Perturb & Observe (P&0), Incremental conductance, Fractional Voltage, Fractional current, fuzzy logic, etc. In this system, P&O algorithm is used to track the Maximum Power Point (MPP). It is relatively simple to implement and can provide

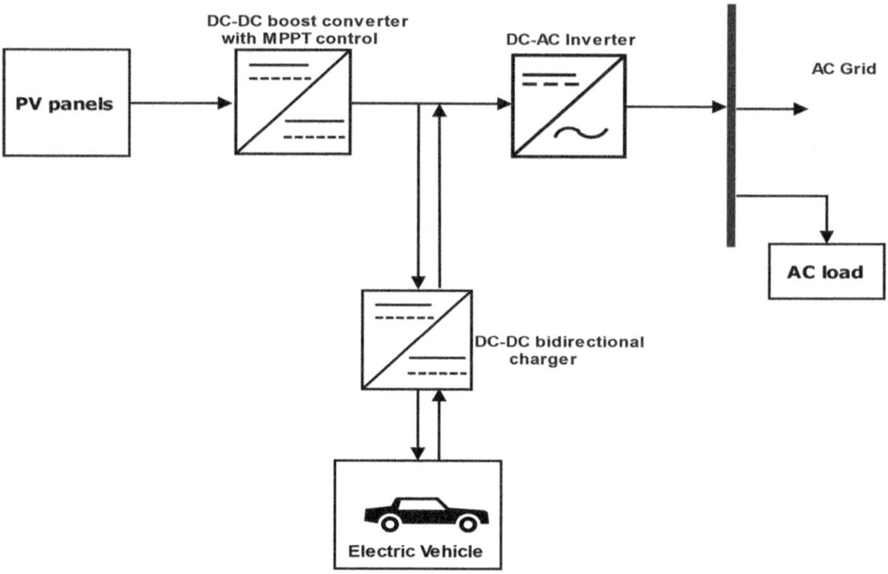

Fig. 30.3 Block diagram of the system

good efficiency. P&O algorithm continuously compares the new and previous values of output power and voltage of PV and then varies the duty cycle according to the direction of MPP tracked on the PV panel.

In the presented system, there are two panels with user-defined input irradiance constant, this system uses irradiance as 1000 constant and temperature as 25 °C. The parameters used for the PV panels are shown in the Table 30.1

DC to DC converter (step up or step down) act as an interface between the load and the PV modules. In this proposed system, a DC–DC boost converter is used for the MPPT control of the PV panels in this proposed system. The circuit diagram of the DC–DC boost converter is shown in Fig. 30.4

Table 30.1 PV parameters

PV parameters	Value
Open circuit voltage Voc	800 V
Maximum voltage	820 V
Current output	6.227 A
Voltage output	750 V
Output power	12 kw watt

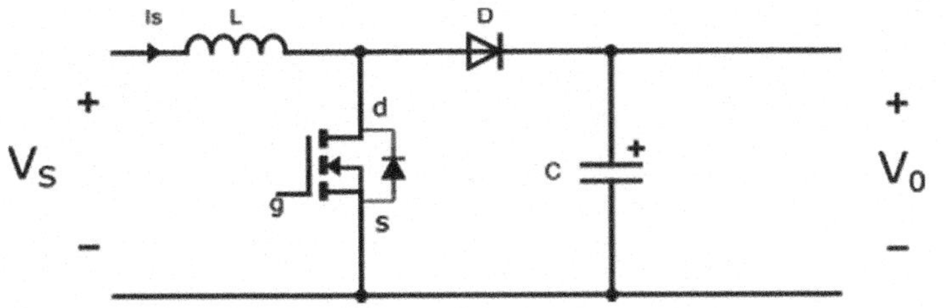

Fig. 30.4 DC–DC Boost converter

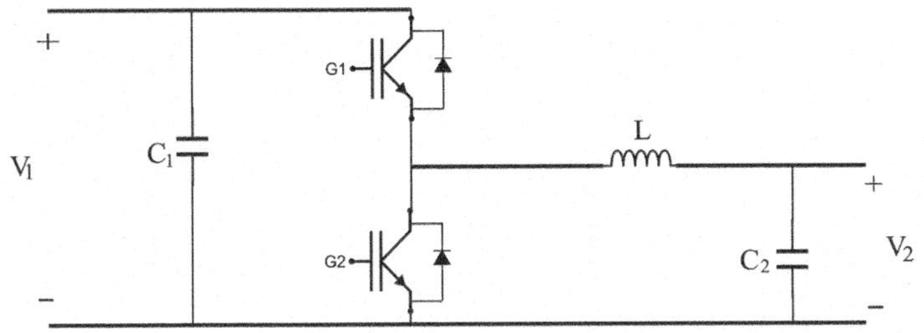

Fig. 30.5 Bidirectional DC–DC converter circuit

30.2.1.2 DC–DC Bidirectional Converter with PI Controller

A bidirectional DC–DC converter can act as both buck and boost converter, i.e., it can step up as well as step down the input DC voltage. Figure 30.5 presents the circuit of the converter circuit. In this system, DC–DC bidirectional converter is used for charging and discharging EV batteries. These bidirectional converters are the reason that EVs can be used for delivering power to the grid. PI controller is used with the DC–DC bidirectional converter in this system which increases the efficiency of the converter. It decides the discharging or charging rate according to the State of Charge (SoC) of EV. The PI controller is used to provide the pulse to the IGBTs used in the converter. PI controller compares the DC output voltage to the desired DC output voltage, it sets the required value and triggers the signal to obtain the desired value of the output voltage from the given input values of the converter (Singh et al. 2020; Yuan et al. 2021; Liu et al. 2020).

30.2.1.3 DC–AC Inverter with SRF–PLL-Based Controller

Inverters are power converters used to convert stable DC power into AC power. The universal bridge is used as PMW IGBT inverter with parameters given in Table 30.2.

The output of the inverter is the Chopped AC voltage with zero DC component consisting of harmonics. Low pass LC filters are added after the inverters to reduce the high-frequency harmonics. In power systems applications, the LC filter with the

Table 30.2 Inverter's parameters

Parameters	Values
Number of Bridge arms	3
Snubber resistance	5000 Ω
Ron	1 mΩ
Forward v [device Vf, diode Vfd(v)]	[0 0] V
Inductance	0.1 mH
Capacitance	500 μF

converters play an important role to obtain good quality voltage regulation. Low pass LC filter is used with the inverter to attenuate the high-frequency harmonics.

The present power system has involved renewable energy sources to feed the grid. This creates an imbalance in the grid due to many reasons such as the uneven nature of supply, fluctuations, etc. For the safe and reliable operation, the grid operators require to maintain power qualities standards and codes related to power quality improvement, active and reactive power, synchronization, stability, fault mitigations, etc. The converters require controlling techniques to control the output of the converter according to the requirements. For synchronization of the grid, the power converters on the grid side require the information of phase voltage, frequency and phase angle from the grid. Phase-Locked Loop (PLL) is the most commonly used method for the determination of grid phase angle, voltage and frequency. PI controller is used with the PLL for comparison and obtaining proper dynamics. Filters are used for reducing the harmonics and imbalances.

Synchronous Reference frame PLL (SRF–PLL) with PI controller is a widely used PLL technique because it is easy to implement. SRF–PLL gives accurate grid phase information under normal grid conditions. Under harmonics and imbalances conditions, SRF–PLL tend to give inaccurate results.

Tuning parameters of the PI controller can improve the results by reducing harmonics, but this method slows the dynamic response of the SRF–PLL. Therefore, various improvements and modifications are done in SRF–PLL in order to improve the dynamic response of SRF–PLL under harmonics conditions.

SRF–PLL uses dq synchronous rotating reference frame for the synchronization process. The basic block diagram of SRF–PLL has been shown in Fig. 30.6.

The grid phase voltages Va, Vb and Vc are measured which are in abc reference frame. Transformation of time-domain abc reference frame of three-phase voltages into αβ stationary reference frame is done by using Clark's transformation. The two-phase system of stationary αβ0 reference is then converted to two-axis system of dq0 rotating reference frame by Park transformation. The rotating direct axis or quadrature axis of dq reference frame can be aligned with the voltage in the αβ stationary reference frame or abc reference frame. PI controller ensures the zero value of Vd in steady-state condition and the alignment of grid voltage vector along the q-axis. PI controller synchronously rotates voltage space vector along q or d axis to obtain phase angle information. This way the DC voltage converted into AC

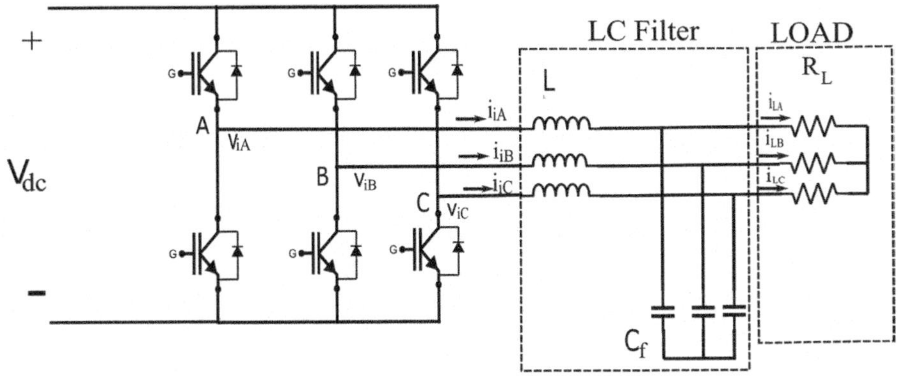

Fig. 30.6 DC–AC inverter with LC filter

three phase voltage by DC to AC inverter can be synchronized with the grid phase parameters in the rotating reference frame.

30.2.2 Implementation in MATLAB/SIMULINK

The simulation model of the proposed energy transfer from solar-to-vehicle and vehicle-to-grid is shown in Fig. 30.7 is developed in MATLAB version 2019b. There are two modes of operations:

Mode 1: S2V

In S2V operation mode, the power from solar panels is used to charge the battery. The DC–DC boost converter steps up the voltage of the solar power, this helps to transfer the power to the grid and to the battery as well. The bidirectional DC–DC converter operates as a buck converter to charge the battery in the S2V operation

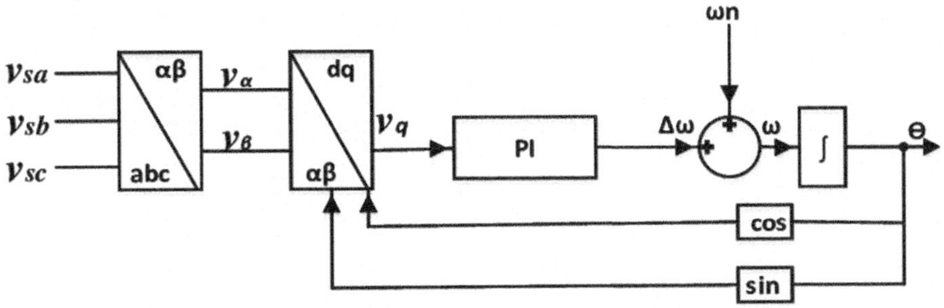

Fig. 30.7 Block diagram of SRF–PLL

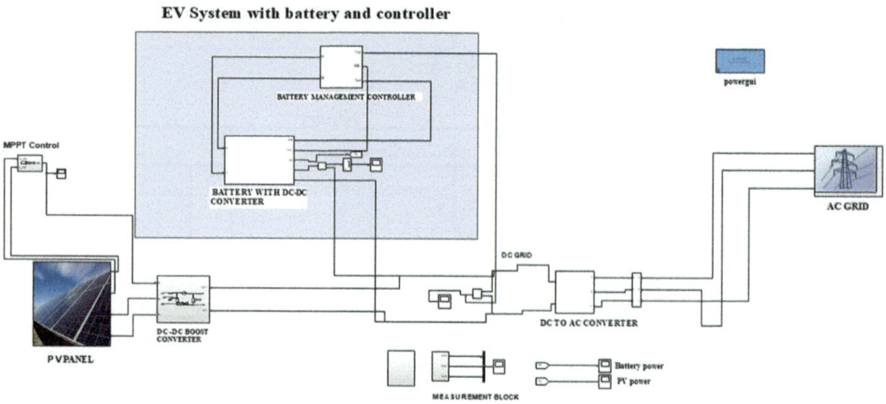

Fig. 30.8 Implementation of the system in MATLAB simulation

mode. The solar power also transfers the excess power to the gird through DC to AC converter. The implementation of grid connected solar system is shown in Fig. 30.8.

Mode 2: V2G

In V2G operation, the DC–DC converter of battery operates as a boost converter and then the output of this boost converter goes into DC to AC inverter to deliver the AC power to the grid (Melo et al. 2018).

A lithium-ion battery has been used in this model. Initial SoC of the battery is 60% and the response time is 30 s which means whenever SoC reaches 60%, it indicates that the battery is fully charged. The battery controller is for voltage regulation of the battery that decides when voltage needs to increase and when to drop, whenever it senses load power is not compensated by other resources, then it provides a pulse to the switches s1 and s2, PI controller compares the value of Vdc and the reference value. PI controller works to achieve the desired value. It provides switching pulses through s1 and s2 which is input to the IGBTs. This way the charging and discharging of the battery is done.

30.2.2.1 System Components in MATLAB/SIMULINK

Figures 30.9, 30.10, 30.11, 30.12 and 30.13 shows the components of the system in MATLAB/SIMULATION.

30.2.3 Results

The proposed S2V [mode 1] is evaluated, then the integration of EV into the grid in V2G [mode 2] operation is verified. The current and voltage waveform of the

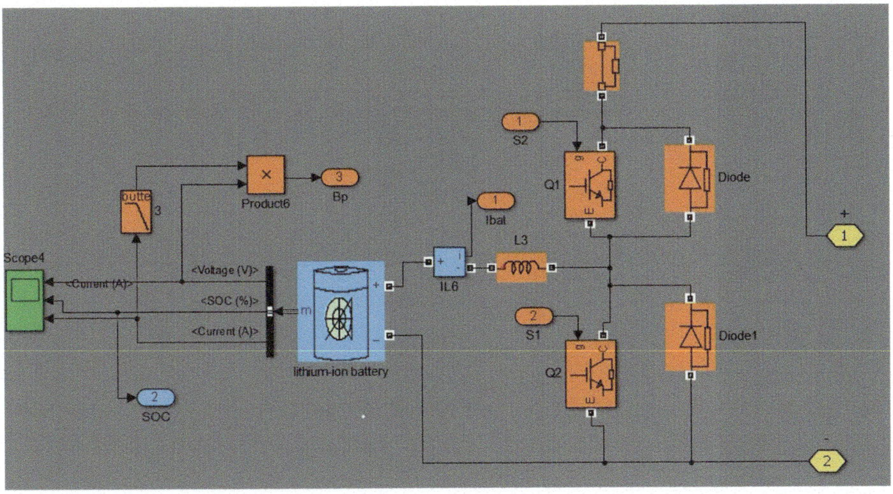

Fig. 30.9 Bidirectional DC–DC converter with EV battery

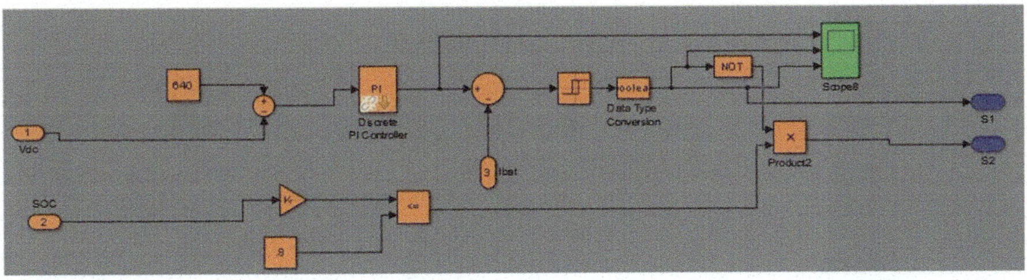

Fig. 30.10 PI controller of bidirectional DC–DC controller

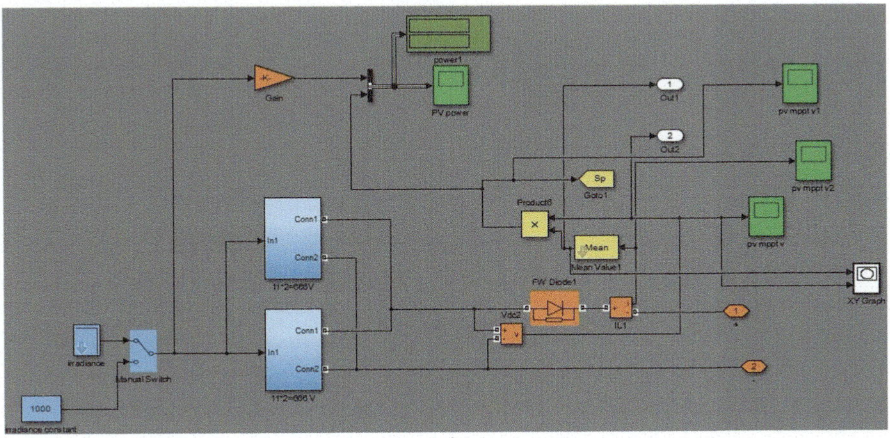

Fig. 30.11 PV panel

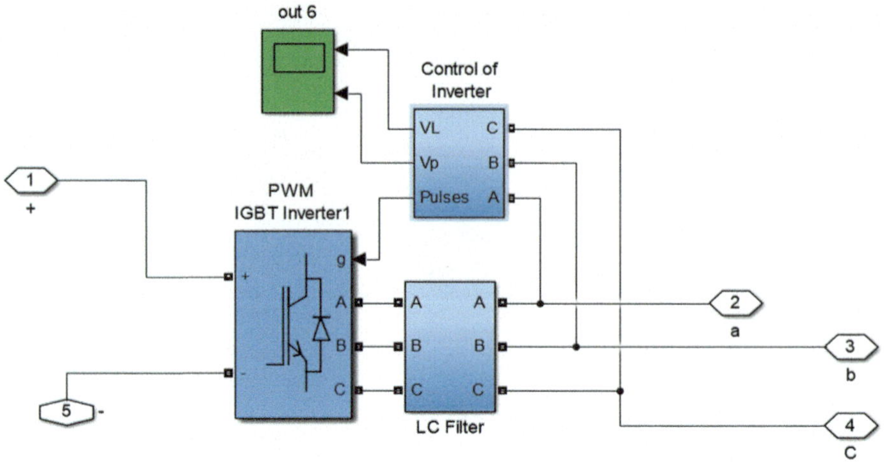

Fig. 30.12 DC–AC inverter

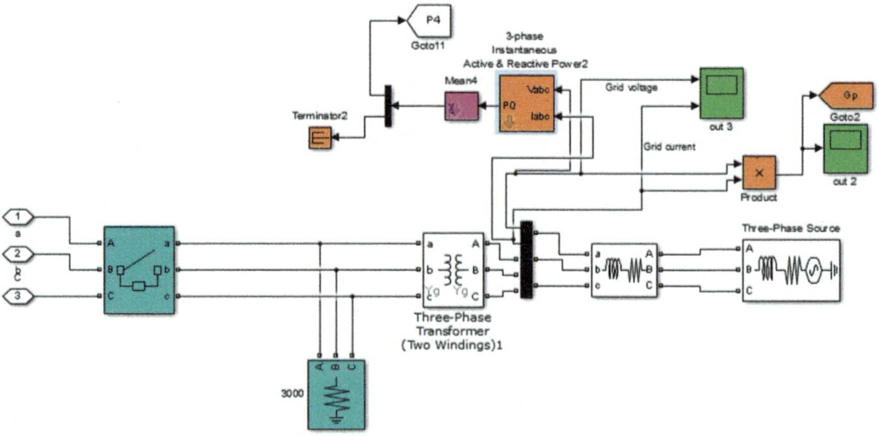

Fig. 30.13 Grid circuit

grid as shown in Fig. 30.14 are sinusoidal and some brief voltage transients occur during abrupt integration of vehicle voltage source to the gird at 2 s can be seen. The battery voltage, current and SoC while charging and discharging process is shown in Fig. 30.15, we observe the drop in voltage and rise in current during discharging process of the battery at 2 s. Figure 30.16 shows the output power of PV, battery and grid.

The irradiance of the PV system is high during the day which is used for charging the EV and access to the grid. As the irradiance level of the PV system reduces, the power injection of the pv power into the grid also reduces. After the PV power generation reaches zero, no pv power is available to charge the EV as well as to inject

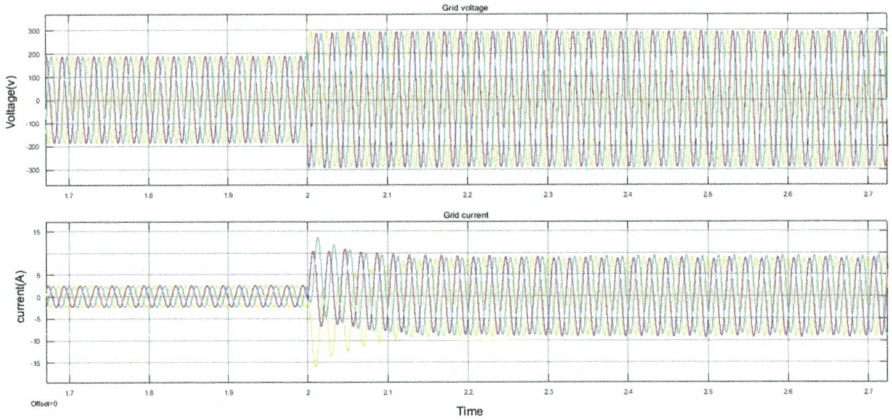

Fig. 30.14 Grid voltage and Grid current profile

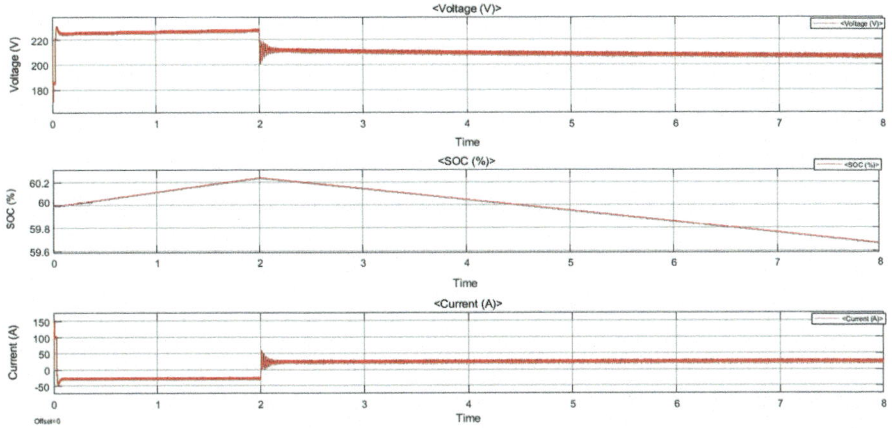

Fig. 30.15 EV battery voltage, SOC and current graph while charging and discharging

into the grid. As the load demand gets high, the EV controller checks the SoC value of the EV, if it is 60%, then it is ready to deliver power to the grid. EV delivers the power to the grid until the SoC value drops to 20% after which the power injected by the EV into the grid is stopped. As the minimum SoC should be available for the convenience of the EV drivers.

The results show the integration of the vehicle to the grid. The waveforms of the battery voltage, SOC and current are appropriate in the charging and discharging process. The grid current is in the opposite phase to the grid voltage as the grid is acting as a load in the V2G operation mode.

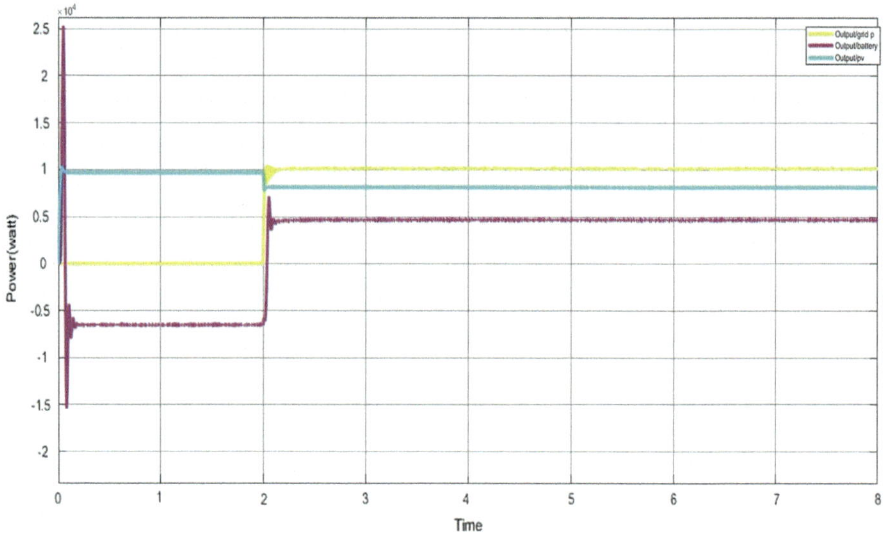

Fig. 30.16 Output power of the grid, battery and PV panel

30.3 Limitations

30.3.1 Challenges Faced by V2G Concept

There are the following challenges to implementing V2G concept such as:

1. In the V2G concept, the bidirectional flow of the EV battery will cause deterioration due to the depth of charging and discharging cycles. Therefore, improving the battery's capabilities and maintenance solutions should be considered as the most important part of this concept.
2. For the V2G concept to be implemented, there is a need for the mass adoption of EVs so that sufficient amount of EVs are available for the V2G operation. Therefore, EVs price needs to be affordable enough for the people.
3. For the convenience of the customers, the EVs should be used in the V2G concept in such a way that they won't face problems in transportation. Therefore, regulations are required about the amount of energy that can be consumed from a single EV for the V2G operation.
4. The EVs integration to smart grid will require a proper secure network in order to avoid any cyberattacks.
5. The installation of V2G components is an expensive process, therefore, the companies and government should provide profitable investment plans and loans so that people will get attracted towards adopting this concept.

30.3.2 Challenges Faced by S2V Concept

The issues related to employing solar-based charging stations are as follows (Huang et al. 2020; Fathabadi 2017).

1. Installation cost: The idea to employ various government schemes to promote the use of renewable energy sources by giving various investment return plans and loans will help the investors and businessmen to invest in the solar charging stations.
2. Availability of rooftop spaces: the houses and buildings in congested areas might face the problems for rooftop spaces but most of the office buildings, parking places and other public places must install solar panels in the way to sufficiently produce the power required to charge the vehicles in the premises.
3. Power storing facility: The storage units are expensive. The idea to use the end of life batteries as the energy storage units can be very profitable in the sense of reducing expenses as well as helping in the recycling of the batteries which will reduce harmful effects on the environment.
4. Since solar power cannot be extracted in bad weather conditions, therefore, the need to plan the solar-based charging unit according to the regional weather pattern and proper storage facility is required.

30.4 Conclusion

This system demonstrates that Electric vehicle (EV) battery is capable of operating in Solar-to-vehicle (S2V) mode and Vehicle-to-grid (V2G) mode. MATLAB/Simulink environment is used to work the proposed system and the results of the system are appropriate.

Grid-tied charging stations cause power stability issues in the grid as well as increase the cost of power generation, therefore, this presented approach of a standalone solar charging station for EV is beneficial to the power system, and also this approach provides controllable solar power injection into the grid. Solar power charging stations can be installed in houses, public areas, as well as isolated areas hence the planning for charging station placement, gets less complicated. V2G operation of EV provides a spinning reserve to the grid, peak-shaving, valley filling, reactive and active power to the grid when required, etc.

Solar charging stations are being installed in many countries but these charging stations require the support of grid and battery storage systems. The focus is on developing standalone charging stations by utilizing only renewable energy sources. EV integration in the smart grid needs advanced communications tools to communicate properly with utilities as well as customers. The EVs will definitely play an important role in the future smart grids and hence efforts are being done in this research area.

Although the integration of EVs in the grid as a power source is still a concept, many agencies around the world are working on developing the technologies to implement this concept of Vehicle- to-Grid in the power sector soon. For V2G to work, we require efficient bidirectional battery chargers that allow power flow in both directions without causing much deterioration.

References

Deb S, Tammi K, Kalita K, Mahanta P (2019) Charging station placement for electric vehicles: a case study of Guwahati City, India. IEEE Access 7:100270–100282. https://doi.org/10.1109/ACCESS.2019.2931055

Fathabadi H (2017) Novel solar powered electric vehicle charging station with the capability of vehicle-to-grid. Solar Energy 142: 136–143. ISSN 0038–092X. https://doi.org/10.1016/j.solener.2016.11.037

Ghazanfari A, Hamzeh M, Mohamed YAI (2018) A resilient plug-and-play decentralized control for DC parking lots. IEEE Trans Smart Grid 9(3):1930–1942. https://doi.org/10.1109/TSG.2016.2602759

Huang P, Zhang X, Copertaro B, Saini PK, Yan D, Wu Y, Chen X (2020) A Technical review of modeling techniques for urban solar mobility: solar to buildings, vehicles, and storage (S2BVS). Sustainability 12:7035. https://doi.org/10.3390/su12177035

Liu S, Xie X, Yang L (2020) Analysis, modeling and implementation of a switching bi-directional buck-boost converter based on electric vehicle hybrid energy storage for V2G system. IEEE Access 8:65868–65879. https://doi.org/10.1109/ACCESS.2020.2985772

de Melo HN, Trovão JPF, Pereirinha PG, Jorge HM, Antunes CH (2018) A controllable bidirectional battery charger for electric vehicles with vehicle-to-grid capability. IEEE Trans Veh Technol 67(1):114–123. https://doi.org/10.1109/TVT.2017.2774189 Jan

Mishra S, Dwivedi G, Upadhyay S, Chauhan A (2021) Modelling of standalone solar photovoltaic based electric bike charging. Materials Today: Proceedings. ISSN 2214–7853. https://doi.org/10.1016/j.matpr.2021.02.738

Mohammed SAQ, Jung J-W (2021) A comprehensive state-of-the-art review of wired/wireless charging technologies for battery electric vehicles: classification/common topologies/future research issues. IEEE Access 9:19572–19585. https://doi.org/10.1109/ACCESS.2021.3055027

Nguyen HV, To D, Lee D (2018) Onboard battery chargers for plug-in electric vehicles with dual functional circuit for low-voltage battery charging and active power decoupling. IEEE Access 6:70212–70222. https://doi.org/10.1109/ACCESS.2018.2876645

Patil H, Kalkhambkar VN (2021) Grid integration of electric vehicles for economic benefits: a review. J Modern Power Syst Clean Energy, 9(1): 13–26, January 2021. https://doi.org/10.35833/MPCE.2019.000326

Preetha Yesheswini B, Jai Iswarya S, Amani B, Prakash P, Sindhu MR (2020) Solar PV charging station for electric vehicles. In: 2020 international conference for emerging technology (INCET), pp 1–7. https://doi.org/10.1109/INCET49848.2020.9154187

Prem P, Sivaraman P, Sakthi Suriya Raj JS, Jagabar Sathik M, Almakhles D (2020) Fast charging converter and control algorithm for solar PV battery and electrical grid integrated electric vehicle charging station. Automatika 61(4): 614-625. https://doi.org/10.1080/00051144.2020.1810506

Qin D, Sun Q, Wang R, Ma D, Liu M (2020) Adaptive bidirectional droop control for electric vehicles parking with vehicle-to-grid service in microgrid. CSEE J Power Energy Syst 6(4): 793–805, Dec. https://doi.org/10.17775/CSEEJPES.2020.00310

Shariff SM, Alam MS, Ahmad F, Rafat Y, Asghar MSJ, Khan S (2020) System design and realization of a solar-powered electric vehicle charging station. IEEE Syst J 14(2):2748–2758. https://doi.org/10.1109/JSYST.2019.2931880.doi:10.1109/JSYST.2019.2931880

Shahrukh Adnan Khan MD, Mahtab Kadir K, Sultan Mahmood K, Ibrahim Ibne Alam Md, Kamal A, Mamoon Al Bashir Md (2019) Technical investigation on V2G, S2V, and V2I for next generation

smart city planning. J Electron Sci Technol 17(4):100010. ISSN1674–862X. https://doi.org/10.1016/j.jnlest.2020.100010
Singh AK, Mishra AK, Gupta KK, Bhatnagar P, Kim T (2020) An integrated converter with reduced components for electric vehicles utilizing solar and grid power sources. IEEE Trans Transport Electrific 6(2):439–452. https://doi.org/10.1109/TTE.2020.2998799
Sneha Angeline PM, Newlin Rajkumar M (2020) Evolution of electric vehicle and its future scope. Materials Today: Proceedings, Volume 33, Part 7, 2020, pp 3930–3936. ISSN 2214–7853. https://doi.org/10.1016/j.matpr.2020.06.266
Taghizadeh S, Hossain MJ, Poursafar N, Lu J, Konstantinou G (2020) A multifunctional single-phase EV on-board charger with a new V2V charging assistance capability. IEEE Access 8:116812–116823. https://doi.org/10.1109/ACCESS.2020.3004931
Tran VT, Islam MR, Muttaqi KM, Sutanto D (2019) An efficient energy management approach for a solar-powered EV battery charging facility to support distribution grids. IEEE Trans Industry Appl 55(6): 6517–6526, Nov.-Dec. https://doi.org/10.1109/TIA.2019.2940923
Yuan J, Dorn-Gomba L, Callegaro AD, Reimers J, Emadi A (2021) A review of bidirectional on-board chargers for electric vehicles. IEEE Access 9:51501–51518. https://doi.org/10.1109/ACCESS.2021.3069448

Chapter 31
Harmonic Reduction for Smart Distribution Network with D-STATCOM and DG Using Gravitational Search Algorithm

Aadesh Kumar Arya, Atma Ram Gupta, Govind Rai Goyal, and Ashwani Kumar Sharma

Abstract The current technological advances of the existing power system offer advantages in almost all aspects of power generation, transmission and distribution at lower cost with an improved voltage profile with minimal power loss to make it a smart power system. Practically, the need for electricity is growing expeditiously all across the world which results in poor voltage profile and an increase in line losses. To reduce the line losses as well as to improve the voltage profile, instead of longer transmission, renewable-based distributed generation (RBDG) can be added locally to the distribution side. Continuous addition of irregular reactive load requires fast-acting variable reactive power compensator like distribution static compensator (D-STATCOM) device to maintain the voltage profile. Consequently, RBDG and D-STATCOM devices are integrated into the distribution system (DS) to fulfill the active and reactive power demand, respectively. Deciding on the distributed generation (DG) and D-STATCOM's position and size is one of the important analyses to be done by the distribution network operator (DNO). In addition to the allocation of devices, harmonic pollution and voltage instability due to nonlinear load are considered to be serious problems in power systems. In the presented chapter, the gravitational search algorithm (GSA) is applied to obtain the accurate feasible position of DG and D-STATCOM by DNO for smart distribution system (SDS). The combined effect of GSA and harmonic load flow is carried out to achieve minimization of total power losses, voltage total harmonic distortion (THDV) and maximization of the annual energy loss reduction (AELR).

Keywords Distributed generation · Gravitational search algorithm · Voltage total harmonic distortion · Distribution network operator · Smart distribution system

A. K. Arya (✉) · G. R. Goyal
College of Engineering Roorkee, Roorkee, India

A. R. Gupta · A. K. Sharma
National Institute of Technology, Kurukshetra, India
e-mail: argupta@nitkkr.ac.in

A. K. Bohre et al. (eds.), *Planning of Hybrid Renewable Energy Systems, Electric Vehicles and Microgrid*, Energy Systems in Electrical Engineering,
https://doi.org/10.1007/978-981-19-0979-5_31

31.1 Introduction

The electricity demand is increasing day-by-day which is a major challenge for the present power grid, but it is an important economic factor for any nation. Electricity generation is expected to increase by 93% during 2010–2040 (Murty and Kumar 2020). The increase in electricity demand can be supplied through the use of micro-grid, renewable-based distributed generation (RBDG), and at the same time devices like D-STATCOM can be added by DNO to fulfill the variable reactive power demand. In addition, the regulatory operation and technological innovations of the smart grid have sparked renewed interest in the future economy. The DNO works to reduce the imbalance between supply and demand with the use of DG and DFACTS devices. With the integration of renewable resources, dependence on fossil fuels is reduced and consequent reduction in the levels of carbon emission in the environment (Murty et al. 2018). Various types of RBDG technologies have been used by the DNO in the electricity market, for renewable-based power generation, namely hydropower generation, PV solar, natural gas engines, fuel cells, micro-turbines and wind energy.

With the integration of RBDG and DFACTS devices, the distribution system behaves as an active distribution system (Saxena and Kumar 2021; Murty and Kumar 2015). The nonlinear components and power electronics devices are increasingly in use due to their characteristics like simple, easy-to-manage and high efficiency. However, the DS experiences harmonics distortion (Bagheri et al. 2015). The nonlinear load, electronic gadgets, thyristor drives, static var compensator (SVC) etc. introduce harmonics into the system. At the point of common coupling harmonic current injection and voltage distortion should be minimum within the acceptable limits as per the standards. Good power quality is a difficult task to keep up with the present load models. Harmonic distortion is a bad factor known to cause certain problems in a system, such as shortening the life of insulation, increasing heat loss, reduction in power factors and reducing power plant output (Mohammadi 2015). Therefore, there is a need to correct the harmonic distortion problem as per IEEE 519–1992 standard to maintain the power loss and power quality in competitive energy market scenario for making conventional DS to SDS (F II, I. 1993).

In the past, harmonic power flow (HPF) methods used the Newton–Raphson and other traditional harmonic methods to analyze harmonic distortion in DS (Au et al. 2006; Mahmoud and Shultz 1982; Pileggi et al. 1981). It is found that HPF methods cannot work effectively. Furthermore, the forward/backward sweep (FBS) is highly effective in the analysis of harmonics in DS (Shirajum et al. 2016). In (Chang and Teng 2002), the authors proposed a fast method for harmonic analysis using backward/forward sweep techniques. The current injection method is one of the most frequently used methods for solving HPF in electrical networks in order to determine the degree of voltage distortion (Archundia-Aranda and Mota-Palomino 2010). In power systems, HPF methods are often used deterministically to predict and solve many of the harmonic problems (Herraiz et al. 2003). Single-phase harmonic analysis methods are generally used in systems where imbalances can be overlooked.

The harmonic analysis methods are categorized by the following techniques: (i) transient-state analysis (time-domain-based) and (ii) steady-state analysis (frequency-domain-based) (Heydt and Galli 1997; Zheng et al. 1999). The steady-state analysis is more accurate than the transient-state analysis. Therefore, the steady-state analysis is employed in HPF programs. Consequently, the steady-state algorithms are the best choice for the analysis of large-scale power systems because of the computational economy (Saini et al. 2012). The harmonic studies play a crucial role in the analysis and design of power systems. However, linear and nonlinear loads and network configurations change randomly and in a probabilistic manner (Zhang and Xu 2008).

In Bonner et al. (1996), the authors discussed two parts of research: first, the overview of the type and modeling of harmonic sources in power systems, and second, modeling of load and investigation of the impact of load as frequency response. In Grady (2006), Grady et al. analyzed the impact of total harmonics distortion (THD) for the various signals and systems. Fangl et al. developed the algorithm for harmonic flow with DG and analyzed it for IEEE 33-bus system using PCLFO software (Zhang et al. 2011). Shokri et al. used a direct approach technique based on BIBC and BCBV matrices for the solution of the HPF method with nonlinear loads for IEEE 34-bus distributions (Shokri et al. 2013). Gupta has discussed that the nonlinear loads have increased in everyday life with the use of home appliances. As a result, the power quality problems in the distribution system have increased.

The advantages of DG and D-STATCOM largely depend on their optimal allocation in DS, and incorrect allocation of these devices may lead to excessive THD, power loss and bad voltage profile. Consequently in Gupta (2017), the author optimally placed several DGs and D-STATCOM in a radial distribution system (RDS) to reduce line loss and THD using variation technique and stability index. D-STATCOM is a compensation device used to absorb reactive power in DS. Further, D-STATCOM is placed in DS to manage the total power loss as a minimum, improvement of power factor and minimal of harmonics pollution produced by nonlinear loads. In Arya et al. (2019), Tagore et al. (2017), the authors have found that if DG and D-STATCOM of optimal size are placed at a proper bus in DS, the system power losses and THD can be minimized with voltage profile improvement.

31.1.1 Contribution of Research Work

The objective of this research work is to improve the voltage profile of the network by minimizing the total power loss as well as harmonics injected by the nonlinear loads. To achieve these objectives, a bi-objective problem is formulated and the objective functions are minimized using gravitational search algorithm (GSA). In this research work optimal allocation and size of DG and D-STATCOM are also found for IEEE 33-bus and 69-bus distribution systems while minimizing the power losses, total harmonic distortion (THD_V) and maximizing the annual energy loss reduction (AELR). In Sect. 31.2, a mathematical model for HPF with DG and D-STATCOM is described. Section 31.3 deals with the outcome and results analysis of IEEE 33-bus and 69-bus distribution systems. The conclusion of the chapter is mentioned in Sect. 31.4.

31.2 Mathematical Model for HPF with DG and D-STATCOM

The electrical power is supplied by the generators and consumed by the load in the power flow model at the distribution side. Furthermore, in the case of a nonlinear load, the load may become a harmonic source that provides a harmonic flow of energy to generators. In the proposed research the current source is taken as a harmonic source and the FBS technique is applied to calculate the HPF in DS with multiple nonlinear loads. In the HPF, the FBS method is used by two processes: (i) the backward sweep process is performed to establish a relationship between the branch current and the injected bus current and (ii) the forward sweep process is performed to determine the bus voltage using the branch current and the injected bus current. The harmonic currents are being generated by the nonlinear load, and D-STATCOM can be expressed by Eq. (31.1):

$$I_{h_i}^{(h),k} = \frac{V_i^{(h),k}}{Z_i^{(h)}} \tag{31.1}$$

It can be expressed in the vector form as Eq. (31.2):

$$I^{(h),k} = \begin{bmatrix} I_h^{(h),k} \\ \dots\dots\dots \\ I_{DST}^{(h),k} \end{bmatrix} \tag{31.2}$$

where $I^{(h)}$ = Harmonic currents at hth order.

$I_h^{(h)}$ = Harmonic current by linear or nonlinear load.

$I_{DST}^{(h)}$ = Absorbed or reactive current by D-STATCOM.

The coefficient vector of harmonic currents, given in Eq. (31.3), is used to obtain the power flow between bus i and bus j (Arya et al. 2019).

$$A_{ij}^{(h)} = \begin{bmatrix} Ah_{ij}^{(h)} \\ \dots\dots\dots \\ As_{ij}^{(h)} \end{bmatrix} \tag{31.3}$$

$Ah_{ij}^{(h)}$ = Coefficient vector of the nonlinear or linear load impedances.

$As_{ij}^{(h)}$ = Absorbed or reactive harmonics coefficient vector by D-STATCOM.

Branch current ($B_{ij}^{(h),k}$) and voltage drop ($\Delta V_{ij}^{(h),k}$) due to harmonic currents can also be expressed by Eqs. (31.4) and (31.5), respectively.

$$B_{ij}^{(h),k} = \left[A_{ij}^{(h)}\right]^T I^{(h),k} \tag{31.4}$$

$$\Delta V_{ij}^{(h),k} = Z_{ij}^{(h)}\left[A_{ij}^{(h)}\right]^T I^{(h),k} \tag{31.5}$$

where $Z_{ij}^{(h)}$ = Harmonic branch impedance of hth order.

31.2.1 Forward Sweep Process

The forward sweep process is used to determine the system matrix [HA] between harmonic current vector and bus voltage vector ($V^{(h)}$) as in Eq. (31.6) (Colmenar-Santos et al. 2016).

$$V^{(h),k} = \left[HA^{(h),k}\right]I^{(h),k} \tag{31.6}$$

The D-STATCOM bus voltage ($V_{DST}{}^{(h)}$) can be obtained by Eq. (31.7):

$$\left[V_{DST}{}^{(h),k}\right] = \left[HAs^{(h),k}\right]\left[I^{(h),k}\right] \tag{31.7}$$

where $\left[HAs^{(h),k}\right]$ is composed of the row vectors with respect to the buses of D-STATCOM.

The D-STATCOM voltage at bus i can be expressed in terms of harmonic load impedance ($Z_{DST)}$) as given in Eq. (31.8):

$$V_{DST}{}^{(h),k} = -I_{DST}{}_i^{(h,k)} \times Z_{DST}{}_i^{(h)} \tag{31.8}$$

Equation (31.7) can be rewritten as Eq. (31.9) or Eq. (31.10) by separating series and shunt terms

$$-\begin{bmatrix} I_{DST\,i}^{(h,k)} Zs_i^{(h)} \\ \cdot \\ \cdot \\ \cdot \\ I_{DST\,n-1}^{(h,k)} Zs_{n-1}^{(h)} \\ I_{DST\,n}^{(h,k)} Zs_n^{(h)} \end{bmatrix} = \begin{bmatrix} HAsh^{(h),k} & \vdots & HAss^{(h),k} \end{bmatrix} \begin{bmatrix} I^{(h),k} \\ \ldots\ldots \\ I_{DST}{}^{(h),k} \end{bmatrix} \tag{31.9}$$

or

$$\left(\left[HAsh^{(h),k} \right] + \begin{bmatrix} Z_{DST\,1}^{(h)} & \cdots & 0 \\ \vdots & \ddots & \vdots \\ 0 & \cdots & Z_{DST\,n}^{(h)} \end{bmatrix} I_{DST}{}^{(h),k} \right) = -\left[HAsh^{(h),k} \right] \left[Ih^{(h),k} \right] \tag{31.10}$$

The $[HPF]$ matrix is determined by taking Z_{DST}. Hence [HPF] is obtained by Eq. (31.11):

$$[HPF] = [HAss] + [Zs] \tag{31.11}$$

Equation (31.12) can be written using Eqs. (31.10) and (31.11):

$$\left[HPF^{(h),k} \right] \left[Is^{(h),k} \right] = -\left[HAsh^{(h),k} \right] \left[Ih^{(h),k} \right] \tag{31.12}$$

31.2.2 Backward Sweep Process

The bus voltage $\left[V^{(h),k} \right]$ calculated after kth iteration is given by Eqs. (31.13), and (31.14) as the stopping criterion for iterations (Al-Rashidi and Al-Hajri 2011).

$$\left[V^{(h),k+1} \right] = \left[V^{(h),0} \right] + \left[V^{(h),k} \right] \tag{31.13}$$

$$\left| V^{(h),k+1} - V^{(h),k} \right| \le \varepsilon fori = 1 \ldots N \tag{31.14}$$

where N is the bus no. and ε is a predetermined tolerance.

31.2.3 Objective Function

In this research work mainly two objective functions are considered for maximization of the annual energy loss reduction (ALER) and superior voltage profile, i.e., (i) minimization of power loss and (ii) minimization of total harmonic distortion of voltage (*THD*v) with optimal allocation of the DGs and D-STATCOM (Arya et al. 2017). A combined composite function of both the objectives is given by Eq. (31.15):

$$F = \omega_1 Min(F_1) + \omega_2 Min(F_2) \tag{31.15}$$

Here, F1 represents the first objective function of minimization of power loss (P_{loss}) and F2 represents the second objective function of minimization of THDv given by Eqs. (31.15a) and (31.15b), respectively.

$$F_1 = MinP_{lossi} = \left[\sum_{i=1,j=j+1}^{n_b} Re(I_i^2 \times Z_{ij}) + \sum_{i=1,j=j+1}^{n_b} \sum_{h=h_o}^{h_{max}} Re(I_i^2 \times Z_{ij})^{(h)} \right] \tag{31.15a}$$

$$F_2 = MinTHD_v \tag{31.15b}$$

The value of THD_v^{bus} is obtained by Eq. (31.16):

$$THD_v^{bus}(\%) = \frac{\sqrt{\sum_{h=h_{min}}^{h_{max}} \left|V_i^h\right|^2}}{\left|V_i^{(1)}\right|^2} \tag{31.16}$$

The annual energy loss reduction (ALER) can be represented by Eq. (31.17):

$$\text{AELR} = Max\left[\sum_{i=1}^{n_b} P_{loss,1} - \sum_{i=1}^{n_b} P_{loss,2} \right] \times t \tag{31.17}$$

where $P_{loss,1}$ = Active power loss with DGs and D-STATCOM.

$P_{loss,2}$ = Active power loss without DGs and D-STATCOM.

t = Time = 1 year = 8760 days.

h = Harmonic order.

Z_{ij} = Line impedance between the ith *and* jth bus.

The weight factor $\omega_1 = 0.6$ for power loss and $\omega_2 = 0.4$ for THD is taken.

31.2.4 Operational Constraints

In the optimization process, the THD constraint given by Eq. (31.18) is imposed in addition to voltage constraint, power balance constraint (Arya et al. 2019; Singhal et al. 2018), DG and D-STATCOM limits as given in Eq. (31.18):

$$THD_v^{bus} \leq THD_v^{max} \tag{31.18}$$

31.3 Result Analysis and Discussions

In the present work, the first normal load flow with the FBS method is examined for the determination of power loss and the voltage profile. Next, two nonlinear loads (six-pulse converter) are connected and harmonic power flow with BFS method is examined, and voltage profile, power losses and THDv at each bus are obtained. For the result analysis, the (MVA) base and (kV) base are taken as 100 MVA and 12.66 kV, respectively. Various cases are taken to verify the usefulness of harmonic power flow with BFS and GSA for the DS of IEEE 33- and 69-bus systems.

31.3.1 Analysis of IEEE 33-Bus DS

In this work, the FBS load flow method is applied to the DS illustrated in Fig. 31.1 and the obtained voltage profile is shown in Fig. 31.2. The corresponding value of the

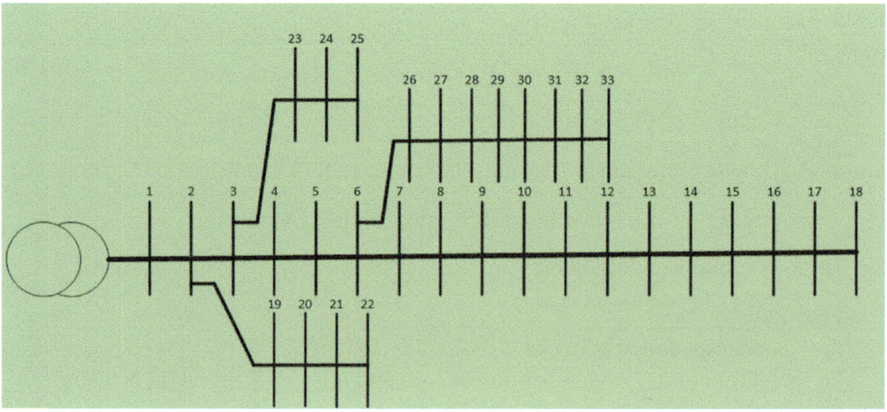

Fig. 31.1 Network of 33-bus DS

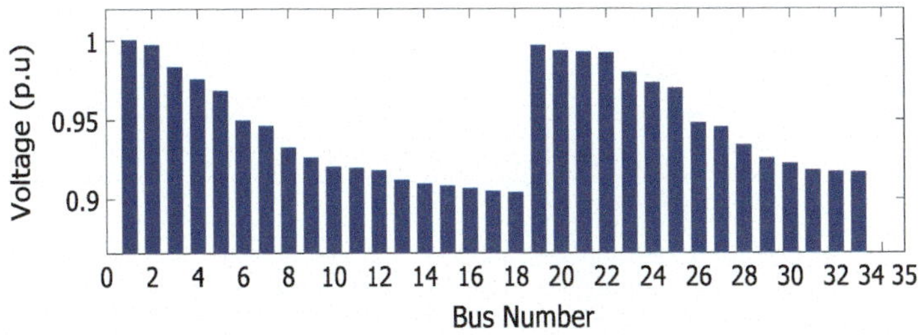

Fig. 31.2 Voltage profile for base case for the network of 33-bus DS

losses of active and reactive power is found as 210.98 kW and 143.02 kVAr, respectively. For the proposed study, three cases are considered: (i) single DG placement (2510 kW @ 6th bus), (ii) double DGs placement (2510 kW and 470 kW @ 6th and 15th bus) and (iii) placement of a combination of single DG and single D-STATCOM (2570 kW and 1240 kVAr @ 6th and 30th bus). GSA is applied for the optimized allocation of DG and D-STATCOM in three cases. The active power loss is found as 111.0349 kW, 95.8335 kW and 58.4815 kW and the reactive power losses as 81.6646 kVAr, 68.4680 kVAr and 47.1354 kVAr in the first, second and third cases, respectively. The voltage profile is improved with the proper allocation of DG and D-STATCOM, which is shown in Fig. 31.3.

For the harmonic analysis, two nonlinear loads are placed at bus 5 and 26 in IEEE 33-bus distribution systems, as shown in Fig. 31.4. The six-pulse converters are considered as nonlinear loads of 1000 kW and 750 kVAr. The harmonic voltage levels corresponding to the 5th, 7th, 11th, 13th and 17th harmonics are shown in Fig. 31.5. The harmonic voltage levels are seen to decrease with increasing order of harmonic. For higher-order harmonics, the voltage levels are seen to become constant. In order to reduce the THD and improve the voltage profile, the HPF based on forward/backward method is incorporated with the gravitational search algorithm, as shown in Fig. 31.6, and the active and reactive power losses of various cases are shown in Table 31.1.

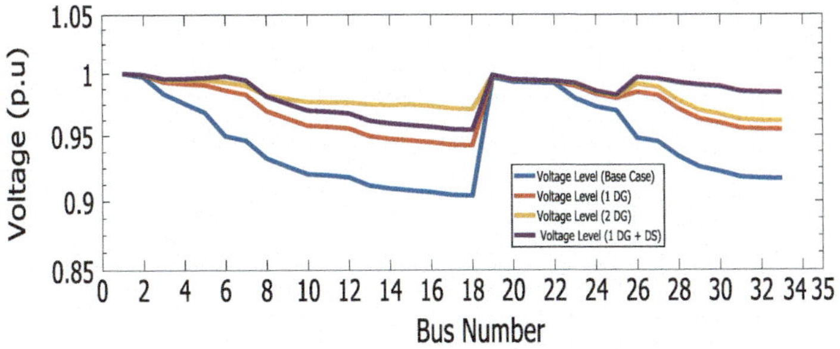

Fig. 31.3 Comparison of voltage profile with and without DG and D-STATCOM for 33 buses system

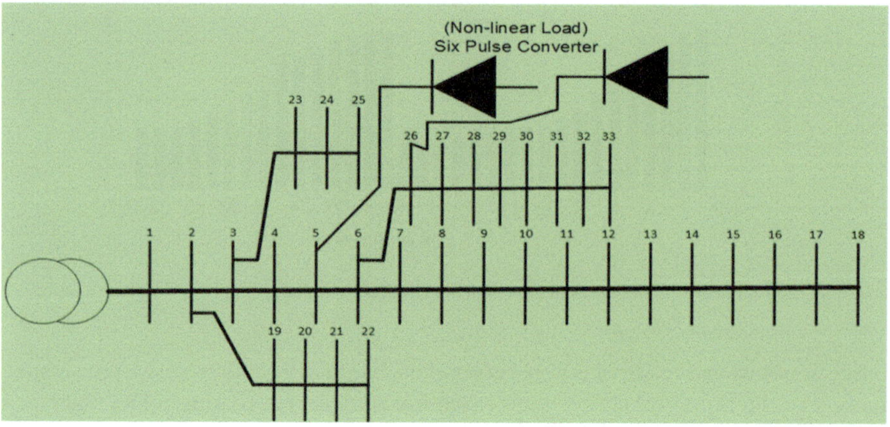

Fig. 31.4 Network of 33-bus DS with nonlinear load

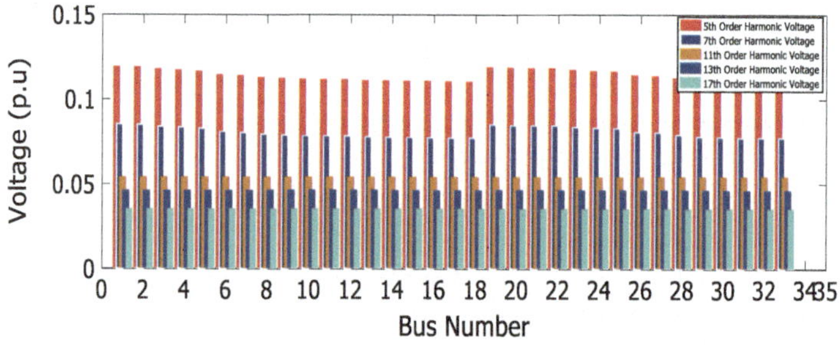

Fig. 31.5 Voltage level for 5th, 7th, 11th, 13th and 17th harmonic order

The population size for the GS algorithm was taken as 100. With the same allocation of DG and D-STATCOM, the THD in each of the three cases is reduced and the voltage profile has improved. The effect of proper location and optimal capacity of DG and D-STATCOM on THD is shown in Fig. 31.7. In addition, the voltage profile is improved, as shown in Fig. 31.8.

The bi-objective function was to minimize the total power loss, THDv and maximize the annual energy loss reduction. Table 31.1 illustrates that with the optimal placing of DG and D-STATCOM, the THD is reduced from 17.12% to 16.71%.

31.3.2 Analysis of IEEE-69 Bus DS

In this work, the FBS load flow method is applied to the IEEE-69 bus system, which is shown in Fig. 31.9, and the voltage profile obtained for the base case is illustrated in Fig. 31.10. The corresponding values of the losses of active and reactive power

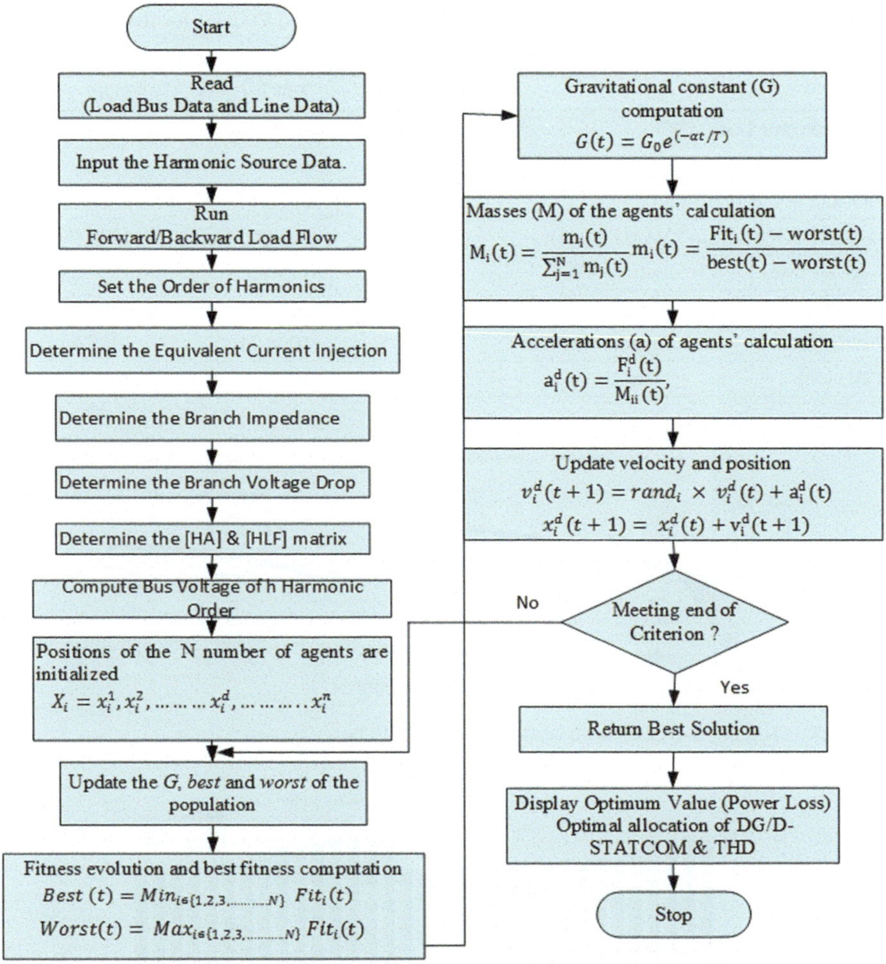

Fig. 31.6 Harmonic load flow incorporated with GSA

are found as 186.62 kW and 81.52 kVAr, respectively. For the proposed study, three cases are considered: (i) Single DG placement (1870 kW @ 60th bus), (ii) double DGs placement (1870 kW @60th bus, 550 kW @ 66th bus) and (iii) placement of a combination of single DG and single D-STATCOM (1870 kW @ 60th bus, 1320 KVAr @ 66th bus). GSA is applied for the optimization of allocation of DG and D-STATCOM in three cases. The active power loss is found as 63.10 kW, 59.24 kW and 8.46 kW and reactive power loss as 29.10 kVAr, 27.33 kVAr and 6.50 kVAr for the first, second and third cases, respectively. The voltage profile is improved with the proper allocation of DG and D-STATCOM, which is shown in Fig. 31.11.

For harmonic analysis, the nonlinear load is placed at bus 19, 30, 38 and 57 in IEEE 69-bus DS, which is shown in Fig. 31.12. The six-pulse converter is considered as a

Table 31.1 Analysis of results with an allocation of DG and D-STATCOM for IEEE 33-bus RDS

Parameters	Base case	With an allocation of DG and D-STATCOM		
		1 DG	2 DG	1DG + 1DST
Active Power Loss (kW)	210.98	111.03	95.83	51.18
Reactive Power Loss (kVAr)	143.02	81.66	68.46	47.13
DG size in kW (Location)	–	2570 (6)	2570(6), 470(15)	2570 (6)
D-STATCOM size in kVAr (Location)	–	–	–	1240 (30)
Active Power Loss Reduction (kW)	–	99.94	115.15	159.8
Active Power Loss Reduction (%)	–	47.37	54.57	75.74
Annual Energy Loss Reduction (MWh)	–	875.56	1008.71	1399.84
THD (%)	17.12	16.96	16.87	16.71

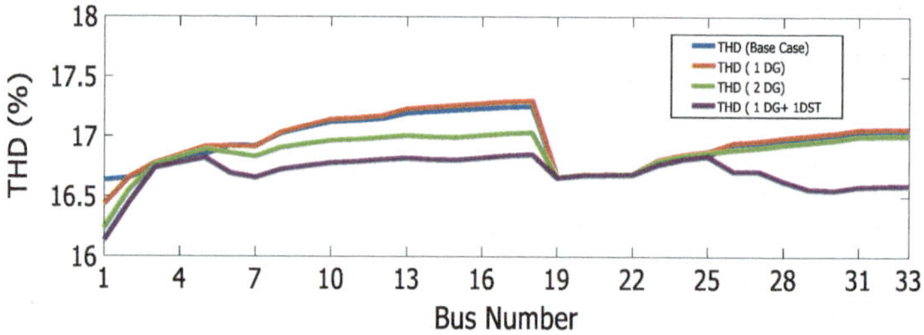

Fig. 31.7 THD in three cases with nonlinear load for 33 buses system

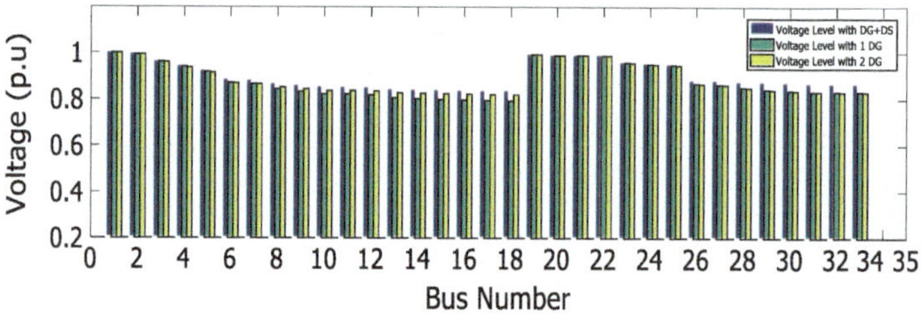

Fig. 31.8 Voltage profile with nonlinear load and DG and D-STATCOM

nonlinear load whose size is 3800 kW and 2690 kVAr. Therefore, 5th, 7th, 11th, 13th and 17th harmonic orders were generated. Figure 31.13 illustrates that by increasing the order of harmonics the voltage level is decreased, and it has been found that the voltage becomes constant for higher-order harmonics.

In order to reduce the THD and increase the voltage profile, the harmonic load flow based on forward/backward method is incorporated with the gravitational search algorithm as shown in Fig. 31.14. The population size for the GS algorithm was taken as 100. The same allocation of DG and D-STATCOM which were discussed above was taken and the THD and voltage profile were obtained in the same three cases

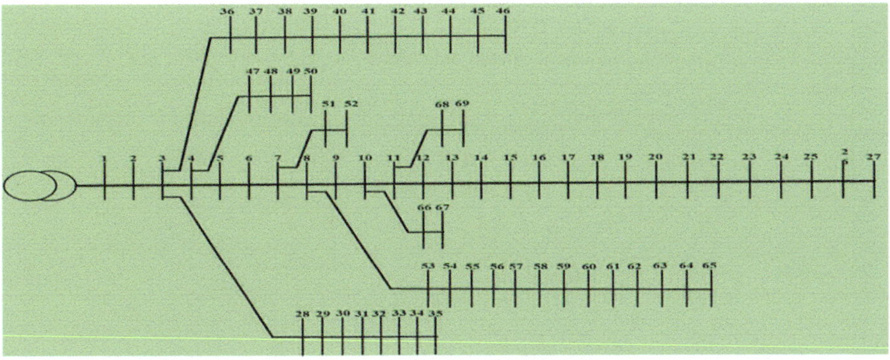

Fig. 31.9 Network of 69-bus RDS

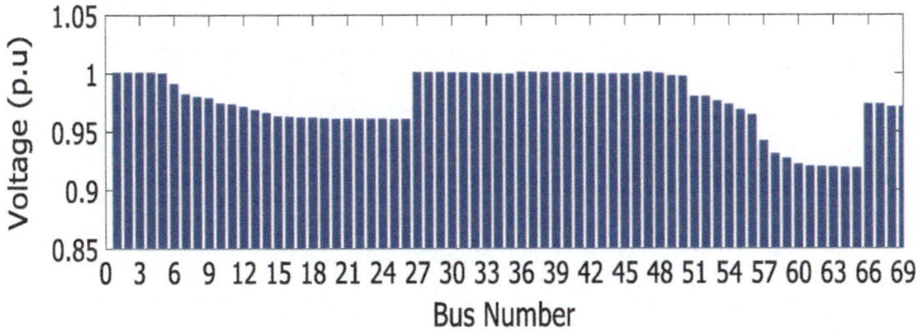

Fig. 31.10 Voltage profile for the base case of IEEE 69-bus RDS

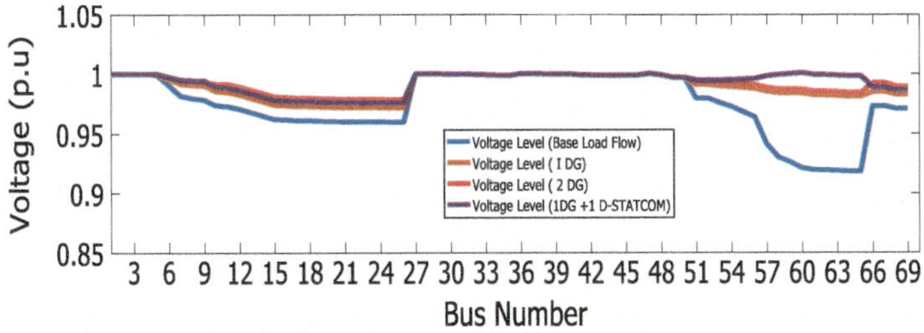

Fig. 31.11 Comparison of voltage profile with and without DG and D-STATCOM

and it was observed that THD is reduced in each case. The effect of proper location and optimal capacity of DG and D-STATCOM on THD is shown in Fig. 31.15. In addition, the voltage profile is improved, as shown in Fig. 31.15. Figure 31.16 illustrates the comparison of voltage profile of with and without nonlinear load, and the active and reactive power losses in various cases are given in Table 31.2.

Table 31.2 illustrates that the THD is 14.62% when the four nonlinear loads placed at various buses are reduced. It is also observed that with the optimal placing of DG and D-STATCOM, the THD is reduced from 14.62% to 12.52%.

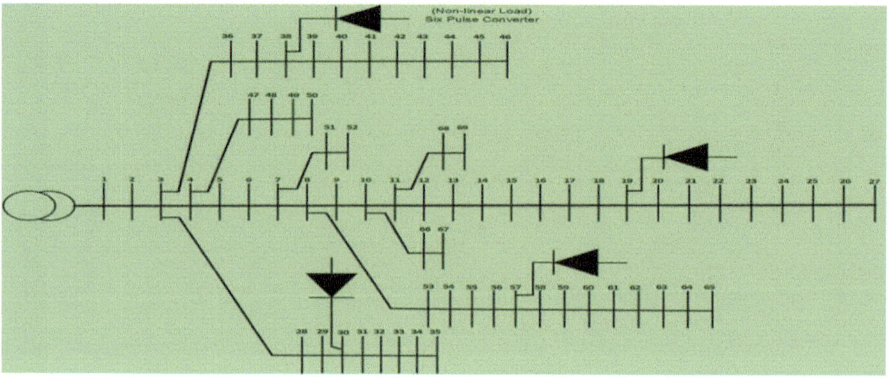

Fig. 31.12 Network for 69-bus RDS with nonlinear load

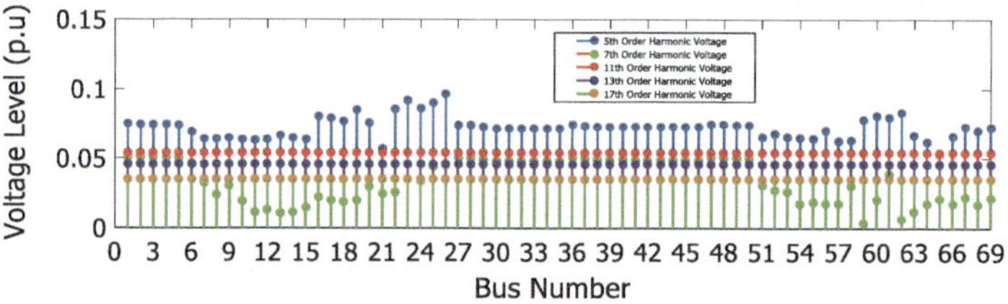

Fig. 31.13 Voltage level for harmonic order 5th, 7th, 11th, 13th and 17th

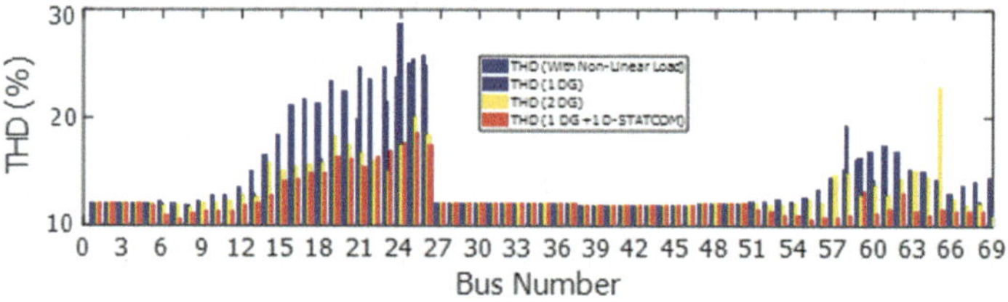

Fig. 31.14 THD in three cases with nonlinear load

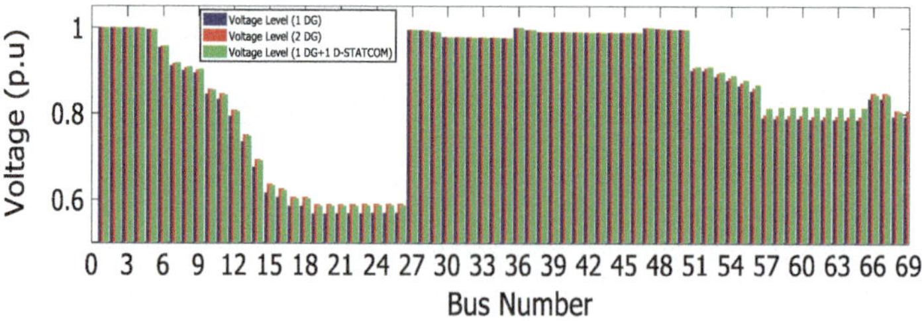

Fig. 31.15 Voltage profile with nonlinear load and DGDSTAT

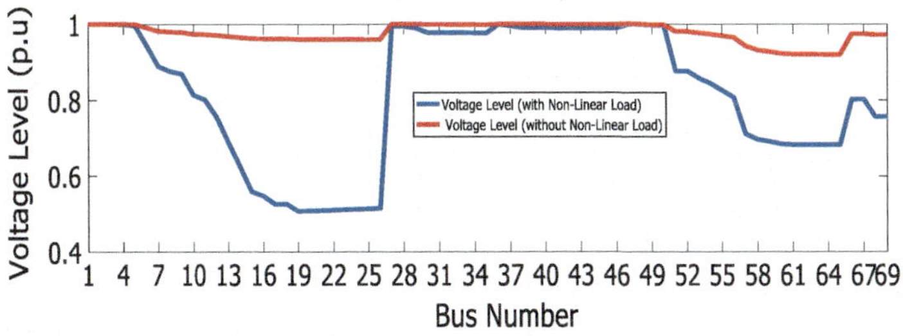

Fig. 31.16 Comparison of voltage profile with and without nonlinear load

Table 31.2 Analysis with DG and D-STATCOM for IEEE 69-bus systems

Parameters	Base case	With DG and D-STATCOM		
		1 DG	2 DG	1 DG + 1DS
Active power loss (kW)	186.62	63.10	59.24	8.46
Reactive power loss (kVAr)	81.52	29.10	27.33	6.50
DG size in kW (Location)	–	1870 (60)	1870 (60) 550 (66)	1870 (60)
D-STATCOM size in kVAr (Location)	–	–	–	1320 (60)
Active power loss reduction (kW)	–	123.52	127.38	178.16
Active power loss reduction (%)	–	66.18	68.25	95.46
Annual energy loss reduction (MWh)	–	1082. 03	1115.84	1560.68
THD with nonlinear load (%)	14.62	13.67	13.36	12.52

31.4 Conclusions

In this work, an HPF with forward/backward method is carried out for the analysis of DS in the presence of nonlinear loads. The bi-objective function is used to minimize the total power loss, THDv and maximize the annual energy loss reduction. It is observed that the applied GSA is very effective in optimally allocating the DG and D-STATCOM. Consequently, after optimal deployment of the DG and D-STATCOM, the results show that GSA with the integration of harmonic load flow based on forward/backward method can be used by DNO to reduce THD, active and reactive power losses along with the enhancement of voltage profile and for maximization of annual energy loss reduction. The reduction of harmonics will result in a reduction of losses, overheating, mal-operation of relays and malfunctioning of the control circuit.

References

Al-Rashidi MR, Al-Hajri MF (2011) Optimal planning of multiple distributed generation sources in distribution networks: a new approach. Energy Convers Manag 52(11):3301–3308

Archundia-Aranda I, Mota-Palomino RO (2010) Harmonic load flow method for radial distribution networks. In Proceedings of 14th international conference on harmonics and quality of power-ICHQP 2010. IEEE

Arya AK, Kumar A, Yadav A (2017) Optimal placement and sizing of distributed generators by using voltage indexing & heuristic method. Int J Recent Innov Trends Comput Commun 5:945–949

Arya AK, Kumar A, Chanana S (2019a) Analysis of distribution system with D-STATCOM by Gravitational Search Algorithm (GSA). J. Inst. Eng. India Ser. B 100:207–215. https://doi.org/10.1007/s40031-019-00383-2

Au MT, Milanovic JV (2006) Development of stochastic aggregate harmonic load model based on field measurements. IEEE Trans Power Delivery 22(1):323–330

Bagheri P, Xu W, Ding T (2015) A distributed filtering scheme to mitigate harmonics in residential distribution systems. IEEE Trans Power Delivery 31(2):648–656

Bonner A et al (1996) Modeling and simulation of the propagation of harmonics in electric power networks. 2. sample systems and examples. IEEE Trans Power Delivery 11(1):466–474

Chang C-Y, Teng J-H (2002) Three-phase harmonic load flow method. In 2002 IEEE international conference on industrial technology, 2002. IEEE ICIT'02.. Vol. 2. IEEE

Colmenar-Santos, A et al (2016) Distributed generation: a review of factors that can contribute most to achieve a scenario of DG units embedded in the new distribution networks. Renew Sustain Energy Rev 59:1130–1148

F II, I (1993) IEEE recommended practices and requirements for harmonic control in electrical power systems. New York, NY, USA:1–1

Grady M (2006) Understanding power system harmonics. University of Texas, Austin, TX

Gupta AR (2017) Effect of optimal allocation of multiple DG and D-STATCOM in radial distribution system for minimizing losses and THD. In 2017 7th international symposium on embedded computing and system design (ISED). IEEE

Herraiz S, Sainz L, Clua J (2003) Review of harmonic load flow formulations. IEEE Trans Power Delivery 18(3):1079–1087

Heydt GT, Galli AW (1997) Transient power quality problems analyzed using wavelets. IEEE Trans Power Delivery 12(2):908–915

Mahmoud AA, Shultz RD (1982) A method for analyzing harmonic distribution in AC power systems. IEEE Trans Power Appar Syst 6:1815–1824

Mohammadi M (2015) Bacterial foraging optimization and adaptive version for economically optimum sitting, sizing and harmonic tuning orders setting of LC harmonic passive power filters in radial distribution systems with linear and nonlinear loads. Appl Soft Comput 29:345–356

Murty VVSN, Kumar A (2015) Optimal placement of DG in radial distribution systems based on new voltage stability index under load growth. Int J Electr Power Energy Syst 69:246–256

Murty VVSN, Kumar A (2018) Impact of D-STATCOM in distribution systems with load growth on stability margin enhancement and energy savings using PSO and GAMS. Int Trans Electr Energy Syst 28(11):e2624

Munir MS, Wei Li Y, Tian H (2016) Improved residential distribution system harmonic compensation scheme using power electronics interfaced DGs. IEEE Trans Smart Grid 7(3):1191–1203

Pileggi DJ, Harish Chandra N, Emanuel AE (1981) Prediction of harmonic voltages in distribution systems. IEEE Trans Power Appar Syst 3:1307–1315

Saini P et al (2012) A new approach of harmonic load flow for radial distribution networks. In 2012 seventeenth national power systems conference (NPSC), IIT BHU

Shokri SMM et al (2013) A direct approach used for solving the distribution system and harmonic load flow solutions. In 2013 IEEE 7th international power engineering and optimization conference (PEOCO). IEEE

Saxena NK, Kumar A (20211) Estimation of dynamic compensation for renewable-based hybrid DG in radial distribution system using least error iterative method. Iranian J Sci Technol Trans Electr Eng 45(1):15–28
Singhal K, Rai Goyal G (2018) Comparative study of power consumption minimization in analog electronic circuit using AI techniques. Eur J Electr Eng 20(4):427
Tagore, Kumar A, Ram Gupta A (2017) Impact of DG and D-STATCOM allocation in radial distribution system for reducing harmonics. In: 2017 8th international conference on computing, communication and networking technologies (ICCCNT). IEEE
Vallem, Murty VVSN, Kumar A (2020) Retracted: Optimal energy dispatch in microgrids with renewable energy sources and demand response. Int Trans Electr Energy Syst 30(5): e12328
Zhang JF et al (2011) The harmonic load flow calculation of distribution network with distributed power. In 2011 international conference on advanced power system automation and protection. Vol. 3. IEEE
Zhang G, Xu W (2008) Estimating harmonic distortion levels for systems with random-varying distributed harmonic-producing loads. IET Gener Transm Distrib 2(6):847–855
Zheng T, Makram EB, Girgis AA (1999) Power system transient and harmonic studies using wavelet transform. IEEE Trans Power Deliv 14(4):1461–1468

Chapter 32
Enhanced Loadability and Inapt Locations Investigation in the Renewable Energy Resource Enriched Power System

Shilpa R. Kalambe, Sanjay Jain, and Bhojraj N. Kale

Abstract The exact location and design parameter investigation of Renewable Energy Resources (RER) prior to its actual installation is an important step in network planning. Many RER allocation methods are available in the literature but rarely define any unsuitability of location for RER installation that may reduce system performance. For closing the gap of those published methods this chapter outlines two auxiliary services of the proposed method for the installation of renewable energy resources in Transmission and Distribution Systems. These services highlights the most important fields of energy systems, which will be carefully analyzed to make the use of RER more profitable. The first service states the enhanced load capacity which can be achieved after appropriate installation of RER as per the proposed method. While the other one reveals the number of buses removed from the competition to be the ideal location for RER insertion. A thorough study of these often underestimated effects of RER or Distributed Generation installation can help to improve the systems ability of load expansion without network congestion. It also investigates the areas where students will be guided to remove the buses for DG inclusion, which can be described as 'Inapt Locations'. This will provide a tool to reduce researchers' efforts to find a suitable location on large networks by reducing the size of the candidates locations.

Keywords Renewable Energy Resources · Transmission Parameters · Distributed Generator · Inapt locations · Loadability

---

S. R. Kalambe (✉)
Electrical Engineering Department, Dr. Babasaheb Ambedkar College of Engineering and Research, Nagpur, Maharashtra, India

S. Jain
Electrical and Electronics Engineering Department, R.K.D.F. University, Bhopal, Madhya Pradesh, India

B. N. Kale
Mechanical Engineering Department, Dr. Babasaheb Ambedkar College of Engineering and Research, Nagpur, Maharashtra, India

A. K. Bohre et al. (eds.), *Planning of Hybrid Renewable Energy Systems, Electric Vehicles and Microgrid*, Energy Systems in Electrical Engineering,
https://doi.org/10.1007/978-981-19-0979-5_32

Nomenclature

nl	Total load buses
br	Total system branches
m	Total generator buses
P_{SS}	Active Power from substation in MW
Q_{SS}	Reactive power from substation in MVAr
P_{RER}	Active power delivered by RER
Q_{RER}	Reactive power delivered by RER
P_{loss}	Power loss in network, MW
P_{di}	Active power demand at ith bus
Q_{di}	Reactive power demand at ith bus
P_{Li}	Active power loss of ith branch
Q_{Li}	Reactive power loss of ith branch
P_{load}	Active power demand fed by RER
Q_{load}	Reactive power demand fed by RER
S_{DG}	DG capacity in MVA
OPF	Optimal power factor
OPF_{DG}	Optimal power factor of DG
$O.L.$	Optimal location
$O.S.$	Optimal size
PF_{load}	Power factor of load
$P_{Load,i}$	Operating value of active load at bus i, p.u.
$Q_{Load,i}$	Operating value of reactive load at bus i, p.u
P_{L0}	Initial operating value of active load at bus i, p.u.
Q_{L0}	Initial operating value of reactive load at bus i, p.u.
α	Loading factor
ΔP_{Load}	Increment in the active load, p.u.
ΔQ_{Load}	Increment in the reactive load, p.u.
OL_{DG}	Optimal DG location
OS_{DG}	Optimal DG size
RER	Renewable Energy Resources
DG	Decentralized generator
DNO	Distributed Network Operator
$Lmax$	Maximum Load
RER	Renewable Energy Resource

32.1 Introduction

The power system with RER and DGs sets a different set of network performance issues, which may be desirable or worse depending on network structure (Pathomthat Chiradeja and Ramakumar 2004; Rezk et al. 2021; Vatani et al. 2016; Kashem et al.

2000; Baran and Felix 1989; Hadisaid et al. 1999). Desirable working conditions enrich the energy system with an improved operating environment and cost-effective system management. On the other hand, unfavorable operating conditions reduce system performance and reduce the reliability levels that place penalties on DNOs in terms of a system with low power output, bad power supply, systemic loss, reduced stability, etc. (Pathomthat Chiradeja and Ramakumar 2004; Rezk et al. 2021; Vatani et al. 2016) These unfavorable conditions require hard workers to minimize the effects of the DG's implementation on the efficiency of the system (CIREDWG04 1999; Rau and Wan 1994; Caisheng Wang and Nehrir 2004; Gadomkar et al. 2005). This chapter outlines a multifaceted approach using the Multiple Indexed Performance Evaluation Function (MIPEF) to ensure the optimal performance of the limited RER parameters that it can incorporate into any power network with advanced performance. This approach focuses on assessing the impacts of RER allocation on the power distribution system, power profile development, and critical bus system and MVA power output due to residual power installation. The Multiple Index Performance Evaluation (MIPE) simplifies the evaluation of the appropriate DG design parameters to be included in the DS. It also enables the system to receive advanced uploads and Inapt locations. These helpful resources help the way to expand the system with a rich solution environment.

32.1.1 Outline of the Work

This chapter includes:

- Details of various technical and economic implications that need to be addressed when considering the installation of generators in the power system.
- Definition of Distributed Generation and Performance Indices used in Multiple Indexed Performance Evaluation (MIPE) Function.
- Description of Multiple Indexed Performance Evaluation (MIPE) Function and the studies performed To establish the utility of proposed MIPEF following studies are performed on 33 bus and 69 bus systems which includes (1) Examination of various indices and (2) Investigation of Inapt locations for DG insertion.
- Investigation of optimal access location for DG by using MIPEF and comparison of the obtained results with conventional Load Flow Analysis.
- Investigation of impact of MIPE function to improve system loadability.
- Definition of Inapt locations and impact of MIPE function in investigation of them.
- Results and Discussion.
- Concluding comments.

32.1.2 Flow of Work

In this work Inapt Locations and the enhanced loadability at an investigated optimal location obtained by application of MIPEF explored in this research has been observed. The proposed approach has been shown in the flowcharts shown in Figs. 32.1 and 32.2. Figure 32.1 shows the methodology used for investigation of optimal location and the Inapt locations whereas Fig. 32.2 explains the line of action required for searching the enhanced loadability which can be ensured after installing the RER at the optimal location investigated by the proposed methodology.

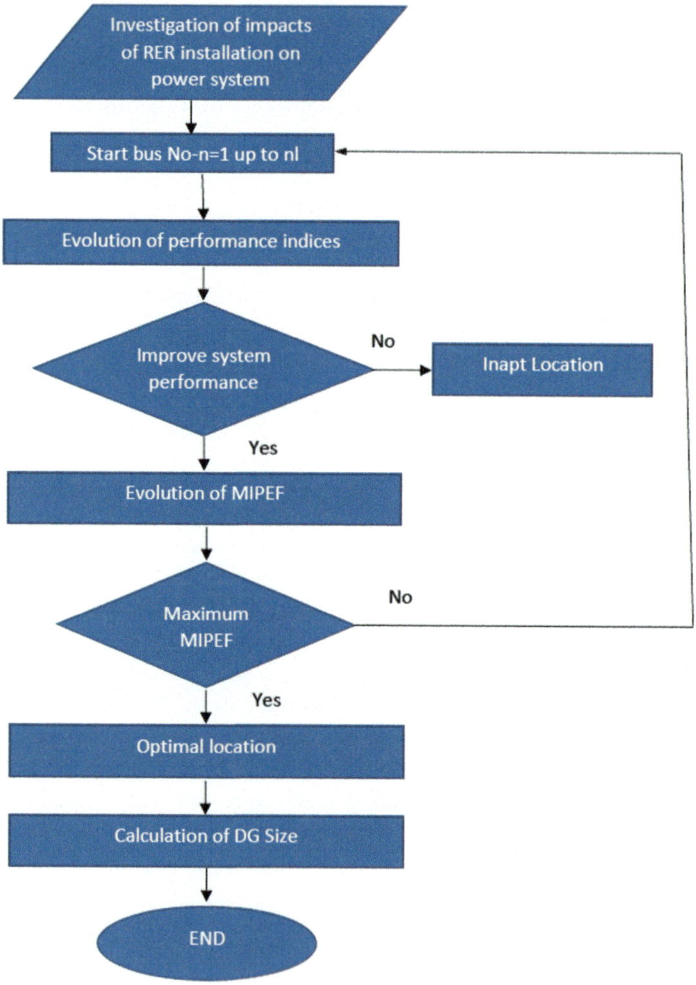

Fig. 32.1 Flowchart of MIPEF for investigation of optimal location and the inapt locations

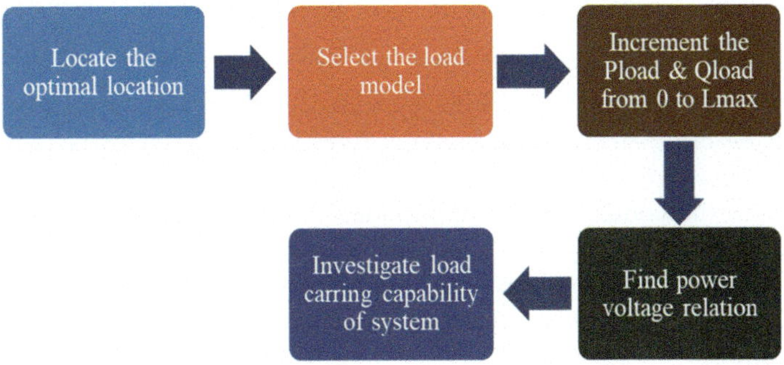

Fig. 32.2 Line of action required for searching the enhanced loadability

32.2 Overview

There are various technical and economic implications that need to be addressed when considering the installation of generators in the power system. The significant technical impacts are (Pathomthat Chiradeja and Ramakumar 2004; Rezk et al. 2021; Vatani et al. 2016; Kashem et al. 2000; Baran and Felix 1989; Hadisaid et al. 1999; CIREDWG04 1999; Rau and Wan 1994; Caisheng Wang and Nehrir 2004; Gadomkar et al. 2005; Quezada et al. 2006; Thukaram et al. 2009; Pinnarelli et al. 2021; Mahmoud et al. 2010; Akorede et al. 2011; Fahad and Abu-Mouti 2011; Kumar and Gao 2010; Failed 2020; Kalambe and Agnihotri 2013; Acharya et al. 2006).

- Power loss reduction
- Energy loss reduction
- Voltage profile correction
- Environmental impacts
- Efficiency improvement
- Reliability improvement
- Security requirement
- Transmission and Distribution congestion management (CIREDWG04 1999)
- The key economic impacts are:
- Deferred investments for line up gradation (CIREDWG04 1999)
- Negligible operational and maintenance cost of some renewable DGs
- Peak shaving
- Reserve supply requirement reduction.

In this work, a multiple indexed function is developed to appraise system performance by considering technical as well as economic benefits availed after RER insertion. The indices are derived by associating and taking the ratio of a measure of a trait with and without RER for the same network combination (CIREDWG04 1999). The maximum value of the index will represent the highest degree of trait and in case of technical impacts the positive and negative sign of the indices will show its favorable or unfavorable impact respectively on network.

32.2.1 *Definitions*

This chapter demonstrates two ancillary utilities of the proposed method. The first application describes the system loading done after DG deployment is arranged according to the proposed method, while the other specifies the number of buses removed from the competition to be the ideal location for DG distribution (Pathomthat Chiradeja and Ramakumar 2004). Therefore, it can be said that the proposed approach facilitates the improved capacity of the system to maintain load expansion without network renewal and testing to reduce the DG's designated areas called 'Inapt Locations'.

32.3 Performance Indices

To illustrate the performance of system after DG insertion, four major technical and two minor economical indices are proposed by strategically giving credence to each index. These credence factors are intended to give the relevant weight to each impact indices for the DG insertion. The use of credence factors enables the emphasis of certain trait depending on the location of the DG and types of loads served by the distribution system (CIREDWG04 1999). With this formulation, the positive highest value of MIPE function obtained will indicate the optimum sizing parameters for the DG which can be installed in system. It also facilitates the investigation of apt and inapt location for DG addition. These indices can be defined as follows:

(1) Voltage Profile Correction Index (V_{PCI})

Addition of DG at an appropriate location may improve the voltage profile of the system because DG shares some part of the power required by load diminishing the proportional current through some part of system which may result in boosting the voltage magnitude of relevant of aligned load buses (Rezk et al. 2021; CIREDWG04 1999; Rau and Wan 1994; Caisheng Wang and Nehrir 2004). Whereas voltage profile may demean as a result of reverse power flows caused with DG insertion at an inapt location. Apt locations and inapt locations for DG insertion with respect to voltage correction of the system can be identified by positive and negative values of V_{PCI} respectively calculated by using Eq. (32.1).

Definition V_{PCI} for the bus i at which DG is installed can be defined as the summation of Voltage Regulation (VR) of all the buses. The voltage regulation of jth can be defined as the difference of voltage of jth bus without DG and with DG (installed at bus i) to the voltage of jth bus without DG. The general expression for V_{PCI} is given as.

$$V_{PCI,i} = \sum_{j=1}^{nl} VR_{j,i} \tag{32.1}$$

where

$$VR_{j,i} = \frac{V_{j,w/oDG} - \mathrm{V_{j,i}}}{V_{j,w/oDG}} \tag{32.2}$$

$V_{j,i}$	Voltage of bus j when DG is installed at bus i.
$V_{j,w/o\ DG}$	Voltage of bus j without DG.
i	load bus at which DG is installed one at a time.

Thus,

$$i \forall L_1, L_2 \ldots L_{nl}$$

$$j \forall L_1, L_2 \ldots L_{nl}$$

Highest value of V_{PCI} obtained after DG installed at certain ith location will bring all node voltages within permissible limits which ensures the most stable system operation with respect to voltage profile.

Location identification with respect to voltage profile correction and aided system stability.

Optimal location	Maximum V_{PCI}.
Apt location	Positive V_{PCI}.
Inapt location	Negative V_{PCI}.

(2) **Critical Bus Voltage Improvement Index(VCRII)**.

In the system without DG, many bus voltages are not in permissible limits. A bus at which the lowest value of voltage is obtained is generally defined as the Critical Bus. Voltage profile improvement of Critical Bus is the most effective trail to enhance the system loadability. As it is reiterated in previous part of this work, DG allocation at an appropriate location can improve the voltage of a major portion of the system to a satisfactory level including critical bus. This attribute of system with DG instigate the investigation of V_{CRII}.

Definition V_{CRII} for bus i at which DG is installed can be defined as the ratio of the difference of critical bus voltage of the system with, and without DG (installed at bus i) to the critical bus voltage without DG. The value of V_{CRII} obtained after DG installed at each load bus, one at a time is found to be very less as compared to other indices. To bring it to the scale of other indices, it can be amplified by using a Multiplication Factor (MF). Thus amplified value of VCRII may provide a significant contribution for inspection DG parameters. The general expression for VCRII is given as

$$V_{CRII,i} = M.F. \frac{V_{CRj,i} - V_{CRj,W/ODG}}{V_{CRj,W/ODG}} \tag{32.3}$$

where

$$i \forall L_1, L_2 ... L_{nl}$$

$$j \forall L_1, L_2 ... L_{nl}$$

$V_{CRj,i}$	critical bus voltage of bus j with DG installed at bus i.
$V_{CRj,W/O\ DG}$	critical bus voltage of bus j without DG.
M.F.Multiplication factor	(in this work M.F. = 5).

Location identification with respect to critical bus voltage improvement.

Optimal location	Maximum VCRII.
Apt location	Positive VCRII.
Inapt location	Negative VCRII.

(3) **Active Power Loss Index (PLOSSI)**

Distribution system is the closing end of power system where high voltage network is converted to low voltage feeders resulting high line currents. Therefore due to high values of current in distribution system, I^2R and I^2X losses are significant in distribution systems. Thus to improve the efficiency of the network considerable efforts are required to reduce these losses. The principal potential benefit accessible by DG is the power loss reduction. It is observed that, with an appropriate DG insertion due to reductions in power flows of a major portion of system, line losses are reduced to a considerable extent. However, depending on the rating and location of DG units, it is also possible that the system will lead to higher power losses at very high (and unrealistic) penetration levels. This provides the key element to decide the criterion for sizing and placing of DG to be installed in the system. Many researchers have developed various methods for distribution system planning for DG insertion by considering active power loss minimization as the foremost criterion. In this work also loss minimization is considered as the prime objective for DG allocation and sizing but to ensure its optimality its relevance is verified by using MIPE approach. Thus in MIPE function explored in this work, active power loss reduction plays a vital role since one of the indices incorporated in the function is Active Power Loss Index which provides the exact measure of the active power loss reduction occurred due to DG installed at each load bus individually.

Definition P_{LOSSI} for the bus i at which DG is installed can be defined as the ratio of the difference of active power loss occurred in the system (with same configuration) without and with DG (installed at bus i) to the loss without DG.

$$P_{LOSSI,i} = \frac{P_{LOSS,w/oDG} - P_{LOSS,withDG,i}}{P_{LOSS,w/oDG}} \tag{32.4}$$

where

$$i \forall L_1, L_2, \cdots L_{nl}$$

$P_{LOSS,\ withDG,i}$	system active power loss with DG installed at bus i.
$P_{LOSS,\ w/oDG}$	system active power loss without DG.

Location identification with respect to active power loss reduction.

Optimal location	Maximum PLOSSI.
Apt location	Positive PLOSSI.
Inapt location	Negative PLOSSI.

(4) **Reactive Power Loss Index (QLOSSI)**.

Another potential benefit of DG installation is reactive power loss reduction. Researcher has shown keen interest in active power loss reduction while working in the area of DG allocation. Accompanied to that reactive power loss should also be considered as an aided advantage of DG installation at an appropriate location. Reactive power support offered by DG improves the system power factor which in turn improves efficiency.

Definition Q_{LOSSI} for the bus i at which DG is installed can be defined as the ratio of the difference of reactive power loss that occurred in the system lines (with same configuration) without and with DG (installed at bus i) to the loss without DG.

$$Q_{LOSSI,i} = \frac{Q_{LOSS,w/oDG} - Q_{LOSS,withDG,i}}{Q_{LOSS,w/oDG}} \tag{32.5}$$

where

$$i \forall L_1, L_2, \cdots L_{nl}$$

$Q_{LOSS,\ withDG,i}$	system reactive power loss with DG installed at bus i.
$Q_{LOSS,\ w/oDG}$	system reactive power loss without DG.

Location identification with respect to reactive power loss reduction.

Optimal location	Maximum QLOSSI.
Apt location	Positive QLOSSI.
Inapt location	Negative QLOSSI.

After investigating the performance indices by considering key technical impacts like active and reactive power loss reduction and voltage profile correction of system as well as critical bus, two additional indices are developed which portray economic impacts such as.

1. The substation capacity release due to the addition of DG,
2. Help to investigate the location at which if DG is installed maximum power loss reduction is possible with an economic DG size.

(5) **Main_MVA Release Index (M_MVARI).**

Urbanization and day-to-day enhanced social progress results in a prolonged gap between electricity generation and never ending demand where unfortunately demand is found to be always higher than generation. This drought of supply evolves the undesirable or complicated requirements like load scheduling, load balancing, demand management, etc. The key economic advantage of DG insertion in the distribution system is reduction of power requirement from sub-station (S.S.). This relieves the grid in terms of power requirement from the conventional plants which in turn reduces the bottlenecks at fuel such as coal, nuclear elements, etc. The reduction of capacity needs of conventional plants is possible due to two reasons: (1) the real power generated by DG units will directly reduce share of conventional plants and (2) reduced line losses will further decrease the power requirements from conventional plants (Pathomthat Chiradeja and Ramakumar 2004). Apart from this more the reduction in power requirement from grid higher the ability of the system expansion with deferred investments in line upgrades. Main_MVA release index provides the major of reduced share of power from substation.

M_MVARI for the bus i at which DG is installed can be defined as the ratio of difference of MVA share of S.S. with DG (installed at i) and without DG to the supply of S.S. without DG (with same configuration).

$$M\_MVA_{RI,i} = \frac{S.S._{Share,w/oDG} - S.S._{Share,i}}{S.S._{Share,w/oDG}} \tag{32.6}$$

S.S.share w/o DG	substation share of system load without DG.
S.S.share, i	substation share of system load with DG installed at bus i.

Therefore,

$$M\_MVA_{RI,i} \leq 1$$

Location identification with respect to Mian_MVA capacity release.

Optimal location	M_MVARI = 0.6–0.3 whatever that may give lowest power loss.
Apt location	M_MVARI < unity.
Inapt location	M_MVARI $\approx$ 1.

This index provides superficial information about location suitability according to power loss reduction percentage. Individual information of the index may misguide the decision of optimal location but in combination with the power loss reduction and voltage stability improvement, Main_MVA capacity release percentage supports the optimality of the solution. If more than one option is available as an optimal DG location as approximately same power loss reduction and improved voltage stability can be achieved from more than one location then the bus with higher value of M_MVA_{RI} can be recommended as a better optimal location for DG installation. Thus this index is counted as one of the minor rather than major performance evaluation indices since it can be considered as a supporting factor but not the prime decisive factor for optimal solution evaluation.

(6) **DG Capacity Index (DGMVAI).**

This index can also be considered as the supporting factor to recommend the optimality of solution in uncertain situations. For example, if the number of locations as a viable candidate to install DG is more at which DG insertion will provide approximately same power loss then the location at which DG of smaller capacity is required is considered as the optimum location. It indicates that in 16 bus system 67.93% of power loss reduction is possible with DG addition of 11.78 MVA at bus 9 whereas approximately same percentage loss reduction, i.e., of 61.67% is possible with DG of only 6.55MVA installed at 12th bus. Thus on the basis of DG size, 12th bus may be considered as the economic solution.

Definition DG_{MVAI} for the bus i at which DG is installed can be defined as the inverse of optimal capacity of DG calculated for bus i with which minimum power loss for that bus can be obtained.

$$\text{DG}_{\text{MVAI,i}} = 1/\text{DG}_{\text{capacity,i}} \tag{32.7}$$

where

$$i \forall L_1, L_2, \cdots L_{nl}$$

Location identification with respect to DG capacity requirement.

Optimal location	DGMVAI = 0.9–0.3 whatever that may give lowest power loss.
Apt location	DGcapacity < S.S.share,w/oDG.
Inapt location	DGcapacity ≥ S.S.share,w/oDG.

32.4 Multiple Indexed Performance Evaluation (MIPE) Function

MIPE function proposed in this work offers the optimal solution for the system by examining the overall performance of each load bus as a viable optimal access location for DG insertion through various technical and economical indices mentioned in the previous section. For emphasizing the impacts in terms of its degree of importance, a specific credence (cr) has been assigned to each index. Selection of credence depends on the degree of significance of respective index. For DG insertion planning is thru then to improve the voltage stability the higher credence can be assigned to V_{PCI}, the system loadability can be enhanced by assigning higher credence to V_{CRII}, lower loss can be achieved by giving higher credence to P_{LOSSI} or Q_{LOSSI}. Similarly relative credence can be assigned according to the requirement. In this work, four major technical and two minor economical indices are proposed. Equal credence is assigned to each major index and a combined credence is assigned to two minor indices in order to plan the DG insertion to improve the overall performance of system. Therefore 1/Nth credence is assigned to each major index and to the combination of two minor indices so that the overall sum of total credence will be unity.

Therefore the Multiple Indexed Performance Evaluation Function is given by

$$\begin{aligned} \mathrm{MIPEF_i} = cr_1 V_{PCI,i} + cr_2 V_{PCII,i} + cr_3 P_{LOSSI,i} + cr_4 Q_{LOSSI,i} \\ + cr_5 (M_{MVA_{RI,i}} + DG_{MVA,i}) \end{aligned} \tag{32.8}$$

where

$$i \forall L_1, L_2, \cdots L_{nl}$$

And,

$$\sum_{i=1}^{5} cr_i = 1 \wedge cr_i \in [0, 1]. \tag{32.9}$$

In general, allocation of credence to relevant impact is a complicated task (Correia et al. 2020). Furthermore, the credence should be flexible and fare enough to cope up with ambiguous concerns of DNO's about loss minimization, voltage stability, load expansion, other associated economic issues, etc. (Correia et al. 2020; Bindeshwar Singh 2015; Ullah et al. 2019; Ghasemi et al. 2019; Graditi and Somma 2021; Kifle et al. 2018; Poornazaryan et al. 2016; Roche and Courtney 2020; Alotaibi and Salama 2018; Handbook of Distribute Generation 2017; Bawazir and Cetin 2020; Choi et al. 2018). The resiliency and fare allocation attained by the proposed MIPEF function presents it as an effective method to investigate the most beneficial optimal location at which DGs can be placed. In this work equal credence is assigned to all the indices to achieve overall improvement of system performance.

32.5 Studies Performed

To establish the utility of proposed MIPEF ensuing studies are executed on 33 bus (Kashem et al. 2000) and 69 bus (Baran and Felix 1989) systems.

Case 1 Examination of various indices.

Case 2 Investigation of optimal access location for DG by using MIPEF and comparison of the obtained results with conventional Load Flow Analysis.

32.5.1 Examination of Various Indices

Figures 32.3, 32.4 show the active-reactive power loss reduction indices evaluated for 69 bus and 33 bus systems respectively. These figures depict information regarding feasible and unfeasible candidate locations for DG insertion. The values of both the power loss indices for buses 2–6 and 28–50 are very small showing their unfeasibility for DG addition in terms of power loss reduction. Whereas a significant loss reduction can be observed from buses 7–27 to 51–69 which indicate their feasibility for DG addition. But comparative analysis recommends buses 58–65 as the striking options showing approximately 70–80% of loss reduction. Similarly Fig. 32.2 shows that in 33 bus system no feasible solution is achieved for buses 2 and 19–25. Whereas a

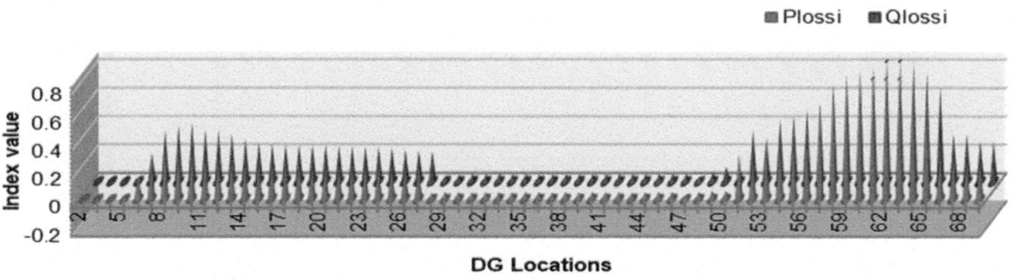

Fig. 32.3 Power loss indices of 69 bus system

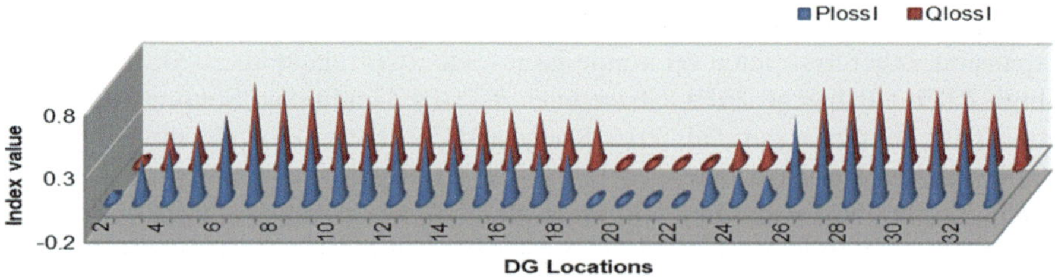

Fig. 32.4 Power loss indices of 33 bus system

feasible solution is obtained for buses 3–18 and 26–33 showing significant power loss reduction percentage. But buses 6, 7, and 26–30 indicate highest percentage of loss reduction, i.e., approximately above 65% thus can be recommended as prominent options for DG allocation.

Figures 32.5 and 32.6 depict system voltage and critical bus voltage correction indices evaluated for 69 bus and 33 bus systems respectively. The specific formulation of indices enables the identification of optimum, apt as well as inapt locations for DG allocations. This significant trait of the indices is demonstrated by these figures. Figure 32.3 indicates that in 69 bus system, a significant voltage profile correction is possible from buses 10–27 and 54–69 showing its higher degree of feasibility

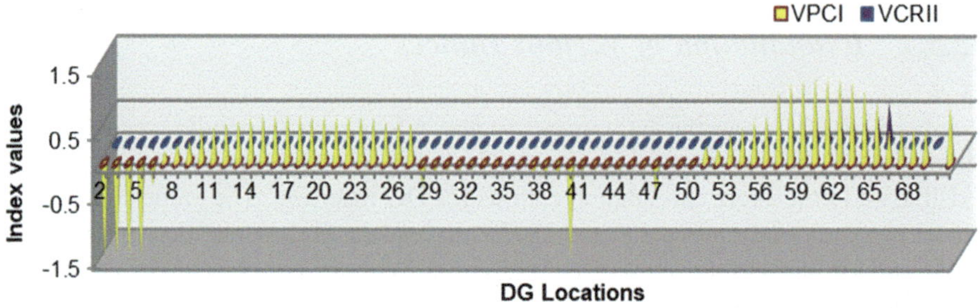

Fig. 32.5 System voltage and critical voltage correction indices of 69 bus system

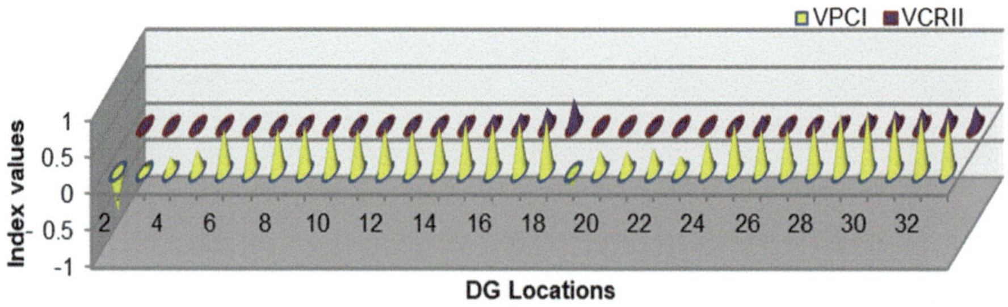

Fig. 32.6 System voltage profile and critical voltage improvement indices of 33 bus system

whereas negative or near zero value of buses 2–9 and 28–53 indicates their failure to provide feasible solution. However buses 14–24 and 56–64 presents the striking options of being best locations for DG addition showing index value approximately above 70%. Similarly for 33 bus system Fig. 32.6, indicates that buses 2 and 19 fails to provide any possible feasible solution whereas buses 3–18 and 19–33 provide feasible solution. However, buses 14–18 and 25–33 show highest degree, i.e., above 60% of the voltage profile improvement attribute.

32.5.2 *Investigation of Optimal Access Location for DG by Using MIPEF and Comparison of the Obtained Results with Load Flow Analysis Method*

Figures 32.7, 32.8 and Table 32.1, present the comparative analysis of optimal access location of DG by using MIPEF and Load Flow Analysis method.

Figure 32.7 shows that for 33-bus system maximum value of MIPEF is obtained at bus 29 indicating it as the optimal access location. Whereas as per the results obtained by implementing conventional Load Flow Analysis method bus 6 is the viable solution. Conversely Fig. 32.8 indicates that in 69 bus system, maximum value of MIPEF is obtained at bus 61 and Load Flow Analysis method also recommends bus

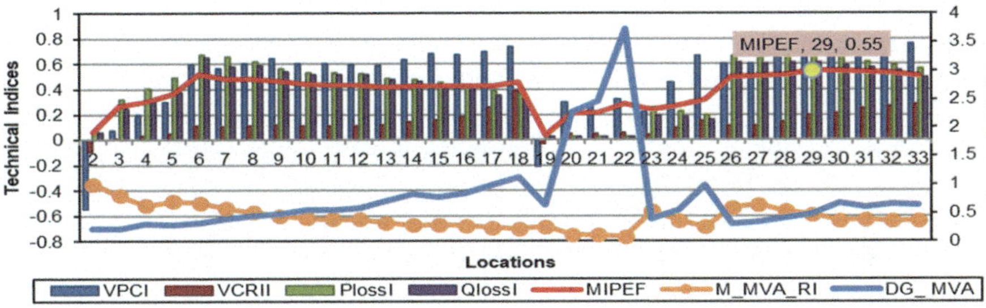

Fig. 32.7 Performance evaluation indices and MIPEF for 33 bus system

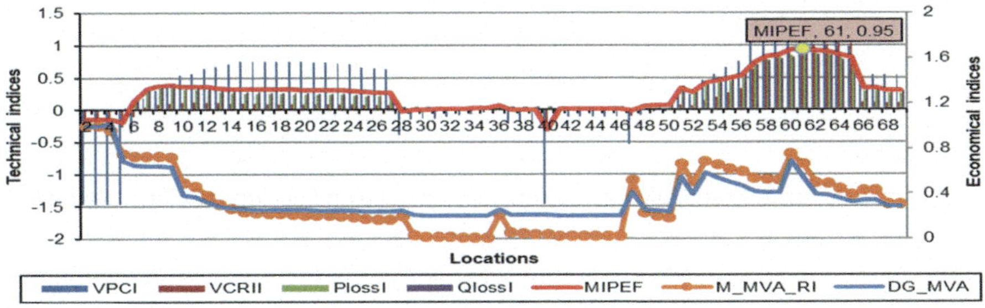

Fig. 32.8 Performance evaluation indices and MIPEF for 69 bus system

Table 32.1 Investigation of optimal access location for RER by using MIPE approach and load flow analysis method

System	Method	Optimal location	DG size	OPFDG	Type
33 bus	Load Flow Analysis	6	3.14	0.83	3
	MIPEF	29	2.12	0.77	3
69 bus	Load Flow Analysis	61	2.37	0.81	3
	MIPEF	61	2.37	0.81	3

Table 32.2 Comparison of optimal access location evaluated by load flow analysis and MIPEF

Indices	bus 6 (Load Flow Analysis)	bus 29 (MIPEF)
V_{PCI}	0.593	0.804
V_{CRII}	0.107	0.195
P_{LOSSI}	0.675	0.660
Q_{LOSSI}	0.648	0.608
M_MVA_{RI}	0.670	0.466
DG_{MVAI}	0.318	0.4712
max(MIPEF)	0.486	0.547

61 as a viable solution. Thus the DG addition impact on the performance of system as per the selected optimal bus differs for 33 bus system. Table 32.1 shows that for 33 bus system, with evaluated optimal location, maximum parameters for DG design also differ. At bus 6 DG of 3.14 with OPF_{DG} 0.83 is required whereas at bus 29 DG of 2.12 with OPF_{DG} 0.77 is required. Variations with the selected optimal location for 33-bus are given in Table 32.2 which indicates power loss reduction indices are slightly higher whereas voltage profile correction indices are significantly lower for bus 6 as compared to bus 29. Similarly in case of economical indices, since DG capacity required at bus 29 is lower the grid capacity relief is also comparatively lower than that from bus 6 location. Thus subsequently the overall MIPEF obtained for bus 29 is higher than the bus 6 indicating enhanced degree of performance improvement by DG addition at it.

This section reveals that though MIPEF provides better optimal location. In this part of study it can be seen that the same optimal solution is obtained in 69-bus system by using both methods (MIPE and load flow analysis) whereas the solution differs little bit for 33 bus system.

32.6 System Loadability

Voltage stability is the ability of a power system to maintain its voltage magnitude between the specified limits at system nodes after being subjected to a static or dynamic disturbance. Voltage collapse is a major consequence of it, which may lead to a blackout or unusual low voltage magnitude in a major portion of the system. It occurs in a configuration that is heavily loaded, faulted, and/or has reactive power shortages, which are not being met due to the generation, transmission, and distribution constrictions of reactive power. Voltage stability of transmission or distribution systems can be defined by their loadability. It may be expressed as the static loading margin or the additional load that a system can withstand over and above the base case load before the collapse point. The location and appropriate size of a RER that alters reactive power level and flows in a radial distribution network can have a substantial influence on it.

Voltage stabilization is the ability of a power system to store its electrical power within the limits set on the system nodes after experiencing static or dynamic interruptions (Baran and Felix 1989). A power outage is a major consequence of it, which can lead to the extinguishing or extinguishing of abnormal electricity for a large part of the system (Baran and Felix 1989). It occurs in highly loaded, faulty, and/or ineffective configurations, which can be achieved due to the production, transmission, and distribution of working capacity. Voltage strength of transmission or distribution systems can be explained by its carrying capacity. It can be displayed as a fixed load line or additional load that the system can withstand in addition to the base case load before the fall point. The exact location and size of the RER that converts the active energy level and flows into a radial distribution network can have a significant impact on it (Baran and Felix 1989).

32.6.1 Enhanced Loadability

Enhancement of loading capability of the system depends on the active power support provided to the system. Active power support is compatible with system voltage and bus power adjustment. Placing the RER at the right size in the right place can reduce the flow of energy on an important part of the system. As a result, it reduces I^2R losses associated with major improvements to the system's power profile and critical bus. The proposed method incorporates indices, indicating the percentage of electricity and the improvement of bus power through the installation of RER (Pathomthat Chiradeja and Ramakumar 2004). Therefore, the installation of RER in the appropriate location with the appropriate size and type proposed in the proposed manner can provide a high voltage profile adjustment of the system and a sensitive bus, which ensures the advanced power support provided in the system. Therefore, system loading is increasing with the increase in operational capacity obtained with the addition of DG. Solid electrical power in the energy system can be analyzed using a PV curve (Pathomthat Chiradeja and Ramakumar 2004). The power drop point on this curve represents the maximum loading of the system. To investigate improvements in high-performance system loading with the addition of RER, the PV curve is obtained by gradually increasing the operating and updating load of the system as given the following statistics:

$$P_{Load,i} = P_{L0,i}(1 + \alpha \Delta P_{Load}) \quad (32.10)$$

$$Q_{Load,i} = Q_{L0,i}(1 + \alpha \Delta Q_{Load}) \quad (32.11)$$

where,

α 0, 1, 2, 3, …
i critical bus.

The α value is increased until a power outage point is detected. Adding a DG to the right size and size can delay the point of a power outage due to the system's fixed power limit (Pathomthat Chiradeja and Ramakumar 2004).

32.6.2 Studies Performed

As mentioned in the previous section, the allocation of DGs according to the proposed method improves the system and profile of critical buses. Therefore, enhanced uploads without a distance acquisition program are possible. Demonstrating this feature of the proposed methods follows studies conducted on 33-bus and 69-bus system.

Case 1 Enhanced Loadability Investigation after DG installation at an optimal location as per Load Flow Analysis method.

Case 2 Enhanced loadability investigation after DG installation at an optimal location as per MIPE approach and investigation of impact of load models.

Table 32.3 indicates in 33-bus system Load Flow Analysis method indicates bus 6 whereas MIPE approach suggests bus 29 as OLDG. So to authenticate the enhanced loadability accomplished by both the approaches in case 1, RER is placed according to

Table 32.3 Optimal DG design as per proposed method and corresponding voltage profile correction

System	Approach	Status	OL_{DG}	OS_{DG}, MVA	OPF	Critical bus	Voltage
33 bus	Basic	Without DG	–	–	–	18	0.83
	Load Flow Analysis	With DG	6	3.22	0.82	18	0.92
	MIPEF	With DG	29	2.12	0.77	18	0.90
69 bus	Basic	Without DG	–	–	–	64	0.84
	Load Flow Analysis	With DG	61	2.37	0.86	65	0.94
	MIPEF	With DG	61	2.37	0.86	65	0.94

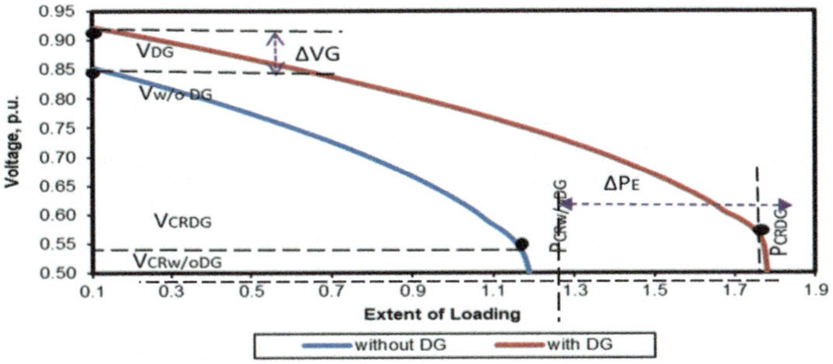

Fig. 32.9 Load application and respective power-voltage curve for 33-bus system

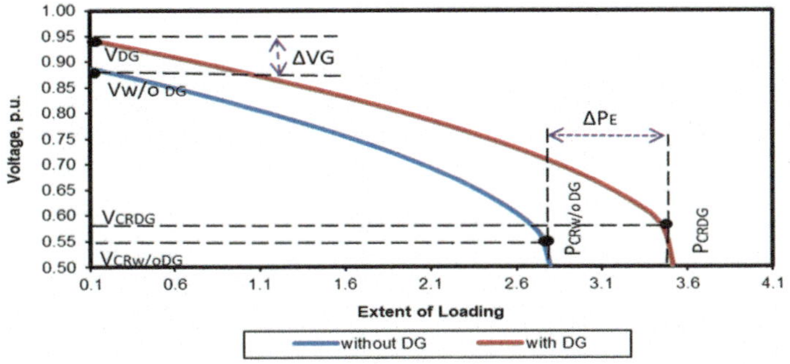

Fig. 32.10 Load application and respective power-voltage Curve in case of 69-bus system

design suggested by Load Flow Analysis method and in case 2, according to the MIPE approach. In case 1 along with investigation of enhanced loadability attained via DG allocation by using Load Flow Analysis method, various terms used to express the stability are defined. However, in case 2 investigation of efficacy of DG allocation by MIPE approach for enhanced loadability is accompanied by going-over the impacts of various load-patterns on it. From Figs. 32.9, 32.10, 32.11, 32.12, it is observed that DG planning by using both the methods enabls the enhanced loadability.

32.6.3 *Enhanced Loadability Investigated After DG Placed at Optimal Parameters Suggested by Load Flow Analysis Method*

In this case, initially the critical bus load of the system without the DG gradually increases until the volume capacity reaches the critical value called $V_{CRw/oDG}$, where the additional load will make the system unstable. Sensitive voltage, V_{CR} can be

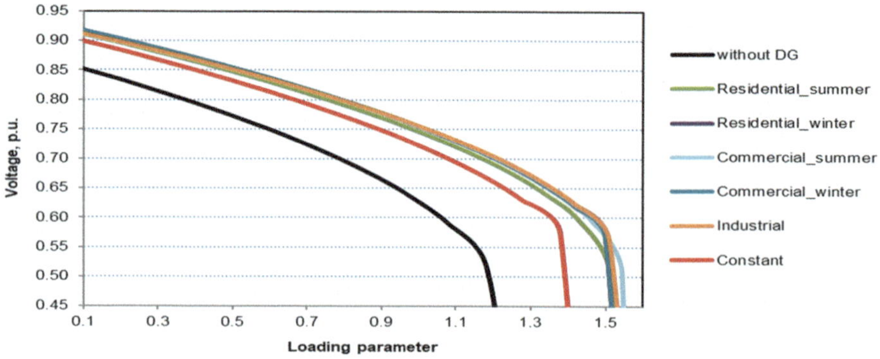

Fig. 32.11 Power-voltage curve in 33-bus with various load models

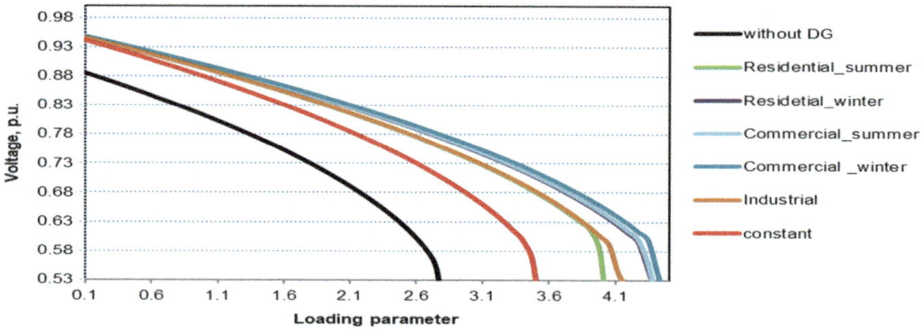

Fig. 32.12 Power-voltage curve in 69-Bus with various load models

defined as a small allowable force above where the system can fall. A load where the system is unstable or a power failure is called a critical load $P_{CRw/oDG}$.

In this case, according to the Load Flow Analysis method, Type 3 DG of 3.22 MVA operating at 0.82 power factor is installed on bus 6 of the 33 system buses and Type 1 DG of 2.37 MVA operating at 0.86 power factor is installed on bus 61 of 69 bus system. Thereafter the load of the critical bus is gradually increased from its original value to the critical value, the P_{CRDG} where the voltage reaches the critical area, the VCRDG. Subsequently the difference between $P_{CRw/oDG}$ and P_{CRDG} indicates the maximum load expansion indicated by ΔPE. Similarly the difference between $V_{w/oDG}$ (working voltage of PV curve without DG) and VDG (operating voltage of PV curve and DG), indicates Voltage Gradation, ΔVG (in that working area) obtained after the effective and beneficial addition of DG to the system.

Figures 32.9 and 32.10 indicate that in both systems a significant increase in load is possible with the DG placed in the correct position with the design parameters to be proposed by the known method of analyzing the load flow as shown in Table 32.3.

Table 32.4 shows the various parameters indicating the enhanced loading of both programs received after the DG installation plan proposed by the System Uploading

Table 32.4 Summarized parameters showing enhanced system loadability

System	$P_{CRw/oDG}$	P_{CRDG}	ΔP_E	% Load expansion	$V_{CRw/oDG}$	$V_{CRw/oDG}$	$V_{w/oDG}$	V_{DG}	ΔVG	% Voltage gradation
33 bus	1.15	1.78	0.63	54.78	0.54	0.57	0.85	0.92	0.07	8.24
69 bus	2.78	3.49	0.71	25.54	0.55	0.58	0.88	0.94	0.06	6.81

method. In the 33-bus system, 54.78% of the load increase of 8.24% of voltage gradation is possible with the installation of the DG and in the 69-bus system 25.54% of the load expansion is obtained by 6.81% of the gradation voltage at the first value of the burden.

32.6.4 Investigation Optimal Design of DG Recommended by MIPE Approach and Examination of Impact of Load Models

In this case, the effectiveness of the MIPE method in the DG's local spatial assessment in terms of improved uploads is demonstrated. The impact of different types of power loads was then considered as the system downtime was investigated. As described in 1, in this case and the critical bus load is gradually increased from its original value to $P_{CRw/oDG}$ to volume capacity, $V_{CRw/oDG}$ after which the voltage drops. The DG is then introduced into the system with permanent loading models, with DG design parameters proposed by the MIPE method as described in Table 32.3, and the critical bus loading and gradually increases from its original value to P_{CRDG} until volume capacity, V_{CRDG} after which the voltage drops. Next, ΔP_E, ΔVG, and the corresponding load increases and the percentage of voltage gradation are tested. The process is repeated throughout the configuration process with the various loading models mentioned and the corresponding changes in the loading improvement are studied.

Figure 32.11 shows that the dependence of the electrical load on loads shows an increase in the percentage increase in load compared to the regular loading model. All power loading models show the same pattern of Power-Voltage curves and the power loading model always shows significant changes in the stability parameters and the Power-Voltage curve pattern. Figure 32.12 shows the same variation on the power-voltage curve in terms of reliability and continuous loading of the system on the 69 bus system.

The results given in Figs. 32.11 and 32.12 show that significant differences in PV curves can be seen in the speculation of a larger loading model and electrical power. A summary of the results obtained in this case from the PV curves given in these figures for the different loading models is presented in Table 32.5.

The result concludes that the loading model plays a major role in the development and demonstration of the operating conditions of the distribution system after the introduction of the DG (Pathomthat Chiradeja and Ramakumar 2004). The results presented in Table 32.5 confirm this conclusion in terms of further improvements.

33 bus system: Figure 32.11 shows that enhanced loadability obtained after DG addition in the system with constant load model configuration is considerably lower than the system with voltage dependent load models. Table 32.5 presents various load expansion parameters evaluated from these curves. Compared to regular loading models on reliability-dependent loading models reduce system loading rates which

Table 32.5 Summary parameters show advanced system loading with permanent and electronic loading models

System	Loadability parameters	Constant	Residential		Commercial		Industrial	
			Summer	Winter	Summer	Winter	Summer	Winter
33 bus	$P_{crw/oDG}$,	1.3	1.3	1.3	1.3	1.32	1.32	1.33
	P_{crDG},	1.135	1.45	1.46	1.45	1.38	1.39	1.51
	ΔP_E	0.19	0.29	0.25	0.31	0.22	0.24	0.26
	$V_{CRw/oDG}$,	0.52	0.52	0.51	0.51	0.52	0.53	0.52
	V_{CRDG},	0.59	0.54	0.52	0.55	0.58	0.58	0.6
	$V_{w/oDG}$, Volts	0.81	0.75	0.81	0.79	0.83	0.83	0.79
	V_{DG}, Volts	0.89	0.89	0.89	0.89	0.89	0.88	0.88
	ΔVG	0.03	0.04	0.04	0.04	0.08	0.05	0.053
69 bus	$P_{crw/oDG}$, MVA	1.83	1.79	1.79	1.78	1.78	1.78	1.78
	P_{crDG}, MVA	2.46	3	4.1	4.1	4.1	3.9	3.9
	ΔP_E	1.3	1.08	1.35	1.23	1.22	1.22	1.23
	$V_{CRw/oDG}$, Volts	0.51	0.51	0.52	0.52	0.51	0.51	0.52
	V_{CRDG}, Volts	0.59	0.55	0.53	0.53	0.56	0.55	0.55
	$V_{w/oDG}$, Volts	0.78	0.78	0.78	0.78	0.78	0.78	0.78
	V_{DG}, Volts	0.82	0.82	0.82	0.82	0.82	0.82	0.82
	ΔVG	0.053	0.053	0.053	0.053	0.053	0.053	0.053

in turn reduce the currents taken from the grid and DG which resulted in line losses and reduced bus collisions. This phenomenon is well emphasized in the results depicted in Table 32.5. In the system with constant load model the DG insertion at bus 29 recommended by MIPE approach provides 16.67% of enhanced loadability and 5.88% voltage gradation at initial value of load. Whereas in the system with all other voltage dependent load models, DG addition at bus 30 as recommended by MIPE approach provides the enhanced loadability in the range of 22–25% with corresponding voltage gradation of 7.3–8.23% range at an initial load.

69 bus system: Figure 32.12 reiterated the same phenomenon of increased voltages due to voltage dependency of loads as related to constant load models. Here also RER addition in the system with constant loads provides considerably lower proportion of load extension paralleled to the voltage dependent load models. Table 32.5 reveals that constant power load model shows only 22.34% of possible load expansion with 7.15% of voltage gradation after DG addition at bus 61. However voltage dependent load models show possible load expansion in the range of 50.3–54.07% with voltage gradation of 7.61–7.84% after DG addition at the same bus.

These variations in the results with an assumption of different load models indicate that for accurate and precise planning of DG insertion proper load model proposition is mandatory otherwise the constant load model may mislead the practicable solution.

32.7 Inapt Locations

The selection of the best installation sites and the ideal size of DG units in large distribution systems is a complex integration problem. (Pathomthat Chiradeja and Ramakumar, 2004; Baran and Felix 1989; Hadisaid et al. 1999; CIREDWG04 1999). Many methods and approaches to best practice are found in the literature (Rau and Wan 1994; Caisheng Wang and Nehrir, 2004; Gadomkar et al. 2005; Quezada et al. 2006; Thukaram et al. 1999; Pinnarelli et al. 2021; Mahmoud et al. 2010; Akorede et al. 2011; Fahad and Abu-Mouti 2011; Kumar and Gao 2010; Failed 2020) to address this complex problem. Most methods require procedures for repetition or over-calculation, which ultimately result in time-consuming and tedious problems. Apart from these many methods (Caisheng Wang and Nehrir, 2004; Thukaram et al. 1999; Kumar and Gao 2010; Bindeshwar Singh 2015; Ullah et al. 2019; Ghasemi et al. 2019; Kifle et al. 2018; Poornazaryan et al. 2016; Roche and Courtney 2020; Alotaibi and Salama 2018; Bawazir and Cetin 2020; Choi et al. 2018) provide sound effects for small or medium power systems but complex systems create problems in integration or precision. Such scenarios indicate the search location for new, functional tools, which can reduce system complexity, or the number of selected locations where the method can be used, i.e., the number of candidate nodes in the system. This proposed method not only investigates the appropriate access point but also provides information about unsuitable locations called Inapt Locations to perform DG. This useful information can help researchers narrow down the search area of this site.

32.7.1 Prominence of Inapt Locations

Inapt environments are defined as system areas where the installation of a DG can create unwanted operating conditions such as repetitive power flow, reverse energy profile, etc. In this study such areas are divided into two categories such as:

- **Adverse Impact Areas (AILs)**—buses in which the DG is installed, lead to a decrease in system performance due to adverse effects imposed on the system by increasing system losses, reduced power intensity and low percentage output of the main source, i.e., location.
- **Concurrent bus (CB)** Kalambe and Agnihotri (2013)—a bus near the main station.

32.7.1.1 Adversative Impact Locations (AIL_S)

In the distribution system, the DG cannot be installed on all buses. Some buses can be found as bad DG inputs as the inconsistent choice of location and DG size can lead to greater system loss or reduced power stabilization than outside the DG (Acharya et al. 2006; Correia et al. 2020)—DG. The inclusion of DG in Inapt areas puts negative impacts on the distribution system. The MIPEF conducted in this study is used to investigate such areas. Task can identify incompatible locations from complete system nodes. The negative or zero number of indices found in candidate areas indicates inconsistencies in the DG's inclusion in that direction. The high negative rating indicates the severity of the negative impact placed on the placement of the DG in the appropriate areas.

32.7.1.2 Sequential Bus (CB)

If a DG is installed on a bus near a small sub-station the power will be distributed to zero power at the station and the full power supply will be sent by the DG alone. Therefore a high dose of DG is required to provide power when installed on the Sequential Bus even though there is a very small amount of potential reduction losses from this area.

Main_MVA release indicator (M_MVARI) and usable decommissioning indicators are used to ensure the correct representation of the consecutive bus to enter the DG. Unity M_MVARI shows zero power distributed from the scale and low or bad values indices for effective performance reduction and support their inadequacy for DG inclusion.

32.7.2 Results and Analysis

After the implementation of the Load Analysis method and the MIPE method, the appropriate access point for DG installation and inapt locations can be investigated. Demonstrating this quality is used in the 33 bus systems and 69 bus systems used in this study and the summary results are given in Tables 32.6 and 32.7.

Table 32.6 and Fig. 32.13 show that in 33 bus system, bus 2 is located next to main sub-station and unit value of M_MVA_{RI} attained after DG installed at this bus indicates that zero power will be shared from the substation and the total power will be delivered by the DG alone. Apart from this, zero value of MIPEF indicates lowest benefits of DG allocation on the system. Thus, bus 2 can be considered as an Adverse Impact location as well as Sequential Bus. Due to the zero value of MIPEF evaluated at bus 19, it is considered as AIL.

The results presented in Tables 32.7 and Fig. 32.12 show that in the 69 bus system, bus 2 can be referred to as a consecutive bus. Apart from this many buses (bus numbers 3, 4, 5, 28, 29, 30, 31, 32, 33, 34, 35, 45, 46, 47, and 69) show a negative impact on the DG allocation in the system and as a result it has been said AIL.

Figures 32.13 and 32.14 show the power supply data for each source that creates a load on the system from the corresponding area showing the bus near the main station, i.e., the second bus can be called the Sequential Bus. It can be seen that the installation of RER on this bus will lead to zero sharing from the station because the full power is provided by RER alone. High Capacity RER is therefore required

Table 32.6 Investigation of inapt locations in 33 bus system

Location	Active power loss, MW	Reactive power loss, MVAr	M_MVARI	MIPEF	AIL	CB
w/o DG	222.5	146	–	–	–	–
2	192.5	142.3	1	0.06	√	√
3	141.4	104.6	0.8	0.26	–	–
4	123.2	95.3	0.63	0.30	–	–
5	105.6	86.4	07	0.36	–	–
6	68.2	53.8	0.67	0.52	–	–
7	71.4	57.9	0.57	0.49	–	–
8	78.8	57.7	0.50	0.48	–	–
9	89.9	63.5	0.45	0.47	–	–
10	96.2	66.6	0.40	0.44	–	–
11	97.2	66.8	0.39	0.44	–	–
12	99.3	67.3	0.39	0.43	–	–
13	106.7	72.5	0.33	0.42	–	–
14	109	75.1	0.29	0.43	–	–
15	113.3	78.2	0.29	0.43	–	–
16	118	80.9	0.28	0.42	–	–
17	126.3	89.3	0.24	0.43	–	–
18	130.3	91.4	0.22	0.46	–	–
19	199.5	134.9	0.26	0.04	√	–
20	202.1	135.3	0.11	0.21	–	–
21	202.2	135.3	0.11	0.22	–	–
22	202.6	135.6	0.08	0.28	–	–
23	154.2	112.1	0.53	0.24	–	–
24	160.3	113.9	0.36	0.27	–	–
25	167.4	117.3	0.25	0.32	–	–
26	69.3	54.2	0.58	0.50	–	–
27	70.4	54.7	0.64	0.51	–	–
28	70.8	53.8	0.52	0.51	–	–
29	68.6	50.5	0.47	0.55	–	–
30	68.8	53.3	0.36	0.55	–	–
31	79.5	58.9	0.38	0.53	–	–
32	83.5	62.4	0.35	0.53	–	–
33	90.1	70.1	0.36	0.51	–	–

Table 32.7 Investigation of inapt locations in 69 bus system

Locations	Active power loss, MW	Reactive power loss, MVAr	M_MVA_{RI}	MIPEF	AIL	CB
Without DG	211.72	96.11	–	–	–	–
2	232.7	101.72	1.00	−0.18	√	√
3	225.63	101.76	0.43	−0.27	√	–
4	225.44	101.3	0.85	−0.20	√	–
5	223.5	99.05	0.56	−0.23	√	–
6	188.93	81.87	0.72	0.10	–	–
7	155.7	65.5	0.72	0.28	–	–
8	148.22	61.77	0.72	0.32	–	–
9	144.54	59.96	0.72	0.34	–	–
10	155.16	64.93	0.49	0.33	–	–
11	155.9	65.53	0.46	0.34	–	–
12	160.03	68.17	0.38	0.34	–	–
13	166.24	71.77	0.31	0.32	–	–
14	169.3	73.81	0.26	0.32	–	–
15	170.92	75.09	0.23	0.32	–	–
16	171.11	75.26	0.23	0.33	–	–
17	171.65	75.66	0.22	0.33	–	–
18	171.66	75.66	0.22	0.33	–	–
19	172.6	76.18	0.21	0.32	–	–
20	173.16	76.49	0.21	0.32	–	–
21	173.98	76.96	0.20	0.32	–	–
22	174.03	76.98	0.20	0.32	–	–
23	174.75	77.33	0.20	0.31	–	–
24	176.21	78.04	0.19	0.31	–	–
25	179.22	79.49	0.17	0.29	–	–
26	180.31	80.02	0.17	0.29	–	–
27	180.92	80.31	0.17	0.29	–	–
28	216.63	97.96	−0.07	−0.03	√	–
29	214.86	97.21	0.02	−0.03	√	–
30	214.79	97.19	0.02	−0.03	√	–
31	214.78	97.18	0.01	−0.03	√	–
32	214.76	97.17	0.01	−0.02	√	–
33	214.72	97.16	0.01	−0.02	√	–
34	214.7	97.15	0.01	−0.01	√	–
35	214.7	97.16	0.01	−0.01	√	–

(continued)

Table 32.7 (continued)

Locations	Active power loss, MW	Reactive power loss, MVAr	M_MVA_{RI}	MIPEF	AIL	CB
36	214.68	97.08	0.19	0.08	–	–
37	214.84	97.17	0.06	0.14	√	–
38	214.77	97.13	0.04	0.18	√	–
39	214.75	97.12	0.04	0.23	√	–
40	214.75	97.12	0.04	−0.04	√	–
41	214.66	97.06	0.03	0.36	–	–
42	214.63	97.04	0.02	0.36	–	–
43	214.63	97.03	0.02	0.39	–	–
44	214.63	97.03	0.02	0.39	–	–
45	214.62	97.03	0.02	−0.02	√	–
46	214.62	97.02	0.02	−0.02	√	–
47	218.99	98.72	0.45	−0.04	√	–
48	214.36	95.77	0.23	0.07	–	–
49	212.64	91.99	0.20	0.13	–	–
50	212.62	91.98	0.18	0.15	–	–
51	154.34	64.91	0.66	0.30	–	–
52	169.56	70.81	0.50	0.24	–	–
53	139.46	57.53	0.68	0.37	–	–
54	133.29	54.55	0.65	0.41	–	–
55	124.76	50.44	0.62	0.45	–	–
56	116.53	46.46	0.59	0.48	–	–
57	71.71	31.53	0.53	0.66	–	–
58	51.02	24.63	0.52	0.73	–	–
59	43.25	22.07	0.52	0.76	–	–
60	34.89	19.57	0.51	0.79	–	–
61	24.17	14.17	0.50	0.86	–	–
62	26.01	15.16	0.50	0.81	–	–
63	28.81	16.66	0.49	0.79	–	–
64	40.99	22.99	0.44	0.74	–	–
65	61.25	33.05	0.38	0.68	–	–
66	161.08	67.74	0.42	0.32	–	–
67	161.19	67.78	0.42	0.32	–	–
68	170.77	73.32	0.30	0.29	–	–
69	170.82	73.35	0.30	−0.18	√	–

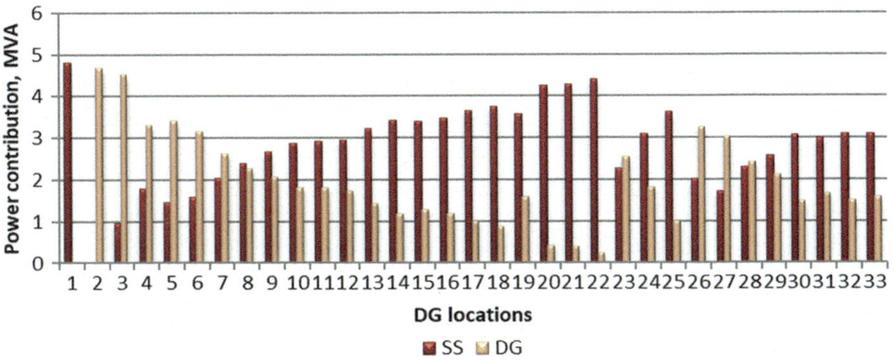

Fig. 32.13 Power contribution from grid and the DG in 33 bus system

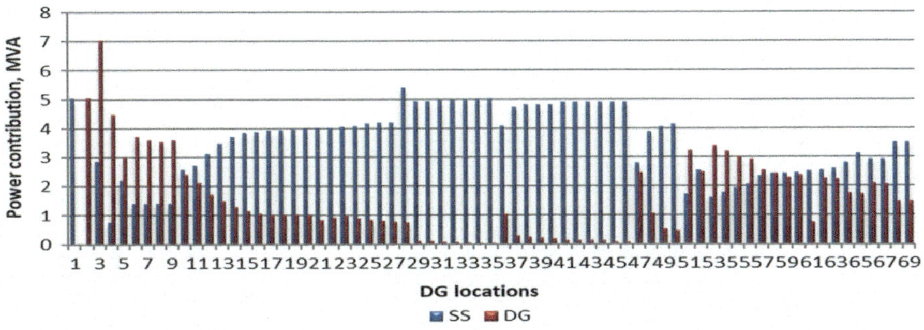

Fig. 32.14 Power contribution from grid and the DG in 69-bus system

to supply power when placed on this bus. Except for that very small amount of loss reduction is found in this area. Therefore, it can be suggested that Sequential Bus is the preferred location for RER installation. In the 33 bus system with a total load of 4.36 MVA, bus 2 is CB where a RER of 4.65 MVA is required which will result in a 7.07% reduction in losses. While in the 69 bus system with a load of 4.26 MVA, bus 2 is CB where the DG of 4.23 MVA is required which will result in a 2.15% reduction in loss this leads to the conclusion that Sequential Bus can be considered the worst RER installation site.

32.7.3 Investigation of Inapt Locations by Using Performance Evaluation Indices

Various performance evaluation indices explored in this chapter can be used for the investigation of Inapt Locations for DG insertion in distribution system. The negative or lower values of indices indicate the unsuitability of the location for DG installation.

33 bus system: Figure 32.15 shows both power loss reduction indices evaluated in case of 33-bus system. It shows that DG addition at bus 2, 19, 20, 23, and 24 can attain loss reduction to a very less extent so indicates their unsuitability for DG insertion in terms of loss reduction. Figure 32.16 shows system and critical bus voltage correction indices evaluated after DG addition at various locations. At bus 2 and bus 18 the negative values of Voltage Contour improvement Index (V_{PCI}), Critical Bus Voltage Improvement Index (V_{CRII}) depict adverse influence of DG addition on the system so these buses might be considered as AILs for voltage profile improvement.

69 bus system: Figure 32.17 indicates both power loss reduction indices assessed after DG placement many locations in 69-bus system. Negative values at 2-5th and 28-48th locations indicate their unsuitability for DG insertion in terms of loss reduction. Similarly DG insertion at any of the buses such as 49, 50, or 51 is de-motivated due to the lower values of power loss reduction indices. Figure 32.18 shows V_{PCI} and V_{CRII} evaluated after DG insertion at various locations. The negative values of indices evaluated at buses 2–6 and 28–50 indicate their unsuitability for DG insertion in terms of voltage contour upgrading of system and critical bus.

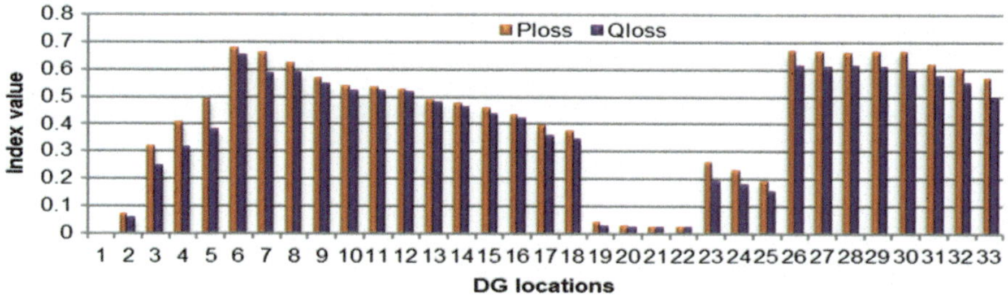

Fig. 32.15 Active-Reactive power loss reduction indices estimated for 33-bus system

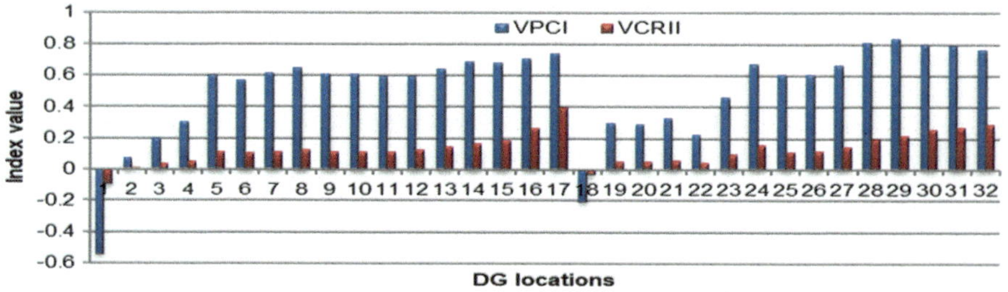

Fig. 32.16 System and critical bus voltage correction indices estimated for 33-bus system

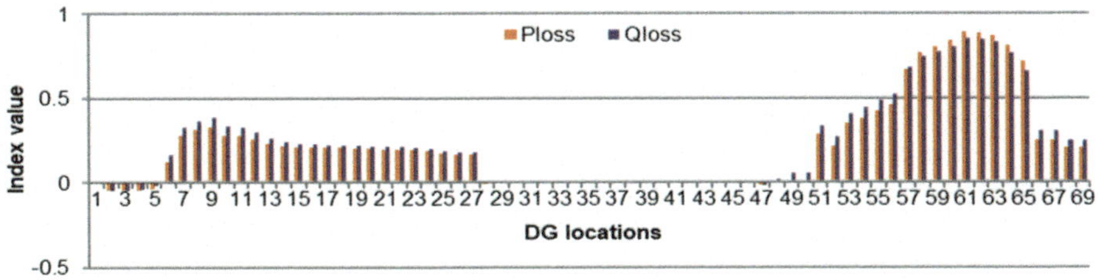

Fig. 32.17 Active-reactive power loss reduction indices assessed for 69-bus system

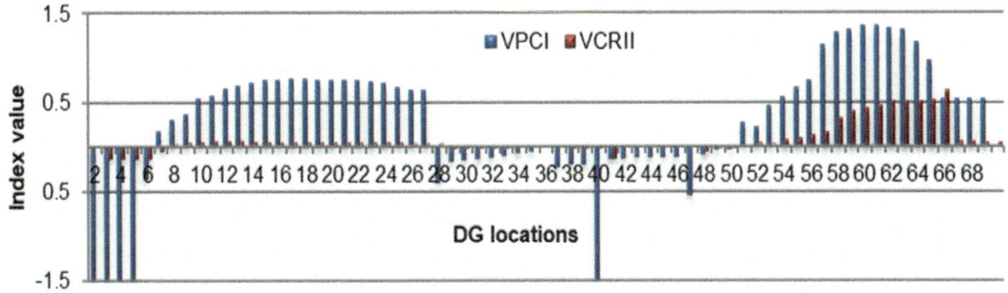

Fig. 32.18 System and critical bus voltage correction indices evaluated for 69-bus system

32.8 Conclusion

This chapter outlines two auxiliary services for the proposed method, enhanced uploads obtained after DG installation and Inapt Location investigations. It also shows the impact of loading model consideration on download load expectations. It is evident that the maximum expansion of the system loading is possible with the installation of the DG in the appropriate location and the design recommended for the proposed method. However, accurate predictions of load expansion are mandatory.

References

Acharya N, Mahat P, Mithulananthan N (2006) An analytical approach for DG allocation in primary distribution network. Int J Electr Power Energy Syst 28(10):669–678. https://doi.org/10.1016/j.ijepes.2006.02.013

Akorede MF, Hizam H, Aris I, Ab Kadir MZA (2011) Effective method for optimal allocation of distributed generation units in meshed electric power systems. IET Gener Transm Distrib 5(2):276–287. https://doi.org/10.1049/iet-gtd.2010.0199

Alotaibi MA, Salama MMA (2018) An Incentive-Based Multistage Expansion Planning Model for Smart Distribution Systems. IEEE Trans Power Syst 33(5):5469–5485. https://doi.org/10.1109/TPWRS.2018.2805322

Baran ME, Wu FF (1989) Optimal sizing of capacitors placed on a radial distribution system. IEEE Trans Power Delivery 4(1):735–743. https://doi.org/10.1109/61.19266

Bawazir RO, Cetin NS (2020) Comprehensive overview of optimizing PV-DG allocation in power system and solar energy resource potential assessments. Energy Rep 6:173–208, ISSN 2352-4847. https://doi.org/10.1016/j.egyr.2019.12.010

Choi DG, Min D, Ryu J-H (2018) Economic value assessment and optimal sizing of an energy storage system in a grid-connected wind farm. Energies 11:591–600. https://doi.org/10.3390/en11030591

CIREDWG04 (1999) Dispersed generation—preliminary report. CIRED-99. Nizza (Fr), June 1999

Correia ET, Calili R, Louzada D (2020) Regulatory model for micro and mini-generation distributed in Brazil: conditions for updating the current rules.In: 2020 47th IEEE Photovoltaic Specialists Conference (PVSC) pp 1952–1956. https://doi.org/10.1109/PVSC45281.2020.9300473

Chauhan J, Surjan BS (2020) Impact of distributed generation in single area load frequency control on system frequency.In: 2020 IEEE International Students' Conference on Electrical, Electronics and Computer Science (SCEECS) pp 1–5. https://doi.org/10.1109/SCEECS48394.2020.77

Chiradeja P, Ramakumar R (2004) An approach to quantify the technical benefits of distributed generation. IEEE Trans Energy Convers 19(4):764–773. https://doi.org/10.1109/TEC.2004.827704

Fahad S, Abu-Mouti, El-Hawary ME (2011) Optimal distributed generation allocation and sizing in distribution systems via artificial bee colony algorithm. IEEE Trans Power Deliv 26(4):2090–2101. https://doi.org/10.1109/TPWRD.2011.2158246

Gadomkar M, Vakillan M, Ehsan M (2005) Optimal distributed generation allocation in distributed network using Hereford Ranch algorithm. In: Proceedings of the 8th international conference on electrical machines and systems—ICEMS 2005, pp 916–918. https://doi.org/10.1109/ICEMS.2005.202678

Ghasemi M, Akbari E, Rahimnejad A et al (2019) Phasor particle swarm optimization: a simple and efficient variant of PSO. Soft Comput 23:9701–9718. https://doi.org/10.1007/s00500-018-3536-8

Graditi G, Di Somma M (2021) Distributed energy resources in local integrated energy systems, 1st edn. Elsevier, 27th February 2021

Hadisaid N, Canard JF, Dumas F (1999) Dispersed generation impact on distribution networks. IEEE Comput Appl Power 12(2):22–28. https://doi.org/10.1109/67.755642

Handbook of Distributed Generation (2017) Springer Science and Business Media LLC

Kalambe S,Agnihotri G (2013) Extraction of transmission parameters for siting and sizing of distributed energy sources in distribution network. J Energy 9:2013. https://doi.org/10.1155/2013/938958

Kashem MA, Ganapathy V, Jasmon GB, Buhari MI (2000) A novel method for loss minimization in distribution networks. In: Proceedings of the international conference on electric utility deregulation and restructuring and power technologies, DRPT 2000. IEEE, pp 251–256. https://doi.org/10.1109/DRPT.2000.855672

Kifle Y, Khan B, Singh P (2018) Assessment and enhancement of distribution system reliability by renewable energy sources and energy storage. J Green Eng 8(3):219262. https://doi.org/10.13052/jge1904-4720.832

Kumar A, Gao W (2010) Optimal distributed generation location using mixed integer non-linear programming in hybrid electricity markets. IET Gener, Transm & Distrib 4(2):281–298. https://doi.org/10.1049/iet-gtd.2009.0026

Mahmoud K, Abdel-Akher M, Ahmed AA (2010) Sizing and locating distributed generations for losses minimization and voltage stability improvement.In: IEEE Internal Conference on Power and Energy (PECon2010) pp 600–604 https://doi.org/10.1109/PECON.2010.5697652

Poornazaryan B, Karimyan P, Gharehpetian GB, Abedi M (2016) Optimal allocation and sizing of DG units considering voltage stability, losses and load variations. Int J Electr Power Energy Syst 79:42–52, ISSN 0142–0615. https://doi.org/10.1016/j.ijepes.2015.12.034.

Pinnarelli A, Menniti D, Sorrentino N, Bayod-Rújula AA, (2021) Optimal management of energy storage systems integrated in nanogrids for virtual "nonsumer" community. In: Distributed energy resources in local integrated energy systems. Elsevier, pp 231–278, ISBN

9780128238998. https://doi.org/10.1016/B978-0-12-823899-8.00004-2.;Atwa YM, El-Saadany EF, Salama MMA, Seethapathy R (2010) Optimal renewable resources mix for distribution system energy loss minimization. IEEE Trans Power Syst 25:360–370 https://doi.org/10.1109/TPWRS.2009.2030276

Quezada VHM, Rivier Abbad J, Gomez San Roman T (2006) Assessment of energy distribution losses for increasing penetration of distributed generation. IEEE Trans Power Syst 21(2):533–540. https://doi.org/10.1109/TPWRS.2006.873115

Rau NS, Wan YH (1994) Optimum location of resources in distributed planning. IEEE Trans Power Syst 9:2014–2020. https://doi.org/10.1109/59.331463

Rezk H, Babu TS, Al-Dhaifallah M, Ziedan HA (2021) A robust parameter estimation approach based on stochastic fractal search optimization algorithm applied to solar PV parameters. Energy Rep 7:620–640. https://doi.org/10.1016/j.egyr.2021.01.024

Roche N, Courtney J (2020) Optimal sizing of energy storage with embedded wind power generation. In: 2020 55th International Universities Power Engineering Conference (UPEC) pp 1–6. https://doi.org/10.1109/UPEC49904.2020.9209807

Singh B, Mukherjee V, Tiwari P (2015) A survey on impact assessment of DG and FACTS controllers in power systems. Renew Sustain Energy Rev 42:846-882. https://doi.org/10.1016/j.rser.2014.10.057

Thukaram D, Vyjayanthi C, Surendra S (2009) Optimal placement of distributed generation for a projected load increase using relative electrical distance approach. In: Third international conference on power systems, Kharagpur, India, 27–29 December 2009. https://doi.org/10.1109/ICPWS.2009.5442774

Ullah Z, Wang S, Radosavljević J (2019) A novel method based on PPSO for optimal placement and sizing of distributed generation. IEEE Transm Electr Electron Eng. https://doi.org/10.1002/tee.23001

Vatani M, Alkaran DS, Sanjari MJ, Gharehpetian GB (2016) Multiple distributed generation units allocation in distribution network for loss reduction based on a combination of analytical and genetic algorithm methods. Gener Trans Distrib IET 10:66–72. https://doi.org/10.1049/iet-gtd.2015.0041

Wang C, Nehrir MH (2004) Analytical approaches for optimal placement of distributed generation sources in power systems. IEEE Trans Power Syst 19(4):2068–2076. https://doi.org/10.1109/PES.2005.1489285

Chapter 33
Analysis of Mesh Distribution Systems with Multiple Wind Turbine and Multiple D-STATCOM Allocation Using Artificial Bee Colony Algorithm

Nandola Maitrey Bharatbhai and Atma Ram Gupta

Abstract The electrical power is transmitted through a high-voltage transmission system to low-voltage consumers on the distribution side. Because of the high R/X ratio, high current, and low voltage in distribution systems, active power loss (TPL) is much higher as compared to transmission systems. The economic incentive for distribution power utilities is to reduce losses in their networks. This inducement is generally the cost differential between actual and standard losses. As a result, when actual losses exceed standard losses, the utilities are economically penalized, while when the reverse happens, they are getting benefitted. As a result, the loss minimization problem in distribution systems is an excellent research subject for academics. All the methods for the loss minimization are different from each other by the means of formulation of the problem, tools used to minimize the losses, constraints taken, solution methods, and results are obtained. In most of the cases, the distribution system is interconnected and due to increment in load demand and limitations of loading capacity of distribution lines, the idea of distribution side generation came to maintain the system parameters like voltage profile, power loss, cost of operation, etc. To minimize losses, several methods are available like distributed generation (DG) using renewable energy sources and combined heat and power systems, capacitor allocation, reconfiguration of the network, and D-FACTS devices like dynamic voltage restorer (DVR), distribution static compensator (D-STATCOM), unified power quality conditioner (UPQC) allocation, etc. In this chapter, 33 and 69 bus mesh distribution system analysis is proposed with implementation of multiple wind turbines (WT) and multiple D-STATCOM. Advantages of WT are cost-effective generation, clean fuel source, sustainability, and less land requirement. Here, the Artificial Bee Colony (ABC) algorithm is used for finding optimal size and location of WG and D-STATCOM. ABC is an optimization technique that offers population-based search procedure in which artificial bees change

N. M. Bharatbhai (✉) · A. R. Gupta
Electrical Department, NIT Kurukshetra, Kurukshetra, India
e-mail: nandola_31904219@nitkkr.ac.in

A. R. Gupta
e-mail: argupta@nitkkr.ac.in

A. K. Bohre et al. (eds.), *Planning of Hybrid Renewable Energy Systems, Electric Vehicles and Microgrid*, Energy Systems in Electrical Engineering,
https://doi.org/10.1007/978-981-19-0979-5_33

individual food locations over time, with the aim of the bees being to locate the places of food supply with the most nectar and recognize it as the one with the greatest nectar. Here direct approach-based load flow method is used to find load flow parameters like bus voltage, branch currents, and power loss. The outcome verifies effective improvement in the system voltage profile, reduction in distribution feeder line loss, and increment in cost savings.

Keywords Distributed generation · Wind turbine · D-STATCOM · Artificial bee colony algorithm · Voltage profile improvement · Loss minimization · Mesh distribution · Load flow analysis

Nomenclature

DS	Distribution system
ES	Energy saving
N-R	NewtonRaphson
AECS	Annual energy cost saving
G-S	GaussSeidel
ACS	Annual cost saving
DG	Distributed generation
ABC	Artificial ee olony
WT	Wind turbine
LSF	Loss ensitivity actor
D-STATCOM	Distribution static compensator
TPL	Total active ower oss
DVR	Dynamic oltage estorer
UPQC	Unified ower uality onditioner

33.1 Introduction

Wind energy is the most attractive clean energy source for a number of reasons, including increased grid efficiency, energy flexibility, lower carbon dioxide emissions, etc. As a result, several countries are pursuing policies to increase wind energy use through incentives and financial options. If the number of Wind Turbines (WTs) connected to distribution networks increases, DNOs face a number of problems, including, voltage volatility, voltage stability, power shortages, and reliability (Mokryani et al. 2013; Banos et al. 2011). These all power quality issues are because of WT installation at the wrong place. Hence, before installing WT in the system, suitable location and suitable size should be investigated. D-STATCOM (D-FACTS) is also used in distribution networks to address the issues (like voltage instability, voltage sag, and voltage fluctuation) and compensate for reactive capacity (Apoorva et al. 2018). Low harmonic production, low power losses, high regulatory capability, low cost, and compact size differentiate D-STATCOM from other reactive power compensation devices.

The DNO's primary responsibility is to deliver loads at the proper voltage and load level; it must implement a fair operational plan that takes into account buying power from the wholesale market while ensuring device safety, dispatching WTs and D-STATCOM and interrupting loads. DNOs act as brokers, buying bulk electricity at volatile prices and reselling it to small consumers at fixed rates. Retailers and DNOs are two distinct market entities with distinct networks, goals, and sizes (Hasani-Marzooni and Hosseini 2012). DNOs can deal with power quality-related issues and its solutions. It uses optimization techniques to find a reliable and economic solution of the problems to maintain the distribution system healthy.

The installation of DGs in the distribution network is the cost-effective way for the grid to benefit from renewable energy. For most cases, careful analysis of DG and D-STATCOM deployment would result in a substantial improvement in the reduction of electrical power network faults, voltage control, and reliability (Hasani-Marzooni and Hosseini 2012; Ackermann and Knyazkin 2002). Excessive losses and feeder overloading result from non-optimal integration of DGs and D-STATCOMs deployed in non-optimal locations with non-optimal ratings (Mohandas et al. 2016; Mendez Quezada et al. 2006). Renewable energy sources' (RES) future availability is obvious, and they are successfully used to provide a stable electricity supply to islanded areas. A grid which uses RES to sustain the main grid and meet load demand is called a micro-grid (Oureilidis et al. 2016).

Many experiments have been performed in the past to assess the most suitable capacities and positions for D-STATCOMs and WTs. From the literature survey, there are many methods like multi-objective optimal power flow (OPF) (Harrison et al. 2007a), stochastic optimization algorithm (Chen et al. 2010a), genetic algorithm (GA) (Harrison et al. 2007b, 2008), particle swarm optimization (PSO) (El-Zonkoly 2011), a mechanism for optimum positioning of WTs to minimize energy loss (Atwa and El-Saadany 2011), hybrid optimization approach (Chen et al. 2010b), biologically based immune algorithm (Taher and Afsari 2014), modified particle swarm optimization algorithm (Devi and Geethanjali 2014), voltage stability index, and power loss index techniques (Gupta and Kumar 2015) used to find most suitable location and size of DG and D-STATCOM.

In this chapter, ABC algorithm is used to assess optimum position and rating of WT and D-STATCOM to get minimum TPL. The ABC algorithm is a meta-heuristics technique based on swarms. It focuses on swarm optimization of honey bees. It was introduced by Karaboga (2005). Here, WTs and D-STATCOMs are implemented in 33 and 69 mesh DS and bus data are taken from 2005. After implementation of WTs and D-STATCOM, system analysis is done based on voltage profile, minimum voltage of the system, TPL, annual energy loss cost savings, etc.

The chapter is organized as follows: Sect. 33.2 contains the problem formulation, Sect. 33.3 contains the ABC algorithm, Sect. 33.4 contains the results and discussions, and Sect. 33.5 contains the conclusions.

33.2 Problem Formulation

33.2.1 Load Flow Analysis

Several planning and operations issues, such as network optimization, loss reduction, load balancing, renewable integration, location and size of DG and D-STATCOM, and service restoration necessitate load flow analysis. To analyze electrical parameters such as node voltages, branch currents, system voltage profiles, etc., load flow analysis is a must. Without breaking the network constraints, the load flow analysis should be completed. Traditional algorithms such as N-R and G-S can be used to solve distribution systems, but these programmers are typically constructed with a high X/R ratio in mind. Since the X/R ratio in distribution systems is low, convergence time for standard algorithms is large. So, a program that needs to be designed specifically for distribution systems would be more effective and simpler than one designed for high-voltage systems. In general, the distribution system is unbalanced, with a radial or weakly meshed arrangement. As a result, load flow studies for distribution systems can be performed for both balanced and unbalanced systems, as well as radial and weakly meshed systems.

The load flow problem of radial and weakly meshed distribution systems was solved by authors in Prakash and Sydulu (2011) using a primitive impedance-based technique. In this form, a converged load flow solution is obtained by computing diagonal elements of the DLF matrix without using the bus admittance matrix. The authors of Ranjan et al. (2004) suggest a procedure in which the branch currents are determined using the receiving bus voltage and the power transfer is calculated using the backward load flow. In Siano et al. (2009), the author proposes a ladder network-based hypothesis for the load flow approach of delivery networks. S. Ghosh and D. Das use the same forward–backward approach in Ghosh and Das (1999), with the bus voltage mismatch serving as the convergence criterion. The authors suggest a new algorithm for power flow in radial networks in Babu et al. (2010), which uses Kirchhoff's laws for forward sweep iteration and a special numbering system. The authors of Zhu and Tomsovic (2002) have suggested an algorithm for solving the radial portion of networks using backward current and forward voltage. The author of Asfari et al. (2002) suggested a backward sweep algorithm for obtaining nodal voltages. The correct estimation of the terminal bus voltage is critical to the feasibility and convergence of this proposed process. In Thukaram et al. (1999) a modified prototype of the forward/backward sweep method has been proposed. In Cheng and Shirmohammadi (1995) the same method is used for both unbalanced radial and mesh distribution systems. In Liu et al. (2002) the author proposed an algorithm where the voltage is updated in the forward/backward process and this method can be applied for both balanced radial and weakly meshed distribution networks. Voltage-based static load models with a high convergence rate are used in Satyanarayana et al. (2007). Voltage-based load models are used for a backward ladder approach in Eminoglu and Hocaoglu (2005), and the effect of line shunt capacitance is also used, resulting in successful results.

In this chapter, a direct approach-based load flow analysis method is applied on meshed DS considering five loops following the same strategy proposed in Teng (2000), Kumar Sharma and Murty (2014), the voltage profile at each node and power loss at each branch are computed for base case.

33.2.2 Objective Function

OF is a mathematical equation which gives minimum or maximum value of the function with satisfying conditional constraints. In this chapter, WT and D-STATCOM are installed in the system with the best size and location which satisfy the objective function with some constraints. The objective function is minimal TPL, and the constraints are maximum and minimum ratings of WT and D-STATCOM. Equation (33.1) is OF and Eq. (33.2), (33.3) are constraints.

$$\mathrm{OF} = \min\left[\sum_{j=1}^{B} \mathrm{Ibr}^2(j) * r(j)\right] \tag{33.1}$$

$$\min(\mathrm{Pdgi}) < \mathrm{Pdgi} < \max(\mathrm{Pdgi}) \tag{33.2}$$

$$\min(Q_{D-STATCOMi}) < Q_{D-STATCOMi} < \max(Q_{D-STATCOMi}) \tag{33.3}$$

where Ibr is branch current, B is number of branches, r is branch resistance, $\min(P_{wti})$ and $\max(P_{wti})$ are minimum and maximum size of WT at bus i. Here, $\max(Q_{D\text{-}STATCOMi})$ and $\min(Q_{D\text{-}STATCOMi})$ are maximum and minimum size of D-STATCOM at bus I.

33.2.3 Wind Turbine Modeling

When WTs are mounted, they normally provide continuous power with less investment than other sources in the distribution system (Yuvaraj and Ravi 2018). The variation in wind speed and air density causes the uncertainty in wind turbine performance at any location. In the equation of radial power transfer, there is a random variable because it varies randomly. WTs are environmentally sustainable, renewable energy source. They provide benefits such as improved power quality, cost-effective operation, reduced land requirements, and more efficient energy solutions than other generation methods. For the analysis of implementation of wind turbine in the system, 1-year data of wind speed is taken from NREL (www.nrel.gov). For calculating wind turbines output, per hour average speed is taken and per hour output power is calculated using Eq. (33.4).

The wind turbine power output (Pwt) is given by Faruk and Engin (2013)

$$Pwt = \begin{cases} 0 & 0 < K < Kci \\ \frac{\text{Kci}-\text{K}}{\text{Kci}-\text{Kr}} * \text{Prated Kci} < \text{K} < \text{Kr} \\ Prated\ \text{Kr} < \text{K} < \text{Kco} \\ 0 & \text{Kco} < \text{K} \end{cases} \quad (33.4)$$

where

- Pwt: power output of wind turbine in KW,
- Prated: rated power output in KW,
- K: wind velocity in m/s,
- Kci: cut-in speed,
- Kco: cut-out speed, and
- Kr: rated speed.

Here, power generated by WT is supplied to the distribution system so wind generation is called distributed generation. There are many types of DGs based on their characteristics of supplying active and reactive power. In this chapter, let's assume that WT supplies only active power with unity power factor. Outputs of WT for each hour are given in Table 33.1. Parameters specified to wind turbines (wind speed (K) and WT output (Pwt)) are given in Table 33.2. Here, the best rating of WT as DG is found using the ABC algorithm. The best size of DG is the one with the lowest TPL, measured in kW or MW.

33.2.4 D-STATCOM

A STATic COMpensator (STATCOM) (which is power electronics-based voltage-source converter) can be operated as either source or sink of a reactive power. In some operations, it can also supply active power if connected with energy storage. This device was initially developed for transmission systems for operations like faster and continuous inductive or capacitive compensation. Similarly, for the distribution system, a tool D-STATCOM is employed. D-STATCOM can inject leading or lagging current as per the requirement for compensation. Here the benefit is the total demand of specified load meets utility connection specification because D-STATCOM is related to a specified load (Chen and Hsu 2008; Devabalaji and Ravi 2014; Failed 2020).

Power quality issues like voltage instability and voltage sag result in less power flow limits, increased power depletion, and longer response times. Shunt capacitors have traditionally been used in distribution systems for reactive power compensation; however, shunt capacitors have the disadvantage of being unable to produce continuous variable reactive power. As a result, utilities must incur additional costs for capacitor installation. It has a few operational issues related to resonance. Furthermore, load balancing is not feasible (Gupta and Kumar 2015).

Table 33.1 Average wind speed per hour and wind turbine output per hour

Time in h	K in m/s	Pwt in KW
1	4.4603	153.7190
2	4.3541	142.5366
3	4.2648	133.1339
4	4.1484	120.8812
5	4.0796	113.6407
6	4.0392	109.3884
7	3.9645	101.5247
8	3.9294	97.8342
9	3.9004	94.7782
10	3.9459	99.5720
11	3.9707	102.1834
12	3.9608	101.1393
13	3.9907	104.2851
14	4.0670	112.3201
15	4.1324	119.2042
16	4.1956	125.8563
17	4.2661	133.2703
18	4.2725	133.9507
19	4.2985	136.6831
20	4.4347	151.0192
21	4.5450	162.6310
22	4.5946	167.8500
23	4.4888	156.7140
24	4.4159	149.0424

Table 33.2 Specifications for WT (www.suzlon.com)

Parameter	Value
Prated (rated power in KW)	1000
Kci (cut-in speed in m/s)	2.5
Kr (rated speed in m/s)	12.5
Kco (cut-out speed in m/s)	20

D-STATCOM is used to compensate for reactive power in the distribution network. It has advantages like lower harmonic production, high speed of operation, less space requirement, low initial cost, etc. D-STATCOM can supply and consume reactive power and it has no operational issues like transient and resonance harmonics (Samal 2020; Yuvaraj et al. 2017). It belongs to the D-FACTS device family.

D-STATCOM has the ability to compensate for capacitive and inductive modes quickly and continuously. As D-STATCOM is connected to a specific load, it can

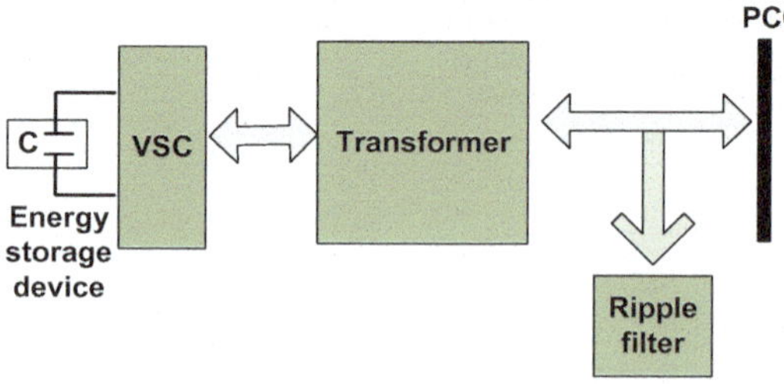

Fig. 33.1 Block diagram of D-STATCOM (Chen and Hsu 2008).VSC is voltage stability control, PCC is point of common control

inject enough leading or lagging compensating current to ensure that the overall demand satisfies the utility connection criteria. Main components of D-STATCOM are DC capacitor, PWM control, inverter modules, AC filter, and transformer (Bapaiah 2013). Figure 33.1 shows a block diagram of D-STATCOM.

In this chapter, D-STATCOM is implemented in mesh distribution systems for reactive power support. Best location and KVAr rating of D-STATCOM are found using ABC algorithm. Here minimum KVAr rating is 50 and maximum KVAr rating is 2000 taken.

33.2.5 Cost Analysis

After installing WT and D-STATCOM in the system, power loss decreases so energy savings become more and indirectly cost savings also. Power distribution utilities and end consumers always want the lower electricity charges along with better power quality. In this chapter, cost analyses like annual energy loss cost savings and annual cost savings have been carried out. For finding investment cost (IC) of DG, consider the DG cost characteristics equation (33.5), and for investment cost of D-STATCOM, consider the cost characteristics equation (33.6).

$$\text{Cost of DG} = (\text{aPdg}^2 + \text{bPdg} + \text{c})\$/\text{MWh} \tag{33.5}$$

$$\text{C(DS)/year} = \text{C(DS)}\frac{(1+\text{B})^{\text{n}} * \text{B}}{(1+\text{B})^{\text{n}} - 1} \tag{33.6}$$

where Pdg is active power supplied by DG in MW. C(DS)/year is annual cost of D-STATCOM in \$ and C(DS) is the investment cost of D-STATCOM in \$/kVAr at the year of installation. B and n are the asset rate of return and longevity of D-STATCOM, respectively. Here for the cost analysis, values are taken as a = 0, b = 20, c = 0.25, C(D-STATCOM) = 50 \$/kVAr, B = 0.1, n = 30.

The price of wind turbines varies depending on the manufacturer. Other costs, with the exception of wind turbine costs, are wind speed, taxation, installation, and maintenance. Wind turbines cost 1150 \$/kW for rated capacity of more than 200 kW

(Adaramola et al. 2011). The energy cost produced by wind turbines is calculated in this analysis using Eq. (33.7) (Mathew 2006).

$$\mathrm{WTc} = \frac{\mathrm{ICt}}{8760 * \mathrm{n}}\left(\frac{1}{\mathrm{PrCf}}\right)[1 + \mathrm{w}\left(\frac{(1+\mathrm{u})^{\mathrm{n}} - 1}{\mathrm{u}(1+\mathrm{u})^{\mathrm{n}}}\right)] \tag{33.7}$$

where ICt is total investment cost, n is lifetime of turbine in year, Cf is capacity factor, Pr is rated power, w is annual maintenance cost, and u is discount rate. Here for the cost analysis, ICt is 40% of turbine cost, n = 20 years, w and u is 4% of turbine cost are taken (Dincan et al. 2020).

Now, for finding Annual Energy loss Cost Savings (AECS), consider Eq. (33.8). Energy saving (ES) is difference between energy loss before and after installation of DG and D-STATCOM (Ashwani Kumar et al. 2017). Annual Cost Savings (ACS) is difference between AECS and IC. It can be calculated using (33.9).

$$\mathrm{AECS} = \mathrm{Ke}\left(\mathrm{TPL} - \mathrm{TPL}'\right)\mathrm{T}\$ \tag{33.8}$$

$$\mathrm{ACS} = \mathrm{AECS} - \mathrm{IC}\$ \tag{33.9}$$

where TPL/TPL' are TPL in kW before/after installation, T is time duration in hours, and Ke is the energy rate in per kWh. Here the values used Ke = 0.06 /KWh, T = 8760 (hours).

33.3 Artificial Bee Colony (ABC) Algorithm

In 2005, Dervis Karaboga proposed the ABC algorithm, inspired by the intelligent behavior of honey bees. The ABC algorithm and its applications to real-world problems have been studied by Karaboga and his research group. The ABC algorithm is a meta-heuristics technique based on swarms. Scots, onlookers, and employed bees are the three classes in the ABC (Karaboga 2005). The ABC developed an initial population (N) of food source solutions that were randomly distributed. Each Z(i) (i = 1, 2,…, N) solution is a d-dimensional vector where d is the size of constraints. The population of the positions is treated to repeated loops of employed bees, onlooker bees, and scout bees after initialization. The swarm size, limit, and number of iterations are the algorithm's control parameters.

A working bee adjusts the position in her mind depending on the nectar volume of the new source (new solution). If the amount of nectar in the current position is larger than the previous one, the bee remembers the new one and forgets the existing one. When all employed bees have concluded their search, they tell the onlooker bees about the dancing area, nectar details, and their location. An onlooker bee evaluates the nectar data collected from all employed bees and chooses a food supply based on a chance equal to nectar volume. Like the employed bee, it adjusts the position in its memory and checks the nectar amount of the applicant source. As long as the nectar is greater than the previous one, the bee remembers the current spot and forgets about the former one (Karaboga and Akay 2011).

The flow chart of ABC algorithm is given in Fig. 33.2. From that, the algorithm can be separated into four parts, initialization, employed phase, onlooker phase, and scout phase. The initial food source or populations can be found by Eq. (33.10) (Karaboga 2005), and Eq. (33.11) is used to calculate the location of the nearest food source (Karaboga 2005):

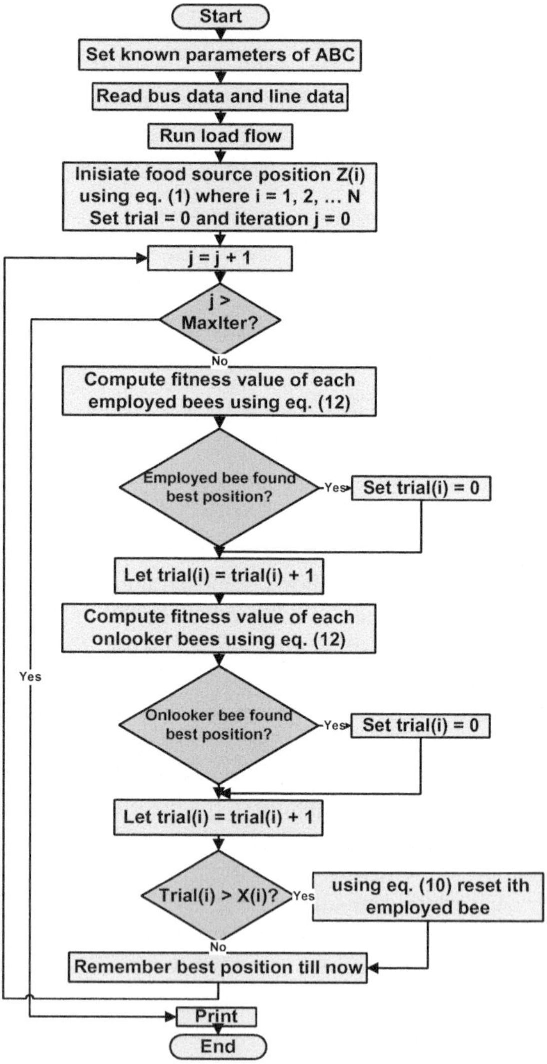

Fig. 33.2 Flow chart of ABC (Bapaiah 2013)

$$\mathrm{Zi} = \mathrm{Lb} + \mathrm{rand}(-1, 1)(\mathrm{Ub} - \mathrm{Lb}) \quad (33.10)$$

$$\mathrm{vi, j} = \mathrm{Zi, j} + \mathrm{U}(\mathrm{Zi, j} - \mathrm{Zk, j}) \quad (33.11)$$

where Ub and Lb are upper and lower bounds of the solution Z, j is the randomly selected parameters index, U is the random choice that exists among [-1,1], and Zk,j is a randomly selected food source. This formula is used to measure fitness, after which greedy selection is added between Zi and vi,j. Fitness value of the objective function is found using Eq. (33.12) (Karaboga 2005).

$$\mathrm{Fitnessm(Zm)} = \begin{cases} \frac{1}{1+fm(Zm)} iffm(Zm) > 0 \\ 1 + fm(Zm)iffm(Zm) < 0 \end{cases} \quad (33.12)$$

where fm(Zm) is the objective function value of Zm.

Probability value (Pi) of every food source can be calculated using Eq. (33.13). Using probability value, onlooker bee selects the food source.

$$\mathrm{Pi} = \frac{(\text{fitness value of the solution i})}{\sum_{i=0}^{N}(\text{fitness value of the solutioni})} \quad (33.13)$$

where N is the number of food source (employed bees).

The ABC employs Eq. (33.11) to create a candidate food location from the old in mind. The scouts use Eq. (33.10) to supplement the food source with a new food source in which the bees have rejected. If a location can't be changed any more after a certain number of cycles, which is called a limit, the food source is considered to be abandoned in ABC. Scout bees are the pioneers but haven't any direction when trying to find food, so it generates probability of sources which have lowest and highest nectar quantity. The parameter "limit" controls scout bee's selection. When the number of trails reaches a certain limit, the employed bee leaves the source and becomes a scout (Ayse Aybike Seker 2013).

33.4 Results and Discussions

In this section, 33 and 69 bus mesh distribution system analysis is done before and after implementation of WT and D-STATCOM. The line data and load data are taken from https://core.ac.uk/download/pdf/53189751.pdf. Total active power load on 33 buses is 3715 kW and reactive power load is 2300 KVAr. In 69 buses, the active power load is 3802 kW and reactive power load is 2694.6 KVAr. In this article, to find load flow parameters like TPL, bus voltage, and branch currents, direct approach-based load flow analysis method is used. For calculation in per unit, here base voltage is

12.66 kV and base MVA is 100 MVA taken. From the load flow calculation, active power loss of 33 bus mesh DS is 121.82 kW and for 69 bus mesh DS, it is 82.71 kW. Minimum system voltage of 33 bus is 0.9526 pu and for 69 bus is 0.9653 pu.

The objective of the work is to reduce TPL and improve the voltage profile of the both systems with improving minimum system voltage. Here, assumption is that WT supplies active power and D-STATCOM supplies reactive power, so after implementing these devices in the system at particular bus, changes in distribution system are shown here. For finding the best size and best location to implement these devices, ABC algorithm is used and the problem formulation and coding for the algorithm are done in MATLAB 2015. The solution of the ABC algorithm depends on the digestive tracts of actual honey bees. This algorithm has three control parameters, the population size (Np), limit value, and maximum cycle number MAXCYCLE. For the calculation, parameter values are Np = 20, limit = 3 and MAXCYCLE = 40. For the precise analysis, implementation results of WT and D-STATCOM are separated with different cases as described in Table 33.3.

Along with the ABC algorithm, loss sensitivity factor (LSF) at each bus for both the systems is also calculated, which shows the weakest bus of the system. If D-STATCOM is installed at the weakest bus of the system, it improves system performance effectively. The result of the ABC algorithm is verified by LSF for single D-STATCOM installation. LSF at each bus is plotted in Figs. 33.3 and 33.4 for 33

Table 33.3 Different cases

Analysis with	Case 1	Case 2	Case 3	Case 4	Case 5
WT	Without WT	Single WT	Two WT	Three WT	Four WT
D-STATCOM	Without D-STATCOM	Single D-STATCOM	Two D-STATCOM	Three D-STATCOM	Four D-STATCOM
WT + D-STATCOM	Without WT + D-STATCOM	Multiple WT	Multiple D-STATCOM	Multiple WT and multiple D-STATCOM	–

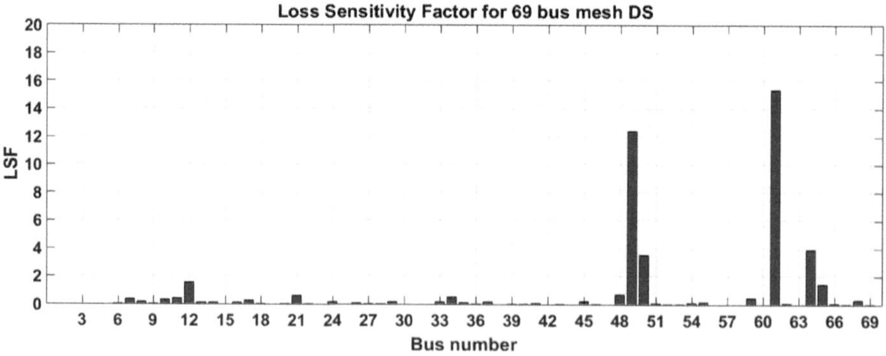

Fig. 33.3 LSF at each bus for 33 mesh DS

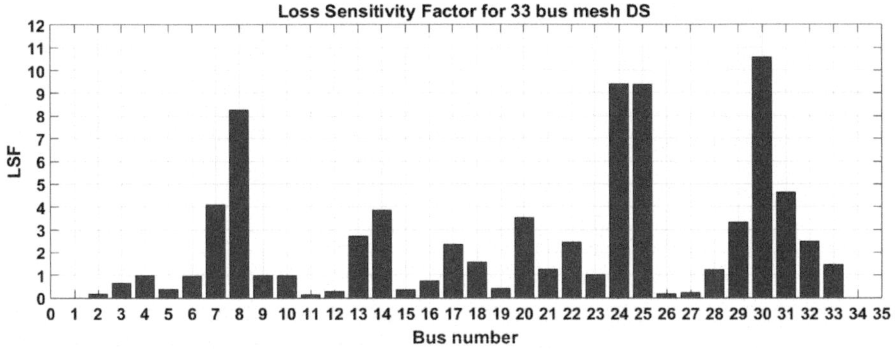

Fig. 33.4 LSF at each bus for IEEE 69 mesh DS

and 69 buses, respectively.

As shown in Fig. 33.3, LSF of 30th bus is highest so it is marked as the weakest bus of the system. In Fig. 33.4, LSF of 61st bus is highest so it is marked as the system's weakest bus. So to implement D-STATCOM in 33 buses, the location of the device is preferred at 30th bus and for 69 bus, preferred location is 61st bus. If there is need to install multiple D-STATCOM then compute LSF at each bus after implementing D-STATCOM at previously preferred locations, so we get a new preferred location to install one more D-STATCOM. This method to implement multiple D-STATCOM and its analysis are difficult and time-consuming. So for multiple device implementations, optimization methods are used. Although for single device installation like capacitor placement, D-STATCOM placement, etc.

Here, first analysis is with implementing WTs in both 33 and 69 bus systems which gives system parameters like TPL, voltage profile, minimum system voltage, energy loss, energy savings, annual energy loss cost, annual cost savings, etc. After that second analysis is with D-STATCOMs in both the systems. At last, analysis of the systems with both WT and D-STATCOM is done.

Tables 33.4 and 33.5 show distribution system results using ABC algorithm after installing WTs in the 33 and 69 bus systems, respectively. The results are separated with various cases including base case.

Here the results show size and locations of WTs for all cases. As we install more WTs, all the system parameters like TPL, % TPL, Vmin, annual energy loss, annual cost savings changes. Here from Table 33.4, TPL reduces 8.83% for case 2, 16.69% for case 3, 23.64% for case 4, and 29.71% for case 5. Vmin of the system improves from 0.9526 pu. to 9645 pu. Results for 69 bus mesh DS are show in Table 33.5. From Table 33.5, as we install WT, TPL reduces by 8.7% for case 2, 16.6% for case 3, 23.9% for case 4, and 29.2% for case 5. Vmin of the system improves from 0.9653 p.u. to 0.9727 p.u.

Figures 33.5 and 33.6 show voltage profile of 33 and 69 bus mesh DS, respectively, and it can be observed that after installation of WTs in the system, voltage profile tries to become flat. The reason for different voltage values at each bus is due to line

Table 33.4 Results for 33 mesh DS with multiple WT

Performance parameters	Base case	Single WT	Two WT	Three WT	Four WT
Location	–	32	32, 18	17, 32, 32	17, 32, 32, 32
Size in KW	–	167.85	167.85	167.85	167.85
TPL in KW	121.85	111.09	101.51	93.04	85.64
% reduction in TPL	–	8.83	16.69	23.64	29.71
Min. system voltage (Vmin)	0.9526	0.9554	0.9583	0.9614	0.9645
Annual energy loss (KWh)	1,067,423	973,148	889,227	815,030	750,197
ES in KWh	–	94,275.12	178,195.82	252,393.12	317,225.88
AECS in $	–	5656.51	10,691.75	15,143.59	19,033.56
ACS in $	–	2299.26	3991.25	5092.84	5632.56

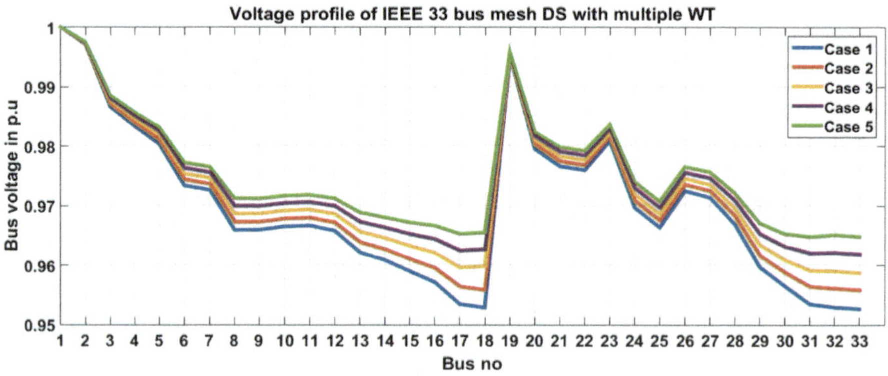

Fig. 33.5 Voltage at each node in 33 bus

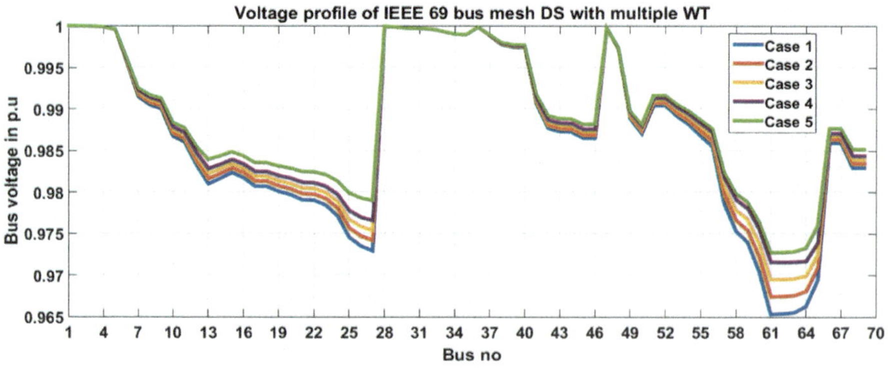

Fig. 33.6 Voltage at each node in 69 bus

losses. If load at any bus is high, then voltage at that bus is lower because of higher current flows which results in higher losses. In Figs. 33.7 and 33.8, per day average TPL for 33 and 69 bus mesh DS is shown and seen that hourly TPL is different for all cases, and it is due to different wind velocities. Here from all cases, it can be observed that more number of WT installation results reduced TPL.

Annual energy loss cost savings are also shown in Tables 33.4 and 33.5. Here AECS in 33 bus are \$ 5656.51 for case 2, \$ 10,691.75 for case 3, \$ 15,143.59 for case 4, and \$ 19,033.56 for case 5. From Table 33.5, AECS in 69 bus are \$ 3782.76 for case 2, \$ 7245.30 for case 3, \$ 10,391.46 for case 4, and \$ 13,679.00 for case 5.

Tables 33.6 and 33.7 show results for 33 and 69 bus mesh DS after installing multiple D-STATCOM, respectively. Best size and location of D-STATCOM give minimum TPL and improved minimum voltage of the system. As D-STATCOM installed in the system, we can see that % reduction of TPL increased and Vmin of the system improves. Also annual energy cost savings increase as power loss reduces but annual cost savings decrease as number of D-STATCOM increases in the system.

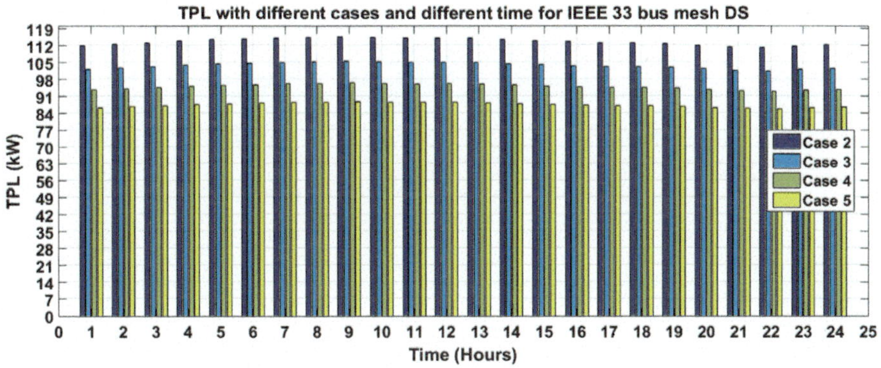

Fig. 33.7 Hourly TPL with different cases for WT in 33 bus mesh DS

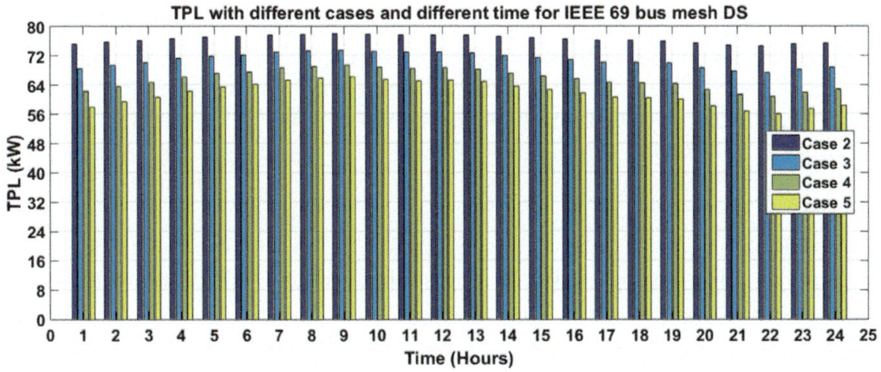

Fig. 33.8 Hourly TPL with different cases for WT in 69 bus mesh DS

Table 33.5 Results for 69 mesh DS with multiple WT

Performance parameters	Base case	Single WT	Two WT	Three WT	Four WT
Location	–	61	61, 61	61, 61, 61	61, 61, 61, 27
Size in KW	–	167.85	167.85	167.85	167.85
TPL in KW	82.7029	75.5059	68.9180	62.9322	56.6774
% reduction in TPL	–	8.7	16.6	23.9	29.2
Min. system voltage (Vmin)	0.9653	0.9674	0.9694	0.9715	0.9727
Annual energy loss (KWh)	724,477	661,432	889,227.6	5,512,861	496,494
ES in KWh	–	63,046	120,755	173,191	227,983.38
AECS in $	–	3782.76	7245.30	10,391.46	13,679.00
ACS in $	–	425.75	530.50	319.25	250.00

Table 33.6 Results for 33 mesh DS with multiple D-STATCOM

Performance parameters	Base case	Single D-STATCOM	Two D-STATCOM	Three D-STATCOM	Four D-STATCOM
Location	–	30	30, 8	30, 8, 25	30, 8, 25, 21
size in KVAr	–	1421	1421, 445	1421, 445, 228	1421, 445, 228, 115
TPL in KW	121.8520	89.5641	86.8952	86.1099	85.8904
% reduction in TPL	–	26.49	28.69	29.33	29.44
Min. bus voltage (Vmin)	0.9526	0.9660	0.9686	0.9694	0.9700
Annual energy loss (KWh)	1,067,423.5	784,581.5	761,201.9	754,322.7	752,399.9
ES in KWh	–	282,842.004	306,221.568	313,100.796	315,023.616
AECS in $	–	16,970.52	18,373.30	18,786.05	18,901.42
ACS in $	–	9433	8476	7680	7185

Here from Table 33.6, TPL reduces 26.49% for case 2, 28.69% for case 3, 29.33% for case 4, and 29.44% for case 5. Vmin of the system improves from 0.9526 pu. to 0.9700 pu. Results for 69 bus mesh DS are shown in Table 33.7. From Table 33.7, as we install D-STATCOM, TPL reduces by 29.68% for case 2, 32% for case 3, 32.88% for case 4, and 33.32% for case 5. Vmin of the system improves from 0.9653 p.u. to 0.9777 p.u.

Figures 33.9 and 33.10 show voltage profile of 33 and 69 bus mesh DS, respectively, and it can be observed that with the installation of D-STATCOMs in the system, voltage profile tries to become flat.

Annual energy loss cost savings are also shown in Tables 33.6 and 33.7. Here AECS in 33 bus are $ 16,970.52 for case 2, $ 18,373.30 for case 3, $ 18,786.05 for

Table 33.7 Results for 69 mesh DS with multiple D-STATCOM

Performance parameters	Base case	Single D-STATCOM	Two D-STATCOM	Three D-STATCOM	Four D-STATCOM
Location	–	61	61, 11	61, 11, 49	61, 11, 49, 21
Size in KVAr	–	1390	1390, 468	1390, 468, 550	1390, 468, 550, 188
TPL in KW	82.7029	58.1511	56.2373	55.5087	55.1416
% reduction in TPL	–	29.68	32	32.88	33.32
Min. bus voltage (Vmin)	0.9653	0.9747	0.9758	0.9770	0.9777
Annual energy loss (KWh)	724,477.4	509,403.6	492,638.7	486,256.2	483,040.4
ES in KWh	–	215,073.768	231,838.656	238,221.192	241,436.98
AECS in $	–	12,904.43	13,910.32	14,293.30	14,486.23
ACS in $	–	5532	4056	1521.3	717.1319

case 4, and $ 18,901.42 for case 5. From Table 33.7, AECS in 69 bus are $ 12,904.43 for case 2, $ 13,910.32 for case 3, $ 14,293.30 for case 4, and $ 14,486.23 for case 5.

From Figs. 33.7, 33.8, 33.9, 33.10, it can be seen that implementation of D-STATCOM affects voltage profile of the system more effectively as compared to WT. Among five cases of WT and D-STATCOM implementation, it can be observed that with D-STATCOM, voltage profile of the system improves more effectively. Also it can be seen that minimum system voltage for both WTs and D-STATCOMs gets improved as we install multiple devices in the system. Figures 33.11 and 33.12 show minimum system voltage of 33 and 69 bus, respectively, and it can be observed that minimum system voltage improves more effectively using D-STATCOM. Here

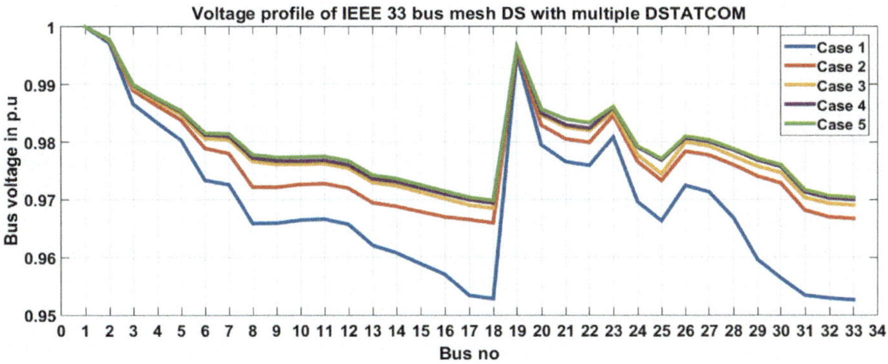

Fig. 33.9 Voltage at each node in 33 bus

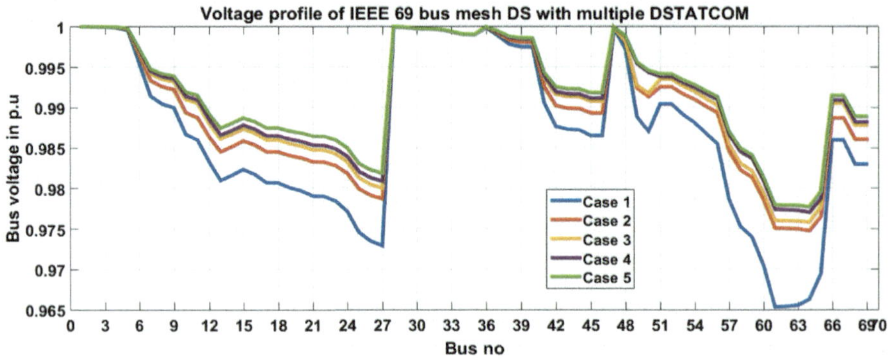

Fig. 33.10 Voltage at each node in 69 bus

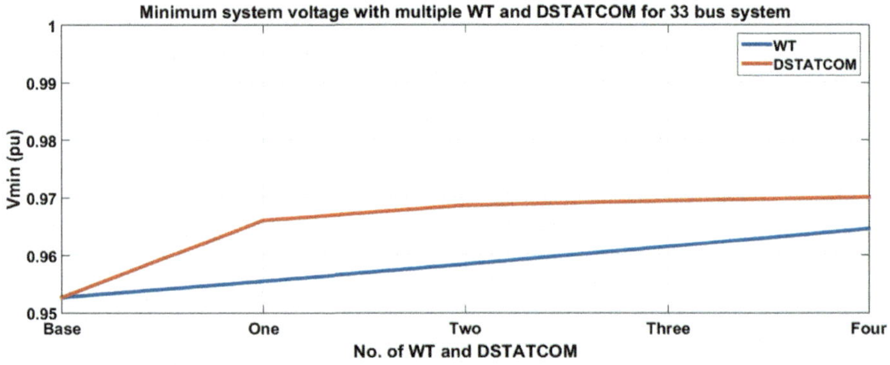

Fig. 33.11 Minimum system voltage with different cases for 33 bus

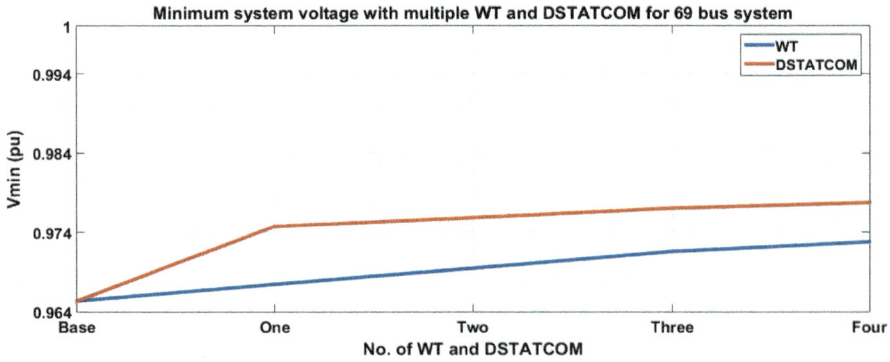

Fig. 33.12 Minimum system voltage with different cases for 69 bus

Table 33.8 Results for 33 mesh DS with multiple WT and multiple D-STATCOM

Performance parameters	Base case	Multiple WT	Multiple D-STATCOM	WT + D-STATCOM
Location	–	17, 32, 32, 32	30, 8, 25, 21	WT (33, 32, 8, 31), D-STATCOM (13, 24, 7, 30)
Size (KW, KVAr)	–	167.85	1421, 445, 228, 115	168, 285, 314, 542, 1014
TPL in KW	121.852	85.639	85.890	49.988
% reduction in TPL	–	29.71	29.44	58.97
Min. bus voltage (Vmin)	0.9526	0.9645	0.9700	0.9779
Annual energy loss (KWh)	1,067,423	750,197	752,399	437,901
ES in KWh	–	317,225.88	315,023.62	629,521.63
AECS in $	–	19,033.56	18,901.42	37,771.30
ACS in $	–	5632	7185	13,485

reason is the size of D-STATCOM. As output power from WT and D-STATCOM is higher, the results of our system improve.

Comparison of results for 33 and 69 bus mesh DS for multiple WT + D-STATCOM is given in Tables 33.8 and 33.9, respectively.

Best size and location of WT + D-STATCOM give minimum TPL and improved minimum voltage of the system. As WT + D-STATCOM installed in the system, % reduction of TPL increased and Vmin of the system improves. Also annual energy cost savings increase as power loss reduces but annual cost savings decrease as more number of D-STATCOM are installed in the system.

Here from Table 33.8, TPL reduces 29.71% for case 2, 29.44% for case 3, and 58.97% for case 4. Vmin of the system improves from 0.9526 pu. to 0.9779 pu. Results for 69 bus mesh DS are shown in Table 33.9. From Table 33.9, as we install multiple WT + D-STATCOM, TPL reduces by 29.2% for case 2, 33.32% for case 3, and 65.80% for case 4. Vmin of the system improves from 0.9653 pu. to 0.9825 pu. Figures 33.13 and 33.14 show voltage profile of 33 and 69 bus mesh DS with multiple WT + D-STATCOM, respectively, and observed that as WTs and D-STATCOMs installed together in the system, voltage profile improves more effectively.

Annual energy loss cost savings are also shown in Tables 33.8 and 33.9. Here AECS in 33 bus are $ 19,033.56 for case 2, $ 18,901.42 for case 3, and $ 37,771.30 for case 4. From Table 33.9, AECS in 69 bus are $ 12,714.06 for case 2, $ 14,486.22 for case 3, and $ 28,604.26 for case 4. Hourly TPL of both the systems with multiple WT and multiple D-STATCOM are given in Figs. 33.15 and 33.16, and from the figure it can be seen that at 22 o'clock power loss is minimum so it means that wind speed at 22 o'clock is higher. It can also be seen that there is a remarkable decrement in annual energy loss and annual energy loss cost.

Table 33.9 Results for 69 mesh DS with multiple WT and multiple D-STATCOM

Performance parameters	Base case	Multiple WT	Multiple D-STATCOM	WT + D-STATCOM
Location	–	61, 61, 61, 27	61, 11, 49, 21	WT (61, 61, 62, 61), D-STATCOM (49, 69, 16, 61)
Size (kW, KVAr)	–	167.85	1390, 468, 550, 188	168, 392, 189, 332 1215
TPL in KW	82.7029	56.6774	55.1416	28.2808
% reduction in TPL	–	29.2	33.32	65.80
Min. bus voltage (Vmin)	0.9653	0.9727	0.9777	0.9825
Annual energy loss (KWh)	724,477	496,494	483,040	247,739
ES in KWh	–	227,983.38	241,436.98	476,737.59
AECS in $	–	13,679.00	14,486.22	28,604.26
ACS in $	–	250	717	27,637

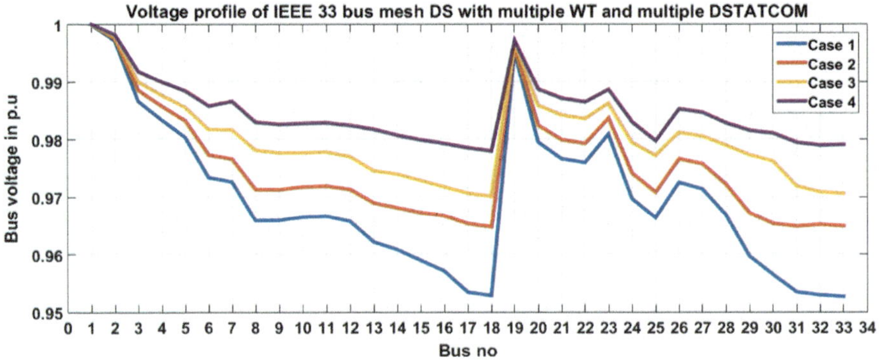

Fig. 33.13 Voltage at each node in 33 bus

Also it can be seen that minimum system voltage for both WTs and D-STATCOMs gets improved as we install multiple devices in the system. Figure 33.17 shows minimum system voltage of 33 and 69 bus with multiple WT and multiple D-STATCOM, and from the figure it can be observed that minimum system voltage improves more effectively using D-STATCOM.

After installing multiple WT and multiple D-STATCOM, active power loss reduces, and branch currents also reduce. As energy loss is given by power loss × time, energy loss also decreases. Figure 33.18 shows annual energy cost savings for all three cases and Fig. 33.19 shows annual cost savings for all three cases.

As shown in Fig. 33.18, annual energy loss cost savings is more in case of WT + D-STATCOM because % reduction in TPL is high. ACS is difference between annual

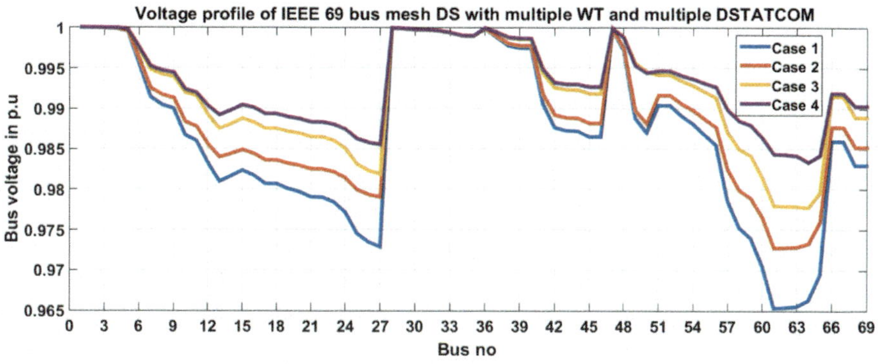

Fig. 33.14 Voltage at each node in 69 bus

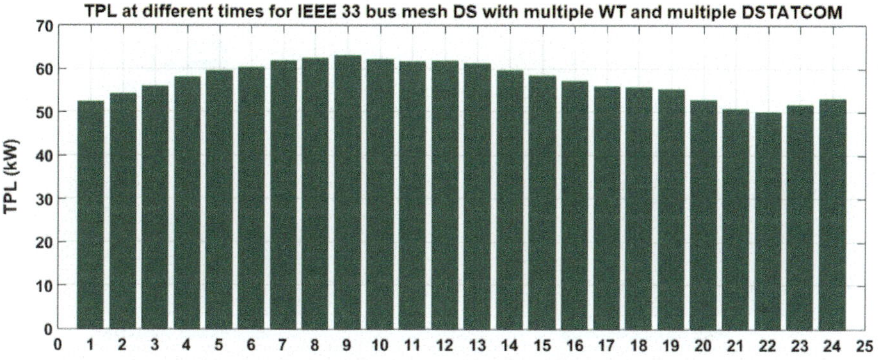

Fig. 33.15 Hourly TPL with multiple WT + D-STATCOM in 33 bus mesh DS

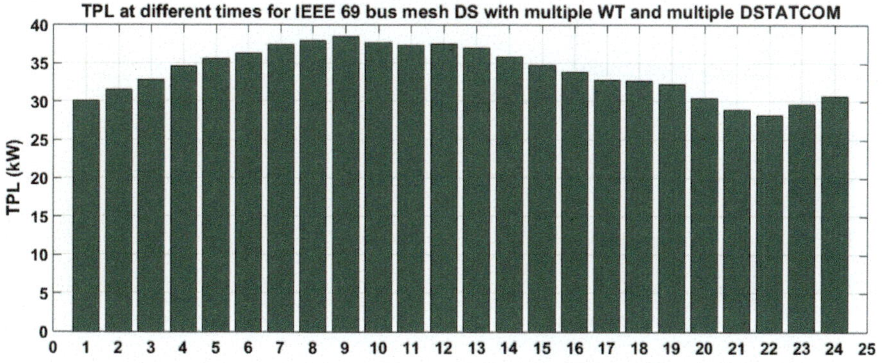

Fig. 33.16 Hourly TPL with multiple WT + D-STATCOM in 69 bus mesh DS

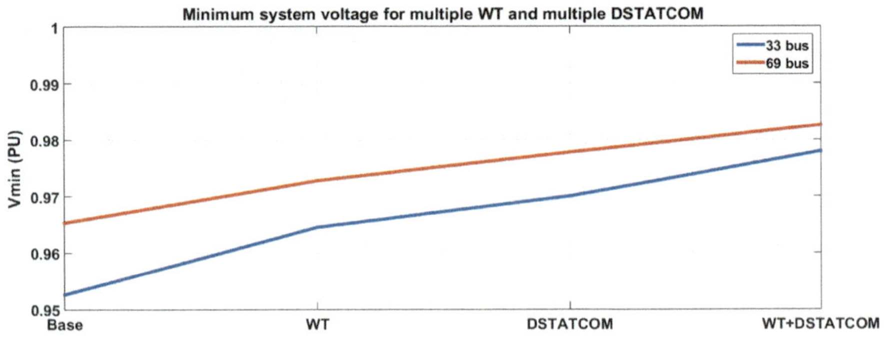

Fig. 33.17 Minimum system voltage with multiple WT and multiple D-STATCOM for 33 and 69 bus systems

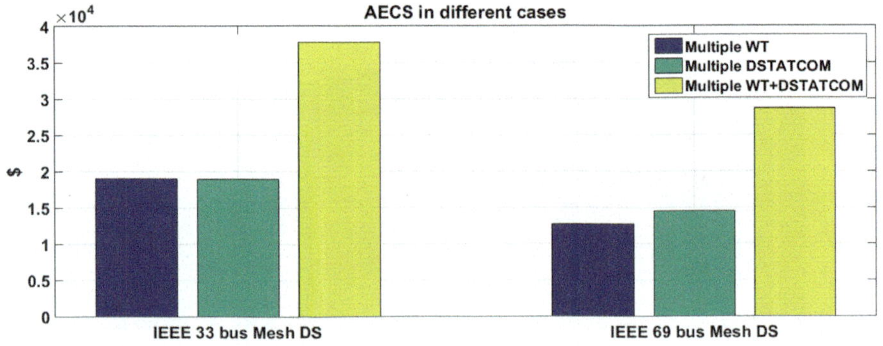

Fig. 33.18 AECS for multiple WT and multiple D-STATCOM

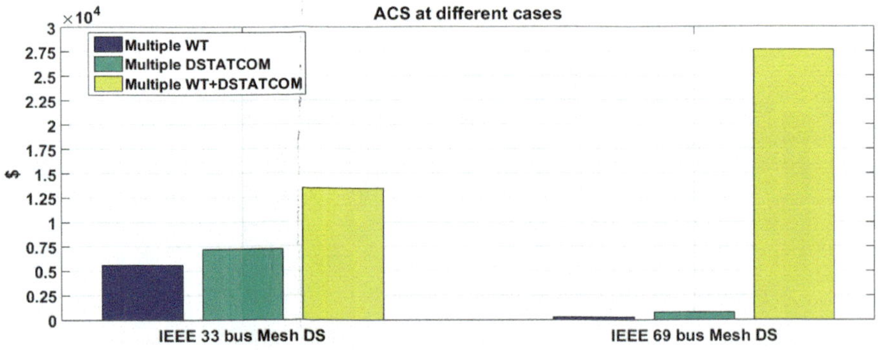

Fig. 33.19 ACS for multiple WT and multiple D-STATCOM

energy loss cost savings and investment cost of a device. In Fig. 33.19, annual cost savings after implementing WT and D-STATCOM are shown, and from the figure, we can see that annual cost savings are higher in case of multiple WT + D-STATCOM.

33.5 Conclusions

In this chapter, 33 and 69 bus meshed distribution system analysis is carried out with allocation of multiple WT and multiple D-STATCOM. Here artificial bee colony algorithm is used to find best location and best size of WT and D-STATCOM to get minimum TPL of the system. Here, analysis is done with various aspects like system voltage profile, total active power loss, annual energy loss, annual energy loss cost, and annual energy savings. Conclusions form the analysis are described below.

- % Reduction in TPL increased from 8.83% to 29.71% for 33 bus and from 8.7% to 29.2% for 69 bus system in case of multiple WT. For multiple D-STATCOM, it improves to 29.44% from 26.49% in 33 bus and 33.32% from 29.68% in 69 bus system. After installing multiple WT and multiple D-STATCOM together in the system, % reduction of TPL for 33 bus is 58.97% and for 69 bus is 65.80%.
- Minimum voltage of the system after implementing multiple WT reaches to 0.9645 p.u from 0.9526 p.u for 33 bus and 0.9727 p.u from 0.9653 p.u for 69 bus system. After installing multiple D-STATCOM, it becomes 0.9700 p.u for 33 bus and 0.9777 p.u for 69 bus system. For multiple WT and multiple D-STATCOM together, it becomes 0.9779 p.u for 33 bus and 0.9825 p.u for 69 bus system.
- Annual energy loss cost savings (AECS) in 33 bus are $ 19,033.56 for multiple WT, $ 18,901.42 for multiple D-STATCOM, and $ 37,771.30 for multiple WT + D-STATCOM. AECS in 69 bus are $ 12,714.06 for multiple WT, $ 14,486.22 for multiple D-STATCOM, and $ 28,604.26 for multiple WT + D-STATCOM.
- Annual cost savings (ACS) for multiple WT in 33 and 69 buses are 5604.55 $ and 250 $, respectively. ACS in case of multiple D-STATCOM is 7185 $ and 717 $ for 33 and 69 bus systems, respectively. For multiple WT + D-STATCOM, annual cost savings are 13,485 $ and 27,637 $ for 33 and 69 bus systems, respectively.

This analysis will help to distribution system utilities for planning distribution network with active and reactive power support devices or systems. Also we can extend this work with installing electric vehicles in the distribution system along with its charging and discharging schedule.

References

Ackermann T, Knyazkin V (2022) Interaction between distributed generation and the distribution network: operation aspects. IEEE/PES Trans Distribut Conf. Exhibit 1357–1362. https://doi.org/10.1109/TDC.2002.1177677.

Adaramola MS, Paul SS, Oyedepo SO (2011) Assessment of electricity generation and energy cost of wind energy conversion systems in North-central Nigeria. Energy Convers Manage 52(12):3363–3368. https://doi.org/10.1016/j.enconman.2011.07.007

Apoorva Y, Balamurugan P, Yuvaraj T, Muthukannan P (2018) Review on optimal allocation of D-STATCOM in radial distribution system. Int J Pure Appl Math. https://doi.org/10.1155/2017/2857926

Asfari M, Singh SP, Raju GS, Rao GK (2002) A fast power flow solution of radial distribution networks. Elect. Power Compon. Syst. 30(3):1065–1074. https://doi.org/10.1080/15325000290085361

Kumar A, Vijay Babu P, Murty VVSN (2017) Distributed generators allocation in radial distribution systems with load growth using loss sensitivity approach. J Inst Eng India Ser B 98(3): 275–287. https://doi.org/10.1007/s40031-016-0242-8.

Atwa YM, El-Saadany EF (2011) Probabilistic approach for optimal allocation of wind-based distributed generation in distribution systems. IET Renew Power Gener 5(1):79–88. https://doi.org/10.1049/iet-rpg.2009.0011

Ayse Aybike Seker, Mehmet Hakan Hocaoglu (2013) Artificial Bee Colony Algorithm for optimal placement and sizing of distributed generation. IEEE ELECO. https://doi.org/10.1109/ELECO.2013.6713817

Babu PR, Kumar MPVVR, Hemachandra VS, Vanamali MPR (2010) A novel power flow solution methodology for radial distribution systems. In: 2010 IEEE Region 8 International Conference on Computational Technologies in Electrical and Electronics Engineering (SIBIRCON), pp 507–512. https://doi.org/10.1109/SIBIRCON.2010.5555377.

Banos R, Manzano-Agugliar F, Montoy FG, Gil G, Alcayd A, Góme J (2011) Optimization methods applied to renewable and sustainable energy: a review. Rrenew Sustain Energy Rev 15: 1753–1766. https://ideas.repec.org/a/eee/rensus/v15y2011i4p1753-1766.html

Bapaiah P (2013) Power quality improvement by using D-STATCOM. Int J Emerg Trends Electr Electron. https://doi.org/10.1109/i-PACT44901.2019.8960234

Chen B, Hsu Y (2008) A minimal harmonic controller for a STATCOM. IEEE Trans Industr Electron 55(2):655–664. https://doi.org/10.1109/TIE.2007.896266

Chen P, Siano P, Bak-Jensen B, Chen Z (2010a) Stochastic optimization of wind turbine power factor using stochastic model of wind power. IEEE Trans. Sustain. Energy 1(1):19–29. https://doi.org/10.1109/TSTE.2010.2044900

Chen P, Siano P, Chen Z, Bak-Jensen B (2010b) Optimal allocation of power-electronic interfaced wind turbines using a genetic algorithm Monte Carlo hybrid optimization method. In: Wind Power Systems. Springer, Berlin, Germany, pp 1–23. https://doi.org/10.1007/978-3-642-13250-6_1

Cheng CS, Shirmohammadi D (1995) A three-phase power flow method for real-time distribution system analysis. IEEE Trans Power Syst 10(2):671–679. https://doi.org/10.1109/59.387902

Devabalaji KR, Ravi K (2014) Power quality improvement in wind farm connected to grid using STATCOM. Int Conf Adv Electr Eng (ICAEE) 1–5. https://doi.org/10.1109/ICAEE.2014.6838508

Devi S, Geethanjali M (2014) Optimal location and sizing determination of Distributed Generation and D-STATCOM using Particle Swarm Optimization algorithm. Electr Power Energy Syst 62:562–570. https://doi.org/10.1016/j.ijepes.2014.05.015

Dincan C, Kjær P, Helle L (2020) Cost of energy assessment of wind turbine configurations. In: 22nd European Conference on Power Electronics and Applications (EPE'20 ECCE Europe), 1–8. https://doi.org/10.23919/EPE20ECCEEurope43536.2020.9215590.

El-Zonkoly M (2011) Optimal placement of multi-distributed generation units including different load models using particle swarm optimization. Swarm Evol Comput 1(1):50–59. https://doi.org/10.1016/j.swevo.2011.02.003

Eminoglu U, Hocaoglu MH (2005) A new power flow method for radial distribution systems including voltage dependent load models. Elect Power Syst Res 76(1–3):106–114. https://doi.org/10.1016/j.epsr.2005.05.008

Faruk U, Engin K (2013) Optimal wind turbine sizing to minimize energy loss. Electr Power Energy Syst 53: 656–663. https://doi.org/10.1016/j.ijepes.2013.05.035
Ghosh S, Das D (1999) Method for load-flow solution of radial distribution networks. IEEE Proc Generation Transm Distrib 146(6):641–648. https://doi.org/10.1049/ip-gtd:19990464
Gupta R, Kumar A (2015) Energy savings using D-STATCOM placement in radial distribution system. In: 4th International Conference on Eco-friendly Computing and Communication Systems, ICECCS. https://doi.org/10.1016/j.procs.2015.10.100
Harrison GP, Piccolo A, Siano P, Wallace AR (2007a) Exploring the trade-offs between incentives for distributed generation developers and DNOs. IEEE Trans Power Syst 22(2):821–828. https://doi.org/10.1109/TPWRS.2007.895176
Harrison GP, Piccolo A, Siano P, Wallace AR (2007b) Distributed generation capacity evaluation using combined genetic algorithm and OPF. Int J Emerg Electr Power Syst 8(2):1–13. https://doi.org/10.2202/1553-779X.1517
Harrison GP, Piccolo A, Siano P, Wallace AR (2008) Hybrid GA and OPF evaluation of network capacity for distributed generation connections. Electr Power Syst Res 78(3):392–398. https://doi.org/10.1016/j.epsr.2007.03.008
Hasani-Marzooni M, Hosseini SH (2012) Dynamic interactions of TGC and electricity markets to promote wind capacity investment. IEEE Syst J 6(1):46–57. https://doi.org/10.1109/JSYST.2011.2162891
Karaboga D (2005) An idea based on honey bee swarm for numerical optimization. Technical Report TR06, Computer Engineering Department, Erciyes University, Turkey
Karaboga D, Akay B (2011) A Modified artificial bee colony (ABC) algorithm for constrained optimization problems. Appl Soft Comput 11:3021–3031
Kumar Sharma A, Murty VVSN (2014) Analysis of mesh distribution systems considering load models and load growth impact with loops on system performance. J Inst Eng India Ser B 95(4): 295–318. https://doi.org/10.1007/s40031-014-0108-x
Liu J, Salama MMA, Mansour RR (2002) An efficient power flow algorithm for distribution systems with polynomial load. Int J Elect Eng 371–386. https://doi.org/10.7227/IJEEE.39.4.7
Mathew S (2006) Wind energy: fundamentals, resource analysis and economics. Springer, Berlin, Germany
Mendez Quezada VH, Rivier Abbad J, Gomez San Roman T. (2006) Assessment of energy distribution losses for increasing penetration of generation. IEEE Trans Power Syst 21(2): 533–540. https://doi.org/10.1109/TPWRS.2006.873115
Mohandas N, Balamurugan R, Lakshminarasimman L (2016) Optimal location and sizing of real power DG units to improve the voltage stability in the distribution system using ABC algorithm united with chaos. Int J Electr Power Energy Syst 66:41–52. https://doi.org/10.1016/j.ijepes.2014.10.033
Mokryani G, Siano P, Piccolo A, Chen Z (2013) Improving fault ride through capability of variable speed wind turbines in distribution networks. IEEE Syst J 7(4):713–722. https://doi.org/10.1109/JSYST.2012.2211812
Oureilidis KO, Bakirtzis EA, Demoulias CS (2016) Frequency-based control of islanded microgrid with renewable energy sources and energy storage. J Mod Power Syst Clean Energy 4(1):54–62. https://doi.org/10.1007/s40565-015-0178-z
Prakash K, Sydulu M (2011) Topological and primitive impedance based load flow method for radial and weakly meshed distribution systems. Iran J Electr Comput Eng 10(1)
Ranjan R, Venkatesh B, Chaturvedi A, Das D (2004) Power flow solution of three phase unbalanced radial distribution network. Elect Power Compon Syst 32(4):421–433. https://doi.org/10.1080/15325000490217452
Samal P (2020) Optimal D-STATCOM allocation using a hybrid seagull-differential evolution algorithm. In: 2020 IEEE 17th India Council International Conference (INDICON), pp 1–6. https://doi.org/10.1109/INDICON49873.2020.9342524

Satyanarayana S, Ramana T, Sivanagaraju S, Rao GK (2007) An efficient load flow solution for radial distribution network including voltage dependent load models. Elect Power Compon Syst 35(5):539–551. https://doi.org/10.1080/15325000601078179

Sallam AA, Malik OP (2019) Distribution system structure. In Electric distribution systems, IEEE, 9–22. https://doi.org/10.1002/9781119509332.ch2

Shadangi P, Swain SD, Ray PK (2021) Experimental realization of D-STATCOM for Power Quality Enhancement Under Various Load Perturbations. In: 2020 3rd International Conference on Energy, Power and Environment: Towards Clean Energy Technologies, pp 1–5. https://doi.org/10.1109/ICEPE50861.2021.9404488

Siano P, Ochoa LF, Hamson GP, Piccolo A (2009) Assessing the strategic benefits of distributed generation ownership for DNOs. Gener Transm Distrib, IET, 3(3): 225–236. https://doi.org/10.1049/iet-gtd:20080235

Taher SA, Afsari SA (2014) Optimal location and sizing of D-STATCOM in distribution systems by immune algorithm. Electr Power Energy Syst 60:34–44. https://doi.org/10.1016/j.ijepes.2014.02.020

Teng JH (2000) A network-topology based three-phase load flow for distribution systems. Proc Natl Sci Council ROC (A) 24(4):259–264

Thukaram D, Banda HMW, Jerome J (1999) A robust three-phase power flow algorithm for radial distribution systems. Elect Power Syst Res 50(3):227–236. https://doi.org/10.1016/S0378-7796(98)00150-3

Yuvaraj T, Ravi K (2018) Multi-objective simultaneous DG and D-STATCOM allocation in radial distribution networks using cuckoo searching algorithm. Alex Eng J 57:2729–2742. https://doi.org/10.1016/j.aej.2018.01.001

Yuvaraj T, Ravi K, Devabalaji K R (2017) "Optimal allocation of DG and D-STATCOM in radial distribution system using cuckoo search optimization algorithm. Model Simul Eng. https://doi.org/10.1155/2017/2857926

Zhu Y, Tomsovic K (2002) Adaptive power flow method for distribution systems with dispersed generation. IEEE Trans Power Delivery 17(3):822–827. https://doi.org/10.1109/TPWRD.2002.1022810

https://core.ac.uk/download/pdf/53189751.pdf

https://www.nrel.gov/gis/wind.html

https://www.suzlon.com/in-en/energy-solutions/s128-wind-turbine-generator

Chapter 34
Magnetohydrodynamic (MHD) Power Generation Systems

Tushar Kanti Bera, Aashish Kumar Bohre, Irfan Ahmed, Aniruddha Bhattacharya, and Partha Sarathee Bhowmik

Abstract A magnetohydrodynamic (MHD) power generation system is an electrical power generating system which generates the electricity utilizing the MHD principle. MHD power generation technique generates the electric power directly from a moving stream of ionized fluid flowing through a magnetic field. Therefore, the MHD power generation systems are found as the non-conventional electric power generation modality which is considered as the green energy harvesting procedures. The MHD generators utilizes the electromagnetic interaction of an ionized fluid flow and a magnetic field. The ionized fluids in MHD generators work as the moving electrical conductor and hence the electromotive force (e.m.f.) could be generated across the ionized conductor due to the Faraday's electromagnetic principle. An MHD system, therefore, can act as a fluid dynamo or MHD power converter. In MHD, as the flow (motion) of the conducting fluid (conductor) under a magnetic field causes an induced voltage across the fluid, the e.m.f. would be found at the perpendicular direction to both the magnetic field and the fluid flow according to Fleming's right-hand rule. The concept of MHD power generation technique was first introduced by Michael Faraday in 1832 during his lecture at the Royal Society, UK. Since then, the MHD systems have been developed and studied by several research groups. Different types of MHD generator geometries have been proposed with different channel geometries, different electrode configurations, different magnetic coil structures, and different working fluids or plasmas. Though a typical coal-fired MHD generator converts about 20% of the thermal input power to the output electricity but, using the combined MHD/steam cycle systems, an energy conversion efficiency up to 60% of the coal's energy can be converted into the electrical energy. In recent time, the green energy harvesting processes are found extremely important to reduce the pollution and to save the fossil fuel in the world for its sustainable development. In this direction, the MHD power generation technique could be utilized for green energy generation without any environmental pollution. In this chapter, The MHD technology has been discussed in detail followed by a discussion on its components, system design issues, and crucial design aspects. A detail review on the historical

T. K. Bera (✉) · A. K. Bohre · I. Ahmed · A. Bhattacharya · P. S. Bhowmik
Department of Electrical Engineering, National Institute of Technology Durgapur (NITDgp), A-Zone, Mahatma Gandhi Avenue, Durgapur, West Bengal 713209, India

A. K. Bohre et al. (eds.), *Planning of Hybrid Renewable Energy Systems, Electric Vehicles and Microgrid*, Energy Systems in Electrical Engineering,
https://doi.org/10.1007/978-981-19-0979-5_34

developments and the associated research works conducted on the MHD power generation process has been presented highlighting the major developments. Along with the limitations and challenges of the MHD power generation method, the present scenario and the future trends are also discussed.

Keywords Electric power · Electric power generation · Non-conventional power generating systems · Green energy · Magnetohydrodynamic (MHD) power generation · MHD Generators · Plasma

34.1 Introduction

Electric power generating stations or electric power plants (Drbal Westra Boston 2012) are developed to generate the electrical energy to fulfil the energy demands in our day-to-day life. Though we need different kinds of energies such as light, heat, sound, etc., in our daily life, the electrical form of the energy is preferred to generate in large amount and the other types of the energies are converted from it when required. Electrical power generating machines called electrical generators are used to generate the electricity in electrical power stations. In thermal power stations, the energy stored inside the fossil fuel such as coal and oil, is extracted by burning process of the fuels inside a heat generation chamber and the steam is generated from the water utilizing the generated heat energy using the boiler system. The steam is used to rotate the turbine (Drbal Westra Boston 2012) which is a mechanical machine coupled with the electric electrical generators which converts the mechanical energy into the electrical energy utilizing electromechanical energy conversion (Begamudre 2007) principle. A magnetohydrodynamic (MHD) power generation technique (Messerle and Messerle 1995; Sheindlin et al. 1979; Dhareppagol and Saurav 2013; Petrick and Shumi͡at͡skiĭ 1978; Rosa 1961; Sarkar 2016) is a non-conventional electric power generation modality (Bansal 2014) which can generate the electrical energy directly from an electrically conducting fluid stream passing through a magnetic field applied. The MHD fluid stream is produced by ionizing a fluid and sending through a pipe-like MHD channel kept within a magnetic field. The ionized fluid in the MHD generators works as the moving electrical conductors (Branover 1978) and hence the e.m.f. is generated according to the electromagnetic principle of Michael Faraday. The MHD power generators (MHD generator) or the MHD power converter (MHD converter) acts as a fluid-based dynamo in which the flow (motion) of the conducting fluid (conductor) under a magnetic field generates an electric potential (e.m.f.) across the fluid stream which is found to be perpendicular to both the magnetic field direction and the fluid flow direction and according to Fleming's right-hand rule.

In 1832, the concept of MHD power generation technique was first time practically studied by Michael Faraday (Jackson and Ralph Strohl 2016) during his lecture delivered to the Royal Society. In 1832, Michael Faraday conducted a practical experimentation with the Thames River water flow near the Waterloo Bridge in London, England. During his experiment, Faraday submerged two electrodes made up of the copper plates and connected them by a wire to observe the electricity generated by MHD process. But, due to some practical issue associated with this experimental setup developed with the river water, electric current generation or the electric current conduction through the electrode-lead wires were not very satisfactory due to the electrochemically polarization of the electrodes (Development of MHD power generators, Internet Article, Encyclopædia Britannica Inc xxxx). After that, MHD principle-based electrical power generation procedures have been explored by several research groups. The MHD-based electrical power generation techniques have been implemented in varieties of practical applications [12 = 13–15 = 16], and hence these techniques have been utilized for electrical energy harvesting. Also, in recent time when an alternating source of green-energy-harvesting processes (Onar 2017) are sought, the MHD-based power generation modality seems to be an important and effective means as the renewable energy sources for the sustainable development. In this chapter, the MHD power generation techniques have been presented in detail and the principle of operation, system components, system classifications, and practical applications have been discussed. A brief review on the historical developments and the associated research works conducted in the domain of MHD-based electrical power generation systems has also been presented. The advantages, limitations and critical issues, and future trends of the MHD power generation process are also discussed.

34.2 Methods

34.2.1 Electricity and Electrical Power Generation

Electricity (Maxwell 1888) represents the state of the present of electrons or charge particles either in static form or in a dynamic form. The static electricity (Maxwell 1888) is created due to accumulation or storage of the electrical charge particles, whereas the electrical current if produced due to the flow or movement of the electrons or other charge particles such as holes or ions. The electrical energy is possible to be converted in many types of other energies such as heat, light, sound, and all and hence the energy conversion (Karady and Holbert 2013), when required in large scale, is done by generating the electrical energy and then the electrical energy is converted into the required form. The generation one electrical energy or electrical power is conducted using an electrical machine called electrical generator which is found as an electromechanical energy conversion device. The electrical generator (Say 1976) generates electrical power by converting the mechanical energy fed into the generator as its input. The mechanical energy is generated and supplied to the generator by a mechanical power generation machine called prime mover which can be a steam engine, internal combustion (IC) engine, water turbine, wind turbine, electrical motors, or else. The electrical power is generated by electromagnetic induction process invented by Michael Faraday. As per the electromagnetic induction process

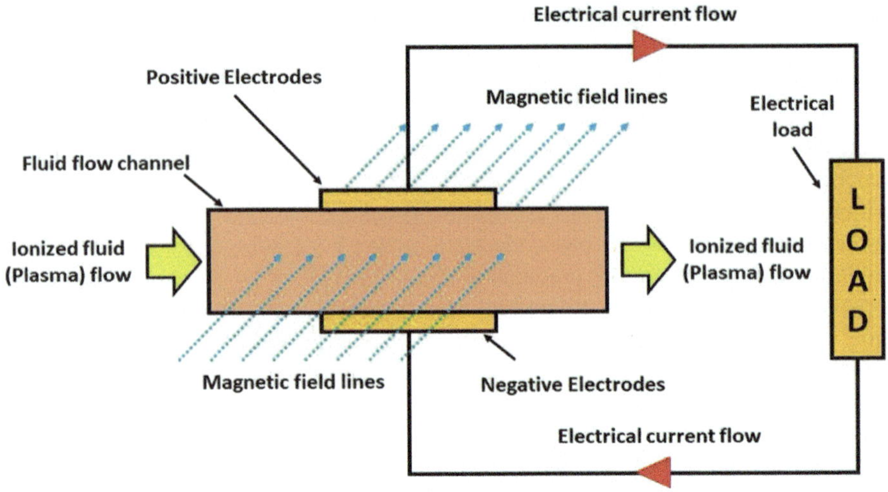

Fig. 34.1 An MHD power generation system schematic

of Faraday, as shown in the Fig. 34.1a, the conductor PQ moving with a velocity, v through a magnetic field, B will generate an e.m.f., E across the conductor PQ. The direction of the e.m.f. will be determined by the Fleming's right-hand rule as show in Fig. 34.2b in the direction from P to Q.

Generally, the electrical power is conventionally generated in large scale by driving the alternating current generators called alternators (Say 1976) by using the turbines called prime movers. In thermal power stations (Nag 2002) the alternators are driven by the steam turbine as (Drbal Westra Boston 2012; Tanuma 2017) the prime mover. In thermal power generating stations, the pulverized coal (dust or power like coal particles of very small sizes) (Pratt et al. 1979) is burnt, and the heat energy generated is used to produce the steam at high temperature and high pressure. The steam is fed to the turbine and the mechanical or rotation motion is obtained at the turbine shaft which is mechanically coupled with the alternator shaft. Thus, the heat energy collected form the coal burning process is used to drive the turbine and in turn to drive the alternator. The electrical conductors, (generally made up of copper), housed in the alternator cuts the magnetic field and produces the electrical power. In small-scale power generators, the conductus is housed on the rotor (rotating part of the alternator) rotating in a magnetic field produced by the magnets or the magnetic poles kept static in space in the alternator. On the contrary, the large alternators which are generally used in the power stations to generate the electricity in large scale, are developed with the electrical conductors housed in the stator, whereas the magnetic poles are housed in the rotor part to rotate for producing the variable magnetic flux linkage over time.

The electrical energy sources which are used to generate electricity can be reobtained either as a renewable or non-renewable type (Mukherjee and Chakrabarti 2004). Electrical energy harvesting systems such as solar photovoltaic energy

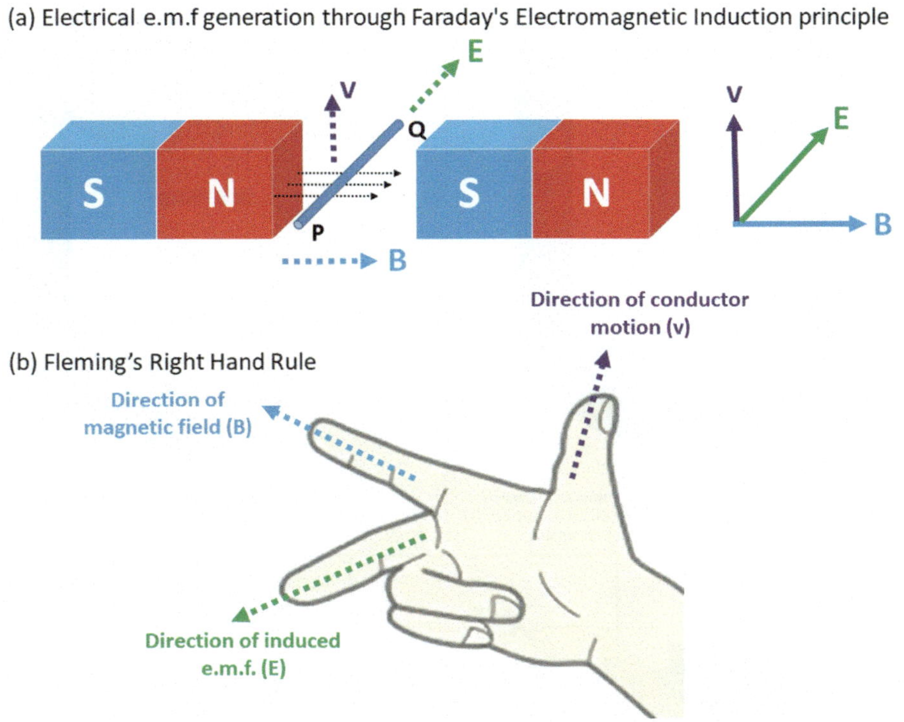

Fig. 34.2 Understanding the generation of e.m.f. for an electrical conductor moving in a magnetic field using Faraday's Electromagnetic Induction principle **a** Electrical e.m.f. generation through Faraday's Electromagnetic Induction principle, **b** Fleming's Right-Hand Rule

generating systems (Sukhatme and Nayak 2017), wind energy conversion system (Mukherjee and Chakrabarti 2004), tidal wave energy harvesting system (Abo-Khalil and Alghamdi 2021), geothermal energy system (Soltani et al. 2021), bioenergy- or biomass (Mukherjee and Chakrabarti 2004) based electrical energy generation system, etc., are the examples of renewable type of energy sources. The electrical energy generation system which are generally used for large-scale power generation using the non-renewable energy sources are called as the conventional energy generating systems. On the contrary, the renewable energy sources also sometimes called as the non-conventional energy sources, are used to generate the electrical power with lesser efficiency and amount. MHD-based electrical energy generation system is also one kind of non-conventional energy (Khan 2006) systems that are used for electricity generation using the conducting fluid or ionized gas or plasma (Daybelge et al. 1968) as the electrical conductors. In an MHD power generating system, the solid metal conductors are replaced by the electrically conducting fluid moving through a pipe like structure called MHD channel kept inside a magnetic field (Fig. 34.3a). In the conventional alternators the rotational motion is obtained for the rotor which may contain either the conductors or the magnetic field circuit, whereas

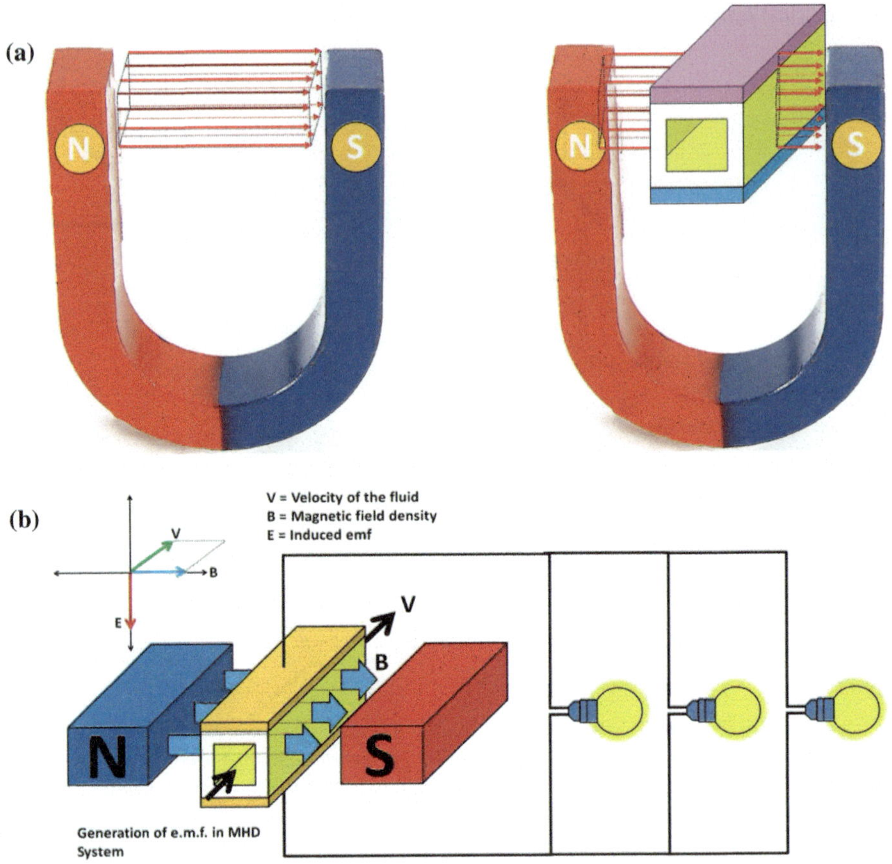

Fig. 34.3 Direction of the e.m.f. generated in an MHD power generation system **a** magnetic field of a permanent magnet, **b** a fluid pipe within a magnetic field, **c** MHD power generation schematic

in the MHD generators, the motion is provided to the conducting fluids keeping the magnetic field fixed in space. According to the Faraday's electromagnetic induction principle, the electrical potential is induced across the conducting fluid which can be utilized for the electric power harvesting using conducting electrodes placed on the surfaces of the conducting fluid stream or the fluid channel (Fig. 34.3b).

34.2.2 History of MHD Power Generation

Interest in MHD power generation was moved forward when it was observed that a plasma can interact with a magnetic field at much higher temperatures compared to the rotating mechanical turbine. The performance limitations in the heat engines and their associated efficiency were studied and established by the French physicist Sadi Carnot in the early nineteenth century (Jackson and Ralph Strohl 2016). In 1824, Sadi Carnot proposed a "theoretical ideal thermodynamic cycle" called "Carnot cycle" which provides an upper limit on the efficiency that can be achieve in any classical thermodynamic engine during the conversion process of heat into work. It is not an actual thermodynamic cycle but is a theoretical construct. The Carnot cycle is obtained from the difference between the hot source-temperature (T_{source}) and the cold sink-temperature (T_{sink}), divided by the source-temperature (T_{source}) (Jackson and Ralph Strohl 2016,) and thus, the Carnot cycle establishes the maximum theoretical efficiency of a heat engine. Considering the limitations and inefficiencies introduced by finite heat transfer rates (Jackson and Ralph Strohl 2016) as well as the system component limitations and inefficiencies in the real heat-engines (Jackson and Ralph Strohl 2016), a thermal power generating system coupled with an MHD generation system provides the energy conversion efficiency (η_{ece}) in the range of 60%–65% (Jackson and Ralph Strohl 2016).

The research activity or the scientific investigation on the MHD process was first conducted by the English chemist Humphry Davy in 1821 when he demonstrated that a magnetic field could deflect an arc (Jackson and Ralph Strohl 2016). After that, when Michael Faraday observed the motional electromagnetic induction process for an electrical conductor moving through the geomagnetic field of the Earth, he got the idea of MHD effect to generate the electricity using an electrically conducting fluid moving in a magnetic field. In 1832, he developed a very basic open-circuit MHD generator using the water of the River Thames in London (Jackson and Ralph Strohl 2016). Though the MHD system was proposed by Faraday in 1832, but MHD power generation received little attention until the early twentieth century, when Bela Karlovitz first proposed a gaseous MHD system (Jackson and Ralph Strohl 2016). In 1938, Bela Karlovitz and D. Halász set up an experimental MHD facility at the Westinghouse Electric Corporation research laboratories (Jackson and Ralph Strohl 2016). In 1946, Karlovitz and Halász demonstrated that small amounts of electric power could be extracted through seeding the working gas (Jackson and Ralph Strohl 2016). But till 1950, the power generation with MHD process was not so successful due to the less of understanding the procedures to convert the working gas into an effective and efficient electrical conducting medium (Jackson and Ralph Strohl 2016).

During the late 1950s, the information about the procedures to convert the working gas into an effective and efficient electrical conducting medium are obtained (Jackson and Ralph Strohl 2016) through the extensive studies on ionized gases and plasma systems. Accordingly, the interest in MHD generating systems resume again (Jackson and Ralph Strohl 2016) and grew quickly in the world along with the other research activities associated with applications of the plasma and ionized gases. In 1959, the American engineer Richard J. Rosa worked to develop the first truly successful MHD generator which was capable to produce an output of about 10 kilowatts of electric power (Jackson and Ralph Strohl 2016). After a few years of time, the American physicist Arthur R. Kantrowitz constructed and operated a 33-megawatt

MHD generator in the Avco Research Laboratory, in 1963 (Jackson and Ralph Strohl 2016).

34.2.3 MHD Power Generation Principle

The MHD generator generates electric power by converting the kinetic energy associated with the hot plasma directly into electricity. Though for both the conventional alternator and the MHD generator the electromagnetic induction process is the original reason behind the electrical power generation process, but the major constructional difference of an MHD generator from a conventional electric generator is required to be noted. The constructional difference between the conventional alternator and the MHD generator lies in the type of conductor/conductors used and the motion of the conductor/conductors. In conventional alternator, the conductors are solid metal conductors (preferably copper), but the MHD generator uses the ionized fluids (which may be a gas or liquid or plasma) as the electrical conductors. In an MHD generator, an ionized fluid is forced to move (at a particular velocity, v) through a pipeline called MHD channel which is placed within a powerful magnetic field (B). As soon as the conducting fluid moves through the MHD channel, the fluid flow interacts with the magnetic field lines and hence the variation in flux linkage ($d\psi/dt$, where the ψ represents the flux linkage) is obtained. According to the principle of Faraday's electromagnetic induction, an e.m.f. is generated within the conducting fluid, which can be suitably utilized to harvest the electric energy by placing two electrodes across fluid stream (Fig. 34.3). If the direction of v and B are known then, by using the Fleming's right-hand rule the direction of the generated e.m.f. could be found. If the conducting fluid or plasma moves towards a direction perpendicular to the magnetic field, the e.m.f. will be found to be developed mutually perpendicular to the direction of fluid motion (v) and that of the magnetic field (B). According to the principle of Faraday's electromagnetic induction, the amplitude of the generated e.m.f. will be proportional to the magnitude of the velocity of the fluid and to the magnitude of the magnetic flux density. Now, if the velocity of the conducting fluid and the magnetic field density are for an MHD generator are represented by v and B, respectively, the e.m.f. induced across the fluid conductor will be given by

$$E_{ind} = v \times B \tag{34.1}$$

Therefore, the induced current density (J_{ind}) will be given by

$$J_{ind} = \sigma \times E_{ind} \tag{34.2}$$

where σ = electric conductivity and E_{ind} is the electric field intensity.

Now combining the Eq. 34.1 and Eq. 34.2, we get e.m.f. with the following form:

$$E_{ind} = \sigma(v \times B) \tag{34.3}$$

Therefore, the induced current density (J_{ind}) will be given by

The retarding force on the conductor is the Lorentz force which will be represented by

$$F_{ind} = J_{ind} \times B \tag{34.4}$$

As in conventional electricity generators, the prime movers (mechanical engines like steam turbine, IC engine, water turbine, or electrical motors) are used to rotate the solid copper conductors within a magnetic field the similar relationship is applicable. In MHD-based electricity generation systems, only difference is that the generators dot not use a solid metal conductor but an ionized gas as the electrical conductors moving across a strong magnetic field created by a powerful magnet.

A basic structural construction of an MHD generator is shown in the Fig. 34.3. In an MHD generator, the electrically conducting hot gas is passed through a channel kept within a magnetic field (Fig. 3a). The practical design and development of the plasma channel is obtained by a nozzle to inject the plasma into the MHD channel. The strength of the magnetic field could be made as powerful as possible as the induced e.m.f. and hence the electrical power generated will be of higher amount if the magnitude of the magnetic field strength is kept high. As shown by the Fig. 34.3b, according to the Faraday's law of induction, the electric field (e.m.f.) is developed across the conducting fluid in a direction which is mutually perpendicular to velocity (v) of the ionized gas as well as the applied magnetic field (B).

The generated electrical e.m.f. can be utilized to harvest the electrical energy by connecting the electrical loads with the surfaces (perpendicular to the direction of e.m.f. and parallel to the direction of v and B) of the ionized conductors. The electrodes can be placed on the internal surface of the fluid channel to get the contact of the conducting fluid. The electrodes developed with metals which can withstand the temperature of the plasma can be suitably used for the MHD system to harvest the electrical power generated in forms of the induced e.m.f. through an electric circuit with electrical loads to be driven.

As shown below (Fig. 34.4), in a practical MHD-based electrical power generation system the fuel is burnt in a combustion chamber to generate heat energy. The heat energy generated in the combustion chamber is used to convert the non-conducting gas into the ionized gas or plasma. The combustion chamber takes the fuel and air to produce the heat energy required to ionize the gas to produce the MHD fluid plasma state. The ionized gas or the plasma is passed through an expansion nozzle to decrease the gas pressure in order to increases the velocity of fluid through the MHD channel. As described by the Eq. 34.1, with the increase in the fluid velocity, the amount of the electrical power output is also increased. After passing through the MHD chamber, the fluid obtained at the gas exhaust of the MHD chamber is then passed through a heat exchanger. The heat exchanger produces hot air by supplying the heat energy (collected from the exhaust gas) to the air taken as an input. The heat exchanger then feed the hot air to the compressor after raising its temperature.

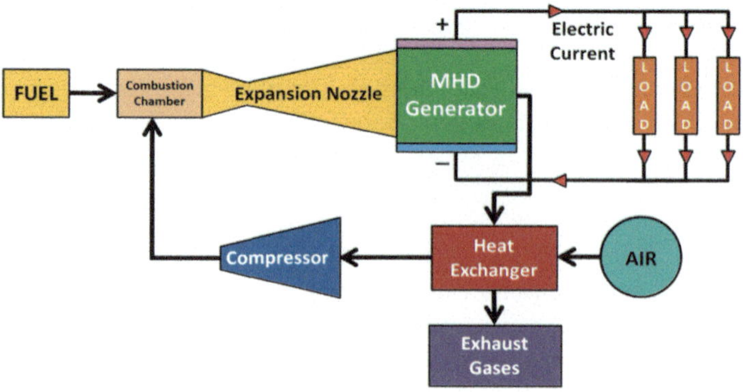

Fig. 34.4 Block diagram of the MHD power generation process showing the various components of an MHD-based electrical power generation system

Compressor takes the hot air from the heat exchanger and feeds the hot air into the combustion chamber again for producing plasma which is used to send to the MHD channel again.

34.2.4 MHD Generator Components

34.2.4.1 MHD Fluid Channel

A pipe like structure is always needed to make a controlled flow in a particular direction without leakage. The fluid flow in an MHD generator is also obtained by using a pipe like structure called MHD channel. The MHD channel guide the fluid and keep the fluid confined within the MHD chamber or MHD duct to make the flow controlled and directed in a required direction. MHD chambers or MHD ducts are made up of insulating materials which can't conduct the electricity. The shape or geometry of the MHD chambers or MHD ducts may be different to obtain a specific MHG system geometry and required operational performances. MHD chambers or MHD ducts may be found with rectangular cross sections or circular cross sections. Generally, the inlet diameter of the MHD chamber is found with smaller diameter (or cross-sectional area) cross-sectional area compared to the outlet diameter (or cross-sectional area) of the MHD duct. This difference between inlet duct diameter and outlet duct diameter is obtained to decrease the pressure for increasing the speed of the fluid flow to generate higher amount of electricity. The performance of the MHD generator depends on the properties and the design parameters of the fluid flow, MHD duct (Zengyu et al. 2009) and other components used in the MHD systems.

34.2.4.2 Magnetic Field

The magnetic field is the domain within which the magnetic flux or the magnetic field lines produced (Fig. 34.5) by the source of the magnetic field such as permanent magnet (Fig. 34.5a), electromagnet, earth magnetic field (Fig. 34.5b), etc. Magnetic flux density could be realized as the number of magnetic field lines passing through a unit area imagined within the magnetic field (Fig. 34.5). In the MHD systems, the voltage generation (and hence the electrical power) is found dependent on the of the magnetic field properties (Kayukawa 1985). The magnetic field density of an MHD generating system should be very high which can be obtained ideally by a superconducting magnet (Kirillin et al. 1985). The superconducting magnets are developed as the electromagnets which are developed with the current-carrying conductors/coils made up of superconducting wires. The superconducting wires can be considered as special electrical conductors which are developed from superconducting materials having negligible Ohmic resistance at their superconducting state. At their superconducting state, which is generally obtained at very low temperature, the superconducting materials exhibit negligible electrical resistance (Schmidt et al. 1997) and hence they are found capable of carrying a huge amount of electric currents compared to the conventional electrical conductors. Therefor, a huge amount of magnetic field strength can be obtained using the superconducting magnets which can be developed by sending a huge amount of electrical current through the electrical conductors (superconductors) used to develop the electromagnets (superconducting magnets). A large amount of magnetic fields produced by the superconducting magnets are found very useful in many practical applications where a strong magnetic field is required

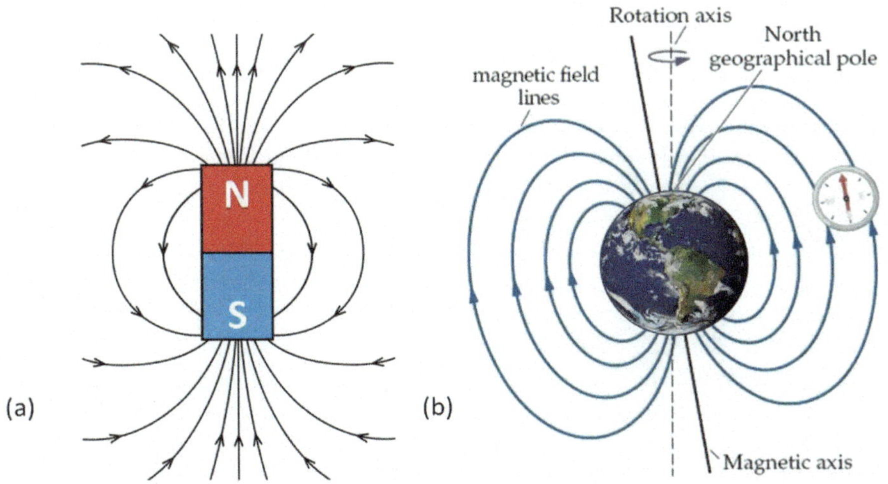

Fig. 34.5 magnetic field lines generated by the magnetic field source **a** magnetic field lines created by the bar magnet **b** magnetic field lines created by the earth magnetic field

such as MHD generating system, magnetic resonance imaging system (MRI) (Brown and Semelka 2011) etc.

34.2.4.3 Conducting Fluids in MHD Generators

In MHD-based electric power generating systems, the electrical conductors or the electrical conducting mediums can be obtained as a conducting fluid which may be either a conducting liquid, or an ionized gas or a plasma. The plasma which is often called as the "fourth state of matter" which is obtained when a gas is ionized gas. When water (or any other liquid) is heated up, it starts boiling and it then converts into the steam (or the vaporized form of that liquid). In similar way, when we heat up some gas, it absorbs energy and change itself into an ionized gas called a plasma.

The plasma used in MHD generators (Article et al. 2020a) can be considered as the superheated gaseous matter which is so hot that the electrons of the atoms are ripped away forming a medium containing the electrons and the positively charged particles or ions and this gaseous mixture of the electrons and ions is called the ionized gas or the plasma. The plasma can be realized as the gas or gaseous fluid containing of positively charged particles (ions) and negatively charged particles (electrons). The process of the ionization of the gas is conducted either by heating up a gas to its plasma state or to add some materials which helps the gas to become ionized. The materials added to the gas to convert it into the ionized gas (electrically conducting medium) are called the seeds or the seeding materials such as salts of alkali metals. The salts of alkali metals are the materials which are ionized easily to enhance the electrical conductivity of the entire fluid medium.

34.2.4.4 MHD Generator Electrodes

The electrodes used in the MHD generators, called MHD electrodes (Rosa et al. 1991a; Celinski and Fischer 1966; Scannell 1980; Witalis 1965; Kessler and Eustis 1968) are used to harvest the electrical power generated in the MHD system. Generally, the MHD electrodes are developed with metals and metal components in MHD systems though the other materials are also possible to be used (Stiglich and Addington 1978). The electrode geometry and electrode position in the MHD-based electricity generation systems are very important as the performance of the MHD systems depends on the electrode geometry and positions and other MHD design parameters. The multi-electrode configurations as well as interconnected multi-electrode configurations also (Stewart 1966) are also used in MHD generators. Hruby et al. (1986) studied the possibility of the platinum clad electrodes may dramatically enhance the generator lifetime, which may in turn help us to make it possible to move forward to the increase the applicability of the MHD energy generators. Authors reported that platinum clad electrode-based MHD system can provide approximately 1300 h of electrode operation.

34.2.4.5 MHD Magnet Coils

High magnetic strength is preferred in MHD generation and hence superconducting magnets are sometimes used. The high magnetic field can be generated by using the superconducting magnet technology. A superconducting magnet is an electromagnet which is developed with the magnetic coils of superconducting wire. The superconducting wire-based magnetic coils are required to be cooled to very low temperatures (cryogenic temperatures) during the operation.

At the lower temperature when the conducting wires used go into their superconducting state, they don't have any electrical resistance and therefore the conductors become superconductors because they can conduct much larger electric currents compared to their ordinary state. The huge amount of current conducted by the conductors in their superconduction state creates a huge amount of magnetic fields and hence the electromagnet is called as the superconducting magnets. Superconducting magnets, therefore, are capable of producing greater magnetic fields with lower energy loss than all the non-superconducting electromagnets as Ohmic loss is found negligible during the superconduction. Superconducting electromagnets are used in mass spectrometers, NMR spectrometers, MRI machines, fusion reactors, particle accelerators, and many other instruments and practical applications.

The magnetic coils used for MHD generators (Fig. 34.6a) have been described by Kayukawa (2004). Kayukawa reported that the saddle-type coil shown as in Fig. 34.3a is the principal design for the linear channels including the Faraday-type MHD generators (Kayukawa 2004), the diagonal type (Kayukawa 2004) and the linear Hall geometries (Kayukawa 2004). In an MHD generator, generally, it is found as a basic need to minimize the eddy current losses at the entrance of the channel and at the exit of the channel where no load current is extracted.

The saddle-type coil structure as shown in Fig. 34.6a helps us to achieve as the field lines are found parallel to the plasma flow at both ends for this type of coils. The magnetic coil for the disc-type channel-based MHD system is generally found as a race track type as shown in the Fig. 34.6b. The plasma in a disc-type MHD system enters the disc-type channel parallel to it and expands radially across the magnetic field line. The plasma leaves the channel also across the magnetic field (Kayukawa 2004). Though the required field magnetic strength of 5–6 T in an MHD system can be obtained using the superconducting magnet technology for both the saddle-type and race track-type magnetic coil geometries (Kayukawa 2004), but the race track-type magnetic coil geometries are beneficial due to lower cost and easy fabrication facilities (Kayukawa 2004). The cost of the saddle-type magnet strongly depends on the channel size and the magnetic field required to be generated. Depending on the magnetic field also other components are designed for optimal performance.

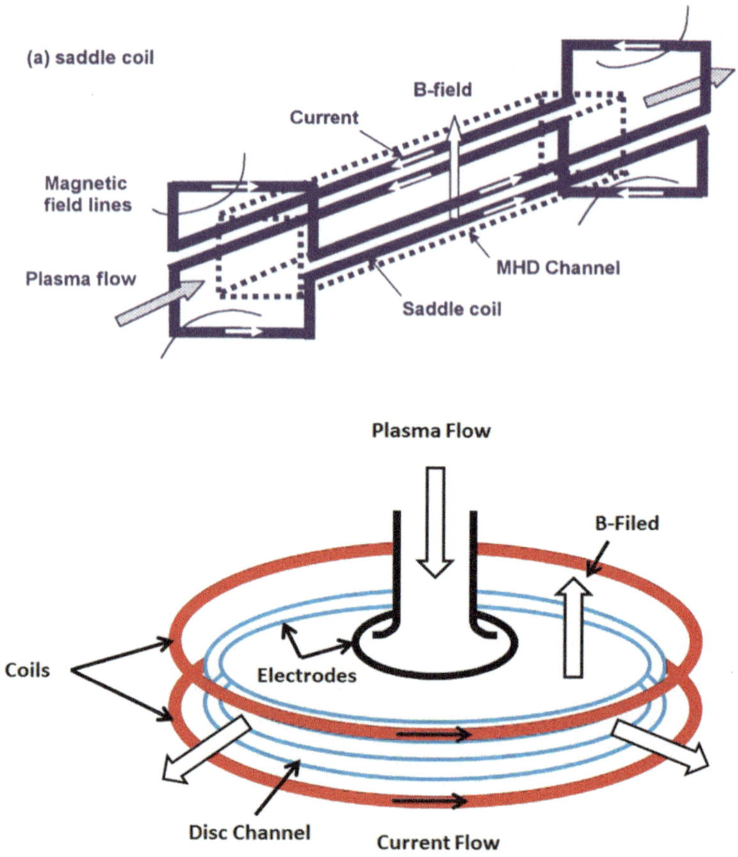

Fig. 34.6 Different coil shapes for different MHD channel geometries **a** A saddle-type magnetic circuit coil suitable for linear channels MHD generators **b** a race track-type magnet coil for disc-type MHD system [Picture Coutesy: Karady and Holbert (2013)]

34.2.5 MHD Generator Classifications

34.2.5.1 Faraday MHD Generator

A Faraday-type MHD generator (Fig. 34.7) or Faraday MHD generator (Rosa 1976; Ishikawa and Itoh 2003; Rankin et al. 1980; Hardianto et al. 2008) is developed using a plasma pipe or plasma duct kept within a strong magnetic field created either by a permanent magnet or by an electromagnet. The plasma pipe used in the MHD system called MHD channel or MHD chamber should be developed with an electrically insulating material to allow the electrically conducting fluid to flow through the duct. Under a strong magnetic field (perpendicular to the MHD channel) when a conductive fluid (hot plasma) flows through the MHD channel, an e.m.f. is induced

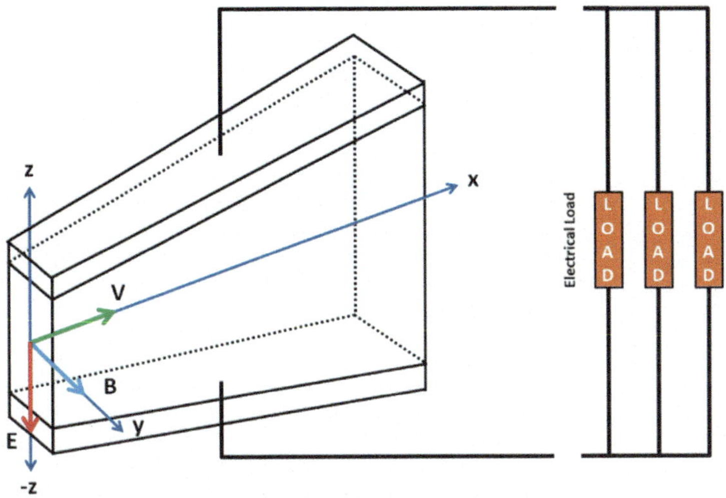

Fig. 34.7 Schematic Diagram of a Faraday MHD generator indicating the direction of the electrical current flows

across the fluid conductor (hot plasma). The direction of the e.m.f. generated will be obtained along the perpendicular direction to both the direction of the magnetic field applied and the direction of the fluid flow. The generated e.m.f. can be probed by using two electrical conducting materials called electrodes (MHD electrodes). The electrical power could be collected by placing two electrodes placed on the fluid flow surfaces (along the direction of e.m.f.). In the Faraday MHD generator, the magnitude of the electrical power collected through the MHD electrodes is found proportional not only to the cross-sectional area of the tube but also to the speed of the conductor (conductive fluid flow).

The Faraday generating system suffer from the short-circuit problems as these types of the MHD generators may provide short-circuit paths through the electrodes on the sides of the MHD duct used. Moreover, the MHD generators design also imposed some limitations on the conducting fluid density as well as the type of magnetic field used in the MHD generators. Due to the flow of the ionized fluid through the MHD duct, the temperature and the velocity of the conducting fluid are reduced (Rosa 1976; Ishikawa and Itoh 2003). The Faraday MHD generator systems also suffer from the problem created by the Hall effect current (Kholshchevnikova 1966; Teno et al. 1966) which makes the Faraday generator very inefficient. A large Faraday generator needs an extremely powerful magnetic field (superconducting magnets) which is quite expensive to develop and hence increases the manufacturing and installation cost can be.

A simple Faraday MHD generator or Faraday generator consists of a wedge-shaped pipe or tube called MHD channel which is generally made up of some electrical insulator or non-conductive material. In a Faraday MHD generator, when an electrically conductive fluid or the plasma flows through the MHD channel, with the

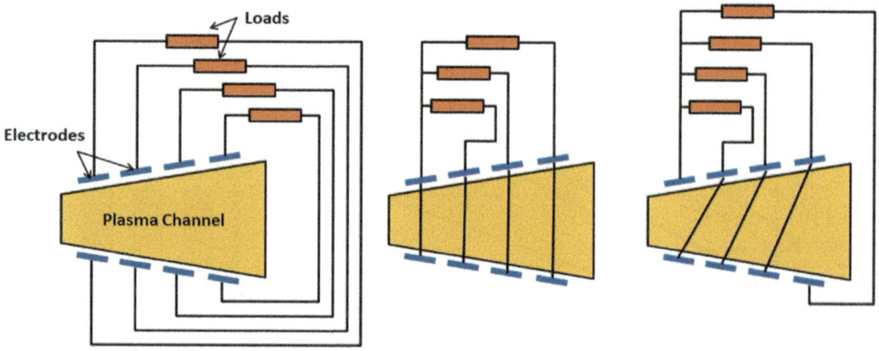

Fig. 34.8 Types of MHD generator: **a** Segmented electrode Faraday MHD system, **b** Hall MHD system, **c** Skewed electrode MHD system type [Photo Courtesy: Kayukawa 2004; Bera 2020]

interaction of the conducting plasma the magnetic field present, a voltage is induced across the plasma stream which can be harvested as the electrical power by placing two electrodes on the plasma surfaces at 90-degree angles to the magnetic field and the plasma flow direction.

One of the major practical problems in a Faraday generator is found as the short-circuit problem through the electrodes on the sides of the duct. The Hall effect current also makes the Faraday duct very inefficient. The optimal magnetic field inside the channel of the duct-shaped MHD generators is found as a sort of saddle shape which requires an extremely powerful magnet. The high magnetic field have been achieved by many research groups by using superconducting magnets as described earlier. Also, there are some other limitations on the density and type of magnetic field used. The conductive plasma gets gradually cooled and slowed down by this process. Faraday MHD generators are also developed with the electrodes segmented into small rectangular electrode geometries (Fig. 34.8a).

34.2.5.2 Hall MHD Generator

Hall effect is found to occur in the Faraday MHD generation systems. In Faraday MHD generation systems, when a large amount of electrical current is produced at the generator output side the current interacts with the magnetic field present which is also very large. The interaction between the electrical current and the in the system and magnetic field produces a charge particle displacement towards the perpendicular direction of the conducting fluid flow path. This phenomenon is known as the Hall effect (Kholshchevnikova 1966; Teno et al. 1966). As a result of the Hall effect, a transverse current is produced perpendicular to the direction of the conducting fluid. The total current produced in the MHD system is then obtained as the vector sum of the traverse current components (I_T) and axial current component (I_A). To overcome this problem created by the Hall effect and also to reduce the energy loss another MHD configuration is developed which is known as the Hall-MHD generator (Fig. 34.8). In Hall MHD generator, the efficiency is improved by reducing the energy loss. In Hall MHD generator, electrode configuration is modified. Each of the rectangular electrodes in the Hall MHD generator are split into an array of segments (Fig. 34.8b) and the segmented electrodes are placed side by side on both the sides of the fluid surface (Fig. 34.8b). The electrical power is collected at a higher voltage with a lower current amplitude. In order to do so, all the electrode segments of same side of the channel are kept insulated from each other but all the segments of one side are connected in series with their corresponding opposite electrode segments (Fig. 34.8b).

As shown in the Fig. 34.8b, the Hall generator is developed with the arrays of segmented electrodes (electrode segments of one side of the channel are kept insulated from each other) which are shorted or connected in series with their corresponding electrode segments of the opposite side of the fluid stream. In Hall generator, this type of segmented electrode replaces the concept of single rectangular electrodes of single electrode Faraday generator. Though, the losses of the Hall generators are found to be less compared to a Faraday generator, but, the generator's efficiency is found very sensitive to the electrical load connected for this type of electrode design.

Sometimes, skewed electrode Hall-MHD systems are also developed as shown in the Fig. 34.8c (Bera 2020). In the skewed electrode Hall MHD systems, the electrode segments of the Hall MHD system are shorted with the diagonal electrode segments of the opposite side of the fluid stream providing a skewed arrangement of the electrodes segments. In this skewed electrode Hall-MHD systems, the electrode segments are connected in a skewed pattern to align the electrode axes with the direction of the vector sum of the Faraday current and Hall Effect currents. Also, the skewed structure of the electrode segments allows us to extract the maximum amount of electrical energy from the ionized fluid used (Jackson and Ralph Strohl 2016).

34.2.5.3 Hall Effect Disc-Type MHD Generator

The Hall effect disc-type MHD generator (Kayukawa 2004; Bera 2020). A Magnetohydrodynamic (MHD) is developed with an MHD chamber having a disc-type geometry through which the conducting fluid can flow between the centers of the disc. The Hall effect disc-type MHD Generator is found as the most efficient design for MHD generators. In a Hall effect disc-type MHD generator the plasma or the ionized fluid flows between the center of a disc, and a duct wrapped around the edge of the disc. The magnetic field excitation is provided by a pair of circular Helmholtz coils (as shown in the Fig. 34.6b) above and below the disc. In the Hall effect disc-type MHD generators, the Faraday currents conducts in a perfect dead short around the disc periphery, whereas the Hall effect currents is found conducting between the ring electrode situated near the center duct and ring electrode situated near the periphery duct. The efficiency of the Hall effect disc-type MHD generator increases due to the reduction in resistance of the conducting fluid as the distance of the fluid flow is reduced for wide flat gas flow configuration achieved in the disc-type MHD generators. The magnets used in the Hall effect disc-type MHD generators are more efficient and produces simple parallel lines of the magnetic flux. In the Hall effect disc-type MHD generators, as the fluid is processed in a disc, the fluid can flow very closer to the magnet which increases the magnetic interaction.

The magnetic field in the Hall effect disc-type MHD generators is produced by a two circular Helmholtz coils above and below the disc. To provide the circulation paths to flow the Hall effect current, two pairs of the ring electrodes (RE) are placed inside the disc chamber. One pair of ring electrodes with smaller radius ($RE_{SmallRadius}$) is placed near the inlet duct at the disc center, whereas the other pair of ring electrodes with larger radius ($RE_{LargeRadius}$) is placed at near the periphery of the disc. In Hall effect disc-type MHD generators, the Faraday currents are passed through the disc periphery, whereas the Hall effect currents are conducted between the RE_{Small} and the RE_{Large}. The efficiency of the Hall effect disc-type MHD generators is enhanced due to the wide flat gas flow, parallel magnetic field lines, and increased magnetic field strength.

34.2.5.4 Plasma MHD

The MHD-based electric power generation systems can be developed either using gas plasma or the liquid metal flow or similar fluid with high electrical conductivity (Article et al. 2020b). The plasma-based MHD-based electrical power generators are the MHD generators which use the hot plasma as its conducting fluid (Article et al. 2020b). Plasma (Rosa et al. 1991a), which is often called "the fourth state of matter", is an ionized gas comprising of electrons (negatively charged particles) and the ions (positively charged particles). The plasma can be obtained by superheating the gas and by ionizing it to produce an ionized gaseous or fluidic matter (Article et al. 2020b) in which freely moving electrons are available. The plasma is produced by ripping the electrons away from the atoms by applying extra energy (heat). The ripping of electrons not only increase the number of free electron in the medium but also it produces positively charged ions making the gas ionized and electrically conducting in nature (Article et al. 2020b). In thermal ionization process of the plasma production, the gas-temperature is raised up to a certain point at which the electrons from the gas atoms are ripped away by applying heat energy. The electrons as well as the ionized atoms (ionized after losing the electrons) are found freely moving and hence the gas becomes gas plasma (electrically conductive). In order to make the gas plasma, some seeding materials are also mixed with the gas to be ionized which practically brings down the temperature at which the gas becomes ionized (Article et al. 2020b). On the contrary, to convert the gas into the gas plasma by only applying the heat energy, a much higher temperature is required. If some chemical agents are sued to ionize the gas at reduced temperature the chemical are called as the seeding materials. The seeding materials like alkali metal, salts, etc., are mixed with the gas to get ionized easily even at the lower temperatures.

As a conducting wire moving through a magnetic field can generate e.m.f. within it, in a plasma–MHD system, the plasma is passed through the MHD channel which is housed within a strong magnetic field. The interaction between the moving plasma with this magnetic field produces an e.m.f. across the plasma surfaces. The direction of the e.m.f. is found by using Fleming's right-hand rule. Using the Fleming's right-hand rule or the right-hand cork-screw rule, the e.m.f. direction is obtained just perpendicular to the direction of the plasma flow (v) as well as the direction of the magnetic field (B). Utilizing thermal ionization process gas plasma is produced for the MHD generators.

34.2.5.5 Coal-Fired MHD Systems

The coal-fired MHD systems (Jackson and Ralph Strohl 2016) use coal as the fuel to generate the heat energy required to produce the plasma. Coal-fired MHD systems burns the coal at a sufficiently high temperature to make the gas ionized. However, as the conducting fluid (plasma) passes through the MHD channel the plasma expands along the length of the chamber. With the expansion of the conducting fluid, the electrical conductivity as well as the fluid temperature both decreases. An MHD generator has low efficiency, around 20% (Hruby et al. 1986) and hence a pure MHD system is not much efficient for energy generation. Also, the conventional coalfired power thermal station has a 30%–40% efficiency (Hruby et al. 1986). To enhance the efficiency of the conventional coal fired power thermal station, a coal-fired MHD system is sometimes coupled with the conventional thermal power system to increase the power generation efficiency. In an MHD-coupled coal-fired thermal power plant, the hot conduction fluid (plasma) is first passed through the MHD generator to generate electrical power through MHD principle and then the plasma collected from the MHD channel exhaust is used to produce the steam to generate electrical power by driving the alternators coupled with the steam turbine.

34.2.5.6 Liquid Metal MHD Systems

Liquid metal MHD generators (Jackson and Ralph Strohl 2016; Branover et al. 1983) are the MHD generators which use the liquid metals as their electrically conducting fluids. As the liquid metals are used as the conducting fluid, the MHD generators are known as the liquid metal MHD generators. The electrical conductivity of the metals is extremely high and hence if the metals are found in a form of liquid, the metal in its liquid form is found to be one of the best conducting fluids for the MHD generators. As the liquid metals can be converted into a fluid flow stream, the liquid metals can suitably passed through the MHD channels to generate electrical power. If the liquid state of the metal is available at a lower temperature, the liquid metal MHD generators can be operated at lower temperature. The high temperature will not be required in these liquid metal MHD generating systems for the producing any plasma. In liquid metal MHD generators, the liquid metals are first combined with a

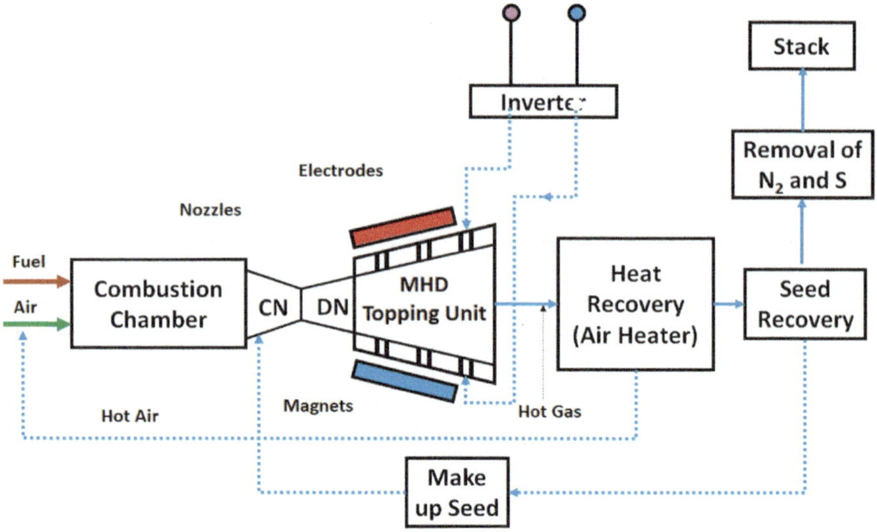

Fig. 34.9 Block diagram of the Open-cycle MHD system (Kayukawa 2004; Bera 2020; Magnetohydrodynamic (MHD) 2020; Ayeleso and Kahn 2018). The figure has been redrawn taking the reference from Ayeleso and Kahn (2018)

driving fluid or driving gas and then accelerated using a thermodynamic pump. After the acceleration obtained then the liquid metals could be separated from the driving gas before it passes through the MHD channel (Jackson and Ralph Strohl 2016). Therefore, the liquid metal MHD system can achieve a conducting fluid with high electrical conductivity at lower temperature which may be below 1,400 K (a much lower temperature compared to the plasma converters) (Jackson and Ralph Strohl 2016).

34.2.5.7 Open-Cycle MHD Systems

Depending on the conducting fluid paths, the MHD systems can be classified as the open-cycle MHD system (Kayukawa 2004; Bera 2020; A Magnetohydrodynamic (MHD) 2020; Ayeleso and Kahn 2018) or the closed-cycle MHD systems (Kayukawa 2004; Bera 2020; Ayeleso and Kahn 2018; Harada et al. 1984). In open-cycle MHD systems (Fig. 34.9), the conducting fluid is passed through the MHD channel and then discharged to the atmosphere after the generation of the electricity.

34.2.5.8 Closed-Cycle MHD Generators

In closed-cycle MHD systems (Ayeleso and Kahn 2018) the working fluid (conducting plasma) is recycled to utilize the heat energy left within the exhausted

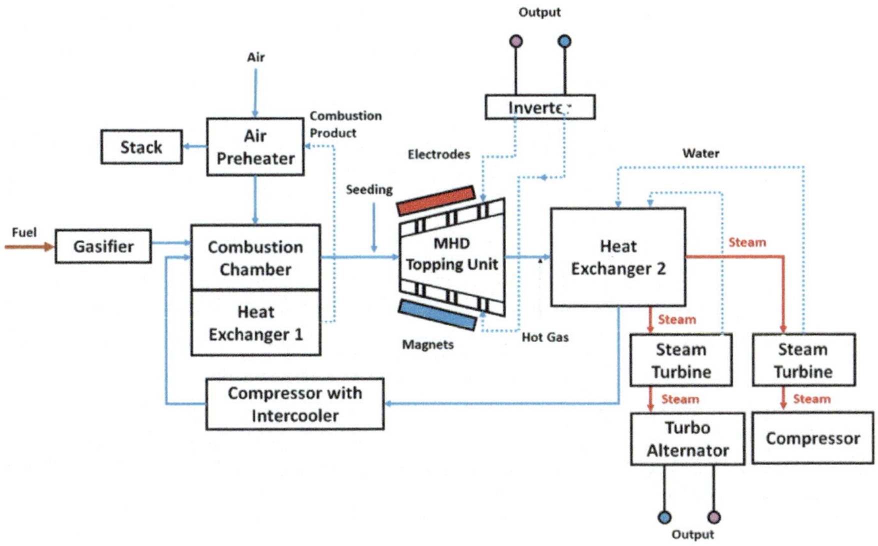

Fig. 34.10 Block diagram of the Closed-cycle MHD system (Ayeleso and Kahn 2018). The figure has been redrawn taking the reference from Ayeleso and Kahn (2018)

plasma after it comes out from the MHD channel. As shown the Fig. 34.10, the closed-cycle MHD system the heat is utilized after collecting it from the heat exchanger 2 and utilized for producing the steam to generate the electrical power using the turbo-alternator.

34.3 Advantages and Limitations of MHD Systems

MHD systems provide around 20% of energy conversion efficiency as a lot of heat remain inside the hot gas exhausted from the MHD channel outlet. Conventional coal-fired thermal power stations can achieve a maximum efficiency of about 35–40%. Coupling with an MHD generator, the efficiency of the coal-fired thermal plant can be enhanced up to 50–60% (Kayukawa 2004) by utilizing the heat energy of the exhausted hot gas collected at the MHD channel outlet. The hot gas collected from the MHD channel outlet is utilized to generate the steam in the conventional thermal power plant unit and hence the efficiency increases (Rosa et al. 1991b). The MHD generator couple thermal power plants generate electrical energy by harvesting the heat energy from the exhausted hot plasma which remains sufficiently hot to boil water evet at the MHD channel outlet. Thus, employing the MHD generators with the thermal power generation units the electricity are generated not only by both the MHD and thermal power units but also the overall efficiency increases. Moreover, as there are no solid moving parts in the MHD-based electrical power generators, the frictional or mechanical losses ere comparatively less with negligible wear and tear in the system during operation. Also, the effect on environmental pollution is much less for an MHD generating system compared to the conventional thermal power generation unit. During electrical power generation in MHD systems, the CO_2 emission is negligible compared to the coal-fired thermal power generation units power generation schemes (Harada et al. 2017). Though the installation cost for MHD generation goes high if the superconducting magnets are used, but the running cost is found low. A large amount of magnetic field required for MHD-based electrical power generation imposes few limitations and challenges in MHD generator design process. For high magnetic field, special system design, with proper magnetic shielding are required to be provided. The higher installation cost is one of the challenges for the MHD-based electrical power generation systems. Plasma generation needs high temperature and hence the MHD channel, magnetic coils and electrodes all are required to be properly designed with special materials and specifications. Velocity of the ionized fluid must be preferably high to generate a large amount of electrical energy.

34.4 Discussion and Conclusions

A MHD power generation system is an electrical power generating system which generates the electricity utilizing the MHD principle. MHD power generation technique generates the electric power directly from a moving stream of ionized fluid flowing through a magnetic field. Therefore, the MHD power generation systems are found as the non-conventional electric power generation modality which is considered as the green energy harvesting procedures. The MHD generators, utilizes the electromagnetic interaction of an ionized fluid flow and a magnetic field. The ionized fluids in MHD generators work as the moving electrical conductor and hence the electromotive force could be generated across the ionized conductor due to the Faraday's electromagnetic principle. An MHD system, therefore, can act as a fluid dynamo or the MHD power converter. In MHD, as the flow (motion) of the conducting fluid (conductor) under a magnetic field causes an induced voltage across the fluid, the e.m.f. would be found at the perpendicular direction to both the magnetic field and the fluid flow according to Fleming's right-hand rule. The concept of MHD power generation technique was first introduced by Michael Faraday in 1832 during his lecture at the Royal Society, UK. Since then, the MHD systems have been developed and studied by several research groups. In recent time, the green energy harvesting processes are found extremely important to reduce the pollution and to save the fossil fuel in the world for its sustainable development. In this direction, the MHD power generation technique could be utilized for green energy generation without any environmental pollution. In this chapter, the MHD technology has been discussed in detail followed by a discussion on its components, instrumentation, and design aspects. A detail review on the research works conducted on the MHD power generation process has been presented highlighting the major components of the MHD systems. Along with the limitations and challenges of the MHD power generation method, the present scenario and the future trends are also discussed. Different types of MHD generator geometries have been proposed with different channel geometries, different electrode configurations, different magnetic coil structures, and different working fluids or plasmas. A typical coal-fired MHD generator converts about 20% (Hruby et al. 1986) of the thermal input power to the output electricity and hence almost 80% of the thermal energy is still in the hot fluid stream which is found to be no longer usable for further MHD-based power production due to the low electrical conductivity of the fluid. As the plasma temperature at the MHD channel output is more than adequate to produce steam, the plasma obtained at the MHD channel exhaust is sometimes used to generate steam to generate the electrical power using conventional turbo-generators. Therefore, using the combined MHD/steam cycle systems (Hruby et al. 1986), which is found to be thermodynamically in series and electrically in parallel, an energy conversion system with a conversion efficiency up to 60% (Hruby et al. 1986) of the coal's energy can be converted into the electrical energy, whereas around 35 to 40% (Hruby et al. 1986) of the coal's energy is possible to be converted into the electrical energy for the conventional coal-fired thermal power plants. Therefore, the combination of a coal-fired thermal power generation system and a MHD generator promises greater efficiency, cheaper electricity and lower environmental pollution compared to the conventional fossil fuel-based thermal power stations (Hruby et al. 1986). Moreover, MHD generators produce fewer pollutants than conventional plants and are found as more environment-friendly. However, the higher construction costs of the MHD-based electrical power generation systems have, sometimes, limited their application in practical fields.

Acknowledgements The author acknowledges the National Institute of Technology Durgapur (NITDgp) for providing the research opportunity and infrastructure to conduct this study.

Conflict of Interest The author hereby confirms that there is no conflict of interest for this research work.

References

Abo-Khalil AG, Alghamdi AS (2021) MPPT of permanent magnet synchronous generator in tidal energy systems using support vector regression. Sustainability 13(4):2223

Ayeleso AO, Kahn MT (2018) Modelling of a combustible ionised gas in thermal power plants using MHD conversion system in South Africa. J King Saud Univ-Sci 30(3):367–374

Bansal NK (2014) Non-conventional energy resources. Vikas Publ House

Begamudre RD (2007) Electromechanical energy conversion with dynamics of machines. New Age International

Bera TK (2020) A magnetohydrodynamic (MHD) power generating system: a technical review. In: IOP Conference Series: Materials Science and Engineering, vol 955, 1st edn. IOP Publishing, p 012075

Bityurin VA, Bocharov AN, Lineberry JT (1999) In: 13th International conference on MHD power generation and high temperature technologies, vol 3. IEE CAS, Beijing, pp 12–15

Branover H (1978) Magnetohydrodynamic flow in ducts. hp
Branover H, Lykoudis PS, Yakhot A (eds) (1983) Liquid-metal flows and magnetohydrodynamics. American Institute of Aeronautics and Astronautics
Brown MA, Semelka RC (2011) MRI: basic principles and applications. John Wiley & Sons
Celinski ZN, Fischer FW (1966) AIAA J 4(3):421–428
Daybelge U, Kruger CH, Mitchner M (1968) Transport properties of MHD-generator plasmas. AIAA J 6(9):1712–1723
Development of MHD power generators, Internet Article. Encycl Br, Inc., https://www.britannica.com/technology/magnetohydrodynamic-power-generator, Accessed Date: July 2021
Dhareppagol VD, Saurav A (2013) Int J Adv Electl Electrn Engg 2:2278–8948
Drbal L, Westra K, Boston P (eds) (2012) Power plant engineering. Springer Science & Business Media
Harada N, Yamasaki H, Oyake T, Watanabe M, Shimizu E, Osogai T, Shioda S (1984) In: Proc. 22nd Symp. on Eng. Asp. of MHD, vol 3, 1st edn
Harada N, Takahashi K, Sasaki T, Kikuchi T (2017) In: international conference on system science and engineering (ICSSE), IEEE pp 191–195
Hardianto T, Sakamoto N, Harada N (2008) IEEE Trans Ind Appl 44(4):1116–1123
Hruby V, Petty S, Kessler R (1986) Platinum-clad electrodes for magnetohydrodynamic generators. Platin Met Rev 30(1):2–11
Hruby V, Petty S, Kessler R (1986) Platinum-clad electrodes for magnetohydrodynamic generators. Platin Met Rev 30(1):2–11
Internet Article, Plasma Science and Fusion Center, Massachusetts Institute of Technology (MIT). 77 Massachusetts Avenue, NW17, Cambridge, MA 02139, USA https://www.psfc.mit.edu/vision/what_is_plasma, Accessed Date: Sept 07, 2020a
Internet Article, Magnetohydrodynamic Electrical Power Generators, https://en.wikipedia.org/wiki/Magnetohydrodynamic_generator, Accessed Date: Sept09, 2020b
Ishikawa M, Itoh K (2003) Energy Convers Manage 44(13):2111–2124
Kantrowitz AR, Brogan TR, Rosa RJ, Louis JF (1962) IRE Trans Mil Electron 1:78–83
Karady GG, Holbert KE (2013) Electrical energy conversion and transport: an interactive computer-based approach, vol 64. John Wiley & Sons
Kayukawa N (1985) J Propul Power 1(4):309–310
Kayukawa N (2004) Open-cycle magnetohydrodynamic electrical power generation: a review and future perspectives. Prog Energy Combust Sci 30(1):33–60
Kessler R, Eustis RH (1968) Proc IEEE 56(9):1502–1510
Khan BH (2006) Non-conventional energy resources. Tata McGraw-Hill Educ
Kholshchevnikova EK (1966) J Appl Mech Tech Phys 7(4):48–54
Kirillin VA, Sheyndlin AY, Asinovskiy EI, Sychev VV, Zenkevich VB (1985) Foreign technology div wright-patterson afb oh
Maxwell JC (1888). An elementary treatise on electricity. At the Clarendon Press
Messerle HK, Messerle HK (1995) Magnetohydrodynamic electrical power generation. Wiley, Chichester, UK
Mukherjee D, Chakrabarti S (2004) Fundamentals of renewable energy systems. New Age Int
Nag PK (2002). Power plant engineering. Tata McGraw-Hill Educ
Onar OC (2017) Energy harvesting: solar, wind, and ocean energy conversion systems. CRC Press
Petrick M, Shumi͡atskiĭ BI, Shumi͡atskiĭ BI (1978) Open-cycle magnetohydrodynamic electrical power generation. Argonne Natl Lab
Pratt DT, Smoot L, Pratt D (1979) Pulverized coal combustion and gasification. Springer, Berlin
Rankin RR, Self SA, Eustis RH (1980) AIAA J 18(9):1094–1100
Rosa RJ (1961) Phys Fluids 4(2):182–194
Rosa RJ, Krueger CH, Shioda S (1991a) Plasmas in MHD power generation. IEEE Trans Plasma Sci 19(6):1180–1190
Rosa RJ, Krueger CH, Shioda S (1991b) IEEE Trans Plasma Sci 19(6):1180–1190

Rosa RJ (1976) Voltage consolidation and control circuits for multiple-electrode MHD generators. eamh, VII-5
Sarkar D (2016) Thermal power plant: pre-operational activities. Elsevier
Say MG (1976) Alternating current machines. Pitman
Scannell EP (1980) US Patent 4,185,213. US Patent and Trademark Office, Washington, DC
Schmidt VV, Schmidt VV, Müller P, Ustinov AV (1997) The physics of superconductors: introduction to fundamentals and applications. Springer Sci & Bus Media
Sheindlin AE, Jackson WD, Brzozowski WS, Rietjens LT (1979) In: Natural Resources Forum, vol 3, 2nd edn. Blackwell Publishing Ltd., Oxford, UK, pp 133–145
Soltani M, Kashkooli FM, Souri M, Rafiei B, Jabarifar M, Gharali K, Nathwani JS (2021) Environmental, economic, and social impacts of geothermal energy systems. Renew Sustain Energy Rev 140:110750
Steg L, Sutton GW. (1960) Astronautics 5
Stewart W (1966) U.S. Patent 3,275,860. U.S. Patent and Trademark Office, Washington, DC
Stewart W (1966) US Patent 3, pp 275–860. U.S. Patent and Trademark Office, Washington, DC
Stiglich JJ, Addington LA (1978) Hot pressed composite ceramic MHD electrode development. In: Process Cryst Ceram. Springer, Boston, MA, pp 493–503
Sukhatme SP, Nayak JK (2017) Solar energy. McGraw-Hill Educ
Tanuma T (2017). Introduction to steam turbines for power plants. In: Advances in steam turbines for modern power plants. Woodhead publishing, pp 3–9
Teno J, Brogan TR, DiNanno LR (1966) Hall configuration MHD generator studies. In: Electricity from MHD, vol III. Proceedings of a Symposium on Magnetohydrodynamic Electrical Power Generation
William D, Jackson, Strohl GR (2020) Magnetohydrodynamic power generator. In: Encyclopædia britannica, encyclopædia britannica, inc. Date Published: March 21, 2016. https://www.britannica.com/technology/magnetohydrodynamic-power-generator, Accessed Sept 08
Witalis EA (1965) J Nucl Energy Part C, Plasma Phys, Accel, Thermonucl Res 7(3):235
Zengyu X, Chuanjie P, Xiujie Z, Li Z, Xuru D, Yong L (2009) Plasma Sci Technol 11(4): 499

Chapter 35
Recent Advancement in Battery Energy Storage System for Launch Vehicle

Kiran H. Raut, Asha Shendge, and Jagdish Chaudhari

Abstract NASA is exploring a host of exciting planetary science exploration ideas for the next decade. The energy storage systems are required for the outer planet, inner planet, Mars, and small body missions. In space missions on energy storage systems place various essential performance conditions. It must be made to specifications and reviewed to maintain trust and to fulfill a wide variety of needs. The energy storage systems used in planetary science missions include main batteries (Non-rechargeable), secondary batteries (rechargeable), and condensers. Fuel cells have been used in human space missions but not in planetary science missions. Therefore, due to this limitation of the fuel cell, it is necessary to develop strong rechargeable batteries to increase the life of the launch vehicle. This chapter offers an overview of energy storage systems that are widely used in the launch vehicle. Storage technologies differ in terms of cost, cycle life, energy density, performance, power output, and discharge time. The benefits and drawbacks of various commercially developed battery chemistries are investigated. The chapter concludes with a discussion on lithium-ion battery recovery and reuse best practices. Advanced technologies are described in this study as those that have not yet been used in space missions and are still in progress. Main batteries, rechargeable batteries, fuel cells, capacitors, and flywheels are among the advanced technologies discussed in this chapter.

Keywords Types of batteries · Fuel cell · Ultracapacitor · Compress air energy storage

---

K. H. Raut (✉)
G H Raisoni Institute of Engineering and Technology, Nagpur, India

A. Shendge
G H Raisoni Institute of Engineering and Technology, Pune, India
e-mail: asha.shendge@raisoni.net

J. Chaudhari
Nagpur Institute of Technology, Nagpur, India

A. K. Bohre et al. (eds.), *Planning of Hybrid Renewable Energy Systems, Electric Vehicles and Microgrid*, Energy Systems in Electrical Engineering,
https://doi.org/10.1007/978-981-19-0979-5_35

Nomenclature

Pb–acid	Lead–acid
RFB	Redox Flow Battery
Li-ion	Lithium-ion
VRB	Vanadium Redox Battery
Li–air	Lithium–air
PSB	Polysulfide–Bromine Battery
Ni–Cd	Nickel–Cadmium
FC	Fuel Cell
Li–S	Lithium–sulfur
PEMFC	Proton Exchange Membrane Fuel cell
Ni–MH	Nickel–metal hydride
AFC	Alkaline Fuel Cell
Zn–air	Zn–air
PAFC	Phosphoric Acid Fuel Cell
IT	Information Technology
MCFC	Melting Carbonate Fuel Cell
UPS	Uninterruptible power supply
SOFC	Solid Oxide Fuel Cell
T & D	Transmission and Distribution
CAES	Compress air energy storage
PMDA	Power management and distribution
ESD	Energy Storage Devices

35.1 Introduction

The purpose of the chapter is to evaluate space power and energy storage technologies' current practice such that advanced energy and energy storage solutions for future space missions are developed and delivered in a timely manner. The major power subsystems are as follows:

1. Power generation,
2. Energy storage, and
3. Power management and distribution (PMDA).

Sub-systems for electricity generation/conversion are solar panels, radioisotope generating systems, power reactor systems, and fuel cells. Batteries, regenerating fuel cells, and condensers include the energy systems used in space missions. PMAD consists of power and transmission distribution, conversion and control, and load management and control. A variety of output parameters characterize power systems.

- The basic power (W/kg) which indicates how much power can be supplied per unit mass of power system is a highly important parameter.
- Basic energy (Wh/kg) and energy density (Wh/m^3) are other related parameters. Power structures cannot always be characterized by a single variable, such as a particular power.

Similarly significant may be other additional functionality. This may include exposure to the temperature, storage capacity, durability, resistance to radiation, etc. These other considerations are becoming more and more relevant as space missions are shifting from orbital missions to in situ missions with their harsh climate.

When viewing power systems, substantial mass and volume reduction (3 to 4x), improved reliability (2 to 3x), and low- and high-temperature operations as well as intense radiation conditions can be achieved. These advanced technologies allow for modern research and exploratory tasks such as power and energy storage: robotic missions, polar Mars missions and Moon missions, and distributed constellations of micro-spacecraft. Space power infrastructure comprises domestic defense systems, including aircraft unmanned (fuel cell, battery, and wireless) equipment, unmanned submarine (AUV) (fuel cell and battery), and mobile power soldier (PV, batteries, wireless power, and PMAD). Terrestrial energy benefit includes all-electric and fuel-driven automobiles, smart grid (PMAD), solar power systems (high-performance solar cells, advanced arrays, PV tuning, and solar concentrations), advanced nuclear systems, and hydroelectric systems, which are the only components of the Earth's energy sector.

The organization of this chapter is as follows. In the first section, classify different energy storage systems and compare the energy efficiency and energy consumption for different energy storage technologies. The fundamental knowledge of batteries and their benefits and limitations on space missions may be included in Section II. The application of the fuel cell using in space missions is discussed in Section III. The ultracapacitor role of energy storage devices is discussed in Section IV. In Section V, the flywheel in launch vehicles is of importance. Section VI discusses the role of compressed air systems.

The proposed work in the form of graphical representation is mentioned in Fig. 35.1.

35.2 Storage Type

Energy storage devices can be categorized as mechanical, electrochemical, chemical, electrical, or thermal devices, depending on the storage technology used as shown in Fig. 35.2.

The oldest technology is mechanical technology, which includes pumped hydropower generation. However, this technology has a drawback in that it needs a lot

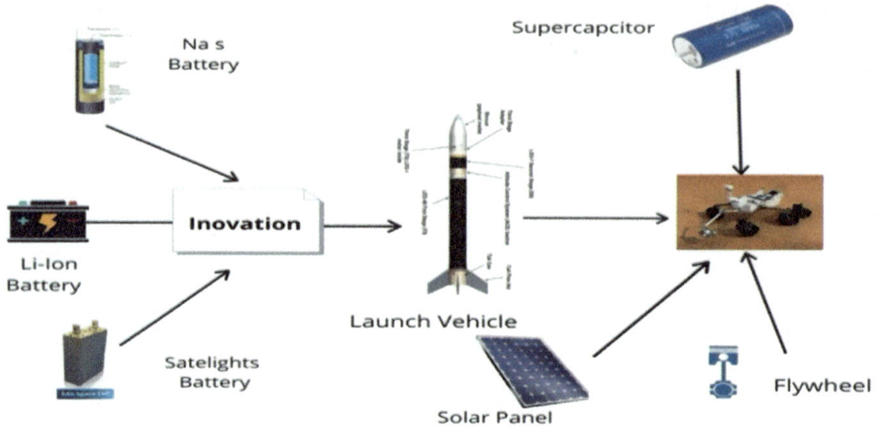

Fig. 35.1 Schematic representation of the proposed work

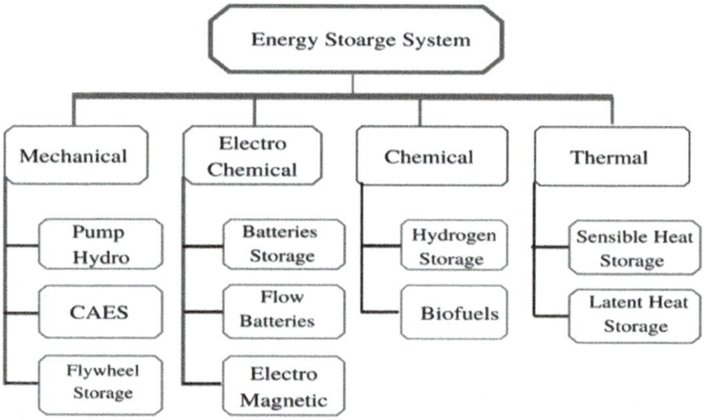

Fig. 35.2 Classification of Energy Storage Technology

of water and a different geographical elevation, as well as the building of power transmission lines to reach households that use electricity. The cost of constructing a transmission line has recently exceeded the cost of constructing a pumped hydropower plant.

Aside from the recent advent of mobile information technology (IT) devices and electric vehicles, expanded mass manufacturing of lithium secondary batteries and reduced prices have bolstered demand for energy storage devices that use those batteries. Using electrochemical technology, lithium secondary batteries transform electric energy to chemical energy and vice versa. Lead storage batteries and sodium–sulfur batteries are examples of such developments. Chemical technology, such as fuel cells, and mechanical technologies, such as electric double-layer capacitors, are examples of chemical technologies. The production and energy density of energy storage devices can be used to determine their efficiency. Based on the equipment used and the storage space, energy storage systems can be used for uninterruptible power supply (UPS), transmission and distribution (T&D) system service, or large-scale generation (Fig. 35.3).

Energy storage system battery technologies can be classified based on their energy capacity, charge and discharge (round trip) performance, life cycle, and environmental friendliness (Table 35.1). The sum of energy that can be contained in a single device per unit volume or weight is known as energy density. Lithium secondary batteries have a storage capacity of 150–250 W-hours per kilogram (kg), which is 1.5–2 times that of Na–S batteries, two to three times that of redox flow batteries, and about five times that of lead storage batteries (Fig. 35.4).

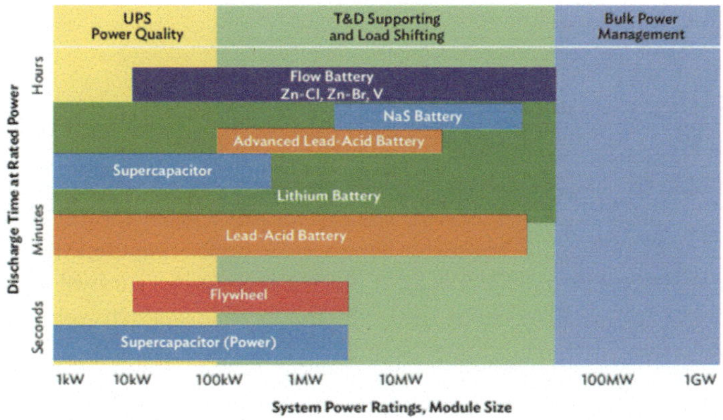

Fig. 35.3 Comparison of Power Output (in watts) and Energy Consumption (in watt-hours) for Various Energy Storage Technologies (Zhai 2018). *Source* Korea Battery Industry Association 2017 "Energy storage system technology and business model"

Table 35.1 Characteristics of various energy storage devices (Zhai 2018)

	Energy Density (Kw/Kg)	Round Trip Efficiency (%)	Life Span (Years)	Eco-friendliness
Li-ion	150	95	10–15	Yes
Na–S	125–150	78–85	10–15	No
Flow	60–80	70–75	5–10	No
Ni–Cd	40–60	60–80	10–15	No
Lead–Acid	30–50	60–70	3–6	No

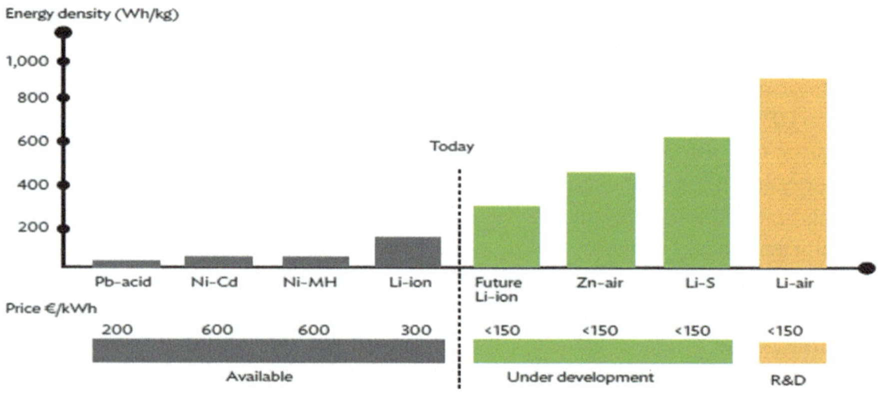

Fig. 35.4 Present and future of batteries technology (Zhai 2018). *Source* Second Life-Batteries as Flexible Storage for Renewable Energies, 2016

35.3 Batteries Chemistry Type

In space missions, batteries are used for a number of purposes. Primary batteries (single discharge batteries) are used in missions that need just a few minutes to several hours of electrical power. Rechargeable batteries (secondary batteries) have been used mainly for load-leveling and for providing electrical power for survival during eclipse periods on solar-powered missions and as the source of power extravehicular activity suits (Chin et al 2018). They have been used in orbital missions (TOPEX, Mars Global Surveyor, and Mars Reconnaissance Observer) as well as Mars Landers (Mars Pathfinder) and Mars Rovers (Spirit and Opportunity). Primary and rechargeable batteries are heavy, bulky, and have limited capability to function in extreme space environments such as high and low temperatures and radiation. Safety concerns exist with some of the primary lithium and rechargeable Li-Ion batteries.

35.3.1 Lithium-Ion (Li-Ion) Battery

Li-ion battery chemicals are considered stable with the most energy density. Based on the desirable performance of electrochemical properties, lithium-ion batteries (LIBs) are commonly used as electrochemical sources in portable electronic applications like smartphone, calculators, and personal and video cameras; they can even play an important role as hybrid vehicles' power supplies in the future. Lithium-ion batteries can be classified mainly into three categories, i.e., anode, cathode, and electrolyte components. The commercially most used anode element is graphite. Lithium iron phosphate (LiFePO4) and lithium manganese (LiMnO2) are being used for the use of cathode products such as sheet structured lithium cobalt oxide (LiCoO2) and titanium disulfide (TiS2) (Fig. 35.5 and Tables 35.2 and 35.3).

Fig. 35.5 Li-Ion batteries MER Rover (Chin et al 2018)

Table 35.2 Advantages and disadvantages of Li-Ion batteries

S. No	Advantages	Disadvantages
1	High energy density (120 W h/kg)	Required protection circuit for thermal runaway
2	High voltage (up to 3–6 V)	High-temperature degradation and processed at high voltage
3	Long Cycle (500–1000 cycles)	Freezing temperature failure to charge quickly
4	Wide temperature range (–20 to 60◦C)	
5	Minimum memory effect	

Table 35.3 Advantages and disadvantages of lead–acid batteries

S. No	Advantages	Disadvantages
1	Simple to manufacture	Full charge takes 14–16 h
2	Low cost	Cycle period is limited
3	High specific power	Low specific Energy
4	Low self-discharge	Repeated deep-cycling reduces battery life
5	Good low- and high-temperature performance	Poor weight-to-energy ratio

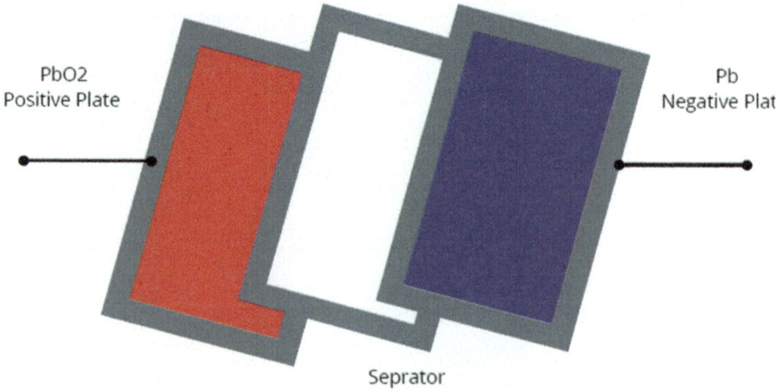

Fig. 35.6 Principle components and chemistry of lead–acid batteries

35.3.2 Lead–Acid (PbA) Battery

The secondary cell type is often found in automobiles and other applications with high load current values. Its key advantages are low cost of capital, technological sophistication, and effective recycling (Mahlia et al. 2014).

The voltage of the nominal cell at 2.05 V is very high. The positive active substance is intensely soluble lead dioxide and the negative active substance is thinly separated. The electrolyte is a diluted sulfuric acid used in the discharge process. As HSO4 ions are discharged, they move to the negative electrode and produce H + ions. At the positive electrode, lead dioxide reacts with the electrolyte to form lead sulfate crystals and water. Both electrodes are discharged to lead sulfate which is a poor conductor, and the electrolyte is progressively diluted as the discharge proceeds (Fig. 35.6).

35.3.3 Nickel–Cadmium (Ni–Cd) Battery

Nickel–Cadmium (Ni–Cd) Battery is the only energy storage device used since the start of the space program on lingerie-life devices. Nickel–Cadmium batteries were used for the early explorer satellites. Nickel oxides electrodes are typical to batteries of Nickel–Cadmium or Nickel–Hydrogen. Only the alkaline Nickel device used in space has a capability of thousands of cycles and standing lives in geosynchronous equatorial orbits in low- to medium-sized altitudes for more than 10 years (Fig. 35.7 and Table 35.4).

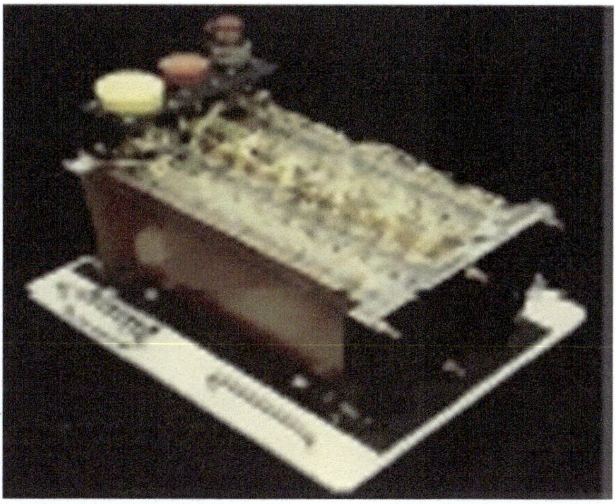

Fig. 35.7 Ni–Cd solar max battery (Luo et al. 2015)

Table 35.4 Advantages and disadvantages of Ni–Cd battery

S. No	Advantages	Disadvantages
1	Lowest cost per cycle	Cadmium is a poisonous metal; not disposable in sites with waste
2	Availability in a wide range of sizes and performance options	High self-discharge
3	Long cycle life	Low specific energy compared with newer systems
4	Load performance is good	Low cell voltage of 1.20 V requires many cells to achieve high voltage
5	Only battery that can be ultra-fast-charged with little stress	

35.3.4 Nickel–Metal Hydride (Ni–MH) Battery

Since their introduction in 1991, the efficiency of Ni–MH batteries has steadily improved according to a range of innovative approaches such as high-density negative electrodes, thinner separators, revamped positive electrodes, and improved packaging efficiencies. For portable power applications, Ni–MH batteries are currently manufactured in large quantities. Nickel–metal hydride batteries have now become a proven power source for HEVs (Luo et al. 2015).

Other rechargeable batteries are outperformed by Ni–MH batteries, which have a higher power and less voltage depression (Fig. 35.8 and Table 35.5).

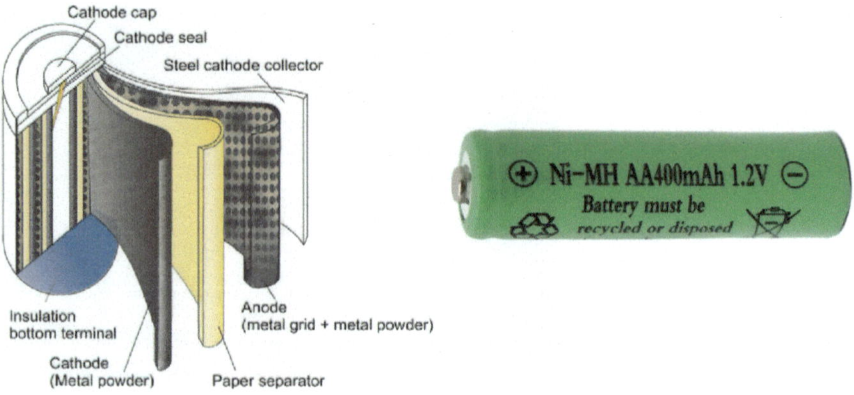

Fig. 35.8 Physical structure of Ni–MH Battery (Pickett 1990)

Table 35.5 Advantages and disadvantages of Ni–MH battery

S. No	Advantages	Disadvantages
1	At low temperature gives greater service advantages over other primary types of battery	Need for a more complex charge algorithm
2	Battery manufacturing, use, and recycling restrictions due to concerns about Cadmium toxicity should be removed	High self-discharge (50% higher self-discharge compared with the Ni–Cd)
3	Longer run times or less space used for the battery can be achieved by increasing energy density	Limited service life (deteriorate after 200–300 cycles)
4	Because of the many structural parallels between the two chemistries, it is easier to incorporate into devices that already use Nickel–Cadmium batteries	Limited discharge current (Good results are achieved with load currents of one-fifth to one-half of the rated capacity)

35.3.5 Sodium–Sulfur (Na–S) Battery

The Na–S battery, also known as a liquid metal battery, is a molten metal battery made of sodium and sulfur. It has a high energy density, high charge and discharge quality (89–92%), and a long cycle life, and it is made of low-cost materials. However, due to their high operating temperatures of 300 °C–350 °C and the extremely corrosive composition of sodium polysulfides, such cells are mainly used for large-scale nonmobile applications such as energy storage for the electricity grid (Fig. 35.9 and Table 35.6).

Fig. 35.9 Structure of Na–S Battery (Schledde, et al. 2018)

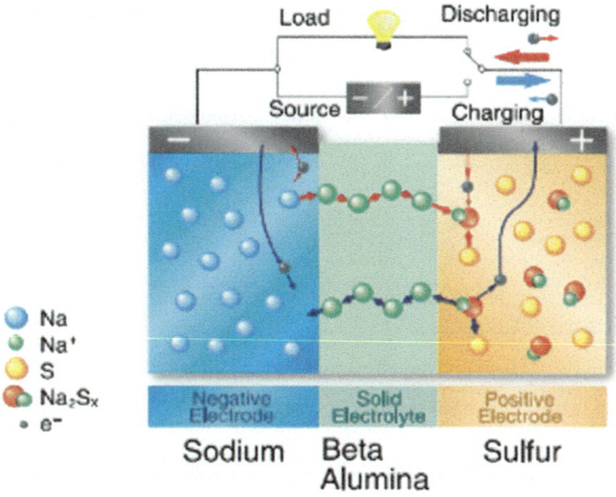

$$2Na + xS \underset{\text{Charge}}{\overset{\text{Discharge}}{\rightleftharpoons}} Na_2S_x$$

Table 35.6 Advantages and disadvantages of Na–S battery

S. No	Advantages	Disadvantages
1	High cycle life	Maintenance required
2	High Efficiency (75 to 90%)	Extra cost of constructing the enclosing structure to prevent leakage
3	Flexible in operation	Need to be operated above 300 °C
4	Good energy density(150–240 Wh/kg)	Highly reactive nature of metallic sodium which is combustible when exposed to water
5	High power density (150–240 Wh/kg)	

35.3.6 Silver Zinc Battery

The main Ag–Zn battery used in the Russian spacecraft Sputnik, launched on October 4, 1956, was the first use of a battery in an orbital spacecraft. This primary battery was used to fuel the spacecraft's communication and navigation systems. Since there were no solar panels available for charging, connectivity was cut off until the electricity ran out. The Ag–Zn primary was designed to last three weeks and supply fuel to the 84 kg spacecraft. The spacecraft stayed in orbit for three months in total (Salkind et al. 2009) (Fig. 35.10 and Table 35.7).

Fig. 35.10 Ag–Zn Battery [Mars pathfinder Lander] (Pickett 1990)

Table 35.7 Advantages and disadvantages of Na–S battery

S. No	Advantages	Disadvantages
1	High charge efficiency, typically >99%	High cost
2	Very high specific energy (up to 250 W h kg^{1})	Limited cycle life (typically no more than 100 cycles)
3	Highest specific power (up to 600 W kg^{1})	Relatively slow recharge
4	Safer to operate than most other systems	If used repeatedly, accelerate the rate of capacity fade of the cells
5		Higher rates reduce the charge efficiency

35.3.7 Redox Flow Battery (RFB)

Both electrochemical energy storage schemes transform electrical energy into chemical energy when charged. Conversion and recycling were carried out in closed cells of traditional batteries. The conversion and storing of energy are, therefore, distinguished by redox flow batteries. Redox flow batteries are different from other conventional batteries in that the energy storage material is conveyed by an energy converter. It requires the energy storage material to be in a flowable form. This structure is like that of fuel cells, whereby in redox flow batteries, charging and discharging processes can take place in the same cell. Different vanadium redox batteries are mentioned in Table 35.8 (Fig. 35.11 and Table 35.9).

Table 35.8 Types of Vanadium Redox Batteries

S. No	Types	Description
1	Vanadium Redox Battery (VRB)	Electrolytes: In VRBs use two vanadium electrolytes (V2 + /V3 + and V4 + V5 +) hich exchange hydrogen ions (H +) through a membrane
2	Polysulfide–Bromine Battery (PSB)	Electrolytes: Sodium sulfide (Na2S2) and sodium tribromide (NaBr3). The sodium ions (Na + pass through the membrane during the charging or discharging process
3	Zinc–bromine (Zn–Br) Battery	Electrodes: Solutions of zinc and a complex bromine

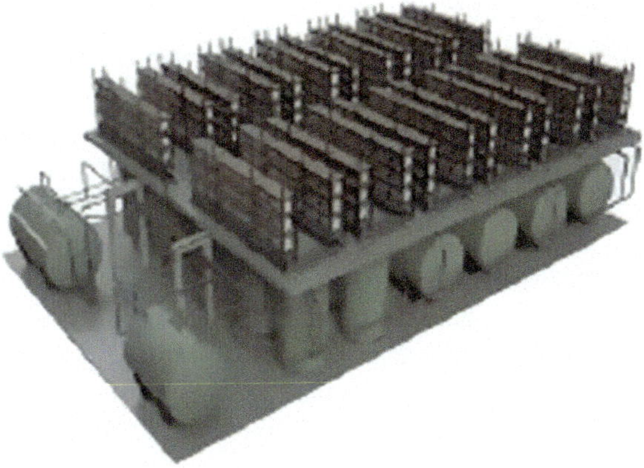

Fig. 35.11 Structure of redox flow battery (Fischer and Tuebke 2018)

Table 35.9 Advantages and disadvantages of RFB battery

S. No	Advantages	Disadvantages
1	Longer service life (around 20 years)	Complexity RFB systems require additional equipment
2	Flexible in design	Low energy density
3	The possibility of a fire with the batteries is extremely low	

Below Fig. 35.12 shows various energy storage batteries in MW scale battery energy storage projects in China and in the world.

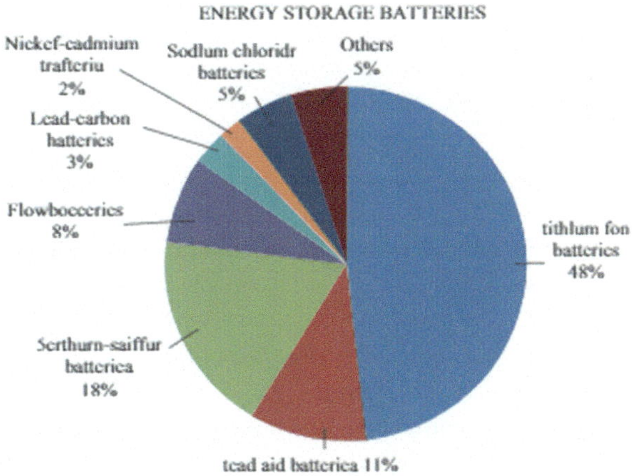

Fig. 35.12 various energy storage batteries in MW scale battery energy storage (Guo and Niu 2018)

35.4 Fuel Cell

A fuel cell (FC) integrates an oxidizer (oxygen or air) with the fuel (hydrogen or hydrogen source) to provide electricity. Similar to batteries, these devices do not fall off or need to be charged. Like a battery, a fuel cell is divided into two electrodes—one cathode and one anode. However, at least one solid metal electrode is used by the batteries and is absorbed with the generation of electricity. In a cell, electricity is not absorbed, so far that more fuel and oxidizer are injected into the electrode and the cell will generate electricity.

For human space missions (e.g., crews, recycled launch vehicles, or human lunar precursor missions), the fuel cells are especially desirable and require several kilowatts of power over long stretches of up to 10 days. Conventional batteries in terms of their much lower real energy and manageability problems are not ideal for such applications. For fractions of an hour to certain hours, planetary missions need a few watts to 100 s of watts.

For a variety of commercial and military uses, many types of fuel cells have been developed. They are as follows: a) Proton Exchange Membrane (PEM) Fuel cell running at a temperature of 80 °C. (b) Alkaline; 175 °C system. (c) Phosphoric acid at 1750 °C. (d) Melting carbonate at a maximum temperature of 650 °C. (e) Solid Oxide; 900–1000 °C. The most promising for future planetary missions include the fuel cells of H2–O2 PEM as well as the regenerative fuel cells, which are listed below (Table 35.10).

For large stationary applications, many varieties of fuel cells are suitable for power generation. Worldwide distributed generation is used by AFC, PAFC, PEMFC, SOFC, and MCFC networks. Figure 35.13 shows the relative weight of the different technology built for high-capacity fuel cells by 2018 (Nadeem et al. 2019; Jose Bellosta von Colbeet al 2019; Tixador 2008) (Fig. 35.14).

35.5 UltraCapacitor

Capacitors are normally used to meet peak power requirements on spacecraft. In the deep space missions of Galileo and Cassini, tantalum condensers (solid and electrolytic designs) were used. The main benefit of condensers is the potential for hundreds of thousands of cycles to supply high pulses over short times. Condensers have low energy and low energy density as their main drawbacks.

NASA also created a new technique to produce supercapacitors to create the energy storage medium by combining electrochemical condenser systems and batteries. Not only can this invention be used to transport electrodes properly, but it also suppresses electrical shortages between electrodes, has a comparatively low interface resistance between any electrode and every material that divides the electrodes physically, and reduces its power (Fig. 35.15 and Table 35.11).

Table 35.10 Advantages, disadvantages, and application of different fuel cells (Felseghi and Carcadea 2019)

S. No	Fuel cell type	Advantages	Disadvantages	Application
1	Proton Exchange Membrane Fuel cell (PEMFC)	(1) Low temperature (2) Quickly start-up (3) Solid electrolyte reduces corrosion and electrolyte management problems	(1) Sensitive to fuel impurities (2) Expensive	(1) Grid support (2) Portable power; Power to power (P2P)
2	Alkaline Fuel Cell (AFC)	(1) Stable materials allow lower cost components (2) Low temperature (3) Quickly start-up	(1) Sensitive to CO2 in fuel and air 2) Electrolyte conductivity (3) Electrolyte management (aqueous)	(1) Military (2) Space
3	Phosphoric Acid Fuel Cell (PAFC)	(1) Suitable for CHP (2) Increased tolerance to fuel impurities	(1) Expensive catalysts (2) Long start-up time (3) Sulfur sensitivity	(1) Distributed generation
4	Melting Carbonate Fuel Cell (MCFC)	(1) Suitable for CHP (2) High Efficiency (3) Suitable for gas turbine cycle (4) Fuel Flexibility	(1) High-temperature corrosion and breakdown of cell components (2) Long start-up time (3) Low power density	(1) Electric utility (2) Distributed generation
5	Solid Oxide Fuel Cell	(1) Potential for reversible operation (2) Suitable for CHP (3) Suitable for gas turbine cycle (4) High Efficiency	(1) High-temperature corrosion and breakdown of cell components (2) Long start-up time (3) Limited number of	(1) Auxiliary power (2) Electric utility

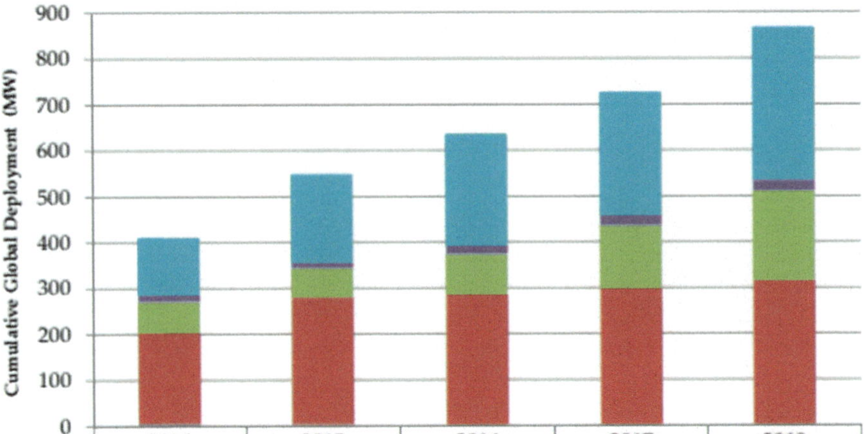

	2014	2015	2016	2017	2018
SOFC	126	193	244	270	334
PEMFC	14	12	18	21	23
PAFC	68	64	88	137	196
MCFC	198	276	283	297	312
AFC	4	3	2	1	2

Fig. 35.13 Large-scale stationary fuel cells

		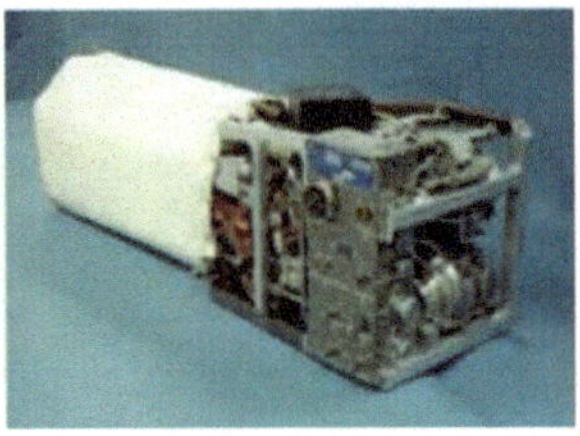
Polymer Electrolyte Membrane (PEM) Fuel Cell(Gemini)	Alkaline Fuel Cell (Apollo)	Alkaline Fuel Cell (Space Shuttle)

Fig. 35.14 Different fuel cells used in space (Felseghi and Carcadea 2019)

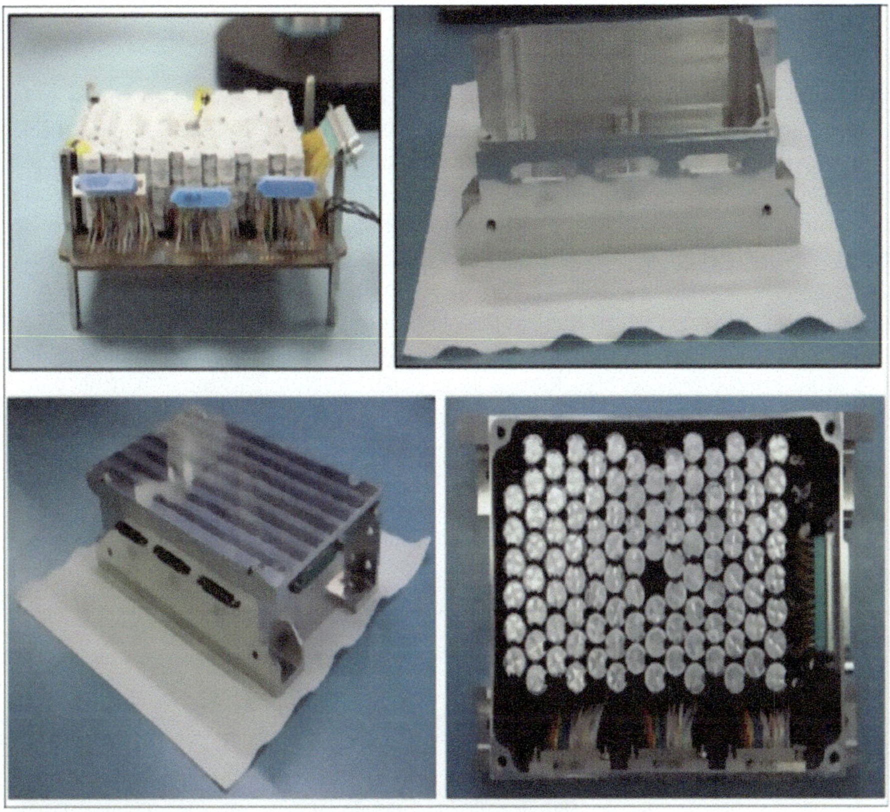

Fig. 35.15 Bank of supercapacitor for space application (Rose et al. 1995)

Table 35.11 Advantages and disadvantages of Ultracapacitor (IRENA 2020)

S. No	Advantages	Disadvantages
1	A longer cycle life under multiple charge/discharge cycles, hence a longer satellite lifetime	Very low specific/energy density compared with conventional rechargeable batteries
2	High charge/discharge efficiency, thus reducing the number of solar cells required	Voltage Interface Compatibility
3	The ability to operate high power demanding payloads while keeping mass and volume of the energy store to a minimum	Space Environment Survivability
4	Insensitivity to temperature changes which leads to a simpler thermal system and mass saving	

35.6 FlyWheels

NASA aims to improve the flywheel's capacity, a wheel that converts energy to power a motor, so that chemical batteries can be replaced on future spacecraft and millions of dollars can be saved. The flywheels are similar to the revolving toy, but they rotate and are much bigger at higher speeds. Before the batteries were commonly used, they were more popular, but they are still used in pottery wheels, clock gears, and treadle sewing machines.

Flywheels could play a crucial role in the driving of the vehicles which take us there in the future as NASA's vision takes people to and beyond the Moon. Flywheels are rotating wheels used to control or accumulate equipment. In a quickly rotating wheel, they store kinetic energy, also known as potential energy, and move it if necessary to use(Nadeem et al. 2019). They are somehow like non-chemical batteries, with a couple of additional benefits: They have no acids or other dangerous material, and like certain batteries are, they are not harmed by high temperatures.

Flywheels were mostly huge, low-speed steel wheels spinning. A new technology allows the use of a more advanced flywheel to store electrical energy. New flywheels use a generator for rotating and transforming kinetic energy into electricity. Flywheels today provide hospitals and production facilities with backup control. Their use of hybrid electric cars is also considered (Fig. 35.16 and Table 35.12).

Fig. 35.16 Flywheel energy storage (Santiago 2014)

Table 35.12 Advantages and disadvantages of Flywheel (Qian 2010)

S. No	Advantages	Disadvantages
1	High power density	Metallic structures are speed limited due to fatigue
2	High energy density	The flywheel rim cannot benefit from the use of high strain composite materials
3	Service life of flywheel systems (15 + years) as compared to chemicals batteries	High radial compressive stresses occur in the composite rim, leading to shortened performance life due to creep
4	The temperature range for flywheels is wider	Short discharge time
5	Large energy storage capacity	

35.7 Compress Air Energy Storage

CAES refers to the energy that is collected as high-pressure air and absorbed in a distinct way than compressed air in energy conversion. In favor of electrical network service, compressed air energy storage operates in compressors using air at high pressure during times of low energy demand, and then the stored compressed air is released to supply the electricity generation expander to satisfy the high demand in peak periods of time as shown in Fig. 35.17.

There are two separate operational phases of the CAES plant, namely compression and extension, as shown in Fig. 35.18. The machine performance (48–54%) is better than the conventional Gas-Turbine systems because the two phases do not function simulcast. Two major commercial CAES plants are currently invoked (Venkataramani et al. 2016).

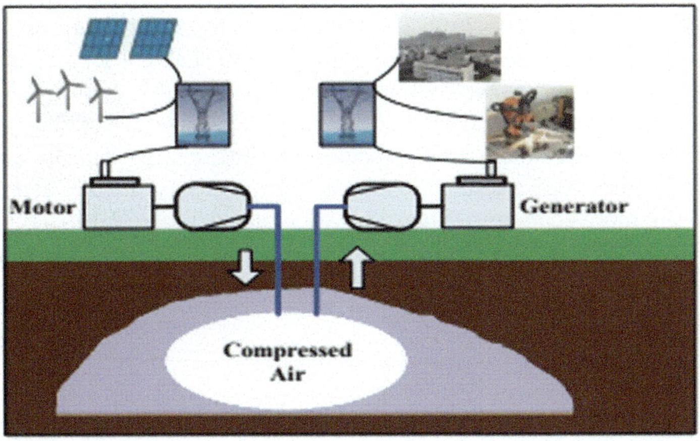

Fig. 35.17 Process of compress energy storage

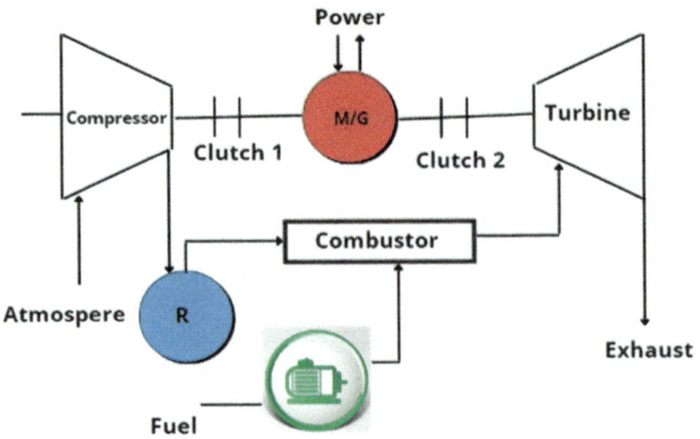

Fig. 35.18 Schematic diagram of CAES system (C—Compressor, G-T—Gas turbine, M/G—Motor/Generator, P—Pump, and R—Reservoir)

Table 35.13 Advantages and disadvantages of compress air energy storage (Schledde et al. 2018)

S. No	Advantages	Disadvantages
1	Free from fire hazards	Damage may occur due to higher pressure in the storage vessel
2	Does not cause any pollution	Energy losses may occur due to poor maintenance in transmission and distribution
3	Easier maintenance	Insufficient control and leakage due to loss of compress power
4	Less cost	Less power output
5	Reliability in operation and remote controlling	Probability leakage of air

In 1978, the first CAES plant was set up in Huntorf to operate (Hariprakash et al. 2009). Its capacity for load-subsequent operation and peak demand while keeping the constant nuclear power plant's capacity factor is rated to 290 MW. In 1991, McIntosh began its second large-scale CAES plant (Hariprakash et al. 2009). This power plant will deliver its maximum power output for up to 26 h, with a capacity of 110 MW with a storage capacity of 2700 MWh (Table 35.13).

Comparing the investment cost, capacity, lifetime, energy density, and storage duration, PHS and CAES are suitable for use in large-scale commercial applications where they are more economic (Figs. 35.19, 35.20 and Table 35.14.

35.8 Conclusion

A small selection of battery technology is the focus of current scientific research. Each technology, however, offers various application areas as well as challenges and problems.

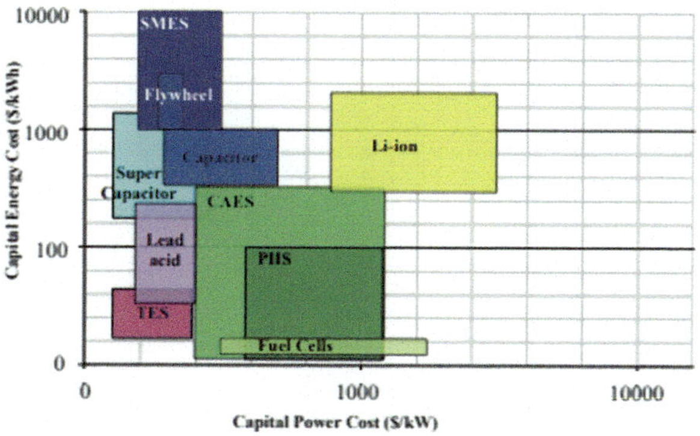

Fig. 35.19 Capital energy cost versus capital power cost (Luo et al. 2014, 2015; Sameer and Johannes 2015; Kousksou et al. 2014; Mahlia et al. 2014; Venkataramani et al. 2016; Wang et al. 2017)

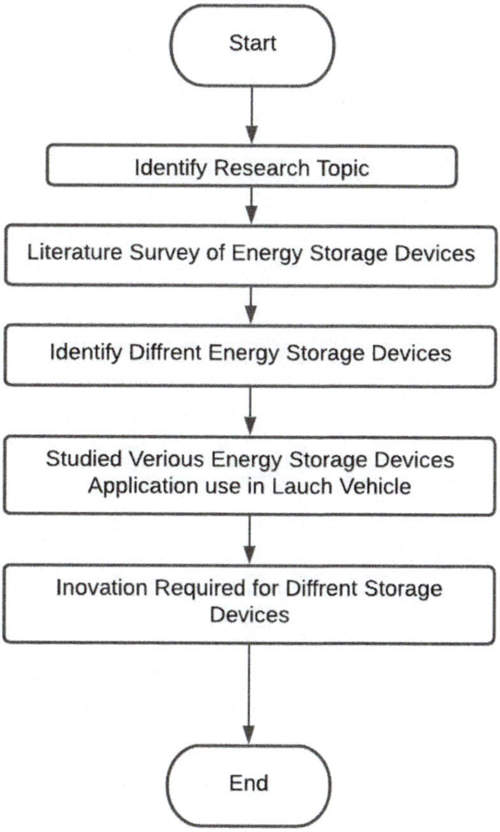

Fig. 35.20 Flow chart of the proposed work

Table 35.14 Comparison of different energy storage technologies

	Lead–acid	lithium-ion	Na–S	Redox flow	Super capacitor	flywheel	CAES
Energy density in Wh/l	50–100	200–350	150–250	20–70	2–10	80–200	3–6
Installation costs In €/kw	150-200	150–200	150–200	1000–1500	150- 200	300	700–1000
Installation costs In €/kwh	100-250	300–800	500–700	300–500	10,000-20,000	1000	40–80
Reaction time	3-5 ms	3-5 ms	3-5 ms	>1 s	<10 ms	>10 ms	3–10 min
Self-discharge rate	0.1–0.4 % / day	5% month	10%/ day	0.1–0.4%/day	up to 25% in first 48 h	5–15 % h	05–1 % / day
Cycle l ife time	500-2000	2000-7000	5000–10,000	>10,000	>1Mill	>1 Mill	–
Lifetime In years	5–15	5–20	15–20	10–15	15	15	ca.25
System efficiency in %	70–75	80–85	68–75	70–80	77–83	80–95	60–70

Due to a lack of reversibility, high cycle numbers, and theoretically projected high capacities, novel technologies with promising high energy densities, such as lithium–sulfur and lithium–air, are now failing to realize and commercialize. Electrolyte stability, cell toxicity, clogging phenomena, and the use of active species and electrodes are all contributing factors. As a result, the solution can only be found at the material level and will necessitate the development of new materials and designs as well as a high level of expertise.

Further innovative technologies, such as redox flow and molten-salt systems, which already give sufficient reversibility and require less intense effort on the material level, naturally find their way into larger testing systems and integration concepts more quickly. Nonetheless, both technologies, as well as the well-established lithium-ion technology, have a lot of room for improvement in terms of electrode design and electrolytic solutions at the material level.

In the scientific review, lithium-ion technology is the main focus at the system level, similar to the project analysis. The high degree of maturity, as well as the high demands on safety, maintaining a tight thermal range, and complex battery state estimation techniques of lithium-ion batteries are all reasons for this. Furthermore, widespread usage of lithium-ion technology in other industries (like transportation and consumer electronics) may have spillover effects on the energy industry.

This book chapter is a guide to implementations, innovations, benefits, and inconveniences which should be taken into account in the assessment of the viability of a battery power storage system. Increase the flexibility of the grid such that sometimes cycles of unsustainable renewable energy production are not limited, or so that major network expansion costs leading to high market prices are less necessary.

35.9 Future Scope

A huge improvement in battery performance with new designs that meet specific purposes is key to meet the future needs of our society. Although initially batteries were simple, their development in comparison with other fields of electronics was quite slow. One major challenge in research is to find the right materials for electrodes and electrolytes, which actually work well together with other aspects of a battery design without undue compromise. In choosing the best combination of design parameters, there is much trial and error.

A combination of battery technologies is likely to be in use in the future, although it is not yet possible to say which these will be or to identify which type will be the most environmentally sustainable. However, it is possible to highlight which areas of battery design need attention in order to ensure that future battery technologies are as sustainable as possible, while continuing to fulfill their valuable purpose. This report draws attention to those areas.

References

Zhai Y (2018) Handbook on battery energy storage system. Asian Dev Bank, ISBN 978-92-9261-470-6 (print), 978-92-9261-471-6 (electronic). https://doi.org/10.22617/TCS189791-2

Ye K, Li P, Li H (2020) Optimization of hybrid energy storage system control strategy for pure electric vehicle based on typical driving cycle. Hindawi Math Prob Eng 29. https://doi.org/10.1155/2020/1365195

Sankarkumar RS, Natarajan R (2021) Energy management techniques and topologies suitable for hybrid energy storage system powered electric vehicles: an overview. Int Trans Electr Energ Syst. John Wiley & Sons Ltd. https://doi.org/10.1002/2050-7038.12819

Zhang Q, Li G (2019) A predictive energy management system for hybrid energy storage systems in electric vehicles. Electr Eng 101:759–770. https://doi.org/10.1007/s00202-019-00822-9

Chin et al (2018) Energy storage technologies for small satellite applications. IEEE Trans. https://doi.org/10.1109/JPROC.2018.2793158

Nadeem F, Hussain SMS et al (2019) Comparative review of energy storage systems, their roles and impacts on future power systems. IEEE Trans 7:11. https://doi.org/10.1109/ACCESS.2018.2888497

Jose Bellosta von Colbeet al (2019) Application of hydrides in hydrogen storage and compression: achievements, outlook and perspectives. Int J Hydrogen Energy 44(15):7780–7808. https://doi.org/10.1016/j.ijhydene.2019.01.104

Tixador P (2008) Superconducting magnetic energy storage: status and perspective. IEEE/CSC & ESAS European Superconductivity News Forum

Chaudhary K , Kumar D, (2018) Satellite solar wireless power transfer for base load ground supply: clean energy for the future. Eur J Futur Res 6–9. https://doi.org/10.1186/s40309-018-0139-7

IRENA (2020), Electricity storage valuation framework: assessing system value and ensuring project viability. Int Renew Energy Agency, Abu Dhabi.www.irena.com

Hybrid Capacitor Market_ Global Industry Analysis, Size, Share, Growth, Trends, and Forecast 2017–2025. Transparency Market Res. https://www.transparencymarketresearch.com/hybridcapacitor

Schledde D, Dabrowski T et al (2018) Support to R&D strategy for battery based energy storage technical analysis of ongoing projects. European Commission Directorate General Energy, 15 August Project number: POWNL16059

Posielek T (2018) An energy management approach for satellites. In: 69th international astronautical congress, Bremen, Germany, pp 1–5

Shimizu NT, Underwood C (2013) Super-capacitor energy storage for micro-satellites: feasibility and potential mission applications. Acta Astronaut 85:138–154. https://doi.org/10.1016/j.actaastro.2012.12.005

Kalair A, Abas N (2021) Role of energy storage systems in energy transition from fossil fuels to renewables. Energy Storage J Wiley Online Publ 3(1). https://doi.org/10.1002/est2.135. http://batteryuniversity.com/learn/article/lead_based_batteries

Geoffrey J (2018) May, Alistair Davidson, Boris Monahov. Lead batteries for utility energy storage: a review. J Energy Storage 15:145–157. https://doi.org/10.1016/j.est.2017.11.008

.Pickett DF (1990) Nickel alkaline batteries for space electrical power systems. SAE Trans Section 1 J Aerospace Part 1 (1990), 99:298–306. https://www.jstor.org/stable/44472501

Hariprakash B, Shukla AK Venugoplan S (2009) Nickel–Metal hydride: overview. Encycl Electrochem Power Sources, Elsevier 494–501. https://doi.org/10.1016/B978-044452745-5.00158-1

Salkind AJ, Karpinski AP, Serenyi JR (2009) Secondary batteries – Zinc–Silver. Encycl Electrochem Power Sources. 513–523. https://doi.org/10.1016/b978-044452745-5.00172-6

E4tech, The Fuel Cell Industry Review (2017) http://www.fuelcellindustryreview.com/archive/TheFuelCellIndustryReview2017.pdf. Accessed 24 Aug 2019

E4tech, The Fuel Cell Industry Review 2018. https://www.californiahydrogen.org/wpcontent/uploads/2019/01/TheFuelCellIndustryReview201. Accessed on 24 August 2019. Energies 2019, 12, 4593 25 of 28

European Hydrogen and Fuel Cell Technology Platform, Strategic Research Agenda. https://www.fch.europa.eu/sites/default/files/documents/hfp-sra004_v9-2004_sra-report-final_22jul2005.pdf (accessed on 23 July 2019)

Felseghi R, Carcadea E et al, (2019) Hydrogen fuel cell technology for the sustainable future of stationary applications. Energies 12:4593. https://doi.org/10.3390/en12234593

Rose MF, Merryman SA, Owens WT (1995) Chemical double layer capacitor technology for space applications, In: 33rd Aerosp Sci Meet Exhib, Reno, USA

Merryman SA, Hall DK (1996) Chemical double-layer capacitor power source for electromechanical thrust vector control actuator. J Propul Power 12(1):89–94

Qian X (2010) Application research of flywheel battery in the wind and solar complementary power generation. 2010 International Conference on Computer Application and System Modeling (ICCASM 2010). https://doi.org/10.1109/iccasm.2010.5622895

Luo X, Wang J, Ma Z (2014) Overview of energy storage technologies and their application prospects in Smart Grid. Smart Grid 2:7–12

Sameer H, Johannes L (2015) A review of large-scale electrical energy storage. Int. J. Energy Storage 39:1179–1195

Kousksou T, Bruel P, Jamil A, Rhafiki T, Zeraouli Y (2014) Energy storage: Applications and challenges. Sol Energy Mater Sol Cell 120:59–80

Mahlia TMI, Saktisahdan TJ, Jannifar A, Hasan MH, Matseelar HSC (2014) A review of available methods and development on energy storage; technology update. Renew Sustain Energy Rev 33:532–545

Venkataramani G, Parankusam P, Ramalingam V, Wang J (2016) A review on compressed air energy storage—A pathway for smart grid and polygeneration. Renew Sustain Energy Rev 62:895–907

Wang J, Lu K et al, (2017) Overview of Compressed Air Energy Storage and Technology Development. Energies, 13. https://doi.org/10.3390/en10070991

Luo X, Wang J, Dooner M, Clarke J (2015) Overview of current development on electrical energy storage technologies and application potential in power system operation. Appl Energy 137:511–536

Guo B, Niu M (2018) Application research on large-scale battery energy storage system. Glob Energy Interconnect Fram 1:(1). https://doi.org/10.14171/j.2096-5117.gei.2018.01.010

Juan de Santiago (2014) A power buffer in an electric driveline: two batteries are better than one. Hindawi Publ Corp, ISRN Automotive Eng 2014, Article ID 525630, 9. https://doi.org/10.1155/2014/525630

Chapter 36
Policies and Prospects to Promote Microgrids for Rural Electrification in Present Indian Scenario: A Comprehensive Review

Anoop Arya, Karuna Nikum, Abhay Wagh, Shweta Mehroliya, and Prateek Mundra

Abstract In India, the growing population increases energy demand and requires more attention to fulfill current energy needs. Renewable energy sources (RES) appear to cope with current energy demand. Coal is majorly utilized for energy generation whereas renewable energy still requires some effort to reach a significant contribution level. A very high number of Indian rural villages still need electrification. Sometimes, power transfer to these villages is not viable because of cost, large distance, and high power losses by a centralized grid. To solve the problem of rural electrification, microgrids (MG) are a potential clean energy solution isolated or in conjunction with the utility grid at present. This study explores various prospects of MG in rural electrification, policies and their initiation, government, planning to improve the economy linked with current challenges, and some recommendations based on the study. Besides, the condition of MG markets with existing situations faces new challenges with opportunities also discussed. According to the International Energy Agency (IEA), the power demand for global needs around two-thirds will be fulfilled by RES. Today, RES contributes around 25% and is expected to contribute 50% by 2040. IEA forecast MG with a standalone system would fulfill 60% of global electricity demand and influence the market positively by an estimated period of 2019–2025. The Paris Agreement formed upon the UN Framework Convention based on climate change in 2015 aimed to lessen the effects of increasing worldwide temperatures from more than 2 °C above, and it is consented to by 144 countries. According to Climatescope, India ranks second in the list of over 100 countries based on readiness for RE investment. After China and the United States, India is in the third position in the largest solar power. Bloomberg New Energy Finance

A. Arya · S. Mehroliya · P. Mundra
Electrical Engineering Department, MANIT, Bhopal, India
e-mail: shwetamehroliya12@gmail.com

K. Nikum (✉)
Department of Engineering Sciences and Humanities, TCET, Mumbai, India

A. Wagh
Directorate of Technical Education, Mumbai, India

A. K. Bohre et al. (eds.), *Planning of Hybrid Renewable Energy Systems, Electric Vehicles and Microgrid*, Energy Systems in Electrical Engineering,
https://doi.org/10.1007/978-981-19-0979-5_36

has estimated the opportunities in a rooftop solar alone investment of around $23 billion in India. India has the world's largest auction for RE and has in recent times embarked on crucial incentives for RE, together with MG growth. In this chapter, various prospects, economics, challenges, market conditions, and government policies for MG related to rural electrification have been discussed. The main challenges of MGs like intermittent power, storage system cost, energy cost, power quality, tariff plans, and subsidy have been discussed. The policies by central and state levels should consider all challenges for rural electrification market planning and its implementation with the consideration of political, environmental, technical, and economic factors. It is required to consider the MG costs, cost recovery models and regulatory bottlenecks are weighing down before implementing any MG project. Finally, RES and its contribution in today's scenario is also reviewed with potential and its requirement for the future.

Keywords Microgrid · Renewable energy source · Rural electrification · Government policies · Economic challenges · MG market

Nomenclature

AT&C	Aggregate Technical and Commercial
BPL	Below poverty line
CGHG	Greenhouse gas
DISCOM	Distribution company
EES	Energy storage system
ESCO	Energy service company
FiT	Feed-in Tariff
HMG	Hybrid microgrid
IEA	International Energy Agency
IRENA	International Renewable Energy Agency
MG	Microgrid
MNRE	Ministry of New and Renewable Energy
PEI	Power electronics interfaces
PV	Photovoltaic
RE	Renewable energy
REC	Rural Electrification Corporation
RES	Renewable energy sources
RPO	Renewable purchase obligation
SERC	State Electricity Regulatory Commission
SHS	Solar home system
SMG	Smart microgrids
UNFCCC	United Nations Framework Convention on Climate Change

36.1 Introduction

The market of MG and mini-grid is promptly emerging due to low carbon emission, cost-effectiveness, and diversification of energy sources (Understanding microgrid and What are the Benefits of the Smart Microgrid Approach Galvin Electricity Initiative 2015). MG is a new idea to connect various sources to a common bus via power electronics control (Zeng et al. 2011; Parhizi et al. 2015; Jiayi et al. 2008). Generally, the distributed generation up to 10 kW is treated as MG, whereas mini-grids have larger than 10 kW capacity. Many government programs run to expand electricity access especially for rural villages by adopting renewable energy sources (RES). To make renewable energy (RE) more efficient and reliable, MGs come into existence (Cabraal et al. 2005; Wang et al. 2009). In today's world, switching topology reaches its mature stage and now it is easy to control power electronics interfaces (PEI) to control the power (Cai and Mitra 2010; Sao and Lehn 2010). MG is a blend of RE sources, grid, storage unit, and loads located in a local distribution (Fu et al. 2012). The whole integration of multiple generation sources, loads, storage, electronic switches, protective and communication devices as well as automation and control systems is the heart of what makes a system reliable (Kroposki 2009; Vaccaro et al. 2011; Che et al. 2014). An MG is a small-scale localized distribution network and accomplished power in island mode or in grid-connected mode (Korres et al. 2011; Amin 2019). Due to the variable nature of renewables, an energy storage system (EES) to compensate fluctuations is essential. In islanded MG, a power supply is needed to ensure the continuous transfer of power. Therefore, the incorporation of EES with MG enhances the load profile with flexibility in power consumption and production (Ross et al. 2011; Chen et al. 2012; Vasiljevska et al. 2013). ESS supports utility grids with several plus points like large savings on cost, improve efficiency, upgrading the reliability of power reduce operation cost and emission (Morris et al. 2012; Farzan et al. 2013; Augustine et al. 2012). Many RE technologies are now well developed, reliable, and cost-competitive as compared to conventional sources. The positive impacts of rural electricity access support economic development and reduce poverty gaps. A new approach to access electricity in rural areas by MG installations pave the way for new business and accelerate economic activity. Electricity not only provides power but also gives facilities connectivity to the Internet, medical, drinking water, educational, and skills programs. Departments of Ministries can emphasize and speed up economic activity related to MG for rural growth throughout the access to health care, education, and employment (www.renewableenergyworld.com; Su et al. 2010; Khodaei 2015). As demand increases, production increases, and simultaneously cost is reduced. Broadly, MGs and RES are classifications as shown in Fig. 36.1.

Based on supply, MG is divided into AC, DC, and hybrid types (Xu and Chen 2011; Liu et al. 2011).

- DC Microgrid (DCMG): All energy sources with storage are directly connected to the DC bus.

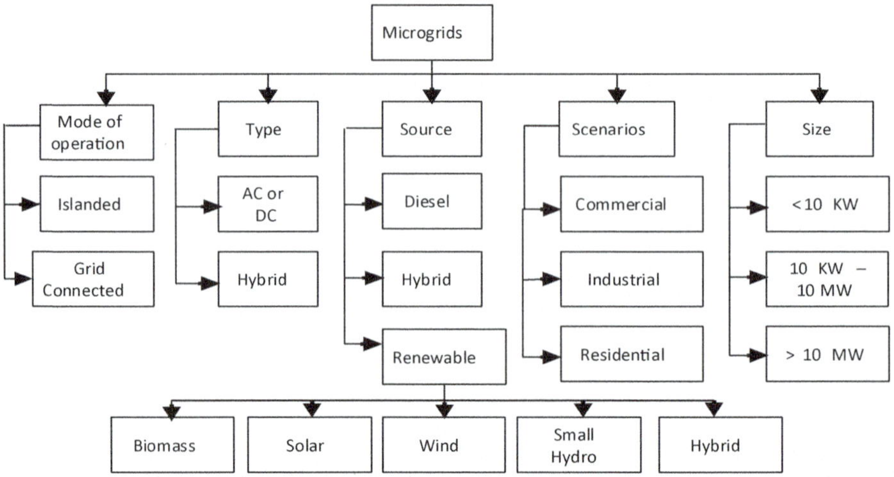

Fig. 36.1 Classification of MG

- AC Microgrid (ACMG): When all energy resources are associated with the AC bus.
- Hybrid Microgrid (HMG): When a system is having DC as well as AC bus via PEI as per need as shown in Fig. 36.2.

Before modernizing the energy supply by just replacing old equipment with new ones, large investment is required to integrate RE resources in isolation or conjunction with the grid by ensuring system security, reliability, and controllability of the system (Zhang et al. 2013; Che and Chen 2012; Kanchev et al. 2011). MG has the additional advantage of no power outages because it uses power from a combination of sources and the capability to support the main grid by sending surplus power. When MG

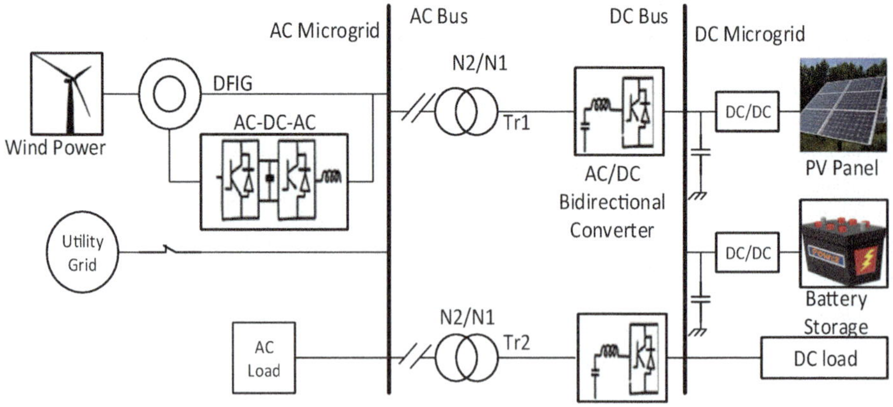

Fig. 36.2 Hybrid microgrid

gives surplus power to the grid, it means MG can help to reduce the greenhouse gas (GHG) effect by aiding low energy sources like solar and wind (Suryanarayanan and Kyriakides 2012; Brown et al. 2012; Khodayar et al. 2012; India 2020).

In this chapter, Sect. 36.2 provides a brief overview of the MG scenario at the world level, and Sect. 36.3 discusses the MG market and its strategies. In Sect. 36.4, various Government schemes to provide a continuous and reliable power supply for rural electrification have been discussed. Section 36.5 introduces the different policies and their impact on MG in rural areas, and Sect. 36.6 discusses the challenges in the deployment of MG in rural parts. Section 36.7 presents the future directions needed in order to mitigate the challenges and finally, the conclusion and future scope have been given in Sects. 36.8 and 36.9, respectively.

36.2 MG's Scenario in the World

At the world level, 14% of the global population is not having electricity and around 700 million people will go under the burden of power in 2030 (Urbanization 2016). Therefore, the world is in severe need of electricity.

The MG already has a value of $20 billion in 2018 as per a world research report. Simultaneously, all the countries will have to transmute their existing power sector, in a direction to use more RE (Kopanos et al. 2013; Millar et al. 2012; International Renewable Energy Agency (IRENA), 2019). The total renewable energy installed worldwide and patent growth are as shown in Figs. 36.3 and 36.4, respectively.

The ranking of countries in deployment of total RE installed capacity and technologies used for electricity generation in MW (https://www.energy.gov/sites/default/files/oeprod/DocumentsandMedia/DOE_Benefits_of_Demand_Response_in_Electricity_Markets_and_Recommendations_for_Achieving_Them_Report_to_Congress.pdf) and the employability development at world level as given in Table 36.1.

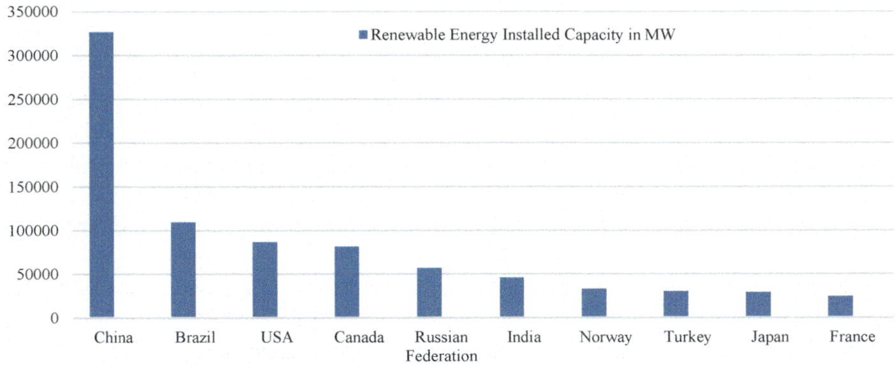

Fig. 36.3 RE installed capacity in MW. *Source* IRENA, 2019

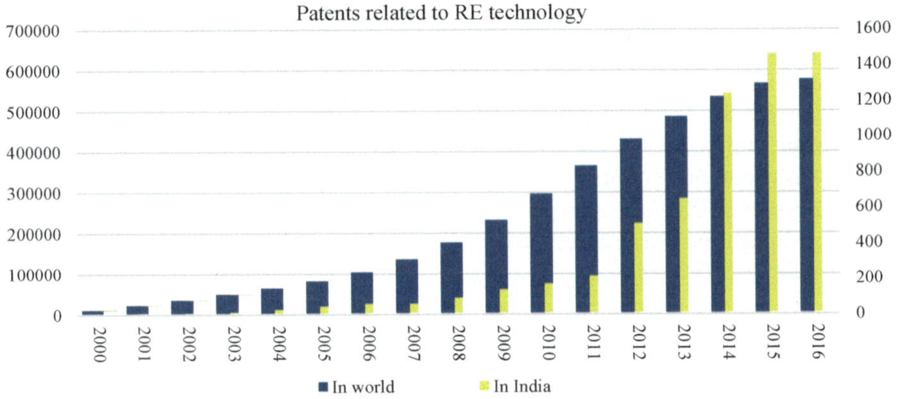

Fig. 36.4 Total number of patents at world versus Indian level. *Source* IRENA, 2019

Table 36.1 Jobs creation and RE technologies used at world level

Type of RES	Total no. of jobs due to RE in thousands	RET used in MW
Solar PV	3605	580,159
Liquid Biofuel	2063	3211.38
Hydro Power	2054	1,189,448
Wind Energy	1160	622,702.3
Solar Thermal	801.4	6274.93
Solid Biomass	787.1	86,623.68
Biogas	334	19,452.79
Geothermal	93.5	13,930.58
Municipal Waste	41.1	14,517.55

MGs are providing a solution to the distribution system and applications in the community, industries, organizations, utilities, and military. The United Nations declared some goals in 2015 in the direction of economics and research expansion while ensuring natural resources for future generations (Schnitzer et al. 2014; Krapels 2013). The world-level research and innovation in RE have continued to grow the most in-depth assessment to overcome carbon emission, cut the cost of energy, support the energy transition, improve the reliability, and make it more flexible (Demonstration and for Energy Reliability and Security (SPIDERS) 2015). The US military world also has an initiative of Smart Power Infrastructure Demonstration for Energy and Security to provide power to critical facilities in the event of an attack. Now, Americans bear 36% of the regional MG market share (www.who.int). IndustryARC's research analyst estimates the application of MG in the healthcare sector to be 14% through to 2025 (IEA, India 2015). According to the IEA, RES will make a bright opportunity in the MG market in the coming years (International Smart Grid Action Network (ISGAN) 2019; The Paris Agreement, the United Nations Framework Convention on Climate Change 2015).

The Paris Agreement aimed to lessen the effects of increasing worldwide temperatures from more than 2 °C above (Climatescope 2019). According to Climatescope, India ranks second in the list of over 100 countries based on readiness for RE investment (Bloomberg 2017). Bloomberg New Energy Finance has estimated the opportunities in a rooftop solar alone investment of around $23 billion in India (Cullen et al. 2019).

36.2.1 MG Market Challenges

The market of MG suffers from the reluctance of the present industries because of the still preference for the conventional grid. The conventional grid gives a better return on investment to the consumer. Also, a more strict law for the penetration of MG in the industrial sector is inevitable (Chaudhuri et al. 2013; Casillas and Kammen 2012).

36.2.2 The Need for Reducing Carbon Footprints:

The conventional generation of electricity leads to hazardous emissions of carbon footprints to prevent global warming. Many governments have integrated programs for reducing carbon footprints and in which MG is one of the initiatives and plays a key role to minimize carbon footprint (He et al. 2012; Casillas and Kammen 2012; Bansal et al. 2019; UNFCCC 2008). UNFCCC implemented a Kyoto protocol to reduce GHGs and global warming. As per the Kyoto protocol, there are six GHGs of which are Methane, Carbon Dioxide, and Nitrous oxide whereas less predominant but very influential GHGs are Sulfur hexafluoride, Perfluorocarbons, and Hydrofluorocarbons. As per the Kyoto protocol, carbon emission reduction is not obligatory by developed countries whereas, in the Paris agreement, it is obligatory to undertake and publish targets every 5 years from nations for reducing carbon emission footprint. By 2020, developed countries are supposed to add $100 billion to back developing countries. India is the third-largest GHGs emitter that has assured to follow a low-carbon footprint model and reduces 33–35% from its 2005 levels by 2030 (2018).

36.2.3 Industrialization and Population Growth

As per the estimation of the Population Reference Bureau, the global population will be 9.9 billion by the year 2050, which is an increased 29% growth as compared to today (Industrial Development Report 2018). According to Industrial Development Organization and the UN, there is a huge demand for industrial development to provide the world with reasonable and justifiable products (Market 2019). Now, these two factors of population growth and huge industrial development need a sustainable power source, which will be achieved by MG.

36.3 MG's Market Trends

The global MG market size is projected to reach USD 47.4 billion by 2025. ABB Ltd. is one of the market leaders in MG that had annual revenue of $34,312 million in the year 2017. Another key player in the MG market is Eaton Corporation PLC, which has seen growth in the MG segment owing to the growing demand for the same. Some other companies crusading from MG market share are General Electric, Siemens AG, Exelon Corporation, Honeywell Corporation, Homer Energy, S&C Electric, Schneider Electric, Power Analytics, and Exelon Corporation (Exelon, US) (Seetharaman et al. 2019). In May 2018, "SHELL" made strategic investments in MG company—GI Energy. SHELL specifically mentioned the vision of a low-carbon future as a reason for the investment. In October 2018, Siemens had announced its plans to acquire Russelectric, especially to strengthen its position in the MG market. In January 2019, CleanSpark—an MG company—completed the acquisition of Pioneer Power Solutions Inc. to meet future power demand. Tata Power partnership with Rockefeller Foundation launches new venture "TP Renewable MG" implemented in collaboration with Smart Power India. This new venture has 11,000 MW of installed generation capacity and will be powered 5 million families which influence 25 million people over the next decade. SELCO foundation also supported many RES-based MG projects in India.

36.3.1 MG's Scenario in India and Its Economics

India is a developing economy and the population has led to an increase in demand for electricity and at the same time has very huge opportunities and availability of RES as given in Table 36.2.

In India, the MG establishment has some important issues related to economic operation (Daneshi and Khorashadi-Zadeh 2012; Department of Energy Office of Electricity Delivery and Energy Reliability 2016) as follows:

i. Optimal technological investment.
ii. Optimal mix and utilization of RE sources.
iii. Optimization of demand and supply.
iv. Optimization of heat and electric supply.
v. Tariff management: penalties and incentives.

Table 36.2 Installed capacity of various resources and electricity consumption in TW

Type of source	%	Installed capacity (MW)	Year	Total RE Consumption (TW)
Coal	55.6	204,724.50	2010–11	811.143
Large Hydro	12.3	45,399.22	2011–12	876.887
Small Hydro	1.3	4,671.56	2012–13	912.056
Wind Power	1.2	37,505.18	2013–14	967.15
Solar Power	9.2	33,730.56	2014–15	1048.673
Biomass	2.7	10,001.11	2015–16	1055.5
Nuclear	1.8	6,780	2016–17	1160.14
Gas	6.8	24,955.36	2017–18	1200.5
Diesel	0.1	509.71	2018–19	1249.337

Source powermin.nic.in

vi. Social, technical, and environmental perspective.

In the existing scenario, MG is one of the best options and can fulfill the exceeded demand of customer load at optimum cost by reducing the grid crowding (Zhang et al. 2010; Ramana et al. 2014). At present India has the major problem of low-income groups, and the government is to implement reforms and boost foreign investment to improve the living conditions (Akinyele et al. 2018; Hossain and Mahmud 2014; Wang et al. 2013; MNRE 2018). At present, around 60–70% of the population is dependent on agriculture and it added 16–17% of GDP. So, energy from agricultural waste material in the form of biomass can be used to generate electricity. The Ministry of New and Renewable Energy (MNRE) has recognized the significance of MG and mini-grids for reaching the objective of power to all for 24 × 7 with the plan of 10,000 MG and mini-grids of 500 MW capacity announced in June 2019.

DISCOM loses its major portion of the money due to electricity theft and therefore, it is mandatory to build a strong and secure power infrastructure, and this can be achieved through smart microgrids (SMG). Expenditure on HMG with smart technologies has greater potential to provide electricity to the customer at affordable prices. HMG is also a new way to reach a remote area for power applications or where the grid is not reached (Shah et al. 2011a). Despite all other sources, solar energy resources got greater attention because it is easily available all over the world. Till February 29, 2020, the solar installation capacity reached up to 34.404 GW and in addition to that, the capital cost per MW is the lowest in India as compared to worldwide cost per MW. The initial target of 20 GW which was scheduled for 2022 is achieved four years ahead of schedule. Meanwhile, in 2015 the target was raised from 20 to 100 GW plus 40 GW solar power from rooftop panels by 2022. The investment amount is the US $100 billion (Shah et al. 2011b).

Nowadays, some MG projects in rural parts have been installed by many small enterprises (Anastasiadis et al. 2010). The present MG in India supported by SELCO foundation is "“Baikampady Mangalore Microgrid” of 1.2 kW with 1600 Ah storage,

" "Neelakantarayanagaddi Village Microgrid" of 1.2 kW, 1600 Ah storage, " "Mendare Village Microgrid" of 1 kW, 800 Ah storage, and Kalkeri Sangeet Vidyalaya DC Microgrid is a solar microgrid in Karnataka of 4 kW with 17 kWh storage. Other notable projects are in Chief Minister's residence in Bihar run on solar of 125 kW, "Sundarbans Village Microgrids" in Bengal 120 kW, Rampura Village Microgrid in Uttar Pradesh of 8.7 kW, India, Dharnai Microgrid in Bihar, 100 kW, KPCL Mandya Karnataka Microgrid in Karnataka of 100 kW, and Indian Coast Guard Microgrid, Andaman, 75 kW.

36.4 Government Schemes

In India, most of the electricity DISCOM runs in a financially stressed condition because of huge power losses and electricity theft. The economic growth and development of the nation require a continuous, reliable, and quality supply 24X7. Also, to achieve 100% of village electrification, DISCOM should perform without losses. The outage of power adversely affects national priorities like "Make in India", "Digital India" and digitalization in other fields of manufacturing, medical, and education. The central government public–private partnerships boost to expand the MG sector in the long run. The ministry meticulously works with public or private sectors like National Bank for Agriculture and Rural Development (NABARD), Solar Energy Corporation of India (SECI), Indian Renewable Energy Development Agency (IREDA), and Panchayat (local governance) to accomplish the target. All schemes by state and central governments give a push to integration RES with the enforcement of environmental agenda to reduce GHG by the exhaustion of fossil fuel and fight climate change resulting in RE proliferation. Rural Electrification Corporation (REC) Ltd. has the following schemes to provide a continuous and reliable power supply.

36.4.1 "24 × 7 Power for All"

It is an initiative of all State and Union Territories (UTs) as well as Central Government to provide electricity by "24 × 7 power for all" scheme by 2019. To accomplish the target of providing affordable, reliable quality power to consumers by 2018–19, the roadmap had been signed by 29 states and seven UTs. The underlying principles of the policy are to provide reliable and adequate power to all sectors of residential, commercial, and industrial as well as agriculture consumers in the next 5 years, i.e. by FY 2018–19.

36.4.2 Pradhan Mantri Sahaj Bijli Har Ghar Yojana— "Saubhagya"

In this scheme, free electricity connection is given to all unelectrified households in rural and poor families of the urban area of around Rs. 4.0 crore by 2018. The use of solar PV-based standalone systems for remote and inaccessible villages, where grid extension is not feasible.

36.4.3 Deen Dayal Upadhyaya Gram Jyoti Yojana (DDUGJY)

This scheme solely provides rural electrification with electricity distribution infrastructure. The previous scheme of "Rajiv Gandhi Grameen Vidyutikaran Yojana" (RGGVY) has been incorporated in the DDUGJY scheme. In this combined scheme of "DDUGJY-RE", total 921 projects to electrify 1,21,225 un-electrified villages, 5,92,979 partially electrified villages, and provide free electricity connections to 397.45 lakh BPL rural households have been sanctioned.

36.4.4 National Electricity Fund (NEF)

This NEF scheme provides interest subsidy to promote capital investment in the distribution sector. It has a provision for interest subsidy on loans availed by distribution utilities in both the public and private sectors.

36.4.5 Ujwal DISCOM Assurance Yojana (Uday)

The government in 2015 for the financial turn-around of power DISCOM in consultation with the various stakeholders ensured a permanent solution to the problem. The scheme UDAY envisions reform in all sectors of generation to distribution with energy efficiency. The underlying principles of the policy empower DISCOMs, and reduce Aggregate Technical and Commercial (AT&C) losses from 22% to 15% by improving operational efficiency and reduction in the cost of power through various measures.

36.5 Different Policies on MGs

The MNRE released a draft policy in 2017 aims to accomplish a total capacity of 500 MW target over the next 5 years. The aim is to set up nearly 10,000 solar, wind, and biomass-based power projects in collaboration with MGs in rural areas. The primary ideology of this policy is to improve affordable energy services and rationalize for Energy Service Companies (ESCOs), and run with the DISCOM of the grid.

36.5.1 Rural Energy Service Providers (RESPs)

It is a special category of service providers receiving special incentives to promote this program including capital and streamlined approvals by the government. Government launches many programs to endorse RES into the electricity market from time to time with energy providers as given in Tables 36.3 and 36.4.

Table 36.3 Different Government policies to endorse RES

Policies	Outcomes
Rural Electricity Supply Technology (REST) Mission: Year Enacted—2001	• Power for all by 2012
Accelerated Rural Electrification Program (AREP): Year Enacted—2003	• The interest subsidy of 4% on loans is offered by government to extend rural electrification
Pradhan Mantri Gramodaya Yojana: Year Enacted—2003	• Under this policy, electrified villages by 2012 through decentralized RES and introduction of multiyear tariffs plus open access
Electricity Act of 2003	• Promote and encourage private MG projects by eliminating licensing requirement for ESCOs • Exempts from tariff 11 to all privately owned MG companies
National Electrification Policy: Rajiv Gandhi Grameen Vidyutikaran Yojana (RGGVY) Year Enacted—2005	• Connect small distribution networks and electrified low-income groups in rural areas, where extension of grids is expensive • With consolidated existing electrification programs, this scheme provides unrestricted connections to households under the poverty line
Rural Electrification Policy: Year Enacted—2006	• Formed a detailed framework of RGGVY • Directed the provision of 1 unit of electricity per day for all households by 2012
2018 Amendments to the Electricity Act of 2003	• 20% initial reduction in cross-subsidies and by the time of 3 years eventual elimination of cross-subsidies • Penalties/removal of a license, if fail to supply quality power

Table 36.4 Different energy providers

Energy Providers	Work implementation
State Electricity Regulatory Commission (SERC): State-Level Agency	• Supervise and create secure payment methods for exit options and exit procedure • Established energy tariffs under the National Tariff Policy • Formation of Services to settle ESCO disputes and established clear and quantifiable metrics of performance
State Nodal Agencies (SNAs): State-Level Agency	• Work in partnership with ESCO supervisors to the identification of areas, where energy projects are required • In case of any non-operational, poorly functioning scenario then replacement of ESCO management has been allowed • Provision of technical and operational backing to SERCs settling the disputes of ESCO • Build, operate, manage, monitor microgrid facilities, and collaborate with the Ministry of Power to generate data for project implementation agencies (PIA)
DISCOMs: Energy Provider	• Buy excess power of MG from ESCOs • Avoid over-competition and interference with MG market • Sometimes, DISCOM is not economically feasible in low-income rural areas
ESCOs: Energy Provider	• Abide by state-specific policy and tariffs • Can charge a tariff to the local community and often eligible for state subsidies • ESCOs operate with local/rural areas that cannot connect with the grid
Project Implementation System (PIA): State-Level System	• Developed by the state's Investment Promotion Board • Data is publicly available and categorizes MG based on kW size • Database registered like number of households connected and metrics of performance to the location of grids details should regularly be uploaded by the ESCOs
Village Energy Committee (VEC) and Panchayat:	• Determine energy tariffs and their recalculation, if surplus energy is sold • Collaborate with energy providers to set tariffs and promote community participation in projects • Give legal permission to new projects implementation and certify NOC • Organize VEC and address issues and disputes of community-related to companies

36.5.2 MG's Strategies for Rural Electrification

Electricity consumption and demand are increased by increasing consumer base, lifestyle, and pattern of consumption. MGs will enhance the overall reliability and fulfill the future requirements of all sectors (Hatziargyriou et al. 2005; Logenthiran and Srinivasan 2009). Rural electrification requires continual changes and reinforcement in new infrastructures of generation, transmission, and distribution at affordable prices by making DISCOM with less financial losses (Guan et al. 2010; Matamoros et al. 2012; Su and Wang 2012; Jain and Arya 2015; Priyadarshi et al. 2019). The rural electrification in India has extensions of electricity through various means of access as given in Table 36.5.

For rural electrification, the following point should be noted:

- Villages located far from DISCOM are supplied with MG and SHS gives high reliability.
- For medium type load, MG is a better solution and for a small load, SHS will be preferred.

Table 36.5 Comparison of extensions of electricity through various means access

Characteristic	Grid Extension	Solar Home System (SHS)	Microgrids
Reliability	Low reliability, especially for rural areas	Reliable for required load within the capacity	Overall, more consistent than SHS
Generation Cost	Rs.3.18/kWh to Rs. 231/kWh	About Rs. 37/kWh	Rs.23 to 33/kWh
Electricity Price	Rs. 3/kWh	Rs. 45 k for a SHS	Rs. 100 to 200/month
Capacity	Unlimited capacity	Limited capacity; small loads only	Limited capacity but greater than SHS
Losses	About 23.97% in 2012	Losses exhibited within the SHS	Overall fewer losses than with SHS
Generation Type	Varies depending on source type	Solar energy	Determined by local DG resources
Constraints based on Geographic Location	Cost of supply increases, if lines extended in remote villages, hilly or forested areas	Suitable for areas with high solar irradiation	Location itself is flexible for MG
Operation and Maintenance (O&M)	Often takes days by DISCOM to fix a problem	Relatively low maintenance	Maintenance varies from low to high

- MG is more reliable than SHS and economically viable than the grid.
- If the capacity or demand for electricity will be increased, then SHS will not be expected to use because of fixed capacity, therefore MG is a better and exact option.
- To recover the initial and capital value of MG and SHS typically require 10–25 years.
- Off-grid is preferred only when villages are not located nearby the central grid.

36.6 Challenges in the Deployment of MG in Rural India

Power outage and load shedding is still a common problem in rural parts of India. Many MG have a higher cost/kWh than the traditional grid and are not scaled to meet the expectation of rural electrification. There are many challenges involved in the implementation of MGs and some of them are as follows.

36.6.1 Absence of Promotion Strategies

- In the field of RE promotion, a strong regulatory framework is absent. The shortage of policies, lack of standards, certification, insufficient incentives, bureaucratic and administrative hurdles, and sometimes impractical government targets to integrate RE technologies with the global market have prevented RE to expand.
- Awareness is an important factor for social recognition of RE, still not very promising in the rural area. It is needed to spread awareness about renewable energy technology (RET) and its environmental benefits.
- Due to lack of awareness, people believe the replacement of the storage system and its maintenance cost is very high and are not ready to use in general.
- Due to inadequate knowledge and no wakefulness, RE plans are not available in the public domain and RE still faces setting up and investment issues in rural areas.
- The understanding of the benefit of RET is not clear to people and they have negative perceptions regarding RE technologies.

36.6.2 Lack of Synchronization, Coordination, and Delays in Policies

- There is a lack of synchronization between different authorities that leads to elongated time in obtaining authorization that delays in planning and puts boundaries on the development phase. Obtaining permission requires a high cost due to lobbying, and it demotivates the RE investors. Sometimes, declared policies do not match with the plans and target of RE deployment.

- The progress of RE is restricted by a lack of support, synchronization, and delays in executing policies for acquiring land for constructing the RE plant. This leads to a decrease in the interest of investors whereas institutes, agencies, stakeholders, and key market players work with poor inter-institutional coordination by MNRE.
- No proper customer care centers to guide developers as regards RE projects and policies. The workforce and experts for RE developments in agencies, organizations, institutes, and ministries are also not enough in numbers.
- The single window project approval process delays the receiving of clearance for projects.
- High taxation, a slowing economy, low tariffs for fossil fuel, and negative sentiments of investors have also been responsible for slow RE market growth.

36.6.3 State-Wise Different Regulatory Framework

- Every state has its defined Renewable Purchase Obligation (RPOs) and different regulatory frameworks and procedure. Additionally, all policies are only applicable for just 5 years, so greater jeopardy for investment is oblivious.
- The percentage of RPO specified for RES is not precise. The RPO target of the central government is complying with only six states out of 29 and 7 UTs.
- Still, RPO goals are hazy and its cell compliance is taking place in May 2018 to collect, check, and deal monthly information with compliance and non-compliance issues with authorities. The biomass factor still does not have any established framework.

36.6.4 Uncertainty Around Tariff, Penalties, and Incentives

- To accelerate wind farms, the government provides an incentive for accelerated depreciation. Many companies only take advantage by building such projects only for tax benefit. It is observed that wind farms more than 10 years ago show not optimally maintained. This policy framework does not take into account maintenance.
- Suppliers make buyers pay a higher cost for equipment which increases the burden on buyers whereas no control over suppliers for activities like commissioning, operation, and maintenance. Sometimes, ready-made projects are sold and customers are inclined to this trap to save tax.
- Only two states (Maharashtra and Rajasthan) have some actions for penalty mechanism and other states are still unclear.
- Ministry should reconsider subsidies because of non-transparency in subsidies and incentives as in 2018.

- The power purchase agreements (PPA) sign up on a pre-determined fixed rate of tariffs, which are usually higher than present bids between power generators and purchasers.

36.6.5 Government Plans About Grid Extension

- Mostly RE plants constructed and installed in remote places require a further extension in lines to connect the utility grid. The further extension seems uneconomical for small-size projects.
- Since the existing grids are not designed to directly integrate with RES, these grids require upgradation.
- Therefore, SERCs are unable to utilize all generated power to accomplish the demands and enforce purchasing power from neighbor states at a higher cost. Therefore, the required infrastructure for the integration with RES is again a leading problem and restricts the capability of RE.

36.6.6 Limited Practical-Based Research, Innovation, and Development

- Non-availability of deep-rooted research centers for the advancement of RE organization. There are not enough research services; quality infrastructure centers are yet to be established to check the quality, reliability of manufactured and imported equipment.
- There is no clear-cut reassuring document for referral institutes, laboratories, standards, review mechanisms, assessment, inspection, and monitoring. For spare parts, equipment, and the latest technologies, the country is still dependent on international suppliers.
- Insufficient investment by the government side in research and development (R&D) both in government organizations and in energy firms. Therefore, RETs and their development stage for practical viability to the manufacturer's point of view are lagging and manufacturers are just replicating the already available technologies.
- Quality control and standards are allotted and issued just in 2018 and 2019. Not enough research institutions and labs to provide standards or certification for validating the quality and suitability of using RETs is still a problem.

36.6.7 Majority Dependence on Conventional Energy Source and Subsidized Kerosene

- The major portion of India is still dependent on fossil fuels for electricity generation. As India is the third largest GHG emitter in the world, therefore, it needs to shift on RES. Accordingly, a significant amount of contribution is required to set up a large scale of RE plants.
- At present, coal remains a dominant player, and investment in fossil fuel is around 55%.
- The falling prices of fossil fuel is still a challenge and offers very tough competition for RE projects. Also, the government delivered higher subsidies to conventional energy sources than RES, and this makes overshadowing the wide use of low emission technologies.
- Quality assurance processes are still at the elementary level when it is about international practices. MNRE issued policies regarding standardization including testing and certification of RE projects on December 11, 2017. Therefore, RE projects are still far behind conventional fuel projects.
- The government continuing the subsidy in kerosene reflects its failure to provide basic service of electricity in rural areas, while the government continuously tries to improve electricity network transition toward other cost-effective alternatives compared to kerosene.

36.6.8 Technological Barricade

- There are some technological barriers is insufficient knowledge of new RETs in the operations and maintenance of such industries. Since RE technology is newly developed, so there is a lack of knowledge on maintenance and operation.
- The remote area maintenance, monitoring, and operation require a skilled or trained person which becomes a challenge for electricity DISCOM. However, challenges present are related to finance and sustainability of such projects, existing price differences with state DISCOM supplied power to city residents, municipalities, and businesses.
- Different forms of power conversion—AC/DC or DC/AC—make the system complicated and simultaneously required high maintenance with trained engineers all time in remote areas.
- As limited advanced technologies and high cost of procuring and importing especially by developing country like India is a big problem, and to survive and dispatch power economically in the market require well-developed indigenous technologies and research and trained technicians, developing countries import all items from other countries and procuring at a very high cost leads to an increase in the overall cost.

36.6.9 Environmental, Land Acquisition, and Installation Problem

- A huge area is required for solar and wind plants to produce the same amount of energy, which can be produced by conventional plants in a lesser area than RE plants. Lack of land for wind or solar plants in many states and penetrating RE projects in India require enormous parts of rural side connectivity via roads or transport.
- The financers, RE producers, and developers face difficulty in securing projects at the rate, which is made available for fossil fuel. Sometimes, RE proposal faces resistance from citizens, political interest, environmental and national groups because of landscape impact and environmental degradation. The RE projects face challenges in planning, profit-loss mechanism, environmental uncertainties, and natural calamities.
- As the high initial cost and low working efficiency in RES make its payback period long which in turn pushes financiers on to the back foot, the situation also worsens by very stringent lending procedures that limit access to funding. Such installation of a plant requires high capital cost, which often becomes unviable due to such a process.
- The dependence of RE on the changeable climate reduces the popularity among the public and investors.
- Installations at offshore affect ocean activities like gravel/oil/gas/sand extraction, navigation, fishing, aqua, and wildlife of marine.
- By making WTs, immobile during low wind can guard the birds but it is not practiced.
- In wind generation, the flicker effect and shadow of turbines are also not considered as severe environmental impact, whereas no use of proper sound absorbent material. Sound by WTs through aerodynamic, mechanical, and visual impacts are associated.
- The manufacturing process of PV cells contains hazardous chemicals and if its wastes are not correctly disposed off, it produces severe environmental threats. The workers face health issues resulting from inhaling silicon dust.
- The dumping of a huge quantity of waste from coal-based power sectors is responsible for 70% of freshwater withdrawal. Each year 160 million tons of ashes are added and over 1 billion tons of ash are lying unused, which is alarming for us (Centre of Science and Environment 2020).
- The outdated argument to backup coal is a low-priced and good choice as a key source of power, but it is a direct burden on green taxes.
- After the record growth in RE capacity in the bulk installation of over 4 years to 2017 and to enhance manufacturing at the domestic level, the government imposes anti-dumping duty on imported material due to which the RE capacity addition is slackened in 2018.
- As per the Ministry of Power, allotted energy generated by conventional sources has a 63.2% share whereas only 22% share to RE source.

36.6.9.1 Inadequate and Delay in Fund Allocation and Investment

- The release of funds and budgets for RE sectors are not within the timeframe. The required finance from land acquisition to final setup before initializing the project is very high and creates a burden on developers.
- Fewer financing organizations for RE projects come forward because less assurance in the future seems high-risk perceptions and this will demotivate investors and developers.
- The government and private financial institutions/banks do not have much understanding in the field of RE market and project. The delayed fund by SERCs imposes debt burden on developers.
- Moneylenders always work on a credit system with a guaranteed bond between developers and moneylenders. Sometimes, it comes to be a financial barrier for developers if there are delays in the fund.
- The allocation for the Ministry of Coal stood at Rs. 20,121 crores, while the MNRE was granted Rs. 12,353.81 crore. Existed such gaps in all budgets since 2009–10 shown in Figs. 36.5 and 36.6 (www.indianbudget.gov.in).
- The RE sectors allocate 63% less funds as compared to coal-based power in 2019–20. Despite a worldwide assurance to increase the consumption of RE and reduce GHG, coal is the sole largest entity of total temperature increasing and despite that total rose to Rs. 600 crores for the allocation of coal and lignite exploration in the last 10 years.
- In the latest budget, the allocation on R&D in renewable and fund including international cooperation 90% less, i.e. Rs. 60 crores in the current budget. Investment still requires more than $250 over the next decade; it indicates an annual investment of more than 1.77 lakh crore for 10 years although the 2019–20 budget allocates 93% less than the target.

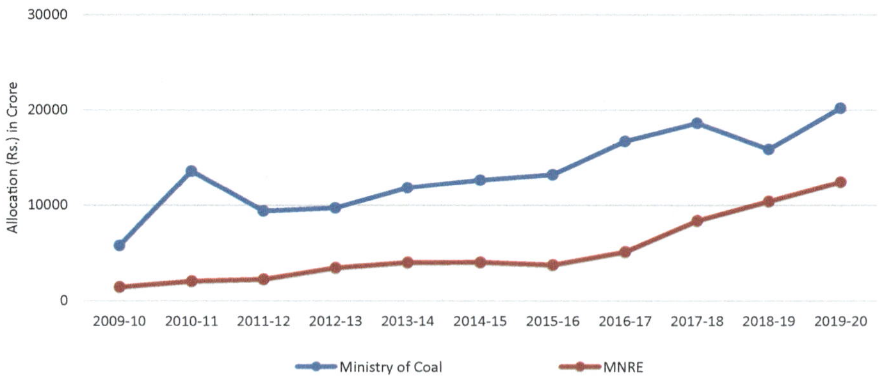

Fig. 36.5 Ministry-wise budgetary allocation

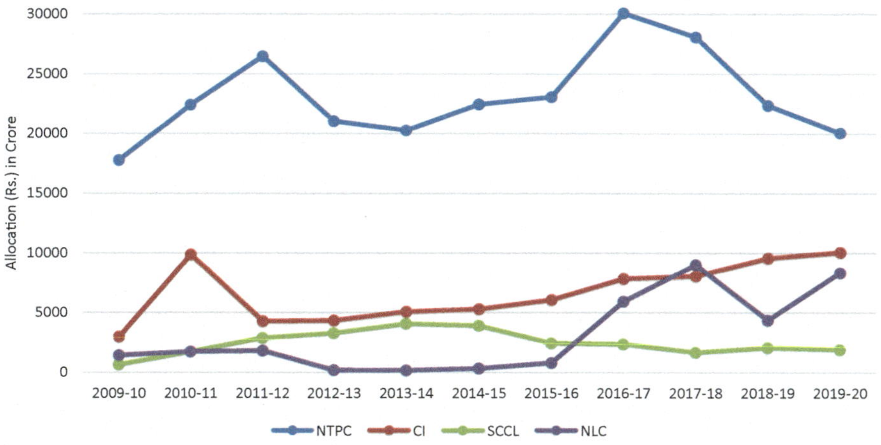

Fig. 36.6 Allocation to public sector enterprises

- Public sector enterprises also have investments in RE such as NTPC and Coal India, even though a rising share of RE at the same time continues to add to coal capacity about 86% of overall installed capacity.
- The tax incentives for R&D have seen a reduction from 200 to 150% at the starting of April 2017 in the weighted tax deduction (WTD). In 2020–21, it further lowered from 150 to 100%.

36.6.9.2 Market-based barricades

At present, the main RE markets classified in India are

i. Government-driven market: Purchase the output of some RE-based project and provide budgetary support to promote RE.
ii. Loan and cash market: Taking a loan to fund and buying RE-based applications needed by individuals.

  - Due to the unfair structure of subsidies in between conventional fossil fuel and RE, i.e. more subsidy is given to conventional power than RE power which leads to wrong information about the RE market.
  - The cost of RE power is based on direct material and labor cost, product overhead cost, and excluding environmental cost. Therefore, the ecological benefits of clean energy are not countable.
  - The biomass market is considerably unorganized and having very high demand–supply gaps creates a continuous and dramatic fluctuation in biomass rates.
  - Usually, investors think that there is uncertainty in the RE sector and exclusively look for developers with well-established suppliers, and operators while newly established developers needed to develop the projects

not getting financial support. Even moneylenders are ambiguous and not prepared to provide funding to new RE market developers.

- MG is at its early stage and its market is fragmented. The estimated market of MG is $ 35 billion by 2020. The local generated and utilized energy system developed as to how urban and rural area is developed or planned. In the future, with MG the following points should be achieved (Gomes et al. 2019; Canziani and Melgarejo 2019):

i. Low GHG, clean, and affordable electric power.
ii. High reliability, security, and control with huge potential which reduces financial loss.
iii. Low stress on the transmission and distribution system.
iv. Guarantees continuity and is easily controllable during disturbances.
v. Integration with less complexity of a large number of RES with or without the grid.

36.6.9.3 Absence of Large Commercial Loads in Villages

- As the large commercial loads are not present in remote areas, it makes the cost of energy and maintenance very high.
- Generally, exporters join their hands to limit the competition and control the prices, hence, the quoted capital cost is very high by the developers and providers of equipment.

In the current scenario, many villages have access to the grid, but in fact, they cannot afford to pay for electricity. Meanwhile, power theft is common in villages and leads to financial instability to utilities as a resulting inability in infrastructure upgrade investments. The important link between generating plants to end-users is a distribution line that requires many infrastructural changes for better control, monitoring, and security. India has set to add 300 million population by the year 2050, and the ambition of the government of achieving "24X7 power" requires investment in smart technologies to deliver power. To resolve this issue, R&D on large-capacity batteries is required. The overall initial and maintenance cost of MGs solely depends on the energy storage system. Due to PEI used in ACMG, DCMG, and HMG power flow control, and conversion is very easy whereas it suffers from the problem of power quality and harmonics into the system. The advancement in control switching techniques in the PEI makes converters usage in RES convenient. These challenges and their solution will not just support India to go ahead, but also have positive effects on India's future through education, infrastructure development, the industrial revolution, and employment. The key areas of focus include MG markets being political, economic, environmental, and technical for implementation in all sectors. Some of the main factors to be assessed and considered before implementing India's energy development initiatives are as given in Table 36.6.

Table 36.6 Factors required in the deployment of MG

Factors	Implementation and Operation
Political Factors • Clarity on government policies • Process of land acquisition • Process of procurement • Role of cooperative societies	• Uniformity of design and ownership • Readiness to accept policies and standards • Reliable, secure power, and economic rate of power • Maintenance, safety Inspection, and quality check • Penalties and permitting process requirement
Economic Factor • Funding process • Financially viable • Consumer subsidies	• Local employment, skill-based training program • Awareness and social perspective • Theft of power, power losses, and AT&C losses • R&D Funding, investment, loan and market condition
Environmental factor • Project locations • Accurate solar/wind data • Varying resource profile (Solar, biomass, and hydro)	• Permitting process and land acquisition • Distributed habitat and monitoring the health of system • Pollution and effect on the environment
Technological Factor • Metering technologies implementation • Role of RES • Load sharing and unbalancing • Security and reliability	• Integration with grid and extension of lines • R&D on MG and Hybrid technology • Smart meters and supply–demand energy management • Local control of PQ, reliability, and security • Tariff mechanism and connection charges

36.7 Future Directions Need to be Taken in Order to Mitigate Challenges

The most influencing factor is economics which affect consumer and manufacturer directly or indirectly. Here, some glitches and complications are discussed with some recommendations.

36.7.1 Awareness, Education, and Training Mechanism

- The promotion and awareness program required immediate actions and was regularly organized in rural areas throughout the country. The MNRE should prepare transparent and achievable promotion plans in consultation with SERC.
- All promotion strategies should be state-wise and region-wise with the consideration of geographical and environmental conditions. All these action plans should be executed within a timeframe. For example, most of the rural areas benefitted from biomass energy, whereas coastal areas are benefitted from offshore wind energy promotional plans.
- Generally, the external skilled and trained workforce does not show interest to work in rural areas. Therefore, developing skills and training of local people through promotional strategies and training programs is an urgent need of time.
- The electricity generation and its technologies should be cost-effective and low maintenance type. It also assures rural people to get benefitted in terms of growth, jobs, more small-scale industry through promotional plans so they also contribute and take interest as well (Bihari et al. 2021; Shafiullah et al. 2021).
- To scale new heights for RES requires teaching RE-related courses and training to be conducted to improve the skills required to set up and function RE projects. The transformation of energy from fossil fuel to RES requires a concrete arrangement of skilled workers and professionals to fulfill the demand for design, operation, and maintenance RE plants.
- A huge investment should be required in the field of training and awareness programs. This will reduce the dependence on foreign technology and ease to trained people. As everyone knows, if technology maintenance is provided by in-house workers or engineers without getting delay, it will reduce costs drastically.

36.7.2 Market-Based Technology, Innovation, and Research

- The government put money and resources into R&D of new cost-effective technologies and its related services such as battery storage.
- Not enough attention is received in R&D activities in India because the country has focused largely on applications limited to conventional fuels and irrigation solar pumps. As per the financial year report, 2016–17, renewable and electrical vehicles get only one-third of subsidies than coal, oil, and gas.
- A high rate of tax is charged for the purchase of land and loan for RE sectors. Goods and Services Tax (GST) on Coal is 5%, whereas solar power is taxed based on valuation methodology. As per valuation methodology, 70% of the cost is considered as goods and is taxed at the rate of 5%, whereas the remaining 30% of the cost is considered to be a service and hence taxed at the rate of 18%.
- Higher rates of taxation would limit the capability of nations to meet its promise of solar mission—2022.

- All trends are inclined toward making coal a continuously dominant player in the energy sector whereas renewable and its innovation still struggle to its establishment with financial restrictions.
- It is expected in upcoming years the capacity of coal power generation is increased by 22.4%, whereas NTPC increased its capacity by 47.3 to 85 GW denoting an 80% increase by 2032.
- Out of 250 companies in the global list, NTPC ranked at 54th position and Coal India ranked first in the year 2016–17 which were responsible for 33% of annual emissions of GHG. India is responsible for more than 80% of mercury emissions, 50% of SO_2 emissions, and over 60% emissions of particulate matter as per the CSE report in 2015.
- To become energy independent and make a stable and reliable grid, support policies should be implemented for battery energy storage technologies in India itself.
- National Labs and research institutions in India should focus more on developing lithium-ion and other advanced technologies for batteries.
- For Intellectual Property Right (IPR) and patent application that normally take 3 years in India which is a very long and tedious process, the government should layout simplification and time-effective measures to improve R&D enthusiasm.
- Promoting R&D in organizations is directly investing in developing advanced technologies to make RE appear more profitable.
- Encourage MGs to use modern analytics for its power control and focus on achieving the cost-effectiveness, reduction of GHG without sacrificing the important qualities, and improvement in reliability for consumers.

36.7.3 Financing the Renewable Sector

- The government should make authoritative and operational structured finance mechanisms and capital market tools should be developed to ensure capital recycling to facilitate successful scaling up of projects. A dedicated government fund needs to be provisioned in the budget for next year to meet these requirements.
- For the growth of the RE sector, the government should allocate appropriate funds to boost "“Make in India”, energy storage revolution, and for proposed subsidy for NITI Aayog’s Giga factory manufacturing plan.
- Address the critical matters of DISCOM such as financial data, difficulties in cost analysis of transmission and distribution line extensions, secure and on-time payments guarantee, and generation benefits by the government to ensure RE growth.

36.7.4 Policy and Regulation Advancements

- If policies and regulatory framework are rationalized then some renewable developers feel that the government focuses on accelerating open-access policies, privatization of DISCOM, addressing many challenges by implementing the amendments in the Electricity Act of 2003.
- Use of outdated equipment, poor coal supply, import of coal from long distances, and lack of PPA added financial stress on coal-based plants. As per the observation based on recent auctions by the International Institute for Sustainable Development (IISD), grid-scale wind and solar plants are now giving competitive prices with coal, without air pollution and reduced GHG.
- As per industry expectation, there should be some clarity and rethinking on the GST for solar projects. There should be an incentive for developing next-generation solar panel manufacturing technology in India as well as investing in storage and hybrid projects.
- The incentives to the renewable power sector against the payment from the DISCOM for giving extra power can be secured as per the timeline specified in the agreement.
- All policies and regulatory frameworks should be rationalized to lessen the dependence on long-term power purchase contracts. Therefore, all processes should move toward flexible and competitive contracts through market-based mechanisms. These steps would bring significant benefits to power sectoral transformation is cost-effective and energy access at competitively priced. This will lift industry profitability, which will manifest in increased funds, investments, and employability.
- Provide tax benefits on loans taken to purchase rooftop solar systems with storage to the consumer. Rooftop solar and energy storage can accelerate the process of the adoption of advanced energy storage technologies even in rural areas.
- The RE sector witnessed several setbacks in 2019–20 and fund flows for developing new projects has slowed down due to various reasons:

    i. Due to the bank's reluctance in money lending and liquidity crunch.
    ii. DISCOM delayed procedure in payments.
    iii. Due to lock-down for COVID-19 in the international market, which hit the economy very badly.

- For the overall and practical development in the infrastructure of the RE sector, the government should ease and provide an exemption to the Indian cement sector for RE plants.
- As the rooftop projects are seriously behind the set target and there have been serious issues in the implementation of rooftop projects, or this condition, DISCOMs are very much responsible due to bad net metering policy and slow response. Therefore, DISCOMS should be levied a heavy penalty if they fail to comply with set targets and this action will be a major boost.

- A concrete action plan and its regulatory framework for the RE sector as well as a consistent check for the running status and monthly report for upgradation.
- A reduction of additional subsidy as well as cross-subsidy surcharges by state and central governments. Therefore, the RE sector rolled out services easily.
- To accelerate the R&D projects, weighted tax deduction back to 200% on R&D and again re-introduce to 200% for in-house R&D projects, which was reduced to 100%.
- The government should formulate incentives for domestic manufacturing in the heavily import-dependent renewable industry and safeguard duty should be imposed on imported products. Therefore, it acts as a safeguard for the local manufacturer to make domestic products cost-effective.
- The health of DISCOMs should be addressed because the market expects a declaration from government schemes to reduce AT&C losses, provision of smart-meters, tariff plans, etc. Set up other projects to end the state monopoly of DISCOMs.
- The budget should provide the facility to free-market forces of the RE sector to operate with DISCOM. So, consumers are also free from the monopoly of being enjoyed by obligatory DISCOM.
- SERCs should be allowed the amendment to moderate regulatory and policies that create ambiguity in RE also allowed to find the drive areas of RE development. Additionally, strong initiatives at the municipality or local level for RE-based projects approvals.
- Appropriate standards and codes make higher market penetration possible if adaptation and implementation are proper. Minimum performance standards should be guided by MNRE which results in higher performance, durability, and performance.
- The necessary RE ecosystem at the developing stage requires efficient funding mechanisms and RE certificate policies for various projects.
- The RPOs need to be upgraded in an efficient manner by offering rewards or penalties for obligated entities by regulatory administration.
- The RE sector decays in terms of growth because of penalty mechanism, obligations, RPO regulations, and implementation still not satisfactory.
- Most states have defined RPO targets but several states' DISCOMs obey completely with their RPO targets. States should adhere to their RPO set by SERCs.
- It is strongly recommended REC should only be traded under recognized exchange because the exchange is intermediate with absolutely no default risk. On the other hand, trading through off-exchange will result in lack of transparency, poor price discovery and there are no risk management phenomena.
- The investment will need to address the integration of RES with an existing grid or in isolated mode. Also, make a provision for DISCOM of giving performance-based incentives as compensation for every MW capacity.
- The tariff orders/Feed-in tariff (FiT) should be stable, consistent, and not limited to a few years.

- Central and state governments should ease the process of getting the subsidy and cross-subsidy, transmission charges and assure the investors of their backup if there is any financial loss as these hinder the development making RE more expensive.
- The government should resolve and find out the inconsistency between all policies and their implementation because such policies create confusion over the implementation of the subsidies.
- The formation of policies and their correct implementation with enabling regulation by easing the process of the promotion for RES is also very important. Therefore, the capacity-building program with awareness of all types of policies, information, and financial benefit of RES should be ready to act among rural as well as DISCOM with RE installation projects people.
- Enough actions to reduce the import tax of the equipment and on the parts required by RE plants. The aim is to create viable competitiveness between RES and fossil-fuel-based technologies by subsidizing RE concerning FiT. Without FiT results, the financial incentive is high and hinders the RE projects.
- The set target of policies should be realistic and sometimes the target has a wide gap and loopholes in the implemented process. The formation of policies should be a clear insight into important legislation and regulations. So, RE projects-based industries can be promoted as steady and proposing growth.
- Strict standards and certificates are required for importing companies to make sure that these companies operate in obedience to local laws. Absenteeism of such standard builds confusion in manufacturing and energy producer companies and faced unnecessary difficulties.

36.8 Conclusion

In this paper, various prospects, economics, challenges, market conditions, and government policies for MG related to rural electrification have been discussed. RE is one of the most important sectors and plays a solution to improve the economy by bringing in new investments, a clean environment, and creating jobs. MG can act as a complete solution to deliver electricity to rural and remote areas at affordable prices. The concept of different MGs is adapted for rural electrification due to its various advantages over conventional sources that's why government and private market key holders should take initiative to promote MG. To make MG operational in rural areas requires the upright scheme to achieve 100% rural electrification then the government should deal with challenges and opportunities in the deployment of MGs. The main challenges of MGs like intermittent power, storage system cost, energy cost, power quality, tariff plans, and subsidy have been discussed. The policies by central and state levels should consider all challenges for rural electrification market planning and its implementation with the consideration of political, environmental, technical, and economic factors.

36.9 Future Scope

Renewable will be India's next IT, so the government formulates incentivize the developers, investors, domestic manufacturing and allocate a separate budget for R&D and further proliferation of RE across the nation. Reasonable exemptions must be budgeted toward the payments on account of high open-access charges, cross-subsidy charges, additional surcharges, transmission charges to make RE more viable. It is required to consider the MG costs, cost recovery models, and regulatory bottlenecks are weighing down before implementing any project. Finally, RES and its contribution in today's scenario is also studied with potential and its requirement for the future and ensures their implementation in all private and public sectors with dedication will also provide by selling power for neighboring countries like Bhutan, Bangladesh, and Nepal on a loan basis.

References

Akinyele D, Belikov J, Levron Y (2018) Challenges of microgrids in remote communities: A STEEP model application, Energies, pp 1–35. https://doi.org/10.3390/en11020432

Adnan ZA (2019) Off-grid renewable energy solutions to expand electricity access: an opportunity not to be missed. IRENA. https://www.irena.org/-/media/Files/IRENA/Agency/Publication/2019/Jan/IRENA_Off-grid_RE_Access_2019.pdf

Anastasiadis AG, Tsikalakis AG, Hatziargyriou ND, (2010) Operational and environmental benefits due to significant penetration of microgrids and topology sensitivity. In proceedings IEEE PES General Meeting, pp 1–8. https://doi.org/10.1109/PES.2010.5590069

Augustine N, Suresh S, Moghe P, Sheikh K (2012) Economic dispatch for a microgrid considering renewable energy cost functions. In: Proceedings IEEE PES Innovative Smart Grid Technol. (ISGT), pp 1–7. https://doi.org/10.1109/ISGT.2012.6175747

Bansal N, Srivastava VK, Kheraluwala J Renewable Energy in India: Policies to Reduce Greenhouse Gas Emissions. Springer Nature Singapore Pte Ltd. 2019. IRENA (International Renewable Energy Agency) (2018a), Renewable Capacity Statistics 2018, International Renewable Energy Agency, Abu Dhabi. www.irena.org/publications/2018/Mar/Renewable-CapacityStatistics-2018

Benefits of demand response in electricity markets and recommendations for achieving them. A report to the United States Congress pursuant to section 1252 of the energy policy act of 2005. U.S. Dept. Energy, Washington, DC, USA, Tech. Rep., 2006. https://www.energy.gov/sites/default/files/oeprod/DocumentsandMedia/DOE_Benefits_of_Demand_Response_in_Electricity_Markets_and_Recommendations_for_Achieving_Them_Report_to_Congress.pdf

Bihari SP et al (2021) A comprehensive review of microgrid control mechanism and impact assessment for hybrid renewable energy integration. IEEE Access 9:88942–88958. https://doi.org/10.1109/ACCESS.2021.3090266

Bloomberg NEF (2017) The carbon brief profile: India. Carbon Brief 2019–03–14. Retrieved 2019–09–25.. https://about.bnef.com/blog/accelerating-indias-clean-energy-transition

Bloomberg NEF (2017) The carbon brief profile: India. Carbon Brief 2019–03–14. Retrieved 2019–09–25. https://about.bnef.com/blog/accelerating-indias-clean-energy-transition

Brown HE, Suryanarayanan S, Natarajan SA, Rajopadhye S (2012) Improving reliability of islanded distribution systems with distributed renewable energy resources. IEEE Trans. Smart Grid 3(4):pp 2028–2038.https://doi.org/10.1109/2FTSG.2012.2200703

Cabraal RA, Barnes DF, Agarwal DG (2005) Productive uses of energy for rural development. Ann Rev Environ Resour 117–44. https://doi.org/10.1146/annurev.energy.30.050504.144228

Cai N, Mitra J (2010) A decentralized control architecture for a microgrid with power electronic interfaces. In: proceedings of the north american power symposium, pp 1–8
Canziani F, Melgarejo Ó (2019) Design and Implementation of Rural Microgrids. In: Zambroni de Souza A., Castilla M. (eds) Microgrids Design and Implementation. Springer, Cham. https://doi.org/10.1007/978-3-319-98687-6_17
Casillas C, Kammen DM (2010) The energy-poverty-climate nexus. Science 330: 1182 – 1182 https://doi.org/10.1126/science.1197412.
Casillas C, Kammen DM (2012) Quantifying the social equity of carbon mitigation strategies. Climate Policy 10:1080–1094. https://doi.org/10.1080/14693062.2012.669097
Centre of Science and Environment. https://www.cseindia.org/page/waste.
Chaudhuri DG et al (2013) Sustainable development of renewable energy mini-grids for energy access: A framework for policy design. https://www.prayaspune.org/peg/publications/item/download/479_831445f88d532d77255b30add84342c4.html
Che L, Khodayar M, Shahidehpour M (2014) Only connect: microgrids for distribution system restoration. IEEE Power Energy Mag 12(1):70–81. https://doi.org/10.1109/MPE.2013.2286317
Che Y, Chen J (2012) Research on design and control of microgrid system. Elect Rev 88(5b):8386. https://doi.org/10.15199/48
Chen S, Gooi HB, Wang M (2012) Sizing of energy storage for microgrids. In: Proceedings IEEE power and energy society general meeting, pp 142–151. https://doi.org/10.1109/TSG.2011.2160745
Climatescope (2019). http://global-climatescope.org/library
Cullen W et al (2019) policy and financial barriers to micro-grid development in india. Roberts Environ Cent Claremont McKenna CollE Sustain Dev Policy Financ Team, May 16. https://cpb-us-w2.wpmucdn.com/sites.cmc.edu/dist/c/13/files/2019/12/Barriers-to-Micro-Grid-Development-in-India.pdf
Daneshi H, Khorashadi-Zadeh H, (2012) Microgrid energy management system: A study of reliability and economic issues. In proceedings IEEE power and energy society general meeting, pp 1–5. https://doi.org/10.1109/PESGM.2012.6344957
Department of Energy Office of Electricity Delivery and Energy Reliability, (2016). Summary Report: 2016. https://fas.org/sgp/crs/misc/R44357.pdf
Farzan F, Lahiri S, Kleinberg M, Gharieh K, Farzan F, Jafari M (2013) Microgrids for fun and profit: the economics of installation investments and operations. IEEE Power Energy Mag 11(4):52–58. https://doi.org/10.1109/MPE.2013.2258282
Fu Q et al (2012) Microgrid generation capacity design with renewables and energy storage addressing power quality and surety. IEEE Trans Smart Grid 3(4):2019–2027. https://doi.org/10.1155/2015/493560
Gomes M, Coelho P, Fernandes J (2019) Electricity Markets and Their Implications. In: Zambroni de Souza A., Castilla M. (eds) Microgrids Design and Implementation. Springer, Cham. https://doi.org/10.1007/978-3-319-98687-6_14
Guan X, Xu Z, Jia QS (2010) Energy-efficient buildings facilitated by microgrid. IEEE Trans Smart Grid 1(3):243–252. https://doi.org/10.1109/TSG.2010.2083705
Hatziargyriou ND et al. (2005) Management of microgrids in market environment. In proceedings of the **international conference** future power system pp 1–7. https://doi.org/10.1109/FPS.2005.204225
He Y, Sharma R, Zhang X, (2012) Microgrid operator's capacity and storage investment strategies under environmental regulations. In: **proceedings of IEEE PES innovative smart grid technologies**. (ISGT), pp 1–7. https://doi.org/10.1109/ISGT.2012.6175635
Hossain J, Mahmud A (2014) Renewable energy integration: Challenges and solutions. Singapore: Springer-Verlag. https://doi.org/10.1007/978-981-4585-27-9.
IEA, India Energy Outlook - World Energy Outlook Special Report 2015. https://www.gita.org.in/Attachments/Reports/IndiaEnergyOutlook_WEO2015.pdf
India 2020 Energy Policy Review, International Energy Agency (IEA). https://niti.gov.in/sites/default/files/2020-01/IEA-India%202020-In-depth-EnergyPolicy 0.pdf

Industrial Development Report, 2018, Demand for Manufacturing:Driving Inclusive and Sustainable Indus Devl. https://www.unido.org/sites/default/files/files/2017-11/IDR2018_FULL%20REPORT.pdf

IndustryArc, Microgrid Market - Forecast (2020 to 2025). https://www.industryarc.com/Report/16322/microgrid-market.html

International Renewable Energy Agency. (IRENA), 2019. https://www.irena.org/Statistics/View-Data-by-Topic/Capacity-and-Generation/Technologies.

International smart grid action network (ISGAN). An initiat clean energy minist, (2019). http://www.cleanenergyministerial.org/sites/default/files/201906/ISGAN%20fact%20sheet%20%28June%202019%29.pdf

Jain R, Arya A, (2015) A Comprehensive Review on Micro Grid Operation, Challenges and Control Strategies. In: Proceedings of the 2015 ACM Sixth International Conference on Future Energy Systems, Bangalore, pp 295–300. https://doi.org/10.1145/2768510.2768514

Jiayi H, Chuanwen J, Rong X (2008) A review on distributed energy resources and MicroGrid. Renew Sustain Energy Rev 12(9):2472–2483. https://doi.org/10.1016/j.rser.2007.06.004

Kanchev H, Lu D, Colas F, Lazarov V, Francois B (2011) Energy management and operational planning of a microgrid with a PV-based active generator for smart grid applications. IEEE Trans Ind Elec 58(10):4583–4592. https://doi.org/10.1109/TIE.2011.2119451

Khodaei A (2015) Provisional microgrids. IEEE Trans Smart Grid 6(3):1107–1115. https://doi.org/10.1109/PES.2010.5589391

Khodayar ME, Barati M, Shahidehpour M (2012) Integration of high reliability distribution system in microgrid operation. IEEE Trans Smart Grid 3(4):1997–2006. https://doi.org/10.1109/TSG.2012.2213348

Kopanos GM, Georgiadis MC, Pistikopoulos EN, (2013) Operational planning in energy networks based on microgeneration. In proceedings of the american control conference, pp 2940–2945. https://doi.org/10.1021/ie404165c

Korres GN, Hatziargyriou ND, Katsikas PJ (2011) State estimation in multi-microgrids. Eur Trans. Elect Power 21(2):1178–1199. https://doi.org/10.1002/etep.442

Krapels EN (2013) Microgrid development: Good for society and utilities [in my view]. IEEE Power Energy Mag 11(4):94–96. https://doi.org/10.1109/MPE.2013.2254081

Kroposki B (2009) An integration facility to accelerate deployment of distributed energy resources in microgrids. In: proceedings **IEEE power and energy society general meeting**, pp 1–4. https://doi.org/10.1109/JPROC.2017.2696878

Liu X, Wang P, Loh PC (2011) A hybrid AC/DC microgrid and its coordination control. IEEE Trans Smart Grid 2(2):278–286. https://doi.org/10.1109/TSG.2011.2116162

Logenthiran T, Srinivasan D (2009) Short term generation scheduling of a microgrid. In proceedings IEEE Region 10 **conference** (TENCON), pp 1–6. https://doi.org/10.1109/2FTENCON.2009.5396184

Microgrid Market, 2019. https://www.marketsandmarkets.com/Market-Reports/micro-grid-electronics-market-917.html

Matamoros J, Gregoratti D, Dohler M, (2012) Microgrids energy trading in islanding mode. In proceedings IEEE 3rd international conference Smart Grid Communication (SmartGridComm), pp 49–54. https://doi.org/10.1109/SmartGridComm.2012.6485958

Millar RJ, Kazemi S, Lehtonen M, Saarijarvi E (2012) Impact of MV connected microgrids on MV distribution planning. IEEE Trans. Smart Grid 3(4):2100–2108. https://doi.org/10.1109/TSG.2012.2212922

MNRE (Ministry of New and Renewable Energy) (2018), 4-Year Achievement Booklet, Government of India, New Delhi. https://mnre.gov.in/sites/default/files/uploads/MNRE-4- Year-Achievement-Booklet.pdf

MNRE, Overview - Annual Report 2018–19. https://mnre.gov.in

Morris GY, Abbey C, Wong S, Joos G (2012) Evaluation of the costs and benefits of microgrids with consideration of services beyond energy supply. In: Proceedings IEEE power and energy society general meeting, pp 1–9. https://doi.org/10.1109/PESGM.2012.6345380

Parhizi S, Lotfi H, Khodaei A, Bahramirad S (2015) State of the art in research on microgrids: a review. IEEE Access, Open Access J 3:890–925. https://doi.org/10.1109/ACCESS.2015.2443119
Priyadarshi R, Thakur T, Arya A (2019) Performance Evaluation of Rural Electrification sector of India. In: 8th International Conference on Power Systems 2019, (ICPS 2019) MNIT Jaipur, December 20 – 22. https://doi.org/10.1109/ICPS48983.2019.9067627.
Ramana MV et al, (2014) Rural Energy Alternatives in India: Opportunities in Financing and Community Engagement for Renewable Energy Microgrid Projects, pp 1–66. https://spia.princeton.edu/sites/default/files/content/591f%20Rural%20Energy%20Alternatives%20in%20India.pdf
Renewable energy policies in a time of transition. https://www.irena.org/-/media/Files/IRENA/Agency/Publication/2018/Apr/IRENA_IEA_REN21_Policies_2018.pdf
Renewable Energy, Forms and Types of Renewable Energy. http://www.altenergy.org/renewables/renewables.html. Accessed 13 Feb 2015
Ross M, Hidalgo R, Abbey C, Joós G (2011) Energy storage system scheduling for an isolated microgrid. IET Renew Power Gener 5(2):117–123. https://doi.org/10.1049/iet-rpg.2009.0204
Sao C, Lehn P (2010) Control and power management of converter fed microgrids. In: IEEE PES general meeting. https://doi.org/10.1109/ICAGE.2014.7050138
Schnitzer D, Lounsbury DS, Carvallo JPR, Deshmukh R, Apt J, Kammen DM (2014) Microgrids for Rural Electrification: A critical review of best practices based on seven case studies. U N Found. https://doi.org/10.13140/RG.2.1.1399.9600
Seetharaman K, Moorthy K, Patwa N, Saravanan, Gupta Y, (2019) Breaking barriers in deployment of renewable energy, Heliyon 5(1). https://doi.org/10.1016/j.heliyon.2019.e01166
Shafiullah G et al (2021) Prospects of hybrid renewable energy-based power system: a case study, post analysis of chipendeke micro-hydro, zimbabwe. IEEE Access 9:73433–73452. https://doi.org/10.1109/ACCESS.2021.3078713
Shah J, Wollenberg BF, Mohan N (2011a) Decentralized power flow control for a smart micro-grid. In: proceedings IEEE power and energy society general meeting, pp 1–6. 978-1-4577-1002-5/11
Shah J, Wollenberg BF, Mohan N (2011b) Decentrali978-1-4577-1002-5/11zed power flow control for a smart micro-grid. In: proceedings IEEE power and energy society general meeting, pp 1–6. 978-1-4577-1002-5/11
Su W, Yuan Z, Chow MY, (2010) Microgrid planning and operation: solar energy and wind energy. In: proceedings. IEEE PES General Meeting pp1–7. https://doi.org/10.1109/PES.2010.5589391
Su W, Wang J (2012) Energy management systems in microgrid operations. Elect J 25(8):45–60. https://doi.org/10.1016/j.tej.2012.09.010
Suryanarayanan S, Kyriakides E (2012) Microgrids: An emerging technology to enhance power system reliability. IEEE Trans. Smart Grid. http://smartgrid.ieee.org/march2012/527-microgrids-an-emerging-technology-to-enhance-powersystem-reliability. Accessed 13 Nov 2014
The Paris Agreement, the United Nations Framework Convention on Climate Change 2015. https://unfccc.int/files/essential_background/convention/application/pdf/english_paris_agreement.pdf
Technology Transition Final Public Report Smart Power Infrastructure Demonstration for Energy Reliability and Security (SPIDERS) 2015. https://www.energy.gov/sites/prod/files/2016/03/f30/spiders_final_report.pdf
U.S. Environmental Protection Agency. State and Local Climate and Energy Program. http://www.epa.gov/statelocalclimate/state/topics/renewable.html. Accessed 13 Feb 2015
Understanding microgrid and what are the benefits of the smart microgrid approach? (2015) Galvin electricity initiative. http://www.galvinpower.org/resources. Accessed 13 Feb 2015
United nation framework convention on climate change (UNFCCC). Kyoto protocol reference manual on accounting of emission and assigned amount, 2008. https://unfccc.int/resource/docs/publications/08_unfccc_kp_ref_manual.pdf
Urbanization and Development: Emerging Futures, World Cities Report 2016, May 21. https://wcr.unhabitat.org/wp-content/uploads/2017/02/WCR-2016_FULL_updated.pdf

Vaccaro A, Popov M, Villacci D, Terzija V (2011) An integrated framework for smart microgrids modeling, monitoring, control, communication, and verification. Proc IEEE 99(1):119–132. https://doi.org/10.1109/JPROC.2010.2081651

Vasiljevska J, Lopes JAP, Matos MA (2013) Integrated microgeneration, load and energy storage control functionality under the multi micro-grid concept. Electr Power Syst Res 95:292–301. https://doi.org/10.1016/j.epsr.2012.09.014

Wang S, Li Z, Wu L, Shahidehpour M, Li Z (2013) New metrics for assessing the reliability and economics of microgrids in distribution system. IEEE Trans Power Syst 28(3):2852–2861. https://doi.org/10.1109/TPWRS.2013.2249539

Wang L et al (2009) A micro hydro power generation system for sustainable microgrid development in rural electrification of Africa. In: proceedings of the IEEE power energy **Society** general meeting, pp 1–8. https://doi.org/10.1109/PES.2009.5275244

Why is Renewable Energy Important? http://www.renewableenergyworld.com/rea/tech/home. Accessed 13 Feb 2015

World bank, mini grids have potential to bring electricity to half a billion people. Accord new world bank study. https://www.worldbank.org/en/news/press-release/2019/06/25/mini-grids-have-potential-to-bring-electricity-to-half-a-billion-people.

World health organization (WHO), Health sustain dev: Energy access resil. www.who.int/sustainabledevelopment/health-sector/health-risks/energy-access/en/

World Population Data Sheet 2018. https://www.prb.org/2018-world-population-data-sheet-with-focus-on-changing-age-structures/

Xu L, Chen D (2011) Control and operation of a DC microgrid with variable generation and energy storage. IEEE Trans Power Del 26(4):2513–2522. https://doi.org/10.1109/TPWRD.2011.2158456

Zeng Z, Yang H, Zhao R (2011) Study on small signal stability of microgrids: a review and a new approach. Renew Sustain Energy Rev 15(9):4818–4828. https://doi.org/10.1016/j.rser.2011.07.069

Zhang Y, Gatsis N, Giannakis GB (2013) Robust energy management for microgrids with high-penetration renewables. IEEE Trans Sustain Energy 4(4):944–953. https://doi.org/10.1109/TSTE.2013.2255135

Zhang Z, Li G, Zhou M, (2010) Application of microgrid in distributed generation together with the benefit research. In proceedings IEEE PES general meeting, pp 1–5. https://doi.org/10.1109/PES.2010.5589277

www.indianbudget.gov.in and https://infogram.com/ministry-wise-budgetary-allocations-2009-10-to-2019-20-1hxj48jgkjr52vg

Chapter 37
Challenges and Issues in Solar PV Development in India

A. S. Werulkar and Prakash S. Kulkarni

Abstract The population of India is growing day by day. This is resulting in more pollution facing the problem of shorter non-rechargeable resources leftover for the succeeding generations. Naturally, it is certainly concluded that India's energy demand will increase in near future. Renewable energy is very useful in playing a significant role for providing pollution-free energy. This will certainly contribute to climate change. There is a target of generating 100 GW of solar energy by 2022. This goal is very pioneering for India under the agreement of Climate Change in Paris. Progress of the Indian solar energy division depends on the mixture of law framework, financial structure, and local production sector and eco-friendly sound technology. Although there is enhancing potential of sun energy in India, today also there are shortcomings in its administration. In this chapter, the authors focus on the present status, challenges, and upcoming prospects of sun power progress in India. It further summarizes the way ahead with recommendations to overcome some of the shortcomings.

Keywords PV scenario · Global energy · Indian PV · Motivational policies · Solar panels · JNNSM · Solar mission achievements

Nomenclature

PV	Photovoltaic
JNNSM	Jawaharlal Nehru National Solar Mission
LPG	Liquefied petroleum gas
UV	Ultraviolet
CNG	Compressed natural gas

A. S. Werulkar (✉)
Department of Electronics and Telecommunication Engineering, SVPCET, Nagpur 441108, India

P. S. Kulkarni
Department of Electrical Engineering, VNIT, Nagpur 440010, India
e-mail: pskulkarni@eee.vnit.ac.in

A. K. Bohre et al. (eds.), *Planning of Hybrid Renewable Energy Systems, Electric Vehicles and Microgrid*, Energy Systems in Electrical Engineering,
https://doi.org/10.1007/978-981-19-0979-5_37

SHLS	Solar home lighting system
RE	Renewable energy
W-LED	White lighting emitting diode
CFL	Compact fluorescent lamp
GSP	Generalized system of preferences

37.1 Introduction

There is a need of providing enough continuous electrical supply and this is the most important and rigorous catch of the world. The emerging economies of the globe are facing the acute problem of electricity. It is the basic need for the growth and progress of any country. India is a developing country and its population is 17% of the world's total residents. Hence, it is facing a severe problem of insufficient electric supply. An approximately more than 80% of the total utilization of fuel is brought in. Due to this, the nation has to spend its valuable foreign money. India is unable to manage the electrical energy requirements of the country though they are bringing in high quantities. A lot of the countryside, as well as suburban areas, are facing power cut-off problems. Govt. of India has aimed to achieve 100% uninterrupted power supply in the country by the year 2022. Because of the above circumstances, Govt. of India and Govt. of some states have a plan for extending facilities like a subsidized supply of materials required for energy production from renewable energy resources, viz., solar, wind, biogas, tidal, etc.

India has ample scope for generating solar energy. The reason is its geographical location and it receives solar radiation almost throughout the year. This amounts to 3,000 h of sunshine. So, it is a preferred location to utilize inexhaustible and clean solar energy for the production of electricity through PV technology. Though PV technology is costlier, in long term, it is beneficial, profitable, maintenance free as well as environmental friendly technology. The use of solar energy will definitely reduce the effect the greenhouse gases. Thus, it can prove as an effective source in saving the environment. India is a developing country, and power is the main constituent factor in the financial growth of any economy especially in an evolving nation like India with 17% of the world's population. However, only 0.4% of the world's oil, 0.4% of natural gas, and 6% of coal reserves are used. This fact of access to energy resources can be considered a major shortfall because India is the fourth largest consumer of energy in the world after the USA, China, and Russia. Recently, Govt. of India has announced that it will provide 24 × 7 power to all by 2022. It is very important for India to assess the trade that it will need to do in order to achieve a balanced goal of energy security, access to energy, and sustainability (http://mnre.gov.in). To fulfill the above objectives, it is very important to examine the components of the Indian ecosystem, their availability, and demand and the interconnected networks. The connection of these parts also needs to be judged. Fortunately, India has great potential for renewable energy. Increasing domestic supply and exploiting

India's vast renewable energy potential could be a major step in reducing India's dependence on imported energy sources. It can also address its growing concerns about climate change. It is also important that the required sectors play an important role in reducing the amount of energy required to deliver the same number of services. Energy efficiency and power supply demand are two key interventions that should be considered and considered under the same lens as supply delivery interventions. Transmission and Distribution networks need to be strengthened and Smart Grids and storage solutions should be implemented and this could be an intervention that would help maintain a more secure energy environment in India (Karakoti et al. 2013).

Almost all activities in the universe require energy for living and performance. Energy is the capacity to do work. This work can be calculated as the transfer of energy from one form to another. It is an integral part of the day-to-day life. As the health of a person is apprehended from his diet, the state of development of the country is recognized from the quantity of energy it utilizes. The countries consuming more energy are experiencing a better standard of living compared to the countries having comparatively less energy consumption. The per capita annual energy consumption in the US is more than 12,000 kWh/year. On the contrary, this figure for INDIA is just 600 kWh/year (Solanki 2009, 2013). This reflects the fact of the developmental status of India. Therefore, for the growth of the developing countries and to upgrade their standard of living, the energy supply should be enhanced. Generally, energy is used in the form of diesel, petrol, coal, LPG, CNG, and electricity. Mostly the electricity is generated from other fuels like coal and petroleum. These natural sources of energy are limited and may exhaust in due course of time. All these sources of energy are known as conventional energy sources. They are also called commercial energy sources, as they are available in the market at a price decided by the suppliers. Regarding conventional energy sources, the fact is to be noted that they cause environmental pollution which is most disadvantageous. There is consistent growth in CO_2 gas levels which is said to be the cause of global warming.

37.2 Renewable Energy Resources

Renewable energy sources are inexhaustible and pollution free. Various forms of renewable energy are solar Energy, wind energy, bioenergy, hydro energy, geothermal energy, wave, and tidal Energy (Sukhatme and Nayak 2010; Khan 2006).

Figure 37.1 depicts global installed Energy Capacity of the world. Solar PV Renewable Energy capacity stands 3.77% of total capacity of world. It is creditable in Indian perspective that it has achieved 11th position in the world's PV installation with the achievement of 36.9 GW installation.

Figure 37.2 presents bar diagram in respect of source-wise RE installation for India from the year 2007–2015. It is apparent from the figure that wind power stands 1st, small hydropower 2nd, whereas solar power has gained 3rd position in RE installation in India.

2012	
Solar PV	0.64
Geothermal	0.28
Biogas	0.16
Solar Thermal	0.04
Gas	41.1
Coal	29.32
Nuclear	8.59
Hydro	5.59
Oil	5.2
Wind	5.08
Biomass	1

2020	
Solar PV	3.77
Geothermal	0.34
Biogas	0.23
Solar Thermal	0.34
Gas	3.21
Coal	25.23
Nuclear	9.41
Hydro	8.58
Oil	3.63
Wind	8.24
Biomass	1.02

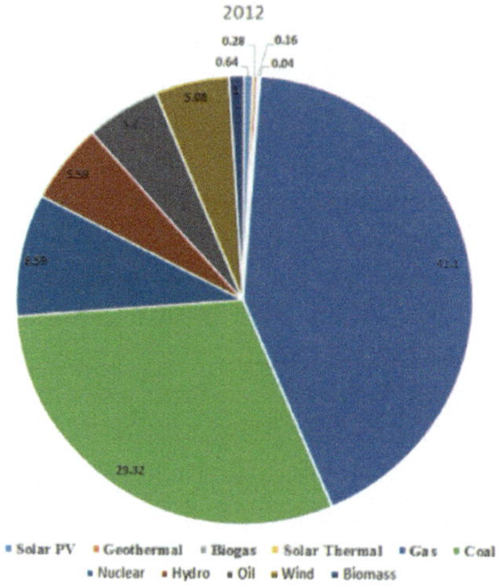

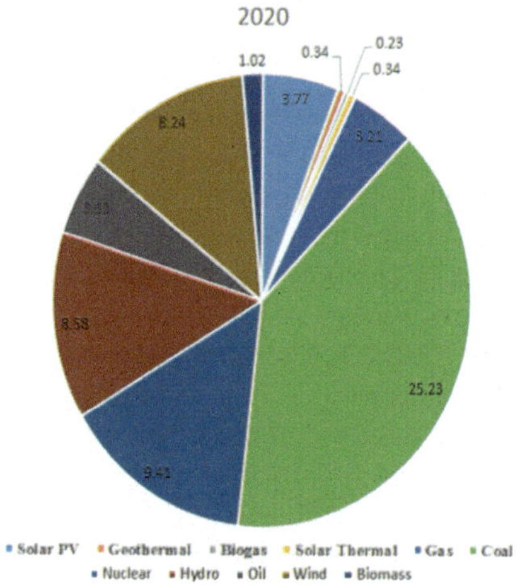

Fig. 37.1 Global installed energy capacity with a comparison of 2012 and 2020

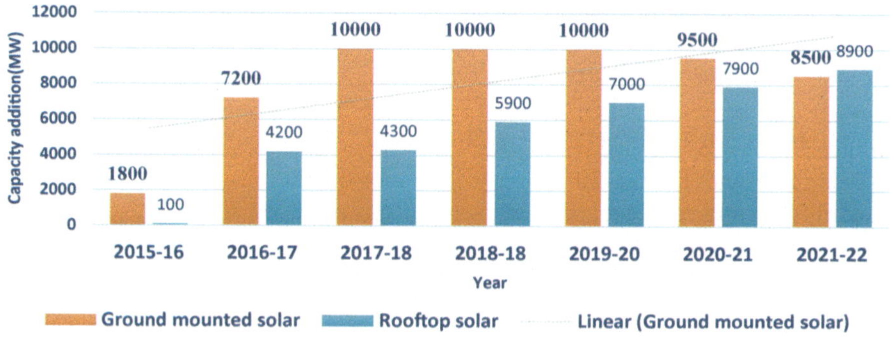

Fig. 37.2 Bar diagram in respect of source wise renewable energy installation in India

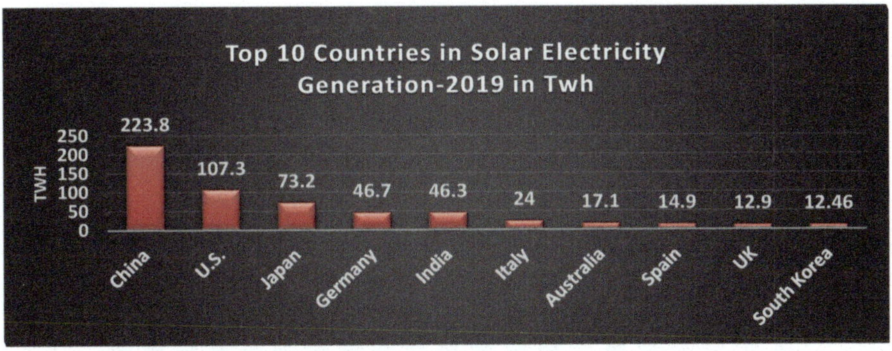

Fig. 37.3 Top ten countries in Global PV installation

The time is not far away when solar energy will stand 1st in RE installation due to Govt. of India's positive approach and policies. The most easily available and free source of energy is solar energy and it is available since the beginning of life on the earth. Solar PV route uses solar energy to generate DC power for illuminating homes and buildings, driving motors, pumps, electric gadgets, etc. (Messenger and Ventre 2000; Patel 2006; Markvart 2000; www.wiki-solar.org).

37.3 Current PV Scenario

In the following paragraphs along with analytical bar diagrams and tables, global and Indian scenario in respect of PV installation is presented.

37.3.1 Global PV Scenario

Figure 37.3 displays global scenario of solar PV installation (European photovoltaic industry association 2014–2018). It shows top ten countries showing PV installation globally. From Fig 37.3, it is concluded that in China PV installation is maximum.

Figure 37.4 shows cumulative global PV capacity by 2030 end.

37.3.2 Indian PV Scenario

Solar Energy provides great energy in a tropical country like India where about 45% of households, especially in rural areas can't get electricity. India has not utilized its maximum capacity as compared to other countries. There is a vast scope for

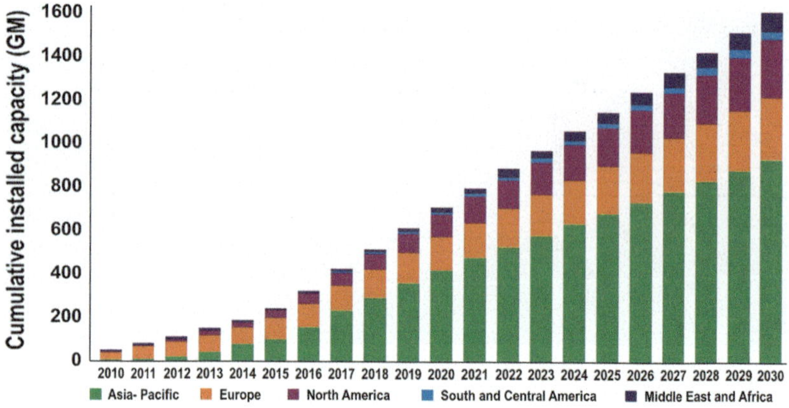

Fig. 37.4 Global PV capacity by 2030 end

enhancement of RE solar capacity considering its favorable topographical location for production of solar energy. It will definitely accelerate India's process of development of solar energy which is lagging behind due to insufficient supply of financial resources. Table 37.1 shows India's Renewable Energy Sector at a glance.

Solar electricity is one of the most important sources of renewable energy. The electricity can be generated from photovoltaic panels in many ways. Since photovoltaic energy is the clean energy, it is gaining a lot of importance in India. The cost of solar panels is decreasing day by day and this in turn is reducing the PV electricity cost. Some of these products are Solar Home Lighting Systems, Solar power LED lights, Solar Fans, Solar lanterns, Solar water heaters, and Solar cookers, etc. (Solanki 2009).

Table 37.1 India's Renewable Energy Sector at a glance

Year	Installed RE capacity (in GW)	% Share of RE in total installed capacity	Generation from renewable sources (in BU)	Total generation from all sources (in BU)	%Share of RE in generation
2014–15	39.55	14.36	61.78	1,110.1	5.56
2015–16	46.58	15.23	65.78	1,172.98	5.6
2016–17	57.9	17.68	81.54	1,241.38	6.56
2017–18	69.77	20.24	101.83	1,303.37	7.81
2018–19	78.31	21.95	126.76	1,375.96	9.21
2019–20	87.07	23.52	138.32	1,390.93	9.95
2020–21	92.54	24.53	111.92	1,017.81	11
	(Up to Jan, 2021)	(Up to Jan, 2021)	(Up to Dec, 2021)	(Up to Dec, 2021)	(Up to Dec, 2021)

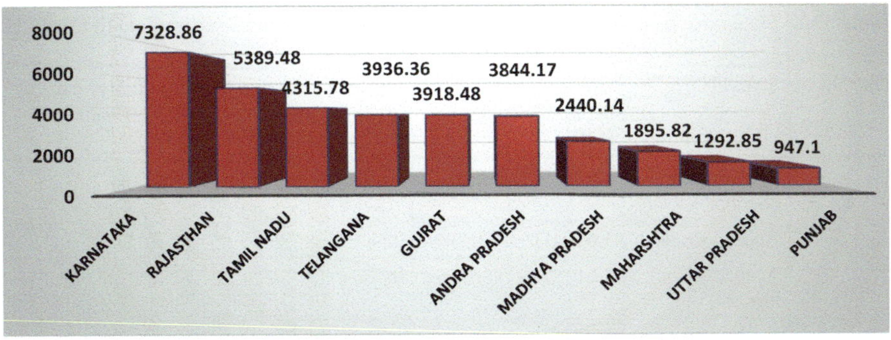

Fig. 37.5 Top 10 states in solar PV installation

There is frequent power cut observed especially in rural areas. The state of Karnataka has maximum installation of solar projects in India with a capacity of 7,328.8 MW. The graphical trend of top 10 states in Solar PV installation is shown in Fig. 37.5.

Though India is trying to enhance its capacity of solar energy, the major constraint in PV installation is the cost of PV panels. Due to high price of PV panels, the PV system installation is not economically viable. However due to continuous research in the technology of solar cell manufacturing since the last 20 years, the cost of PV panels is significantly reduced.

The cost of thin film silicon is low and that of crystalline silicon is high.

Table 37.2 demonstrates state-wise solar installed projects in India.

Figure 37.6 displays the graphical representation of reducing cost of solar panel in US $. The cost of thin film silicon is low and that of crystalline silicon is high. However, since the last 20 years, it has reduced gradually from 7.5 US $/W to approximately 2 US $/W. Due to reduction in the cost, installation of PV systems has indicated remarkable rise in recent years.

37.4 Motivational Policies of Government of India

Introducing India's National Action Plan on Climate Change on June 30, 2008, former Indian Prime Minister Dr. Manmohan Singh said: "Our vision is to make India's economic development energy-efficient. Over a period of time, we must pioneer a graduated shift from economic activity based on fossil fuels to one based on non-fossil fuels and from reliance on non-renewable and depleting sources of energy to renewable sources of energy. Table 37.3 shows phase-wise targets of JNNSM.

In this strategy, the sun occupies Center-stage, as it should, being literally the original source of all energy. We will pool our scientific, technical and managerial talents, with sufficient financial resources, to develop solar energy as a source of abundant energy to power our economy and to transform the lives of our people. Our

Table 37.2 State-wise installed capacity of solar projects under various schemes up to December 2020

SR. No	State/UT	Cumulative capacity till 31-03-2020 (MW)	Capacity added in 2020–21 till 31-12-2020 (MW)	Cumulative capacity till 31-12-2020 (MW)
1	Andaman & Nicobar	12.19	17.03	29.22
2	Andhra Pradesh	3,610.02	234.15	3,844.17
3	Arunachal Pradesh	5.61	0	5.61
4	Assam	41.23	1.76	42.99
5	Bihar	151.57	5.84	157.41
6	Chandigarh	40.55	4.61	45.16
7	Chhattisgarh	231.35	8.65	240
8	Dadra & Nagar	5.46	0	5.46
9	Daman & Diu	19.86	20.1	39.97
10	Delhi	165.16	11.3	176.46
11	Goa	4.78	0	4.78
12	Gujrat	2,948.37	970.11	3,918.48
13	Haryana	252.14	10.28	262.42
14	Himachal Pradesh	32.93	9.8	42.73
15	Jammu & Kashmir	19.3	1.43	20.73
16	Jharkhand	38.4	0.66	39.06
17	Karnataka	7,277.92	50.94	7,328.68
18	Kerala	142.22	4.69	146.92
19	Lakshadweep	0.75	0	0.75
20	Madhya Pradesh	2,258.45	181.69	2,440.14
21	Maharashtra	1,801.8	94.02	1,895.82
22	Manipur	5.16	1.2	6.36
23	Meghalaya	0.12	0	0.12
24	Mizoram	1.52	0.01	1.53
25	Nagaland	1	0	1
26	Odisha	397.84	1.76	399.6
27	Pondicherry	5.51	2.03	7.54
28	Punjab	947.1	0	947.1

(continued)

Table 37.2 (continued)

SR. No	State/UT	Cumulative capacity till 31-03-2020 (MW)	Capacity added in 2020–21 till 31-12-2020 (MW)	Cumulative capacity till 31-12-2020 (MW)
29	Rajasthan	5,137.91	251.57	5,389.48
30	Sikkim	0.07	0	0.07
31	Tamil Nadu	3,915.88	399.9	4,315.78
32	Telangana	3,620.75	315.61	3,936.36
33	Tripura	9.41	0	9.41
34	Uttar Pradesh	1,095.1	197.75	1,292.85
35	Uttarakhand	315.9	4.54	320.44
36	West Bengal	114.46	35.38	149.84
		34,627.79	2,836.81	37,464.44

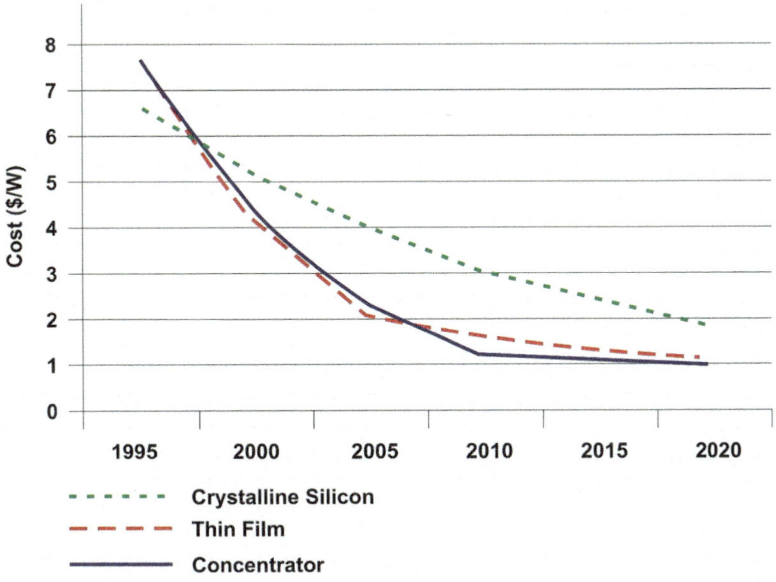

Fig. 37.6 Cost of PV panel from 1995 to 2020

Table 37.3 Phase-wise targets of JNNSM

	Till 2009	Phase 1	Phase 2	Phase 3	Total
Grid-connected (MW)	6	1,100	3,000	16,000	20,000
Off-grid (MW)	2.4	200	800	1,000	2,000
Thermal collectors (million m^2)	3.1	7	8	5	20
Solar lighting systems (million)	1.3				20

success in this endeavor will change the face of India. It would also enable India to help change the destinies of people around the world."

The mission has adopted a 3 phase approach. Phase 1 comprises the remaining period of the 11th Plan (until 2012) and first year of the 12th Plan (up to 2012–2013). Phase 2 consists of the remaining 4 years of the 12th Plan (2013–2017) and phase 3 comprises the 13th Plan (2017–2022). At the end of each plan, and mid-term during the 12th and 13th Plan, progress of each phase is being evaluated and capacities as targets of subsequent phases are being reviewed.

37.5 Solar Resource (Sukhatme and Nayak 2010)

Figure 37.7 shows state-wise solar potential in India. From this figure, it can be apparently seen that Rajasthan state in India has the highest solar potential. Figure 37.8 shows the earth's solar radiation budget. The radiation of the sun reaches to earth in various wavelengths from 300 nm to 4 μm approximately. It is partly reflected by the atmosphere and partly transmitted to the earth's surface. The actual solar radiation reaching to earth's surface is only 48%. This energy is absorbed by the land and ocean.

Sun's spectral distributions can be divided into two categories:

(a) Air mass zero (AM_0) spectrum outside the atmosphere and
(b) $AM_{1.5G}$ spectrum at sea level. They are defined at certain standard conditions.

Figure 37.9 indicates PSpice 9.1 simulated output of air mass 1.5 G standard spectrum of sun (Castaner and Silvestre 2002).

As shown in this spectrum, the peak irradiance is about 1,600 $W/m^2/\mu m$. This solar spectrum at the earth's surface mainly consists of visible and infrared radiation. The UV component in the spectrum is small. The dispersion of solar spectrum on the earth's surface is shown in Table 37.4.

From Table 37.4, it is clear that maximum energy from the sun is available in the visible radiation range. Hence this range of radiation is used to generate electricity from PV panels.

Most of the states of India are also functioning through participation in the captioned scheme for enhancement of solar energy production. They are actively involved in solar energy production through their individual solar state policies. Karnataka is the most vibrant state which has been leading in India in solar power generation through GSSP 2009. This state has developed many individual solar projects as well as public–private partnership-based large-scale solar parks of about 7,328 MW up to December 2020. Though it contributes towards maximum solar power energy production, it is apparently seen that though strong implementation mechanism is the key strength of the policy [A-06]. As the greenhouse gases emission is the main hurdle in installation of biopower generation units to meet energy requirement, their installation will have adverse environmental as well as social impact. As the decentralized solar rooftop PV system overcomes the adverse effects

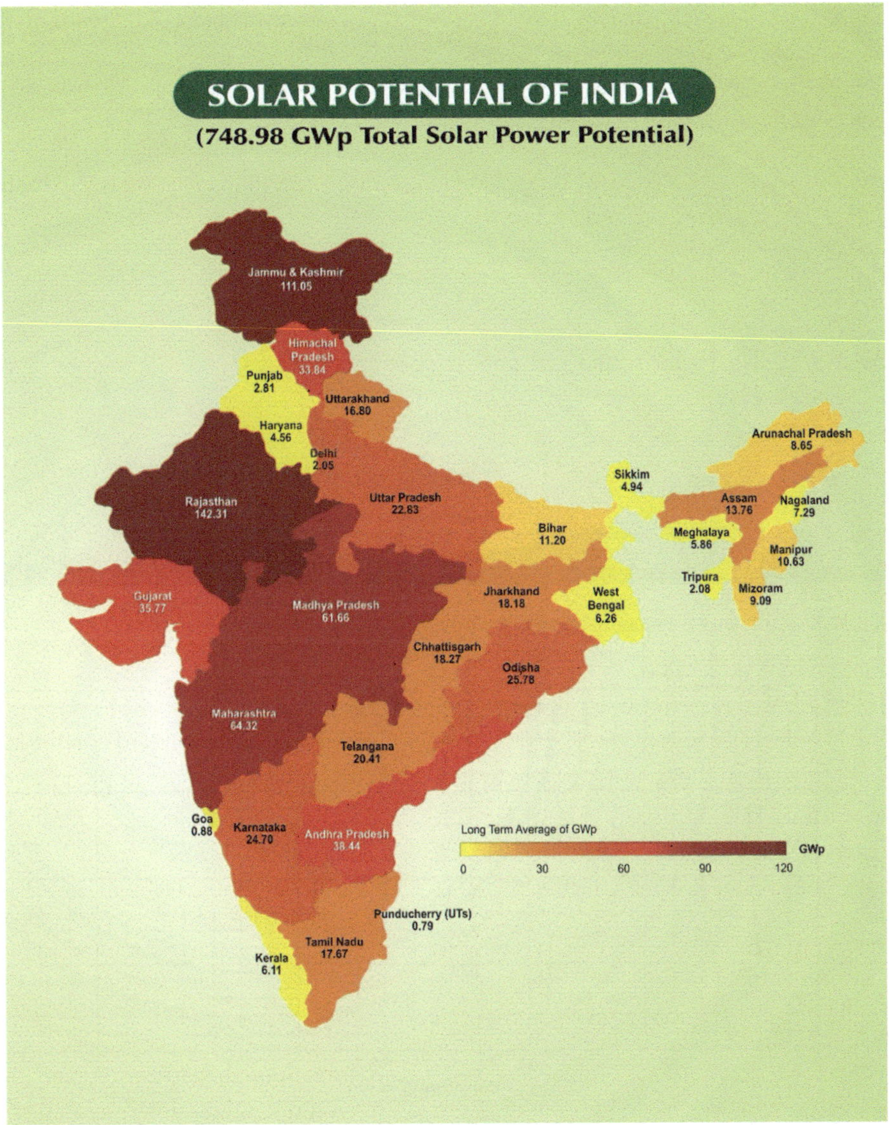

Fig. 37.7 India's state-wise solar potential (www.wiki-solar.org)

of fossil-based power plants in a sustainable way through its successful systems and environmental advantages, it will be beneficial in providing clean energy to the public who has still no access to electricity. Due to the unforeseen corona pandemic, since March 2020 in India, it is not possible to achieve the upward target of achieving installation of 40 GW decentralized solar rooftop PV up to 2022 (Rakesh Kumar Tarai and Paresh Kale 2018).

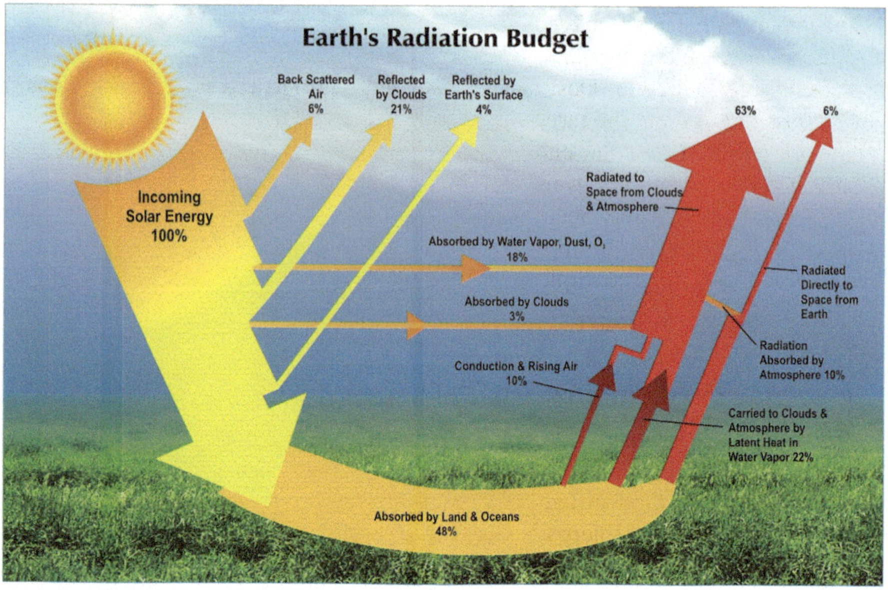

Fig. 37.8 Earth's solar radiation budget (www.wiki-solar.org)

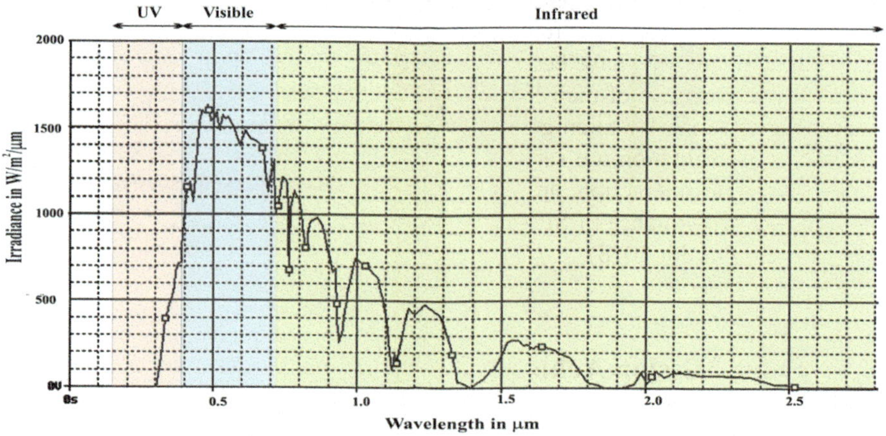

Fig. 37.9 Solar spectrum using Pspice 9.1

Table 37.4 Distribution of spectral content of solar insolation on Earth's surface

Type of radiation	Range of wavelengths (micrometers)	Percentage of energy carried
Ultraviolet	0.15–0.38	7.6
Visible	0.38–0.72	48.4
Infrared	0.72–4	43
Other	> 4	1

The geographical position of India in the globe is such that more than three fourth of India has favorable atmospheric conditions to receive abundant sunlight required for PV energy production. Despite these encouraging environmental conditions, it will be possible to meet the target if the bureaucracy plans meticulously the policy framework required for solar energy production. Though the solar industries in the country had started blooming rapidly with the implementation of solar policy 36 states, a great setback has been experienced during 20–21 due to pandemic. All states are trying their best to implement solar energy plants by using their wasteland and thereby decreasing the implementation cost (Kapoor et al. 2014).

37.6 Solar Performance Indicators

India is the peninsula and has the benefits of its geographical conditions due to which it has a tremendous God gifted potential for generating clean electricity through Hydro, Wind, and Solar Energy Sources, which are renewable. Since 2014, as the policies for the reduction of carbon footprints are implemented meticulously, India has been recognized for its consciousness and efforts to save earth. [A-09]Grid proves to be unreliable in remote areas of India and so the sustainable energy solution is the need of time. The Solar photovoltaic microgrids have proven a great relief in energy supply. Not only in the remote areas but also in the towns and cities the solar photovoltaic microgrids have become popular due to their production capacity and availability irrespective of their higher cost. The outlook for the assessment of the functioning of a solar PV microgrid is being constantly evaluated by the researchers. The means of evaluation are self-consumption, capacity utilization factor, performance ratio, grid interaction, and self-sufficiency. It is noticed that such research improves the performance of SPV microgrids, particularly in grid-connected mode (Purohita and Purohit 2018).

37.7 Solar Mission Achievement

Incompetent and unstable political scenario in India was the major hurdle and significant challenge in the implementation of solar power sector constantly trying to enhance its solar power production. Government of India uplifted the solar energy production projects by subsidizing them on individual as well as corporate basis (Rathore and Rathore 2021).

It is anticipated that a cumulative capacity of around 10,000 MW has been installed up to March 31, 2020–21. It is shown in the trend of Fig. 37.10.

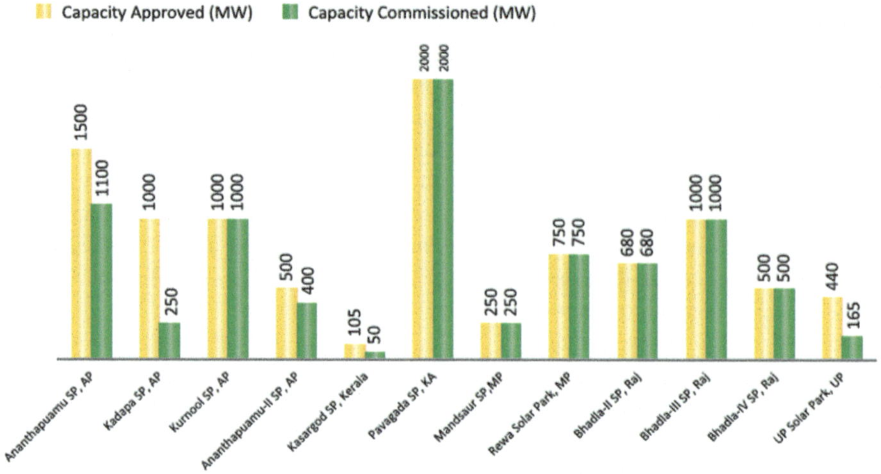

Fig. 37.10 Solar parks: capacity approved and capacity commissioned

37.8 Solar Operation Challenges

India is still a developing country, and there is a vast potential to develop eco-friendly industries. There will be a high demand of power. The target of the Government up to 2022 is generation of 100 GW under the National Solar Mission. The main obstacles and gridlocks faced by solar energy suppliers to achieve their target are being studied on various levels. There are different suggested motivational factors of solar energy to attract large solar producers around the globe to invest in India (Jaina et al. 2018).

Rapid urbanization and constant increase in population are the factors responsible for India's continuous growing electricity requirements. It is apparent that with growth of population and flow of people towards suburban areas India's electricity demand is definitely going to increase in near future. The reason for shifting towards non-conventional energy resources are changed climate, increased carbon emissions, and the speedily exhausting conventional energy resources (Sannigrahi 2000).

The exploitation of solar energy is mostly done by (a) Electricity generation by PV systems, (b) heat and electricity generation with solar thermal power plant, and use of solar thermal appliances to consume thermal energy. The target of 100 GW generation has many obstacles on the ground as there are issues relating to achieving success in a factual manner. Some of the hurdles are requirement of huge amount of land space for installing solar power gadgets and power grid framework. The challenge of using less cultivable land can be overcome by preferring rooftop desert areas rooftops wastelands and inland water bodies. The forthcoming challenge is to protect our environment from the harmful effects of the discarding of damaged or dead panels and batteries of a solar PV system. Serious attention should be provided for the use of recyclable or more efficient material. In the cyclonic or coastal areas, light weight solar PV modules are the need of time to make it sustainable. The projected life span of 30 years of a solar panel remains only on paper shrinking to 7 years in India as it practically faces several challenges of positioning and emplacement, intense environmental conditions, and also on kind and recurrence of maintenance. Long life of the panels will definitely attract the investors and users. It is claimed that solar technology is maintenance free, but the factors like dust accumulation, bird dejection, or piling of leaves from tree on solar panels drop down the performance efficiency up to 50%. Solar panels require a high maintenance such as washing them with filtered water to avoid water patches on the panels and rusting in the aluminum structures. The solution to the above catch is by covering the panels with an industry processed translucent indium tin oxide electrodes (self-dusting Martian technology). Power grid integration faces many difficulties. There is a need of tremendous in-depth research and development of a solar technology which will tolerate and sustain Indian dynamic environment. A product which is developed with the collaborative efforts of scientists, industry, and society is sustainable in long term. The only way out from the emission of greenhouse and toxic gases in the environment is the more and more usage of solar technology (Yaduvir Singh and Nitai Pal 2020).

Energy has a noteworthy role in economic, national equity and improving quality of life. Reducing fuel resources and the high risk of associated pollution have necessitated the consideration of other renewable and renewable resources. Due to the development and need for non-conventional energy sources, like solar or wind energy, tidal, wave energy, etc., are gaining popularity. Both the government and the central government of India have also supported research on solar photovoltaic power over the past decade. The current study discusses non-conventional energy and tends to solar photovoltaic energy trends in India as well as comparative studies of current and future developments. This chapter also reviews the policies and promotional measures taken by the Indian government. Finally, this review chapter will help identify current trends, challenges, and barriers to better planning and management in establishing and implementing other energy resources (Bhatt and Jani 2019).

37.9 Case Studies and Analysis

Production, distribution, and use of power include various employees at different levels of management. Permitting easy-to-use, dependable, well organized, prompt, and smooth communication to sensitive facts and figures for all such employees is a highly desirable characteristics and a Smart City measure.

In the current scenario, most of the cities are developing e-Governance to improve state services and to improve citizens. Gandhinagar already exists as an Indian city taking the lead in the use of high-tech communications technology and thus modifying itself into Smart City. In the past Gandhinagar has often experienced power outages and declining demand for energy which has caused not only disruption, but also leading to significant job losses and product losses and turbulent situations. A network that promotes citizen participation has played a major role in making Gandhinagar a world-class city which is alive and well today (Khan and Rana 2019).

Current research shows technical testing and the saving of a 100 kW frame on the corresponding table roof based on operational efficiency by the homer system. Both annual vigor is done and system devouring is resolved. Examination of a PV-matched PV frame includes Current Temporary Theme Cost, Estimated Energy Cost (Rs/kWh), sustainability rate energy (solar energy) used in construction, lowering the cost of bills for this presentation parameters (Behura et al. 2021).

Solar home lighting systems (SHLS) are categorized under decentralized energy systems. They can be treated as an replacing option to grid extension or as a supplement to grid-provided power. SHLS is the best alternative for rural areas.

The characteristic of any solar PV system is that it does not supply consistent power throughout the day when sun moves from east to west. The power which is gained from sunlight requires an efficient management for its fullest possible utilization. In addition, the energy generated from the solar panels needs to be utilized directly either to the load or transferred to the storage device like battery. It is very important to match the source and the sink energy capacities in a PV system.

37.9.1 Solar Home Lighting System (SHLS)

The SHLS has been designed keeping in view the objective of rural electrification. Generally, it consists of lighting and fan either DC or AC type with/without inverter. The SHLS consists of the following types.

(a) **White LED-based SHLS**

It is a solid-state gadget, and it emits light due to electric current. It can also be used to operate a small DC fan or a 12-V DC television and so also W-LED Lamps.

(b) **CFL-based solar home lighting system**

In this SHLS, W-LED is replaced with CFL. Performance specifications of a CFL-based SHLS are given in Table 37.5.

(c) **LED-based solar lantern**

This is useful for either all types of lighting areas including indoor or outdoor, covering a full range of circular illumination around it. Its luminary consists of W-LED.

Performance specifications of a W-LED light source-based solar lantern system are given in Table 37.6.

(d) **CFL-based solar lantern**

In this lantern, W-LED is replaced with CFL.

Performance specifications of a CFL light source-based solar lantern system are given in Table 37.7.

Table 37.5 Performance specifications of a CFL-based SHLS

Accessories	Specification
PV module	18–74 W peak under STC
Battery	Lead acid sealed maintenance free, lead acid tubular flooded or Gel/VRLA or NiMH or lithium-ion
Light source	9–11 W CFL
Mounting of light	Wall or ceiling
Electronics	Min 85% efficiency
Inverter	Quasi sine wave or sine wave type
Average duty cycle	5 h a day under average daily solar radiation of 5.5 kWh/m^2 on a horizontal surface
Autonomy	3 days or minimum 15 operating hours per permissible discharge

Table 37.6 Performance specifications of a W-LED-based solar lantern

Accessories	Specification
PV module	5 Wp under STC
Battery	Sealed maintenance free (SMF) lead acid battery or NiMH battery or lithium ion battery
Light source	W-LED luminary, dispersed beam, soothing to eyes with the use of proper optics
Mounting of light	Wall or ceiling
Electronics	Approximately 85%
Average duty cycle	4 h a day under average daily solar radiation of 5.5 kWh/m^2 on a horizontal surface
Autonomy	Minimum of 3 days or 12 operating hours per permissible discharge

Table 37.7 Performance specifications of a CFL-based solar lantern

Accessories	Specification
PV module	10 Wp under STC
Battery	Sealed maintenance free (SMF) lead acid battery or NiMH battery or lithium ion battery
Light source	7 W CFL luminary with 4 pins only along with proper pre-heating circuit
Electronics	Approximately 85%
Average duty cycle	4 h a day under average daily solar radiation of 5.5 kWh/m^2 on a horizontal surface
Autonomy	Minimum of 3 days or 12 operating hours per permissible discharge

37.10 Conclusion

The rising demand for energy supplies worldwide has led to the acquisition of another source of energy other than conventional sources. Solar radiation is abundant and sought after as a useful source of energy for developing nations such as India. Although PV energy is a fruitful source of energy, it is at the forefront of the PV sector with proper research and development and the efficient use of energy conversion. The Indian government has introduced various goals and initiatives to disseminate greater PV energy in the nation. Also, the success of various projects and various programs implemented by the Indian Government such as JNNSM in 2009, in the use of PV energy as a source of renewable energy has earned India a reputation and global reputation. However, it still has to maintain a steady pace. The Indian government still faces major challenges in giving solar energy a key role in generating electricity. If India wants its renewable energy goals to be achieved successfully in its time, it must be more involved in the country's ability to turn its environment into its own.

References

Bhatt JG, Jani OK, E-Governance for Solar Photo Voltaic Powergrid: Solar City Gandhinagar, Gujarat, India. E-Governance for Smart Cities, pp 177–230

Behura AK, Kumar A, Rajak DK, Pruncu CI, Luciano Lamberti (2021) Towards better performances for a novel rooftop solar PV system, Solar Energy 216:518–529

Castaner L, Silvestre S (2002) Modeling of photovoltaic systems using PSpice.J. Willey & Sons, Ltd.

Chandel SS, Aggarwal RK (2011) Estimation of hourly solar radiation on horizontal and inclined surfaces in western Himalayas. Smart Grid Renew Energy 2(151):45–55

Indira K, Prasun Kumar Das, Bandyopadhyay B (2013) An analytical study on daily solar radiation data Research Communications, Curr Sci 105(2):215–224

Jaina S, Jain NK, Jamie Vaughnd W (2018) Challenges in meeting all of India's electricity from solar: an energetic Approach. Renew Sustain Energy Rev 82:1006–1013

Kapoor K, Pandey KK, Jain AK, Nandan A (2014) Evolution of solar energy in India: a review. Renew Sustain Energy Rev 40:475–487

Kazimier MK (2008) Pulse width modulated DC–DC power converters. John Wiley and Sons Limited

Khan BH (2006) Non-conventional energy resources. Tata McGraw-Hill Publishing Company Limited, New Delhi, India

Khan A, Rana KB (2019) Techno-economic analysis of 100 kW rooftop photovoltaic grid—type system based on actual performance through Homer software. In: 2nd International Conference on Issues and Challenges in Intelligent Computing Techniques (ICICT) 978-1-7281-1772-0 ©2019 IEEE

Muzathik AM, Ibrahim MZ, Samo KB, Wan Nik WB (2011) Estimation of global solar irradiation on horizontal and inclined surfaces based on horizontal measurements. Energy 36:812–818

Masters GM (2004) Renewable and efficient electric power systems. Wiley Interscience, New Jersey

Messenger R, Ventre J (2000) Photovoltaic systems engineering. CRC Press, New York

Markvart T (2000) Solar electricity, 2nd edn. John Wiley & Sons Ltd., Chichester

Mukerjee AK, Nivedita Thakur (2011) Photovoltaic systems, analysis and design. PHI Learning Pvt. Ltd.

Olmo FJ, Vida J, Foyo I, Castro-Diez Y, Alados-Arboledas L (1999) Prediction of global irradiance on inclined surfaces from horizontal global irradiance. Energy 24:689–704

Patel MR (2006) Wind and solar power systems: design, analysis and operation, 2nd edn. Taylor & Francis, Boca Raton

Pritam Satsangi K, Sailesh Babu GS, Bhagwan Das D, Saxena AK, Performance indicators for assessing solar photovoltaic microgrids in grid-connected mode, applications of computing, automation and wireless systems in electrical engineering, Lect Notes Electr Eng 553

Purohita I, Purohit P (2018) Performance assessment of grid-interactive solar photovoltaic projects under India's national solar mission. Appl Energy 222:25–41

Ravindra M, Kulkarni PS (2010) Reliability analysis of solar photovoltaic system using hourly mean solar radiation data. Solar Energy 84:691–702

Rathore PKS, Chauhan DS, Singh RP (2018) Decentralized solar rooftop photovoltaic in India: on the path of sustainable energy security. Renew Energy. https://doi.org/10.1016/j.renene.2018.07.049

Rathore PKS, Rathore S, Singh RP, Agnihotri S Solar power utility sector in India: challenges and opportunities. Renew Sustain Energy Rev

Solanki CS (2009) Solar photovoltaics fundamentals, technologies and applications, vol 150. PHI Learning Private Limited, New Delhi, India

Solanki CS (2013) Solar photovoltaic technology and systems-a manual for technicians, trainers and engineers. PHI Learning Pvt. Limited, Delhi, India

Sukhatme SP, Nayak JK (2010) Solar energy, 3rd edn. Tata Mc-Graw Hill Education Pvt. Ltd., New Delhi

Sannigrahi D, Major issues on beneficial utilization of solar energy in India. In: Handbook of Environmental Materials Management. https://doi.org/10.1007/978-3-319-58538-3_6-1

Singh Y, Pal N (2020) Obstacles and comparative analysis in the advancement of photovoltaic power stations in India. Sustain Comput: Inform Syst 25:100372

Tarai RK, Kale P (2018) Solar PV policy framework of Indian States: overview, pitfalls, challenges, and improvements. Renew Energy Focus 26(00):46–57

Werulkar AS, Kulkarni PS (2014) Energy analysis of solar home lighting system with microcontroller-based charge controller. J Sol Energy Eng Trans ASME 136(031010-1 to 031010-9)

Werulkar AS, Kulkarni PS (2015) A case study of residential solar photovoltaic system with utility backup in Nagpur, India. Renew Sustain Energy Rev 52:1809–1822

Werulkar A, Shankar D, Kulkarni PS (2013) A soft switching boost converter with simulation of maximum power point tracking for solar home lighting system. Int J Chem Tech Res IJCRGG 5(2):935–946. CODEN(USA). ISSN: 0974-4290

Werulkar AS, Kulkarni PS (2016) Microcontroller based ZVS resonant buck converter for solar home lighting system. Int J Electr Electron Commun Eng 1(1):1–17. ManTech Publications

Werulkar AS, Kulkarni PS (2018) Simulation of utility-tied micro inverter for solar photovoltaic system. Rashtrasant Tukadoji Maharaj Sci J RTMNU 14:114–120

Yenneti K (2016) Industry perceptions on feed in tariff (FiT) based solar power policies—a case of Gujarat, India. Renew Sustain Energy Rev 57:988–998

www.powermin.nic.in

http://mnre.gov.in/

www.wiki-solar.org

https://en.wikipedia.org/wiki/solarpower_Europe_Association (2014–2018) European photovoltaic industry association (Report of Global market outlook for photovoltaics)

http://www.rbi.org.in, Handbook of Statistics on Indian Economy (2013–14). Reserve Bank of India

www.indiasmartgrid.org (2015) India's energy story,100 years from its independence. Smart Grid Bullet 2(8)

Printed by Printforce, the Netherlands